P9-DBI-881

MATHEMATICAL METHODS IN THE PHYSICAL SCIENCES

MATHEMATICAL METHODS IN THE PHYSICAL SCIENCES

Third Edition

MARY L. BOAS

DePaul University

PUBLISHER	Kaye Pace
SENIOR ACQUISITIONS Editor	Stuart Johnson
PRODUCTION MANAGER	Pam Kennedy
PRODUCTION EDITOR	Sarah Wolfman-Robichaud
MARKETING MANAGER	Amanda Wygal
SENIOR DESIGNER	Dawn Stanley
EDITORIAL ASSISTANT	Krista Jarmas/Alyson Rentrop
PRODUCTION MANAGER	Jan Fisher/Publication Services

This book was set in 10/12 Computer Modern by Publication Services and manufactured in Singapore by Markono Print Media Pte Ltd.The cover was manufactured in Singapore by Markono Print Media Pte Ltd.

This book is printed on acid free paper.

ISBN-13 978-0-471-19826-0
ISBN-WIE-13 978-0-471-36580-8

Printed in Singapore

24

To the memory of RPB

PREFACE

This book is particularly intended for the student with a year (or a year and a half) of calculus who wants to develop, in a short time, a basic competence in each of the many areas of mathematics needed in junior to senior-graduate courses in physics, chemistry, and engineering. Thus it is intended to be accessible to sophomores (or freshmen with AP calculus from high school). It may also be used effectively by a more advanced student to review half-forgotten topics or learn new ones, either by independent study or in a class. Although the book was written especially for students of the physical sciences, students in any field (say mathematics or mathematics for teaching) may find it useful to survey many topics or to obtain some knowledge of areas they do not have time to study in depth. Since theorems are stated carefully, such students should not need to unlearn anything in their later work.

The question of proper mathematical training for students in the physical sciences is of concern to both mathematicians and those who use mathematics in applications. Some instructors may feel that if students are going to study mathematics at all, they should study it in careful and thorough detail. For the undergraduate physics, chemistry, or engineering student, this means either (1) learning more mathematics than a mathematics major or (2) learning a few areas of mathematics thoroughly and the others only from snatches in science courses. The second alternative is often advocated; let me say why I think it is unsatisfactory. It is certainly true that motivation is increased by the immediate application of a mathematical technique, but there are a number of disadvantages:

1. The discussion of the mathematics is apt to be sketchy since that is not the primary concern.
2. Students are faced simultaneously with learning a new mathematical method and applying it to an area of science that is also new to them. Frequently the

difficulty in comprehending the new scientific area lies more in the distraction caused by poorly understood mathematics than it does in the new scientific ideas.

3. Students may meet what is actually the same mathematical principle in two different science courses without recognizing the connection, or even learn apparently contradictory theorems in the two courses! For example, in thermodynamics students learn that the integral of an exact differential around a closed path is always zero. In electricity or hydrodynamics, they run into $\int_0^{2\pi} d\theta$, which is certainly the integral of an exact differential around a closed path but is not equal to zero!

Now it would be fine if every science student could take the separate mathematics courses in differential equations (ordinary and partial), advanced calculus, linear algebra, vector and tensor analysis, complex variables, Fourier series, probability, calculus of variations, special functions, and so on. However, most science students have neither the time nor the inclination to study that much mathematics, yet they are constantly hampered in their science courses for lack of the basic techniques of these subjects. It is the intent of this book to give these students enough background in each of the needed areas so that they can cope successfully with junior, senior, and beginning graduate courses in the physical sciences. I hope, also, that some students will be sufficiently intrigued by one or more of the fields of mathematics to pursue it futher.

It is clear that something must be omitted if so many topics are to be compressed into one course. I believe that two things can be left out without serious harm at this stage of a student's work: generality, and detailed proofs. Stating and proving a theorem in its most general form is important to the mathematician and to the advanced student, but it is often unnecessary and may be confusing to the more elementary student. This is not in the least to say that science students have no use for careful mathematics. Scientists, even more than pure mathematicians, need careful statements of the limits of applicability of mathematical processes so that they can use them with confidence without having to supply proof of their validity. Consequently I have endeavored to give accurate statements of the needed theorems, although often for special cases or without proof. Interested students can easily find more detail in textbooks in the special fields.

Mathematical physics texts at the senior-graduate level are able to assume a degree of mathematical sophistication and knowledge of advanced physics not yet attained by students at the sophomore level. Yet such students, if given simple and clear explanations, can readily master the techniques we cover in this text. (They not only *can*, but will *have to* in one way or another, if they are going to pass their junior and senior physics courses!) These students are not ready for detailed applications—these they will get in their science courses—but they do need and want to be given some idea of the use of the methods they are studying, and some simple applications. This I have tried to do for each new topic.

For those of you familiar with the second edition, let me outline the changes for the third:

1. Prompted by several requests for matrix diagonalization in Chapter 3, I have moved the first part of Chapter 10 to Chapter 3 and then have amplified the treatment of tensors in Chapter 10. I have also changed Chapter 3 to include more detail about linear vector spaces and then have continued the discussion of basis functions in Chapter 7 (Fourier series), Chapter 8 (Differential equations),

Chapter 12 (Series solutions) and Chapter 13 (Partial differential equations).

2. Again, prompted by several requests, I have moved Fourier integrals back to the Fourier series Chapter 7. Since this breaks up the integral transforms chapter (old Chapter 15), I decided to abandon that chapter and move the Laplace transform and Dirac delta function material back to the ordinary differential equations Chapter 8. I have also amplified the treatment of the delta function.
3. The Probability chapter (old Chapter 16) now becomes Chapter 15. Here I have changed the title to Probability and Statistics, and have revised the latter part of the chapter to emphasize its purpose, namely to clarify for students the theory behind the rules they learn for handling experimental data.
4. The very rapid development of technological aids to computation poses a steady question for instructors as to their best use. Without selecting any particular Computer Algebra System, I have simply tried for each topic to point out to students both the usefulness and the pitfalls of computer use. (Please see my comments at the end of "To the Student" just ahead.)

The material in the text is so arranged that students who study the chapters in order will have the necessary background at each stage. However, it is not always either necessary or desirable to follow the text order. Let me suggest some rearrangements I have found useful. If students have previously studied the material in any of chapters 1, 3, 4, 5, 6, or 8 (in such courses as second-year calculus, differential equations, linear algebra), then the corresponding chapter(s) could be omitted, used for reference, or, preferably, be reviewed briefly with emphasis on problem solving. Students may know Taylor's theorem, for example, but have little skill in using series approximations; they may know the theory of multiple integrals, but find it difficult to set up a double integral for the moment of inertia of a spherical shell; they may know existence theorems for differential equations, but have little skill in solving, say, $y'' + y = x \sin x$. Problem solving is the essential core of a course on Mathematical Methods.

After Chapters 7 (Fourier Series) and 8 (Ordinary Differential Equations) I like to cover the first four sections of Chapter 13 (Partial Differential Equations). This gives students an introduction to Partial Differential Equations but requires only the use of Fourier series expansions. Later on, after studying Chapter 12, students can return to complete Chapter 13. Chapter 15 (Probability and Statistics) is almost independent of the rest of the text; I have covered this material anywhere from the beginning to the end of a one-year course.

It has been gratifying to hear the enthusiastic responses to the first two editions, and I hope that this third edition will prove even more useful. I want to thank many readers for helpful suggestions and I will appreciate any further comments. If you find misprints, please send them to me at MLBoas@aol.com. I also want to thank the University of Washington physics students who were my LaTeX typists: Toshiko Asai, Jeff Sherman, and Jeffrey Frasca. And I especially want to thank my son, Harold P. Boas, both for mathematical consultations, and for his expert help with LaTeX problems.

Instructors who have adopted the book for a class should consult the publisher about an Instructor's Answer Book, and about a list correlating 2nd and 3rd edition problem numbers for problems which appear in both editions.

Mary L. Boas

TO THE STUDENT

As you start each topic in this book, you will no doubt wonder and ask "Just why should I study this subject and what use does it have in applications?" There is a story about a young mathematics instructor who asked an older professor "What do you say when students ask about the practical applications of some mathematical topic?" The experienced professor said "I tell them!" This text tries to follow that advice. However, you must on your part be reasonable in your request. It is not possible in one book or course to cover both the mathematical methods and very many detailed applications of them. You will have to be content with some information as to the areas of application of each topic and some of the simpler applications. In your later courses, you will then use these techniques in more advanced applications. At that point you can concentrate on the physical application instead of being distracted by learning new mathematical methods.

One point about your study of this material cannot be emphasized too strongly: To use mathematics effectively in applications, you need not just knowledge but *skill*. Skill can be obtained only through practice. You can obtain a certain superficial *knowledge* of mathematics by listening to lectures, but you cannot obtain *skill* this way. How many students have I heard say "It looks so easy when you do it," or "I understand it but I can't do the problems!" Such statements show lack of practice and consequent lack of skill. The *only* way to develop the skill necessary to use this material in your later courses is to practice by solving many problems. Always study with pencil and paper at hand. Don't just read through a solved problem—try to do it yourself! Then solve some similar ones from the problem set for that section, trying to choose the most appropriate method from the solved examples. See the Answers to Selected Problems and check your answers to any problems listed there. If you meet an unfamiliar term, look for it in the Index (or in a dictionary if it is nontechnical).

My students tell me that one of my most frequent comments to them is "You're working too hard." There is no merit in spending hours producing a solution to a problem that can be done by a better method in a few minutes. Please ignore anyone who disparages problem-solving techniques as "tricks" or "shortcuts." You will find that the more able you are to choose effective methods of solving problems in your science courses, the easier it will be for you to master new material. But this means practice, practice, practice! The *only* way to learn to solve problems is to solve problems. In this text, you will find both drill problems and harder, more challenging problems. You should not feel satisfied with your study of a chapter until you can solve a reasonable number of these problems.

You may be thinking "I don't really need to study this—my computer will solve all these problems for me." Now Computer Algebra Systems are wonderful—as you know, they save you a lot of laborious calculation and quickly plot graphs which clarify a problem. But a computer is a tool; *you* are the one in charge. A very perceptive student recently said to me (about the use of a computer for a special project): "*First* you learn how to do it; *then* you see what the computer can do to make it easier." Quite so! A very effective way to study a new technique is to do some simple problems by hand in order to understand the process, and compare your results with a computer solution. You will then be better able to use the method to set up and solve similar more complicated applied problems in your advanced courses. So, in one problem set after another, I will remind you that the point of solving some simple problems is not to get an answer (which a computer will easily supply) but rather to learn the ideas and techniques which will be so useful in your later courses.

M. L. B.

CONTENTS

MATHEMATICAL METHODS IN THE PHYSICAL SCIENCES

CHAPTER 1

Infinite Series, Power Series

1. THE GEOMETRIC SERIES

As a simple example of many of the ideas involved in series, we are going to consider the geometric series. You may recall that in a geometric progression we multiply each term by some fixed number to get the next term. For example, the *sequences*

$$2,\ 4,\ 8,\ 16,\ 32, \ldots, \tag{1.1a}$$

$$1,\ \tfrac{2}{3},\ \tfrac{4}{9},\ \tfrac{8}{27},\ \tfrac{16}{81}, \ldots, \tag{1.1b}$$

$$a,\ ar,\ ar^2,\ ar^3, \ldots, \tag{1.1c}$$

are geometric progressions. It is easy to think of examples of such progressions. Suppose the number of bacteria in a culture doubles every hour. Then the terms of (1.1a) represent the number by which the bacteria population has been multiplied after 1 hr, 2 hr, and so on. Or suppose a bouncing ball rises each time to $\frac{2}{3}$ of the height of the previous bounce. Then (1.1b) would represent the heights of the successive bounces in yards if the ball is originally dropped from a height of 1 yd.

In our first example it is clear that the bacteria population would increase without limit as time went on (mathematically, anyway; that is, assuming that nothing like lack of food prevented the assumed doubling each hour). In the second example, however, the height of bounce of the ball decreases with successive bounces, and we might ask for the total distance the ball goes. The ball falls a distance 1 yd, rises a distance $\frac{2}{3}$ yd and falls a distance $\frac{2}{3}$ yd, rises a distance $\frac{4}{9}$ yd and falls a distance $\frac{4}{9}$ yd, and so on. Thus it seems reasonable to write the following expression for the total distance the ball goes:

$$1 + 2\cdot\tfrac{2}{3} + 2\cdot\tfrac{4}{9} + 2\cdot\tfrac{8}{27} + \cdots = 1 + 2\left(\tfrac{2}{3} + \tfrac{4}{9} + \tfrac{8}{27} + \cdots\right), \tag{1.2}$$

where the three dots mean that the terms continue as they have started (each one being $\frac{2}{3}$ the preceding one), and there is never a last term. Let us consider the expression in parentheses in (1.2), namely

$$\frac{2}{3} + \frac{4}{9} + \frac{8}{27} + \cdots. \tag{1.3}$$

This expression is an example of an *infinite series*, and we are asked to find its sum. Not all infinite series have sums; you can see that the series formed by adding the terms in (1.1a) does not have a finite sum. However, even when an infinite series does have a finite sum, we cannot find it by adding the terms because no matter how many we add there are always more. Thus we must find another method. (It is actually deeper than this; what we really have to do is to *define* what we mean by the sum of the series.)

Let us first find the sum of n terms in (1.3). The formula (Problem 2) for the sum of n terms of the geometric progression (1.1c) is

$$S_n = \frac{a(1-r^n)}{1-r}. \tag{1.4}$$

Using (1.4) in (1.3), we find

$$S_n = \frac{2}{3} + \frac{4}{9} + \cdots + \left(\frac{2}{3}\right)^n = \frac{\frac{2}{3}[1-(\frac{2}{3})^n]}{1-\frac{2}{3}} = 2\left[1-\left(\frac{2}{3}\right)^n\right]. \tag{1.5}$$

As n increases, $(\frac{2}{3})^n$ decreases and approaches zero. Then the sum of n terms approaches 2 as n increases, and we say that the sum of the series is 2. (This is really a definition: The sum of an infinite series is the limit of the sum of n terms as $n \to \infty$.) Then from (1.2), the total distance traveled by the ball is $1 + 2 \cdot 2 = 5$. This is an answer to a mathematical problem. A physicist might well object that a bounce the size of an atom is nonsense! However, after a number of bounces, the remaining infinite number of small terms contribute very little to the final answer (see Problem 1). Thus it makes little difference (in our answer for the total distance) whether we insist that the ball rolls after a certain number of bounces or whether we include the entire series, and it is easier to find the sum of the series than to find the sum of, say, twenty terms.

Series such as (1.3) whose terms form a geometric progression are called *geometric series*. We can write a geometric series in the form

$$a + ar + ar^2 + \cdots + ar^{n-1} + \cdots. \tag{1.6}$$

The sum of the geometric series (if it has one) is by definition

$$S = \lim_{n\to\infty} S_n, \tag{1.7}$$

where S_n is the sum of n terms of the series. By following the method of the example above, you can show (Problem 2) that a geometric series has a sum if and only if $|r| < 1$, and in this case the sum is

$$S = \frac{a}{1-r}. \tag{1.8}$$

The series is then called *convergent.*

Here is an interesting use of (1.8). We can write $0.3333\cdots = \frac{3}{10} + \frac{3}{100} + \frac{3}{1000} + \cdots = \frac{3/10}{1-1/10} = \frac{1}{3}$ by (1.8). Now of course you knew that, but how about $0.785714285714\cdots$? We can write this as $0.5 + 0.285714285714\cdots = \frac{1}{2} + \frac{0.285714}{1-10^{-6}} = \frac{1}{2} + \frac{285714}{999999} = \frac{1}{2} + \frac{2}{7} = \frac{11}{14}$. (Note that any repeating decimal is equivalent to a fraction which can be found by this method.) If you want to use a computer to do the arithmetic, be sure to tell it to give you an exact answer or it may hand you back the decimal you started with! You can also use a computer to sum the series, but using (1.8) may be simpler. (Also see Problem 14.)

PROBLEMS, SECTION 1

1. In the bouncing ball example above, find the height of the tenth rebound, and the distance traveled by the ball after it touches the ground the tenth time. Compare this distance with the total distance traveled.

2. Derive the formula (1.4) for the sum S_n of the geometric progression $S_n = a + ar + ar^2 + \cdots + ar^{n-1}$. *Hint:* Multiply S_n by r and subtract the result from S_n; then solve for S_n. Show that the geometric series (1.6) converges if and only if $|r| < 1$; also show that if $|r| < 1$, the sum is given by equation (1.8).

Use equation (1.8) to find the fractions that are equivalent to the following repeating decimals:

3. $0.55555\cdots$ **4.** $0.818181\cdots$ **5.** $0.583333\cdots$

6. $0.61111\cdots$ **7.** $0.185185\cdots$ **8.** $0.694444\cdots$

9. $0.857142857142\cdots$ **10.** $0.576923076923076923\cdots$

11. $0.678571428571428571\cdots$

12. In a water purification process, one-nth of the impurity is removed in the first stage. In each succeeding stage, the amount of impurity removed is one-nth of that removed in the preceding stage. Show that if $n = 2$, the water can be made as pure as you like, but that if $n = 3$, at least one-half of the impurity will remain no matter how many stages are used.

13. If you invest a dollar at "6% interest compounded monthly," it amounts to $(1.005)^n$ dollars after n months. If you invest \$10 at the beginning of each month for 10 years (120 months), how much will you have at the end of the 10 years?

14. A computer program gives the result 1/6 for the sum of the series $\sum_{n=0}^{\infty}(-5)^n$. Show that this series is divergent. Do you see what happened? *Warning hint:* Always consider whether an answer is reasonable, whether it's a computer answer or your work by hand.

15. Connect the midpoints of the sides of an equilateral triangle to form 4 smaller equilateral triangles. Leave the middle small triangle blank, but for each of the other 3 small triangles, draw lines connecting the midpoints of the sides to create 4 tiny triangles. Again leave each middle tiny triangle blank and draw the lines to divide the others into 4 parts. Find the infinite series for the total area left blank if this process is continued indefinitely. (Suggestion: Let the area of the original triangle be 1; then the area of the first blank triangle is 1/4.) Sum the series to find the total area left blank. Is the answer what you expect? *Hint:* What is the "area" of a straight line? (Comment: You have constructed a *fractal* called the Sierpiński gasket. A fractal has the property that a magnified view of a small part of it looks very much like the original.)

16. Suppose a large number of particles are bouncing back and forth between $x = 0$ and $x = 1$, except that at each endpoint some escape. Let r be the fraction reflected each time; then $(1 - r)$ is the fraction escaping. Suppose the particles start at $x = 0$ heading toward $x = 1$; eventually all particles will escape. Write an infinite series for the fraction which escape at $x = 1$ and similarly for the fraction which escape at $x = 0$. Sum both the series. What is the largest fraction of the particles which can escape at $x = 0$? (Remember that r must be between 0 and 1.)

2. DEFINITIONS AND NOTATION

There are many other infinite series besides geometric series. Here are some examples:

$$1^2 + 2^2 + 3^2 + 4^2 + \cdots, \tag{2.1a}$$

$$\frac{1}{2} + \frac{2}{2^2} + \frac{3}{2^3} + \frac{4}{2^4} + \cdots, \tag{2.1b}$$

$$x - \frac{x^2}{2} + \frac{x^3}{3} - \frac{x^4}{4} + \cdots. \tag{2.1c}$$

In general, an infinite series means an expression of the form

$$a_1 + a_2 + a_3 + \cdots + a_n + \cdots, \tag{2.2}$$

where the a_n's (one for each positive integer n) are numbers or functions given by some formula or rule. The three dots in each case mean that the series never ends. The terms continue according to the law of formation, which is supposed to be evident to you by the time you reach the three dots. If there is apt to be doubt about how the terms are formed, a general or nth term is written like this:

$$1^2 + 2^2 + 3^2 + \cdots + n^2 + \cdots, \tag{2.3a}$$

$$x - x^2 + \frac{x^3}{2} + \cdots + \frac{(-1)^{n-1}x^n}{(n-1)!} + \cdots. \tag{2.3b}$$

(The quantity $n!$, read n factorial, means, for integral n, the product of all integers from 1 to n; for example, $5! = 5 \cdot 4 \cdot 3 \cdot 2 \cdot 1 = 120$. The quantity $0!$ is defined to be 1.) In (2.3a), it is easy to see without the general term that each term is just the square of the number of the term, that is, n^2. However, in (2.3b), if the formula for the general term were missing, you could probably make several reasonable guesses for the next term. To be sure of the law of formation, we must either know a good many more terms or have the formula for the general term. You should verify that the fourth term in (2.3b) is $-x^4/6$.

We can also write series in a shorter abbreviated form using a summation sign $\sum$ followed by the formula for the nth term. For example, (2.3a) would be written

$$1^2 + 2^2 + 3^2 + 4^2 + \cdots = \sum_{n=1}^{\infty} n^2 \tag{2.4}$$

(read "the sum of n^2 from $n = 1$ to ∞"). The series (2.3b) would be written

$$x - x^2 + \frac{x^3}{2} - \frac{x^4}{6} + \cdots = \sum_{n=1}^{\infty} \frac{(-1)^{n-1}x^n}{(n-1)!}$$

For printing convenience, sums like (2.4) are often written $\sum_{n=1}^{\infty} n^2$.

In Section 1, we have mentioned both sequences and series. The lists in (1.1) are sequences; a *sequence* is simply a set of quantities, one for each n. A *series* is an indicated sum of such quantities, as in (1.3) or (1.6). We will be interested in various sequences related to a series: for example, the sequence a_n of terms of the series, the sequence S_n of partial sums [see (1.5) and (4.5)], the sequence R_n [see (4.7)], and the sequence ρ_n [see (6.2)]. In all these examples, we want to find the limit of a sequence as $n \to \infty$ (if the sequence has a limit). Although limits can be found by computer, many simple limits can be done faster by hand.

Example 1. Find the limit as $n \to \infty$ of the sequence

$$\frac{(2n-1)^4 + \sqrt{1+9n^8}}{1-n^3-7n^4}.$$

We divide numerator and denominator by n^4 and take the limit as $n \to \infty$. Then all terms go to zero except

$$\frac{2^4+\sqrt{9}}{-7} = -\frac{19}{7}.$$

Example 2. Find $\lim_{n\to\infty} \frac{\ln n}{n}$. By L'Hôpital's rule (see Section 15)

$$\lim_{n\to\infty} \frac{\ln n}{n} = \lim_{n\to\infty} \frac{1/n}{1} = 0.$$

Comment: Strictly speaking, we can't differentiate a function of n if n is an integer, but we can consider $f(x) = (\ln x)/x$, and the limit of the sequence is the same as the limit of $f(x)$.

Example 3. Find $\lim_{n\to\infty} \left(\frac{1}{n}\right)^{1/n}$. We first find

$$\ln\left(\frac{1}{n}\right)^{1/n} = -\frac{1}{n}\ln n.$$

Then by Example 2, the limit of $(\ln n)/n$ is 0, so the original limit is $e^0 = 1$.

PROBLEMS, SECTION 2

In the following problems, find the limit of the given sequence as $n \to \infty$.

1. $\dfrac{n^2+5n^3}{2n^3+3\sqrt{4+n^6}}$

2. $\dfrac{(n+1)^2}{\sqrt{3+5n^2+4n^4}}$

3. $\dfrac{(-1)^n\sqrt{n+1}}{n}$

4. $\dfrac{2^n}{n^2}$

5. $\dfrac{10^n}{n!}$

6. $\dfrac{n^n}{n!}$

7. $(1+n^2)^{1/\ln n}$

8. $\dfrac{(n!)^2}{(2n)!}$

9. $n\sin(1/n)$

3. APPLICATIONS OF SERIES

In the example of the bouncing ball in Section 1, we saw that it is possible for the sum of an infinite series to be nearly the same as the sum of a fairly small number of terms at the beginning of the series (also see Problem 1.1). Many applied problems cannot be solved exactly, but we may be able to find an answer in terms of an infinite series, and then use only as many terms as necessary to obtain the needed accuracy. We shall see many examples of this both in this chapter and in later chapters. Differential equations (see Chapters 8 and 12) and partial differential equations (see Chapter 13) are frequently solved by using series. We will learn how to find series that represent functions; often a complicated function can be approximated by a few terms of its series (see Section 15).

But there is more to the subject of infinite series than making approximations. We will see (Chapter 2, Section 8) how we can use power series (that is, series whose terms are powers of x) to give meaning to functions of complex numbers, and (Chapter 3, Section 6) how to define a function of a matrix using the power series of the function. Also power series are just a first example of infinite series. In Chapter 7 we will learn about Fourier series (whose terms are sines and cosines). In Chapter 12, we will use power series to solve differential equations, and in Chapters 12 and 13, we will discuss other series such as Legendre and Bessel. Finally, in Chapter 14, we will discover how a study of power series clarifies our understanding of the mathematical functions we use in applications.

4. CONVERGENT AND DIVERGENT SERIES

We have been talking about series which have a finite sum. We have also seen that there are series which do not have finite sums, for example (2.1a). If a series has a finite sum, it is called *convergent.* Otherwise it is called *divergent.* It is important to know whether a series is convergent or divergent. Some weird things can happen if you try to apply ordinary algebra to a divergent series. Suppose we try it with the following series:

$$S = 1 + 2 + 4 + 8 + 16 + \cdots . \tag{4.1}$$

Then,

$$\begin{aligned} 2S &= 2 + 4 + 8 + 16 + \cdots = S - 1, \\ S &= -1. \end{aligned}$$

This is obvious nonsense, and you may laugh at the idea of trying to operate with such a violently divergent series as (4.1). But the same sort of thing can happen in more concealed fashion, and has happened and given wrong answers to people who were not careful enough about the way they used infinite series. At this point you probably would not recognize that the series

$$1 + \frac{1}{2} + \frac{1}{3} + \frac{1}{4} + \frac{1}{5} + \cdots \tag{4.2}$$

is divergent, but it is; and the series

$$1 - \frac{1}{2} + \frac{1}{3} - \frac{1}{4} + \frac{1}{5} - \cdots \tag{4.3}$$

is convergent as it stands, but can be made to have *any* sum you like by combining the terms in a different order! (See Section 8.) You can see from these examples how essential it is to know whether a series converges, and also to know how to apply algebra to series correctly. There are even cases in which some divergent series can be used (see Chapter 11), but in this chapter we shall be concerned with convergent series.

Before we consider some tests for convergence, let us repeat the definition of convergence more carefully. Let us call the terms of the series a_n so that the series is

$$a_1 + a_2 + a_3 + a_4 + \cdots + a_n + \cdots. \tag{4.4}$$

Remember that the three dots mean that there is never a last term; the series goes on without end. Now consider the sums S_n that we obtain by adding more and more terms of the series. We define

$$\begin{aligned} S_1 &= a_1, \\ S_2 &= a_1 + a_2, \\ S_3 &= a_1 + a_2 + a_3, \\ &\cdots \\ S_n &= a_1 + a_2 + a_3 + \cdots + a_n. \end{aligned} \tag{4.5}$$

Each S_n is called a *partial sum*; it is the sum of the first n terms of the series. We had an example of this for a geometric progression in (1.4). The letter n can be any integer; for each n, S_n stops with the nth term. (Since S_n is not an infinite series, there is no question of convergence for it.) As n increases, the partial sums may increase without any limit as in the series (2.1a). They may oscillate as in the series $1-2+3-4+5-\cdots$ (which has partial sums $1, -1, 2, -2, 3, \cdots$) or they may have some more complicated behavior. One possibility is that the S_n's may, after a while, not change very much any more; the a_n's may become very small, and the S_n's come closer and closer to some value S. We are particularly interested in this case in which the S_n's approach a limiting value, say

$$\lim_{n\to\infty} S_n = S. \tag{4.6}$$

(It is understood that S is a finite number.) If this happens, we make the following definitions.

a. If the partial sums S_n of an infinite series tend to a limit S, the series is called *convergent.* Otherwise it is called *divergent.*

b. The limiting value S is called the *sum of the series.*

c. The difference $R_n = S - S_n$ is called the *remainder* (or the remainder after n terms). From (4.6), we see that

$$\lim_{n\to\infty} R_n = \lim_{n\to\infty} (S - S_n) = S - S = 0. \tag{4.7}$$

Example 1. We have already (Section 1) found S_n and S for a geometric series. From (1.8) and (1.4), we have for a geometric series, $R_n = \frac{ar^n}{1-r}$ which $\to 0$ as $n \to \infty$ if $|r| < 1$.

Example 2. By partial fractions, we can write $\frac{2}{n^2-1} = \frac{1}{n-1} - \frac{1}{n+1}$. Let's write out a number of terms of the series

$$\sum_2^\infty \frac{2}{n^2-1} = \sum_2^\infty \left(\frac{1}{n-1} - \frac{1}{n+1}\right) = \sum_1^\infty \left(\frac{1}{n} - \frac{1}{n+2}\right)$$

$$= 1 - \frac{1}{3} + \frac{1}{2} - \frac{1}{4} + \frac{1}{3} - \frac{1}{5} + \frac{1}{4} - \frac{1}{6} + \frac{1}{5} - \frac{1}{7} + \frac{1}{6} - \frac{1}{8} + \cdots$$

$$+ \frac{1}{n-2} - \frac{1}{n} + \frac{1}{n-1} - \frac{1}{n+1} + \frac{1}{n} - \frac{1}{n+2} + \cdots.$$

Note the cancellation of terms; this kind of series is called a telescoping series. Satisfy yourself that when we have added the nth term $(\frac{1}{n} - \frac{1}{n+2})$, the only terms which have not cancelled are $1, \frac{1}{2}, \frac{-1}{n+1}$, and $\frac{-1}{n+2}$, so we have

$$S_n = \frac{3}{2} - \frac{1}{n+1} - \frac{1}{n+2}, \qquad S = \frac{3}{2}, \qquad R_n = \frac{1}{n+1} + \frac{1}{n+2}.$$

Example 3. Another interesting series is

$$\sum_1^\infty \ln\left(\frac{n}{n+1}\right) = \sum_1^\infty [\ln n - \ln(n+1)]$$

$$= \ln 1 - \ln 2 + \ln 2 - \ln 3 + \ln 3 - \ln 4 + \cdots + \ln n - \ln(n+1) \cdots.$$

Then $S_n = -\ln(n+1)$ which $\to -\infty$ as $n \to \infty$, so the series diverges. However, note that $a_n = \ln \frac{n}{n+1} \to \ln 1 = 0$ as $n \to \infty$, so we see that even if the terms tend to zero, a series may diverge.

PROBLEMS, SECTION 4

For the following series, write formulas for the sequences a_n, S_n, and R_n, and find the limits of the sequences as $n \to \infty$ (if the limits exist).

1. $\displaystyle\sum_1^\infty \frac{1}{2^n}$

2. $\displaystyle\sum_0^\infty \frac{1}{5^n}$

3. $1 - \frac{1}{2} + \frac{1}{4} - \frac{1}{8} + \frac{1}{16} \cdots$

4. $\displaystyle\sum_1^\infty e^{-n\ln 3}$ *Hint:* What is $e^{-\ln 3}$?

5. $\displaystyle\sum_0^\infty e^{2n\ln\sin(\pi/3)}$ *Hint:* Simplify this.

6. $\displaystyle\sum_1^\infty \frac{1}{n(n+1)}$ *Hint:* $\displaystyle\frac{1}{n(n+1)} = \frac{1}{n} - \frac{1}{n+1}$.

7. $\displaystyle\frac{3}{1\cdot 2} - \frac{5}{2\cdot 3} + \frac{7}{3\cdot 4} - \frac{9}{4\cdot 5} + \cdots$

5. TESTING SERIES FOR CONVERGENCE; THE PRELIMINARY TEST

It is not in general possible to write a simple formula for S_n and find its limit as $n \to \infty$ (as we have done for a few special series), so we need some other way to find out whether a given series converges. Here we shall consider a few simple tests for convergence. These tests will illustrate some of the ideas involved in testing series for convergence and will work for a good many, but not all, cases. There are more complicated tests which you can find in other books. In some cases it may be quite a difficult mathematical problem to investigate the convergence of a complicated series. However, for our purposes the simple tests we give here will be sufficient.

First we discuss a useful *preliminary test.* In most cases you should apply this to a series before you use other tests.

Preliminary test. If the terms of an infinite series do *not* tend to zero (that is, if $\lim_{n\to\infty} a_n \neq 0$), the series diverges. If $\lim_{n\to\infty} a_n = 0$, we must test further.

This is *not* a test for convergence; what it does is to weed out some very badly divergent series which you then do not have to spend time testing by more complicated methods. *Note carefully:* The preliminary test can *never* tell you that a series converges. It does *not* say that series converge if $a_n \to 0$ and, in fact, often they do not. A simple example is the harmonic series (4.2); the nth term certainly tends to zero, but we shall soon show that the series $\sum_{n=1}^{\infty} 1/n$ is divergent. On the other hand, in the series

$$\frac{1}{2} + \frac{2}{3} + \frac{3}{4} + \frac{4}{5} + \cdots$$

the terms are tending to 1, so by the preliminary test, this series diverges and no further testing is needed.

PROBLEMS, SECTION 5

Use the preliminary test to decide whether the following series are divergent or require further testing. *Careful:* Do *not* say that a series is convergent; the preliminary test cannot decide this.

1. $\frac{1}{2} - \frac{4}{5} + \frac{9}{10} - \frac{16}{17} + \frac{25}{26} - \frac{36}{37} + \cdots$

2. $\sqrt{2} + \frac{\sqrt{3}}{2} + \frac{\sqrt{4}}{3} + \frac{\sqrt{5}}{4} + \frac{\sqrt{6}}{5} + \cdots$

3. $\sum_{n=1}^{\infty} \frac{n+3}{n^2+10n}$

4. $\sum_{n=1}^{\infty} \frac{(-1)^n n^2}{(n+1)^2}$

5. $\sum_{n=1}^{\infty} \frac{n!}{n!+1}$

6. $\sum_{n=1}^{\infty} \frac{n!}{(n+1)!}$

7. $\sum_{n=1}^{\infty} \frac{(-1)^n n}{\sqrt{n^3+1}}$

8. $\sum_{n=1}^{\infty} \frac{\ln n}{n}$

9. $\sum_{n=1}^{\infty} \frac{3^n}{2^n+3^n}$

10. $\sum_{n=2}^{\infty} \left(1 - \frac{1}{n^2}\right)$

11. Using (4.6), give a proof of the preliminary test. *Hint:* $S_n - S_{n-1} = a_n$.

6. CONVERGENCE TESTS FOR SERIES OF POSITIVE TERMS; ABSOLUTE CONVERGENCE

We are now going to consider four useful tests for series whose terms are all positive. If some of the terms of a series are negative, we may still want to consider the related series which we get by making all the terms positive; that is, we may consider the series whose terms are the absolute values of the terms of our original series. If this new series converges, we call the original series *absolutely convergent.* It can be proved that if a series converges absolutely, then it converges (Problem 7.9). This means that if the series of absolute values converges, the series is still convergent when you put back the original minus signs. (The sum is different, of course.) The following four tests may be used, then, either for testing series of positive terms, or for testing any series for absolute convergence.

A. The Comparison Test

This test has two parts, (a) and (b).

(a) Let

$$m_1 + m_2 + m_3 + m_4 + \cdots$$

be a series of positive terms which you know converges. Then the series you are testing, namely

$$a_1 + a_2 + a_3 + a_4 + \cdots$$

is absolutely convergent if $|a_n| \leq m_n$ (that is, if the absolute value of each term of the a series is no larger than the corresponding term of the m series) for all n from some point on, say after the third term (or the millionth term). See the example and discussion below.

(b) Let

$$d_1 + d_2 + d_3 + d_4 + \cdots$$

be a series of positive terms which you know diverges. Then the series

$$|a_1| + |a_2| + |a_3| + |a_4| + \cdots$$

diverges if $|a_n| \geq d_n$ for all n from some point on.

Warning: Note carefully that neither $|a_n| \geq m_n$ nor $|a_n| \leq d_n$ tells us anything. That is, if a series has terms larger than those of a convergent series, it may still converge or it may diverge—we must test it further. Similarly, if a series has terms smaller than those of a divergent series, it may still diverge, or it may converge.

Example. Test $\displaystyle\sum_{n=1}^{\infty} \frac{1}{n!} = 1 + \frac{1}{2} + \frac{1}{6} + \frac{1}{24} + \cdots$ for convergence.

As a comparison series, we choose the geometric series

$$\sum_{n=1}^{\infty} \frac{1}{2^n} = \frac{1}{2} + \frac{1}{4} + \frac{1}{8} + \frac{1}{16} + \cdots .$$

Notice that we do not care about the first few terms (or, in fact, any finite number of terms) in a series, because they can affect the sum of the series but *not* whether it converges. When we ask whether a series converges or not, we are asking what happens as we add more and more terms for larger and larger n. Does the sum increase indefinitely, or does it approach a limit? What the first five or hundred or million terms are has no effect on whether the sum eventually increases indefinitely or approaches a limit. Consequently we frequently ignore some of the early terms in testing series for convergence.

In our example, the terms of $\sum_{n=1}^{\infty} 1/n!$ are smaller than the corresponding terms of $\sum_{n=1}^{\infty} 1/2^n$ for all $n > 3$ (Problem 1). We know that the geometric series converges because its ratio is $\frac{1}{2}$. Therefore $\sum_{n=1}^{\infty} 1/n!$ converges also.

PROBLEMS, SECTION 6

1. Show that $n! > 2^n$ for all $n > 3$. *Hint:* Write out a few terms; then consider what you multiply by to go from, say, $5!$ to $6!$ and from 2^5 to 2^6.

2. Prove that the harmonic series $\sum_{n=1}^{\infty} 1/n$ is divergent by comparing it with the series

$$1+\frac{1}{2}+\left(\frac{1}{4}+\frac{1}{4}\right)+\left(\frac{1}{8}+\frac{1}{8}+\frac{1}{8}+\frac{1}{8}\right)+\left(8 \text{ terms each equal to } \frac{1}{16}\right)+\cdots,$$

$$\text{which is} \quad 1+\frac{1}{2}+\frac{1}{2}+\frac{1}{2}+\frac{1}{2}+\cdots.$$

3. Prove the convergence of $\sum_{n=1}^{\infty} 1/n^2$ by grouping terms somewhat as in Problem 2.

4. Use the comparison test to prove the convergence of the following series:

(a) $\displaystyle\sum_{n=1}^{\infty} \frac{1}{2^n+3^n}$ (b) $\displaystyle\sum_{n=1}^{\infty} \frac{1}{n\,2^n}$

5. Test the following series for convergence using the comparison test.

(a) $\displaystyle\sum_{n=1}^{\infty} \frac{1}{\sqrt{n}}$ *Hint:* Which is larger, n or $\sqrt{n}$? (b) $\displaystyle\sum_{n=2}^{\infty} \frac{1}{\ln n}$

6. There are 9 one-digit numbers (1 to 9), 90 two-digit numbers (10 to 99). How many three-digit, four-digit, etc., numbers are there? The first 9 terms of the harmonic series $1+\frac{1}{2}+\frac{1}{3}+\cdots+\frac{1}{9}$ are all greater than $\frac{1}{10}$; similarly consider the next 90 terms, and so on. Thus prove the divergence of the harmonic series by comparison with the series

$$\left[\tfrac{1}{10}+\tfrac{1}{10}+\cdots(9 \text{ terms each} = \tfrac{1}{10})\right]+\left[\,90 \text{ terms each} = \tfrac{1}{100}\right]+\cdots$$
$$= \tfrac{9}{10}+\tfrac{90}{100}+\cdots = \tfrac{9}{10}+\tfrac{9}{10}+\cdots.$$

The comparison test is really the basic test from which other tests are derived. It is probably the most useful test of all for the experienced mathematician but it is often hard to think of a satisfactory m series until you have had a good deal of experience with series. Consequently, you will probably not use it as often as the next three tests.

B. The Integral Test

We can use this test when the terms of the series are positive and not increasing, that is, when $a_{n+1} \le a_n$. (Again remember that we can ignore any finite number of terms of the series; thus the test can still be used even if the condition $a_{n+1} \le a_n$ does not hold for a finite number of terms.) To apply the test we think of a_n as a function of the variable n, and, forgetting our previous meaning of n, we allow it to take all values, not just integral ones. The test states that:

> If $0 < a_{n+1} \leq a_n$ for $n > N$, then $\sum^{\infty} a_n$ converges if $\int^{\infty} a_n \, dn$ is finite and diverges if the integral is infinite. (The integral is to be evaluated *only* at the upper limit; no lower limit is needed.)

To understand this test, imagine a graph sketched of a_n as a function of n. For example, in testing the harmonic series $\sum_{n=1}^{\infty} 1/n$, we consider the graph of the function $y = 1/n$ (similar to Figures 6.1 and 6.2) letting n have all values, not just integral ones. Then the values of y on the graph at $n = 1, 2, 3, \cdots$, are the terms of the series. In Figures 6.1 and 6.2, the areas of the rectangles are just the terms of the series. Notice that in Figure 6.1 the top edge of each rectangle is above the curve, so that the area of the rectangles is greater than the corresponding area under the curve. On the other hand, in Figure 6.2 the rectangles lie below the curve, so their area is less than the corresponding area under the curve. Now the areas of the rectangles are just the terms of the series, and the area under the curve is an integral of $y \, dn$ or $a_n \, dn$. The upper limit on the integrals is ∞ and the lower limit could be made to correspond to any term of the series we wanted to start with. For example (see Figure 6.1), $\int_3^{\infty} a_n \, dn$ is less than the sum of the series from a_3 on, but (see Figure 6.2) greater than the sum of the series from a_4 on. If the integral is finite, then the sum of the series from a_4 on is finite, that is, the series converges. Note again that the terms at the beginning of a series have nothing to do with convergence. On the other hand, if the integral is infinite, then the sum of the series from a_3 on is infinite and the series diverges. Since the beginning terms are of no interest, you should simply evaluate $\int^{\infty} a_n \, dn$. (Also see Problem 16.)

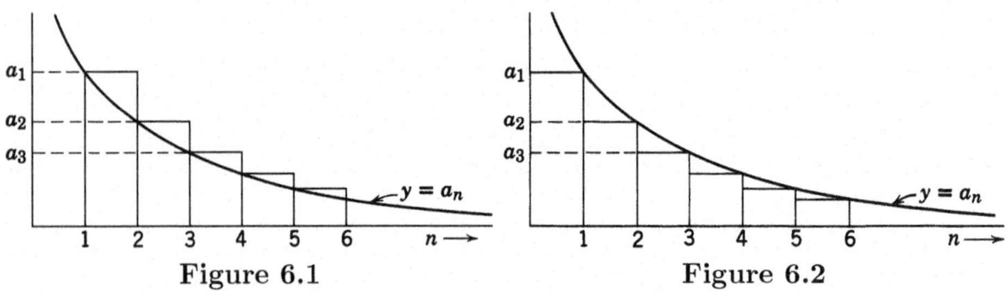

Figure 6.1 Figure 6.2

▶ **Example.** Test for convergence the harmonic series

$$1 + \frac{1}{2} + \frac{1}{3} + \frac{1}{4} + \cdots. \tag{6.1}$$

Using the integral test, we evaluate

$$\int^{\infty} \frac{1}{n} \, dn = \ln n \Big|^{\infty} = \infty.$$

(We use the symbol ln to mean a natural logarithm, that is, a logarithm to the base e.) Since the integral is infinite, the series diverges.

PROBLEMS, SECTION 6

Use the integral test to find whether the following series converge or diverge. *Hint* and *warning:* Do *not* use lower limits on your integrals (see Problem 16).

7. $\sum_{n=2}^{\infty} \frac{1}{n \ln n}$ **8.** $\sum_{n=1}^{\infty} \frac{n}{n^2+4}$ **9.** $\sum_{n=3}^{\infty} \frac{1}{n^2-4}$

10. $\sum_{n=1}^{\infty} \frac{e^n}{e^{2n}+9}$ **11.** $\sum_{1}^{\infty} \frac{1}{n(1+\ln n)^{3/2}}$ **12.** $\sum_{1}^{\infty} \frac{n}{(n^2+1)^2}$

13. $\sum_{1}^{\infty} \frac{n^2}{n^3+1}$ **14.** $\sum_{1}^{\infty} \frac{1}{\sqrt{n^2+9}}$

15. Use the integral test to prove the following so-called p-series test. The series

$$\sum_{n=1}^{\infty} \frac{1}{n^p} \text{ is } \begin{cases} \text{convergent} & \text{if } p > 1, \\ \text{divergent} & \text{if } p \le 1. \end{cases}$$

Caution: Do $p = 1$ separately.

16. In testing $\sum 1/n^2$ for convergence, a student evaluates $\int_0^{\infty} n^{-2} dn = -n^{-1}\big|_0^{\infty} = 0 + \infty = \infty$ and concludes (erroneously) that the series diverges. What is wrong? *Hint:* Consider the area under the curve in a diagram such as Figure 6.1 or 6.2. This example shows the danger of using a lower limit in the integral test.

17. Use the integral test to show that $\sum_{n=0}^{\infty} e^{-n^2}$ converges. *Hint:* Although you cannot *evaluate* the integral, you can show that it is finite (which is all that is necessary) by comparing it with $\int^{\infty} e^{-n} dn$.

C. The Ratio Test

The integral test depends on your being able to integrate $a_n dn$; this is not always easy! We consider another test which will handle many cases in which we cannot evaluate the integral. Recall that in the geometric series each term could be obtained by multiplying the one before it by the ratio r, that is, $a_{n+1} = ra_n$ or $a_{n+1}/a_n = r$. For other series the ratio a_{n+1}/a_n is not constant but depends on n; let us call the absolute value of this ratio ρ_n. Let us also find the limit (if there is one) of the sequence ρ_n as $n \to \infty$ and call this limit ρ. Thus we define ρ_n and ρ by the equations

$$\rho_n = \left|\frac{a_{n+1}}{a_n}\right|, \qquad \rho = \lim_{n\to\infty} \rho_n. \tag{6.2}$$

If you recall that a geometric series converges if $|r| < 1$, it may seem plausible that a series with $\rho < 1$ should converge and this is true. This statement can be proved (Problem 30) by comparing the series to be tested with a geometric series. Like a geometric series with $|r| > 1$, a series with $\rho > 1$ also diverges (Problem 30). However, if $\rho = 1$, the ratio test does not tell us anything; some series with $\rho = 1$ converge and some diverge, so we must find another test (say one of the two preceding tests). To summarize the ratio test:

$$\text{(6.3)} \qquad \text{If } \begin{cases} \rho < 1, & \text{the series converges;} \\ \rho = 1, & \text{use a different test;} \\ \rho > 1, & \text{the series diverges.} \end{cases}$$

Example 1. Test for convergence the series

$$1 + \frac{1}{2!} + \frac{1}{3!} + \cdots + \frac{1}{n!} + \cdots.$$

Using (6.2), we have

$$\begin{aligned} \rho_n &= \left| \frac{1}{(n+1)!} \div \frac{1}{n!} \right| \\ &= \frac{n!}{(n+1)!} = \frac{n(n-1)\cdots 3\cdot 2\cdot 1}{(n+1)(n)(n-1)\cdots 3\cdot 2\cdot 1} = \frac{1}{n+1}, \\ \rho &= \lim_{n\to\infty} \rho_n = \lim_{n\to\infty} \frac{1}{n+1} = 0. \end{aligned}$$

Since $\rho < 1$, the series converges.

Example 2. Test for convergence the harmonic series

$$1 + \frac{1}{2} + \frac{1}{3} + \cdots + \frac{1}{n} + \cdots.$$

We find

$$\begin{aligned} \rho_n &= \left| \frac{1}{n+1} \div \frac{1}{n} \right| = \frac{n}{n+1}, \\ \rho &= \lim_{n\to\infty} \frac{n}{n+1} = \lim_{n\to\infty} \frac{1}{1+\frac{1}{n}} = 1. \end{aligned}$$

Here the test tells us nothing and we must use some different test. A word of warning from this example: Notice that $\rho_n = n/(n+1)$ is always less than 1. Be careful not to confuse this ratio with ρ and conclude incorrectly that this series converges. (It is actually divergent as we proved by the integral test.) Remember that ρ is *not* the same as the ratio $\rho_n = |a_{n+1}/a_n|$, but is the *limit* of this ratio as $n \to \infty$.

PROBLEMS, SECTION 6

Use the ratio test to find whether the following series converge or diverge:

18. $\displaystyle\sum_{n=1}^{\infty} \frac{2^n}{n^2}$ **19.** $\displaystyle\sum_{n=0}^{\infty} \frac{3^n}{2^{2n}}$ **20.** $\displaystyle\sum_{n=0}^{\infty} \frac{n!}{(2n)!}$

21. $\displaystyle\sum_{n=0}^{\infty}\frac{5^n(n!)^2}{(2n)!}$ **22.** $\displaystyle\sum_{n=1}^{\infty}\frac{10^n}{(n!)^2}$ **23.** $\displaystyle\sum_{n=1}^{\infty}\frac{n!}{100^n}$

24. $\displaystyle\sum_{n=0}^{\infty}\frac{3^{2n}}{2^{3n}}$ **25.** $\displaystyle\sum_{n=0}^{\infty}\frac{e^n}{\sqrt{n!}}$ **26.** $\displaystyle\sum_{n=0}^{\infty}\frac{(n!)^3e^{3n}}{(3n)!}$

27. $\displaystyle\sum_{n=0}^{\infty}\frac{100^n}{n^{200}}$ **28.** $\displaystyle\sum_{n=0}^{\infty}\frac{n!(2n)!}{(3n)!}$ **29.** $\displaystyle\sum_{n=0}^{\infty}\frac{\sqrt{(2n)!}}{n!}$

30. Prove the ratio test. *Hint:* If $|a_{n+1}/a_n| \to \rho < 1$, take σ so that $\rho < \sigma < 1$. Then $|a_{n+1}/a_n| < \sigma$ if n is large, say $n \geq N$. This means that we have $|a_{N+1}| < \sigma|a_N|$, $|a_{N+2}| < \sigma|a_{N+1}| < \sigma^2|a_N|$, and so on. Compare with the geometric series

$$\sum_{n=1}^{\infty}\sigma^n|a_N|.$$

Also prove that a series with $\rho > 1$ diverges. *Hint:* Take $\rho > \sigma > 1$, and use the preliminary test.

D. A Special Comparison Test

This test has two parts: (a) a convergence test, and (b) a divergence test. (See Problem 37.)

> (a) If $\sum_{n=1}^{\infty} b_n$ is a convergent series of positive terms and $a_n \geq 0$ and a_n/b_n tends to a (finite) limit, then $\sum_{n=1}^{\infty} a_n$ converges.
> (b) If $\sum_{n=1}^{\infty} d_n$ is a divergent series of positive terms and $a_n \geq 0$ and a_n/d_n tends to a limit greater than 0 (or tends to $+\infty$), then $\sum_{n=1}^{\infty} a_n$ diverges.

There are really two steps in using either of these tests, namely, to decide on a comparison series, and then to compute the required limit. The first part is the most important; given a good comparison series it is a routine process to find the needed limit. The method of finding the comparison series is best shown by examples.

Example 1. Test for convergence

$$\sum_{n=3}^{\infty}\frac{\sqrt{2n^2-5n+1}}{4n^3-7n^2+2}.$$

Remember that whether a series converges or diverges depends on what the terms are as n becomes larger and larger. We are interested in the nth term as $n \to \infty$. Think of $n = 10^{10}$ or 10^{100}, say; a little calculation should convince you that as n increases, $2n^2 - 5n + 1$ is $2n^2$ to quite high accuracy. Similarly, the denominator in our example is nearly $4n^3$ for large n. By Section 9, fact 1, we see that the factor $\sqrt{2}/4$ in every term does not affect convergence. So we consider as a comparison series just

$$\sum_{n=3}^{\infty}\frac{\sqrt{n^2}}{n^3} = \sum_{n=3}^{\infty}\frac{1}{n^2}$$

which we recognize (say by integral test) as a convergent series. Hence we use test (a) to try to show that the given series converges. We have:

$$\lim_{n\to\infty}\frac{a_n}{b_n} = \lim_{n\to\infty}\left(\frac{\sqrt{2n^2-5n+1}}{4n^3-7n^2+2}\div\frac{1}{n^2}\right)$$

$$= \lim_{n\to\infty}\frac{n^2\sqrt{2n^2-5n+1}}{4n^3-7n^2+2}$$

$$= \lim_{n\to\infty}\frac{\sqrt{2-\frac{5}{n}+\frac{1}{n^2}}}{4-\frac{7}{n}+\frac{2}{n^3}} = \frac{\sqrt{2}}{4}.$$

Since this is a finite limit, the given series converges. (With practice, you won't need to do all this algebra! You should be able to look at the original problem and see that, for large n, the terms are essentially $1/n^2$, so the series converges.)

Example 2. Test for convergence

$$\sum_{n=2}^{\infty}\frac{3^n-n^3}{n^5-5n^2}.$$

Here we must first decide which is the important term as $n\to\infty$; is it 3^n or n^3? We can find out by comparing their logarithms since $\ln N$ and N increase or decrease together. We have $\ln 3^n = n\ln 3$, and $\ln n^3 = 3\ln n$. Now $\ln n$ is much smaller than n, so for large n we have $n\ln 3 > 3\ln n$, and $3^n > n^3$. (You might like to compute $100^3 = 10^6$, and $3^{100} > 5\times 10^{47}$.) The denominator of the given series is approximately n^5. Thus the comparison series is $\sum_{n=2}^{\infty} 3^n/n^5$. It is easy to prove this divergent by the ratio test. Now by test (b)

$$\lim_{n\to\infty}\left(\frac{3^n-n^3}{n^5-5n^2}\div\frac{3^n}{n^5}\right) = \lim_{n\to\infty}\frac{1-\frac{n^3}{3^n}}{1-\frac{5}{n^3}} = 1$$

which is greater than zero, so the given series diverges.

PROBLEMS, SECTION 6

Use the special comparison test to find whether the following series converge or diverge.

31. $\displaystyle\sum_{n=9}^{\infty}\frac{(2n+1)(3n-5)}{\sqrt{n^2-73}}$

32. $\displaystyle\sum_{n=0}^{\infty}\frac{n(n+1)}{(n+2)^2(n+3)}$

33. $\displaystyle\sum_{n=5}^{\infty}\frac{1}{2^n-n^2}$

34. $\displaystyle\sum_{n=1}^{\infty}\frac{n^2+3n+4}{n^4+7n^3+6n-3}$

35. $\displaystyle\sum_{n=3}^{\infty}\frac{(n-\ln n)^2}{5n^4-3n^2+1}$

36. $\displaystyle\sum_{n=1}^{\infty}\frac{\sqrt{n^3+5n-1}}{n^2-\sin n^3}$

37. Prove the special comparison test. *Hint* (part a): If $a_n/b_n\to L$ and $M>L$, then $a_n < Mb_n$ for large n. Compare $\sum_{n=1}^{\infty} a_n$ with $\sum_{n=1}^{\infty} Mb_n$.

7. ALTERNATING SERIES

So far we have been talking about series of positive terms (including series of absolute values). Now we want to consider one important case of a series whose terms have mixed signs. An *alternating series* is a series whose terms are alternately plus and minus; for example,

$$1-\frac{1}{2}+\frac{1}{3}-\frac{1}{4}+\frac{1}{5}-\cdots+\frac{(-1)^{n+1}}{n}+\cdots \tag{7.1}$$

is an alternating series. We ask two questions about an alternating series. Does it converge? Does it converge absolutely (that is, when we make all signs positive)? Let us consider the second question first. In this example the series of absolute values

$$1+\frac{1}{2}+\frac{1}{3}+\frac{1}{4}+\cdots+\frac{1}{n}+\cdots$$

is the harmonic series (6.1), which diverges. We say that the series (7.1) is not absolutely convergent. Next we must ask whether (7.1) converges as it stands. If it had turned out to be absolutely convergent, we would not have to ask this question since an absolutely convergent series is also convergent (Problem 9). However, a series which is not absolutely convergent may converge or it may diverge; we must test it further. For alternating series the test is very simple:

Test for alternating series. An alternating series converges if the absolute value of the terms decreases steadily to zero, that is, if $|a_{n+1}| \leq |a_n|$ and $\lim_{n\to\infty} a_n = 0$.

In our example $\dfrac{1}{n+1} < \dfrac{1}{n}$, and $\displaystyle\lim_{n\to\infty}\frac{1}{n} = 0$, so (7.1) converges.

PROBLEMS, SECTION 7

Test the following series for convergence.

1. $\displaystyle\sum_{n=1}^{\infty}\frac{(-1)^n}{\sqrt{n}}$

2. $\displaystyle\sum_{n=1}^{\infty}\frac{(-2)^n}{n^2}$

3. $\displaystyle\sum_{n=1}^{\infty}\frac{(-1)^n}{n^2}$

4. $\displaystyle\sum_{n=1}^{\infty}\frac{(-3)^n}{n!}$

5. $\displaystyle\sum_{n=2}^{\infty}\frac{(-1)^n}{\ln n}$

6. $\displaystyle\sum_{n=1}^{\infty}\frac{(-1)^n n}{n+5}$

7. $\displaystyle\sum_{n=0}^{\infty}\frac{(-1)^n n}{1+n^2}$

8. $\displaystyle\sum_{n=1}^{\infty}\frac{(-1)^n\sqrt{10n}}{n+2}$

9. Prove that an absolutely convergent series $\sum_{n=1}^{\infty} a_n$ is convergent. *Hint:* Put $b_n = a_n + |a_n|$. Then the b_n are nonnegative; we have $|b_n| \leq 2|a_n|$ and $a_n = b_n - |a_n|$.

10. The following alternating series are divergent (but you are not asked to prove this). Show that $a_n \to 0$. Why doesn't the alternating series test prove (incorrectly) that these series converge?

(a) $2-\frac{1}{2}+\frac{2}{3}-\frac{1}{4}+\frac{2}{5}-\frac{1}{6}+\frac{2}{7}-\frac{1}{8}\cdots$

(b) $\frac{1}{\sqrt{2}}-\frac{1}{2}+\frac{1}{\sqrt{3}}-\frac{1}{3}+\frac{1}{\sqrt{4}}-\frac{1}{4}+\frac{1}{\sqrt{5}}-\frac{1}{5}\cdots$

8. CONDITIONALLY CONVERGENT SERIES

A series like (7.1) which converges, but does not converge absolutely, is called *conditionally convergent*. You have to use special care in handling conditionally convergent series because the positive terms alone form a divergent series and so do the negative terms alone. If you rearrange the terms, you will probably change the sum of the series, and you may even make it diverge! It is possible to rearrange the terms to make the sum any number you wish. Let us do this with the alternating harmonic series $1 - \frac{1}{2} + \frac{1}{3} - \frac{1}{4} + \cdots$. Suppose we want to make the sum equal to 1.5. First we take enough positive terms to add to just over 1.5. The first three positive terms do this:

$$1 + \frac{1}{3} + \frac{1}{5} = 1\frac{8}{15} > 1.5.$$

Then we take enough negative terms to bring the partial sum back under 1.5; the one term $-\frac{1}{2}$ does this. Again we add positive terms until we have a little more than 1.5, and so on. Since the terms of the series are decreasing in absolute value, we are able (as we continue this process) to get partial sums just a little more or a little less than 1.5 but always nearer and nearer to 1.5. But this is what convergence of the series to the sum 1.5 means: that the partial sums should approach 1.5. You should see that we could pick in advance *any* sum that we want, and rearrange the terms of this series to get it. Thus, we must not rearrange the terms of a conditionally convergent series since its convergence and its sum depend on the fact that the terms are added in a particular order.

Here is a physical example of such a series which emphasizes the care needed in applying mathematical approximations in physical problems. Coulomb's law in electricity says that the force between two charges is equal to the product of the charges divided by the square of the distance between them (in electrostatic units; to use other units, say SI, we need only multiply by a numerical constant). Suppose there are unit positive charges at $x = 0, \sqrt{2}, \sqrt{4}, \sqrt{6}, \sqrt{8}, \cdots$, and unit negative charges at $x = 1, \sqrt{3}, \sqrt{5}, \sqrt{7}, \cdots$. We want to know the total force acting on the unit positive charge at $x = 0$ due to all the other charges. The negative charges attract the charge at $x = 0$ and try to pull it to the right; we call the forces exerted by them positive, since they are in the direction of the positive x axis. The forces due to the positive charges are in the negative x direction, and we call them negative. For example, the force due to the positive charge at $x = \sqrt{2}$ is $-(1 \cdot 1)/(\sqrt{2})^2 = -1/2$. The total force on the charge at $x = 0$ is, then,

$$F = 1 - \frac{1}{2} + \frac{1}{3} - \frac{1}{4} + \frac{1}{5} - \frac{1}{6} + \cdots. \tag{8.1}$$

Now we know that this series converges as it stands (Section 7). But we have also seen that its sum (even the fact that it converges) can be changed by rearranging the terms. Physically this means that the force on the charge at the origin depends not only on the size and position of the charges, but also on the *order* in which we place them in their positions! This may very well go strongly against your physical intuition. You feel that a physical problem like this should have a definite answer. Think of it this way. Suppose there are two crews of workers, one crew placing the positive charges and one placing the negative. If one crew works faster than the other, it is clear that the force at any stage may be far from the F of equation (8.1) because there are many extra charges of one sign. The crews can never place *all* the charges because there are an infinite number of them. At any stage the forces which would arise from the positive charges that are not yet in place, form a divergent series; similarly, the forces due to the unplaced negative charges form a divergent series of the opposite sign. We cannot then stop at some point and say that the rest of the series is negligible as we could in the bouncing ball problem in Section 1. But if we specify the *order* in which the charges are to be placed, then the sum S of the series is determined (S is probably different from F in (8.1) unless the charges are placed alternately). Physically this means that the value of the force as the crews proceed comes closer and closer to S, and we can use the sum of the (properly arranged) *infinite* series as a good approximation to the force.

9. USEFUL FACTS ABOUT SERIES

We state the following facts for reference:

1. The convergence or divergence of a series is not affected by multiplying every term of the series by the same nonzero constant. Neither is it affected by changing a finite number of terms (for example, omitting the first few terms).

2. Two convergent series $\sum_{n=1}^{\infty} a_n$ and $\sum_{n=1}^{\infty} b_n$ may be added (or subtracted) term by term. (Adding "term by term" means that the nth term of the sum is $a_n + b_n$.) The resulting series is convergent, and its sum is obtained by adding (subtracting) the sums of the two given series.

3. The terms of an *absolutely convergent series* may be rearranged in any order without affecting either the convergence or the sum. This is *not true* of conditionally convergent series as we have seen in Section 8.

PROBLEMS, SECTION 9

Test the following series for convergence or divergence. Decide for yourself which test is easiest to use, but don't forget the preliminary test. Use the facts stated above when they apply.

1. $\displaystyle\sum_{n=1}^{\infty} \frac{n-1}{(n+2)(n+3)}$

2. $\displaystyle\sum_{n=1}^{\infty} \frac{n^2-1}{n^2+1}$

3. $\displaystyle\sum_{n=1}^{\infty} \frac{1}{n^{\ln 3}}$

4. $\displaystyle\sum_{n=0}^{\infty} \frac{n^2}{n^3+4}$

5. $\displaystyle\sum_{n=1}^{\infty} \frac{n}{n^3-4}$

6. $\displaystyle\sum_{n=0}^{\infty} \frac{(n!)^2}{(2n)!}$

7. $\displaystyle\sum_{n=0}^{\infty} \frac{(2n)!}{3^n (n!)^2}$

8. $\displaystyle\sum_{n=1}^{\infty} \frac{n^5}{5^n}$

9. $\displaystyle\sum_{n=1}^{\infty} \frac{n^n}{n!}$

10. $\displaystyle\sum_{n=2}^{\infty} (-1)^n \frac{n}{n-1}$

11. $\displaystyle\sum_{n=4}^{\infty} \frac{2n}{n^2-9}$

12. $\displaystyle\sum_{n=2}^{\infty} \frac{1}{n^2-n}$

13. $\displaystyle\sum_{n=0}^{\infty} \frac{n}{(n^2+4)^{3/2}}$

14. $\displaystyle\sum_{n=2}^{\infty} \frac{(-1)^n}{n^2-n}$

15. $\displaystyle\sum_{n=1}^{\infty} \frac{(-1)^n n!}{10^n}$

16. $\displaystyle\sum_{n=0}^{\infty} \frac{2+(-1)^n}{n^2+7}$

17. $\displaystyle\sum_{n=1}^{\infty} \frac{(n!)^3}{(3n)!}$

18. $\displaystyle\sum_{n=1}^{\infty} \frac{(-1)^n}{2^{\ln n}}$

19. $\dfrac{1}{2^2} - \dfrac{1}{3^2} + \dfrac{1}{2^3} - \dfrac{1}{3^3} + \dfrac{1}{2^4} - \dfrac{1}{3^4} + \cdots$

20. $\dfrac{1}{2} + \dfrac{1}{2^2} - \dfrac{1}{3} - \dfrac{1}{3^2} + \dfrac{1}{4} + \dfrac{1}{4^2} - \dfrac{1}{5} - \dfrac{1}{5^2} + \cdots$

21. $\displaystyle\sum_{n=1}^{\infty} a_n$ if $a_{n+1} = \dfrac{n}{2n+3} a_n$

22. (a) $\displaystyle\sum_{n=1}^{\infty} \frac{1}{3^{\ln n}}$ (b) $\displaystyle\sum_{n=1}^{\infty} \frac{1}{2^{\ln n}}$

(c) For what values of k is $\displaystyle\sum_{n=1}^{\infty} \frac{1}{k^{\ln n}}$ convergent?

10. POWER SERIES; INTERVAL OF CONVERGENCE

We have been discussing series whose terms were constants. Even more important and useful are series whose terms are functions of x. There are many such series, but in this chapter we shall consider series in which the nth term is a constant times x^n or a constant times $(x-a)^n$ where a is a constant. These are called *power series*, because the terms are multiples of powers of x or of $(x-a)$. In later chapters we shall consider Fourier series whose terms involve sines and cosines, and other series (Legendre, Bessel, etc.) in which the terms may be polynomials or other functions.

By definition, a power series is of the form

$$\sum_{n=0}^{\infty} a_n x^n = a_0 + a_1 x + a_2 x^2 + a_3 x^3 + \cdots \quad \text{or} \tag{10.1}$$
$$\sum_{n=0}^{\infty} a_n (x-a)^n = a_0 + a_1(x-a) + a_2(x-a)^2 + a_3(x-a)^3 + \cdots,$$

where the coefficients a_n are constants. Here are some examples:

$$1 - \frac{x}{2} + \frac{x^2}{4} - \frac{x^3}{8} + \cdots + \frac{(-x)^n}{2^n} + \cdots, \tag{10.2a}$$

$$x - \frac{x^2}{2} + \frac{x^3}{3} - \frac{x^4}{4} + \cdots + \frac{(-1)^{n+1}x^n}{n} + \cdots, \tag{10.2b}$$

$$x - \frac{x^3}{3!} + \frac{x^5}{5!} - \frac{x^7}{7!} + \cdots + \frac{(-1)^{n+1}x^{2n-1}}{(2n-1)!} + \cdots, \tag{10.2c}$$

$$1 + \frac{(x+2)}{\sqrt{2}} + \frac{(x+2)^2}{\sqrt{3}} + \cdots + \frac{(x+2)^n}{\sqrt{n+1}} + \cdots. \tag{10.2d}$$

Whether a power series converges or not depends on the value of x we are considering. We often use the ratio test to find the values of x for which a series converges. We illustrate this by testing each of the four series (10.2). Recall that in the ratio test we divide term $n+1$ by term n and take the absolute value of this ratio to get ρ_n, and then take the limit of ρ_n as $n \to \infty$ to get ρ.

Example 1. For (10.2a), we have

$$\rho_n = \left| \frac{(-x)^{n+1}}{2^{n+1}} \div \frac{(-x)^n}{2^n} \right| = \left| \frac{x}{2} \right|,$$
$$\rho = \left| \frac{x}{2} \right|.$$

The series converges for $\rho < 1$, that is, for $|x/2| < 1$ or $|x| < 2$, and it diverges for $|x| > 2$ (see Problem 6.30). Graphically we consider the interval on the x axis between $x = -2$ and $x = 2$; for any x in this interval the series (10.2a) converges. The endpoints of the interval, $x = 2$ and $x = -2$, must be considered separately. When $x = 2$, (10.2a) is

$$1 - 1 + 1 - 1 + \cdots,$$

which is divergent; when $x = -2$, (10.2a) is $1 + 1 + 1 + 1 + \cdots$, which is divergent. Then the interval of convergence of (10.2a) is stated as $-2 < x < 2$.

Example 2. For (10.2b) we find

$$\rho_n = \left|\frac{x^{n+1}}{n+1} \div \frac{x^n}{n}\right| = \left|\frac{nx}{n+1}\right|,$$

$$\rho = \lim_{n\to\infty}\left|\frac{nx}{n+1}\right| = |x|.$$

The series converges for $|x| < 1$. Again we must consider the endpoints of the interval of convergence, $x = 1$ and $x = -1$. For $x = 1$, the series (10.2b) is $1 - \frac{1}{2} + \frac{1}{3} - \frac{1}{4} + \cdots$; this is the alternating harmonic series and is convergent. For $x = -1$, (10.2b) is $-1 - \frac{1}{2} - \frac{1}{3} - \frac{1}{4} - \cdots$; this is the harmonic series (times -1) and is divergent. Then we state the interval of convergence of (10.2b) as $-1 < x \le 1$. Notice carefully how this differs from our result for (10.2a). Series (10.2a) did not converge at either endpoint and we used only $<$ signs in stating its interval of convergence. Series (10.2b) converges at $x = 1$, so we use the sign $\le$ to include $x = 1$. You must always test a series at its endpoints and include the results in your statement of the interval of convergence. A series may converge at neither, either one, or both of the endpoints.

Example 3. In (10.2c), the absolute value of the nth term is $|x^{2n-1}/(2n-1)!|$. To get term $n+1$ we replace n by $n+1$; then $2n-1$ is replaced by $2(n+1)-1 = 2n+1$, and the absolute value of term $n+1$ is

$$\left|\frac{x^{2n+1}}{(2n+1)!}\right|.$$

Thus we get

$$\rho_n = \left|\frac{x^{2n+1}}{(2n+1)!} \div \frac{x^{2n-1}}{(2n-1)!}\right| = \left|\frac{x^2}{(2n+1)(2n)}\right|,$$

$$\rho = \lim_{n\to\infty}\left|\frac{x^2}{(2n+1)(2n)}\right| = 0.$$

Since $\rho < 1$ for all values of x, this series converges for all x.

Example 4. In (10.2d), we find

$$\rho_n = \left|\frac{(x+2)^{n+1}}{\sqrt{n+2}} \div \frac{(x+2)^n}{\sqrt{n+1}}\right|,$$

$$\rho = \lim_{n\to\infty}\left|(x+2)\frac{\sqrt{n+1}}{\sqrt{n+2}}\right| = |x+2|.$$

The series converges for $|x+2| < 1$; that is, for $-1 < x+2 < 1$, or $-3 < x < -1$. If $x = -3$, (10.2d) is

$$1 - \frac{1}{\sqrt{2}} + \frac{1}{\sqrt{3}} - \frac{1}{\sqrt{4}} + \cdots$$

which is convergent by the alternating series test. For $x = -1$, the series is

$$1 + \frac{1}{\sqrt{2}} + \frac{1}{\sqrt{3}} + \cdots = \sum_{n=0}^{\infty} \frac{1}{\sqrt{n+1}}$$

which is divergent by the integral test. Thus, the series converges for $-3 \le x < 1$.

PROBLEMS, SECTION 10

Find the interval of convergence of each of the following power series; be sure to investigate the endpoints of the interval in each case.

1. $\sum_{n=0}^{\infty}(-1)^n x^n$

2. $\sum_{n=0}^{\infty}\frac{(2x)^n}{3^n}$

3. $\sum_{n=1}^{\infty}\frac{(-1)^n x^n}{n(n+1)}$

4. $\sum_{n=1}^{\infty}\frac{x^{2n}}{2^n n^2}$

5. $\sum_{n=1}^{\infty}\frac{x^n}{(n!)^2}$

6. $\sum_{n=1}^{\infty}\frac{(-1)^n x^n}{(2n)!}$

7. $\sum_{n=1}^{\infty}\frac{x^{3n}}{n}$

8. $\sum_{n=1}^{\infty}\frac{(-1)^n x^n}{\sqrt{n}}$

9. $\sum_{n=1}^{\infty}(-1)^n n^3 x^n$

10. $\sum_{n=1}^{\infty}\frac{(-1)^n x^{2n}}{(2n)^{3/2}}$

11. $\sum_{n=1}^{\infty}\frac{1}{n}\left(\frac{x}{5}\right)^n$

12. $\sum_{n=1}^{\infty} n(-2x)^n$

13. $\sum_{n=1}^{\infty}\frac{n(-x)^n}{n^2+1}$

14. $\sum_{n=1}^{\infty}\frac{n}{n+1}\left(\frac{x}{3}\right)^n$

15. $\sum_{n=1}^{\infty}\frac{(x-2)^n}{3^n}$

16. $\sum_{n=1}^{\infty}\frac{(x-1)^n}{2^n}$

17. $\sum_{n=1}^{\infty}\frac{(-1)^n(x+1)^n}{n}$

18. $\sum_{n=1}^{\infty}\frac{(-2)^n(2x+1)^n}{n^2}$

The following series are *not* power series, but you can transform each one into a power series by a change of variable and so find out where it converges.

19. $\sum_0^\infty 8^{-n}(x^2-1)^n$ *Method:* Let $y = x^2 - 1$. The power series $\sum_0^\infty 8^{-n}y^n$ converges for $|y| < 8$, so the original series converges for $|x^2 - 1| < 8$, which means $|x| < 3$.

20. $\sum_{0}^{\infty}(-1)^n\frac{2^n}{n!}(x^2+1)^{2n}$

21. $\sum_{2}^{\infty}\frac{(-1)^n x^{n/2}}{n\ln n}$

22. $\sum_{0}^{\infty}\frac{n!(-1)^n}{x^n}$

23. $\sum_{0}^{\infty}\frac{3^n(n+1)}{(x+1)^n}$

24. $\sum_{0}^{\infty}\left(\sqrt{x^2+1}\right)^n\frac{2^n}{3^n+n^3}$

25. $\sum_{0}^{\infty}(\sin x)^n(-1)^n 2^n$

11. THEOREMS ABOUT POWER SERIES

We have seen that a power series $\sum_{n=0}^{\infty} a_n x^n$ converges in some interval with center at the origin. For each value of x (in the interval of convergence) the series has a finite sum whose value depends, of course, on the value of x. Thus we can write the sum of the series as $S(x) = \sum_{n=0}^{\infty} a_n x^n$. We see then that a power series (within its interval of convergence) defines a function of x, namely $S(x)$. In describing the relation of the series and the function $S(x)$, we may say that the series converges to the function $S(x)$, or that the function $S(x)$ is represented by the series, or that the series is the power series of the function. Here we have thought of obtaining the function from a given series. We shall also (Section 12) be interested in finding a power series that converges to a given function. When we are working with power series and the functions they represent, it is useful to know the following theorems (which we state without proof; see advanced calculus texts). Power series are very useful and convenient because within their interval of convergence they can be handled much like polynomials.

1. A power series may be differentiated or integrated term by term; the resulting series converges to the derivative or integral of the function represented by the original series within the same interval of convergence as the original series (that is, not necessarily at the endpoints of the interval).

2. Two power series may be added, subtracted, or multiplied; the resultant series converges at least in the common interval of convergence. You may divide two series if the denominator series is not zero at $x = 0$, or if it is and the zero is canceled by the numerator [as, for example, in $(\sin x)/x$; see (13.1)]. The resulting series will have *some* interval of convergence (which can be found by the ratio test or more simply by complex variable theory—see Chapter 2, Section 7).

3. One series may be substituted in another provided that the values of the substituted series are in the interval of convergence of the other series.

4. The power series of a function is unique, that is, there is just one power series of the form $\sum_{n=0}^{\infty} a_n x^n$ which converges to a given function.

12. EXPANDING FUNCTIONS IN POWER SERIES

Very often in applied work, it is useful to find power series that represent given functions. We illustrate one method of obtaining such series by finding the series for $\sin x$. In this method we *assume* that there *is* such a series (see Section 14 for discussion of this point) and set out to find what the coefficients in the series must be. Thus we write

$$\sin x = a_0 + a_1 x + a_2 x^2 + \cdots + a_n x^n + \cdots \tag{12.1}$$

and try to find numerical values of the coefficients a_n to make (12.1) an identity (within the interval of convergence of the series). Since the interval of convergence of a power series contains the origin, (12.1) must hold when $x = 0$. If we substitute $x = 0$ into (12.1), we get $0 = a_0$ since $\sin 0 = 0$ and all the terms except a_0 on the right-hand side of the equation contain the factor x. Then to make (12.1) valid at $x = 0$, we must have $a_0 = 0$. Next we differentiate (12.1) term by term to get

$$\cos x = a_1 + 2a_2 x + 3a_3 x^2 + \cdots . \tag{12.2}$$

(This is justified by Theorem 1 of Section 11.) Again putting $x = 0$, we get $1 = a_1$. We differentiate again, and put $x = 0$ to get

$$\begin{aligned} -\sin x &= 2a_2 + 3\cdot 2a_3 x + 4\cdot 3a_4 x^2 + \cdots , \\ 0 &= 2a_2. \end{aligned} \tag{12.3}$$

Continuing the process of taking successive derivatives of (12.1) and putting $x = 0$, we get

$$\begin{aligned} -\cos x &= 3\cdot 2a_3 + 4\cdot 3\cdot 2a_4 x + \cdots , \\ -1 &= 3!\, a_3, \qquad a_3 = -\frac{1}{3!}; \\ \sin x &= 4\cdot 3\cdot 2\cdot a_4 + 5\cdot 4\cdot 3\cdot 2a_5 x + \cdots , \\ 0 &= a_4; \\ \cos x &= 5\cdot 4\cdot 3\cdot 2a_5 + \cdots , \\ 1 &= 5!\, a_5, \cdots . \end{aligned} \tag{12.4}$$

We substitute these values back into (12.1) and get

$$\sin x = x - \frac{x^3}{3!} + \frac{x^5}{5!} - \cdots . \tag{12.5}$$

You can probably see how to write more terms of this series without further computation. The $\sin x$ series converges for all x; see Example 3, Section 10.

Series obtained in this way are called *Maclaurin series* or *Taylor series about the origin.* A Taylor series in general means a series of powers of $(x - a)$, where a is some constant. It is found by writing $(x - a)$ instead of x on the right-hand side of an equation like (12.1), differentiating just as we have done, but substituting $x = a$ instead of $x = 0$ at each step. Let us carry out this process in general for a function $f(x)$. As above, we assume that there is a Taylor series for $f(x)$, and write

$$\begin{aligned} f(x) &= a_0 + a_1(x-a) + a_2(x-a)^2 + a_3(x-a)^3 + a_4(x-a)^4 + \cdots \\ &\quad + a_n(x-a)^n + \cdots , \\ f'(x) &= a_1 + 2a_2(x-a) + 3a_3(x-a)^2 + 4a_4(x-a)^3 + \cdots \\ &\quad + na_n(x-a)^{n-1} + \cdots , \\ f''(x) &= 2a_2 + 3\cdot 2a_3(x-a) + 4\cdot 3a_4(x-a)^2 + \cdots \\ &\quad + n(n-1)a_n(x-a)^{n-2} + \cdots , \\ f'''(x) &= 3!\, a_3 + 4\cdot 3\cdot 2a_4(x-a) + \cdots \\ &\quad + n(n-1)(n-2)a_n(x-a)^{n-3} + \cdots , \\ &\ \vdots \\ f^{(n)}(x) &= n(n-1)(n-2)\cdots 1\cdot a_n + \text{terms containing powers of } (x-a). \end{aligned} \tag{12.6}$$

[The symbol $f^{(n)}(x)$ means the nth derivative of $f(x)$.] We now put $x = a$ in each equation of (12.6) and obtain

$$(12.7)\qquad \begin{gathered} f(a) = a_0, \quad f'(a) = a_1, \quad f''(a) = 2a_2, \\ f'''(a) = 3!\, a_3, \quad \cdots, \quad f^{(n)}(a) = n!\, a_n. \end{gathered}$$

[Remember that $f'(a)$ means to differentiate $f(x)$ and then put $x = a$; $f''(a)$ means to find $f''(x)$ and then put $x = a$, and so on.]

We can then write the Taylor series for $f(x)$ about $x = a$:

$$(12.8)\quad f(x) = f(a) + (x-a)f'(a) + \frac{1}{2!}(x-a)^2 f''(a) + \cdots + \frac{1}{n!}(x-a)^n f^{(n)}(a) + \cdots.$$

The Maclaurin series for $f(x)$ is the Taylor series about the origin. Putting $a = 0$ in (12.8), we obtain the Maclaurin series for $f(x)$:

$$(12.9)\qquad f(x) = f(0) + xf'(0) + \frac{x^2}{2!}f''(0) + \frac{x^3}{3!}f'''(0) + \cdots + \frac{x^n}{n!}f^{(n)}(0) + \cdots.$$

We have written this in general because it is sometimes convenient to have the formulas for the coefficients. However, finding the higher order derivatives in (12.9) for any but the simplest functions is unnecessarily complicated (try it for, say, $e^{\tan x}$). In Section 13, we shall discuss much easier ways of getting Maclaurin and Taylor series by combining a few basic series. Meanwhile, you should verify (Problem 1, below) the basic series (13.1) to (13.5) and memorize them.

PROBLEMS, SECTION 12

1. By the method used to obtain (12.5) [which is the series (13.1) below], verify each of the other series (13.2) to (13.5) below.

13. TECHNIQUES FOR OBTAINING POWER SERIES EXPANSIONS

There are often simpler ways for finding the power series of a function than the successive differentiation process in Section 12. Theorem 4 in Section 11 tells us that for a given function there is *just one* power series, that is, just one series of the form $\sum_{n=0}^{\infty} a_n x^n$. Therefore we can obtain it by any correct method and be sure that it is the same Maclaurin series we would get by using the method of Section 12. We shall illustrate a variety of methods for obtaining power series. First of all, it is a great timesaver for you to verify (Problem 12.1) and then memorize the basic series (13.1) to (13.5). We shall use these series without further derivation when we need them.

convergent for

(13.1) $$\sin x = \sum_{n=0}^{\infty} \frac{(-1)^n x^{2n+1}}{(2n+1)!} = x - \frac{x^3}{3!} + \frac{x^5}{5!} - \frac{x^7}{7!} + \cdots, \qquad \text{all } x;$$

(13.2) $$\cos x = \sum_{n=0}^{\infty} \frac{(-1)^n x^{2n}}{(2n)!} = 1 - \frac{x^2}{2!} + \frac{x^4}{4!} - \frac{x^6}{6!} + \cdots, \qquad \text{all } x;$$

(13.3) $$e^x = \sum_{n=0}^{\infty} \frac{x^n}{n!} = 1 + x + \frac{x^2}{2!} + \frac{x^3}{3!} + \frac{x^4}{4!} + \cdots, \qquad \text{all } x;$$

(13.4) $$\ln(1+x) = \sum_{n=1}^{\infty} \frac{(-1)^{n+1} x^n}{n} = x - \frac{x^2}{2} + \frac{x^3}{3} - \frac{x^4}{4} + \cdots, \qquad -1 < x \le 1;$$

(13.5) $$(1+x)^p = \sum_{n=0}^{\infty} \binom{p}{n} x^n = 1 + px + \frac{p(p-1)}{2!}x^2 + \frac{p(p-1)(p-2)}{3!}x^3 + \cdots, \qquad |x| < 1,$$

(binomial series; p is any real number, positive or negative and $\binom{p}{n}$ is called a binomial coefficient—see method C below.)

When we use a series to approximate a function, we may want only the first few terms, but in derivations, we may want the formula for the general term so that we can write the series in summation form. Let's look at some methods of obtaining either or both of these results.

A. Multiplying a Series by a Polynomial or by Another Series

▶ **Example 1.** To find the series for $(x+1)\sin x$, we multiply $(x+1)$ times the series (13.1) and collect terms:

$$\begin{aligned}(x+1)\sin x &= (x+1)\left(x - \frac{x^3}{3!} + \frac{x^5}{5!} - \cdots\right)\\ &= x + x^2 - \frac{x^3}{3!} - \frac{x^4}{3!} + \cdots.\end{aligned}$$

You can see that this is easier to do than taking the successive derivatives of the product $(x+1)\sin x$, and Theorem 4 assures us that the results are the same.

Example 2. To find the series for $e^x \cos x$, we multiply (13.2) by (13.3):

$$\begin{aligned}
e^x \cos x &= \left(1 + x + \frac{x^2}{2!} + \frac{x^3}{3!} + \frac{x^4}{4!} + \cdots\right)\left(1 - \frac{x^2}{2!} + \frac{x^4}{4!} - \cdots\right) \\
&= 1 \quad + x + \frac{x^2}{2!} + \frac{x^3}{3!} + \frac{x^4}{4!} \cdots \\
&\qquad\qquad - \frac{x^2}{2!} - \frac{x^3}{2!} - \frac{x^4}{2!\,2!} \cdots \\
&\qquad\qquad\qquad\qquad + \frac{x^4}{4!} \cdots \\
\hline
&= 1 \quad + x + 0x^2 - \frac{x^3}{3} - \frac{x^4}{6} \cdots = 1 + x - \frac{x^3}{3} - \frac{x^4}{6} \cdots .
\end{aligned}$$

There are two points to note here. First, as you multiply, line up the terms involving each power of x in a column; this makes it easier to combine them. Second, be careful to include *all* the terms in the product out to the power you intend to stop with, but don't include *any* higher powers. In the above example, note that we did not include the $x^3 \cdot x^2$ terms; if we wanted the x^5 term in the answer, we would have to include *all* products giving x^5 (namely, $x \cdot x^4, x^3 \cdot x^2$, and $x^5 \cdot 1$).

Also see Chapter 2, Problem 17.30, for a simple way of getting the general term of this series.

B. Division of Two Series or of a Series by a Polynomial

Example 1. To find the series for $(1/x)\ln(1 + x)$, we divide (13.4) by x. You should be able to do this in your head and just write down the answer.

$$\frac{1}{x}\ln(1 + x) = 1 - \frac{x}{2} + \frac{x^2}{3} - \frac{x^3}{4} + \cdots .$$

To obtain the summation form, we again just divide (13.4) by x. We can simplify the result by changing the limits to start at $n = 0$, that is, replace n by $n + 1$.

$$\frac{1}{x}\ln(1 + x) = \sum_{n=1}^{\infty} \frac{(-1)^{n+1}x^{n-1}}{n} = \sum_{n=0}^{\infty} \frac{(-1)^n x^n}{n + 1}.$$

▶ **Example 2.** To find the series for $\tan x$, we divide the series for $\sin x$ by the series for $\cos x$ by long division:

$$
\begin{array}{r|l}
 & x+\frac{x^3}{3}+\frac{2}{15}x^5\cdots \\
\hline
1-\frac{x^2}{2!}+\frac{x^4}{4!}\cdots & x-\frac{x^3}{3!}+\frac{x^5}{5!}\cdots \\
 & x-\frac{x^3}{2!}+\frac{x^5}{4!}\cdots \\
\hline
 & \frac{x^3}{3}-\frac{x^5}{30}\cdots \\
 & \frac{x^3}{3}-\frac{x^5}{6}\cdots \\
\hline
 & \frac{2x^5}{15}\cdots, \text{ etc.}
\end{array}
$$

C. Binomial Series

If you recall the binomial theorem, you may see that (13.5) looks just like the beginning of the binomial theorem for the expansion of $(a+b)^n$ if we put $a=1$, $b=x$, and $n=p$. The difference here is that we allow p to be negative or fractional, and in these cases the expansion is an infinite series. The series converges for $|x|<1$ as you can verify by the ratio test. (See Problem 1.)

From (13.5), we see that the binomial coefficients are:

$$
\begin{aligned}
\binom{p}{0} &= 1, \\
\binom{p}{1} &= p, \\
\binom{p}{2} &= \frac{p(p-1)}{2!}, \\
\binom{p}{3} &= \frac{p(p-1)(p-2)}{3!}, \cdots, \\
\binom{p}{n} &= \frac{p(p-1)(p-2)\cdots(p-n+1)}{n!}.
\end{aligned}
\tag{13.6}
$$

▶ **Example 1.** To find the series for $1/(1+x)$, we use the binomial series (13.5) to write

$$
\begin{aligned}
\frac{1}{1+x} = (1+x)^{-1} &= 1 - x + \frac{(-1)(-2)}{2!}x^2 + \frac{(-1)(-2)(-3)}{3!}x^3 + \cdots \\
&= 1 - x + x^2 - x^3 + \cdots = \sum_{n=0}^{\infty}(-x)^n.
\end{aligned}
$$

Example 2. The series for $\sqrt{1+x}$ is (13.5) with $p = 1/2$.

$$\begin{aligned}\sqrt{1+x} = (1+x)^{1/2} &= \sum_{n=0}^{\infty} \binom{1/2}{n} x^n \\ &= 1 + \frac{1}{2}x + \frac{\frac{1}{2}(-\frac{1}{2})}{2!}x^2 + \frac{\frac{1}{2}(-\frac{1}{2})(-\frac{3}{2})}{3!}x^3 + \frac{\frac{1}{2}(-\frac{1}{2})(-\frac{3}{2})(-\frac{5}{2})}{4!}x^4 + \cdots \\ &= 1 + \frac{1}{2}x - \frac{1}{8}x^2 + \frac{1}{16}x^3 - \frac{5}{128}x^4 \cdots .\end{aligned}$$

From (13.6) we can see that the binomial coefficients when $n = 0$ and $n = 1$ are $\binom{1/2}{0} = 1$ and $\binom{1/2}{1} = 1/2$. For $n \geq 2$, we can write

$$\begin{aligned}\binom{\frac{1}{2}}{n} &= \frac{(\frac{1}{2})(-\frac{1}{2})(-\frac{3}{2})\cdots(\frac{1}{2}-n+1)}{n!} = \frac{(-1)^{n-1}3\cdot5\cdot7\cdots(2n-3)}{n!\,2^n} \\ &= \frac{(-1)^{n-1}(2n-3)!!}{(2n)!!}\end{aligned}$$

where the double factorial of an odd number means the product of that number times all smaller odd numbers, and a similar definition for even numbers. For example, $7!! = 7\cdot5\cdot3$, and $8!! = 8\cdot6\cdot4\cdot2$.

PROBLEMS, SECTION 13

1. Use the ratio test to show that a binomial series converges for $|x| < 1$.

2. Show that the binomial coefficients $\binom{-1}{n} = (-1)^n$.

3. Show that if p is a positive integer, then $\binom{p}{n} = 0$ when $n > p$, so $(1+x)^p = \sum \binom{p}{n} x^n$ is just a sum of $p+1$ terms, from $n = 0$ to $n = p$. For example, $(1+x)^2$ has 3 terms, $(1+x)^3$ has 4 terms, etc. This is just the familiar binomial theorem.

4. Write the Maclaurin series for $1/\sqrt{1+x}$ in $\sum$ form using the binomial coefficient notation. Then find a formula for the binomial coefficients in terms of n as we did in Example 2 above.

D. Substitution of a Polynomial or a Series for the Variable in Another Series

Example 1. Find the series for e^{-x^2}. Since we know the series (13.3) for e^x, we simply replace the x there by $-x^2$ to get

$$\begin{aligned}e^{-x^2} &= 1 - x^2 + \frac{(-x^2)^2}{2!} + \frac{(-x^2)^3}{3!} + \cdots \\ &= 1 - x^2 + \frac{(x^4)}{2!} - \frac{x^6}{3!} + \cdots .\end{aligned}$$

Example 2. Find the series for $e^{\tan x}$. Here we must replace the x in (13.3) by the series of Example 2 in method B. Let us agree in advance to keep terms only as far as x^4; we then write only terms which can give rise to powers of x up to 4, and neglect any higher powers:

$$
\begin{aligned}
e^{\tan x} &= 1+\left(x+\frac{x^3}{3}+\cdots\right)+\frac{1}{2!}\left(x+\frac{x^3}{3}+\cdots\right)^2 \\
&\quad +\frac{1}{3!}\left(x+\frac{x^3}{3}+\cdots\right)^3+\frac{1}{4!}(x+\cdots)^4+\cdots \\
&= 1+x \qquad +\frac{x^3}{3} \qquad +\cdots \\
&\quad +\frac{x^2}{2!} \qquad +\frac{2x^4}{3\cdot 2!}+\cdots \\
&\quad +\frac{x^3}{3!}+\frac{x^4}{4!}+\cdots \\
\hline
&= 1+x+\frac{x^2}{2}+\frac{x^3}{2}+\frac{3}{8}x^4+\cdots.
\end{aligned}
$$

E. Combination of Methods

▶ **Example.** Find the series for $\operatorname{arc}\tan x$. Since

$$\int_0^x \frac{dt}{1+t^2} = \operatorname{arc}\tan t\Big|_0^x = \operatorname{arc}\tan x,$$

we first write out (as a binomial series) $(1+t^2)^{-1}$ and then integrate term by term:

$$
\begin{aligned}
&(1+t^2)^{-1} = 1-t^2+t^4-t^6+\cdots; \\
&\int_0^x \frac{dt}{1+t^2} = t-\frac{t^3}{3}+\frac{t^5}{5}-\frac{t^7}{7}+\cdots\Big|_0^x.
\end{aligned}
$$

Thus, we have

$$\operatorname{arc}\tan x = x-\frac{x^3}{3}+\frac{x^5}{5}-\frac{x^7}{7}+\cdots. \tag{13.7}$$

Compare this simple way of getting the series with the method in Section 12 of finding successive derivatives of $\operatorname{arc}\tan x$.

F. Taylor Series Using the Basic Maclaurin Series

In many simple cases it is possible to obtain a Taylor series using the basic memorized Maclaurin series instead of the formulas or method of Section 12.

▶ **Example 1.** Find the first few terms of the Taylor series for $\ln x$ about $x = 1$. [This means a series of powers of $(x-1)$ rather than powers of x.] We write

$$\ln x = \ln[1+(x-1)]$$

and use (13.4) with x replaced by $(x-1)$:

$$\ln x = \ln[1+(x-1)] = (x-1)-\frac{1}{2}(x-1)^2+\frac{1}{3}(x-1)^3-\frac{1}{4}(x-1)^4\cdots.$$

Example 2. Expand $\cos x$ about $x = 3\pi/2$. We write

$$\cos x = \cos\left[\frac{3\pi}{2} + \left(x - \frac{3\pi}{2}\right)\right] = \sin\left(x - \frac{3\pi}{2}\right)$$
$$= \left(x - \frac{3\pi}{2}\right) - \frac{1}{3!}\left(x - \frac{3\pi}{2}\right)^3 + \frac{1}{5!}\left(x - \frac{3\pi}{2}\right)^5 \cdots$$

using (13.1) with x replaced by $(x - 3\pi/2)$.

G. Using a Computer

You can also do problems like these using a computer. This is a good method for complicated functions where it saves you a lot of algebra. However, you're not saving time if it takes longer to type a problem into the computer than to do it in your head! For example, you should be able to just write down the first few terms of $(\sin x)/x$ or $(1 - \cos x)/x^2$. A good method of study is to practice doing problems by hand and also check your results using the computer. This will turn up errors you are making by hand, and also let you discover what the computer will do and what it won't do! It is very illuminating to computer plot the function you are expanding, along with several partial sums of the series, in order to see how accurately the partial sums represent the function—see the following example.

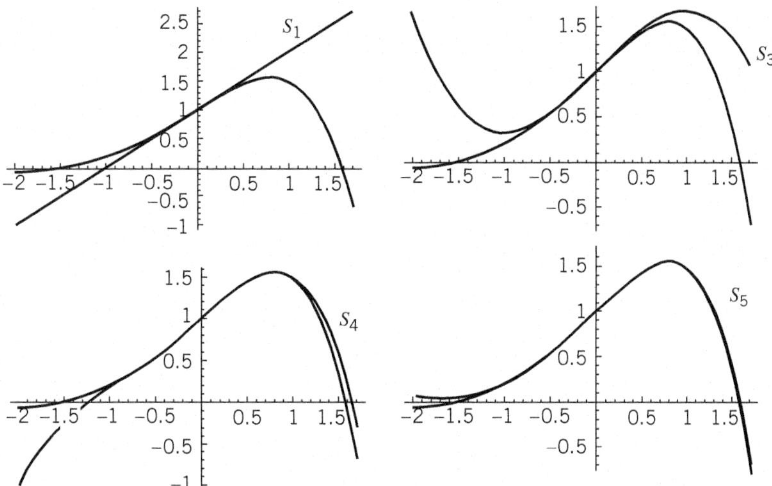

Figure 13.1

Example. Plot the function $e^x \cos x$ together with several partial sums of its Maclaurin series. Using Example 2 in 13A or a computer, we have

$$e^x \cos x = 1 + x - \frac{x^3}{3} - \frac{x^4}{6} - \frac{x^5}{30} \cdots .$$

Figure 13.1 shows plots of the function along with each of the partial sums $S_1 = 1+x$, $S_3 = 1 + x - \frac{x^3}{3}$, $S_4 = 1 + x - \frac{x^3}{3} - \frac{x^4}{6}$, $S_5 = 1 + x - \frac{x^3}{3} - \frac{x^4}{6} - \frac{x^5}{30}$. We can see from the graphs the values of x for which an approximation is fairly good. Also see Section 14.

PROBLEMS, SECTION 13

Using the methods of this section:

(a) Find the first few terms of the Maclaurin series for each of the following functions.

(b) Find the general term and write the series in summation form.

(c) Check your results in (a) by computer.

(d) Use a computer to plot the function and several approximating partial sums of the series.

5. $x^2 \ln(1-x)$ **6.** $x\sqrt{1+x}$ **7.** $\frac{1}{x}\sin x$

8. $\frac{1}{\sqrt{1-x^2}}$ **9.** $\frac{1+x}{1-x}$ **10.** $\sin x^2$

11. $\frac{\sin\sqrt{x}}{\sqrt{x}}, \quad x>0$ **12.** $\int_0^x \cos t^2\,dt$ **13.** $\int_0^x e^{-t^2}\,dt$

14. $\ln\sqrt{\frac{1+x}{1-x}} = \int_0^x \frac{dt}{1-t^2}$ **15.** $\arcsin x = \int_0^x \frac{dt}{\sqrt{1-t^2}}$

16. $\cosh x = \frac{e^x+e^{-x}}{2}$ **17.** $\ln\frac{1+x}{1-x}$

18. $\int_0^x \frac{\sin t\,dt}{t}$ **19.** $\ln(x+\sqrt{1+x^2}) = \int_0^x \frac{dt}{\sqrt{1+t^2}}$

Find the first few terms of the Maclaurin series for each of the following functions and check your results by computer.

20. $e^x \sin x$ **21.** $\tan^2 x$ **22.** $\frac{e^x}{1-x}$

23. $\frac{1}{1+x+x^2}$ **24.** $\sec x = \frac{1}{\cos x}$ **25.** $\frac{2x}{e^{2x}-1}$

26. $\frac{1}{\sqrt{\cos x}}$ **27.** $e^{\sin x}$ **28.** $\sin[\ln(1+x)]$

29. $\sqrt{1+\ln(1+x)}$ **30.** $\sqrt{\frac{1-x}{1+x}}$ **31.** $\cos(e^x-1)$

32. $\ln(1+xe^x)$ **33.** $\frac{1-\sin x}{1-x}$ **34.** $\ln(2-e^{-x})$

35. $\frac{x}{\sin x}$ **36.** $\int_0^u \frac{\sin x\,dx}{\sqrt{1-x^2}}$

37. $\ln\cos x$ *Hints:* Method 1: Write $\cos x = 1+(\cos x-1) = 1+u$; use the series you know for $\ln(1+u)$; replace u by the Maclaurin series for $(\cos x-1)$. Method 2: $\ln\cos x = -\int_0^x \tan u\,du$. Use the series of Example 2 in method B.

38. $e^{\cos x}$ *Hint:* $e^{\cos x} = e\cdot e^{\cos x-1}$.

Using method F above, find the first few terms of the Taylor series for the following functions about the given points.

39. $f(x)=\sin x, \quad a=\pi/2$ **40.** $f(x)=\frac{1}{x}, \quad a=1$

41. $f(x)=e^x, \quad a=3$ **42.** $f(x)=\cos x, \quad a=\pi$

43. $f(x)=\cot x, \quad a=\pi/2$ **44.** $f(x)=\sqrt{x} \quad a=25$

14. ACCURACY OF SERIES APPROXIMATIONS

The thoughtful student might well be disturbed about the mathematical manipulations we have been doing. How do we know whether these processes we have shown really give us series that approximate the functions being expanded? Certainly *some* functions cannot be expanded in a power series; since a power series becomes just a_0 when $x = 0$, it cannot be equal to any function (like $1/x$ or $\ln x$) which is infinite at the origin. So we might ask whether there are other functions (besides those that become infinite at the origin) which cannot be expanded in a power series. All we have done so far is to show methods of finding the power series for a function *if it has one.* Now is there a chance that there might be some functions which do not have series expansions, but for which our formal methods would give us a spurious series? Unfortunately, the answer is "Yes"; fortunately, this is not a very common difficulty in practice. However, you should know of the possibility and what to do about it. You may first think of the fact that, say, the equation

$$\frac{1}{1+x} = 1 - x + x^2 - x^3 + \cdots$$

is not valid for $|x| \geq 1$. This is a fairly easy restriction to determine; from the beginning we recognized that we could use our series expansions only when they converged. But there is another difficulty which can arise. It is possible for a series found by the above methods to converge and still not represent the function being expanded! A simple example of this is $e^{-(1/x^2)}$ for which the formal series is $0+0+0+\cdots$ because $e^{-(1/x^2)}$ and all its derivatives are zero at the origin (Problem 15.26). It is clear that $e^{-(1/x^2)}$ is not zero for $x^2 > 0$, so the series is certainly not correct. You can startle your friends with the following physical interpretation of this. Suppose that at $t = 0$ a car is at rest (zero velocity), and has zero acceleration, zero rate of change of acceleration, etc. (all derivatives of distance with respect to time are zero at $t = 0$). Then according to Newton's second law (force equals mass times acceleration), the instantaneous force acting on the car is also zero (and, in fact, so are all the derivatives of the force). Now we ask "Is it possible for the car to be moving immediately after $t = 0$?" The answer is "Yes"! For example, let its distance from the origin as a function of time be $e^{-(1/t^2)}$.

This strange behavior is really the fault of the function itself and not of our method of finding series. The most satisfactory way of avoiding the difficulty is to recognize, by complex variable theory, when functions can or cannot have power series. We shall consider this in Chapter 14, Section 2. Meanwhile, let us consider two important questions: (1) Does the Taylor or Maclaurin series in (12.8) or (12.9) actually converge to the function being expanded? (2) In a computation problem, if we know that a series converges to a given function, how rapidly does it converge? That is, how many terms must we use to get the accuracy we require? We take up these questions in order.

The *remainder* $R_n(x)$ in a Taylor series is the difference between the value of the function and the sum of $n + 1$ terms of the series:

$$(14.1)\quad R_n(x) = f(x) - \left[f(a) + (x-a)f'(a) + \frac{1}{2!}(x-a)^2 f''(a) + \cdots + \frac{1}{n!}(x-a)^n f^{(n)}(a)\right].$$

Saying that the series converges to the function means that $\lim_{n\to\infty} |R_n(x)| = 0$. There are many different formulas for $R_n(x)$ which are useful for special purposes; you can find these in calculus books. One such formula is

$$R_n(x) = \frac{(x-a)^{n+1} f^{(n+1)}(c)}{(n+1)!} \tag{14.2}$$

where c is some point between a and x. You can use this formula in some simple cases to prove that the Taylor or Maclaurin series for a function does converge to the function (Problems 11 to 13).

Error in Series Approximations Now suppose that we know in advance that the power series of a function does converge to the function (within the interval of convergence), and we want to use a series approximation for the function. We would like to estimate the error caused by using only a few terms of the series.

There is an easy way to estimate this error when the series is alternating and meets the alternating series test for convergence (Section 7). In this case the error is (in absolute value) less than the absolute value of the first neglected term (see Problem 1).

(14.3) If $S = \sum_{n=1}^{\infty} a_n$ is an alternating series with $|a_{n+1}| < |a_n|$, and $\lim_{n\to\infty} a_n = 0$, then $|S - (a_1 + a_2 + \cdots + a_n)| \le |a_{n+1}|$.

▶ **Example 1.** Consider the series

$$1 - \frac{1}{2} + \frac{1}{4} - \frac{1}{8} + \frac{1}{16} - \frac{1}{32} + \frac{1}{64} \cdots .$$

The sum of this series [see (1.8), $a = 1, r = -\frac{1}{2}$] is $S = \frac{2}{3} = 0.666\cdots$. The sum of the terms through $-\frac{1}{32}$ is 0.656+, which differs from S by about 0.01. This is less than the next term $= \frac{1}{64} = 0.015+$.

Estimating the error by the first neglected term may be quite misleading for convergent series that are not alternating.

▶ **Example 2.** Suppose we approximate $\sum_{n=1}^{\infty} 1/n^2$ by the sum of the first five terms; the error is then about 0.18 [see problem 2(a)]. But the first neglected term is $1/6^2 = 0.028$ which is much less than the error. However, note that we are finding the sum of the power series $\sum_{n=1}^{\infty} x^n/n^2$ when $x = 1$, which is the largest x for which the series converges. If, instead, we ask for the sum of the series when $x = 1/2$, we find [see Problem 2(b)]:

$$S = \sum_{n=1}^{\infty} \frac{1}{n^2}\left(\frac{1}{2}\right)^n = 0.5822+.$$

The sum of the first five terms of the series is 0.5815+, so the error is about 0.0007. The next term is $(\frac{1}{6})^2/6^2 = 0.0004$, which is less than the error but still of the same order of magnitude. We can state the following theorem [Problem 2(c)] which covers many practical problems.

$$\text{If } S = \sum_{n=0}^{\infty} a_n x^n \text{ converges for } |x| < 1, \text{ and if}$$
$$|a_{n+1}| < |a_n| \text{ for } n > N, \text{ then} \tag{14.4}$$
$$\left| S - \sum_{n=0}^{N} a_n x^n \right| < |a_{N+1} x^{N+1}| \div (1 - |x|).$$

That is, as in (14.3), the error may be estimated by the first neglected term, but here the error may be a few times as large as the first neglected term instead of smaller. In the example of $\sum x^n/n^2$ with $x = \frac{1}{2}$, we have $1 - x = \frac{1}{2}$, so (14.4) says that the error is less than two times the next term. We observe that the error 0.0007 is less than 2(0.0004) as (14.4) says.

For values of $|x|$ much less than 1, $1 - |x|$ is about 1, so the next term gives a good error estimate in this case. If the interval of convergence is not $|x| < 1$, but, for example, $|x| < 2$ as in

$$\sum_{n=1}^{\infty} \frac{1}{n^2} \left(\frac{x}{2}\right)^n,$$

we can easily let $x/2 = y$, and apply the theorem in terms of y.

PROBLEMS, SECTION 14

1. Prove theorem (14.3). *Hint:* Group the terms in the error as $(a_{n+1}+a_{n+2})+(a_{n+3}+a_{n+4})+\cdots$ to show that the error has the same sign as a_{n+1}. Then group them as $a_{n+1}+(a_{n+2}+a_{n+3})+(a_{n+4}+a_{n+5})+\cdots$ to show that the error has magnitude less than $|a_{n+1}|$.

2. (a) Using computer or tables (or see Chapter 7, Section 11), verify that $\sum_{n=1}^{\infty} 1/n^2 = \pi^2/6 = 1.6449+$, and also verify that the error in approximating the sum of the series by the first five terms is approximately 0.1813.

(b) By computer or tables verify that $\sum_{n=1}^{\infty}(1/n^2)(1/2)^n = \pi^2/12 - (1/2)(\ln 2)^2 = 0.5822+$, and that the sum of the first five terms is 0.5815+.

(c) Prove theorem (14.4). *Hint:* The error is $\left|\sum_{N+1}^{\infty} a_n x^n\right|$. Use the fact that the absolute value of a sum is less than or equal to the sum of the absolute values. Then use the fact that $|a_{n+1}| \le |a_n|$ to replace all a_n by a_{N+1}, and write the appropriate inequality. Sum the geometric series to get the result.

In Problems 3 to 7, assume that the Maclaurin series converges to the function.

3. If $0 < x < \frac{1}{2}$, show [using theorem (14.3)] that $\sqrt{1+x} = 1 + \frac{1}{2}x$ with an error less than 0.032. *Hint:* Note that the series is alternating after the first term.

4. Show that $\sin x = x$ with an error less than 0.021 for $0 < x < \frac{1}{2}$, and with an error less than 0.0002 for $0 < x < 0.1$. *Hint:* Use theorem (14.3) and note that the "next" term is the x^3 term.

5. Show that $1 - \cos x = x^2/2$ with an error less than 0.003 for $|x| < \frac{1}{2}$.

6. Show that $\ln(1-x) = -x$ with an error less than 0.0056 for $|x| < 0.1$. *Hint:* Use theorem (14.4).

7. Show that $2/\sqrt{4-x} = 1 + \frac{1}{8}x$ with an error less than $\frac{1}{32}$ for $0 < x < 1$. *Hint:* Let $x = 4y$, and use theorem (14.4).

8. Estimate the error if $\sum_{n=1}^{\infty} x^n/n^3$ is approximated by the sum of its first three terms for $|x| < \frac{1}{2}$.

9. Consider the series in Problem 4.6 and show that the remainder after n terms is $R_n = 1/(n+1)$. Compare the value of term $n+1$ with R_n for $n = 3, n = 10, n = 100, n = 500$ to see that the first neglected term is not a useful estimate of the error.

10. Show that the interval of convergence of the series $\sum_{n=1}^{\infty} x^n/(n^2+n)$ is $|x| \le 1$. (For $x = 1$, this is the series of Problem 9.) Using theorem (14.4), show that for $x = \frac{1}{2}$, four terms will give two decimal place accuracy.

11. Show that the Maclaurin series for $\sin x$ converges to $\sin x$. *Hint:* If $f(x) = \sin x$, $f^{(n+1)}(x) = \pm\sin x$ or $\pm\cos x$, and so $|f^{(n+1)}(x)| \le 1$ for all x and all n. Let $n \to \infty$ in (14.2).

12. Show as in Problem 11 that the Maclaurin series for e^x converges to e^x.

13. Show that the Maclaurin series for $(1+x)^p$ converges to $(1+x)^p$ when $0 < x < 1$.

15. SOME USES OF SERIES

In this chapter we are going to consider a few rather straightforward uses of series. In later chapters there will also be many other cases where we need them.

Numerical Computation With computers and calculators so available, you may wonder why we would ever want to use series for numerical computation. Here is an example to warn you of the pitfalls of blind computation.

Example 1. Evaluate $f(x) = \ln\sqrt{(1+x)/(1-x)} - \tan x$ at $x = 0.0015$.

Here are answers from several calculators and computers: -9×10^{-16}, 3×10^{-10}, 6.06×10^{-16}, 5.5×10^{-16}. All of these are wrong! Let's use series to see what's going on. By Section 13 methods we find, for $x = 0.0015$:

$$\begin{aligned}
\ln\sqrt{(1+x)/(1-x)} &= x + \frac{x^3}{3} + \frac{x^5}{5} + \frac{x^7}{7}\cdots &&= 0.001500001125001518752441,\\
\tan x &= x + \frac{x^3}{3} + \frac{2x^5}{15} + \frac{17x^7}{315}\cdots &&= 0.001500001125001012500922,\\
f(x) &= \frac{x^5}{15} + \frac{4x^7}{45}\cdots &&= 5.0625 \times 10^{-16}
\end{aligned}$$

with an error of the order of x^7 or 10^{-21}. Now we see that the answer is the difference of two numbers which are identical until the 16th decimal place, so any computer carrying fewer digits will lose all accuracy in the subtraction. It may also be necessary to tell your computer that the value of x is an exact number and not a 4 decimal place approximation. The moral here is that a computer is a tool—a very useful tool, yes—but you need to be constantly aware of whether an answer is reasonable when you are doing problems either by hand or by computer. A final point is that in an applied problem you may want, not a numerical value, but a simple approximation for a complicated function. Here we might approximate $f(x)$ by $x^5/15$ for small x.

Example 2. Evaluate

$$\frac{d^5}{dx^5}\left(\frac{1}{x}\sin x^2\right)\bigg|_{x=0}.$$

We can do this by computer, but it's probably faster to use $\sin x^2 = x^2 - (x^2)^3/3! \cdots$, and observe that when we divide this by x and take 5 derivatives, the x^2 term is gone. The second term divided by x is an x^5 term and the fifth derivative of x^5 is 5!. Any further terms will have a power of x which is zero at $x = 0$. Thus we have

$$\frac{d^5}{dx^5}\left(\frac{1}{x}\cdot\frac{-(x^2)^3}{3!}\right)_{x=0} = -\frac{5!}{3!} = -20.$$

Summing series We have seen a few numerical series which we could sum exactly (see Sections 1 and 4) and we will see some others later (see Chapter 7, Section 11). Here it is interesting to note that if $f(x) = \sum a_n x^n$, and we let x have a particular value (within the interval of convergence), then we get a numerical series whose sum is the value of the function for that x. For example, if we substitute $x = 1$ in (13.4), we get

$$\ln(1+1) = \ln 2 = 1 - \frac{1}{2} + \frac{1}{3} - \frac{1}{4}\cdots$$

so the sum of the alternating harmonic series is $\ln 2$.

We can also find sums of series from tables or computer, either the exact sum if that is known, or a numerical approximation (see Problems 20 to 22, and also Problems 14.2, 16.1, 16.30, and 16.31).

Integrals By Theorem 1 of Section 11, we may integrate a power series term by term. Then we can find an approximation for an integral when the indefinite integral cannot be found in terms of elementary functions. As an example, consider the Fresnel integrals (integrals of $\sin x^2$ and $\cos x^2$) which occur in the problem of Fresnel diffraction in optics. We find

$$\begin{aligned}\int_0^t \sin x^2\,dx &= \int_0^t \left(x^2 - \frac{x^6}{3!} + \frac{x^{10}}{5!} - \cdots\right)dx\\ &= \frac{t^3}{3} - \frac{t^7}{7\cdot 3!} + \frac{t^{11}}{11\cdot 5!} - \cdots\end{aligned}$$

so for $t < 1$, the integral is approximately $\frac{t^3}{3} - \frac{t^7}{42}$ with an error < 0.00076 since this is an alternating series (see (14.3)).

Evaluation of Indeterminate Forms Suppose we want to find

$$\lim_{x\to 0}\frac{1-e^x}{x}.$$

If we try to substitute $x = 0$, we get $0/0$. Expressions that lead us to such meaningless results when we substitute are called indeterminate forms. You can evaluate these by computer, but simple ones can often be done quickly by series. For example,

$$\begin{aligned}\lim_{x\to 0}\frac{1-e^x}{x} &= \lim_{x\to 0}\frac{1-(1+x+(x^2/2!)+\cdots)}{x}\\ &= \lim_{x\to 0}\left(-1-\frac{x}{2!}-\cdots\right) = -1.\end{aligned}$$

You may recall L'Hôpital's rule which says that

$$\lim_{x\to a}\frac{f(x)}{\phi(x)} = \lim_{x\to a}\frac{f'(x)}{\phi'(x)}$$

when $f(a)$ and $\phi(a)$ are both zero, and f'/ϕ' approaches a limit or tends to infinity (that is, does not oscillate) as $x \to a$. Let's use power series to see why this is true. We consider functions $f(x)$ and $\phi(x)$ which are expandable in a Taylor series about $x = a$, and assume that $\phi'(a) \neq 0$. Using (12.8), we have

$$\lim_{x\to a}\frac{f(x)}{\phi(x)} = \lim_{x\to a}\frac{f(a) + (x-a)f'(a) + (x-a)^2 f''(a)/2! + \cdots}{\phi(a) + (x-a)\phi'(a) + (x-a)^2\phi''(a)/2! + \cdots}.$$

If $f(a) = 0$ and $\phi(a) = 0$, and we cancel one $(x - a)$ factor, this becomes

$$\lim_{x\to a}\frac{f'(a) + (x-a)f''(a)/2! + \cdots}{\phi'(a) + (x-a)\phi''(a)/2! + \cdots} = \frac{f'(a)}{\phi'(a)} = \lim_{x\to a}\frac{f'(x)}{\phi'(x)}$$

as L'Hôpital's rule says. If $f'(a) = 0$ and $\phi'(a) = 0$, and $\phi''(a) \neq 0$, then a repetition of the rule gives the limit as $f''(a)/\phi''(a)$, and so on.

There are other indeterminate forms besides $0/0$, for example, ∞/∞, $0 \cdot \infty$, etc. L'Hôpital's rule holds for the ∞/∞ form as well as the $0/0$ form. Series are most useful for the $0/0$ form or others which can easily be put into the $0/0$ form. For example, the limit $\lim_{x\to 0}(1/x)\sin x$ is an $\infty \cdot 0$ form, but is easily written as $\lim_{x\to 0}(\sin x)/x$ which is a $0/0$ form. Also *note carefully:* Series (of powers of x) are useful mainly in finding limits as $x \to 0$, because for $x = 0$ such a series collapses to the constant term; for any other value of x we have an infinite series whose sum we probably do not know (see Problem 25, however).

Series Approximations When a problem in, say, differential equations or physics is too difficult in its exact form, we often can get an approximate answer by replacing one or more of the functions in the problem by a few terms of its infinite series. We shall illustrate this idea by two examples.

▶ **Example 3.** In elementary physics we find that the equation of motion of a simple pendulum is (see Chapter 11, Section 8, or a physics textbook):

$$\frac{d^2\theta}{dt^2} = -\frac{g}{l}\sin\theta.$$

This differential equation cannot be solved for θ in terms of elementary functions (see Chapter 11, Section 8), and you may recall that what is usually done is to approximate $\sin\theta$ by θ. Recall the infinite series (13.1) for $\sin\theta$; θ is simply the first term of the series for $\sin\theta$. (Remember that θ is in radians; see discussion in Chapter 2, end of Section 3.) For small values of θ (say $\theta < \frac{1}{2}$ radian or about 30°), this series converges rapidly, and using the first term gives a good approximation (see Problem 14.4). The solutions of the differential equation are then $\theta = A\sin\sqrt{g/l}\,t$ and $\theta = B\cos\sqrt{g/l}\,t$ (A and B constants) as you can verify; we say that the pendulum is executing simple harmonic motion (see Chapter 7, Section 2).

Example 4. Let us consider a radioactive substance containing N_0 atoms at $t = 0$. It is known that the number of atoms remaining at a later time t is given by the formula (see Chapter 8, Section 3):

$$N = N_0 e^{-\lambda t} \tag{15.1}$$

where λ is a constant which is characteristic of the radioactive substance. To find λ for a given substance, a physicist measures in the laboratory the number of decays ΔN during the time interval Δt for a succession of Δt intervals. It is customary to plot each value of $\Delta N/\Delta t$ at the midpoint of the corresponding time interval Δt. If $\lambda\,\Delta t$ is small, this graph is a good approximation to the exact dN/dt graph. A better approximation can be obtained by plotting $\Delta N/\Delta t$ a little to the left of the midpoint. Let us show that the midpoint *does* give a good approximation and also find the more accurate t value. (An approximate value of λ, good enough for calculating the correction, is assumed known from a rough preliminary graph.)

What we should *like* to plot is the graph of dN/dt, that is, the graph of the slope of the curve in Figure 15.1. What we *measure* is the value of $\Delta N/\Delta t$ for each Δt interval. Consider one such Δt interval in Figure 15.1, from t_1 to t_2. To get an accurate graph we should plot the measured value of the quotient $\Delta N/\Delta t$ at the point between t_1 and t_2 where $\Delta N/\Delta t = dN/dt$. Let us write this condition and find the t which satisfies it. The quantity ΔN is the change in N, that is, $N(t_2) - N(t_1)$; the value of dN/dt we get from (15.1). Then $dN/dt = \Delta N/\Delta t$ becomes

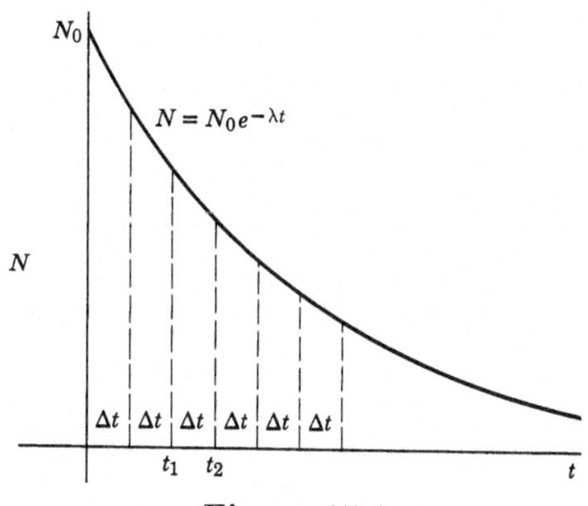

Figure 15.1

$$-\lambda N_0 e^{-\lambda t} = \frac{N_0 e^{-\lambda t_2} - N_0 e^{-\lambda t_1}}{\Delta t}. \tag{15.2}$$

Multiplying this equation by $(\Delta t/N_0)e^{\lambda(t_1+t_2)/2}$, we get

$$-\lambda\,\Delta t\, e^{-\lambda[t-(t_1+t_2)/2]} = e^{-\lambda(t_2-t_1)/2} - e^{\lambda(t_2-t_1)/2} = e^{-\lambda\,\Delta t/2} - e^{\lambda\,\Delta t/2} \tag{15.3}$$

since $t_2 - t_1 = \Delta t$. Since we assumed $\lambda\,\Delta t$ to be small, we can expand the exponentials on the right-hand side of (15.3) in power series; this gives

$$-\lambda\,\Delta t\, e^{-\lambda[t-(t_1+t_2)/2]} = -\lambda\,\Delta t - \frac{1}{3}\left(\frac{\lambda\,\Delta t}{2}\right)^3 \cdots \tag{15.4}$$

or, canceling $(-\lambda\,\Delta t)$,

$$e^{-\lambda[t-(t_1+t_2)/2]} = 1 + \frac{1}{24}(\lambda\,\Delta t)^2 \cdots. \tag{15.5}$$

Suppose $\lambda\,\Delta t$ is small enough so that we can neglect the term $\frac{1}{24}(\lambda\,\Delta t)^2$. Then

(15.5) reduces to

$$e^{-\lambda[t-(t_1+t_2)/2]} = 1,$$

$$-\lambda\left(t - \frac{t_1+t_2}{2}\right) = 0,$$

$$t = \frac{t_1+t_2}{2}.$$

Thus we have justified the usual practice of plotting $\Delta N/\Delta t$ at the midpoint of the interval Δt.

Next consider a more accurate approximation. From (15.5) we get

$$-\lambda\left(t - \frac{t_1+t_2}{2}\right) = \ln\left(1 + \frac{1}{24}(\lambda\,\Delta t)^2 \cdots\right).$$

Since $\frac{1}{24}(\lambda\,\Delta t)^2 \ll 1$, we can expand the logarithm by (13.4) to get

$$-\lambda\left(t - \frac{t_1+t_2}{2}\right) = \frac{1}{24}(\lambda\,\Delta t)^2 \cdots.$$

Then we have

$$t = \frac{t_1+t_2}{2} - \frac{1}{24\lambda}(\lambda\,\Delta t)^2 \cdots.$$

Thus the measured $\Delta N/\Delta t$ should be plotted a little to the left of the midpoint of Δt, as we claimed.

▶ PROBLEMS, SECTION 15

In Problems 1 to 4, use power series to evaluate the function at the given point. Compare with computer results, using the computer to find the series, and also to do the problem without series. Resolve any disagreement in results (see Example 1).

1. $e^{\arcsin x} + \ln\left(\dfrac{1-x}{e}\right)$ at $x = 0.0003$

2. $\dfrac{1}{\sqrt{1+x^4}} - \cos x^2$ at $x = 0.012$

3. $\ln\left(x + \sqrt{1+x^2}\right) - \sin x$ at $x = 0.001$

4. $e^{\sin x} - (1/x^3)\ln(1+x^3e^x)$ at $x = 0.00035$

Use Maclaurin series to evaluate each of the following. Although you could do them by computer, you can probably do them in your head faster than you can type them into the computer. So use these to practice quick and skillful use of basic series to make simple calculations.

5. $\dfrac{d^4}{dx^4}\ln(1+x^3)$ at $x = 0$

6. $\dfrac{d^3}{dx^3}\left(\dfrac{x^2e^x}{1-x}\right)$ at $x = 0$

7. $\dfrac{d^{10}}{dx^{10}}(x^8\tan^2 x)$ at $x = 0$

8. $\lim_{x\to0}\dfrac{1-\cos x}{x^2}$ **9.** $\lim_{x\to0}\dfrac{\sin x - x}{x^3}$ **10.** $\lim_{x\to0}\dfrac{1-e^{x^3}}{x^3}$

11. $\lim_{x\to0}\dfrac{\sin^2 2x}{x^2}$ **12.** $\lim_{x\to0}\dfrac{\tan x - x}{x^3}$ **13.** $\lim_{x\to0}\dfrac{\ln(1-x)}{x}$

Find a two term approximation for each of the following integrals and an error bound for the given t interval.

14. $\int_0^t e^{-x^2}\,dx,\quad 0<t<0.1$ **15.** $\int_0^t \sqrt{x}\,e^{-x}\,dx,\quad 0<t<0.01$

Find the sum of each of the following series by recognizing it as the Maclaurin series for a function evaluated at a point.

16. $\sum_{n=1}^{\infty}\dfrac{2^n}{n!}$ **17.** $\sum_{n=0}^{\infty}\dfrac{(-1)^n}{(2n)!}\left(\dfrac{\pi}{2}\right)^{2n}$

18. $\sum_{n=1}^{\infty}\dfrac{1}{n\,2^n}$ **19.** $\sum_{n=0}^{\infty}\binom{-1/2}{n}\left(-\dfrac{1}{2}\right)^n$

20. By computer or tables, find the exact sum of each of the following series.

(a) $\sum_{n=1}^{\infty}\dfrac{n}{(4n^2-1)^2}$ (b) $\sum_{n=1}^{\infty}\dfrac{n^3}{n!}$ (c) $\sum_{n=1}^{\infty}\dfrac{n(n+1)}{3^n}$

21. By computer, find a numerical approximation for the sum of each of the following series.

(a) $\sum_{n=1}^{\infty}\dfrac{n}{(n^2+1)^2}$ (b) $\sum_{n=2}^{\infty}\dfrac{\ln n}{n^2}$ (c) $\sum_{n=1}^{\infty}\dfrac{1}{n^n}$

22. The series $\sum_{n=1}^{\infty}1/n^s$, $s>1$, is called the Riemann Zeta function, $\zeta(s)$. (In Problem 14.2(a) you found $\zeta(2)=\pi^2/6$. When n is an even integer, these series can be summed exactly in terms of π.) By computer or tables, find

(a) $\zeta(4)=\sum_{n=1}^{\infty}\dfrac{1}{n^4}$ (b) $\zeta(3)=\sum_{n=1}^{\infty}\dfrac{1}{n^3}$ (c) $\zeta\left(\dfrac{3}{2}\right)=\sum_{n=1}^{\infty}\dfrac{1}{n^{3/2}}$

23. Find the following limits using Maclaurin series and check your results by computer. *Hint:* First combine the fractions. Then find the first term of the denominator series and the first term of the numerator series.

(a) $\lim_{x\to0}\left(\dfrac{1}{x}-\dfrac{1}{e^x-1}\right)$ (b) $\lim_{x\to0}\left(\dfrac{1}{x^2}-\dfrac{\cos x}{\sin^2 x}\right)$

(c) $\lim_{x\to0}\left(\csc^2 x-\dfrac{1}{x^2}\right)$ (d) $\lim_{x\to0}\left(\dfrac{\ln(1+x)}{x^2}-\dfrac{1}{x}\right)$

24. Evaluate the following indeterminate forms by using L'Hôpital's rule and check your results by computer. (Note that Maclaurin series would not be useful here because x does not tend to zero, or because a function ($\ln x$, for example) is not expandable in a Maclaurin series.)

(a) $\lim_{x\to\pi}\dfrac{x\sin x}{x-\pi}$ (b) $\lim_{x\to\pi/2}\dfrac{\ln(2-\sin x)}{\ln(1+\cos x)}$

(c) $\lim_{x\to1}\dfrac{\ln(2-x)}{x-1}$ (d) $\lim_{x\to\infty}\dfrac{\ln x}{\sqrt{x}}$

(e) $\lim_{x\to0}x\ln 2x$

(f) $\lim_{x\to\infty} x^n e^{-x}$ (n not necessarily integral)

25. In general, we do not expect Maclaurin series to be useful in evaluating indeterminate forms except when x tends to zero (see Problem 24). Show, however, that Problem 24(f) can be done by writing $x^n e^{-x} = x^n/e^x$ and using the series (13.3) for e^x. *Hint:* Divide numerator and denominator by x^n before you take the limit. What is special about the e^x series which makes it possible to know what the limit of the infinite series is?

26. Find the values of several derivatives of e^{-1/t^2} at $t = 0$. *Hint:* Calculate a few derivatives (as functions of t); then make the substitution $x = 1/t^2$, and use the result of Problem 24(f) or 25.

27. The velocity v of electrons from a high energy accelerator is very near the velocity c of light. Given the voltage V of the accelerator, we often want to calculate the ratio v/c. The relativistic formula for this calculation is (approximately, for $V \gg 1$)

$$\frac{v}{c} = \sqrt{1 - \left(\frac{0.511}{V}\right)^2}, \quad V = \text{number of million volts.}$$

Use two terms of the binomial series (13.5) to find $1 - v/c$ in terms of V. Use your result to find $1 - v/c$ for the following values of V. *Caution:* V = the number of *million* volts.

(a) $V = 100$ million volts
(b) $V = 500$ million volts
(c) $V = 25,000$ million volts
(d) $V = 100$ gigavolts (100×10^9 volts $= 10^5$ million volts)

28. The energy of an electron at speed v in special relativity theory is $mc^2(1-v^2/c^2)^{-1/2}$, where m is the electron mass, and c is the speed of light. The factor mc^2 is called the rest mass energy (energy when $v = 0$). Find two terms of the series expansion of $(1 - v^2/c^2)^{-1/2}$, and multiply by mc^2 to get the energy at speed v. What is the second term in the energy series? (If v/c is very small, the rest of the series can be neglected; this is true for everyday speeds.)

29. The figure shows a heavy weight suspended by a cable and pulled to one side by a force F. We want to know how much force F is required to hold the weight in equilibrium at a given distance x to one side (say to place a cornerstone correctly). From elementary physics, $T\cos\theta = W$, and $T\sin\theta = F$.

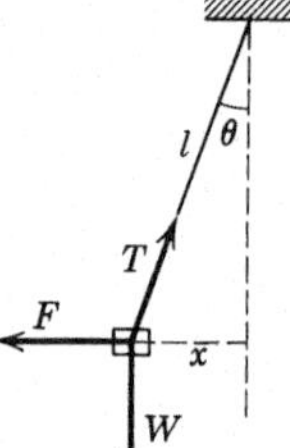

(a) Find F/W as a series of powers of θ.
(b) Usually in a problem like this, what we know is not θ, but x and l in the diagram. Find F/W as a series of powers of x/l.

30. Given a strong chain and a convenient tree, could you pull your car out of a ditch in the following way? Fasten the chain to the car and to the tree. Pull with a force F at the center of the chain as shown in the figure. From mechanics, we have $F = 2T\sin\theta$, or $T = F/(2\sin\theta)$, where T is the tension in the chain, that is, the force exerted on the car.

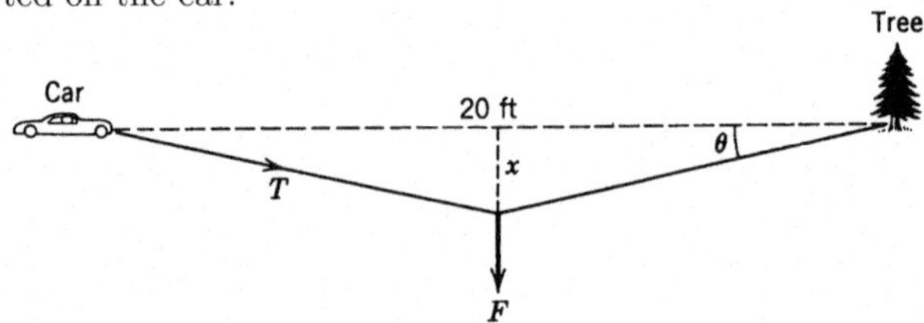

(a) Find T as x^{-1} times a series of powers of x.

(b) Find T as θ^{-1} times a series of powers of θ.

31. A tall tower of circular cross section is reinforced by horizontal circular disks (like large coins), one meter apart and of negligible thickness. The radius of the disk at height n is $1/(n \ln n)$ $(n \geq 2)$.

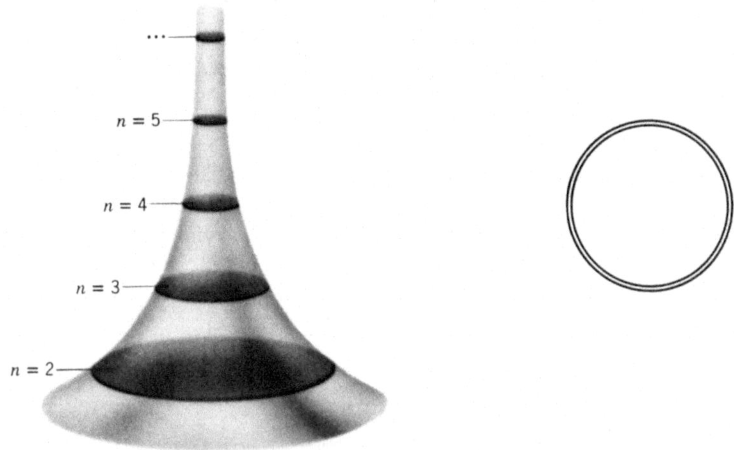

Assuming that the tower is of infinite height:

(a) Will the total area of the disks be finite or not? *Hint:* Can you compare the series with a simpler one?

(b) If the disks are strengthened by wires going around their circumferences like tires, will the total length of wire required be finite or not?

(c) Explain why there is not a contradiction between your answers in (a) and (b). That is, how is it possible to start with a set of disks of finite area, remove a little strip around the circumference of each, and get an infinite total length of these strips? *Hint:* Think about units—you can't compare area and length. Consider two cases: (1) Make the width of each strip equal to one percent of the radius of the disk from which you cut it. Now the total length is infinite but what about the total area? (2) Try to make the strips all the same width; what happens? Also see Chapter 5, Problem 3.31(b).

32. Show that the "doubling time" (time for your money to double) is n periods at interest rate $i\%$ per period with $ni = 69$, approximately. Show that the error in the approximation is less than 10% if $i\% \leq 20\%$. (Note that n does not have to be the number of years; it can be the number of months with $i =$ interest rate per month, etc.) *Hint:* You want $(1 + i/100)^n = 2$; take ln of both sides of this equation and use equation (13.4). Also see theorem (14.3).

33. If you are at the top of a tower of height h above the surface of the earth, show that the distance you can see along the surface of the earth is approximately $s = \sqrt{2Rh}$, where R is the radius of the earth. *Hints:* See figure. Show that $h/R = \sec\theta - 1$; find two terms of the series for $\sec\theta = 1/\cos\theta$, and use $s = R\theta$. Thus show that the distance in miles is approximately $\sqrt{3h/2}$ with h in feet.

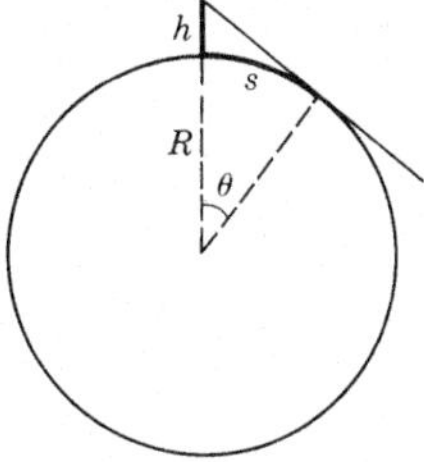

16. MISCELLANEOUS PROBLEMS

1. (a) Show that it is possible to stack a pile of identical books so that the top book is as far as you like to the right of the bottom book. Start at the top and each time place the pile already completed on top of another book so that the pile is just at the point of tipping. (In practice, of course, you can't let them overhang quite this much without having the stack topple. Try it with

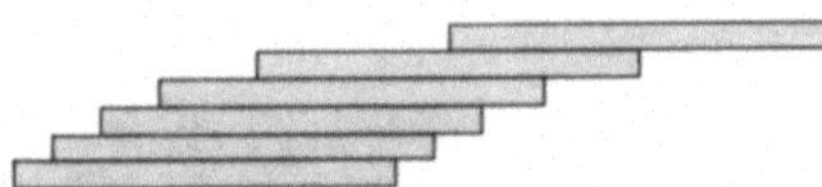

a deck of cards.) Find the distance from the right-hand end of each book to the right-hand end of the one beneath it. To find a general formula for this distance, consider the three forces acting on book n, and write the equation for the torque about its right-hand end. Show that the sum of these setbacks is a divergent series (proportional to the harmonic series). [See "Leaning Tower of *The Physical Reviews*," Am. J. Phys. **27**, 121–122 (1959).]

   (b) By computer, find the sum of N terms of the harmonic series with $N = 25$, 100, 200, 1000, 10^6, 10^{100}.

   (c) From the diagram in (a), you can see that with 5 books (count down from the top) the top book is completely to the right of the bottom book, that is, the overhang is slightly over one book. Use your series in (a) to verify this. Then using parts (a) and (b) and a computer as needed, find the number of books needed for an overhang of 2 books, 3 books, 10 books, 100 books.

2. The picture is a mobile constructed of dowels (or soda straws) connected by thin threads. Each thread goes from the left-hand end of a rod to a point on the rod below. Number the rods from the bottom and find, for rod n, the distance from its left end to the thread so that all rods of the mobile will be horizontal. *Hint:* Can you see the relation between this problem and Problem 1?

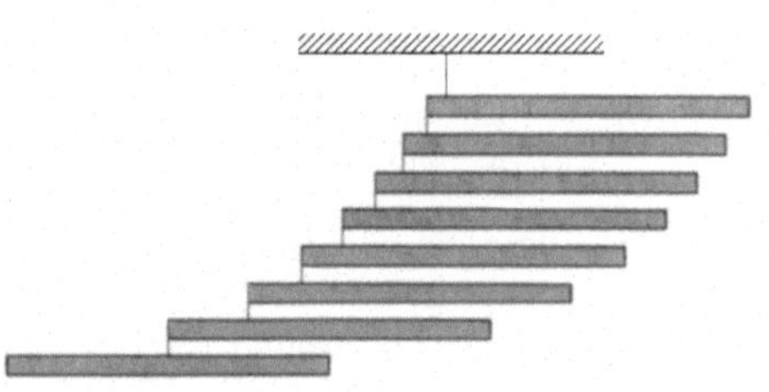

3. Show that $\sum_{n=2}^{\infty} 1/n^{3/2}$ is convergent. What is wrong with the following "proof" that it diverges?

$$\frac{1}{\sqrt{8}}+\frac{1}{\sqrt{27}}+\frac{1}{\sqrt{64}}+\frac{1}{\sqrt{125}}+\cdots>\frac{1}{\sqrt{9}}+\frac{1}{\sqrt{36}}+\frac{1}{\sqrt{81}}+\frac{1}{\sqrt{144}}+\cdots$$

which is

$$\frac{1}{3}+\frac{1}{6}+\frac{1}{9}+\frac{1}{12}+\cdots=\frac{1}{3}\left(1+\frac{1}{2}+\frac{1}{3}+\frac{1}{4}+\cdots\right).$$

Since the harmonic series diverges, the original series diverges. *Hint:* Compare $3n$ and $n\sqrt{n}$.

Test for convergence:

4. $\displaystyle\sum_{n=1}^{\infty}\frac{2^n}{n!}$

5. $\displaystyle\sum_{n=2}^{\infty}\frac{(n-1)^2}{1+n^2}$

6. $\displaystyle\sum_{n=2}^{\infty}\frac{\sqrt{n-1}}{(n+1)^2-1}$

7. $\displaystyle\sum_{n=2}^{\infty}\frac{1}{n\ln(n^3)}$

8. $\displaystyle\sum_{n=2}^{\infty}\frac{2n^3}{n^4-2}$

Find the interval of convergence, including end-point tests:

9. $\sum_{n=1}^{\infty} \frac{x^n}{\ln(n+1)}$ **10.** $\sum_{n=1}^{\infty} \frac{(n!)^2 x^n}{(2n)!}$ **11.** $\sum_{n=1}^{\infty} \frac{(-1)^n x^{2n-1}}{2n-1}$

12. $\sum_{n=1}^{\infty} \frac{x^n n^2}{5^n(n^2+1)}$ **13.** $\sum_{n=1}^{\infty} \frac{(x+2)^n}{(-3)^n \sqrt{n}}$

Find the Maclaurin series for the following functions.

14. $\cos[\ln(1+x)]$ **15.** $\ln\left(\frac{\sin x}{x}\right)$ **16.** $\frac{1}{\sqrt{1+\sin x}}$

17. $e^{1-\sqrt{1-x^2}}$ **18.** $\arctan x = \int_0^x \frac{du}{1+u^2}$

Find the first few terms of the Taylor series for the following functions about the given points.

19. $\sin x,\ a = \pi$ **20.** $\sqrt[3]{x},\ a = 8$ **21.** $e^x,\ a = 1$

Use series you know to show that:

22. $1 - \frac{1}{3} + \frac{1}{5} - \frac{1}{7} + \cdots = \frac{\pi}{4}$. *Hint:* See Problem 18.

23. $\frac{\pi^2}{3!} - \frac{\pi^4}{5!} + \frac{\pi^6}{7!} - \cdots = 1$ **24.** $\ln 3 + \frac{(\ln 3)^2}{2!} + \frac{(\ln 3)^3}{3!} + \cdots = 2$

25. Evaluate the limit $\lim_{x\to 0} x^2/\ln\cos x$ by series (in your head), by L'Hôpital's rule, and by computer.

Use Maclaurin series to do Problems 26 to 29 and check your results by computer.

26. $\lim_{x\to 0}\left(\frac{1}{x^2} - \frac{1}{1-\cos^2 x}\right)$ **27.** $\lim_{x\to 0}\left(\frac{1}{x^2} - \cot^2 x\right)$

28. $\lim_{x\to 0}\left(\frac{1+x}{x} - \frac{1}{\sin x}\right)$ **29.** $\left.\frac{d^6}{dx^6}(x^4 e^{x^2})\right|_{x=0}$

30. (a) It is clear that you (or your computer) can't find the sum of an infinite series just by adding up the terms one by one. For example, to get $\zeta(1.1) = \sum_{n=1}^{\infty} 1/n^{1.1}$ (see Problem 15.22) with error < 0.005 takes about 10^{33} terms. To see a simple alternative (for a series of positive decreasing terms) look at Figures 6.1 and 6.2. Show that when you have summed N terms, the sum R_N of the rest of the series is between $I_N = \int_N^\infty a_n\,dn$ and $I_{N+1} = \int_{N+1}^\infty a_n\,dn$.

(b) Find the integrals in (a) for the $\zeta(1.1)$ series and verify the claimed number of terms needed for error < 0.005. *Hint*: Find N such that $I_N = 0.005$. Also find upper and lower bounds for $\zeta(1.1)$ by computing $\sum_{n=1}^{N} 1/n^{1.1} + \int_N^\infty n^{-1.1}\,dn$ and $\sum_{n=1}^{N} 1/n^{1.1} + \int_{N+1}^\infty n^{-1.1}\,dn$ where N is far less than 10^{33}. *Hint*: You want the difference between the upper and lower limits to be about 0.005; find N so that term $a_N = 0.005$.

31. As in Problem 30, for each of the following series, find the number of terms required to find the sum with error < 0.005, and find upper and lower bounds for the sum using a much smaller number of terms.

(a) $\sum_{1}^{\infty} \frac{1}{n^{1.01}}$ (b) $\sum_{1}^{\infty} \frac{1}{n(1+\ln n)^2}$ (c) $\sum_{3}^{\infty} \frac{1}{n \ln n (\ln\ln n)^2}$

CHAPTER 2

Complex Numbers

1. INTRODUCTION

You will probably recall using imaginary and complex numbers in algebra. The general solution of the quadratic equation

$$az^2 + bz + c = 0 \tag{1.1}$$

for the unknown z, is given by the *quadratic formula*

$$z = \frac{-b \pm \sqrt{b^2 - 4ac}}{2a}. \tag{1.2}$$

If the *discriminant* $d = (b^2 - 4ac)$ is negative, we must take the square root of a negative number in order to find z. Since only non-negative numbers have real square roots, it is impossible to use (1.2) when $d < 0$ unless we introduce a new kind of number, called an imaginary number. We use the symbol $i = \sqrt{-1}$ with the understanding that $i^2 = -1$. Then

$$\sqrt{-16} = 4i, \quad \sqrt{-3} = i\sqrt{3}, \quad i^3 = -i$$

are imaginary numbers, but

$$i^2 = -1, \quad \sqrt{-2}\sqrt{-8} = i\sqrt{2} \cdot i\sqrt{8} = -4, \quad i^{4n} = 1$$

are real. In (1.2) we also need combinations of real and imaginary numbers.

Example. The solution of

$$z^2 - 2z + 2 = 0$$

is

$$z = \frac{2 \pm \sqrt{4-8}}{2} = \frac{2 \pm \sqrt{-4}}{2} = 1 \pm i.$$

We use the term *complex number* to mean any one of the whole set of numbers, real, imaginary, or combinations of the two like $1 \pm i$. Thus, $i + 5$, $17i$, 4, $3 + i\sqrt{5}$ are all examples of complex numbers.

Once the new kind of number is admitted into our number system, fascinating possibilities open up. Can we attach any meaning to marks like $\sin i$, $e^{i\pi}$, $\ln(1+i)$? We'll see later that we can and that, in fact, such expressions may turn up in problems in physics, chemistry, and engineering, as well as mathematics.

When people first considered taking square roots of negative numbers, they felt very uneasy about the problem. They thought that such numbers could not have any meaning or any connection with reality (hence the term "imaginary"). They certainly would not have believed that the new numbers could be of any practical use. Yet complex numbers are of great importance in a variety of applied fields; for example, the electrical engineer would, to say the least, be severely handicapped without them. The complex notation often simplifies setting up and solving vibration problems in either dynamical or electrical systems, and is useful in solving many differential equations which arise from problems in various branches of physics. (See Chapters 7 and 8.) In addition, there is a highly developed field of mathematics dealing with functions of a complex variable (see Chapter 14) which yields many useful methods for solving problems about fluid flow, elasticity, quantum mechanics, and other applied problems. Almost every field of either pure or applied mathematics makes some use of complex numbers.

2. REAL AND IMAGINARY PARTS OF A COMPLEX NUMBER

A complex number such as $5 + 3i$ is the sum of two terms. The real term (not containing i) is called the *real part* of the complex number. The *coefficient* of i in the other term is called the *imaginary part* of the complex number. In $5 + 3i$, 5 is the real part and 3 is the imaginary part. Notice carefully that the *imaginary part* of a complex number is *not imaginary*!

Either the real part or the imaginary part of a complex number may be zero. If the real part is zero, the complex number is called imaginary (or, for emphasis, *pure* imaginary). The zero real part is usually omitted; thus $0 + 5i$ is written just $5i$. If the imaginary part of the complex number is zero, the number is real. We write $7 + 0i$ as just 7. Complex numbers then include both real numbers and pure imaginary numbers as special cases.

In algebra a complex number is ordinarily written (as we have been doing) as a sum like $5 + 3i$. There is another very useful way of thinking of a complex number. As we have said, every complex number has a real part and an imaginary part (either of which may be zero). These are two *real* numbers, and we could, if we liked, agree to write $5+3i$ as $(5, 3)$. Any complex number could be written this way as a pair of real numbers, the real part first and then the imaginary part (which, you must remember, is real). This would not be a very convenient form for computation, but it suggests a very useful geometrical representation of a complex number which we shall now consider.

3. THE COMPLEX PLANE

In analytic geometry we plot the point $(5, 3)$ as shown in Figure 3.1. As we have seen, the symbol $(5, 3)$ could also mean the complex number $5+3i$. The point $(5, 3)$ may then be labeled either $(5, 3)$ or $5+3i$. Similarly, any complex number $x+iy$ (x and y real) can be represented by a point (x, y) in the (x, y) plane. Also any point (x, y) in the (x, y) plane can be labeled $x + iy$ as well as (x, y). When the (x, y)

plane is used in this way to plot complex numbers, it is called the *complex plane.* It is also sometimes called an *Argand diagram.* The x axis is called the real axis, and the y axis is called the imaginary axis (note, however, that you plot y and *not* iy).

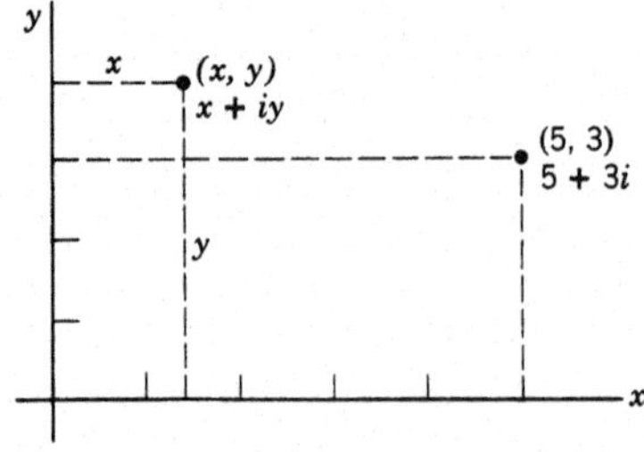

Figure 3.1

When a complex number is written in the form $x + iy$, we say that it is in *rectangular form* because x and y are the rectangular coordinates of the point representing the number in the complex plane. In analytic geometry, we can locate a point by giving its polar coordinates (r, θ) instead of its rectangular coordinates (x, y). There is a corresponding way to write any complex number. In Figure 3.2,

$$x = r\cos\theta, \quad y = r\sin\theta. \tag{3.1}$$

Then we have

$$\begin{aligned} x + iy &= r\,\cos\theta + ir\,\sin\theta \\ &= r\,(\cos\theta + i\,\sin\theta). \end{aligned} \tag{3.2}$$

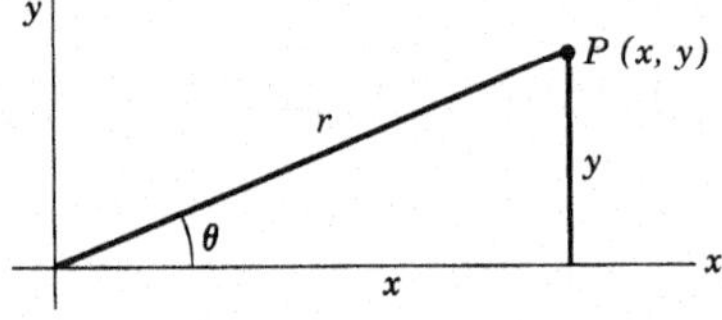

Figure 3.2

This last expression is called the *polar form* of the complex number. As we shall see (Sections 9 to 16), the expression $(\cos\theta + i\,\sin\theta)$ can be written as $e^{i\theta}$, so a convenient way to write the polar form of a complex number is

$$x + iy = r(\cos\theta + i\,\sin\theta) = re^{i\theta}. \tag{3.3}$$

The polar form $re^{i\theta}$ of a complex number is often simpler to use than the rectangular form.

▶ **Example.** In Figure 3.3 the point A could be labeled as $(1, \sqrt{3})$ or as $1 + i\sqrt{3}$. Similarly, using polar coordinates, the point A could be labeled with its (r, θ) values as $(2, \pi/3)$. Notice that r is always taken positive. Using (3.3) we have

$$1 + i\sqrt{3} = 2\left(\cos\frac{\pi}{3} + i\,\sin\frac{\pi}{3}\right) = 2e^{i\pi/3}.$$

This gives two more ways to label point A in Figure 3.3.

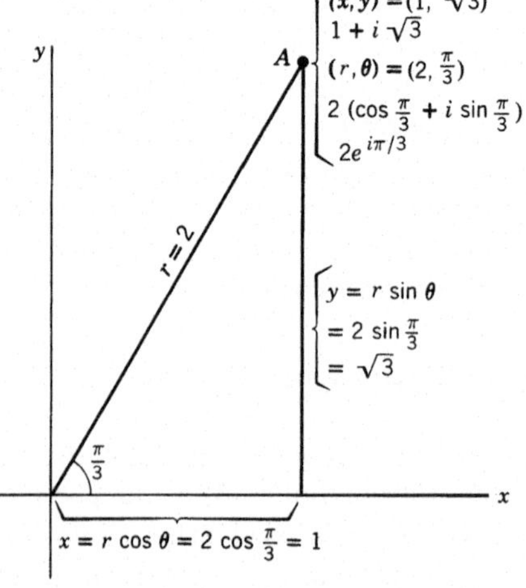

Figure 3.3

Radians and Degrees In Figure 3.3, the angle $\pi/3$ is in radians. Ever since you studied calculus, you have been expected to measure angles in radians and not degrees. Do you know why? You have learned that $(d/dx)\sin x = \cos x$. This formula is **not** correct—**unless** x is in radians. (Look up the derivation in your calculus book!) Many of the formulas you now know and use are correct *only* if you use radian measure; consequently that is what you are usually advised to do. However, it is sometimes convenient to do computations with complex numbers using degrees, so it is important to know when you can and when you cannot use degrees. You can use degrees to measure an angle and to add and subtract angles as long as the final step is to find the sine, cosine, or tangent of the resulting angle (with your calculator in degree mode). For example, in Figure 3.3, we can, if we like, say that $\theta = 60°$ instead of $\theta = \pi/3$. If we want to find $\sin(\pi/3 - \pi/4) = \sin(\pi/12) = 0.2588$ (calculator in radian mode), we can instead find $\sin(60° - 45°) = \sin 15° = 0.2588$ (calculator in degree mode). Note carefully that an angle is in radians unless the degree symbol is used; for example, in $\sin 2$, the 2 is 2 radians or about 115°.

In formulas, however, use radians. For example, in using infinite series, we say that $\sin\theta \cong \theta$ for very small θ. Try this on your calculator; you will find that it is true in radian mode but not in degree mode. As another example, consider $\int_0^1 dx/(1+x^2) = \arctan 1 = \pi/4 = 0.785$. Here $\arctan 1$ is *not* an angle; it is the numerical value of the integral, so the answer 45 (obtained from a calculator in degree mode) is wrong! Do not use degree mode in reading an arc tan (or arc sin or arc cos) *unless* you are finding an *angle* [for example, in Figure 3.2, $\theta = \arctan(y/x)$, and in Figure 3.3, $\theta = \arctan\sqrt{3} = \pi/3$ or 60°].

4. TERMINOLOGY AND NOTATION

Both i and j are used to represent $\sqrt{-1}$, j usually in any problem dealing with electricity since i is needed there for current. A physicist should be able to work with ease using either symbol. We shall for consistency use i throughout this book.

We often label a point with a single letter (for example, P in Figure 3.2 and A in Figure 3.3) even though it requires two coordinates to locate the point. If you have studied vectors, you will recall that a vector is represented by a single letter, say $\mathbf{v}$, although it has (in two dimensions) two components. It is customary to use a single letter for a complex number even though we realize that it is actually a pair of real numbers. Thus we write

$$z = x + iy = r(\cos\theta + i\sin\theta) = re^{i\theta}. \tag{4.1}$$

Here z is a complex number; x is the *real part* of the complex number z, and y is the *imaginary part* of z. The quantity r is called the *modulus* or *absolute value* of z, and θ is called the *angle* of z (or the *phase*, or the *argument*, or the *amplitude* of z). In symbols:

$$\begin{aligned} &\operatorname{Re} z = x, & &|z| = \operatorname{mod} z = r = \sqrt{x^2+y^2}, \\ &\operatorname{Im} z = y\ (\textbf{not } iy), & &\text{angle of } z = \theta. \end{aligned} \tag{4.2}$$

The values of θ should be found from a diagram rather than a formula, although we do sometimes write $\theta = \arctan(y/x)$. An example shows this clearly.

▶ **Example.** Write $z = -1 - i$ in polar form. Here we have $x = -1$, $y = -1$, $r = \sqrt{2}$ (Figure 4.1). There are an infinite number of values of θ,

$$\theta = \frac{5\pi}{4} + 2n\pi, \tag{4.3}$$

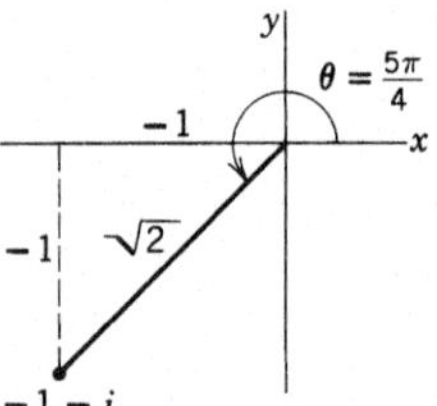

Figure 4.1

where n is any integer, positive or negative. The value $\theta = 5\pi/4$ is sometimes called the *principal angle* of the complex number $z = -1 - i$. Notice carefully, however, that this is not the same as the principal value $\pi/4$ of arc tan 1 as defined in calculus. The angle of a complex number must be in the same quadrant as the point representing the number. For our present work, any one of the values in (4.3) will do; here we would probably use either $5\pi/4$ or $-3\pi/4$. Then we have in our example

$$\begin{aligned} z = -1 - i &= \sqrt{2}\left[\cos\left(\frac{5\pi}{4} + 2n\pi\right) + i\sin\left(\frac{5\pi}{4} + 2n\pi\right)\right] \\ &= \sqrt{2}\left(\cos\frac{5\pi}{4} + i\sin\frac{5\pi}{4}\right) = \sqrt{2}\,e^{5i\pi/4}. \end{aligned}$$

[We could also write $z = \sqrt{2}\,(\cos 225° + i \sin 225°)$.]

The complex number $x - iy$, obtained by changing the sign of i in $z = x + iy$, is called the *complex conjugate* or simply the *conjugate* of z. We usually write the conjugate of $z = x + iy$ as $\bar{z} = x - iy$. Sometimes we use z^* instead of $\bar{z}$ (in fields such as statistics or quantum mechanics where the bar may be used to mean an average value). Notice carefully that the conjugate of $7i - 5$ is $-7i - 5$; that is, it is the i term whose sign is changed.

Complex numbers come in conjugate pairs; for example, the conjugate of $2 + 3i$ is $2 - 3i$ and the conjugate of $2-3i$ is $2+3i$. Such a pair of points in the complex plane are mirror images of each other with the x axis as the mirror (Figure 4.2). Then in polar form, z and $\bar{z}$ have the same r value, but their θ values are negatives of each other. If we write $z = r(\cos\theta + i\sin\theta)$, then

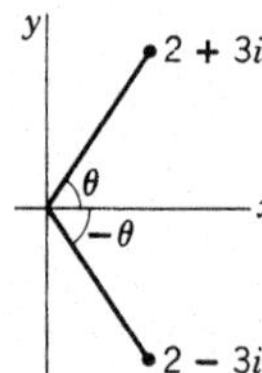

Figure 4.2

$$\bar{z} = r[\cos(-\theta) + i\sin(-\theta)] = r(\cos\theta - i\sin\theta) = re^{-i\theta}. \tag{4.4}$$

▶ PROBLEMS, SECTION 4

For each of the following numbers, first visualize where it is in the complex plane. With a little practice you can quickly find x, y, r, θ in your head for these simple problems. Then plot the number and label it in five ways as in Figure 3.3. Also plot the complex conjugate of the number.

1. $1+i$ **2.** $i-1$ **3.** $1-i\sqrt{3}$

4. $-\sqrt{3}+i$ **5.** $2i$ **6.** $-4i$

7. -1 **8.** 3 **9.** $2i-2$

10. $2-2i$ **11.** $2\left(\cos\frac{\pi}{6}+i\sin\frac{\pi}{6}\right)$

12. $4\left(\cos\frac{2\pi}{3}-i\sin\frac{2\pi}{3}\right)$ **13.** $\cos\frac{3\pi}{2}+i\sin\frac{3\pi}{2}$

14. $2\left(\cos\frac{\pi}{4}+i\sin\frac{\pi}{4}\right)$ **15.** $\cos\pi-i\sin\pi$

16. $5(\cos 0+i\sin 0)$ **17.** $\sqrt{2}\,e^{-i\pi/4}$

18. $3\,e^{i\pi/2}$ **19.** $5(\cos 20^\circ+i\sin 20^\circ)$

20. $7(\cos 110^\circ-i\sin 110^0)$

5. COMPLEX ALGEBRA

A. Simplifying to $x+iy$ form

Any complex number can be written in the rectangular form $x+iy$. To add, subtract, and multiply complex numbers, remember that they follow the ordinary rules of algebra and that $i^2=-1$.

Example 1.

$$(1+i)^2 = 1+2i+i^2 = 1+2i-1 = 2i$$

To divide one complex number by another, first write the quotient as a fraction. Then reduce the fraction to rectangular form by multiplying numerator and denominator by the conjugate of the denominator; this makes the denominator real.

Example 2.

$$\frac{2+i}{3-i} = \frac{2+i}{3-i}\cdot\frac{3+i}{3+i} = \frac{6+5i+i^2}{9-i^2} = \frac{5+5i}{10} = \frac{1}{2}+\frac{1}{2}i.$$

It is sometimes easier to multiply or divide complex numbers in polar form.

Example 3. To find $(1+i)^2$ in polar form, we first sketch (or picture mentally) the point $(1,1)$. From Figure 5.1, we see that $r=\sqrt{2}$, and $\theta=\pi/4$, so $(1+i)=\sqrt{2}\,e^{i\pi/4}$. Then from Figure 5.2 we find the same result as in Example 1.

$$(1+i)^2 = (\sqrt{2}\,e^{i\pi/4})^2 = 2\,e^{i\pi/2} = 2i.$$

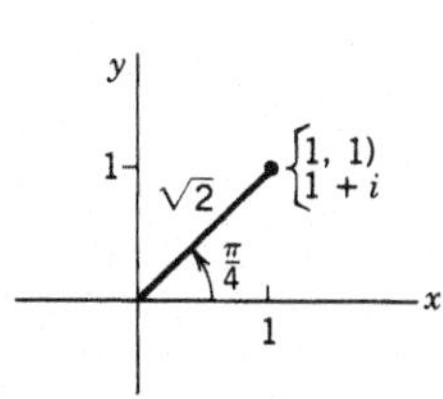

Figure 5.1

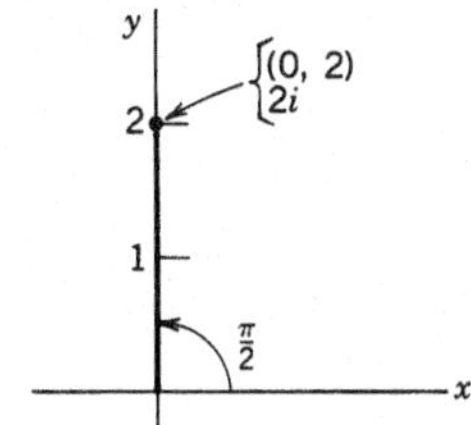

Figure 5.2

Example 4. Write $1/[2(\cos 20° + i\sin 20°]$ in $x+iy$ form. Since $20° = \pi/9$ radians,

$$\frac{1}{2(\cos 20° + i\sin 20°)} = \frac{1}{2(\cos\pi/9 + i\sin\pi/9)} = \frac{1}{2\,e^{i\pi/9}} = 0.5\,e^{-i\pi/9}$$
$$= 0.5(\cos\pi/9 - i\sin\pi/9) = 0.47 - 0.17i,$$

by calculator in radian mode. We obtain the same result leaving the angle in degrees and using a calculator in degree mode: $0.5(\cos 20° - i\sin 20°) = 0.47 - 0.17i$.

PROBLEMS, SECTION 5

First simplify each of the following numbers to the $x+iy$ form or to the $re^{i\theta}$ form. Then plot the number in the complex plane.

1. $\dfrac{1}{1+i}$ **2.** $\dfrac{1}{i-1}$ **3.** i^4

4. i^2+2i+1 **5.** $\left(i+\sqrt{3}\right)^2$ **6.** $\left(\dfrac{1+i}{1-i}\right)^2$

7. $\dfrac{3+i}{2+i}$ **8.** $1.6-2.7i$

9. $25\,e^{2i}$ *Careful!* The angle is 2 radians.

10. $\dfrac{3i-7}{i+4}$ *Careful! Not* $3-7i$

11. $17-12i$ **12.** $3(\cos 28° + i\sin 28°)$

13. $5\left(\cos\dfrac{2\pi}{5} + i\sin\dfrac{2\pi}{5}\right)$ **14.** $2.8e^{-i(1.1)}$

15. $\dfrac{5-2i}{5+2i}$ **16.** $\dfrac{1}{0.5(\cos 40° + i\sin 40°)}$

17. $(1.7-3.2i)^2$ **18.** $(0.64+0.77i)^4$

Find each of the following in rectangular $(a+bi)$ form if $z = 2-3i$; if $z = x+iy$.

19. z^{-1} **20.** $\dfrac{1}{z^2}$ **21.** $\dfrac{1}{z+1}$

22. $\dfrac{1}{z-i}$ **23.** $\dfrac{1+z}{1-z}$ **24.** $z/\bar{z}$

B. Complex Conjugate of a Complex Expression

It is easy to see that the conjugate of the sum of two complex numbers is the sum of the conjugates of the numbers. If

$$z_1 = x_1 + iy_1 \qquad \text{and} \qquad z_2 = x_2 + iy_2,$$

then

$$\bar{z}_1 + \bar{z}_2 = x_1 - iy_1 + x_2 - iy_2 = x_1 + x_2 - i(y_1 + y_2).$$

The conjugate of $(z_1 + z_2)$ is

$$\overline{(x_1+x_2) + i(y_1+y_2)} = (x_1+x_2) - i(y_1+y_2).$$

Similarly, you can show that the conjugate of the difference (or product or quotient) of two complex numbers is equal to the difference (or product or quotient) of the conjugates of the numbers (Problem 25). In other words, you can get the conjugate of an expression containing i's by just changing the signs of all the i terms. We must watch out for hidden i's, however.

Example. If

$$z = \frac{2-3i}{i+4}, \qquad \text{then} \qquad \bar{z} = \frac{2+3i}{-i+4}.$$

But if $z = f+ig$, where f and g are themselves complex, then the complex conjugate of z is $\bar{z} = \bar{f} - i\bar{g}$ (*not* $f - ig$).

PROBLEMS, SECTION 5

25. Prove that the conjugate of the quotient of two complex numbers is the quotient of the conjugates. Also prove the corresponding statements for difference and product. *Hint:* It is easier to prove the statements about product and quotient using the polar coordinate $re^{i\theta}$ form; for the difference, it is easier to use the rectangular form $x + iy$.

C. Finding the Absolute Value of z

Recall that the definition of $|z|$ is $|z| = r = \sqrt{x^2+y^2}$ (positive square root!). Since $z\bar{z} = (x+iy)(x-iy) = x^2+y^2$, or, in polar coordinates, $z\bar{z} = (re^{i\theta})(re^{-i\theta}) = r^2$, we see that $|z|^2 = z\bar{z}$, or $|z| = \sqrt{z\bar{z}}$. Note that $z\bar{z}$ is always real and ≥ 0, since x, y, and r are real. We have

$$|z| = r = \sqrt{x^2+y^2} = \sqrt{z\bar{z}}. \tag{5.1}$$

By Problem 25 and (5.1), the absolute value of a quotient of two complex numbers is the quotient of the absolute values (and a similar statement for product).

Example.

$$\left|\frac{\sqrt{5}+3i}{1-i}\right| = \frac{|\sqrt{5}+3i|}{|1-i|} = \frac{\sqrt{14}}{\sqrt{2}} = \sqrt{7}.$$

PROBLEMS, SECTION 5

Find the absolute value of each of the following using the discussion above. Try to do simple problems like these in your head—it saves time.

26. $\dfrac{2i-1}{i-2}$ **27.** $\dfrac{2+3i}{1-i}$ **28.** $\dfrac{z}{\bar{z}}$

29. $(1+2i)^3$ **30.** $\dfrac{3i}{i-\sqrt{3}}$ **31.** $\dfrac{5-2i}{5+2i}$

32. $(2-3i)^4$ **33.** $\dfrac{25}{3+4i}$ **34.** $\left(\dfrac{1+i}{1-i}\right)^5$

D. Complex Equations

In working with equations involving complex quantities, we must always remember that a complex number is actually a pair of real numbers. Two complex numbers are equal if and only if their real parts are equal and their imaginary parts are equal. For example, $x + iy = 2 + 3i$ means $x = 2$ and $y = 3$. In other words, any equation involving complex numbers is really two equations involving real numbers.

Example. Find x and y if

$$(x + iy)^2 = 2i. \tag{5.2}$$

Since $(x + iy)^2 = x^2 + 2ixy - y^2$, (5.2) is equivalent to the two real equations

$$x^2 - y^2 = 0,$$
$$2xy = 2.$$

From the first equation $y^2 = x^2$, we find $y = x$ or $y = -x$. Substituting these into the second equation gives

$$2x^2 = 2 \qquad \text{or} \qquad -2x^2 = 2.$$

Since x is real, x^2 cannot be negative. Thus we find only

$$x^2 = 1 \qquad \text{and} \qquad y = x,$$

that is,

$$x = y = 1 \qquad \text{and} \qquad x = y = -1.$$

PROBLEMS, SECTION 5

Solve for all possible values of the real numbers x and y in the following equations.

35. $x + iy = 3i - 4$

36. $2ix + 3 = y - i$

37. $x + iy = 0$

38. $x + iy = 2i - 7$

39. $x + iy = y + ix$

40. $x + iy = 3i - ix$

41. $(2x - 3y - 5) + i(x + 2y + 1) = 0$

42. $(x + 2y + 3) + i(3x - y - 1) = 0$

43. $(x + iy)^2 = 2ix$

44. $x + iy = (1 - i)^2$

45. $(x + iy)^2 = (x - iy)^2$

46. $\dfrac{x + iy}{x - iy} = -i$

47. $(x + iy)^3 = -1$

48. $\dfrac{x + iy + 2 + 3i}{2x + 2iy - 3} = i + 2$

49. $|1 - (x + iy)| = x + iy$

50. $|x + iy| = y - ix$

E. Graphs

Using the graphical representation of the complex number z as the point (x, y) in a plane, we can give geometrical meaning to equations and inequalities involving z.

Example 1. What is the curve made up of the points in the (x, y) plane satisfying the equation $|z| = 3$?

Since

$$|z| = \sqrt{x^2 + y^2},$$

the given equation is

$$\sqrt{x^2 + y^2} = 3 \quad \text{or} \quad x^2 + y^2 = 9.$$

Thus $|z| = 3$ is the equation of a circle of radius 3 with center at the origin. Such an equation might describe, for example, the path of an electron or of a satellite. (See Section F below.)

Example 2.

(a) $|z - 1| = 2$. This is the circle $(x - 1)^2 + y^2 = 4$.

(b) $|z - 1| \leq 2$. This is the disk whose boundary is the circle in (a).

Note that we use "circle" to mean a curve and "disk" to mean an area. The interior of the disk is given by $|z - 1| < 2$.

Example 3. (Angle of z) $= \pi/4$. This is the half-line $y = x$ with $x > 0$; this might be the path of a light ray starting at the origin.

Example 4. $\operatorname{Re} z > \frac{1}{2}$. This is the half-plane $x > \frac{1}{2}$.

PROBLEMS, SECTION 5

Describe geometrically the set of points in the complex plane satisfying the following equations.

51. $|z| = 2$

52. $\operatorname{Re} z = 0$

53. $|z - 1| = 1$

54. $|z - 1| < 1$

55. $z - \bar{z} = 5i$

56. angle of $z = \dfrac{\pi}{2}$

57. $\operatorname{Re}(z^2) = 4$

58. $\operatorname{Re} z > 2$

59. $|z + 3i| = 4$

60. $|z - 1 + i| = 2$

61. $\operatorname{Im} z < 0$

62. $|z + 1| + |z - 1| = 8$

63. $z^2 = \bar{z}^2$

64. $z^2 = -\bar{z}^2$

65. Show that $|z_1 - z_2|$ is the distance between the points z_1 and z_2 in the complex plane. Use this result to identify the graphs in Problems 53, 54, 59, and 60 without computation.

F. Physical Applications

Problems in physics as well as geometry may often be simplified by using one complex equation instead of two real equations. See the following example and also Section 16.

Example. A particle moves in the (x, y) plane so that its position (x, y) as a function of time t is given by

$$z = x + iy = \frac{i + 2t}{t - i}.$$

Find the magnitudes of its velocity and its acceleration as functions of t.

We *could* write z in $x + iy$ form and so find x and y as functions of t. It is easier to do the problem as follows. We define the complex velocity and complex acceleration by

$$\frac{dz}{dt} = \frac{dx}{dt} + i\frac{dy}{dt} \quad \text{and} \quad \frac{d^2z}{dt^2} = \frac{d^2x}{dt^2} + i\frac{d^2y}{dt^2}.$$

Then the magnitude v of the velocity is $v = \sqrt{(dx/dt)^2 + (dy/dt)^2} = |dz/dt|$, and similarly the magnitude a of the acceleration is $a = |d^2z/dt^2|$. Thus we have

$$\frac{dz}{dt} = \frac{2(t - i) - (i + 2t)}{(t - i)^2} = \frac{-3i}{(t - i)^2}.$$

$$v = \left|\frac{dz}{dt}\right| = \sqrt{\frac{-3i}{(t - i)^2} \cdot \frac{+3i}{(t + i)^2}} = \frac{3}{t^2 + 1},$$

$$\frac{d^2z}{dt^2} = \frac{(-3i)(-2)}{(t - i)^3} = \frac{6i}{(t - i)^3},$$

$$a = \left|\frac{d^2z}{dt^2}\right| = \frac{6}{(t^2 + 1)^{3/2}}.$$

Note carefully that all physical quantities (x, y, v, and a) are real; the complex expressions are used just for convenience in calculation.

PROBLEMS, SECTION 5

66. Find x and y as functions of t for the example above, and verify for this case that v and a are correctly given by the method of the example.

67. Find v and a if $z = (1 - it)/(2t + i)$.

68. Find v and a if $z = \cos 2t + i \sin 2t$. Can you describe the motion?

6. COMPLEX INFINITE SERIES

In Chapter 1 we considered infinite series whose terms were real. We shall be very much interested in series with complex terms; let us reconsider our definitions and theorems for this case. The partial sums of a series of complex numbers will be complex numbers, say $S_n = X_n + iY_n$, where X_n and Y_n are real. Convergence is defined just as for real series: If S_n approaches a limit $S = X + iY$ as $n \to \infty$, we call the series convergent and call S its sum. This means that $X_n \to X$ and $Y_n \to Y$; in other words, the real and the imaginary parts of the series are each convergent series.

It is useful, just as for real series, to discuss absolute convergence first. It can be proved (Problem 1) that an absolutely convergent series converges. Absolute convergence means here, just as for real series, that the series of absolute values of the terms is a convergent series. Remember that $|z| = r = \sqrt{x^2 + y^2}$ is a positive number. Thus any of the tests given in Chapter 1 for convergence of series of positive terms may be used here to test a complex series for absolute convergence.

Example 1. Test for convergence

$$1+\frac{1+i}{2}+\frac{(1+i)^2}{4}+\frac{(1+i)^3}{8}+\cdots+\frac{(1+i)^n}{2^n}+\cdots.$$

Using the ratio test, we find

$$\rho=\lim_{n\to\infty}\left|\frac{(1+i)^{n+1}}{2^{n+1}}\div\frac{(1+i)^n}{2^n}\right|=\lim_{n\to\infty}\left|\frac{1+i}{2}\right|=\left|\frac{1+i}{2}\right|=\frac{\sqrt{2}}{2}<1.$$

. Since $\rho<1$, the series is absolutely convergent and therefore convergent.

Example 2. Test for convergence $\sum_1^\infty i^n/\sqrt{n}$. Here the ratio test gives 1 so we must try a different test. Let's write out a few terms of the series:

$$i-\frac{1}{\sqrt{2}}-\frac{i}{\sqrt{3}}+\frac{1}{\sqrt{4}}+\frac{i}{\sqrt{5}}-\frac{1}{\sqrt{6}}\cdots.$$

We see that the real part of the series is

$$-\frac{1}{\sqrt{2}}+\frac{1}{\sqrt{4}}-\frac{1}{\sqrt{6}}+\cdots=\sum_1^\infty\frac{(-1)^n}{\sqrt{2n}},$$

and the imaginary part of the series is

$$1-\frac{1}{\sqrt{3}}+\frac{1}{\sqrt{5}}\cdots=\sum_0^\infty\frac{(-1)^n}{\sqrt{2n+1}}.$$

Verify that both these series satisfy the alternating series test for convergence. Thus, the original series converges.

Example 3. Test for convergence $\sum_0^\infty z^n=\sum_0^\infty(re^{i\theta})^n=\sum_0^\infty r^ne^{in\theta}$. This is a geometric series with ratio $=z=re^{i\theta}$; it converges if and only if $|z|<1$. Recall that $|z|=r$. Thus, $\sum_0^\infty r^ne^{in\theta}$ converges if and only if $r<1$.

PROBLEMS, SECTION 6

1. Prove that an absolutely convergent series of complex numbers converges. This means to prove that $\sum(a_n+ib_n)$ converges (a_n and b_n real) if $\sum\sqrt{a_n^2+b_n^2}$ converges. *Hint:* Convergence of $\sum(a_n+ib_n)$ means that $\sum a_n$ and $\sum b_n$ *both* converge. Compare $\sum|a_n|$ and $\sum|b_n|$ with $\sum\sqrt{a_n^2+b_n^2}$, and use Problem 7.9 of Chapter 1.

Test each of the following series for convergence.

2. $\sum(1+i)^n$

3. $\sum\frac{1}{(1+i)^n}$

4. $\sum\left(\frac{1-i}{1+i}\right)^n$

5. $\sum\left(\frac{1}{n^2}+\frac{i}{n}\right)$

6. $\sum\frac{1+i}{n^2}$

7. $\sum\frac{(i-1)^n}{n}$

8. $\sum e^{in\pi/6}$

9. $\sum\frac{i^n}{n}$

10. $\sum\left(\frac{1+i}{1-i\sqrt{3}}\right)^n$

11. $\sum\left(\frac{2+i}{3-4i}\right)^{2n}$

12. $\sum\frac{(3+2i)^n}{n!}$

13. $\sum\left(\frac{1+i}{2-i}\right)^n$

14. Prove that a series of complex terms diverges if $\rho>1$ ($\rho=$ ratio test limit). *Hint:* The nth term of a convergent series tends to zero.

7. COMPLEX POWER SERIES; DISK OF CONVERGENCE

In Chapter 1 we considered series of powers of x, $\sum a_n x^n$. We are now interested in series of powers of z,

$$\sum a_n z^n, \tag{7.1}$$

where $z = x + iy$, and the a_n are complex numbers. [Notice that (7.1) includes real series as a special case since $z = x$ if $y = 0$.] Here are some examples.

$$1 - z + \frac{z^2}{2} - \frac{z^3}{3} + \frac{z^4}{4} + \cdots, \tag{7.2a}$$

$$1 + iz + \frac{(iz)^2}{2!} + \frac{(iz)^3}{3!} + \cdots = 1 + iz - \frac{z^2}{2!} - \frac{iz^3}{3!} + \cdots, \tag{7.2b}$$

$$\sum_{n=0}^{\infty} \frac{(z+1-i)^n}{3^n\, n^2}. \tag{7.2c}$$

Let us use the ratio test to find for what z these series are absolutely convergent. For (7.2a), we have

$$\rho = \lim_{n\to\infty} \left| \frac{z \cdot n}{n+1} \right| = |z|.$$

The series converges if $\rho < 1$, that is, if $|z| < 1$, or $\sqrt{x^2 + y^2} < 1$. This is the interior of a disk of radius 1 with center at the origin in the complex plane. This disk is called the *disk of convergence* of the infinite series and the radius of the disk is called the *radius of convergence.* The disk of convergence replaces the interval of convergence which we had for real series. In fact (see Figure 7.1), the interval of convergence for the series $\sum(-x)^n/n$ is just the interval $(-1, 1)$ on the x axis contained within the disk of convergence of $\sum(-z)^n/n$, as it must be since x is the value of z when $y = 0$. For this reason we sometimes speak of the *radius* of convergence of a power series even though we are considering only real values of z. (Also see Chapter 14, Equations (2.5) and (2.6) and Figure 2.4.)

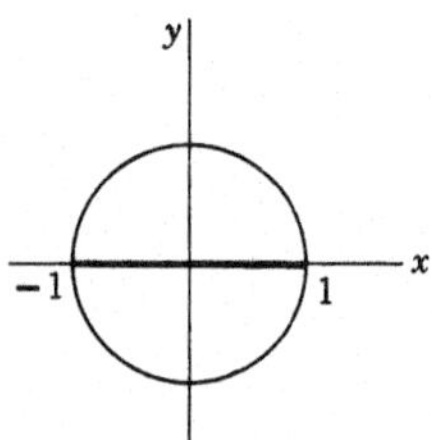

Figure 7.1

Next consider series (7.2b); here we have

$$\rho = \lim_{n\to\infty} \left| \frac{(iz)^{n+1}}{(n+1)!} \div \frac{(iz)^n}{n!} \right| = \lim_{n\to\infty} \left| \frac{iz}{n+1} \right| = 0.$$

This is an example of a series which converges for all values of z. For series (7.2c), we have

$$\rho = \lim_{n\to\infty} \left| \frac{(z+1-i)}{3} \frac{n^2}{(n+1)^2} \right| = \left| \frac{z+1-i}{3} \right|.$$

Thus, this series converges for

$$|z + 1 - i| < 3, \text{ or } |z - (-1 + i)| < 3.$$

This is the interior of a disk (Figure 7.2) of radius 3 and center at $z = -1 + i$ (see Problem 5.65).

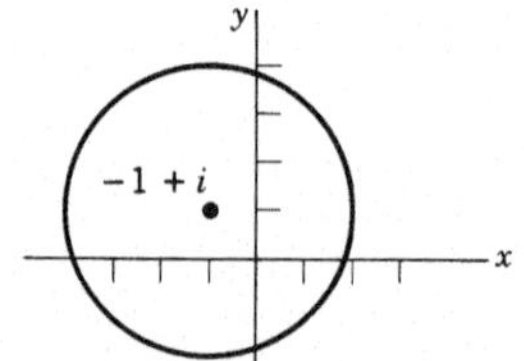

Figure 7.2

Just as for real series, if $\rho > 1$, the series diverges (Problem 6.14). For $\rho = 1$ (that is, on the boundary of the disk of convergence) the series may either converge or diverge. It may be difficult to find out which and we shall not in general need to consider the question.

The four theorems about power series (Chapter 1, Section 11) are true also for complex series (replace *interval* by *disk* of convergence). Also we can now state for Theorem 2 what the disk of convergence is for the quotient of two series of powers of z. Assume to start with that any common factor z has been cancelled. Let r_1 and r_2 be the radii of convergence of the numerator and denominator series. Find the closest point to the origin in the complex plane where the denominator is zero; call the distance from the origin to this point s. Then the quotient series converges at least inside the smallest of the three disks of radii r_1, r_2, and s, with center at the origin. (See Chapter 14, Section 2.)

Example. Find the disk of convergence of the Maclaurin series for $(\sin z)/[z(1+z^2)]$.

We shall soon see that the series for $\sin z$ has the same form as the real series for $\sin x$ in Chapter 1. Using this fact we find (Problem 17)

$$\frac{\sin z}{z(1+z^2)} = 1 - \frac{7z^2}{6} + \frac{47z^4}{40} - \frac{5923z^6}{5040} + \cdots. \tag{7.3}$$

From (7.3) we can't find the radius of convergence, but let's use the theorem above. Let the numerator series be $(\sin z)/z$. By ratio test, the series for $(\sin z)/z$ converges for all z (if you like, $r_1 = \infty$). There is no r_2 since the denominator is not an infinite series. The denominator $1 + z^2$ is zero when $z = \pm i$, so $s = 1$. Then the series (7.3) converges inside a disk of radius 1 with center at the origin.

PROBLEMS, SECTION 7

Find the disk of convergence for each of the following complex power series.

1. $e^z = 1 + z + \dfrac{z^2}{2!} + \dfrac{z^3}{3!} \cdots$ [equation (8.1)]

2. $z - \dfrac{z^2}{2} + \dfrac{z^3}{3} - \dfrac{z^4}{4} + \cdots$

3. $1 - \dfrac{z^2}{3!} + \dfrac{z^4}{5!} - \cdots$

4. $\displaystyle\sum_{n=0}^{\infty} z^n$

5. $\displaystyle\sum_{n=0}^{\infty} \left(\frac{z}{2}\right)^n$

6. $\displaystyle\sum_{n=1}^{\infty} n^2(3iz)^n$

7. $\displaystyle\sum_{n=0}^{\infty} \frac{(-1)^n z^{2n}}{(2n)!}$

8. $\displaystyle\sum_{n=1}^{\infty} \frac{z^{2n}}{(2n+1)!}$

9. $\displaystyle\sum_{n=1}^{\infty} \frac{z^n}{\sqrt{n}}$

10. $\displaystyle\sum_{n=1}^{\infty} \frac{(iz)^n}{n^2}$

11. $\displaystyle\sum_{n=0}^{\infty} \frac{(n!)^3 z^n}{(3n)!}$

12. $\displaystyle\sum_{n=0}^{\infty} \frac{(n!)^2 z^n}{(2n)!}$

13. $\displaystyle\sum_{n=1}^{\infty} \frac{(z-i)^n}{n}$

14. $\displaystyle\sum_{n=0}^{\infty} n(n+1)(z-2i)^n$

15. $\displaystyle\sum_{n=0}^{\infty} \frac{(z-2+i)^n}{2^n}$

16. $\displaystyle\sum_{n=1}^{\infty} 2^n(z+i-3)^{2n}$

17. Verify the series in (7.3) by computer. Also show that it can be written in the form

$$\sum_{n=0}^{\infty}(-1)^n z^{2n}\sum_{k=0}^{n}\frac{1}{(2k+1)!}.$$

Use this form to show by ratio test that the series converges in the disk $|z|<1$.

8. ELEMENTARY FUNCTIONS OF COMPLEX NUMBERS

The so-called elementary functions are powers and roots, trigonometric and inverse trigonometric functions, logarithmic and exponential functions, and combinations of these. All these you can compute or find in tables, as long as you want them as functions of real numbers. Now we want to find things like i^i, $\sin(1+i)$, or $\ln i$. These are not just curiosities for the amusement of the mathematically inclined, but may turn up to be evaluated in applied problems. To be sure, the values of experimental measurements are not imaginary. But the values of $\operatorname{Re} z$, $\operatorname{Im} z$, $|z|$, angle of z, are real, and these are the quantities which have experimental meaning. Meanwhile, mathematical solutions of problems may involve manipulations of complex numbers before we arrive finally at a real answer to compare with experiment.

Polynomials and rational functions (quotients of polynomials) of z are easily evaluated.

Example. If $f(z)=(z^2+1)/(z-3)$, we find $f(i-2)$ by substituting $z=i-2$:

$$f(i-2)=\frac{(i-2)^2+1}{i-2-3}=\frac{-4i+4}{i-5}\cdot\frac{-i-5}{-i-5}=\frac{8i-12}{13}.$$

Next we want to investigate the possible meaning of other functions of complex numbers. We should like to define expressions like e^z or $\sin z$ so that they will obey the familiar laws we know for the corresponding real expressions [for example, $\sin 2x=2\sin x\cos x$, or $(d/dx)e^x=e^x$]. We must, for consistency, define functions of complex numbers so that any equations involving them reduce to correct real equations when $z=x+iy$ becomes $z=x$, that is, when $y=0$. These requirements will be met if we define e^z by the power series

$$e^z=\sum_{0}^{\infty}\frac{z^n}{n!}=1+z+\frac{z^2}{2!}+\frac{z^3}{3!}+\cdots. \tag{8.1}$$

This series converges for all values of the complex number z (Problem 7.1) and therefore gives us the value of e^z for any z. If we put $z=x$ (x real), we get the familiar series for e^x.

It is easy to show, by multiplying the series (Problem 1), that

$$e^{z_1}\cdot e^{z_2}=e^{z_1+z_2}. \tag{8.2}$$

In Chapter 14 we shall consider in detail the meaning of derivatives with respect to a complex z. However, it is worth while for you to know that $(d/dz)z^n=nz^{n-1}$, and that, in fact, the other differentiation and integration formulas which you know from elementary calculus hold also with x replaced by z. You can verify that $(d/dz)e^z = e^z$ when e^z is defined by (8.1) by differentiating (8.1) term by term (Problem 2). It can be shown that (8.1) is the only definition of e^z which preserves these familiar formulas. We now want to consider the consequences of this definition.

PROBLEMS, SECTION 8

Show from the power series (8.1) that

1. $e^{z_1} \cdot e^{z_2} = e^{z_1+z_2}$

2. $\dfrac{d}{dz}e^z = e^z$

3. Find the power series for $e^x \cos x$ and for $e^x \sin x$ from the series for e^z in the following way: Write the series for e^z; put $z = x + iy$. Show that $e^z = e^x(\cos y + i \sin y)$; take real and imaginary parts of the equation, and put $y = x$.

9. EULER'S FORMULA

For real θ, we know from Chapter 1 the power series for $\sin\theta$ and $\cos\theta$:

$$\begin{aligned} \sin\theta &= \theta - \frac{\theta^3}{3!} + \frac{\theta^5}{5!} - \cdots, \\ \cos\theta &= 1 - \frac{\theta^2}{2!} + \frac{\theta^4}{4!} - \cdots. \end{aligned} \tag{9.1}$$

From our definition (8.1), we can write the series for e to any power, real or imaginary. We write the series for $e^{i\theta}$, where θ is real:

$$\begin{aligned} e^{i\theta} &= 1 + i\theta + \frac{(i\theta)^2}{2!} + \frac{(i\theta)^3}{3!} + \frac{(i\theta)^4}{4!} + \frac{(i\theta)^5}{5!} + \cdots \\ &= 1 + i\theta - \frac{\theta^2}{2!} - i\frac{\theta^3}{3!} + \frac{\theta^4}{4!} + i\frac{\theta^5}{5!} \cdots \\ &= 1 - \frac{\theta^2}{2!} + \frac{\theta^4}{4!} + \cdots + i\left(\theta - \frac{\theta^3}{3!} + \frac{\theta^5}{5!} \cdots\right). \end{aligned} \tag{9.2}$$

(The rearrangement of terms is justified because the series is absolutely convergent.) Now compare (9.1) and (9.2); the last line in (9.2) is just $\cos\theta + i\sin\theta$. We then have the very useful result we introduced in Section 3, known as Euler's formula:

$$e^{i\theta} = \cos\theta + i\sin\theta. \tag{9.3}$$

Thus we have justified writing any complex number as we did in (4.1), namely

$$z = x + iy = r(\cos\theta + i\sin\theta) = re^{i\theta}. \tag{9.4}$$

Here are some examples of the use of (9.3) and (9.4). These problems can be done very quickly graphically or just by picturing them in your mind.

Examples. Find the values of $2\,e^{i\pi/6}$, $e^{i\pi}$, $3\,e^{-i\pi/2}$, $e^{2n\pi i}$.

$2\,e^{i\pi/6}$ is $re^{i\theta}$ with $r=2$, $\theta=\pi/6$. From Figure 9.1, $x=\sqrt{3}$, $y=1$, $x+iy=\sqrt{3}+i$, so $2\,e^{i\pi/6}=\sqrt{3}+i$.

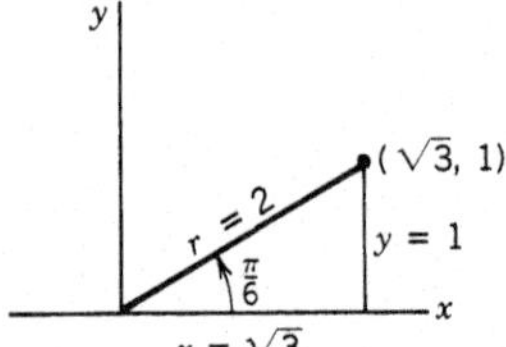

Figure 9.1

$e^{i\pi}$ is $re^{i\theta}$ with $r=1$, $\theta=\pi$. From Figure 9.2, $x=-1$, $y=0$, $x+iy=-1+0i$, so $e^{i\pi}=-1$. Note that $r=1$ and $\theta=-\pi, \pm 3\pi, \pm 5\pi, \cdots$, give the same point, so $e^{-i\pi}=-1$, $e^{3\pi i}=-1$, and so on.

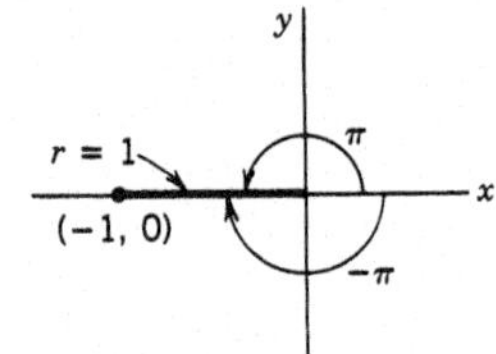

Figure 9.2

$3e^{-i\pi/2}$ is $re^{i\theta}$ with $r=3$, $\theta=-\pi/2$. From Figure 9.3, $x=0$, $y=-3$, so $3e^{-i\pi/2}=x+iy=0-3i=-3i$.

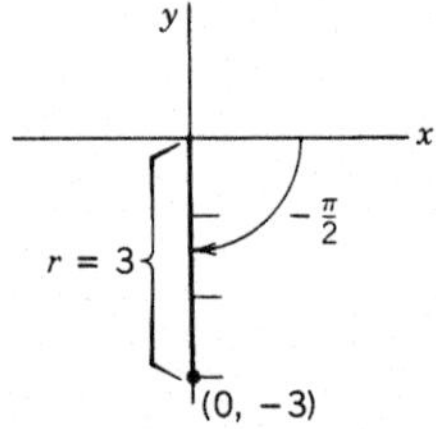

Figure 9.3

$e^{2n\pi i}$ is $re^{i\theta}$ with $r=1$ and $\theta=2n\pi=n(2\pi)$; that is, θ is an integral multiple of 2π. From Figure 9.4, $x=1$, $y=0$, so $e^{2n\pi i}=1+0i=1$.

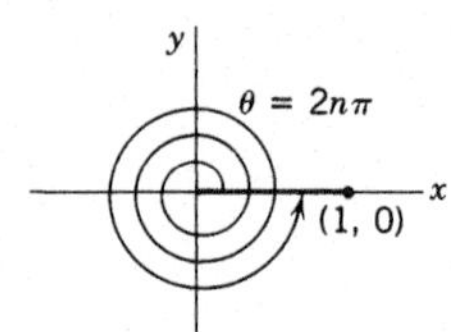

Figure 9.4

It is often convenient to use Euler's formula when we want to multiply or divide complex numbers. From (8.2) we obtain two familiar looking laws of exponents which are now valid for imaginary exponents:

$$e^{i\theta_1}\cdot e^{i\theta_2}=e^{i(\theta_1+\theta_2)},$$
$$e^{i\theta_1}\div e^{i\theta_2}=e^{i(\theta_1-\theta_2)}. \tag{9.5}$$

Remembering that *any* complex number can be written in the form $re^{i\theta}$ by (9.4), we get

(9.6)
$$z_1 \cdot z_2 = r_1\, e^{i\theta_1} \cdot r_2\, e^{i\theta_2} = r_1\, r_2\, e^{i(\theta_1+\theta_2)},$$
$$z_1 \div z_2 = \frac{r_1}{r_2}\, e^{i(\theta_1-\theta_2)}.$$

In words, to multiply two complex numbers, we multiply their absolute values and add their angles. To divide two complex numbers, we divide the absolute values and subtract the angles.

Example. Evaluate $(1+i)^2/(1-i)$. From Figure 5.1 we have $1+i = \sqrt{2}\, e^{i\pi/4}$. We plot $1-i$ in Figure 9.5 and find $r = \sqrt{2}$, $\theta = -\pi/4$ (or $+7\pi/4$), so $1-i = \sqrt{2}\, e^{-i\pi/4}$. Then

$$\frac{(1+i)^2}{1-i} = \frac{(\sqrt{2}\, e^{i\pi/4})^2}{\sqrt{2}\, e^{-i\pi/4}} = \frac{2\, e^{i\pi/2}}{\sqrt{2}\, e^{-i\pi/4}} = \sqrt{2}\, e^{3i\pi/4}.$$

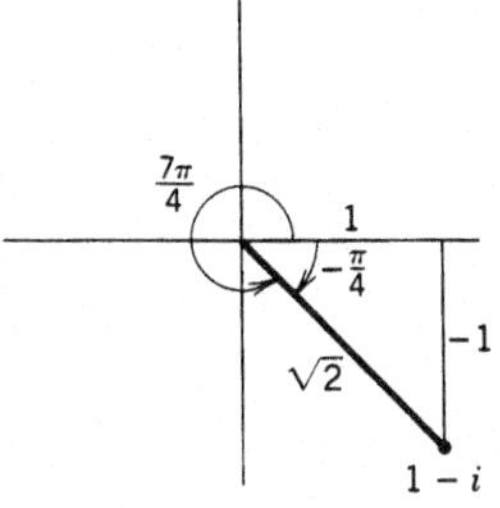

Figure 9.5

From Figure 9.6, we find $x = -1$, $y = 1$, so

$$\frac{(1+i)^2}{1-i} = x + iy = -1 + i.$$

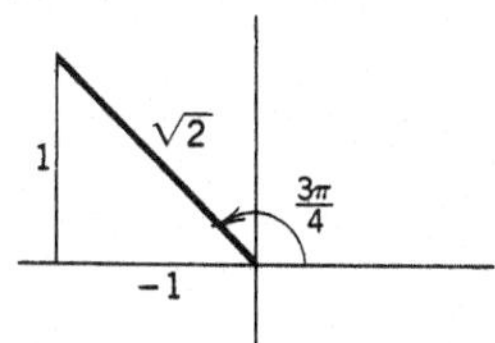

Figure 9.6

We could use degrees in this problem. By (9.6), we find that the angle of $(1+i)^2/(1-i)$ is $2(45°) - (-45°) = 135°$ as in Figure 9.6.

PROBLEMS, SECTION 9

Express the following complex numbers in the $x+iy$ form. Try to visualize each complex number, using sketches as in the examples if necessary. The first twelve problems you should be able to do in your head (and maybe some of the others—try it!) Doing a problem quickly in your head saves time over using a computer. Remember that the point in doing problems like this is to gain skill in manipulating complex expressions, so a good study method is to do the problems by hand and use a computer to check your answers.

1. $e^{-i\pi/4}$ **2.** $e^{i\pi/2}$ **3.** $9\, e^{3\pi i/2}$

4. $e^{(1/3)(3+4\pi i)}$ **5.** $e^{5\pi i}$ **6.** $e^{-2\pi i} - e^{-4\pi i} + e^{-6\pi i}$

7. $3\, e^{2(1+i\pi)}$ **8.** $2\, e^{5\pi i/6}$ **9.** $2\, e^{-i\pi/2}$

10. $e^{i\pi} + e^{-i\pi}$ **11.** $\sqrt{2}\, e^{5i\pi/4}$ **12.** $4\, e^{-8i\pi/3}$

13. $\dfrac{(i-\sqrt{3})^3}{1-i}$ **14.** $\left(1+i\sqrt{3}\right)^6$ **15.** $(1+i)^2 + (1+i)^4$

16. $\left(i-\sqrt{3}\right)\left(1+i\sqrt{3}\right)$ **17.** $\dfrac{1}{(1+i)^3}$ **18.** $\left(\dfrac{1+i}{1-i}\right)^4$

19. $(1-i)^8$ **20.** $\left(\dfrac{\sqrt{2}}{i-1}\right)^{10}$ **21.** $\left(\dfrac{1-i}{\sqrt{2}}\right)^{40}$

22. $\left(\dfrac{1-i}{\sqrt{2}}\right)^{42}$ **23.** $\dfrac{(1+i)^{48}}{(\sqrt{3}-i)^{25}}$ **24.** $\dfrac{(1-i\sqrt{3})^{21}}{(i-1)^{38}}$

25. $\left(\dfrac{i\sqrt{2}}{1+i}\right)^{12}$ **26.** $\left(\dfrac{2i}{i+\sqrt{3}}\right)^{19}$

27. Show that for any real y, $|e^{iy}| = 1$. Hence show that $|e^z| = e^x$ for every complex z.

28. Show that the absolute value of a product of two complex numbers is equal to the product of the absolute values. Also show that the absolute value of the quotient of two complex numbers is the quotient of the absolute values. *Hint:* Write the numbers in the $re^{i\theta}$ form.

Use Problems 27 and 28 to find the following absolute values. If you understand Problems 27 and 28 and equation (5.1), you should be able to do these in your head.

29. $|e^{i\pi/2}|$ **30.** $|e^{\sqrt{3}-i}|$ **31.** $|5\,e^{2\pi i/3}|$ **32.** $|3e^{2+4i}|$

33. $|2\,e^{3+i\pi}|$ **34.** $|4\,e^{2i-1}|$ **35.** $|3\,e^{5i}\cdot 7\,e^{-2i}|$ **36.** $|2\,e^{i\pi/6}|^2$

37. $\left|\dfrac{1+i}{1-i}\right|$ **38.** $\left|\dfrac{e^{i\pi}}{1+i}\right|$

10. POWERS AND ROOTS OF COMPLEX NUMBERS

Using the rules (9.6) for multiplication and division of complex numbers, we have

$$z^n = (re^{i\theta})^n = r^n e^{in\theta} \tag{10.1}$$

for any integral n. In words, to obtain the nth power of a complex number, we take the nth power of the modulus and multiply the angle by n. The case $r = 1$ is of particular interest. Then (10.1) becomes DeMoivre's theorem:

$$(e^{i\theta})^n = (\cos\theta + i\sin\theta)^n = \cos n\theta + i\sin n\theta. \tag{10.2}$$

You can use this equation to find the formulas for $\sin 2\theta$, $\cos 2\theta$, $\sin 3\theta$, etc. (Problems 27 and 28).

The nth root of z, $z^{1/n}$, means a complex number whose nth power is z. From (10.1) you can see that this is

$$z^{1/n} = (re^{i\theta})^{1/n} = r^{1/n}e^{i\theta/n} = \sqrt[n]{r}\left(\cos\frac{\theta}{n} + i\sin\frac{\theta}{n}\right). \tag{10.3}$$

This formula must be used with care (see Examples 2 to 4 below).

Some examples will show how useful these formulas are.

Example 1.

$$[\cos(\pi/10) + i\sin(\pi/10)]^{25} = (e^{i\pi/10})^{25} = e^{2\pi i}e^{i\pi/2} = 1\cdot i = i.$$

Example 2. Find the cube roots of 8. We know that 2 is a cube root of 8, but there are also two complex cube roots of 8; let us see why. Plot the complex number 8 (that is, $x = 8$, $y = 0$) in the complex plane; the polar coordinates of the point are $r = 8$, and $\theta = 0$, or $360°, 720°, 1080°$, etc. (We can use either degrees or radians here; read the end of Section 3.) Now by equation (10.3), $z^{1/3} = r^{1/3}e^{i\theta/3}$; that is, to find the polar coordinates of the cube root of a number $re^{i\theta}$, we find the cube root of r and divide the angle by 3. Then the polar coordinates of $\sqrt[3]{8}$ are

$$
\begin{aligned}
r = 2, \quad \theta &= 0°, \quad 360°/3, \quad 720°/3, \quad 1080°/3 \cdots \\
&= 0°, \quad 120°, \quad 240°, \quad 360° \cdots .
\end{aligned}
\tag{10.4}
$$

We plot these points in Figure 10.1. Observe that the point (2, 0°) and the point (2, 360°) are the same. The points in (10.4) are all on a circle of radius 2 and are equally spaced $360°/3 = 120°$ apart. Starting with $\theta = 0$, if we add 120° repeatedly, we just repeat the three angles shown. Thus, there are exactly three cube roots for any number z, always on a circle of radius $\sqrt[3]{|z|}$ and spaced 120° apart.

Now to find the values of $\sqrt[3]{8}$ in rectangular form, we can read them from Figure 10.1, or we can calculate them from $z = r(\cos\theta + i\sin\theta)$ with $r = 2$ and $\theta = 0$, $120° = 2\pi/3$, $240° = 4\pi/3$. We can also use a computer to solve the equation $z^3 = 8$. By any of these methods we find

$$\sqrt[3]{8} = \{2,\ -1 + i\sqrt{3},\ -1 - i\sqrt{3}\}.$$

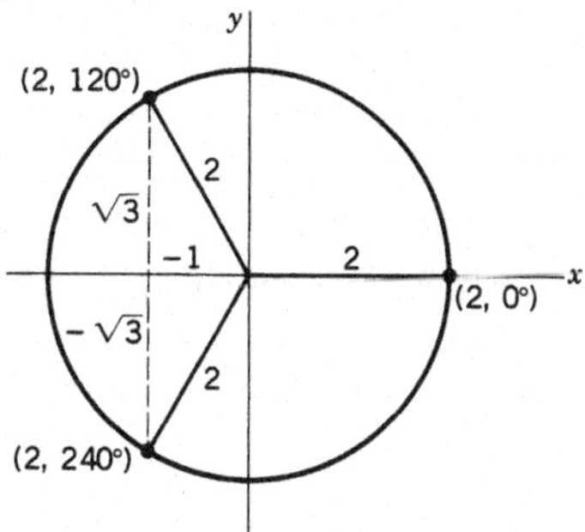

Figure 10.1

Example 3. Find and plot all values of $\sqrt[4]{-64}$. From Figure 10.2 (or by visualizing a plot of -64), we see that the polar coordinates of -64 are $r = 64$, $\theta = \pi + 2k\pi$

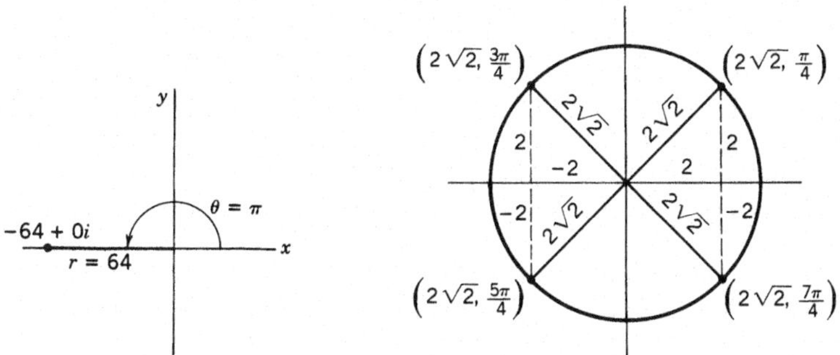

Figure 10.2 Figure 10.3

(where $k = 0, 1, 2, \cdots$). Then since $z^{1/4} = r^{1/4}e^{i\theta/4}$, the polar coordinates of $\sqrt[4]{-64}$ are

$$
\begin{aligned}
r &= \sqrt[4]{64} = 2\sqrt{2}, \\
\theta &= \frac{\pi}{4}, \frac{\pi + 2\pi}{4}, \frac{\pi + 4\pi}{4}, \frac{\pi + 6\pi}{4}, \cdots = \frac{\pi}{4}, \frac{3\pi}{4}, \frac{5\pi}{4}, \frac{7\pi}{4}.
\end{aligned}
$$

We plot these points in Figure 10.3. Observe that they are all on a circle of radius $2\sqrt{2}$, equally spaced $2\pi/4 = \pi/2$ apart. Starting with $\theta = \pi/4$, we add $\pi/2$

repeatedly, and find exactly 4 fourth roots. We can read the values of $\sqrt[4]{-64}$ in rectangular form from Figure 10.3:

$$\sqrt[4]{-64} = \pm 2 \pm 2i \quad \text{(all four combinations of } \pm \text{ signs)}$$

or we can calculate them as in Example 2, or we can solve the equation $z^4 = -64$ by computer.

Example 4. Find and plot all values of $\sqrt[6]{-8i}$. The polar coordinates of $-8i$ are $r = 8$, $\theta = 270^\circ + 360^\circ k = 3\pi/2 + 2\pi k$. Then the polar coordinates of $\sqrt[6]{-8i}$ are

$$(10.5) \qquad r = \sqrt{2}, \quad \theta = \frac{270^\circ + 360^\circ k}{6} = 45^\circ + 60^\circ k \quad \text{or} \quad \theta = \frac{\pi}{4} + \frac{\pi}{3}k.$$

In Figure 10.4, we sketch a circle of radius $\sqrt{2}$. On it we plot the point at 45° and then plot the rest of the 6 equally spaced points 60° apart. To find the roots in rectangular coordinates, we need to find all the values of $r(\cos\theta + i\sin\theta)$ with r and θ given by (10.5). We can do this one root at a time or more simply by using a computer to solve the equation $z^6 = -8i$. We find (see Problem 33)

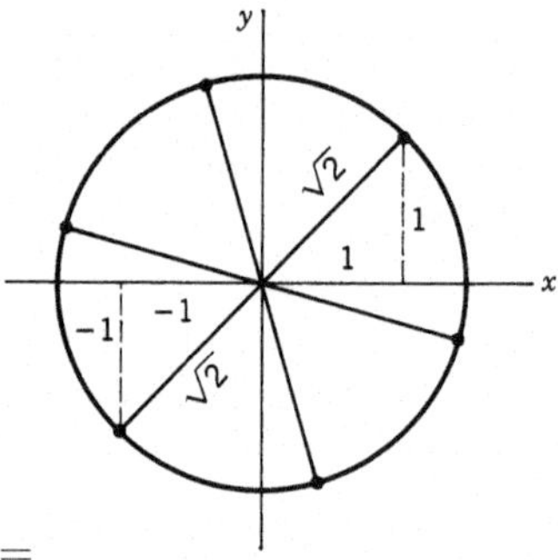

Figure 10.4

$$\pm\left\{1+i, \quad \frac{\sqrt{3}+1}{2} - \frac{\sqrt{3}-1}{2}i, \quad \frac{\sqrt{3}-1}{2} - \frac{\sqrt{3}+1}{2}i\right\} =$$

$$\pm\{1+i, \quad 1.366 - 0.366i, \quad 0.366 - 1.366i\}.$$

Summary In each of the preceding examples, our steps in finding $\sqrt[n]{re^{i\theta}}$ were:

(a) Find the polar coordinates of the roots: Take the nth root of r and divide $\theta + 2k\pi$ by n.

(b) Make a sketch: Draw a circle of radius $\sqrt[n]{r}$, plot the root with angle θ/n, and then plot the rest of the n roots around the circle equally spaced $2\pi/n$ apart. Note that we have now essentially solved the problem. From the sketch you can see the approximate rectangular coordinates of the roots and check your answers in (c). Since this sketch is quick and easy to do, it is worthwhile even if you use a computer to do part (c).

(c) Find the $x + iy$ coordinates of the roots by one of the methods in the examples. If you are using a computer, you may want to make a computer plot of the roots which should be a perfected copy of your sketch in (b).

PROBLEMS, SECTION 10

Follow steps (a), (b), (c) above to find all the values of the indicated roots.

1. $\sqrt[3]{1}$	**2.** $\sqrt[3]{27}$	**3.** $\sqrt[4]{1}$
4. $\sqrt[4]{16}$	**5.** $\sqrt[6]{1}$	**6.** $\sqrt[6]{64}$
7. $\sqrt[8]{16}$	**8.** $\sqrt[8]{1}$	**9.** $\sqrt[5]{1}$
10. $\sqrt[5]{32}$	**11.** $\sqrt[3]{-8}$	**12.** $\sqrt[3]{-1}$

13. $\sqrt[4]{-4}$ **14.** $\sqrt[4]{-1}$ **15.** $\sqrt[6]{64}$

16. $\sqrt[6]{-1}$ **17.** $\sqrt[5]{-1}$ **18.** $\sqrt{i}$

19. $\sqrt[3]{i}$ **20.** $\sqrt[3]{-8i}$ **21.** $\sqrt{2+2i\sqrt{3}}$

22. $\sqrt[3]{2i-2}$ **23.** $\sqrt[4]{8i\sqrt{3}-8}$ **24.** $\sqrt[8]{\dfrac{-1-i\sqrt{3}}{2}}$

25. $\sqrt[5]{-1-i}$ **26.** $\sqrt[5]{i}$

27. Using the fact that a complex equation is really two real equations, find the double angle formulas (for $\sin 2\theta$, $\cos 2\theta$) by using equation (10.2).

28. As in Problem 27, find the formulas for $\sin 3\theta$ and $\cos 3\theta$.

29. Show that the center of mass of three identical particles situated at the points z_1, z_2, z_3 is $(z_1 + z_2 + z_3)/3$.

30. Show that the sum of the three cube roots of 8 is zero.

31. Show that the sum of the n nth roots of any complex number is zero.

32. The three cube roots of $+1$ are often called 1, ω, and ω^2. Show that this is reasonable, that is, show that the cube roots of $+1$ are $+1$ and two other numbers, each of which is the square of the other.

33. Verify the results given for the roots in Example 4. You can find the exact values in terms of $\sqrt{3}$ by using trigonometric addition formulas or more easily by using a computer to solve $z^6 = -8i$. (You still may have to do a little work by hand to put the computer's solution into the given form.)

11. THE EXPONENTIAL AND TRIGONOMETRIC FUNCTIONS

Although we have already defined e^z by a power series (8.1), it is worth while to write it in another form. By (8.2) we can write

$$e^z = e^{x+iy} = e^x e^{iy} = e^x(\cos y + i \sin y). \tag{11.1}$$

This is more convenient to use than the infinite series if we want values of e^z for given z. For example,

$$e^{2-i\pi} = e^2 e^{-i\pi} = e^2 \cdot (-1) = -e^2$$

from Figure 9.2.

We have already seen that there is a close relationship [Euler's formula (9.3)] between complex exponentials and trigonometric functions of real angles. It is useful to write this relation in another form. We write Euler's formula (9.3) as it is, and also write it with θ replaced by $-\theta$. Remember that $\cos(-\theta) = \cos\theta$ and $\sin(-\theta) = -\sin\theta$. Then we have

$$\begin{aligned} e^{i\theta} &= \cos\theta + i\sin\theta, \\ e^{-i\theta} &= \cos\theta - i\sin\theta. \end{aligned} \tag{11.2}$$

These two equations can be solved for $\sin\theta$ and $\cos\theta$. We get (Problem 2)

$$\begin{aligned} \sin\theta &= \frac{e^{i\theta} - e^{-i\theta}}{2i}, \\ \cos\theta &= \frac{e^{i\theta} + e^{-i\theta}}{2}. \end{aligned} \tag{11.3}$$

These formulas are useful in evaluating integrals since products of exponentials are easier to integrate than products of sines and cosines. (See Problems 11 to 16, and Chapter 7, Section 5.)

So far we have discussed only trigonometric functions of real angles. We could define $\sin z$ and $\cos z$ for complex z by their power series as we did for e^z. We could then compare these series with the series for e^{iz} and derive Euler's formula and (11.3) with θ replaced by z. However, it is simpler to use the complex equations corresponding to (11.3) as our definitions for $\sin z$ and $\cos z$. We define

$$\begin{aligned} \sin z &= \frac{e^{iz} - e^{-iz}}{2i}, \\ \cos z &= \frac{e^{iz} + e^{-iz}}{2}. \end{aligned} \tag{11.4}$$

The rest of the trigonometric functions of z are defined in the usual way in terms of these; for example, $\tan z = \sin z / \cos z$.

▶ **Example 1.** $\cos i = \dfrac{e^{i\cdot i} + e^{-i\cdot i}}{2} = \dfrac{e^{-1} + e}{2} = 1.543\cdots$. (We will see in Section 15 that this expression is called the hyperbolic cosine of 1.)

▶ **Example 2.**

$$\begin{aligned} \sin\left(\frac{\pi}{2} + i\ln 2\right) &= \frac{e^{i(\pi/2 + i\ln 2)} - e^{-i(\pi/2 + i\ln 2)}}{2i} \\ &= \frac{e^{i\pi/2}e^{-\ln 2} - e^{-i\pi/2}e^{\ln 2}}{2i} \qquad \text{by (8.2).} \end{aligned}$$

From Figures 5.2 and 9.3, $e^{i\pi/2} = i$, and $e^{-i\pi/2} = -i$. By the definition of $\ln x$ [or see equations (13.1) and (13.2)], $e^{\ln 2} = 2$, so $e^{-\ln 2} = 1/e^{\ln 2} = 1/2$. Then

$$\sin\left(\frac{\pi}{2} + i\ln 2\right) = \frac{(i)(1/2) - (-i)(2)}{2i} = \frac{5}{4}.$$

Notice from both these examples that sines and cosines of complex numbers may be greater than 1. As we shall see (Section 15), although $|\sin x| \le 1$ and $|\cos x| \le 1$ for *real* x, when z is a complex number, $\sin z$ and $\cos z$ can have *any* value we like.

Using the definitions (11.4) of $\sin z$ and $\cos z$, you can show that the familiar trigonometric identities and calculus formulas hold when θ is replaced by z.

Example 3. Prove that $\sin^2 z + \cos^2 z = 1$.

$$\sin^2 z = \left(\frac{e^{iz} - e^{-iz}}{2i}\right)^2 = \frac{e^{2iz} - 2 + e^{-2iz}}{-4},$$

$$\cos^2 z = \left(\frac{e^{iz} + e^{-iz}}{2}\right)^2 = \frac{e^{2iz} + 2 + e^{-2iz}}{4},$$

$$\sin^2 z + \cos^2 z = \frac{2}{4} + \frac{2}{4} = 1.$$

Example 4. Using the definitions (11.4), verify that $(d/dz)\sin z = \cos z$.

$$\sin z = \frac{e^{iz} - e^{-iz}}{2i},$$

$$\frac{d}{dz}\sin z = \frac{1}{2i}(ie^{iz} + ie^{-iz}) = \frac{e^{iz} + e^{-iz}}{2} = \cos z.$$

PROBLEMS, SECTION 11

1. Define $\sin z$ and $\cos z$ by their power series. Write the power series for e^{iz}. By comparing these series obtain the definition (11.4) of $\sin z$ and $\cos z$.
2. Solve the equations $e^{i\theta} = \cos\theta + i\sin\theta$, $e^{-i\theta} = \cos\theta - i\sin\theta$, for $\cos\theta$ and $\sin\theta$ and so obtain equations (11.3).

Find each of the following in rectangular form $x + iy$ and check your results by computer. Remember to save time by doing as much as you can in your head.

3. $e^{-(i\pi/4)+\ln 3}$
4. $e^{3\ln 2 - i\pi}$
5. $e^{(i\pi/4)+(\ln 2)/2}$
6. $\cos(i\ln 5)$
7. $\tan(i\ln 2)$
8. $\cos(\pi - 2i\ln 3)$
9. $\sin(\pi - i\ln 3)$
10. $\sin(i\ln i)$

In the following integrals express the sines and cosines in exponential form and then integrate to show that:

11. $\displaystyle\int_{-\pi}^{\pi} \cos 2x \cos 3x\, dx = 0$
12. $\displaystyle\int_{-\pi}^{\pi} \cos^2 3x\, dx = \pi$
13. $\displaystyle\int_{-\pi}^{\pi} \sin 2x \sin 3x\, dx = 0$
14. $\displaystyle\int_{0}^{2\pi} \sin^2 4x\, dx = \pi$
15. $\displaystyle\int_{-\pi}^{\pi} \sin 2x \cos 3x\, dx = 0$
16. $\displaystyle\int_{-\pi}^{\pi} \sin 3x \cos 4x\, dx = 0$

Evaluate $\int e^{(a+ib)x}dx$ and take real and imaginary parts to show that:

17. $\displaystyle\int e^{ax}\cos bx\, dx = \frac{e^{ax}(a\cos bx + b\sin bx)}{a^2 + b^2}$
18. $\displaystyle\int e^{ax}\sin bx\, dx = \frac{e^{ax}(a\sin bx - b\cos bx)}{a^2 + b^2}$

12. HYPERBOLIC FUNCTIONS

Let us look at $\sin z$ and $\cos z$ for pure imaginary z, that is, $z = iy$:

$$(12.1)\qquad \begin{aligned} \sin iy &= \frac{e^{-y} - e^{y}}{2i} = i\,\frac{e^{y} - e^{-y}}{2}, \\ \cos iy &= \frac{e^{-y} + e^{y}}{2} = \frac{e^{y} + e^{-y}}{2}. \end{aligned}$$

The real functions on the right have special names because these particular combinations of exponentials arise frequently in problems. They are called the hyperbolic sine (abbreviated sinh) and the hyperbolic cosine (abbreviated cosh). Their definitions for all z are

$$(12.2)\qquad \begin{aligned} \sinh z &= \frac{e^{z} - e^{-z}}{2}, \\ \cosh z &= \frac{e^{z} + e^{-z}}{2}. \end{aligned}$$

The other hyperbolic functions are named and defined in a similar way to parallel the trigonometric functions:

$$(12.3)\qquad \begin{aligned} \tanh z &= \frac{\sinh z}{\cosh z}, &\qquad \coth z &= \frac{1}{\tanh z}, \\ \operatorname{sech} z &= \frac{1}{\cosh z}, &\qquad \operatorname{csch} z &= \frac{1}{\sinh z}. \end{aligned}$$

(See Problem 38 for the reason behind the term "hyperbolic" functions.)

We can write (12.1) as

$$(12.4)\qquad \begin{aligned} \sin iy &= i\,\sinh y, \\ \cos iy &= \cosh y. \end{aligned}$$

Then we see that the hyperbolic functions of y are (except for one i factor) the trigonometric functions of iy. From (12.2) we can show that (12.4) holds with y replaced by z. Because of this relation between hyperbolic and trigonometric functions, the formulas for hyperbolic functions look very much like the corresponding trigonometric identities and calculus formulas. They are not identical, however.

Example. You can prove the following formulas (see Problems 9, 10, 11 and 38).

$$\begin{aligned} &\cosh^2 z - \sinh^2 z = 1 &\qquad &(\text{compare } \sin^2 z + \cos^2 z = 1), \\ &\frac{d}{dz}\cosh z = \sinh z &\qquad &\left(\text{compare } \frac{d}{dz}\cos z = -\sin z\right). \end{aligned}$$

PROBLEMS, SECTION 12

Verify each of the following by using equations (11.4), (12.2), and (12.3).

1. $\sin z = \sin(x + iy) = \sin x\, \cosh y + i\, \cos x\, \sinh y$

2. $\cos z = \cos x \cosh y - i \sin x \sinh y$

3. $\sinh z = \sinh x \cos y + i \cosh x \sin y$

4. $\cosh z = \cosh x \cos y + i \sinh x \sin y$

5. $\sin 2z = 2 \sin z \cos z$

6. $\cos 2z = \cos^2 z - \sin^2 z$

7. $\sinh 2z = 2 \sinh z \cosh z$

8. $\cosh 2z = \cosh^2 z + \sinh^2 z$

9. $\dfrac{d}{dz} \cos z = -\sin z$

10. $\dfrac{d}{dz} \cosh z = \sinh z$

11. $\cosh^2 z - \sinh^2 z = 1$

12. $\cos^4 z + \sin^4 z = 1 - \dfrac{1}{2} \sin^2 2z$

13. $\cos 3z = 4 \cos^3 z - 3 \cos z$

14. $\sin iz = i \sinh z$

15. $\sinh iz = i \sin z$

16. $\tan iz = i \tanh z$

17. $\tanh iz = i \tan z$

18. $\tan z = \tan(x + iy) = \dfrac{\tan x + i \tanh y}{1 - i \tan x \tanh y}$

19. $\tanh z = \dfrac{\tanh x + i \tan y}{1 + i \tanh x \tan y}$

20. Show that $e^{nz} = (\cosh z + \sinh z)^n = \cosh nz + \sinh nz$. Use this and a similar equation for e^{-nz} to find formulas for $\cosh 3z$ and $\sinh 3z$ in terms of $\sinh z$ and $\cosh z$.

21. Use a computer to plot graphs of $\sinh x$, $\cosh x$, and $\tanh x$.

22. Using (12.2) and (8.1), find, in summation form, the power series for $\sinh x$ and $\cosh x$. Check the first few terms of your series by computer.

Find the real part, the imaginary part, and the absolute value of

23. $\cosh(ix)$

24. $\cos(ix)$

25. $\sin(x - iy)$

26. $\cosh(2 - 3i)$

27. $\sin(4 + 3i)$

28. $\tanh(1 - i\pi)$

Find each of the following in the $x + iy$ form and check your answers by computer.

29. $\cosh 2\pi i$

30. $\tanh \dfrac{3\pi i}{4}$

31. $\sinh \left(\ln 2 + \dfrac{i\pi}{3} \right)$

32. $\cosh \left(\dfrac{i\pi}{2} - \ln 3 \right)$

33. $\tan i$

34. $\sin \dfrac{i\pi}{2}$

35. $\cosh(i\pi + 2)$

36. $\sinh \left(1 + \dfrac{i\pi}{2} \right)$

37. $\cos(i\pi)$

38. The functions $\sin t$, $\cos t$, $\cdots$, are called "circular functions" and the functions $\sinh t$, $\cosh t$, $\cdots$, are called "hyperbolic functions". To see a reason for this, show that $x = \cos t$, $y = \sin t$, satisfy the equation of a circle $x^2 + y^2 = 1$, while $x = \cosh t$, $y = \sinh t$, satisfy the equation of a hyperbola $x^2 - y^2 = 1$.

13. LOGARITHMS

In elementary mathematics you learned to find logarithms of positive numbers only; in fact, you may have been told that there were no logarithms of negative numbers. This is true if you use only real numbers, but it is not true when we allow complex numbers as answers. We shall now see how to find the logarithm of any complex number $z \neq 0$ (including negative real numbers as a special case). If

$$z = e^w, \tag{13.1}$$

then by definition

$$w = \ln z. \tag{13.2}$$

(We use ln for natural logarithms to avoid the cumbersome $\log_e$ and to avoid confusion with logarithms to the base 10.)

We can write the law of exponents (8.2), using the letters of (13.1), as

$$z_1 z_2 = e^{w_1} \cdot e^{w_2} = e^{w_1 + w_2}. \tag{13.3}$$

Taking logarithms of this equation, that is, using (13.1) and (13.2), we get

$$\ln z_1 z_2 = w_1 + w_2 = \ln z_1 + \ln z_2. \tag{13.4}$$

This is the familiar law for the logarithm of a product, justified now for complex numbers. We can then find the real and imaginary parts of the logarithm of a complex number from the equation

$$w = \ln z = \ln(re^{i\theta}) = \operatorname{Ln} r + \ln e^{i\theta} = \operatorname{Ln} r + i\theta, \tag{13.5}$$

where $\operatorname{Ln} r$ means the ordinary real logarithm to the base e of the real positive number r.

Since θ has an infinite number of values (all differing by multiples of 2π), a complex number has infinitely many logarithms, differing from each other by multiples of $2\pi i$. The *principal value* of $\ln z$ (often written as $\operatorname{Ln} z$) is the one using the principal value of θ, that is $0 \leq \theta < 2\pi$. (Some references use $-\pi < \theta \leq \pi$.)

Example 1. Find $\ln(-1)$. From Figure 9.2, we see that the polar coordinates of the point $z = -1$ are $r = 1$ and $\theta = \pi,\ -\pi,\ 3\pi,\ \cdots$. Then,

$$\ln(-1) = \operatorname{Ln}(1) + i(\pi \pm 2n\pi) = i\pi, -i\pi, 3\pi i, \cdots.$$

Example 2. Find $\ln(1+i)$. From Figure 5.1, for $z = 1+i$, we find $r = \sqrt{2}$, and $\theta = \pi/4 \pm 2n\pi$. Then

$$\ln(1+i) = \operatorname{Ln}\sqrt{2} + i\left(\frac{\pi}{4} \pm 2n\pi\right) = 0.347\cdots + i\left(\frac{\pi}{4} \pm 2n\pi\right).$$

Even a positive real number now has infinitely many logarithms, since its angle can be taken as $0,\ 2\pi,\ -2\pi$, etc. Only one of these logarithms is real, namely the principal value $\operatorname{Ln} r$ using the angle $\theta = 0$.

14. COMPLEX ROOTS AND POWERS

For real positive numbers, the equation $\ln a^b = b \ln a$ is equivalent to $a^b = e^{b \ln a}$. We define complex powers by the same formula with complex a and b. By definition, for complex a and b $(a \neq e)$,

$$(14.1) \qquad a^b = e^{b \ln a}.$$

[The case $a = e$ is excluded because we have already defined powers of e by (8.1).] Since $\ln a$ is multiple valued (because of the infinite number of values of θ), powers a^b are usually multiple valued, and unless you want just the principal value of $\ln z$ or of a^b you must use all values of θ. In the following examples we find all values of each complex power and write the answers in the $x + iy$ form.

Example 1. Find all values of i^{-2i}. From Figure 5.2, and equation (13.5) we find $\ln i = \operatorname{Ln} 1 + i(\pi/2 \pm 2n\pi) = i(\pi/2 \pm 2n\pi)$ since $\operatorname{Ln} 1 = 0$. Then, by equation (14.1),

$$i^{-2i} = e^{-2i \ln i} = e^{-2i \cdot i(\pi/2 \pm 2n\pi)} = e^{\pi \pm 4n\pi} = e^{\pi}, e^{5\pi}, e^{-3\pi}, \cdots,$$

where $e^{\pi} = 23.14\cdots$. Note the infinite set of values of i^{-2i}, all real! Also read the end of Section 3, and note that here the final step is *not* to find sine or cosine of $\pi \pm 4n\pi$; thus, in finding $\ln i = i\theta$, we must *not* write θ in degrees.

Example 2. Find all values of $i^{1/2}$. Using $\ln i$ from Example 1 we have $i^{1/2} = e^{(1/2)\ln i} = e^{i(\pi/4 + n\pi)} = e^{i\pi/4} e^{in\pi}$. Now $e^{in\pi} = +1$ when n is even (Fig. 9.4), and $e^{in\pi} = -1$ when n is odd (Fig. 9.2). Thus,

$$i^{1/2} = \pm e^{i\pi/4} = \pm \frac{1+i}{\sqrt{2}}$$

using Figure 5.1. Notice that although $\ln i$ has an infinite set of values, we find just two values for $i^{1/2}$ as we should for a square root. (Compare the method of Section 10 which is easier for this problem.)

Example 3. Find all values of $(1+i)^{1-i}$. Using (14.1) and the value of $\ln(1+i)$ from Example 2, Section 13, we have

$$\begin{aligned}
(1+i)^{1-i} &= e^{(1-i)\ln(1+i)} = e^{(1-i)[\operatorname{Ln}\sqrt{2} + i(\pi/4 \pm 2n\pi)]} \\
&= e^{\operatorname{Ln}\sqrt{2}}\, e^{-i \operatorname{Ln}\sqrt{2}}\, e^{i\pi/4}\, e^{\pm 2n\pi i}\, e^{\pi/4}\, e^{\pm 2n\pi} \\
&= \sqrt{2}\, e^{i(\pi/4 - \operatorname{Ln}\sqrt{2})}\, e^{\pi/4}\, e^{\pm 2n\pi} \qquad (\text{since } e^{\pm 2n\pi i} = 1) \\
&= \sqrt{2}\, e^{\pi/4}\, e^{\pm 2n\pi}[\cos(\pi/4 - \operatorname{Ln}\sqrt{2}) + i \sin(\pi/4 - \operatorname{Ln}\sqrt{2})] \\
&\cong e^{\pm 2n\pi}(2.808 + 1.318i).
\end{aligned}$$

Now you may be wondering why not just do these problems by computer. The most important point is that it is useful for advanced work to have skill in manipulating complex expressions. A second point is that there may be several forms for an answer (see Section 15, Example 2) or there may be many answers (see examples above), and your computer may not give you the one you want (see Problem 25). So to obtain needed skills, a good study method is to do problems by hand and compare with computer solutions.

PROBLEMS, SECTION 14

Evaluate each of the following in $x + iy$ form, and compare with a computer solution.

1. $\ln(-e)$ **2.** $\ln(-i)$ **3.** $\ln(i + \sqrt{3})$

4. $\ln(i - 1)$ **5.** $\ln(-\sqrt{2} - i\sqrt{2})$ **6.** $\ln\left(\frac{1-i}{\sqrt{2}}\right)$

7. $\ln\left(\frac{1+i}{1-i}\right)$ **8.** $i^{2/3}$ **9.** $(-1)^i$

10. $i^{\ln i}$ **11.** 2^i **12.** i^{3+i}

13. $i^{2i/\pi}$ **14.** $(2i)^{1+i}$ **15.** $(-1)^{\sin i}$

16. $\left(\frac{1+i\sqrt{3}}{2}\right)^i$ **17.** $(i-1)^{i+1}$ **18.** $\cos(2i \ln i)$

19. $\cos(\pi + i \ln 2)$ **20.** $\sin\left(i \ln \frac{1-i}{1+i}\right)$ **21.** $\cos[i \ln(-1)]$

22. $\sin\left[i \ln\left(\frac{\sqrt{3}+i}{2}\right)\right]$ **23.** $\left(1 - \sqrt{2i}\right)^i$. *Hint:* Find $\sqrt{2i}$ first.

24. Show that $(a^b)^c$ can have more values than a^{bc}. As examples compare

(a) $[(-i)^{2+i}]^{2-i}$ and $(-i)^{(2+i)(2-i)} = (-i)^5$;

(b) $(i^i)^i$ and i^{-1}.

25. Use a computer to find the three solutions of the equation $x^3 - 3x - 1 = 0$. Find a way to show that the solutions can be written as $2\cos(\pi/9)$, $-2\cos(2\pi/9)$, $-2\cos(4\pi/9)$.

15. INVERSE TRIGONOMETRIC AND HYPERBOLIC FUNCTIONS

We have already defined the trigonometric and hyperbolic functions of a complex number z. For example,

$$w = \cos z = \frac{e^{iz} + e^{-iz}}{2} \tag{15.1}$$

defines $w = \cos z$; that is, for each complex number z, (15.1) gives us the complex number w. We now define the inverse cosine or arc cos w by

$$z = \operatorname{arc\,cos} w \qquad \text{if} \qquad w = \cos z. \tag{15.2}$$

The other inverse trigonometric and hyperbolic functions are defined similarly.

In dealing with real numbers, you know that $\sin x$ and $\cos x$ are never greater than 1. This is no longer true for $\sin z$ and $\cos z$ with z complex. To illustrate the method of finding inverse trigonometric (or inverse hyperbolic) functions, let's find arc cos 2.

Example 1. We want z, where

$$z = \arccos 2 \quad \text{or} \quad \cos z = 2.$$

Then we have

$$\frac{e^{iz} + e^{-iz}}{2} = 2.$$

To simplify the algebra, let $u = e^{iz}$. Then $e^{-iz} = u^{-1}$, and the equation becomes

$$\frac{u + u^{-1}}{2} = 2.$$

Multiply by 2 and by u to get $u^2 + 1 = 4u$ or $u^2 - 4u + 1 = 0$. Solve this equation by the quadratic formula to find

$$u = \frac{4 \pm \sqrt{16 - 4}}{2} = 2 \pm \sqrt{3}, \quad \text{or} \quad e^{iz} = u = 2 \pm \sqrt{3}.$$

Take logarithms of both sides of this equation, and solve for z:

$$iz = \ln(2 \pm \sqrt{3}) = \text{Ln}(2 \pm \sqrt{3}) + 2n\pi i,$$
$$\arccos 2 = z = 2n\pi - i\,\text{Ln}(2 \pm \sqrt{3}\,) = 2n\pi \pm i\,\text{Ln}(2 + \sqrt{3}\,)$$

since $\text{Ln}(2 - \sqrt{3}) = -\text{Ln}(2 + \sqrt{3})$.

It is instructive now to find $\cos z$ and see that it *is* 2. For $iz = \ln(2 \pm \sqrt{3})$, we have

$$e^{iz} = e^{\ln(2 \pm \sqrt{3})} = 2 \pm \sqrt{3},$$
$$e^{-iz} = \frac{1}{e^{iz}} = \frac{1}{2 \pm \sqrt{3}} = \frac{2 \mp \sqrt{3}}{4 - 3} = 2 \mp \sqrt{3}.$$

Then

$$\cos z = \frac{e^{iz} + e^{-iz}}{2} = \frac{2 \pm \sqrt{3} + 2 \mp \sqrt{3}}{2} = \frac{4}{2} = 2,$$

as claimed.

By the same method, we can find all the inverse trigonometric and hyperbolic functions in terms of logarithms. (See Problems, Section 17.) Here is one more example.

Example 2. In integral tables or from your computer you may find for the indefinite integral

$$\int \frac{dx}{\sqrt{x^2 + a^2}} \tag{15.3}$$

either

$$\sinh^{-1}\frac{x}{a} \quad \text{or} \quad \ln(x + \sqrt{x^2 + a^2}\,). \tag{15.4}$$

How are these related? Put

$$z = \sinh^{-1}\frac{x}{a} \quad \text{or} \quad \frac{x}{a} = \sinh z = \frac{e^z - e^{-z}}{2}. \tag{15.5}$$

We solve for z as in the previous example. Let $e^z = u$, $e^{-z} = 1/u$. Then

$$u - \frac{1}{u} = \frac{2x}{a},$$
$$au^2 - 2xu - a = 0,$$
$$e^z = u = \frac{2x \pm \sqrt{4x^2 + 4a^2}}{2a} = \frac{x \pm \sqrt{x^2 + a^2}}{a}. \tag{15.6}$$

For real integrals, that is, for real z, $e^z > 0$, so we must use the positive sign. Then, taking the logarithm of (15.6) we have

$$z = \ln(x + \sqrt{x^2 + a^2}\,) - \ln a. \tag{15.7}$$

Comparing (15.5) and (15.7) we see that the two answers in (15.4) differ only by the constant $\ln a$, which is a constant of integration.

PROBLEMS, SECTION 15

Find each of the following in the $x + iy$ form and compare a computer solution.

1. $\arcsin 2$

2. $\arctan 2i$

3. $\cosh^{-1}(1/2)$

4. $\sinh^{-1}(i/2)$

5. $\arccos\left(i\sqrt{8}\right)$

6. $\tanh^{-1}(-i)$

7. $\arctan\left(i\sqrt{2}\right)$

8. $\arcsin(5/3)$

9. $\tanh^{-1}\left(i\sqrt{3}\right)$

10. $\arccos(5/4)$

11. $\sinh^{-1}\left(i/\sqrt{2}\right)$

12. $\cosh^{-1}\left(\sqrt{3}/2\right)$

13. $\cosh^{-1}(-1)$

14. $\arcsin(3i/4)$

15. $\arctan(2 + i)$

16. $\tanh^{-1}(1 - 2i)$

17. Show that $\tan z$ never takes the values $\pm i$. *Hint:* Try to solve the equation $\tan z = i$ and find that it leads to a contradiction.

18. Show that $\tanh z$ never takes the values ± 1.

16. SOME APPLICATIONS

Motion of a Particle We have already seen (end of Section 5) that the path of a particle in the (x, y) plane is given by $z = z(t)$. As another example of this, suppose $z = 1 + 3e^{2it}$. We see that

$$|z - 1| = |3\,e^{2it}| = 3. \tag{16.1}$$

Recall that $|z - 1|$ is the distance between the points z and 1; (16.1) says that this distance is 3. Thus the particle traverses a circle of radius 3, with center at $(1, 0)$. The magnitude of its velocity is $|dz/dt| = |6i\,e^{2it}| = 6$, so it moves around the circle at constant speed. (Also see Problem 2).

PROBLEMS, SECTION 16

1. Show that if the line through the origin and the point z is rotated $90°$ about the origin, it becomes the line through the origin and the point iz. This fact is sometimes expressed by saying that multiplying a complex number by i rotates it through $90°$. Use this idea in the following problem. Let $z = a\,e^{i\omega t}$ be the displacement of a particle from the origin at time t. Show that the particle travels in a circle of radius a at velocity $v = a\omega$ and with acceleration of magnitude v^2/a directed toward the center of the circle.

In each of the following problems, z represents the displacement of a particle from the origin. Find (as functions of t) its speed and the magnitude of its acceleration, and describe the motion.

2. $z = 5e^{i\omega t}$, $\omega = \text{const.}$ *Hint:* See Problem 1.

3. $z = (1+i)e^{it}$.

4. $z = (1+i)t - (2+i)(1-t)$. *Hint:* Show that the particle moves along a straight line through the points $(1+i)$ and $(-2-i)$.

5. $z = z_1 t + z_2(1-t)$. *Hint:* See Problem 4; the straight line here is through the points z_1 and z_2.

Electricity In the theory of electric circuits, it is shown that if V_R is the voltage across a resistance R, and I is the current flowing through the resistor, then

$$V_R = IR \quad \text{(Ohm's law)}. \tag{16.2}$$

It is also known that the current and voltage across an inductance L are related by

$$V_L = L\frac{dI}{dt} \tag{16.3}$$

and the current and voltage across a capacitor are related by

$$\frac{dV_C}{dt} = \frac{I}{C}, \tag{16.4}$$

where C is the capacitance. Suppose the current I and voltage V in the circuit of Figure 16.1 vary with time so that I is given by

$$I = I_0 \sin\omega t. \tag{16.5}$$

You can verify that the following voltages across R, L, and C are consistent with (16.2), (16.3), and (16.4):

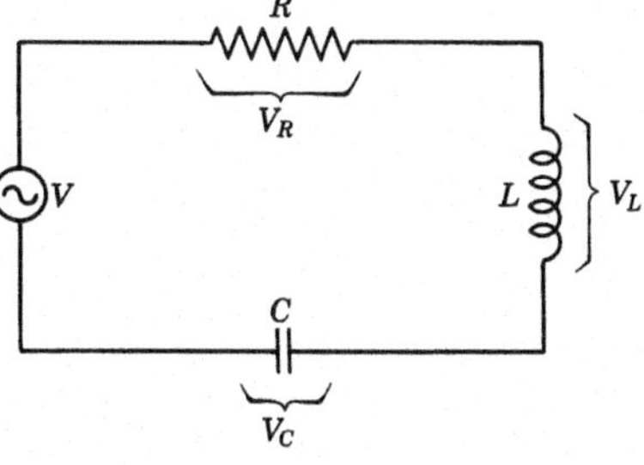

Figure 16.1

$$V_R = RI_0 \sin\omega t, \tag{16.6}$$

$$V_L = \omega L I_0 \cos\omega t, \tag{16.7}$$

$$V_C = -\frac{1}{\omega C} I_0 \cos\omega t. \tag{16.8}$$

The total voltage

$$V = V_R + V_L + V_C \tag{16.9}$$

is then a complicated function. A simpler method of discussing a-c circuits uses complex quantities as follows. Instead of (16.5) we write

$$I = I_0 \, e^{i\omega t}, \tag{16.10}$$

where it is understood that the actual physical current is given by the imaginary part of I in (16.10), that is, by (16.5). Note, by comparing (16.5) and (16.10), that the maximum value of I, namely I_0, is given in (16.10) by $|I|$. Now equations (16.6) to (16.9) become

$$V_R = RI_0 \, e^{i\omega t} = RI, \tag{16.11}$$

$$V_L = i\omega L I_0 \, e^{i\omega t} = i\omega L I, \tag{16.12}$$

$$V_C = \frac{1}{i\omega C} I_0 \, e^{i\omega t} = \frac{1}{i\omega C} I, \tag{16.13}$$

$$V = V_R + V_L + V_C = \left[R + i\left(\omega L - \frac{1}{\omega C}\right)\right] I. \tag{16.14}$$

The complex quantity Z defined by

$$Z = R + i\left(\omega L - \frac{1}{\omega C}\right) \tag{16.15}$$

is called the (complex) impedance. Using it we can write (16.14) as

$$V = ZI \tag{16.16}$$

which looks much like Ohm's law. In fact, Z for an a-c circuit corresponds to R for a d-c circuit. The more complicated a-c circuit equations now take the same simple form as the d-c equations except that all quantities are complex. For example, the rules for combining resistances in series and in parallel hold for combining complex impedances (see Problems below).

PROBLEMS, SECTION 16

In electricity we learn that the resistance of two resistors in series is $R_1 + R_2$ and the resistance of two resistors in parallel is $(R_1^{-1} + R_2^{-1})^{-1}$. Corresponding formulas hold for complex impedances. Find the impedance of Z_1 and Z_2 in series, and in parallel, given:

6. (a) $Z_1 = 2 + 3i$, $Z_2 = 1 - 5i$ (b) $Z_1 = 2\sqrt{3}\, e^{i\pi/6}$, $Z_2 = 2\, e^{2i\pi/3}$

7. (a) $Z_1 = 1 - i$, $Z_2 = 3i$ (b) $|Z_1| = 3.16$, $\theta_1 = 18.4°$; $|Z_2| = 4.47$, $\theta_2 = 63.4°$

8. Find the impedance of the circuit in Figure 16.2 (R and L in series, and then C in parallel with them). A circuit is said to be *in resonance* if Z is real; find ω in terms of R, L, and C at resonance.

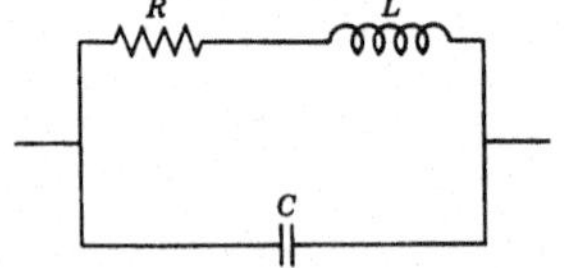

Figure 16.2

9. For the circuit in Figure 16.1:

(a) Find ω in terms of R, L, and C if the angle of Z is 45°.

(b) Find the resonant frequency ω (see Problem 8).

10. Repeat Problem 9 for a circuit consisting of R, L, and C, all in parallel.

Optics In optics we frequently need to combine a number of light waves (which can be represented by sine functions). Often each wave is "out of phase" with the preceding one by a fixed amount; this means that the waves can be written as $\sin t$, $\sin(t+\delta)$, $\sin(t+2\delta)$, and so on. Suppose we want to add all these sine functions together. An easy way to do it is to see that each sine is the imaginary part of a complex number, so what we want is the imaginary part of the series

$$e^{it} + e^{i(t+\delta)} + e^{i(t+2\delta)} + \cdots. \tag{16.17}$$

This is a geometric progression with first term e^{it} and ratio $e^{i\delta}$. If there are n waves to be combined, we want the sum of n terms of this progression, which is

$$\frac{e^{it}(1-e^{in\delta})}{1-e^{i\delta}}. \tag{16.18}$$

We can simplify this expression by writing

$$1-e^{i\delta} = e^{i\delta/2}(e^{-i\delta/2}-e^{i\delta/2}) = -e^{i\delta/2}\cdot 2i\sin\frac{\delta}{2} \tag{16.19}$$

by (11.3). Substituting (16.19) and a similar formula for $(1-e^{in\delta})$ into (16.18), we get

$$\frac{e^{it}e^{in\delta/2}}{e^{i\delta/2}}\,\frac{\sin(n\delta/2)}{\sin(\delta/2)} = e^{i\{t+[(n-1)/2]\delta\}}\frac{\sin(n\delta/2)}{\sin(\delta/2)}. \tag{16.20}$$

The imaginary part of the series (16.17) which we wanted is then the imaginary part of (16.20), namely

$$\sin\left(t+\frac{n-1}{2}\delta\right)\sin\frac{n\delta}{2}\Big/\sin\frac{\delta}{2}.$$

PROBLEMS, SECTION 16

11. Prove that

$$\cos\theta+\cos 3\theta+\cos 5\theta+\cdots+\cos(2n-1)\theta = \frac{\sin 2n\theta}{2\sin\theta},$$
$$\sin\theta+\sin 3\theta+\sin 5\theta+\cdots+\sin(2n-1)\theta = \frac{\sin^2 n\theta}{\sin\theta}.$$

Hint: Use Euler's formula and the geometric progression formula.

12. In optics, the following expression needs to be evaluated in calculating the intensity of light transmitted through a film after multiple reflections at the surfaces of the film:

$$\left(\sum_{n=0}^{\infty} r^{2n}\cos n\theta\right)^2 + \left(\sum_{n=0}^{\infty} r^{2n}\sin n\theta\right)^2.$$

Show that this is equal to $|\sum_{n=0}^{\infty} r^{2n}e^{in\theta}|^2$ and so evaluate it assuming $|r|<1$ (r is the fraction of light reflected each time).

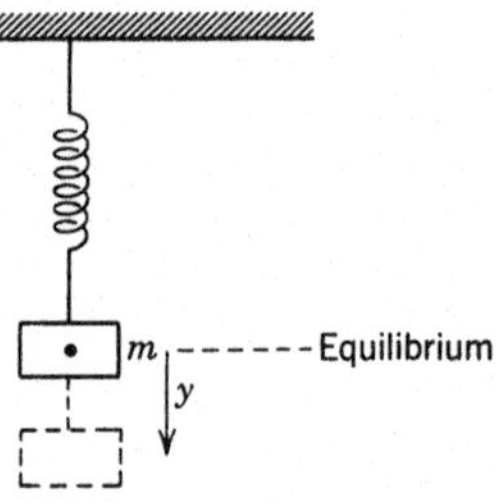

Figure 16.3

Simple Harmonic Motion It is very convenient to use complex notation even for motion along a straight line. Think of a mass m attached to a spring and oscillating up and down (see Figure 16.3). Let y be the vertical displacement of the mass from its equilibrium position (the point at which it would hang at rest). Recall that the force on m due to the stretched or compressed spring is then $-ky$, where k is the spring constant, and the minus sign indicates that the force and displacement are in opposite directions. Then Newton's second law (force = mass times acceleration) gives

$$m\frac{d^2y}{dt^2} = -ky \quad \text{or} \quad \frac{d^2y}{dt^2} = -\frac{k}{m}y = -\omega^2 y \quad \text{if} \quad \omega^2 = \frac{k}{m}. \tag{16.21}$$

Now we want a function $y(t)$ with the property that differentiating it twice just multiplies it by a constant. You can easily verify that this is true for exponentials, sines, and cosines (see problem 13). Just as in discussing electric circuits (see (16.10)), we may write a solution of (16.21) as

$$y = y_0\, e^{i\omega t} \tag{16.22}$$

with the understanding that the actual physical displacement is either the real or the imaginary part of (16.22). The constant $\omega = \sqrt{k/m}$ is called the angular frequency (see Chapter 7, Section 2). We will use this notation in Chapter 3, Section 12.

PROBLEMS, SECTION 16

13. Verify that $e^{i\omega t}$, $e^{-i\omega t}$, $\cos\omega t$, and $\sin\omega t$ satisfy equation (16.21).

17. MISCELLANEOUS PROBLEMS

Find one or more values of each of the following complex expressions and compare with a computer solution.

1. $\left(\dfrac{1+i}{1-i}\right)^{2718}$ **2.** $\left(\dfrac{1+i\sqrt{3}}{\sqrt{2}+i\sqrt{2}}\right)^{50}$ **3.** $\sqrt[5]{-4-4i}$

4. $\sinh(1+i\pi/2)$ **5.** $\tanh(i\pi/4)$ **6.** $(-e)^{i\pi}$

7. $(-i)^i$ **8.** $\cos\left[2i\ln\dfrac{1-i}{1+i}\right]$ **9.** $\arcsin\left[\left(\dfrac{\sqrt{3}+i}{\sqrt{3}-i}\right)^{12}\right]$

10. $e^{2i\arctan(i\sqrt{3})}$ **11.** $e^{2\tanh^{-1} i}$ **12.** $e^{i\arcsin i}$

13. Find *real* x and y for which $|z+3| = 1 - iz$, where $z = x + iy$.

14. Find the disk of convergence of the series $\sum (z-2i)^n/n$.

15. For what z is the series $\sum z^{\ln n}$ absolutely convergent? *Hints:* Use equation (14.1). Also see Chapter 1, Problem 6.15.

16. Describe the set of points z for which $\mathrm{Re}(e^{i\pi/2}z) > 2$.

Verify the formulas in Problems 17 to 24.

17. $\arcsin z = -i \ln(iz \pm \sqrt{1 - z^2})$

18. $\arccos z = i \ln(z \pm \sqrt{z^2 - 1})$

19. $\arctan z = \dfrac{1}{2i} \ln \dfrac{1 + iz}{1 - iz}$

20. $\sinh^{-1} z = \ln(z \pm \sqrt{z^2 + 1})$

21. $\cosh^{-1} z = \ln(z \pm \sqrt{z^2 - 1}) = \pm \ln(z + \sqrt{z^2 - 1})$

22. $\tanh^{-1} z = \dfrac{1}{2} \ln \dfrac{1 + z}{1 - z}$

23. $\cos iz = \cosh z$

24. $\cosh iz = \cos z$

25. (a) Show that $\overline{\cos z} = \cos \bar{z}$.

(b) Is $\overline{\sin z} = \sin \bar{z}$?

(c) If $f(z) = 1 + iz$, is $\overline{f(z)} = f(\bar{z})$?

(d) If $f(z)$ is expanded in a power series with *real* coefficients, show that $\overline{f(z)} = f(\bar{z})$.

(e) Using part (d), verify, *without computing its value*, that $i[\sinh(1 + i) - \sinh(1 - i)]$ is real.

26. Find $\left|\dfrac{2e^{i\theta} - i}{ie^{i\theta} + 2}\right|$. *Hint:* See equation (5.1).

27. (a) Show that $\operatorname{Re} z = \frac{1}{2}(z + \bar{z})$ and that $\operatorname{Im} z = (1/2i)(z - \bar{z})$.

(b) Show that $|e^z|^2 = e^{2 \operatorname{Re} z}$.

(c) Use (b) to evaluate $|e^{(1+ix)^2(1-it)-|1+it|^2}|^2$ which occurs in quantum mechanics.

28. Evaluate the following absolute square of a complex number (which arises in a problem in quantum mechanics). Assume a and b are real. Express your answer in terms of a hyperbolic function.

$$\left|\frac{(a + bi)^2 e^b - (a - bi)^2 e^{-b}}{4abie^{-ia}}\right|^2$$

29. If $z = \dfrac{a}{b}$ and $\dfrac{1}{a + b} = \dfrac{1}{a} + \dfrac{1}{b}$, find z.

30. Write the series for $e^{x(1+i)}$. Write $1 + i$ in the $re^{i\theta}$ form and so obtain (easily) the powers of $(1 + i)$. Thus show, for example, that the $e^x \cos x$ series has no x^2 term, no x^6 term, etc., and a similar result for the $e^x \sin x$ series. Find (easily) a formula for the general term for each series.

31. Show that if a sequence of complex numbers tends to zero, then the sequence of absolute values tends to zero too, and vice versa. *Hint:* $a_n + ib_n \to 0$ means $a_n \to 0$ and $b_n \to 0$.

32. Use a series you know to show that $\displaystyle\sum_{n=0}^{\infty} \frac{(1 + i\pi)^n}{n!} = -e$.

CHAPTER 3

Linear Algebra

1. INTRODUCTION

In this chapter, we are going to discuss a combination of algebra and geometry which is important in many applications. As you know, problems in various fields of science and mathematics involve the solution of sets of linear equations. This sounds like algebra, but it has a useful geometric interpretation. Suppose you have solved two simultaneous linear equations and have found $x = 2$ and $y = -3$. We can think of $x = 2$, $y = -3$ as the point $(2, -3)$ in the (x, y) plane. Since two linear equations represent two straight lines, the solution is then the point of intersection of the lines. The geometry helps us to understand that sometimes there is no solution (parallel lines) and sometimes there are infinitely many solutions (both equations represent the same line).

The language of vectors is very useful in studying sets of simultaneous equations. You are familiar with quantities such as the velocity of an object, the force acting on it, or the magnetic field at a point, which have both magnitude and direction. Such quantities are called *vectors*; contrast them with such quantities as mass, time, or temperature, which have magnitude only and are called *scalars*. A vector can be represented by an arrow and labeled by a boldface letter (**A** in Figure 1.1; also see Section 4). The length of the arrow tells us the magnitude of the vector and the direction of the arrow tells us the direction of the vector. It is not necessary to use coordinate axes as in Figure 1.1; we can, for example, point a finger to tell someone which way it is to town without knowing the direction of north. This is the geometric method of discussing vectors (see Section 4). However, if we do use a coordinate system as in Figure 1.1, we can specify the vector by giving its *components* A_x and A_y which are the projections of the vector on the x axis and the y axis. Thus we have two distinct methods of defining and working with vectors. A vector may be a geometric entity (arrow), or it may be a set of numbers (components relative to a coordinate system) which we use algebraically. As we shall see, this double interpretation of everything we do makes the use of vectors a very powerful tool in applications.

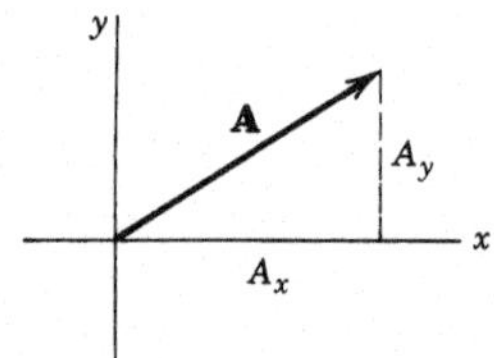

Figure 1.1

One of the great advantages of vector formulas is that they are independent of the choice of coordinate system. For example, suppose we are discussing the motion of a mass m sliding down an inclined plane. Newton's second law $\mathbf{F} = m\mathbf{a}$ is then a correct equation no matter how we choose our axes. We might, say, take the x axis horizontal and the y axis vertical, or alternatively we might take the x axis along the inclined plane and the y axis perpendicular to the plane. F_x would, of course, be different in the two cases, but for either case it would be true that $F_x = ma_x$ and $F_y = ma_y$, that is, the *vector* equation $\mathbf{F} = m\mathbf{a}$ would be true.

As we have just seen, a vector equation in two dimensions is equivalent to two component equations. In three dimensions, a vector equation is equivalent to three component equations. We will find it useful to generalize this to n dimensions and think of a set of n equations in n unknowns as the component equations for a vector equation in an n dimensional space (Section 10).

We shall also be interested in sets of linear equations which you can think of as changes of variable, say

$$\begin{cases} x' = ax + by, \\ y' = cx + dy, \end{cases} \tag{1.1}$$

where a, b, c, d, are constants. Alternatively, we can think of (1.1) geometrically as telling us to move each point (x, y) to another point (x', y'), an operation we will refer to as a transformation of the plane. Or if we think of (x, y) and (x', y') as being components of vectors from the origin to the given points, then (1.1) tells us how to change each vector in the plane to another vector. Equations (1.1) could also correspond to a change of axes (say a rotation of axes around the origin) where (x, y) and (x', y') are the coordinates of the same point relative to different axes. We will learn (Sections 11 and 12) how to choose the best coordinate system or set of variables to use in solving various problems. The same methods and tools (such as matrices and determinants) which can be used to solve sets of numerical equations are what we need to work with transformations and changes of coordinate system. After we have considered 2- and 3-dimensional space, we will extend these ideas to n-dimensional space and finally to a space in which the "vectors" are functions. This generalization is of great importance in applications.

2. MATRICES; ROW REDUCTION

A matrix (plural: matrices) is just a rectangular array of quantities, usually inclosed in large parentheses, such as

$$\mathrm{A} = \begin{pmatrix} 1 & 5 & -2 \\ -3 & 0 & 6 \end{pmatrix}. \tag{2.1}$$

We will ordinarily indicate a matrix by a roman letter such as A (or B, C, M, r, etc.), but the letter does not have a numerical value; it simply stands for the array. To indicate a number in the array, we will write A_{ij} where i is the row number and j is the column number. For example, in (2.1), $A_{11} = 1$, $A_{12} = 5$, $A_{13} = -2$, $A_{21} = -3$, $A_{22} = 0$, $A_{23} = 6$. We will call a matrix with m rows and n columns an m by n matrix. Thus the matrix in (2.1) is a 2 by 3 matrix, and the matrix in (2.2) below is a 3 by 2 matrix.

Transpose of a Matrix We write

$$\mathrm{A}^{\mathrm{T}} = \begin{pmatrix} 1 & -3 \\ 5 & 0 \\ -2 & 6 \end{pmatrix}, \tag{2.2}$$

and call A^{T} the transpose of the matrix A in (2.1). To transpose a matrix, we simply write the rows as columns, that is, we interchange rows and columns. Note that, using index notation, we have $(\mathrm{A}^{\mathrm{T}})_{ij} = A_{ji}$. You will find a summary of matrix notation in Section 9.

Sets of Linear Equations Historically linear algebra grew out of efforts to find efficient methods for solving sets of linear equations. As we have said, the subject has developed far beyond the solution of sets of numerical equations (which are easily solved by computer), but the ideas and methods developed for that purpose are needed in later work. A simple way to learn these techniques is to use them to solve some numerical problems by hand. In this section and the next we will develop methods of working with sets of linear equations, and introduce definitions and notation which will be useful later. Also, as you will see, we will discover how to tell whether a given set of equations has a solution or not.

▶ **Example 1.** Consider the set of equations

$$\begin{cases} 2x & & - & z = 2, \\ 6x & + 5y & + & 3z = 7, \\ 2x & - y & & = 4. \end{cases} \tag{2.3}$$

Let's agree always to write sets of equations in this *standard form* with the x terms lined up in a column (and similarly for the other variables), and with the constants on the right hand sides of the equations. Then there are several matrices of interest connected with these equations. First is the *matrix of the coefficients* which we will call M:

$$\mathrm{M} = \begin{pmatrix} 2 & 0 & -1 \\ 6 & 5 & 3 \\ 2 & -1 & 0 \end{pmatrix}. \tag{2.4}$$

Then there are two 3 by 1 matrices which we will call r and k:

$$\mathrm{r} = \begin{pmatrix} x \\ y \\ z \end{pmatrix}, \qquad \mathrm{k} = \begin{pmatrix} 2 \\ 7 \\ 4 \end{pmatrix}. \tag{2.5}$$

If we use index notation and replace x, y, z, by x_1, x_2, x_3, and call the constants k_1, k_2, k_3, then we could write the equations (2.3) in the form (Problem 1)

$$\sum_{j=1}^{3} M_{ij} x_j = k_i, \quad i = 1, 2, 3. \tag{2.6}$$

It is interesting to note that, as we will see in Section 6, this is exactly how matrices are multiplied, so we will learn to write sets of equations like (2.3) as $\mathrm{Mr} = \mathrm{k}$.

For right now we are interested in the fact that we can display all the essential numbers in equations (2.3) as a matrix known as the *augmented matrix* which we call A. Note that the first three columns of A are just the columns of M, and the fourth column is the column of constants on the right hand sides of the equations.

$$\text{(2.7)} \qquad \mathrm{A} = \begin{pmatrix} 2 & 0 & -1 & 2 \\ 6 & 5 & 3 & 7 \\ 2 & -1 & 0 & 4 \end{pmatrix}.$$

Instead of working with a set of equations and writing all the variables, we can just work with the matrix (2.7). The process which we are going to show is called *row reduction* and is essentially the way your computer solves a set of linear equations. Row reduction is just a systematic way of taking linear combinations of the given equations to produce a simpler but equivalent set of equations. We will show the process, writing side-by-side the equations and the matrix corresponding to them.

(a) The first step is to use the first equation in (2.3) to eliminate the x terms in the other two equations. The corresponding matrix operation on (2.7) is to subtract 3 times the first row from the second row and subtract the first row from the third row. This gives:

$$\begin{cases} 2x \qquad\quad - \;z = 2, \\ \qquad 5y + 6z = 1, \\ \quad\;\; - \;y + \;\;z = 2. \end{cases} \qquad \begin{pmatrix} 2 & 0 & -1 & 2 \\ 0 & 5 & 6 & 1 \\ 0 & -1 & 1 & 2 \end{pmatrix}$$

(b) Now it is convenient to interchange the second and third equations to get:

$$\begin{cases} 2x \qquad\quad - \;z = 2, \\ \quad\;\; - \;y + \;\;z = 2, \\ \qquad 5y + 6z = 1. \end{cases} \qquad \begin{pmatrix} 2 & 0 & -1 & 2 \\ 0 & -1 & 1 & 2 \\ 0 & 5 & 6 & 1 \end{pmatrix}$$

(c) Next we use the second equation to eliminate the y terms from the other equations:

$$\begin{cases} 2x \qquad\quad - \;\;\; z = 2, \\ \quad\;\; - \;y + \;\;\; z = 2, \\ \qquad\qquad 11z = 11. \end{cases} \qquad \begin{pmatrix} 2 & 0 & -1 & 2 \\ 0 & -1 & 1 & 2 \\ 0 & 0 & 11 & 11 \end{pmatrix}$$

(d) Finally, we divide the third equation by 11 and then use it to eliminate the z terms from the other equations:

$$\begin{cases} 2x \qquad\qquad = 3, \\ \quad\;\; - \;y \qquad = 1, \\ \qquad\qquad z = 1. \end{cases} \qquad \begin{pmatrix} 2 & 0 & 0 & 3 \\ 0 & -1 & 0 & 1 \\ 0 & 0 & 1 & 1 \end{pmatrix}$$

It is customary to divide each equation by the leading coefficient so that the equations read $x = 3/2$, $y = -1$, $z = 1$. The row reduced matrix is then:

$$\begin{pmatrix} 1 & 0 & 0 & 3/2 \\ 0 & 1 & 0 & -1 \\ 0 & 0 & 1 & 1 \end{pmatrix}.$$

The important thing to understand here is that in finding a row reduced matrix we have just taken linear combinations of the original equations. This process is reversible, so the final simple equations are equivalent to the original ones. Let's summarize the allowed operations in row reducing a matrix (called *elementary row operations*).

(2.8)
i. Interchange two rows [see step (b)];
ii. Multiply (or divide) a row by a (nonzero) constant [see step (d)];
iii. Add a multiple of one row to another; this includes subtracting, that is, using a negative multiple [see steps (a) and (c)].

▶ **Example 2.** Write and row reduce the augmented matrix for the equations:

$$\left\{\begin{aligned} x - y + 4z &= 5,\\ 2x - 3y + 8z &= 4,\\ x - 2y + 4z &= 9.\end{aligned}\right. \tag{2.9}$$

This time we won't write the equations, just the augmented matrix. Remember the routine: Use the first row to clear the rest of the first column; use the new second row to clear the rest of the second column; etc. Also, since matrices are equal only if they are identical, we will not use equal signs between them. Let's use arrows.

$$\begin{pmatrix} 1 & -1 & 4 & 5\\ 2 & -3 & 8 & 4\\ 1 & -2 & 4 & 9\end{pmatrix} \to \begin{pmatrix} 1 & -1 & 4 & 5\\ 0 & -1 & 0 & -6\\ 0 & -1 & 0 & 4\end{pmatrix} \to \begin{pmatrix} 1 & 0 & 4 & 11\\ 0 & -1 & 0 & -6\\ 0 & 0 & 0 & -20\end{pmatrix}$$

We don't need to go any farther! The last row says $0 \cdot z = -20$ which isn't true for any finite value of z. Now you see why your computer doesn't give an answer—there isn't any. We say that the equations are *inconsistent*. If this happens for a set of equations you have written for a physics problem, you know to look for a mistake.

Rank of a Matrix There is another way to discuss Example 2 using the following definition: The number of nonzero rows remaining when a matrix has been row reduced is called the *rank* of the matrix. (It is a theorem that the rank of A^{T} is the same as the rank of A.) Now look at the reduced augmented matrix for Example 2; it has 3 nonzero rows so its rank is 3. But the matrix M (matrix of the coefficients = first three columns of A) has only 2 nonzero rows so its rank is 2. Note that (rank of M) $<$ (rank of A) and the equations are inconsistent.

▶ **Example 3.** Consider the equations

$$\left\{\begin{aligned} x + 2y - z &= 4,\\ 2x - z &= 1,\\ x - 2y &= -3.\end{aligned}\right. \tag{2.10}$$

Either by hand or by computer we row reduce the augmented matrix to get:

$$\begin{pmatrix} 1 & 2 & -1 & 4\\ 2 & 0 & -1 & 1\\ 1 & -2 & 0 & -3\end{pmatrix} \to \begin{pmatrix} 1 & 0 & -1/2 & 1/2\\ 0 & 1 & -1/4 & 7/4\\ 0 & 0 & 0 & 0\end{pmatrix}.$$

The last row of zeros tells us that there are infinitely many solutions. For any z we find from the first two rows that $x = (z+1)/2$ and $y = (z+7)/4$. Here we see that the rank of M and the rank of A are both 2 but the number of unknowns is 3, and we are able to find two unknowns in terms of the third.

To make this all very clear, let's look at some simple examples where the results are obvious. We write three sets of equations together with the row reduced matrices:

$$\text{(2.11)} \qquad \begin{cases} x + y = 2, \\ x + y = 5. \end{cases} \qquad \begin{pmatrix} 1 & 1 & 2 \\ 0 & 0 & 3 \end{pmatrix}$$

$$\text{(2.12)} \qquad \begin{cases} x + \ \ y = 2, \\ 2x + 2y = 4. \end{cases} \qquad \begin{pmatrix} 1 & 1 & 2 \\ 0 & 0 & 0 \end{pmatrix}$$

$$\text{(2.13)} \qquad \begin{cases} x + y = 2, \\ x - y = 4. \end{cases} \qquad \begin{pmatrix} 1 & 0 & 3 \\ 0 & 1 & -1 \end{pmatrix}$$

In (2.11), since $x + y$ can't be equal to both 2 and 5, it is clear that there is no solution; the equations are inconsistent. Note that the last row of the reduced matrix is all zeros except for the last entry and so (rank M) $<$ (rank A). In (2.12), the second equation is just twice the first so they are really the same equation; we say that the equations are dependent. There is an infinite set of solutions, namely all points on the line $y = 2 - x$. Note that the last line of the matrix is all zeros; this indicates linear dependence. We have (rank A) $=$ (rank M) $= 1$, and we can solve for one unknown in terms of the other. Finally in (2.13) we have a set of equations with one solution, $x = 3$, $y = -1$, and we see that the row reduced matrix gives this result. Note that (rank A) $=$ (rank M) $=$ number of unknowns $= 2$.

Now let's consider the general problem of solving m equations in n unknowns. Then M has m rows (corresponding to m equations) and n columns (corresponding to n unknowns) and A has one more column (the constants). The following summary outlines the possible cases.

(2.14)

a. If (rank M) $<$ (rank A), the equations are inconsistent and there is no solution.

b. If (rank M) $=$ (rank A) $= n$ (number of unknowns), there is one solution.

c. If (rank M) $=$ (rank A) $= R < n$, then R unknowns can be found in terms of the remaining $n - R$ unknowns.

Example 4. Here is a set of equations and the row reduced matrix:

$$\text{(2.15)} \qquad \begin{cases} x + y - z = 7, \\ 2x - y - 5z = 2, \\ -5x + 4y + 14z = 1, \\ 3x - y - 7z = 5. \end{cases} \qquad \begin{pmatrix} 1 & 0 & -2 & 3 \\ 0 & 1 & 1 & 4 \\ 0 & 0 & 0 & 0 \\ 0 & 0 & 0 & 0 \end{pmatrix}$$

From the reduced matrix, the solution is $x = 3 + 2z$, $y = 4 - z$. We see that this is an example of (2.14c) with $m = 4$ (number of equations), $n = 3$ (number of unknowns), (rank M) = (rank A) = $R = 2 < n = 3$. Then by (2.14c), we solve for $R = 2$ unknowns (x and y) in terms of the $n - R = 1$ unknown (z).

PROBLEMS, SECTION 2

1. The first equation in (2.6) written out in detail is

$$M_{11}x_1 + M_{12}x_2 + M_{13}x_3 = k_1.$$

Write out the other two equations in the same way and then substitute $x_1, x_2, x_3 = x, y, z$ and the values of M_{ij} and k_i from (2.4) and (2.5) to verify that (2.6) is really (2.3).

2. As in Problem 1, write out in detail in terms of M_{ij}, x_j, and k_i, equations like (2.6) for two equations in four unknowns; for four equations in two unknowns.

For each of the following problems write and row reduce the augmented matrix to find out whether the given set of equations has exactly one solution, no solutions, or an infinite set of solutions. Check your results by computer. *Warning hint:* Be sure your equations are written in standard form. *Comment*: Remember that the point of doing these problems is not just to get an answer (which your computer will give you), but to become familiar with the terminology, ideas, and notation we are using.

3. $\begin{cases} x - 2y + 13 = 0 \\ y - 4x = 17 \end{cases}$

4. $\begin{cases} 2x + y - z = 2 \\ 4x + y - 2z = 3 \end{cases}$

5. $\begin{cases} 2x + y - z = 2 \\ 4x + 2y - 2z = 3 \end{cases}$

6. $\begin{cases} x + y - z = 1 \\ 3x + 2y - 2z = 3 \end{cases}$

7. $\begin{cases} 2x + 3y = 1 \\ x + 2y = 2 \\ x + 3y = 5 \end{cases}$

8. $\begin{cases} -x + y - z = 4 \\ x - y + 2z = 3 \\ 2x - 2y + 4z = 6 \end{cases}$

9. $\begin{cases} x - y + 2z = 5 \\ 2x + 3y - z = 4 \\ 2x - 2y + 4z = 6 \end{cases}$

10. $\begin{cases} x + 2y - z = 1 \\ 2x + 3y - 2z = -1 \\ 3x + 4y - 3z = -4 \end{cases}$

11. $\begin{cases} x - 2y = 4 \\ 5x + z = 7 \\ x + 2y - z = 3 \end{cases}$

12. $\begin{cases} 2x + 5y + z = 2 \\ x + y + 2z = 1 \\ x + 5z = 3 \end{cases}$

13. $\begin{cases} 4x + 6y - 12z = 7 \\ 5x - 2y + 4z = -15 \\ 3x + 4y - 8z = 4 \end{cases}$

14. $\begin{cases} 2x + 3y - z = -2 \\ x + 2y - z = 4 \\ 4x + 7y - 3z = 11 \end{cases}$

Find the rank of each of the following matrices.

15. $\begin{pmatrix} 1 & 1 & 2 \\ 2 & 4 & 6 \\ 3 & 2 & 5 \end{pmatrix}$

16. $\begin{pmatrix} 2 & -3 & 5 & 3 \\ 4 & -1 & 1 & 1 \\ 3 & -2 & 3 & 4 \end{pmatrix}$

17. $\begin{pmatrix} 1 & 1 & 4 & 3 \\ 3 & 1 & 10 & 7 \\ 4 & 2 & 14 & 10 \\ 2 & 0 & 6 & 4 \end{pmatrix}$

18. $\begin{pmatrix} 1 & 0 & 1 & 0 \\ -1 & -2 & -1 & 0 \\ 2 & 2 & 5 & 3 \\ 2 & 4 & 8 & 6 \end{pmatrix}$

3. DETERMINANTS; CRAMER'S RULE

We have said that a matrix is simply a display of a set of numbers; it does *not* have a numerical value. For a square matrix, however, there is a useful number called the *determinant* of the matrix. Although a computer will quickly give the value of a determinant, we need to know what this value means in order to use it in applications. [See, for example, equations (4.19), (6.24) and (8.5).] We also need to know how to work with determinants. An easy way to learn these things is to solve some numerical problems by hand. We shall outline some of the facts about determinants without proofs (for more details, see linear algebra texts).

Evaluating Determinants To indicate that we mean the determinant of a square matrix A (written det A), we replace the large parentheses inclosing A by single bars. The value of det A if A is a 1 by 1 matrix is just the value of the single element. For a 2 by 2 matrix,

$$\text{(3.1)} \qquad \text{A} = \begin{pmatrix} a & b \\ c & d \end{pmatrix}, \qquad \det \text{A} = \begin{vmatrix} a & b \\ c & d \end{vmatrix} = ad - bc.$$

Equation (3.1) gives the value of a second order determinant. We shall describe how to evaluate determinants of higher order.

First we need some notation and definitions. It is convenient to write an nth order determinant like this:

$$\text{(3.2)} \qquad \begin{vmatrix} a_{11} & a_{12} & a_{13} & \cdots & a_{1n} \\ a_{21} & a_{22} & a_{23} & \cdots & a_{2n} \\ a_{31} & a_{32} & a_{33} & \cdots & a_{3n} \\ \vdots & & & \ddots & \vdots \\ a_{n1} & a_{n2} & a_{n3} & \cdots & a_{nn} \end{vmatrix}.$$

Notice that a_{23} is the element in the second row and the third column; that is, the first subscript is the number of the row and the second subscript is the number of the column in which the element is. Thus the element a_{ij} is in row i and column j. As an abbreviation for the determinant in (3.2), we sometimes write simply $|a_{ij}|$, that is, the determinant whose elements are a_{ij}. In this form it looks exactly like the absolute value of the element a_{ij} and you have to tell from the context which of these meanings is intended.

If we remove one row and one column from a determinant of order n, we have a determinant of order $n - 1$. Let us remove the row and column containing the element a_{ij} and call the remaining determinant M_{ij}. The determinant M_{ij} is called the *minor* of a_{ij}. For example, in the determinant

$$\text{(3.3)} \qquad \begin{vmatrix} 1 & -5 & 2 \\ 7 & 3 & 4 \\ 2 & 1 & 5 \end{vmatrix},$$

the minor of the element $a_{23} = 4$ is

$$M_{23} = \begin{vmatrix} 1 & -5 \\ 2 & 1 \end{vmatrix},$$

obtained by crossing off the row and column containing 4. The signed minor $(-1)^{i+j}M_{ij}$ is called the *cofactor* of a_{ij}. In (3.3), the element 4 is in the second row ($i = 2$) and third column ($j = 3$), so $i + j = 5$, and the cofactor of 4 is $(-1)^5 M_{23} = -11$. It is very convenient to get the proper sign (plus or minus) for the factor $(-1)^{i+j}$ by thinking of a checkerboard of plus and minus signs like this:

$$(3.4) \qquad \begin{vmatrix} + & - & + & - & & & \\ - & + & - & + & & & \\ + & - & + & - & \text{etc.} & & \\ - & + & - & + & & & \\ & \text{etc.} & & & \ddots & & \\ & & & & & + & - \\ & & & & & - & + \end{vmatrix}.$$

Then the sign $(-1)^{i+j}$ to be attached to M_{ij} is just the checkerboard sign in the same position as a_{ij}. For the element a_{23}, you can see that the checkerboard sign is minus.

Now we can easily say how to find the *value of a determinant: Multiply each element of one row (or one column) by its cofactor and add the results.* It can be shown that we get the same answer whichever row or column we use.

▶ **Example 1.** Let us evaluate the determinant in (3.3) using elements of the third column. We get

$$\begin{vmatrix} 1 & -5 & 2 \\ 7 & 3 & 4 \\ 2 & 1 & 5 \end{vmatrix} = 2\begin{vmatrix} 7 & 3 \\ 2 & 1 \end{vmatrix} - 4\begin{vmatrix} 1 & -5 \\ 2 & 1 \end{vmatrix} + 5\begin{vmatrix} 1 & -5 \\ 7 & 3 \end{vmatrix}$$
$$= 2 \cdot 1 - 4 \cdot 11 + 5 \cdot 38 = 148.$$

As a check, using elements of the first row, we get

$$1\begin{vmatrix} 3 & 4 \\ 1 & 5 \end{vmatrix} + 5\begin{vmatrix} 7 & 4 \\ 2 & 5 \end{vmatrix} + 2\begin{vmatrix} 7 & 3 \\ 2 & 1 \end{vmatrix} = 11 + 135 + 2 = 148.$$

The method of evaluating a determinant which we have described here is one form of Laplace's development of a determinant. If the determinant is of fourth order (or higher), using the Laplace development once gives us a set of determinants of order one less than we started with; then we use the Laplace development all over again to evaluate each of these, and so on until we get determinants of second order which we know how to evaluate. This is obviously a lot of work! We will see below how to simplify the calculation. A word of warning to anyone who has learned a special method of evaluating a third-order determinant by recopying columns to the right and multiplying along diagonals: this method *does not work* for fourth order (and higher).

Useful Facts About Determinants We state these facts without proof. (See algebra books for proofs.)

1. If each element of *one* row (or *one* column) of a determinant is multiplied by a number k, the value of the determinant is multiplied by k.
2. The value of a determinant is zero if
   (a) all elements of one row (or column) are zero; or if
   (b) two rows (or two columns) are identical; or if
   (c) two rows (or two columns) are proportional.
3. If two rows (or two columns) of a determinant are interchanged, the value of the determinant changes sign.
4. The value of a determinant is unchanged if
   (a) rows are written as columns and columns as rows; or if
   (b) we add to each element of one row, k times the corresponding element of another row, where k is any number (and a similar statement for columns).

Let us look at a few examples of the use of these facts.

Example 2. Find the equation of a plane through the three given points $(0,0,0)$, $(1,2,5)$, and $(2,-1,0)$.

We shall verify that the answer in determinant form is

$$\begin{vmatrix} x & y & z & 1 \\ 0 & 0 & 0 & 1 \\ 1 & 2 & 5 & 1 \\ 2 & -1 & 0 & 1 \end{vmatrix} = 0.$$

By a Laplace development using elements of the first row, we would find that this is a linear equation in x, y, z; thus it represents a plane. We need now to show that the three points are in the plane. Suppose $(x,y,z) = (0,0,0)$; then the first two rows of the determinant are identical and by Fact 2b the determinant is zero. Similarly if the point (x,y,z) is either of the other given points, two rows of the determinant are identical and the determinant is zero. Thus all three points lie in the plane.

Example 3. Evaluate the determinant

$$D = \begin{vmatrix} 0 & a & -b \\ -a & 0 & c \\ b & -c & 0 \end{vmatrix}.$$

If we interchange rows and columns in D, then by Facts 4a and 1 we have

$$D = \begin{vmatrix} 0 & -a & b \\ a & 0 & -c \\ -b & c & 0 \end{vmatrix} = (-1)^3 \begin{vmatrix} 0 & a & -b \\ -a & 0 & c \\ b & -c & 0 \end{vmatrix},$$

where in the last step we have factored -1 out of each column by Fact 1. Thus we have $D = -D$, so $D = 0$.

We can use Facts 1 to 4 to simplify finding the value of a determinant. First we check Facts 2a, 2b, 2c, in case the determinant is trivially equal to zero. Then we try to get as many zeros as possible in some row or column in order to have fewer terms in the Laplace development. We look for rows (or columns) which can be combined (using Fact 4b) to give zeros. Although this is something like row reduction, we can operate with columns as well as rows. However, we can't just cancel a number from a row (or column); by Fact 1 we must keep it as a factor in our answer. And we must keep track of any row (or column) interchanges since by Fact 3 each interchange multiplies the determinant by (-1).

▶ **Example 4.** Evaluate the determinant

$$D = \begin{vmatrix} 4 & 3 & 0 & 1 \\ 9 & 7 & 2 & 3 \\ 4 & 0 & 2 & 1 \\ 3 & -1 & 4 & 0 \end{vmatrix}.$$

Subtract 4 times the fourth column from the first column, and subtract 2 times the fourth column from the third column to get:

$$D = \begin{vmatrix} 0 & 3 & -2 & 1 \\ -3 & 7 & -4 & 3 \\ 0 & 0 & 0 & 1 \\ 3 & -1 & 4 & 0 \end{vmatrix}.$$

Do a Laplace development using the third row:

$$D = (-1) \begin{vmatrix} 0 & 3 & -2 \\ -3 & 7 & -4 \\ 3 & -1 & 4 \end{vmatrix}. \tag{3.5}$$

Add the second row to the third row:

$$D = (-1) \begin{vmatrix} 0 & 3 & -2 \\ -3 & 7 & -4 \\ 0 & 6 & 0 \end{vmatrix}.$$

Do a Laplace development using the first column:

$$D = (-1)(-1)(-3) \begin{vmatrix} 3 & -2 \\ 6 & 0 \end{vmatrix} = (-3)[0 - 6(-2)] = -36.$$

This is the answer but you might like to look for some shorter solutions. For example, consider the determinant (3.5) above. If we immediately do another Laplace development using the first row, the minor of 3 in the first row, second column is

$$\begin{vmatrix} -3 & -4 \\ 3 & 4 \end{vmatrix}.$$

Without even evaluating it, we should recognize by Fact 2c that it is zero. Then proceeding with the Laplace development of (3.5) using the first row gives just

$$D = (-1)(-2)\begin{vmatrix} -3 & 7 \\ 3 & -1 \end{vmatrix} = 2(3-21) = -36 \quad \text{as above.}$$

Now you may be wondering why you should learn about this when your computer will do it for you. Suppose you have a determinant with elements which are algebraic expressions, and you want to write it in a different form. Then you need to know what manipulations you can do without changing its value. Also, if you know the rules, you may see that a determinant is zero without evaluating it. An easy way to learn these things is to evaluate some simple numerical determinants by hand.

Cramer's Rule This is a formula in terms of determinants for the solution of n linear equations in n unknowns when there is exactly one solution. As we said for row reduction and for evaluating determinants, your computer will quickly give you the solution of a set of linear equations when there is one. However, for theoretical purposes, we need the Cramer's rule formula, and a simple way to learn about it is to use it to solve sets of linear equations with numerical coefficients.

Let us first show the use of Cramer's rule to solve two equations in two unknowns. Then we will generalize it to n equations in n unknowns. Consider the set of equations

$$\begin{cases} a_1x + b_1y = c_1, \\ a_2x + b_2y = c_2. \end{cases} \tag{3.6}$$

If we multiply the first equation by b_2, the second by b_1, and then subtract the results and solve for x, we get (if $a_1b_2 - a_2b_1 \neq 0$)

$$x = \frac{c_1b_2 - c_2b_1}{a_1b_2 - a_2b_1}. \tag{3.7a}$$

Solving for y in a similar way, we get

$$y = \frac{a_1c_2 - a_2c_1}{a_1b_2 - a_2b_1}. \tag{3.7b}$$

Using the definition (3.1) of a second order determinant, we can write the solutions (3.7) of (3.6) in the form

$$x = \frac{\begin{vmatrix} c_1 & b_1 \\ c_2 & b_2 \end{vmatrix}}{\begin{vmatrix} a_1 & b_1 \\ a_2 & b_2 \end{vmatrix}}, \qquad y = \frac{\begin{vmatrix} a_1 & c_1 \\ a_2 & c_2 \end{vmatrix}}{\begin{vmatrix} a_1 & b_1 \\ a_2 & b_2 \end{vmatrix}}. \tag{3.8}$$

It is helpful in remembering (3.8) to say in words how we find the correct determinants. First, the equations must be written in standard form as for row reduction (Section 2). Then if we simply write the array of coefficients on the left-hand side of (3.6), these form the denominator determinant in (3.8). This determinant (which we shall denote by D) is called the *determinant of the coefficients.* To find the numerator determinant for x, start with D, erase the x coefficients a_1 and a_2, and replace them by the constants c_1 and c_2 from the right-hand sides of the equations. Similarly, we replace the y coefficients in D by the constant terms to find the numerator determinant in y.

Example 5. Use (3.8) to solve the set of equations

$$\begin{cases} 2x + 3y = 3, \\ x - 2y = 5. \end{cases}$$

We find

$$D = \begin{vmatrix} 2 & 3 \\ 1 & -2 \end{vmatrix} = -4 - 3 = -7,$$

$$x = \frac{1}{D}\begin{vmatrix} 3 & 3 \\ 5 & -2 \end{vmatrix} = \frac{-6-15}{-7} = 3,$$

$$y = \frac{1}{D}\begin{vmatrix} 2 & 3 \\ 1 & 5 \end{vmatrix} = \frac{10-3}{-7} = -1.$$

This method of solution of a set of linear equations is called Cramer's rule. It may be used to solve n equations in n unknowns if $D \neq 0$; the solution then consists of one value for each unknown. The denominator determinant D is the n by n determinant of the coefficients when the equations are arranged in standard form. The numerator determinant for each unknown is the determinant obtained by replacing the column of coefficients of that unknown in D by the constant terms from the right-hand sides of the equations. Then to find the unknowns, we must evaluate each of the determinants and divide.

Rank of a Matrix Here is another way to find the rank of a matrix (Section 2). A submatrix means a matrix remaining if we remove some rows and/or remove some columns from the original matrix. To find the rank of a matrix, we look at all the square submatrices and find their determinants. The order of the largest nonzero determinant is the rank of the matrix.

Example 6. Find the rank of the matrix

$$\begin{pmatrix} 1 & -1 & 2 & 3 \\ -2 & 2 & -1 & 0 \\ 4 & -4 & 5 & 6 \end{pmatrix}.$$

We need to look at the four 3 by 3 determinants containing columns 1,2,3 or 1,2,4 or 1,3,4 or 2,3,4. We note that the first two columns are negatives of each other, so by Fact 2c the first two of these determinants are both zero. The last two determinants differ only in the sign of their first column, so we just have to look at one of them, say:

$$\begin{pmatrix} 1 & 2 & 3 \\ -2 & -1 & 0 \\ 4 & 5 & 6 \end{pmatrix}.$$

If we now subtract twice the first row from the third row, we have

$$\begin{pmatrix} 1 & 2 & 3 \\ -2 & -1 & 0 \\ 2 & 1 & 0 \end{pmatrix},$$

and we see by Fact 2c that the determinant is zero. So the rank of the matrix is less than 3. To show that it is 2, we just have to find *one* 2 by 2 submatrix with nonzero determinant. There are several of them; find one. Thus the rank of the matrix is 2. (If we had needed to show that the rank was 1, we would have had to show that *all* the 2 by 2 submatrices had determinants equal to zero.)

PROBLEMS, SECTION 3

Evaluate the determinants in Problems 1 to 6 by the methods shown in Example 4. Remember that the reason for doing this is not just to get the answer (your computer can give you that) but to learn how to manipulate determinants correctly. Check your answers by computer.

1. $\begin{vmatrix} -2 & 3 & 4 \\ 3 & 4 & -2 \\ 5 & 6 & -3 \end{vmatrix}$ **2.** $\begin{vmatrix} 5 & 17 & 3 \\ 2 & 4 & -3 \\ 11 & 0 & 2 \end{vmatrix}$

3. $\begin{vmatrix} 1 & 1 & 1 & 1 \\ 1 & 2 & 3 & 4 \\ 1 & 3 & 6 & 10 \\ 1 & 4 & 10 & 20 \end{vmatrix}$ **4.** $\begin{vmatrix} -2 & 4 & 7 & 3 \\ 8 & 2 & -9 & 5 \\ -4 & 6 & 8 & 4 \\ 2 & -9 & 3 & 8 \end{vmatrix}$

5. $\begin{vmatrix} 7 & 0 & 1 & -3 & 5 \\ 2 & -1 & 0 & 1 & 4 \\ 7 & -3 & 2 & -1 & 4 \\ 8 & 6 & -2 & -7 & 4 \\ 1 & 3 & -5 & 7 & 5 \end{vmatrix}$ **6.** $\begin{vmatrix} 0 & 1 & 1 & 1 & 1 \\ 1 & 0 & 1 & 1 & 1 \\ 1 & 1 & 0 & 1 & 1 \\ 1 & 1 & 1 & 0 & 1 \\ 1 & 1 & 1 & 1 & 0 \end{vmatrix}$

7. Prove the following by appropriate manipulations using Facts 1 to 4; do not just evaluate the determinants.

$$\begin{vmatrix} 1 & a & bc \\ 1 & b & ac \\ 1 & c & ab \end{vmatrix} = \begin{vmatrix} 1 & a & a^2 \\ 1 & b & b^2 \\ 1 & c & c^2 \end{vmatrix} = (c-a)(b-a)(c-b)\begin{vmatrix} 1 & a & a^2 \\ 0 & 1 & b+a \\ 0 & 0 & 1 \end{vmatrix}$$
$$= (c-a)(b-a)(c-b).$$

8. Show that if, in using the Laplace development, you accidentally multiply the elements of one row by the cofactors of another row, you get zero.
Hint: Consider Fact 2b.

9. Show without computation that the following determinant is equal to zero.
Hint: Consider the effect of interchanging rows and columns.

$$\begin{vmatrix} 0 & 2 & -3 \\ -2 & 0 & 4 \\ 3 & -4 & 0 \end{vmatrix}$$

10. A determinant or a square matrix is called skew-symmetric if $a_{ij} = -a_{ji}$. (The determinant in Problem 9 is an example of a skew-symmetric determinant.) Show that a skew-symmetric determinant of odd order is zero.

In Problems 11 and 12 evaluate the determinants.

11. $\begin{vmatrix} 0 & 5 & -3 & -4 & 1 \\ -5 & 0 & 2 & 6 & -2 \\ 3 & -2 & 0 & -3 & 7 \\ 4 & -6 & 3 & 0 & -3 \\ -1 & 2 & -7 & 3 & 0 \end{vmatrix}$ **12.** $\begin{vmatrix} 0 & 1 & 2 & -1 \\ -1 & 0 & -3 & 0 \\ -2 & 3 & 0 & 1 \\ 1 & 0 & -1 & 0 \end{vmatrix}$

13. Show that

$$\begin{vmatrix} \cos\theta & 1 & 0 \\ 1 & 2\cos\theta & 1 \\ 0 & 1 & 2\cos\theta \end{vmatrix} = \cos 3\theta.$$

14. Show that the n-rowed determinant

$$\begin{vmatrix} \cos\theta & 1 & 0 & 0 & & & 0 \\ 1 & 2\cos\theta & 1 & 0 & \cdots & \cdots & 0 \\ 0 & 1 & 2\cos\theta & 1 & & & 0 \\ 0 & 0 & 1 & 2\cos\theta & & & 0 \\ & & \vdots & & \ddots & & \vdots \\ & & \vdots & & & 2\cos\theta & 1 \\ 0 & 0 & 0 & 0 & \cdots & 1 & 2\cos\theta \end{vmatrix} = \cos n\theta.$$

Hint: Expand using elements of the last row or column. Use mathematical induction and the trigonometric addition formulas.

15. Use Cramer's rule to solve Problems 2.3 and 2.11.

16. In the following set of equations (from a quantum mechanics problem), A and B are the unknowns, k and K are given, and $i = \sqrt{-1}$. Use Cramer's rule to find A and show that $|A|^2 = 1$.

$$\begin{cases} A - B = -1 \\ ikA - KB = ik \end{cases}$$

17. Use Cramer's rule to solve for x and t the Lorentz equations of special relativity:

$$\begin{cases} x' = \gamma(x - vt) \\ t' = \gamma(t - vx/c^2) \end{cases} \quad \text{where} \quad \gamma^2(1 - v^2/c^2) = 1$$

Caution: Arrange the equations in standard form.

18. Find z by Cramer's rule:

$$\begin{cases} (a-b)x - (a-b)y + 3b^2z = 3ab \\ (a+2b)x - (a+2b)y - (3ab+3b^2)z = 3b^2 \\ bx + ay - (2b^2+a^2)z = 0 \end{cases}$$

4. VECTORS

Notation We shall indicate a vector by a boldface letter (for example, $\mathbf{A}$) and a component of a vector by a subscript (for example A_x is the x component of $\mathbf{A}$), as in Figure 4.1. Since it is not easy to handwrite boldface letters, you should write a vector with an arrow over it (for example, $\vec{A}$). It is very important to indicate clearly whether a letter represents a vector, since, as we shall see below, the same letter in italics (not boldface) is often used with a different meaning.

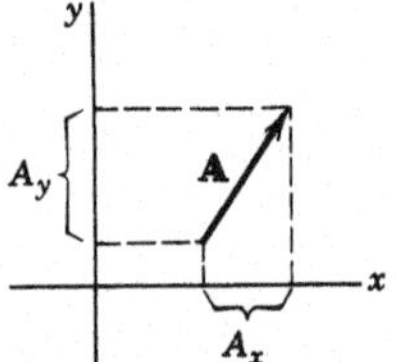

Figure 4.1

Magnitude of a Vector The length of the arrow representing a vector $\mathbf{A}$ is called the *length* or the *magnitude* of $\mathbf{A}$ (written $|\mathbf{A}|$ or A) or (see Section 10) the *norm* of $\mathbf{A}$ (written $||\mathbf{A}||$). Note the use of A to mean the magnitude of $\mathbf{A}$; for this reason it is important to make it clear whether you mean a vector or its magnitude (which is a scalar). By the Pythagorean theorem, we find

(4.1)
$$A = |\mathbf{A}| = \sqrt{A_x^2 + A_y^2} \quad \text{in two dimensions, or}$$
$$A = |\mathbf{A}| = \sqrt{A_x^2 + A_y^2 + A_z^2} \quad \text{in three dimensions.}$$

Example 1. In Figure 4.2 the force $\mathbf{F}$ has an x component of 4 lb and a y component of 3 lb. Then we write

$$F_x = 4 \text{ lb},$$
$$F_y = 3 \text{ lb},$$
$$|\mathbf{F}| = 5 \text{ lb},$$
$$\theta = \arctan \tfrac{3}{4}.$$

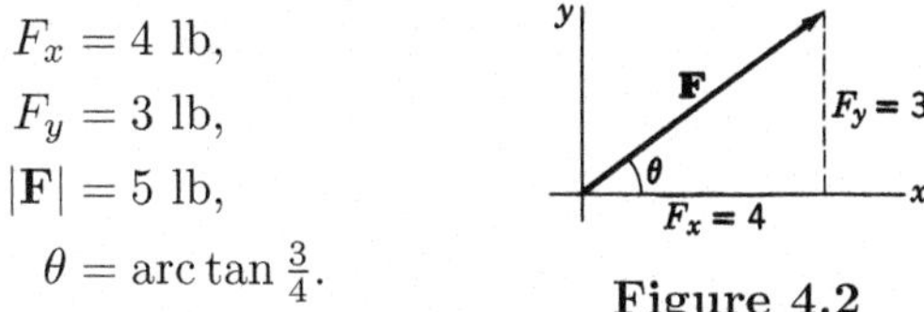

Figure 4.2

Addition of Vectors There are two ways to get the sum of two vectors. One is by the parallelogram law: To find $\mathbf{A} + \mathbf{B}$, place the tail of $\mathbf{B}$ at the head of $\mathbf{A}$ and draw the vector from the tail of $\mathbf{A}$ to the head of $\mathbf{B}$ as shown in Figures 4.3 and 4.4.

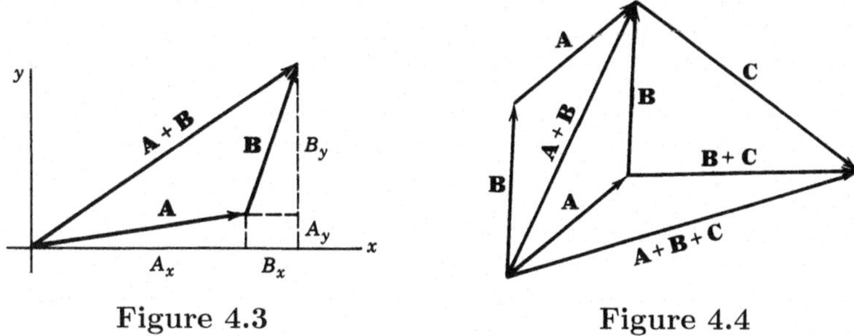

Figure 4.3

Figure 4.4

The second way of finding $\mathbf{A} + \mathbf{B}$ is to add components: $\mathbf{A} + \mathbf{B}$ has components $A_x + B_x$ and $A_y + B_y$. You should satisfy yourself from Figure 4.3 that these two methods of finding $\mathbf{A} + \mathbf{B}$ are equivalent. From Figure 4.4 and either definition of vector addition, it follows that

$$\mathbf{A} + \mathbf{B} = \mathbf{B} + \mathbf{A} \qquad \text{(commutative law for addition);}$$
$$(\mathbf{A} + \mathbf{B}) + \mathbf{C} = \mathbf{A} + (\mathbf{B} + \mathbf{C}) \qquad \text{(associative law for addition).}$$

In other words, vectors may be added together by the usual laws of algebra.

It seems reasonable to use the symbol $3\mathbf{A}$ for the vector $\mathbf{A} + \mathbf{A} + \mathbf{A}$. By the methods of vector addition above, we can say that the vector $\mathbf{A}+\mathbf{A}+\mathbf{A}$ is a vector three times as long as $\mathbf{A}$ and in the same direction as $\mathbf{A}$ and that each component of $3\mathbf{A}$ is three times the corresponding component of $\mathbf{A}$. As a natural extension of these facts we define the vector $c\mathbf{A}$ (where c is any real positive number) to be a vector c times as long as $\mathbf{A}$ and in the same direction as $\mathbf{A}$; each component of $c\mathbf{A}$ is then c times the corresponding component of $\mathbf{A}$ (Figure 4.5).

The negative of a vector is defined as a vector of the same magnitude but in the opposite direction. Then (Figure 4.6) each component of $-\mathbf{B}$ is the negative of the corresponding component of $\mathbf{B}$. We can now define subtraction of vectors by

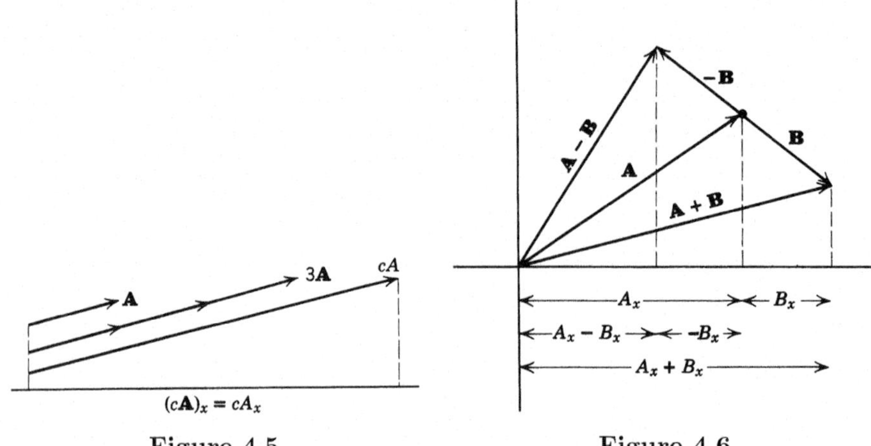

Figure 4.5 Figure 4.6

saying that $\mathbf{A} - \mathbf{B}$ means the sum of the vectors $\mathbf{A}$ and $-\mathbf{B}$. Each component of $\mathbf{A} - \mathbf{B}$ is then obtained by subtracting the corresponding components of $\mathbf{A}$ and $\mathbf{B}$, that is, $(\mathbf{A} - \mathbf{B})_x = A_x - B_x$, etc. Like addition, subtraction of vectors can be done geometrically (by the parallelogram law) or algebraically by subtracting the components (Figure 4.6).

The *zero vector* (which might arise as $\mathbf{A} = \mathbf{B} - \mathbf{B} = \mathbf{0}$, or as $\mathbf{A} = c\mathbf{B}$ with $c = 0$) is a vector of zero magnitude; its components are all zero and it does not have a direction. A vector of length or magnitude 1 is called a *unit vector*. Then for any $\mathbf{A} \neq 0$, the vector $\mathbf{A}/|\mathbf{A}|$ is a unit vector. In Example 1, $\mathbf{F}/5$ is a unit vector.

We have just seen that there are two ways to combine vectors: geometric (head to tail addition), and algebraic (using components). Let us look first at an example of the geometric method; then we shall consider the algebraic method. Example 2 below illustrates the geometric method. By similar proofs, many of the facts of elementary geometry can be easily proved using vectors, with no reference to components or a coordinate system. (See Problems 3 to 8.)

▶ **Example 2.** Prove that the medians of a triangle intersect at a point two-thirds of the way from any vertex to the midpoint of the opposite side.

To prove this, we call two of the sides of the triangle $\mathbf{A}$ and $\mathbf{B}$. The third side of the triangle is then $\mathbf{A} + \mathbf{B}$ by the parallelogram law, with the directions of $\mathbf{A}$, $\mathbf{B}$, and $\mathbf{A} + \mathbf{B}$ as indicated in Figure 4.7. If we add the vector $\frac{1}{2}\mathbf{B}$ to the vector $\mathbf{A}$ (head to tail as in Figure 4.7b), we have a vector from point O to the midpoint of the opposite side of the triangle, that is, we have the median to side $\mathbf{B}$. Next, take two-thirds of this vector; we now have the vector $\frac{2}{3}(\mathbf{A} + \frac{1}{2}\mathbf{B}) = \frac{2}{3}\mathbf{A} + \frac{1}{3}\mathbf{B}$ extending from O to P in Figure 4.7b. We want to show that P is the intersection point of the three medians and also the "$\frac{2}{3}$ point" for each. We prove this by showing that P is the "$\frac{2}{3}$ point" on the median to side $\mathbf{A}$; then since $\mathbf{A}$ and $\mathbf{B}$ represent *any* two sides of the triangle, the proof holds for all three medians. The vector from R to Q (Figure 4.7c) is $\frac{1}{2}\mathbf{A} + \mathbf{B}$; this is the median to $\mathbf{A}$. The "$\frac{2}{3}$ point" on this median is the point P' (Figure 4.7d); the vector from R to P' is equal to $\frac{1}{3}(\frac{1}{2}\mathbf{A} + \mathbf{B})$. Then the vector from O to P' is $\frac{1}{2}\mathbf{A} + \frac{1}{3}(\frac{1}{2}\mathbf{A} + \mathbf{B}) = \frac{2}{3}\mathbf{A} + \frac{1}{3}\mathbf{B}$. Thus P and P' are the same point and all three medians have their "$\frac{2}{3}$ points" there. Note that we have made no reference to a coordinate system or to components in this proof.

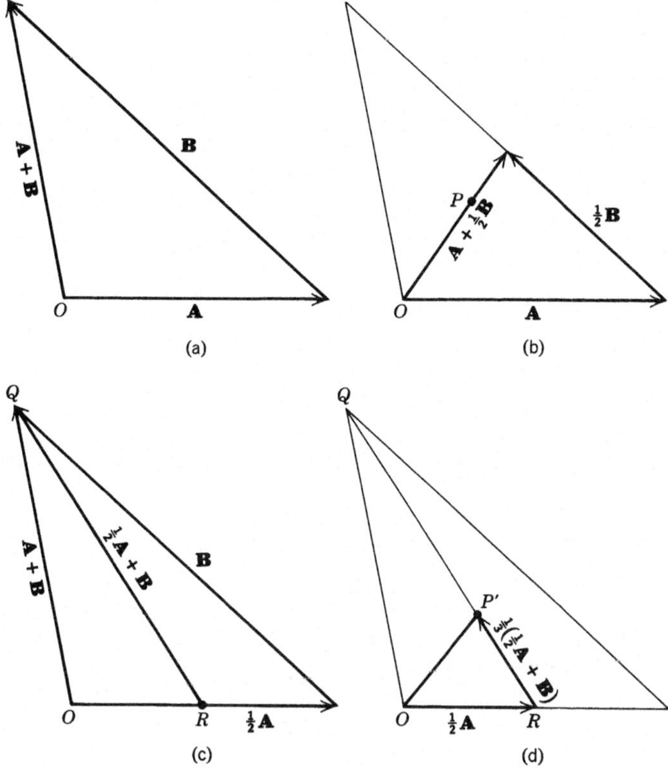

Figure 4.7

PROBLEMS, SECTION 4

1. Draw diagrams and prove (4.1).

2. Given the vectors making the given angles θ with the positive x axis:

$\mathbf{A}$ of magnitude 5, $\theta = 45°$,

$\mathbf{B}$ of magnitude 3, $\theta = -30°$,

$\mathbf{C}$ of magnitude 7, $\theta = 120°$,

(a) Draw diagrams representing $2\mathbf{A}$, $\mathbf{A} - 2\mathbf{B}$, $\mathbf{C} - \mathbf{B}$, $\frac{2}{5}\mathbf{A} - \frac{1}{7}\mathbf{C}$.

(b) Draw diagrams to show that

$$\mathbf{A} + \mathbf{B} = \mathbf{B} + \mathbf{A} \qquad \mathbf{A} - (\mathbf{B} - \mathbf{C}) = (\mathbf{A} - \mathbf{B}) + \mathbf{C},$$
$$(\mathbf{A} + \mathbf{B}) + \mathbf{C} = (\mathbf{A} + \mathbf{C}) + \mathbf{B}, \qquad (\mathbf{A} + \mathbf{B})_x = \mathbf{A}_x + \mathbf{B}_x,$$
$$(\mathbf{B} - \mathbf{C})_x = \mathbf{B}_x - \mathbf{C}_x.$$

Use vectors to prove the following theorems from geometry:

3. The diagonals of a parallelogram bisect each other.

4. The line segment joining the midpoints of two sides of any triangle is parallel to the third side and half its length.

5. In a parallelogram, the two lines from one corner to the midpoints of the two opposite sides trisect the diagonal they cross.

6. In any quadrilateral (four-sided figure with sides of various lengths and—in general—four different angles), the lines joining the midpoints of opposite sides bisect each other. *Hint:* Label three sides **A**, **B**, **C**; what is the vector along the fourth side?

7. A line through the midpoint of one side of a triangle and parallel to a second side bisects the third side. *Hint:* Call parallel vectors **A** and $c\mathbf{A}$.

8. The median of a trapezoid (four-sided figure with just two parallel sides) means the line joining the midpoints of the two nonparallel sides. Prove that the median bisects both diagonals; that the median is parallel to the two parallel bases and equal to half the sum of their lengths.

We have discussed in some detail the geometric method of adding vectors (parallelogram law or head to tail addition) and its importance in stating and proving geometric and physical facts without the intrusion of a special coordinate system. There are, however, many cases in which algebraic methods (using components relative to a particular coordinate system) are better. We shall discuss this next.

Vectors in Terms of Components We consider a set of rectangular axes as in Figure 4.8. Let the vector **i** be a unit vector in the positive x direction (out of the paper toward you), and let **j** and **k** be unit vectors in the positive y and z directions. If A_x and A_y are the scalar components of a vector in the (x, y) plane, then $\mathbf{i}A_x$ and $\mathbf{j}A_y$ are its vector components, and their sum is the vector **A** (Figure 4.9).

$$\mathbf{A} = \mathbf{i}A_x + \mathbf{j}A_y.$$

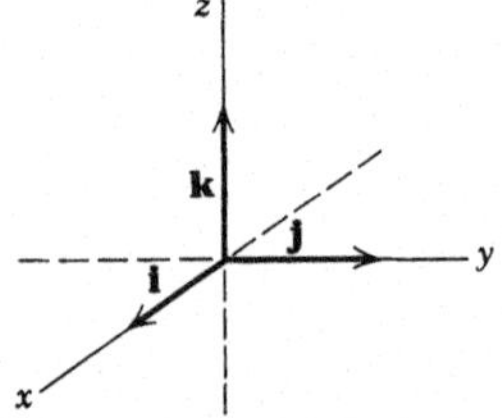

Figure 4.8

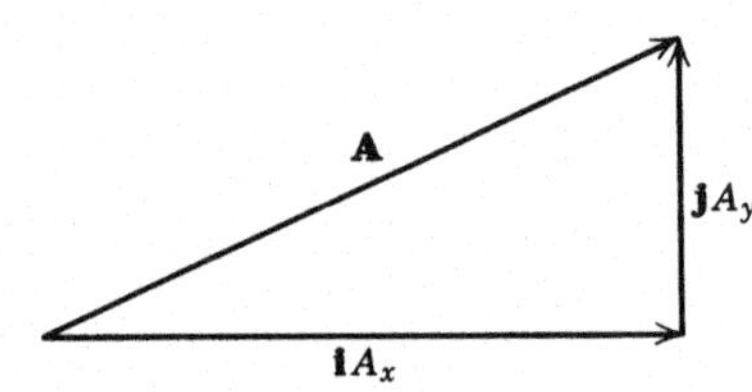

Figure 4.9

Similarly, in three dimensions

$$\mathbf{A} = \mathbf{i}A_x + \mathbf{j}A_y + \mathbf{k}A_z.$$

It is easy to add (or subtract) vectors in this form: If **A** and **B** are vectors in two dimensions, then

$$\mathbf{A} + \mathbf{B} = (\mathbf{i}A_x + \mathbf{j}A_y) + (\mathbf{i}B_x + \mathbf{j}B_y) = \mathbf{i}(A_x + B_x) + \mathbf{j}(A_y + B_y).$$

This is just the familiar result of adding components; the unit vectors **i** and **j** serve to keep track of the separate components and allow us to write **A** as a single algebraic expression. The vectors **i**, **j**, **k** are called *unit basis vectors.*

Multiplication of Vectors There are two kinds of product of two vectors. One, called the *scalar product* (or *dot product* or *inner product*), gives a result which is a scalar; the other, called the *vector product* (or *cross product*), gives a vector answer.

Scalar Product By definition, the scalar product of $\mathbf{A}$ and $\mathbf{B}$ (written $\mathbf{A} \cdot \mathbf{B}$) is a scalar equal to the magnitude of $\mathbf{A}$ times the magnitude of $\mathbf{B}$ times the cosine of the angle θ between $\mathbf{A}$ and $\mathbf{B}$:

$$\mathbf{A} \cdot \mathbf{B} = |\mathbf{A}|\,|\mathbf{B}| \cos\theta. \tag{4.2}$$

You should observe from (4.2) that the commutative law (4.3) holds for scalar multiplication:

$$\mathbf{A} \cdot \mathbf{B} = \mathbf{B} \cdot \mathbf{A}. \tag{4.3}$$

A useful interpretation of the dot product is shown in Figure 4.10.

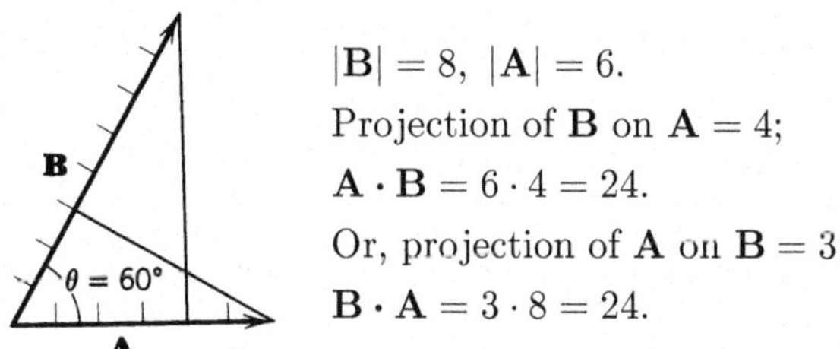

B
θ = 60°
A
$|\mathbf{B}| = 8,\ |\mathbf{A}| = 6.$
Projection of $\mathbf{B}$ on $\mathbf{A} = 4$;
$\mathbf{A} \cdot \mathbf{B} = 6 \cdot 4 = 24.$
Or, projection of $\mathbf{A}$ on $\mathbf{B} = 3$;
$\mathbf{B} \cdot \mathbf{A} = 3 \cdot 8 = 24.$

Figure 4.10

Since $|\mathbf{B}| \cos\theta$ is the projection of $\mathbf{B}$ on $\mathbf{A}$, we can write

$$\mathbf{A} \cdot \mathbf{B} = |\mathbf{A}| \text{ times (projection of } \mathbf{B} \text{ on } \mathbf{A}), \tag{4.4}$$

or, alternatively,

$$\mathbf{A} \cdot \mathbf{B} = |\mathbf{B}| \text{ times (projection of } \mathbf{A} \text{ on } \mathbf{B}).$$

Also we find from (4.2) that

$$\mathbf{A} \cdot \mathbf{A} = |\mathbf{A}|^2 \cos 0^\circ = |\mathbf{A}|^2 = A^2. \tag{4.5}$$

Sometimes $\mathbf{A}^2$ is written instead of $|\mathbf{A}|^2$ or A^2; you should understand that the square of a vector always means the square of its magnitude or its dot product with itself.

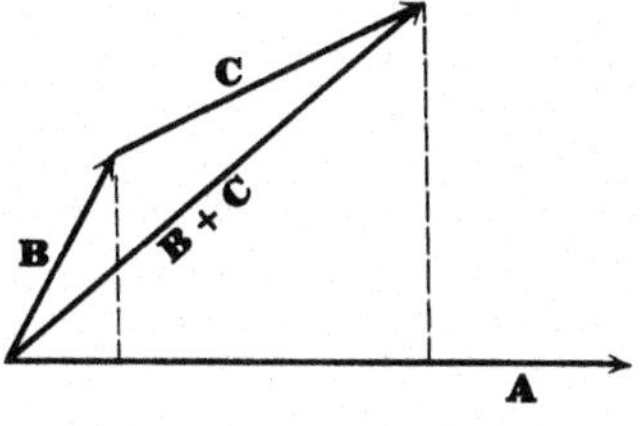

Figure 4.11

From Figure 4.11 we can see that the projection of $\mathbf{B}+\mathbf{C}$ on $\mathbf{A}$ is equal to the projection of $\mathbf{B}$ on $\mathbf{A}$ plus the projection of $\mathbf{C}$ on $\mathbf{A}$. Then by (4.4)

$$\begin{aligned}(4.6)\quad \mathbf{A}\cdot(\mathbf{B}+\mathbf{C}) &= |\mathbf{A}| \text{ times (projection of } (\mathbf{B}+\mathbf{C}) \text{ on } \mathbf{A})\\ &= |\mathbf{A}| \text{ times (projection of } \mathbf{B} \text{ on } \mathbf{A} + \text{ projection of } \mathbf{C} \text{ on } \mathbf{A})\\ &= \mathbf{A}\cdot\mathbf{B}+\mathbf{A}\cdot\mathbf{C}.\end{aligned}$$

This is the distributive law for scalar multiplication. By (4.3) we get also

$$(4.7)\qquad (\mathbf{B}+\mathbf{C})\cdot\mathbf{A} = \mathbf{B}\cdot\mathbf{A}+\mathbf{C}\cdot\mathbf{A} = \mathbf{A}\cdot\mathbf{B}+\mathbf{A}\cdot\mathbf{C}.$$

The component form of $\mathbf{A}\cdot\mathbf{B}$ is very useful. We write

$$(4.8)\qquad \mathbf{A}\cdot\mathbf{B} = (\mathbf{i}A_x+\mathbf{j}A_y+\mathbf{k}A_z)\cdot(\mathbf{i}B_x+\mathbf{j}B_y+\mathbf{k}B_z).$$

By the distributive law we can multiply this out getting nine terms such as $A_xB_x\mathbf{i}\cdot\mathbf{i}$, $A_xB_y\mathbf{i}\cdot\mathbf{j}$, and so on. Using the definition of the scalar product, we find

$$(4.9)\qquad \begin{aligned}&\mathbf{i}\cdot\mathbf{i} = |\mathbf{i}|\cdot|\mathbf{i}|\cos 0^\circ = 1\cdot 1\cdot 1 = 1, \text{ and similarly, } \mathbf{j}\cdot\mathbf{j}=1,\ \mathbf{k}\cdot\mathbf{k}=1;\\ &\mathbf{i}\cdot\mathbf{j} = |\mathbf{i}|\cdot|\mathbf{j}|\cos 90^\circ = 1\cdot 1\cdot 0 = 0, \text{ and similarly, } \mathbf{i}\cdot\mathbf{k}=0,\ \mathbf{j}\cdot\mathbf{k}=0.\end{aligned}$$

Using (4.9) in (4.8), we get

$$(4.10)\qquad \mathbf{A}\cdot\mathbf{B} = A_xB_x+A_yB_y+A_zB_z.$$

Equation (4.10) is an important formula which you should memorize. There are several immediate uses of this formula and of the dot product.

Angle Between Two Vectors Given the vectors, we can find the angle between them by using both (4.2) and (4.10) and solving for $\cos\theta$.

▶ **Example 3.** Find the angle between the vectors $\mathbf{A} = 3\mathbf{i}+6\mathbf{j}+9\mathbf{k}$ and $\mathbf{B} = -2\mathbf{i}+3\mathbf{j}+\mathbf{k}$.
By (4.2) and (4.10) we get

$$(4.11)\qquad \begin{aligned}&\mathbf{A}\cdot\mathbf{B} = |\mathbf{A}|\,|\mathbf{B}|\cos\theta = 3\cdot(-2)+6\cdot 3+9\cdot 1 = 21,\\ &|\mathbf{A}| = \sqrt{3^2+6^2+9^2} = 3\sqrt{14},\quad |\mathbf{B}| = \sqrt{2^2+3^2+1^2} = \sqrt{14},\\ &3\sqrt{14}\sqrt{14}\cos\theta = 21,\quad \cos\theta = \frac{1}{2},\quad \theta = 60^\circ.\end{aligned}$$

Perpendicular and Parallel Vectors If two vectors are perpendicular, then $\cos\theta = 0$; thus

$$(4.12)\qquad A_xB_x+A_yB_y+A_zB_z = 0 \quad \text{if } \mathbf{A} \text{ and } \mathbf{B} \text{ are perpendicular vectors.}$$

If two vectors are parallel, their components are proportional; thus (when no components are zero)

(4.13) $$\frac{A_x}{B_x} = \frac{A_y}{B_y} = \frac{A_z}{B_z} \quad \text{if } \mathbf{A} \text{ and } \mathbf{B} \text{ are parallel vectors.}$$

(Of course, if $B_x = 0$, then $A_x = 0$, etc.)

Vector Product The vector or cross product of **A** and **B** is written $\mathbf{A} \times \mathbf{B}$. By definition, $\mathbf{A} \times \mathbf{B}$ is a vector whose magnitude and direction are given as follows:

The magnitude of $\mathbf{A} \times \mathbf{B}$ is

(4.14) $$|\mathbf{A} \times \mathbf{B}| = |\mathbf{A}|\,|\mathbf{B}| \sin\theta,$$

where θ is the positive angle ($\leq 180°$) between **A** and **B**. The direction of $\mathbf{A} \times \mathbf{B}$ is perpendicular to the plane of **A** and **B** and in the sense **C** of advance of a right-handed screw rotated from **A** to **B** as in Figure 4.12.

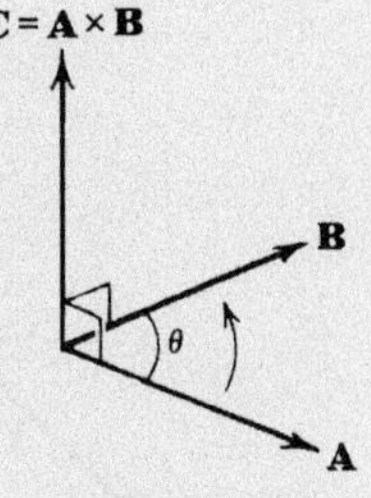

Figure 4.12

It is convenient to find the direction of $\mathbf{C} = \mathbf{A} \times \mathbf{B}$ by the following right-hand rule. Think of grasping the line **C** (or a screwdriver driving a right-handed screw in the direction **C**) with the right hand. The fingers then curl in the direction of rotation of **A** into **B** (arrow in Figure 4.12) and the thumb points along $\mathbf{C} = \mathbf{A} \times \mathbf{B}$.

Perhaps the most startling result of the vector product definition is that $\mathbf{A} \times \mathbf{B}$ and $\mathbf{B} \times \mathbf{A}$ are not equal; in fact, $\mathbf{A} \times \mathbf{B} = -\mathbf{B} \times \mathbf{A}$. In mathematical language, vector multiplication is not commutative.

We find from (4.14) that the cross product of any two parallel (or antiparallel) vectors has magnitude $|\mathbf{A} \times \mathbf{B}| = AB \sin 0° = 0$ (or $AB \sin 180° = 0$). Thus

(4.15) $$\begin{aligned} &\mathbf{A} \times \mathbf{B} = 0 \quad \text{if } \mathbf{A} \text{ and } \mathbf{B} \text{ are parallel or antiparallel,} \\ &\mathbf{A} \times \mathbf{A} = 0 \quad \text{for any } \mathbf{A}. \end{aligned}$$

Then we have the useful results

(4.16) $$\mathbf{i} \times \mathbf{i} = \mathbf{j} \times \mathbf{j} = \mathbf{k} \times \mathbf{k} = 0.$$

Also from (4.14) we find

$$|\mathbf{i} \times \mathbf{j}| = |\mathbf{i}|\,|\mathbf{j}| \sin 90° = 1 \cdot 1 \cdot 1 = 1,$$

and similarly for the magnitude of the cross product of any two different unit vectors **i**, **j**, **k**. From the right-hand rule and Figure 4.13, we see that the direction of $\mathbf{i} \times \mathbf{j}$ is **k**, and since its magnitude is 1, we have $\mathbf{i} \times \mathbf{j} = \mathbf{k}$; however, $\mathbf{j} \times \mathbf{i} = -\mathbf{k}$. Similarly evaluating the other cross products, we find

$$\text{(4.17)} \quad \begin{array}{lll} \mathbf{i}\times\mathbf{j}=\mathbf{k} & \mathbf{j}\times\mathbf{k}=\mathbf{i} & \mathbf{k}\times\mathbf{i}=\mathbf{j}. \\ \mathbf{j}\times\mathbf{i}=-\mathbf{k} & \mathbf{k}\times\mathbf{j}=-\mathbf{i} & \mathbf{i}\times\mathbf{k}=-\mathbf{j}. \end{array}$$

A good way to remember these is to write them cyclically (around a circle as indicated in Figure 4.14). Reading around the circle counterclockwise (positive θ direction), we get the positive products (for example, $\mathbf{i}\times\mathbf{j}=\mathbf{k}$); reading the other way we get the negative products (for example, $\mathbf{i}\times\mathbf{k}=-\mathbf{j}$).

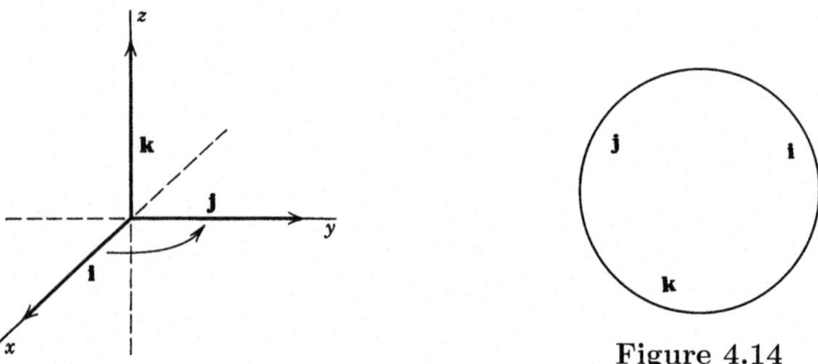

Figure 4.13

Figure 4.14

It is well to note here that the results (4.17) depend upon the way we have labeled the axes in Figure 4.13. We have arranged the (x, y, z) axes so that a rotation of the x into the y axis (through 90°) corresponds to the rotation of a right-handed screw advancing in the positive z direction. Such a coordinate system is called a *right-handed system.* If we used a left-handed system (say exchanging x and y), then all the equations in (4.17) would have their signs changed. This would be confusing; consequently, we practically always use right-handed coordinate systems, and we must be careful about this in drawing diagrams. (See Chapter 10, Section 6.)

To write $\mathbf{A}\times\mathbf{B}$ in component form we need the distributive law, namely

$$\text{(4.18)} \qquad \mathbf{A}\times(\mathbf{B}+\mathbf{C})=\mathbf{A}\times\mathbf{B}+\mathbf{A}\times\mathbf{C}.$$

(see Problem 7.18).

Then we find

$$\text{(4.19)} \quad \begin{aligned} \mathbf{A}\times\mathbf{B} &= (\mathbf{i}A_x+\mathbf{j}A_y+\mathbf{k}A_z)\times(\mathbf{i}B_x+\mathbf{j}B_y+\mathbf{k}B_z) \\ &= \mathbf{i}(A_yB_z-A_zB_y)+\mathbf{j}(A_zB_x-A_xB_z)+\mathbf{k}(A_xB_y-A_yB_x) \\ &= \begin{vmatrix} \mathbf{i} & \mathbf{j} & \mathbf{k} \\ A_x & A_y & A_z \\ B_x & B_y & B_z \end{vmatrix}. \end{aligned}$$

The second line in (4.19) is obtained by multiplying out the first line (getting nine products) and using (4.16) and (4.17). The determinant in (4.19) is the most convenient way to remember the component form of the vector product. You should verify that multiplying out the determinant using the elements of the first row gives the result in the line above it.

Since $\mathbf{A} \times \mathbf{B}$ is a vector perpendicular to $\mathbf{A}$ and to $\mathbf{B}$, we can use (4.19) to find a vector perpendicular to two given vectors.

Example 4. Find a vector perpendicular to both $\mathbf{A} = 2\mathbf{i} + \mathbf{j} - \mathbf{k}$ and $\mathbf{B} = \mathbf{i} + 3\mathbf{j} - 2\mathbf{k}$.

$$\mathbf{A} \times \mathbf{B} = \begin{vmatrix} \mathbf{i} & \mathbf{j} & \mathbf{k} \\ 2 & 1 & -1 \\ 1 & 3 & -2 \end{vmatrix} = \mathbf{i}(-2+3) - \mathbf{j}(-4+1) + \mathbf{k}(6-1)$$
$$= \mathbf{i} + 3\mathbf{j} + 5\mathbf{k}.$$

PROBLEMS, SECTION 4

9. Let $\mathbf{A} = 2\mathbf{i} + 3\mathbf{j}$ and $\mathbf{B} = 4\mathbf{i} - 4\mathbf{j}$. Show graphically, and find algebraically, the vectors $-\mathbf{A}$, $3\mathbf{B}$, $\mathbf{A} - \mathbf{B}$, $\mathbf{B} + 2\mathbf{A}$, $\frac{1}{2}(\mathbf{A} + \mathbf{B})$.

10. If $\mathbf{A} + \mathbf{B} = 4\mathbf{j} - \mathbf{i}$ and $\mathbf{A} - \mathbf{B} = \mathbf{i} + 3\mathbf{j}$, find $\mathbf{A}$ and $\mathbf{B}$ algebraically. Show by a diagram how to find $\mathbf{A}$ and $\mathbf{B}$ geometrically.

11. Let $3\mathbf{i} - \mathbf{j} + 4\mathbf{k}$, $7\mathbf{j} - 2\mathbf{k}$, $\mathbf{i} - 3\mathbf{j} + \mathbf{k}$ be three vectors with tails at the origin. Then their heads determine three points A, B, C in space which form a triangle. Find vectors representing the sides AB, BC, CA in that order and direction (for example, A to B, not B to A) and show that the sum of these vectors is zero.

12. Find the angle between the vectors $\mathbf{A} = -2\mathbf{i} + \mathbf{j} - 2\mathbf{k}$ and $\mathbf{B} = 2\mathbf{i} - 2\mathbf{j}$.

13. If $\mathbf{A} = 4\mathbf{i} - 3\mathbf{k}$ and $\mathbf{B} = -2\mathbf{i} + 2\mathbf{j} - \mathbf{k}$, find the scalar projection of $\mathbf{A}$ on $\mathbf{B}$, the scalar projection of $\mathbf{B}$ on $\mathbf{A}$, and the cosine of the angle between $\mathbf{A}$ and $\mathbf{B}$.

14. Find the angles between (a) the space diagonals of a cube; (b) a space diagonal and an edge; (c) a space diagonal and a diagonal of a face.

15. Let $\mathbf{A} = 2\mathbf{i} - \mathbf{j} + 2\mathbf{k}$. (a) Find a *unit* vector in the same direction as $\mathbf{A}$. *Hint:* Divide $\mathbf{A}$ by $|\mathbf{A}|$. (b) Find a vector in the same direction as $\mathbf{A}$ but of magnitude 12. (c) Find a vector perpendicular to $\mathbf{A}$. *Hint:* There are *many* such vectors; you are to find one of them. (d) Find a unit vector perpendicular to $\mathbf{A}$. See hint in (a).

16. Find a unit vector in the same direction as the vector $\mathbf{A} = 4\mathbf{i} - 2\mathbf{j} + 4\mathbf{k}$, and another unit vector in the same direction as $\mathbf{B} = -4\mathbf{i} + 3\mathbf{k}$. Show that the vector sum of these unit vectors bisects the angle between $\mathbf{A}$ and $\mathbf{B}$. *Hint:* Sketch the rhombus having the two unit vectors as adjacent sides.

17. Find three vectors (none of them parallel to a coordinate axis) which have lengths and directions such that they could be made into a right triangle.

18. Show that $2\mathbf{i} - \mathbf{j} + 4\mathbf{k}$ and $5\mathbf{i} + 2\mathbf{j} - 2\mathbf{k}$ are orthogonal (perpendicular). Find a third vector perpendicular to both.

19. Find a vector perpendicular to both $\mathbf{i} - 3\mathbf{j} + 2\mathbf{k}$ and $5\mathbf{i} - \mathbf{j} - 4\mathbf{k}$.

20. Find a vector perpendicular to both $\mathbf{i} + \mathbf{j}$ and $\mathbf{i} - 2\mathbf{k}$.

21. Show that $\mathbf{B}|\mathbf{A}| + \mathbf{A}|\mathbf{B}|$ and $\mathbf{A}|\mathbf{B}| - \mathbf{B}|\mathbf{A}|$ are orthogonal.

22. Square $(\mathbf{A} + \mathbf{B})$; interpret your result geometrically. *Hint:* Your answer is a law which you learned in trigonometry.

23. If $\mathbf{A} = 2\mathbf{i} - 3\mathbf{j} + \mathbf{k}$ and $\mathbf{A} \cdot \mathbf{B} = 0$, does it follow that $\mathbf{B} = 0$? (Either prove that it does or give a specific example to show that it doesn't.) Answer the same question if $\mathbf{A} \times \mathbf{B} = \mathbf{0}$. And again answer the same question if $\mathbf{A} \cdot \mathbf{B} = 0$ *and* $\mathbf{A} \times \mathbf{B} = \mathbf{0}$.

24. What is the value of $(\mathbf{A} \times \mathbf{B})^2 + (\mathbf{A} \cdot \mathbf{B})^2$? *Comment*: This is a special case of Lagrange's identity. (*See* Chapter 6, Problem 3.12b, page 284.)

Use vectors as in Problems 3 to 8, and also the dot and cross product, to prove the following theorems from geometry.

25. The sum of the squares of the diagonals of a parallelogram is equal to twice the sum of the squares of two adjacent sides of the parallelogram.

26. The median to the base of an isosceles triangle is perpendicular to the base.

27. In a kite (four-sided figure made up of two pairs of equal adjacent sides), the diagonals are perpendicular.

28. The diagonals of a rhombus (four-sided figure with all sides of equal length) are perpendicular and bisect each other.

5. LINES AND PLANES

A great deal of analytic geometry can be simplified by the use of vector notation. Such things as equations of lines and planes, and distances between points or between lines and planes often occur in physics and it is very useful to be able to find them quickly. We shall talk about three-dimensional space most of the time although the ideas apply also to two dimensions. In analytic geometry a point is a set of three coordinates (x, y, z); we shall think of this point as the *head* of a vector $\mathbf{r} = \mathbf{i}x + \mathbf{j}y + \mathbf{k}z$ *with tail at the origin.* Most of the time the *vector* will be in the background of our minds and we shall not draw it; we shall just plot the point (x, y, z) which is the head of the vector. In other words, the point (x, y, z) and the vector $\mathbf{r}$ will be synonymous. We shall also use vectors joining two points. In Figure 5.1 the vector $\mathbf{A}$ from $(1, 2, 3)$ to (x, y, z) is

$$\mathbf{A} = \mathbf{r} - \mathbf{C} = (x, y, z) - (1, 2, 3) = (x - 1, y - 2, z - 3) \quad \text{or}$$
$$\mathbf{A} = \mathbf{i}x + \mathbf{j}y + \mathbf{k}z - (\mathbf{i} + 2\mathbf{j} + 3\mathbf{k}) = \mathbf{i}(x - 1) + \mathbf{j}(y - 2) + \mathbf{k}(z - 3).$$

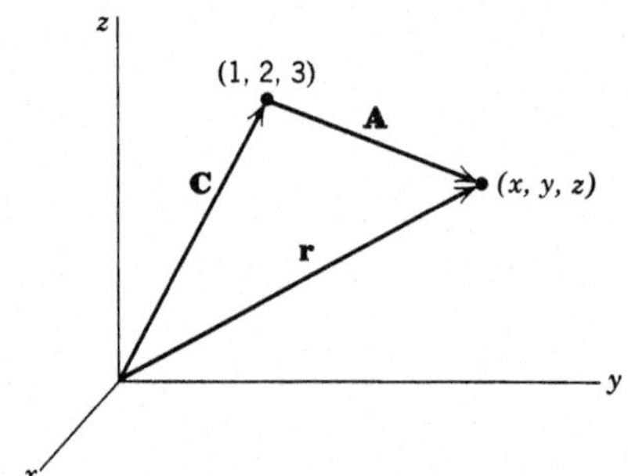

Figure 5.1

Thus we have two ways of writing vector equations; we may choose the one we prefer. Note the possible advantage of writing $(1, 0, -2)$ for $\mathbf{i} - 2\mathbf{k}$; since the zero is explicitly written, there is less chance of accidentally confusing $\mathbf{i} - 2\mathbf{k}$ with $\mathbf{i} - 2\mathbf{j} = (1, -2, 0)$. On the other hand, $5\mathbf{j}$ is simpler than $(0, 5, 0)$.

In two dimensions, we write the equation of a straight line through (x_0, y_0) with slope m as

$$\frac{y - y_0}{x - x_0} = m. \tag{5.1}$$

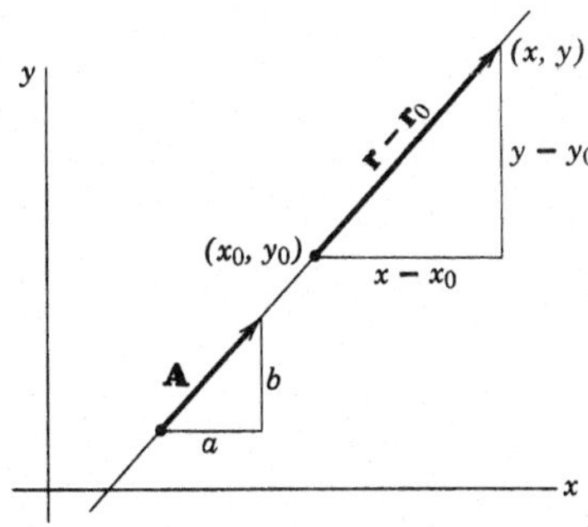

Figure 5.2

Suppose, instead of the slope, we are given a vector in the direction of the line, say $\mathbf{A} = \mathbf{i}a + \mathbf{j}b$ (Figure 5.2). Then the line through (x_0, y_0) and in the direction $\mathbf{A}$ is determined and we should be able to write its equation. The directed line segment from (x_0, y_0) to any point (x, y) on the line is the vector $\mathbf{r} - \mathbf{r}_0$ with components $x - x_0$ and $y - y_0$:

$$\mathbf{r} - \mathbf{r}_0 = \mathbf{i}(x - x_0) + \mathbf{j}(y - y_0). \tag{5.2}$$

This vector is parallel to $\mathbf{A} = \mathbf{i}a + \mathbf{j}b$. Now if two vectors are parallel, their components are proportional. Thus we can write (for $a, b \neq 0$)

$$\frac{x - x_0}{a} = \frac{y - y_0}{b} \quad \text{or} \quad \frac{y - y_0}{x - x_0} = \frac{b}{a}. \tag{5.3}$$

This is the equation of the given straight line. As a check we see that the slope of the line is $m = b/a$, so (5.3) is the same as (5.1).

Another way to write this equation is to say that if $\mathbf{r} - \mathbf{r}_0$ and $\mathbf{A}$ are parallel vectors, one is some scalar multiple of the other, that is,

$$\mathbf{r} - \mathbf{r}_0 = \mathbf{A}t, \quad \text{or} \quad \mathbf{r} = \mathbf{r}_0 + \mathbf{A}t, \tag{5.4}$$

where t is the scalar multiple. We can think of t as a parameter; the component form of (5.4) is a set of parametric equations of the line, namely

$$\begin{aligned} x - x_0 &= at, \\ y - y_0 &= bt, \end{aligned} \quad \text{or} \quad \begin{aligned} x &= x_0 + at, \\ y &= y_0 + bt. \end{aligned} \tag{5.5}$$

Eliminating t yields the equation of the line in (5.3).

In three dimensions, the same ideas can be used. We want the equations of a straight line through a given point (x_0, y_0, z_0) and parallel to a given vector $\mathbf{A} = a\mathbf{i} + b\mathbf{j} + c\mathbf{k}$. If (x, y, z) is any point on the line, the vector joining (x_0, y_0, z_0) and (x, y, z) is parallel to $\mathbf{A}$. Then its components $x - x_0$, $y - y_0$, $z - z_0$ are proportional to the components a, b, c of $\mathbf{A}$ and we have

$$\frac{x - x_0}{a} = \frac{y - y_0}{b} = \frac{z - z_0}{c} \quad \text{(symmetric equations of a straight line, } a, b, c \neq 0\text{).} \tag{5.6}$$

If c, for instance, happens to be zero, we would have to write (5.6) in the form

$$\text{(5.7)}\qquad \frac{x-x_0}{a}=\frac{y-y_0}{b},\quad z=z_0 \qquad \text{(symmetric equations of a straight line when } c=0\text{).}$$

As in the two-dimensional case, equations (5.6) and (5.7) could be written

$$\text{(5.8)}\qquad \mathbf{r}=\mathbf{r}_0+\mathbf{A}t,\quad \text{or}\quad \begin{cases} x=x_0+at,\\ y=y_0+bt,\\ z=z_0+ct,\end{cases} \qquad \text{(parametric equations of a straight line).}$$

The parametric equations (5.8) have a particularly useful interpretation when the parameter t means time. Consider a particle m (electron, billiard ball, or star) moving along the straight line L in Figure 5.3. Position yourself at the origin and watch m move from P_0 to P along L. Your line of sight is the vector $\mathbf{r}$; it swings from $\mathbf{r}_0$ at $t=0$ to $\mathbf{r}=\mathbf{r}_0+\mathbf{A}t$ at time t. Note that the velocity of m is $d\mathbf{r}/dt=\mathbf{A}$; $\mathbf{A}$ is a vector along the line of motion.

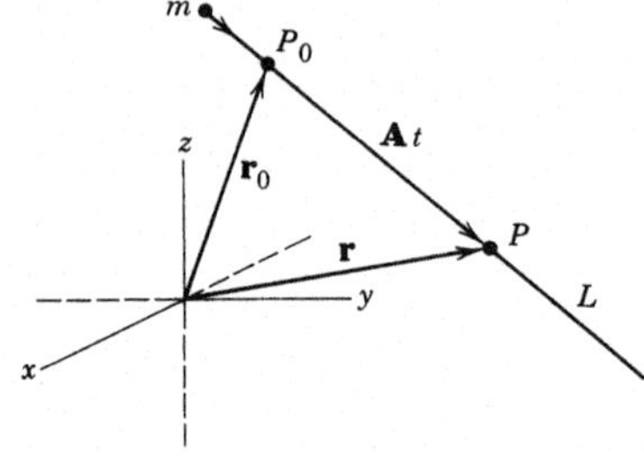

Figure 5.3

Going back to two dimensions, suppose we want the equation of a straight line L through the point (x_0, y_0) and perpendicular to a given vector $\mathbf{N}=a\mathbf{i}+b\mathbf{j}$. As above, the vector

$$\mathbf{r}-\mathbf{r}_0=(x-x_0)\mathbf{i}+(y-y_0)\mathbf{j}$$

lies along the line. This time we want this vector perpendicular to $\mathbf{N}$; recall that two vectors are perpendicular if their dot product is zero. Setting the dot product of $\mathbf{N}$ and $\mathbf{r}-\mathbf{r}_0$ equal to zero gives

$$\text{(5.9)}\qquad a(x-x_0)+b(y-y_0)=0 \quad \text{or} \quad \frac{y-y_0}{x-x_0}=-\frac{a}{b}.$$

This is the desired equation of the straight line L perpendicular to $\mathbf{N}$. As a check, note from Figure 5.4 that the slope of the line L is

$$\tan\theta=-\cot\phi=-a/b.$$

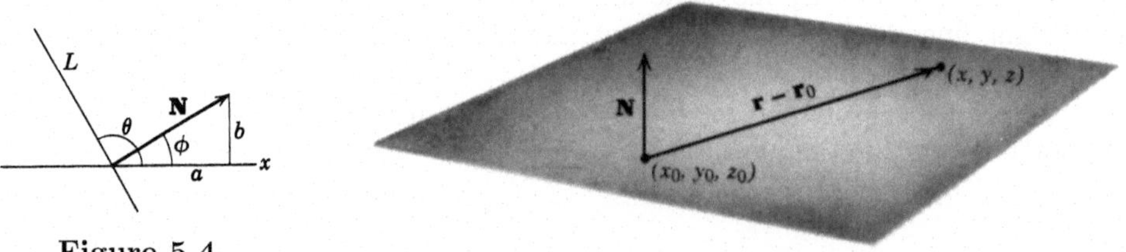

Figure 5.4

Figure 5.5

In three dimensions, we use this method to write the equation of a plane. If (x_0, y_0, z_0) is a given point in the plane and (x, y, z) is any other point in the plane, the vector (Figure 5.5)

$$\mathbf{r} - \mathbf{r}_0 = (x - x_0)\mathbf{i} + (y - y_0)\mathbf{j} + (z - z_0)\mathbf{k}$$

is in the plane. If $\mathbf{N} = a\mathbf{i} + b\mathbf{j} + c\mathbf{k}$ is normal (perpendicular) to the plane, then $\mathbf{N}$ and $\mathbf{r} - \mathbf{r}_0$ are perpendicular, so the equation of the plane is $\mathbf{N} \cdot (\mathbf{r} - \mathbf{r}_0) = 0$, or

(5.10)
$$a(x - x_0) + b(y - y_0) + c(z - z_0) = 0, \quad \text{or} \quad ax + by + cz = d, \qquad \text{(equation of a plane)}$$

where $d = ax_0 + by_0 + cz_0$.

If we are given equations like the ones above, we can read backwards to find $\mathbf{A}$ or $\mathbf{N}$. Thus we can say that the equations (5.6), (5.7), and (5.8) are the equations of a straight line which is parallel to the vector $\mathbf{A} = a\mathbf{i} + b\mathbf{j} + c\mathbf{k}$, and either equation in (5.10) is the equation of a plane perpendicular to the vector $\mathbf{N} = a\mathbf{i} + b\mathbf{j} + c\mathbf{k}$.

Example 1. Find the equation of the plane through the three points $A(-1,1,1)$, $B(2,3,0)$, $C(0,1,-2)$.

A vector joining any pair of the given points lies in the plane. Two such vectors are $\overrightarrow{AB} = (2,3,0) - (-1,1,1) = (3,2,-1)$ and $\overrightarrow{AC} = (1,0,-3)$. The cross product of these two vectors is perpendicular to the plane. This is

$$\mathbf{N} = (\overrightarrow{AB}) \times (\overrightarrow{AC}) = \begin{vmatrix} \mathbf{i} & \mathbf{j} & \mathbf{k} \\ 3 & 2 & -1 \\ 1 & 0 & -3 \end{vmatrix} = -6\mathbf{i} + 8\mathbf{j} - 2\mathbf{k}.$$

Now we write the equation of the plane with normal direction $\mathbf{N}$ through one of the given points, say B, using (5.10):

$$-6(x-2) + 8(y-3) - 2z = 0 \quad \text{or} \quad 3x - 4y + z + 6 = 0.$$

(Note that we could have divided $\mathbf{N}$ by -2 to save arithmetic.)

Example 2. Find the equations of a line through $(1,0,-2)$ and perpendicular to the plane of Example 1.

The vector $3\mathbf{i} - 4\mathbf{j} + \mathbf{k}$ is perpendicular to the plane of Example 1 and so parallel to the desired line. Thus by (5.6) the symmetric equations of the line are

$$\frac{(x-1)}{3} = \frac{y}{-4} = \frac{(z+2)}{1}.$$

By (5.8) the parametric equations of the line are $\mathbf{r} = \mathbf{i} - 2\mathbf{k} + (3\mathbf{i} - 4\mathbf{j} + \mathbf{k})t$ or, if you like, $\mathbf{r} = (1,0,-2) + (3,-4,1)t$.

Vectors give us a very convenient way of finding distances between points and lines or planes. Suppose we want to find the (perpendicular) distance from a point P

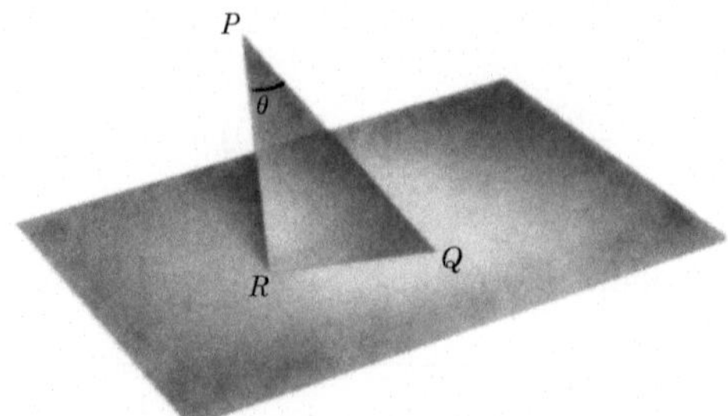

Figure 5.6

to the plane (5.10). (See Figure 5.6.) We pick *any* point Q we like in the plane (just by looking at the equation of the plane and thinking of some simple numbers x, y, z that satisfy it). The distance PR is what we want. Since PR and RQ are perpendicular (because PR is perpendicular to the plane), we have from Figure 5.6

$$PR = PQ\cos\theta. \tag{5.11}$$

From the equation of the plane, we can find a vector $\mathbf{N}$ normal to the plane. If we divide $\mathbf{N}$ by its magnitude, we have a unit vector normal to the plane; we denote this unit vector by $\mathbf{n}$. Then $|\overrightarrow{PQ}\cdot\mathbf{n}| = (PQ)\cos\theta$, which is what we need in (5.11) to find PR. (We have put in absolute value signs because $\overrightarrow{PQ}\cdot\mathbf{n}$ might be negative, whereas $(PQ)\cos\theta$, with θ acute as in Figure 5.6, is positive.)

▶ **Example 3.** Find the distance from the point $P(1,-2,3)$ to the plane $3x-2y+z+1=0$.

One point in the plane is $(1,2,0)$; call this point Q. Then the vector from P to Q is

$$\overrightarrow{PQ} = (1,2,0)-(1,-2,3) = (0,4,-3) = 4\mathbf{j}-3\mathbf{k}.$$

From the equation of the plane we get the normal vector

$$\mathbf{N} = 3\mathbf{i}-2\mathbf{j}+\mathbf{k}.$$

We get $\mathbf{n}$ by dividing $\mathbf{N}$ by $|\mathbf{N}| = \sqrt{14}$. Then we have

$$\begin{aligned}|PR| &= \left|\overrightarrow{PQ}\cdot\mathbf{n}\right| = \left|(4\mathbf{j}-3\mathbf{k})\cdot(3\mathbf{i}-2\mathbf{j}+\mathbf{k})/\sqrt{14}\right| \\ &= \left|(-8-3)/\sqrt{14}\right| = 11/\sqrt{14}.\end{aligned}$$

We can find the distance from a point P to a line in a similar way. In Figure 5.7 we want the perpendicular distance PR. We select any point on the line [that is, we pick any (x,y,z) satisfying the equations of the line]; call this point Q. Then (see Figure 5.7) $PR = PQ\sin\theta$. Let $\mathbf{A}$ be a vector along the line and $\mathbf{u}$ a unit vector along the line (obtained by dividing $\mathbf{A}$ by its magnitude). Then

$$\left|\overrightarrow{PQ}\times\mathbf{u}\right| = |PQ|\sin\theta,$$

so we get

$$|PR| = \left|\overrightarrow{PQ}\times\mathbf{u}\right|.$$

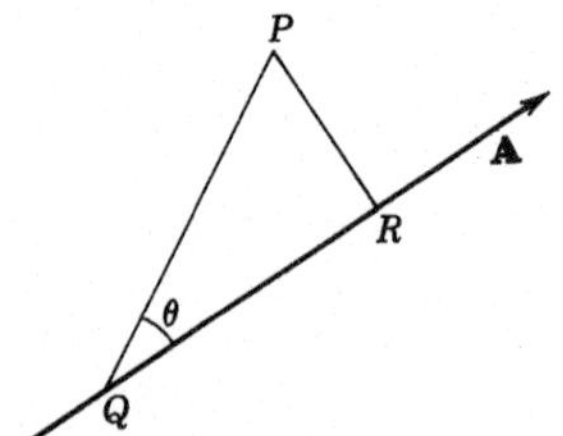

Figure 5.7

Example 4. Find the distance from $P(1,2,-1)$ to the line joining $P_1(0,0,0)$ and $P_2(-1,0,2)$.

Let $\mathbf{A} = \overrightarrow{P_1P_2} = -\mathbf{i} + 2\mathbf{k}$; this is a vector along the line. Then a unit vector along the line is $\mathbf{u} = (1/\sqrt{5})(-\mathbf{i} + 2\mathbf{k})$. Let us take Q to be $P_1(0,0,0)$. Then $\overrightarrow{PQ} = -\mathbf{i} - 2\mathbf{j} + \mathbf{k}$, so we get for the distance $|PR|$:

$$|PR| = \frac{1}{\sqrt{5}}|(-\mathbf{i} - 2\mathbf{j} + \mathbf{k}) \times (-\mathbf{i} + 2\mathbf{k})| = \frac{1}{\sqrt{5}}|-4\mathbf{i} + \mathbf{j} - 2\mathbf{k}| = \sqrt{21/5}.$$

It is also straightforward to find the distance between two skew lines (and if you really want to appreciate vectors, just look up this calculation in an analytic geometry book that doesn't use vectors!). Pick two points P and Q, one on each line (Figure 5.8). Then $|\overrightarrow{PQ} \cdot \mathbf{n}|$, where $\mathbf{n}$ is a unit vector perpendicular to both lines, is the distance we want. Now if $\mathbf{A}$ and $\mathbf{B}$ are vectors along the two lines, then $\mathbf{A} \times \mathbf{B}$ is perpendicular to both, and $\mathbf{n}$ is just $\mathbf{A} \times \mathbf{B}$ divided by $|\mathbf{A} \times \mathbf{B}|$.

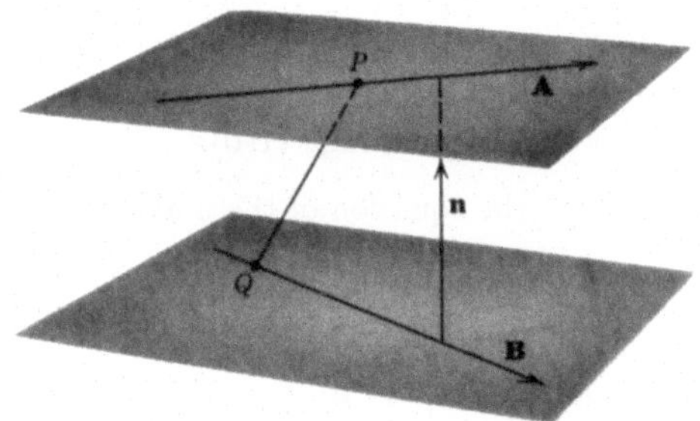

Figure 5.8

Example 5. Find the distance between the lines $\mathbf{r} = \mathbf{i} - 2\mathbf{j} + (\mathbf{i} - \mathbf{k})t$ and $\mathbf{r} = 2\mathbf{j} - \mathbf{k} + (\mathbf{j} - \mathbf{i})t$.

If we write the first line as $\mathbf{r} = \mathbf{r}_0 + \mathbf{A}t$, then (the head of) $\mathbf{r}_0$ is a simple choice for P, so we have

$$P = (1, -2, 0) \quad \text{and} \quad \mathbf{A} = \mathbf{i} - \mathbf{k}.$$

Similarly, from the second line we find

$$Q = (0, 2, -1) \quad \text{and} \quad \mathbf{B} = \mathbf{j} - \mathbf{i}.$$

Then $\mathbf{A} \times \mathbf{B} = \mathbf{i} + \mathbf{j} + \mathbf{k}$ and $\mathbf{n} = \left(1/\sqrt{3}\right)(\mathbf{i} + \mathbf{j} + \mathbf{k})$. Also

$$\overrightarrow{PQ} = (0, 2, -1) - (1, -2, 0) = (-1, 4, -1) = -\mathbf{i} + 4\mathbf{j} - \mathbf{k}.$$

Thus we get for the distance between the lines

$$\left|\overrightarrow{PQ} \cdot \mathbf{n}\right| = \left|(-\mathbf{i} + 4\mathbf{j} - \mathbf{k}) \cdot (\mathbf{i} + \mathbf{j} + \mathbf{k})/\sqrt{3}\right| = |-1 + 4 - 1|/\sqrt{3} = 2/\sqrt{3}.$$

Example 6. Find the direction of the line of intersection of the planes $x - 2y + 3z = 4$ and $2x + y - z = 5$.

The desired line lies in both planes, and so is perpendicular to the two normal vectors to the planes, namely $\mathbf{i} - 2\mathbf{j} + 3\mathbf{k}$ and $2\mathbf{i} + \mathbf{j} - \mathbf{k}$. Then the direction of the line is that of the cross product of these normal vectors; this is $-\mathbf{i} + 7\mathbf{j} + 5\mathbf{k}$.

Example 7. Find the cosine of the angle between the planes of Example 6.

The angle between the planes is the same as the angle between the normals to the planes. Thus our problem is to find the angle between the vectors $\mathbf{A} = \mathbf{i} - 2\mathbf{j} + 3\mathbf{k}$ and $\mathbf{B} = 2\mathbf{i} + \mathbf{j} - \mathbf{k}$. Since $\mathbf{A} \cdot \mathbf{B} = |\mathbf{A}|\,|\mathbf{B}| \cos\theta$, we have $-3 = \sqrt{14}\sqrt{6}\cos\theta$, and so $\cos\theta = -\sqrt{3/28}$. This gives the obtuse angle between the planes; the corresponding acute angle is $\pi - \theta$, or $\arccos\sqrt{3/28}$.

PROBLEMS, SECTION 5

In Problems 1 to 5, all lines are in the (x, y) plane.

1. Write the equation of the straight line through $(2, -3)$ with slope $3/4$, in the parametric form $\mathbf{r} = \mathbf{r}_0 + \mathbf{A}t$.
2. Find the slope of the line whose parametric equation is $\mathbf{r} = (\mathbf{i} - \mathbf{j}) + (2\mathbf{i} + 3\mathbf{j})t$.
3. Write, in parametric form [as in Problem 1], the equation of the straight line that joins $(1, -2)$ and $(3, 0)$.
4. Write, in parametric form, the equation of the straight line that is perpendicular to $\mathbf{r} = (2\mathbf{i} + 4\mathbf{j}) + (\mathbf{i} - 2\mathbf{j})t$ and goes through $(1, 0)$.
5. Write, in parametric form, the equation of the y axis.

Find the symmetric equations (5.6) or (5.7) and the parametric equations (5.8) of a line, and/or the equation (5.10) of the plane satisfying the following given conditions.

6. Line through $(1, -1, -5)$ and $(2, -3, -3)$.
7. Line through $(2, 3, 4)$ and $(5, 1, -2)$.
8. Line through $(0, -2, 4)$ and $(3, -2, -1)$.
9. Line through $(-1, 3, 7)$ and $(-1, -2, 7)$.
10. Line through $(3, 4, -1)$ and parallel to $2\mathbf{i} - 3\mathbf{j} + 6\mathbf{k}$.
11. Line through $(4, -1, 3)$ and parallel to $\mathbf{i} - 2\mathbf{k}$.
12. Line through $(5, -4, 2)$ and parallel to the line $\mathbf{r} = \mathbf{i} - \mathbf{j} + (5\mathbf{i} - 2\mathbf{j} + \mathbf{k})t$.
13. Line through $(3, 0, -5)$ and parallel to the line $\mathbf{r} = (2, 1, -5) + (0, -3, 1)t$.
14. Plane containing the triangle ABC of Problem 4.11.
15. Plane through the origin and the points in Problem 8.
16. Plane through the point and perpendicular to the line in Problem 12.
17. Plane through the point and perpendicular to the line in Problem 13.
18. Plane containing the two parallel lines in Problem 12.
19. Plane containing the two parallel lines in Problem 13.
20. Plane containing the three points $(0, 1, 1)$, $(2, 1, 3)$, and $(4, 2, 1)$.

In Problems 21 to 23, find the angle between the given planes.

21. $2x + 6y - 3z = 10$ and $5x + 2y - z = 12$.
22. $2x - y - z = 4$ and $3x - 2y - 6z = 7$.
23. $2x + y - 2z = 3$ and $3x - 6y - 2z = 4$.

24. Find a point on *both* the planes (that is, on their line of intersection) in Problem 21. Find a vector parallel to the line of intersection. Write the equations of the line of intersection of the planes. Find the distance from the origin to the line.

25. As in Problem 24, find the equations of the line of intersection of the planes in Problem 22. Find the distance from the point $(2, 1, -1)$ to the line.

26. As in Problem 24, find the equations of the line of intersection of the planes in Problem 23. Find the distance from the point $(1, 0, 0)$ to the line.

27. Find the equation of the plane through $(2, 3, -2)$ and perpendicular to both planes in Problem 21.

28. Find the equation of the plane through $(-4, -1, 2)$ and perpendicular to both planes in Problem 22.

29. Find a point on the plane $2x - y - z = 13$. Find the distance from $(7, 1, -2)$ to the plane.

30. Find the distance from the origin to the plane $3x - 2y - 6z = 7$.

31. Find the distance from $(-2, 4, 5)$ to the plane $2x + 6y - 3z = 10$.

32. Find the distance from $(3, -1, 2)$ to the plane $5x - y - z = 4$.

33. Find the perpendicular distance between the two parallel lines in Problem 12.

34. Find the distance (perpendicular is understood) between the two parallel lines in Problem 13.

35. Find the distance from $(2, 5, 1)$ to the line in Problem 10.

36. Find the distance from $(3, 2, 5)$ to the line in Problem 11.

37. Determine whether the lines

$$\frac{x-1}{2} = \frac{y+3}{1} = \frac{z-4}{-3} \quad \text{and} \quad \frac{x+3}{4} = \frac{y+4}{1} = \frac{8-z}{4}$$

intersect. *Two suggestions:* (1) Can you find the intersection point, if any? (2) Consider the distance between the lines.

38. Find the angle between the lines in Problem 37.

In Problems 39 and 40, show that the given lines intersect and find the acute angle between them.

39. $\mathbf{r} = 2\mathbf{j} + \mathbf{k} + (3\mathbf{i} - \mathbf{k})t_1 \quad \text{and} \quad \mathbf{r} = 7\mathbf{i} + 2\mathbf{k} + (2\mathbf{i} - \mathbf{j} + \mathbf{k})t_2.$

40. $\mathbf{r} = (5, -2, 0) + (1, -1, -1)t_1 \quad \text{and} \quad \mathbf{r} = (4, -4, -1) + (0, 3, 2)t_2.$

In Problems 41 to 44, find the distance between the two given lines.

41. $\mathbf{r} = (4, 3, -1) + (1, 1, 1)t \quad \text{and} \quad \mathbf{r} = (4, -1, 1) + (1, -2, -1)t.$

42. The line that joins $(0, 0, 0)$ to $(1, 2, -1)$, and the line that joins $(1, 1, 1)$ to $(2, 3, 4)$.

43. $\dfrac{x-1}{2} = \dfrac{y+2}{3} = \dfrac{2z-1}{4} \quad \text{and} \quad \dfrac{x+2}{-1} = \dfrac{2-y}{2}, \quad z = \dfrac{1}{2}.$

44. The x axis and $\mathbf{r} = \mathbf{j} - \mathbf{k} + (2\mathbf{i} - 3\mathbf{j} + \mathbf{k})t$.

45. A particle is traveling along the line $(x-3)/2 = (y+1)/(-2) = z - 1$. Write the equation of its path in the form $\mathbf{r} = \mathbf{r}_0 + \mathbf{A}t$. Find the distance of closest approach of the particle to the origin (that is, the distance from the origin to the line). If t represents time, show that the time of closest approach is $t = -(\mathbf{r}_0 \cdot \mathbf{A})/|\mathbf{A}|^2$. Use this value to check your answer for the distance of closest approach. *Hint:* See Figure 5.3. If P is the point of closest approach, what is $\mathbf{A} \cdot \mathbf{r}$?

6. MATRIX OPERATIONS

In Section 2 we used matrices simply as arrays of numbers. Now we want to go farther into the subject and discuss the meaning and use of multiplying a matrix by a number and of combining matrices by addition, subtraction, multiplication, and even (in a sense) division. We will see that we may be able to find functions of matrices such as e^{M}. These are, of course, all questions of definition, but we shall show some applications which might suggest reasonable definitions; or alternatively, given the definitions, we shall see what applications we can make of the matrix operations.

Matrix Equations Let us first emphasize again that two matrices are equal *only* if they are identical. Thus the matrix equation

$$\begin{pmatrix} x & r & u \\ y & s & v \end{pmatrix} = \begin{pmatrix} 2 & 1 & -5 \\ 3 & -7i & 1-i \end{pmatrix}$$

is really the set of six equations

$$x = 2, \qquad y = 3, \qquad r = 1, \qquad s = -7i, \qquad u = -5, \qquad v = 1 - i.$$

(Recall similar situations we have met before: The equation $z = x + iy = 2 - 3i$ is equivalent to the two real equations $x = 2$, $y = -3$; a vector equation in three dimensions is equivalent to three component equations.) In complicated problems involving many numbers or variables, it is often possible to save a great deal of writing by using a single matrix equation to replace a whole set of ordinary equations. Any time it is possible to so abbreviate the writing of a mathematical equation (like using a single letter for a complicated parenthesis) it not only saves time but often enables us to think more clearly.

Multiplication of a Matrix by a Number A convenient way to display the components of the vector $\mathbf{A} = 2\mathbf{i} + 3\mathbf{j}$ is to write them as elements of a matrix, either

$$\mathrm{A} = \begin{pmatrix} 2 \\ 3 \end{pmatrix} \quad \text{called a column matrix or column vector,}$$

or

$$\mathrm{A}^{\mathrm{T}} = \begin{pmatrix} 2 & 3 \end{pmatrix} \quad \text{called a row matrix or row vector.}$$

The row matrix A^{T} is the transpose of the column matrix A. Observe the notation we are using here: We will often use the same letter for a vector and its column matrix, but we will usually write the letter representing the matrix as A (roman, not boldface), the vector as boldface **A**, and the length of the vector as italic A.

Now suppose we want a vector of twice the length of **A** and in the same direction; we would write this as $2\mathbf{A} = 4\mathbf{i} + 6\mathbf{j}$. Then we would like to write its matrix representation as

$$2\mathrm{A} = 2\begin{pmatrix} 2 \\ 3 \end{pmatrix} = \begin{pmatrix} 4 \\ 6 \end{pmatrix}, \qquad 2\mathrm{A}^{\mathrm{T}} = 2(2 \quad 3) = (4 \quad 6).$$

This is, in fact, exactly how a matrix is multiplied by a number: *every* element of the matrix is multiplied by the number. Thus

$$k\begin{pmatrix} a & c & e \\ b & d & f \end{pmatrix} = \begin{pmatrix} ka & kc & ke \\ kb & kd & kf \end{pmatrix}$$

and

$$\begin{pmatrix} -\frac{1}{2} & \frac{3}{4} \\ -1 & -\frac{5}{8} \end{pmatrix} = -\frac{1}{8}\begin{pmatrix} 4 & -6 \\ 8 & 5 \end{pmatrix}.$$

Note carefully a difference between determinants and matrices: multiplying a matrix by a number k means multiplying every element by k, but multiplying just *one* row of a determinant by k multiplies the determinant by k. Thus $\det(k\mathrm{A}) = k^2 \det \mathrm{A}$ for a 2 by 2 matrix, $\det(k\mathrm{A}) = k^3 \det \mathrm{A}$ for a 3 by 3 matrix, and so on.

Addition of Matrices When we add vectors algebraically, we add them by components. Matrices are added in the same way, by adding corresponding elements. For example,

$$(6.1)\qquad \begin{pmatrix} 1 & 3 & -2 \\ 4 & 7 & 1 \end{pmatrix} + \begin{pmatrix} 2 & -1 & 4 \\ 3 & -7 & -2 \end{pmatrix} = \begin{pmatrix} 1+2 & 3-1 & -2+4 \\ 4+3 & 7-7 & 1-2 \end{pmatrix} = \begin{pmatrix} 3 & 2 & 2 \\ 7 & 0 & -1 \end{pmatrix}.$$

Note that if we add $\mathrm{A} + \mathrm{A}$ we would get $2\mathrm{A}$ in accord with our definition of twice a matrix above. Suppose we have

$$\mathrm{A} = \begin{pmatrix} 1 & 3 & -2 \\ 4 & 7 & 1 \end{pmatrix} \quad \text{and} \quad \mathrm{B} = \begin{pmatrix} 2 & -1 \\ 3 & 5 \end{pmatrix}.$$

In this case we cannot add A and B; we say that the sum is undefined or meaningless.

In applications, then, matrices are useful in representing things which are added by components. Suppose, for example, that, in (6.1), the columns represent displacements of three particles. The first particle is displaced by $\mathbf{i}+4\mathbf{j}$ (first column of the first matrix) and later by $2\mathbf{i}+3\mathbf{j}$ (first column of the second matrix). The total displacement is then $3\mathbf{i}+7\mathbf{j}$ (first column of the sum of the matrices). Similarly the second and third columns represent displacements of the second and third particles.

Multiplication of Matrices Let us start by defining the product of two matrices and then see what use we can make of the process. Here is a simple example to show what is meant by the product $\mathrm{AB} = \mathrm{C}$ of two matrices A and B:

$$(6.2a)\qquad \mathrm{AB} = \begin{pmatrix} a & b \\ c & d \end{pmatrix}\begin{pmatrix} e & f \\ g & h \end{pmatrix} = \begin{pmatrix} ae+bg & af+bh \\ ce+dg & cf+dh \end{pmatrix} = \mathrm{C}.$$

Observe that in the product matrix C, the element in the first row and first column is obtained by multiplying each element of the first row in A times the corresponding element in the first column of B and adding the results. This is referred to as "row times column" multiplication; when we compute $ae + bg$, we say that we have "multiplied the first row of A times the first column of B." Next examine the element $af + bh$ in the first row and second column of C; it is the "first row of A times the second column of B." Similarly, $ce+dg$ in the second row and first column of C is the "second row of A times the first column of B," and $cf+dh$ in the second row and second column of C is the "second row of A times the second column of B." Thus all the elements of C may be obtained by using the following simple rule:

The element in row i and column j of the product matrix AB is equal to row i of A times column j of B. In index notation

(6.2b)
$$(\mathrm{AB})_{ij} = \sum_k A_{ik} B_{kj}.$$

Here is another useful way of saying this: Think of the elements in a row (or a column) of a matrix as the components of a vector. Then row times column multiplication for the matrix product AB corresponds to finding the dot product of a row vector of A and a column vector of B.

It is not necessary for matrices to be square in order for us to multiply them. Consider the following example.

▶ **Example 1.** Find the product of A and B if

$$\mathrm{A} = \begin{pmatrix} 4 & 2 \\ -3 & 1 \end{pmatrix}, \qquad \mathrm{B} = \begin{pmatrix} 1 & 5 & 3 \\ 2 & 7 & -4 \end{pmatrix}.$$

Following the rule we have stated, we get

$$\begin{aligned}
\mathrm{AB} &= \begin{pmatrix} 4 & 2 \\ -3 & 1 \end{pmatrix} \begin{pmatrix} 1 & 5 & 3 \\ 2 & 7 & -4 \end{pmatrix} \\
&= \begin{pmatrix} 4 \cdot 1 + 2 \cdot 2 & 4 \cdot 5 + 2 \cdot 7 & 4 \cdot 3 + 2(-4) \\ -3 \cdot 1 + 1 \cdot 2 & -3 \cdot 5 + 1 \cdot 7 & -3 \cdot 3 + 1(-4) \end{pmatrix} \\
&= \begin{pmatrix} 8 & 34 & 4 \\ -1 & -8 & -13 \end{pmatrix}.
\end{aligned}$$

Notice that the third column in B caused us no difficulty in following our rule; we simply multiplied each row of A times the third column of B to obtain the elements in the third column of AB. But suppose we tried to find the product BA. In B a row contains 3 elements, while in A a column contains only two; thus we are not able to apply the "row times column" method. Whenever this happens, we say that B is *not conformable* with respect to A, and the product BA is not defined (that is, it is meaningless and we do not use it).

The product AB (in that order) can be found if and only if the number of elements in a row of A equals the number of elements in a column of B; the matrices A, B in that order are then called *conformable*. (Observe that the number of rows in A and of columns in B have nothing to do with the question of whether we can find AB or not.)

Example 2. Find AB and BA, given

$$A = \begin{pmatrix} 3 & -1 \\ -4 & 2 \end{pmatrix}, \qquad B = \begin{pmatrix} 5 & 2 \\ -7 & 3 \end{pmatrix}.$$

Note that here the matrices are conformable in both orders, so we can find both AB and BA.

$$\begin{aligned} AB &= \begin{pmatrix} 3 & -1 \\ -4 & 2 \end{pmatrix}\begin{pmatrix} 5 & 2 \\ -7 & 3 \end{pmatrix} \\ &= \begin{pmatrix} 3\cdot 5 - 1(-7) & 3\cdot 2 - 1\cdot 3 \\ -4\cdot 5 + 2(-7) & -4\cdot 2 + 2\cdot 3 \end{pmatrix} = \begin{pmatrix} 22 & 3 \\ -34 & -2 \end{pmatrix}. \\ BA &= \begin{pmatrix} 5 & 2 \\ -7 & 3 \end{pmatrix}\begin{pmatrix} 3 & -1 \\ -4 & 2 \end{pmatrix} \\ &= \begin{pmatrix} 5\cdot 3 + 2(-4) & 5(-1) + 2\cdot 2 \\ -7\cdot 3 + 3(-4) & -7(-1) + 3\cdot 2 \end{pmatrix} = \begin{pmatrix} 7 & -1 \\ -33 & 13 \end{pmatrix}. \end{aligned}$$

Observe that AB is *not* the same as BA. We say that matrix multiplication is *not commutative*, or that, in general, matrices do not commute under multiplication. (Of course, two particular matrices may happen to commute.) We define the *commutator* of the matrices A and B by

$$[A, B] = AB - BA = \textit{commutator} \text{ of A and B.} \tag{6.3}$$

(Commutators are of interest in classical and quantum mechanics.) Since matrices do not in general commute, be careful not to change the order of factors in a product of matrices unless you know they commute. For example

$$(A - B)(A + B) = A^2 + AB - BA - B^2 = A^2 - B^2 + [A, B].$$

This is not equal to $A^2 - B^2$ when A and B don't commute. Also see the discussion just after (6.17). On the other hand, the associative law is valid, that is, $A(BC) = (AB)C$, so we can write either as simply ABC. Also the distributive law holds: $A(B + C) = AB + AC$ and $(A + B)C = AC + BC$ as we have been assuming above. (See Section 9.)

Zero Matrix The *zero* or *null* matrix means one with all its elements equal to zero. It is often abbreviated by 0, but we must be careful about this. For example:

$$\text{If} \quad M = \begin{pmatrix} 2 & -4 \\ 1 & -2 \end{pmatrix}, \quad \text{then} \quad M^2 = \begin{pmatrix} 0 & 0 \\ 0 & 0 \end{pmatrix} \tag{6.4}$$

so we have $M^2 = 0$, but $M \neq 0$. Also see Problems 9 and 10.

Identity Matrix or Unit Matrix This is a square matrix with every element of the main diagonal (upper left to lower right) equal to 1 and all other elements equal to zero. For example

$$(6.5)\qquad \begin{pmatrix} 1 & 0 & 0 \\ 0 & 1 & 0 \\ 0 & 0 & 1 \end{pmatrix}$$

is a unit or identity matrix of order 3 (that is, three rows and three columns). An identity or unit matrix is called 1 or I or U or E in various references. You should satisfy yourself that in multiplication, a unit matrix acts like the number 1, that is, if A is any matrix and I is the unit matrix conformable with A in the order in which we multiply, then IA = AI = A (Problem 11).

Operations with Determinants We do not define addition for determinants. However, multiplication is useful; we multiply determinants the same way we multiply matrices. It can be shown that if A and B are square matrices of the same order, then

$$(6.6)\qquad \det \mathrm{AB} = \det \mathrm{BA} = (\det \mathrm{A}) \cdot (\det \mathrm{B}).$$

Look at Example 2 above to see that (6.6) is true even when matrices AB and BA are not equal, that is, when A and B do not commute.

Applications of Matrix Multiplication We can now write sets of simultaneous linear equations in a very simple form using matrices. Consider the matrix equation

$$(6.7)\qquad \begin{pmatrix} 1 & 0 & -1 \\ -2 & 3 & 0 \\ 1 & -3 & 2 \end{pmatrix} \begin{pmatrix} x \\ y \\ z \end{pmatrix} = \begin{pmatrix} 5 \\ 1 \\ -10 \end{pmatrix}.$$

If we multiply the first two matrices, we have

$$(6.8)\qquad \begin{pmatrix} x - z \\ -2x + 3y \\ x - 3y + 2z \end{pmatrix} = \begin{pmatrix} 5 \\ 1 \\ -10 \end{pmatrix}.$$

Now recall that two matrices are equal only if they are identical. Thus (6.8) is the set of three equations

$$(6.9)\qquad \begin{cases} x \qquad\quad - z = 5 \\ -2x + 3y \qquad = 1 \\ x - 3y + 2z = -10 \end{cases}.$$

Consequently (6.7) is the matrix form for the set of equations (6.9). In this way we can write any set of linear equations in matrix form. If we use letters to represent the matrices in (6.7),

$$(6.10)\qquad \mathrm{M} = \begin{pmatrix} 1 & 0 & -1 \\ -2 & 3 & 0 \\ 1 & -3 & 2 \end{pmatrix}, \quad \mathrm{r} = \begin{pmatrix} x \\ y \\ z \end{pmatrix}, \quad \mathrm{k} = \begin{pmatrix} 5 \\ 1 \\ -10 \end{pmatrix},$$

then we can write (6.7) or (6.9) as

$$\text{(6.11)} \qquad \mathrm{Mr} = \mathrm{k}.$$

Or, in index notation, we can write $\sum_j M_{ij}x_j = k_i$. [Review Section 2, equations (2.3) to (2.6).] Note that (6.11) could represent any number of equations or unknowns (say 100 equations in 100 unknowns!). Thus we have a great simplification in notation which may help us to think more clearly about a problem. For example, if (6.11) were an ordinary algebraic equation, we would solve it for r to get

$$\text{(6.12)} \qquad \mathrm{r} = \mathrm{M}^{-1}\mathrm{k}.$$

Since M is a matrix, (6.12) only makes sense if we can give a meaning to M^{-1} such that (6.12) gives the solution of (6.7) or (6.9). Let's try to do this.

Inverse of a Matrix The reciprocal or inverse of a number x is x^{-1} such that the product $xx^{-1} = 1$. We define the inverse of a matrix M (if it has one) as the matrix M^{-1} such that MM^{-1} and $\mathrm{M}^{-1}\mathrm{M}$ are both equal to a unit matrix I. Note that only square matrices can have inverses (otherwise we could not multiply both MM^{-1} and $\mathrm{M}^{-1}\mathrm{M}$). Actually, some square matrices do not have inverses either. You can see from (6.6) that if $\mathrm{M}^{-1}\mathrm{M} = \mathrm{I}$, then $(\det \mathrm{M}^{-1})(\det \mathrm{M}) = \det \mathrm{I} = 1$. If two numbers have product $= 1$, then neither of them is zero; thus $\det \mathrm{M} \neq 0$ is a requirement for M to have an inverse.

If a matrix has an inverse we say that it is *invertible*; if it doesn't have an inverse, it is called *singular*. For simple numerical matrices your computer will easily produce the inverse of an invertible matrix. However, for theoretical purposes, we need a formula for the inverse; let's discuss this. The cofactor of an element in a square matrix M means exactly the same thing as the cofactor of that element in det M [see (3.3) and (3.4)]. Thus, the cofactor C_{ij} of the element m_{ij} in row i and column j is a number equal to $(-1)^{i+j}$ times the value of the determinant remaining when we cross off row i and column j. Then to find M^{-1}: Find the cofactors C_{ij} of all elements, write the matrix C whose elements are C_{ij}, transpose it (interchange rows and columns), and divide by det M. (See Problem 23.)

$$\text{(6.13)} \qquad \mathrm{M}^{-1} = \frac{1}{\det \mathrm{M}}\mathrm{C}^{\mathrm{T}} \quad \text{where } C_{ij} = \text{ cofactor of } m_{ij}$$

Although (6.13) is particularly useful in theoretical work, you should practice using it (as we said for Cramer's rule) on simple numerical problems in order to learn what the formula means.

▶ **Example 3.** For the matrix M of the coefficients in equations (6.7) or (6.9), find M^{-1}.

$$\mathrm{M} = \begin{pmatrix} 1 & 0 & -1 \\ -2 & 3 & 0 \\ 1 & -3 & 2 \end{pmatrix}.$$

We find $\det \mathrm{M} = 3$. The cofactors of the elements are:

$$1^{\text{st}} \text{ row}: \quad \begin{vmatrix} 3 & 0 \\ -3 & 2 \end{vmatrix} = 6, \quad -\begin{vmatrix} -2 & 0 \\ 1 & 2 \end{vmatrix} = 4, \quad \begin{vmatrix} -2 & 3 \\ 1 & -3 \end{vmatrix} = 3.$$

$$2^{\text{nd}} \text{ row}: \quad -\begin{vmatrix} 0 & -1 \\ -3 & 2 \end{vmatrix} = 3, \quad \begin{vmatrix} 1 & -1 \\ 1 & 2 \end{vmatrix} = 3, \quad -\begin{vmatrix} 1 & 0 \\ 1 & -3 \end{vmatrix} = 3.$$

$$3^{\text{rd}} \text{ row}: \quad \begin{vmatrix} 0 & -1 \\ 3 & 0 \end{vmatrix} = 3, \quad -\begin{vmatrix} 1 & -1 \\ -2 & 0 \end{vmatrix} = 2, \quad \begin{vmatrix} 1 & 0 \\ -2 & 3 \end{vmatrix} = 3.$$

Then

$$\mathrm{C} = \begin{pmatrix} 6 & 4 & 3 \\ 3 & 3 & 3 \\ 3 & 2 & 3 \end{pmatrix} \quad \text{so} \quad \mathrm{M}^{-1} = \frac{1}{\det \mathrm{M}} \mathrm{C}^{\mathrm{T}} = \frac{1}{3} \begin{pmatrix} 6 & 3 & 3 \\ 4 & 3 & 2 \\ 3 & 3 & 3 \end{pmatrix}.$$

Now we can use M^{-1} to solve equations (6.9). By (6.12), the solution is given by the column matrix $\mathrm{r} = \mathrm{M}^{-1}\mathrm{k}$, so we have

$$\begin{pmatrix} x \\ y \\ z \end{pmatrix} = \frac{1}{3} \begin{pmatrix} 6 & 3 & 3 \\ 4 & 3 & 2 \\ 3 & 3 & 3 \end{pmatrix} \begin{pmatrix} 5 \\ 1 \\ -10 \end{pmatrix} = \begin{pmatrix} 1 \\ 1 \\ -4 \end{pmatrix},$$

or $x = 1$, $y = 1$, $z = -4$. (See Problem 12.)

Rotation Matrices As another example of matrix multiplication, let's consider a case where we know the answer, just to see that our definition of matrix multiplication works the way we want it to. You probably know the rotation equations [for reference, see the next section, equation (7.12) and Figure 7.4]. Equation (7.12) gives the matrix which rotates the vector $\mathbf{r} = \mathbf{i}x + \mathbf{j}y$ through angle θ to become the vector $\mathbf{R} = \mathbf{i}X + \mathbf{j}Y$. Suppose we further rotate $\mathbf{R}$ through angle ϕ to become $\mathbf{R}' = \mathbf{i}X' + \mathbf{j}Y'$. We could write the matrix equations for the rotations in the form $\mathrm{R} = \mathrm{Mr}$ and $\mathrm{R}' = \mathrm{M}'\mathrm{R}$ where M and M′ are the rotation matrices (7.12) for rotation through angles θ and ϕ. Then, solving for R′ in terms of r, we get $\mathrm{R}' = \mathrm{M}'\mathrm{Mr}$. We expect the matrix product M′M to give us the matrix for a rotation through the angle $\theta + \phi$, that is we expect to find

$$\begin{pmatrix} \cos\phi & -\sin\phi \\ \sin\phi & \cos\phi \end{pmatrix} \begin{pmatrix} \cos\theta & -\sin\theta \\ \sin\theta & \cos\theta \end{pmatrix} = \begin{pmatrix} \cos(\theta+\phi) & -\sin(\theta+\phi) \\ \sin(\theta+\phi) & \cos(\theta+\phi) \end{pmatrix}. \tag{6.14}$$

It is straightforward to multiply the two matrices (Problem 25) and verify (by using trigonometric identities) that (6.14) is correct. Also note that these two rotation matrices commute (that is, rotation through angle θ and then through angle ϕ gives the same result as rotation through ϕ followed by rotation through θ). This is true in this problem in two dimensions. As we will see in Section 7, rotation matrices in three dimensions do not in general commute if the two rotation axes are different. (See Problems 7.30 and 7.31.) But all rotations in the (x, y) plane are rotations about the z axis and so they commute.

Functions of Matrices Since we now know how to multiply matrices and how to add them, we can evaluate any power of a matrix A and so evaluate a polynomial in A. The constant term c or $c\text{A}^0$ in a polynomial is defined to mean c times the unit matrix I [see (6.16) below].

Example 4.

$$\text{If}\quad \text{A} = \begin{pmatrix} 1 & \sqrt{2} \\ -\sqrt{2} & -1 \end{pmatrix}, \quad \text{then}\quad \text{A}^2 = \begin{pmatrix} -1 & 0 \\ 0 & -1 \end{pmatrix} = -\text{I}, \tag{6.15}$$
$$\text{A}^3 = -\text{A},\ \text{A}^4 = \text{I}, \quad \text{and so on.}$$

(Verify these powers and the fact that higher powers simply repeat these four results: A, −I, −A, I, over and over.) Then we can find (Problem 28)

$$\begin{aligned} f(\text{A}) &= 3 - 2\text{A}^2 - \text{A}^3 - 5\text{A}^4 + \text{A}^6 \\ &= 3\text{I} + 2\text{I} + \text{A} - 5\text{I} - \text{I} = \text{A} - \text{I} = \begin{pmatrix} 0 & \sqrt{2} \\ -\sqrt{2} & -2 \end{pmatrix}. \end{aligned} \tag{6.16}$$

We can extend this to other functions by expanding a given $f(x)$ in a power series if all the series we need to use happen to converge. For example, the series for e^z converges for all z, so we can find $e^{k\text{A}}$ when A is a given matrix and k is any number, real or complex. Let A be the matrix in (6.15). Then (Problem 28), we find

$$\begin{aligned} e^{k\text{A}} &= 1 + k\text{A} + \frac{k^2\text{A}^2}{2!} + \frac{k^3\text{A}^3}{3!} + \frac{k^4\text{A}^4}{4!} + \frac{k^5\text{A}^5}{5!} + \cdots \\ &= (1 - \frac{k^2}{2!} + \frac{k^4}{4!} + \cdots)\text{I} + (k - \frac{k^3}{3!} + \frac{k^5}{5!})\text{A} \\ &= (\cos k)\text{I} + (\sin k)\text{A} = \begin{pmatrix} \cos k + \sin k & \sqrt{2}\sin k \\ -\sqrt{2}\sin k & \cos k - \sin k \end{pmatrix}. \end{aligned} \tag{6.17}$$

A word of warning about functions of two matrices when A and B don't commute: Familiar formulas may mislead you; see (6.3) and the discussion following it. Be sure to write $(\text{A}+\text{B})^2 = \text{A}^2 + \text{AB} + \text{BA} + \text{B}^2$; don't write 2AB. Similarly, you can show that $e^{\text{A}+\text{B}}$ is not the same as $e^{\text{A}}e^{\text{B}}$ when A and B don't commute (see Problem 29 and Problem 15.34).

PROBLEMS, SECTION 6

In Problems 1 to 3, find AB, BA, A + B, A − B, A^2, B^2, 5A, 3B. Observe that AB ≠ BA. Show that $(\text{A}-\text{B})(\text{A}+\text{B}) \neq (\text{A}+\text{B})(\text{A}-\text{B}) \neq \text{A}^2 - \text{B}^2$. Show that det AB = det BA = (det A)(det B), but that det(A + B) ≠ det A + det B. Show that det(5A) ≠ 5 det A, and find n so that $\det(5\text{A}) = 5^n \det \text{A}$. Find similar results for det(3B). Remember that the point of doing these simple problems by hand is to learn how to manipulate determinants and matrices correctly. Check your answers by computer.

1. $\text{A} = \begin{pmatrix} 3 & 1 \\ 2 & 5 \end{pmatrix}, \quad \text{B} = \begin{pmatrix} -2 & 2 \\ 1 & 4 \end{pmatrix}.$

2. $\text{A} = \begin{pmatrix} 2 & -5 \\ -1 & 3 \end{pmatrix}, \quad \text{B} = \begin{pmatrix} -1 & 4 \\ 0 & 2 \end{pmatrix}.$

3. $A = \begin{pmatrix} 1 & 0 & 2 \\ 3 & -1 & 0 \\ 0 & 5 & 1 \end{pmatrix}, \qquad B = \begin{pmatrix} 1 & 1 & 0 \\ 0 & 2 & 1 \\ 3 & -1 & 0 \end{pmatrix}.$

4. Given the matrices

$$A = \begin{pmatrix} 2 & 3 & 1 & -4 \\ 2 & 1 & 0 & 5 \end{pmatrix}, \quad B = \begin{pmatrix} 2 & 4 \\ 1 & -1 \\ 3 & -1 \end{pmatrix}, \quad C = \begin{pmatrix} 2 & 1 & 3 \\ 4 & -1 & -2 \\ -1 & 0 & 1 \end{pmatrix},$$

compute or mark as meaningless all products of two of these matrices (AB, BA, A^2, etc.); of three of them (ABC, A^2C, A^3, etc.).

5. Compute the product of each of the matrices in Problem 4 with its transpose [see (2.2) or (9.1)] in both orders, that is AA^T and A^TA, etc.

6. The Pauli spin matrices in quantum mechanics are

$$A = \begin{pmatrix} 0 & 1 \\ 1 & 0 \end{pmatrix}, \quad B = \begin{pmatrix} 0 & -i \\ i & 0 \end{pmatrix}, \quad C = \begin{pmatrix} 1 & 0 \\ 0 & -1 \end{pmatrix}.$$

(You will probably find these called σ_x, σ_y, σ_z in your quantum mechanics texts.) Show that $A^2 = B^2 = C^2 =$ a unit matrix. Also show that any two of these matrices anticommute, that is, $AB = -BA$, etc. Show that the commutator of A and B, that is, $AB - BA$, is $2iC$, and similarly for other pairs in cyclic order.

7. Find the matrix product

$$\begin{pmatrix} 2 & 3 \end{pmatrix} \begin{pmatrix} -1 & 4 \\ 2 & -1 \end{pmatrix} \begin{pmatrix} -1 \\ 2 \end{pmatrix}.$$

By evaluating this in two ways, verify the associative law for matrix multiplication, that is, $A(BC) = (AB)C$, which justifies our writing just ABC.

8. Show, by multiplying the matrices, that the following equation represents an ellipse.

$$\begin{pmatrix} x & y \end{pmatrix} \begin{pmatrix} 5 & -7 \\ 7 & 3 \end{pmatrix} \begin{pmatrix} x \\ y \end{pmatrix} = 30.$$

9. Find AB and BA given

$$A = \begin{pmatrix} 1 & 2 \\ 3 & 6 \end{pmatrix}, \quad B = \begin{pmatrix} 10 & 4 \\ -5 & -2 \end{pmatrix}.$$

Observe that AB is the null matrix; if we call it 0, then $AB = 0$, but neither A nor B is 0. Show that A is singular.

10. Given

$$C = \begin{pmatrix} 7 & 6 \\ 2 & 3 \end{pmatrix}, \quad D = \begin{pmatrix} -3 & 2 \\ 7 & 5 \end{pmatrix}$$

and A as in Problem 9, show that $AC = AD$, but $C \neq D$ and $A \neq 0$.

11. Show that the unit matrix I has the property that we associate with the number 1, that is, $IA = A$ and $AI = A$, assuming that the matrices are conformable.

12. For the matrices in Example 3, verify that MM^{-1} and $M^{-1}M$ both equal a unit matrix. Multiply $M^{-1}k$ to verify the solution of equations (6.9).

In Problems 13 to 16, use (6.13) to find the inverse of the given matrix.

13. $\begin{pmatrix} 6 & 9 \\ 3 & 5 \end{pmatrix}$ **14.** $\begin{pmatrix} 2 & 1 \\ 0 & -3 \end{pmatrix}$

15. $\begin{pmatrix} -1 & 2 & 3 \\ 2 & 0 & -4 \\ -1 & -1 & 1 \end{pmatrix}$ **16.** $\begin{pmatrix} -2 & 0 & 1 \\ 1 & -1 & 2 \\ 3 & 1 & 0 \end{pmatrix}$

17. Given the matrices

$$\mathrm{A} = \begin{pmatrix} 1 & -1 & 1 \\ 4 & 0 & -1 \\ 4 & -2 & 0 \end{pmatrix}, \quad \mathrm{B} = \begin{pmatrix} 1 & 0 & 1 \\ 2 & 1 & 1 \\ 2 & 1 & 2 \end{pmatrix}.$$

(a) Find A^{-1}, B^{-1}, $\mathrm{B}^{-1}\mathrm{AB}$, and $\mathrm{B}^{-1}\mathrm{A}^{-1}\mathrm{B}$.

(b) Show that the last two matrices are inverses, that is, that their product is the unit matrix.

18. Problem 17(b) is a special case of the general theorem that the inverse of a product of matrices is the product of the inverses in reverse order. Prove this. *Hint:* Multiply ABCD times $\mathrm{D}^{-1}\mathrm{C}^{-1}\mathrm{B}^{-1}\mathrm{A}^{-1}$ to show that you get a unit matrix.

In Problems 19 to 22, solve each set of equations by the method of finding the inverse of the coefficient matrix. *Hint:* See Example 3.

19. $\begin{cases} x - 2y = 5 \\ 3x + y = 15 \end{cases}$ **20.** $\begin{cases} 2x + 3y = -1 \\ 5x + 4y = 8 \end{cases}$

21. $\begin{cases} x + 2z = 8 \\ 2x - y = -5 \\ x + y + z = 4 \end{cases}$ **22.** $\begin{cases} x - y + z = 4 \\ 2x + y - z = -1 \\ 3x + 2y + 2z = 5 \end{cases}$

23. Verify formula (6.13). *Hint:* Consider the product of the matrices MC^{T}. Use Problem 3.8.

24. Use the method of solving simultaneous equations by finding the inverse of the matrix of coefficients, together with the formula (6.13) for the inverse of a matrix, to obtain Cramer's rule.

25. Verify (6.14) by multiplying the matrices and using trigonometric addition formulas.

26. In (6.14), let $\theta = \phi = \pi/2$ and verify the result numerically.

27. Do Problem 26 if $\theta = \pi/2$, $\phi = \pi/4$.

28. Verify the calculations in (6.15), (6.16), and (6.17).

29. Show that if A and B are matrices which don't commute, then $e^{\mathrm{A}+\mathrm{B}} \neq e^{\mathrm{A}}e^{\mathrm{B}}$, but if they do commute then the relation holds. *Hint:* Write out several terms of the infinite series for e^{A}, e^{B}, and $e^{\mathrm{A}+\mathrm{B}}$ and do the multiplications carefully assuming that A and B don't commute. Then see what happens if they do commute.

30. For the Pauli spin matrix A in Problem 6, find the matrices $\sin k\mathrm{A}$, $\cos k\mathrm{A}$, $e^{k\mathrm{A}}$, and $e^{ik\mathrm{A}}$ where $i = \sqrt{-1}$.

31. Repeat Problem 30 for the Pauli spin matrix C in Problem 6. *Hint*: Show that if a matrix is diagonal, say $\mathrm{D} = \begin{pmatrix} a & 0 \\ 0 & b \end{pmatrix}$, then $f(\mathrm{D}) = \begin{pmatrix} f(a) & 0 \\ 0 & f(b) \end{pmatrix}$.

32. For the Pauli spin matrix B in Problem 6, find $e^{i\theta\mathrm{B}}$ and show that your result is a rotation matrix. Repeat the calculation for $e^{-i\theta\mathrm{B}}$.

7. LINEAR COMBINATIONS, LINEAR FUNCTIONS, LINEAR OPERATORS

Given two vectors $\mathbf{A}$ and $\mathbf{B}$, the vector $3\mathbf{A} - 2\mathbf{B}$ is called a "linear combination" of $\mathbf{A}$ and $\mathbf{B}$. In general, a linear combination of $\mathbf{A}$ and $\mathbf{B}$ means $a\mathbf{A} + b\mathbf{B}$ where a and b are scalars. Geometrically, if $\mathbf{A}$ and $\mathbf{B}$ have the same tail and do not lie along a line, then they determine a plane. You should satisfy yourself that all linear combinations of $\mathbf{A}$ and $\mathbf{B}$ then lie in the plane. It is also true that every vector in the plane can be written as a linear combination of $\mathbf{A}$ and $\mathbf{B}$; we shall consider this in Section 8. The vector $\mathbf{r} = \mathbf{i}x + \mathbf{j}y + \mathbf{k}z$ with tail at the origin (which we used in writing equations of lines and planes) is a linear combination of the unit basis vectors $\mathbf{i}$, $\mathbf{j}$, $\mathbf{k}$.

A function of a vector, say $f(\mathbf{r})$, is called linear if

(7.1) $$f(\mathbf{r}_1 + \mathbf{r}_2) = f(\mathbf{r}_1) + f(\mathbf{r}_2), \quad \text{and} \quad f(a\mathbf{r}) = af(\mathbf{r}),$$

where a is a scalar.

For example, if $\mathbf{A} = 2\mathbf{i} + 3\mathbf{j} - \mathbf{k}$ is a given vector, then $f(\mathbf{r}) = \mathbf{A} \cdot \mathbf{r} = 2x + 3y - z$ is a linear function because

$$f(\mathbf{r}_1 + \mathbf{r}_2) = \mathbf{A} \cdot (\mathbf{r}_1 + \mathbf{r}_2) = \mathbf{A} \cdot \mathbf{r}_1 + \mathbf{A} \cdot \mathbf{r}_2 = f(\mathbf{r}_1) + f(\mathbf{r}_2), \quad \text{and}$$

$$f(a\mathbf{r}) = \mathbf{A} \cdot (a\mathbf{r}) = a\mathbf{A} \cdot \mathbf{r} = af(\mathbf{r}).$$

On the other hand, $f(\mathbf{r}) = |\mathbf{r}|$ is *not* a linear function, because the length of the sum of two vectors is not in general the sum of their lengths. That is,

$$f(\mathbf{r}_1 + \mathbf{r}_2) = |\mathbf{r}_1 + \mathbf{r}_2| \neq |\mathbf{r}_1| + |\mathbf{r}_2| = f(\mathbf{r}_1) + f(\mathbf{r}_2),$$

as you can see from Figure 7.1. Also note that although we call $y = mx + b$ a linear equation (it is the equation of a straight line), the function $f(x) = mx + b$ is not linear (unless $b = 0$) because

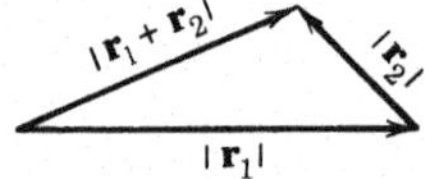

Figure 7.1

$$f(x_1 + x_2) = m(x_1 + x_2) + b \neq (mx_1 + b) + (mx_2 + b) = f(x_1) + f(x_2).$$

We can also consider vector functions of a vector $\mathbf{r}$. The magnetic field at each point (x, y, z), that is, at the head of the vector $\mathbf{r}$, is a vector $\mathbf{B} = \mathbf{i}B_x + \mathbf{j}B_y + \mathbf{k}B_z$. The components B_x, B_y, B_z may vary from point to pint, that is, they are functions of (x, y, z) or $\mathbf{r}$. Then

$\mathbf{F}(\mathbf{r})$ is a linear vector function if

(7.2) $$\mathbf{F}(\mathbf{r}_1 + \mathbf{r}_2) = \mathbf{F}(\mathbf{r}_1) + \mathbf{F}(\mathbf{r}_2) \quad \text{and} \quad \mathbf{F}(a\mathbf{r}) = a\mathbf{F}(\mathbf{r}),$$

where a is a scalar.

For example, $\mathbf{F}(\mathbf{r}) = b\mathbf{r}$ (where b is a scalar) is a linear vector function of $\mathbf{r}$.

You know from calculus that

$$(7.3)\qquad \frac{d}{dx}[f(x)+g(x)] = \frac{d}{dx}f(x) + \frac{d}{dx}g(x) \quad \text{and}$$
$$\frac{d}{dx}[kf(x)] = k\frac{d}{dx}f(x),$$

where k is a constant. We say that d/dx is a "linear operator" [compare (7.3) with (7.1) and (7.2)]. An "operator" or "operation" simply means a rule or some kind of instruction telling us what to do with whatever follows it. In other words, a linear operator is a linear function. Then

> O is a linear operator if
>
> $$(7.4)\qquad O(A+B) = O(A) + O(B) \quad \text{and} \quad O(kA) = kO(A),$$

where k is a number, and A and B are numbers, functions, vectors, and so on. Many of the errors people make happen because they assume that operators are linear when they are not (see problems).

Example 1. Is square root a linear operator? We are asking, is $\sqrt{A+B}$ the same as $\sqrt{A}+\sqrt{B}$? The answer is no; taking the square root is not a linear operation.

Example 2. Is taking the complex conjugate a linear operation? We want to know whether $\overline{A+B} = \bar{A}+\bar{B}$ and $\overline{kA} = k\bar{A}$. The first equation is true; the second equation is true if we restrict k to real numbers.

Matrix Operators, Linear Transformations Consider the set of equations

$$(7.5)\qquad \begin{cases} X = ax + by, \\ Y = cx + dy, \end{cases} \quad \text{or} \quad \begin{pmatrix} X \\ Y \end{pmatrix} = \begin{pmatrix} a & b \\ c & d \end{pmatrix}\begin{pmatrix} x \\ y \end{pmatrix}, \quad \text{or} \quad \mathrm{R} = \mathrm{Mr},$$

where a, b, c, d, are constants. For every point (x, y), these equations give us a point (X, Y). If we think of each point of the (x, y) plane being moved to some other point (with some points like the origin not being moved), we can call this process a *mapping* or *transformation* of the plane into itself. All the information about this transformation is contained in the matrix M. We say that this matrix is an operator which maps the plane into itself. Any matrix can be thought of as an operator on (conformable) column matrices r. Since

$$(7.6)\qquad \mathrm{M}(\mathrm{r}_1 + \mathrm{r}_2) = \mathrm{Mr}_1 + \mathrm{Mr}_2 \quad \text{and} \quad \mathrm{M}(k\mathrm{r}) = k(\mathrm{Mr}),$$

the matrix M is a linear operator.

Equations (7.5) can be interpreted geometrically in two ways. In Figure 7.2, we have one set of coordinate axes and the vector **r** has been changed to the vector **R** by the transformation (7.5). In Figure 7.3, we have *two* sets of coordinate axes,

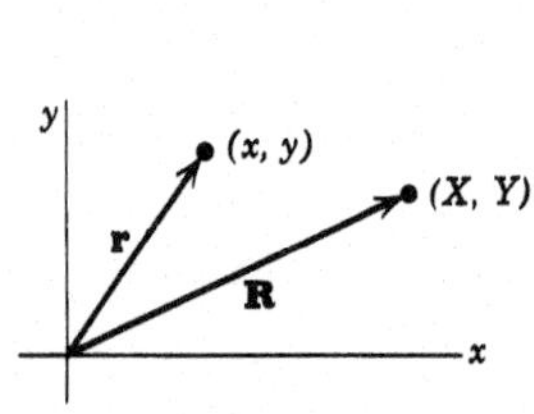

Figure 7.2

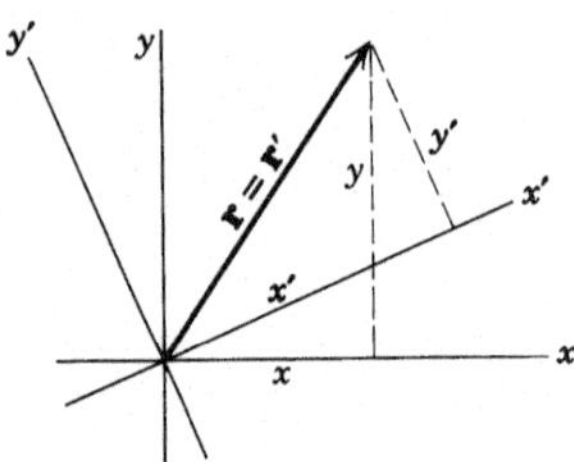

Figure 7.3

(x, y) and (x', y'), and *one* vector $\mathbf{r} = \mathbf{r}'$ with coordinates relative to each set of axes. This time the transformation

$$(7.7) \qquad \begin{cases} x' = ax + by, \\ y' = cx + dy, \end{cases} \quad \text{or} \quad \begin{pmatrix} x' \\ y' \end{pmatrix} = \begin{pmatrix} a & b \\ c & d \end{pmatrix} \begin{pmatrix} x \\ y \end{pmatrix}, \quad \text{or} \quad \mathrm{r}' = \mathrm{Mr},$$

tells us how to get the components of the vector $\mathbf{r} = \mathbf{r}'$ relative to axes (x', y') when we know its components relative to axes (x, y).

Orthogonal Transformations We shall be particularly interested in the special case of a linear transformation which preserves the length of a vector. We call (7.7) an *orthogonal transformation* if

$$(7.8) \qquad x'^2 + y'^2 = x^2 + y^2,$$

and similarly for (7.5). You can see from the figures that this requirement says that the length of a vector is not changed by an orthogonal transformation. In Figure 7.2, the vector would be rotated (or perhaps reflected) with its length held fixed (that is $R = r$ for an orthogonal transformation). In Figure 7.3, the axes are rotated (or reflected), while the vector stays fixed. The matrix M of an orthogonal transformation is called an *orthogonal matrix.* Let's show that the inverse of an orthogonal matrix equals its transpose; in symbols

$$(7.9) \qquad \mathrm{M}^{-1} = \mathrm{M}^{\mathrm{T}}, \quad \mathrm{M} \text{ orthogonal.}$$

From (7.8) and (7.7) we have

$$\begin{aligned} x'^2 + y'^2 &= (ax + by)^2 + (cx + dy)^2 \\ &= (a^2 + c^2)x^2 + 2(ab + cd)xy + (b^2 + d^2)y^2 \equiv x^2 + y^2. \end{aligned}$$

Thus we must have $a^2 + c^2 = 1$, $b^2 + d^2 = 1$, $ab + cd = 0$. Then

$$(7.10) \qquad \begin{aligned} \mathrm{M}^{\mathrm{T}}\mathrm{M} &= \begin{pmatrix} a & c \\ b & d \end{pmatrix} \begin{pmatrix} a & b \\ c & d \end{pmatrix} \\ &= \begin{pmatrix} a^2 + c^2 & ab + cd \\ ab + cd & b^2 + d^2 \end{pmatrix} \equiv \begin{pmatrix} 1 & 0 \\ 0 & 1 \end{pmatrix}. \end{aligned}$$

Since M^TM is the unit matrix, M and M^T are inverse matrices as we claimed in (7.9). We have defined an orthogonal transformation in two dimensions and we have proved (7.9) for the 2-dimensional case. However, a square matrix of any order is called orthogonal if it satisfies (7.9), and you can easily show that the corresponding transformation preserves the lengths of vectors (Problem 9.24).

Now if we write (7.9) as $M^TM = I$ and use the facts from Section 3 that $\det(M^TM) = (\det M^T)(\det M)$ and $\det M^T = \det M$, we have $(\det M)^2 = \det(M^TM) = \det I = 1$, so

$$\det M = \pm 1, \quad \text{M orthogonal.} \tag{7.11}$$

This is true for M of any order since we have used only the definition (7.9) of an orthogonal matrix and some properties of determinants. As we shall see, $\det M = 1$ corresponds geometrically to a rotation, and $\det M = -1$ means that a reflection is involved.

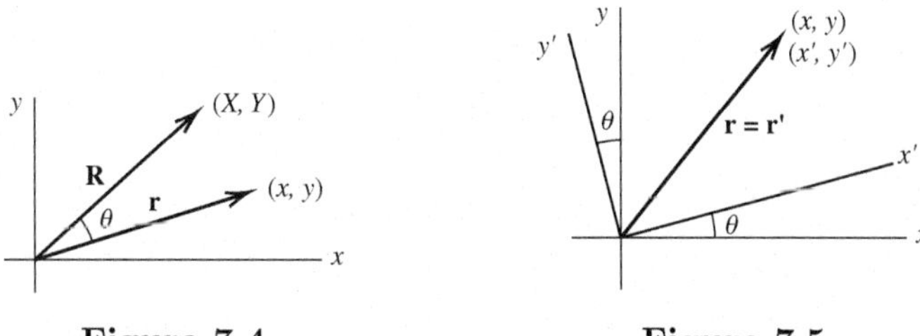

Figure 7.4 **Figure 7.5**

Rotations in 2 Dimensions In Figure 7.4, we have sketched the vector $\mathbf{r} = (x, y)$, and the vector $\mathbf{R} = (X, Y)$ which is the vector $\mathbf{r}$ rotated by angle θ. We write in matrix form the equations relating the components of $\mathbf{r}$ and $\mathbf{R}$ (Problem 19).

$$\begin{pmatrix} X \\ Y \end{pmatrix} = \begin{pmatrix} \cos\theta & -\sin\theta \\ \sin\theta & \cos\theta \end{pmatrix} \begin{pmatrix} x \\ y \end{pmatrix}, \quad \text{vector rotated.} \tag{7.12}$$

In Figure 7.5, we have sketched two sets of axes with the primed axes rotated by angle θ with respect to the unprimed axes. The vector $\mathbf{r} = (x, y)$, and the vector $\mathbf{r}' = (x', y')$ are the same vector, but with components relative to different axes. These components are related by the equations (Problem 20).

$$\begin{pmatrix} x' \\ y' \end{pmatrix} = \begin{pmatrix} \cos\theta & \sin\theta \\ -\sin\theta & \cos\theta \end{pmatrix} \begin{pmatrix} x \\ y \end{pmatrix}, \quad \text{axes rotated.} \tag{7.13}$$

Both equations (7.12) and equations (7.13) are referred to as "rotation equations" and the θ matrices are called "rotation matrices". To distinguish them, we refer to the rotation (7.12) as an "active" transformation (vectors rotated), and to (7.13) as a "passive" transformation (vectors not moved but their components changed because the axes are rotated). Equations (7.7) or (7.13) are also referred to as a "change of basis". (Remember that we called $\mathbf{i}$, $\mathbf{j}$, $\mathbf{k}$ unit basis vectors; here we have changed from the $\mathbf{i}$, $\mathbf{j}$, $\mathbf{k}$ basis to the $\mathbf{i}'$, $\mathbf{j}'$, $\mathbf{k}'$ basis. Also see Section 10.) Observe that the matrices in (7.12) and (7.13) are inverses of each other. You can see from the figures why this must be so. The rotation of a vector in, say, the counterclockwise direction produces the same result as the rotation of the axes in the opposite (clockwise) direction.

We note that $\det M = \cos^2\theta + \sin^2\theta = 1$ for a rotation matrix. Any 2 by 2 orthogonal matrix with determinant 1 corresponds to a rotation, and any 2 by 2 orthogonal matrix with determinant $= -1$ corresponds to a reflection through a line.

▶ **Example 3.** Find what transformation corresponds to each of the following matrices.

$$\text{(7.14)} \qquad \mathrm{A} = \frac{1}{2}\begin{pmatrix} -1 & \sqrt{3} \\ -\sqrt{3} & -1 \end{pmatrix}, \quad \mathrm{B} = \begin{pmatrix} 1 & 0 \\ 0 & -1 \end{pmatrix}, \quad \mathrm{C} = \mathrm{AB}, \quad \mathrm{D} = \mathrm{BA}.$$

First we can show that all these matrices are orthogonal, and that $\det \mathrm{A} = 1$, but the determinants of the other three are -1 (Problem 21). Thus A is a rotation and B, C and D are reflections. Let's view these as active transformations (fixed axes, vectors rotated or reflected). Then by comparing A with (7.12), we have $\cos\theta = -1/2$, $\sin\theta = -\frac{1}{2}\sqrt{3}$, so this is a rotation of 240° (or −120°). Alternatively, we could ask what happens to the vector $\mathbf{i}$. We multiply matrix A times the column matrix $\binom{1}{0}$ and get

$$\frac{1}{2}\begin{pmatrix} -1 & \sqrt{3} \\ -\sqrt{3} & -1 \end{pmatrix}\begin{pmatrix} 1 \\ 0 \end{pmatrix} = \frac{1}{2}\begin{pmatrix} -1 \\ -\sqrt{3} \end{pmatrix} \quad \text{or} \quad -\frac{1}{2}(\mathbf{i} + \mathbf{j}\sqrt{3}),$$

which is $\mathbf{i}$ rotated by 240° as we had before.

Now B operating on $\binom{x}{y}$ leaves x fixed and changes the sign of y (check this); that is, B corresponds to a reflection through the x axis.

We find C = AB and D = BA by multiplying the matrices (Problem 21).

$$\text{(7.15)} \qquad \mathrm{C} = \mathrm{AB} = \frac{1}{2}\begin{pmatrix} -1 & -\sqrt{3} \\ -\sqrt{3} & 1 \end{pmatrix}, \quad \mathrm{D} = \mathrm{BA} = \frac{1}{2}\begin{pmatrix} -1 & \sqrt{3} \\ \sqrt{3} & 1 \end{pmatrix}.$$

We know that these are reflections since they have determinant $= -1$. To find the line through which the plane is reflected, we realize that the vectors along that line are unchanged by the reflection, so we want to find x and y, that is vector $\mathbf{r}$, which is mapped to itself by the transformation. For matrix C we write $\mathrm{Cr} = \mathrm{r}$.

$$\text{(7.16)} \qquad \frac{1}{2}\begin{pmatrix} -1 & -\sqrt{3} \\ -\sqrt{3} & 1 \end{pmatrix}\begin{pmatrix} x \\ y \end{pmatrix} = \begin{pmatrix} x \\ y \end{pmatrix}.$$

You can verify (Problem 21) that the two equations in (7.16) are really the same equation, namely $y = -x\sqrt{3}$. Vectors along this line, say $\mathbf{i} - \mathbf{j}\sqrt{3}$, are not changed by the reflection [see (7.17)] so this is the reflection line. As further verification we can show [see (7.17)] that a vector perpendicular to this line, say $\mathbf{i}\sqrt{3} + \mathbf{j}$, is changed into its negative, that is, it is reflected through the line.

$$\text{(7.17)} \qquad \begin{aligned} \frac{1}{2}\begin{pmatrix} -1 & -\sqrt{3} \\ -\sqrt{3} & 1 \end{pmatrix}\begin{pmatrix} 1 \\ -\sqrt{3} \end{pmatrix} &= \begin{pmatrix} 1 \\ -\sqrt{3} \end{pmatrix}, \\ \frac{1}{2}\begin{pmatrix} -1 & -\sqrt{3} \\ -\sqrt{3} & 1 \end{pmatrix}\begin{pmatrix} \sqrt{3} \\ 1 \end{pmatrix} &= \begin{pmatrix} -\sqrt{3} \\ -1 \end{pmatrix}. \end{aligned}$$

Comment: The solution of the equation $\mathrm{Cr} = \mathrm{r}$ is an example of an eigenvalue, eigenvector problem. We shall discuss such problems in detail in Section 11.

We can analyze the transformation D in the same way we did C to find (Problem 21) that the reflection line is $y = x\sqrt{3}$. Note that matrices A and B do not commute and the transformations C and D are different.

Rotations and Reflections in 3 Dimensions Let's consider 3 by 3 orthogonal matrices as active transformations rotating or reflecting vectors $\mathbf{r} = (x, y, z)$. A simple form for a rotation matrix is

$$\text{(7.18)} \qquad \mathrm{A} = \begin{pmatrix} \cos\theta & -\sin\theta & 0 \\ \sin\theta & \cos\theta & 0 \\ 0 & 0 & 1 \end{pmatrix}.$$

You should satisfy yourself that this transformation produces a rotation of vectors about the z axis through angle θ. We can then find the rotation angle from (7.12) as we did in 2 dimensions. Similarly the matrix

$$\text{(7.19)} \qquad \mathrm{B} = \begin{pmatrix} \cos\theta & -\sin\theta & 0 \\ \sin\theta & \cos\theta & 0 \\ 0 & 0 & -1 \end{pmatrix}$$

produces a rotation about the z axis of angle θ together with a reflection through the (x, y) plane, and again we can find the rotation angle as in 2 dimensions.

We will show in Section 11 that any 3 by 3 orthogonal matrix with determinant $= 1$ can be written in the form (7.18) by choosing the z axis as the rotation axis, and any 3 by 3 orthogonal matrix with determinant $= -1$ can be written in the form (7.19). For now, let's look at a few simple problems we can do just by considering how the matrix maps certain vectors.

Example 4. The matrix for a rotation about the y axis is

$$\text{(7.20)} \qquad \mathrm{F} = \begin{pmatrix} \cos\theta & 0 & \sin\theta \\ 0 & 1 & 0 \\ -\sin\theta & 0 & \cos\theta \end{pmatrix}.$$

You should satisfy yourself that the entry $-\sin\theta$ is in the right place for an active transformation. Let $\theta = 90°$; then the matrix F in (7.20) maps the vector $\mathbf{i} = (1, 0, 0)$ to the vector $-\mathbf{k} = (0, 0, -1)$; this is correct for a $90°$ rotation around the y axis. Check that $(0, 0, 1)$ is mapped to $(1, 0, 0)$.

Example 5. Find the mappings produced by the matrices

$$\text{(7.21)} \qquad \mathrm{G} = \begin{pmatrix} 0 & 0 & 1 \\ 0 & -1 & 0 \\ 1 & 0 & 0 \end{pmatrix}, \quad \mathrm{K} = \begin{pmatrix} 0 & 0 & 1 \\ -1 & 0 & 0 \\ 0 & -1 & 0 \end{pmatrix}.$$

First we find that the determinants are 1 so these are rotations. For G, either by inspection or by solving $\mathrm{Gr} = \mathrm{r}$ as in (7.16), we find that the vector $(1, 0, 1)$ is unchanged and so $\mathbf{i} + \mathbf{k}$ is the rotation axis. Now G^2 is the identity matrix (corresponding to a $360°$ rotation); thus the rotation angle for G is $180°$.

Similarly for K, we find that the vector $(1, -1, 1)$ is unchanged by the transformation so $\mathbf{i} - \mathbf{j} + \mathbf{k}$ is the rotation axis. Now verify that K maps $\mathbf{i}$ to $-\mathbf{j}$, and $-\mathbf{j}$ to $\mathbf{k}$, and $\mathbf{k}$ to $\mathbf{i}$ (or, alternatively that K^3 is the identity matrix) so the rotation angle for K^3 is $\pm 360°$. From the geometry we see that the rotation $\mathbf{i} \to -\mathbf{j} \to \mathbf{k} \to \mathbf{i}$ is a rotation of $-120°$ about $\mathbf{i} - \mathbf{j} + \mathbf{k}$. (Also see Section 11.)

Example 6. Find the mapping produced by the matrix

$$\mathrm{L} = \begin{pmatrix} 0 & -1 & 0 \\ -1 & 0 & 0 \\ 0 & 0 & 1 \end{pmatrix}.$$

Since $\det \mathrm{L} = -1$, this is a reflection through some plane. The vector perpendicular to the reflection plane is reversed by the reflection, so we ask for a vector satisfying $\mathrm{Lr} = -\mathrm{r}$. Either by solving these equations or by inspection we find $\mathbf{r} = (1,1,0) = \mathbf{i} + \mathbf{j}$. The reflecting plane is the plane through the origin perpendicular to this vector, that is, the plane $x + y = 0$ (see Section 5).

PROBLEMS, SECTION 7

Are the following linear functions? Prove your conclusions by showing that $f(\mathbf{r})$ satisfies both of the equations (7.1) or that it does not satisfy at least one of them.

1. $f(\mathbf{r}) = \mathbf{A} \cdot \mathbf{r} + 3$, where $\mathbf{A}$ is a given vector.

2. $f(\mathbf{r}) = \mathbf{A} \cdot (\mathbf{r} - \mathbf{k}z)$.

3. $\mathbf{r} \cdot \mathbf{r}$.

Are the following linear vector functions? Prove your conclusions using (7.2).

4. $\mathbf{F}(\mathbf{r}) = \mathbf{r} - \mathbf{i}x = \mathbf{j}y + \mathbf{k}z$.

5. $\mathbf{F}(\mathbf{r}) = \mathbf{A} \times \mathbf{r}$, where $\mathbf{A}$ is a given vector.

6. $\mathbf{F}(\mathbf{r}) = \mathbf{r} + \mathbf{A}$, where $\mathbf{A}$ is a given vector.

Are the following operators linear?

7. Definite integral with respect to x from 0 to 1; the objects being operated on are functions of x.

8. Find the logarithm; operate on positive real numbers.

9. Find the square; operate on numbers or on functions.

10. Find the reciprocal; operate on numbers or on functions.

11. Find the absolute value; operate on complex numbers.

12. Let D stand for $\dfrac{d}{dx}$, D^2 for $\dfrac{d^2}{dx^2}$, $D^3 = \dfrac{d^3}{dx^3}$, and so on. Are D, D^2, D^3 linear? Operate on functions of x which can be differentiated as many times as needed.

13. (a) As in Problem 12, is $D^2 + 2D + 1$ linear?

(b) Is $x^2D^2 - 2xD + 7$ a linear operator?

14. Find the maximum; operate on functions of x.

15. Find the transpose; operate on matrices.

16. Find the inverse; operate on square matrices.

17. Find the determinant; operate on square matrices.

18. With the cross product of two vectors defined by (4.14), show that finding the cross product is a linear operation, that is, show that (4.18) is valid. *Warning hint:* Don't try to prove it by writing out components: Writing, for example, $\mathbf{i}A_x \times (\mathbf{j}B_y + \mathbf{k}B_z) = \mathbf{i}A_x \times \mathbf{j}B_y + \mathbf{i}A_x \times \mathbf{k}B_z$ would be assuming what you're trying to prove. *Further hints:* First show that (4.18) is valid if $\mathbf{B}$ and $\mathbf{C}$ are both perpendicular to $\mathbf{A}$ by sketching (in the plane perpendicular to $\mathbf{A}$) the vectors $\mathbf{B}$, $\mathbf{C}$, $\mathbf{B}+\mathbf{C}$, and their vector products with $\mathbf{A}$. Then do the general case by first showing that $\mathbf{A}\times\mathbf{B}$ and $\mathbf{A}\times\mathbf{B}_\perp$ (where $\mathbf{B}_\perp$ is the vector component of $\mathbf{B}$ perpendicular to $\mathbf{A}$) have the same magnitude and the same direction.

19. If we multiply a complex number $z = re^{i\phi}$ by $e^{i\theta}$, we get $e^{i\theta}z = re^{i(\phi+\theta)}$, that is, a complex number with the same r but with its angle increased by θ. We can say that the vector $\mathbf{r}$ from the origin to the point $z = x+iy$ has been rotated by angle θ as in Figure 7.4 to become the vector $\mathbf{R}$ from the origin to the point $Z = X+iY$. Then we can write $X+iY = e^{i\theta}z = e^{i\theta}(x+iy)$. Take real and imaginary parts of this equation to obtain equations (7.12).

20. Verify equations (7.13) using Figure 7.5. *Hints*: Write $\mathbf{r}' = \mathbf{r}$ as $\mathbf{i}'x' + \mathbf{j}'y' = \mathbf{i}x + \mathbf{j}y$ and take the dot product of this equation with $\mathbf{i}'$ and with $\mathbf{j}'$ to get x' and y'. Evaluate the dot products of the unit vectors in terms of θ using Figure 7.5. For example, $\mathbf{i}'\cdot\mathbf{j}$ is the cosine of the angle between the x' axis and the y axis.

21. Do the details of Example 3 as follows:

(a) Verify that the four matrices in (7.14) are all orthogonal and verify the stated values of their determinants.

(b) Verify the products C = AB and D = BA in (7.15).

(c) Solve (7.16) to find the reflection line.

(d) Analyze the transformation D as we did C.

Let each of the following matrices represent an active transformation of vectors in the (x, y) plane (axes fixed, vectors rotated or reflected). As in Example 3, show that each matrix is orthogonal, find its determinant, and find the rotation angle, or find the line of reflection.

22. $\dfrac{1}{\sqrt{2}}\begin{pmatrix} 1 & 1 \\ -1 & 1 \end{pmatrix}$

23. $\dfrac{1}{2}\begin{pmatrix} -\sqrt{3} & 1 \\ -1 & -\sqrt{3} \end{pmatrix}$

24. $\begin{pmatrix} 0 & -1 \\ -1 & 0 \end{pmatrix}$

25. $\dfrac{1}{3}\begin{pmatrix} -1 & 2\sqrt{2} \\ 2\sqrt{2} & 1 \end{pmatrix}$

26. $\dfrac{1}{5}\begin{pmatrix} 3 & 4 \\ 4 & -3 \end{pmatrix}$

27. $\dfrac{1}{\sqrt{2}}\begin{pmatrix} -1 & -1 \\ 1 & -1 \end{pmatrix}$

28. Write the matrices which produce a rotation θ about the x axis, or that rotation combined with a reflection through the (y, z) plane. [Compare (7.18) and (7.19) for rotation about the z axis.]

29. Construct the matrix corresponding to a rotation of 90° about the y axis together with a reflection through the (x, z) plane.

30. For the matrices G and K in (7.21), find the matrices R = GK and S = KG. Note that R $\neq S$. (In 3 dimensions, rotations about two different axes do not in general commute.) Find what geometric transformations are produced by R and S.

31. To see a physical example of non-commuting rotations, do the following experiment. Put a book on your desk and imagine a set of rectangular axes with the x and y axes in the plane of the desk with the z axis vertical. Place the book in the first quadrant with the x and y axes along the edges of the book. Rotate the book 90° about the x axis and then 90° about the z axis; note its position. Now repeat the experiment, this time rotating 90° about the z axis first, and then 90° about the x axis; note the different result. Write the matrices representing the 90° rotations and multiply them in both orders. In each case, find the axis and angle of rotation.

For each of the following matrices, find its determinant to see whether it produces a rotation or a reflection. If a rotation, find the axis and angle of rotation. If a reflection, find the reflecting plane and the rotation (if any) about the normal to this plane.

32. $\begin{pmatrix} 0 & 0 & -1 \\ 0 & -1 & 0 \\ -1 & 0 & 0 \end{pmatrix}$ **33.** $\begin{pmatrix} 0 & 0 & -1 \\ -1 & 0 & 0 \\ 0 & 1 & 0 \end{pmatrix}$

34. $\begin{pmatrix} 1 & 0 & 0 \\ 0 & 0 & -1 \\ 0 & -1 & 0 \end{pmatrix}$ **35.** $\begin{pmatrix} 0 & -1 & 0 \\ 1 & 0 & 0 \\ 0 & 0 & -1 \end{pmatrix}$

8. LINEAR DEPENDENCE AND INDEPENDENCE

We say that the three vectors $\mathbf{A} = \mathbf{i} + \mathbf{j}$, $\mathbf{B} = \mathbf{i} + \mathbf{k}$, and $\mathbf{C} = 2\mathbf{i} + \mathbf{j} + \mathbf{k}$ are *linearly dependent* because $\mathbf{A} + \mathbf{B} - \mathbf{C} = \mathbf{0}$. The two vectors $\mathbf{i}$ and $\mathbf{j}$ are *linearly independent* because there are no numbers a and b (not *both* zero) such that the linear combination $a\mathbf{i} + b\mathbf{j}$ is zero. In general, a set of vectors is linearly dependent if some linear combination of them is zero (with not *all* the coefficients equal to zero). In the simple examples above, it was easy to see by inspection whether the vectors were linearly independent or not. In more complicated cases, we need a method of determining linear dependence. Consider the set of vectors

$$(1, 4, -5), (5, 2, 1), (2, -1, 3), \text{ and } (3, -6, 11); \tag{8.1}$$

We want to know whether they are linearly dependent, and if so, we want to find a smaller linearly independent set. Let us row reduce the matrix whose rows are the given vectors (see Section 2):

$$\begin{pmatrix} 1 & 4 & -5 \\ 5 & 2 & 1 \\ 2 & -1 & 3 \\ 3 & -6 & 11 \end{pmatrix} \rightarrow \begin{pmatrix} 9 & 0 & 7 \\ 0 & -9 & 13 \\ 0 & 0 & 0 \\ 0 & 0 & 0 \end{pmatrix}. \tag{8.2}$$

In row reduction, we are forming linear combinations of the rows by elementary row operations [see (2.8)]. All these operations are reversible, so we could, if we liked, reverse our calculations and combine the two vectors $(9, 0, 7)$ and $(0, -9, 13)$ to obtain each of the four original vectors (Problem 1). Thus there are only two independent vectors in (8.1); we refer to these independent vectors as basis vectors since all the original vectors can be written in terms of them (see Section 10). Note that the rank (see Section 2) of the matrix in (8.2) is equal to the number of independent or basis vectors.

Linear Independence of Functions By a definition similar to that for vectors, we say that the functions $f_1(x)$, $f_2(x)$, $\cdots$, $f_n(x)$ are linearly dependent if some linear combination of them is identically zero, that is, if there are constants k_1, k_2, $\cdots$, k_n, not all zero, such that

$$k_1 f_1(x) + k_2 f_2(x) + \cdots + k_n f_n(x) \equiv 0. \tag{8.3}$$

For example, $\sin^2 x$ and $(1 - \cos^2 x)$ are linearly dependent since

$$\sin^2 x - (1 - \cos^2 x) \equiv 0.$$

But $\sin x$ and $\cos x$ are linearly independent since there are no numbers k_1 and k_2, not both zero, such that

$$k_1 \sin x + k_2 \cos x \tag{8.4}$$

is zero for *all* x (Problem 8).

We shall be particularly interested in knowing that a given set of functions is linearly independent. For this purpose the following theorem is useful (Problems 8 to 16, and Chapter 8, Section 5).

> If $f_1(x)$, $f_2(x)$, $\cdots$, $f_n(x)$ have derivatives of order $n-1$, and if the determinant
>
> $$W = \begin{vmatrix} f_1(x) & f_2(x) & \cdots & f_n(x) \\ f_1'(x) & f_2'(x) & \cdots & f_n'(x) \\ f_1''(x) & f_2''(x) & \cdots & f_n''(x) \\ \vdots & \vdots & \ddots & \vdots \\ f_1^{(n-1)}(x) & f_2^{(n-1)}(x) & \cdots & f_n^{(n-1)}(x) \end{vmatrix} \not\equiv 0, \tag{8.5}$$
>
> then the functions are linearly independent. (See Problem 16.) The determinant W is called the *Wronskian* of the functions.

Example 1. Using (8.5), show that the functions 1, x, $\sin x$ are linearly independent.

We write and evaluate the Wronskian,

$$W = \begin{vmatrix} 1 & x & \sin x \\ 0 & 1 & \cos x \\ 0 & 0 & -\sin x \end{vmatrix} = -\sin x.$$

Since $-\sin x$ is not identically equal to zero, the functions are linearly independent.

Example 2. Now let's compute the Wronskian for a case when the functions are linearly dependent.

$$W = \begin{vmatrix} x & \sin x & 2x - 3\sin x \\ 1 & \cos x & 2 - 3\cos x \\ 0 & -\sin x & 3\sin x \end{vmatrix} = \begin{vmatrix} x & \sin x & 2x \\ 1 & \cos x & 2 \\ 0 & -\sin x & 0 \end{vmatrix} = (\sin x)(2x - 2x) \equiv 0,$$

as we expected. However, note that "functions dependent" implies $W \equiv 0$, but $W \equiv 0$ does not necessarily imply "functions dependent". (See Problem 16.)

Homogeneous Equations In Section 2 we considered sets of linear equations. Here we want to consider the special case of such equations when the constants on the right hand sides are all zero; these are called homogeneous equations. We write the homogeneous equations corresponding to (2.12) and (2.13) together with the row reduced matrices:

$$(8.6) \qquad \begin{cases} x + y = 0 \\ x - y = 0 \end{cases} \qquad \begin{pmatrix} 1 & 0 & 0 \\ 0 & 1 & 0 \end{pmatrix}$$

$$(8.7) \qquad \begin{cases} x + \ \ y = 0 \\ 2x + 2y = 0 \end{cases} \qquad \begin{pmatrix} 1 & 1 & 0 \\ 0 & 0 & 0 \end{pmatrix}.$$

We can draw several conclusions from these examples. Note that in (8.6) the only solution is $x = y = 0$; the rank of the matrix is 2, the same as the number of unknowns. In (8.7), the rank of the matrix is 1; this is less than the number of unknowns. This reflects what we could see in (8.7), that we really have just one equation in two unknowns; all the points on a line satisfy $x + y = 0$. In (8.8) we summarize the facts for homogeneous equations:

(8.8) Homogeneous equations are never inconsistent; they always have the solution "all unknowns = 0" (often called the "trivial solution"). If the number of independent equations (that is, the rank of the matrix) is the same as the number of unknowns, this is the only solution. If the rank of the matrix is less than the number of unknowns, there are infinitely many solutions.

A very important special case is a set of n homogeneous equations in n unknowns. By (8.8), these equations have only the trivial solution unless the rank of the matrix is less than n. This means that at least one row of the row reduced n by n matrix of the coefficients is a zero row. But then the determinant D of the coefficients is zero. Thus we have an important result (see Problems 21 to 25; also see Section 11):

(8.9) A system of n homogeneous equations in n unknowns has solutions other than the trivial solution if and only if the determinant of the coefficients is zero.

Solutions in Vector Form Geometrically, solutions of sets of linear equations may be points or lines or planes.

▶ **Example 3.** In Section 2, Example 4, we solved equations (2.15):

$$(8.10) \qquad x = 3 + 2z, \qquad y = 4 - z.$$

This solution set consists of all points on the line which is the intersection of these two planes. An interesting way to write the solution is the vector form

$$(8.11) \qquad \mathbf{r} = (x, y, z) = (3 + 2z, 4 - z, z) = (3, 4, 0) + (2, -1, 1)z.$$

If we put $z = t$, this is the parametric form of the equations of a straight line, $\mathbf{r} = \mathbf{r}_0 + \mathbf{A}t$ [see (5.8)].

Now let's consider the homogeneous equations (zero right hand sides) corresponding to equations (2.15). The equations and the row reduced matrix are:

$$\text{(8.12)} \qquad \begin{pmatrix} 1 & 1 & -1 \\ 2 & -1 & -5 \\ -5 & 4 & 14 \\ 3 & -1 & -7 \end{pmatrix} \begin{pmatrix} x \\ y \\ z \end{pmatrix} = \begin{pmatrix} 0 \\ 0 \\ 0 \\ 0 \end{pmatrix}, \qquad \begin{pmatrix} 1 & 0 & -2 & 0 \\ 0 & 1 & 1 & 0 \\ 0 & 0 & 0 & 0 \\ 0 & 0 & 0 & 0 \end{pmatrix},$$

so the solutions are

$$\text{(8.13)} \qquad x = 2z, \quad y = -z, \quad \text{or} \quad \mathbf{r} = (2, -1, 1)z.$$

Comparing (8.11) and (8.13), we see that the solution of the homogeneous equations Mr = 0 is a straight line through the origin; the solution of the equations Mr = k is a parallel straight line through the point $(3, 4, 0)$. We could say that the solution of Mr = k is the solution of the corresponding homogeneous equations plus the particular solution $\mathbf{r} = (3, 4, 0)$.

Here is an example of an important use of (8.9).

Example 4. For what values of λ does the following set of equations have nontrivial solutions for x and y? For each value of λ find the corresponding relation between x and y. This is an example of an *eigenvalue* problem; we shall discuss such problems in detail in Sections 11 and 12. The values of λ are called eigenvalues and the corresponding vectors (x, y) are called eigenvectors.

$$\text{(8.14)} \qquad \begin{cases} (1-\lambda)x + 2y = 0, \\ 2x + (4-\lambda)y = 0. \end{cases}$$

By (8.9), we set the determinant M of the coefficients equal to zero. Then we solve for λ, and for each value of λ we solve for x and y.

$$\begin{vmatrix} 1-\lambda & 2 \\ 2 & 4-\lambda \end{vmatrix} = \lambda^2 - 5\lambda + 4 - 4 = \lambda(\lambda - 5) = 0, \quad \lambda = 0, 5.$$

For $\lambda = 0$, we find $x + 2y = 0$. For $\lambda = 5$, we find $2x - y = 0$. In vector notation the eigenvectors are: For $\lambda = 0$, $\mathbf{r} = (2, -1)s$, and for $\lambda = 5$, $\mathbf{r} = (1, 2)t$, where s and t are parameters in these vector equations of straight lines through the origin.

PROBLEMS, SECTION 8

1. Write each of the vectors (8.1) as a linear combination of the vectors $(9, 0, 7)$ and $(0, -9, 13)$. *Hint:* To get the right x component in $(1, 4, -5)$, you have to use $(1/9)(9, 0, 7)$. How do you get the right y component? Is the z component now correct?

In Problems 2 to 4, find out whether the given vectors are dependent or independent; if they are dependent, find a linearly independent subset. Write each of the given vectors as a linear combination of the independent vectors.

2. $(1, -2, 3)$, $(1, 1, 1)$, $(-2, 1, -4)$, $(3, 0, 5)$

3. $(0,1,1),\ (-1,5,3),\ (1,0,2),\ (2,-15,1)$

4. $(3,5,-1),\ (1,4,2),\ (-1,0,5),\ (6,14,5)$

5. Show that any vector $\mathbf{V}$ in a plane can be written as a linear combination of two non-parallel vectors $\mathbf{A}$ and $\mathbf{B}$ in the plane; that is, find a and b so that $\mathbf{V} = a\mathbf{A} + b\mathbf{B}$. *Hint:* Find the cross products $\mathbf{A} \times \mathbf{V}$ and $\mathbf{B} \times \mathbf{V}$; what are $\mathbf{A} \times \mathbf{A}$ and $\mathbf{B} \times \mathbf{B}$? Take components perpendicular to the plane to show that

$$a = \frac{(\mathbf{B} \times \mathbf{V}) \cdot \mathbf{n}}{(\mathbf{B} \times \mathbf{A}) \cdot \mathbf{n}}$$

where $\mathbf{n}$ is normal to the plane, and a similar formula for b.

6. Use Problem 5 to write $\mathbf{V} = 3\mathbf{i} + 5\mathbf{j}$ as a linear combination of $\mathbf{A} = 2\mathbf{i} + \mathbf{j}$ and $\mathbf{B} = 3\mathbf{i} - 2\mathbf{j}$. Show that the formulas in Problem 5, written as a quotient of 2 by 2 determinants, are just the Cramer's rule solution of simultaneous equations for a and b.

7. As in Problem 6, write $\mathbf{V} = 4\mathbf{i} - 5\mathbf{j}$ in terms of the basis vectors $\mathbf{i} - 4\mathbf{j}$ and $5\mathbf{i} + 2\mathbf{j}$.

In Problems 8 to 15, use (8.5) to show that the given functions are linearly independent.

8. $\sin x,\ \cos x$

9. $e^{ix},\ \sin x$

10. $x,\ e^x,\ xe^x$

11. $\sin x,\ \cos x,\ x\sin x,\ x\cos x$

12. $1,\ x^2,\ x^4,\ x^6$

13. $\sin x,\ \sin 2x$

14. $e^{ix},\ e^{-ix}$

15. $e^x,\ e^{ix},\ \cosh x$

16. (a) Prove that if the Wronskian (8.5) is not identically zero, then the functions f_1, $f_2, \ldots, f_n$ are linearly independent. Note that this is equivalent to proving that if the functions are linearly dependent, then W is identically zero. *Hints:* Suppose (8.3) were true; you want to find the k's. Differentiate (8.3) repeatedly until you have a set of n equations for the n unknown k's. Then use (8.9).

(b) In part (a) you proved that if $W \not\equiv 0$, then the functions are linearly independent. You might think that if $W \equiv 0$, the functions would be linearly dependent. This is not necessarily true; if $W \equiv 0$, the functions might be either dependent or independent. For example, consider the functions x^3 and $|x^3|$ on the interval $(-1, 1)$. Show that $W \equiv 0$, but the functions are not linearly dependent on $(-1, 1)$. (Sketch them.) On the other hand, they are linearly dependent (in fact identical) on $(0, 1)$.

In Problems 17 to 20, solve the sets of homogeneous equations by row reducing the matrix.

17. $$\begin{cases} x - 2y + 3z = 0 \\ x + 4y - 6z = 0 \\ 2x + 2y - 3z = 0 \end{cases}$$

18. $$\begin{cases} 2x + 3z = 0 \\ 4x + 2y + 5z = 0 \\ x - y + 2z = 0 \end{cases}$$

19. $$\begin{cases} 3x + y + 3z + 6w = 0 \\ 4x - 7y - 3z + 5w = 0 \\ x + 3y + 4z - 3w = 0 \\ 3x + 2z + 7w = 0 \end{cases}$$

20. $$\begin{cases} 2x - 3y + 5z = 0 \\ x + 2y - z = 0 \\ x - 5y + 6z = 0 \\ 4x + y + 3z = 0 \end{cases}$$

21. Find a condition for four points in space to lie in a plane. Your answer should be in the form a determinant which must be equal to zero. *Hint:* The equation of a plane is of the form $ax + by + cz = d$, where a, b, c, d are constants. The four points (x_1, y_1, z_1), (x_2, y_2, z_2), etc., are all to satisfy this equation. When can you find a, b, c, d not all zero?

22. Find a condition for three lines in a plane to intersect in one point. *Hint:* See Problem 21. Write the equation of a line as $ax + by = c$. Assume that no two of the lines are parallel.

Using (8.9), find the values of λ such that the following equations have nontrivial solutions, and for each λ, solve the equations. (See Example 4.)

23. $\begin{cases} (4-\lambda)x - 2y = 0 \\ -2x + (7-\lambda)y = 0 \end{cases}$

24. $\begin{cases} (6-\lambda)x + 3y = 0 \\ 3x - (2+\lambda)y = 0 \end{cases}$

25. $\begin{cases} -(1+\lambda)x + y + 3z = 0, \\ x + (2-\lambda)y = 0, \\ 3x + (2-\lambda)z = 0. \end{cases}$

For each of the following, write the solution in vector form [see (8.11) and (8.13)].

26. $\begin{cases} 2x - 3y + 5z = 3 \\ x + 2y - z = 5 \\ x - 5y + 6z = -2 \\ 4x + y + 3z = 13 \end{cases}$

27. $\begin{cases} x - y + 2z = 3 \\ -2x + 2y - z = 0 \\ 4x - 4y + 5z = 6 \end{cases}$

28. $\begin{cases} 2x + y - 5z = 7 \\ x - 2y = 1 \\ 3x - 5y - z = 4 \end{cases}$

9. SPECIAL MATRICES AND FORMULAS

In this section we want to discuss various terms used in work with matrices, and prove some important formulas. First we list for reference needed definitions and facts about matrices.

There are several special matrices which are related to a given matrix A. We outline in (9.1) what these matrices are called, what notations are used for them, and how we get them from A.

(9.1)	*Name of Matrix*	*Notations for it*	*How to get it from* A
	Transpose of A, or A transpose	A^T or $\widetilde{A}$ or A' or A^t	Interchange rows and columns in A.
	Complex conjugate of A	$\bar{A}$ or A^*	Take the complex conjugate of each element.
	Transpose conjugate, Hermitian conjugate, adjoint (Problem 9), Hermitian adjoint.	$A^\dagger$ (A dagger)	Take the complex conjugate of each element and transpose.
	Inverse of A	A^{-1}	See Formula (6.13).

There is another set of names for special types of matrices. In (9.2), we list these and their definitions for reference.

(9.2)	A matrix is called	if it satisfies the condition(s)
	real	$A = \bar{A}$
	symmetric	$A = A^T$, A real (matrix = its transpose)
	skew-symmetric or antisymmetric	$A = -A^T$, A real
	orthogonal	$A^{-1} = A^T$, A real (inverse = transpose)
	pure imaginary	$A = -\bar{A}$
	Hermitian	$A = A^\dagger$ (matrix = its transpose conjugate)
	anti-Hermitian	$A = -A^\dagger$
	unitary	$A^{-1} = A^\dagger$ (inverse = transpose conjugate)
	normal	$AA^\dagger = A^\dagger A$ (A and $A^\dagger$ commute)

Now let's consider some examples and proofs using these terms.

Index Notation We are going to need index notation in some of our work below, so for reference we restate the rule in (6.2b) for matrix multiplication.

(9.3)
$$(AB)_{ij} = \sum_k A_{ik}B_{kj}.$$

Study carefully the index notation for "row times column" multiplication. To find the element in row i and column j of the product matrix AB, we multiply row i of A times column j of B. Note that the k's (the sum is over k) are next to each other in (9.3). If we should happen to have $\sum_k B_{kj}A_{ik}$, we should rewrite it as $\sum_k A_{ik}B_{kj}$ (with the k's next to each other) to recognize it as an element of the matrix AB (not BA). We will see an example of this in (9.10) below.

Kronecker δ The *Kronecker* δ is defined by

(9.4)
$$\delta_{ij} = \begin{cases} 1, & \text{if } i = j, \\ 0, & \text{if } i \neq j. \end{cases}$$

For example, $\delta_{11} = 1$, $\delta_{12} = 0$, $\delta_{22} = 1$, $\delta_{31} = 0$, and so on. In this notation a unit matrix is one whose elements are δ_{ij} and we can write

(9.5)
$$I = (\delta_{ij}).$$

(*Also see* Chapter 10, Section 5.) The Kronecker δ notation is useful for other purposes. For example, since (for positive integers m and n)

$$\int_{-\pi}^{\pi} \cos nx \cos mx \, dx = \begin{cases} \pi, & \text{if } m = n, \\ 0, & \text{if } m \neq n, \end{cases} \tag{9.6a}$$

we can write

$$\int_{-\pi}^{\pi} \cos nx \cos mx \, dx = \pi \cdot \delta_{nm}. \tag{9.6b}$$

This is the same as (9.6a) because $\delta_{nm} = 0$ if $m \neq n$, and $\delta_{nm} = 1$ if $m = n$.

Using the Kronecker δ, we can give a formal proof that for any matrix M and a conformable unit matrix I, the product of I and M is just M. Using index notation and equations (9.3) and (9.4), we have

$$(\mathrm{IM})_{ij} = \sum_k \delta_{ik} M_{kj} = M_{ij} \quad \text{or} \quad \mathrm{IM} = \mathrm{M} \tag{9.7}$$

since $\delta_{ik} = 0$ unless $k = i$.

More Useful Theorems Let's use index notation to prove the associative law for matrix multiplication, that is

$$\mathrm{A(BC)} = \mathrm{(AB)C} = \mathrm{ABC}. \tag{9.8}$$

First we write $(\mathrm{BC})_{kj} = \sum_l B_{kl} C_{lj}$. Then we have

$$\begin{aligned} [\mathrm{A(BC)}]_{ij} &= \sum_k A_{ik} (\mathrm{BC})_{kj} = \sum_k A_{ik} \sum_l B_{kl} C_{lj} \\ &= \sum_k \sum_l A_{ik} B_{kl} C_{lj} = (\mathrm{ABC})_{ij} \end{aligned} \tag{9.9}$$

which is the index notation for $\mathrm{A(BC)} = \mathrm{ABC}$ as in (9.8). We can prove $\mathrm{(AB)C} = \mathrm{ABC}$ in a similar way (Problem 1).

In formulas we may want the transpose of the product of two matrices. First note that $A^{\mathrm{T}}_{ik} = A_{ki}$ [see (2.1) or (9.1)]. Then

$$\begin{aligned} (\mathrm{AB})^{\mathrm{T}}_{ik} &= (\mathrm{AB})_{ki} = \sum_j A_{kj} B_{ji} = \sum_j A^{\mathrm{T}}_{jk} B^{\mathrm{T}}_{ij} \\ &= \sum_j B^{\mathrm{T}}_{ij} A^{\mathrm{T}}_{jk} = (\mathrm{B^T A^T})_{ik}, \quad \text{or,} \\ (\mathrm{AB})^{\mathrm{T}} &= \mathrm{B^T A^T}. \end{aligned} \tag{9.10}$$

The theorem applies to a product of any number of matrices (see Problem 8b). For example

$$(\mathrm{ABCD})^{\mathrm{T}} = \mathrm{D^T C^T B^T A^T}. \tag{9.11}$$

The transpose of a product of matrices is equal to the product of the transposes in reverse order.

A similar theorem is true for the inverse of a product (see Section 6, Problem 18).

$$(9.12)\qquad (\mathrm{ABCD})^{-1} = \mathrm{D}^{-1}\mathrm{C}^{-1}\mathrm{B}^{-1}\mathrm{A}^{-1}.$$

The inverse of a product of matrices is equal to the product of the inverses in reverse order.

Trace of a Matrix The *trace* (or *spur*) or a square matrix A (written Tr A) is the sum of the elements on the main diagonal. Thus the trace of a unit n by n matrix is n, and the trace of the matrix M in (6.10) is 6. It is a theorem that the trace of a product of matrices is not changed by permuting them in cyclic order. For example

$$(9.13)\qquad \mathrm{Tr(ABC)} = \mathrm{Tr(BCA)} = \mathrm{Tr(CAB)}.$$

We can prove this as follows:

$$\begin{aligned}\mathrm{Tr(ABC)} &= \sum_i (\mathrm{ABC})_{ii} = \sum_i\sum_j\sum_k A_{ij}B_{jk}C_{ki}\\ &= \sum_i\sum_j\sum_k B_{jk}C_{ki}A_{ij} = \mathrm{Tr(BCA)}\\ &= \sum_i\sum_j\sum_k C_{ki}A_{ij}B_{jk} = \mathrm{Tr(CAB)}.\end{aligned}$$

Warning: Tr(ABC) is *not* equal to Tr(ACB) in general.

Theorem: If H is a Hermitian matrix, then $\mathrm{U} = e^{i\mathrm{H}}$ is a unitary matrix. (This is an important relation in quantum mechanics.) By (9.2) we need to prove that $\mathrm{U}^\dagger = \mathrm{U}^{-1}$ if $\mathrm{H}^\dagger = \mathrm{H}$. First, $e^{i\mathrm{H}}e^{-i\mathrm{H}} = e^{i\mathrm{H}-i\mathrm{H}}$ since H commutes with itself—see Problem 6.29. But this is e^0 which is the unit matrix [see Section 6] so $\mathrm{U}^{-1} = e^{-i\mathrm{H}}$. To find $\mathrm{U}^\dagger = (e^{i\mathrm{H}})^\dagger$, we expand $\mathrm{U} = e^{i\mathrm{H}}$ in a power series to get $\mathrm{U} = \sum_k (i\mathrm{H})^k/k!$ and then take the transpose conjugate. To do this we just need to realize that the transpose of a sum of matrices is the sum of the transposes, and that the transpose of a power of a matrix, say $(\mathrm{M}^n)^\mathrm{T}$ is equal to $(\mathrm{M}^\mathrm{T})^n$ (Problem 9.21). Also recall from Chapter 2 that you find the complex conjugate of an expression by changing the signs of all the i's. This means that $(i\mathrm{H})^\dagger = -i\mathrm{H}^\dagger = -i\mathrm{H}$ since H is Hermitian. Then summing the series we get $\mathrm{U}^\dagger = e^{-i\mathrm{H}}$, which is just what we found for U^{-1} above. Thus $\mathrm{U}^\dagger = \mathrm{U}^{-1}$, so U is a unitary matrix. (Also see Problem 11.61.)

PROBLEMS, SECTION 9

1. Use index notation as in (9.9) to prove the second part of the associative law for matrix multiplication: (AB)C = ABC.

2. Use index notation to prove the distributive law for matrix multiplication, namely: A(B + C) = AB + AC.

3. Given the following matrix, find the transpose, the inverse, the complex conjugate, and the transpose conjugate of A. Verify that $AA^{-1} = A^{-1}A =$ the unit matrix.

$$A = \begin{pmatrix} 1 & 0 & 5i \\ -2i & 2 & 0 \\ 1 & 1+i & 0 \end{pmatrix},$$

4. Repeat Problem 3 given

$$A = \begin{pmatrix} 0 & 2i & -1 \\ -i & 2 & 0 \\ 3 & 0 & 0 \end{pmatrix}.$$

5. Show that the product AA^T is a symmetric matrix.

6. Give numerical examples of: a symmetric matrix; a skew-symmetric matrix; a real matrix; a pure imaginary matrix.

7. Write each of the items in the second column of (9.2) in index notation.

8. (a) Prove that $(AB)^\dagger = B^\dagger A^\dagger$. *Hint:* See (9.10).

(b) Verify (9.11), that is, show that (9.10) applies to a product of any number of matrices. *Hint:* Use (9.10) and (9.8).

9. In (9.1) we have defined the adjoint of a matrix as the transpose conjugate. This is the usual definition except in algebra where the adjoint is defined as the transposed matrix of cofactors [see (6.13)]. Show that the two definitions are the same for a unitary matrix with determinant $= +1$.

10. Show that if a matrix is orthogonal and its determinant is $+1$, then each element of the matrix is equal to its own cofactor. *Hint:* Use (6.13) and the definition of an orthogonal matrix.

11. Show that a real Hermitian matrix is symmetric. Show that a real unitary matrix is orthogonal. *Note:* Thus we see that Hermitian is the complex analogue of symmetric, and unitary is the complex analogue of orthogonal. (See Section 11.)

12. Show that the definition of a Hermitian matrix $(A = A^\dagger)$ can be written $a_{ij} = \bar{a}_{ji}$ (that is, the diagonal elements are real and the other elements have the property that $a_{12} = \bar{a}_{21}$, etc.). Construct an example of a Hermitian matrix.

13. Show that the following matrix is a unitary matrix.

$$\begin{pmatrix} (1+i\sqrt{3})/4 & \dfrac{\sqrt{3}}{2\sqrt{2}}(1+i) \\ \dfrac{-\sqrt{3}}{2\sqrt{2}}(1+i) & (\sqrt{3}+i)/4 \end{pmatrix}$$

14. Use (9.11) and (9.12) to simplify $(AB^TC)^T$, $(C^{-1}MC)^{-1}$, $(AH)^{-1}(AHA^{-1})^3(HA^{-1})^{-1}$.

15. (a) Show that the Pauli spin matrices (Problem 6.6) are Hermitian.

(b) Show that the Pauli spin matrices satisfy the Jacobi identity $[A,[B,C]] + [B,[C,A]] + [C,[A,B]] = 0$ where $[A,B]$ is the commutator of A, B [*see* (6.3)].

(c) Generalize (b) to prove the Jacobi identity for any (conformable) matrices A, B, C. *Also see* Chapter 6, Problem 3.14.

16. Let $C_{ij} = (-1)^{i+j}M_{ij}$ be the cofactor of element a_{ij} in the determinant A. Show that the statement of Laplace's development and the statement of Problem 3.8 can be combined in the equations

$$\sum_j a_{ij}C_{kj} = \delta_{ik} \cdot \det A, \quad \text{or} \quad \sum_i a_{ij}C_{ik} = \delta_{jk} \cdot \det A.$$

17. (a) Show that if A and B are symmetric, then AB is not symmetric unless A and B commute.

(b) Show that a product of orthogonal matrices is orthogonal.

(c) Show that if A and B are Hermitian, then AB is not Hermitian unless A and B commute.

(d) Show that a product of unitary matrices is unitary.

18. If A and B are symmetric matrices, show that their commutator is antisymmetric [see equation (6.3)].

19. (a) Prove that $\mathrm{Tr(AB)} = \mathrm{Tr(BA)}$. *Hint:* See proof of (9.13).

(b) Construct matrices A, B, C for which $\mathrm{Tr(ABC)} \neq \mathrm{Tr(CBA)}$, but verify that $\mathrm{Tr(ABC)} = \mathrm{Tr(CAB)}$.

(c) If S is a symmetric matrix and A is an antisymmetric matrix, show that $\mathrm{Tr(SA)} = 0$. *Hint:* Consider $\mathrm{Tr(SA)^T}$ and prove that $\mathrm{Tr(SA)} = -\,\mathrm{Tr(SA)}$.

20. Show that the determinant of a unitary matrix is a complex number with absolute value $= 1$. *Hint:* See proof of equation (7.11).

21. Show that the transpose of a sum of matrices is equal to the sum of the transposes. Also show that $(\mathrm{M}^n)^{\mathrm{T}} = (\mathrm{M}^{\mathrm{T}})^n$. *Hint:* Use (9.11) and (9.8).

22. Show that a unitary matrix is a normal matrix, that is, that it commutes with its transpose conjugate [see (9.2)]. Also show that orthogonal, symmetric, antisymmetric, Hermitian, and anti-Hermitian matrices are normal.

23. Show that the following matrices are Hermitian whether A is Hermitian or not: $\mathrm{AA}^\dagger$, $\mathrm{A}+\mathrm{A}^\dagger$, $i(\mathrm{A}-\mathrm{A}^\dagger)$.

24. Show that an orthogonal transformation preserves the length of vectors. *Hint:* If r is the column matrix of vector **r** [see (6.10)], write out $\mathrm{r^Tr}$ to show that it is the square of the length of **r**. Similarly $\mathrm{R^TR} = |\mathbf{R}|^2$ and you want to show that $|\mathbf{R}|^2 = |\mathbf{r}|^2$, that is, $\mathrm{R^TR} = \mathrm{r^Tr}$ if $\mathrm{R} = \mathrm{Mr}$ and M is orthogonal. Use (9.11).

25. (a) Show that the inverse of an orthogonal matrix is orthogonal. *Hint:* Let $\mathrm{A} = \mathrm{O}^{-1}$; from (9.2), write the condition for O to be orthogonal and show that A satisfies it.

(b) Show that the inverse of a unitary matrix is unitary. See hint in (a).

(c) If H is Hermitian and U is unitary, show that $\mathrm{U^{-1}HU}$ is Hermitian.

10. LINEAR VECTOR SPACES

We have used extensively the vector $\mathbf{r} = \mathbf{i}x+\mathbf{j}y+\mathbf{k}z$ to mean a vector from the origin to the point (x, y, z). There is a one-to-one correspondence between the vectors **r** and the points (x, y, z); the collection of all such points or all such vectors makes up the 3-dimensional space often called R_3 (R for real) or V_3 (V for vector) or E_3 (E for Euclidean). Similarly, we can consider a 2-dimensional space V_2 of vectors $\mathbf{r} = \mathbf{i}x + \mathbf{j}y$ or points (x, y) making up the (x, y) plane. V_2 might also mean *any* plane through the origin. And V_1 means all the vectors from the origin to points on some line through the origin.

We also use x, y, z to mean the variables or unknowns in a problem. Now applied problems often involve more than three variables. By extension of the idea of V_3, it is convenient to call an ordered set of n numbers a point or vector in the n-dimensional space V_n. For example, the 4-vectors of special relativity are ordered sets of four numbers; we say that space-time is 4-dimensional. A point of the *phase space* used in classical and quantum mechanics is an ordered set of six numbers, the three components of the position of a particle and the three components of its momentum; thus the phase space of a particle is the 6-dimensional space V_6.

In such cases, we can't represent the variables as coordinates of a point in *physical* space since physical space has only three dimensions. But it is convenient and customary to extend our geometrical *terminology* anyway. Thus we use the terms *variables* and *coordinates* interchangeably and speak, for example, of a "point in 5-dimensional space," meaning an ordered set of values of five variables, and similarly for any number of variables. In three dimensions, we think of the coordinates of a point as the components of a vector from the origin to the point. By analogy, we call an ordered set of five numbers a "vector in 5-dimensional space" or an ordered set of n numbers a "vector in n-dimensional space."

Much of the geometrical terminology which is familiar in two and three dimensions can be extended to problems in n dimensions (that is, n variables) by using the algebra which parallels the geometry. For example, the distance from the origin to the point (x, y, z) is $\sqrt{x^2 + y^2 + z^2}$. By analogy in a problem in the five variables x, y, z, u, v, we define the distance from the origin $(0, 0, 0, 0, 0)$ to the point (x, y, z, u, v) as $\sqrt{x^2 + y^2 + z^2 + u^2 + v^2}$. By using the algebra which goes with the geometry, we can easily extend such ideas as the length of a vector, the dot product of two vectors, and therefore the angle between the vectors and the idea of orthogonality, etc. We saw in Section 7, that an orthogonal transformation in two or three dimensions corresponds to a rotation. Thus we might say, in a problem in n variables, that a linear transformation (that is a linear change of variables) satisfying "sum of squares of new variables = sum of squares of old variables" [compare (7.8)] corresponds to a "rotation in *n-dimensional space.*"

Example 1. Find the distance between the points $(3, 0, 5, -2, 1)$ and $(0, 1, -2, 3, 0)$.

Generalizing what we would do in three dimensions, we find $d^2 = (3-0)^2 + (0-1)^2 + (5+2)^2 + (-2-3)^2 + (1-0)^2 = 9 + 1 + 49 + 25 + 1 = 85$, $d = \sqrt{85}$.

If we start with several vectors, and find linear combinations of them in the algebraic way (by components), then we say that the original set of vectors and all their linear combinations form a *linear vector space* (or just *vector space* or *linear space* or *space*). Note that if $\mathbf{r}$ is one of our original vectors, then $\mathbf{r} - \mathbf{r} = \mathbf{0}$ is one of the linear combinations; thus the zero vector (that is, the origin) must be a point in every vector space. A line or plane not passing through the origin is not a vector space.

Subspace, Span, Basis, Dimension Suppose we start with the four vectors in (8.1). We showed in (8.2) that they are all linear combinations of the two vectors $(9, 0, 7)$ and $(0, -9, 13)$. Now two linearly independent vectors (remember their tails are at the origin) determine a plane; all linear combinations of the two vectors lie in the plane. [The plane we are talking about in this example is the plane through the three points $(9, 0, 7)$, $(0, -9, 13)$, and the origin.] Since all the vectors making up this plane V_2 are also part of 3-dimensional space V_3, we call V_2 a *subspace* of V_3. Similarly any line lying in this plane and passing through the origin is a subspace of V_2 and of V_3. We say that either the original four vectors or the two independent ones *span* the space V_2; a set of vectors spans a space if all the vectors in the space can be written as linear combinations of the spanning set. A set of *linearly independent* vectors which span a vector space is called a *basis.* Here the vectors $(9,0,7)$ and $(0,-9,13)$ are one possible choice as a basis for the space V_2; another choice would be any two of the original vectors since in (8.2) no two of the vectors are dependent.

The *dimension* of a vector space is equal to the number of basis vectors. Note that this statement implies (correctly—see Problem 8) that no matter how you pick the basis vectors for a given vector space, there will always be the same number of them. This number is the dimension of the space. In 3 dimensions, we have frequently used the unit basis vectors $\mathbf{i}$, $\mathbf{j}$, $\mathbf{k}$ which can also be written as $(1,0,0)$, $(0,1,0)$, $(0,0,1)$. Then in, say 5 dimensions, a corresponding set of unit basis vectors would be $(1,0,0,0,0)$, $(0,1,0,0,0)$, $(0,0,1,0,0)$, $(0,0,0,1,0)$, $(0,0,0,0,1)$. You should satisfy yourself that these five vectors are linearly independent and span a 5 dimensional space.

Example 2. Find the dimension of the space spanned by the following vectors, and a basis for the space: $(1,0,1,5,-2)$, $(0,1,0,6,-3)$, $(2,-1,2,4,1)$, $(3,0,3,15,-6)$.

We write the matrix whose rows are the components of the vectors and row reduce it to find that there are three linearly independent vectors: $(1,0,1,5,0)$, $(0,1,0,6,0)$, $(0,0,0,0,1)$. These three vectors are a basis for the space which is therefore 3-dimensional.

Inner Product, Norm, Orthogonality Recall from (4.10) that the scalar (or dot or inner) product of two vectors $\mathbf{A} = (A_1, A_2, A_3)$ and $\mathbf{B} = (B_1, B_2, B_3)$ is $A_1B_1 + A_2B_2 + A_3B_3 = \sum_{i=1}^{3} A_iB_i$. This is very easy to generalize to n dimensions. By definition, the inner product of two vectors in n dimensions is given by

$$\text{(10.1)} \qquad \mathbf{A}\cdot\mathbf{B} = (\text{Inner product of } \mathbf{A} \text{ and } \mathbf{B}) = \sum_{i=1}^{n} A_iB_i.$$

Similarly, generalizing (4.1), we can define the length or *norm* of a vector in n dimensions by the formula:

$$\text{(10.2)} \qquad A = \text{Norm of } \mathbf{A} = ||\mathbf{A}|| = \sqrt{\mathbf{A}\cdot\mathbf{A}} = \sqrt{\sum_{i=1}^{n} A_i^2}.$$

In 3 dimensions, we also write the scalar product as $AB\cos\theta$ [see (4.2)] so if two vectors are orthogonal (perpendicular) their scalar product is $AB\cos\pi/2 = 0$. We generalize this to n dimensions by saying that two vectors in n dimensions are orthogonal if their inner product is zero.

$$\text{(10.3)} \qquad \mathbf{A} \text{ and } \mathbf{B} \text{ are orthogonal if } \sum_{i=1}^{n} A_iB_i = 0.$$

Schwarz Inequality In 2 or 3 dimensions we can find the angle between two vectors [see (4.11)] from the formula $\mathbf{A}\cdot\mathbf{B} = AB\cos\theta$. It is tempting to use the same formula in n dimensions, but before we do we should be sure that the resulting value of $\cos\theta$ will satisfy $|\cos\theta| \le 1$, that is

$$|\mathbf{A}\cdot\mathbf{B}| \le AB, \quad \text{or} \quad \left|\sum_{i=1}^{n} A_iB_i\right| \le \sqrt{\sum_{i=1}^{n} A_i^2}\sqrt{\sum_{i=1}^{n} B_i^2}. \tag{10.4}$$

This is called the Schwarz inequality (for n-dimensional Euclidean space). We can prove it as follows. First note that if $\mathbf{B} = \mathbf{0}$, (10.4) just says $0 \le 0$ which is certainly true. For $\mathbf{B} \ne \mathbf{0}$, we consider the vector $\mathbf{C} = B\mathbf{A} - (\mathbf{A}\cdot\mathbf{B})\mathbf{B}/B$, and find $\mathbf{C}\cdot\mathbf{C}$. Now $\mathbf{C}\cdot\mathbf{C} = \sum C_i^2 \ge 0$, so we have

$$\begin{aligned}\mathbf{C}\cdot\mathbf{C} &= B^2(\mathbf{A}\cdot\mathbf{A}) - 2B(\mathbf{A}\cdot\mathbf{B})(\mathbf{A}\cdot\mathbf{B})/B + (\mathbf{A}\cdot\mathbf{B})^2(\mathbf{B}\cdot\mathbf{B})/B^2 \\ &= A^2B^2 - 2(\mathbf{A}\cdot\mathbf{B})^2 + (\mathbf{A}\cdot\mathbf{B})^2 = A^2B^2 - (\mathbf{A}\cdot\mathbf{B})^2 = C^2 \ge 0,\end{aligned} \tag{10.5}$$

which gives (10.4). Thus, if we like, we can define the cosine of the angle between two vectors in n dimensions by $\cos\theta = \mathbf{A}\cdot\mathbf{B}/(AB)$. Note that equality holds in Schwarz's inequality if and only if $\cos\theta = \pm 1$, that is, when $\mathbf{A}$ and $\mathbf{B}$ are parallel or antiparallel, say $\mathbf{B} = k\mathbf{A}$.

Example 3. Find the cosine of the angle between each pair of the 3 basis vectors we found in Example 2.

By (10.2) we find that the norms of the first two basis vectors are $\sqrt{1+1+25} = \sqrt{27}$ and $\sqrt{1+36} = \sqrt{37}$. By (10.1), the inner product of these two vectors is $1\cdot 0 + 0\cdot 1 + 1\cdot 0 + 5\cdot 6 + 0\cdot 0 = 30$. Thus $\cos\theta = 30/(\sqrt{27\cdot 37}) \simeq 0.949$, which, we note, is < 1 as Schwarz's inequality says. The third basis vector in Example 2 is orthogonal to the other two since the inner products are zero, that is, $\cos\theta = 0$.

Orthonormal Basis; Gram-Schmidt Method We call a set of vectors *orthonormal* if they are all mutually *ortho*gonal (perpendicular), and each vector is *normal*ized (that is, its norm is one—it has unit length). For example, the vectors **i**, **j**, **k**, form an orthonormal set. If we have a set of basis vectors for a space, it is often convenient to take combinations of them to form an orthonormal basis. The Gram-Schmidt method is a systematic process for doing this. It is very simple in idea although the details of carrying it out can get messy. Suppose we have basis vectors **A**, **B**, **C**. Normalize **A** to get the first vector of a set of orthonormal basis vectors. To get a second basis vector, subtract from **B** its component along **A**; what remains is orthogonal to **A**. [See equation (4.4) and Figure 4.10.] Normalize this remainder to find the second vector of an orthonormal basis. Similarly, subtract from **C** its components along **A** and **B** to find a third vector orthogonal to both **A** and **B** and normalize this third vector. We now have 3 mutually orthogonal unit vectors; this is the desired set of orthonormal basis vectors. In a space of higher dimension, this process can be continued. (We will see a use for this method in Section 11; see degeneracy, pages 152–153.)

▶ **Example 4.** Given the basis vectors **A**, **B**, **C**, below, use the Gram-Schmidt method to find an orthonormal set of basis vectors $\mathbf{e}_1$, $\mathbf{e}_2$, $\mathbf{e}_3$. Following the outline above, we find

$$\begin{aligned}
&\mathbf{A} = (0,0,5,0); & \mathbf{e}_1 &= \mathbf{A}/A = (0,0,1,0);\\
&\mathbf{B} = (2,0,3,0); & \mathbf{B} - (\mathbf{e}_1\cdot\mathbf{B})\mathbf{e}_1 &= \mathbf{B} - 3\mathbf{e}_1 = (2,0,0,0);\\
& & \mathbf{e}_2 &= (1,0,0,0);\\
&\mathbf{C} = (7,1,-5,3); & \mathbf{C} - (\mathbf{e}_1\cdot\mathbf{C})\mathbf{e}_1 - (\mathbf{e}_2\cdot\mathbf{C})\mathbf{e}_2 &= \mathbf{C} - (-5)\mathbf{e}_1 - 7\mathbf{e}_2\\
& & &= (0,1,0,3);\\
& & \mathbf{e}_3 &= (0,1,0,3)/\sqrt{10}.
\end{aligned}$$

Complex Euclidean Space In applications it is useful to allow vector components to be complex. For example, in three dimensions we might consider vectors like $(5+2i, 3-i, 1+i)$. Let's go back and see what modifications are needed in this case. In (10.2), we want the quantity under the square root sign to be positive. To assure this, we replace the square of A_i by the absolute square of A_i, that is by $|A_i|^2 = A_i^* A_i$ where A_i^* is the complex conjugate of A_i (see Chapter 2). Similarly, in (10.1) and (10.3), we replace $A_i B_i$ by $A_i^* B_i$. Thus we define

$$\text{(Inner product of } \mathbf{A} \text{ and } \mathbf{B}) = \sum_{i=1}^{n} A_i^* B_i \tag{10.6}$$

$$\text{(Norm of } \mathbf{A}) = ||\mathbf{A}|| = \sqrt{\sum_{i=1}^{n} A_i^* A_i} \tag{10.7}$$

$$\mathbf{A} \text{ and } \mathbf{B} \text{ are orthogonal if } \sum_{i=1}^{n} A_i^* B_i = 0. \tag{10.8}$$

The Schwarz inequality becomes (see Problem 6)

$$\left|\sum_{i=1}^{n} A_i^* B_i\right| \le \sqrt{\sum_{i=1}^{n} A_i^* A_i}\sqrt{\sum_{i=1}^{n} B_i^* B_i}. \tag{10.9}$$

Note that we can write the inner product in matrix form. If A is a column matrix with elements A_i, then the transpose conjugate matrix $\mathrm{A}^\dagger$ is a row matrix with elements A_i^*. Using this notation we can write $\sum A_i^* B_i = \mathrm{A}^\dagger\mathrm{B}$ (Problem 9).

▶ **Example 5.** Given $\mathbf{A} = (3i, 1-i, 2+3i, 1+2i)$, $\mathbf{B} = (-1, 1+2i, 3-i, i)$, $\mathbf{C} = (4-2i, 2-i, 1, i-2)$, we find by (10.6) to (10.8):

$$\begin{aligned}
\text{(Inner product of } \mathbf{A} \text{ and } \mathbf{B}) &= (-3i)(-1) + (1+i)(1+2i)\\
&\quad + (2-3i)(3-i) + (1-2i)i = 4-4i.
\end{aligned}$$

$$\begin{aligned}(\text{Norm of } \mathbf{A})^2 &= (-3i)(3i) + (1+i)(1-i) + (2-3i)(2+3i) + (1-2i)(1+2i)\\ &= 9 + 2 + 13 + 5 = 29, \quad \|\mathbf{A}\| = \sqrt{29}.\\ (\text{Norm of } \mathbf{B})^2 &= 1 + 5 + 10 + 1 = 17, \quad \|\mathbf{B}\| = \sqrt{17}.\end{aligned}$$

Note that $|4-4i| = 4\sqrt{2} < \sqrt{29}\sqrt{17}$ in accord with the Schwarz inequality (10.9).

$$\begin{aligned}(\text{Inner product of } \mathbf{B} \text{ and } \mathbf{C}) &= (-1)(4-2i) + (1-2i)(2-i) + (3+i)(1)\\ &\quad + (-i)(i-2) = -4 + 2i - 5i + 3 + i + 1 + 2i = 0.\end{aligned}$$

Thus by (10.8), **B** and **C** are orthogonal.

PROBLEMS, SECTION 10

1. Find the distance between the points

(a) $(4, -1, 2, 7)$ and $(2, 3, 1, 9)$;

(b) $(-1, 5, -3, 2, 4)$ and $(2, 6, 2, 7, 6)$;

(c) $(5, -2, 3, 3, 1, 0)$ and $(0, 1, 5, 7, 2, 1)$.

2. For the given sets of vectors, find the dimension of the space spanned by them and a basis for this space.

(a) $(1, -1, 0, 0)$, $(0, -2, 5, 1)$, $(1, -3, 5, 1)$, $(2, -4, 5, 1)$;

(b) $(0, 1, 2, 0, 0, 4)$, $(1, 1, 3, 5, -3, 5)$, $(1, 0, 0, 5, 0, 1)$, $(-1, 1, 3, -5, -3, 3)$, $(0, 0, 1, 0, -3, 0)$;

(c) $(0, 10, -1, 1, 10)$, $(2, -2, -4, 0, -3)$, $(4, 2, 0, 4, 5)$, $(3, 2, 0, 3, 4)$, $(5, -4, 5, 6, 2)$.

3. (a) Find the cosines of the angles between pairs of vectors in Problem 2(a).

(b) Find two orthogonal vectors in Problem 2(b).

4. For each given set of basis vectors, use the Gram-Schmidt method to find an orthonormal set.

(a) $\mathbf{A} = (0, 2, 0, 0)$, $\mathbf{B} = (3, -4, 0, 0)$, $\mathbf{C} = (1, 2, 3, 4)$.

(b) $\mathbf{A} = (0, 0, 0, 7)$, $\mathbf{B} = (2, 0, 0, 5)$, $\mathbf{C} = (3, 1, 1, 4)$.

(c) $\mathbf{A} = (6, 0, 0, 0)$, $\mathbf{B} = (1, 0, 2, 0)$, $\mathbf{C} = (4, 1, 9, 2)$.

5. By (10.6) and (10.7), find the norms of **A** and **B** and the inner product of **A** and **B**, and note that the Schwarz inequality (10.9) is satisfied:

(a) $\mathbf{A} = (3+i, 1, 2-i, -5i, i+1)$, $\mathbf{B} = (2i, 4-3i, 1+i, 3i, 1)$;

(b) $\mathbf{A} = (2, 2i-3, 1+i, 5i, i-2)$, $\mathbf{B} = (5i-2, 1, 3+i, 2i, 4)$.

6. Write out the proof of the Schwarz inequality (10.9) for a complex Euclidean space. *Hint:* Follow the proof of (10.4) in (10.5), replacing the definitions of norm and inner product in (10.1) and (10.2) by the definitions in (10.6) and (10.7). Remember that norms are real and ≥ 0.

7. Show that, in n-dimensional space, any $n+1$ vectors are linearly dependent. *Hint:* See Section 8.

8. Show that two different sets of basis vectors for the same vector space must contain the same number of vectors. *Hint:* Suppose a basis for a given vector space contains n vectors. Use Problem 7 to show that there cannot be more than n vectors in a basis for this space. Conversely, if there were a correct basis with less than n vectors, what can you say about the claimed n-vector basis?

9. Write equations (10.6) to (10.9) in matrix form as discussed just after (10.9).

10. Prove that $||\mathbf{A}+\mathbf{B}|| \le ||\mathbf{A}|| + ||\mathbf{B}||$. This is called the triangle inequality; in two or three dimensions, it simply says that the length of one side of a triangle $\le$ sum of the lengths of the other 2 sides. *Hint:* To prove it in n-dimensional space, write the square of the desired inequality using (10.2) and also use the Schwarz inequality (10.4). Generalize the theorem to complex Euclidean space by using (10.7) and (10.9).

11. EIGENVALUES AND EIGENVECTORS; DIAGONALIZING MATRICES

We can give the following physical interpretation to Figure 7.2 and equations (7.5). Suppose the (x, y) plane is covered by an elastic membrane which can be stretched, shrunk, or rotated (with the origin fixed). Then any point (x, y) of the membrane becomes some point (X, Y) after the deformation, and we can say that the matrix M describes the deformation. Let us now ask whether there are any vectors such that $\mathbf{R} = \lambda\mathbf{r}$ where $\lambda =$ const. Such vectors are called *eigenvectors* (or *characteristic* vectors) of the transformation, and the values of λ are called the *eigenvalues* (or *characteristic* values) of the matrix M of the transformation.

Eigenvalues To illustrate finding eigenvalues, let's consider the transformation

$$\begin{pmatrix} X \\ Y \end{pmatrix} = \begin{pmatrix} 5 & -2 \\ -2 & 2 \end{pmatrix} \begin{pmatrix} x \\ y \end{pmatrix}. \tag{11.1}$$

The eigenvector condition $\mathbf{R} = \lambda\mathbf{r}$ is, in matrix notation,

$$\begin{pmatrix} X \\ Y \end{pmatrix} = \begin{pmatrix} 5 & -2 \\ -2 & 2 \end{pmatrix} \begin{pmatrix} x \\ y \end{pmatrix} = \lambda \begin{pmatrix} x \\ y \end{pmatrix} = \begin{pmatrix} \lambda x \\ \lambda y \end{pmatrix},$$

or written out in equation form:

$$\begin{aligned} 5x - 2y &= \lambda x, \\ -2x + 2y &= \lambda y, \end{aligned} \qquad \text{or} \qquad \begin{aligned} (5-\lambda)x - 2y &= 0, \\ -2x + (2-\lambda)y &= 0. \end{aligned} \tag{11.2}$$

These equations are homogeneous. Recall from (8.9) that a set of homogeneous equations has solutions other than $x = y = 0$ only if the determinant of the coefficients is zero. Thus we want

$$\begin{vmatrix} 5-\lambda & -2 \\ -2 & 2-\lambda \end{vmatrix} = 0. \tag{11.3}$$

This is called the *characteristic equation* of the matrix M, and the determinant in (11.3) is called the *secular determinant.*

> To obtain the characteristic equation of a matrix M, we subtract λ from the elements on the main diagonal of M, and then set the determinant of the resulting matrix equal to zero.

We solve (11.3) for λ to find the characteristic values of M:

$$\begin{aligned} &(5-\lambda)(2-\lambda) - 4 = \lambda^2 - 7\lambda + 6 = 0, \\ &\lambda = 1 \quad \text{or} \quad \lambda = 6. \end{aligned} \tag{11.4}$$

Eigenvectors Substituting the λ values from (11.4) into (11.2), we get:

(11.5) $\begin{aligned} 2x - y &= 0 \quad \text{from either of the equations (11.2) when } \lambda = 1; \\ x + 2y &= 0 \quad \text{from either of the equations (11.2) when } \lambda = 6. \end{aligned}$

We were looking for vectors $\mathbf{r} = \mathbf{i}x + \mathbf{j}y$ such that the transformation (11.1) would give an $\mathbf{R}$ parallel to $\mathbf{r}$. What we have found is that *any* vector $\mathbf{r}$ with x and y components satisfying either of the equations (11.5) has this property. Since equations (11.5) are equations of straight lines through the origin, such vectors lie along these lines (Figure 11.1). Then equations (11.5) show that any vector $\mathbf{r}$ from the origin to a point on $x + 2y = 0$ is changed by the transformation (11.1) to a vector in the same direction but six times as long, and any vector from the origin to a point on $2x - y = 0$ is unchanged by the transformation (11.1). These vectors (along $x + 2y = 0$ and $2x - y = 0$) are the eigenvectors of the transformation. Along these two directions (and only these), the deformation of the elastic membrane was a pure stretch with no shear (rotation).

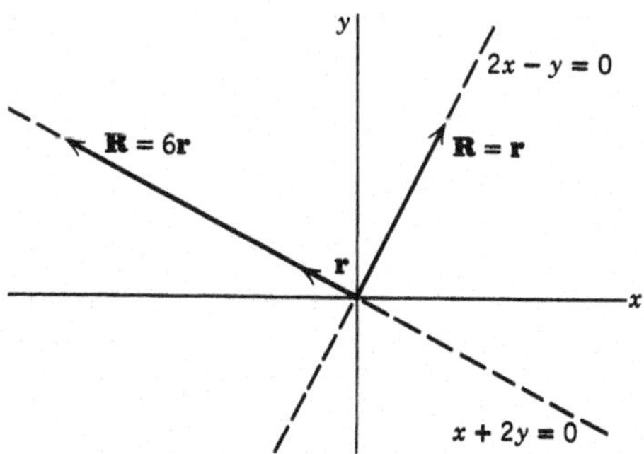

Figure 11.1

Diagonalizing a Matrix We next write (11.2) once with $\lambda = 1$, and again with $\lambda = 6$, using subscripts 1 and 2 to identify the corresponding eigenvectors:

(11.6) $$\begin{aligned} 5x_1 - 2y_1 &= x_1, & 5x_2 - 2y_2 &= 6x_2, \\ -2x_1 + 2y_1 &= y_1, & -2x_2 + 2y_2 &= 6y_2. \end{aligned}$$

These four equations can be written as one matrix equation, as you can easily verify by multiplying out both sides (Problem 1):

(11.7) $$\begin{pmatrix} 5 & -2 \\ -2 & 2 \end{pmatrix} \begin{pmatrix} x_1 & x_2 \\ y_1 & y_2 \end{pmatrix} = \begin{pmatrix} x_1 & x_2 \\ y_1 & y_2 \end{pmatrix} \begin{pmatrix} 1 & 0 \\ 0 & 6 \end{pmatrix}.$$

All we really can say about (x_1, y_1) is that $2x_1 - y_1 = 0$; however, it is convenient to pick numerical values of x_1 and y_1 to make $\mathbf{r}_1 = (x_1, y_1)$ a unit vector, and similarly for $\mathbf{r}_2 = (x_2, y_2)$. Then we have

(11.8) $$x_1 = \frac{1}{\sqrt{5}}, \quad y_1 = \frac{2}{\sqrt{5}}, \quad x_2 = \frac{-2}{\sqrt{5}}, \quad y_2 = \frac{1}{\sqrt{5}},$$

and (11.7) becomes

$$\begin{pmatrix} 5 & -2 \\ -2 & 2 \end{pmatrix} \begin{pmatrix} \frac{1}{\sqrt{5}} & \frac{-2}{\sqrt{5}} \\ \frac{2}{\sqrt{5}} & \frac{1}{\sqrt{5}} \end{pmatrix} = \begin{pmatrix} \frac{1}{\sqrt{5}} & \frac{-2}{\sqrt{5}} \\ \frac{2}{\sqrt{5}} & \frac{1}{\sqrt{5}} \end{pmatrix} \begin{pmatrix} 1 & 0 \\ 0 & 6 \end{pmatrix}. \tag{11.9}$$

Representing these matrices by letters we can write

$$\mathrm{MC} = \mathrm{CD}, \qquad \text{where}$$

$$\mathrm{M} = \begin{pmatrix} 5 & -2 \\ -2 & 2 \end{pmatrix}, \quad \mathrm{C} = \begin{pmatrix} \frac{1}{\sqrt{5}} & \frac{-2}{\sqrt{5}} \\ \frac{2}{\sqrt{5}} & \frac{1}{\sqrt{5}} \end{pmatrix}, \quad \mathrm{D} = \begin{pmatrix} 1 & 0 \\ 0 & 6 \end{pmatrix}. \tag{11.10}$$

If, as here, the determinant of C is not zero, then C has an inverse C^{-1}; let us multiply (11.10) by C^{-1} and remember that $\mathrm{C}^{-1}\mathrm{C}$ is the unit matrix; then $\mathrm{C}^{-1}\mathrm{MC} = \mathrm{C}^{-1}\mathrm{CD} = \mathrm{D}$.

$$\mathrm{C}^{-1}\mathrm{MC} = \mathrm{D}. \tag{11.11}$$

The matrix D has elements different from zero only down the main diagonal; it is called a *diagonal matrix.* The matrix D is called *similar* to M, and when we obtain D given M, we say that we have *diagonalized* M *by a similarity transformation.*

We shall see shortly that this amounts physically to a simplification of the problem by a better choice of variables. For example, in the problem of the membrane, it is simpler to describe the deformation if we use axes along the eigenvectors. Later we shall see more examples of the use of the diagonalization process.

Observe that it is easy to find D; we need only solve the characteristic equation of M. Then D is a matrix with these characteristic values down the main diagonal and zeros elsewhere. We can also find C (with more work), but for many purposes only D is needed.

Note that the order of the eigenvalues down the main diagonal of D is arbitrary; for example we could write (11.6) as

$$\begin{pmatrix} 5 & -2 \\ -2 & 2 \end{pmatrix} \begin{pmatrix} x_2 & x_1 \\ y_2 & y_1 \end{pmatrix} = \begin{pmatrix} x_2 & x_1 \\ y_2 & y_1 \end{pmatrix} \begin{pmatrix} 6 & 0 \\ 0 & 1 \end{pmatrix} \tag{11.12}$$

instead of (11.7). Then (11.11) still holds, with a different C, of course, and with

$$\mathrm{D} = \begin{pmatrix} 6 & 0 \\ 0 & 1 \end{pmatrix}$$

instead of as in (11.10) (Problem 1).

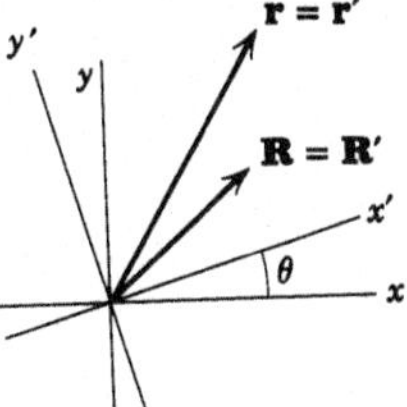

Figure 11.2

Meaning of C and D To see more clearly the meaning of (11.11) let us find what the matrices C and D mean physically. We consider two sets of axes (x, y) and (x', y') with (x', y') rotated through θ from (x, y) (Figure 11.2). The (x, y) and (x', y') coordinates of *one* point (or components of one vector $\mathbf{r} = \mathbf{r}'$) relative to the two systems are related by (7.13). Solving (7.13) for x and y, we have

$$\begin{aligned} x &= x' \cos\theta - y' \sin\theta, \\ y &= x' \sin\theta + y' \cos\theta, \end{aligned} \tag{11.13}$$

or in matrix notation

$$\mathrm{r} = \mathrm{Cr}' \qquad \text{where} \qquad \mathrm{C} = \begin{pmatrix} \cos\theta & -\sin\theta \\ \sin\theta & \cos\theta \end{pmatrix}. \tag{11.14}$$

This equation is true for *any* single vector with components given in the two systems. Suppose we have another vector $\mathbf{R} = \mathbf{R}'$ (Figure 11.2) with components X, Y and X', Y'; these components are related by

$$\mathrm{R} = \mathrm{CR}'. \tag{11.15}$$

Now let M be a matrix which describes a deformation of the plane in the (x, y) system. Then the equation

$$\mathrm{R} = \mathrm{Mr} \tag{11.16}$$

says that the vector $\mathbf{r}$ becomes the vector $\mathbf{R}$ after the deformation, both vectors given relative to the (x, y) axes. Let us ask how we can describe the deformation in the (x', y') system, that is, what matrix carries $\mathbf{r}'$ into $\mathbf{R}'$? We substitute (11.14) and (11.15) into (11.16) and find $\mathrm{CR}' = \mathrm{MCr}'$ or

$$\mathrm{R}' = \mathrm{C}^{-1}\mathrm{MCr}'. \tag{11.17}$$

Thus the answer to our question is that

> $\mathrm{D} = \mathrm{C}^{-1}\mathrm{MC}$ is the matrix which describes in the (x', y') system the same deformation that M describes in the (x, y) system.

Next we want to show that if the matrix C is chosen to make $\mathrm{D} = \mathrm{C}^{-1}\mathrm{MC}$ a diagonal matrix, then the new axes (x', y') are along the directions of the eigenvectors of M. Recall from (11.10) that the columns of C are the components of the unit eigenvectors. If the eigenvectors are perpendicular, as they are in our example (see Problem 2) then the new axes (x', y') along the eigenvector directions are a set of perpendicular axes rotated

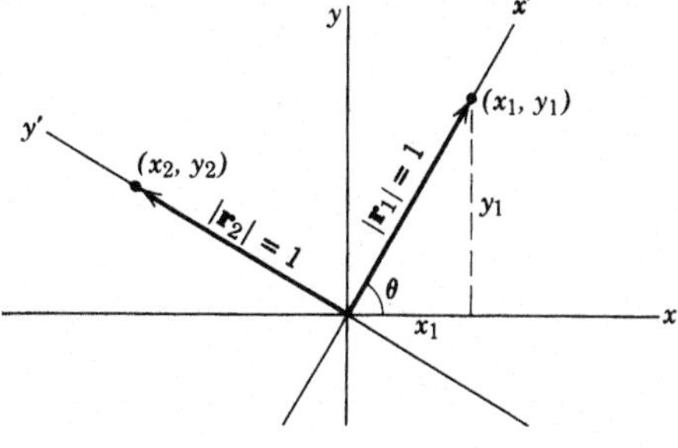

Figure 11.3

from axes (x, y) by some angle θ (Figure 11.3). The unit eigenvectors $\mathbf{r}_1$ and $\mathbf{r}_2$ are shown in Figure 11.3; from the figure we find

$$\begin{aligned} &x_1 = |\mathbf{r}_1|\cos\theta = \cos\theta, \quad x_2 = -|\mathbf{r}_2|\sin\theta = -\sin\theta \\ &y_1 = |\mathbf{r}_1|\sin\theta = \sin\theta, \quad y_2 = |\mathbf{r}_2|\cos\theta = \cos\theta; \\ &\mathrm{C} = \begin{pmatrix} x_1 & x_2 \\ y_1 & y_2 \end{pmatrix} = \begin{pmatrix} \cos\theta & -\sin\theta \\ \sin\theta & \cos\theta \end{pmatrix}. \end{aligned} \tag{11.18}$$

Thus, the matrix C which diagonalizes M is the rotation matrix C in (11.14) when the (x', y') axes are along the directions of the eigenvectors of M.

Relative to these new axes, the diagonal matrix D describes the deformation. For our example we have

$$\begin{aligned} &\mathrm{R}' = \mathrm{D}\mathrm{r}' \quad \text{or} \quad \begin{pmatrix} X' \\ Y' \end{pmatrix} = \begin{pmatrix} 1 & 0 \\ 0 & 6 \end{pmatrix} \begin{pmatrix} x' \\ y' \end{pmatrix} \quad \text{or} \\ &X' = x', \qquad Y' = 6y'. \end{aligned} \tag{11.19}$$

In words, (11.19) says that [in the (x', y') system] each point (x', y') has its x' coordinate unchanged by the deformation and its y' coordinate multiplied by 6, that is, the deformation is simply a stretch in the y' direction. This is a simpler description of the deformation and clearer physically than the description given by (11.1).

You can see now why the order of eigenvalues down the main diagonal in D is arbitrary and why (11.12) is just as satisfactory as (11.7). The new axes (x', y') are along the eigenvectors, but it is unimportant which eigenvector we call x' and which we call y'. In doing a problem we simply select a D with the eigenvalues of M in some (arbitrary) order down the main diagonal. Our choice of D then determines which eigenvector direction is called the x' axis and which is called y'.

It was unnecessary in the above discussion to have the x' and y' axes perpendicular, although this is the most useful case. If $\mathrm{r} = \mathrm{C}\mathrm{r}'$ but C is just any (nonsingular) matrix [not necessarily the orthogonal rotation matrix as in (11.14)], then (11.17) still follows. That is, $\mathrm{C}^{-1}\mathrm{M}\mathrm{C}$ describes the deformation using (x', y') axes. But if C is not an orthogonal matrix, then the (x', y') axes are not perpendicular (Figure 11.4) and $x^2 + y^2 \neq x'^2 + y'^2$, that is, the transformation is not a rotation of axes. Recall that C is the matrix of unit eigenvectors; if these are perpendicular, then C is an orthogonal matrix (Problem 6). It can be shown that this will be the case if and only if the matrix M is symmetric. [See equation (11.27) and the discussion just before it. Also see Problems 33 to 35, and Problem 15.25.]

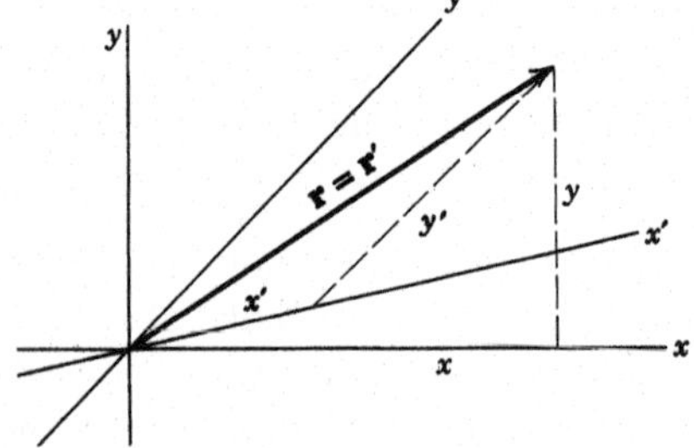

Figure 11.4

Degeneracy For a symmetric matrix, we have seen that the eigenvectors corresponding to different eigenvalues are orthogonal. If two (or more) eigenvalues are the same, then that eigenvalue is called *degenerate*. Degeneracy means that two (or more) independent eigenvectors correspond to the same eigenvalue.

Example 1. Consider the following matrix:

$$\text{(11.20)} \qquad \mathrm{M} = \begin{pmatrix} 1 & -4 & 2 \\ -4 & 1 & -2 \\ 2 & -2 & -2 \end{pmatrix}.$$

The eigenvalues of M are $\lambda = 6, -3, -3$, and the eigenvector corresponding to $\lambda = 6$ is $(2, -2, 1)$ (Problem 36). For $\lambda = -3$, the eigenvector condition is $2x - 2y + z = 0$. This is a plane orthogonal to the $\lambda = 6$ eigenvector, and any vector in this plane is an eigenvector corresponding to $\lambda = -3$. That is, the $\lambda = -3$ eigenspace is a plane. It is convenient to choose two orthogonal eigenvectors as basis vectors in this $\lambda = -3$ eigenplane, for example $(1, 1, 0)$ and $(-1, 1, 4)$. (See Problem 36.)

You may ask how you find these orthogonal eigenvectors except by inspection. Recall that the cross product of two vectors is perpendicular to both of them. Thus in the present case we could pick one vector in the $\lambda = -3$ eigenplane and then take its cross product with the $\lambda = 6$ eigenvector. This gives a second vector in the $\lambda = -3$ eigenplane, perpendicular to the first one we picked. However, this only works in three dimensions; if we are dealing with spaces of higher dimension (see Section 10), then we need another method. Suppose we first write down just any two (different) vectors in the eigenplane not trying to make them orthogonal. Then we can use the Gram-Schmidt method (see Section 10) to find an orthogonal set. For example, in the problem above, suppose you had thought of (or your computer had given you) the vectors $\mathbf{A} = (1, 1, 0)$ and $\mathbf{B} = (-1, 0, 2)$ which are vectors in the $\lambda = -3$ eigenplane but not orthogonal to each other. Following the Gram-Schmidt method, we find

$$\mathbf{A} = (1, 1, 0), \qquad \mathbf{e} = \mathbf{A}/A = (1, 1, 0)/\sqrt{2},$$
$$\mathbf{B} - (\mathbf{e} \cdot \mathbf{B})\mathbf{e} = (-1, 0, 2) - \frac{-1}{2}(1, 1, 0) = \left(\frac{-1}{2}, \frac{1}{2}, 2\right),$$

or $(-1, 1, 4)$ as we had above. For a degenerate subspace of dimension $m > 2$, we just need to write down m linearly independent eigenvectors, and then find an orthogonal set by the Gram-Schmidt method.

Diagonalizing Hermitian Matrices We have seen how to diagonalize symmetric matrices by orthogonal similarity transformations. The complex analogue of a symmetric matrix ($\mathrm{S}^{\mathrm{T}} = \mathrm{S}$) is a Hermitian matrix ($\mathrm{H}^\dagger = \mathrm{H}$) and the complex analogue of an orthogonal matrix ($\mathrm{O}^{\mathrm{T}} = \mathrm{O}^{-1}$) is a unitary matrix ($\mathrm{U}^\dagger = \mathrm{U}^{-1}$). So let's discuss diagonalizing Hermitian matrices by unitary similarity transformations. This is of great importance in quantum mechanics.

Although Hermitian matrices may have complex off-diagonal elements, the eigenvalues of a Hermitian matrix are always real. Let's prove this. (Refer to Section 9 for definitions and theorems as needed.) Let H be a Hermitian matrix, and let r be the column matrix of a non-zero eigenvector of H corresponding to the eigenvalue λ. Then the eigenvector condition is $\mathrm{Hr} = \lambda\mathrm{r}$. We want to take the transpose conjugate (dagger) of this equation. Using the complex conjugate of equation (9.10), we get $(\mathrm{Hr})^\dagger = \mathrm{r}^\dagger\mathrm{H}^\dagger = \mathrm{r}^\dagger\mathrm{H}$ since $\mathrm{H}^\dagger = \mathrm{H}$ for a Hermitian matrix. The transpose conjugate of $\lambda\mathrm{r}$ is $\lambda^*\mathrm{r}^\dagger$ (since λ is a number, we just need to take its complex conjugate). Now we have the two equations

$$\text{(11.21)} \qquad \mathrm{Hr} = \lambda\mathrm{r} \qquad \text{and} \qquad \mathrm{r}^\dagger\mathrm{H} = \lambda^*\mathrm{r}^\dagger.$$

Multiply the first equation in (11.21) on the left [see discussion following (10.9)] by the row matrix $\mathrm{r}^\dagger$ and the second equation on the right by the column matrix r to get

$$\mathrm{r}^\dagger \mathrm{H}\mathrm{r} = \lambda \mathrm{r}^\dagger \mathrm{r} \qquad \text{and} \qquad \mathrm{r}^\dagger \mathrm{H}\mathrm{r} = \lambda^* \mathrm{r}^\dagger \mathrm{r}. \tag{11.22}$$

Subtracting the two equations we find $(\lambda - \lambda^*)\mathrm{r}^\dagger \mathrm{r} = 0$. Since we assumed $\mathrm{r} \neq 0$, we have $\lambda^* = \lambda$, that is, λ is real.

We can also show that for a Hermitian matrix the eigenvectors corresponding to two different eigenvalues are orthogonal. Start with the two eigenvector conditions,

$$\mathrm{H}\mathrm{r}_1 = \lambda_1 \mathrm{r}_1 \qquad \text{and} \qquad \mathrm{H}\mathrm{r}_2 = \lambda_2 \mathrm{r}_2. \tag{11.23}$$

From these we can show (Problem 37)

$$\mathrm{r}_1^\dagger \mathrm{H}\mathrm{r}_2 = \lambda_1 \mathrm{r}_1^\dagger \mathrm{r}_2 = \lambda_2 \mathrm{r}_1^\dagger \mathrm{r}_2, \quad \text{or} \quad (\lambda_1 - \lambda_2)\mathrm{r}_1^\dagger \mathrm{r}_2 = 0. \tag{11.24}$$

Thus if $\lambda_1 \neq \lambda_2$, then the inner product of r_1 and r_2 is zero, that is, they are orthogonal [see (10.8)].

We can also prove that if a matrix M has real eigenvalues and can be diagonalized by a unitary similarity transformation, then it is Hermitian. In symbols, we write $\mathrm{U}^{-1}\mathrm{M}\mathrm{U} = \mathrm{D}$, and find the transpose conjugate of this equation to get (Problem 38)

$$(\mathrm{U}^{-1}\mathrm{M}\mathrm{U})^\dagger = \mathrm{U}^{-1}\mathrm{M}^\dagger \mathrm{U} = \mathrm{D}^\dagger = \mathrm{D}. \tag{11.25}$$

Thus $\mathrm{U}^{-1}\mathrm{M}\mathrm{U} = \mathrm{D} = \mathrm{U}^{-1}\mathrm{M}^\dagger\mathrm{U}$, so $\mathrm{M} = \mathrm{M}^\dagger$, which says that M is Hermitian. So we have proved that

(11.26) A matrix has real eigenvalues and can be diagonalized by a unitary similarity transformation if and only if it is Hermitian.

Since a real Hermitian matrix is a symmetric matrix and a real unitary matrix is an orthogonal matrix, the corresponding statement for symmetric matrices is (Problem 39).

(11.27) A matrix has real eigenvalues and can be diagonalized by an orthogonal similarity transformation if and only if it is symmetric.

Recall from (9.2) and Problem 9.22 that normal matrices include symmetric, Hermitian, orthogonal, and unitary matrices (as well as some others). It may be useful to know the following general theorem which we state without proof [see, for example, Am. J. Phys. **52**, 513–515 (1984)].

(11.28) A matrix can be diagonalized by a unitary similarity transformation if and only if it is normal.

Example 2. To illustrate diagonalizing a Hermitian matrix by a unitary similarity transformation, we consider the matrix

$$\text{(11.29)} \qquad \mathrm{H} = \begin{pmatrix} 2 & 3-i \\ 3+i & -1 \end{pmatrix}.$$

(Verify that H is Hermitian.) We follow the same routine we used to find the eigenvalues and eigenvectors of a symmetric matrix. The eigenvalues are given by

$$(2-\lambda)(-1-\lambda) - (3+i)(3-i) = 0,$$
$$\lambda^2 - \lambda - 12 = 0, \quad \lambda = -3, 4.$$

For $\lambda = -3$, an eigenvector satisfies the equations

$$\begin{pmatrix} 5 & 3-i \\ 3+i & 2 \end{pmatrix}\begin{pmatrix} x \\ y \end{pmatrix} = 0, \quad \text{or}$$
$$5x + (3-i)y = 0, \quad (3+i)x + 2y = 0.$$

These equations are satisfied by $x = 2$, $y = (-3-i)$. A choice for the unit eigenvector is $(2, -3-i)/\sqrt{14}$. For $\lambda = 4$, we find similarly the equations

$$-2x + (3-i)y = 0, \quad (3+i)x - 5y = 0,$$

which are satisfied by $y = 2$, $x = 3-i$, so a unit eigenvector is $(3-i, 2)/\sqrt{14}$. We can verify that the two eigenvectors are orthogonal (as we proved above that they must be) by finding that their inner product [see (10.8)] is $(2, -3-i)^* \cdot (3-i, 2) = 2(3-i) + 2(-3+i) = 0$. As in (11.10) we write the unit eigenvectors as the columns of a matrix U which diagonalizes H by a similarity transformation.

$$\mathrm{U} = \frac{1}{\sqrt{14}}\begin{pmatrix} 2 & 3-i \\ -3-i & 2 \end{pmatrix}, \quad \mathrm{U}^\dagger = \frac{1}{\sqrt{14}}\begin{pmatrix} 2 & -3+i \\ 3+i & 2 \end{pmatrix}$$

You can easily verify that $\mathrm{U}^\dagger\mathrm{U}$ = the unit matrix, so $\mathrm{U}^{-1} = \mathrm{U}^\dagger$. Then (Problem 40)

$$\text{(11.30)} \qquad \mathrm{U}^{-1}\mathrm{HU} = \mathrm{U}^\dagger\mathrm{HU} = \begin{pmatrix} -3 & 0 \\ 0 & 4 \end{pmatrix},$$

that is, H is diagonalized by a unitary similarity transformation.

Orthogonal Transformations in 3 Dimensions In Section 7, we considered the active rotation and/or reflection of vectors $\mathbf{r}$ which was produced by a given 3 by 3 orthogonal matrix. Study Equations (7.18) and (7.19) carefully to see that, acting on a column vector $\mathbf{r}$, they rotate the vector by angle θ around the z axis and/or reflect it through the (x, y) plane. We would now like to see how to find the effect of more complicated orthogonal matrices. We can do this by using an orthogonal similarity transformation to write a given orthogonal matrix relative to a new coordinate system in which the rotation axis is the z axis, and/or the (x, y) plane is the reflecting plane (in vector space language, this is a change of basis). Then a comparison with (7.18) or (7.19) gives the rotation angle. Recall how we construct a C matrix so that $\mathrm{C}^{-1}\mathrm{MC}$ describes the same transformation relative to a new set of axes that M described relative to the original axes: The columns of the C matrix are the components of unit vectors along the new axes [see (11.18) and Figure 11.3].

▶ **Example 3.** Consider the following matrices.

$$(11.31)\qquad A=\frac{1}{2}\begin{pmatrix}1 & \sqrt{2} & 1\\ -\sqrt{2} & 0 & \sqrt{2}\\ 1 & -\sqrt{2} & 1\end{pmatrix},\quad B=\frac{1}{3}\begin{pmatrix}-2 & -1 & -2\\ 2 & -2 & -1\\ 1 & 2 & -2\end{pmatrix}$$

You can verify that A and B are both orthogonal, and that $\det A = 1$, $\det B = -1$ (Problem 45). Thus A is a rotation matrix while B involves a reflection (and perhaps also a rotation). For A, a vector along the rotation axis is not affected by the transformation so we find the rotation axis by solving the equation $Ar = r$. We did this in Section 7, but now you should recognize this as an eigenvector equation. We want the eigenvector corresponding to the eigenvalue 1. By hand or by computer (Problem 45) we find that the eigenvector of A corresponding to $\lambda = 1$ is $(1, 0, 1)$ or $\mathbf{i}+\mathbf{k}$; this is the rotation axis. We want the new z axis to lie along this direction, so we take the elements of the third column of matrix C to be the components of the unit vector $\mathbf{u} = (1, 0, 1)/\sqrt{2}$. For the first column (new x axis) we choose a unit vector perpendicular to the rotation axis, say $\mathbf{v} = (1, 0, -1)/\sqrt{2}$, and for the second column (new y axis), we use the cross product $\mathbf{u}\times\mathbf{v} = (0, 1, 0)$ (so that the new axes form a right-handed orthogonal triad). This gives (Problem 45)

$$(11.32)\qquad C=\frac{1}{\sqrt{2}}\begin{pmatrix}1 & 0 & 1\\ 0 & \sqrt{2} & 0\\ -1 & 0 & 1\end{pmatrix},\quad C^{-1}AC=\begin{pmatrix}0 & 1 & 0\\ -1 & 0 & 0\\ 0 & 0 & 1\end{pmatrix}.$$

Comparing this result with (7.18), we see that $\cos\theta = 0$ and $\sin\theta = -1$, so the rotation is $-90°$ around the axis $\mathbf{i}+\mathbf{k}$ (or, if you prefer, $+90°$ around $-\mathbf{i}-\mathbf{k}$).

▶ **Example 4.** For the matrix B, a vector perpendicular to the reflection plane is reversed in direction by the reflection. Thus we want to solve the equation $Br = -r$, that is, to find the eigenvector corresponding to $\lambda = -1$. You can verify (Problem 45) that this is the vector $(1, -1, 1)$ or $\mathbf{i}-\mathbf{j}+\mathbf{k}$. The reflection is through the plane $x - y + z = 0$, and the rotation (if any) is about the vector $\mathbf{i}-\mathbf{j}+\mathbf{k}$. As we did for matrix A, we construct a matrix C from this vector and two perpendicular vectors, to get (Problem 45)

$$(11.33)\qquad C=\begin{pmatrix}\frac{1}{\sqrt{6}} & \frac{1}{\sqrt{2}} & \frac{1}{\sqrt{3}}\\ \frac{-1}{\sqrt{6}} & \frac{1}{\sqrt{2}} & \frac{-1}{\sqrt{3}}\\ \frac{-2}{\sqrt{6}} & 0 & \frac{1}{\sqrt{3}}\end{pmatrix},\quad C^{-1}BC=\begin{pmatrix}\frac{-1}{2} & \frac{-\sqrt{3}}{2} & 0\\ \frac{\sqrt{3}}{2} & \frac{-1}{2} & 0\\ 0 & 0 & -1\end{pmatrix}.$$

Compare this with (7.19) to get $\cos\theta = -\frac{1}{2}$, $\sin\theta = \frac{\sqrt{3}}{2}$, so matrix B produces a rotation of $120°$ around $\mathbf{i}-\mathbf{j}+\mathbf{k}$ and a reflection through the plane $x - y + z = 0$.

You may have discovered that matrices A and B have two complex eigenvalues (see Problem 46). The corresponding eigenvectors are also complex, and we didn't use them because this would take us into complex vector space (see Section 10, and Problem 47) and our rotation and reflection problems are in ordinary real 3-dimensional space. (Note also that we did not diagonalize A and B, but just used similarity transformations to display them relative to rotated axes.) However, when all the eigenvalues of an orthogonal matrix are real (see Problem 48), then this process produces a diagonalized matrix with the eigenvalues down the main diagonal.

Example 5. Consider the matrix

$$\text{F} = \frac{1}{7}\begin{pmatrix} 2 & 6 & 3 \\ 6 & -3 & 2 \\ 3 & 2 & -6 \end{pmatrix}. \tag{11.34}$$

You can verify (Problem 49) that $\det \text{F} = 1$, that the rotation axis (eigenvector corresponding to the eigenvalue $\lambda = 1$) is $3\mathbf{i} + 2\mathbf{j} + \mathbf{k}$, and that the other two eigenvalues are $-1, -1$. Then the diagonalized F (relative to axes with the new z axis along the rotation axis) is

$$\begin{pmatrix} -1 & 0 & 0 \\ 0 & -1 & 0 \\ 0 & 0 & 1 \end{pmatrix}. \tag{11.35}$$

Comparing this with equation (7.18), we see that $\cos\theta = -1$, $\sin\theta = 0$, so F produces a rotation of 180° about $3\mathbf{i} + 2\mathbf{j} + \mathbf{k}$.

An even easier way to find the rotation angle in this problem is to use the trace of F (Problem 50). From (7.18) and (11.34) we have $2\cos\theta + 1 = -1$. Thus $\cos\theta = -1$, $\theta = 180°$ as before. This method gives $\cos\theta$ for any rotation or reflection matrix, but unless $\cos\theta = \pm 1$, we also need more information (say the value of $\sin\theta$) to determine whether θ is positive or negative.

Powers and Functions of Matrices In Section 6 we found functions of some matrices A for which it was easy to find the powers because they repeated periodically [see equations (6.15) to (6.17)]. When this doesn't happen, it isn't so easy to find powers directly (Problem 58). But it is easy to find powers of a diagonal matrix, and you can also show that (Problem 57)

$$\text{M}^n = \text{CD}^n\text{C}^{-1}, \quad \text{where} \quad \text{C}^{-1}\text{MC} = \text{D}, \quad \text{D diagonal}. \tag{11.36}$$

This result is useful not just for evaluating powers and functions of numerical matrices but also for proving theorems (Problem 60).

Example 6. We can show that if, as above, $\text{C}^{-1}\text{MC} = \text{D}$, then

$$\det e^{\text{M}} = e^{\text{Tr(M)}}. \tag{11.37}$$

As in (6.17) we define e^{M} by its power series. For each term of the series $\text{M}^n = \text{CD}^n\text{C}^{-1}$ by (11.36), so $e^{\text{M}} = \text{C}e^{\text{D}}\text{C}^{-1}$. By (6.6), the determinant of a product = the product of the determinants, and $\det \text{CC}^{-1} = 1$, so we have $\det e^{\text{M}} = \det e^{\text{D}}$. Now the matrix e^{D} is diagonal and the diagonal elements are e^{λ_i} where λ_i are the eigenvalues of M. Thus $\det e^{\text{D}} = e^{\lambda_1}e^{\lambda_2}e^{\lambda_3}\cdots = e^{\text{Tr D}}$. But by (9.13), $\text{Tr D} = \text{Tr}(\text{CC}^{-1}\text{M}) = \text{Tr M}$, so we have (11.37).

Simultaneous Diagonalization Can we diagonalize two (or more) matrices using the same similarity transformation? Sometimes we can, namely if, and only if, they commute. Let's see why this is true. Recall that the diagonalizing C matrix has columns which are mutually orthogonal unit eigenvectors of the matrix being diagonalized. Suppose we can find the same set of eigenvectors for two matrices F and G; then the same C will diagonalize both. So the problem amounts to showing how to find a common set of eigenvectors for F and G if they commute.

▶ **Example 7.** Let's start by diagonalizing F. Suppose r (a column matrix) is the eigenvector corresponding to the eigenvalue λ, that is, $\mathrm{Fr} = \lambda \mathrm{r}$. Multiply this on the left by G and use GF = FG (matrices commute) to get

$$\mathrm{GFr} = \lambda \mathrm{Gr}, \qquad \text{or} \qquad \mathrm{F(Gr)} = \lambda(\mathrm{Gr}). \tag{11.38}$$

This says that Gr is an eigenvector of F corresponding to the eigenvalue λ. If λ is not degenerate (that is if there is just one eigenvector corresponding to λ) then Gr must be the same vector as r (except maybe for length), that is, Gr is a multiple of r, or $\mathrm{Gr} = \lambda' \mathrm{r}$. This is the eigenvector equation for G; it says that r is an eigenvector of G. If all eigenvalues of F are non-degenerate, then F and G have the same set of eigenvectors, and so can be diagonalized by the same C matrix.

▶ **Example 8.** Now suppose that there are two (or more) linearly independent eigenvectors corresponding to the eigenvalue λ of F. Then every vector in the degenerate eigenspace corresponding to λ is an eigenvector of matrix F (see discussion of degeneracy above). Next consider matrix G. Corresponding to all non-degenerate F eigenvalues we already have the same set of eigenvectors for G as for F. So we just have to find the eigenvectors of G in the degenerate eigenspace of F. Since all vectors in this subspace are eigenvectors of F, we are free to choose ones which are eigenvectors of G. Thus we now have the same set of eigenvectors for both matrices, and so we can construct a C matrix which will diagonalize both F and G. For the converse, see Problem 62.

▶ PROBLEMS, SECTION 11

1. Verify (11.7). Also verify (11.12) and find the corresponding different C in (11.11). *Hint:* To find C, start with (11.12) instead of (11.7) and follow through the method of getting (11.10) from (11.7).

2. Verify that the two eigenvectors in (11.8) are perpendicular, and that C in (11.10) satisfies the condition (7.9) for an orthogonal matrix.

3. (a) If C is orthogonal and M is symmetric, show that $\mathrm{C}^{-1}\mathrm{MC}$ is symmetric.

   (b) If C is orthogonal and M antisymmetric, show that $\mathrm{C}^{-1}\mathrm{MC}$ is antisymmetric.

4. Find the inverse of the rotation matrix in (7.13); you should get C in (11.14). Replace θ by $-\theta$ in (7.13) to see that the matrix C corresponds to a rotation through $-\theta$.

5. Show that the C matrix in (11.10) does represent a rotation by finding the rotation angle. Write equations (7.13) and (11.13) for this rotation.

6. Show that if C is a matrix whose columns are the components (x_1, y_1) and (x_2, y_2) of two perpendicular vectors each of unit length, then C is an orthogonal matrix. *Hint:* Find $\mathrm{C}^{\mathrm{T}}\mathrm{C}$.

7. Generalize Problem 6 to three dimensions; to n dimensions.

8. Show that under the transformation (11.1), all points (x, y) on a given straight line through the origin go into points (X, Y) on another straight line through the origin. *Hint:* Solve (11.1) for x and y in terms of X and Y and substitute into the equation $y = mx$ to get an equation $Y = kX$, where k is a constant. *Further hint:* If $\mathrm{R} = \mathrm{Mr}$, then $\mathrm{r} = \mathrm{M}^{-1}\mathrm{R}$.

9. Show that $\det(\mathrm{C}^{-1}\mathrm{MC}) = \det \mathrm{M}$. *Hints:* See (6.6). What is the product of $\det(\mathrm{C}^{-1})$ and $\det \mathrm{C}$? Thus show that the product of the eigenvalues of M is equal to $\det \mathrm{M}$.

10. Show that $\mathrm{Tr}(\mathrm{C}^{-1}\mathrm{MC}) = \mathrm{Tr}\,\mathrm{M}$. *Hint:* See (9.13). Thus show that the sum of the eigenvalues of M is equal to $\mathrm{Tr}\,\mathrm{M}$.

11. Find the inverse of the transformation $x' = 2x - 3y$, $y' = x + y$, that is, find x, y in terms of x', y'. (*Hint:* Use matrices.) Is the transformation orthogonal?

Find the eigenvalues and eigenvectors of the following matrices. Do some problems by hand to be sure you understand what the process means. Then check your results by computer.

12. $\begin{pmatrix} 1 & 3 \\ 2 & 2 \end{pmatrix}$ **13.** $\begin{pmatrix} 2 & 2 \\ 2 & -1 \end{pmatrix}$ **14.** $\begin{pmatrix} 3 & -2 \\ -2 & 0 \end{pmatrix}$

15. $\begin{pmatrix} 2 & 3 & 0 \\ 3 & 2 & 0 \\ 0 & 0 & 1 \end{pmatrix}$ **16.** $\begin{pmatrix} 2 & 0 & 2 \\ 0 & 2 & 0 \\ 2 & 0 & -1 \end{pmatrix}$ **17.** $\begin{pmatrix} 5 & 0 & 2 \\ 0 & 3 & 0 \\ 2 & 0 & 5 \end{pmatrix}$

18. $\begin{pmatrix} -1 & 1 & 3 \\ 1 & 2 & 0 \\ 3 & 0 & 2 \end{pmatrix}$ **19.** $\begin{pmatrix} 1 & 2 & 2 \\ 2 & 3 & 0 \\ 2 & 0 & 3 \end{pmatrix}$ **20.** $\begin{pmatrix} -1 & 2 & 1 \\ 2 & 3 & 0 \\ 1 & 0 & 3 \end{pmatrix}$

21. $\begin{pmatrix} 1 & 1 & 1 \\ 1 & -1 & 1 \\ 1 & 1 & -1 \end{pmatrix}$ **22.** $\begin{pmatrix} -3 & 2 & 2 \\ 2 & 1 & 3 \\ 2 & 3 & 1 \end{pmatrix}$ **23.** $\begin{pmatrix} 13 & 4 & -2 \\ 4 & 13 & -2 \\ -2 & -2 & 10 \end{pmatrix}$

24. $\begin{pmatrix} 3 & 2 & 4 \\ 2 & 0 & 2 \\ 4 & 2 & 3 \end{pmatrix}$ **25.** $\begin{pmatrix} 1 & 1 & -1 \\ 1 & 1 & 1 \\ -1 & 1 & -1 \end{pmatrix}$ **26.** $\begin{pmatrix} 2 & 1 & 1 \\ 1 & 2 & 1 \\ 1 & 1 & 2 \end{pmatrix}$

Let each of the following matrices M describe a deformation of the (x, y) plane. For each given M find: the eigenvalues and eigenvectors of the transformation, the matrix C which diagonalizes M and specifies the rotation to new axes (x', y') along the eigenvectors, and the matrix D which gives the deformation relative to the new axes. Describe the deformation relative to the new axes.

27. $\begin{pmatrix} 2 & -1 \\ -1 & 2 \end{pmatrix}$ **28.** $\begin{pmatrix} 5 & 2 \\ 2 & 2 \end{pmatrix}$ **29.** $\begin{pmatrix} 3 & 4 \\ 4 & 9 \end{pmatrix}$

30. $\begin{pmatrix} 3 & 1 \\ 1 & 3 \end{pmatrix}$ **31.** $\begin{pmatrix} 3 & 2 \\ 2 & 3 \end{pmatrix}$ **32.** $\begin{pmatrix} 6 & -2 \\ -2 & 3 \end{pmatrix}$

33. Find the eigenvalues and eigenvectors of the real symmetric matrix

$$\mathrm{M} = \begin{pmatrix} A & H \\ H & B \end{pmatrix}.$$

Show that the eigenvalues are real and the eigenvectors are perpendicular.

34. By multiplying out $M = CDC^{-1}$ where C is the rotation matrix (11.14) and D is the diagonal matrix

$$\begin{pmatrix} \lambda_1 & 0 \\ 0 & \lambda_2 \end{pmatrix},$$

show that if M can be diagonalized by a rotation, then M is symmetric.

35. The characteristic equation for a second-order matrix M is a quadratic equation. We have considered in detail the case in which M is a real symmetric matrix and the roots of the characteristic equation (eigenvalues) are real, positive, and unequal. Discuss some other possibilities as follows:

(a) M real and symmetric, eigenvalues real, one positive and one negative. Show that the plane is reflected in one of the eigenvector lines (as well as stretched or shrunk). Consider as a simple special case

$$M = \begin{pmatrix} 1 & 0 \\ 0 & -1 \end{pmatrix}.$$

(b) M real and symmetric, eigenvalues equal (and therefore real). Show that M must be a multiple of the unit matrix. Thus show that the deformation consists of dilation or shrinkage in the radial direction (the same in all directions) with no rotation (and reflection in the origin if the root is negative).

(c) M real, *not* symmetric, eigenvalues real and not equal. Show that in this case the eigenvectors are not orthogonal. *Hint:* Find their dot product.

(d) M real, *not* symmetric, eigenvalues complex. Show that all vectors are rotated, that is, there are no (real) eigenvectors which are unchanged in direction by the transformation. Consider the characteristic equation of a rotation matrix as a special case.

36. Verify the eigenvalues and eigenvectors of matrix M in (11.20). Find some other pairs of orthogonal eigenvectors in the $\lambda = -3$ eigenplane.

37. Starting with (11.23), obtain (11.24). *Hints:* Take the transpose conjugate (dagger) of the first equation in (11.23), (remember that H is Hermitian and the λ's are real) and multiply on the right by r_2. Multiply the second equation in (11.23) on the left by $r_1^\dagger$.

38. Verify equation (11.25). *Hint:* Remember from Section 9 that the transpose conjugate (dagger) of a product of matrices is the product of the transpose conjugates in reverse order and that $U^\dagger = U^{-1}$. Also remember that we have assumed real eigenvalues, so D is a real diagonal matrix.

39. Write out the detailed proof of (11.27). *Hint:* Follow the proof of (11.26) in equations (11.21) to (11.25), replacing the Hermitian matrix H by a symmetric matrix M which is real. However, don't assume that the eigenvalues λ are real until you prove it.

40. Verify the details as indicated in diagonalizing H in (11.29).

Verify that each of the following matrices is Hermitian. Find its eigenvalues and eigenvectors, write a unitary matrix U which diagonalizes H by a similarity transformation, and show that $U^{-1}HU$ is the diagonal matrix of eigenvalues.

41. $\begin{pmatrix} 2 & i \\ -i & 2 \end{pmatrix}$

42. $\begin{pmatrix} 3 & 1-i \\ 1+i & 2 \end{pmatrix}$

43. $\begin{pmatrix} 1 & 2i \\ -2i & -2 \end{pmatrix}$

44. $\begin{pmatrix} -2 & 3+4i \\ 3-4i & -2 \end{pmatrix}$

45. Verify the details in the discussion of the matrices in (11.31).

46. We have seen that an orthogonal matrix with determinant 1 has at least one eigenvalue = 1, and an orthogonal matrix with determinant = −1 has at least one eigenvalue = −1. Show that the other two eigenvalues in both cases are $e^{i\theta}$, $e^{-i\theta}$, which, of course, includes the real values 1 (when $\theta = 0$), and −1 (when $\theta = \pi$). *Hint:* See Problem 9, and remember that rotations and reflections do not change the length of vectors so eigenvalues must have absolute value = 1.

47. Find a unitary matrix U which diagonalizes A in (11.31) and verify that $\mathrm{U}^{-1}\mathrm{AU}$ is diagonal with the eigenvalues down the main diagonal.

48. Show that an orthogonal matrix M with all real eigenvalues is symmetric. *Hints*: Method 1. When the eigenvalues are real, so are the eigenvectors, and the unitary matrix which diagonalizes M is orthogonal. Use (11.27). Method 2. From Problem 46, note that the only real eigenvalues of an orthogonal M are ±1. Thus show that $\mathrm{M} = \mathrm{M}^{-1}$. Remember that M is orthogonal to show that $\mathrm{M} = \mathrm{M}^{\mathrm{T}}$.

49. Verify the results for F in the discussion of (11.34).

50. Show that the trace of a rotation matrix equals $2\cos\theta + 1$ where θ is the rotation angle, and the trace of a reflection matrix equals $2\cos\theta - 1$. *Hint*: See equations (7.18) and (7.19), and Problem 10.

Show that each of the following matrices is orthogonal and find the rotation and/or reflection it produces as an operator acting on vectors. If a rotation, find the axis and angle; if a reflection, find the reflecting plane and the rotation, if any, about the normal to that plane.

51. $\dfrac{1}{11}\begin{pmatrix} 2 & 6 & 9 \\ 6 & 7 & -6 \\ 9 & -6 & 2 \end{pmatrix}$

52. $\dfrac{1}{2}\begin{pmatrix} -1 & -1 & \sqrt{2} \\ 1 & 1 & \sqrt{2} \\ \sqrt{2} & -\sqrt{2} & 0 \end{pmatrix}$

53. $\dfrac{1}{3}\begin{pmatrix} -1 & 2 & 2 \\ 2 & -1 & 2 \\ 2 & 2 & -1 \end{pmatrix}$

54. $\dfrac{1}{2}\begin{pmatrix} 1 & \sqrt{2} & -1 \\ \sqrt{2} & 0 & \sqrt{2} \\ 1 & -\sqrt{2} & -1 \end{pmatrix}$

55. $\dfrac{1}{9}\begin{pmatrix} -1 & 8 & 4 \\ -4 & -4 & 7 \\ -8 & 1 & -4 \end{pmatrix}$

56. $\dfrac{1}{2\sqrt{2}}\begin{pmatrix} 2 & \sqrt{2} & \sqrt{2} \\ -\sqrt{2} & 1+\sqrt{2} & 1-\sqrt{2} \\ -\sqrt{2} & 1-\sqrt{2} & 1+\sqrt{2} \end{pmatrix}$

57. Show that if D is a diagonal matrix, then D^n is the diagonal matrix with elements equal to the nth power of the elements of D. Also show that if $\mathrm{D} = \mathrm{C}^{-1}\mathrm{MC}$, then $\mathrm{D}^n = \mathrm{C}^{-1}\mathrm{M}^n\mathrm{C}$, so $\mathrm{M}^n = \mathrm{CD}^n\mathrm{C}^{-1}$. *Hint*: For $n = 2$, $(\mathrm{C}^{-1}\mathrm{MC})^2 = \mathrm{C}^{-1}\mathrm{MCC}^{-1}\mathrm{MC}$; what is CC^{-1}?

58. Note in Section 6 [see (6.15)] that, for the given matrix A, we found $\mathrm{A}^2 = -\mathrm{I}$, so it was easy to find all the powers of A. It is not usually this easy to find high powers of a matrix directly. Try it for the square matrix M in equation (11.1). Then use the method outlined in Problem 57 to find M^4, M^{10}, e^{M}.

59. Repeat the last part of Problem 58 for the matrix $\mathrm{M} = \begin{pmatrix} 3 & -1 \\ -1 & 3 \end{pmatrix}$.

60. The Caley-Hamilton theorem states that "A matrix satisfies its own characteristic equation." Verify this theorem for the matrix M in equation (11.1). *Hint*: Substitute the matrix M for λ in the characteristic equation (11.4) and verify that you have a correct matrix equation. *Further hint*: Don't do all the arithmetic. Use (11.36) to write the left side of your equation as $\mathrm{C}(\mathrm{D}^2 - 7\mathrm{D} + 6)\mathrm{C}^{-1}$ and show that the parenthesis = 0. Remember that, by definition, the eigenvalues satisfy the characteristic equation.

61. At the end of Section 9 we proved that if H is a Hermitian matrix, then the matrix $e^{i\mathrm{H}}$ is unitary. Give another proof by writing $\mathrm{H} = \mathrm{CDC}^{-1}$, remembering that now C is unitary and the eigenvalues in D are real. Show that $e^{i\mathrm{D}}$ is unitary and that $e^{i\mathrm{H}}$ is a product of three unitary matrices. See Problem 9.17d.

62. Show that if matrices F and G can be diagonalized by the same C matrix, then they commute. *Hint*: Do diagonal matrices commute?

12. APPLICATIONS OF DIAGONALIZATION

We next consider some examples of the use of the diagonalization process. A central conic section (ellipse or hyperbola) with center at the origin has the equation

$$Ax^2 + 2Hxy + By^2 = K, \tag{12.1}$$

where A, B, H and K are constants. In matrix form this can be written

$$\begin{pmatrix} x & y \end{pmatrix} \begin{pmatrix} A & H \\ H & B \end{pmatrix} \begin{pmatrix} x \\ y \end{pmatrix} = K \quad \text{or} \quad \begin{pmatrix} x & y \end{pmatrix} \mathrm{M} \begin{pmatrix} x \\ y \end{pmatrix} = K \tag{12.2}$$

if we call

$$\begin{pmatrix} A & H \\ H & B \end{pmatrix} = \mathrm{M}$$

(as you can verify by multiplying out the matrices). We want to choose the principal axes of the conic as our reference axes in order to write the equation in simpler form. Consider Figure 11.2; let the axes (x', y') be rotated by some angle θ from (x, y). Then the (x, y) and (x', y') coordinates of a point are related by (11.13) or (11.14):

$$\begin{pmatrix} x \\ y \end{pmatrix} = \begin{pmatrix} \cos\theta & -\sin\theta \\ \sin\theta & \cos\theta \end{pmatrix} \begin{pmatrix} x' \\ y' \end{pmatrix} \quad \text{or} \quad \begin{pmatrix} x \\ y \end{pmatrix} = \mathrm{C} \begin{pmatrix} x' \\ y' \end{pmatrix}. \tag{12.3}$$

By (9.11) the transpose of (12.3) is

$$\begin{aligned} &\begin{pmatrix} x & y \end{pmatrix} = \begin{pmatrix} x' & y' \end{pmatrix} \begin{pmatrix} \cos\theta & \sin\theta \\ -\sin\theta & \cos\theta \end{pmatrix} \quad \text{or} \\ &\begin{pmatrix} x & y \end{pmatrix} = \begin{pmatrix} x' & y' \end{pmatrix} \mathrm{C}^{\mathrm{T}} = \begin{pmatrix} x' & y' \end{pmatrix} \mathrm{C}^{-1} \end{aligned} \tag{12.4}$$

since C is an orthogonal matrix. Substituting (12.3) and (12.4) into (12.2), we get

$$\begin{pmatrix} x' & y' \end{pmatrix} \mathrm{C}^{-1}\mathrm{MC} \begin{pmatrix} x' \\ y' \end{pmatrix} = K. \tag{12.5}$$

If C is the matrix which diagonalizes M, then (12.5) is the equation of the conic relative to its principal axes.

Example 1. Consider the conic

$$5x^2 - 4xy + 2y^2 = 30. \tag{12.6}$$

In matrix form this can be written

$$\begin{pmatrix} x & y \end{pmatrix} \begin{pmatrix} 5 & -2 \\ -2 & 2 \end{pmatrix} \begin{pmatrix} x \\ y \end{pmatrix} = 30. \tag{12.7}$$

We have here the same matrix,

$$\mathrm{M}=\begin{pmatrix}5 & -2\\ -2 & 2\end{pmatrix},$$

whose eigenvalues we found in Section 11. In that section we found a C such that

$$\mathrm{C}^{-1}\mathrm{MC}=\mathrm{D}=\begin{pmatrix}1 & 0\\ 0 & 6\end{pmatrix}.$$

Then the equation (12.5) of the conic relative to principal axes is

$$(12.8)\qquad \begin{pmatrix}x' & y'\end{pmatrix}\begin{pmatrix}1 & 0\\ 0 & 6\end{pmatrix}\begin{pmatrix}x'\\ y'\end{pmatrix}=x'^2+6y'^2=30.$$

Observe that changing the order of 1 and 6 in D would give $6x'^2+y'^2=30$ as the new equation of the ellipse instead of (12.8). This amounts simply to interchanging the x' and y' axes.

By comparing the matrix C of the unit eigenvectors in (11.10) with the rotation matrix in (11.14), we see that the rotation angle θ (Figure 11.3) from the original axes (x, y) to the principal axes (x', y') is

$$(12.9)\qquad \theta=\arccos\frac{1}{\sqrt{5}}.$$

Notice that in writing the conic section equation in matrix form (12.2) and (12.7), we split the xy term evenly between the two nondiagonal elements of the matrix; this made M symmetric. Recall (end of Section 11) that M can be diagonalized by a similarity transformation $\mathrm{C}^{-1}\mathrm{MC}$ with C an orthogonal matrix (that is, by a rotation of axes) if and only if M is symmetric. We choose M symmetric (by splitting the xy term in half) to make our process work.

Although for simplicity we have been working in two dimensions, the same ideas apply to three (or more) dimensions (that is, three or more variables). As we have said (Section 10), although we can represent only three coordinates in physical space, it is very convenient to use the same geometrical terminology even though the number of variables is greater than three. Thus if we diagonalize a matrix of any order, we still use the terms eigenvalues, eigenvectors, principal axes, rotation to principal axes, etc.

Example 2. Rotate to principal axes the quadric surface

$$x^2+6xy-2y^2-2yz+z^2=24.$$

In matrix form this equation is

$$\begin{pmatrix}x & y & z\end{pmatrix}\begin{pmatrix}1 & 3 & 0\\ 3 & -2 & -1\\ 0 & -1 & 1\end{pmatrix}\begin{pmatrix}x\\ y\\ z\end{pmatrix}=24.$$

The characteristic equation of this matrix is

$$\begin{vmatrix}1-\lambda & 3 & 0\\ 3 & -2-\lambda & -1\\ 0 & -1 & 1-\lambda\end{vmatrix}=0=-\lambda^3+13\lambda-12$$

$$=-(\lambda-1)(\lambda+4)(\lambda-3).$$

The characteristic values are

$$\lambda = 1, \qquad \lambda = -4, \qquad \lambda = 3.$$

Relative to the principal axes (x', y', z') the quadric surface equation becomes

$$\begin{pmatrix} x' & y' & z' \end{pmatrix} \begin{pmatrix} 1 & 0 & 0 \\ 0 & -4 & 0 \\ 0 & 0 & 3 \end{pmatrix} \begin{pmatrix} x' \\ y' \\ z' \end{pmatrix} = 24$$

or

$$x'^2 - 4y'^2 + 3z'^2 = 24.$$

From this equation we can identify the quadric surface (hyperboloid of one sheet) and sketch its size and shape using (x', y', z') axes without finding their relation to the original (x, y, z) axes. However, if we do want to know the relation between the two sets of axes, we find the C matrix in the following way. Recall from Section 11 that C is the matrix whose columns are the components of the unit eigenvectors. One of the eigenvectors can be found by substituting the eigenvalue $\lambda = 1$ into the equations

$$\begin{pmatrix} 1 & 3 & 0 \\ 3 & -2 & -1 \\ 0 & -1 & 1 \end{pmatrix} \begin{pmatrix} x \\ y \\ z \end{pmatrix} = \begin{pmatrix} \lambda x \\ \lambda y \\ \lambda z \end{pmatrix}$$

and solving for x, y, z. Then $\mathbf{i}x + \mathbf{j}y + \mathbf{k}z$ is an eigenvector corresponding to $\lambda = 1$, and by dividing it by its magnitude we get a *unit* eigenvector (Problem 8). Repeating this process for each of the other values of λ, we get the following three unit eigenvectors:

$$\left(\frac{1}{\sqrt{10}}, 0, \frac{3}{\sqrt{10}}\right) \qquad \text{when} \quad \lambda = 1;$$

$$\left(\frac{-3}{\sqrt{35}}, \frac{5}{\sqrt{35}}, \frac{1}{\sqrt{35}}\right) \qquad \text{when} \quad \lambda = -4;$$

$$\left(\frac{-3}{\sqrt{14}}, \frac{-2}{\sqrt{14}}, \frac{1}{\sqrt{14}}\right) \qquad \text{when} \quad \lambda = 3.$$

Then the rotation matrix C is

$$\mathrm{C} = \begin{pmatrix} \frac{1}{\sqrt{10}} & \frac{-3}{\sqrt{35}} & \frac{-3}{\sqrt{14}} \\ 0 & \frac{5}{\sqrt{35}} & \frac{-2}{\sqrt{14}} \\ \frac{3}{\sqrt{10}} & \frac{1}{\sqrt{35}} & \frac{1}{\sqrt{14}} \end{pmatrix}$$

The numbers in C are the cosines of the nine angles between the (x, y, z) and (x', y', z') axes. (Compare Figure 11.3 and the discussion of it.)

A useful physical application of this method occurs in discussing vibrations. We illustrate this with a simple problem.

Example 3. Find the characteristic vibration frequencies for the system of masses and springs shown in Figure 12.1.

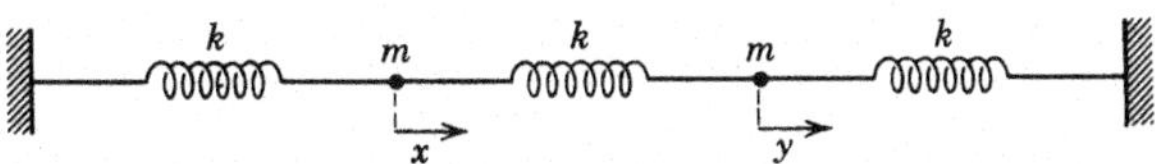

Figure 12.1

Let x and y be the coordinates of the two masses at time t relative to their equilibrium positions, as shown in Figure 12.1. We want to write the equations of motion (mass times acceleration = force) for the two masses (see Chapter 2, end of Section 16). We *can* just write the forces by inspection as we did in Chapter 2, but for more complicated problems it is useful to have a systematic method. First write the potential energy; for a spring this is $V = \frac{1}{2}ky^2$ where y is the compression or extension of the spring from its equilibrium length. Then the force exerted on a mass attached to the spring is $-ky = -dV/dy$. If V is a function of two (or more) variables, say x and y as in Figure 12.1, then the forces on the two masses are $-\partial V/\partial x$ and $-\partial V/\partial y$ (and so on for more variables). For Figure 12.1, the extension or compression of the middle spring is $x - y$ so its potential energy is $\frac{1}{2}k(x-y)^2$. For the other two springs, the potential energies are $\frac{1}{2}kx^2$ and $\frac{1}{2}ky^2$ so the total potential energy is

$$V = \frac{1}{2}kx^2 + \frac{1}{2}k(x-y)^2 + \frac{1}{2}ky^2 = k(x^2 - xy + y^2). \tag{12.10}$$

In writing the equations of motion it is convenient to use a dot to indicate a time derivative (as we often use a prime to mean an x derivative). Thus $\dot{x} = dx/dt$, $\ddot{x} = d^2x/dt^2$, etc. Then the equations of motion are

$$\begin{cases} m\ddot{x} = -\partial V/\partial x = -2kx + ky, \\ m\ddot{y} = -\partial V/\partial y = kx - 2ky. \end{cases} \tag{12.11}$$

In a *normal* or *characteristic* mode of vibration, the x and y vibrations have the same frequency. As in Chapter 2, equations (16.22), we assume solutions $x = x_0e^{i\omega t}$, $y = y_0e^{i\omega t}$, with the same frequency ω for both x and y. [Or, if you prefer, we could replace $e^{i\omega t}$ by $\sin\omega t$ or $\cos\omega t$ or $\sin(\omega t + \alpha)$, etc.] Note that (for any of these solutions),

$$\ddot{x} = -\omega^2 x, \quad \text{and} \quad \ddot{y} = -\omega^2 y. \tag{12.12}$$

Substituting (12.12) into (12.11) we get (Problem 10)

$$\begin{cases} -m\omega^2 x = -2kx + ky, \\ -m\omega^2 y = kx - 2ky. \end{cases} \tag{12.13}$$

In matrix form these equations are

$$\lambda\begin{pmatrix} x \\ y \end{pmatrix} = \begin{pmatrix} 2 & -1 \\ -1 & 2 \end{pmatrix}\begin{pmatrix} x \\ y \end{pmatrix} \quad \text{with} \quad \lambda = \frac{m\omega^2}{k}. \tag{12.14}$$

Note that this is an eigenvalue problem (see Section 11). To find the eigenvalues λ, we write

$$\begin{vmatrix} 2-\lambda & -1 \\ -1 & 2-\lambda \end{vmatrix} = 0 \tag{12.15}$$

and solve for λ to find $\lambda = 1$ or $\lambda = 3$. Thus [by the definition of λ in (12.14)] the characteristic frequencies are

$$\omega_1 = \sqrt{\frac{k}{m}} \quad \text{and} \quad \omega_2 = \sqrt{\frac{3k}{m}}. \tag{12.16}$$

The eigenvectors (not normalized) corresponding to these eigenvalues are:

$$\text{For } \lambda = 1\text{: } y = x \text{ or } \mathbf{r} = (1, 1)\text{; for } \lambda = 3\text{: } y = -x \text{ or } \mathbf{r} = (1, -1). \tag{12.17}$$

Thus at frequency ω_1 (with $y = x$), the two masses oscillate back and forth together like this $\rightarrow\rightarrow$ and then like this $\leftarrow\leftarrow$. At frequency ω_2 (with $y = -x$), they oscillate in opposite directions like this $\longleftarrow\!\!\longrightarrow$ and then like this $\rightarrow\leftarrow$. These two especially simple ways in which the system can vibrate, each involving just one vibration frequency, are called the characteristic (or normal) modes of vibration; the corresponding frequencies are called the characteristic (or normal) frequencies of the system.

The problem we have just done shows an important method which can be used in many different applications. There are numerous examples of vibration problems in physics—in acoustics: the vibrations of strings of musical instruments, of drumheads, of the air in organ pipes or in a room; in mechanics and its engineering applications: vibrations of mechanical systems all the way from the simple pendulum to complicated structures like bridges and airplanes; in electricity: the vibrations of radio waves, of electric currents and voltages as in a tuned radio; and so on. In such problems, it is often useful to find the characteristic vibration frequencies of the system under consideration and the characteristic modes of vibration. More complicated vibrations can then be discussed as combinations of these simpler normal modes of vibration.

▶ **Example 4.** In Example 3 and Figure 12.1, the two masses were equal and all the spring constants were the same. Changing the spring constants to different values doesn't cause any problems but when the masses are different, there is a possible difficulty which we want to discuss. Consider an array of masses and springs as in Figure 12.1 but with the following masses and spring constants: $2k$, $2m$, $6k$, $3m$, $3k$. We want to find the characteristic frequencies and modes of vibration. Following our work in Example 3, we write the potential energy V, find the forces, write the equations of motion, and substitute $\ddot{x} = -\omega^2 x$, and $\ddot{y} = -\omega^2 y$, in order to find the characteristic frequencies. (*Do the details*: Problem 11.)

$$V = \frac{1}{2}2kx^2 + \frac{1}{2}6k(x-y)^2 + \frac{1}{2}3ky^2 = \frac{1}{2}k(8x^2 - 12xy + 9y^2) \tag{12.18}$$

$$\begin{cases} 2m\ddot{x} = -\partial V/\partial x, \\ 3m\ddot{y} = -\partial V/\partial y, \end{cases} \quad \text{or} \quad \begin{cases} -2m\omega^2 x = -k(8x - 6y), \\ -3m\omega^2 y = -k(-6x + 9y). \end{cases} \tag{12.19}$$

Next divide each equation by its mass and write the equations in matrix form.

$$\omega^2 \begin{pmatrix} x \\ y \end{pmatrix} = \frac{k}{m} \begin{pmatrix} 4 & -3 \\ -2 & 3 \end{pmatrix} \begin{pmatrix} x \\ y \end{pmatrix}. \tag{12.20}$$

With $\lambda = m\omega^2/k$, the eigenvalues of the square matrix are $\lambda = 1$ and $\lambda = 6$. Thus the characteristic frequencies of vibration are

$$\omega_1 = \sqrt{\frac{k}{m}} \quad \text{and} \quad \omega_2 = \sqrt{\frac{6k}{m}}. \tag{12.21}$$

The corresponding eigenvectors are:

(12.22) For $\lambda = 1$: $y = x$ or $\mathbf{r} = (1, 1)$; for $\lambda = 6$: $3y = -2x$ or $\mathbf{r} = (3, -2)$.

Thus at frequency ω_1 the two masses oscillate back and forth together with equal amplitudes like this $\leftarrow\leftarrow$ and then like this $\rightarrow\rightarrow$. At frequency ω_2 the two masses oscillate in opposite directions with amplitudes in the ratio 3 to 2 like this $\longleftarrow\longrightarrow$ and then like this $\rightarrow\leftarrow$.

Now we seem to have solved the problem; where is the difficulty? Note that the square matrix in (12.20) is not symmetric [and compare (12.14) where the square matrix was symmetric]. In Section 11 we discussed the fact that (for real matrices) only symmetric matrices have orthogonal eigenvectors and can be diagonalized by an orthogonal transformation. Here note that the eigenvectors in Example 3 were orthogonal [dot product of $(1, 1)$ and $(1, -1)$ is zero] but the eigenvectors for (12.20) are not orthogonal [dot product of $(1, 1)$ and $(3, -2)$ is not zero]. If we want orthogonal eigenvectors, we can make the change of variables (also see Example 6)

$$X = x\sqrt{2}, \quad Y = y\sqrt{3}, \tag{12.23}$$

where the constants are the square roots of the numerical factors in the masses $2m$ and $3m$. (Note that geometrically this just amounts to different changes in scale along the two axes, not to a rotation.) Then (12.20) becomes

$$\omega^2 \begin{pmatrix} X \\ Y \end{pmatrix} = \frac{k}{m} \begin{pmatrix} 4 & -\sqrt{6} \\ -\sqrt{6} & 3 \end{pmatrix} \begin{pmatrix} X \\ Y \end{pmatrix}. \tag{12.24}$$

By inspection we see that the characteristic equation for the square matrix in (12.24) is the same as the characteristic equation for (12.20) so the eigenvalues and the characteristic frequencies are the same as before (as they must be by physical reasoning). However the (12.24) matrix is symmetric and so we know that its eigenvectors are orthogonal. By direct substitution of (12.23) into (12.22), [or by solving for the eigenvectors in the (12.24) matrix] we find the eigenvectors in the X, Y coordinates:

(12.25) For $\lambda = 1$: $\mathbf{R} = (X, Y) = (\sqrt{2}, -\sqrt{3})$; for $\lambda = 6$: $\mathbf{R} = (3\sqrt{2}, 2\sqrt{3})$.

As expected, these eigenvectors are orthogonal.

Example 5. Let's consider a model of a linear triatomic molecule in which we approximate the forces between the atoms by forces due to springs (Figure 12.2).

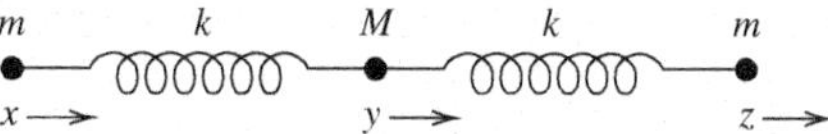

Figure 12.2

As in Example 3, let x, y, z be the coordinates of the three masses relative to their equilibrium positions. We want to find the characteristic vibration frequencies of the molecule. Following our work in Examples 3 and 4, we find (Problem 12)

$$V = \tfrac{1}{2}k(x-y)^2 + \tfrac{1}{2}k(y-z)^2 = \tfrac{1}{2}k(x^2+2y^2+z^2-2xy-2yz), \tag{12.26}$$

$$\begin{cases} m\ddot{x} = -\partial V/\partial x = -k(x-y), \\ M\ddot{y} = -\partial V/\partial y = -k(2y-x-z), \\ m\ddot{z} = -\partial V/\partial z = -k(z-y), \end{cases}$$

(12.27) or

$$\begin{cases} -m\omega^2 x = -k(x-y), \\ -M\omega^2 y = -k(2y-x-z), \\ -m\omega^2 z = -k(z-y). \end{cases}$$

We are going to consider several different ways of solving this problem in order to learn some useful techniques. First of all, if we add the three equations we get

$$m\ddot{x} + M\ddot{y} + m\ddot{z} = 0. \tag{12.28}$$

Physically (12.28) says that the center of mass is at rest or moving at constant speed (that is, has zero acceleration). Since we are just interested in vibrational motion, let's assume that the center of mass is at rest at the origin. Then we have $mx + My + mz = 0$. Solving this equation for y gives

$$y = -\frac{m}{M}(x+z). \tag{12.29}$$

Substitute (12.29) into the second set of equations in (12.27) to get the x and z equations

$$\begin{aligned} -m\omega^2 x &= -k(1+\frac{m}{M})x - k\frac{m}{M}z, \\ -m\omega^2 z &= -k\frac{m}{M}x - k(1+\frac{m}{M})z. \end{aligned} \tag{12.30}$$

In matrix form equations (12.30) become [compare (12.14)]

$$\lambda\begin{pmatrix} x \\ y \end{pmatrix} = \begin{pmatrix} 1+\frac{m}{M} & \frac{m}{M} \\ \frac{m}{M} & 1+\frac{m}{M} \end{pmatrix}\begin{pmatrix} x \\ y \end{pmatrix} \quad \text{with} \quad \lambda = \frac{m\omega^2}{k}. \tag{12.31}$$

We solve this eigenvalue problem to find

$$\omega_1 = \sqrt{\frac{k}{m}}, \qquad \omega_2 = \sqrt{\frac{k}{m}\left(1+\frac{2m}{M}\right)}. \tag{12.32}$$

For ω_1 we find $z = -x$, and consequently by (12.29), $y = 0$. For ω_2, we find $z = x$ and so $y = -\frac{2m}{M}x$. Thus at frequency ω_1, the central mass M is at rest and the two masses m vibrate in opposite directions like this $\leftarrow m \;\; M \;\; m\rightarrow$ and then like this $m\rightarrow \;\; M \;\; \leftarrow m$. At the higher frequency ω_2, the central mass M moves in one direction while the two masses m move in the opposite direction, first like this $m\rightarrow \;\; \leftarrow M \;\; m\rightarrow$ and then like this $\leftarrow m \;\; M\rightarrow \;\; \leftarrow m$.

Now suppose that we had not thought about eliminating the translational motion and had set this problem up as a 3 variable problem. Let's go back to the second set of equations in (12.27), and divide the x and z equations by m and the y equation by M. Then in matrix form these equations can be written as

$$\omega^2 \begin{pmatrix} x \\ y \\ z \end{pmatrix} = \frac{k}{m} \begin{pmatrix} 1 & -1 & 0 \\ \frac{-m}{M} & \frac{2m}{M} & \frac{-m}{M} \\ 0 & -1 & 1 \end{pmatrix} \begin{pmatrix} x \\ y \\ z \end{pmatrix}. \tag{12.33}$$

With $\lambda = m\omega^2/k$, the eigenvalues of the square matrix are $\lambda = 0, 1, 1 + \frac{2m}{M}$, and the corresponding eigenvectors are (check these)

$$\begin{aligned} &\text{For } \lambda = 0, \ \mathbf{r} = (1,1,1); \\ &\text{for } \lambda = 1, \ \mathbf{r} = (1,0,-1); \\ &\text{for } \lambda = 1 + \frac{2m}{M}, \ \mathbf{r} = (1, -\frac{2m}{M}, 1). \end{aligned} \tag{12.34}$$

We recognize the $\lambda = 0$ solution as corresponding to translation both because $\omega = 0$ (so there is no vibration), and because $\mathbf{r} = (1,1,1)$ says that any motion is the same for all three masses. The other two modes of vibration are the same ones we had above. We note that the square matrix in (12.33) is not symmetric and so, as expected, the eigenvectors in (12.34) are not an orthogonal set. However, the last two (which correspond to vibrations) are orthogonal so if we are just interested in modes of vibration we can ignore the translation eigenvector. If we want to consider all motion of the molecule along its axis (both translation and vibration), and want an orthogonal set of eigenvectors, we can make the change of variables discussed in Example 4, namely

$$X = x, \qquad Y = y\sqrt{\frac{M}{m}}, \qquad Z = z. \tag{12.35}$$

Then the eigenvectors become

$$(1, \sqrt{M/m}, 1), \quad (1, 0, -1), \quad (1, -2\sqrt{m/M}, 1) \tag{12.36}$$

which are an orthogonal set. The first eigenvector (corresponding to translation) may seem confusing, looking as if the central mass M doesn't move with the others (as it must for pure translation). But remember from Example 4 that changes of variable like (12.23) and (12.35) correspond to changes of scale, so in the XYZ system we are not using the same measuring stick to find the position of the central mass as for the other two masses. Their physical displacements are actually all the same.

Example 6. Let's consider Example 4 again in order to illustrate a very compact form for the eigenvalue equation. Satisfy yourself (Problem 13) that we can write the potential energy V in (12.18) as

$$V = \frac{1}{2} k \mathrm{r}^{\mathrm{T}} \mathrm{V} \mathrm{r} \quad \text{where} \quad \mathrm{V} = \begin{pmatrix} 8 & -6 \\ -6 & 9 \end{pmatrix}, \quad \mathrm{r} = \begin{pmatrix} x \\ y \end{pmatrix}, \quad \mathrm{r}^{\mathrm{T}} = \begin{pmatrix} x & y \end{pmatrix}. \tag{12.37}$$

Similarly the kinetic energy $T = \frac{1}{2}(2m\dot{x}^2 + 3m\dot{y}^2)$ can be written as

$$T = \frac{1}{2} m \dot{\mathrm{r}}^{\mathrm{T}} \mathrm{T} \dot{\mathrm{r}} \quad \text{where} \quad \mathrm{T} = \begin{pmatrix} 2 & 0 \\ 0 & 3 \end{pmatrix}, \quad \dot{\mathrm{r}} = \begin{pmatrix} \dot{x} \\ \dot{y} \end{pmatrix}, \quad \dot{\mathrm{r}}^{\mathrm{T}} = \begin{pmatrix} \dot{x} & \dot{y} \end{pmatrix}. \tag{12.38}$$

(Notice that the T matrix is diagonal and is a unit matrix when the masses are equal; otherwise T has the mass factors along the main diagonal and zeros elsewhere.) Now using the matrices T and V, we can write the equations of motion (12.19) as

$$m\omega^2 \begin{pmatrix} 2 & 0 \\ 0 & 3 \end{pmatrix} \begin{pmatrix} x \\ y \end{pmatrix} = k \begin{pmatrix} 8 & -6 \\ -6 & 9 \end{pmatrix} \begin{pmatrix} x \\ y \end{pmatrix} \quad \text{or}$$

$$\lambda \mathrm{Tr} = \mathrm{Vr} \quad \text{where} \quad \lambda = \frac{m\omega^2}{k}. \tag{12.39}$$

We can think of (12.39) as the basic eigenvalue equation. If T is a unit matrix, then we just have $\lambda \mathrm{r} = \mathrm{Vr}$ as in (12.14). If not, then we can multiply (12.39) by T^{-1} to get

$$\lambda \mathrm{r} = \mathrm{T}^{-1}\mathrm{Vr} = \begin{pmatrix} 1/2 & 0 \\ 0 & 1/3 \end{pmatrix} \begin{pmatrix} 8 & -6 \\ -6 & 9 \end{pmatrix} \mathrm{r} = \begin{pmatrix} 4 & -3 \\ -2 & 3 \end{pmatrix} \begin{pmatrix} x \\ y \end{pmatrix} \tag{12.40}$$

as in (12.20). However, we see that this matrix is not symmetric and so the eigenvectors will not be orthogonal. If we want the eigenvectors to be orthogonal as in (12.23), we choose new variables so that the T matrix is the unit matrix, that is variables X and Y so that

$$T = \tfrac{1}{2}(2m\dot{x}^2 + 3m\dot{y}^2) = \tfrac{1}{2}m(\dot{X}^2 + \dot{Y}^2). \tag{12.41}$$

But this means that we want $X^2 = 2x^2$ and $Y^2 = 3y^2$ as in (12.23), or in matrix form,

$$\mathrm{R} = \begin{pmatrix} X \\ Y \end{pmatrix} = \begin{pmatrix} x\sqrt{2} \\ y\sqrt{3} \end{pmatrix} = \begin{pmatrix} \sqrt{2} & 0 \\ 0 & \sqrt{3} \end{pmatrix} \begin{pmatrix} x \\ y \end{pmatrix} = \mathrm{T}^{1/2}\mathrm{r} \quad \text{or}$$

$$\mathrm{r} = \mathrm{T}^{-1/2}\mathrm{R} = \begin{pmatrix} 1/\sqrt{2} & 0 \\ 0 & 1/\sqrt{3} \end{pmatrix} \begin{pmatrix} X \\ Y \end{pmatrix}. \tag{12.42}$$

Substituting (12.42) into (12.39), we get $\lambda \mathrm{TT}^{-1/2}\mathrm{R} = \mathrm{VT}^{-1/2}\mathrm{R}$. Then multiplying on the left by $\mathrm{T}^{-1/2}$ and noting that $\mathrm{T}^{-1/2}\mathrm{TT}^{-1/2} = \mathrm{I}$, we have

$$\lambda \mathrm{R} = \mathrm{T}^{-1/2}\mathrm{VT}^{-1/2}\mathrm{R} \tag{12.43}$$

as the eigenvalue equation in terms of the new variables X and Y. Substituting the numerical $\mathrm{T}^{-1/2}$ from (12.42) into (12.43) gives the result we had in (12.24).

We have simply demonstrated that (12.39) and (12.43) give compact forms of the eigenvalue equations for Example 4. However, it is straightforward to show that these equations are just a compact summary of the equations of motion for any similar vibrations problem, in any number of variables, just by writing the potential and kinetic energy matrices and comparing the equations of motion in matrix form.

▶ **Example 7.** Find the characteristic frequencies and the characteristic modes of vibration for the system of masses and springs shown in Figure 12.3, where the motion is along a vertical line.

Let's use the simplified method of Example 6 for this problem. We first write the expressions for the kinetic energy and the potential energy as in previous examples.

(Note carefully that we measure x and y from the equilibrium positions of the masses when they are hanging at rest; then the gravitational forces are already balanced and gravitational potential energy does not come into the expression for V.)

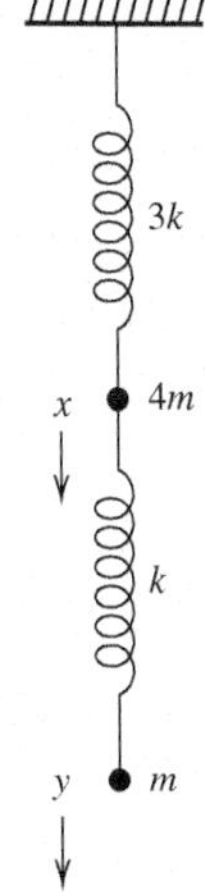

Figure 12.3

$$T = \tfrac{1}{2}m(4\dot{x}^2 + \dot{y}^2),$$
$$V = \tfrac{1}{2}k[3x^2 + (x-y)^2] = \tfrac{1}{2}k(4x^2 - 2xy + y^2). \tag{12.44}$$

The corresponding matrices are [see equations (12.37) and (12.38)]:

$$\mathrm{T} = \begin{pmatrix} 4 & 0 \\ 0 & 1 \end{pmatrix}, \qquad \mathrm{V} = \begin{pmatrix} 4 & -1 \\ -1 & 1 \end{pmatrix}. \tag{12.45}$$

As in equation (12.40), we find $\mathrm{T}^{-1}\mathrm{V}$ and its eigenvalues and eigenvectors.

$$\mathrm{T}^{-1}\mathrm{V} = \begin{pmatrix} 1/4 & 0 \\ 0 & 1 \end{pmatrix}\begin{pmatrix} 4 & -1 \\ -1 & 1 \end{pmatrix} = \begin{pmatrix} 1 & -1/4 \\ -1 & 1 \end{pmatrix}, \quad \lambda = \frac{m\omega^2}{k} = \frac{1}{2}, \frac{3}{2}.$$

$$\text{For } \omega = \sqrt{\frac{k}{2m}}, \ \mathbf{r} = (1,2); \quad \text{for } \omega = \sqrt{\frac{3k}{2m}}, \ \mathbf{r} = (1,-2). \tag{12.46}$$

As expected (since $\mathrm{T}^{-1}\mathrm{V}$ is not symmetric), the eigenvectors are not orthogonal. If we want orthogonal eigenvectors, we make the change of variables $X = 2x$, $Y = y$, to find the eigenvectors $\mathbf{R} = (1,1)$ and $\mathbf{R} = (1,-1)$ which are orthogonal. Alternatively, we can find the matrix $\mathrm{T}^{-1/2}\mathrm{V}\mathrm{T}^{-1/2}$

$$\begin{pmatrix} 1/2 & 0 \\ 0 & 1 \end{pmatrix}\begin{pmatrix} 4 & -1 \\ -1 & 1 \end{pmatrix}\begin{pmatrix} 1/2 & 0 \\ 0 & 1 \end{pmatrix} = \begin{pmatrix} 1 & -1/2 \\ -1/2 & 1 \end{pmatrix}, \tag{12.47}$$

and find its eigenvalues and eigenvectors.

PROBLEMS, SECTION 12

1. Verify that (12.2) multiplied out is (12.1).

Find the equations of the following conics and quadric surfaces relative to principal axes.

2. $2x^2 + 4xy - y^2 = 24$

3. $8x^2 + 8xy + 2y^2 = 35$

4. $3x^2 + 8xy - 3y^2 = 8$

5. $5x^2 + 3y^2 + 2z^2 + 4xz = 14$

6. $x^2 + y^2 + z^2 + 4xy + 2xz - 2yz = 12$

7. $x^2 + 3y^2 + 3z^2 + 4xy + 4xz = 60$

8. Carry through the details of Example 2 to find the unit eigenvectors. Show that the resulting rotation matrix C is orthogonal. *Hint:* Find CC^{T}.

9. For Problems 2 to 7, find the rotation matrix C which relates the principal axes and the original axes. See Example 2.

10. Verify equations (12.13) and (12.14). Solve (12.15) to find the eigenvalues and verify (12.16). Find the corresponding eigenvectors as stated in (12.17).

11. Verify the details of Example 4, equations (12.18) to (12.25).

12. Verify the details of Example 5, equations (12.26) to (12.36).

13. Verify the details of Example 6, equations (12.37) to (12.43).

Find the characteristic frequencies and the characteristic modes of vibration for systems of masses and springs as in Figure 12.1 and Examples 3, 4, and 6 for the following arrays.

14. $k, m, 2k, m, k$

15. $5k, m, 2k, m, 2k$

16. $4k, m, 2k, m, k$

17. $3k, 3m, 2k, 4m, 2k$

18. $2k, m, k, 5m, 10k$

19. $4k, 2m, k, m, k$

20. Carry through the details of Example 7.

Find the characteristic frequencies and the characteristic modes of vibration as in Example 7 for the following arrays of masses and springs, reading from top to bottom in a diagram like Figure 12.3.

21. $3k, m, 2k, m$

22. $4k, 3m, k, m$

23. $2k, 4m, k, 2m$

13. A BRIEF INTRODUCTION TO GROUPS

We will not go very far into group theory—there are whole books on the subject as well as on its applications in physics. But since so many of the ideas we are discussing in this chapter are involved, it is interesting to have a quick look at groups.

Example 1. Think about the four numbers ± 1, $\pm i$. Notice that no matter what products and powers of them we compute, we never get any numbers besides these four. This property of a set of elements with a law of combination is called *closure*. Now think about these numbers written in polar form: $e^{i\pi/2}$, $e^{i\pi}$, $e^{3i\pi/2}$, $e^{2i\pi} = 1$, or the corresponding rotations of a vector (in the xy plane with tail at the origin), or the set of rotation matrices corresponding to these successive $90°$ rotations of a vector (Problem 1). Note also that these numbers are the four fourth roots of 1, so we could write them as A, A^2, A^3, $A^4 = 1$. All these sets are examples of groups, or more precisely, they are all *representations* of the same group known as the *cyclic group of order* 4. We will be particularly interested in groups of matrices, that is, in matrix representations of groups, since this is very important in applications. Now just what is a group?

Definition of a Group A group is a set $\{A, B, C, \cdots\}$ of elements—which may be numbers, matrices, operations (such as the rotations above)—together with a law of combination of two elements (often called the "product" and written as AB—see discussion below) subject to the following four conditions.

1. Closure: The combination of any two elements is an element of the group.

2. Associative law: The law of combination satisfies the associative law: $(AB)C = A(BC)$.

3. Unit element: There is a unit element I with the property that $IA = AI = A$ for every element of the group.

4. Inverses: Every element of the group has an inverse in the group; that is, for any element A there is an element B such that $AB = BA = I$.

We can easily verify that these four conditions are satisfied for the set ± 1, $\pm i$ under multiplication.

1. We have already discussed closure.

2. Multiplication of numbers is associative.

3. The unit element is 1.

4. The numbers i and $-i$ are inverses since their product is 1; -1 is its own inverse, and 1 is its own inverse.

Thus the set ± 1, $\pm i$, under the operation of multiplication, is a group. The *order of a finite group* is the number of elements in the group. When the elements of a group of order n are of the form A, A^2, A^3, $\cdots$, $A^n = 1$, it is called a *cyclic group*. Thus the group ± 1, $\pm i$, under multiplication, is a cyclic group of order 4 as we claimed above.

A *subgroup* is a subset which is itself a group. The whole group, or the unit element, are called *trivial subgroups*; any other subgroup is called a *proper subgroup*. The group ± 1, $\pm i$ has the proper subgroup ± 1.

Product, Multiplication Table In the definition of a group and in the discussion so far, we have used the term "product" and have written AB for the combination of two elements. However, terms like "product" or "multiplication" are used here in a generalized sense to refer to whatever the operation is for combining group elements. In applications, group elements are often matrices and the operation is matrix multiplication. In general mathematical group theory, the operation might be, for example, addition of two elements, and that sounds confusing to say "product" when we mean sum! Look at one of the first examples we discussed, namely the rotation of a vector by angles $\pi/2$, π, $3\pi/2$, 2π or 0. If the group elements are rotation matrices, then we multiply them, but if the group elements are the angles, then we add them. But the physical problem is exactly the same in both cases. So remember that group multiplication refers to the law of combination for the group rather than just to ordinary multiplication in arithmetic.

Multiplication tables for groups are very useful; equations (13.1), (13.2), and (13.4) show some examples. Look at (13.1) for the group ± 1, $\pm i$. The first column and the top row (set off by lines) list the group elements. The sixteen possible products of these elements are in the body of the table. Note that each element of the group appears exactly once in each row and in each column (Problem 3). At the intersection of the row starting with i and the column headed by $-i$, you find the product $(i)(-i) = 1$, and similarly for the other products.

(13.1)

	1	i	-1	$-i$
1	1	i	-1	$-i$
i	i	-1	$-i$	1
-1	-1	$-i$	1	i
$-i$	$-i$	1	i	-1

In (13.2) below, note that you add the angles as we discussed above. However, it's not quite just adding—it's really the familiar process of adding angles until you get to 2π and then starting over again at zero. In mathematical language this is called adding (mod 2π) and we write $\pi/2 + 3\pi/2 \equiv 0 \pmod{2\pi}$. Hours on an ordinary clock add in a similar way. If it's 10 o'clock and then 4 hours elapse, the clock says it's 2 o'clock. We write $10 + 4 \equiv 2 \pmod{12}$. (See Problems 6 and 7 for more examples.)

(13.2)

	0	$\pi/2$	π	$3\pi/2$
0	0	$\pi/2$	π	$3\pi/2$
$\pi/2$	$\pi/2$	π	$3\pi/2$	0
π	π	$3\pi/2$	0	$\pi/2$
$3\pi/2$	$3\pi/2$	0	$\pi/2$	π

Two groups are called *isomorphic* if their multiplication tables are identical except for the names we attach to the elements [compare (13.1) and (13.2)]. Thus all the 4-element groups we have discussed so far are isomorphic to each other, that is, they are really all the same group. However, there are two different groups of order 4, the cyclic group we have discussed, and another group called the 4's group (see Problem 4).

Symmetry Group of the Equilateral Triangle Consider three identical atoms at the corners of an equilateral triangle in the xy plane, with the center of the triangle at the origin as shown in Figure 13.1. What rotations and reflections of vectors in the xy plane (as in Section 7) will produce an identical array of atoms? By considering Figure 13.1, we see that there are three possible rotations: 0°, 120°, 240°, and three possible reflections, through the three lines F, G, H (lines along the altitudes of the triangle). Think of moving just the triangle (that is, the atoms), leaving the axes and the lines F, G, H fixed in the background. As in Section 7, we can write a 2 by 2 rotation or reflection matrix for each of these six transformations and set up a multiplication table to show that they do form a group of order 6. This group is called the symmetry group of the equilateral triangle. We find (Problem 8)

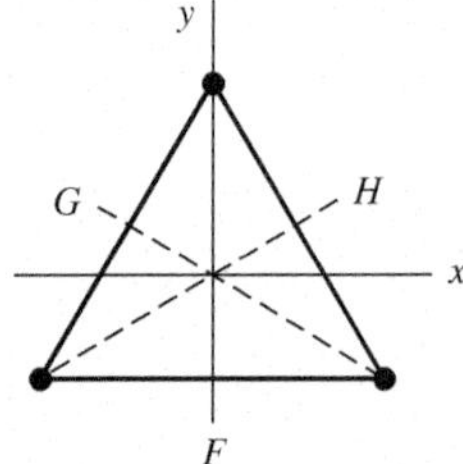

Figure 13.1

(13.3)

Identity, 0° rotation $\quad \mathrm{I} = \begin{pmatrix} 1 & 0 \\ 0 & 1 \end{pmatrix}$

120° rotation $\quad \mathrm{A} = \frac{1}{2}\begin{pmatrix} -1 & -\sqrt{3} \\ \sqrt{3} & -1 \end{pmatrix}$

240° rotation $\quad \mathrm{B} = \frac{1}{2}\begin{pmatrix} -1 & \sqrt{3} \\ -\sqrt{3} & -1 \end{pmatrix}$

Reflection through line F (y axis) $\quad \mathrm{F} = \begin{pmatrix} -1 & 0 \\ 0 & 1 \end{pmatrix}$

Reflection through line G $\quad \mathrm{G} = \frac{1}{2}\begin{pmatrix} 1 & -\sqrt{3} \\ -\sqrt{3} & -1 \end{pmatrix}$

Reflection through line H $\quad \mathrm{H} = \frac{1}{2}\begin{pmatrix} 1 & \sqrt{3} \\ \sqrt{3} & -1 \end{pmatrix}$

The group multiplication table is:

(13.4)

	I	A	B	F	G	H
I	I	A	B	F	G	H
A	A	B	I	G	H	F
B	B	I	A	H	F	G
F	F	H	G	I	B	A
G	G	F	H	A	I	B
H	H	G	F	B	A	I

Note here that GF = A, but FG = B, not surprising since we know that matrices don't always commute. In group theory, if every two group elements commute, the group is called *Abelian*. Our previous group examples have all been Abelian, but the group in (13.4) is not Abelian.

This is just one example of a symmetry group. Group theory is so important in applications because it offers a systematic way of using the symmetry of a physical problem to simplify the solution. As we have seen, groups can be represented by sets of matrices, and this is widely used in applications.

Conjugate Elements, Class, Character Two group elements A and B are called *conjugate* elements if there is a group element C such that $C^{-1}AC = B$. By letting C be successively one group element after another, we can find all the group elements conjugate to A. This set of conjugate elements is called a *class*. Recall from Section 11 that if A is a matrix describing a transformation (such as a rotation or some sort of mapping of a space onto itself), then $B = C^{-1}AC$ describes the same mapping but relative to a different set of axes (different basis). Thus all the elements of a class really describe the same mapping, just relative to different bases.

Example 2. Find the classes for the group in (13.3) and (13.4). We find the elements conjugate to F as follows [use (13.4) to find inverses and products]:

$$
\begin{aligned}
&I^{-1}FI = F;\\
&A^{-1}FA = BFA = BH = G;\\
&B^{-1}FB = AFB = AG = H;\\
&F^{-1}FF = F;\\
&G^{-1}FG = GFG = GB = H;\\
&H^{-1}FH = HFH = HA = G.
\end{aligned}
\tag{13.5}
$$

Thus the elements F, G, and H are conjugate to each other and form one class. You can easily show (Problem 12) that elements A and B are another class, and the unit element I is a class by itself. Now notice what we observed above. The elements F, G, and H all just interchange two atoms, that is, all of them do the same thing, just seen from a different viewpoint. The elements A and B rotate the atoms, A by 120° and B by 240° which is the same as 120° looked at upside down. And finally the unit element I leaves things unchanged so it is a class by itself. Notice that a class is not a group (except for the class consisting of I) since a group must contain the unit element. So a class is a subset of a group, but not a subgroup.

Recall from (9.13) and Problem 11.10 that the trace of a matrix (sum of diagonal elements) is not changed by a similarity transformation. Thus all the matrices of a class have the same trace. Observe that this is true for the group (13.3): Matrix I has trace = 2, A and B have trace $= -\frac{1}{2} - \frac{1}{2} = -1$, and F, G, and H have trace = 0. In this connection, the trace of a matrix is called its *character*, so we see that all matrices of a class have the same character. Also note that we could write the matrices (13.3) in (infinitely) many other ways by rotating the reference axes, that is, by performing similarity transformations. But since similarity transformations do not change the trace, that is, the character, we now have a number attached to each class which is independent of the particular choice of coordinate system (basis). Classes and their associated character are very important in applications of group theory.

One more number is important here, and that is the dimension of a representation. In (13.3), we used 2 by 2 matrices (2 dimensions), but it would be possible to work in 3 dimensions. Then, for example, the A matrix would describe a 120° rotation around the z axis and would be

$$\text{(13.6)} \qquad \mathrm{A} = \begin{pmatrix} -1/2 & -\sqrt{3}/2 & 0 \\ \sqrt{3}/2 & -1/2 & 0 \\ 0 & 0 & 1 \end{pmatrix},$$

and the other matrices in (13.3) would have a similar form, called *block diagonalized.* But now the traces of all the matrices are increased by 1. To avoid having any ambiguity about character, we use what are called "irreducible representations" in finding character; let's discuss this.

Irreducible Representations A 2-dimensional representation is called *reducible* if all the group matrices can be diagonalized by the same unitary similarity transformation (that is, the same change of basis). For example, the matrices in Problem 1 and the matrices in Problem 4 both give 2-dimensional reducible representations of their groups (see Problems 13, 15, and 16). On the other hand, the matrices in (13.3) cannot be simultaneously diagonalized (see Problem 13), so (13.3) is called a 2-dimensional *irreducible representation* of the equilateral triangle symmetry group. If a group of 3 by 3 matrices can all be either diagonalized or put in the form of (13.6) (block diagonalized) by the same unitary similarity transformation, then the representation is called reducible; if not, it is a 3-dimensional irreducible representation. For still larger matrices, imagine the matrices block diagonalized with blocks along the main diagonal which are the matrices of irreducible representations.

Thus we see that any representation is made up of irreducible representations. For each irreducible representation, we find the character of each class. Such lists are known as character tables, but their construction is beyond our scope.

Infinite Groups Here we survey some examples of infinite groups as well as some sets which are not groups.

(13.7)

(a) The set of all integers, positive, negative, and zero, under ordinary addition, is a group. *Proof*: The sum of two integers is an integer. Ordinary addition obeys the associative law. The unit element is 0. The inverse of the integer N is $-\mathrm{N}$ since $\mathrm{N} + (-\mathrm{N}) = 0$.

(b) The same set under ordinary multiplication is not a group because 0 has no inverse. But even if we omit 0, the inverses of the other integers are fractions which are not in the set.

(c) Under ordinary multiplication, the set of all rational numbers except zero, is a group. *Proof*: The product of two rational numbers is a rational number. Ordinary multiplication is associative. The unit element is 1, and the inverse of a rational number is just its reciprocal.

Similarly, you can show that the following sets are groups under ordinary multiplication (Problem 17): All real numbers except zero, all complex numbers except zero, all complex numbers $re^{i\theta}$ with $r = 1$.

(d) Ordinary subtraction or division cannot be group operations because they don't satisfy the associative law; for example, $x-(y-z) \neq (x-y)-z$. (Problem 18.)

(e) The set of all orthogonal 2 by 2 matrices under matrix multiplication is a group called O(2). If the matrices are required to be rotation matrices, that is, have determinant +1, the set is a group called SO(2) (the S stands for special). Similarly, the following sets of matrices are groups under matrix multiplication: The set of all orthogonal 3 by 3 matrices, called O(3); its subgroup SO(3) with determinant = 1; or the corresponding sets of orthogonal matrices of any dimension n, called O(n) and SO(n). (Problem 19.)

(f) The set of all unitary n by n matrices, $n = 1, 2, 3, \cdots$, called U(n), is a group under matrix multiplication, and its subgroup SU(n) of unitary matrices with determinant = 1 is also a group. *Proof*: We have repeatedly noted that matrix multiplication is associative and that the unit matrix is the unit element of a group of matrices. So we just need to check closure and inverses. The product of two unitary matrices is unitary (see Section 9). If two matrices have determinant = 1, their product has determinant = 1 [see equation (6.6)]. The inverse of a unitary matrix is unitary (see Problem 9.25).

PROBLEMS, SECTION 13

1. Write the four rotation matrices for rotations of vectors in the xy plane through angles 90°, 180°, 270°, 360° (or 0°) [see equation (7.12)]. Verify that these 4 matrices under matrix multiplication satisfy the four group requirements and are a matrix representation of the cyclic group of order 4. Write their multiplication table and compare with Equations (13.1) and (13.2).

2. Following the text discussion of the cyclic group of order 4, and Problem 1, discuss

(a) the cyclic group of order 3 (see Chapter 2, Problem 10.32);

(b) the cyclic group of order 6.

3. Show that, in a group multiplication table, each element appears exactly once in each row and in each column. *Hint:* Suppose that an element appears twice, and show that this leads to a contradiction, namely that two elements assumed different are the same element.

4. Show that the matrices

$$\mathrm{I}=\begin{pmatrix}1&0\\0&1\end{pmatrix},\quad \mathrm{A}=\begin{pmatrix}0&1\\1&0\end{pmatrix},\quad \mathrm{B}=\begin{pmatrix}0&-1\\-1&0\end{pmatrix},\quad \mathrm{C}=\begin{pmatrix}-1&0\\0&-1\end{pmatrix},$$

under matrix multiplication, form a group. Write the group multiplication table to see that this group (called the 4's group) is not isomorphic to the cyclic group of order 4 in Problem 1. Show that the 4's group is Abelian but not cyclic.

5. Consider the group of order 4 with unit element I and other elements A, B, C, where $AB = BA = C$, and $A^2 = B^2 = I$. Write the group multiplication table and verify that it is a group. There are two groups of order 4 (discussed in Problems 1 and 4). To which is this one isomorphic? *Hint:* Compare the multiplication tables.

6. Consider the integers 0, 1, 2, 3 under addition (mod 4). Write the group "multiplication" table and show that you have a group of order 4. Is this group isomorphic to the cyclic group of order 4 or to the 4's group?

7. Consider the set of numbers 1, 3, 5, 7 with multiplication (mod 8) as the law of combination. Write the multiplication table to show that this is a group. [To multiply two numbers (mod 8), you multiply them and then take the remainder after dividing by 8. For example, $5 \times 7 = 35 \equiv 3 (\text{mod } 8)$.] Is this group isomorphic to the cyclic group of order 4 or to the 4's group?

8. Verify (13.3) and (13.4). *Hints:* For the rotation and reflection matrices, see Section 7. In checking the multiplication table, be sure you are multiplying the matrices in the right order. Remember that matrices are operators on the vectors in the plane (Section 7), and matrices may not commute. GFA means apply A, then F, then G.

9. Show that any cyclic group is Abelian. *Hint:* Does a matrix commute with itself?

10. As we did for the equilateral triangle, find the symmetry group of the square. *Hints:* Draw the square with its center at the origin and its sides parallel to the x and y axes. Find a set of eight 2 by 2 matrices (4 rotation and 4 reflection) which map the square onto itself, and write the multiplication table to show that you have a group.

11. Do Problem 10 for a rectangle. Note that now only two rotations and 2 reflections leave the rectangle unchanged. So you have a group of order 4. To which is it isomorphic, the cyclic group or the 4's group?

12. Verify (13.5) and then also show that A, B are the elements of a class, and that I is a class by itself. Show that it will always be true in any group that I is a class by itself. *Hint:* What is C^{-1}IC for any element C of a group?

13. Using the discussion of simultaneous diagonalization at the end of Section 11, show that the 2-dimensional matrices in Problems 1 and 4 are reducible representations of their groups, and the matrices in (13.5) give an irreducible representation of the equilateral triangle symmetry group. *Hint:* Look at the multiplication tables to see which matrices commute.

14. Use the multiplication table you found in Problem 10 to find the classes in the symmetry group of a square. Show that the 2 by 2 matrices you found are an irreducible representation of the group (see Problem 13), and find the character of each class for that representation. Note that it is possible for the character to be the same for two classes, but it is not possible for the character of two elements of the same class to be different.

15. By Problem 13, you know that the matrices in Problem 4 are a reducible representation of the 4's group, that is they can all be diagonalized by the same unitary similarity transformation (in this case orthogonal since the matrices are symmetric). Demonstrate this directly by finding the matrix C and diagonalizing all 4 matrices.

16. Do Problem 15 for the group of matrices you found in Problem 1. Be careful here—you are working in a complex vector space and your C matrix will be unitary but not orthogonal (see Sections 10 and 11). *Comment:* Not surprisingly, the numbers 1, i, -1, $-i$ give a 1-dimensional representation—note that a single number can be thought of as a 1-dimensional matrix.

17. Verify that the sets listed in (13.7c) are groups.

18. Show that division cannot be a group operation. *Hint:* See (13.7d).

19. Verify that the sets listed in (13.7e) are groups. *Hint:* See the proofs in (13.7f).

20. Is the set of all orthogonal 3-by-3 matrices with determinant $= -1$ a group? If so, what is the unit element?

21. Is the group SO(2) Abelian? What about SO(3)? *Hint:* See the discussion following equation (6.14).

14. GENERAL VECTOR SPACES

In this section we are going to introduce a generalization of our picture of vector spaces which is of great importance in applications. This will be merely an introduction because the ideas here will be used in many of the following chapters as you will discover. The basic idea will be to set up an outline of the requirements for 3-dimensional vector spaces (as we listed the requirements for a group), and then show that these familiar 3-dimensional vector space requirements are satisfied by sets of things like functions or matrices which we would not ordinarily think of as vectors.

Definition of a Vector Space A vector space is a set of elements $\{\mathbf{U}, \mathbf{V}, \mathbf{W}, \cdots\}$ called vectors, together with two operations: addition of vectors, and multiplication of a vector by a scalar (which for our purposes will be a real or a complex number), and subject to the following requirements:

1. Closure: The sum of any two vectors is a vector in the space.
2. Vector addition is:
   (a) commutative: $\mathbf{U} + \mathbf{V} = \mathbf{V} + \mathbf{U}$,
   (b) associative: $(\mathbf{U} + \mathbf{V}) + \mathbf{W} = \mathbf{U} + (\mathbf{V} + \mathbf{W})$.
3. (a) There is a zero vector $\mathbf{0}$ such that $\mathbf{0} + \mathbf{V} = \mathbf{V} + \mathbf{0} = \mathbf{V}$ for every element $\mathbf{V}$ in the space.
   (b) Every element $\mathbf{V}$ has an additive inverse $(-\mathbf{V})$ such that $\mathbf{V} + (-\mathbf{V}) = \mathbf{0}$.
4. Multiplication of vectors by scalars has the expected properties:
   (a) $k(\mathbf{U} + \mathbf{V}) = k\mathbf{U} + k\mathbf{V}$;
   (b) $(k_1 + k_2)\mathbf{V} = k_1\mathbf{V} + k_2\mathbf{V}$;
   (c) $(k_1k_2)\mathbf{V} = k_1(k_2\mathbf{V})$;
   (d) $0 \cdot \mathbf{V} = \mathbf{0}$, and $1 \cdot \mathbf{V} = \mathbf{V}$.

You should go over these and satisfy yourself that they are all true for ordinary two and three dimensional vector spaces. Now let's look at some examples of things we don't usually think of as vectors which, nevertheless, satisfy the above requirements.

Example 1. Consider the set of polynomials of the third degree or less, namely functions of the form $f(x) = a_0 + a_1x + a_2x^2 + a_3x^3$. Is this a vector space? If so, find a basis. What is the dimension of the space?

We go over the requirements listed above:

1. The sum of two polynomials of degree ≤ 3 is a polynomial of degree ≤ 3 and so is a member of the set.

2. Addition of algebraic expressions is commutative and associative.

3. The "zero vector" is the polynomial with all coefficients a_i equal to 0, and adding it to any other polynomial just gives that other polynomial. The additive inverse of a function $f(x)$ is just $-f(x)$, and $-f(x) + f(x) = 0$ as required for a vector space.

4. All the listed familiar rules are just what we do every time we work with algebraic expressions.

So we have a vector space! Now let's try to find a basis for it. Consider the set of functions: $\{1, x, x^2, x^3\}$. They span the space since any polynomial of degree ≤ 3 is a linear combination of them. You can easily show (Problem 1) by computing the Wronskian [equation (8.5)] that they are linearly independent. Therefore they are a basis, and since there are 4 basis vectors, the dimension of the space is 4.

Example 2. Consider the set of linear combinations of the functions

$$\{e^{ix},\ e^{-ix},\ \sin x,\ \cos x,\ x \sin x\}.$$

It is straightforward to verify that all our requirements above are met (Problem 1). To find a basis, we must find a linearly independent set of functions which spans the space. We note that the given functions are not linearly independent since e^{ix} and e^{-ix} are linear combinations of $\sin x$ and $\cos x$ (Chapter 2, Section 4). However, the set $\{\sin x,\ \cos x,\ x \sin x\}$ is a linearly independent set and it spans the space. So this is a possible basis and the dimension of the space is 3. Another possible basis would be $\{e^{ix}, e^{-ix}, x \sin x\}$. You will meet sets of functions like these as solutions of differential equations (*see* Chapter 8, Problems 5.13 to 5.18).

Example 3. Modify Example 1 to consider the set of polynomials of degree ≤ 3 with $f(1) = 1$. Is this a vector space? Suppose we add two of the polynomials; then the value of the sum at $x = 1$ is 2, so it is not an element of the set. Thus requirement 1 is not satisfied so this is not a vector space. Note that a subset of the vectors of a vector space is not necessarily a subspace. On the other hand, if we consider polynomials of degree ≤ 3 with $f(1) = 0$, then the sum of two of them is zero at $x = 1$; this is a vector space. You can easily verify (Problem 1) that it is a subspace of dimension 3 and a possible basis is $\{x - 1,\ x^2 - 1,\ x^3 - 1\}$.

Example 4. Consider the set of all polynomials of any degree $\leq N$. The sum of two polynomials of degree $\leq N$ is another such polynomial, and you can easily verify (Problem 1) that the rest of the requirements are met, so this is a vector space. A simple choice of basis is the set of powers of x from $x^0 = 1$ to x^N. Thus we see that the dimension of this space is $N + 1$.

Example 5. Consider the set of all 2 by 3 matrices with matrix addition as the law of combination, and multiplication by scalars defined as in Section 6. Recall that you add matrices by adding corresponding elements. Thus a sum of two 2 by 3 matrices is another 2 by 3 matrix. For matrix addition and multiplication by scalars, it is straightforward to show that the other requirements listed above are satisfied (Problem 1). As a basis, we could use the six matrices:

$$\begin{pmatrix} 1 & 0 & 0 \\ 0 & 0 & 0 \end{pmatrix}, \quad \begin{pmatrix} 0 & 1 & 0 \\ 0 & 0 & 0 \end{pmatrix}, \quad \begin{pmatrix} 0 & 0 & 1 \\ 0 & 0 & 0 \end{pmatrix},$$
$$\begin{pmatrix} 0 & 0 & 0 \\ 1 & 0 & 0 \end{pmatrix}, \quad \begin{pmatrix} 0 & 0 & 0 \\ 0 & 1 & 0 \end{pmatrix}, \quad \begin{pmatrix} 0 & 0 & 0 \\ 0 & 0 & 1 \end{pmatrix}.$$

Satisfy yourself that these are linearly independent and that they span the space (that is, that you could write any 2 by 3 matrix as a linear combination of these six). Since there are 6 basis vectors, the dimension of this space is 6.

Inner Product, Norm, Orthogonality The definitions of these terms need to be generalized when our "vectors" are functions, that is, we want to generalize equations (10.1) to (10.3). A natural generalization of a sum is an integral, so we might reasonably replace $\sum A_i B_i$ by $\int A(x)B(x)\,dx$, and $\sum A_i^2$ by $\int [A(x)]^2\,dx$. However, in applications we frequently want to consider complex functions of the real variable x (for example, e^{ix} as in Example 2). Thus, given functions $A(x)$ and $B(x)$ on $a \le x \le b$, we define

$$\text{(14.1)} \qquad [\text{Inner Product of } A(x) \text{ and } B(x)] = \int_a^b A^*(x)B(x)\,dx,$$

$$\text{(14.2)} \qquad [\text{Norm of } A(x)] = ||A(x)|| = \sqrt{\int_a^b A^*(x)A(x)\,dx},$$

$$\text{(14.3)} \qquad A(x) \text{ and } B(x) \text{ are orthogonal on } (a,b) \quad \text{if} \quad \int_a^b A^*(x)B(x)\,dx = 0.$$

Let's now generalize our definition (14.1) of inner product still further. Let A, B, C, $\cdots$ be elements of a vector space, and let a, b, c, $\cdots$ be scalars. We will use the bracket $\langle A|B\rangle$ to mean the inner product of A and B. This vector space is called an *inner product space* if an inner product is defined subject to the conditions:

$$\text{(14.4a)} \qquad \langle A|B\rangle^* = \langle B|A\rangle;$$
$$\text{(14.4b)} \qquad \langle A|A\rangle \ge 0, \quad \langle A|A\rangle = 0 \text{ if and only if } A = 0;$$
$$\text{(14.4c)} \qquad \langle C|aA + bB\rangle = a\langle C|A\rangle + b\langle C|B\rangle.$$

(See Problem 11.) It follows from (14.4) that (Problem 12)

$$\text{(14.5a)} \qquad \langle aA + bB|C\rangle = a^*\langle A|C\rangle + b^*\langle B|C\rangle, \quad \text{and}$$
$$\text{(14.5b)} \qquad \langle aA|bB\rangle = a^*b\langle A|B\rangle.$$

You will find various other notations for the inner product, such as (A, B) or $[A, B]$ or $\langle A, B\rangle$. The notation $\langle A|B\rangle$ is used in quantum mechanics. Most mathematics books put the complex conjugate on the second factor in (14.1) and make the corresponding changes in (14.4) and (14.5). Most physics and mathematical methods books handle the complex conjugate as we have. If you are confused by this notation and equations (14.4) and (14.5), keep going back to (14.1) where $\langle A|B\rangle = \int A^*B$ until you get used to the bracket notation. Also study carefully our use of the bracket notation in the next section and do Problems 11 to 14.

Schwarz's Inequality In Section 10 we proved the Schwarz inequality for n-dimensional Euclidean space. For an inner product space satisfying (14.4), it becomes [compare (10.9)]

$$|\langle A|B\rangle|^2 \leq \langle A|A\rangle\langle B|B\rangle. \tag{14.6}$$

To prove this, we first note that it is true if $B = 0$. For $B \neq 0$, let $C = A - \mu B$, where $\mu = \langle B|A\rangle/\langle B|B\rangle$, and find $\langle C|C\rangle$ which is ≥ 0 by (14.4b). Using (14.4) and (14.5), we write

$$\langle A - \mu B|A - \mu B\rangle = \langle A|A\rangle - \mu^*\langle B|A\rangle - \mu\langle A|B\rangle + \mu^*\mu\langle B|B\rangle \geq 0. \tag{14.7}$$

Now substitute the values of μ and μ^* to get (see Problem 13)

$$\begin{aligned} &\langle A|A\rangle - \frac{\langle A|B\rangle}{\langle B|B\rangle}\langle B|A\rangle - \frac{\langle B|A\rangle}{\langle B|B\rangle}\langle A|B\rangle + \frac{\langle A|B\rangle}{\langle B|B\rangle}\frac{\langle B|A\rangle}{\langle B|B\rangle}\langle B|B\rangle \\ &= \langle A|A\rangle - \frac{\langle A|B\rangle\langle A|B\rangle^*}{\langle B|B\rangle} = \langle A|A\rangle - \frac{|\langle A|B\rangle|^2}{\langle B|B\rangle} \geq 0 \end{aligned} \tag{14.8}$$

which gives (14.6).

For a function space as in (14.1) to (14.3), Schwarz's inequality becomes (see Problem 14):

$$\left|\int_a^b A^*(x)B(x)\,dx\right|^2 \leq \left(\int_a^b A^*(x)A(x)\,dx\right)\left(\int_a^b B^*(x)B(x)\,dx\right). \tag{14.9}$$

Orthonormal Basis; Gram-Schmidt Method Two functions are called *orthogonal* if they satisfy (14.3); a function is *normalized* if its norm in (14.2) is 1. By a combination of the two words, we call a set of functions *orthonormal* if they are all mutually orthogonal and they all have norm 1. It is often convenient to write the functions of a vector space in terms of an orthonormal basis (compare writing ordinary vectors in three dimensions in terms of **i**, **j**, **k**). Let's see how the Gram-Schmidt method applies to a vector space of functions with inner product, norm, and orthogonality defined by (14.1) to (14.3). (Compare Section 10, Example 4 and the paragraph before it.)

▶ **Example 6.** In Example 1, we found that the set of all polynomials of degree ≤ 3 is a vector space of dimension 4 with basis 1, x, x^2, x^3. Let's consider these polynomials on the interval $-1 \leq x \leq 1$ and construct an orthonormal basis. To keep track of what we're doing, let $f_0, f_1, f_2, f_3 = 1, x, x^2, x^3$; let p_0, p_1, p_2, p_3 be a corresponding orthogonal basis (which we find by the Gram-Schmidt method); and let e_0, e_1, e_2, e_3, be the orthonormal basis (which we get by normalizing the functions p_i). Recall the Gram-Schmidt routine (see Section 10, Example 4): Normalize the first function to get e_0. Then for the rest of the functions, subtract from f_i each preceding e_j multiplied by the inner product of e_j and f_i, that is, find

$$p_i = f_i - \sum_{j<i} e_j \langle e_j | f_i \rangle = f_i - \sum_{j<i} e_j \int_{-1}^{1} e_j f_i \, dx. \tag{14.10}$$

Finally, normalize p_i to get e_i.

We can save effort by noting in advance that many of the inner products we need are going to be zero. You can easily show (Problem 15) that the integral of an odd power of x from $x = -1$ to 1 is zero, and consequently any even power of x is orthogonal to any odd power. Observe that the f_i are alternately even and odd powers of x. Then you can show that the corresponding p_i and e_i will also involve just even or just odd powers of x. The Gram-Schmidt method gives the following results (Problem 16).

$$f_0 = 1 = p_0, \quad ||p_0||^2 = \int_{-1}^{1} 1^2 \, dx = 2, \quad e_0 = \frac{1}{\sqrt{2}}.$$

$f_1 = x; \quad p_1 = x$ because x is orthogonal to e_0.

$$||p_1||^2 = \int_{-1}^{1} x^2 \, dx = \frac{2}{3}, \quad e_1 = x\sqrt{\frac{3}{2}}.$$

$f_2 = x^2$. Since x^2 is orthogonal to e_1 but not to e_0,

$$p_2 = x^2 - \frac{1}{\sqrt{2}} \int_{-1}^{1} \frac{1}{\sqrt{2}} x^2 \, dx = x^2 - \frac{1}{3}.$$

$$||p_2||^2 = \int_{-1}^{1} \left(x^2 - \frac{1}{3} \right)^2 dx = \frac{8}{45}, \quad e_2 = (3x^2 - 1)\sqrt{\frac{5}{8}}.$$

$f_3 = x^3$. Since x^3 is orthogonal to e_0 and e_2,

$$p_3 = x^3 - x\sqrt{\frac{3}{2}} \int_{-1}^{1} x\sqrt{\frac{3}{2}} x^3 \, dx = x^3 - \frac{3}{5}x,$$

$$||p_3||^2 = \int_{-1}^{1} \left(x^3 - \frac{3}{5}x \right)^2 dx = \frac{8}{175}, \quad e_3 = (5x^3 - 3x)\sqrt{\frac{7}{8}}.$$

This process could be continued for a vector space with basis 1, x, x^2, $\cdots$, x^N (but it is not very efficient). The orthonormal functions e_i are well-known functions called (normalized) *Legendre polynomials.* In Chapters 12 and 13, we will discover these functions as solutions of differential equations and see their applications in physics problems.

Infinite Dimensional Spaces If a vector space does not have a finite basis, it is called an infinite dimensional vector space. It is beyond our scope to go into a detailed mathematical study of such spaces. However, you should know that, by analogy with finite dimensional vector spaces, we still use the term basis functions for sets of functions (like x^n or $\sin nx$) in terms of which we can expand suitably restricted functions in infinite series. So far we have discussed only power series (Chapter 1). In later chapters you will discover many other sets of functions which provide useful bases in applications: sines and cosines in Chapter 7, various special functions in Chapters 12 and 13. When we introduce them, we will discuss questions of convergence of the infinite series, and of completeness of sets of basis functions.

PROBLEMS, SECTION 14

1. Verify the statements indicated in Examples 1 to 5 above.

For each of the following sets, either verify (as in Example 1) that it is a vector space, or show which requirements are not satisfied. If it is a vector space, find a basis and the dimension of the space.

2. Linear combinations of the set of functions $\{e^x, \sinh x, xe^x\}$.

3. Linear combinations of the set of functions $\{x, \cos x, x\cos x, e^x\cos x, (2-3e^x)\cos x, x(1+5\cos x)\}$.

4. Polynomials of degree ≤ 3 with $a_2 = 0$.

5. Polynomials of degree ≤ 5 with $a_1 = a_3$.

6. Polynomials of degree ≤ 6 with $a_3 = 3$.

7. Polynomials of degree ≤ 7 with all the even coefficients equal to each other and all the odd coefficients equal to each other.

8. Polynomials of degree ≤ 7 but with all odd powers missing.

9. Polynomials of degree ≤ 10 but with all even powers having positive coefficients.

10. Polynomials of degree ≤ 13, but with the coefficient of each odd power equal to half the preceding coefficient of an even power.

11. Verify that the definitions in (14.1) and (14.2) satisfy the requirements for an inner product listed in (14.4) and (14.5). *Hint:* Write out all the equations (14.4) and (14.5) in the integral notation of (14.1) and (14.2).

12. Verify that the relations in (14.5) follow from (14.4). *Hints:* For (14.5a), take the complex conjugate of (14.4c). To take the complex conjugate of a bracket, use (14.4a).

13. Verify (14.7) and (14.8) *Hints:* Remember that a norm squared, like $\langle B|B\rangle$, is a real and non-negative.scalar, so its complex conjugate is just itself. But $\langle B|A\rangle$ is a complex scalar and $\langle B|A\rangle = \langle A|B\rangle^*$ by (14.4). Show that $\mu^* = \langle A|B\rangle/\langle B|B\rangle$.

14. Verify that (14.9) is (14.6) with the definition of scalar product as in (14.1).

15. For Example 6, verify the claimed orthogonality on $(-1,1)$ of an even power of x and an odd power of x. *Hint:* For example, consider $\int_{-1}^{1} x^2x^3\,dx$.

16. For Example 6, verify the details of the terms omitted in the functions p_i because of orthogonality. *Hint:* See Problem 15. Also verify the calculations of inner products and norms and the orthonormal set e_i.

15. MISCELLANEOUS PROBLEMS

1. Show that if each element of one row (or column) of a determinant is the sum of two terms, the determinant can be written as a sum of two determinants; for example,

$$\begin{vmatrix} a_{11} & a_{12}+b_{12} & a_{13} \\ a_{21} & a_{22}+b_{22} & a_{23} \\ a_{31} & a_{32}+b_{32} & a_{33} \end{vmatrix} = \begin{vmatrix} a_{11} & a_{12} & a_{13} \\ a_{21} & a_{22} & a_{23} \\ a_{31} & a_{32} & a_{33} \end{vmatrix} + \begin{vmatrix} a_{11} & b_{12} & a_{13} \\ a_{21} & b_{22} & a_{23} \\ a_{31} & b_{32} & a_{33} \end{vmatrix}.$$

Use this result to verify Fact 4b of Section 3.

2. What is wrong with the following argument? "If we add the first row of a determinant to the second row and the second row to the first row, then the first two rows of the determinant are identical, and the value of the determinant is zero. Therefore all determinants have the value zero."

3. (a) Find the equations of the line through the points $(4, -1, 2)$ and $(3, 1, 4)$.

(b) Find the equation of the plane through the points $(0, 0, 0)$, $(1, 2, 3)$ and $(2, 1, 1)$.

(c) Find the distance from the point $(1, 1, 1)$ to the plane $3x - 2y + 6z = 12$.

(d) Find the distance from the point $(1, 0, 2)$ to the line $\mathbf{r} = 2\mathbf{i}+\mathbf{j}-\mathbf{k}+(\mathbf{i}-2\mathbf{j}+2\mathbf{k})t$.

(e) Find the angle between the plane in (c) and the line in (d).

4. Given the line $\mathbf{r} = 3\mathbf{i} - \mathbf{j} + (2\mathbf{i} + \mathbf{j} - 2\mathbf{k})t$:

(a) Find the equation of the plane containing the line and the point $(2, 1, 0)$.

(b) Find the angle between the line and the (y, z) plane.

(c) Find the perpendicular distance between the line and the x axis.

(d) Find the equation of the plane through the point $(2, 1, 0)$ and perpendicular to the line.

(e) Find the equations of the line of intersection of the plane in (d) and the plane $y = 2z$.

5. (a) Write the equations of a straight line through the points $(2, 7, -1)$ and $(5, 7, 3)$.

(b) Find the equation of the plane determined by the two lines $\mathbf{r} = (\mathbf{i} - 2\mathbf{j} + \mathbf{k})t$ and $\mathbf{r} = (6\mathbf{i} - 3\mathbf{j} + 2\mathbf{k})t$.

(c) Find the angle which the line in (a) makes with the plane in (b).

(d) Find the distance from $(1, 1, 1)$ to the plane in (b).

(e) Find the distance from $(1, 6, -3)$ to the line in (a).

6. Derive the formula

$$D = \frac{|ax_0 + by_0 + cz_0 - d|}{\sqrt{a^2 + b^2 + c^2}}$$

for the distance from (x_0, y_0, z_0) to $ax + by + cz = d$.

7. Given the matrices A, B, C below, find or mark as meaningless the matrices: A^{T}, A^{-1}, AB, $\bar{\mathrm{A}}$, $\mathrm{A}^{\mathrm{T}}\mathrm{B}^{\mathrm{T}}$, $\mathrm{B}^{\mathrm{T}}\mathrm{A}^{\mathrm{T}}$, BA^{T}, ABC, $\mathrm{AB}^{\mathrm{T}}\mathrm{C}$, $\mathrm{B}^{\mathrm{T}}\mathrm{AC}$, $\mathrm{A}^{\dagger}$, $\mathrm{B}^{\mathrm{T}}\mathrm{C}$, $\mathrm{B}^{-1}\mathrm{C}$, $\mathrm{C}^{-1}\mathrm{A}$, CB^{T}.

$$\mathrm{A} = \begin{pmatrix} 1 & -1 \\ 0 & i \end{pmatrix}, \quad \mathrm{B} = \begin{pmatrix} 2 & 1 & -1 \\ 0 & 3 & 5 \end{pmatrix}, \quad \mathrm{C} = \begin{pmatrix} 0 & 1 \\ -1 & 0 \end{pmatrix}.$$

8. Given

$$\mathrm{A} = \begin{pmatrix} 1 & 0 & 2i \\ i & -3 & 0 \\ 1 & 0 & i \end{pmatrix}, \quad \text{find } \mathrm{A}^{\mathrm{T}}, \bar{\mathrm{A}}, \mathrm{A}^{\dagger}, \mathrm{A}^{-1}.$$

9. The following matrix product is used in discussing a thick lens in air:

$$\mathrm{A} = \begin{pmatrix} 1 & (n-1)/R_2 \\ 0 & 1 \end{pmatrix} \begin{pmatrix} 1 & 0 \\ d/n & 1 \end{pmatrix} \begin{pmatrix} 1 & -(n-1)/R_1 \\ 0 & 1 \end{pmatrix},$$

where d is the thickness of the lens, n is its index of refraction, and R_1 and R_2 are the radii of curvature of the lens surfaces. It can be shown that element A_{12} of A is $-1/f$ where f is the focal length of the lens. Evaluate A and det A (which should equal 1) and find $1/f$. [See Am. J. Phys. **48**, 397–399 (1980).]

10. The following matrix product is used in discussing two thin lenses in air:

$$\mathrm{M} = \begin{pmatrix} 1 & -1/f_2 \\ 0 & 1 \end{pmatrix} \begin{pmatrix} 1 & 0 \\ d & 1 \end{pmatrix} \begin{pmatrix} 1 & -1/f_1 \\ 0 & 1 \end{pmatrix},$$

where f_1 and f_2 are the focal lengths of the lenses and d is the distance between them. As in Problem 9, element M_{12} is $-1/f$ where f is the focal length of the combination. Find M, det M, and $1/f$.

11. There is a one-to-one correspondence between two-dimensional vectors and complex numbers. Show that the real and imaginary parts of the product $z_1 z_2^*$ (the star denotes complex conjugate) are respectively the scalar product and $\pm$ the magnitude of the vector product of the vectors corresponding to z_1 and z_2.

12. The vectors $\mathbf{A} = a\mathbf{i} + b\mathbf{j}$ and $\mathbf{B} = c\mathbf{i} + d\mathbf{j}$ form two sides of a parallelogram. Show that the area of the parallelogram is given by the absolute value of the following determinant. (Also see Chapter 6, Section 3.)

$$\begin{vmatrix} a & b \\ c & d \end{vmatrix}.$$

13. The plane $2x+3y+6z = 6$ intersects the coordinate axes at points P, Q, R, forming a triangle. Find the vectors $\overrightarrow{PQ}$ and $\overrightarrow{PR}$. Write a vector formula for the area of the triangle PQR, and find the area.

In Problems 14 to 17, multiply matrices to find the resultant transformation. *Caution:* Be sure you are multiplying the matrices in the right order.

14. $\begin{cases} x' = (x + y\sqrt{3})/2 \\ y' = (-x\sqrt{3} + y)/2 \end{cases}$ $\quad\begin{cases} x'' = (-x' + y'\sqrt{3})/2 \\ y'' = -(x'\sqrt{3} + y')/2 \end{cases}$

15. $\begin{cases} x' = 2x + 5y \\ y' = x + 3y \end{cases}$ $\quad\begin{cases} x'' = x' - 2y' \\ y'' = 3x' - 5y' \end{cases}$

16. $\begin{cases} x' = (x + y\sqrt{2} + z)/2 \\ y' = (x\sqrt{2} - z\sqrt{2})/2 \\ z' = (-x + y\sqrt{2} - z)/2 \end{cases}$ $\quad\begin{cases} x'' = (x'\sqrt{2} + z'\sqrt{2})/2 \\ y'' = (-x' - y'\sqrt{2} + z')/2 \\ z'' = (x' - y'\sqrt{2} - z')/2 \end{cases}$

17. $\begin{cases} x' = (2x + y + 2z)/3 \\ y' = (x + 2y - 2z)/3 \\ z' = (2x - 2y - z)/3 \end{cases}$ $\quad\begin{cases} x'' = (2x' + y' + 2z')/3 \\ y'' = (-x' - 2y' + 2z')/3 \\ z'' = (-2x' + 2y' + z')/3 \end{cases}$

Find the eigenvalues and eigenvectors of the matrices in the following problems.

18. $\begin{pmatrix} 1 & 0 \\ 3 & -2 \end{pmatrix}$ **19.** $\begin{pmatrix} 5 & 1 \\ 4 & 2 \end{pmatrix}$ **20.** $\begin{pmatrix} 5 & -4 \\ -4 & 5 \end{pmatrix}$ **21.** $\begin{pmatrix} 4 & 2 \\ 2 & 1 \end{pmatrix}$

22. $\begin{pmatrix} 3 & 0 & -2 \\ 0 & 4 & 0 \\ -2 & 0 & 3 \end{pmatrix}$ **23.** $\begin{pmatrix} 3 & 0 & 1 \\ 0 & 3 & 1 \\ 1 & 1 & 2 \end{pmatrix}$ **24.** $\begin{pmatrix} 2 & -3 & 4 \\ -3 & 2 & 0 \\ 4 & 0 & 2 \end{pmatrix}$

25. Find the C matrix which diagonalizes the matrix M of Problem 18. Observe that M is not symmetric, and C is not orthogonal (see Section 11). However, C does have an inverse; find C^{-1} and show that $\mathrm{C}^{-1}\mathrm{MC} = \mathrm{D}$.

26. Repeat Problem 25 for Problem 19.

In Problems 27 to 30, rotate the given quadric surface to principal axes. What is the name of the surface? What is the shortest distance from the origin to the surface?

27. $x^2 + y^2 - 5z^2 + 4xy = 15$

28. $7x^2 + 4y^2 + z^2 - 8xz = 36$

29. $3x^2 + 5y^2 - 3z^2 + 6yz = 54$

30. $7x^2 + 7y^2 + 7z^2 + 10xz - 24yz = 20$

31. Find the characteristic vibration frequencies of a system of masses and springs as in Figure 12.1 if the spring constants are k, $3k$, k.

32. Do Problem 31 if the spring constants are $6k$, $2k$, $3k$.

33. Prove the Caley-Hamilton theorem (Problem 11.60) for any matrix M for which $\mathrm{D} = \mathrm{C}^{-1}\mathrm{MC}$ is diagonal. See hints in Problem 11.60.

34. In problems 6.30 and 6.31, you found the matrices e^{A} and e^{C} (put $k = 1$) where A and C are the Pauli matrices from Problem 6.6. Now find the matrix $(\mathrm{A}+\mathrm{C})$ and its powers and so find the matrix $e^{\mathrm{A}+\mathrm{C}}$ to show that $e^{\mathrm{A}+\mathrm{C}} \neq e^{\mathrm{A}}e^{\mathrm{C}}$. See Problem 6.29.

35. Show that a square matrix A has an inverse if and only if $\lambda = 0$ is not an eigenvalue of A. *Hint:* Write the condition for A to have an inverse (Section 6), and the condition for A to have the eigenvalue $\lambda = 0$ (Section 11).

36. Write the three 3 by 3 matrices for 180° rotations about the x, y, z axes. Show that these three matrices commute (contrary to what we usually expect—see Problems 7.30 and 7.31). By writing the multiplication table, show that these three matrices with the unit matrix form a group. To which order 4 group is it isomorphic? *Hint:* See Problem 13.5.

37. Show that for a given irreducible representation of a group, the character of the class consisting of the identity is always the dimension of the irreducible representation. *Hint:* What is the trace of a unit n-by-n matrix?

38. For a cyclic group, show that every element is a class by itself. Show this also for an Abelian group.

CHAPTER 4

Partial Differentiation

1. INTRODUCTION AND NOTATION

If $y = f(x)$, then dy/dx can be thought of either as the slope of the curve $y = f(x)$ or as the rate of change of y with respect to x. Rates occur frequently in physics; time rates such as velocity, acceleration, and rate of cooling of a hot body are obvious examples. There are also other rates: rate of change of volume of a gas with applied pressure, rate of decrease of the fuel in your automobile tank with distance traveled, and so on. Equations involving rates (differential equations) often need to be solved in applied problems. Derivatives are also used in finding maximum and minimum points of a curve and in finding the power series of a function. All these applications, and more, occur also when we consider a function of several variables.

Let z be a function of two variables x and y; we write $z = f(x, y)$. Just as we think of $y = f(x)$ as a curve in two dimensions, so it is useful to interpret $z = f(x, y)$ geometrically. If x, y, z are rectangular coordinates, then for each x, y the equation gives us a value of z, and so determines a point (x, y, z) in three dimensions. All the points satisfying the equation ordinarily form a surface in three-dimensional space (see Figure 1.1). (It might happen that an equation would not be satisfied by any real points, for example $x^2 + y^2 + z^2 = -1$, but we shall be interested in equations whose graphs are real surfaces.) Now suppose x is constant; think of a plane $x =$ const. intersecting the surface (see Figure 1.1). The points satisfying $z = f(x, y)$ *and* $x =$ const. then lie on a curve (the curve of intersection of the surface and the $x =$ const. plane; this is AB in Figure 1.1). We might want the slope, maximum and minimum points, etc., of this curve. Since z is a function of y (on this curve), we might write dz/dy for the slope. However, to show that z is actually a function of two variables x and y with one of them (x) temporarily a constant, we write $\partial z/\partial y$; we call $\partial z/\partial y$ the partial derivative of z with respect to y. Similarly, we can

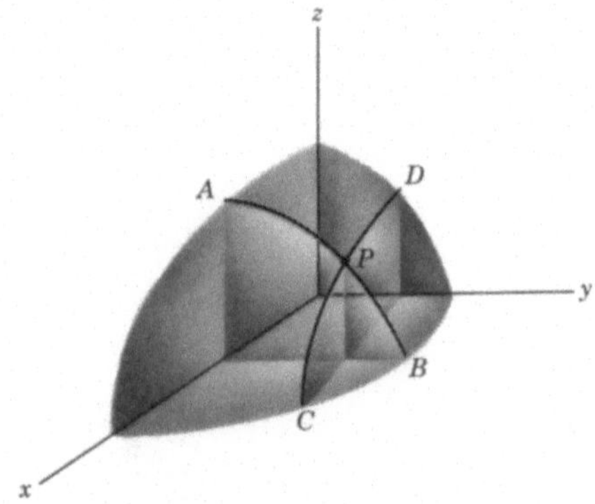

Figure 1.1

hold y constant and find $\partial z/\partial x$, the partial derivative of z with respect to x. If these partial derivatives are differentiated further, we write

$$\frac{\partial}{\partial x}\frac{\partial z}{\partial x}=\frac{\partial^2 z}{\partial x^2},\qquad \frac{\partial}{\partial x}\frac{\partial z}{\partial y}=\frac{\partial^2 z}{\partial x\partial y},\qquad \frac{\partial}{\partial x}\frac{\partial^2 z}{\partial x\partial y}=\frac{\partial^3 z}{\partial x^2\partial y},\qquad \text{etc.}$$

Other notations are often useful. If $z = f(x, y)$, we may use z_x or f_x or f_1 for $\partial f/\partial x$, and corresponding notations for the higher derivatives.

Example. Given $z = f(x, y) = x^3y - e^{xy}$, then

$$\begin{aligned}
\frac{\partial f}{\partial x} &\equiv \frac{\partial z}{\partial x} \equiv f_x \equiv z_x \equiv f_1 = 3x^2y - ye^{xy},\\
\frac{\partial f}{\partial y} &\equiv \frac{\partial z}{\partial y} \equiv f_y \equiv z_y \equiv f_2 = x^3 - xe^{xy},\\
\frac{\partial^2 f}{\partial x\,\partial y} &\equiv \frac{\partial^2 z}{\partial x\,\partial y} \equiv f_{yx} \equiv z_{yx} \equiv f_{21} = 3x^2 - e^{xy} - xye^{xy},\\
\frac{\partial^2 f}{\partial x^2} &\equiv \frac{\partial^2 z}{\partial x^2} \equiv f_{xx} \equiv z_{xx} \equiv f_{11} = 6xy - y^2e^{xy},\\
\frac{\partial^3 f}{\partial y^3} &\equiv \frac{\partial^3 z}{\partial y^3} \equiv f_{yyy} \equiv z_{yyy} \equiv f_{222} = -x^3e^{xy},\\
\frac{\partial^3 f}{\partial x^2\,\partial y} &\equiv \frac{\partial^3 z}{\partial x^2\,\partial y} \equiv f_{yxx} \equiv z_{yxx} \equiv f_{211} = 6x - 2ye^{xy} - xy^2e^{xy}.
\end{aligned}$$

We can also consider functions of more variables than two, although in this case it is not so easy to give a geometrical interpretation. For example, the temperature T of the air in a room might depend on the point (x, y, z) at which we measured it and on the time t; we would write $T = T(x, y, z, t)$. We could then find, say, $\partial T/\partial y$, meaning the rate at which T is changing with y for fixed x and z at one instant of time t.

A notation which is frequently used in applications (particularly thermodynamics) is $(\partial z/\partial x)_y$, meaning $\partial z/\partial x$ when z is expressed as a function of x and y. (Note two different uses of the subscript y; in the example above, f_y meant $\partial f/\partial y$. A subscript *on a partial derivative*, however, does *not* mean another derivative, but just indicates the variable being held constant in the indicated partial differentiation.) For example, let $z = x^2 - y^2$. Then using polar coordinates r and θ, (recall that $x = r\cos\theta$, $y = r\sin\theta$, $x^2 + y^2 = r^2$), we can write z in several other ways. For each new expression let us find $\partial z/\partial r$.

$$\begin{aligned}
&z = x^2 - y^2,\\
&z = r^2\cos^2\theta - r^2\sin^2\theta, &&\left(\frac{\partial z}{\partial r}\right)_\theta = 2r(\cos^2\theta - \sin^2\theta),\\
&z = 2x^2 - x^2 - y^2 = 2x^2 - r^2, &&\left(\frac{\partial z}{\partial r}\right)_x = -2r,\\
&z = x^2 + y^2 - 2y^2 = r^2 - 2y^2, &&\left(\frac{\partial z}{\partial r}\right)_y = +2r.
\end{aligned}$$

These three expressions for $\partial z/\partial r$ have different values and are derivatives of three different functions, so we distinguish them as indicated by writing the second independent variable as a subscript. Note that we do *not* write $z(x, y)$ or $z(r, \theta)$; z is *one* variable, but it is equal to several *different* functions. Pure mathematics books usually avoid the subscript notation by writing, say, $z = f(r,\theta) = g(r,x) = h(r,y)$, etc.; then $(\partial z/\partial r)_\theta$ can be written as just $\partial f/\partial r$, and similarly

$$\left(\frac{\partial z}{\partial r}\right)_x = \frac{\partial g}{\partial r} \quad \text{and} \quad \left(\frac{\partial z}{\partial r}\right)_y = \frac{\partial h}{\partial r}.$$

However, this multiplicity of notation ($z = f = g = h$, etc.) would be inconvenient and confusing in applications where the letters have *physical* meanings. For example, in thermodynamics, we might need

$$\left(\frac{\partial T}{\partial p}\right)_v, \quad \left(\frac{\partial T}{\partial v}\right)_s, \quad \left(\frac{\partial T}{\partial p}\right)_u, \quad \left(\frac{\partial T}{\partial s}\right)_p, \quad \text{etc.,}$$

as well as many other similar partial derivatives. Now T means temperature (and the other letters similarly have physical meanings which must be recognized). If we wrote $T = A(p,v) = B(v,s) = C(p,u) = D(s,p)$ and similar formulas for the eight commonly used quantities in thermodynamics, each as functions of pairs from the other seven, we would not only have an unwieldy system, but the physical meaning of equations would be lost until we translated them back to standard letters. Thus the subscript notation is essential.

The symbol $(\partial z/\partial r)_x$ is usually read "the partial of z with respect to r, with x held constant." However, the important point to understand is that the notation means that z has been written as a function of the variables r and x *only*, and then differentiated with respect to r.

A little experimenting with various functions $f(x,y)$ will probably convince you that $(\partial/\partial x)(\partial f/\partial y) = (\partial/\partial y)(\partial f/\partial x)$; this is usually (but not always) true in applied problems. It can be proved (see advanced calculus texts) that if the first and second order partial derivatives of f are continuous, then $\partial^2 f/\partial x\,\partial y$ and $\partial^2 f/\partial y\,\partial x$ *are* equal. In many applied problems, these conditions are met; for example, in thermodynamics they are normally assumed and are called the reciprocity relations.

PROBLEMS, SECTION 1

1. If $u = x^2/(x^2+y^2)$, find $\partial u/\partial x$, $\partial u/\partial y$.
2. If $s = t^u$, find $\partial s/\partial t$, $\partial s/\partial u$.
3. If $z = \ln\sqrt{u^2+v^2+w^2}$, find $\partial z/\partial u$, $\partial z/\partial v$, $\partial z/\partial w$.
4. For $w = x^3 - y^3 - 2xy + 6$, find $\partial^2 w/\partial x^2$ and $\partial^2 w/\partial y^2$ at the points where $\partial w/\partial x = \partial w/\partial y = 0$.
5. For $w = 8x^4 + y^4 - 2xy^2$, find $\partial^2 w/\partial x^2$ and $\partial^2 w/\partial y^2$ at the points where $\partial w/\partial x = \partial w/\partial y = 0$.
6. For $u = e^x \cos y$,

   (a) verify that $\partial^2 u/\partial x\partial y = \partial^2 u/\partial y\partial x$;

   (b) verify that $\partial^2 u/\partial x^2 + \partial^2 u/\partial y^2 = 0$.

If $z = x^2 + 2y^2$, $x = r\cos\theta$, $y = r\sin\theta$, find the following partial derivatives.

7. $\left(\frac{\partial z}{\partial x}\right)_y$ **8.** $\left(\frac{\partial z}{\partial x}\right)_r$ **9.** $\left(\frac{\partial z}{\partial x}\right)_\theta$ **10.** $\left(\frac{\partial z}{\partial y}\right)_x$ **11.** $\left(\frac{\partial z}{\partial y}\right)_r$ **12.** $\left(\frac{\partial z}{\partial y}\right)_\theta$

13. $\left(\frac{\partial z}{\partial \theta}\right)_x$ **14.** $\left(\frac{\partial z}{\partial \theta}\right)_y$ **15.** $\left(\frac{\partial z}{\partial \theta}\right)_r$ **16.** $\left(\frac{\partial z}{\partial r}\right)_\theta$ **17.** $\left(\frac{\partial z}{\partial r}\right)_x$ **18.** $\left(\frac{\partial z}{\partial r}\right)_y$

19. $\frac{\partial^2 z}{\partial r\,\partial y}$ **20.** $\frac{\partial^2 z}{\partial x\,\partial \theta}$ **21.** $\frac{\partial^2 z}{\partial y\,\partial \theta}$ **22.** $\frac{\partial^2 z}{\partial r\,\partial x}$ **23.** $\frac{\partial^2 z}{\partial r\,\partial \theta}$ **24.** $\frac{\partial^2 z}{\partial x\,\partial y}$

7′ to 24′. Repeat Problems 7 to 24 if $z = r^2\tan^2\theta$.

2. POWER SERIES IN TWO VARIABLES

Just as in the one-variable case discussed in Chapter 1, the power series (about a given point) for a function of two variables is unique, and we may use any convenient method of finding it (see Chapter 1 for methods).

Example 1. Expand $f(x, y) = \sin x \cos y$ in a two-variable Maclaurin series. We write and multiply the series for $\sin x$ and $\cos y$. This gives

$$\sin x \cos y = \left(x - \frac{x^3}{3!} + \cdots\right)\left(1 - \frac{y^2}{2!} + \cdots\right) = x - \frac{x^3}{3!} - \frac{xy^2}{2!} + \cdots.$$

Example 2. Find the two-variable Maclaurin series for $\ln(1 + x - y)$. We replace x in equation (13.4) of Chapter 1 by $x - y$ to get

$$\begin{aligned}\ln(1 + x - y) &= (x - y) - (x - y)^2/2 + (x - y)^3/3 + \cdots \\ &= x - y - x^2/2 + xy - y^2/2 + x^3/3 - x^2y + xy^2 - y^3/3 + \cdots.\end{aligned}$$

The methods of Chapter 1, used as we have just shown, provide an easy way of obtaining the power series for many simple functions $f(x, y)$. However, it is also convenient, for theoretical purposes, to have formulas for the coefficients in the Taylor series or the Maclaurin series for $f(x, y)$; see, for example, Problem 8.2. Following a process similar to that used in Chapter 1, Section 12, we can find the coefficients of the power series for a function of two variables $f(x, y)$ (assuming that it can be expanded in a power series). To find the series expansion of $f(x, y)$ about the point (a, b) we write $f(x, y)$ as a series of powers of $(x - a)$ and $(y - b)$ and then differentiate this equation repeatedly as follows.

$$\begin{aligned}f(x, y) &= a_{00} + a_{10}(x - a) + a_{01}(y - b) + a_{20}(x - a)^2 + a_{11}(x - a)(y - b) \\ &\quad + a_{02}(y - b)^2 + a_{30}(x - a)^3 + a_{21}(x - a)^2(y - b) \\ &\quad + a_{12}(x - a)(y - b)^2 + a_{03}(y - b)^3 + \cdots. \\ f_x &= a_{10} + 2a_{20}(x - a) + a_{11}(y - b) + \cdots, \\ f_y &= a_{01} + a_{11}(x - a) + 2a_{02}(y - b) + \cdots, \\ f_{xx} &= 2a_{20} + \text{ terms containing } (x - a) \text{ and/or } (y - b), \\ f_{xy} &= a_{11} + \text{ terms containing } (x - a) \text{ and/or } (y - b).\end{aligned} \tag{2.1}$$

[We have written only a few derivatives to show the idea. You should be able to calculate others in the same way (Problem 7).] Now putting $x = a$, $y = b$ in (2.1), we get

$$(2.2)\qquad \begin{aligned} &f(a,b)=a_{00}, \quad f_x(a,b)=a_{10}, \quad f_y(a,b)=a_{01},\\ &f_{xx}(a,b)=2a_{20}, \quad f_{xy}(a,b)=a_{11}, \quad \text{etc.} \end{aligned}$$

[Remember that $f_x(a,b)$ means that we are to find the partial derivative of f with respect to x and then put $x=a$, $y=b$, and similarly for the other derivatives.] Substituting the values for the coefficients into (2.1), we find

(2.3)

$$\begin{aligned} f(x,y) &= f(a,b)+f_x(a,b)(x-a)+f_y(a,b)(y-b)\\ &+\frac{1}{2!}[f_{xx}(a,b)(x-a)^2+2f_{xy}(a,b)(x-a)(y-b)+f_{yy}(a,b)(y-b)^2]\cdots. \end{aligned}$$

This can be written in a simpler form if we put $x-a=h$ and $y-b=k$. Then the second-order terms (for example) become

$$(2.4)\qquad \frac{1}{2!}[f_{xx}(a,b)h^2+2f_{xy}(a,b)hk+f_{yy}(a,b)k^2].$$

We can write this in the form

$$(2.5)\qquad \frac{1}{2!}\left(h\frac{\partial}{\partial x}+k\frac{\partial}{\partial y}\right)^2 f(a,b)$$

if we understand that the parenthesis is to be squared and then a term of the form $h(\partial/\partial x)k(\partial/\partial y)f(a,b)$ is to mean $hkf_{xy}(a,b)$. It can be shown (Problem 7) that the third-order terms can be written in this notation as

$$(2.6)\qquad \frac{1}{3!}\left(h\frac{\partial}{\partial x}+k\frac{\partial}{\partial y}\right)^3 f(a,b)=\frac{1}{3!}[h^3f_{xxx}(a,b)+3h^2kf_{xxy}(a,b)+\cdots]$$

and so on for terms of any order. Thus we can write the series (2.3) in the form

$$(2.7)\qquad f(x,y)=\sum_{n=0}^{\infty}\frac{1}{n!}\left(h\frac{\partial}{\partial x}+k\frac{\partial}{\partial y}\right)^n f(a,b).$$

The numbers appearing in the nth order terms are the familiar binomial coefficients [in the expansion of $(p+q)^n$] divided by $(n!)$. (See Chapter 1, Section 13C.)

PROBLEMS, SECTION 2

Find the two-variable Maclaurin series for the following functions.

1. $\cos x \sinh y$ **2.** $\cos(x+y)$ **3.** $\dfrac{\ln(1+x)}{1+y}$

4. e^{xy} **5.** $\sqrt{1+xy}$ **6.** e^{x+y}

7. Verify the coefficients of the third-order terms [(2.6) or $n=3$ in (2.7)] of the power series for $f(x,y)$ by finding the third-order partial derivatives in (2.1) and substituting $x=a$, $y=b$.

8. Find the two-variable Maclaurin series for $e^x\cos y$ and $e^x\sin y$ by finding the series for $e^z=e^{x+iy}$ and taking real and imaginary parts. (See Chapter 2.)

3. TOTAL DIFFERENTIALS

The graph (Figure 3.1) of the equation $y = f(x)$ is a curve in the (x, y) plane and

$$y' = \frac{dy}{dx} = \frac{d}{dx}f(x) \tag{3.1}$$

is the slope of the tangent to the curve at the point (x, y). In calculus, we use Δx to mean a change in x, and Δy means the corresponding change in y (see Figure 3.1). By definition

$$\frac{dy}{dx} = \lim_{\Delta x \to 0} \frac{\Delta y}{\Delta x}. \tag{3.2}$$

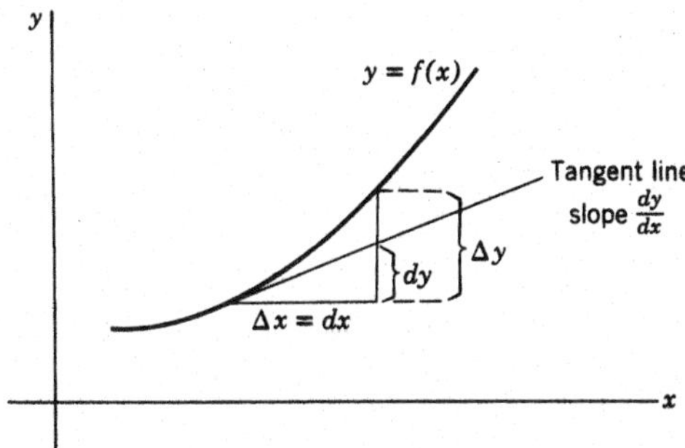

Figure 3.1

We shall now define the differential dx of the independent variable as

$$dx = \Delta x. \tag{3.3}$$

However, dy is not the same as Δy. From Figure 3.1 and equation (3.1), we can see that Δy is the change in y along the curve, but $dy = y'dx$ is the change in y along the tangent line. We say that dy is the tangent approximation (or linear approximation) to Δy.

Example. If $y = f(t)$ represents the distance a particle has gone as a function of t, then dy/dt is the speed. The actual distance the particle has gone between time t and time $t + dt$ is Δy. The tangent approximation $dy = (dy/dt)dt$ is the distance it would have gone if it had continued with the same speed dy/dt which it had at time t.

You can see from the graph (Figure 3.1) that dy is a good approximation to Δy if dx is small. We can say this more exactly using (3.2). Saying that dy/dx is the limit of $\Delta y/\Delta x$ as $\Delta x \to 0$ means that the difference $\Delta y/\Delta x - dy/dx \to 0$ as $\Delta x \to 0$. Let us call this difference ϵ; then we can say

$$\frac{\Delta y}{\Delta x} = \frac{dy}{dx} + \epsilon, \qquad \text{where } \epsilon \to 0 \quad \text{as} \quad \Delta x \to 0, \tag{3.4}$$

or since $dx = \Delta x$

$$\Delta y = (y' + \epsilon)dx, \qquad \text{where } \epsilon \to 0 \quad \text{as} \quad \Delta x \to 0. \tag{3.5}$$

The differential $dy = y'dx$ is called the principal part of Δy; since ϵ is small for small dx, you can see from (3.5) that dy is then a good approximation to Δy.

In our example, suppose $y = t^2$, $t = 1$, $dt = 0.1$. Then

$$\Delta y = (1.1)^2 - 1^2 = 0.21,$$
$$dy = \frac{dy}{dt}dt = 2 \cdot 1 \cdot (0.1) = 0.2,$$
$$\epsilon = \frac{\Delta y}{\Delta t} - \frac{dy}{dt} = 2.1 - 2 = 0.1,$$
$$\Delta y = (y' + \epsilon)dt = (2 + 0.1)(0.1) = dy + \epsilon dt = 0.2 + 0.01.$$

Thus dy is a good approximation to Δy.

For a function of two variables, $z = f(x, y)$, we want to do something similar to this. We have said that this equation represents a surface and that the derivatives $\partial f/\partial x$, $\partial f/\partial y$, at a point, are the slopes of the two tangent lines to the surface in the x and y directions at that point. The symbols $\Delta x = dx$ and $\Delta y = dy$ represent changes in the independent variables x and y. The quantity Δz means the corresponding change in z along the surface. We define dz by the equation

$$dz = \frac{\partial z}{\partial x}dx + \frac{\partial z}{\partial y}dy. \tag{3.6}$$

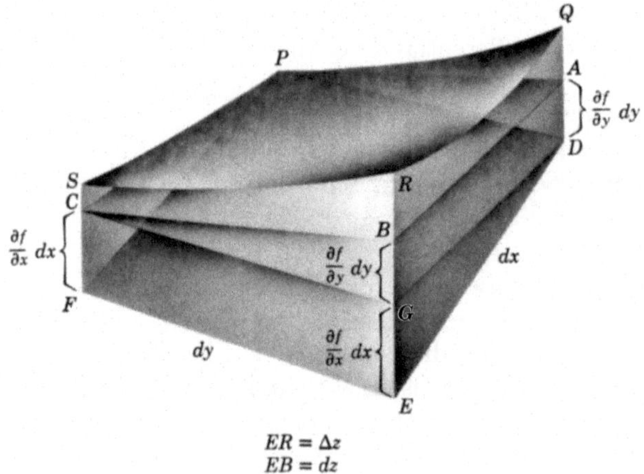

Figure 3.2

The differential dz is called the *total differential* of z. Let us consider the geometrical meaning of dz. Recall (Figure 3.1) that for $y = f(x)$, dy was the change in y along the tangent line; here we shall see that dz is the change in z along the tangent plane. In Figure 3.2, $PQRS$ is a surface, $PABC$ is the plane tangent to the surface at P, and $PDEF$ is a horizontal plane through P. Thus $PSCF$ is the plane $y =$ const. (through P), PS is the curve of intersection of this plane with the surface, and PC is the tangent line to this curve and so has the slope $\partial f/\partial x$; then (just as in Figure 3.1), if $PF = dx$, we have $CF = (\partial f/\partial x)\,dx$. Similarly, $PQAD$ is a plane $x =$ const., intersecting the surface in the curve PQ, whose tangent is PA; with

$PD = dy$, we have $DA = (\partial f/\partial y)\,dy$. From the figure, $GE = CF$, and $BG = AD$, so

$$EB = CF + DA = \frac{\partial f}{\partial x}\,dx + \frac{\partial f}{\partial y}\,dy = dz.$$

Thus, as we said, dz is the change in z along the tangent plane when x changes by dx and y by dy. In the figure, $ER = \Delta z$, the change in z along the surface.

From the geometry, we can reasonably expect dz to be a good approximation to Δz if dx and dy are small. However, we should like to say this more accurately in an equation corresponding to (3.5). We can do this if $\partial f/\partial x$ and $\partial f/\partial y$ are continuous functions. By definition

$$\Delta z = f(x + \Delta x, y + \Delta y) - f(x, y). \tag{3.7}$$

By adding and subtracting a term, we get

$$\Delta z = f(x + \Delta x, y) - f(x, y) + f(x + \Delta x, y + \Delta y) - f(x + \Delta x, y). \tag{3.8}$$

Recall from calculus that the mean value theorem (law of the mean) says that for a differentiable function $f(x)$,

$$f(x + \Delta x) - f(x) = (\Delta x) f'(x_1), \tag{3.9}$$

where x_1 is between x and $x + \Delta x$. Geometrically this says (Figure 3.3) that there is a tangent line somewhere between x and $x + \Delta x$ which has the same slope as the

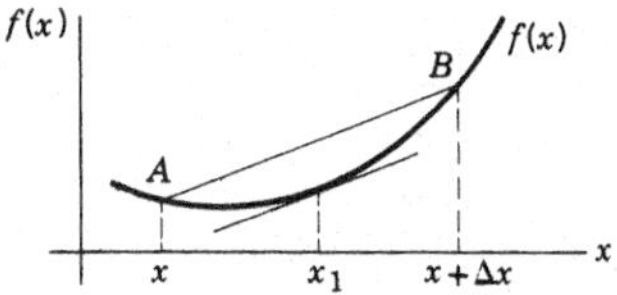

Figure 3.3

line AB. In the first two terms of the right side of (3.8), y is constant, and we can use (3.9) if we write $\partial f/\partial x$ for f'. In the last two terms of (3.8), x is constant and we can use an equation like (3.9) with y as the variable; y_1 will mean a value of y between y and $y + \Delta y$. Then (3.8) becomes

$$\Delta z = \frac{\partial f(x_1, y)}{\partial x}\,\Delta x + \frac{\partial f(x + \Delta x, y_1)}{\partial y}\,\Delta y. \tag{3.10}$$

If the partial derivatives of f are continuous, then their values in (3.10) at points *near* (x, y) differ from their values *at* (x, y) by quantities which approach zero as Δx and Δy approach zero. Let us call these quantities ϵ_1 and ϵ_2. Then we can write

(3.11)

$$\Delta z = \left(\frac{\partial f}{\partial x} + \epsilon_1\right)\Delta x + \left(\frac{\partial f}{\partial y} + \epsilon_2\right)\Delta y = dz + \epsilon_1\,\Delta x + \epsilon_2\,\Delta y$$

$$(\epsilon_1 \text{ and } \epsilon_2 \to 0 \text{ as } \Delta x \text{ and } \Delta y \to 0),$$

where $\partial f/\partial x$ and $\partial f/\partial y$ in (3.11) are evaluated at (x, y). Equation (3.11) [like (3.5) for the $y = f(x)$ case] tells us algebraically what we suspected from the geometry, that (if $\partial f/\partial x$ and $\partial f/\partial y$ are continuous) dz is a good approximation to Δz for small dx and dy. The differential dz is called the principal part of Δz.

Everything we have said about functions of two variables works just as well for functions of any number of variables. if $u = f(x, y, z, \cdots)$, then by definition

$$du = \frac{\partial f}{\partial x}\,dx + \frac{\partial f}{\partial y}\,dy + \frac{\partial f}{\partial z}\,dz + \cdots \tag{3.12}$$

and du is a good approximation to Δu if the partial derivatives of f are continuous and dx, dy, dz, etc., are small.

PROBLEMS, SECTION 3

1. Consider a function $f(x, y)$ which can be expanded in a two-variable power series, (2.3) or (2.7). Let $x - a = h = \Delta x$, $y - b = k = \Delta y$; then $x = a + \Delta x$, $y = b + \Delta y$ so that $f(x, y)$ becomes $f(a + \Delta x, b + \Delta y)$. The change Δz in $z = f(x, y)$ when x changes from a to $a + \Delta x$ and y changes from b to $b + \Delta y$ is then

$$\Delta z = f(a + \Delta x, b + \Delta y) - f(a, b).$$

Use the series (2.7) to obtain (3.11) and to see explicitly what ϵ_1 and ϵ_2 are and that they approach zero as Δx and $\Delta y \to 0$.

4. APPROXIMATIONS USING DIFFERENTIALS

Let's consider some examples.

Example 1. Find approximately the value of

$$\frac{1}{\sqrt{0.25 - 10^{-20}}} - \frac{1}{\sqrt{0.25}}.$$

If $f(x) = 1/\sqrt{x}$, the desired difference is $\Delta f = f(0.25 - 10^{-20}) - f(0.25)$. But Δf is approximately $df = d(1/\sqrt{x})$ with $x = 0.25$ and $dx = -10^{-20}$.

$$d(1/\sqrt{x}) = (-1/2)x^{-3/2}dx = (-1/2)(0.25)^{-3/2}(-10^{-20}) = 4 \times 10^{-20}.$$

Now why not just use a computer or calculator for a problem like this? First note that we are subtracting two numbers which are almost equal to each other. If your calculator or computer isn't carrying enough digits, you may lose all accuracy in the subtraction (see Chapter 1, Section 15, Example 1). So it may take you more time to check on this and to type the problem into the computer than to find df which you can probably do in your head! However, there is another important point here which is shown in the next example. For theoretical purposes, we may want a *formula* rather than a numerical result.

Example 2. Show that when n is very large

$$\frac{1}{n^2} - \frac{1}{(n+1)^2} \cong \frac{2}{n^3}$$

($\cong$ means "approximately equal to"). If $f(x) = 1/x^2$, the desired difference is $\Delta f = f(n) - f(n+1)$. But Δf is approximately $df = d(1/x^2)$ with $x = n$ and $dx = -1$.

$$d\left(\frac{1}{x^2}\right) = -\frac{2}{x^3}dx = -\frac{2}{n^3}(-1) = \frac{2}{n^3}.$$

(This result is used in obtaining the "correspondence principle" in quantum mechanics; see texts on quantum physics.) Also see Problem 17.

Example 3. The *reduced mass* μ of a system of two masses m_1 and m_2 is defined by $\mu^{-1} = m_1^{-1} + m_2^{-1}$. If m_1 is increased by 1%, what fractional change in m_2 leaves μ unchanged? Taking differentials of the equation and substituting $dm_1 = 0.01m_1$, we find

$$0 = -m_1^{-2}\,dm_1 - m_2^{-2}\,dm_2,$$

$$\frac{dm_2}{m_2^2} = -\frac{dm_1}{m_1^2} = -\frac{0.01m_1}{m_1^2} \qquad \text{or} \qquad \frac{dm_2}{m_2} = -0.01m_2/m_1.$$

For example, if $m_1 = m_2$, m_2 should be decreased by 1%; if $m_2 = 3m_1$, m_2 should be decreased by 3%; and so on.

Example 4. The electrical resistance R of a wire is proportional to its length and inversely proportional to the square of its radius, that is, $R = kl/r^2$. If the relative error in length measurement is 5% and the relative error in radius measurement is 10%, find the relative error in R in the worst possible case.

The relative error in l means the actual error in measuring l divided by the length measured. Since we might measure l either too large or too small, the relative error dl/l might be either $+0.05$ or -0.05 in the worst cases. Similarly $|dr/r|$ might be as large as 0.10. We want the largest value which $|dR/R|$ could have; we can find dR/R by differentiating $\ln R$. From $R = kl/r^2$ we find

$$\ln R = \ln k + \ln l - 2\ln r.$$

Then

$$\frac{dR}{R} = \frac{dl}{l} - 2\frac{dr}{r}.$$

In the worst case (that is, largest value of $|dR/R|$), dl/l and dr/r might have opposite signs so the two terms would add. Then we would have:

$$\text{Largest } \left|\frac{dR}{R}\right| = \left|\frac{dl}{l}\right| + 2\left|\frac{dr}{r}\right| = 0.05 + 2(0.10) = 0.25 \qquad \text{or} \qquad 25\%.$$

Example 5. Estimate the change in

$$f(x) = \int_0^x \frac{\sin t}{t} dt$$

when x changes from $\pi/2$ to $(1+\epsilon)\pi/2$ where $\epsilon << 1/10$. Recall from calculus that $df/dx = (\sin x)/x$. Then we want $df = (df/dx)\,dx$ with $x = \pi/2$ and $dx = \epsilon\pi/2$. Thus

$$df = \frac{\sin \pi/2}{\pi/2}(\epsilon\pi/2) = \epsilon.$$

Note that the approximations we have been making correspond to using a Taylor series through the f' term. We can write Chapter 1 equation (12.8) with the replacements $x \to x + \Delta x$, $a \to x$, $x - a \to \Delta x$, to get

$$f(x + \Delta x) = f(x) + f'(x)\Delta x + f''(x)(\Delta x)^2/2! + \cdots.$$

Dropping the $(\Delta x)^2$ and higher terms we have the approximation we have been using:

$$df \cong \Delta f = f(x + \Delta x) - f(x) \cong f'(x)\Delta x = f'(x)dx.$$

PROBLEMS, SECTION 4

1. Use differentials to show that, for very large n, $\dfrac{1}{(n+1)^3} - \dfrac{1}{n^3} \cong -\dfrac{3}{n^4}$.

2. Use differentials to show that, for large n and small a, $\sqrt{n+a} - \sqrt{n} \cong \dfrac{a}{2\sqrt{n}}$.

   Find the approximate value of $\sqrt{10^{26}+5} - \sqrt{10^{26}}$.

3. The thin lens formula is
$$\frac{1}{i} + \frac{1}{o} = \frac{1}{f},$$
where f is the focal length of the lens and o and i are the distances from the lens to the object and image. If $i = 15$ when $o = 10$, use differentials to find i when $o = 10.1$.

4. Do Problem 3 if $i = 12$ when $o = 18$, to find i if $o = 17.5$.

5. Let R be the resistance of $R_1 = 25$ ohms and $R_2 = 15$ ohms in parallel. (See Chapter 2, Problem 16.6.) If R_1 is changed to 25.1 ohms, find R_2 so that R is not changed.

6. The acceleration of gravity can be found from the length l and period T of a pendulum; the formula is $g = 4\pi^2 l/T^2$. Find the relative error in g in the worst case if the relative error in l is 5%, and the relative error in T is 2%.

7. Coulomb's law for the force between two charges q_1 and q_2 at distance r apart is $F = kq_1q_2/r^2$. Find the relative error in q_2 in the worst case if the relative error in q_1 is 3%; in r, 5%; and in F, 2%.

8. About how much (in percent) does an error of 1% in a and b affect a^2b^3?

9. Show that the approximate relative error $(df)/f$ of a product $f = gh$ is the sum of the approximate relative errors of the factors.

10. A force of 500 nt is measured with a possible error of 1 nt. Its component in a direction 60° away from its line of action is required, where the angle is subject to an error of 0.5°. What is (approximately) the largest possible error in the component?

11. Show how to make a quick estimate (to two decimal places) of $\sqrt{(4.98)^2 - (3.03)^2}$ without using a computer or a calculator. *Hint:* Consider $f(x, y) = \sqrt{x^2 - y^2}$.

12. As in Problem 11, estimate $\sqrt[3]{(2.05)^2 + (1.98)^2}$.

13. Without using a computer or a calculator, estimate the change in length of a space diagonal of a box whose dimensions are changed from $200\times200\times100$ to $201\times202\times99$.

14. Estimate the change in

$$f(x) = \int_0^x \frac{e^{-t}}{t^2 + 0.51}\,dt$$

if x changes from 0.7 to 0.71.

15. For an ideal gas of N molecules, the number of molecules with speeds $\le v$ is given by the formula

$$n(v) = \frac{4a^3 N}{\sqrt{\pi}} \int_0^v x^2 e^{-a^2x^2}\,dx,$$

where a is a constant and N is the total number of molecules. If $N = 10^{26}$, estimate the number of molecules with speeds between $v = 1/a$ and $1.01/a$.

16. The operating equation for a synchrotron in the relativistic range is

$$qB = \omega m[1 - (\omega R)^2/c^2]^{-1/2},$$

where q and m are the charge and rest mass of the particle being accelerated, B is the magnetic field strength, R is the orbit radius, ω is the angular frequency, and c is the speed of light. If ω and B are varied (all other quantities constant), show that the relation between $d\omega$ and dB can be written as

$$\frac{dB}{B^3} = \left(\frac{q}{m}\right)^2 \frac{d\omega}{\omega^3}, \qquad \text{or as} \qquad \frac{dB}{B} = \frac{d\omega}{\omega}[1 - (\omega R/c)^2]^{-1}.$$

17. Here are some other ways of obtaining the formula in Example 2.
(a) Combine the two fractions to get $(2n+1)/[n^2(n+1)^2]$. Then note that for large n, $2n + 1 \cong 2n$ and $n + 1 \cong n$.
(b) Factor the expression as $\left(\frac{1}{n^2}\right)\left(1 - \frac{1}{(1+\frac{1}{n})^2}\right)$, expand $\left(1 + \frac{1}{n}\right)^{-2}$ by binomial series to two terms, and then simplify.

5. CHAIN RULE OR DIFFERENTIATING A FUNCTION OF A FUNCTION

You already know about the chain rule whether you have called it that or not. Look at this example.

Example 1. Find dy/dx if $y = \ln \sin 2x$.

You would say

$$\frac{dy}{dx} = \frac{1}{\sin 2x} \cdot \frac{d}{dx}(\sin 2x) = \frac{1}{\sin 2x} \cdot \cos 2x \cdot \frac{d}{dx}(2x) = 2 \cot 2x.$$

We *could* write this problem as

$$y = \ln u, \qquad \text{where} \qquad u = \sin v \qquad \text{and} \qquad v = 2x.$$

Then we would say

$$\frac{dy}{dx} = \frac{dy}{du}\frac{du}{dv}\frac{dv}{dx}.$$

This is an example of the chain rule. We shall want a similar equation for a function of several variables. Consider another example.

▶ **Example 2.** Find dz/dt if $z = 2t^2 \sin t$.

Differentiating the product, we get

$$\frac{dz}{dt} = 4t \sin t + 2t^2 \cos t.$$

We *could* have written this problem as

$$z = xy, \qquad \text{where } x = 2t^2 \qquad \text{and } y = \sin t,$$
$$\frac{dz}{dt} = y\frac{dx}{dt} + x\frac{dy}{dt}.$$

But since x is $\partial z/\partial y$ and y is $\partial z/\partial x$, we could also write

$$\frac{dz}{dt} = \frac{\partial z}{\partial x}\frac{dx}{dt} + \frac{\partial z}{\partial y}\frac{dy}{dt}. \tag{5.1}$$

We would like to be sure that (5.1) is a correct formula in general, when we are given any function $z(x, y)$ with continuous partial derivatives and x and y are differentiable functions of t. To see this, recall from our discussion of differentials that we had

$$\Delta z = \frac{\partial z}{\partial x}\Delta x + \frac{\partial z}{\partial y}\Delta y + \epsilon_1 \Delta x + \epsilon_2 \Delta y, \tag{5.2}$$

where ϵ_1 and $\epsilon_2 \to 0$ with Δx and Δy. Divide this equation by Δt and let $\Delta t \to 0$; since Δx and $\Delta y \to 0$, ϵ_1 and $\epsilon_2 \to 0$ also, and we get (5.1).

It is often convenient to use differentials rather than derivatives as in (5.1). We would like to be able to use (3.6), but in (3.6) x and y were independent variables and now they are functions of t. However, it is possible to show (Problem 8) that dz as defined in (3.6) is a good approximation to Δz even though x and y are related. We may then write

$$dz = \frac{\partial z}{\partial x}dx + \frac{\partial z}{\partial y}dy \tag{5.3}$$

whether or not x and y are independent variables, and we may think of getting (5.1) by dividing (5.3) by dt. This is very convenient in doing problems. Thus we could do Example 2 in the following way:

$$dz = x\,dy + y\,dx = x \cos t\,dt + y \cdot 4t\,dt = (2t^2 \cos t + 4t \sin t)dt,$$
$$\frac{dz}{dt} = 2t^2 \cos t + 4t \sin t.$$

In doing problems, we may then use either differentials or derivatives. Here is another example.

Example 3. Find dz/dt given $z = x^y$, where $y = \tan^{-1} t$, $x = \sin t$.
Using differentiais, we find

$$dz = yx^{y-1}\,dx + x^y \ln x\,dy = yx^{y-1}\cos t\,dt + x^y \ln x \cdot \frac{dt}{1+t^2},$$

$$\frac{dz}{dt} = yx^{y-1}\cos t + x^y \ln x \cdot \frac{1}{1+t^2}.$$

You may wonder in a problem like this why we don't just substitute x and y as functions of t into $z = x^y$ to get z as a function of t and then differentiate. Sometimes this may be the best thing to do but not always. For example, the resulting formula may be very complicated and it may save a lot of algebra to use (5.1) or (5.3). This is especially true if we want dz/dt for a numerical value of t. Then there are cases when we *cannot* substitute; for example, if x as a function of t is given by $x + e^x = t$, we cannot solve for x as a function of t in terms of elementary functions. But we *can* find dx/dt, and so we can find dz/dt by (5.1). Finding dx/dt from such an equation is called implicit differentiation; we shall discuss this process in the next section.

Computers can find derivatives, so why should we learn the methods shown here and in the following sections? Perhaps the most important reason is that the techniques are needed in theoretical derivations. However, there is also a practical reason: When a problem involves a number of variables, there may be many ways to express the answer. (You might like to verify that $dz/dt = z(y \cot t + \ln x \cos^2 y)$ is another form for the answer in Example 3 above). Your computer may not give you the form you want and it may be as easy to do the problem by hand as to convert the computer result. But in problems involving a lot of algebra, a computer can save time, so a good study method is to do problems both by hand and by computer and compare results.

PROBLEMS, SECTION 5

1. Given $z = xe^{-y}$, $x = \cosh t$, $y = \cos t$, find dz/dt.

2. Given $w = \sqrt{u^2 + v^2}$, $u = \cos[\ln\tan(p + \frac{1}{4}\pi)]$, $v = \sin[\ln\tan(p + \frac{1}{4}\pi)]$, find dw/dp.

3. Given $r = e^{-p^2-q^2}$, $p = e^s$, $q = e^{-s}$, find dr/ds.

4. Given $x = \ln(u^2 - v^2)$, $u = t^2$, $v = \cos t$, find dx/dt.

5. If we are given $z = z(x, y)$ and $y = y(x)$, show that the chain rule (5.1) gives

$$\frac{dz}{dx} = \frac{\partial z}{\partial x} + \frac{\partial z}{\partial y}\frac{dy}{dx}.$$

6. Given $z = (x + y)^5$, $y = \sin 10x$, find dz/dx.

7. Given $c = \sin(a - b)$, $b = ae^{2a}$, find dc/da.

8. Prove the statement just after (5.2), that dz given by (3.6) is a good approximation to Δz even though dx and dy are not independent. *Hint:* let x and y be functions of t; then (5.2) is correct, but $\Delta x \neq dx$ and $\Delta y \neq dy$ (because x and y are not independent variables). However, $\Delta x/\Delta t$ is nearly dx/dt for small dt and $dt = \Delta t$ since t is the independent variable. You can then show that

$$\Delta x = \left(\frac{dx}{dt} + \epsilon_x\right) dt = dx + \epsilon_x dt,$$

and a similar formula for Δy , and get

$$\Delta z = \frac{\partial z}{\partial x}dx + \frac{\partial z}{\partial y}dy + (\text{terms containing } \epsilon\text{'s}) \cdot dt = dz + \epsilon dt,$$

where $\epsilon \to 0$ as $\Delta t \to 0$.

6. IMPLICIT DIFFERENTIATION

Some examples will show the use of implicit differentiation.

Example 1. Given $x + e^x = t$, find dx/dt and d^2x/dt^2.

If we give values to x, find the corresponding t values, and plot x against t, we have a graph whose slope is dx/dt. In other words, x *is* a function of t even though we cannot solve the equation for x in terms of elementary functions of t. To find dx/dt, we realize that x *is* a function of t and just differentiate each term of the equation with respect to t (this is called implicit differentiation). We get

$$\frac{dx}{dt} + e^x \frac{dx}{dt} = 1. \tag{6.1}$$

Solving for dx/dt, we get

$$\frac{dx}{dt} = \frac{1}{1+e^x}.$$

Alternatively, we could use differentials here, and write first $dx + e^x dx = dt$; dividing by dt then gives (6.1).

We can also find higher derivatives by implicit differentiation (but do *not* use differentials for this since we have not given any meaning to the derivative or differential of a differential). Let us differentiate each term of (6.1) with respect to t; we get

$$\frac{d^2x}{dt^2} + e^x \frac{d^2x}{dt^2} + e^x \left(\frac{dx}{dt}\right)^2 = 0. \tag{6.2}$$

Solving for d^2x/dt^2 and substituting the value already found for dx/dt, we get

$$\frac{d^2x}{dt^2} = \frac{-e^x \left(\frac{dx}{dt}\right)^2}{1+e^x} = \frac{-e^x}{(1+e^x)^3}. \tag{6.3}$$

This problem is even easier if we want only the numerical values of the derivatives at a point. For $x = 0$ and $t = 1$, (6.1) gives

$$\frac{dx}{dt} + 1 \cdot \frac{dx}{dt} = 1 \qquad \text{or} \qquad \frac{dx}{dt} = \frac{1}{2},$$

and (6.2) gives

$$\frac{d^2x}{dt^2} + 1 \cdot \frac{d^2x}{dt^2} + 1 \cdot \left(\frac{1}{2}\right)^2 = 0 \qquad \text{or} \qquad \frac{d^2x}{dt^2} = -\frac{1}{8}.$$

Implicit differentiation is the best method to use in finding slopes of curves with complicated equations.

Example 2. Find the equation of the tangent line to the curve $x^3 - 3y^3 + xy + 21 = 0$ at the point $(1, 2)$.

We differentiate the given equation implicitly with respect to x to get

$$3x^2 - 9y^2\frac{dy}{dx} + x\frac{dy}{dx} + y = 0.$$

Substitute $x = 1, y = 2$:

$$3 - 36\frac{dy}{dx} + \frac{dy}{dx} + 2 = 0, \qquad \frac{dy}{dx} = \frac{5}{35} = \frac{1}{7}.$$

Then the equation of the tangent line is

$$\frac{y-2}{x-1} = \frac{1}{7} \qquad \text{or} \qquad x - 7y + 13 = 0.$$

By computer plotting the curve and the tangent line on the same axes, you can check to be sure that the line appears tangent to the curve.

PROBLEMS, SECTION 6

1. If $pv^a = C$ (where a and C are constants), find dv/dp and d^2v/dp^2.
2. If $ye^{xy} = \sin x$ find dy/dx and d^2y/dx^2 at $(0, 0)$.
3. If $x^y = y^x$, find dy/dx at $(2, 4)$.
4. If $xe^y = ye^x$, find dy/dx and d^2y/dx^2 for $y \neq 1$.
5. If $xy^3 - yx^3 = 6$ is the equation of a curve, find the slope and the equation of the tangent line at the point $(1, 2)$. Computer plot the curve and the tangent line on the same axes.
6. In Problem 5 find d^2y/dx^2 at $(1, 2)$.
7. If $y^3 - x^2y = 8$ is the equation of a curve, find the slope and the equation of the tangent line at the point $(3, -1)$. Computer plot the curve and the tangent line on the same axes.
8. In Problem 7 find d^2y/dx^2 at $(3, -1)$.
9. For the curve $x^{2/3} + y^{2/3} = 4$, find the equations of the tangent lines at $(2\sqrt{2}, -2\sqrt{2})$, at $(8, 0)$, and at $(0, 8)$. Computer plot the curve and the tangent lines on the same axes.
10. For the curve $xe^y + ye^x = 0$, find the equation of the tangent line at the origin. *Caution:* Substitute $x = y = 0$ as soon as you have differentiated. Computer plot the curve and the tangent line on the same axes.
11. In Problem 10, find y'' at the origin.

7. MORE CHAIN RULE

Above we have considered $z = f(x, y)$, where x and y are functions of t. Now suppose $z = f(x, y)$ as before, but x and y are each functions of two variables s and t. Then z is a function of s and t and we want to find $\partial z/\partial s$ and $\partial z/\partial t$. We show by some examples how to do problems like this.

Example 1. Find $\partial z/\partial s$ and $\partial z/\partial t$ given

$$z = xy, \qquad x = \sin(s+t), \qquad y = s - t.$$

We take differentials of each of the three equations to get

$$dz = y\,dx + x\,dy, \qquad dx = \cos(s+t)(ds+dt), \qquad dy = ds - dt.$$

Substituting dx and dy into dz, we get

$$\begin{aligned} dz &= y\cos(s+t)(ds+dt) + x(ds-dt) \\ &= [y\cos(s+t)+x]\,ds + [y\cos(s+t)-x]\,dt. \end{aligned} \tag{7.1}$$

Now if s is constant, $ds = 0$, z is a function of one variable t, and we can divide (7.1) by dt [see (5.1) and the discussion following it]. For $dz \div dt$ on the left we write $\partial z/\partial t$ because that is the notation which properly describes what we are finding, namely, the rate of change of z with t when s is constant. Thus we have

$$\frac{\partial z}{\partial t} = y\cos(s+t) - x$$

and similarly

$$\frac{\partial z}{\partial s} = y\cos(s+t) + x.$$

Notice that in (7.1) the coefficient of ds is $\partial z/\partial s$ and the coefficient of dt is $\partial z/\partial t$ [also compare (5.3)]. If you realize this, you can simply read off $\partial z/\partial s$ and $\partial z/\partial t$ from (7.1).

We can do problems with more variables in the same way.

Example 2. Find $\partial u/\partial s$, $\partial u/\partial t$, given $u = x^2 + 2xy - y\ln z$ and $x = s + t^2$, $y = s - t^2$, $z = 2t$.

We find

$$\begin{aligned} du &= 2x\,dx + 2x\,dy + 2y\,dx - \frac{y}{z}dz - \ln z\,dy \\ &= (2x+2y)(ds + 2t\,dt) + (2x - \ln z)(ds - 2t\,dt) - \frac{y}{z}(2\,dt) \\ &= (4x + 2y - \ln z)\,ds + \left(4yt + 2t\ln z - \frac{2y}{z}\right)dt. \end{aligned}$$

Then

$$\frac{\partial u}{\partial s} = 4x + 2y - \ln z, \qquad \frac{\partial u}{\partial t} = 4yt + 2t\ln z - \frac{2y}{z}.$$

If we want just one derivative, say $\partial u/\partial t$, we can save some work by letting $ds = 0$ to start with. To make it clear that we have done this, we write

$$\begin{aligned} du_s &= (2x+2y)(2t\,dt) + (2x - \ln z)(-2t\,dt) - \frac{y}{z}(2\,dt) \\ &= \left(4yt + 2t\,\ln z - \frac{2y}{z}\right)dt. \end{aligned}$$

The subscript s indicates that s is being held constant. Then dividing by dt, we have $\partial u/\partial t$ as before. We could also use derivatives instead of differentials. By an equation like (5.1), we have

$$\frac{\partial u}{\partial t} = \frac{\partial u}{\partial x}\frac{\partial x}{\partial t} + \frac{\partial u}{\partial y}\frac{\partial y}{\partial t} + \frac{\partial u}{\partial z}\frac{\partial z}{\partial t}, \tag{7.2}$$

where we have written all the t derivatives as partials since u, x, y, and z depend on both s and t. Using (7.2), we get

$$\frac{\partial u}{\partial t} = (2x+2y)(2t) + (2x - \ln z)(-2t) + \left(-\frac{y}{z}\right)(2) = 4yt + 2t\ln z - \frac{2y}{z}.$$

It is sometimes useful to write chain rule formulas in matrix form (for matrix multiplication, see Chapter 3, Section 6). Given, as above, $u = f(x, y, z)$, $x(s, t)$, $y(s, t)$, $z(s, t)$, we can write equations like (7.2) in the following matrix form:

$$\begin{pmatrix} \dfrac{\partial u}{\partial s} & \dfrac{\partial u}{\partial t} \end{pmatrix} = \begin{pmatrix} \dfrac{\partial u}{\partial x} & \dfrac{\partial u}{\partial y} & \dfrac{\partial u}{\partial z} \end{pmatrix} \begin{pmatrix} \dfrac{\partial x}{\partial s} & \dfrac{\partial x}{\partial t} \\ \dfrac{\partial y}{\partial s} & \dfrac{\partial y}{\partial t} \\ \dfrac{\partial z}{\partial s} & \dfrac{\partial z}{\partial t} \end{pmatrix}. \tag{7.3}$$

[Sometimes (7.3) is written in the abbreviated form

$$\frac{\partial(u)}{\partial(s,t)} = \frac{\partial(u)}{\partial(x,y,z)}\frac{\partial(x,y,z)}{\partial(s,t)}$$

which is reminiscent of

$$\frac{dy}{dt} = \frac{dy}{dx}\frac{dx}{dt};$$

but be careful of this for two reasons: (a) It may be helpful in remembering the formula but to use it you must understand that it means the matrix product (7.3). (b) The symbol $\partial(u, v)/\partial(x, y)$ usually means a determinant rather than a matrix of partial derivatives—see Chapter 5, Section 4].

Again in these problems, you may say, why not just substitute? Look at the following problem.

Example 3. Find dz/dt given $z = x - y$ and

$$x^2 + y^2 = t^2,$$
$$x \sin t = ye^y.$$

From the z equation, we have

$$dz = dx - dy.$$

We need dx and dy; here we cannot solve for x and y in terms of t. But we *can* find dx and dy in terms of dt from the other two equations and this is all we need. Take differentials of both equations to get

$$2x\,dx + 2y\,dy = 2t\,dt,$$
$$\sin t\,dx + x\cos t\,dt = (ye^y + e^y)dy.$$

Rearrange terms:

$$x\,dx + y\,dy = t\,dt,$$
$$\sin t\,dx - (y+1)e^y dy = -x\cos t\,dt.$$

Solve for dx and dy (in terms of dt) by determinants:

$$dx = \frac{\begin{vmatrix} t\,dt & y \\ -x\cos t\,dt & -(y+1)e^y \end{vmatrix}}{\begin{vmatrix} x & y \\ \sin t & -(y+1)e^y \end{vmatrix}} = \frac{-t(y+1)e^y + xy\cos t}{-x(y+1)e^y - y\sin t}\,dt,$$

and similarly for dy. Substituting dx and dy into the formula for dz and dividing by dt, we get dz/dt. A computer may save us some time with the algebra.

We can also do problems like this when x and y are given implicitly as functions of two variables s and t.

▸ **Example 4.** Find $\partial z/\partial s$ and $\partial z/\partial t$ given

$$z = x^2 + xy,$$
$$x^2 + y^3 = st + 5,$$
$$x^3 - y^2 = s^2 + t^2.$$

We have $dz = 2x\,dx + x\,dy + y\,dx$. To find dx and dy from the other two equations, we take differentials of each equation:

$$\begin{aligned} 2x\,dx + 3y^2\,dy &= s\,dt + t\,ds, \\ 3x^2\,dx - 2y\,dy &= 2s\,ds + 2t\,dt. \end{aligned} \tag{7.4}$$

We can solve these two equations for dx and dy in terms of ds and dt to get

$$dx = \frac{\begin{vmatrix} s\,dt + t\,ds & 3y^2 \\ 2s\,ds + 2t\,dt & -2y \end{vmatrix}}{\begin{vmatrix} 2x & 3y^2 \\ 3x^2 & -2y \end{vmatrix}} = \frac{(-2ys - 6ty^2)\,dt + (-2yt - 6sy^2)\,ds}{-4xy - 9x^2y^2}$$

and a similar expression for dy. We substitute these values of dx and dy into dz and find dz in terms of ds and dt just as in Example 1; we can then write $\partial z/\partial s$ and $\partial z/\partial t$ just as we did there (Problem 11). Notice that if we want only one derivative, say $\partial z/\partial t$, we could save some algebra by putting $ds = 0$ in (7.4). Also note that we can save some algebra if we want the derivatives only at one point. Suppose we were asked for $\partial z/\partial s$ and $\partial z/\partial t$ at $x = 3$, $y = 1$, $s = 1$, $t = 5$. We substitute these values into (7.4) to get

$$6\,dx + 3\,dy = dt + 5\,ds,$$
$$27\,dx - 2\,dy = 10\,dt + 2\,ds.$$

We solve these equations for dx and dy and substitute into dz just as before, but the algebra is easier with the numerical coefficients (Problem 11).

So far, we have been assuming that the independent variables were "natural" pairs like x and y, or s and t. For example, we wrote $\partial x/\partial s$ above, taking it for granted that the variable held constant was t. In some applications (particularly thermodynamics), it is not at all clear what the other independent variable is and we have to be more explicit. We write $(\partial x/\partial s)_t$; this means that s and t are the two independent variables, that x is thought of as a function of them, and then x is differentiated partially with respect to s. Suppose we try to find from the three equations of Example 4 a rather peculiar looking derivative.

Example 5. Given the equations of Example 4, find $(\partial s/\partial z)_x$.

First, let us see that the question makes sense. There are five variables in the three equations. If we give values to two of them, we can solve for the other three; that is, there are *two independent* variables, and the other three are functions of these two. If z and x are the independent ones, then s, t, and y are functions of z and x; we should be able to find their partial derivatives, for example $(\partial s/\partial z)_x$ which we wanted. To carry out the necessary work, we first rearrange equations (7.4) and the dz equation to get

$$\begin{aligned} -x\,dy &= (2x+y)\,dx - dz, \\ t\,ds + s\,dt - 3y^2\,dy &= 2x\,dx, \\ 2s\,ds + 2t\,dt + 2y\,dy &= 3x^2\,dx. \end{aligned}$$

From these three equations we could solve for ds, dt, and dy in terms of dx and dz (by determinants or by elimination—the same methods you use to solve any set of linear equations). Then we could find any partial derivative of $s(x,z)$, $t(x,z)$, or $y(x,z)$ with respect to x to z. For example, to find $(\partial y/\partial z)_x$, we get from the first equation

$$dy = \frac{1}{x}\,dz - \frac{2x+y}{x}\,dx,$$
$$\left(\frac{\partial y}{\partial z}\right)_x = \frac{1}{x}.$$

Note that we would not need to differentiate all three equations if we wanted only this derivative; you should always look ahead to see how much differentiation is necessary! To find $(\partial s/\partial z)_x$, we must solve the three equations for ds in terms of dx and dz; we can save ourselves some work [if we want only $(\partial s/\partial z)_x$] by putting $dx = 0$ to start with. To make it clear that we have done this we write ds_x and dz_x. Then we get

$$ds_x = \frac{\begin{vmatrix} -dz_x & 0 & -x \\ 0 & s & -3y^2 \\ 0 & 2t & 2y \end{vmatrix}}{\begin{vmatrix} 0 & 0 & -x \\ t & s & -3y^2 \\ 2s & 2t & 2y \end{vmatrix}} = \frac{-(2sy+6ty^2)dz_x}{-x(2t^2-2s^2)},$$

$$\left(\frac{\partial s}{\partial z}\right)_x = \frac{sy+3ty^2}{x(t^2-s^2)}.$$

We could use a computer to save us some algebra in this problem.

Example 6. Let x, y be rectangular coordinates and r, θ be polar coordinates in a plane. Then the equations relating them are

$$x = r\cos\theta, \qquad y = r\sin\theta, \tag{7.5}$$

or

$$r = \sqrt{x^2 + y^2}, \qquad \theta = \tan^{-1}\frac{y}{x}. \tag{7.6}$$

Suppose we want to find $\partial\theta/\partial x$. Remembering that if $y = f(x)$, dy/dx and dx/dy are reciprocals, you might be tempted to find $\partial\theta/\partial x$ by taking the reciprocal of $\partial x/\partial\theta$, which is easier to find than $\partial\theta/\partial x$. *This is wrong.* From (7.6) we get

$$\frac{\partial\theta}{\partial x} = \frac{-y/x^2}{1 + (y^2/x^2)} = -\frac{y}{r^2}. \tag{7.7}$$

From $x = r\cos\theta$ we get

$$\frac{\partial x}{\partial\theta} = -r\sin\theta = -y. \tag{7.8}$$

These are not reciprocals. You should think carefully about the reason for this; $\partial\theta/\partial x$ means $(\partial\theta/\partial x)_y$, whereas $\partial x/\partial\theta$ means $(\partial x/\partial\theta)_r$. In one case y is held constant and in the other case r is held constant; this is why the two derivatives are not reciprocals. It *is* true that $(\partial\theta/\partial x)_y$ and $(\partial x/\partial\theta)_y$ are reciprocals. But to find $(\partial x/\partial\theta)_y$ directly, we have to express x as a function of θ and y. We find $x = y\cot\theta$, so we get

$$\left(\frac{\partial x}{\partial\theta}\right)_y = y(-\csc^2\theta) = \frac{-y}{\sin^2\theta} = \frac{-y}{y^2/r^2} = -\frac{r^2}{y}, \tag{7.9}$$

which *is* the reciprocal of $\partial\theta/\partial x$ in (7.7).

> This is a general rule: $\partial u/\partial v$ and $\partial v/\partial u$ are *not* usually reciprocals; they *are* reciprocals if the other independent variables (besides u or v) are the same in both cases.

You can see this clearly from the equations involving differentials. From the equation $\theta = \arctan(y/x)$, we can find

$$d\theta = \frac{x\,dy - y\,dx}{x^2} \bigg/ \left(1 + \frac{y^2}{x^2}\right) = \frac{x\,dy - y\,dx}{r^2}. \tag{7.10}$$

From $x = r\cos\theta$, we get

$$dx = \cos\theta\,dr - r\sin\theta\,d\theta = \frac{x}{r}\,dr - y\,d\theta. \tag{7.11}$$

From (7.10), if y is constant, $dy = 0$, and we can write

$$d\theta_y = -\frac{y}{r^2}\,dx_y, \tag{7.12}$$

where the y subscript indicates that y is constant. From (7.12), we then find either

$$\left(\frac{\partial\theta}{\partial x}\right)_y = \frac{d\theta_y}{dx_y} \qquad \text{or} \qquad \left(\frac{\partial x}{\partial\theta}\right)_y = \frac{dx_y}{d\theta_y},$$

and these are reciprocals. From (7.11), however, we can find $(\partial x/\partial\theta)_r$ or $(\partial\theta/\partial x)_r$; these are again reciprocals of each other, but are different from the derivatives found from (7.12).

It is interesting to write equations like (7.11) in matrix notation:

$$\text{(7.13)} \qquad \begin{pmatrix} dx \\ dy \end{pmatrix} = \begin{pmatrix} \dfrac{\partial x}{\partial r} & \dfrac{\partial x}{\partial\theta} \\ \dfrac{\partial y}{\partial r} & \dfrac{\partial y}{\partial\theta} \end{pmatrix} \begin{pmatrix} dr \\ d\theta \end{pmatrix} = \begin{pmatrix} \cos\theta & -r\sin\theta \\ \sin\theta & r\cos\theta \end{pmatrix} \begin{pmatrix} dr \\ d\theta \end{pmatrix} = \mathrm{A}\begin{pmatrix} dr \\ d\theta \end{pmatrix},$$

where A stands for the square matrix in (7.13). Similarly, we can write

$$\text{(7.14)} \qquad \begin{pmatrix} dr \\ d\theta \end{pmatrix} = \begin{pmatrix} \dfrac{\partial r}{\partial x} & \dfrac{\partial r}{\partial y} \\ \dfrac{\partial\theta}{\partial x} & \dfrac{\partial\theta}{\partial y} \end{pmatrix} \begin{pmatrix} dx \\ dy \end{pmatrix} = \mathrm{A}^{-1}\begin{pmatrix} dx \\ dy \end{pmatrix};$$

we have written the square matrix as A^{-1} since by (7.13),

$$\mathrm{A}^{-1}\begin{pmatrix} dx \\ dy \end{pmatrix} = \mathrm{A}^{-1}\mathrm{A}\begin{pmatrix} dr \\ d\theta \end{pmatrix} = \begin{pmatrix} dr \\ d\theta \end{pmatrix}.$$

Then, finding A^{-1} (Problem 9) and using (7.14), we have

$$\text{(7.15)} \qquad \begin{pmatrix} \dfrac{\partial r}{\partial x} & \dfrac{\partial r}{\partial y} \\ \dfrac{\partial\theta}{\partial x} & \dfrac{\partial\theta}{\partial y} \end{pmatrix} = \begin{pmatrix} \cos\theta & \sin\theta \\ -\dfrac{1}{r}\sin\theta & \dfrac{1}{r}\cos\theta. \end{pmatrix}$$

We can simply read off the four partial derivatives of r, θ, with respect to x, y, from equation (7.15). (Also see Problem 9.) Also using (7.5), and specifically noting that x and y are independent variables, we have:

$$\text{(7.16)} \qquad \begin{aligned} \frac{\partial r}{\partial x} &= \left(\frac{\partial r}{\partial x}\right)_y = \cos\theta = \frac{x}{r}, & \frac{\partial r}{\partial y} &= \left(\frac{\partial r}{\partial y}\right)_x = \sin\theta = \frac{y}{r}, \\ \frac{\partial\theta}{\partial x} &= \left(\frac{\partial\theta}{\partial x}\right)_y = -\frac{1}{r}\sin\theta = -\frac{y}{r^2}, & \frac{\partial\theta}{\partial y} &= \left(\frac{\partial\theta}{\partial y}\right)_x = \frac{1}{r}\cos\theta = \frac{x}{r^2}. \end{aligned}$$

[In the notation mentioned just after (7.3), we could write

$$\mathrm{AA}^{-1} = \frac{\partial(x,y)}{\partial(r,\theta)}\frac{\partial(r,\theta)}{\partial(x,y)} = \text{unit matrix};$$

thus, although the individual pairs of partial derivatives discussed above are not reciprocals, the two matrices of partial derivatives are inverses.]

PROBLEMS, SECTION 7

1. If $x = yz$ and $y = 2\sin(y+z)$, find dx/dy and d^2x/dy^2.

2. If $P = r\cos t$ and $r\sin t - 2te^r = 0$, find dP/dt.

3. If $z = xe^{-y}$ and $x = \cosh t$, $y = \cos s$, find $\partial z/\partial s$ and $\partial z/\partial t$.

4. If $w = e^{-r^2-s^2}$, $r = uv$, $s = u + 2v$, find $\partial w/\partial u$ and $\partial w/\partial v$.

5. If $u = x^2y^3z$ and $x = \sin(s+t)$, $y = \cos(s+t)$, $z = e^{st}$, find $\partial u/\partial s$ and $\partial u/\partial t$.

6. If $w = f(x,y)$ and $x = r\cos\theta$, $y = r\sin\theta$, find formulas for $\partial w/\partial r$, $\partial w/\partial \theta$, and $\partial^2 w/\partial r^2$.

7. If $x = r\cos\theta$ and $y = r\sin\theta$, find $(\partial y/\partial\theta)_r$ and $(\partial y/\partial\theta)_x$. Also find $(\partial\theta/\partial y)_x$ in two ways (by eliminating r from the given equations and then differentiating, or by taking differentials in both equations and then eliminating dr). When are $\partial y/\partial\theta$ and $\partial\theta/\partial y$ reciprocals?

8. If $xs^2 + yt^2 = 1$ and $x^2s + y^2t = xy - 4$, find $\partial x/\partial s$, $\partial x/\partial t$, $\partial y/\partial s$, $\partial y/\partial t$, at $(x,y,s,t) = (1,-3,2,-1)$. *Hint:* To simplify the work, substitute the numerical values just after you have taken differentials.

9. Verify (7.16) in three ways:

(a) Differentiate equations (7.6).

(b) Take differentials of (7.5) and solve for dr and $d\theta$.

(c) Find A^{-1} in (7.15) from A in (7.13); note that this is (b) in matrix notation.

10. If $x^2 + y^2 = 2st - 10$ and $2xy = s^2 - t^2$, find $\partial x/\partial s$, $\partial x/\partial t$, $\partial y/\partial s$, $\partial y/\partial t$ at $(x,y,s,t) = (4,2,5,3)$.

11. Finish Example 4 above, both for the general case and for the given numerical values. Substitute the numerical values into your general formulas to check your answers.

12. If $w = x + y$ with $x^3 + xy + y^3 = s$ and $x^2y + xy^2 = t$, find $\partial w/\partial s$, $\partial w/\partial t$.

13. If $m = pq$ with $a\sin p - p = q$ and $b\cos q + q = p$, find $(\partial p/\partial q)_m$, $(\partial p/\partial q)_a$, $(\partial p/\partial q)_b$, $(\partial b/\partial a)_p$, $(\partial a/\partial q)_m$.

14. If $u = x^2 + y^2 + xyz$ and $x^4 + y^4 + z^4 = 2x^2y^2z^2 + 10$, find $(\partial u/\partial x)_z$ at the point $(x,y,z) = (2,1,1)$.

15. Given $x^2u - y^2v = 1$, and $x + y = uv$. Find $(\partial x/\partial u)_v$, $(\partial x/\partial u)_y$.

16. Let $w = x^2 + xy + z^2$.

(a) If $x^3 + x = 3t$, $y^4 + y = 4t$, $z^5 + z = 5t$, find dw/dt.

(b) If $y^3 + xy = 1$ and $z^3 - xz = 2$, find dw/dx.

(c) If $x^3z + z^3y + y^3x = 0$, find $(\partial w/\partial x)_y$.

17. If $p^3 + sq = t$, and $q^3 + tp = s$, find $(\partial p/\partial s)_t$, $(\partial p/\partial s)_q$ at $(p,q,s,t) = (-1,2,3,5)$.

18. If $m = a + b$ and $n = a^2 + b^2$ find $(\partial b/\partial m)_n$ and $(\partial m/\partial b)_a$.

19. If $z = r + s^2$, $x + y = s^3 + r^3 - 3$, $xy = s^2 - r^2$, find $(\partial x/\partial z)_s$, $(\partial x/\partial z)_r$, $(\partial x/\partial z)_y$ at $(r,s,x,y,z) = (-1,2,3,1,3)$.

20. If $u^2 + v^2 = x^3 - y^3 + 4$, $u^2 - v^2 = x^2y^2 + 1$, find $(\partial u/\partial x)_y$, $(\partial u/\partial x)_v$, $(\partial x/\partial u)_y$, $(\partial x/\partial u)_v$ at $(x,y,u,v) = (2,-1,3,2)$.

21. Given $x^2+y^2+z^2=6$, and $w^3+z^3=5xy+12$, find the following partial derivatives at the point $(x,y,z,w)=(1,-2,1,1)$.

$$\left(\frac{\partial z}{\partial x}\right)_y, \quad \left(\frac{\partial z}{\partial x}\right)_w, \quad \left(\frac{\partial z}{\partial y}\right)_x, \quad \left(\frac{\partial z}{\partial y}\right)_w, \quad \left(\frac{\partial w}{\partial x}\right)_z, \quad \left(\frac{\partial x}{\partial w}\right)_z.$$

22. If $w=f(ax+by)$, show that $b\dfrac{\partial w}{\partial x}-a\dfrac{\partial w}{\partial y}=0$.

Hint: Let $ax+by=z$.

23. If $u=f(x-ct)+g(x+ct)$, show that $\dfrac{\partial^2 u}{\partial x^2}=\dfrac{1}{c^2}\dfrac{\partial^2 u}{\partial t^2}$.

24. If $z=\cos(xy)$, show that $x\dfrac{\partial z}{\partial x}-y\dfrac{\partial z}{\partial y}=0$.

25. The formulas of this problem are useful in thermodynamics.

(a) Given $f(x,y,z)=0$, find formulas for

$$\left(\frac{\partial y}{\partial x}\right)_z, \quad \left(\frac{\partial x}{\partial y}\right)_z, \quad \left(\frac{\partial y}{\partial z}\right)_x, \quad \text{and} \quad \left(\frac{\partial z}{\partial x}\right)_y.$$

(b) Show that

$$\left(\frac{\partial x}{\partial y}\right)_z\left(\frac{\partial y}{\partial x}\right)_z=1$$

and

$$\left(\frac{\partial x}{\partial y}\right)_z\left(\frac{\partial y}{\partial z}\right)_x\left(\frac{\partial z}{\partial x}\right)_y=-1.$$

(c) If x,y,z are each functions of t, show that $\left(\dfrac{\partial y}{\partial z}\right)_x=\left(\dfrac{\partial y}{\partial t}\right)_x\Big/\left(\dfrac{\partial z}{\partial t}\right)_x$ and corresponding formulas for $\left(\dfrac{\partial z}{\partial x}\right)_y$ and $\left(\dfrac{\partial x}{\partial y}\right)_z$.

26. Given $f(x,y,z)=0$ and $g(x,y,z)=0$, find a formula for dy/dx.

27. Given $u(x,y)$ and $y(x,z)$, show that

$$\left(\frac{\partial u}{\partial x}\right)_z=\left(\frac{\partial u}{\partial x}\right)_y+\left(\frac{\partial u}{\partial y}\right)_x\left(\frac{\partial y}{\partial x}\right)_z.$$

28. Given $s(v,T)$ and $v(p,T)$, we define $c_p=T(\partial s/\partial T)_p$, $c_v=T(\partial s/\partial T)_v$. (The c's are specific heats in thermodynamics.) Show that

$$c_p-c_v=T\left(\frac{\partial s}{\partial v}\right)_T\left(\frac{\partial v}{\partial T}\right)_p.$$

8. APPLICATION OF PARTIAL DIFFERENTIATION TO MAXIMUM AND MINIMUM PROBLEMS

You will recall that derivatives give slopes as well as rates and that you find maximum and minimum points of $y=f(x)$ by setting $dy/dx=0$. Often in applied problems we want to find maxima or minima of functions of more than one variable. Think of $z=f(x,y)$ which represents a surface. If there is a maximum point on it (like the top of a hill), then the curves for $x=$ const. and $y=$ const. which pass through the maximum point also have maxima at the same point. That is, $\partial z/\partial x$

and $\partial z/\partial y$ are zero at the maximum point. Recall that $dy/dx = 0$ was a necessary condition for a maximum point of $y = f(x)$, but not sufficient; the point might have been a minimum or perhaps a point of inflection with a horizontal tangent. Something similar can happen for $z = f(x, y)$. The point where $\partial z/\partial x = 0$ and $\partial z/\partial y = 0$ may be a maximum point, a minimum point, or neither. (An interesting example of neither is a "saddle point"—a curve from front to back on a saddle has a minimum; one from side to side has a maximum. See Figure 8.1.) In finding maxima of $y = f(x)$, it is sometimes possible to tell from the geometry or physics that you have a maximum. If necessary you can find d^2y/dx^2; if it is negative, then you know you have a maximum point. There is a similar (rather complicated) second derivative test for functions of two variables (see Problems 1 to 7), but we use it only if we have to; usually we can tell from the problem whether we have a maximum, a minimum, or neither. Let us consider some examples of maximum or minimum problems.

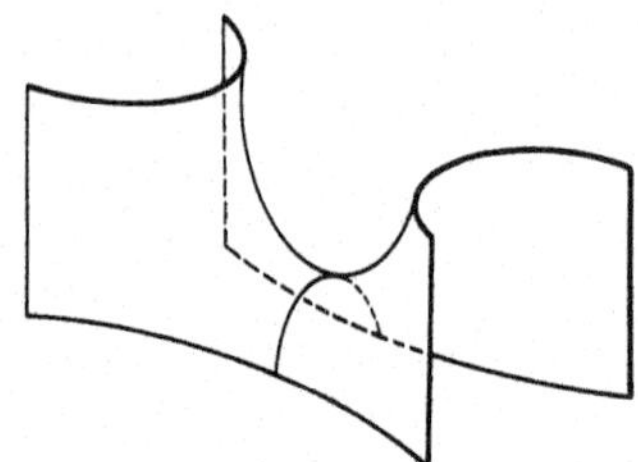

Figure 8.1

▶ **Example.** A pup tent (Figure 8.2) of given volume V, with ends but no floor, is to be made using the least possible material. Find the proportions.

Using the letters indicated in the figure, we find the volume V and the area A.

$$V = \frac{1}{2} \cdot 2w \cdot l \cdot w \tan\theta = w^2 l \tan\theta,$$

$$A = 2w^2 \tan\theta + \frac{2lw}{\cos\theta}.$$

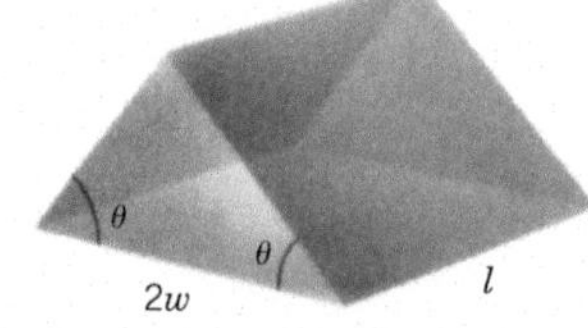

Figure 8.2

Since V is given, only two of the three variables w, l, and θ are independent, and we must eliminate one of them from A before we try to minimize A. Solving the V equation for l and substituting into A, we get

$$A = 2w^2 \tan\theta + \frac{2w}{\cos\theta} \frac{V}{w^2 \tan\theta} = 2w^2 \tan\theta + \frac{2V}{w} \csc\theta.$$

We now have A as a function of two independent variables w and θ. To minimize A we find $\partial A/\partial w$ and $\partial A/\partial \theta$ and set them equal to zero.

$$\frac{\partial A}{\partial w} = 4w \tan\theta - \frac{2V \csc\theta}{w^2} = 0,$$

$$\frac{\partial A}{\partial \theta} = 2w^2 \sec^2\theta - \frac{2V}{w} \csc\theta \cot\theta = 0.$$

Solving each of these equations for w^3 and setting the results equal, we get

$$w^3 = \frac{V \csc\theta}{2\tan\theta} = \frac{V \csc\theta \cot\theta}{\sec^2\theta} \quad \text{or} \quad \frac{\cos\theta}{2\sin^2\theta} = \frac{\cos\theta \cos^2\theta}{\sin^2\theta}.$$

You should convince yourself that neither $\sin\theta = 0$ nor $\cos\theta = 0$ is possible (the tent collapses to zero volume in both cases). Therefore we may assume $\sin\theta \neq 0$ and $\cos\theta \neq 0$ and cancel these factors, getting $\cos^2\theta = \frac{1}{2}$ or $\theta = 45°$. Then $\tan\theta = 1$, $V = w^2 l$, and from the $\partial A/\partial w$ equation we have $2w = l\sqrt{2}$. Then the height of the tent (at the peak) is $w \tan\theta = w = 1/\sqrt{2}$.

PROBLEMS, SECTION 8

1. Use the Taylor series about $x = a$ to verify the familiar "second derivative test" for a maximum or minimum point. That is, show that if $f'(a) = 0$, then $f''(a) > 0$ implies a minimum point at $x = a$ and $f''(a) < 0$ implies a maximum point at $x = a$. *Hint:* For a minimum point, say, you must show that $f(x) > f(a)$ for all x near enough to a.

2. Using the two-variable Taylor series [say (2.7)] prove the following "second derivative tests" for maximum or minimum points of functions of two variables. If $f_x = f_y = 0$ at (a, b), then

(a, b) is a minimum point if at (a, b), $f_{xx} > 0, f_{yy} > 0$, and $f_{xx}f_{yy} > f_{xy}^2$;

(a, b) is a maximum point if at (a, b), $f_{xx} < 0, f_{yy} < 0$, and $f_{xx}f_{yy} > f_{xy}^2$;

(a, b) is neither a maximum nor a minimum point if $f_{xx}f_{yy} < f_{xy}^2$. (Note that this includes $f_{xx}f_{yy} < 0$, that is, f_{xx} and f_{yy} of opposite sign.)

Hint: Let $f_{xx} = A$, $f_{xy} = B$, $f_{yy} = C$; then the second derivative terms in the Taylor series are $Ah^2 + 2Bhk + Ck^2$; this can be written $A(h + Bk/A)^2 + (C - B^2/A)k^2$. Find out when this expression is positive for *all* small h, k [that is, all (x, y) near (a, b)]; also find out when it is negative for all small h, k, and when it has both positive and negative values for small h, k.

Use the facts stated in Problem 2 to find the maximum and minimum points of the functions in Problems 3 to 6.

3. $x^2 + y^2 + 2x - 4y + 10$

4. $x^2 - y^2 + 2x - 4y + 10$

5. $4 + x + y - x^2 - xy - \frac{1}{2}y^2$

6. $x^3 - y^3 - 2xy + 2$

7. Given $z = (y - x^2)(y - 2x^2)$, show that z has neither a maximum nor a minimum at $(0, 0)$, although z has a minimum on every straight line through $(0, 0)$.

8. A roof gutter is to be made from a long strip of sheet metal, 24 cm wide, by bending up equal amounts at each side through equal angles. Find the angle and the dimensions that will make the carrying capacity of the gutter as large as possible.

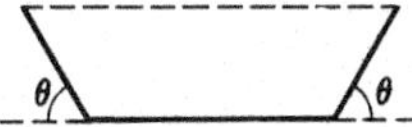

9. An aquarium with rectangular sides and bottom (and no top) is to hold 5 gal. Find its proportions so that it will use the least amount of material.

10. Repeat Problem 9 if the bottom is to be three times as thick as the sides.

11. Find the most economical proportions for a tent as in the figure, with no floor.

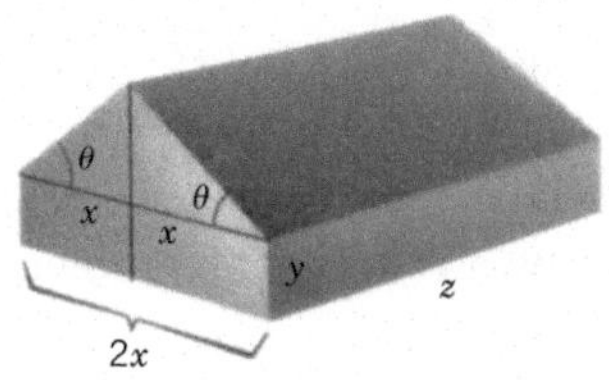

12. Find the shortest distance from the origin to the surface $z = xy + 5$.

13. Given particles of masses m, $2m$, and $3m$ at the points $(0, 1)$, $(1, 0)$, and $(2, 3)$, find the point P about which their total moment of inertia will be least. (Recall that to find the moment of inertia of m about P, you multiply m by the square of its distance from P.)

14. Repeat Problem 13 for masses m_1, m_2, m_3 at (x_1, y_1), (x_2, y_2), (x_3, y_3). Show that the point you find is the center of mass.

15. Find the point on the line through $(1, 0, 0)$ and $(0, 1, 0)$ that is closest to the line $x = y = z$. Also find the point on the line $x = y = z$ that is closest to the line through $(1, 0, 0)$ and $(0, 1, 0)$.

16. To find the best straight line fit to a set of data points (x_n, y_n) in the "least squares" sense means the following: Assume that the equation of the line is $y = mx + b$ and verify that the vertical deviation of the line from the point (x_n, y_n) is $y_n - (mx_n + b)$. Write $S =$ sum of the squares of the deviations, substitute the given values of x_n, y_n to give S as a function of m and b, and then find m and b to minimize S.

Carry through this routine for the set of points: $(-1, -2)$, $(0, 0)$, $(1, 3)$. Check your results by computer, and also computer plot (on the same axes) the given points and the approximating line.

17. Repeat Problem 16 for each of the following sets of data points.

(a) $(1, 0)$, $(2, -1)$, $(3, -8)$

(b) $(-2, -6)$, $(-1, -3)$, $(0, 0)$, $(1, 9/2)$, $(2, 7)$

(c) $(-2, 4)$, $(-1, 0)$, $(0, -1)$, $(1, -8)$, $(2, -10)$

9. MAXIMUM AND MINIMUM PROBLEMS WITH CONSTRAINTS; LAGRANGE MULTIPLIERS

Some examples will illustrate these methods.

Example 1. A wire is bent to fit the curve $y = 1 - x^2$ (Figure 9.1). A string is stretched from the origin to a point (x, y) on the curve. Find (x, y) to minimize the length of the string.

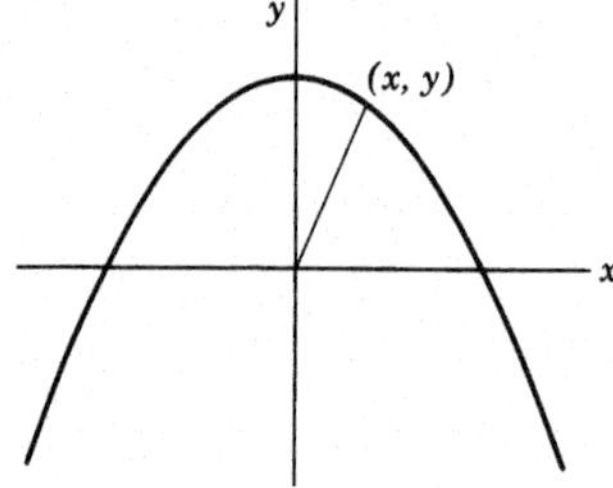

Figure 9.1

We want to minimize the distance $d = \sqrt{x^2 + y^2}$ from the origin to the point (x, y); this is equivalent to minimizing $f = d^2 = x^2 + y^2$. But x and y are not independent; they are related by the equation of the curve. This extra relation between the variables is what we mean by a *constraint.* Problems involving constraints occur frequently in applications.

There are several ways to do a problem like this. We shall discuss the following methods: (a) elimination, (b) implicit differentiation, (c) Lagrange multipliers.

(a) Elimination The most obvious method is to eliminate y. Then we want to minimize

$$f = x^2 + (1 - x^2)^2 = x^2 + 1 - 2x^2 + x^4 = x^4 - x^2 + 1.$$

This is just an ordinary calculus problem:

$$\frac{df}{dx} = 4x^3 - 2x = 0, \quad x = 0, \quad \text{or} \quad x = \pm\sqrt{\frac{1}{2}}.$$

It is not immediately obvious which of these points is a maximum and which is a minimum, so in this simple problem it is worth while to find the second derivative:

$$\frac{d^2 f}{dx^2} = 12x^2 - 2 = \begin{cases} -2 & \text{at} \quad x = 0 \qquad \text{(relative maximum)}, \\ 4 & \text{at} \quad x = \pm\sqrt{1/2} \quad \text{(minimum)}. \end{cases}$$

The minimum we wanted then occurs at $x = \pm\sqrt{1/2}$, $y = 1/2$.

(b) Implicit Differentiation Suppose it had not been possible to solve for y and substitute; we could still do the problem. From $f = x^2 + y^2$, we find

$$df = 2x\,dx + 2y\,dy \quad \text{or} \quad \frac{df}{dx} = 2x + 2y\frac{dy}{dx}. \tag{9.1}$$

From an equation like $y = 1 - x^2$ relating x and y, we could find dy in terms of dx even if the equation were not solvable for y. Here we get

$$dy = -2x\,dx.$$

Eliminating dy from df, we have

$$df = (2x - 4xy)dx \quad \text{or} \quad \frac{df}{dx} = 2x - 4xy.$$

To minimize f, we set $df/dx = 0$ (or in the differential notation we set $df = 0$ for arbitrary dx). This gives

$$2x - 4xy = 0.$$

This equation must now be solved simultaneously with the equation of the curve $y = 1 - x^2$. We get $2x - 4x(1 - x^2) = 0$, $x = 0$ or $\pm\sqrt{1/2}$ as before.

To test for maxima or minima we need d^2f/dx^2. Differentiating df/dx in (9.1) with respect to x, we get

$$\frac{d^2f}{dx^2} = 2 + 2\left(\frac{dy}{dx}\right)^2 + 2y\frac{d^2y}{dx^2}.$$

At $x = 0$, we find $y = 1$, $dy/dx = 0$, $d^2y/dx^2 = -2$, so

$$\frac{d^2f}{dx^2} = 2 - 4 = -2;$$

this is a maximum point. At $x = \pm\sqrt{1/2}$, we find

$$y = \frac{1}{2}, \quad \frac{dy}{dx} = \mp\sqrt{2}, \quad \frac{d^2y}{dx^2} = -2,$$

so

$$\frac{d^2f}{dx^2} = 2 + 4 - 2 = 4;$$

this point is the required minimum. Notice particularly here that you could do every step of (b) even if the equation of the curve could not be solved for y.

We can do problems with several independent variables by methods similar to those we have just used in Example 1. Consider this problem.

Example 2. Find the shortest distance from the origin to the plane $x - 2y - 2z = 3$.

We want to minimize the distance $d = \sqrt{x^2 + y^2 + z^2}$ from the origin to a point (x, y, z) on the plane. This is equivalent to minimizing $f = d^2 = x^2 + y^2 + z^2$ if $x - 2y - 2z = 3$. We can eliminate one variable, say x, from f using the equation of the plane. Then we have

$$f = (3 + 2y + 2z)^2 + y^2 + z^2.$$

Here f is a function of the two independent variables y and z, so to minimize f we set $\partial f/\partial y = 0, \partial f/\partial z = 0$.

$$\frac{\partial f}{\partial y} = 2(3 + 2y + 2z) \cdot 2 + 2y = 0,$$
$$\frac{\partial f}{\partial z} = 2(3 + 2y + 2z) \cdot 2 + 2z = 0.$$

Solving these equations for y and z, we get $y = z = -2/3$, so from the equation of the plane we get $x = 1/3$. Then

$$f_{\text{min}} = \left(\frac{1}{3}\right)^2 + \left(\frac{2}{3}\right)^2 + \left(\frac{2}{3}\right)^2 = 1, \quad d_{\text{min}} = 1.$$

It is clear from the geometry that there *is* a minimum distance from the origin to a plane; therefore this is it without a second-derivative test. (Also see Chapter 3, Section 5 for another way to do this problem.)

Problems with any number of variables *can* be done this way, or by method (b) if the equations are implicit.

(c) Lagrange Multipliers However, methods (a) and (b) can involve an enormous amount of algebra. We can shortcut this algebra by a process known as the method of *Lagrange multipliers* or undetermined multipliers. We want to consider a problem like the one we discussed in (a) or (b). In general, we want to find the maximum or minimum of a function $f(x, y)$, where x and y are related by an equation $\phi(x, y) = \text{const.}$ Then f is really a function of one variable (say x). To find the maximum or minimum points of f, we set $df/dx = 0$ or $df = 0$ as in (9.1). Since $\phi = \text{const.}$, we get $d\phi = 0$.

$$(9.2) \qquad \begin{aligned} df &= \frac{\partial f}{\partial x}dx + \frac{\partial f}{\partial y}dy = 0, \\ d\phi &= \frac{\partial \phi}{\partial x}dx + \frac{\partial \phi}{\partial y}dy = 0. \end{aligned}$$

In method (b) we solved the $d\phi$ equation for dy in terms of dx and substituted it into df; this often involves messy algebra. Instead, we shall multiply the $d\phi$ equation by λ (this is the undetermined multiplier—we shall find its value later) and add it to the df equation; then we have

$$\left(\frac{\partial f}{\partial x} + \lambda\frac{\partial \phi}{\partial x}\right) dx + \left(\frac{\partial f}{\partial y} + \lambda\frac{\partial \phi}{\partial y}\right) dy = 0. \tag{9.3}$$

We now pick λ so that

$$\frac{\partial f}{\partial y} + \lambda\frac{\partial \phi}{\partial y} = 0. \tag{9.4}$$

[That is, we pick $\lambda = -(\partial f/\partial y)/(\partial\phi/\partial y)$, but it isn't necessary to write it in this complicated form! In fact, this is exactly the point of the Lagrange multiplier λ; by using the abbreviation λ for a complicated expression, we avoid some algebra.] Then from (9.3) and (9.4) we have

$$\frac{\partial f}{\partial x} + \lambda\frac{\partial \phi}{\partial x} = 0. \tag{9.5}$$

Equations (9.4), (9.5), and $\phi(x, y) =$ const. can now be solved for the three unknowns x, y, λ. We don't actually want the value of λ, but often the algebra is simpler if we do find it and use it in finding x and y which we do want. Note that equations (9.4) and (9.5) are exactly the equations we would write if we had a function

$$F(x, y) = f(x, y) + \lambda\phi(x, y) \tag{9.6}$$

of two independent variables x and y and we wanted to find its maximum and minimum values. Actually, of course, x and y are not independent; they are related by the ϕ equation. However, (9.6) gives us a simple way of stating and remembering how to get equations (9.4) and (9.5). Thus we can state the method of Lagrange multipliers in the following way:

(9.7) To find the maximum or minimum values of $f(x, y)$ when x and y are related by the equation $\phi(x, y) =$ const., form the function $F(x, y)$ as in (9.6) and set the two partial derivatives of F equal to zero [equations (9.4) and (9.5)]. Then solve these two equations and the equation $\phi(x, y) =$ const. for the three unknowns x, y, and λ.

As a simple illustration of the method we shall do the problem of Example 1 by Lagrange multipliers. Here

$$f(x, y) = x^2 + y^2, \qquad \phi(x, y) = y + x^2 = 1,$$

and we write the equations to minimize

$$F(x, y) = f + \lambda\phi = x^2 + y^2 + \lambda(y + x^2),$$

namely

$$
\begin{aligned}
\frac{\partial F}{\partial x} &= 2x + \lambda \cdot 2x = 0, \\
\frac{\partial F}{\partial y} &= 2y + \lambda = 0.
\end{aligned} \tag{9.8}
$$

We solve these simultaneously with the ϕ equation $y + x^2 = 1$. From the first equation in (9.8), either $x = 0$ or $\lambda = -1$. If $x = 0$, $y = 1$ from the ϕ equation (and $\lambda = -2$). If $\lambda = -1$, the second equation gives $y = \frac{1}{2}$, and then the ϕ equation gives $x^2 = \frac{1}{2}$. These are the same values we had before. The method offers nothing new in testing whether we have found a maximum or a minimum, so we shall not repeat that work; if it is possible to see from the geometry or the physics what we have found, we don't bother to test.

Lagrange multipliers simplify the work enormously in more complicated problems. Consider this problem.

▶ **Example 3.** Find the volume of the largest rectangular parallelepiped (that is, box), with edges parallel to the axes, inscribed in the ellipsoid

$$\frac{x^2}{a^2} + \frac{y^2}{b^2} + \frac{z^2}{c^2} = 1.$$

Let the point (x, y, z) be the corner in the first octant where the box touches the ellipsoid. Then (x, y, z) satisfies the ellipsoid equation and the volume of the box is $8xyz$ (since there are 8 octants). Our problem is to maximize $f(x, y, z) = 8xyz$, where x, y, z are related by the ellipsoid equation

$$\phi(x, y, z) = \frac{x^2}{a^2} + \frac{y^2}{b^2} + \frac{z^2}{c^2} = 1.$$

By the method of Lagrange multipliers we write

$$F(x, y, z) = f + \lambda\phi = 8xyz + \lambda\left(\frac{x^2}{a^2} + \frac{y^2}{b^2} + \frac{z^2}{c^2}\right)$$

and set the three partial derivatives of F equal to 0:

$$
\begin{aligned}
\frac{\partial F}{\partial x} &= 8yz + \lambda \cdot \frac{2x}{a^2} = 0, \\
\frac{\partial F}{\partial y} &= 8xz + \lambda \cdot \frac{2y}{b^2} = 0, \\
\frac{\partial F}{\partial z} &= 8xy + \lambda \cdot \frac{2z}{c^2} = 0.
\end{aligned}
$$

We solve these three equations and the equation $\phi = 1$ simultaneously for x, y, z, and λ. (Although we don't *have* to find λ, it may be simpler to find it first.) Multiply the first equation by x, the second by y, and the third by z, and add to get

$$3 \cdot 8xyz + 2\lambda\left(\frac{x^2}{a^2} + \frac{y^2}{b^2} + \frac{z^2}{c^2}\right) = 0.$$

Using the equation of the ellipsoid, we can simplify this to

$$24xyz + 2\lambda = 0 \qquad \text{or} \qquad \lambda = -12xyz.$$

Substituting λ into the $\partial F/\partial x$ equation, we find that

$$8yz - 12xyz \cdot \frac{2x}{a^2} = 0.$$

From the geometry it is clear that the corner of the box should not be where y or z is equal to zero, so we divide by yz and solve for x, getting

$$x^2 = \frac{1}{3}a^2.$$

The other two equations could be solved in the same way. However, it is pretty clear from symmetry that the solutions will be $y^2 = \frac{1}{3}b^2$ and $z^2 = \frac{1}{3}c^2$. Then the maximum volume is

$$8xyz = \frac{8abc}{3\sqrt{3}}.$$

You might contrast this fairly simple algebra with what would be involved in method (a). There you would have to solve the ellipsoid equation for, say, z, substitute this into the volume formula, and then differentiate the square root. Even by method (b) you would have to find $\partial z/\partial x$ or similar expressions from the ellipsoid equation.

We should show that the Lagrange multiplier method is justified for problems involving several independent variables. We want to find maximum or minimum values of $f(x, y, z)$ if $\phi(x, y, z) = \text{const.}$ (You might note at each step that the proof could easily be extended to more variables.) We take differentials of both the f and the ϕ equations. Since $\phi = \text{const.}$, we have $d\phi = 0$. We *put* $df = 0$ because we want maximum and minimum values of f. Thus we write

$$\text{(9.9)} \qquad \begin{aligned} df &= \frac{\partial f}{\partial x}dx + \frac{\partial f}{\partial y}dy + \frac{\partial f}{\partial z}dz = 0, \\ d\phi &= \frac{\partial \phi}{\partial x}dx + \frac{\partial \phi}{\partial y}dy + \frac{\partial \phi}{\partial z}dz = 0. \end{aligned}$$

We *could* find dz from the $d\phi$ equation and substitute it into the df equation; this corresponds to method (b) and may involve complicated algebra. Instead, we form the sum $F = f + \lambda\phi$ and find, using (9.9),

$$\text{(9.10)} \qquad \begin{aligned} dF &= df + \lambda\, d\phi \\ &= \left(\frac{\partial f}{\partial x} + \lambda\frac{\partial \phi}{\partial x}\right)dx + \left(\frac{\partial f}{\partial y} + \lambda\frac{\partial \phi}{\partial y}\right)dy + \left(\frac{\partial f}{\partial z} + \lambda\frac{\partial \phi}{\partial z}\right)dz. \end{aligned}$$

There are two independent variables in this problem (since x, y, and z are related by $\phi = \text{const.}$). Suppose x and y are the independent ones; then z is determined from the ϕ equation. Similarly, dx and dy may have any values we choose, and dz is determined. Let us select λ so that

$$\text{(9.11)} \qquad \frac{\partial f}{\partial z} + \lambda\frac{\partial \phi}{\partial z} = 0.$$

Then from (9.10), for $dy = 0$, we get

$$\frac{\partial f}{\partial x} + \lambda \frac{\partial \phi}{\partial x} = 0 \tag{9.12}$$

and for $dx = 0$ we get

$$\frac{\partial f}{\partial y} + \lambda \frac{\partial \phi}{\partial y} = 0. \tag{9.13}$$

We can state a rule similar to (9.7) for obtaining equations (9.11), (9.12), and (9.13).

(9.14) To find the maximum and minimum values of $f(x, y, z)$ if $\phi(x, y, z) = \text{const.}$, we form the function $F = f + \lambda\phi$ and set the three partial derivatives of F equal to zero. We solve these equations and the equation $\phi = \text{const.}$ for x, y, z, and λ. (For a problem with still more variables there are more equations, but no change in method.)

It is interesting to consider the geometric meaning of equations (9.9) to (9.13). Recall that x, y, z are related by the equation $\phi(x, y, z) = \text{const.}$ We might, for example, think of solving the ϕ equation for $z = z(x, y)$. Then x and y are independent variables, and z is a function of them. Geometrically, $z = z(x, y)$ is a surface as in Figure 3.2. If we start at the point P of this surface (see Figure 3.2) and increase x by dx, y by dy, and z by dz as given in equation (3.6), we are at a point on the plane tangent to the surface at P. That is, the vector $d\mathbf{r} = \mathbf{i}dx + \mathbf{j}dy + \mathbf{k}dz$ ($\overrightarrow{PB}$ in Figure 3.2) lies in the tangent plane of the surface. Now the second equation in (9.9) is a dot product [see Chapter 3, equation (4.10)] of $d\mathbf{r}$ with the vector

$$\mathbf{i}\frac{\partial \phi}{\partial x} + \mathbf{j}\frac{\partial \phi}{\partial y} + \mathbf{k}\frac{\partial \phi}{\partial z}$$

(called the gradient of ϕ and written grad ϕ; see Chapter 6, Section 6ff.). We could write the second equation in (9.9) as $d\phi = (\text{grad } \phi) \cdot d\mathbf{r} = 0$. Recall [Chapter 3, equation (4.12)] that if the dot product of two vectors is zero, the vectors are perpendicular. Thus since $d\mathbf{r}$ lies anywhere in the plane tangent to the surface $\phi = \text{const.}$, (9.9) says that grad ϕ is perpendicular to this plane, or perpendicular to the surface $\phi = \text{const.}$ at P. The first of equations (9.9) says that grad f is also perpendicular to this plane. Thus grad ϕ and grad f are in the same direction, so their components are proportional; this is what equations (9.11), (9.12), and (9.13) say. We can also say that the surfaces $\phi = \text{const.}$ and $f = \text{const.}$ are tangent to each other at P; that is, they have the same tangent plane and their normals, grad ϕ and grad f, are in the same direction.

We can also use the method of Lagrange multipliers if there are several conditions (ϕ equations). Suppose we want to find the maximum or minimum of $f(x, y, z, w)$ if $\phi_1(x, y, z, w) = \text{const.}$ and $\phi_2(x, y, z, w) = \text{const.}$ There are two independent variables, say x and y. We write

$$
\begin{aligned}
df &= \frac{\partial f}{\partial x}\,dx + \frac{\partial f}{\partial y}\,dy + \frac{\partial f}{\partial z}\,dz + \frac{\partial f}{\partial w}\,dw = 0,\\
d\phi_1 &= \frac{\partial \phi_1}{\partial x}\,dx + \frac{\partial \phi_1}{\partial y}\,dy + \frac{\partial \phi_1}{\partial z}\,dz + \frac{\partial \phi_1}{\partial w}\,dw = 0, \qquad (9.15)\\
d\phi_2 &= \frac{\partial \phi_2}{\partial x}\,dx + \frac{\partial \phi_2}{\partial y}\,dy + \frac{\partial \phi_2}{\partial z}\,dz + \frac{\partial \phi_2}{\partial w}\,dw = 0.
\end{aligned}
$$

Again we *could* use the $d\phi_1$ and $d\phi_2$ equations to eliminate dz and dw from df (method b), but the algebra is forbidding! Instead, by the Lagrange multiplier method we form the function $F = f + \lambda_1\phi_1 + \lambda_2\phi_2$ and write, using (9.15),

$$
\begin{aligned}
(9.16)\qquad dF &= df + \lambda_1\,d\phi_1 + \lambda_2\,d\phi_2\\
&= \left(\frac{\partial f}{\partial x} + \lambda_1\frac{\partial \phi_1}{\partial x} + \lambda_2\frac{\partial \phi_2}{\partial x}\right)dx + \left(\frac{\partial f}{\partial y} + \lambda_1\frac{\partial \phi_1}{\partial y} + \lambda_2\frac{\partial \phi_2}{\partial y}\right)dy\\
&+ \left(\frac{\partial f}{\partial z} + \lambda_1\frac{\partial \phi_1}{\partial z} + \lambda_2\frac{\partial \phi_2}{\partial z}\right)dz + \left(\frac{\partial f}{\partial w} + \lambda_1\frac{\partial \phi_1}{\partial w} + \lambda_2\frac{\partial \phi_2}{\partial w}\right)dw.
\end{aligned}
$$

We determine λ_1 and λ_2 from the two equations

$$
(9.17)\qquad
\begin{aligned}
\frac{\partial f}{\partial z} + \lambda_1\frac{\partial \phi_1}{\partial z} + \lambda_2\frac{\partial \phi_2}{\partial z} &= 0,\\
\frac{\partial f}{\partial w} + \lambda_1\frac{\partial \phi_1}{\partial w} + \lambda_2\frac{\partial \phi_2}{\partial w} &= 0.
\end{aligned}
$$

Then for $dy = 0$, we have

$$
(9.18)\qquad \frac{\partial f}{\partial x} + \lambda_1\frac{\partial \phi_1}{\partial x} + \lambda_2\frac{\partial \phi_2}{\partial x} = 0
$$

and for $dx = 0$, we have

$$
(9.19)\qquad \frac{\partial f}{\partial y} + \lambda_1\frac{\partial \phi_1}{\partial y} + \lambda_2\frac{\partial \phi_2}{\partial y} = 0.
$$

As before, we can remember the method of finding (9.17), (9.18), and (9.19) by thinking:

(9.20) To find the maximum or minimum of f subject to the conditions $\phi_1 =$ const. and $\phi_2 =$ const., define $F = f + \lambda_1\phi_1 + \lambda_2\phi_2$ and set each of the partial derivatives of F equal to zero. Solve these equations and the ϕ equations for the variables and the λ's.

Example 4. Find the minimum distance from the origin to the intersection of $xy = 6$ with $7x + 24z = 0$.

We are to minimize $x^2 + y^2 + z^2$ subject to the two conditions $xy = 6$ and $7x + 24z = 0$. By the Lagrange multiplier method, we find the three partial derivatives of

$$
F = x^2 + y^2 + z^2 + \lambda_1(7x + 24z) + \lambda_2 xy
$$

and set each of them equal to zero. We get

$$
\begin{aligned}
2x + 7\lambda_1 + \lambda_2 y &= 0,\\
2y + \lambda_2 x &= 0, \qquad (9.21)\\
2z + 24\lambda_1 &= 0.
\end{aligned}
$$

These equations can be solved with $xy = 6$ and $7x + 24z = 0$ to get (Problem 10)

$$x = \pm 12/5, \qquad y = \pm 5/2, \qquad z = \mp 7/10.$$

Then the required minimum distance is (Problem 10)

$$d = \sqrt{x^2 + y^2 + z^2} = 5/\sqrt{2} = 3.54.$$

PROBLEMS, SECTION 9

1. What proportions will maximize the area shown in the figure (rectangle with isosceles triangles at its ends) if the perimeter is given?

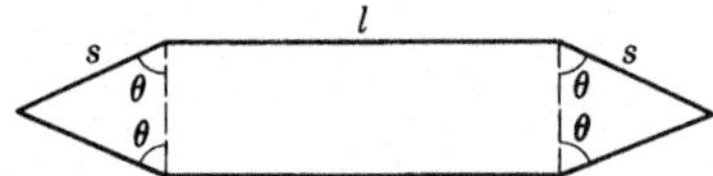

2. What proportions will maximize the volume of a projectile in the form of a circular cylinder with one conical end and one flat end, if the surface area is given?

3. Find the largest rectangular parallelepiped (box) that can be shipped by parcel post (length plus girth = 108 in).

4. Find the largest box (with faces parallel to the coordinate axes) that can be inscribed in
$$\frac{x^2}{4} + \frac{y^2}{9} + \frac{z^2}{25} = 1.$$

5. Find the point on $2x + 3y + z - 11 = 0$ for which $4x^2 + y^2 + z^2$ is a minimum.

6. A box has three of its faces in the coordinate planes and one vertex on the plane $2x + 3y + 4z = 6$. Find the maximum volume for the box.

7. Repeat Problem 6 if the plane is $ax + by + cz = d$.

8. A point moves in the (x, y) plane on the line $2x + 3y - 4 = 0$. Where will it be when the sum of the squares of its distances from $(1, 0)$ and $(-1, 0)$ is smallest?

9. Find the largest triangle that can be inscribed in the ellipse $(x^2/a^2) + (y^2/b^2) = 1$ (assume the triangle symmetric about one axis of the ellipse with one side perpendicular to this axis).

10. Complete Example 4 above.

11. Find the shortest distance from the origin to the line of intersection of the planes $2x + y - z = 1$ and $x - y + z = 2$.

12. Find the right triangular prism of given volume and least area if the base is required to be a right triangle.

10. ENDPOINT OR BOUNDARY POINT PROBLEMS

So far we have been assuming that if there is a maximum or minimum point, calculus will find it. Some simple examples (see Figures 10.1 to 10.4) show that this may not be true. Suppose, in a given problem, x can have values only between 0 and 1; this sort of restriction occurs frequently in applications. For example the *graph* of $f(x) = 2 - x^2$ exists for all real x, but if $x = |\cos\theta|$, θ real, the graph has no meaning except for $0 \le x \le 1$. As another example, suppose x is the length of a rectangle whose perimeter is 2; then $x < 0$ is meaningless in this problem since x is a length, and $x > 1$ is impossible because the perimeter is 2. Let us ask for the largest and smallest values of each of the functions in Figures 10.1 to 10.4 for $0 \le x \le 1$. In Figure 10.1, calculus will give us the minimum point, but the maximum of $f(x)$ *for x between* 0 *and* 1 occurs at $x = 1$ and cannot be obtained by calculus, since $f'(x) \neq 0$ there. In Figure 10.2, both the maximum and the minimum of $f(x)$ are at endpoints, the maximum at $x = 0$ and the minimum at $x = 1$. In Figure 10.3 a relative maximum at P and a relative minimum at Q are given by calculus, but the absolute minimum between 0 and 1 occurs at $x = 0$, and the absolute maximum at $x = 1$. Here is a practical example of this sort of function. It is said that geographers used to give as the highest point in Florida the top of the highest hill; then it was found that the highest point is on the Alabama border! [See H. A. Thurston, *American Mathematical Monthly*, vol. 68 (1961), pp. 650-652. A later paper, same journal, vol. 98, (1991), pp. 752-3, reports that the high point is actually just south of the Alabama border, but gives another example of a geographic boundary point maximum.] Figure 10.4 illustrates another way in which calculus may fail to give us a desired maximum or minimum point; here the derivative is discontinuous at the maximum point.

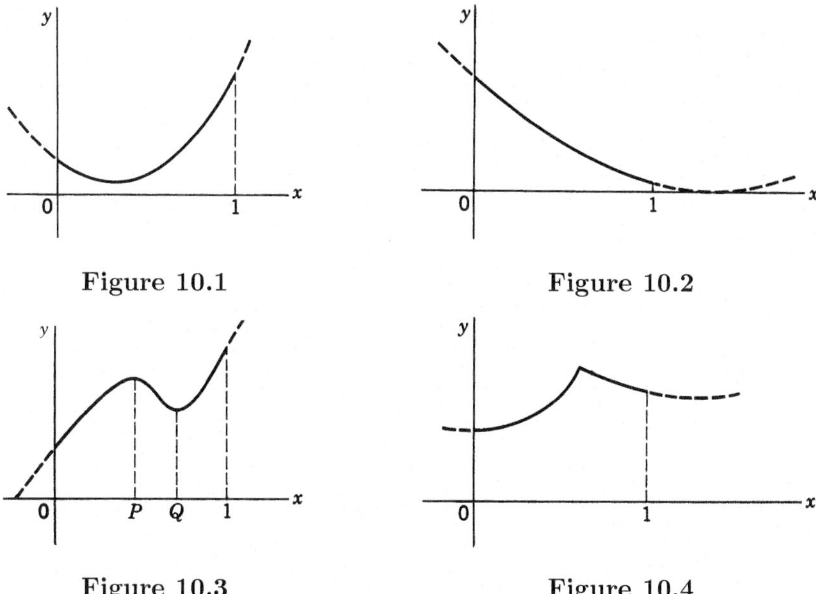

Figure 10.1

Figure 10.2

Figure 10.3

Figure 10.4

These are difficulties we must watch out for whenever there is any restriction on the values any of the variables may take (or any discontinuity in the functions or their derivatives). These restrictions are not usually stated in so many words;

you have to see them for yourself. For example, if $x^2 + y^2 = 25$, x and y are both between -5 and $+5$. If $y^2 = x^2 - 1$, then $|x|$ must be greater than or equal to 1. If $x = \csc\theta$, where θ is a first-quadrant angle, then $x \geq 1$. If $y = \sqrt{x}$, y' is discontinuous at the origin.

▶ **Example 1.** A piece of wire 40 cm long is to be used to form the perimeters of a square and a circle in such a way as to make the total area (of square and circle) a maximum. Call the radius of the circle r; then the circumference of the circle is $2\pi r$. A length $40 - 2\pi r$ is left for the four sides of the square, so one side is $10 - \frac{1}{2}\pi r$. The total area is

$$A = \pi r^2 + (10 - \tfrac{1}{2}\pi r)^2.$$

Then

$$\frac{dA}{dr} = 2\pi r + 2(10 - \tfrac{1}{2}\pi r)(-\tfrac{1}{2}\pi) = 2\pi r\left(1 + \frac{\pi}{4}\right) - 10\pi.$$

If $dA/dr = 0$, we get

$$r\left(1 + \frac{\pi}{4}\right) = 5, \qquad r = 2.8, \qquad A = 56 + .$$

Now we might think that this is the maximum area. But let us apply the second derivative test to see whether we have a maximum. We find

$$\frac{d^2A}{dr^2} = 2\pi\left(1 + \frac{\pi}{4}\right) > 0;$$

we have found the *minimum* area! The problem asks for a maximum. One way to find it would be to sketch A as a function of r and look at the graph to see where A has its largest value. A simpler way is this. A is a continuous function of r with a continuous derivative. If there were an interior maximum (that is, one between $r = 0$ and $2\pi r = 40$), calculus would find it. Therefore the maximum must be at one end or the other.

$$\text{At} \quad r = 0, \qquad A = 100.$$

$$\text{At} \quad 2\pi r = 40, \qquad r = 20/\pi, \qquad A = 400/\pi = 127 + .$$

We see that A takes its largest value at $r = 20/\pi$; $A = 400/\pi = 127+$ is then the desired maximum. It corresponds to using all the wire to make a circle; the side of the square is zero.

A similar difficulty can arise in problems with more variables.

▶ **Example 2.** The temperature in a rectangular plate bounded by the lines $x = 0$, $y = 0$, $x = 3$, and $y = 5$ is

$$T = xy^2 - x^2y + 100.$$

Find the hottest and coldest points of the plate.

We first set the partial derivatives of T equal to zero to find any interior maxima and minima. We get

$$\frac{\partial T}{\partial x} = y^2 - 2xy = 0,$$

$$\frac{\partial T}{\partial y} = 2xy - x^2 = 0.$$

The only solution of these equations is $x = y = 0$, for which $T = 100$.

We must next ask whether there are points around the boundary of the plate where T has a value larger or smaller than 100. To see that this might happen, think of a graph of T plotted as a function of x and y; this is a surface above the (x, y) plane. The mathematical surface does not have to stop at $x = 3$ and $y = 5$, but it has no meaning for our problem beyond these values. Just as for the curves in Figures 10.1 to 10.4, the graph of the temperature may be increasing or decreasing as we cross a boundary; calculus will not then give us a zero derivative even though the temperature at the boundary may be larger (or smaller) than at other points of the plate. Thus we must consider the complete boundary of the plate (*not* just the corners!). The lines $x = 0$, $y = 0$, $x = 3$, and $y = 5$ are the boundaries; we consider each of them in turn. On $x = 0$ and $y = 0$ the temperature is 100. On the line $x = 3$, we have

$$T = 3y^2 - 9y + 100.$$

We can use calculus to see whether T has maxima or minima as a function of y along this line. We have

$$\frac{dT}{dy} = 6y - 9 = 0,$$
$$y = \tfrac{3}{2}, \qquad T = 93\tfrac{1}{4}.$$

Similarly, along the line $y = 5$, we find

$$T = 25x - 5x^2 + 100,$$
$$\frac{dT}{dx} = 25 - 10x = 0,$$
$$x = \tfrac{5}{2}, \qquad T = 131\tfrac{1}{4}.$$

Finally, we must find T at the corners.

$$\text{At} \quad (0,0), (0,5), \text{ and } (3,0), \quad T = 100.$$
$$\text{At} \quad (3,5), \quad T = 130.$$

Putting all our results together, we see that the hottest point is $(\frac{5}{2}, 5)$ with $T = 131\frac{1}{4}$, and the coldest point is $(3, \frac{3}{2})$ with $T = 93\frac{1}{4}$.

Example 3. Find the point or points closest to the origin on the surfaces

$$x^2 - 4yz = 8, \tag{10.1a}$$
$$z^2 - x^2 = 1. \tag{10.1b}$$

We want to minimize $f = x^2 + y^2 + z^2$ subject to a condition [(a) or (b)]. If we eliminate x^2 in each case, we have

$$f = 8 + 4yz + y^2 + z^2, \tag{10.2a}$$
$$f = z^2 - 1 + y^2 + z^2 = 2z^2 + y^2 - 1. \tag{10.2b}$$

In both problems (a) and (b) the *mathematical function* $f(y, z)$ is defined for all y and z. *For our problems*, however, this is not true. In (a), since $x^2 \geq 0$, we have

$x^2 = 8 + 4yz \geq 0$ so we are interested in minimum values of $f(y, z)$ in (a) only in the region $yz \geq -2$. [Compare Example 2 where $T(x, y)$ was of interest only inside a rectangle.] Thus we look for "interior" minima in (a) satisfying $yz \geq -2$; then we substitute $z = -2/y$ into (10.2a) and find any minima on the boundary of the region of interest. In (b), since $x^2 = z^2 - 1 \geq 0$, we must have $z^2 \geq 1$. Again we try to find "interior" minima satisfying $z^2 \geq 1$; then we set $z^2 = 1$ and look for boundary minima. We now carry out these steps.

From (10.2a), we find

$$\text{(10.3a)} \qquad \left.\begin{aligned} \frac{\partial f}{\partial y} &= z + 2y = 0, \\ \frac{\partial f}{\partial z} &= y + 2z = 0, \end{aligned}\right\} \quad y = z = 0.$$

These values satisfy the condition $yz > -2$ and so give points inside the region of interest. We find from (10.1a), $x^2 = 8$, $x = \pm 2\sqrt{2}$; the points are $(\pm 2\sqrt{2}, 0, 0)$ at distance $2\sqrt{2}$ from the origin. Next we consider the boundary $x = 0$, $z = -2/y$; from (10.2a),

$$f = 0 + y^2 + \frac{4}{y^2}, \qquad \frac{df}{dy} = 2y - \frac{8}{y^3} = 0,$$
$$y^4 = 4, \qquad y = \pm\sqrt{2}, \qquad z = -2/y = \mp\sqrt{2}.$$

Remembering that $x = 0$, we have the points $(0, \sqrt{2}, -\sqrt{2})$ and $(0, -\sqrt{2}, \sqrt{2})$ at distance 2 from the origin. Since $2 < 2\sqrt{2}$, these boundary points are closest to the origin.

$$\text{(10.4a)} \qquad \text{Answer to (a): } (0, \sqrt{2}, -\sqrt{2}), (0, -\sqrt{2}, \sqrt{2}).$$

From (10.2b) we find

$$\text{(10.3b)} \qquad \left.\begin{aligned} \frac{\partial f}{\partial y} &= 2y = 0, \\ \frac{\partial f}{\partial z} &= 4z = 0. \end{aligned}\right\} \quad y = z = 0.$$

Since $z = 0$ does not satisfy $z^2 \geq 1$, there is no minimum point *inside the region of interest*, so we look at the boundary $z^2 = 1$. From (10.1b), $x = 0$, and from (10.2b)

$$f = y^2 + 1, \qquad \frac{df}{dy} = 2y = 0, \qquad y = 0.$$

Thus we find the points $(0, 0, \pm 1)$ at distance 1 from the origin. Since the geometry tells us that there must be a point or points closest to the origin, and calculus tells us that these are the only possible minimum points, these must be the desired points.

$$\text{(10.4b)} \qquad \text{Answer to (b): } (0, 0, \pm 1).$$

In both these problems, we could have avoided having to consider the boundary of the region of interest by eliminating z to obtain f as a function of x and y. Since

x and y are allowed by (10.1a) or (10.1b) to take *any* values, there are no boundaries to the region of interest. In (b) this is a satisfactory method; in (a) the algebra is complicated. In both problems, Lagrange multipliers offer a more routine method. For example, in (a) we write

$$F = x^2 + y^2 + z^2 + \lambda(x^2 - 4yz);$$

$$\frac{\partial F}{\partial x} = 2x(1+\lambda) = 0, \qquad x = 0 \text{ or } \lambda = -1;$$

$$\frac{\partial F}{\partial y} = 2y - 4\lambda z = 0; \qquad \text{if } \lambda = -1,\ y = z = 0,\ x^2 = 8;$$

$$\frac{\partial F}{\partial z} = 2z - 4\lambda y = 0; \qquad \text{if } x = 0,\ \lambda = \frac{y}{2z} = \frac{z}{2y},\ y^2 = z^2 = 2.$$

We obtain the same results as above, namely, the points $(\pm 2\sqrt{2}, 0, 0)$, $(0, \pm\sqrt{2}, \mp\sqrt{2})$; the points $(0, \sqrt{2}, -\sqrt{2})$, $(0, -\sqrt{2}, \sqrt{2})$ are closer to the origin by inspection. Part (b) can be done similarly (Problem 14).

We see that using Lagrange multipliers may simplify maximum and minimum problems. However, the Lagrange multiplier method still relies on calculus; consequently, it can work only if the maximum and minimum can be found by calculus using *some* set of variables (x and y, *not* y and z, in Example 3). For example, a problem in which the maximum or minimum occurs at endpoints in all variables cannot be done by *any* method that depends on setting derivatives equal to zero.

Example 4. Find the maximum value of $y - x$ for nonnegative x and y if $x^2 + y^2 = 1$.

Here we must have both x and y between 0 and 1. Then the values $y = 1$ and $x = 0$ give $y - x$ its largest value; these are both endpoint values which cannot be found by calculus.

PROBLEMS, SECTION 10

1. Find the shortest distance from the origin to $x^2 - y^2 = 1$.

2. Find the largest and smallest distances from the origin to the conic whose equation is $5x^2 - 6xy + 5y^2 - 32 = 0$ and hence determine the lengths of the semiaxes of this conic.

3. Repeat Problem 2 for the conic $6x^2 + 4xy + 3y^2 = 28$.

Find the shortest distance from the origin to each of the following quadric surfaces. *Hint:* See Example 3 above.

4. $3x^2 + y^2 - 4xz = 4$.

5. $2z^2 + 6xy = 3$.

6. $4y^2 + 2z^2 + 3xy = 18$.

7. Find the largest z for which $2x + 4y = 5$ and $x^2 + z^2 = 2y$.

8. If the temperature at the point (x, y, z) is $T = xyz$, find the hottest point (or points) on the surface of the sphere $x^2 + y^2 + z^2 = 12$, and find the temperature there.

9. The temperature T of the disk $x^2 + y^2 \le 1$ is given by $T = 2x^2 - 3y^2 - 2x$. Find the hottest and coldest points of the disk.

10. The temperature at a point (x, y, z) in the ball $x^2+y^2+z^2 \le 1$ is given by $T = y^2+xz$. Find the largest and smallest values which T takes

(a) on the circle $y = 0$, $x^2 + z^2 = 1$,

(b) on the surface $x^2 + y^2 + z^2 = 1$,

(c) in the whole ball.

11. The temperature of a rectangular plate bounded by the lines $x = \pm 1$, $y = \pm 1$, is given by $T = 2x^2 - 3y^2 - 2x + 10$. Find the hottest and coldest points of the plate.

12. Find the largest and smallest values of the sum of the acute angles that a line through the origin makes with the three coordinate axes.

13. Find the largest and smallest values of the sum of the acute angles that a line through the origin makes with the three coordinate planes.

14. Do Example 3b using Lagrange multipliers.

11. CHANGE OF VARIABLES

One important use of partial differentiation is in making changes of variables (for example, from rectangular to polar coordinates). This may give a simpler expression or a simpler differential equation or one more suited to the physical problem one is doing. For example, if you are working with the vibration of a circular membrane, or the flow of heat in a circular cylinder, polar coordinates are better; for a problem about sound waves in a room, rectangular coordinates are better. Consider the following problems.

Example 1. Make the change of variables $r = x + vt$, $s = x - vt$ in the *wave equation*

$$\frac{\partial^2 F}{\partial x^2} - \frac{1}{v^2}\frac{\partial^2 F}{\partial t^2} = 0, \tag{11.1}$$

and solve the equation. (Also see Chapter 13, Sections 1, 4, and 6.)

We use the equations

$$\begin{aligned} r &= x + vt, \\ s &= x - vt, \end{aligned} \tag{11.2}$$

and equations like (7.2) to find

$$\begin{aligned} \frac{\partial F}{\partial x} &= \frac{\partial F}{\partial r}\frac{\partial r}{\partial x} + \frac{\partial F}{\partial s}\frac{\partial s}{\partial x} = \frac{\partial F}{\partial r} + \frac{\partial F}{\partial s} = \left(\frac{\partial}{\partial r} + \frac{\partial}{\partial s}\right)F, \\ \frac{\partial F}{\partial t} &= \frac{\partial F}{\partial r}\frac{\partial r}{\partial t} + \frac{\partial F}{\partial s}\frac{\partial s}{\partial t} = v\frac{\partial F}{\partial r} - v\frac{\partial F}{\partial s} = v\left(\frac{\partial}{\partial r} - \frac{\partial}{\partial s}\right)F. \end{aligned} \tag{11.3}$$

It is helpful to say in words what we have written in (11.3): To find the partial of a function with respect to x, we find its partial with respect to r plus its partial with respect to s; to find the partial with respect to t, we find the partial with respect to r minus the partial with respect to s and multiply by the constant v. It is useful to write this in *operator* notation (see Chapter 3, Section 7):

$$\frac{\partial}{\partial x} = \frac{\partial}{\partial r} + \frac{\partial}{\partial s}, \qquad \frac{\partial}{\partial t} = v\left(\frac{\partial}{\partial r} - \frac{\partial}{\partial s}\right). \tag{11.4}$$

Then from (11.3) and (11.4) we find

(11.5)
$$\frac{\partial^2 F}{\partial x^2} = \frac{\partial}{\partial x}\left(\frac{\partial F}{\partial x}\right) = \left(\frac{\partial}{\partial r} + \frac{\partial}{\partial s}\right)\left(\frac{\partial F}{\partial r} + \frac{\partial F}{\partial s}\right) = \frac{\partial^2 F}{\partial r^2} + 2\frac{\partial^2 F}{\partial r\partial s} + \frac{\partial^2 F}{\partial s^2},$$
$$\frac{\partial^2 F}{\partial t^2} = \frac{\partial}{\partial t}\left(\frac{\partial F}{\partial t}\right) = v\left(\frac{\partial}{\partial r} - \frac{\partial}{\partial s}\right)\left(v\frac{\partial F}{\partial r} - v\frac{\partial F}{\partial s}\right) = v^2\left(\frac{\partial^2 F}{\partial r^2} - 2\frac{\partial^2 F}{\partial r\partial s} + \frac{\partial^2 F}{\partial s^2}\right).$$

Substitute (11.5) into (11.1) to get

(11.6)
$$\frac{\partial^2 F}{\partial x^2} - \frac{1}{v^2}\frac{\partial^2 F}{\partial t^2} = 4\frac{\partial^2 F}{\partial r\partial s} = 0.$$

We can easily solve (11.6). We have

$$\frac{\partial^2 F}{\partial r\partial s} = \frac{\partial}{\partial r}\left(\frac{\partial F}{\partial s}\right) = 0,$$

that is, the r derivative of $\partial F/\partial s$ is zero. Then $\partial F/\partial s$ must be independent of r, so $\partial F/\partial s$ = some function of s alone. We integrate with respect to s to find $F = f(s) +$ "const."; the "constant" is a constant as far as s is concerned, but it may be any function of r, say $g(r)$, since $(\partial/\partial s)g(r) = 0$. Thus we find that the solution of (11.6) is

(11.7)
$$F = f(s) + g(r).$$

Then, using (11.2), we find the solution of (11.1):

(11.8)
$$F = f(x - vt) + g(x + vt),$$

where f and g are arbitrary functions. This is known as d'Alembert's solution of the wave equation. Also see Problem 7.23 and Chapter 13, Problem 1.2.

Example 2. Write the Laplace equation

(11.9)
$$\frac{\partial^2 F}{\partial x^2} + \frac{\partial^2 F}{\partial y^2} = 0$$

in terms of polar coordinates r, θ, where

(11.10)
$$x = r\cos\theta,$$
$$y = r\sin\theta.$$

Note that equations (11.10) give the old variables x and y in terms of the new ones, r and θ, whereas (11.2) gave the new variables r and s in terms of the old ones. In this situation, there are several ways to get equations like (11.3). One way is to write

(11.11)
$$\frac{\partial F}{\partial r} = \frac{\partial F}{\partial x}\frac{\partial x}{\partial r} + \frac{\partial F}{\partial y}\frac{\partial y}{\partial r} = \cos\theta\,\frac{\partial F}{\partial x} + \sin\theta\,\frac{\partial F}{\partial y},$$
$$\frac{\partial F}{\partial \theta} = \frac{\partial F}{\partial x}\frac{\partial x}{\partial \theta} + \frac{\partial F}{\partial y}\frac{\partial y}{\partial \theta} = -r\sin\theta\,\frac{\partial F}{\partial x} + r\cos\theta\,\frac{\partial F}{\partial y},$$

and then solve (11.11) for $\partial F/\partial x$ and $\partial F/\partial y$ (Problem 5). Another way is to find the needed partial derivatives of r and θ with respect to x and y [for methods and results, see Section 7, Example 6, equation (7.16) and Problem 7.9] and then write as in (11.3), using (7.16),

$$\begin{aligned}\frac{\partial F}{\partial x} &= \frac{\partial F}{\partial r}\frac{\partial r}{\partial x} + \frac{\partial F}{\partial \theta}\frac{\partial \theta}{\partial x} = \cos\theta\,\frac{\partial F}{\partial r} - \frac{\sin\theta}{r}\frac{\partial F}{\partial \theta},\\ \frac{\partial F}{\partial y} &= \frac{\partial F}{\partial r}\frac{\partial r}{\partial y} + \frac{\partial F}{\partial \theta}\frac{\partial \theta}{\partial y} = \sin\theta\,\frac{\partial F}{\partial r} + \frac{\cos\theta}{r}\frac{\partial F}{\partial \theta}.\end{aligned} \tag{11.12}$$

In finding the second derivatives, it will be convenient to use the abbreviations $G = \partial F/\partial x$ and $H = \partial F/\partial y$. Thus,

$$\begin{aligned}G &= \frac{\partial F}{\partial x} = \cos\theta\,\frac{\partial F}{\partial r} - \frac{\sin\theta}{r}\frac{\partial F}{\partial \theta},\\ H &= \frac{\partial F}{\partial y} = \sin\theta\,\frac{\partial F}{\partial r} + \frac{\cos\theta}{r}\frac{\partial F}{\partial \theta}.\end{aligned} \tag{11.13}$$

Then

$$\frac{\partial^2 F}{\partial x^2} = \frac{\partial G}{\partial x}, \qquad \frac{\partial^2 F}{\partial y^2} = \frac{\partial H}{\partial y}, \qquad \text{so} \qquad \frac{\partial^2 F}{\partial x^2} + \frac{\partial^2 F}{\partial y^2} = \frac{\partial G}{\partial x} + \frac{\partial H}{\partial y}. \tag{11.14}$$

Now equations (11.12) are correct for *any* function F; in particular they are correct if we replace F by G or by H. Let us replace F by G in the first equation (11.12) and replace F by H in the second equation. Then we have

$$\begin{aligned}\frac{\partial G}{\partial x} &= \cos\theta\,\frac{\partial G}{\partial r} - \frac{\sin\theta}{r}\frac{\partial G}{\partial \theta},\\ \frac{\partial H}{\partial y} &= \sin\theta\,\frac{\partial H}{\partial r} + \frac{\cos\theta}{r}\frac{\partial H}{\partial \theta}.\end{aligned} \tag{11.15}$$

Substituting (11.15) into (11.14), we get

$$\frac{\partial^2 F}{\partial x^2} + \frac{\partial^2 F}{\partial y^2} = \cos\theta\,\frac{\partial G}{\partial r} + \sin\theta\,\frac{\partial H}{\partial r} + \frac{1}{r}\left(\cos\theta\,\frac{\partial H}{\partial \theta} - \sin\theta\,\frac{\partial G}{\partial \theta}\right). \tag{11.16}$$

We find the four partial derivatives of G and H which we need in (11.16), by differentiating the right-hand sides of equations (11.13).

$$\begin{aligned}\frac{\partial G}{\partial r} &= \cos\theta\,\frac{\partial^2 F}{\partial r^2} - \frac{\sin\theta}{r}\frac{\partial^2 F}{\partial r\partial\theta} + \frac{\sin\theta}{r^2}\frac{\partial F}{\partial \theta},\\ \frac{\partial H}{\partial r} &= \sin\theta\,\frac{\partial^2 F}{\partial r^2} + \frac{\cos\theta}{r}\frac{\partial^2 F}{\partial r\partial\theta} - \frac{\cos\theta}{r^2}\frac{\partial F}{\partial \theta},\\ \frac{\partial H}{\partial \theta} &= \sin\theta\,\frac{\partial^2 F}{\partial \theta\partial r} + \cos\theta\frac{\partial F}{\partial r} + \frac{\cos\theta}{r}\frac{\partial^2 F}{\partial \theta^2} - \frac{\sin\theta}{r}\frac{\partial F}{\partial \theta},\\ \frac{\partial G}{\partial \theta} &= \cos\theta\,\frac{\partial^2 F}{\partial \theta\partial r} - \sin\theta\frac{\partial F}{\partial r} - \frac{\sin\theta}{r}\frac{\partial^2 F}{\partial \theta^2} - \frac{\cos\theta}{r}\frac{\partial F}{\partial \theta}.\end{aligned} \tag{11.17}$$

We combine these to obtain the expressions needed in (11.16):

$$\begin{aligned}\cos\theta\,\frac{\partial G}{\partial r} + \sin\theta\,\frac{\partial H}{\partial r} &= \frac{\partial^2 F}{\partial r^2},\\ \frac{1}{r}\left(\cos\theta\,\frac{\partial H}{\partial \theta} - \sin\theta\,\frac{\partial G}{\partial \theta}\right) &= \frac{1}{r}\left(\frac{\partial F}{\partial r} + \frac{1}{r}\frac{\partial^2 F}{\partial \theta^2}\right).\end{aligned} \tag{11.18}$$

Finally, substituting (11.18) into (11.16) gives

$$\begin{aligned}\frac{\partial^2 F}{\partial x^2}+\frac{\partial^2 F}{\partial y^2}&=\frac{\partial^2 F}{\partial r^2}+\frac{1}{r}\frac{\partial F}{\partial r}+\frac{1}{r^2}\frac{\partial^2 F}{\partial \theta^2}\\&=\frac{1}{r}\frac{\partial}{\partial r}\left(r\frac{\partial F}{\partial r}\right)+\frac{1}{r^2}\frac{\partial^2 F}{\partial \theta^2}.\end{aligned} \tag{11.19}$$

We next discuss a simple kind of change of variables which is very useful in thermodynamics and mechanics. This process is sometimes known as a *Legendre transformation.* Suppose we are given a function $f(x,y)$; then we can write

$$df=\frac{\partial f}{\partial x}dx+\frac{\partial f}{\partial y}dy. \tag{11.20}$$

Let us call $\partial f/\partial x = p$, and $\partial f/\partial y = q$; then we have

$$df = p\,dx + q\,dy. \tag{11.21}$$

If we now subtract from df the quantity $d(qy)$, we have

$$\begin{aligned}df-d(qy)&=p\,dx+q\,dy-q\,dy-y\,dq \qquad \text{or}\\ d(f-qy)&=p\,dx-y\,dq.\end{aligned} \tag{11.22}$$

If we define the function g by

$$g=f-qy, \tag{11.23}$$

then by (11.22)

$$dg = p\,dx - y\,dq. \tag{11.24}$$

Because dx and dq appear in (11.24), it is convenient to think of g as a function of x and q. The partial derivatives of g are then of simple form, namely,

$$\frac{\partial g}{\partial x}=p, \qquad \frac{\partial g}{\partial q}=-y. \tag{11.25}$$

Similarly, we could replace the $p\,dx$ term in df by $-x\,dp$ by considering the function $f-xp$. This sort of change of independent variables is called a Legendre transformation. (For applications, see Problems 10 to 13.) For a discussion of Legendre transformations, see Callen, Chapter 5.

From the equations above, we can find useful relations between partial derivatives. For example, from equations (11.24) and (11.25) we can write

$$\frac{\partial^2 g}{\partial q\partial x}=\left(\frac{\partial p}{\partial q}\right)_x \qquad \text{and} \qquad \frac{\partial^2 g}{\partial x\partial q}=-\left(\frac{\partial y}{\partial x}\right)_q. \tag{11.26}$$

Assuming $\dfrac{\partial^2 g}{\partial q\partial x}=\dfrac{\partial^2 g}{\partial x\partial q}$ (reciprocity relations, see end of Section 1), then we have

$$\left(\frac{\partial p}{\partial q}\right)_x=-\left(\frac{\partial y}{\partial x}\right)_q. \tag{11.27}$$

Many equations like these appear in thermodynamics (see Problems 12 and 13).

PROBLEMS, SECTION 11

1. In the partial differential equation

$$\frac{\partial^2 z}{\partial x^2} - 5\frac{\partial^2 z}{\partial x \partial y} + 6\frac{\partial^2 z}{\partial y^2} = 0$$

put $s = y + 2x$, $t = y + 3x$ and show that the equation becomes $\partial^2 z/\partial s \partial t = 0$. Following the method of solving (11.6), solve the equation.

2. As in Problem 1, solve

$$2\frac{\partial^2 z}{\partial x^2} + \frac{\partial^2 z}{\partial x \partial y} - 10\frac{\partial^2 z}{\partial y^2} = 0$$

by making the change of variables $u = 5x - 2y$, $v = 2x + y$.

3. Suppose that $w = f(x, y)$ satisfies

$$\frac{\partial^2 w}{\partial x^2} - \frac{\partial^2 w}{\partial y^2} = 1.$$

Put $x = u + v$, $y = u - v$, and show that w satisfies $\partial^2 w/\partial u \partial v = 1$. Hence solve the equation.

4. Verify the chain rule formulas

$$\frac{\partial F}{\partial x} = \frac{\partial F}{\partial r}\frac{\partial r}{\partial x} + \frac{\partial F}{\partial \theta}\frac{\partial \theta}{\partial x},$$

and similar formulas for

$$\frac{\partial F}{\partial y}, \qquad \frac{\partial F}{\partial r}, \qquad \frac{\partial F}{\partial \theta},$$

using differentials. For example, write

$$dF = \frac{\partial F}{\partial r}dr + \frac{\partial F}{\partial \theta}d\theta$$

and substitute for dr and $d\theta$:

$$dr = \frac{\partial r}{\partial x}dx + \frac{\partial r}{\partial y}dy \qquad \text{(and similarly } d\theta).$$

Collect coefficients of dx and dy; these are the values of $\partial F/\partial x$ and $\partial F/\partial y$.

5. Solve equations (11.11) to get equations (11.12).

6. Reduce the equation

$$x^2\left(\frac{d^2 y}{dx^2}\right) + 2x\left(\frac{dy}{dx}\right) - 5y = 0$$

to a differential equation with constant coefficients in d^2y/dz^2, dy/dz, and y by the change of variable $x = e^z$. (See Chapter 8, Section 7d.)

7. Change the independent variable from x to θ by $x = \cos\theta$ and show that the Legendre equation

$$(1 - x^2)\frac{d^2 y}{dx^2} - 2x\frac{dy}{dx} + 2y = 0$$

becomes

$$\frac{d^2 y}{d\theta^2} + \cot\theta\frac{dy}{d\theta} + 2y = 0.$$

8. Change the independent variable from x to $u = 2\sqrt{x}$ in the Bessel equation

$$x^2\frac{d^2y}{dx^2} + x\frac{dy}{dx} - (1-x)y = 0$$

and show that the equation becomes

$$u^2\frac{d^2y}{du^2} + u\frac{dy}{du} + (u^2-4)y = 0.$$

9. If $x = e^s \cos t$, $y = e^s \sin t$, show that

$$\frac{\partial^2 u}{\partial x^2} + \frac{\partial^2 u}{\partial y^2} = e^{-2s}\left(\frac{\partial^2 u}{\partial s^2} + \frac{\partial^2 u}{\partial t^2}\right).$$

10. Given $du = T\,ds - p\,dv$, find a Legendre transformation giving

(a) a function $f(T,v)$;

(b) a function $h(s,p)$;

(c) a function $g(T,p)$.

Hint for (c): Perform a Legendre transformation on both terms in du.

11. Given $L(q,\dot q)$ such that $dL = \dot p\,dq + p\,d\dot q$, find $H(p,q)$ so that $dH = \dot q\,dp - \dot p\,dq$. *Comments:* L and H are functions used in mechanics called the Lagrangian and the Hamiltonian. The quantities $\dot q$ and $\dot p$ are actually time derivatives of p and q, but you make no use of the fact in this problem. Treat $\dot p$ and $\dot q$ as if they were two more variables having nothing to do with p and q. *Hint:* Use a Legendre transformation. On your first try you will probably get $-H$. Look at the text discussion of Legendre transformations and satisfy yourself that $g = qy - f$ would have been just as satisfactory as $g = f - qy$ in (11.23).

12. Using du in Problem 10, and the text method of obtaining (11.27), show that $\left(\frac{\partial T}{\partial v}\right)_s = -\left(\frac{\partial p}{\partial s}\right)_v$. (This is one of the Maxwell relations in thermodynamics.)

13. As in Problem 12, find three more Maxwell relations by using your results in Problem 10, parts (a), (b), (c).

12. DIFFERENTIATION OF INTEGRALS; LEIBNIZ' RULE

According to the definition of an integral as an antiderivative, if

$$f(x) = \frac{dF(x)}{dx}, \tag{12.1}$$

then

$$\int_a^x f(t)\,dt = F(t)\Big|_a^x = F(x) - F(a), \tag{12.2}$$

where a is a constant. If we differentiate (12.2) with respect to x, we have

$$\frac{d}{dx}\int_a^x f(t)\,dt = \frac{d}{dx}[F(x) - F(a)] = \frac{dF(x)}{dx} = f(x) \tag{12.3}$$

by (12.1). Similarly,

$$\int_x^a f(t)\,dt = F(a) - F(x),$$

so

$$\frac{d}{dx}\int_x^a f(t)\,dt = -\frac{dF(x)}{dx} = -f(x). \tag{12.4}$$

▶ **Example 1.** Find $\dfrac{d}{dx}\displaystyle\int_{\pi/4}^{x} \sin t\,dt$.

By (12.3), we find immediately that the answer is $\sin x$. We can check this by finding the integral and then differentiating. We get

$$\int_{\pi/4}^{x} \sin t\,dt = -\cos t\Big|_{\pi/4}^{x} = -\cos x + \frac{1}{2}\sqrt{2}$$

and the derivative of this is $\sin x$ as before.

By replacing x in (12.3) by v, and replacing x in (12.4) by u, we can then write

$$\frac{d}{dv}\int_{a}^{v} f(t)\,dt = f(v) \tag{12.5}$$

and

$$\frac{d}{du}\int_{u}^{b} f(t)\,dt = -f(u). \tag{12.6}$$

Suppose u and v are functions of x and we want dI/dx where

$$I = \int_{u}^{v} f(t)\,dt.$$

When the integral is evaluated, the answer depends on the limits u and v. Finding dI/dx is then a partial differentiation problem; I is a function of u and v, which are functions of x. We can write

$$\frac{dI}{dx} = \frac{\partial I}{\partial u}\frac{du}{dx} + \frac{\partial I}{\partial v}\frac{dv}{dx}. \tag{12.7}$$

But $\partial I/\partial v$ means to differentiate I with respect to v when u is a constant; this is just (12.5), so $\partial I/\partial v = f(v)$. Similarly, $\partial I/\partial u$ means that v is constant and we can use (12.6) to get $\partial I/\partial u = -f(u)$. Then we have

$$\frac{d}{dx}\int_{u(x)}^{v(x)} f(t)\,dt = f(v)\,\frac{dv}{dx} - f(u)\,\frac{du}{dx}. \tag{12.8}$$

▶ **Example 2.** Find dI/dx if $I = \displaystyle\int_{0}^{x^{1/3}} t^2\,dt$.

By (12.8) we get

$$\frac{dI}{dx} = (x^{1/3})^2\frac{d}{dx}(x^{1/3}) = x^{2/3}\cdot\frac{1}{3}x^{-2/3} = \frac{1}{3}.$$

We *could* also integrate first and then differentiate with respect to x:

$$I = \int_{0}^{x^{1/3}} t^2 dt = \frac{t^3}{3}\Big|_{0}^{x^{1/3}} = \frac{x}{3}, \qquad \frac{dI}{dx} = \frac{1}{3}.$$

This last method seems so simple you may wonder why we need (12.8). Look at another example.

Example 3. Find dI/dx if

$$I = \int_{x^2}^{\sin^{-1} x} \frac{\sin t}{t}\, dt.$$

Here the indefinite integral cannot be evaluated in terms of elementary functions; however, we can find dI/dx by using (12.8). We get

$$\begin{aligned}\frac{dI}{dx} &= \frac{\sin(\sin^{-1} x)}{\sin^{-1} x}\frac{1}{\sqrt{1-x^2}} - \frac{\sin x^2}{x^2}\cdot 2x\\ &= \frac{x}{\sqrt{1-x^2}\,\sin^{-1} x} - \frac{2}{x}\sin x^2.\end{aligned}$$

Finally, we may want to find dI/dx when $I = \int_a^b f(x,t)\,dt$, where a and b are constants. Under not too restrictive conditions,

$$\text{(12.9)} \qquad \frac{d}{dx}\int_a^b f(x,t)\,dt = \int_a^b \frac{\partial f(x,t)}{\partial x}\,dt;$$

that is, we can differentiate under the integral sign. [A set of sufficient conditions for this to be correct would be that $\int_a^b f(x,t)\,dt$ exists, $\partial f/\partial x$ is continuous and $|\partial f(x,t)/\partial x| \le g(t)$, where $\int_a^b g(t)\,dt$ exists. For most practical purposes this means that if both integrals in (12.9) exist, then (12.9) is correct.] Equation (12.9) is often useful in evaluating definite integrals.

Example 4. Find $\int_0^\infty t^n e^{-kt^2}\,dt$ for odd n, $k > 0$.

First we evaluate the integral

$$I = \int_0^\infty te^{-kt^2}\,dt = -\frac{1}{2k}e^{-kt^2}\Big|_0^\infty = \frac{1}{2k}.$$

Now we calculate successive derivatives of I with respect to k.

$$\frac{dI}{dk} = \int_0^\infty -t^2te^{-kt^2}\,dt = -\frac{1}{2k^2} \qquad\text{or}\qquad \int_0^\infty t^3e^{-kt^2}\,dt = \frac{1}{2k^2}.$$

Repeating the differentiation with respect to k, we get

$$\int_0^\infty -t^2t^3e^{-kt^2}\,dt = -\frac{2}{2k^3} \qquad\text{or}\qquad \int_0^\infty t^5e^{-kt^2}\,dt = \frac{1}{k^3}.$$

$$\int_0^\infty -t^2t^5e^{-kt^2}\,dt = -\frac{3}{k^4} \qquad\text{or}\qquad \int_0^\infty t^7e^{-kt^2} = \frac{3}{k^4}.$$

Continuing in this way (Problem 17), we can find the integral of any odd power of t times e^{-kt^2}:

$$\text{(12.10)} \qquad \int_0^\infty t^{2n+1}e^{-kt^2}\,dt = \frac{n!}{2k^{n+1}}.$$

Your computer may give you this result in terms of the gamma function (see Chapter 11, Sections 1 to 5). The relation is $n! = \Gamma(n+1)$.

Example 5. Evaluate

$$I = \int_0^1 \frac{t^a - 1}{\ln t}\, dt, \qquad a > -1. \tag{12.11}$$

First we differentiate I with respect to a, and evaluate the resulting integral.

$$\frac{dI}{da} = \int_0^1 \frac{t^a \ln t}{\ln t}\, dt = \int_0^1 t^a\, dt = \left.\frac{t^{a+1}}{a+1}\right|_0^1 = \frac{1}{a+1}.$$

Now we integrate dI/da with respect to a to get I back again (plus an integration constant):

$$I = \int \frac{da}{a+1} = \ln(a+1) + C. \tag{12.12}$$

If $a = 0$, (12.11) gives $I = 0$ and (12.12) gives $I = C$, so $C = 0$ and we have from (12.12), $I = \ln(a+1)$.

It is convenient to collect formulas (12.8) and (12.9) into one formula known as *Leibniz' rule*:

$$\frac{d}{dx}\int_{u(x)}^{v(x)} f(x,t)\, dt = f(x,v)\frac{dv}{dx} - f(x,u)\frac{du}{dx} + \int_u^v \frac{\partial f}{\partial x}\, dt. \tag{12.13}$$

Example 6. Find dI/dx if

$$I = \int_x^{2x} \frac{e^{xt}}{t}\, dt.$$

By (12.13) we get

$$\begin{aligned}\frac{dI}{dx} &= \frac{e^{x\cdot 2x}}{2x}\cdot 2 - \frac{e^{x\cdot x}}{x}\cdot 1 + \int_x^{2x} \frac{te^{xt}}{t}\, dt \\ &= \frac{1}{x}(e^{2x^2} - e^{x^2}) + \left[\frac{e^{xt}}{x}\right]_x^{2x} \\ &= \frac{1}{x}(e^{2x^2} - e^{x^2} + e^{2x^2} - e^{x^2}) = \frac{2}{x}(e^{2x^2} - e^{x^2}).\end{aligned}$$

Although you can do problems like this by computer, in many cases you can just write down the answer using (12.13) in less time than it takes to type the problem into the computer.

PROBLEMS, SECTION 12

1. If $y = \int_0^{\sqrt{x}} \sin t^2 dt$, find dy/dx.

2. If $s = \int_u^v \frac{1-e^t}{t} dt$, find $\partial s/\partial v$ and $\partial s/\partial u$ and also their limits as u and v tend to zero.

3. If $z = \int_{\sin x}^{\cos x} \frac{\sin t}{t}\,dt$, find $\frac{dz}{dx}$.

4. Use L'Hôpital's rule to evaluate $\lim_{x\to 2} \frac{1}{x-2}\int_2^x \frac{\sin t}{t}\,dt$.

5. If $u = \int_x^{y-x} \frac{\sin t}{t}\,dt$, find $\frac{\partial u}{\partial x}, \frac{\partial u}{\partial y}$, and $\frac{\partial y}{\partial x}$ at $x = \pi/2$, $y = \pi$.
Hint: Use differentials.

6. If $w = \int_{xy}^{2x-3y} \frac{du}{\ln u}$, find $\frac{\partial w}{\partial x}, \frac{\partial w}{\partial y}$, and $\frac{\partial y}{\partial x}$ at $x = 3$, $y = 1$.

7. If $\int_u^v e^{-t^2}\,dt = x$ and $u^v = y$, find $\left(\frac{\partial u}{\partial x}\right)_y, \left(\frac{\partial u}{\partial y}\right)_x$, and $\left(\frac{\partial y}{\partial x}\right)_u$ at $u = 2$, $v = 0$.

8. If $\int_0^x e^{-s^2}\,ds = u$, find $\frac{dx}{du}$.

9. If $y = \int_0^\pi \sin xt\,dt$, find dy/dx (a) by evaluating the integral and then differentiating, (b) by differentiating first and then evaluating the integral.

10. Find dy/dx explicitly if $y = \int_0^1 \frac{e^{xu}-1}{u}\,du$.

11. Find $\frac{d}{dx}\int_{3-x}^{x^2}(x-t)\,dt$ by evaluating the integral first, and by differentiating first.

12. Find $\frac{d}{dx}\int_x^{x^2}\frac{du}{\ln(x+u)}$.

13. Find $\frac{d}{dx}\int_{1/x}^{2/x}\frac{\sin xt}{t}\,dt$.

14. Given that $\int_0^\infty \frac{dx}{y^2+x^2} = \frac{\pi}{2y}$, differentiate with respect to y and so evaluate

$$\int_0^\infty \frac{dx}{(y^2+x^2)^2}.$$

15. Given that

$$\int_0^\infty e^{-ax}\sin kx\,dx = \frac{k}{a^2+k^2},$$

differentiate with respect to a to show that

$$\int_0^\infty xe^{-ax}\sin kx\,dx = \frac{2ka}{(a^2+k^2)^2}$$

and differentiate with respect to k to show that

$$\int_0^\infty xe^{-ax}\cos kx\,dx = \frac{a^2-k^2}{(a^2+k^2)^2}.$$

16. In kinetic theory we have to evaluate integrals of the form $I = \int_0^\infty t^n e^{-at^2}\,dt$. Given that $\int_0^\infty e^{-at^2}\,dt = \frac{1}{2}\sqrt{\pi/a}$, evaluate I for $n = 2, 4, 6, \cdots, 2m$.

17. Complete Example 4 to obtain (12.10).

18. Show that $u(x,y) = \dfrac{y}{\pi}\displaystyle\int_{-\infty}^{\infty} \frac{f(t)\,dt}{(x-t)^2+y^2}$ satisfies $u_{xx}+u_{yy}=0$.

19. Show that $y = \displaystyle\int_0^x f(u)\sin(x-u)\,du$ satisfies $y''+y=f(x)$.

20. (a) Show that $y = \displaystyle\int_0^x f(x-t)\,dt$ satisfies $(dy/dx)=f(x)$. (*Hint:* It is helpful to make the change of variable $x-t=u$ in the integral.)

(b) Show that $y = \displaystyle\int_0^x (x-u)f(u)\,du$ satisfies $y''=f(x)$.

(c) Show that $y = \dfrac{1}{(n-1)!}\displaystyle\int_0^x (x-u)^{n-1}f(u)\,du$ satisfies $y^{(n)}=f(x)$.

13. MISCELLANEOUS PROBLEMS

1. A function $f(x,y,z)$ is called homogeneous of degree n if $f(tx,ty,tz)=t^n f(x,y,z)$. For example, $z^2\ln(x/y)$ is homogeneous of degree 2 since

$$(tz)^2\ln\frac{tx}{ty} = t^2\left(z^2\ln\frac{x}{y}\right).$$

Euler's theorem on homogeneous functions says that if f is homogeneous of degree n, then

$$x\frac{\partial f}{\partial x}+y\frac{\partial f}{\partial y}+z\frac{\partial f}{\partial z}=nf.$$

Prove this theorem. *Hints:* Differentiate $f(tx,ty,tz)=t^n f(x,y,z)$ with respect to t, and then let $t=1$. It is convenient to call $\partial f/\partial(tx)=f_1$ (that is, the partial derivative of f with respect to its first variable), $f_2=\partial f/\partial(ty)$, and so on. Or, you can at first call $tx=u$, $ty=v$, $tz=w$. (Both the definition and the theorem can be extended to any number of variables.)

2. (a) Given the point $(2,1)$ in the (x,y) plane and the line $3x+2y=4$, find the distance from the point to the line by using the method of Chapter 3, Section 5.

(b) Solve part (a) by writing a formula for the distance from $(2,1)$ to (x,y) and minimizing the distance (use Lagrange multipliers).

(c) Derive the formula

$$D=\left|\frac{ax_0+by_0-c}{\sqrt{a^2+b^2}}\right|$$

for the distance from (x_0,y_0) to $ax+by=c$ by the methods suggested in parts (a) and (b).

In Problems 3 to 6, assume that x, y and r, θ are rectangular and polar coordinates.

3. Find $\dfrac{\partial^2 y}{\partial x\,\partial\theta}$.

4. Find $\dfrac{\partial^2 r}{\partial\theta\,\partial y}$.

5. Given $z=y^2-2x^2$, find $\left(\dfrac{\partial z}{\partial x}\right)_r$, $\left(\dfrac{\partial z}{\partial\theta}\right)_x$, $\dfrac{\partial^2 z}{\partial x\,\partial\theta}$.

6. If $z=r^2-x^2$, find $\left(\dfrac{\partial z}{\partial r}\right)_\theta$, $\left(\dfrac{\partial z}{\partial\theta}\right)_r$, $\dfrac{\partial^2 z}{\partial r\,\partial\theta}$, $\left(\dfrac{\partial z}{\partial x}\right)_y$.

7. About how much (in percent) does an error of 1% in x and y affect x^3y^2?

8. Assume that the earth is a perfect sphere. Suppose that a rope lies along the equator with its ends fastened so that it fits exactly. Now let the rope be made 2 ft longer, and let it be held up the same distance above the surface of the Earth at all points of the equator. About how high up is it? (For example, could you crawl under? Could a fly?) Answer the same questions for the moon.

9. If $z = xy$ and $\begin{cases} 2x^3 + 2y^3 = 3t^2, \\ 3x^2 + 3y^2 = 6t, \end{cases}$ find dz/dt.

10. If $w = (r\cos\theta)^{r\sin\theta}$, find $\partial w/\partial\theta$.

11. If $\dfrac{x^2}{a^2} + \dfrac{y^2}{b^2} = 1$, find $\dfrac{dy}{dx}$ and $\dfrac{d^2y}{dx^2}$ by implicit differentiation.

12. Given $z = r^2 + s^2 + rst$, $r^4 + s^4 + t^4 = 2r^2s^2t^2 + 10$, find $(\partial z/\partial r)_t$ when $r = 2$, $s = t = 1$.

13. Given $\begin{cases} 2t + e^x = s - \cos y - 2, \\ 2s - t = \sin y + x - 1, \end{cases}$ find $\left(\dfrac{\partial s}{\partial t}\right)_y$ at $(x, y, s, t) = (0, \pi/2, -1, -2)$.

14. If $w = f(x, s, t)$, $s = 2x + y$, $t = 2x - y$, find $(\partial w/\partial x)_y$ in terms of f and its derivatives.

15. If $w = f(x, x^2 + y^2, 2xy)$, find $(\partial w/\partial x)_y$ (compare Problem 14).

16. If $z = \dfrac{1}{x} f\left(\dfrac{y}{x}\right)$, prove that $x\dfrac{\partial z}{\partial x} + y\dfrac{\partial z}{\partial y} + z = 0$.

17. Find the shortest distance from the origin to the surface $x = yz + 10$.

18. Find the shortest distance from the origin to the line of intersection of the planes

$$2x - 3y + z = 5,$$
$$3x - y - 2z = 11,$$

(a) using vector methods (see Chapter 3, Section 5);

(b) using Lagrange multipliers.

19. Find by the Lagrange multiplier method the largest value of the product of three positive numbers if their sum is 1.

20. Find the largest and smallest values of $y = 4x^3 + 9x^2 - 12x + 3$ if $x = \cos\theta$.

21. Find the hottest and coldest points on a bar of length 5 if $T = 4x - x^2$, where x is the distance measured from the left end.

22. Find the hottest and coldest points of the region $y^2 \le x < 5$ if $T = x^2 - y^2 - 3x$.

23. Find $\dfrac{d}{dt}\displaystyle\int_0^{\sin t} \frac{\sin^{-1} x}{x}\, dx$.

24. Find $\dfrac{d}{dx}\displaystyle\int_{t=1/x}^{t=2/x} \frac{\cosh xt}{t}\, dt$.

25. Find $\dfrac{d}{dx}\displaystyle\int_1^{1/x} \frac{e^{xt}}{t}\, dt$.

26. Find $\dfrac{d}{dx}\displaystyle\int_0^{x^2} \frac{\sin xt}{t}\, dt$.

27. Show that $\dfrac{d}{dx}\displaystyle\int_{\cos x}^{\sin x} \sqrt{1 - t^2}\, dt = 1$.

28. In discussing the velocity distribution of molecules of an ideal gas, a function $F(x, y, z) = f(x)f(y)f(z)$ is needed such that $d(\ln F) = 0$ when $\phi = x^2 + y^2 + z^2 =$ const. Then by the Lagrange multiplier method $d(\ln F + \lambda\phi) = 0$. Use this to show that

$$F(x, y, z) = Ae^{-(\lambda/2)(x^2+y^2+z^2)}.$$

29. The time dependent temperature at a point of a long bar is given by

$$T(t) = 100^\circ \left(1 - \frac{2}{\sqrt{\pi}} \int_0^{8/\sqrt{t}} e^{-\tau^2}\, d\tau\right).$$

When $t = 64$, $T = 15.73^\circ$. Use differentials to estimate how long it will be until $T = 17^\circ$.

30. Evaluate $\dfrac{d^2}{dx^2} \displaystyle\int_0^x \int_0^x f(s,t)\, ds\, dt$.

CHAPTER 5

Multiple Integrals; Applications of Integration

1. INTRODUCTION

In calculus and elementary physics, you have seen a number of uses for integration such as finding area, volume, mass, moment of inertia, and so on. In this chapter we want to consider these and other applications of both single and multiple integrals. We shall discuss both how to set up integrals to represent physical quantities and methods of evaluating them. In later chapters we will need to use both single and multiple integrals.

Computers and integral tables are very useful in evaluating integrals. But to use these tools efficiently, you need to understand the notation and meaning of integrals which we will discuss in this chapter. There is another important point here. A computer will give you an answer for a definite integral, but an indefinite integral has many possible answers (differing from each other by a constant of integration), and your computer or integral tables may not give you the form you need. (See problems below.) If this happens, here are some ideas you can try:

(a) Look in other integral tables, or try to induce your computer to change the form.

(b) See if some algebra will give the form you want (see Problem 1 below; also see Chapter 2, Section 15, Example 2).

(c) A simple substitution may give the desired result (see Problem 2 below).

(d) To check a claimed answer, differentiate it (by hand or computer) to see whether you get the integrand.

PROBLEMS, SECTION 1

Verify each of the following answers for an indefinite integral by one or more of the methods suggested above.

1. $\int 2\sin\theta\cos\theta\,d\theta = \sin^2\theta$ or $-\cos^2\theta$ or $-\frac{1}{2}\cos 2\theta$. *Hint*: Use trig identities.

2. $\int \frac{dx}{\sqrt{x^2+a^2}} = \sinh^{-1}\frac{x}{a}$ or $\ln\left(x+\sqrt{x^2+a^2}\right)$. *Hint*: To find the $\sinh^{-1}$ form, make the substitution $x = a\sinh u$. Or see Chapter 2, Sections 15 and 17.

3. $\int \frac{dy}{\sqrt{y^2-a^2}} = \cosh^{-1}\frac{y}{a}$ or $\ln\left(y+\sqrt{y^2-a^2}\right)$. *Hint*: See Problem 2 hints.

4. $\int \sqrt{1+a^2x^2}\,dx = \frac{x}{2}\sqrt{1+a^2x^2} + \frac{1}{2a}\sinh^{-1} ax$ or
$$\frac{x}{2}\sqrt{1+a^2x^2} + \frac{1}{2a}\ln\left(ax+\sqrt{1+a^2x^2}\right).$$

5. $\int \frac{K\,dr}{\sqrt{1-K^2r^2}} = \sin^{-1}Kr$ or $-\cos^{-1}Kr$ or $\tan^{-1}\frac{Kr}{\sqrt{1-K^2r^2}}$.
Hints: Sketch a right triangle with acute angles u and v and label the sides so that $\sin u = Kr$. Also note that $u+v=\pi/2$; then if u is an indefinite integral, so is $-v$ since they differ by a constant of integration.

6. $\int \frac{K\,dr}{r\sqrt{r^2-K^2}} = \cos^{-1}\frac{K}{r}$ or $\sec^{-1}\frac{r}{K}$ or $-\sin^{-1}\frac{K}{r}$ or $-\tan^{-1}\frac{K}{\sqrt{r^2-K^2}}$.

2. DOUBLE AND TRIPLE INTEGRALS

Recall from calculus that $\int_a^b y\,dx = \int_a^b f(x)\,dx$ gives the area "under the curve" in Figure 2.1. Recall also the definition of the integral as the limit of a sum: We approximate the area by a sum of rectangles as in Figure 2.1; a representative rectangle (shaded) has width Δx. The geometry indicates that if we increase the number of rectangles and let all the widths $\Delta x \to 0$, the sum of the areas of the rectangles will tend to the area under the curve. We *define* $\int_a^b f(x)\,dx$ as the limit of the sum of the areas of the rectangles; then we *evaluate* the integral as an antiderivative, and *use* $\int_a^b f(x)\,dx$ to calculate the area under the curve.

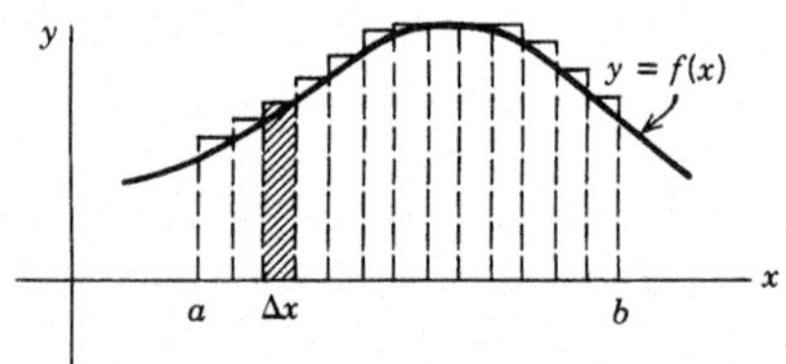

Figure 2.1

We are going to do something very similar in order to find the volume of the cylinder in Figure 2.2 under the surface $z = f(x,y)$. We cut the (x,y) plane into little rectangles of area $\Delta A = (\Delta x)(\Delta y)$ as shown in Figure 2.2; above each $\Delta x\,\Delta y$ is a tall slender box reaching up to the surface. We can approximate the desired volume by a sum of these boxes just as we approximated the area in Figure 2.1 by a sum of rectangles. As the number of boxes increases and all Δx and $\Delta y \to 0$, the geometry indicates that the sum of the volumes of the boxes will tend to the desired volume. We *define* the

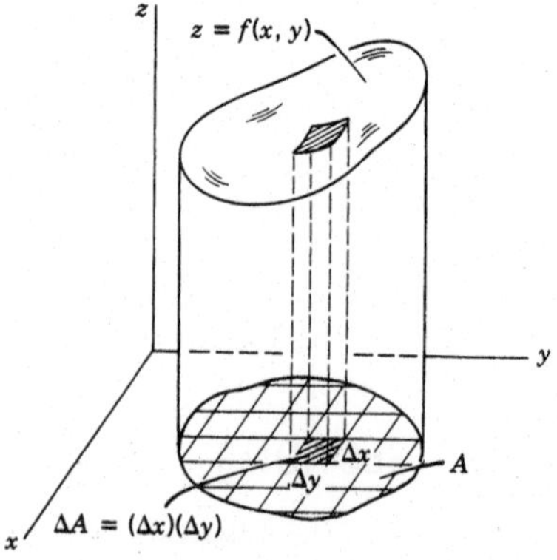

Figure 2.2

double integral of $f(x, y)$ over the area A in the (x, y) plane (Figure 2.2) as the limit of this sum, and we write it as $\iint_A f(x, y)\,dx\,dy$. Before we can use the double integral to compute volumes, however, we need to see how double integrals are evaluated. Even though we may use a computer to do the work, we need to understand the process in order to set up integrals correctly and find and correct errors. Doing some hand evaluation is a good way to learn this.

Iterated Integrals We now show by some examples the details of evaluating double integrals.

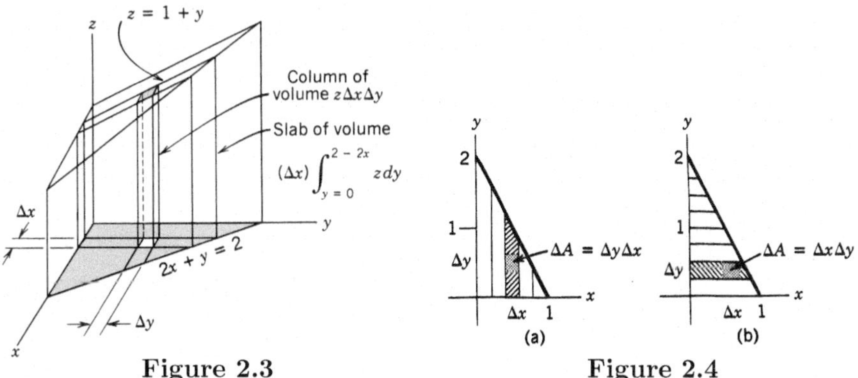

Figure 2.3

Figure 2.4

Example 1. Find the volume of the solid (Figure 2.3) below the plane $z = 1 + y$, bounded by the coordinate planes and the vertical plane $2x + y = 2$. From our discussion above, this is $\iint_A z\,dx\,dy = \iint_A (1 + y)\,dx\,dy$, where A is the shaded triangle in the (x, y) plane [A is shown also in Figure 2.4 (a and b)]. We are going to consider two ways of evaluating this double integral. We think of the triangle A cut up into little rectangles $\Delta A = \Delta x\,\Delta y$ (Figure 2.4) and the whole solid cut into vertical columns of height z and base ΔA (Figure 2.3). We want the (limit of the) sum of the volumes of these columns. First add up the columns (Figure 2.4a) for a fixed value of x producing the volume of a slab (Figure 2.3) of thickness Δx. This corresponds to integrating with respect to y (holding x constant, Figure 2.4a) from $y = 0$ to y on the line $2x + y = 2$, that is $y = 2 - 2x$; we find

$$\text{(2.1)} \qquad \int_{y=0}^{2-2x} z\,dy = \int_{y=0}^{2-2x} (1 + y)\,dy = \left(y + \frac{y^2}{2}\right)\Bigg|_0^{2-2x}$$
$$= (2 - 2x) + (2 - 2x)^2/2 = 4 - 6x + 2x^2.$$

(What we have found is the area of the slab in Figure 2.3; its volume is the area times Δx.) Now we add up the volumes of the slabs; this corresponds to integrating (2.1) with respect to x from $x = 0$ to $x = 1$:

$$\text{(2.2)} \qquad \int_{x=0}^{1} (4 - 6x + 2x^2)\,dx = \frac{5}{3}.$$

We could summarize (2.1) and (2.2) by writing

$$(2.3)\qquad \int_{x=0}^{1}\left(\int_{y=0}^{2-2x}(1+y)\,dy\right)dx \quad\text{or}\quad \int_{x=0}^{1}\int_{y=0}^{2-2x}(1+y)\,dy\,dx$$
$$\text{or}\quad \int_{x=0}^{1}dx\int_{y=0}^{2-2x}dy\,(1+y).$$

We call (2.3) an *iterated* (repeated) integral. Multiple integrals are usually evaluated by using iterated integrals. Note that the large parentheses in (2.3) are not really necessary if we are always careful to state the variable in giving the limits on an integral; that is, always write $\int_{x=0}^{1}$, *not* just $\int_0^1$.

Now we could also add up the volume $z(\Delta A)$ by first integrating with respect to x (for fixed y, Figure 2.4b) from $x = 0$ to $x = 1 - y/2$ giving the volume of a slab perpendicular to the y axis in Figure 2.3, and then add up the volumes of the slabs by integrating with respect to y from $y = 0$ to $y = 2$ (Figure 2.4b). We write

$$(2.4)\qquad \int_{y=0}^{2}\left(\int_{x=0}^{1-y/2}(1+y)\,dx\right)dy = \int_{y=0}^{2}(1+y)x\Big|_{x=0}^{1-y/2}dy$$
$$= \int_{y=0}^{2}(1+y)(1-y/2)\,dy$$
$$= \int_{y=0}^{2}(1+y/2-y^2/2)\,dy = \frac{5}{3}.$$

As the geometry would indicate, the results in (2.2) and (2.4) are the same; we have two methods of evaluating the double integral by using iterated integrals.

Often one of these two methods is more convenient than the other; we choose whichever method is easier. To see how to decide, study the following sketches of areas A over which we want to find $\iint_A f(x,y)\,dx\,dy$. In each case we think of combining little rectangles $dx\,dy$ to form strips (as shown) and then combining the strips to cover the whole area.

Areas shown in Figure 2.5: Integrate with respect to y first. Note that the top and bottom of area A are curves whose equations we know; the boundaries at $x = a$ and $x = b$ are either vertical straight lines or else points.

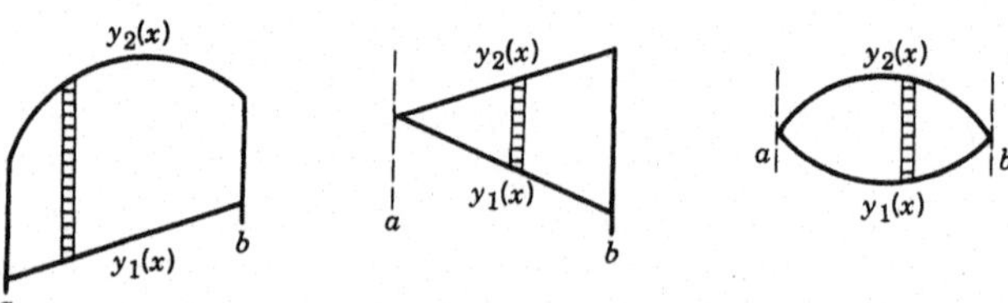

Figure 2.5

We find

$$(2.5)\qquad \iint_A f(x,y)\,dx\,dy = \int_{x=a}^{b}\left(\int_{y=y_1(x)}^{y_2(x)} f(x,y)\,dy\right)dx.$$

Areas shown in Figure 2.6: Integrate with respect to x first. Note that the sides of area A are curves whose equations we know; the boundaries at $y = c$ and $y = d$ are either horizontal straight lines or else points.

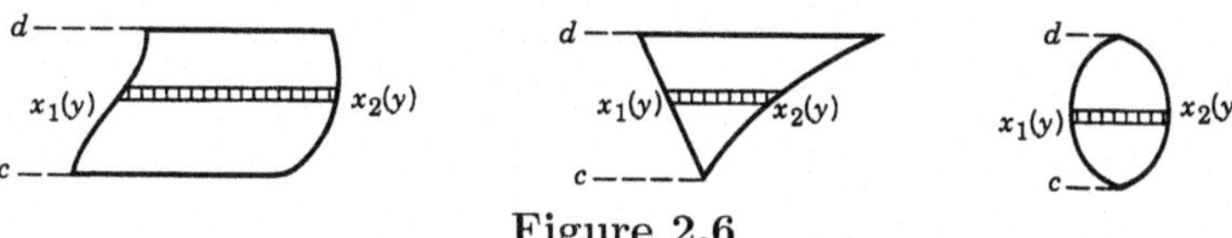

Figure 2.6

We find

$$\iint_A f(x,y)\,dx\,dy = \int_{y=c}^{d}\left(\int_{x=x_1(y)}^{x_2(y)} f(x,y)\,dx\right)dy. \tag{2.6}$$

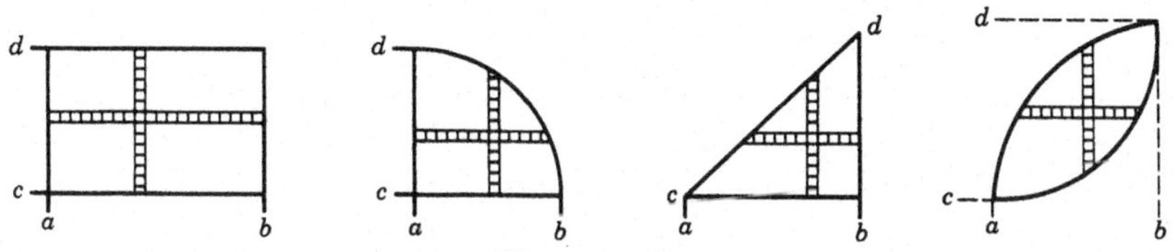

Figure 2.7

Areas shown in Figure 2.7: Integrate in either order. Note that these areas all satisfy the requirements for both (2.5) and (2.6).

We find

$$\begin{aligned}\iint_A f(x,y)\,dx\,dy &= \int_{x=a}^{b}\int_{y=y_1(x)}^{y_2(x)} f(x,y)\,dy\,dx \\ &= \int_{y=c}^{d}\int_{x=x_1(y)}^{x_2(y)} f(x,y)\,dx\,dy.\end{aligned} \tag{2.7}$$

An important special case is a double integral over a rectangle (both x and y limits are constants) when $f(x,y)$ is a product, $f(x,y) = g(x)h(y)$. Then

$$\begin{aligned}\iint_A f(x,y)\,dx\,dy &= \int_{x=a}^{b}\int_{y=c}^{d} g(x)h(y)\,dy\,dx \\ &= \left(\int_a^b g(x)\,dx\right)\left(\int_c^d h(y)\,dy\right).\end{aligned} \tag{2.8}$$

When areas are more complicated than those shown, we may break them into two or more simpler areas (Problems 9 and 10).

We have seen how to set up and evaluate double integrals to find areas and volumes. Recall, however, that we use single integrals for other purposes than finding areas. Similarly, now that we know how to evaluate a double integral, we can use it to find other quantities besides areas and volumes.

Example 2. Find the mass of a rectangular plate bounded by $x = 0$, $x = 2$, $y = 0$, $y = 1$, if its density (mass per unit area) is $f(x, y) = xy$. The mass of a tiny rectangle $\Delta A = \Delta x\,\Delta y$ is approximately $f(x, y)\,\Delta x\,\Delta y$, where $f(x, y)$ is evaluated at some point in ΔA. We want to add up the masses of all the ΔA's; this is what we find by evaluating the double integral of $dM = xy\,dx\,dy$. We call dM an *element* of mass and think of adding up all the dM's to get M.

$$M = \iint_A xy\,dx\,dy = \int_{x=0}^{2}\int_{y=0}^{1} xy\,dx\,dy \tag{2.9}$$
$$= \left(\int_0^2 x\,dx\right)\left(\int_0^1 y\,dy\right) = 2 \cdot \frac{1}{2} = 1.$$

A triple integral of $f(x, y, z)$ over a volume V, written $\iiint_V f(x, y, z)\,dx\,dy\,dz$, is also defined as the limit of a sum and is evaluated by an iterated integral. If the integral is over a box, that is, all limits are constants, then we can do the x, y, z integrations in any order. If the volume is complicated, then we have to consider the geometry as we did for double integrals to decide on the best order and find the limits. This process can best be learned from examples (below and Section 3) and practice (see problems).

Example 3. Find the volume of the solid in Figure 2.3 by using a triple integral. Here we imagine the whole solid cut into tiny boxes of volume $\Delta x\,\Delta y\,\Delta z$; an element of volume is $dx\,dy\,dz$. We first add up the volumes of the tiny boxes to get the volume of a column; this means integrating with respect to z from 0 to $1 + y$ with x and y constant. Then we add up the columns to get a slab and the slabs to get the whole volume just as we did in Example 1. Thus:

$$V = \iiint_V dx\,dy\,dz \tag{2.10}$$
$$= \int_{x=0}^{1}\int_{y=0}^{2-2x}\left(\int_{z=0}^{1+y} dz\right) dy\,dx \quad \text{or} \quad \int_{x=0}^{1}\int_{y=0}^{2-2x}\int_{z=0}^{1+y} dz\,dy\,dx$$
$$= \int_{x=0}^{1}\int_{y=0}^{2-2x}(1 + y)\,dy\,dx = \frac{5}{3},$$

as in (2.1) and (2.2). Or, we could have used (2.4).

Example 4. Find the mass of the solid in Figure 2.3 if the density (mass per unit volume) is $x + z$. An element of mass is $dM = (x + z)\,dx\,dy\,dz$. We add up elements of mass just as we add up elements of volume; that is, the limits are the same as in Example 3.

$$M = \int_{x=0}^{1}\int_{y=0}^{2-2x}\int_{z=0}^{1+y}(x + z)\,dz\,dy\,dx = 2 \tag{2.11}$$

where we evaluate the integrals as we did (2.1) to (2.4). (Check the result by hand and by computer.)

PROBLEMS, SECTION 2

In the problems of this section, set up and evaluate the integrals by hand and check your results by computer.

1. $\int_{x=0}^{1}\int_{y=2}^{4} 3x\,dy\,dx$

2. $\int_{y=-2}^{1}\int_{x=1}^{2} 8xy\,dx\,dy$

3. $\int_{y=0}^{2}\int_{x=2y}^{4} dx\,dy$

4. $\int_{x=0}^{4}\int_{y=0}^{x/2} y\,dy\,dx$

5. $\int_{x=0}^{1}\int_{y=x}^{e^x} y\,dy\,dx$

6. $\int_{y=1}^{2}\int_{x=\sqrt{y}}^{y^2} x\,dx\,dy$

In Problems 7 to 18 evaluate the double integrals over the areas described. To find the limits, sketch the area and compare Figures 2.5 to 2.7.

7. $\iint_A (2x-3y)\,dx\,dy$, where A is the triangle with vertices $(0,0)$, $(2,1)$, $(2,0)$.

8. $\iint_A 6y^2\cos x\,dx\,dy$, where A is the area inclosed by the curves $y=\sin x$, the x axis, and the line $x=\pi/2$.

9. $\iint_A \sin x\,dx\,dy$ where A is the area shown in Figure 2.8.

10. $\iint_A y\,dx\,dy$ where A is the area in Figure 2.8.

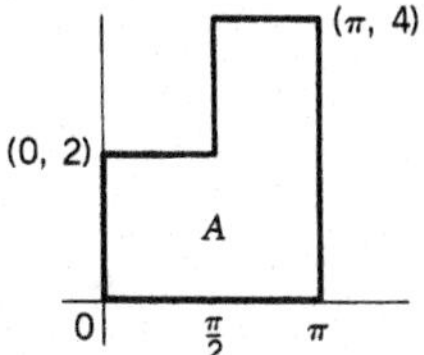

Figure 2.8

11. $\iint_A x\,dx\,dy$, where A is the area between the parabola $y=x^2$ and the straight line $2x-y+8=0$.

12. $\iint y\,dx\,dy$ over the triangle with vertices $(-1,0)$, $(0,2)$, and $(2,0)$.

13. $\iint 2xy\,dx\,dy$ over the triangle with vertices $(0,0)$, $(2,1)$, $(3,0)$.

14. $\iint x^2 e^{x^2 y}\,dx\,dy$ over the area bounded by $y=x^{-1}$, $y=x^{-2}$, and $x=\ln 4$.

15. $\iint dx\,dy$ over the area bounded by $y=\ln x$, $y=e+1-x$, and the x axis.

16. $\iint (9+2y^2)^{-1}\,dx\,dy$ over the quadrilateral with vertices $(1,3)$, $(3,3)$, $(2,6)$, $(6,6)$.

17. $\iint (x/y)\,dx\,dy$ over the triangle with vertices $(0,0)$, $(1,1)$, $(1,2)$.

18. $\iint y^{-1/2}\,dx\,dy$ over the area bounded by $y=x^2$, $x+y=2$, and the y axis.

In Problems 19 to 24, use double integrals to find the indicated volumes.

19. Above the square with vertices at $(0,0)$, $(2,0)$, $(0,2)$, and $(2,2)$, and under the plane $z=8-x+y$.

20. Above the rectangle with vertices $(0,0)$, $(0,1)$, $(2,0)$, and $(2,1)$, and below the surface $z^2=36x^2(4-x^2)$.

21. Above the triangle with vertices $(0,0)$, $(2,0)$, and $(2,1)$, and below the paraboloid $z=24-x^2-y^2$.

22. Above the triangle with vertices $(0,2)$, $(1,1)$, and $(2,2)$, and under the surface $z=xy$.

23. Under the surface $z=y(x+2)$, and over the area bounded by $x+y=0$, $y=1$, $y=\sqrt{x}$.

24. Under the surface $z=1/(y+2)$, and over the area bounded by $y=x$ and $y^2+x=2$.

In Problems 25 to 28, sketch the area of integration, observe that it is like the areas in Figure 2.7, and so write an equivalent integral with the integration in the opposite order. Check your work by evaluating the double integral both ways. Also check that your computer gives the same answer for both orders of integration.

25. $\int_{x=0}^{1}\int_{y=0}^{3-3x} dy\,dx$

26. $\int_{y=0}^{2}\int_{x=y/2}^{1} (x+y)\,dx\,dy$

27. $\int_{x=0}^{4}\int_{y=0}^{\sqrt{x}} y\sqrt{x}\,dy\,dx$

28. $\int_{y=0}^{1}\int_{x=0}^{\sqrt{1-y^2}} y\,dx\,dy$

In Problems 29 to 32, observe that the inside integral cannot be expressed in terms of elementary functions. As in Problems 25 to 28, change the order of integration and so evaluate the double integral. Also try using your computer to evaluate these for both orders of integration.

29. $\int_{y=0}^{\pi}\int_{x=y}^{\pi} \frac{\sin x}{x}\,dx\,dy$

30. $\int_{x=0}^{2}\int_{y=x}^{2} e^{-y^2/2}\,dy\,dx$

31. $\int_{x=0}^{\ln 16}\int_{y=e^{x/2}}^{4} \frac{dy\,dx}{\ln y}$

32. $\int_{y=0}^{1}\int_{x=y^2}^{1} \frac{e^x}{\sqrt{x}}\,dx\,dy$

33. A lamina covering the quarter disk $x^2+y^2 \le 4$, $x > 0$, $y > 0$, has (area) density $x+y$. Find the mass of the lamina.

34. A dielectric lamina with charge density proportional to y covers the area between the parabola $y = 16 - x^2$ and the x axis. Find the total charge.

35. A triangular lamina is bounded by the coordinate axes and the line $x + y = 6$. Find its mass if its density at each point P is proportional to the square of the distance from the origin to P.

36. A partially silvered mirror covers the square area with vertices at $(\pm 1, \pm 1)$. The fraction of incident light which it reflects at (x, y) is $(x-y)^2/4$. Assuming a uniform intensity of incident light, find the fraction reflected.

In Problems 37 to 40, evaluate the triple integrals.

37. $\int_{x=1}^{2}\int_{y=x}^{2x}\int_{z=0}^{y-x} dz\,dy\,dx$

38. $\int_{z=0}^{2}\int_{x=z}^{2}\int_{y=8x}^{z} dy\,dx\,dz$

39. $\int_{y=-2}^{3}\int_{z=1}^{2}\int_{x=y+z}^{2y+z} 6y\,dx\,dz\,dy$

40. $\int_{x=1}^{2}\int_{z=x}^{2x}\int_{y=0}^{1/z} z\,dy\,dz\,dx.$

41. Find the volume between the planes $z = 2x + 3y + 6$ and $z = 2x + 7y + 8$, and over the triangle with vertices $(0,0)$, $(3,0)$, and $(2,1)$.

42. Find the volume between the planes $z = 2x + 3y + 6$ and $z = 2x + 7y + 8$, and over the square in the (x, y) plane with vertices $(0,0)$, $(1,0)$, $(0,1)$, $(1,1)$.

43. Find the volume between the surfaces $z = 2x^2 + y^2 + 12$ and $z = x^2 + y^2 + 8$, and over the triangle with vertices $(0,0)$, $(1,0)$, and $(1,2)$.

44. Find the mass of the solid in Problem 42 if the density is proportional to y.

45. Find the mass of the solid in Problem 43 if the density is proportional to x.

46. Find the mass of a cube of side 2 if the density is proportional to the square of the distance from the center of the cube.

47. Find the volume in the first octant bounded by the coordinate planes and the plane $x + 2y + z = 4$.

48. Find the volume in the first octant bounded by the cone $z^2 = x^2 - y^2$ and the plane $x = 4$.

49. Find the volume in the first octant bounded by the paraboloid $z = 1 - x^2 - y^2$, the plane $x + y = 1$, and all three coordinate planes.

50. Find the mass of the solid in Problem 48 if the density is z.

3. APPLICATIONS OF INTEGRATION; SINGLE AND MULTIPLE INTEGRALS

Many different physical quantities are given by integrals; let us do some problems to illustrate setting up and evaluating these integrals. The basic idea which we use in setting up the integrals in these problems is that an integral is the "limit of a sum." Thus we imagine the physical object (whose volume, moment of inertia, etc., we are trying to find) cut into a large number of small pieces called *elements*. We write an approximate formula for the volume, moment of inertia, etc., of an element and then sum over all elements of the object. The limit of this sum (as the number of elements tends to infinity and the size of each element tends to zero) is what we find by integration and is what we want in the physical problem.

Using a computer to evaluate the integrals saves time and we will concentrate mainly on setting up integrals. However, in order to do a skillful job of finding limits, deciding order of integration, detecting and correcting errors, making useful changes of variables, and understanding the meaning of the symbols used, it is important to learn to evaluate multiple integrals by hand. So a good study method is to do some integrals both by hand and by computer. A computer is also very useful to plot graphs of curves and surfaces to help you find the limits in a multiple integral.

Example 1. Given the curve $y = x^2$ from $x = 0$ to $x = 1$, find

(a) the area under the curve (that is, the area bounded by the curve, the x axis, and the line $x = 1$; see Figure 3.1);

(b) the mass of a plane sheet of material cut in the shape of this area if its density (mass per unit area) is xy;

(c) the arc length of the curve;

(d) the centroid of the area;

(e) the centroid of the arc;

(f) the moments of inertia about the x, y, and z axes of the lamina in (b).

(a) The area is

$$A = \int_{x=0}^{1} y\,dx = \int_0^1 x^2\,dx = \left.\frac{x^3}{3}\right|_0^1 = \frac{1}{3}.$$

We could also find the area as a double integral of $dA = dy\,dx$ (see Figure 3.1). We have then

$$A = \int_{x=0}^{1}\int_{y=0}^{x^2} dy\,dx = \int_0^1 x^2\,dx$$

as before. Although the double integral is entirely unnecessary in finding the area in this problem, we shall need to use a double integral to find the mass in part (b).

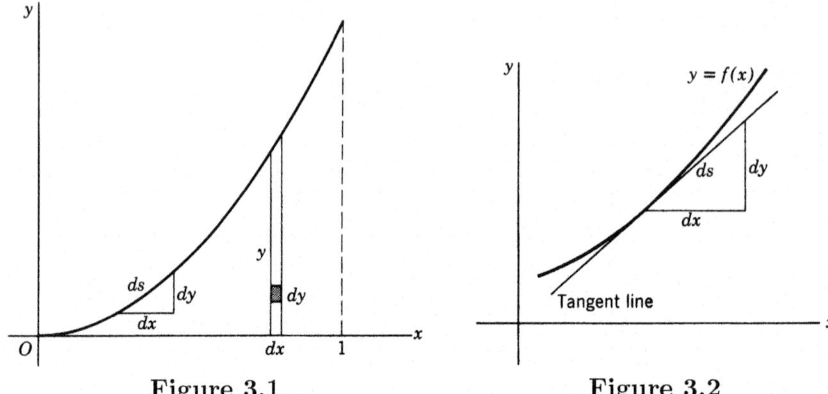

Figure 3.1 **Figure 3.2**

(b) The element of area, as in the double integral method in (a), is $dA = dy\,dx$. Since the density is $\rho = xy$, the element of mass is $dM = xy\,dy\,dx$, and the total mass is

$$M = \int_{x=0}^{1}\int_{y=0}^{x^2} xy\,dy\,dx = \int_0^1 x\,dx\left[\frac{y^2}{2}\right]_0^{x^2} = \int_0^1 \frac{x^5}{2}\,dx = \frac{1}{12}.$$

Observe that we could not do this problem as a single integral because the density depends on both x and y.

(c) The element of arc length ds is defined as indicated in Figures 3.1 and 3.2. Thus we have

$$\text{(3.1)}\qquad \begin{aligned} ds^2 &= dx^2 + dy^2,\\ ds &= \sqrt{dx^2+dy^2} = \sqrt{1+(dy/dx)^2}\,dx = \sqrt{(dx/dy)^2+1}\,dy. \end{aligned}$$

If $y = f(x)$ has a continuous first derivative $dy/\,dx$ (except possibly at a finite number of points), we can find the arc length of the curve $y = f(x)$ between a and b by calculating $\int_a^b ds$. For our example, we have

$$\text{(3.2)}\qquad \begin{aligned} \frac{dy}{dx} &= 2x, \qquad ds = \sqrt{1+4x^2}\,dx,\\ s &= \int_0^1 \sqrt{1+4x^2}\,dx = \frac{2\sqrt{5}+\ln(2+\sqrt{5})}{4} \end{aligned}$$

(see Problem 32).

(d) Recall from elementary physics that:

The *center of mass* of a body has coordinates $\bar{x}$, $\bar{y}$, $\bar{z}$ given by the equations

$$\int \bar{x}\,dM = \int x\,dM, \qquad \int \bar{y}\,dM = \int y\,dM, \qquad \int \bar{z}\,dM = \int z\,dM, \tag{3.3}$$

where dM is an element of mass and the integrals are over the whole body.

Although we have written single integrals in (3.3), they may be single, double, or triple integrals depending on the problem and the method of evaluation. Since $\bar{x}$, $\bar{y}$, and $\bar{z}$ are constants, we *can* take them outside the integrals in (3.3) and solve for them. However, you may find it easier to remember the definitions in the form (3.3). For the example we are doing, $\bar{z} = 0$ since the body is a sheet of material in the (x, y) plane. The element of mass is $dM = \rho\,dA = \rho\,dx\,dy$, where ρ is the density (mass per unit area in this problem). For a variable density as in (b), we would substitute the value of ρ into (3.3) and integrate both sides of each equation to find the coordinates of the center of mass. However, let us suppose the density is a constant. Then the first integral in (3.3) is

$$\int \bar{x}\rho\,dA = \int x\rho\,dA \quad \text{or} \quad \int \bar{x}\,dA = \int x\,dA. \tag{3.4}$$

Similarly, a *constant* density ρ can be canceled from all the equations in (3.3). The quantities $\bar{x}$, $\bar{y}$, $\bar{z}$, are then called the coordinates of the *centroid* of the area (or volume or arc).

The centroid of a body is the center of mass when we assume constant density.

In our example, we have

$$\begin{aligned} \int_{x=0}^{1}\int_{y=0}^{x^2} \bar{x}\,dy\,dx &= \int_{x=0}^{1}\int_{y=0}^{x^2} x\,dy\,dx \quad \text{or} \quad \bar{x}A = \left.\frac{x^4}{4}\right|_0^1 = \frac{1}{4}, \\ \int_{x=0}^{1}\int_{y=0}^{x^2} \bar{y}\,dy\,dx &= \int_{x=0}^{1}\int_{y=0}^{x^2} y\,dy\,dx \quad \text{or} \quad \bar{y}A = \left.\frac{x^5}{10}\right|_0^1 = \frac{1}{10}. \end{aligned} \tag{3.5}$$

(Double integrals are not really necessary for any of these but the last.) Using the value of A from part(a), we find $\bar{x} = \frac{3}{4}$, $\bar{y} = \frac{3}{10}$.

(e) The center of mass $(\bar{x}, \bar{y})$ of a wire bent in the shape of the curve $y = f(x)$ is given by

$$\int \bar{x}\rho\,ds = \int x\rho\,ds, \qquad \int \bar{y}\rho\,ds = \int y\rho\,ds, \tag{3.6}$$

where ρ is the density (mass per unit length), and the integrals are *single* integrals with ds given by (3.1). If ρ is constant, (3.6) defines the coordinates of the centroid.

In our example we have

$$(3.7)\qquad \begin{aligned} &\int_0^1 \bar{x}\sqrt{1+4x^2}\,dx = \int_0^1 x\sqrt{1+4x^2}\,dx,\\ &\int_0^1 \bar{y}\sqrt{1+4x^2}\,dx = \int_0^1 y\sqrt{1+4x^2}\,dx = \int_0^1 x^2\sqrt{1+4x^2}\,dx. \end{aligned}$$

Note carefully here that it is correct to put $y = x^2$ in the last integral of (3.7), but it would *not* have been correct to do this in the last integral of (3.5); the reason is that over the *area*, y could take values from zero to x^2, but *on* the arc, y takes only the value x^2. By calculating the integrals in (3.7) we can find $\bar{x}$ and $\bar{y}$.

(f) We need the following definition:

> The *moment of inertia* I of a point mass m about an axis is by definition the product ml^2 of m times the square of the distance l from m to the axis. For an extended object we must integrate $l^2 dM$ over the whole object, where l is the distance from dM to the axis.

In our example with variable density $\rho = xy$, we have $dM = xy\,dy\,dx$. The distance from dM to the x axis is y (Figure 3.3); similarly, the distance from dM to the y axis is x. The distance from dM to the z axis (the z axis is perpendicular to the paper in Figure 3.3) is $\sqrt{x^2+y^2}$. Then the three moments of inertia about the three coordinate axes are:

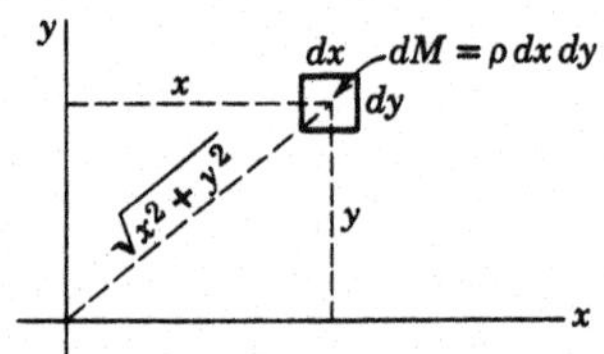

Figure 3.3

$$\begin{aligned} I_x &= \int_{x=0}^{1}\int_{y=0}^{x^2} y^2 xy\,dy\,dx = \int_0^1 \frac{x^9}{4}\,dx = \frac{1}{40},\\ I_y &= \int_{x=0}^{1}\int_{y=0}^{x^2} x^2 xy\,dy\,dx = \int_0^1 \frac{x^7}{2}\,dx = \frac{1}{16},\\ I_z &= \int_{x=0}^{1}\int_{y=0}^{x^2} (x^2+y^2)xy\,dy\,dx = I_x + I_y = \frac{7}{80}. \end{aligned}$$

> The fact that $I_x + I_y = I_z$ for a plane lamina in the (x, y) plane is known as the perpendicular axis theorem.

It is customary to write moments of inertia as multiples of the mass; using $M = \frac{1}{12}$ from (b), we write

$$I_x = \frac{12}{40}M = \frac{3}{10}M,\qquad I_y = \frac{12}{16}M = \frac{3}{4}M,\qquad I_z = \frac{7\cdot 12}{80}M = \frac{21}{20}M.$$

▶ **Example 2.** Rotate the area of Example 1 about the x axis to form a volume and surface of revolution, and find

(a) the volume;

(b) the moment of inertia about the x axis of a solid of constant density occupying the given volume;

(c) the area of the curved surface;

(d) the centroid of the curved surface.

(a) We want to find the given volume.

> The easiest way to find a volume of revolution is to take as volume element a thin slab of the solid as shown in Figure 3.4. The slab has circular cross section of radius y and thickness dx; thus the volume element is $\pi y^2\,dx$.

Then the volume in our example is

$$V = \int_0^1 \pi y^2\,dx = \int_0^1 \pi x^4\,dx = \frac{\pi}{5}. \tag{3.8}$$

We have really avoided part of the integration here because we knew the formula for the area of a circle. In finding volumes of solids which are not solids of revolution, we may have to use double or triple integrals. Even for a solid of revolution we might need multiple integrals to find the mass if the density is variable.

To illustrate setting up such integrals, let us do the above problem using triple integrals. For this we need the equation of the surface which is (see Problem 16)

$$y^2 + z^2 = x^4, \qquad x > 0. \tag{3.9}$$

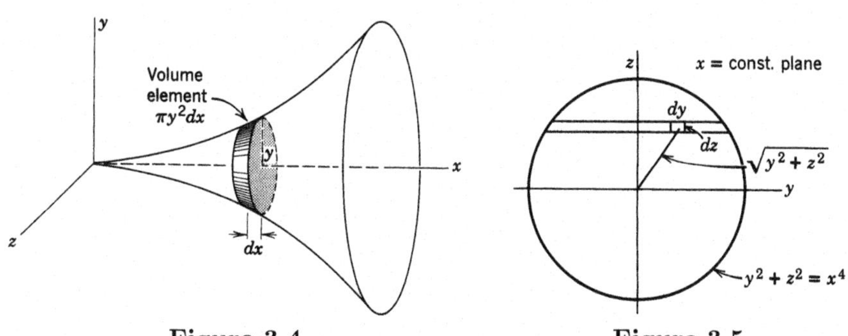

Figure 3.4 Figure 3.5

To set up a multiple integral for the volume of a solid, we cut the solid into slabs as in Figure 3.4 (not necessarily circular slabs, although they are in our example) and then as in Figure 3.5 we cut each slab into strips and each strip into tiny boxes of volume $dx\,dy\,dz$. The volume is

$$V = \iiint dx\,dy\,dz;$$

the only problem is to find the limits! To do this, we start by adding up tiny boxes to get a strip; as we have drawn Figure 3.5, this means to integrate with respect to y from one side of the circle $y^2 + z^2 = x^4$ to the other, that is, from

$$y = -\sqrt{x^4 - z^2} \qquad \text{to} \quad y = +\sqrt{x^4 - z^2}.$$

Next we add all the strips in a slab. This means that, in Figure 3.5, we integrate with respect to z from the bottom to the top of the circle $y^2 + z^2 = x^4$; thus the z limits are $z = \pm$ radius of circle $= \pm x^2$. And finally we add all the slabs to obtain the solid. This means to integrate in Figure 3.4 from $x = 0$ to $x = 1$; this is just what we did in our first simple method. The final integral is then

$$V = \int_{x=0}^{1} \int_{z=-x^2}^{x^2} \int_{y=-\sqrt{x^4-z^2}}^{\sqrt{x^4-z^2}} dy\, dz\, dx. \tag{3.10}$$

(See Problem 33).

Although the triple integral is an unnecessarily complicated way of finding a volume of revolution, this simple problem illustrates the general method of setting up an integral for any kind of volume. Once we have the volume as a triple integral, it is easy to write the integrals for the mass with a given variable density, for the coordinates of the centroid, for the moments of inertia, and so on. The limits of integration are the same as for the volume; we need only insert the proper expressions (density, etc.) in the integrand to get the mass, centroid, and so on.

(b) To find the moment of inertia of the solid about the x axis, we must integrate the quantity $l^2 dM$, where l is the distance from dM to the x axis; from Figure 3.5, since the x axis is perpendicular to the paper, $l^2 = y^2 + z^2$. The limits on the integrals are the same as in (3.10). We are assuming constant density, so the factor ρ can be written outside the integrals. Then we have

$$I_x = \rho \int_{x=0}^{1} \int_{z=-x^2}^{x^2} \int_{y=-\sqrt{x^4-z^2}}^{\sqrt{x^4-z^2}} (y^2 + z^2)\, dy\, dz\, dx = \frac{\pi}{18}\rho.$$

Since from (3.8) the mass of the solid is

$$M = \rho V = \frac{\pi}{5}\rho$$

we can write I_x (as is customary) as a multiple of M:

$$I_x = \frac{\pi}{18}\frac{5}{\pi}M = \frac{5}{18}M.$$

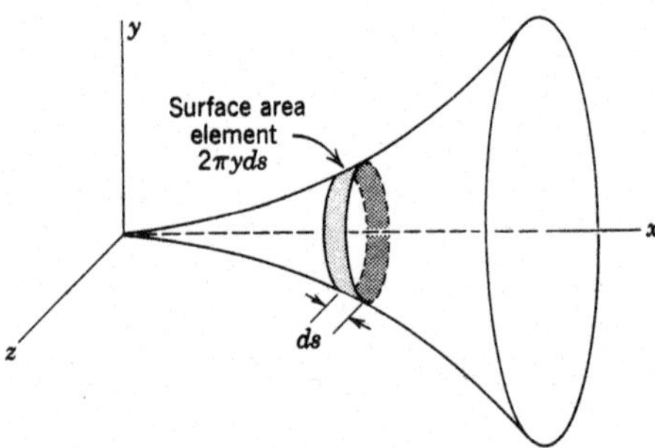

Figure 3.6

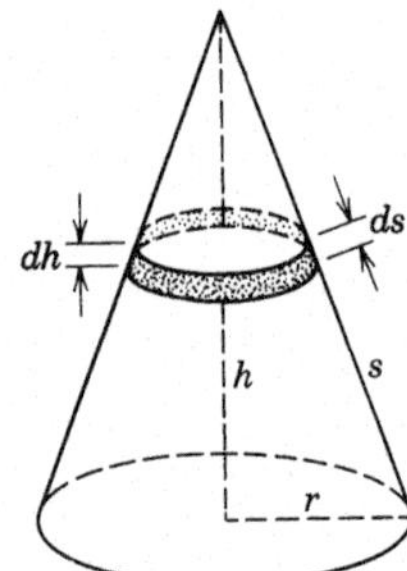

Figure 3.7

(c) We find the area of the surface of revolution by using as element the curved surface of a thin slab as in Figure 3.6. This is a strip of circumference $2\pi y$ and width ds. To see this clearly and to understand why we use ds here but dx in the volume element in (3.8), think of the slab as a thin section of a cone (Figure 3.7) between planes perpendicular to the axis of the cone. If you wanted to find the total volume $V = \frac{1}{3}\pi r^2 h$ of the cone, you would use the height h perpendicular to the base, but in finding the total curved surface area $S = \frac{1}{2} \cdot 2\pi r \cdot s$, you would use the *slant* height s. The same ideas hold in finding the volume and surface elements. The approximate volume of the thin slab is the area of a face of the slab times its thickness (dh in Figure 3.7, dx in Figure 3.4). But if you think of a narrow strip of paper just covering the curved surface of the thin slab, the width of the strip of paper is ds, and its length is the circumference of the thin slab.

The element of surface area (in Figure 3.6) is then

(3.11) $$dA = 2\pi y\, ds.$$

The total area is [using ds from (3.2)]

$$A = \int_{x=0}^{1} 2\pi y\, ds = \int_0^1 2\pi x^2 \sqrt{1+4x^2}\, dx.$$

(For more general surfaces, there is a way to calculate areas by double integration; we shall take this up in Section 5.)

(d) The y and z coordinates of the centroid of the surface area are zero by symmetry. For the x coordinate, we have by (3.4)

$$\int \bar{x}\, dA = \int x\, dA,$$

or, using $dA = 2\pi y\, ds$ and the total area A from (c), we have

$$\bar{x}A = \int_{x=0}^{1} x \cdot 2\pi y\, ds = \int_0^1 x \cdot 2\pi x^2 \sqrt{1+4x^2}\, dx.$$

PROBLEMS, SECTION 3

The following notation is used in the problems:

M = mass,

$\bar{x}$, $\bar{y}$, $\bar{z}$ = coordinates of center of mass (or centroid if the density is constant),

I = moment of inertia (about axis stated),

I_x, I_y, I_z = moments of inertia about x, y, z axes,

I_m = moment of inertia (about axis stated) through the center of mass.

Note: It is customary to give answers for I, I_m, I_x, etc., as multiples of M (for example, $I = \frac{1}{3}Ml^2$).

1. Prove the "parallel axis theorem": The moment of inertia I of a body about a given axis is $I = I_m + Md^2$, where M is the mass of the body, I_m is the moment of inertia of the body about an axis through the center of mass and parallel to the given axis, and d is the distance between the two axes.

2. For a thin rod of length l and uniform density ρ find
   (a) M,
   (b) I_m about an axis perpendicular to the rod,
   (c) I about an axis perpendicular to the rod and passing through one end (see Problem 1).

3. A thin rod 10 ft long has a density which varies uniformly from 4 to 24 lb/ft. Find
   (a) M,
   (b) $\bar{x}$,
   (c) I_m about an axis perpendicular to the rod,
   (d) I about an axis perpendicular to the rod passing through the heavy end.

4. Repeat Problem 3 for a rod of length l with density varying uniformly from 2 to 1.

5. For a square lamina of uniform density, find I about
   (a) a side,
   (b) a diagonal,
   (c) an axis through a corner and perpendicular to the plane of the lamina. *Hint:* See the perpendicular axis theorem, Example 1f.

6. A triangular lamina has vertices $(0, 0)$, $(0, 6)$ and $(6, 0)$, and uniform density. Find:
   (a) $\bar{x}$, $\bar{y}$,
   (b) I_x,
   (c) I_m about an axis parallel to the x axis. *Hint:* Use Problem 1 *carefully.*

7. A rectangular lamina has vertices $(0, 0)$, $(0, 2)$, $(3, 0)$, $(3, 2)$ and density xy. Find
   (a) M,
   (b) $\bar{x}$, $\bar{y}$,
   (c) I_x, I_y,
   (d) I_m about an axis parallel to the z axis. *Hint:* Use the parallel axis theorem and the perpendicular axis theorem.

8. For a uniform cube, find I about one edge.

9. For the pyramid inclosed by the coordinate planes and the plane $x + y + z = 1$:
   (a) Find its volume.
   (b) Find the coordinates of its centroid.
   (c) If the density is z, find M and $\bar{z}$.

10. A uniform chain hangs in the shape of the catenary $y = \cosh x$ between $x = -1$ and $x = 1$. Find (a) its length, (b) $\bar{y}$.

11. A chain in the shape $y = x^2$ between $x = -1$ and $x = 1$ has density $|x|$. Find (a) M, (b) $\bar{x}$, $\bar{y}$.

Prove the following two theorems of Pappus:

12. The area A inside a closed curve in the (x, y) plane, $y \geq 0$, is revolved about the x axis. The volume of the solid generated is equal to A times the circumference of the circle traced by the centroid of A. *Hint:* Write the integrals for the volume and for the centroid.

13. An arc in the (x, y) plane, $y \geq 0$, is revolved about the x axis. The surface area generated is equal to the length of the arc times the circumference of the circle traced by the centroid of the arc.

14. Use Problems 12 and 13 to find the volume and surface area of a torus (doughnut).

15. Use Problems 12 and 13 to find the centroids of a semicircular area and of a semicircular arc. *Hint:* Assume the formulas $A = 4\pi r^2$, $V = \frac{4}{3}\pi r^3$ for a sphere.

16. Let a curve $y = f(x)$ be revolved about the x axis, thus forming a surface of revolution. Show that the cross sections of this surface in any plane $x =$ const. [that is, parallel to the (y, z) plane] are circles of radius $f(x)$. Thus write the general equation of a surface of revolution and verify the special case $f(x) = x^2$ in (3.9).

In Problems 17 to 30, for the curve $y = \sqrt{x}$, between $x = 0$ and $x = 2$, find:

17. The area under the curve.

18. The arc length.

19. The volume of the solid generated when the area is revolved about the x axis.

20. The curved area of this solid.

21, 22, 23. The centroids of the arc, the volume, and the surface area.

24, 25, 26, 27. The moments of inertia about the x axis of a lamina in the shape of the plane area under the curve; of a wire bent along the arc of the curve; of the solid of revolution; and of a thin shell whose shape is the curved surface of the solid (assuming constant density for all these problems).

28. The mass of a wire bent in the shape of the arc if its density (mass per unit length) is $\sqrt{x}$.

29. The mass of the solid of revolution if the density (mass per unit volume) is $|xyz|$.

30. The moment of inertia about the y axis of the solid of revolution if the density is $|xyz|$.

31. (a) Revolve the curve $y = x^{-1}$, from $x = 1$ to $x = \infty$, about the x axis to create a surface and a volume. Write integrals for the surface area and the volume. Find the volume, and show that the surface area is infinite. *Hint:* The surface area integral is not easy to evaluate, but you can easily show that it is greater than $\int_1^\infty x^{-1}\,dx$ which you can evaluate.

(b) The following question is a challenge to your ability to fit together your mathematical calculations and physical facts: In (a) you found a finite volume and an infinite area. Suppose you fill the finite volume with a finite amount of paint and then pour off the excess leaving what sticks to the surface. Apparently you have painted an infinite area with a finite amount of paint! What is wrong? (Compare Problem 15.31c of Chapter 1.)

32. Use a computer or tables to evaluate the integral in (3.2) and verify that the answer is equivalent to the text answer. *Hint:* See Problem 1.4 and also Chapter 2, Sections 15 and 17.

33. Verify that (3.10) gives the same result as (3.8).

4. CHANGE OF VARIABLES IN INTEGRALS; JACOBIANS

In many applied problems, it is more convenient to use other coordinate systems instead of the rectangular coordinates we have been using. For example, in the plane we often use polar coordinates, and in three dimensions we often use cylindrical coordinates or spherical coordinates. It is important to know how to set up multiple integrals directly in these coordinate systems which occur so frequently in practice. That is, we need to know what the area, volume, and arc length elements are, what the variables r, θ, etc., mean geometrically, and how they are related to the rectangular coordinates. We are going to discuss finding elements of area, etc., geometrically for several important coordinate systems. However, if we are given equations like $x = r\cos\theta$, $y = r\sin\theta$, relating new variables to the rectangular ones, it is useful to know how to find the elements of area, etc., *algebraically*, without having to rely on the geometry. We are going to discuss this and illustrate it by verifying the results which we can get geometrically for several of the familiar coordinate systems.

In the plane, the polar coordinates r, θ are related to the rectangular coordinates x, y by the equations

$$x = r\cos\theta, \quad y = r\sin\theta. \tag{4.1}$$

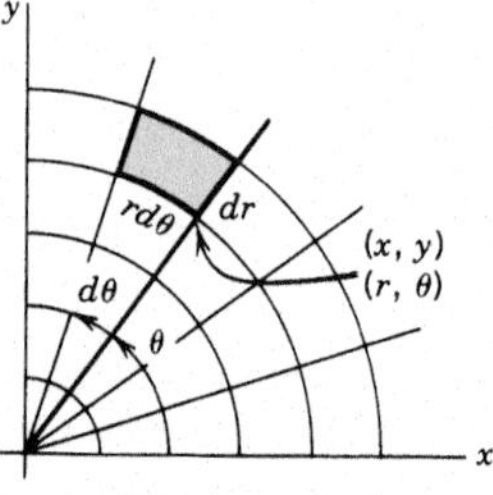

Figure 4.1

Recall that we found the area element $dy\,dx$ by drawing a grid of lines $x =$ const., $y =$ const., which cut the plane into little rectangles dx by dy; the area of one rectangle was then $dy\,dx$. We can make a similar construction for polar coordinates by drawing lines $\theta =$ const. and circles $r =$ const.; we then obtain the grid shown in Figure 4.1. Observe that the sides of the area element are not dr and $d\theta$, but dr and $r\,d\theta$, and its area is then

$$dA = dr \cdot r\,d\theta = r\,dr\,d\theta. \tag{4.2}$$

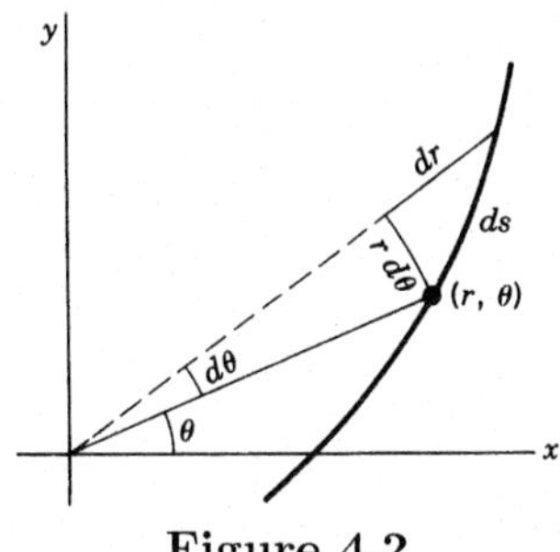

Figure 4.2

Similarly, we can see from Figure 4.2 that the arc length element ds is given by

(4.3)
$$ds^2 = dr^2 + r^2\,d\theta^2,$$
$$ds = \sqrt{\left(\frac{dr}{d\theta}\right)^2 + r^2}\;d\theta = \sqrt{1 + r^2\left(\frac{d\theta}{dr}\right)^2}\;dr.$$

Example 1. Given a semicircular sheet of material of radius a and constant density ρ, find

(a) the centroid of the semicircular area;

(b) the moment of inertia of the sheet of material about the diameter forming the straight side of the semicircle.

(a) In Figure 4.3, we see by symmetry that $\bar{y} = 0$. We want to find $\bar{x}$. By (3.4), we have

$$\int \bar{x} r\,dr\,d\theta = \int x r\,dr\,d\theta.$$

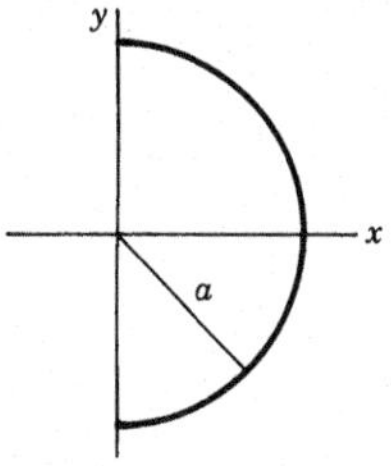

Changing the x to polar coordinates and putting in the limits, we get

Figure 4.3

$$\bar{x}\int_{r=0}^{a}\int_{\theta=-\pi/2}^{\pi/2} r\,dr d\theta = \int_{r=0}^{a}\int_{\theta=-\pi/2}^{\pi/2} r\cos\theta\, r\,dr\,d\theta.$$

We calculate the integrals and find $\bar{x}$:

$$\bar{x}\,\frac{a^2}{2}\,\pi = \frac{a^3}{3}\sin\theta\Big|_{-\pi/2}^{\pi/2} = \frac{a^3}{3}\cdot 2,$$
$$\bar{x} = \frac{4a}{3\pi}.$$

(b) We want the moment of inertia about the y axis in Figure 4.3; by definition this is $\int x^2\,dM$. In polar coordinates, $dM = \rho\,dA = \rho r\,dr\,d\theta$. We are given that the density ρ is constant. Then we have

$$I_y = \rho\int x^2 r\,dr\,d\theta = \rho\int_{r=0}^{a}\int_{\theta=-\pi/2}^{\pi/2} r^2\cos^2\theta\, r\,dr\,d\theta = \rho\frac{\pi a^4}{8}.$$

The mass of the semicircular object is

$$M = \rho \int r\, dr\, d\theta = \rho \int_{r=0}^{a} \int_{\theta=-\pi/2}^{\pi/2} r\, dr\, d\theta = \rho \frac{\pi a^2}{2}.$$

We write I_y in terms of M to get

$$I_y = \frac{2M}{\pi a^2} \frac{\pi a^4}{8} = \frac{Ma^2}{4}.$$

Spherical and Cylindrical Coordinates The two most important coordinate systems (besides rectangular) in three dimensions are the spherical and the cylindrical coordinate systems. Figures 4.4 and 4.5 and equations (4.4) and (4.5) show the geometrical meaning of the variables, their algebraic relation to x, y, z, the appearance of the volume elements, and the formulas for the volume, arc length, and surface area elements.

Cylindrical coordinates are just polar coordinates in the (x, y) plane with z for the third variable. Note that the spherical coordinates r and θ in Figure 4.5 are different from the cylindrical or polar coordinates r and θ in Figures 4.4 and 4.1. Since we seldom use both systems in the same problem, this should cause no confusion. (If necessary, use ρ or R for one of the r's and use ϕ instead of θ in cylindrical coordinates.) Watch out, however, for the discrepancy in notation for spherical coordinates in various texts. Most calculus books interchange θ and ϕ. This can be confusing later since the notation of Figure 4.5 is almost universal in applications to the physical sciences, and is often used in advanced mathematics (partial differential equations, special functions), in computer programs, and in reference books of formulas and tables. You will need to learn a number of useful formulas involving spherical coordinates [for example, (4.7), (4.19), and (4.20) below; also see Chapter 10, Section 9 and Chapter 13, Section 7]. It is best to learn these formulas in the notation that you will use in applications.

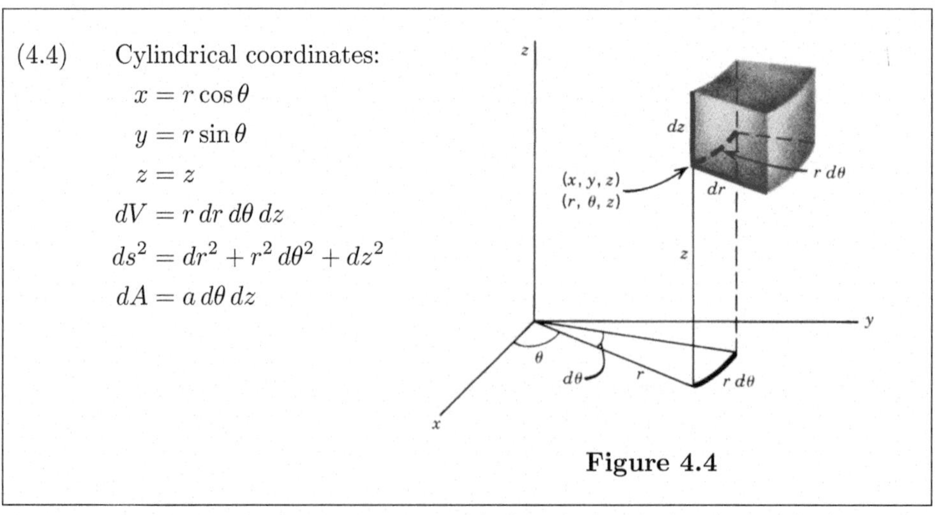

(4.4) Cylindrical coordinates:

$$x = r\cos\theta$$
$$y = r\sin\theta$$
$$z = z$$
$$dV = r\, dr\, d\theta\, dz$$
$$ds^2 = dr^2 + r^2\, d\theta^2 + dz^2$$
$$dA = a\, d\theta\, dz$$

Figure 4.4

(4.5) Spherical coordinates:

$$x = r\sin\theta\cos\phi$$
$$y = r\sin\theta\sin\phi$$
$$z = r\cos\theta$$
$$dV = r^2\sin\theta\,dr\,d\theta\,d\phi$$
$$ds^2 = dr^2 + r^2\,d\theta^2 + r^2\sin^2\theta\,d\phi^2$$
$$dA = a^2\sin\theta\,d\theta\,d\phi$$

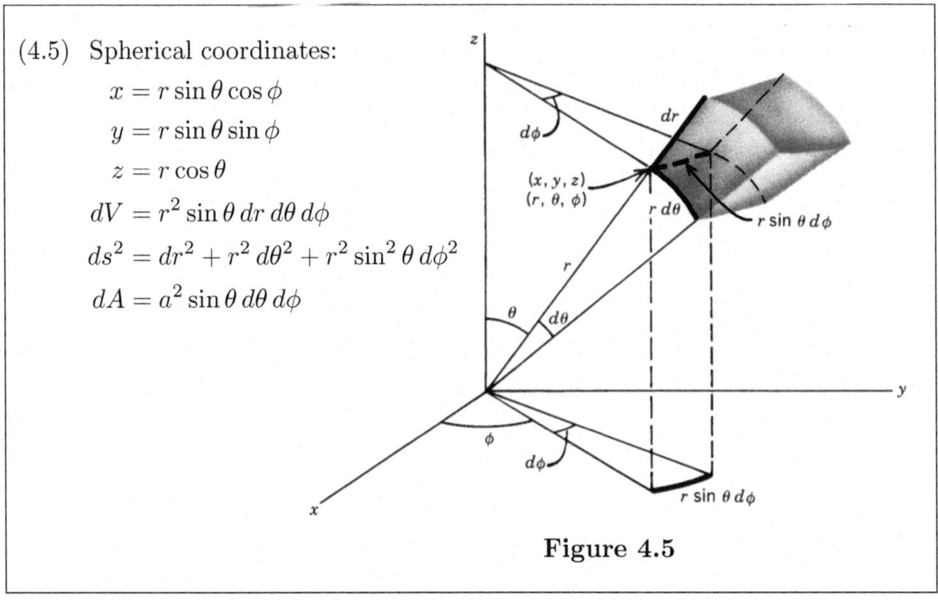

Figure 4.5

We will need the volume and surface area elements in these two systems [and also the arc length elements—see equations (4.18) and (4.19)]. To find the polar coordinate area element in Figure 4.1, we drew a grid of curves $r =$ const., $\theta =$ const. In three dimensions we want to draw a grid of surfaces. In cylindrical coordinates these surfaces are the cylinders $r =$ const., the half-planes $\theta =$ const. (through the z axis), and the planes $z =$ const. [parallel to the (x, y) plane]. One of the elements formed by this grid of surfaces is sketched in Figure 4.4. From the geometry we see that three edges of the element are dr, $r\,d\theta$, and dz, giving the volume element

(4.6) $$dV = r\,dr\,d\theta\,dz \qquad \text{(cylindrical coordinates).}$$

If r is constant, then the surface area element on the cylinder $r = a$ has edges $a\,d\theta$ and dz, so $dA = a\,d\theta\,dz$. Similarly for the spherical coordinate case, we draw the spheres $r =$ const., the cones $\theta =$ const., and the half-planes $\phi =$ const. The volume elements formed by this grid (Figure 4.5) have edges dr, $r\,d\theta$, and r sin $d\theta\,d\phi$; thus we get

(4.7) $$dV = r^2\sin\theta\,dr\,d\theta\,d\phi \qquad \text{(spherical coordinates).}$$

If r is constant, then the surface area element on the sphere $r = a$ has edges $a\,d\theta$ and $a\sin\theta\,d\phi$, so $dA = a^2\sin\theta\,d\theta\,d\phi$.

Jacobians. For polar, cylindrical, and spherical coordinates, we have seen how to find area and volume elements from the geometry. However, it is convenient to know an algebraic way of finding them which we can use for unfamiliar coordinate systems

(Problems 16 and 17) or for any change of variables in a multiple integral (Problems 19 and 20). Here we state without proof (see Chapter 6, Section 3, Example 2) some theorems which tell us how to do this. First, in two dimensions, suppose x and y are given as functions of two new variables s and t. The *Jacobian* of x, y, with respect to s, t, is the determinant in (4.8) below; we also show abbreviations used for it.

$$J = J\left(\frac{x,y}{s,t}\right) = \frac{\partial(x,y)}{\partial(s,t)} = \begin{vmatrix} \dfrac{\partial x}{\partial s} & \dfrac{\partial x}{\partial t} \\ \dfrac{\partial y}{\partial s} & \dfrac{\partial y}{\partial t} \end{vmatrix}. \tag{4.8}$$

Then the area element $dy\,dx$ is replaced in the s, t system by the area element

$$dA = |J|\,ds\,dt \tag{4.9}$$

where $|J|$ is the absolute value of the Jacobian in (4.8).

Let us find the Jacobian of x, y with respect to the polar coordinates r, θ, and thus verify that (4.8) and our geometric method give the same result (4.2) for the polar coordinate area element. We have

$$\frac{\partial(x,y)}{\partial(r,\theta)} = \begin{vmatrix} \dfrac{\partial x}{\partial r} & \dfrac{\partial x}{\partial \theta} \\ \dfrac{\partial y}{\partial r} & \dfrac{\partial y}{\partial \theta} \end{vmatrix} = \begin{vmatrix} \cos\theta & -r\sin\theta \\ \sin\theta & r\cos\theta \end{vmatrix} = r. \tag{4.10}$$

Thus by (4.9) the area element is $r\,dr\,d\theta$ as in (4.2).

The use of Jacobians extends to more variables. Also, it is not necessary to start with rectangular coordinates; let us state a more general theorem.

Suppose we have a triple integral

$$\iiint f(u,v,w)\,du\,dv\,dw \tag{4.11}$$

in some set of variables u, v, w. Let r, s, t be another set of variables, related to u, v, w by given equations

$$u = u(r,s,t), \qquad v = v(r,s,t), \qquad w = w(r,s,t).$$

Then if the determinant

$$J = \frac{\partial(u,v,w)}{\partial(r,s,t)} = \begin{vmatrix} \frac{\partial u}{\partial r} & \frac{\partial u}{\partial s} & \frac{\partial u}{\partial t} \\ \frac{\partial v}{\partial r} & \frac{\partial v}{\partial s} & \frac{\partial v}{\partial t} \\ \frac{\partial w}{\partial r} & \frac{\partial w}{\partial s} & \frac{\partial w}{\partial t} \end{vmatrix} \tag{4.12}$$

is the Jacobian of u, v, w with respect to r, s, t, then the triple integral in the new variables is

$$\iiint f \cdot |J| \cdot dr\, ds\, dt, \tag{4.13}$$

where, of course, f and J must both be expressed in terms of r, s, t, and the limits must be properly adjusted to correspond to the new variables.

We can use (4.12) to verify the volume element (4.6) for cylindrical coordinates (Problem 15) and the volume element (4.7) for spherical coordinates. Let us do the calculation for spherical coordinates. From (4.5), we have

$$\begin{aligned} \frac{\partial(x,y,z)}{\partial(r,\theta,\phi)} &= \begin{vmatrix} \frac{\partial x}{\partial r} & \frac{\partial x}{\partial \theta} & \frac{\partial x}{\partial \phi} \\ \frac{\partial y}{\partial r} & \frac{\partial y}{\partial \theta} & \frac{\partial y}{\partial \phi} \\ \frac{\partial z}{\partial r} & \frac{\partial z}{\partial \theta} & \frac{\partial z}{\partial \phi} \end{vmatrix} = \begin{vmatrix} \sin\theta\,\cos\phi & r\,\cos\theta\,\cos\phi & -r\,\sin\theta\,\sin\phi \\ \sin\theta\,\sin\phi & r\,\cos\theta\,\sin\phi & r\,\sin\theta\,\cos\phi \\ \cos\theta & -r\,\sin\theta & 0 \end{vmatrix} \\ &= r^2\sin\theta[-\sin^2\phi(-\sin^2\theta-\cos^2\theta)-\cos^2\phi(-\sin^2\theta-\cos^2\theta)] \\ &= r^2\sin\theta. \end{aligned} \tag{4.14}$$

Thus the spherical coordinate volume element is $dV = r^2\,\sin\theta\, dr\, d\theta\, d\phi$ as in (4.7).

Example 2. Find the z coordinate of the centroid of a uniform solid cone (part of one nappe) of height h equal to the radius of the base r. Also find the moment of inertia of the solid about its axis.

If we take the cone as shown in Figure 4.6, its equation in cylindrical coordinates is $r = z$, since at any height z, the cross section is a circle of radius equal to the height. To find the mass we must integrate $dM = \rho r\,dr\,d\theta\,dz$, where ρ is the constant density. The limits of integration are

$$\theta:\quad 0 \text{ to } 2\pi, \qquad r:\quad 0 \text{ to } z, \qquad z:\quad 0 \text{ to } h.$$

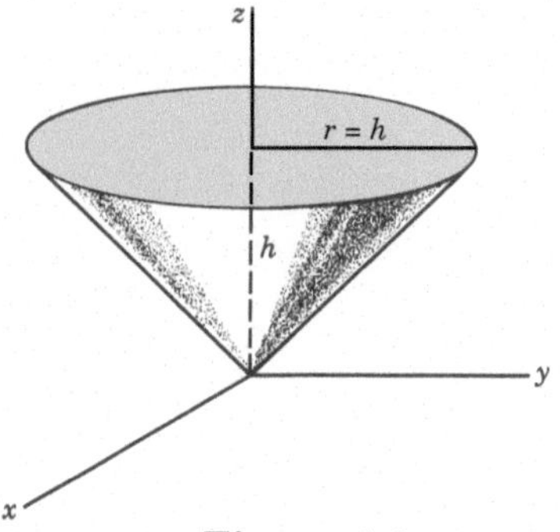

Figure 4.6

Then we have

$$M = \int \rho\,dV = \rho \int_{z=0}^{h} \int_{r=0}^{z} \int_{\theta=0}^{2\pi} r\,dr\,d\theta\,dz = \rho \cdot 2\pi \int_0^h \frac{z^2}{2} dz = \frac{\rho\pi h^3}{3},$$

(4.15)
$$\begin{aligned}\int \bar{z}\,dV = \int z\,dV &= \int_{z=0}^{h} \int_{r=0}^{z} \int_{\theta=0}^{2\pi} zr\,dr\,d\theta\,dz \\ &= 2\pi \int_0^h z \cdot \frac{1}{2} z^2\,dz = \frac{\pi h^4}{4}, \\ \bar{z} \cdot \frac{\pi h^3}{3} &= \frac{\pi h^4}{4}, \\ \bar{z} &= \frac{3}{4} h.\end{aligned}$$

For the moment of inertia about the z axis we have

$$I = \rho \int_{z=0}^{h} \int_{r=0}^{z} \int_{\theta=0}^{2\pi} r^2 r\,dr\,d\theta\,dz = \rho \cdot 2\pi \int_0^h \frac{z^4}{4}\,dz = \rho \frac{\pi h^5}{10}.$$

Using the value of M from (4.15), we write I in the usual form as a multiple of M:

$$I = \frac{3M}{\pi h^3} \frac{\pi h^5}{10} = \frac{3}{10} M h^2.$$

In the following examples and problems, note that we use *sphere* ($r = a$) to mean surface area, and *ball* ($r \leq a$) to mean volume (just as we use *circle* to mean circumference and *disk* to mean area).

▶ **Example 3.** Find the moment of inertia of a solid ball of radius a about a diameter. In spherical coordinates, the equation of the ball is $r \leq a$. Then the mass is

(4.16)
$$\begin{aligned}M = \rho \int dV &= \rho \int_{\phi=0}^{2\pi} \int_{\theta=0}^{\pi} \int_{r=0}^{a} r^2 \sin\theta\,dr\,d\theta\,d\phi \\ &= \rho \frac{a^3}{3} \cdot 2 \cdot 2\pi = \frac{4}{3} \pi a^3 \rho.\end{aligned}$$

(to no one's surprise!). The moment of inertia about the z axis is

$$I = \int (x^2 + y^2)\,dM = \rho \int_{\phi=0}^{2\pi} \int_{\theta=0}^{\pi} \int_{r=0}^{a} (r^2 \sin^2\theta)\, r^2 \sin\theta\,dr\,d\theta\,d\phi$$

$$= \rho \cdot \frac{a^5}{5} \cdot \frac{4}{3} \cdot 2\pi = \frac{8\pi a^5 \rho}{15};$$

or, using the value of M, we get

$$I = \frac{2}{5}Ma^2. \tag{4.17}$$

Example 4. Find the moment of inertia about the z axis of the solid ellipsoid inside

$$\frac{x^2}{a^2} + \frac{y^2}{b^2} + \frac{z^2}{c^2} = 1.$$

We want to evaluate

$$M = \rho \iiint dx\,dy\,dz \quad \text{and} \quad I = \rho \iiint (x^2 + y^2)\,dx\,dy\,dz$$

where the triple integrals are over the volume of the ellipsoid. Make the change of variables $x = ax'$, $y = by'$, $z = cz'$; then $x'^2 + y'^2 + z'^2 = 1$, so in the primed variables we integrate over the volume of a ball of radius 1. Then

$$M = \rho\,abc \iiint dx'\,dy'\,dz' = \rho\,abc \cdot \text{ volume of ball of radius 1.}$$

Using (4.16), we have

$$M = \rho\,abc \cdot \frac{4}{3}\pi \cdot 1^3 = \frac{4}{3}\pi\rho\,abc.$$

Similarly, we find

$$I = \rho\,abc \iiint (a^2x'^2 + b^2y'^2)\,dV'$$

where the triple integral is over the volume of a ball of radius 1. Now, by symmetry,

$$\iiint x'^2 dV' = \iiint y'^2 dV' = \iiint z'^2 dV' = \frac{1}{3}\iiint r'^2 dV'$$

where $r'^2 = x'^2 + y'^2 + z'^2$, and we are integrating over the volume inside the sphere $r' = 1$. Let us use spherical coordinates in the primed system. Then

$$\iiint r'^2 dV' = \int_{\phi=0}^{2\pi}\int_{\theta=0}^{\pi}\int_{r=0}^{1} r'^2(r'^2 \sin\theta'\,dr'\,d\theta'\,d\phi')$$
$$= 4\pi \int_0^1 r'^4 dr' = \frac{4\pi}{5}.$$

Thus,

$$I = \rho\,abc\left[a^2 \iiint x'^2\,dV' + b^2 \iiint y'^2\,dV'\right] = \rho\,abc(a^2 + b^2)\frac{1}{3}\cdot\frac{4\pi}{5},$$

or, in terms of M,

$$I = \frac{1}{5}M(a^2 + b^2).$$

In order to find arc lengths using spherical or cylindrical coordinates, we need the arc length element ds. Recall that we found the polar coordinate arc length element ds (Figure 4.2) as the hypotenuse of the right triangle with sides dr and $r\,d\theta$. From Figure 4.1 you can see that ds can also be thought of as a diagonal of the area element. Similarly, in cylindrical and spherical coordinates (see Figures 4.4 and 4.5), the arc length element ds is a space diagonal of the volume element. In cylindrical coordinates (4.4), the sides of the volume element are dr, $r\,d\theta$, dz, so the arc length element is given by

$$ds^2 = dr^2 + r^2\,d\theta^2 + dz^2 \qquad \text{(cylindrical coordinates)}. \tag{4.18}$$

In spherical coordinates (4.5), the sides of the volume element are dr, $r\,d\theta$, $r\,\sin\theta\,d\phi$, so the arc length element is given by

$$ds^2 = dr^2 + r^2\,d\theta^2 + r^2\,\sin^2\theta\,d\phi^2 \qquad \text{(spherical coordinates)}. \tag{4.19}$$

It is also convenient to be able to find arc lengths algebraically. Let us do this for polar coordinates; the same method can be used in three dimensions. From (4.1) we have

$$\begin{aligned} dx &= \cos\theta\,dr - r\,\sin\theta\,d\theta, \\ dy &= \sin\theta\,dr + r\,\cos\theta\,d\theta. \end{aligned}$$

Squaring and adding these two equations, we get

$$\begin{aligned} ds^2 &= dx^2 + dy^2 \\ &= (\cos^2\theta + \sin^2\theta)\,dr^2 + 0\cdot dr\,d\theta + r^2(\sin^2\theta + \cos^2\theta)\,d\theta^2 \\ &= dr^2 + r^2\,d\theta^2 \end{aligned}$$

as in (4.3). Using the same method for cylindrical and spherical coordinates (Problem 21) you can verify equations (4.18) and (4.19).

▶ **Example 5.** Express the velocity of a moving particle in spherical coordinates.

If s represents the distance the particle has moved along some path, then ds/dt is the velocity of the particle. Dividing (4.19) by dt^2, we find for the square of the velocity

(4.20)

$$v^2 = \left(\frac{ds}{dt}\right)^2 = \left(\frac{dr}{dt}\right)^2 + r^2\left(\frac{d\theta}{dt}\right)^2 + r^2\sin^2\theta\left(\frac{d\phi}{dt}\right)^2 \qquad \text{(spherical coordinates)}.$$

We have just seen how to find the arc length element ds in polar coordinates (or other systems) by calculating $\sqrt{dx^2 + dy^2}$. You might be tempted to try to find the area element by computing $dx\,dy$, but you would discover that this does not work—we must use the Jacobian [or geometry as in (4.2)] to get volume or area elements. You can see why by looking at Figure 4.1. The element of area $r\,dr\,d\theta$ at the point (x, y) is not the same as the element of area $dx\,dy$ at that point. Then consider Figure 4.2; the element of arc ds is the hypotenuse of the triangle with legs dr and $rd\theta$, and it is also the hypotenuse of the triangle with legs dx and dy. Thus ds is the *same element* for both x, y and r, θ and this is why we can compute ds in polar coordinates by calculating $\sqrt{dx^2 + dy^2}$. These comments hold for other coordinate systems, too. We can always find ds by computing $\sqrt{dx^2 + dy^2}$ or $\sqrt{dx^2 + dy^2 + dz^2}$, but we cannot compute area or volume elements directly from the rectangular ones—we must use the Jacobian or else geometrical methods.

PROBLEMS, SECTION 4

As needed, use a computer to plot graphs of figures and to check values of integrals.

1. For the disk $r \le a$, find by integration using polar coordinates:

(a) the area of the disk;

(b) the centroid of one quadrant of the disk;

(c) the moment of inertia of the disk about a diameter;

(d) the circumference of the circle $r = a$;

(e) the centroid of a quarter circle arc.

2. Using polar coordinates:

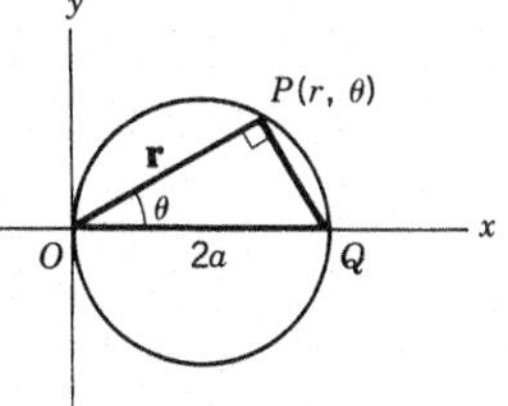

(a) Show that the equation of the circle sketched is $r = 2a\cos\theta$. *Hint:* Use the right triangle OPQ.

(b) By integration, find the area of the disk $r \le 2a\cos\theta$.

(c) Find the centroid of the area of the first quadrant half disk.

(d) Find the moments of inertia of the disk about each of the three coordinate axes, assuming constant area density.

(e) Find the length and the centroid of the semicircular arc in the first quadrant.

(f) Find the center of mass and the moments of inertia of the disk if the density is r.

(g) Find the area common to the disk sketched and the disk $r \le a$.

3. (a) Find the moment of inertia of a circular disk (uniform density) about an axis through its center and perpendicular to the plane of the disk.

(b) Find the moment of inertia of a solid right circular cylinder (uniform density) about its axis.

(c) Do (a) using Problem 1c and the perpendicular axis theorem (Section 3, Example 1f).

4. For the sphere $r = a$, find by integration:

(a) its surface area;

(b) the centroid of the curved surface area of a hemisphere;

(c) the moment of inertia of the whole spherical shell (that is, surface area) about a diameter (assuming constant area density);

(d) the volume of the ball $r \le a$;

(e) the centroid of a solid half ball.

5. (a) Write a triple integral in spherical coordinates for the volume inside the cone $z^2 = x^2 + y^2$ and between the planes $z = 1$ and $z = 2$. Evaluate the integral.

(b) Do (a) in cylindrical coordinates.

6. Find the mass of the solid in Problem 5 if the density is $(x^2 + y^2 + z^2)^{-1}$. Check your work by doing the problem in both spherical and cylindrical coordinates.

7. (a) Using spherical coordinates, find the volume cut from the ball $r \le a$ by the cone $\theta = \alpha < \pi/2$.

(b) Show that the z coordinate of the centroid of the volume in (a) is given by the formula $\bar{z} = 3a(1 + \cos\alpha)/8$.

8. For the solid in Problem 7, find I_z/M if $\alpha = \pi/3$ and the density is constant.

9. Let the solid in Problem 7 have density $= \cos\theta$. Show that then $I_z = \frac{3}{10} M a^2 \sin^2\alpha$.

10. (a) Find the volume inside the cone $3z^2 = x^2 + y^2$, above the plane $z = 2$ and inside the sphere $x^2 + y^2 + z^2 = 36$. *Hint:* Use spherical coordinates.

(b) Find the centroid of the volume in (a).

11. Write a triple integral in cylindrical coordinates for the volume inside the cylinder $x^2 + y^2 = 4$ and between $z = 2x^2 + y^2$ and the (x, y) plane. Evaluate the integral.

12. (a) Write a triple integral in cylindrical coordinates for the volume of the solid cut from a ball of radius 2 by a cylinder of radius 1, one of whose rulings is a diameter of the ball. *Hint:* Take the axis of the cylinder parallel to the z axis; a cross section of the cylinder then looks like the figure in Problem 2.

(b) Write a triple integral for the moment of inertia about the z axis of a uniform solid occupying this volume.

(c) Evaluate the integrals in (a) and (b), and find I as a multiple of the mass.

13. (a) Write a triple integral in cylindrical coordinates for the volume of the part of a ball between two parallel planes which intersect the ball.

(b) Evaluate the integral in (a). *Warning hint:* Do the r and θ integrals first.

(c) Find the centroid of this volume.

14. Express the integral

$$I = \int_0^1 dx \int_0^{\sqrt{1-x^2}} e^{-x^2-y^2}\, dy$$

as an integral in polar coordinates (r, θ) and so evaluate it.

15. Find the cylindrical coordinate volume element by Jacobians.

Find the Jacobians $\partial(x, y)/\partial(u, v)$ of the given transformations from variables x, y to variables u, v:

16. $x = \frac{1}{2}(u^2 - v^2)$,
$y = uv$, (u and v are called parabolic cylinder coordinates).

17. $x = a\cosh u\,\cos v$,
$y = a\sinh u\,\sin v$, (u and v are called elliptic cylinder coordinates).

18. Prove the following theorems about Jacobians.

$$\frac{\partial(u,v)}{\partial(x,y)}\frac{\partial(x,y)}{\partial(u,v)} = 1.$$

$$\frac{\partial(x,y)}{\partial(u,v)}\frac{\partial(u,v)}{\partial(s,t)} = \frac{\partial(x,y)}{\partial(s,t)}.$$

Hint: Multiply the determinants (as you would matrices) and show that each element in the product determinant can be written as a single partial derivative. Also see Chapter 4, Section 7.

19. In the integral

$$I = \int_0^\infty \int_0^\infty \frac{x^2+y^2}{1+(x^2-y^2)^2}\,e^{-2xy}\,dx\,dy$$

make the change of variables

$$u = x^2 - y^2$$
$$v = 2xy$$

and evaluate I. *Hint:* Use (4.8) and the accompanying discussion.

20. In the integral

$$I = \int_{x=0}^{1/2} \int_{y=x}^{1-x} \left(\frac{x-y}{x+y}\right)^2 dy\,dx,$$

make the change of variables

$$x = \frac{1}{2}(r-s),$$
$$y = \frac{1}{2}(r+s),$$

and evaluate I. *Hints:* See Problem 19. To find the r and s limits, sketch the area of integration in the (x, y) plane and sketch the r and s axes. Then show that to cover the same integration area, you may take the r and s limits to be: s from 0 to r, r from 0 to 1.

21. Verify equations (4.18) and (4.19).

22. Use equation (4.18) to set up an integral for the length of wire required to wind a coil spirally about a cylinder of radius 1 in., and length 1 ft, if there are three turns per inch.

23. A loxodrome or rhumb line is a curve on the earth's surface along which a ship sails without changing its course, that is, such that it crosses the meridians at a constant angle α. Show that then $\tan\alpha = \sin\theta\,d\phi/d\theta$ (θ and ϕ are spherical coordinates). Use (4.19) to set up an integral for the distance traveled by a ship along a rhumb line. Show that although a rhumb line winds infinitely many times around either the north or the south pole, its total length is finite.

24. Compute the gravitational attraction on a unit mass at the origin due to the mass (of constant density) occupying the volume inside the sphere $r = 2a$ and above the plane $z = a$. *Hint:* The magnitude of the gravitational force on the unit mass due to the element of mass dM at (r, θ, ϕ) is $(G/r^2)dM$. You want the z component of this since the other components of the total force are zero by symmetry. Use spherical coordinates.

25. The volume inside a sphere of radius r is $V = \frac{4}{3}\pi r^3$. Then $dV = 4\pi r^2\,dr = A\,dr$, where A is the area of the sphere. What is the geometrical meaning of the fact that the derivative of the volume is the area? Could you use this fact to find the volume formula given the area formula?

26. Use the parallel axis theorem (Problem 3.1)

(a) and Example 3, to find the moment of inertia of a solid ball about a line tangent to it;

(b) and Problem 3b to find the moment of inertia of a solid cylinder about a ruling.

27. Use the spherical coordinates θ and ϕ to find the area of a zone of a sphere (that is, the spherical surface area between two parallel planes). *Hint:* See dA in (4.5).

28. Find the center of mass of a hemispherical shell of constant density (mass per unit area) by using double integrals and the area element dA in (4.5). [Compare your result in Problem 4(b).]

5. SURFACE INTEGRALS

In the preceding sections we found surface areas, moments of them, etc., for surfaces of revolution. We now want to consider a way of computing surface integrals in general whether the surface is a surface of revolution or not. Consider a part of a surface as in Figure 5.1 and its projection in the (x, y) plane. We assume that any line parallel to the z axis intersects the surface only once. If this is not true, we must work with part of the surface at a time, or project the surface into a different plane. For example, if the surface is closed, we could find the areas of the upper and lower parts separately. For a cylinder with axis parallel to the z axis we could project the front and back parts separately into the (y, z) plane.

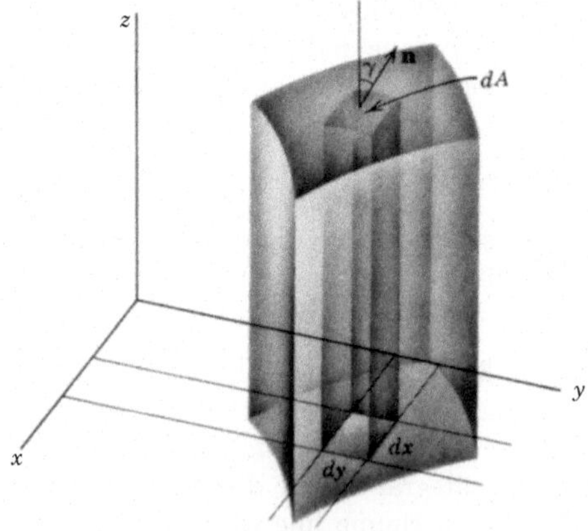

Figure 5.1

Let dA (Figure 5.1) be an element of surface area which projects onto $dx\,dy$ in the (x, y) plane and let γ be the acute angle between dA (that is, the tangent plane at dA) and the (x, y) plane. Then we have

$$(5.1) \qquad dx\,dy = dA\cos\gamma \qquad \text{or} \qquad dA = \sec\gamma\,dx\,dy.$$

The surface area is then

$$\iint dA = \iint \sec\gamma \, dx\, dy \tag{5.2}$$

where the limits on x and y must be such that we integrate over the projected area in the (x, y) plane.

Now we must find $\sec\gamma$. The (acute) angle between two planes is the same as the (acute) angle between the normals to the planes. If $\mathbf{n}$ is a unit vector normal to the surface at dA (Figure 5.1), then γ is the (acute) angle between $\mathbf{n}$ and the z axis, that is, between the vectors $\mathbf{n}$ and $\mathbf{k}$, so $\cos\gamma = |\mathbf{n}\cdot\mathbf{k}|$. Let the equation of the surface be $\phi(x, y, z) = \text{const}$. Recall from Chapter 4 just after equation (9.14) that the vector

$$\operatorname{grad}\phi = \mathbf{i}\frac{\partial\phi}{\partial x} + \mathbf{j}\frac{\partial\phi}{\partial y} + \mathbf{k}\frac{\partial\phi}{\partial z} \tag{5.3}$$

is normal to the surface $\phi(x, y, z) = \text{const}$. (Also see Chapter 6, Section 6.) Then $\mathbf{n}$ is a unit vector in the direction of $\operatorname{grad}\phi$, so

$$\mathbf{n} = (\operatorname{grad}\phi)/|\operatorname{grad}\phi|. \tag{5.4}$$

From (5.3) and (5.4) we find

$$\mathbf{n}\cdot\mathbf{k} = \frac{\mathbf{k}\cdot\operatorname{grad}\phi}{|\operatorname{grad}\phi|} = \frac{\partial\phi/\partial z}{|\operatorname{grad}\phi|},$$

$$\sec\gamma = \frac{1}{\cos\gamma} = \frac{1}{|\mathbf{n}\cdot\mathbf{k}|},$$

so

$$\sec\gamma = \frac{|\operatorname{grad}\phi|}{|\partial\phi/\partial z|} = \frac{\sqrt{\left(\dfrac{\partial\phi}{\partial x}\right)^2 + \left(\dfrac{\partial\phi}{\partial y}\right)^2 + \left(\dfrac{\partial\phi}{\partial z}\right)^2}}{|\partial\phi/\partial z|}. \tag{5.5}$$

Often the equation of a surface is given in the form $z = f(x, y)$. In this case $\phi(x, y, z) = z - f(x, y)$, so $\partial\phi/\partial z = 1$, and (5.5) simplifies to

$$\sec\gamma = \sqrt{(\partial f/\partial x)^2 + (\partial f/\partial y)^2 + 1}. \tag{5.6}$$

We then substitute (5.5) or (5.6) into (5.2) and integrate to find the area. To find centroids, moments of inertia, etc., we insert the proper factor into (5.2) as we have discussed in Section 3.

Example 1. Find the area cut from the upper half of the sphere $x^2 + y^2 + z^2 = 1$ by the cylinder $x^2 + y^2 - y = 0$.

This is the same as the area on the sphere which projects onto the disk $x^2 + y^2 - y \leq 0$ in the (x, y) plane. Thus we want to integrate (5.2) over the area of this disk. Figure 5.2 shows the disk of integration (shaded) and the equatorial circle of the sphere (large circle). We compute $\sec\gamma$ from the equation of the sphere; we could use (5.6), but it is easier in this problem to use (5.5):

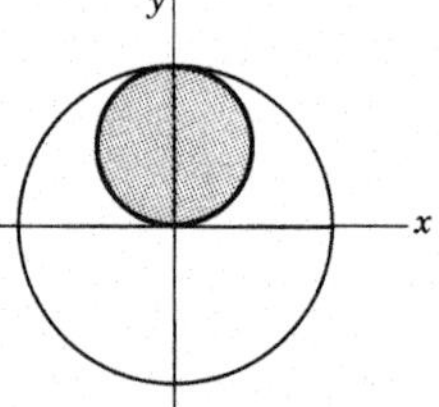

Figure 5.2

$$\phi = x^2 + y^2 + z^2,$$

$$\sec\gamma = \frac{|\operatorname{grad}\phi|}{|\partial\phi/\partial z|} = \frac{1}{2z}\sqrt{(2x)^2 + (2y)^2 + (2z)^2} = \frac{1}{z} = \frac{1}{\sqrt{1 - x^2 - y^2}}.$$

We find the limits of integration from the equation of the shaded disk, $x^2+y^2-y \leq 0$. Because of the symmetry we can integrate over the first-quadrant part of the shaded area and double our result. Then the limits are:

$$x \text{ from } 0 \text{ to } \sqrt{y - y^2},$$
$$y \text{ from } 0 \text{ to } 1.$$

The desired area is

$$A = 2\int_{y=0}^{1}\int_{x=0}^{\sqrt{y-y^2}} \frac{dx\,dy}{\sqrt{1 - x^2 - y^2}}. \tag{5.7}$$

This integral is simpler in polar coordinates. The equation of the cylinder is then $r = \sin\theta$, so the limits are: r from 0 to $\sin\theta$, and θ from 0 to $\pi/2$. Thus (5.7) becomes

$$A = 2\int_{\theta=0}^{\pi/2}\int_{r=0}^{\sin\theta} \frac{r\,dr\,d\theta}{\sqrt{1 - r^2}}. \tag{5.8}$$

This is still simpler if we make the change of variable $z = \sqrt{1 - r^2}$. Then $dz = -r\,dr/\sqrt{1 - r^2}$, and the limits $r = 0$ to $\sin\theta$ become $z = 1$ to $\cos\theta$. Thus (5.8) becomes

$$A = -2\int_{\theta=0}^{\pi/2}\int_{z=1}^{\cos\theta} dz\,d\theta = \pi - 2. \tag{5.9}$$

PROBLEMS, SECTION 5

For these problems, the most important sketch is the projection in the plane of integration, which is easy to do by hand. However, you might like to use your computer to plot the corresponding 3 dimensional picture.

1. Find the area of the plane $x - 2y + 5z = 13$ cut out by the cylinder $x^2 + y^2 = 9$.

2. Find the surface area cut from the cone $2x^2 + 2y^2 = 5z^2$, $z > 0$, by the cylinder $x^2 + y^2 = 2y$.

3. Find the area of the paraboloid $x^2 + y^2 = z$ inside the cylinder $x^2 + y^2 = 9$.

4. Find the area of the part of the cone $2z^2 = x^2 + y^2$ in the first octant cut out by the planes $y = 0$, and $y = x/\sqrt{3}$, and the cylinder $x^2 + y^2 = 4$.

5. Find the area of the part of the cone $z^2 = 3(x^2 + y^2)$ which is inside the sphere $x^2 + y^2 + z^2 = 16$.

6. In Example 1, find the area of the cylinder inside the sphere.

7. Find the area of the part of the cylinder $y^2 + z^2 = 4$ in the first octant, cut out by the planes $x = 0$ and $y = x$.

8. Find the area of the part of the cylinder $z = x + y^2$ that lies below the second-quadrant area bounded by the x axis, $x = -1$, and $y^2 = -x$.

9. Find the area of the part of the cone $x^2+y^2 = z^2$ that is over the disk $(x-1)^2+y^2 \le 1$.

10. Find the area of the part of the sphere of radius a and center at the origin which is above the square in the (x, y) plane bounded by $x = \pm a/\sqrt{2}$ and $y = \pm a/\sqrt{2}$. *Hint for evaluating the integral*: Change to polar coordinates and evaluate the r integral first.

11. The part of the plane $x + y + z = 1$ which is in the first octant is a triangular area (sketch it). Find the area and its centroid by integration. You might like to check your work by geometry.

12. In Problem 11, let the triangle have a density (mass per unit area) equal to x. Find the total mass and the coordinates of the center of mass.

13. For the area of Example 1, find the z coordinate of the centroid.

14. For the area in Example 1, let the mass per unit area be equal to $|x|$. Find the total mass.

15. For a uniform mass distribution over the area of Example 1, find the moment of inertia about the z axis.

16. Find the centroid of the surface area in Problem 2.

6. MISCELLANEOUS PROBLEMS

As needed, use a computer to plot graphs and to check values of integrals.

1. Find the volume inside the cone $z^2 = x^2 + y^2$, above the (x, y) plane, and between the spheres $x^2 + y^2 + z^2 = 1$ and $x^2 + y^2 + z^2 = 4$. *Hint:* Use spherical coordinates.

2. Find the z coordinate of the centroid of the volume in Problem 1.

3. Find the mass of the solid in Problem 1 if the density is equal to z.

4. Find the moment of inertia of a hoop (wire bent to form a circle of radius R)

   (a) about a diameter;

   (b) about a tangent line.

5. The rectangle in the figure has sides $2a$ and $2b$; the curve is an ellipse. If the figure is rotated about the dotted line it generates three solids of revolution: a cone, an ellipsoid, and a cylinder. Show that the volumes are in the ratio 1 : 2 : 3. (See L.H. Lange, American Mathematical Monthly vol. 88 (1981), p. 339.)

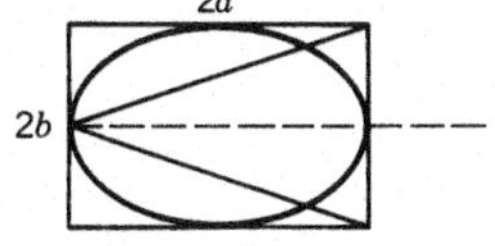

6. (a) Find the area inside the circle $r = 2$, with $x > 0$ and $y > 1$.

(b) Find the centroid of the area in (a).

7. For a lamina of density 1 in the shape of the area of Problem 6, find its moment of inertia about the z axis.

8. For the solid bounded above by the sphere $x^2+y^2+z^2 = 4$ and below by a horizontal plane through $(0, 0, 1)$, find

(a) the volume (see Problem 6 and Problem 3.12);

(b) the z coordinate of the centroid (use cylindrical coordinates).

9. Find the centroid of the area above $y = x^2$ and below $y = c\,(c > 0)$.

10. (a) Find the centroid of the area between the x axis and one arch of $y = \sin x$.

(b) Find the volume formed if the area in (a) is rotated about the x axis.

(c) Find I_x of a mass of constant density occupying the volume in (b).

11. Show that the z coordinate of the centroid of the volume inside the elliptic cone

$$\frac{z^2}{h^2} = \frac{x^2}{a^2} + \frac{y^2}{b^2}, \quad 0 < z < h, \quad \text{is} \quad \bar{z} = \frac{3}{4}h.$$

(Note that the result is independent of a and b.) *Hint:* To evaluate the triple integrals, let $z = hz'$, $x = ax'$, $y = by'$, and then change to cylindrical coordinates in the primed system (see Example 4, Section 4). Compare Example 2, Section 4.

12. Find the mass of the solid inside the ellipsoid

$$\frac{x^2}{a^2} + \frac{y^2}{b^2} + \frac{z^2}{c^2} = 1$$

if the density is $|xyz|$. *Hint:* Evaluate the triple integral as in Example 4, Section 4.

13. Find the surface area of the part of the cylinder $x^2 + z^2 = a^2$ inside the cylinder $x^2 + y^2 = a^2$. Use your computer to graph the two cylinders on the same axes.

14. Find the volume that is inside both cylinders in Problem 13.

15. Find I_x and I_y for a mass distribution of constant density occupying the solid in Problem 14. *Hint:* Do the x integration last.

16. Find the centroid of the first quadrant part of the arc $x^{2/3} + y^{2/3} = a^{2/3}$. *Hint:* Let $x = a\cos^3\theta$, $y = a\sin^3\theta$.

17. Find the moment of inertia about a diagonal of a framework consisting of the four sides of a square of side a.

18. Find the center of mass of the solid right circular cone inside $r^2 = z^2$, $0 < z < h$, if the density is $r^2 = x^2 + y^2$. Use cylindrical coordinates.

19. For the cone in Problem 18, find I_x/M, I_y/M, I_z/M. Also find I/M about a line through the center of mass parallel to the x axis.

20. (a) Find the area of the surface $z = 1 + x^2 + y^2$ inside the cylinder $x^2 + y^2 = 1$.

(b) Find the volume inside the cylinder between the surface and the (x, y) plane. Use cylindrical coordinates.

21. Find the gravitational attraction on a unit mass at the origin due to a mass (of constant density) occupying the volume inside the cone $z^2 = x^2 + y^2$, $0 < z < h$. See Problem 4.24.

22. Find I_x/M, I_y/M, I_z/M, for a lamina in the shape of an ellipse $(x^2/a^2)+(y^2/b^2)=1$. *Hint:* See Problem 11.

23. (a) Find the centroid of the solid paraboloid inside $z = x^2 + y^2$, $0 < z < c$.

(b) Repeat part (a) if the density is $\rho = r = \sqrt{x^2+y^2}$.

24. Repeat Problem 23a for the paraboloid

$$\frac{z}{c} = \frac{x^2}{a^2} + \frac{y^2}{b^2}, \qquad 0 < z < c.$$

25. By changing to polar coordinates, evaluate

$$\int_0^\infty \int_0^\infty e^{-\sqrt{x^2+y^2}}\,dx\,dy.$$

26. Make the change of variables $u = x - y$, $v = x + y$, to evaluate the integral

$$\int_0^1 dy \int_0^{1-y} e^{(x-y)/(x+y)}\,dx.$$

27. Make the change of variables $u = y/x$, $v = x + y$, to evaluate the integral

$$\int_0^1 dx \int_0^x \frac{(x+y)e^{x+y}}{x^2}\,dy.$$

CHAPTER 6

Vector Analysis

1. INTRODUCTION

In Chapter 3, Sections 4 and 5, we have discussed the basic ideas of vector algebra. The principal topic of this chapter will be vector calculus. First (Sections 2 and 3) we shall consider some applications of vector products. Then (Section 4 ff.) we shall discuss differentiation and integration of vector functions. You have probably seen Newton's second law $\mathbf{F} = m\mathbf{a}$ written as $\mathbf{F} = m\, d^2\mathbf{r}/dt^2$. You may have met Gauss's law in electricity which uses a surface integral of the normal component of a vector (Section 10). Derivatives and integrals of vector functions are important in almost every area of applied mathematics. Such diverse fields as mechanics, quantum mechanics, electrodynamics, theory of heat, hydrodynamics, optics, etc., make use of the vector equations and theorems we shall discuss in this chapter.

2. APPLICATIONS OF VECTOR MULTIPLICATION

In Chapter 3, Section 4, we defined the scalar or dot product of vectors $\mathbf{A}$ and $\mathbf{B}$, and the vector or cross product of $\mathbf{A}$ and $\mathbf{B}$ as follows, where θ is the angle ($\leq 180°$) between the vectors:

(2.1) $$\mathbf{A} \cdot \mathbf{B} = AB\cos\theta = A_xB_x + A_yB_y + A_zB_z.$$

(2.2) $\mathbf{A}\times\mathbf{B} = \mathbf{C}$, where $|\mathbf{C}\,| = AB\sin\theta$, and the direction of $\mathbf{C}$ is perpendicular to the plane of $\mathbf{A}$ and $\mathbf{B}$ and in the sense of the rotation of $\mathbf{A}$ to $\mathbf{B}$ through the angle θ (Figure 2.1).

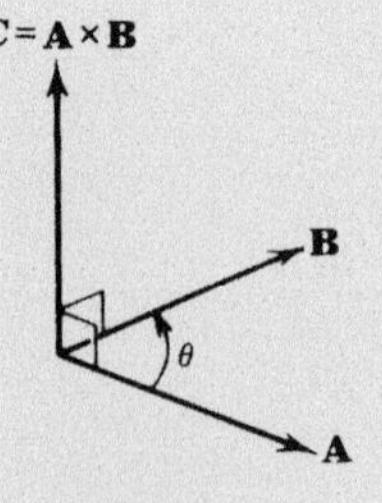

Figure 2.1

Let us consider some applications of these definitions.

Work In elementary physics you learned that work equals force times displacement. If the force and displacement are not parallel, then the component of the force perpendicular to the displacement does no work.

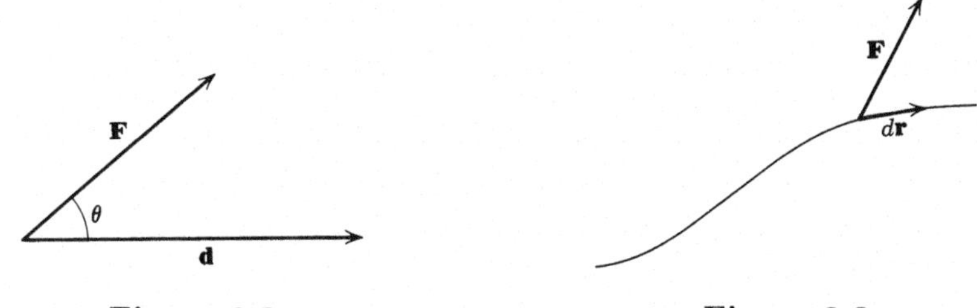

Figure 2.2 **Figure 2.3**

The work in this case is the component of the force parallel to the displacement, multiplied by the displacement; that is $W = (F\cos\theta)\cdot d = Fd\cos\theta$ (Figure 2.2). This can now conveniently be written as

$$W = Fd\cos\theta = \mathbf{F}\cdot\mathbf{d}. \tag{2.3}$$

If the force varies with distance, and perhaps also the direction of motion **d** changes with time, we can write, for an infinitesimal vector displacement $d\mathbf{r}$ (Figure 2.3)

$$dW = \mathbf{F}\cdot d\mathbf{r}. \tag{2.4}$$

We shall see later (Section 8) how to integrate dW in (2.4) to find the total work W done on a particle which is pushed along some path by a variable force **F**.

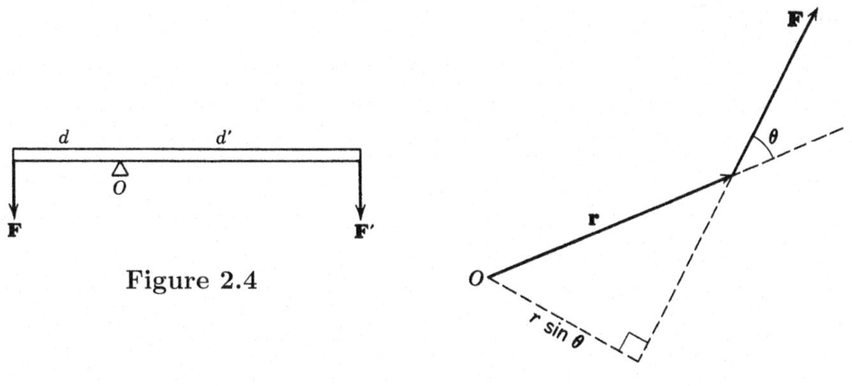

Figure 2.4

Figure 2.5

Torque In doing a seesaw or lever problem (Figure 2.4), you multiply force times distance; the quantity Fd is called the *torque* or *moment** of **F**, and the distance d

*If the force **F** is due to a weight $w = mg$, then the torque about O in Figure 2.4 is $mg\cdot d = g\cdot(md)$; the moment of inertia (Chapter 5, Section 3) of m about O is md^2. The quantity md is called the moment (or *first* moment) of m about O, and the quantity md^2 is called the moment of inertia (or *second* moment) of m about O. By extension, we call mgd the moment of mg, or Fd the moment of **F**. For an object which is not a point mass, the quantities md and md^2 become integrals (Chapter 5, Section 3).

from the fulcrum O to the line of action of $\mathbf{F}$ is the *lever arm* of $\mathbf{F}$. The lever arm is by definition the *perpendicular* distance from O to the line of action of $\mathbf{F}$. Then in general (Figure 2.5) the torque (or moment) of a force about O (really about an axis through O perpendicular to the paper) is defined as the magnitude of the force times its lever arm; in Figure 2.5 this is $Fr\sin\theta$. Now $\mathbf{r}\times\mathbf{F}$ has magnitude $rF\sin\theta$, so the magnitude of the torque is $|\mathbf{r}\times\mathbf{F}|$. We can also use the direction of $\mathbf{r}\times\mathbf{F}$ in describing the torque, in the following way. If you curve the fingers of your right hand in the direction of the rotation produced by applying the torque, then your thumb points in a direction parallel to the rotation axis. It is customary to call this the direction of the torque. By comparing Figures 2.5 and 2.1, we see that this is also the direction of $\mathbf{r}\times\mathbf{F}$. With this agreement, then, $\mathbf{r}\times\mathbf{F}$ is the torque or moment of $\mathbf{F}$ about an axis through O and perpendicular to the plane of the paper in Figure 2.5.

Angular Velocity In a similar way, a vector is used to represent the angular velocity of a rotating body. The direction of the vector is along the axis of rotation in the direction of progression of a right-handed screw turned the way the body is rotating. Suppose P in Figure 2.6 represents a point in a rigid body rotating with angular velocity $\boldsymbol{\omega}$. We can show that the linear velocity $\mathbf{v}$ of point P is $\mathbf{v}=\boldsymbol{\omega}\times\mathbf{r}$. First of all, $\mathbf{v}$ is in the right direction: It is perpendicular to the plane of $\mathbf{r}$ and $\boldsymbol{\omega}$ and in the right sense. Next we want to show that the magnitude of $\mathbf{v}$ is the same as $|\boldsymbol{\omega}\times\mathbf{r}|=\omega r\sin\theta$. But $r\sin\theta$ is the radius of the circle in which P is traveling, and ω is the angular velocity; thus $(r\sin\theta)\omega$ is $|\mathbf{v}|$, as we claimed.

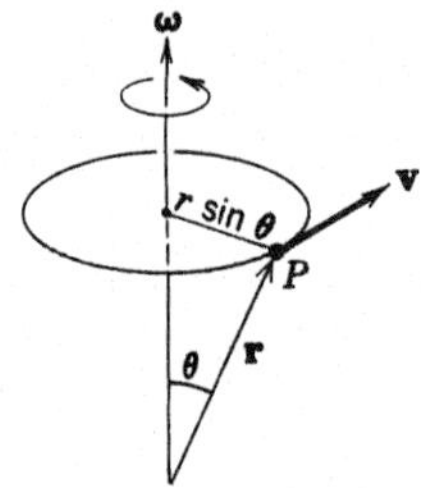

Figure 2.6

3. TRIPLE PRODUCTS

There are two products involving three vectors, one called the triple scalar product (because the answer is a scalar) and the other called the triple vector product (because the answer is a vector).

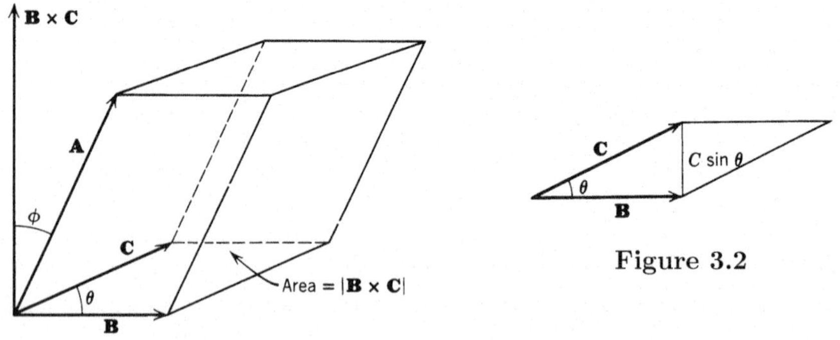

Figure 3.1

Figure 3.2

Triple Scalar Product This is written $\mathbf{A}\cdot(\mathbf{B}\times\mathbf{C})$. There is a useful geometrical interpretation of the triple scalar product (see Figure 3.1). Construct a parallelepiped using $\mathbf{A},\mathbf{B},\mathbf{C}$ as three intersecting edges. Then $|\mathbf{B}\times\mathbf{C}|$ is the area of the base (Figure 3.2) because $|\mathbf{B}\times\mathbf{C}| = |\mathbf{B}|\,|\mathbf{C}|\sin\theta$, which is the area of a parallelogram with sides $|\mathbf{B}|$, $|\mathbf{C}|$, and angle θ. The height of the parallelepiped is $|\mathbf{A}|\cos\phi$ (Figure 3.1). Then the volume of the parallelepiped is

$$|\mathbf{B}|\,|\mathbf{C}|\sin\theta|\mathbf{A}|\cos\phi = |\mathbf{B}\times\mathbf{C}|\,|\mathbf{A}|\cos\phi = \mathbf{A}\cdot(\mathbf{B}\times\mathbf{C}).$$

If $\phi > 90°$, this will come out negative, so in general we should say that the volume is $|\mathbf{A}\cdot(\mathbf{B}\times\mathbf{C})|$. Any side may be used as base, so, for example, $\mathbf{B}\cdot(\mathbf{C}\times\mathbf{A})$ must also be either plus or minus the volume. There are six such triple scalar products, all equal except for sign [or twelve if you count both the type $\mathbf{A}\cdot(\mathbf{B}\times\mathbf{C})$ and the type $(\mathbf{B}\times\mathbf{C})\cdot\mathbf{A}$].

To write the triple scalar product in component form we first write $\mathbf{B}\times\mathbf{C}$ in determinant form [Chapter 3, equation (4.19)]:

$$\mathbf{B}\times\mathbf{C} = \begin{vmatrix} \mathbf{i} & \mathbf{j} & \mathbf{k} \\ B_x & B_y & B_z \\ C_x & C_y & C_z \end{vmatrix}. \tag{3.1}$$

Now $\mathbf{A}\cdot(\mathbf{B}\times\mathbf{C}) = A_x(\mathbf{B}\times\mathbf{C})_x + A_y(\mathbf{B}\times\mathbf{C})_y + A_z(\mathbf{B}\times\mathbf{C})_z$, and this is exactly what we get by expanding, by elements of the first row, the determinant in (3.2) below; this determinant is then equal to $\mathbf{A}\cdot(\mathbf{B}\times\mathbf{C})$.

$$\mathbf{A}\cdot(\mathbf{B}\times\mathbf{C}) = \begin{vmatrix} A_x & A_y & A_z \\ B_x & B_y & B_z \\ C_x & C_y & C_z \end{vmatrix}. \tag{3.2}$$

Recalling that an interchange of rows changes the sign of a determinant, we can now easily write out the six (or twelve) products mentioned above with their proper signs. You should convince yourself of, and then remember, the following facts: The *order* of the factors is all that counts; the dot and cross may be interchanged. If the order of factors is cyclic (one way around the circle in Figure 3.3), all such triple scalar products are equal. If you go the other way, you get another set all equal to each other and the negatives of the first set. For example,

$$\begin{aligned} (\mathbf{A}\times\mathbf{B})\cdot\mathbf{C} &= \mathbf{A}\cdot(\mathbf{B}\times\mathbf{C}) \\ &= \mathbf{C}\cdot(\mathbf{A}\times\mathbf{B}) \\ &= -(\mathbf{A}\times\mathbf{C})\cdot\mathbf{B}, \text{ etc.} \end{aligned} \tag{3.3}$$

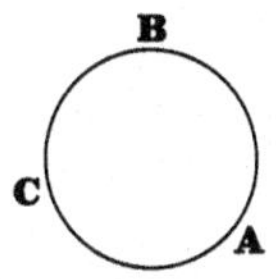

B
C
A

Figure 3.3

Since it doesn't matter where the dot and cross are, the triple scalar product is often written as $(\mathbf{ABC})$, meaning $\mathbf{A}\cdot(\mathbf{B}\times\mathbf{C})$ or $(\mathbf{A}\times\mathbf{B})\cdot\mathbf{C}$.

Triple Vector Product This is written $\mathbf{A}\times(\mathbf{B}\times\mathbf{C})$. Before we try to evaluate it, we can make the following observations. $\mathbf{B}\times\mathbf{C}$ is perpendicular to the plane of $\mathbf{B}$ and $\mathbf{C}$. $\mathbf{A}\times(\mathbf{B}\times\mathbf{C})$ is perpendicular to the plane of $\mathbf{A}$ and $(\mathbf{B}\times\mathbf{C})$; we are particularly interested in the fact that $\mathbf{A}\times(\mathbf{B}\times\mathbf{C})$ is perpendicular to $(\mathbf{B}\times\mathbf{C})$.

Now (see Figure 3.4) *any* vector perpendicular to $\mathbf{B}\times\mathbf{C}$ lies in the plane perpendicular to $\mathbf{B}\times\mathbf{C}$, that is, the plane of $\mathbf{B}$ and $\mathbf{C}$. Thus $\mathbf{A}\times(\mathbf{B}\times\mathbf{C})$ is *some* vector in the plane of $\mathbf{B}$ and $\mathbf{C}$, and can be written as some combination $a\mathbf{B}+b\mathbf{C}$, where a and b are scalars which we want to find. (See Chapter 3, Section 8, Problem 5.) One way to find a and b is to write out $\mathbf{A}\times(\mathbf{B}\times\mathbf{C})$ in component form. We can simplify this work by choosing our coordinate system carefully; recall that a vector equation is true independently of the coordinate system. Given the vectors $\mathbf{A}$, $\mathbf{B}$, $\mathbf{C}$, we take the x axis along $\mathbf{B}$, and the y axis in the plane of $\mathbf{B}$ and $\mathbf{C}$; then $\mathbf{B}\times\mathbf{C}$ is in the z direction. The vectors in component form relative to these axes are:

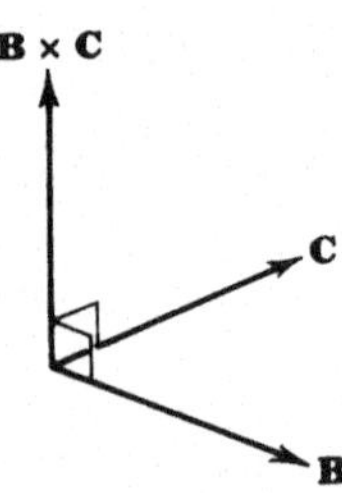

Figure 3.4

$$\begin{aligned} \mathbf{B} &= B_x\mathbf{i}, \\ \mathbf{C} &= C_x\mathbf{i}+C_y\,\mathbf{j}, \\ \mathbf{A} &= A_x\mathbf{i}+A_y\,\mathbf{j}+A_z\mathbf{k}. \end{aligned} \tag{3.4}$$

Using (3.4) we find

$$\begin{aligned} \mathbf{B}\times\mathbf{C} &= B_x\mathbf{i}\times(C_x\mathbf{i}+C_y\,\mathbf{j}) = B_xC_y(\mathbf{i}\times\mathbf{j}) = B_xC_y\mathbf{k}, \\ \mathbf{A}\times(\mathbf{B}\times\mathbf{C}) &= A_xB_xC_y(\mathbf{i}\times\mathbf{k}) + A_yB_xC_y(\mathbf{j}\times\mathbf{k}) \\ &= A_xB_xC_y(-\mathbf{j}) + A_yB_xC_y(\mathbf{i}). \end{aligned} \tag{3.5}$$

We would like to write $\mathbf{A}\times(\mathbf{B}\times\mathbf{C})$ in (3.5) as a combination of $\mathbf{B}$ and $\mathbf{C}$; we can do this by adding and subtracting $A_xB_xC_x\mathbf{i}$:

$$\mathbf{A}\times(\mathbf{B}\times\mathbf{C}) = -A_xB_x(C_x\mathbf{i}+C_y\,\mathbf{j}) + (A_yC_y+A_xC_x)B_x\mathbf{i}. \tag{3.6}$$

Each of these expressions is something simple in terms of the vectors in (3.4):

$$\begin{aligned} A_xB_x &= \mathbf{A}\cdot\mathbf{B}, \qquad & A_yC_y+A_xC_x &= \mathbf{A}\cdot\mathbf{C}, \\ C_x\mathbf{i}+C_y\mathbf{j} &= \mathbf{C}, & B_x\mathbf{i} &= \mathbf{B}. \end{aligned} \tag{3.7}$$

Using (3.7) in (3.6), we get

$$\mathbf{A}\times(\mathbf{B}\times\mathbf{C}) = (\mathbf{A}\cdot\mathbf{C})\mathbf{B} - (\mathbf{A}\cdot\mathbf{B})\mathbf{C}. \tag{3.8}$$

This important formula should be learned, but not memorized in terms of letters, because that is confusing when you want some other combination of the same letters. Learn instead the following three facts:

(3.9) The value of a triple vector product is a linear combination of the two vectors in the parenthesis [$\mathbf{B}$ and $\mathbf{C}$ in (3.8)]; the coefficient of each vector is the dot product of the other two; the middle vector in the triple product [$\mathbf{B}$ in (3.8)] always has the positive sign.

This method also covers triple vector products with the parenthesis first; by (3.9), the value of $(\mathbf{B}\times\mathbf{C})\times\mathbf{A}$ is $(\mathbf{A}\cdot\mathbf{B})\mathbf{C} - (\mathbf{A}\cdot\mathbf{C})\mathbf{B}$. This is correct since it is just the negative of what we had above for $\mathbf{A}\times(\mathbf{B}\times\mathbf{C})$.

Applications of the Triple Scalar Product We have shown that the torque of a force $\mathbf{F}$ about an axis may be written as $\mathbf{r} \times \mathbf{F}$ in one special case, namely when $\mathbf{r}$ and $\mathbf{F}$ are in a plane perpendicular to the axis. Now let us consider the general case of finding the torque produced by a force $\mathbf{F}$ about *any* given line (axis) L in Figure 3.5. Let $\mathbf{r}$ be a vector from some (that is, any) point on L to the line of action of $\mathbf{F}$; let O be the tail of $\mathbf{r}$. Then we define the torque about the *point* O to be $\mathbf{r} \times \mathbf{F}$. Note that this cannot contradict our previous discussion of torque because we were considering torque about a *line* before, and this definition is of torque about a *point*. However, we shall show how the two notions are connected. Also notice that $\mathbf{r} \times \mathbf{F}$ is not changed if the *head* of $\mathbf{r}$ is moved along $\mathbf{F}$; for this just adds a multiple of $\mathbf{F}$ to $\mathbf{r}$, and $\mathbf{F} \times \mathbf{F} = 0$ (see Problem 10).

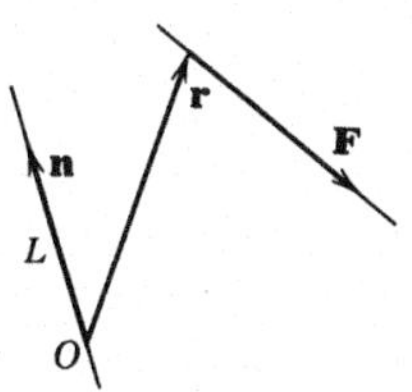

Figure 3.5

We shall now show that the torque of $\mathbf{F}$ about the line L through O is $\mathbf{n} \cdot (\mathbf{r} \times \mathbf{F})$, where $\mathbf{n}$ is a unit vector along L. To simplify the calculation, choose the positive z axis in the direction $\mathbf{n}$; then $\mathbf{n} = \mathbf{k}$. Think of a door hinged to rotate about the z

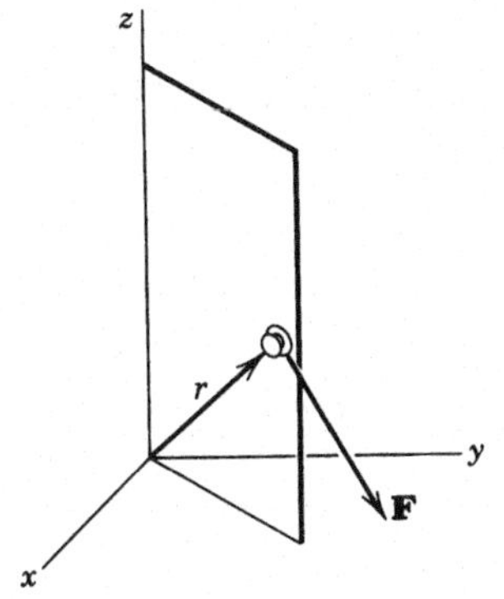

Figure 3.6

Figure 3.7

axis as in Figure 3.6. Let a force $\mathbf{F}$ be applied to it at the head of the vector $\mathbf{r}$. We first find the torque of $\mathbf{F}$ about the z axis by elementary methods and definition. Break $\mathbf{F}$ into its components; the z component is parallel to the rotation axis and produces no torque about it (pulling straight up or down on a door handle does not tend to open or close the door!). The x and y components can be seen better if we draw them in the (x, y) plane (Figure 3.7; note that the x and y axes are rotated 90° clockwise from their usual position in order to compare this figure more easily with Figure 3.6). The torque about the z axis produced by F_x and F_y is $xF_y - yF_x$ by the elementary definition of torque. We want to show that this is the same as $\mathbf{n} \cdot (\mathbf{r} \times \mathbf{F})$ or here $\mathbf{k} \cdot (\mathbf{r} \times \mathbf{F})$. Using (3.2) we find

$$\mathbf{k} \cdot (\mathbf{r} \times \mathbf{F}) = \begin{vmatrix} 0 & 0 & 1 \\ x & y & z \\ F_x & F_y & F_z \end{vmatrix} = xF_y - yF_x.$$

To summarize:

(3.10)	In Figure 3.5, the torque of $\mathbf{F}$ about point O is $\mathbf{r} \times \mathbf{F}$. The torque of $\mathbf{F}$ about the line L through O is $\mathbf{n} \cdot (\mathbf{r} \times \mathbf{F})$ where $\mathbf{n}$ is a unit vector along L.

This proof can easily be given without reference to a coordinate system. Let the symbols $\|$ and $\perp$ stand for parallel and perpendicular to the given rotation axis $\mathbf{n}$. Then any vector ($\mathbf{F}$ or $\mathbf{r}$, say) can be written as the sum of a vector parallel to the axis and a vector perpendicular to the axis (that is, somewhere in the plane perpendicular to $\mathbf{n}$):

$$\mathbf{r} = \mathbf{r}_\perp + \mathbf{r}_\|, \qquad \mathbf{F} = \mathbf{F}_\perp + \mathbf{F}_\|.$$

Then the torque about O produced by $\mathbf{F}$ is

$$\begin{aligned}\mathbf{r} \times \mathbf{F} &= (\mathbf{r}_\perp + \mathbf{r}_\|)\times(\mathbf{F}_\perp + \mathbf{F}_\|)\\ &= \mathbf{r}_\perp\times\mathbf{F}_\perp + \mathbf{r}_\perp\times\mathbf{F}_\| + \mathbf{r}_\|\times\mathbf{F}_\perp + \mathbf{r}_\|\times\mathbf{F}_\|.\end{aligned}$$

The last term is zero (cross product of parallel vectors). Also $\mathbf{r}_\|$ and $\mathbf{F}_\|$ are parallel to $\mathbf{n}$; therefore their cross products with anything are in the plane perpendicular to $\mathbf{n}$, and the dot product of $\mathbf{n}$ with these is zero. Hence we have

$$\mathbf{n} \cdot (\mathbf{r} \times \mathbf{F}) = \mathbf{n} \cdot (\mathbf{r}_\perp\times\mathbf{F}_\perp).$$

Now $\mathbf{r}_\perp$ and $\mathbf{F}_\perp$ are in a plane perpendicular to $\mathbf{n}$; thus the torque about $\mathbf{n}$ produced by $\mathbf{F}_\perp$ is (by Section 2) $\mathbf{r}_\perp\times\mathbf{F}_\perp$. But since only the component of $\mathbf{F}$ perpendicular to $\mathbf{n}$ produces a torque about $\mathbf{n}$, $\mathbf{r}_\perp\times\mathbf{F}_\perp$ is the total torque about $\mathbf{n}$ produced by $\mathbf{F}$. The vector torque $\mathbf{r}_\perp\times\mathbf{F}_\perp$ is in the $\pm\mathbf{n}$ direction since $\mathbf{r}_\perp$ and $\mathbf{F}_\perp$ are perpendicular to $\mathbf{n}$; the dot product of this vector torque with the unit vector $\mathbf{n}$ gives a scalar torque of the same magnitude; the $\pm$ sign indicates whether the torque is in the $+\mathbf{n}$ or the $-\mathbf{n}$ direction.

▶ **Example 1.** If $\mathbf{F} = \mathbf{i} + 3\mathbf{j} - \mathbf{k}$ acts at the point $(1, 1, 1)$, find the torque of $\mathbf{F}$ about the line $\mathbf{r} = 3\mathbf{i} + 2\mathbf{k} + (2\mathbf{i} - 2\mathbf{j} + \mathbf{k})t$.

We first find the vector torque about a point on the line, say the point $(3, 0, 2)$. By (3.10) and Figure 3.5, this is $\mathbf{r} \times \mathbf{F}$ where $\mathbf{r}$ is the vector *from* the point about which we want the torque, *to* the point at which $\mathbf{F}$ acts, that is, from $(3, 0, 2)$ to $(1, 1, 1)$; then $\mathbf{r} = (1, 1, 1) - (3, 0, 2) = (-2, 1, -1)$. The vector torque is

$$\mathbf{r} \times \mathbf{F} = \begin{vmatrix} \mathbf{i} & \mathbf{j} & \mathbf{k} \\ -2 & 1 & -1 \\ 1 & 3 & -1 \end{vmatrix} = 2\mathbf{i} - 3\mathbf{j} - 7\mathbf{k}.$$

The torque about the line is $\mathbf{n} \cdot (\mathbf{r} \times \mathbf{F})$ where $\mathbf{n}$ is a unit vector along the line, namely $\mathbf{n} = \frac{1}{3}(2\mathbf{i} - 2\mathbf{j} + \mathbf{k})$. Then the torque about the line is

$$\mathbf{n} \cdot (\mathbf{r} \times \mathbf{F}) = \frac{1}{3}(2\mathbf{i} - 2\mathbf{j} + \mathbf{k}) \cdot (2\mathbf{i} - 3\mathbf{j} - 7\mathbf{k}) = 1.$$

Example 2. As another application of the triple scalar product, let's find the Jacobian we used in Chapter 5, Section 4 for changing variables in a multiple integral. As you know, in rectangular coordinates the volume element is a rectangular box of volume $dx\,dy\,dz$. In other coordinate systems, the volume element may be approximately a parallelepiped as in Figure 3.1. We want a formula for the volume element in this case. (See, for example, the cylindrical and spherical coordinate volume elements in Chapter 5, Figures 4.4 and 4.5.)

Suppose we are given formulas for x, y, z as functions of new variables u, v, w. Then we want to find the vectors along the edges of the volume element in the u, v, w system. Suppose vector $\mathbf{A}$ in Figure 3.1 is along the direction in which u increases while v and w remain constant. Then if $d\mathbf{r} = \mathbf{i}\,dx + \mathbf{j}\,dy + \mathbf{k}\,dz$ is a vector in this direction, we have

$$\mathbf{A} = \frac{\partial \mathbf{r}}{\partial u}\,du = \left(\mathbf{i}\frac{\partial x}{\partial u} + \mathbf{j}\frac{\partial y}{\partial u} + \mathbf{k}\frac{\partial z}{\partial u}\right) du.$$

Similarly if $\mathbf{B}$ is along the increasing v edge of the volume element and $\mathbf{C}$ is along the increasing w edge, we have

$$\mathbf{B} = \frac{\partial \mathbf{r}}{\partial v}\,dv = \left(\mathbf{i}\frac{\partial x}{\partial v} + \mathbf{j}\frac{\partial y}{\partial v} + \mathbf{k}\frac{\partial z}{\partial v}\right) dv,$$

$$\mathbf{C} = \frac{\partial \mathbf{r}}{\partial w}\,dw = \left(\mathbf{i}\frac{\partial x}{\partial w} + \mathbf{j}\frac{\partial y}{\partial w} + \mathbf{k}\frac{\partial z}{\partial w}\right) dw.$$

Then by (3.2)

$$\mathbf{A}\cdot(\mathbf{B}\times\mathbf{C}) = \begin{vmatrix} \dfrac{\partial x}{\partial u} & \dfrac{\partial y}{\partial u} & \dfrac{\partial z}{\partial u} \\[2ex] \dfrac{\partial x}{\partial v} & \dfrac{\partial y}{\partial v} & \dfrac{\partial z}{\partial v} \\[2ex] \dfrac{\partial x}{\partial w} & \dfrac{\partial y}{\partial w} & \dfrac{\partial z}{\partial w} \end{vmatrix} du\,dv\,dw = J\,du\,dv\,dw$$

where J is the Jacobian of the transformation from x, y, z to u, v, w. Recall from the discussion of (3.2) that the triple scalar product may turn out to be positive or negative. Since we want a volume element to be positive, we use the absolute value of J. Thus the u, v, w volume element is $|J|\,du\,dv\,dw$ as stated in Chapter 5, Section 4.

Applications of the Triple Vector Product In Figure 3.8 (compare Figure 2.6), suppose the particle m is at rest on a rotating rigid body (for example, the earth). Then the *angular momentum* $\mathbf{L}$ of m about point O is defined by the equation $\mathbf{L} = \mathbf{r}\times(m\mathbf{v}) = m\mathbf{r}\times\mathbf{v}$. In the discussion of Figure 2.6, we showed that $\mathbf{v} = \boldsymbol{\omega}\times\mathbf{r}$. Thus, $\mathbf{L} = m\mathbf{r}\times(\boldsymbol{\omega}\times\mathbf{r})$. See Problem 16 and also Chapter 10, Section 4.

As another example, it is shown in mechanics that the centripetal acceleration of m in Figure 3.8 is $\mathbf{a} = \boldsymbol{\omega}\times(\boldsymbol{\omega}\times\mathbf{r})$. See Problem 17.

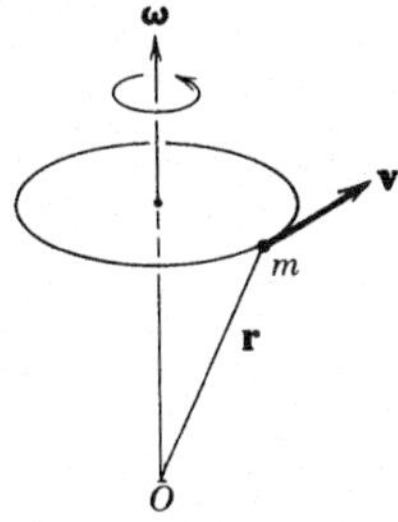

Figure 3.8

PROBLEMS, SECTION 3

1. If $\mathbf{A} = 2\mathbf{i} - \mathbf{j} - \mathbf{k}$, $\mathbf{B} = 2\mathbf{i} - 3\mathbf{j} + \mathbf{k}$, $\mathbf{C} = \mathbf{j} + \mathbf{k}$, find $(\mathbf{A} \cdot \mathbf{B})\mathbf{C}$, $\mathbf{A}(\mathbf{B} \cdot \mathbf{C})$, $(\mathbf{A} \times \mathbf{B}) \cdot \mathbf{C}$, $\mathbf{A} \cdot (\mathbf{B} \times \mathbf{C})$, $(\mathbf{A} \times \mathbf{B}) \times \mathbf{C}$, $\mathbf{A} \times (\mathbf{B} \times \mathbf{C})$.

For Problems 2 to 6, given $\mathbf{A} = \mathbf{i} + \mathbf{j} - 2\mathbf{k}$, $\mathbf{B} = 2\mathbf{i} - \mathbf{j} + 3\mathbf{k}$, $\mathbf{C} = \mathbf{j} - 5\mathbf{k}$:

2. Find the work done by the force **B** acting on an object which undergoes the displacement **C**.

3. Find the total work done by forces **A** and **B** if the object undergoes the displacement **C**. *Hint:* Can you add the two forces first?

4. Let O be the tail of **B** and let **A** be a force acting at the head of **B**. Find the torque of **A** about O; about a line through O perpendicular to the plane of **A** and **B**; about a line through O parallel to **C**.

5. Let **A** and **C** be drawn from a common origin and let **C** rotate about **A** with an angular velocity of 2 rad/sec. Find the velocity of the head of **C**.

6. In Problem 5, draw **B** with its tail at the head of **A**. If the figure is rotating as in Problem 5, find the velocity of the head of **B**. With the same diagram, let **B** be a force; find the torque of **B** about the head of **C**, and about the line **C**.

7. A force $\mathbf{F} = 2\mathbf{i} - 3\mathbf{j} + \mathbf{k}$ acts at the point $(1, 5, 2)$. Find the torque due to **F**

(a) about the origin;

(b) about the y axis;

(c) about the line $x/2 = y/1 = z/(-2)$.

8. A vector force with components $(1, 2, 3)$ acts at the point $(3, 2, 1)$. Find the vector torque about the origin due to this force and find the torque about each of the coordinate axes.

9. The force $\mathbf{F} = 2\mathbf{i} - \mathbf{j} - 5\mathbf{k}$ acts at the point $(-5, 2, 1)$. Find the torque due to **F** about the origin and about the line $2x = -4y = -z$.

10. In Figure 3.5, let $\mathbf{r}'$ be another vector from O to the line of F. Show that $\mathbf{r}' \times \mathbf{F} = \mathbf{r} \times \mathbf{F}$. *Hint:* $\mathbf{r} - \mathbf{r}'$ is a vector along the line of **F** and so is a scalar multiple of **F**. (The scalar has physical units of distance divided by force, but this fact is irrelevant for the vector proof.) Show also that moving the tail of **r** along **n** does not change $\mathbf{n} \cdot \mathbf{r} \times \mathbf{F}$. *Hint:* The triple scalar product is not changed by interchanging the dot and the cross.

11. Write out the twelve triple scalar products involving **A**, **B**, and **C** and verify the facts stated just above (3.3).

12. (a) Simplify $(\mathbf{A} \cdot \mathbf{B})^2 - [(\mathbf{A} \times \mathbf{B}) \times \mathbf{B}] \cdot \mathbf{A}$ by using (3.9).

(b) Prove *Lagrange's identity*: $(\mathbf{A} \times \mathbf{B}) \cdot (\mathbf{C} \times \mathbf{D}) = (\mathbf{A} \cdot \mathbf{C})(\mathbf{B} \cdot \mathbf{D}) - (\mathbf{A} \cdot \mathbf{D})(\mathbf{B} \cdot \mathbf{C})$.

13. Prove that the triple scalar product of $(\mathbf{A} \times \mathbf{B})$, $(\mathbf{B} \times \mathbf{C})$, and $(\mathbf{C} \times \mathbf{A})$, is equal to the square of the triple scalar product of **A**, **B**, and **C**. *Hint:* First let $(\mathbf{B} \times \mathbf{C}) = \mathbf{D}$, and evaluate $(\mathbf{A} \times \mathbf{B}) \times \mathbf{D}$. [See Am. J. Phys. **66**, 739 (1998).]

14. Prove the *Jacobi identity*: $\mathbf{A} \times (\mathbf{B} \times \mathbf{C}) + \mathbf{B} \times (\mathbf{C} \times \mathbf{A}) + \mathbf{C} \times (\mathbf{A} \times \mathbf{B}) = 0$. *Hint:* Expand each triple product as in equations (3.8) and (3.9).

15. In the figure $\mathbf{u}_1$ is a unit vector in the direction of an incident ray of light, and $\mathbf{u}_3$ and $\mathbf{u}_2$ are unit vectors in the directions of the reflected and refracted rays. If $\mathbf{u}$ is a unit vector normal to the surface AB, the laws of optics say that $\theta_1 = \theta_3$ and $n_1 \sin\theta_1 = n_2 \sin\theta_2$, where n_1 and n_2 are constants (indices of refraction). Write these laws in vector form (using dot or cross products).

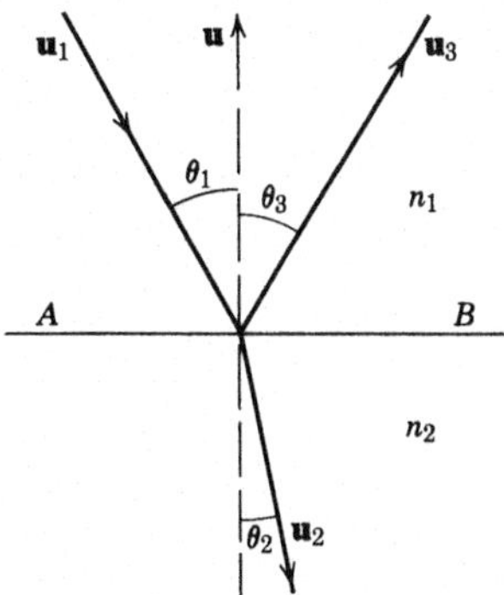

16. In the discussion of Figure 3.8, we found for the angular momentum, the formula $\mathbf{L} = m\mathbf{r} \times (\boldsymbol{\omega} \times \mathbf{r})$. Use (3.9) to expand this triple product. If $\mathbf{r}$ is perpendicular to $\boldsymbol{\omega}$, show that you obtain the elementary formula, angular momentum $= mvr$.

17. Expand the triple product for $\mathbf{a} = \boldsymbol{\omega} \times (\boldsymbol{\omega} \times \mathbf{r})$ given in the discussion of Figure 3.8. If $\mathbf{r}$ is perpendicular to $\boldsymbol{\omega}$ (Problem 16), show that $\mathbf{a} = -\omega^2\mathbf{r}$, and so find the elementary result that the acceleration is toward the center of the circle and of magnitude v^2/r.

18. Two moving charged particles exert forces on each other because each one creates a magnetic field in which the other moves (see Problem 4.6). These two forces are proportional to $\mathbf{v}_1 \times [\mathbf{v}_2 \times \mathbf{r}]$ and $\mathbf{v}_2 \times [\mathbf{v}_1 \times (-\mathbf{r})]$ where $\mathbf{r}$ is the vector joining the particles. By using (3.9), show that these forces are equal and opposite (Newton's third "law") if and only if $\mathbf{r} \times (\mathbf{v}_1 \times \mathbf{v}_2) = 0$. Compare Problem 14.

19. The force $\mathbf{F} = \mathbf{i} + 3\mathbf{j} + 2\mathbf{k}$ acts at the point $(1, 1, 1)$.

(a) Find the torque of the force about the point $(2, -1, 5)$. *Careful!* The vector $\mathbf{r}$ goes from $(2, -1, 5)$ to $(1, 1, 1)$.

(b) Find the torque of the force about the line $\mathbf{r} = 2\mathbf{i} - \mathbf{j} + 5\mathbf{k} + (\mathbf{i} - \mathbf{j} + 2\mathbf{k})t$. Note that the line goes through the point $(2, -1, 5)$.

20. The force $\mathbf{F} = 2\mathbf{i} - 5\mathbf{k}$ acts at the point $(3, -1, 0)$. Find the torque of $\mathbf{F}$ about each of the following lines.

(a) $\mathbf{r} = (2\mathbf{i} - \mathbf{k}) + (3\mathbf{j} - 4\mathbf{k})t$.

(b) $\mathbf{r} = \mathbf{i} + 4\mathbf{j} + 2\mathbf{k} + (2\mathbf{i} + \mathbf{j} - 2\mathbf{k})t$.

4. DIFFERENTIATION OF VECTORS

If $\mathbf{A} = \mathbf{i}A_x + \mathbf{j}A_y + \mathbf{k}A_z$, where $\mathbf{i}$, $\mathbf{j}$, $\mathbf{k}$ are fixed unit vectors and A_x, A_y, A_z are functions of t, then we define the derivative $d\mathbf{A}/dt$ by the equation

$$\frac{d\mathbf{A}}{dt} = \mathbf{i}\frac{dA_x}{dt} + \mathbf{j}\frac{dA_y}{dt} + \mathbf{k}\frac{dA_z}{dt}. \tag{4.1}$$

Thus the derivative of a vector $\mathbf{A}$ means a vector whose components are the derivatives of the components of $\mathbf{A}$.

▶ **Example 1.**

Let (x, y, z) be the coordinates of a moving particle at time t; then x, y, z are functions of t. The vector displacement of the particle from the origin at time t is

$$\mathbf{r} = \mathbf{i}x + \mathbf{j}y + \mathbf{k}z, \tag{4.2}$$

where $\mathbf{r}$ is a vector from the origin to the particle at time t. We say that $\mathbf{r}$ is the position vector or vector coordinate of the particle. The components of the velocity of the particle at time t are dx/dt, dy/dt, dz/dt so the velocity vector is

$$\mathbf{v} = \frac{d\mathbf{r}}{dt} = \mathbf{i}\frac{dx}{dt} + \mathbf{j}\frac{dy}{dt} + \mathbf{k}\frac{dz}{dt}. \tag{4.3}$$

The acceleration vector is

$$\mathbf{a} = \frac{d\mathbf{v}}{dt} = \mathbf{i}\frac{d^2x}{dt^2} + \mathbf{j}\frac{d^2y}{dt^2} + \mathbf{k}\frac{d^2z}{dt^2}. \tag{4.4}$$

The product of a scalar and a vector and the dot and cross products of vectors are differentiated by the ordinary calculus rules for differentiating a product, with one word of caution: The order of the factors must be kept in a cross product. You can easily prove equations (4.5) below by writing out components (Problem 1) and using (4.1).

$$\begin{aligned} \frac{d}{dt}(a\mathbf{A}) &= \frac{da}{dt}\mathbf{A} + a\frac{d\mathbf{A}}{dt}, \\ \frac{d}{dt}(\mathbf{A}\cdot\mathbf{B}) &= \mathbf{A}\cdot\frac{d\mathbf{B}}{dt} + \frac{d\mathbf{A}}{dt}\cdot\mathbf{B}, \\ \frac{d}{dt}(\mathbf{A}\times\mathbf{B}) &= \mathbf{A}\times\frac{d\mathbf{B}}{dt} + \frac{d\mathbf{A}}{dt}\times\mathbf{B}. \end{aligned} \tag{4.5}$$

The second term in $(d/dt)(\mathbf{A}\cdot\mathbf{B})$ *can* be written $\mathbf{B}\cdot d\mathbf{A}/dt$ if you like since $\mathbf{A}\cdot\mathbf{B} = \mathbf{B}\cdot\mathbf{A}$. But the corresponding term in $(d/dt)(\mathbf{A}\times\mathbf{B})$ must *not* be turned around unless you put a minus sign in front of it since $\mathbf{A}\times\mathbf{B} = -\mathbf{B}\times\mathbf{A}$.

▶ **Example 2.** Consider the motion of a particle in a circle at constant speed. We can then write

$$\begin{aligned} r^2 &= \mathbf{r}\cdot\mathbf{r} = \text{ const.}, \\ v^2 &= \mathbf{v}\cdot\mathbf{v} = \text{ const.} \end{aligned} \tag{4.6}$$

If we differentiate these two equations using (4.5), we get

$$\begin{aligned} 2\mathbf{r}\cdot\frac{d\mathbf{r}}{dt} &= 0 \quad \text{or} \quad \mathbf{r}\cdot\mathbf{v} = 0, \\ 2\mathbf{v}\cdot\frac{d\mathbf{v}}{dt} &= 0 \quad \text{or} \quad \mathbf{v}\cdot\mathbf{a} = 0. \end{aligned} \tag{4.7}$$

Also differentiating $\mathbf{r} \cdot \mathbf{v} = 0$, we get

$$\mathbf{r} \cdot \mathbf{a} + \mathbf{v} \cdot \mathbf{v} = 0 \quad \text{or} \quad \mathbf{r} \cdot \mathbf{a} = -v^2. \tag{4.8}$$

The first of equations (4.7) says that $\mathbf{r}$ is perpendicular to $\mathbf{v}$; the second says that $\mathbf{a}$ is perpendicular to $\mathbf{v}$. Therefore $\mathbf{a}$ and $\mathbf{r}$ are either parallel or antiparallel (since the motion is in a plane) and the angle θ between $\mathbf{a}$ and $\mathbf{r}$ is either 0° or 180°. From (4.8) and the definition of scalar product, we have

$$\mathbf{r} \cdot \mathbf{a} = |\mathbf{r}|\,|\mathbf{a}| \cos\theta = -v^2. \tag{4.9}$$

Thus we see that $\cos\theta$ is negative, so $\theta = 180°$. Then from (4.9) we get

$$|\mathbf{r}|\,|\mathbf{a}|(-1) = -v^2 \quad \text{or} \quad a = \frac{v^2}{r}. \tag{4.10}$$

We have just given a vector proof that for motion in a circle at constant speed the acceleration is toward the center of the circle and of magnitude v^2/r.

So far we have written vectors only in terms of their rectangular components using the unit basis vectors $\mathbf{i}$, $\mathbf{j}$, $\mathbf{k}$. It is often convenient to use other coordinate systems, for example polar coordinates in two dimensions and spherical or cylindrical coordinates in three dimensions (see Chapter 5, Section 4, and Chapter 10, Sections 8 and 9). We shall consider using vectors in various coordinate systems in detail in Chapter 10, but it will be useful to discuss briefly here the use of plane polar coordinates. In Figure 4.1, think of starting at the point (x, y) or (r, θ) and moving along the line $\theta =$ const. in the direction of increasing r. We call this the "r direction"; we draw a unit vector (that is, a vector of length 1) in this direction

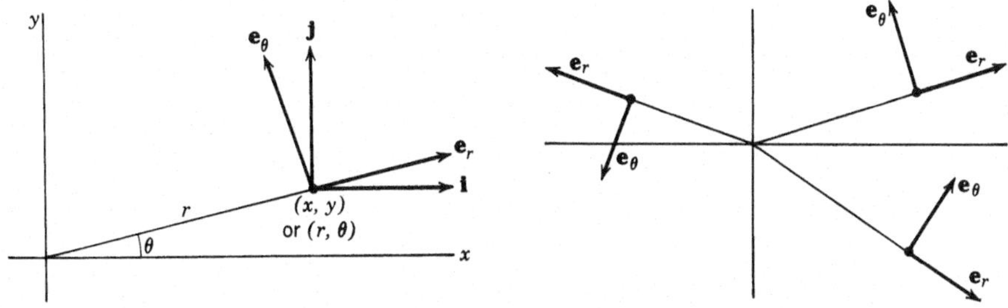

Figure 4.1 Figure 4.2

and label it $\mathbf{e}_r$. Similarly, think of moving along the circle $r =$ const. in the direction of increasing θ. We call this the "θ direction"; we draw a unit vector tangent to the circle and label it $\mathbf{e}_\theta$. These two vectors $\mathbf{e}_r$ and $\mathbf{e}_\theta$ are the polar coordinate unit basis vectors just as $\mathbf{i}$ and $\mathbf{j}$ are the rectangular unit basis vectors. We can now write any given vector in terms of its components in the directions $\mathbf{e}_r$ and $\mathbf{e}_\theta$ (by finding its projections in these directions). There is a complication here, however. In rectangular coordinates, the vectors $\mathbf{i}$ and $\mathbf{j}$ are constant in magnitude *and direction.* The polar coordinate unit basis vectors are constant in magnitude but their directions change from point to point (Figure 4.2). Thus in calculating the derivative of a vector written in polar coordinates, we must differentiate the basis vectors as well as the components [compare (4.1) where we differentiate the components only.] One straightforward way to do this is to express the vectors $\mathbf{e}_r$ and $\mathbf{e}_\theta$ in terms of $\mathbf{i}$ and $\mathbf{j}$. From Figure 4.3, we see that the x and y components of $\mathbf{e}_r$ are $\cos\theta$ and $\sin\theta$. Thus we have

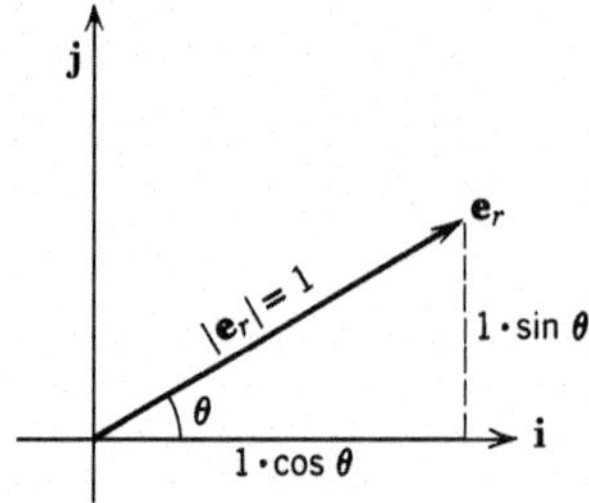

Figure 4.3

$$\mathbf{e}_r = \mathbf{i}\cos\theta + \mathbf{j}\sin\theta. \tag{4.11}$$

Similarly (Problem 7) we find

$$\mathbf{e}_\theta = -\mathbf{i}\sin\theta + \mathbf{j}\cos\theta. \tag{4.12}$$

Differentiating $\mathbf{e}_r$ and $\mathbf{e}_\theta$ with respect to t, we get

$$\begin{aligned} \frac{d\mathbf{e}_r}{dt} &= -\mathbf{i}\sin\theta\,\frac{d\theta}{dt} + \mathbf{j}\cos\theta\,\frac{d\theta}{dt} = \mathbf{e}_\theta\frac{d\theta}{dt}, \\ \frac{d\mathbf{e}_\theta}{dt} &= -\mathbf{i}\cos\theta\,\frac{d\theta}{dt} - \mathbf{j}\sin\theta\,\frac{d\theta}{dt} = -\mathbf{e}_r\frac{d\theta}{dt}. \end{aligned} \tag{4.13}$$

We can now use (4.13) in calculating the derivative of any vector which is written in terms of its polar components.

▶ **Example 3.** Given $\mathbf{A} = A_r\mathbf{e}_r + A_\theta\mathbf{e}_\theta$, where A_r and A_θ are functions of t, find $d\mathbf{A}/dt$. We get

$$\frac{d\mathbf{A}}{dt} = \mathbf{e}_r\frac{dA_r}{dt} + A_r\frac{d\mathbf{e}_r}{dt} + \mathbf{e}_\theta\frac{dA_\theta}{dt} + A_\theta\frac{d\mathbf{e}_\theta}{dt}.$$

Using (4.13), we find

$$\frac{d\mathbf{A}}{dt} = \mathbf{e}_r\frac{dA_r}{dt} + \mathbf{e}_\theta A_r\frac{d\theta}{dt} + \mathbf{e}_\theta\frac{dA_\theta}{dt} - \mathbf{e}_r A_\theta\frac{d\theta}{dt}.$$

We can find higher-order derivatives if we like by differentiating again using (4.13) each time to evaluate the derivatives of $\mathbf{e}_r$ and $\mathbf{e}_\theta$.

PROBLEMS, SECTION 4

1. Verify equations (4.5) by writing out the components.

2. Let the position vector (with its tail at the origin) of a moving particle be $\mathbf{r} = \mathbf{r}(t) = t^2\mathbf{i} - 2t\mathbf{j} + (t^2 + 2t)\mathbf{k}$, where t represents time.

(a) Show that the particle goes through the point $(4, -4, 8)$. At what time does it do this?

(b) Find the velocity vector and the speed of the particle at time t; at the time when it passes though the point $(4, -4, 8)$.

(c) Find the equations of the line tangent to the curve described by the particle and the plane normal to this curve, at the point $(4, -4, 8)$.

3. As in Problem 2, if the position vector of a particle is $\mathbf{r} = (4 + 3t)\mathbf{i} + t^3\mathbf{j} - 5t\mathbf{k}$, at what time does it pass through the point $(1, -1, 5)$? Find its velocity at this time. Find the equations of the line tangent to its path and the plane normal to the path, at $(1, -1, 5)$.

4. Let $\mathbf{r} = \mathbf{r}(t)$ be a vector whose *length* is always 1 (it may vary in direction). Prove that either $\mathbf{r}$ is a constant vector or $d\mathbf{r}/dt$ is perpendicular to $\mathbf{r}$. *Hint:* Differentiate $\mathbf{r} \cdot \mathbf{r}$.

5. The position of a particle at time t is given by $\mathbf{r} = \mathbf{i}\cos t + \mathbf{j}\sin t + \mathbf{k}t$. Show that both the speed and the magnitude of the acceleration are constant. Describe the motion.

6. The force acting on a moving charged particle in a magnetic field $\mathbf{B}$ is $\mathbf{F} = q(\mathbf{v} \times \mathbf{B})$ where q is the electric charge of the particle, and $\mathbf{v}$ is its velocity. Suppose that a particle moves in the (x, y) plane with a uniform $\mathbf{B}$ in the z direction. Assuming Newton's second law, $m\,d\mathbf{v}/dt = \mathbf{F}$, show that the force and velocity are perpendicular and that both have constant magnitude. *Hint:* Find $(d/dt)(\mathbf{v} \cdot \mathbf{v})$.

7. Sketch a figure and verify equation (4.12).

8. In polar coordinates, the position vector of a particle is $\mathbf{r} = r\mathbf{e}_r$. Using (4.13), find the velocity and acceleration of the particle.

9. The angular momentum of a particle m is defined by $\mathbf{L} = m\mathbf{r} \times (d\mathbf{r}/dt)$ (see end of Section 3). Show that

$$\frac{d\mathbf{L}}{dt} = m\mathbf{r} \times \frac{d^2\mathbf{r}}{dt^2}.$$

10. If $\mathbf{V}(t)$ is a vector function of t, find the indefinite integral

$$\int \left(\mathbf{V} \times \frac{d^2\mathbf{V}}{dt^2}\right) dt.$$

5. FIELDS

Many physical quantities have different values at different points in space. For example, the temperature in a room is different at different points: high near a register, low near an open window, and so on. The electric field around a point charge is large near the charge and decreases as we go away from the charge. Similarly, the gravitational force acting on a satellite depends on its distance from the earth. The velocity of flow of water in a stream is large in rapids and in narrow channels and small over flat areas and where the stream is wide. In all these examples there is a particular region of space which is of interest for the problem at hand; at every point of this region some physical quantity has a value. The term *field* is used to mean both the region and the value of the physical quantity in the region (for example, electric field, gravitational field). If the physical quantity is a scalar (for example, temperature), we speak of a *scalar field.* If the quantity is a vector (for example, electric field, force, or velocity), we speak of a *vector field.* Note again a point which we discussed in "endpoint problems" in Chapter 4, Section 10: Physical problems are often restricted to certain regions of space, and our mathematics must take account of this.

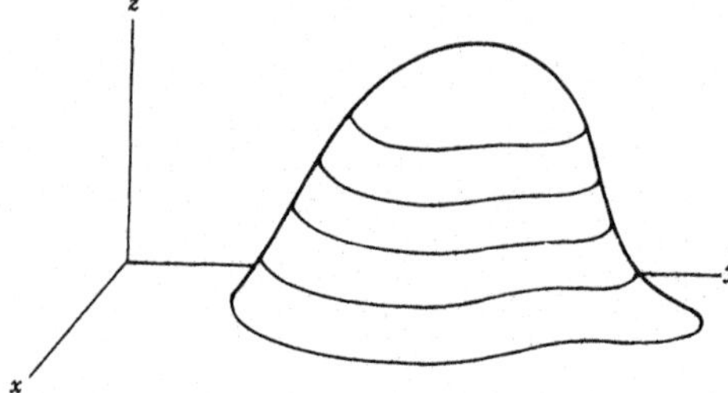

Figure 5.1

A simple example of a scalar field is the gravitational potential energy near the earth; its value is $V = mgz$ at every point of height z above some arbitrary reference level [which we take as the (x, y) plane]. Suppose that on a hill (Figure 5.1) we mark a series of curves each corresponding to some value of z (curves of constant elevation, often called *contour lines* or *level lines*). Any curve or surface on which a potential is constant is called an *equipotential.* Thus these level lines are equipotentials of the gravitational field since along any one curve the value of the gravitational potential energy mgz is constant. The horizontal planes which intersect the hill in these curves are equipotential surfaces (or level surfaces) of the gravitational field. (See Problems, Section 6 for more examples.)

As another example, let us ask for the equipotential surfaces in the field of an electric point charge q. The potential is $V = 9 \cdot 10^9 q/r$ (in SI units) at a point which is a distance r from the charge. The potential V is constant if r is constant; that is, the equipotentials of this electric field are spheres with centers at the charge. Similarly we could imagine drawing a set of surfaces (probably very irregular) in a room so that at every point of a single surface the temperature would be constant. These surfaces would be like equipotentials; they are called *isothermals* when the constant quantity is the temperature.

6. DIRECTIONAL DERIVATIVE; GRADIENT

Suppose that we know the temperature $T(x, y, z)$ at every point of a room, say, or of a metal bar. Starting at a given point we could ask for the rate of change of the temperature with distance (in degrees per centimeter) as we move away from the starting point. The chances are that the temperature increases in some directions and decreases in other directions, and that it increases more rapidly in some directions than others. Thus the rate of change of temperature with distance depends upon the *direction* in which we move; consequently it is called a *directional* derivative. In symbols, we want to find the limiting value of $\Delta T/\Delta s$ where Δs is an element of distance (arc length) in a given direction, and ΔT is the corresponding change in temperature; we write the directional derivative as dT/ds. We could also ask for the direction in which dT/ds has its largest value; this is physically the direction from which heat flows (that is, heat flows from hot to cold, in the opposite direction from the maximum rate of temperature increase).

Before we discuss how to calculate directional derivatives, consider another example. Suppose we are standing at a point on the side of the hill of Figure 5.1 (not at the top), and ask the question "In what direction does the hill slope downward most steeply from this point?" This is the direction in which you would start to slide if you lost your footing; it is the direction most people would probably call "straight" down. We want to make this vague idea more precise. Suppose we move a small distance Δs on the hill; the vertical distance Δz which we have gone may be positive (uphill) or negative (downhill) or zero (around the hill). Then $\Delta z/\Delta s$ and its limit dz/ds depend upon the *direction* in which we go; dz/ds is a directional derivative. The direction of steepest slope is the direction in which dz/ds has its largest absolute value. Notice that since the gravitational potential energy of a mass m is $V = mgz$, maximizing dz/ds is the same as maximizing dV/ds, where the equipotentials on the hill are $V(x, y) = mgz(x, y) = \text{const.}$

Let us now state and solve the general problem of finding a directional derivative. We are given a scalar field, that is, a function $\phi(x, y, z)$ [or $\phi(x, y)$ in a two-variable problem; the following discussion applies to two-variable problems if we simply drop terms and equations containing z]. We want to find $d\phi/ds$, the rate of change of ϕ with distance, at a given point (x_0, y_0, z_0) and in a given direction. Let $\mathbf{u} = \mathbf{i}a + \mathbf{j}b + \mathbf{k}c$ be a unit vector in the given direction. In Figure 6.1, we start at (x_0, y_0, z_0) and go a distance s $(s \geq 0)$ in the direction $\mathbf{u}$ to the point (x, y, z); the vector joining these points is $\mathbf{u}s$ since $\mathbf{u}$ is a unit vector. Then,

$$(x, y, z) - (x_0, y_0, z_0) = \mathbf{u}s = (a\mathbf{i} + b\mathbf{j} + c\mathbf{k})s$$

or

$$\left\{\begin{aligned} x &= x_0 + as, \\ y &= y_0 + bs, \\ z &= z_0 + cs. \end{aligned}\right. \tag{6.1}$$

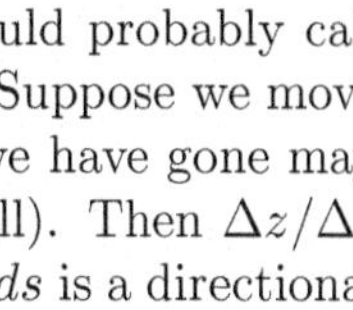

Figure 6.1

Equations (6.1) are the parametric equations of the line through (x_0, y_0, z_0) in the direction $\mathbf{u}$ [see Chapter 3, equation (5.8)] with the distance s (instead of t) as the parameter, and with $\mathbf{u}$ (instead of $\mathbf{A}$) as the vector along the line. From (6.1) we see that along the line, x, y, and z are each functions of a single variable, namely s [all the other letters in (6.1) are given constants]. If we substitute x, y, z in (6.1) into $\phi(x, y, z)$, then ϕ becomes a function of just the one variable s. That is, *along the straight line* (6.1), ϕ is a function of one variable, namely the distance along the line measured from (x_0, y_0, z_0). Since ϕ depends on s alone, we can find $d\phi/ds$:

$$\begin{aligned} \frac{d\phi}{ds} &= \frac{\partial\phi}{\partial x}\frac{dx}{ds} + \frac{\partial\phi}{\partial y}\frac{dy}{ds} + \frac{\partial\phi}{\partial z}\frac{dz}{ds} \\ &= \frac{\partial\phi}{\partial x}a + \frac{\partial\phi}{\partial y}b + \frac{\partial\phi}{\partial z}c. \end{aligned} \tag{6.2}$$

This is the dot product of $\mathbf{u}$ with the vector $\mathbf{i}(\partial\phi/\partial x) + \mathbf{j}(\partial\phi/\partial y) + \mathbf{k}(\partial\phi/\partial z)$. This vector is called the *gradient* of ϕ and is written $\operatorname{grad}\phi$ or $\boldsymbol{\nabla}\phi$ (read "del ϕ"). By definition

(6.3) $$\nabla\phi = \text{grad}\,\phi = \mathbf{i}\frac{\partial\phi}{\partial x} + \mathbf{j}\frac{\partial\phi}{\partial y} + \mathbf{k}\frac{\partial\phi}{\partial z}.$$

Then we can write (6.2) as

(6.4) $$\frac{d\phi}{ds} = \nabla\phi\cdot\mathbf{u} \quad \text{(directional derivative).}$$

Example 1. Find the directional derivative of $\phi = x^2y + xz$ at $(1, 2, -1)$ in the direction $\mathbf{A} = 2\mathbf{i} - 2\mathbf{j} + \mathbf{k}$.

Here $\mathbf{u}$ is a unit vector obtained by dividing $\mathbf{A}$ by $|\mathbf{A}|$. Then we have

$$\mathbf{u} = \frac{1}{3}(2\mathbf{i} - 2\mathbf{j} + \mathbf{k}).$$

Using (6.3) we get

$$\nabla\phi = \mathbf{i}\frac{\partial\phi}{\partial x} + \mathbf{j}\frac{\partial\phi}{\partial y} + \mathbf{k}\frac{\partial\phi}{\partial z} = (2xy + z)\mathbf{i} + x^2\mathbf{j} + x\mathbf{k},$$

$$\nabla\phi \text{ at the point } (1, 2, -1) = 3\mathbf{i} + \mathbf{j} + \mathbf{k}.$$

Then from (6.4) we find

$$\frac{d\phi}{ds} \text{ at } (1, 2, -1) = \nabla\phi\cdot\mathbf{u} = 2 - \frac{2}{3} + \frac{1}{3} = \frac{5}{3}.$$

The gradient of a function has useful geometrical and physical meanings which we shall now investigate. From (6.4), using the definition of a dot product, and the fact that $|\mathbf{u}| = 1$, we have

(6.5) $$\frac{d\phi}{ds} = |\nabla\phi|\cos\theta,$$

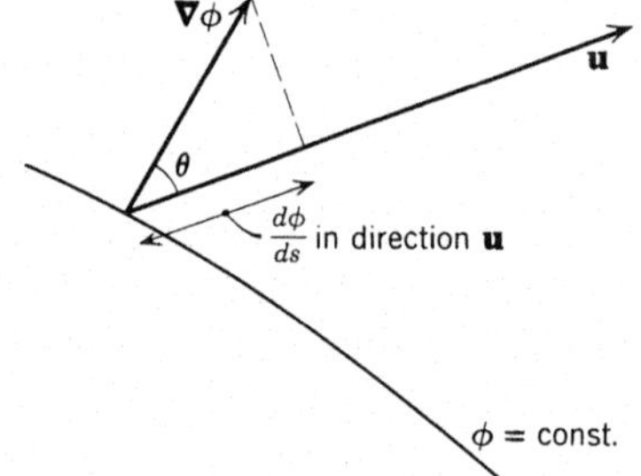

Figure 6.2

where θ is the angle between $\mathbf{u}$ and the vector $\nabla\phi$. Thus $d\phi/ds$ is the projection of $\nabla\phi$ on the direction $\mathbf{u}$ (Figure 6.2). We find the largest value of $d\phi/ds$ (namely $|\nabla\phi|$) if we go in the direction of $\nabla\phi$ (that is, $\theta = 0$ in Figure 6.2). If we go in the opposite direction (that is, $\theta = 180°$ in Figure 6.2) we find the largest rate of decrease of ϕ, namely $d\phi/ds = -|\nabla\phi|$.

Example 2. Suppose that the temperature T at the point (x, y, z) is given by the equation $T = x^2 - y^2 + xyz + 273$. In which direction is the temperature increasing most rapidly at $(-1, 2, 3)$, and at what rate? Here $\nabla T = (2x+yz)\mathbf{i}+(-2y+xz)\mathbf{j}+xy\mathbf{k} = 4\mathbf{i} - 7\mathbf{j} - 2\mathbf{k}$ at $(-1, 2, 3)$, and the increase in temperature is fastest in the direction of this vector. The rate of increase is $dT/ds = |\nabla T| = \sqrt{16 + 49 + 4} = \sqrt{69}$. We can also say that the temperature is decreasing most rapidly in the direction $-\nabla T$; in this direction, $dT/ds = -\sqrt{69}$. Heat flows in the direction $-\nabla T$ (that is, from hot to cold).

Next suppose $\mathbf{u}$ is tangent to the surface $\phi = \text{const.}$ at the point $P(x_0, y_0, z_0)$ (Figure 6.3). We want to show that $d\phi/ds$ in the direction $\mathbf{u}$ is then equal to zero. Consider $\Delta\phi/\Delta s$ for paths PA, PB, PC, etc., approaching the tangent $\mathbf{u}$. Since $\phi = \text{const.}$ on the surface, and P, A, B, C, etc. are all on the surface, $\Delta\phi = 0$, and $\Delta\phi/\Delta s = 0$ for such paths. But $d\phi/ds$ in the tangent direction is the limit of $\Delta\phi/\Delta s$ as $\Delta s \to 0$ (that is, as PA, PB, etc., approach $\mathbf{u}$), so $d\phi/ds$ in the direction $\mathbf{u}$ is zero also. Then for $\mathbf{u}$ along the tangent to $\phi = \text{const.}$, $\nabla\phi \cdot \mathbf{u} = 0$; this means that $\nabla\phi$ is perpendicular to $\mathbf{u}$. Since this is true for any $\mathbf{u}$ tangent to the surface at the point (x_0, y_0, z_0), then at that point:

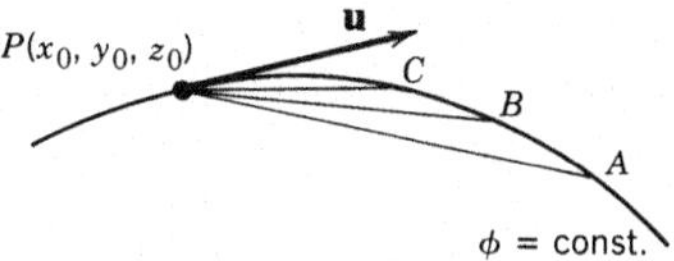

Figure 6.3

> The vector $\nabla\phi$ is perpendicular (normal) to the surface $\phi =$const.

Since $|\nabla\phi|$ is the value of the directional derivative in the direction normal (that is, perpendicular) to the surface, it is often called the *normal derivative* and written $|\nabla\phi| = d\phi/dn$.

We now see that the direction of largest rate of change of a given function ϕ with distance is perpendicular to the equipotentials (or level lines) $\phi = \text{const.}$ In the temperature problem, the direction of maximum dT/ds is then perpendicular to the isothermals. At any point this is the direction of ∇T and is called the direction of the temperature gradient. In the problem of the hill, the direction of steepest slope at any point is perpendicular to the level lines, that is, along ∇z or ∇V.

Example 3. Given the surface $x^3y^2z = 12$, find the equations of the tangent plane and normal line at $(1, -2, 3)$.

This is a level surface of the function $w = x^3y^2z$, so the normal direction is the direction of the gradient

$$\nabla w = 3x^2y^2z\mathbf{i} + 2x^3yz\mathbf{j} + x^3y^2\mathbf{k} = 36\mathbf{i} - 12\mathbf{j} + 4\mathbf{k} \quad \text{at} \quad (1, -2, 3).$$

A simpler vector in the same direction is $9\mathbf{i} - 3\mathbf{j} + \mathbf{k}$. Then (see Chapter 3, Section 5) the equation of the tangent plane is

$$9(x-1) - 3(y+2) + (z-3) = 0,$$

and the equations of the normal line are

$$\frac{x-1}{9} = \frac{y+2}{-3} = \frac{z-3}{1}. \tag{6.6}$$

In (6.3) we have written the gradient in terms of its rectangular components. It is useful to write it in cylindrical and spherical coordinates also. (Note that this includes polar coordinates when $z = 0$). In cylindrical coordinates we want the components of $\nabla\phi$ in the directions $\mathbf{e}_r$, $\mathbf{e}_\theta$, and $\mathbf{e}_z = \mathbf{k}$. According to (6.4), the component of ∇f in any direction $\mathbf{u}$ is the directional derivative df/ds in that direction. (We are changing the function from ϕ to f since ϕ is used as an angle in spherical, and sometimes in cylindrical and polar, coordinates.) The element of arc length ds in the r direction is dr so the directional derivative in the r direction is df/dr (θ and z constant) which we write as $\partial f/\partial r$. In the θ direction, the element of arc length is $r\,d\theta$ (Chapter 5, Section 4) so the directional derivative in the θ direction is $df/(r\,d\theta)$ (with r and z constant) which we write as $(1/r)\partial f/\partial\theta$. Thus we have in cylindrical coordinates (or polar without the z term)

$$(6.7)\qquad \nabla f = \mathbf{e}_r\frac{\partial f}{\partial r} + \mathbf{e}_\theta\frac{1}{r}\frac{\partial f}{\partial \theta} + \mathbf{e}_z\frac{\partial f}{\partial z} \qquad \text{in cylindrical coordinates.}$$

In a similar way we can show (Problem 21) that

$$(6.8)\qquad \nabla f = \mathbf{e}_r\frac{\partial f}{\partial r} + \mathbf{e}_\theta\frac{1}{r}\frac{\partial f}{\partial \theta} + \mathbf{e}_\phi\frac{1}{r\sin\phi}\frac{\partial f}{\partial \phi} \qquad \text{in spherical coordinates.}$$

PROBLEMS, SECTION 6

1. Find the gradient of $w = x^2y^3z$ at $(1, 2, -1)$.

2. Starting from the point $(1, 1)$, in what direction does the function $\phi = x^2 - y^2 + 2xy$ *decrease* most rapidly?

3. Find the derivative of $xy^2 + yz$ at $(1, 1, 2)$ in the direction of the vector $2\mathbf{i} - \mathbf{j} + 2\mathbf{k}$.

4. Find the derivative of $ze^x \cos y$ at $(1, 0, \pi/3)$ in the direction of the vector $\mathbf{i} + 2\mathbf{j}$.

5. Find the gradient of $\phi = z\sin y - xz$ at the point $(2, \pi/2, -1)$. Starting at this point, in what direction is ϕ *decreasing* most rapidly? Find the derivative of ϕ in the direction $2\mathbf{i} + 3\mathbf{j}$.

6. Find a vector normal to the surface $x^2 + y^2 - z = 0$ at the point $(3, 4, 25)$. Find the equations of the tangent plane and normal line to the surface at that point.

7. Find the direction of the line normal to the surface $x^2y + y^2z + z^2x + 1 = 0$ at the point $(1, 2, -1)$. Write the equations of the tangent plane and normal line at this point.

8. (a) Find the directional derivative of $\phi = x^2 + \sin y - xz$ in the direction $\mathbf{i} + 2\mathbf{j} - 2\mathbf{k}$ at the point $(1, \pi/2, -3)$.

(b) Find the equation of the tangent plane and the equations of the normal line to $\phi = 5$ at the point $(1, \pi/2, -3)$.

9. (a) Given $\phi = x^2 - y^2z$, find $\nabla\phi$ at $(1, 1, 1)$.

(b) Find the directional derivative of ϕ at $(1, 1, 1)$ in the direction $\mathbf{i} - 2\mathbf{j} + \mathbf{k}$.

(c) Find the equations of the normal line to the surface $x^2 - y^2z = 0$ at $(1, 1, 1)$.

For Problems 10 to 14, use a computer as needed to make plots of the given surfaces and the isothermal or equipotential curves. Try both 3D graphs and contour plots.

10. If the temperature in the (x, y) plane is given by $T = xy - x$, sketch a few isothermal curves, say for $T = 0, 1, 2, -1, -2$. Find the direction in which the temperature changes most rapidly with distance from the point $(1, 1)$, and the maximum rate of change. Find the directional derivative of T at $(1, 1)$ in the direction of the vector $3\mathbf{i} - 4\mathbf{j}$. Heat flows in the direction $-\nabla T$ (perpendicular to the isothermals). Sketch a few curves along which heat would flow.

11. (a) Given $\phi = x^2 - y^2$, sketch on one graph the curves $\phi = 4$, $\phi = 1$, $\phi = 0$, $\phi = -1$, $\phi = -4$. If ϕ is the electrostatic potential, the curves $\phi =$ const. are equipotentials, and the electric field is given by $\mathbf{E} = -\nabla\phi$. If ϕ is temperature, the curves $\phi =$const. are isothermals and $\nabla\phi$ is the temperature gradient; heat flows in the direction $-\nabla\phi$.

(b) Find and draw on your sketch the vectors $-\nabla\phi$ at the points $(x, y) = (\pm 1, \pm 1)$, $(0, \pm 2)$, $(\pm 2, 0)$. Then, remembering that $\nabla\phi$ is perpendicular to $\phi =$ const., sketch, without computation, several curves along which heat would flow [see(a)].

12. For Problem 11,

(a) Find the magnitude and direction of the electric field at $(2, 1)$.

(b) Find the direction in which the temperature is *decreasing* most rapidly at $(-3, 2)$.

(c) Find the rate of change of temperature with distance at $(1, 2)$ in the direction $3\mathbf{i} - \mathbf{j}$.

13. Let $\phi = e^x \cos y$. Let ϕ represent either temperature or electrostatic potential. Refer to Problem 11 for definitions and find:

(a) The direction in which the temperature is increasing most rapidly at $(1, -\pi/4)$ and the magnitude of the rate of increase.

(b) The rate of change of temperature with distance at $(0, \pi/3)$ in the direction $\mathbf{i} + \mathbf{j}\sqrt{3}$.

(c) The direction and magnitude of the electric field at $(0, \pi)$.

(d) The magnitude of the electric field at $x = -1$, any y.

14. (a) Suppose that a hill (as in Fig. 5.1) has the equation $z = 32 - x^2 - 4y^2$, where $z =$ height measured from some reference level (in hundreds of feet). Sketch a contour map (that is, draw on one graph a set of curves $z =$ const.); use the contours $z = 32, 19, 12, 7, 0$.

(b) If you start at the point $(3, 2)$ and in the direction $\mathbf{i} + \mathbf{j}$, are you going uphill or downhill, and how fast?

15. Repeat Problem 14b for the following points and directions.

(a) $(4, -2), \mathbf{i} + \mathbf{j}$ (b) $(-3, 1), 4\mathbf{i} + 3\mathbf{j}$
(c) $(2, 2), -3\mathbf{i} + \mathbf{j}$ (d) $(-4, -1), 4\mathbf{i} - 3\mathbf{j}$

16. Show by the Lagrange multiplier method that the maximum value of $d\phi/ds$ is $|\nabla\phi|$. That is, maximize $d\phi/ds$ given by (6.3) subject to the condition $a^2 + b^2 + c^2 = 1$. You should get two values ($\pm$) for the Lagrange multiplier λ, and two values (maximum and minimum) for $d\phi/ds$. Which is the maximum and which is the minimum?

17. Find ∇r, where $r = \sqrt{x^2 + y^2}$, using (6.7) and also using (6.3). Show that your results are the same by using (4.11) and (4.12).

As in Problem 17, find the following gradients in two ways and show that your answers are equivalent.

18. ∇x **19.** ∇y **20.** $\nabla(r^2)$

21. Verify equation (6.8); that is, find ∇f in spherical coordinates as we did for cylindrical coordinates. *Hint*: What is ds in the ϕ direction? See Chapter 5, Figure 4.5.

7. SOME OTHER EXPRESSIONS INVOLVING ∇

If we write $\nabla\phi$ as $[\mathbf{i}(\partial/\partial x)+\mathbf{j}(\partial/\partial y)+\mathbf{k}(\partial/\partial z)]\phi$, we can then call the bracket ∇. By itself ∇ has no meaning (just as d/dx alone has no meaning; we must put some function after it to be differentiated). However, it is useful to use ∇ much as we use d/dx to indicate a certain operation.

We call ∇ a *vector operator* and write

$$\nabla = \mathbf{i}\frac{\partial}{\partial x} + \mathbf{j}\frac{\partial}{\partial y} + \mathbf{k}\frac{\partial}{\partial z}. \tag{7.1}$$

It is more complicated than d/dx (which is a *scalar operator*) because ∇ has vector properties too.

So far we have considered $\nabla\phi$ where ϕ is a scalar; we next want to consider whether ∇ can operate on a vector.

Suppose $\mathbf{V}(x,y,z)$ is a vector function, that is, the three components V_x, V_y, V_z of $\mathbf{V}$ are functions of x, y, z:

$$\mathbf{V}(x,y,z) = \mathbf{i}V_x(x,y,z) + \mathbf{j}V_y(x,y,z) + \mathbf{k}V_z(x,y,z).$$

(The subscripts mean components, *not* partial derivatives.) Physically, $\mathbf{V}$ represents a vector field (for example, the electric field about a point charge). At each point of space there is a vector $\mathbf{V}$, but the magnitude and direction of $\mathbf{V}$ may vary from point to point. We can form two useful combinations of ∇ and $\mathbf{V}$. We define the *divergence* of $\mathbf{V}$, abbreviated div $\mathbf{V}$ or $\nabla\cdot\mathbf{V}$, by (7.2):

$$\nabla\cdot\mathbf{V} = \operatorname{div}\mathbf{V} = \frac{\partial V_x}{\partial x} + \frac{\partial V_y}{\partial y} + \frac{\partial V_z}{\partial z}. \tag{7.2}$$

We define the *curl* of $\mathbf{V}$, written $\nabla\times\mathbf{V}$, by (7.3):

$$\begin{aligned}\nabla\times\mathbf{V} &= \operatorname{curl}\mathbf{V} \\ &= \mathbf{i}\left(\frac{\partial V_z}{\partial y} - \frac{\partial V_y}{\partial z}\right) + \mathbf{j}\left(\frac{\partial V_x}{\partial z} - \frac{\partial V_z}{\partial x}\right) + \mathbf{k}\left(\frac{\partial V_y}{\partial x} - \frac{\partial V_x}{\partial y}\right) \\ &= \begin{vmatrix} \mathbf{i} & \mathbf{j} & \mathbf{k} \\ \dfrac{\partial}{\partial x} & \dfrac{\partial}{\partial y} & \dfrac{\partial}{\partial z} \\ V_x & V_y & V_z \end{vmatrix}.\end{aligned} \tag{7.3}$$

You should study these expressions to see how we are using ∇ as "almost" a vector. The *definitions* of divergence and curl are the partial derivative expressions, of course. However, the similarity of the formulas (7.2) and (7.3) to those for $\mathbf{A}\cdot\mathbf{B}$ and $\mathbf{A}\times\mathbf{B}$ helps us to remember $\nabla\cdot\mathbf{V}$ and $\nabla\times\mathbf{V}$. But you must remember to put the partial derivative "components" of ∇ *before* the components of $\mathbf{V}$ in each term [for example, in evaluating the determinant in (7.3)]. Note that $\nabla \cdot \mathbf{V}$ is a scalar and $\nabla \times \mathbf{V}$ is a vector (compare $\mathbf{A} \cdot \mathbf{B}$ and $\mathbf{A} \times \mathbf{B}$). We shall discuss later the meaning and some of the applications of the divergence and the curl of a vector function.

The quantity $\nabla\phi$ in (6.3) is a vector function; we can then let $\mathbf{V} = \nabla\phi$ in (7.2) and find $\nabla \cdot \nabla\phi = \operatorname{div}\operatorname{grad}\phi$. This is a very important expression called the *Laplacian* of ϕ; it is usually written as $\nabla^2\phi$. From (6.3) and (7.2), we have

(7.4) $$\nabla^2\phi = \nabla \cdot \nabla\phi = \text{div grad } \phi = \frac{\partial}{\partial x}\frac{\partial\phi}{\partial x} + \frac{\partial}{\partial y}\frac{\partial\phi}{\partial y} + \frac{\partial}{\partial z}\frac{\partial\phi}{\partial z}$$
$$= \frac{\partial^2\phi}{\partial x^2} + \frac{\partial^2\phi}{\partial y^2} + \frac{\partial^2\phi}{\partial z^2} \quad \text{(the Laplacian).}$$

The Laplacian is part of several important equations in mathematical physics:

$\nabla^2\phi = 0$ Laplace's equation.

$\nabla^2\phi = \frac{1}{a^2}\frac{\partial^2\phi}{\partial t^2}$ wave equation.

$\nabla^2\phi = \frac{1}{a^2}\frac{\partial\phi}{\partial t}$ diffusion, heat conduction, Schrödinger equation.

These equations arise in numerous problems in heat, hydrodynamics, electricity and magnetism, aerodynamics, elasticity, optics, etc.; we shall discuss solving such equations in Chapter 13.

There are many other more complicated expressions involving ∇ and one or more scalar or vector functions, which arise in various applications of vector analysis. For reference we list a table of such expressions at the end of the chapter (page 339). Notice that these are of two kinds: (1) expressions involving two applications of ∇ such as $\nabla \cdot \nabla\phi = \nabla^2\phi$; (2) combinations of ∇ with two functions (vectors or scalars) such as $\nabla \times (\phi\mathbf{V})$. We *can* verify these expressions simply by writing out components. However, it is usually simpler to use the same formulas we would use if ∇ were an ordinary vector, being careful to remember that ∇ is also a differential operator.

Example 1. Evaluate $\nabla \times (\nabla \times \mathbf{V})$. We use (3.8) for $\mathbf{A} \times (\mathbf{B} \times \mathbf{C})$ being careful to write both ∇'s *before* the vector function $\mathbf{V}$ which they must differentiate. Then we get

$$\nabla \times (\nabla \times \mathbf{V}) = \nabla(\nabla \cdot \mathbf{V}) - (\nabla \cdot \nabla)\mathbf{V}$$
$$= \nabla(\nabla \cdot \mathbf{V}) - \nabla^2\mathbf{V}.$$

This is a vector as it should be; the Laplacian of a vector, $\nabla^2\mathbf{V}$, simply means a vector whose components are $\nabla^2 V_x$, $\nabla^2 V_y$, $\nabla^2 V_z$.

Example 2. Find $\nabla \cdot (\phi\mathbf{V})$, where ϕ is a scalar function and $\mathbf{V}$ is a vector function. Here we must differentiate a product, so our result will contain two terms. We could write these as

(7.5) $$\nabla \cdot (\phi\mathbf{V}) = \nabla_\phi \cdot (\phi\mathbf{V}) + \nabla_\mathbf{V} \cdot (\phi\mathbf{V}),$$

where the subscripts on ∇ indicate which function is to be differentiated. Since ϕ is a scalar, it can be moved past the dot. Then

$$\nabla_\phi \cdot (\phi \mathbf{V}) = (\nabla_\phi \phi) \cdot \mathbf{V} = \mathbf{V} \cdot (\nabla \phi),$$

where we have removed the subscript in the last step since $\mathbf{V}$ no longer appears after ∇. Actually you may see in books $(\nabla\phi)\cdot\mathbf{V}$ meaning that only the ϕ is to be differentiated, but it is clearer to write it as $\mathbf{V}\cdot(\nabla\phi)$. [Be careful with $(\nabla\phi)\times\mathbf{V}$, however; assuming that this means that only ϕ is to be differentiated, the clear way to write it is $-\mathbf{V}\times(\nabla\phi)$; note the minus sign.] In the second term of (7.5), ϕ is a scalar and is not differentiated; thus it is just like a constant and we can write this term as $\phi(\nabla\cdot\mathbf{V})$. Collecting our results, we have

$$\nabla \cdot (\phi \mathbf{V}) = \mathbf{V} \cdot \nabla\phi + \phi(\nabla \cdot \mathbf{V}). \tag{7.6}$$

In Chapter 10, Section 9, we will derive the formulas for $\operatorname{div}\mathbf{V} = \nabla\cdot\mathbf{V}$ and $\nabla^2 f$ in cylindrical and spherical coordinates. However, it is useful to have the results for reference, so we state them here. Actually, these can be done as partial differentiation problems (see Chapter 4, Section 11), but the algebra is messy.

In cylindrical coordinates (or polar by omitting the z term):

$$\nabla \cdot \mathbf{V} = \frac{1}{r}\frac{\partial}{\partial r}(rV_r) + \frac{1}{r}\frac{\partial}{\partial \theta}V_\theta + \frac{\partial}{\partial z}V_z \tag{7.7}$$

$$\nabla^2 f = \frac{1}{r}\frac{\partial}{\partial r}\left(r\frac{\partial f}{\partial r}\right) + \frac{1}{r^2}\frac{\partial^2 f}{\partial \theta^2} + \frac{\partial^2 f}{\partial z^2}. \tag{7.8}$$

In spherical coordinates:

$$\nabla \cdot \mathbf{V} = \frac{1}{r^2}\frac{\partial}{\partial r}(r^2 V_r) + \frac{1}{r\sin\theta}\frac{\partial}{\partial\theta}(V_\theta \sin\theta) + \frac{1}{r\sin\theta}\frac{\partial V_\phi}{\partial\phi} \tag{7.9}$$

$$\nabla^2 f = \frac{1}{r^2}\frac{\partial}{\partial r}\left(r^2\frac{\partial f}{\partial r}\right) + \frac{1}{r^2\sin\theta}\frac{\partial}{\partial\theta}\left(\sin\theta\frac{\partial f}{\partial\theta}\right) + \frac{1}{r^2\sin^2\theta}\frac{\partial^2 f}{\partial\phi^2}. \tag{7.10}$$

PROBLEMS, SECTION 7

The purpose in doing the following simple problems is to become familiar with the formulas we have discussed. So a good study method is to do them by hand and then check your results by computer.

Compute the divergence and the curl of each of the following vector fields.

1. $\mathbf{r} = x\mathbf{i} + y\mathbf{j} + z\mathbf{k}$

2. $\mathbf{r} = x\mathbf{i} + y\mathbf{j}$

3. $\mathbf{V} = z\mathbf{i} + y\mathbf{j} + x\mathbf{k}$

4. $\mathbf{V} = y\mathbf{i} + z\mathbf{j} + x\mathbf{k}$

5. $\mathbf{V} = x^2\mathbf{i} + y^2\mathbf{j} + z^2\mathbf{k}$

6. $\mathbf{V} = x^2y\mathbf{i} + y^2x\mathbf{j} + xyz\mathbf{k}$

7. $\mathbf{V} = x\sin y\,\mathbf{i} + \cos y\,\mathbf{j} + xy\mathbf{k}$

8. $\mathbf{V} = \sinh z\,\mathbf{i} + 2y\,\mathbf{j} + x\cosh z\,\mathbf{k}$

Calculate the Laplacian ∇^2 of each of the following scalar fields.

9. $x^3 - 3xy^2 + y^3$

10. $\ln(x^2 + y^2)$

11. $\sqrt{x^2 - y^2}$

12. $(x + y)^{-1}$

13. $xy(x^2 + y^2 - 5z^2)$

14. $(x^2 + y^2 + z^2)^{-1/2}$

15. $xyz(x^2 - 2y^2 + z^2)$

16. $\ln(x^2 + y^2 + z^2)$

17. Verify formulas (b), (c), (d), (g), (h), (i), (j), (k) of the table of vector identities at the end of the chapter. *Hint* for (j): Start by expanding the two triple vector products on the right.

For $\mathbf{r} = x\mathbf{i} + y\mathbf{j} + z\mathbf{k}$, evaluate

18. $\nabla \times (\mathbf{k} \times \mathbf{r})$

19. $\nabla \cdot \left(\frac{\mathbf{r}}{|\mathbf{r}|}\right)$

20. $\nabla \times \left(\frac{\mathbf{r}}{|\mathbf{r}|}\right)$

8. LINE INTEGRALS

In Section 2, we discussed the fact that the work done by a force $\mathbf{F}$ on an object which undergoes an infinitesimal vector displacement $d\mathbf{r}$ can be written as

$$dW = \mathbf{F} \cdot d\mathbf{r}. \tag{8.1}$$

Suppose the object moves along some path (say A to B in Fig. 8.1), with the force $\mathbf{F}$ acting on it varying as it moves. For example, $\mathbf{F}$ might be the force on a charged particle in an electric field; then $\mathbf{F}$ would vary from point to point, that is $\mathbf{F}$ would be a function of x, y, z. However, *on a curve*, x, y, z are related by the equations of the curve. In three dimensions it takes two equations to determine a curve (as an intersection of two surfaces; for example, consider the equations of a straight line in Chapter 3, Section 5). Thus along a curve there is only *one* independent variable;

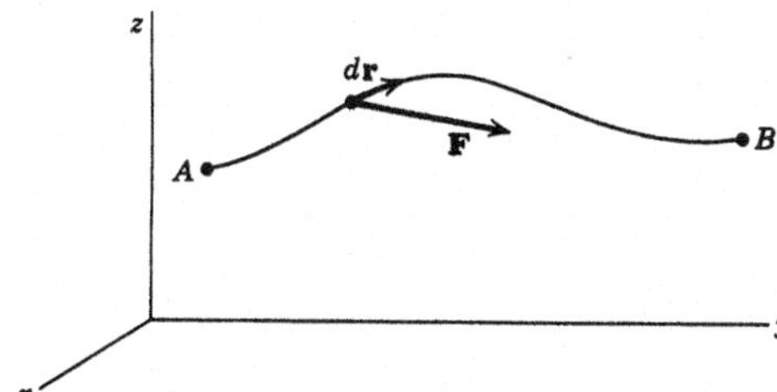

Figure 8.1

we can then write $\mathbf{F}$ and $d\mathbf{r} = \mathbf{i}\,dx + \mathbf{j}\,dy + \mathbf{k}\,dz$ as functions of a single variable. The integral of $dW = \mathbf{F} \cdot d\mathbf{r}$ along the given curve then becomes an ordinary integral of a function of one variable and we can evaluate it to find the total work done by $\mathbf{F}$ in moving an object in Figure 8.1 from A to B. Such an integral is called a *line integral.* A line integral means an integral along a curve (or line), that is, a single integral as contrasted to a double integral over a surface or area, or a triple integral over a volume. The essential point to understand about a line integral is that there is *one* independent variable, because we are required to remain on a *curve*. In two dimensions, the equation of a curve might be written $y = f(x)$, where x is the independent variable. In three dimensions, the equations of a curve (for example, a straight line) can be written either like (6.6) (where we could take x as the independent variable and find y and z as functions of x), or (6.1) (where s is the independent variable and x, y, z are all functions of s). To evaluate a line integral, then, we must write it as a *single* integral using one independent variable.

▶ **Example 1.** Given the force $\mathbf{F} = xy\,\mathbf{i} - y^2\,\mathbf{j}$, find the work done by $\mathbf{F}$ along the paths indicated in Figure 8.2 from $(0,0)$ to $(2,1)$.

Since $\mathbf{r} = x\,\mathbf{i} + y\,\mathbf{j}$ on the (x, y) plane, we have

$$d\mathbf{r} = \mathbf{i}\,dx + \mathbf{j}\,dy,$$
$$\mathbf{F}\cdot d\mathbf{r} = xy\,dx - y^2\,dy.$$

We want to evaluate

$$W = \int (xy\,dx - y^2\,dy). \tag{8.2}$$

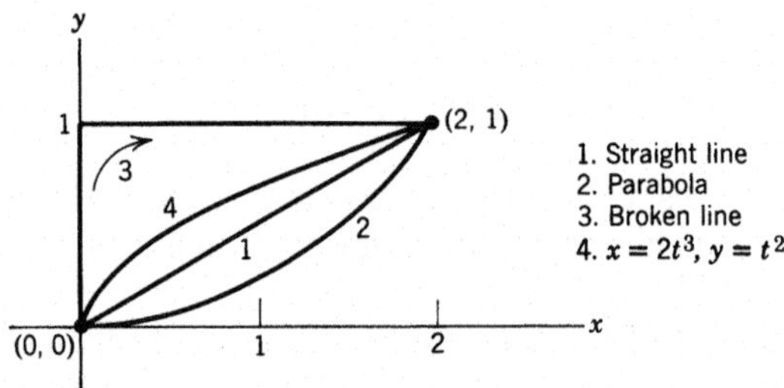

Figure 8.2

First we must write the integrand in terms of *one* variable. Along path 1 (a straight line), $y = \frac{1}{2}x$, $dy = \frac{1}{2}dx$. Substituting these values into (8.2), we obtain an integral in the one variable x. The limits for x (Figure 8.2) are 0 to 2. Thus, we get

$$W_1 = \int_0^2 \left[x \cdot \frac{1}{2}x\,dx - \left(\frac{1}{2}x\right)^2 \cdot \frac{1}{2}\,dx\right] = \int_0^2 \frac{3}{8}x^2\,dx = \frac{x^3}{8}\bigg|_0^2 = 1.$$

We could just as well use y as the independent variable and put $x = 2y$, $dx = 2dy$, and integrate from 0 to 1. (You should verify that the answer is the same.)

Along path 2 in Figure 8.2 (a parabola), $y = \frac{1}{4}x^2$, $dy = \frac{1}{2}x\,dx$. Then we get

$$W_2 = \int_0^2 \left(x \cdot \frac{1}{4}x^2\,dx - \frac{1}{16}x^4 \cdot \frac{1}{2}x\,dx\right) = \int_0^2 \left(\frac{1}{4}x^3 - \frac{1}{32}x^5\right)dx$$
$$= \frac{x^4}{16} - \frac{x^6}{192}\bigg|_0^2 = \frac{2}{3}.$$

Along path 3 (the broken line), we have to use a different method. We integrate first from $(0,0)$ to $(0,1)$ and then from $(0,1)$ to $(2,1)$ and add the results. Along $(0,0)$ to $(0,1)$, $x=0$ and $dx=0$ so we must use y as the variable. Then we have

$$\int_{y=0}^{1}(0\cdot y\cdot 0-y^2dy)=-\frac{y^3}{3}\bigg|_0^1=-\frac{1}{3}.$$

Along $(0,1)$ to $(2,1)$, $y=1$, $dy=0$, so we use x as the variable. We have

$$\int_{x=0}^{2}(x\cdot 1\cdot dx-1\cdot 0)=\frac{x^2}{2}\bigg|_0^2=2.$$

Then the total $W_3=-\frac{1}{3}+2=\frac{5}{3}$.

Path 4 illustrates still another technique. Instead of using either x or y as the integration variable, we can use a parameter t. For $x=2t^3$, $y=t^2$, we have $dx=6t^2\,dt$, $dy=2t\,dt$. At the origin, $t=0$, and at $(2,1)$, $t=1$. Substituting these values into (8.2), we get

$$W_4=\int_0^1(2t^3\cdot t^2\cdot 6t^2\,dt-t^4\cdot 2t\,dt)=\int_0^1(12t^7-2t^5)\,dt=\frac{12}{8}-\frac{2}{6}=\frac{7}{6}.$$

Example 2. Find the value of

$$I=\int\frac{x\,dy-y\,dx}{x^2+y^2}$$

along each of the two paths indicated in Figure 8.3 from $(-1,0)$ to $(1,0)$. [Notice that we *could* have written $I=\int\mathbf{F}\cdot d\mathbf{r}$ with $\mathbf{F}=(-\mathbf{i}y+x\mathbf{j})/(x^2+y^2)$; however, there are also many other kinds of problems in which line integrals may arise.]

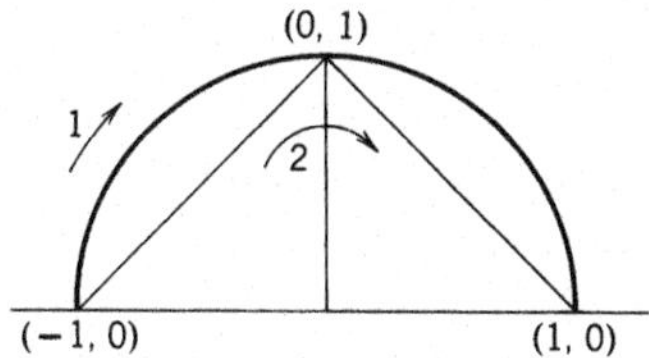

Figure 8.3

Along the circle it is simplest to use polar coordinates; then $r=1$ at all points of the circle and θ is the only variable. We then have

$$\begin{aligned}
x&=\cos\theta, & dx&=-\sin\theta\,d\theta, & &\\
y&=\sin\theta, & dy&=\cos\theta\,d\theta, & x^2+y^2&=1,
\end{aligned}$$

$$\frac{x\,dy-y\,dx}{x^2+y^2}=\frac{\cos^2\theta\,d\theta-\sin\theta(-\sin\theta)\,d\theta}{1}=d\theta.$$

At $(-1,0)$, $\theta=\pi$; at $(1,0)$, $\theta=0$. Then we get

$$I_1=\int_\pi^0 d\theta=-\pi.$$

Along path 2, we integrate from $(-1,0)$ to $(0,1)$ and from $(0,1)$ to $(1,0)$ and add the results. The first straight line has the equation $y = x+1$; then $dy = dx$, and the integral is

$$\begin{aligned}\int_{-1}^{0} \frac{x\,dx - (x+1)\,dx}{x^2+(x+1)^2} = \int_{-1}^{0} \frac{-dx}{2x^2+2x+1} &= \int_{-1}^{0} \frac{-2\,dx}{(2x+1)^2+1} \\ &= -\arctan(2x+1)\Big|_{-1}^{0} \\ &= -\arctan 1 + \arctan(-1) \\ &= -\frac{\pi}{4} + \left(-\frac{\pi}{4}\right) = -\frac{\pi}{2}.\end{aligned}$$

Along the second straight line $y = 1-x$, $dy = -dx$, and the integral is

$$\begin{aligned}-\int_{0}^{1} \frac{x\,dx + (1-x)\,dx}{x^2+(1-x)^2} &= \int_{0}^{1} \frac{-2\,dx}{(2x-1)^2+1} = -\arctan(2x-1)\Big|_{0}^{1} \\ &= -\frac{\pi}{2}.\end{aligned}$$

Adding the results for the integrals along the two parts of path 2, we get $I_2 = -\pi$.

Conservative Fields Notice that in Example 1 the answers were different for different paths, but in Example 2 they are the same. (See Section 11, however.) We can give a physical meaning to these facts if we interpret the integrals in all cases as the work done by a force on an object which moves along the path of integration. Suppose you want to get a heavy box across a sidewalk and up into a truck. Compare the work done in dragging the box across the sidewalk and then lifting it, with the work done in lifting it and then swinging it across in the air. In the first case work is done against friction in addition to the work required to lift the box; in the second case the only work done is that required to lift the box. Thus we see that the work done in moving an object from one point to another *may* depend on the path the object follows; in fact, it usually will when there is friction. Our example 1 was such a case. A force field for which $W = \int \mathbf{F} \cdot d\mathbf{r}$ depends upon the path as well as the endpoints is called *nonconservative*; physically this means that energy has been dissipated, say by friction. There are however, *conservative fields* for which $\int \mathbf{F} \cdot d\mathbf{r}$ is the same between two given points regardless of what path we calculate it along. For example, the work done in raising a mass m to the top of a mountain of height h is $W = mgh$ whether we lift the mass straight up a cliff or carry it up a slope, as long as no friction is involved. Thus the gravitational field is conservative.

It is useful to be able to recognize conservative and nonconservative fields before we do the integration. We shall see later (Section 11) that ordinarily $\operatorname{curl} \mathbf{F} = 0$ [see (7.3) for the definition of curl] is a necessary and sufficient condition for $\int \mathbf{F} \cdot d\mathbf{r}$ to be independent of the path, that is, $\operatorname{curl} \mathbf{F} = 0$ for conservative fields and $\operatorname{curl} \mathbf{F} \neq 0$ for nonconservative fields. (See Section 11 for a more careful discussion of this.) It is not hard to see why this is usually so. Suppose that for a given $\mathbf{F}$ there is a function $W(x, y, z)$ such that

$$(8.3)\qquad \begin{aligned} \mathbf{F} = \nabla W &= \mathbf{i}\frac{\partial W}{\partial x} + \mathbf{j}\frac{\partial W}{\partial y} + \mathbf{k}\frac{\partial W}{\partial z}, \\ F_x = \frac{\partial W}{\partial x}, \quad F_y &= \frac{\partial W}{\partial y}, \quad F_z = \frac{\partial W}{\partial z}. \end{aligned}$$

Then assuming that $\partial^2 W/\partial x \partial y = \partial^2 W/\partial y \partial x$, etc. (see Chapter 4, end of Section 1), we get from (8.3)

$$(8.4)\qquad \frac{\partial F_x}{\partial y} = \frac{\partial^2 W}{\partial y \partial x} = \frac{\partial F_y}{\partial x}, \quad \text{and similarly} \quad \frac{\partial F_y}{\partial z} = \frac{\partial F_z}{\partial y}, \quad \frac{\partial F_x}{\partial z} = \frac{\partial F_z}{\partial x}.$$

Using the definition (7.3) of curl $\mathbf{F}$, we see that equations (8.4) say that the three components of curl $\mathbf{F}$ are equal to zero. Thus if $\mathbf{F} = \nabla W$, then curl $\mathbf{F} = 0$. Conversely (as we shall show later), if curl $\mathbf{F} = 0$, then we can find a function $W(x, y, z)$ for which $\mathbf{F} = \nabla W$. Now if $\mathbf{F} = \nabla W$, we can write

$$(8.5)\qquad \begin{aligned} \mathbf{F} \cdot d\mathbf{r} = \nabla W \cdot d\mathbf{r} &= \frac{\partial W}{\partial x}dx + \frac{\partial W}{\partial y}dy + \frac{\partial W}{\partial z}dz = dW, \\ \int_A^B \mathbf{F} \cdot d\mathbf{r} &= \int_A^B dW = W(B) - W(A), \end{aligned}$$

where $W(B)$ and $W(A)$ mean the values of the function W at the endpoints A and B of the path of integration. Since the value of the integral depends only on the endpoints A and B, it is independent of the path along which we integrate from A to B, that is, $\mathbf{F}$ is conservative.

Potentials In mechanics, if $\mathbf{F} = \nabla W$ (that is, if $\mathbf{F}$ is conservative), then W is the work done by $\mathbf{F}$. For example, if a mass m falls a distance z under gravity, the work done on it is mgz. If, however, we lift the mass a distance z against gravity, the work done *by the force* $\mathbf{F}$ *of gravity* is $W = -mgz$ since the direction of motion is opposite to $\mathbf{F}$. The increase in potential energy of m in this case is $\phi = +mgz$, that is, $W = -\phi$, or $\mathbf{F} = -\nabla\phi$. The function ϕ is called the potential energy or the *scalar potential* of the force $\mathbf{F}$. (Of course, ϕ can be changed by adding any constant; this corresponds to a choice of the zero level of the potential energy and has no effect on $\mathbf{F}$.) More generally for any vector $\mathbf{V}$, if curl $\mathbf{V} = 0$, there is a function ϕ, called the scalar potential of $\mathbf{V}$, such that $\mathbf{V} = -\nabla\phi$. (This is the customary definition of scalar potential in mechanics and electricity; in hydrodynamics many authors define the *velocity potential* so that $\mathbf{V} = +\nabla\phi$.)

Now suppose that we are given $\mathbf{F}$ or $dW = \mathbf{F} \cdot d\mathbf{r}$, and we find by calculation that curl $\mathbf{F} = 0$. We then know that there *is* a function W and we want to know how to find it (up to an arbitrary additive constant of integration). To do this we can calculate the line integral in (8.5) from some reference point A to the variable point B along any convenient path; since the integral is independent of the path when curl $\mathbf{F} = 0$, this process gives the value of W at the point B. (There is, of course, an additive constant in W whose value depends on our choice of the reference point A.)

Example 3. Show that

$$\mathbf{F} = (2xy - z^3)\mathbf{i} + x^2\mathbf{j} - (3xz^2 + 1)\mathbf{k} \tag{8.6}$$

is conservative, and find a scalar potential ϕ such that $\mathbf{F} = -\boldsymbol{\nabla}\phi$.

We find

$$\boldsymbol{\nabla} \times \mathbf{F} = \begin{vmatrix} \mathbf{i} & \mathbf{j} & \mathbf{k} \\ \dfrac{\partial}{\partial x} & \dfrac{\partial}{\partial y} & \dfrac{\partial}{\partial z} \\ 2xy - z^3 & x^2 & -3xz^2 - 1 \end{vmatrix} = 0, \tag{8.7}$$

so $\mathbf{F}$ is conservative. Then

$$W = \int_A^B \mathbf{F} \cdot d\mathbf{r} = \int_A^B (2xy - z^3)\,dx + x^2\,dy - (3xz^2 + 1)\,dz \tag{8.8}$$

is independent of the path. Let us choose the origin as our reference point and integrate (8.8) from the origin to the point (x, y, z). As the path of integration, we choose the broken line from $(0, 0, 0)$ to $(x, 0, 0)$ to $(x, y, 0)$ to (x, y, z). From $(0, 0, 0)$ to $(x, 0, 0)$, we have $y = z = 0$, $dy = dz = 0$, so the integral is zero along this part of the path. From $(x, 0, 0)$ to $(x, y, 0)$, we have $x =$ const., $z = 0$, $dx = dz = 0$, so the integral is

$$\int_0^y x^2\,dy = x^2 \int_0^y dy = x^2 y.$$

From $(x, y, 0)$ to (x, y, z) we have $x =$ const., $y =$ const., $dx = dy = 0$, so the integral is

$$-\int_0^z (3xz^2 + 1)dz = -xz^3 - z.$$

Adding the three results, we get

$$W = x^2 y - xz^3 - z, \tag{8.9}$$

or

$$\phi = -W = -x^2 y + xz^3 + z. \tag{8.10}$$

Example 4. Find the scalar potential for the electric field due to a point charge q at the origin.

Recall that the electric field at a point $\mathbf{r} = \mathbf{i}x + \mathbf{j}y + \mathbf{k}z$ means the force on a unit charge at $\mathbf{r}$ due to q and is (in Gaussian units)

$$\mathbf{E} = \frac{q}{r^2}\mathbf{e}_r = \frac{q}{r^2}\frac{\mathbf{r}}{r} = \frac{q}{r^3}\mathbf{r}. \tag{8.11}$$

(This is Coulomb's law in electricity.) If we take the zero level of the potential energy at infinity, then the scalar potential ϕ means the negative of the work done by the field on the unit charge as the charge moves from infinity to the point $\mathbf{r}$. This is

$$(8.12)\qquad \phi = -\int_{\infty \text{ to } \mathbf{r}} \mathbf{E}\cdot d\mathbf{r} = q\int_{\mathbf{r} \text{ to } \infty} \frac{\mathbf{r}\cdot d\mathbf{r}}{r^3}.$$

It is simplest to evaluate the line integral using the spherical coordinate variable r along a radial line. This is justified by showing that $\operatorname{curl}\mathbf{E} = 0$, that is, that $\mathbf{E}$ is conservative (Problem 19). Since the differential of $(\mathbf{r}\cdot\mathbf{r})$ can be written as either $d(\mathbf{r}\cdot\mathbf{r}) = 2\mathbf{r}\cdot d\mathbf{r}$ or as $d(\mathbf{r}\cdot\mathbf{r}) = d(r^2) = 2r\,dr$, we have $\mathbf{r}\cdot d\mathbf{r} = r\,dr$ and (8.12) gives

$$(8.13)\qquad \phi = q\int_r^\infty \frac{r\,dr}{r^3} = q\int_r^\infty \frac{dr}{r^2} = -\frac{q}{r}\Big|_r^\infty = \frac{q}{r}.$$

It is interesting to obtain $\mathbf{r}\cdot d\mathbf{r} = r\,dr$ geometrically; in fact, for any vector $\mathbf{A}$, let us see that $\mathbf{A}\cdot d\mathbf{A} = A\,dA$. The vector $d\mathbf{A}$ means a change in the vector $\mathbf{A}$; a vector can change in both magnitude and direction (Figure 8.4). The scalar A means $|\mathbf{A}|$; the scalar dA means $d|\mathbf{A}|$. Thus dA is the increase in length of $\mathbf{A}$ and is *not* the same as $|d\mathbf{A}|$. In fact, from Figure 8.4, we see that

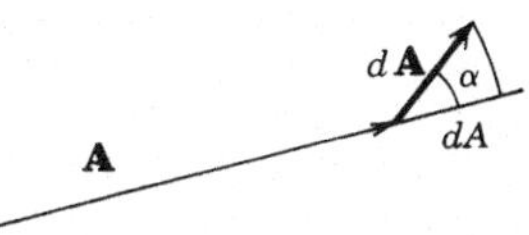

Figure 8.4

$$(8.14)\qquad \mathbf{A}\cdot d\mathbf{A} = |\mathbf{A}|\,|d\mathbf{A}|\cos\alpha = A\,dA$$

since $dA = |d\mathbf{A}|\cos\alpha$. For the vector $\mathbf{r}$, we have

$$(8.15)\qquad \begin{aligned} \mathbf{r} &= \mathbf{i}x + \mathbf{j}y + \mathbf{k}z, \\ d\mathbf{r} &= \mathbf{i}\,dx + \mathbf{j}\,dy + \mathbf{k}\,dz, \\ |d\mathbf{r}| &= \sqrt{dx^2 + dy^2 + dz^2} = ds \qquad \text{(see Chapter 5)}, \\ r &= |\mathbf{r}| = \sqrt{x^2 + y^2 + z^2}, \\ dr &= \frac{1}{2}(x^2 + y^2 + z^2)^{-1/2}(2x\,dx + 2y\,dy + 2z\,dz) \\ &= \frac{1}{r}(\mathbf{r}\cdot d\mathbf{r}), \end{aligned}$$

as above.

Exact Differentials The differential dW in (8.5) of a function $W(x, y, z)$ is called an *exact differential.* We could then say that $\operatorname{curl}\mathbf{F} = 0$ is a necessary and sufficient condition for $\mathbf{F}\cdot d\mathbf{r}$ to be an exact differential (but see Section 11). To make this clear, let us consider some examples in which $\mathbf{F}\cdot d\mathbf{r}$ is, or is not, an exact differential.

Example 5. Consider the function W in (8.9).
Then

$$(8.16)\qquad dW = (2xy - z^3)\,dx + x^2\,dy - (3xz^2 + 1)dz.$$

Here dW is an exact differential by definition since we got it by differentiating a function W. We can easily verify that if we write $dW = \mathbf{F}\cdot d\mathbf{r}$, then equations (8.4) are true:

$$(8.17)\qquad \begin{aligned} \frac{\partial}{\partial x}(x^2) &= 2x = \frac{\partial}{\partial y}(2xy - z^3), \\ \frac{\partial}{\partial x}(-3xz^2 - 1) &= -3z^2 = \frac{\partial}{\partial z}(2xy - z^3), \\ \frac{\partial}{\partial y}(-3xz^2 - 1) &= 0 = \frac{\partial}{\partial z}(x^2). \end{aligned}$$

You should observe carefully how to get (8.17) from (8.16): the equations (8.17) say that, in (8.16), the partial derivative with respect to x of the coefficient of dy equals the partial derivative with respect to y of the coefficient of dx, and similarly for the other pairs of variables. The equations (8.17) are called the *reciprocity relations* in thermodynamics; in mechanics they are the components of $\operatorname{curl} \mathbf{F} = 0$ [see (8.7)]. In both cases they are true assuming that the mixed second partial derivatives are the same in either order, for example $\partial^2 W/\partial x \partial y = \partial^2 W/\partial y \partial x$ (see Chapter 4, end of Section 1).

We obtained dW in (8.16) by taking the differential of (8.9); now suppose we start with a given $dW = \mathbf{F} \cdot d\mathbf{r}$.

Example 6. Let us consider

$$(8.18)\qquad dW = \mathbf{F} \cdot d\mathbf{r} = (2xy - z^3)\,dx + x^2\,dy + (3xz^2 + 1)\,dz.$$

This is almost the same as (8.16); just the sign of the dz term is changed. Then two of the equations corresponding to (8.17) do not hold, so $\operatorname{curl} \mathbf{F} \neq 0$, and dW is not an exact differential. We ask whether there is a function W of which (8.18) is the differential; the answer is "No" because if there were, the mixed second partial derivatives of W *would* be equal, and so $\operatorname{curl} \mathbf{F}$ would be zero. Equations like (8.18) often occur in applications. When dW is not exact, then $\mathbf{F}$ is a nonconservative force, and $\int \mathbf{F} \cdot d\mathbf{r}$, which is the work done by $\mathbf{F}$, depends not only on the points A and B but also upon the path along which the object moves. As we have said, this happens when there are friction forces.

PROBLEMS, SECTION 8

1. Evaluate the line integral $\int (x^2 - y^2)\,dx - 2xy\,dy$ along each of the following paths from $(0, 0)$ to $(1, 2)$.

(a) $y = 2x^2$.

(b) $x = t^2$, $y = 2t$.

(c) $y = 0$ from $x = 0$ to $x = 2$; then along the straight line joining $(2, 0)$ to $(1, 2)$.

2. Evaluate the line integral $\oint (x + 2y)\,dx - 2x\,dy$ along each of the following closed paths, taken counterclockwise:

(a) the circle $x^2 + y^2 = 1$;

(b) the square with corners at $(1, 1)$, $(-1, 1)$, $(-1, -1)$, $(1, -1)$;

(c) the square with corners $(0, 1)$, $(-1, 0)$, $(0, -1)$, $(1, 0)$.

3. Evaluate the line integral $\int xy\,dx + x\,dy$ from $(0, 0)$ to $(1, 2)$ along the paths shown in the sketch.

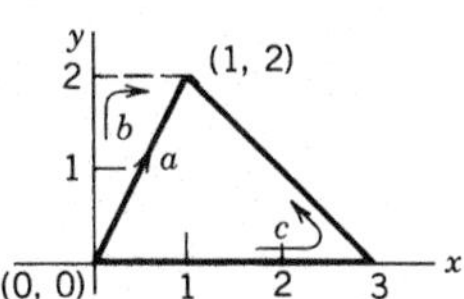

4. Evaluate the line integral $\int_C y^2\,dx + 2x\,dy + dz$, where C connects $(0, 0, 0)$ with $(1, 1, 1)$,

(a) along straight lines from $(0, 0, 0)$ to $(1, 0, 0)$ to $(1, 0, 1)$ to $(1, 1, 1)$;

(b) on the circle $x^2 + y^2 - 2y = 0$ to $(1, 1, 0)$ and then on a vertical line to $(1, 1, 1)$.

5. Find the work done by the force $\mathbf{F} = x^2y\mathbf{i} - xy^2\mathbf{j}$ along the paths shown from $(1, 1)$ to $(4, 2)$.

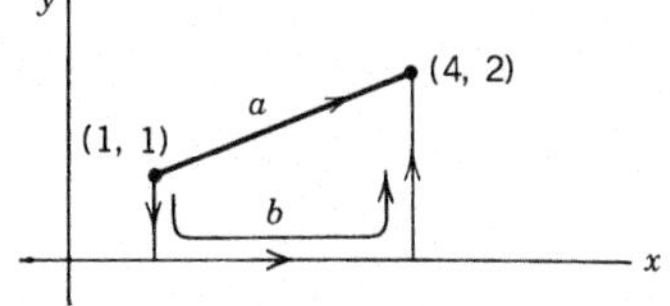

6. Find the work done by the force $\mathbf{F} = (2xy - 3)\mathbf{i} + x^2\mathbf{j}$ in moving an object from $(1, 0)$ to $(0, 1)$ along each of the three paths shown:

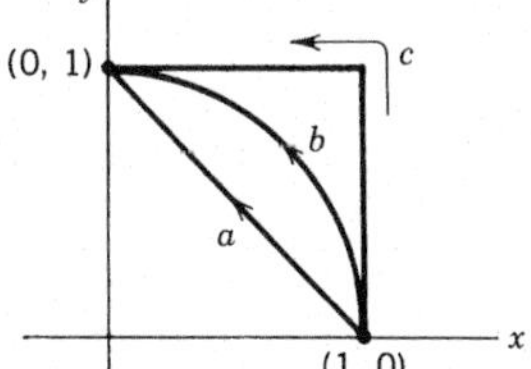

(a) straight line,

(b) circular arc,

(c) along lines parallel to the axes.

7. For the force field $\mathbf{F} = (y + z)\mathbf{i} - (x + z)\mathbf{j} + (x + y)\mathbf{k}$, find the work done in moving a particle around each of the following closed curves:

(a) the circle $x^2 + y^2 = 1$ in the (x, y) plane, taken counterclockwise;

(b) the circle $x^2 + z^2 = 1$ in the (z, x) plane, taken counterclockwise;

(c) the curve starting from the origin and going successively along the x axis to $(1, 0, 0)$, parallel to the z axis to $(1, 0, 1)$, parallel to the (y, z) plane to $(1, 1, 1)$, and back to the origin along $x = y = z$;

(d) from the origin to $(0, 0, 2\pi)$ on the curve $x = 1 - \cos t$, $y = \sin t$, $z = t$, and back to the origin along the z axis.

Verify that each of the following force fields is conservative. Then find, for each, a scalar potential ϕ such that $\mathbf{F} = -\nabla\phi$.

8. $\mathbf{F} = \mathbf{i} - z\mathbf{j} - y\mathbf{k}$.

9. $\mathbf{F} = (3x^2yz - 3y)\mathbf{i} + (x^3z - 3x)\mathbf{j} + (x^3y + 2z)\mathbf{k}$.

10. $\mathbf{F} = -k\mathbf{r}$, $\mathbf{r} = \mathbf{i}x + \mathbf{j}y + \mathbf{k}z$, $\qquad k = \text{const.}$

11. $\mathbf{F} = y \sin 2x\,\mathbf{i} + \sin^2 x\,\mathbf{j}$.

12. $\mathbf{F} = y\mathbf{i} + x\mathbf{j} + \mathbf{k}$.

13. $\mathbf{F} = z^2 \sinh y\,\mathbf{j} + 2z \cosh y\,\mathbf{k}$.

14. $\mathbf{F} = \dfrac{y}{\sqrt{1 - x^2y^2}}\mathbf{i} + \dfrac{x}{\sqrt{1 - x^2y^2}}\mathbf{j}$.

15. $\mathbf{F} = 2x \cos^2 y\,\mathbf{i} - (x^2 + 1) \sin 2y\,\mathbf{j}$.

16. Given $\mathbf{F}_1 = 2x\mathbf{i} - 2yz\mathbf{j} - y^2\mathbf{k}$ and $\mathbf{F}_2 = y\mathbf{i} - x\mathbf{j}$,

(a) Are these forces conservative? Find the potential corresponding to any conservative force.

(b) For any nonconservative force, find the work done if it acts on an object moving from $(-1, -1)$ to $(1, 1)$ along each of the paths shown.

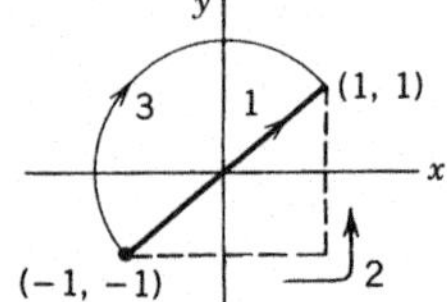

17. Which, if either, of the two force fields

$$\mathbf{F}_1 = -y\mathbf{i} + x\mathbf{j} + z\mathbf{k}, \qquad \mathbf{F}_2 = y\mathbf{i} + x\mathbf{j} + z\mathbf{k}$$

is conservative? Calculate for each field the work done in moving a particle around the circle $x = \cos t$, $y = \sin t$ in the (x, y) plane.

18. For the force field $\mathbf{F} = -y\mathbf{i} + x\mathbf{j} + z\mathbf{k}$, calculate the work done in moving a particle from $(1, 0, 0)$ to $(-1, 0, \pi)$

(a) along the helix $x = \cos t$, $y = \sin t$, $z = t$;

(b) along the straight line joining the points.

Do you expect your answers to be the same? Why or why not?

19. Show that the electric field $\mathbf{E}$ of a point charge [equation (8.11)] is conservative. Write ϕ in (8.13) in rectangular coordinates, and find $\mathbf{E} = -\nabla\phi$ using both rectangular coordinates (6.3) and cylindrical coordinates. Verify that your results are equivalent to (8.11).

20. For motion near the surface of the earth, we usually assume that the gravitational force on a mass m is

$$\mathbf{F} = -mg\mathbf{k},$$

but for motion involving an appreciable variation in distance r from the center of the earth, we must use

$$\mathbf{F} = -\frac{C}{r^2}\mathbf{e}_r = -\frac{C}{r^2}\frac{\mathbf{r}}{|\mathbf{r}|} = -\frac{C}{r^3}\mathbf{r},$$

where C is a constant. Show that both these $\mathbf{F}$'s are conservative, and find the potential for each.

21. Consider a uniform distribution of total mass m' over a spherical shell of radius r'. The potential energy ϕ of a mass m in the gravitational field of the spherical shell is

$$\phi = \begin{cases} \text{const.} & \text{if } m \text{ is inside the spherical shell,} \\ -\dfrac{Cm'}{r} & \text{if } m \text{ is outside the spherical shell, where } r \text{ is the distance} \\ & \text{from the center of the sphere to } m, \text{ and } C \text{ is a constant.} \end{cases}$$

Assuming that the earth is a spherical ball of radius R and constant density, find the potential and the force on a mass m outside and inside the earth. Evaluate the constants in terms of the acceleration of gravity g, to get

$$\mathbf{F} = -\frac{mgR^2}{r^2}\mathbf{e}_r, \quad \text{and} \quad \phi = -\frac{mgR^2}{r}, \qquad m \text{ outside the earth;}$$

$$\mathbf{F} = -\frac{mgr}{R}\mathbf{e}_r, \quad \text{and} \quad \phi = \frac{mg}{2R}(r^2 - 3R^2), \qquad m \text{ inside the earth.}$$

Hint: To find the constants, recall that at the surface of the earth the magnitude of the force on m is mg.

9. GREEN'S THEOREM IN THE PLANE

The fundamental theorem of calculus says that the integral of the derivative of a function is the function, or more precisely:

$$\int_a^b \frac{d}{dt} f(t)\,dt = f(b) - f(a). \tag{9.1}$$

We are going to consider some useful generalizations of this theorem to two and three dimensions. The divergence theorem and Stokes' theorem (Sections 10 and 11) are very important in electrodynamics and other applications; in this section we will find two-dimensional forms of these theorems. First we develop an underlying useful theorem relating an area integral to the line integral around its boundary (see applications in examples and problems and also Chapter 14, Section 3).

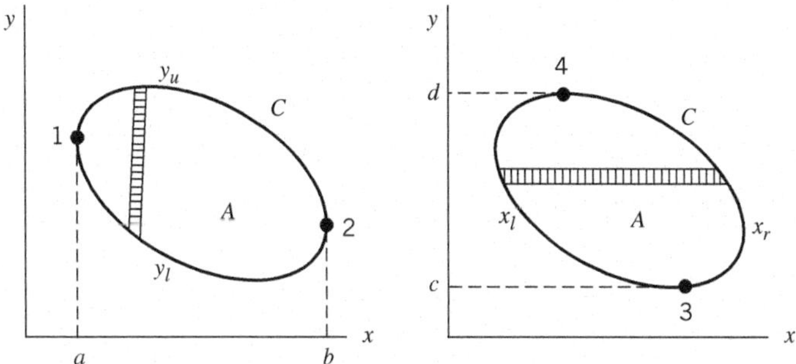

Figure 9.1

Recall that we know how to evaluate line integrals (Section 8), and that we learned in Chapter 5 to evaluate double integrals over areas in the (x, y) plane. We are going to consider areas (such as those in Figure 9.1 or in Chapter 5, Figure 2.7) for which we can evaluate the double integral over the area either with respect to x first or with respect to y first. Look at Figure 9.1. We want to find a relation between a double integral over the area A and a line integral around the curve C, for simple closed curves C. (A simple curve does not cross itself; for example, it is *not* a figure 8.) Now in Figure 9.1, the *upper* part of C between points 1 and 2 is given by an equation $y = y_u(x)$ and the *lower* part by an equation $y = y_l(x)$. (Think of solving the equation of a circle for $y_u(x) = \sqrt{1-x^2}$ and $y_l(x) = -\sqrt{1-x^2}$.) Similarly in Figure 9.1, we can find $x_l(y)$ and $x_r(y)$ for the left and right parts of C between points 3 and 4.

Let $P(x, y)$ and $Q(x, y)$ be continuous functions with continuous first derivatives. We are going to show that the double integral of $\partial P(x, y)/\partial y$ over the area A is equal to the line integral of P around C. We write the double integral using Figure 9.1 to integrate first with respect to y, and do the y integration by equation (9.1) with $t = y$ to get:

$$\iint_A \frac{\partial P(x,y)}{\partial y}\,dy\,dx = \int_a^b dx \int_{y_l}^{y_u} \frac{\partial P(x,y)}{\partial y}\,dy = \int_a^b [P(x, y_u) - P(x, y_l)]\,dx \tag{9.2}$$

$$= -\int_a^b P(x, y_l)\,dx - \int_b^a P(x, y_u)\,dx.$$

Now we have our answer—we just have to recognize it! Think how you would evaluate the line integral of $P(x, y)dx$ along the lower part of C in Figure 9.1 from point 1 to point 2. You would substitute $y = y_l(x)$ into $P(x, y)$ and integrate from $x = a$ to b (see Section 8).

$$\int_a^b P(x, y_l)\,dx = \text{line integral of } P\,dx \tag{9.3}$$

along lower part of C from point 1 to point 2.

This is one of the terms in (9.2). Similarly, to find the line integral of $P(x, y)\,dx$ along the upper part of C from point 2 to point 1, we substitute $y = y_u(x)$ and integrate from b to a.

$$\int_b^a P(x, y_u)\,dx = \text{line integral of } P\,dx \tag{9.4}$$

along upper part of C from point 2 to point 1.

Combining (9.3) and (9.4) gives us the line integral all the way around C in the counterclockwise direction, that is, so that A is always on our left as we go around C. (The symbol $\oint$ means an integral around a closed curve back to the starting point.) Then, from (9.2), we have

$$\oint_C P\,dx = -\iint_A \frac{\partial P(x, y)}{\partial y}\,dx\,dy. \tag{9.5}$$

Repeating the calculation but integrating first with respect to x, we find

$$\iint_A \frac{\partial Q}{\partial x}\,dx\,dy = \int_c^d dy \int_{x_l}^{x_r} \frac{\partial Q}{\partial x}\,dx = \int_c^d [Q(x_r, y) - Q(x_l, y)]dy = \oint_C Q\,dy. \tag{9.6}$$

Adding (9.5) and (9.6) and using the notation ∂A to mean the boundary of A (that is, C) we have

Green's theorem in the plane:

$$\iint_A \left(\frac{\partial Q}{\partial x} - \frac{\partial P}{\partial y}\right) dx\,dy = \oint_{\partial A} (P\,dx + Q\,dy) \tag{9.7}$$

The line integral is counterclockwise around the boundary of area A.

Using Green's theorem we can evaluate either a line integral around a closed path or a double integral over the area inclosed, whichever is easier to do. If the area is not of the simple type we have assumed, it may be possible to cut it into pieces (see Figure 9.2) so that our proof applies to each piece. Then the line integrals along the dotted cuts in Figure 9.2 are in opposite directions for adjacent pieces, and so cancel. Thus the theorem is valid for this more general area and its inclosing curve. In fact, we can even close up Figure 9.2 creating an area with a hole in the middle.

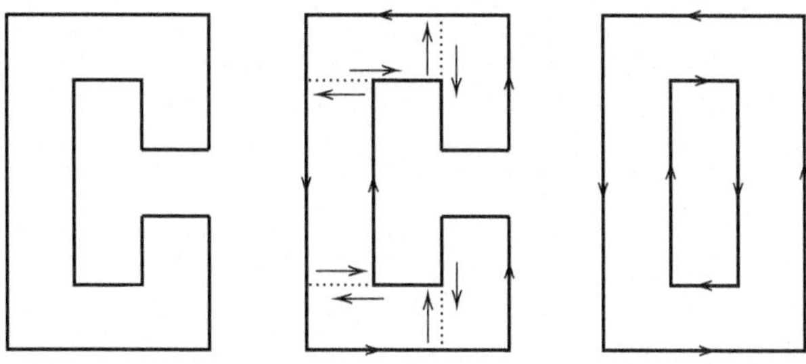

Figure 9.2

Green's theorem still holds, but now the line integral consists of a counterclockwise integral around the outside plus a clockwise integral around the hole as you can see in Figure 9.2. We say that this area is not "simply connected"—see further discussion of this in Section 11.

Example 1. In Example 1, Section 8, we found the line integral (8.2) along several paths (Figure 8.2). Suppose we want the line integral in Figure 8.2 around the closed loop (Figure 9.3) from $(0,0)$ to $(2,1)$ and back as shown. From Section 8, Example 1, this is the work done along path 2 minus the work done along path 3 (since we are now going in the opposite direction); we find $W_2 - W_3 = \frac{2}{3} - \frac{5}{3} = -1$. Let us evaluate this using Green's theorem. From (8.2) and (9.7) we have

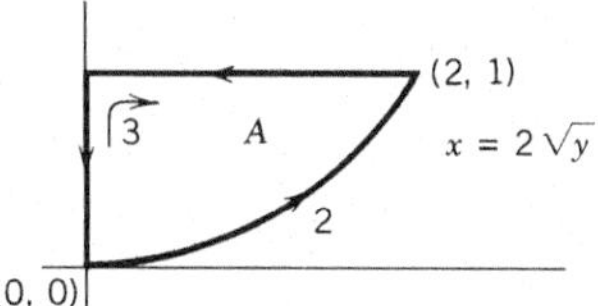

Figure 9.3

$$W = \oint_{\partial A} xy\,dx - y^2\,dy = \iint_A \left[\frac{\partial}{\partial x}(-y^2) - \frac{\partial}{\partial y}(xy)\right] dx\,dy$$

$$= \iint_A -x\,dx\,dy = -\int_{y=0}^{1}\int_{x=0}^{2\sqrt{y}} x\,dx\,dy = -1$$

as before.

Example 2. In Section 8, we discussed conservative forces for which work done is independent of the path. By Green's theorem (9.7), the work done by a force $\mathbf{F}$ around a closed path in the (x,y) plane is

$$W = \oint_{\partial A} (F_x\,dx + F_y\,dy) = \iint_A \left(\frac{\partial F_y}{\partial x} - \frac{\partial F_x}{\partial y}\right) dx\,dy.$$

If $(\partial F_y/\partial x) - (\partial F_x/\partial y) = 0$ (note that this is the z component of $\operatorname{curl} \mathbf{F} = 0$) then W around any closed path is zero, which means that the work from one point to another is independent of the path (also see Section 11).

The functions $P(x,y)$ and $Q(x,y)$ in (9.7) are arbitrary; we may choose them to suit our purposes. Note that a two-dimensional vector function $\mathbf{i}V_x(x,y) + \mathbf{j}V_y(x,y)$

contains two functions, V_x and V_y. In the next two examples, we are going to define P and Q in terms of V_x and V_y in order to obtain two useful results.

▶ **Example 3.** We define:

$$Q = V_x, \quad P = -V_y, \quad \text{where} \quad \mathbf{V} = \mathbf{i}V_x + \mathbf{j}V_y. \tag{9.8}$$

Then

$$\frac{\partial Q}{\partial x} - \frac{\partial P}{\partial y} = \frac{\partial V_x}{\partial x} + \frac{\partial V_y}{\partial y} = \operatorname{div} \mathbf{V} \tag{9.9}$$

by (7.2) with $V_z = 0$. Along the curve bounding an area A (Figure 9.4) the vector

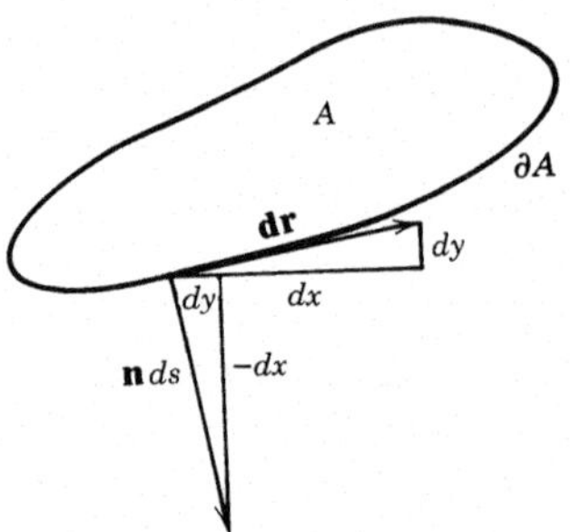

Figure 9.4

$$d\mathbf{r} = \mathbf{i}\,dx + \mathbf{j}\,dy \quad \text{(tangent)} \tag{9.10}$$

is a tangent vector, and the vector

$$\mathbf{n}\,ds = \mathbf{i}\,dy - \mathbf{j}\,dx \quad \text{(outward normal)}, \tag{9.11}$$
where $\mathbf{n}$ is a unit vector and $ds = \sqrt{dx^2 + dy^2}$,

is a normal vector (perpendicular to the tangent) pointing out of area A. Using (9.11) and (9.8), we can write

$$\begin{aligned} P\,dx + Q\,dy &= -V_y\,dx + V_x\,dy = (\mathbf{i}V_x + \mathbf{j}V_y) \cdot (\mathbf{i}\,dy - \mathbf{j}\,x) \\ &= \mathbf{V} \cdot \mathbf{n}\,ds. \end{aligned} \tag{9.12}$$

Then substitute (9.9) and (9.12) into (9.7) to get

$$\iint_A \operatorname{div} \mathbf{V}\,dx\,dy = \int_{\partial A} \mathbf{V} \cdot \mathbf{n}\,ds. \tag{9.13}$$

This is the *divergence theorem* in two dimensions. It can be extended to three dimensions (also see Section 10). Let τ represent a volume; then $\partial\tau$ (read boundary of τ) means the closed surface area of τ. Let $d\tau$ mean a volume element and let $d\sigma$

mean an element of surface area. At each point of the surface, let **n** be a unit vector perpendicular to the surface and pointing outward. Then the divergence theorem in three dimensions says (also see Section 10)

$$\iiint_{\tau} \operatorname{div} \mathbf{V}\, d\tau = \iint_{\partial\tau} \mathbf{V} \cdot \mathbf{n}\, d\sigma. \quad \text{Divergence theorem} \tag{9.14}$$

Example 4. To see another application of (9.7) to vector functions, we let

$$Q = V_y, \quad P = V_x, \quad \text{where} \quad \mathbf{V} = \mathbf{i}V_x + \mathbf{j}V_y. \tag{9.15}$$

Then

$$\frac{\partial Q}{\partial x} - \frac{\partial P}{\partial y} = \frac{\partial V_y}{\partial x} - \frac{\partial V_x}{\partial y} = (\operatorname{curl} \mathbf{V}) \cdot \mathbf{k} \tag{9.16}$$

by (7.3) with $V_z = 0$. Equations (9.10) and (9.15) give

$$P\, dx + Q\, dy = (\mathbf{i}V_x + \mathbf{j}V_y) \cdot (\mathbf{i}\, dx + \mathbf{j}\, dy) = \mathbf{V} \cdot d\mathbf{r}. \tag{9.17}$$

Substituting (9.16) and (9.17) into (9.7), we get

$$\iint_A (\operatorname{curl} \mathbf{V}) \cdot \mathbf{k}\, dx\, dy = \oint_{\partial A} \mathbf{V} \cdot d\mathbf{r}. \tag{9.18}$$

This is *Stokes' theorem* in two dimensions. It can be extended to three dimensions (Section 11). Let σ be an open surface (for example, a hemisphere); then $\partial\sigma$ means the curve bounding the surface (Figure 9.5). Let **n** be a unit vector normal to the surface. Then Stokes' theorem in three dimensions is (also see Section 11)

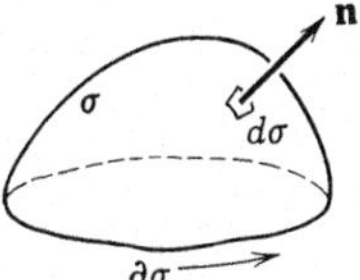

Figure 9.5

$$\iint_{\sigma} (\operatorname{curl} \mathbf{V}) \cdot \mathbf{n}\, d\sigma = \int_{\partial\sigma} \mathbf{V} \cdot d\mathbf{r}. \quad \text{Stokes' theorem.} \tag{9.19}$$

The direction of integration for the line integral is as shown in Figure 9.5 (see also Section 11).

PROBLEMS, SECTION 9

1. Write out the equations corresponding to (9.3) and (9.4) for $\int Q\, dy$ between points 3 and 4 in Figure 9.2, and add them to get (9.6).

In Problems 2 to 5 use Green's theorem [formula (9.7)] to evaluate the given integrals.

2. $\oint 2x\, dy - 3y\, dx$ around the square with vertices $(0, 2)$, $(2, 0)$, $(-2, 0)$, and $(0, -2)$.

3. $\oint_C xy\,dx + x^2\,dy$, where C is as sketched.

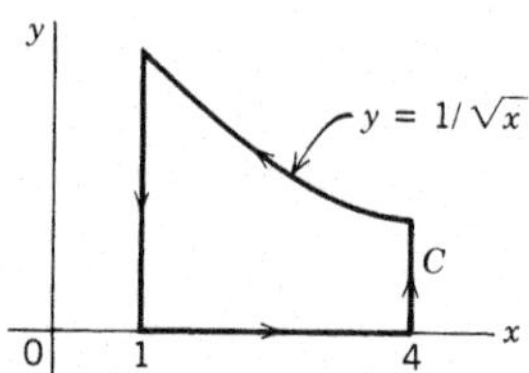

4. $\int_C e^x \cos y\,dx - e^x \sin y\,dy$, where C is the broken line from $A = (\ln 2, 0)$ to $D = (0,1)$ and then from D to $B = (-\ln 2, 0)$. *Hint:* Apply Green's theorem to the integral around the closed curve $ADBA$.

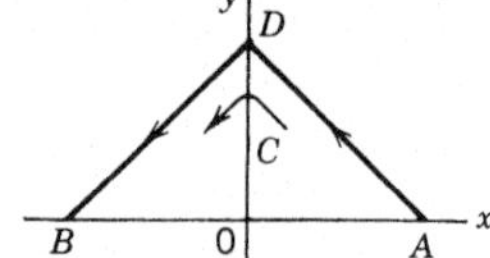

5. $\int_C (ye^x - 1)\,dx + e^x\,dy$, where C is the semicircle through $(0,-10)$, $(10,0)$, and $(0,10)$. (Compare Problem 4.)

6. For a simple closed curve C in the plane show by Green's theorem that the area inclosed is

$$A = \frac{1}{2}\oint_C (x\,dy - y\,dx).$$

7. Use Problem 6 to show that the area inside the ellipse $x = a\cos\theta$, $y = b\sin\theta$, $0 \le \theta \le 2\pi$, is $A = \pi ab$.

8. Use Problem 6 to find the area inside the curve $x^{2/3} + y^{2/3} = 4$.

9. Apply Green's theorem with $P = 0$, $Q = \frac{1}{2}x^2$ to the triangle with vertices $(0,0)$, $(0,3)$, $(3,0)$. You will then have $\iint x\,dx\,dy$ over the triangle expressed as a very simple line integral. Use this to locate the centroid of the triangle. (Compare Chapter 5, Section 3.)

Evaluate each of the following integrals in the easiest way you can.

10. $\oint(2y\,dx - 3x\,dy)$ around the square bounded by $x = 3$, $x = 5$, $y = 1$ and $y = 3$.

11. $\int_C (x\sin x - y)\,dx + (x - y^2)\,dy$, where C is the triangle in the (x,y) plane with vertices $(0,0)$, $(1,1)$, and $(2,0)$.

12. $\int (y^2 - x^2)\,dx + (2xy + 3)\,dy$ along the x axis from $(0,0)$ to $(\sqrt{5}, 0)$ and then along a circular arc from $(\sqrt{5}, 0)$ to $(1,2)$.

10. THE DIVERGENCE AND THE DIVERGENCE THEOREM

We have defined (in Section 7) the *divergence* of a vector function $\mathbf{V}(x,y,z)$ as

$$\text{div}\,\mathbf{V} = \nabla\cdot\mathbf{V} = \frac{\partial V_x}{\partial x} + \frac{\partial V_y}{\partial y} + \frac{\partial V_z}{\partial z}. \tag{10.1}$$

We now want to investigate the meaning and use of the divergence in physical applications.

Consider a region in which water is flowing. We can imagine drawing at every point a vector $\mathbf{v}$ equal to the velocity of the water at that point. The vector function $\mathbf{v}$ then represents a vector field. The curves tangent to $\mathbf{v}$ are called stream lines. We could in the same way discuss the flow of a gas, of heat, of electricity, or of particles (say from a radioactive source). We are going to show that if $\mathbf{v}$ represents the velocity of flow of any of these things, then div $\mathbf{v}$ is related to the amount of the substance which flows out of a given volume. This could be different from zero either because of a change in density (more air flows out than in as a room is heated) or because there is a source or sink in the volume (alpha particles flow out of but not into a box containing an alpha-radioactive source). Exactly the same mathematics applies to the electric and magnetic fields where $\mathbf{v}$ is replaced by $\mathbf{E}$ or $\mathbf{B}$ and the quantity corresponding to outflow of a material substance is called flux.

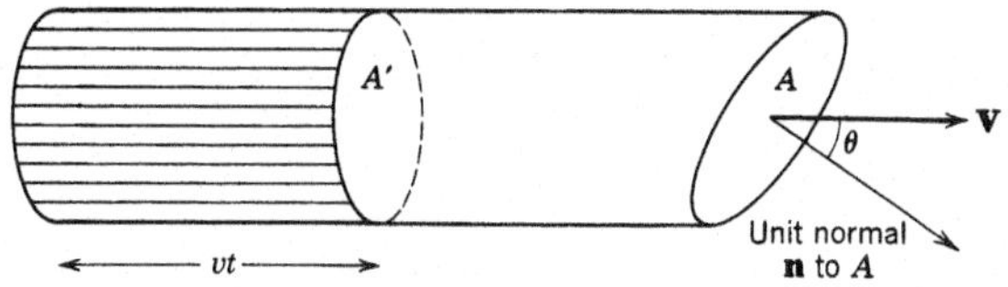

Figure 10.1

For our example of water flow, let $\mathbf{V} = \mathbf{v}\rho$, where ρ is the density of the water. Then the amount of water crossing in time t an area A' which is perpendicular to the direction of flow, is (see Figure 10.1) the amount of water in a cylinder of cross section A' and length vt. This amount of water is

$$(vt)(A')(\rho). \tag{10.2}$$

The same amount of water crosses area A (see Figure 10.1) whose normal is inclined at angle θ to $\mathbf{v}$. Since $A' = A\cos\theta$,

$$vtA'\rho = vt\rho A\cos\theta. \tag{10.3}$$

Then if water is flowing in the direction $\mathbf{v}$ making an angle θ with the normal $\mathbf{n}$ to a surface, the amount of water crossing *unit* area of the surface in *unit* time is

$$v\rho\cos\theta = V\cos\theta = \mathbf{V}\cdot\mathbf{n} \tag{10.4}$$

if $\mathbf{n}$ is a unit vector.

Now consider an element of volume $dx\,dy\,dz$ in the region through which the water is flowing (Figure 10.2). Water is flowing either in or out of the volume $dx\,dy\,dz$ through each of the six surfaces of the volume element; we shall calculate the net outward flow. In Figure 10.2, the rate at which water flows into $dx\,dy\,dz$

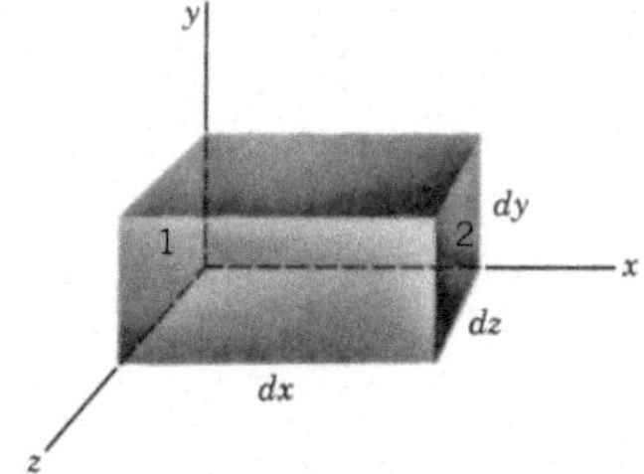

Figure 10.2

through surface 1 is [by (10.4)] $\mathbf{V}\cdot\mathbf{i}$ per unit area, or $(\mathbf{V}\cdot\mathbf{i})\,dy\,dz$ through the area $dy\,dz$ of surface 1. Since $\mathbf{V}\cdot\mathbf{i} = V_x$, we find that the rate at which water flows across surface 1 is $V_x\,dy\,dz$. A similar expression gives the rate at which water flows *out* through surface 2, except that V_x must be the x component of $\mathbf{V}$ at surface 2 instead of at surface 1. We want the difference of the two V_x values at two points, one on surface 1 and one on surface 2, directly opposite each other, that is, for the same y and z. These two values of V_x differ by ΔV_x which can be approximated (as in Chapter 4) by dV_x. For constant y and z, $dV_x = (\partial V_x/\partial x)\,dx$. Then the *net outflow* through these two surfaces is the outflow through surface 2 minus the inflow through surface 1, namely,

$$[(V_x \text{ at surface } 2) - (V_x \text{ at surface } 1)]dy\,dz = \left(\frac{\partial V_x}{\partial x}dx\right)dy\,dz. \tag{10.5}$$

We get similar expressions for the net outflow through the other two pairs of opposite surfaces:

$$\begin{aligned} &\frac{\partial V_y}{\partial y}\,dx\,dy\,dz && \text{through top and bottom, and} \\ &\frac{\partial V_z}{\partial z}\,dx\,dy\,dz && \text{through the other two sides.} \end{aligned} \tag{10.6}$$

Then the total net rate of loss of water from $dx\,dy\,dz$ is

$$\left(\frac{\partial V_x}{\partial x}+\frac{\partial V_y}{\partial y}+\frac{\partial V_z}{\partial z}\right)dx\,dy\,dz = \text{div}\,\mathbf{V}\,dx\,dy\,dz \quad \text{or} \quad \boldsymbol{\nabla}\cdot\mathbf{V}\,dx\,dy\,dz. \tag{10.7}$$

If we divide (10.7) by $dx\,dy\,dz$, we have the rate of loss of water per unit volume. This is the physical meaning of a divergence: It is the net rate of outflow *per unit volume* evaluated at a point (let $dx\,dy\,dz$ shrink to a point). This is outflow of actual substance for liquids, gases, or particles; it is called flux for electric and magnetic fields. You should note that this is somewhat like a density. Density is mass *per unit volume*, but it is evaluated *at a point* and may vary from point to point. Similarly, the divergence is evaluated at each point and may vary from point to point.

As we have said, div $\mathbf{V}$ may be different from zero either because of time variation of the density or because of sources and sinks. Let

ψ = *source density* minus *sink density*
= net mass of fluid being created (or added via something like a minute sprinkler system) per unit time per unit volume;
ρ = density of the fluid = mass per unit volume;
$\partial\rho/\partial t$ = time rate of increase of mass per unit volume.

Then:

Rate of increase of mass in $dx\,dy\,dz$ = rate of creation minus rate of outward flow,

or in symbols

$$\frac{\partial \rho}{\partial t}dx\,dy\,dz = \psi\,dx\,dy\,dz - \boldsymbol{\nabla}\cdot\mathbf{V}\,dx\,dy\,dz.$$

Canceling $dx\,dy\,dz$, we have

$$\frac{\partial \rho}{\partial t} = \psi - \boldsymbol{\nabla}\cdot\mathbf{V}$$

or

$$\boldsymbol{\nabla}\cdot\mathbf{V} = \psi - \frac{\partial \rho}{\partial t}. \tag{10.8}$$

If there are no sources or sinks, then $\psi = 0$; the resulting equation is often called the *equation of continuity*. (See Problem 15.)

$$\text{(10.9)} \qquad \nabla \cdot \mathbf{V} + \frac{\partial \rho}{\partial t} = 0. \qquad \text{Equation of continuity}$$

If $\partial\rho/\partial t = 0$, then

$$\text{(10.10)} \qquad \nabla \cdot \mathbf{V} = \psi.$$

In the case of the electric field, the "sources" and "sinks" are electric charges and the equation corresponding to (10.10) is $\operatorname{div}\mathbf{D} = \psi$, where ψ is the charge density and $\mathbf{D}$ is the electric displacement. For the magnetic field $\mathbf{B}$ you would expect the sources to be magnetic poles; however, there are no free magnetic poles, so $\operatorname{div}\mathbf{B} = 0$ always.

We have shown that the mass of fluid crossing a plane area A per unit time is $A\mathbf{V} \cdot \mathbf{n}$, where $\mathbf{n}$ is a unit vector normal to A, $\mathbf{v}$ and ρ are the velocity and density of the fluid, and $\mathbf{V} = \mathbf{v}\rho$. Consider any closed surface, and let $d\sigma$ represent an area element on the surface (Figure 10.3). For example: for a plane, $d\sigma = dx\,dy$; for a spherical surface,

$$d\sigma = r^2 \sin\theta \, d\theta \, d\phi.$$

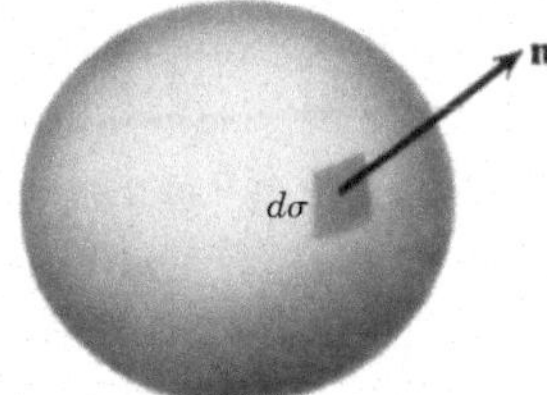

Figure 10.3

Let $\mathbf{n}$ be the unit vector normal to $d\sigma$ and pointing *out* of the surface ($\mathbf{n}$ varies in direction from point to point on the surface). Then the mass of fluid flowing out through $d\sigma$ is $\mathbf{V} \cdot \mathbf{n}\, d\sigma$ by (10.4) and the total outflow from the volume inclosed by the surface is

$$\text{(10.11)} \qquad \iint \mathbf{V} \cdot \mathbf{n}\, d\sigma,$$

where the double integral is evaluated over the closed surface.

We showed previously [see (10.7)] that for the volume element $d\tau = dx\,dy\,dz$:

$$\text{(10.12)} \qquad \text{The outflow from } d\tau \text{ is} \quad \nabla \cdot \mathbf{V}\, d\tau.$$

For simplicity, we proved this for a rectangular coordinate volume element $dx\,dy\,dz$. With extra effort we could prove it more generally, say for volume elements with slanted sides or for spherical coordinate volume elements. From now on we shall assume that $d\tau$ includes more general volume element shapes.

It is worth noticing here another way [besides (7.2)] of defining the divergence. If we write (10.11) for the surface of a volume element $d\tau$, we have two expressions for the total outflow from $d\tau$, and these must be equal. Thus

$$\text{(10.13)} \qquad \nabla \cdot \mathbf{V}\, d\tau = \iint_{\substack{\text{surface}\\ \text{of } d\tau}} \mathbf{V} \cdot \mathbf{n}\, d\sigma.$$

The value of $\nabla \cdot \mathbf{V}$ on the left is, of course, an average value of $\nabla \cdot \mathbf{V}$ in $d\tau$, but if we divide (10.13) by $d\tau$ and let $d\tau$ shrink to a point, we have a definition of $\nabla \cdot \mathbf{V}$ at the point:

$$\nabla \cdot \mathbf{V} = \lim_{d\tau \to 0} \frac{1}{d\tau} \iint\limits_{\substack{\text{surface} \\ \text{of } d\tau}} \mathbf{V} \cdot \mathbf{n}\, d\sigma. \tag{10.14}$$

If we start with (10.14) as the definition of $\nabla \cdot \mathbf{V}$, then the discussion leading to (10.7) is a proof that $\nabla \cdot \mathbf{V}$ as defined in (10.14) is equal to $\nabla \cdot \mathbf{V}$ as defined in (7.2).

The Divergence Theorem See (10.17). The divergence theorem is also called Gauss's theorem, but be careful to distinguish this mathematical theorem from Gauss's law which is a law of physics; see (10.23).

Consider a large volume τ; imagine it cut up into volume elements $d\tau_i$ (a cross section of this is shown in Figure 10.4). The outflow from each $d\tau_i$ is $\nabla \cdot \mathbf{V}\, d\tau_i$; let us add together the outflow from all the $d\tau_i$ to get

$$\sum_i \nabla \cdot \mathbf{V}\, d\tau_i. \tag{10.15}$$

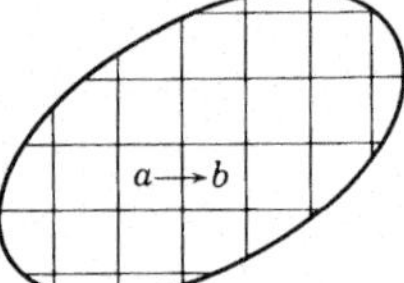

Figure 10.4

We shall show that (10.15) is the outflow from the large volume τ. Consider the flow between the elements marked a and b in Figure 10.4 across their common face. An outflow from a to b is an inflow (negative outflow) from b to a, so that in the sum (10.15) such outflows across interior faces cancel. The total sum in (10.15) then equals just the total outflow from the large volume. As the size of the volume elements tends to zero, this sum approaches a triple integral over the volume,

$$\iiint \nabla \cdot \mathbf{V}\, d\tau. \tag{10.16}$$

We have shown that both (10.11) and (10.16) are equal to the total outflow from the large volume; hence they are equal to each other, and we have the divergence theorem as stated in (9.14):

$$\iiint\limits_{\text{volume } \tau} \nabla \cdot \mathbf{V}\, d\tau = \iint\limits_{\substack{\text{surface} \\ \text{inclosing } \tau}} \mathbf{V} \cdot \mathbf{n}\, d\sigma. \qquad \text{Divergence theorem} \tag{10.17}$$

($\mathbf{n}$ points out of the closed surface σ.)

Notice that the divergence theorem converts a volume integral into an integral over a closed surface or vice versa; we can then evaluate whichever one is the easier to do.

In (10.17) we have carefully written the volume integral with three integral signs and the surface integral with two integral signs. However, it is rather common to write only one integral sign for either case when the volume or area element is indicated by a single differential ($d\tau$, dV, etc., for volume; $d\sigma$, dA, dS, etc., for surface area). Thus we might write $\iiint d\tau$ or $\int d\tau$ or $\iiint dx\,dy\,dz$, all meaning the same thing. When the single integral sign is used to indicate a surface or volume integral, you must see from the notation (τ for volume, σ for area), or the words under the integral, what is really meant. To indicate a surface integral over a *closed* surface or a line integral around a *closed* curve, the symbol $\oint$ is often used. Thus we might write either $\iint d\sigma$ or $\oint d\sigma$ for a surface integral over a closed surface. A different notation for the integrand $\mathbf{V}\cdot\mathbf{n}\,d\sigma$ is often used. Instead of using a unit vector $\mathbf{n}$ and the scalar magnitude $d\sigma$, we may write the vector $d\boldsymbol{\sigma}$ meaning a vector of magnitude $d\sigma$ in the direction $\mathbf{n}$; thus $d\boldsymbol{\sigma}$ means exactly the same thing as $\mathbf{n}\,d\sigma$, and we may replace $\mathbf{V}\cdot\mathbf{n}\,d\sigma$ by $\mathbf{V}\cdot d\boldsymbol{\sigma}$ in (10.17).

Example of the Divergence Theorem Let $\mathbf{V} = \mathbf{i}x + \mathbf{j}y + \mathbf{k}z$ and evaluate $\oint \mathbf{V}\cdot\mathbf{n}\,d\sigma$ over the closed surface of the cylinder shown in Figure 10.5.

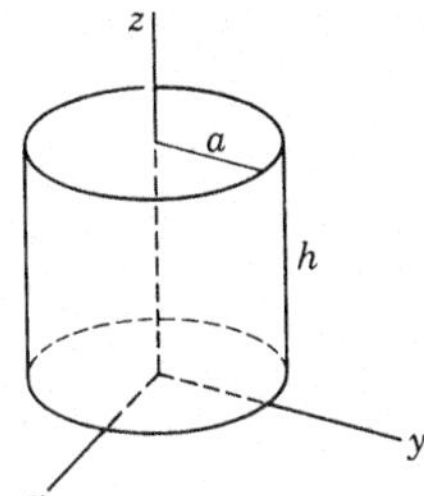

Figure 10.5

By the divergence theorem this is equal to $\int \nabla\cdot\mathbf{V}\,d\tau$ over the volume of the cylinder. (Note that we are using single integral signs, but the notation and words make it clear which integral is a volume integral and which a surface integral.) We find from the definition of divergence

$$\nabla\cdot\mathbf{V} = \frac{\partial x}{\partial x} + \frac{\partial y}{\partial y} + \frac{\partial z}{\partial z} = 3.$$

Then by (10.17)

$$\oint_{\substack{\text{surface of}\\ \text{cylinder}}} \mathbf{V}\cdot\mathbf{n}\,d\sigma = \int_{\substack{\text{volume of}\\ \text{cylinder}}} \nabla\cdot\mathbf{V}\,d\tau = \int 3\,d\tau = 3\int d\tau$$

$$= 3 \text{ times volume of cylinder} = 3\pi a^2 h.$$

It is harder to evaluate $\oint \mathbf{V}\cdot\mathbf{n}\,d\sigma$ directly, but we might do it to show an example of calculating a surface integral and to verify the divergence theorem in a special case. We need the surface normal $\mathbf{n}$. On the top surface (Figure 10.5) $\mathbf{n} = \mathbf{k}$, and there $\mathbf{V}\cdot\mathbf{n} = \mathbf{V}\cdot\mathbf{k} = z = h$. Then

$$\int_{\substack{\text{top surface of}\\ \text{cylinder}}} \mathbf{V}\cdot\mathbf{n}\,d\sigma = h\int d\sigma = h\cdot\pi a^2.$$

On the bottom surface, $\mathbf{n} = -\mathbf{k}$, $\mathbf{V}\cdot\mathbf{n} = -z = 0$; hence the integral over the bottom surface is zero. On the curved surface we might see by inspection that the vector $\mathbf{i}x + \mathbf{j}y$ is normal to the surface, so for the curved surface we have

$$\mathbf{n} = \frac{\mathbf{i}x + \mathbf{j}y}{\sqrt{x^2+y^2}} = \frac{\mathbf{i}x + \mathbf{j}y}{a}.$$

If the vector $\mathbf{n}$ is not obvious by inspection, we can easily find it; recall (Section 6) that if the equation of a surface is $\phi(x,y,z) = \text{const.}$, then $\nabla\phi$ is perpendicular to the surface. In this problem, the equation of the cylinder is $x^2 + y^2 = a^2$; then

$\phi = x^2 + y^2$, $\nabla\phi = 2x\mathbf{i} + 2y\mathbf{j}$, and we get the same unit vector $\mathbf{n}$ as above. Then for the curved surface we find

$$\mathbf{V}\cdot\mathbf{n} = \frac{x^2+y^2}{a} = \frac{a^2}{a} = a,$$

$$\int_{\substack{\text{curved}\\ \text{surface}}} \mathbf{V}\cdot\mathbf{n}\,d\sigma = a\int d\sigma = a\cdot(\text{area of curved surface}) = a\cdot 2\pi ah.$$

The value of $\oint \mathbf{V}\cdot\mathbf{n}\,d\sigma$ over the whole surface of the cylinder is then $\pi a^2h + 2\pi a^2h = 3\pi a^2h$ as before.

Gauss's Law The divergence theorem is very important in electricity. In order to see how it is used, we need a law in electricity known as Gauss's law. Let us derive this law from the more familiar Coulomb's law (8.11). Coulomb's law (written this time in SI units) gives for the electric field at $\mathbf{r}$ due to a point charge q at the origin

(10.18) $$\mathbf{E} = \frac{q}{4\pi\epsilon_0 r^2}\mathbf{e}_r. \qquad \text{Coulomb's law}$$

(ϵ_0 is a constant called the *permittivity of free space* and $\dfrac{1}{4\pi\epsilon_0} = 9\cdot 10^9$ in SI units.) The electric displacement $\mathbf{D}$ is defined (in free space) by $\mathbf{D} = \epsilon_0\mathbf{E}$; then

(10.19) $$\mathbf{D} = \frac{q}{4\pi r^2}\mathbf{e}_r.$$

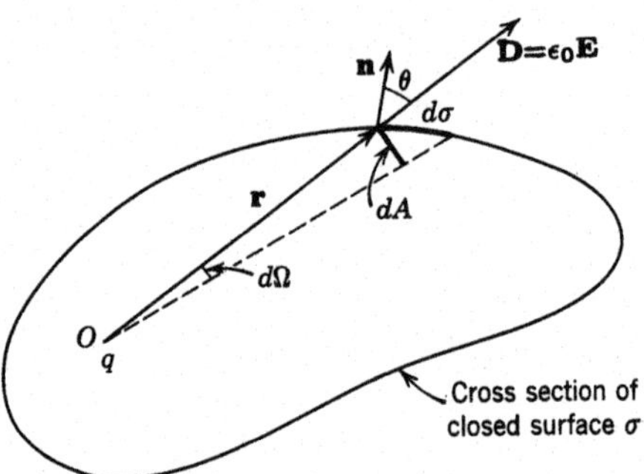

Figure 10.6

Let σ be a closed surface surrounding the point charge q at the origin; let $d\sigma$ be an element of area of the surface at the point $\mathbf{r}$, and let $\mathbf{n}$ be a unit normal to $d\sigma$ (Figures 10.3 and 10.6). Also (Figure 10.6) let dA be the projection of $d\sigma$ onto a sphere of radius r and center at O and let $d\Omega$ be the solid angle subtended by $d\sigma$ (and dA) at O. Then by definition of solid angle

(10.20) $$d\Omega = \frac{1}{r^2}dA.$$

From Figure 10.6 and equations (10.19) and (10.20), we get

(10.21) $$\mathbf{D}\cdot\mathbf{n}\,d\sigma = D\cos\theta\,d\sigma = D\,dA = \frac{q}{4\pi r^2}\cdot r^2\,d\Omega = \frac{1}{4\pi}q\,d\Omega$$

We want to find the surface integral of $\mathbf{D}\cdot\mathbf{n}\,d\sigma$ over the closed surface σ; by (10.21) this is

$$\oint_{\substack{\text{closed}\\ \text{surface } \sigma}} \mathbf{D}\cdot\mathbf{n}\,d\sigma = \frac{q}{4\pi}\int_{\substack{\text{total}\\ \text{solid angle}}} d\Omega = \frac{q}{4\pi}\cdot 4\pi = q \qquad (q \text{ inside } \sigma). \tag{10.22}$$

This is a simple case of Gauss's law when we have only one point charge q; for most purposes we shall want Gauss's law in the forms (10.23) or (10.24) below. Before we derive these, we should note carefully that in (10.22) the charge q is *inside* the closed surface σ. If we repeat the derivation of (10.22) for a point charge q outside the surface (Problem 13), we find that in this case

$$\oint_{\text{closed } \sigma} \mathbf{D}\cdot\mathbf{n}\,d\sigma = 0.$$

Next suppose there are several charges q_i inside the closed surface. For each q_i and the D_i corresponding to it, we could write an equation like (10.22). But the total electric displacement vector $\mathbf{D}$ at a point due to all the q_i is the vector sum of the vectors $\mathbf{D}_i$. Thus we have

$$\oint_{\substack{\text{closed}\\ \text{surface } \sigma}} \mathbf{D}\cdot\mathbf{n}\,d\sigma = \sum_i \oint_{\substack{\text{closed}\\ \text{surface } \sigma}} \mathbf{D}_i\cdot\mathbf{n}\,d\sigma = \sum q_i.$$

Therefore for any charge distribution inside a closed surface

$$\oint_{\substack{\text{closed}\\ \text{surface}}} \mathbf{D}\cdot\mathbf{n}\,d\sigma = \text{total charge inside the closed surface.} \qquad \text{Gauss's law} \tag{10.23}$$

If, instead of isolated charges, we have a charge distribution with charge density ρ (which may vary from point to point), then the total charge is $\int \rho\,d\tau$, so

$$\oint_{\substack{\text{closed}\\ \text{surface } \sigma}} \mathbf{D}\cdot\mathbf{n}\,d\sigma = \int_{\substack{\text{volume}\\ \text{bounded by } \sigma}} \rho\,d\tau. \qquad \text{Gauss's law} \tag{10.24}$$

Since (by Problem 13) charges outside the closed surface σ do not contribute to the integral, (10.23) and (10.24) are correct if $\mathbf{D}$ is the total electric displacement due to all charges inside and outside the surface. The total charge on the right-hand side of these equations is, however, just the charge inside the surface σ. Either (10.23) or (10.24) is called Gauss's law.

We now want to see the use of the divergence theorem in connection with Gauss's law. By the divergence theorem, the surface integral on the left-hand side of (10.23) or (10.24) is equal to

$$\int_{\substack{\text{volume}\\ \text{bounded by } \sigma}} \boldsymbol{\nabla}\cdot\mathbf{D}\,d\tau.$$

Then (10.24) can be written as

$$\int \nabla \cdot \mathbf{D}\, d\tau = \int \rho\, d\tau.$$

Since this is true for *every* volume, we must have $\nabla \cdot \mathbf{D} = \rho$; this is one of the Maxwell equations in electricity. What we have done is to start by assuming Coulomb's law; we have derived Gauss's law from it, and then by use of the divergence theorem, we have derived the Maxwell equation $\nabla \cdot \mathbf{D} = \rho$. From a more sophisticated viewpoint, we might take the Maxwell equation as one of our basic assumptions in electricity. We could then use the divergence theorem to obtain Gauss's law:

$$\oint_{\substack{\text{closed}\\ \text{surface } \sigma}} \mathbf{D} \cdot \mathbf{n}\, d\sigma = \int_{\substack{\text{volume } \tau\\ \text{inside } \sigma}} \nabla \cdot \mathbf{D}\, d\tau = \int_{\text{volume } \tau} \rho\, d\tau \tag{10.25}$$

$$= \text{total charge inclosed by } \sigma.$$

From Gauss's law we could then derive Coulomb's law (Problem 14); more generally we can often use Gauss's law to obtain the electric field produced by a given charge distribution as in the following example.

▶ **Example.** Find $\mathbf{E}$ just above a very large conducting plate carrying a surface charge of C coulombs per square meter on each surface.

The electric field inside a conductor is zero when we are considering an electrostatics problem (otherwise current would flow). From the symmetry of the problem (all horizontal directions are equivalent), we can say that $\mathbf{E}$ (and $\mathbf{D}$) must be vertical as shown in Figure 10.7. We now find $\oint \mathbf{D} \cdot \mathbf{n}\, d\sigma$ over the box whose cross section is shown by the dotted lines. The integral over the bottom surface is zero since $\mathbf{D} = 0$ inside the conductor. The integral over the vertical sides is zero because $\mathbf{D}$ is perpendicular to $\mathbf{n}$ there. On the top surface $\mathbf{D} \cdot \mathbf{n} = |\mathbf{D}|$ and $\int \mathbf{D} \cdot \mathbf{n}\, d\sigma = |\mathbf{D}| \cdot$ (surface area). By (10.25) this is equal to the charge inclosed by the box, which is $C\cdot$ (surface area). Thus we have $|\mathbf{D}| \cdot$ (surface area) $= C \cdot$ (surface area) , or $|\mathbf{D}| = C$ and $|\mathbf{E}| = C/\epsilon_0$.

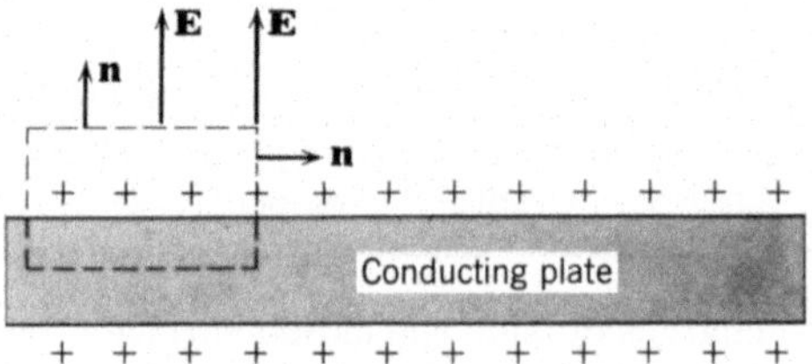

Figure 10.7

▶ PROBLEMS, SECTION 10

1. Evaluate both sides of (10.17) if $\mathbf{V} = \mathbf{r} = \mathbf{i}x + \mathbf{j}y + \mathbf{k}z$, and τ is the volume $x^2 + y^2 + z^2 \le 1$, and so verify the divergence theorem in this case.

2. Given $\mathbf{V} = x^2\mathbf{i} + y^2\mathbf{j} + z^2\mathbf{k}$, integrate $\mathbf{V} \cdot \mathbf{n}\, d\sigma$ over the whole surface of the cube of side 1 with four of its vertices at $(0, 0, 0)$, $(0, 0, 1)$, $(0, 1, 0)$, $(1, 0, 0)$. Evaluate the same integral by means of the divergence theorem.

Evaluate each of the integrals in Problems 3 to 8 as either a volume integral or a surface integral, whichever is easier.

3. $\iint \mathbf{r}\cdot\mathbf{n}\,d\sigma$ over the whole surface of the cylinder bounded by $x^2+y^2=1$, $z=0$, and $z=3$; $\mathbf{r}$ means $\mathbf{i}x+\mathbf{j}y+\mathbf{k}z$.

4. $\iint \mathbf{V}\cdot\mathbf{n}\,d\sigma$ if $\mathbf{V}=x\cos^2 y\,\mathbf{i}+xz\,\mathbf{j}+z\sin^2 y\,\mathbf{k}$ over the surface of a sphere with center at the origin and radius 3.

5. $\iiint(\boldsymbol{\nabla}\cdot\mathbf{F})\,d\tau$ over the region $x^2+y^2+z^2\le 25$, where

$$\mathbf{F}=(x^2+y^2+z^2)(x\mathbf{i}+y\mathbf{j}+z\mathbf{k}).$$

6. $\iiint \boldsymbol{\nabla}\cdot\mathbf{V}\,d\tau$ over the unit cube in the first octant, where

$$\mathbf{V}=(x^3-x^2)y\mathbf{i}+(y^3-2y^2+y)x\mathbf{j}+(z^2-1)\mathbf{k}.$$

7. $\iint \mathbf{r}\cdot\mathbf{n}\,d\sigma$ over the entire surface of the cone with base $x^2+y^2\le 16$, $z=0$, and vertex at $(0,0,3)$, where $\mathbf{r}=\mathbf{i}x+\mathbf{j}y+\mathbf{k}z$.

8. $\iiint \boldsymbol{\nabla}\cdot\mathbf{V}\,d\tau$ over the volume $x^2+y^2\le 4$, $0\le z\le 5$, $\mathbf{V}=(\sqrt{x^2+y^2})(\mathbf{i}x+\mathbf{j}y)$.

9. If $\mathbf{F}=x\mathbf{i}+y\mathbf{j}$, calculate $\iint \mathbf{F}\cdot\mathbf{n}\,d\sigma$ over the part of the surface $z=4-x^2-y^2$ that is above the (x,y) plane, by applying the divergence theorem to the volume bounded by the surface and the piece that it cuts out of the (x,y) plane. *Hint:* What is $\mathbf{F}\cdot\mathbf{n}$ *on* the (x,y) plane?

10. Evaluate $\iint \mathbf{V}\cdot\mathbf{n}\,d\sigma$ over the *curved* surface of the hemisphere $x^2+y^2+z^2=9$, $z\ge 0$, if $\mathbf{V}=y\mathbf{i}+xz\mathbf{j}+(2z-1)\mathbf{k}$. *Careful:* See Problem 9.

11. Given that $\mathbf{B}=\operatorname{curl}\mathbf{A}$, use the divergence theorem to show that $\oint \mathbf{B}\cdot\mathbf{n}\,d\sigma$ over any closed surface is zero.

12. A cylindrical capacitor consists of two long concentric metal cylinders. If there is a charge of k coulombs per meter on the inside cylinder of radius R_1, and $-k$ coulombs per meter on the outside cylinder of radius R_2, find the electric field $\mathbf{E}$ between the cylinders. *Hint:* Use Gauss's law and the method indicated in Figure 10.7. What is $\mathbf{E}$ inside the inner cylinder? Outside the outer cylinder? (Again use Gauss's law.) Find, either by inspection or by direct integration, the potential ϕ such that $\mathbf{E}=-\boldsymbol{\nabla}\phi$ for each of the three regions above. In each case $\mathbf{E}$ is not affected by adding an arbitrary constant to ϕ. Adjust the additive constant to make ϕ a continuous function for all space.

13. Draw a figure similar to Figure 10.6 but with q outside the surface. A vector (like $\mathbf{r}$ in the figure) from q to the surface now intersects it twice, and for each solid angle $d\Omega$ there are two $d\sigma$'s, one where $\mathbf{r}$ enters and one where it leaves the surface. Show that $\mathbf{D}\cdot\mathbf{n}\,d\sigma$ is given by (10.21) for the $d\sigma$ where $\mathbf{r}$ leaves the surface and the negative of (10.21) for the $d\sigma$ where $\mathbf{r}$ enters the surface. Hence show that the total $\oint \mathbf{D}\cdot\mathbf{n}\,d\sigma$ over the closed surface is zero.

14. Obtain Coulomb's law from Gauss's law by considering a spherical surface σ with center at q.

15. Suppose the density ρ of a fluid varies from point to point as well as with time, that is, $\rho=\rho(x,y,z,t)$. If we follow the fluid along a streamline, then x, y, z are functions of t such that the fluid velocity is

$$\mathbf{v}=\mathbf{i}\frac{dx}{dt}+\mathbf{j}\frac{dy}{dt}+\mathbf{k}\frac{dz}{dt}.$$

Show that then $d\rho/dt = \partial\rho/\partial t + \mathbf{v}\cdot\nabla\rho$. Combine this equation with (10.9) to get

$$\rho\nabla\cdot\mathbf{v} + \frac{d\rho}{dt} = 0.$$

(Physically, $d\rho/dt$ is the rate of change of density with time as we follow the fluid along a streamline; $\partial\rho/\partial t$ is the corresponding rate at a fixed point.) For a steady state (that is, time-independent), $\partial\rho/\partial t = 0$, but $d\rho/dt$ is not necessarily zero. For an incompressible fluid, $d\rho/dt = 0$; show that then $\nabla\cdot\mathbf{v} = 0$. (Note that incompressible does not necessarily mean constant density since $d\rho/dt = 0$ does not imply either time or space independence of ρ; consider, for example, a flow of water mixed with blobs of oil.)

16. The following equations are variously known as Green's first and second identities or formulas or theorems. Derive them, as indicated, from the divergence theorem.

$$(1)\quad \int_{\substack{\text{volume } \tau\\ \text{inside } \sigma}} (\phi\nabla^2\psi + \nabla\phi\cdot\nabla\psi)\,d\tau = \oint_{\substack{\text{closed}\\ \text{surface } \sigma}} (\phi\nabla\psi)\cdot\mathbf{n}\,d\sigma.$$

To prove this, let $\mathbf{V} = \phi\nabla\psi$ in the divergence theorem.

$$(2)\quad \int_{\substack{\text{volume } \tau\\ \text{inside } \sigma}} (\phi\nabla^2\psi - \psi\nabla^2\phi)\,d\tau = \oint_{\substack{\text{closed}\\ \text{surface } \sigma}} (\phi\nabla\psi - \psi\nabla\phi)\cdot\mathbf{n}\,d\sigma.$$

To prove this, copy Theorem 1 above as is and also with ϕ and ψ interchanged; then subtract the two equations.

11. THE CURL AND STOKES' THEOREM

We have already defined $\operatorname{curl}\mathbf{V} = \nabla\times\mathbf{V}$ [see (7.3)] and have considered one application of the curl, namely, to determine whether or not a line integral between two points is independent of the path of integration (Section 8). Here is another application of the curl. Suppose a rigid body is rotating with constant angular velocity $\boldsymbol{\omega}$; this means that $|\boldsymbol{\omega}|$ is the magnitude of the angular velocity and $\boldsymbol{\omega}$ is a vector along the axis of rotation (see Figure 2.6). Then we showed in Section 2 that the velocity $\mathbf{v}$ of a particle in the rigid body is $\mathbf{v} = \boldsymbol{\omega}\times\mathbf{r}$, where $\mathbf{r}$ is a radius vector from a point on the rotation axis to the particle. Let us calculate $\nabla\times\mathbf{v} = \nabla\times(\boldsymbol{\omega}\times\mathbf{r})$; we can evaluate this by the method described in Section 7. We use the formula for the triple vector product $\mathbf{A}\times(\mathbf{B}\times\mathbf{C}) = (\mathbf{A}\cdot\mathbf{C})\mathbf{B} - (\mathbf{A}\cdot\mathbf{B})\mathbf{C}$, being careful to remember that ∇ is not an ordinary vector—it has both vector and differential-operator properties, and so must be written before variables that it differentiates. Then

$$\nabla\times(\boldsymbol{\omega}\times\mathbf{r}) = (\nabla\cdot\mathbf{r})\boldsymbol{\omega} - (\boldsymbol{\omega}\cdot\nabla)\mathbf{r}. \tag{11.1}$$

Since $\boldsymbol{\omega}$ is constant, the first term of (11.1) means

$$\boldsymbol{\omega}(\nabla\cdot\mathbf{r}) = \boldsymbol{\omega}\left(\frac{\partial x}{\partial x} + \frac{\partial y}{\partial y} + \frac{\partial z}{\partial z}\right) = 3\boldsymbol{\omega}. \tag{11.2}$$

In the second term of (11.1) we intentionally wrote $\boldsymbol{\omega}\cdot\nabla$ instead of $\nabla\cdot\boldsymbol{\omega}$ since $\boldsymbol{\omega}$ is constant, and ∇ operates only on $\mathbf{r}$; this term means

$$\left(\omega_x\frac{\partial}{\partial x} + \omega_y\frac{\partial}{\partial y} + \omega_z\frac{\partial}{\partial z}\right)(\mathbf{i}x + \mathbf{j}y + \mathbf{k}z) = \mathbf{i}\omega_x + \mathbf{j}\omega_y + \mathbf{k}\omega_z = \boldsymbol{\omega}$$

since $\partial y/\partial x = \partial z/\partial x = 0$, etc. Then

$$(11.3) \qquad \nabla \times \mathbf{v} = \nabla \times (\boldsymbol{\omega} \times \mathbf{r}) = 2\boldsymbol{\omega} \qquad \text{or} \qquad \boldsymbol{\omega} = \frac{1}{2}(\nabla \times \mathbf{v}).$$

This result gives a clue as to the name curl $\mathbf{v}$ (or rotation $\mathbf{v}$ or rot $\mathbf{v}$ as it is sometimes called). For this simple case curl $\mathbf{v}$ gave the angular velocity of rotation. In a more complicated case such as flow of fluid, the value of curl $\mathbf{v}$ at a point is a measure of the angular velocity of the fluid in the neighborhood of the point. When $\nabla \times \mathbf{v} = 0$ everywhere in some region, the velocity field $\mathbf{v}$ is called *irrotational* in that region. Notice that this is the same mathematical condition as for a force $\mathbf{F}$ to be *conservative.*

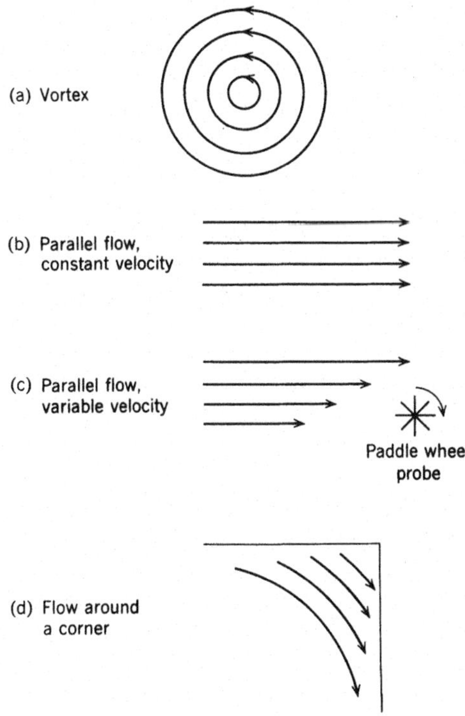

Figure 11.1

Consider a vector field $\mathbf{V}$ (for example, $\mathbf{V} = \mathbf{v}\rho$ for flow of water, or $\mathbf{V} =$ force $\mathbf{F}$). We define the *circulation* as the line integral $\oint \mathbf{V} \cdot d\mathbf{r}$ around a closed plane curve. If $\mathbf{V}$ is a force $\mathbf{F}$, then this integral is equal to the work done by the force. For flow of water, we can get a physical picture of the meaning of the circulation in the following way. Think of placing a tiny paddle-wheel probe (Figure 11.1c) in any of the flow patterns pictured in Figure 11.1. If the velocity of the fluid is greater on one side of the wheel than on the other, for example, as in (c), then the wheel will turn. Suppose we calculate the circulation $\oint \mathbf{V} \cdot d\mathbf{r}$ around the axis of the paddle wheel along a closed curve in a plane perpendicular to the axis (plane of the paper in Figure 11.1). If $\mathbf{V} = \mathbf{v}\rho$ is larger on one side of the wheel than the other, then the circulation is different from zero, but if [as in (b)] $\mathbf{V}$ is the same on both sides, then the circulation is zero. We shall show that the component of curl $\mathbf{V}$ along the axis of the paddle wheel equals

$$\lim_{d\sigma\to 0}\frac{1}{d\sigma}\oint \mathbf{V}\cdot d\mathbf{r} \tag{11.4}$$

where $d\sigma$ is the area inclosed by the curve along which we calculate the circulation. The paddle wheel then acts as a "curl meter" to measure $\operatorname{curl}\mathbf{V}$; if it does not rotate, $\operatorname{curl}\mathbf{V}=0$; if it does, then $\operatorname{curl}\mathbf{V}\neq 0$. In (a), $\operatorname{curl}\mathbf{V}\neq 0$ at the center of the vortex. In (b), $\operatorname{curl}\mathbf{V}=0$. In (c), $\operatorname{curl}\mathbf{V}\neq 0$ in spite of the fact that the flow lines are parallel. In (d), it is possible to have $\operatorname{curl}\mathbf{V}=0$ even though the stream lines go around a corner; in fact, for the flow of water around a corner, $\operatorname{curl}\mathbf{V}=0$. What you should realize is that the value of $\operatorname{curl}\mathbf{V}$ at a point depends upon the circulation in the neighborhood of the point and not on the overall flow pattern.

We want to show the relation between the circulation $\oint \mathbf{V}\cdot d\mathbf{r}$ and $\operatorname{curl}\mathbf{V}$ for a given vector field $\mathbf{V}$. Given a point P and a direction $\mathbf{n}$, let us find the component of $\operatorname{curl}\mathbf{V}$ in the direction $\mathbf{n}$ at P. Draw a plane through P perpendicular to $\mathbf{n}$ and choose axes so that it is the (x,y) plane with $\mathbf{n}$ parallel to $\mathbf{k}$. Find the circulation around an element of area $d\sigma$ centered on P. (See Figures 9.5 and 11.2.) By (9.18) with area A replaced by the *element of area* $d\sigma$, and with $\mathbf{n}=\mathbf{k}$

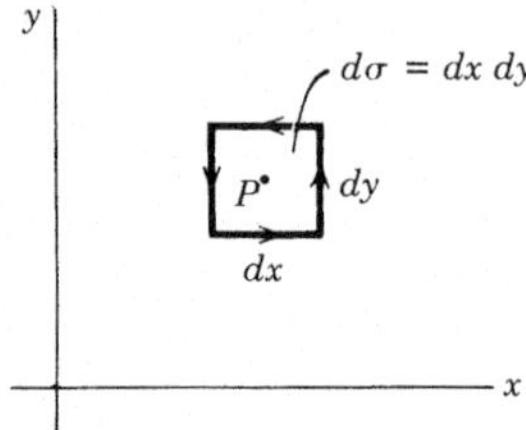

Figure 11.2

$$\oint_{\text{around } d\sigma} \mathbf{V}\cdot d\mathbf{r} = \iint_{d\sigma} (\operatorname{curl}\mathbf{V})\cdot\mathbf{k}\,dx\,dy = \iint_{d\sigma} (\operatorname{curl}\mathbf{V})\cdot\mathbf{n}\,d\sigma \tag{11.5}$$

Note that, since we proved (9.7) and so (9.18) for non-rectangular areas A (see Section 9), $d\sigma$ here may be more general than $dx\,dy$, say with curved or slanted sides.

We assume that the components of $\mathbf{V}$ have continuous first derivatives; then $\operatorname{curl}\mathbf{V}$ is continuous. Thus the value of $(\operatorname{curl}\mathbf{V})\cdot\mathbf{n}$ over $d\sigma$ is nearly the same as $(\operatorname{curl}\mathbf{V})\cdot\mathbf{n}$ at P, so the double integral in (11.5) is approximately the value of $(\operatorname{curl}\mathbf{V})\cdot\mathbf{n}$ at P multiplied by $d\sigma$. If we divide (11.5) by $d\sigma$ and take the limit as $d\sigma\to 0$, we have an exact equation

$$(\nabla\times\mathbf{V})\cdot\mathbf{n} = \lim_{d\sigma\to 0}\frac{1}{d\sigma}\oint_{\text{around } d\sigma} \mathbf{V}\cdot d\mathbf{r}. \tag{11.6}$$

This equation can be used as a definition of $\operatorname{curl}\mathbf{V}$; then the discussion above shows that [see equation (9.16)] the components of $\operatorname{curl}\mathbf{V}$ are those given in our previous definition (7.3).

In evaluating the line integral we must go around the area element $d\sigma$ as in Figure 11.2 keeping the area to our left. Another way of saying this is that we go around $d\sigma$ in the direction indicated by $\mathbf{n}$ and the right-hand rule; that is, if the thumb of your right hand points in the direction $\mathbf{n}$, your fingers curve in the direction you must go around the boundary of $d\sigma$ in evaluating the line integral. (See Figure 11.2 with $\mathbf{n}=\mathbf{k}$.)

Stokes' Theorem This theorem relates an integral over an open surface to the line integral around the curve bounding the surface (Figure 11.3). A butterfly net is a good example of what we are talking about; the net is the surface and the supporting rim is the curve bounding the surface. The surfaces we consider here (and which arise in applications) will be surfaces which could be obtained by deforming a hemisphere (or the butterfly net of Figure 11.3). In particular, the surfaces we consider must be *two*-sided. You can easily construct a *one*-sided surface by taking a long strip of paper, giving it a half twist, and joining the ends (Figure 11.4). A belt of this shape is sometimes used for driving machinery. This surface is called a Moebius strip, and you can verify that it has only one side by tracing your finger around it or imagining trying to paint one side. Stokes' theorem does not apply to such surfaces because we cannot define the sense of the normal vector $\mathbf{n}$ to such a surface. We require the bounding curve to be simple (that is, it must not cross itself) and closed.

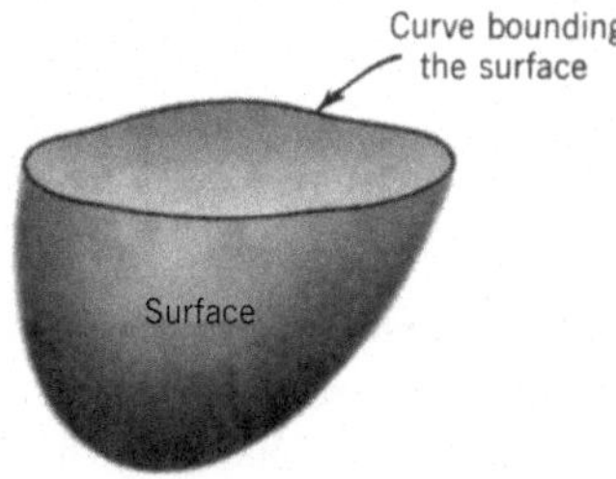

Figure 11.3

Figure 11.4

Consider the kind of surface we have described and imagine it divided into area elements $d\sigma$ by a network of curves as in Figure 11.5. Draw a unit vector $\mathbf{n}$ perpendicular to each area element; $\mathbf{n}$, of course, varies from element to element, but all $\mathbf{n}$'s must be on the same side of the two-sided surface. Each area element

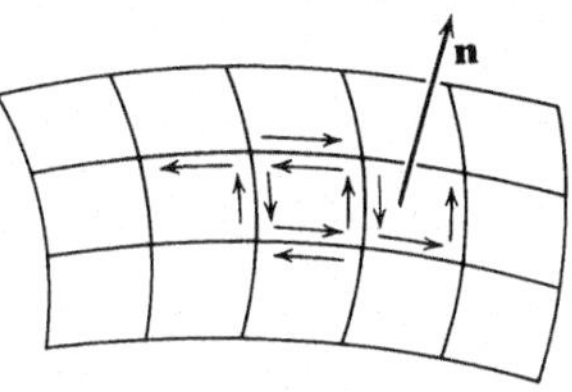

Figure 11.5

is approximately an element of the tangent plane to the surface at a point in $d\sigma$. Then, as in (11.5), we have

$$\oint_{\text{around } d\sigma} \mathbf{V} \cdot d\mathbf{r} = \iint_{d\sigma} (\boldsymbol{\nabla} \times \mathbf{V}) \cdot \mathbf{n}\, d\sigma \tag{11.7}$$

for each element. Recall from Section 9 and the comment just after equation (11.5), that $d\sigma$ includes area elements such as those along the edges in Figure 11.5. Then if we sum the equations in (11.7) for all the area elements of the whole surface area, we get

$$\sum_{\text{all } d\sigma} \oint \mathbf{V} \cdot d\mathbf{r} = \iint_{\text{surface } \sigma} (\nabla \times \mathbf{V}) \cdot \mathbf{n}\, d\sigma. \tag{11.8}$$

From Figure 11.5 we see that all the interior line integrals cancel because along a border between two $d\sigma$'s the two integrals are in opposite directions. Then the left side of (11.8) becomes simply the line integral around the outside curve bounding the surface. Thus we have Stokes' theorem as stated in (9.19):

$$\oint_{\substack{\text{curve} \\ \text{bounding } \sigma}} \mathbf{V} \cdot d\mathbf{r} = \iint_{\text{surface } \sigma} (\nabla \times \mathbf{V}) \cdot \mathbf{n}\, d\sigma. \qquad \text{Stokes' theorem} \tag{11.9}$$

You should have it clearly in mind that this is for an open surface bounded by a simple closed curve. Recall the example of a butterfly net. Notice that Stokes' theorem says that the line integral $\oint \mathbf{V} \cdot d\mathbf{r}$ is equal to the surface integral of $(\nabla \times \mathbf{V}) \cdot \mathbf{n}$ over *any* surface of which the curve is a boundary; in other words, you don't change the value of the integral by deforming the butterfly net! An easy way to determine the direction of integration for the line integral is to imagine collapsing the surface and its bounding curve into a plane; then the "surface" is just the plane area inside the curve and $\mathbf{n}$ is normal to the plane. The direction of integration is then given by the right-hand rule as discussed just after equation (11.6).

▶ **Example 1.** Given $\mathbf{V} = 4y\mathbf{i} + x\mathbf{j} + 2z\mathbf{k}$, find $\int (\nabla \times \mathbf{V}) \cdot \mathbf{n}\, d\sigma$ over the hemisphere $x^2 + y^2 + z^2 = a^2$, $z \geq 0$.

Using (7.3), we find that $\nabla \times \mathbf{V} = -3\mathbf{k}$. There are several ways we could do the problem: (a) integrate the expression as it stands; (b) use Stokes' theorem and evaluate $\oint \mathbf{V} \cdot d\mathbf{r}$ around the circle $x^2 + y^2 = a^2$ in the (x, y) plane; (c) use Stokes' theorem to say that the integral is the same over *any* surface bounded by this circle, for example, the plane area inside the circle! Since this plane area is in the (x, y) plane, we have

$$\mathbf{n} = \mathbf{k}, \qquad (\nabla \times \mathbf{V}) \cdot \mathbf{n} = -3\mathbf{k} \cdot \mathbf{k} = -3,$$

so the integral is

$$-3 \int d\sigma = -3 \cdot \pi a^2 = -3\pi a^2.$$

This is the easiest way to do the problem; however, for this simple case it is not too hard by the other methods. We shall leave (b) for you to do and do (a). Since the surface is a sphere with center at the origin, $\mathbf{r}$ is normal to it (but for any surface we could get the normal from the gradient). Then on the surface

$$\mathbf{n} = \frac{\mathbf{r}}{|\mathbf{r}|} = \frac{\mathbf{r}}{a} = \frac{\mathbf{i}x + \mathbf{j}y + \mathbf{k}z}{a},$$

$$(\nabla \times \mathbf{V}) \cdot \mathbf{n} = -3\mathbf{k} \cdot \frac{\mathbf{r}}{a} = -3\frac{z}{a}.$$

We want to evaluate $\int -3(z/a)\,d\sigma$ over the hemisphere. In spherical coordinates (see Chapter 5, Section 4) we have

$$z = r\cos\theta,$$
$$d\sigma = r^2 \sin\theta\, d\theta\, d\phi.$$

For our surface $r = a$. Then the integral is

$$\int_{\phi=0}^{2\pi}\int_{\theta=0}^{\pi/2} -3\frac{a\cos\theta}{a}a^2\sin\theta\,d\theta\,d\phi = -3a^2\int_0^{2\pi} d\phi\int_0^{\pi/2}\sin\theta\cos\theta\,d\theta$$
$$= -3a^2\cdot 2\pi\cdot\frac{1}{2} = -3\pi a^2$$

(as before).

Ampère's Law Stokes' theorem is of interest in electromagnetic theory. (Compare the use of the divergence theorem in connection with Gauss's law in Section 10.) Ampère's circuital law (in SI units) says that

$$\oint_C \mathbf{H}\cdot d\mathbf{r} = I,$$

where $\mathbf{H} = \mathbf{B}/\mu_0$, $\mathbf{B}$ is the magnetic field, μ_0 is a constant (called the *permeability of free space*), C is a closed curve, and I is the current "linking" C, that is crossing any surface area bounded by C. The surface area and the curve C are related just as in Stokes' theorem (butterfly net and its rim). If we think of a bundle of wires linking a closed curve C (Figure 11.6) and then spreading out, we can see that the same current crosses any surface whose bounding curve is C.

Just as Gauss's law (10.23) is useful in computing electric fields, so Ampère's law is useful in computing magnetic fields. Consider, for example, a long straight wire carrying a current I (Figure 11.7). At a distance r from the wire, $\mathbf{H}$ is tangent

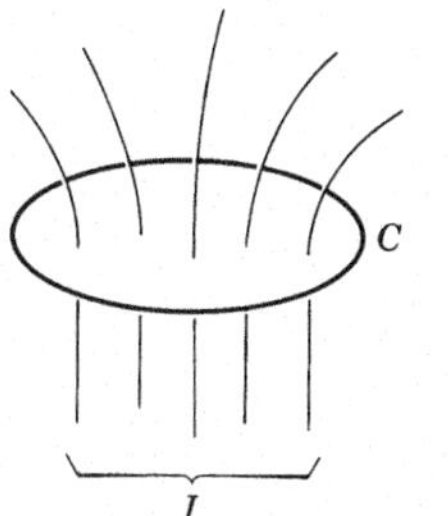

Figure 11.6

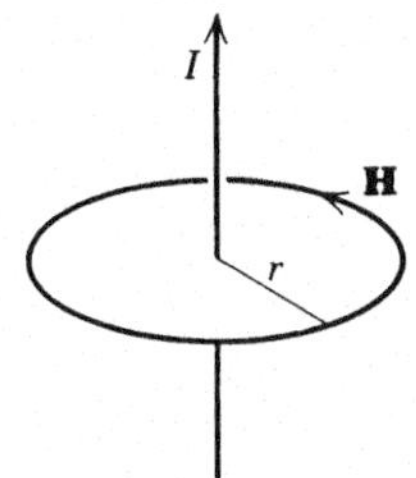

Figure 11.7

to a circle of radius r in a plane perpendicular to the wire. By symmetry, $|\mathbf{H}|$ same at all points of the circle. We can then find $|\mathbf{H}|$ by Ampère's law. Taking C to be the circle of radius r, we have

$$\oint_C \mathbf{H}\cdot d\mathbf{r} = \int_0^{2\pi} |\mathbf{H}|r\,d\theta = |\mathbf{H}|r\cdot 2\pi = I$$

or

$$|\mathbf{H}| = \frac{I}{2\pi r}.$$

If, in Figure 11.6, $\mathbf{J}$ is the current density (current crossing unit area perpendicular to $\mathbf{J}$), then $\mathbf{J}\cdot\mathbf{n}\,d\sigma$ is the current across a surface element $d\sigma$ [compare (10.4)] and $\iint_\sigma \mathbf{J}\cdot\mathbf{n}\,d\sigma$, over any surface σ bounded by C, is the total current I linking C. Then by Ampère's law

$$\oint_C \mathbf{H}\cdot d\mathbf{r} = \iint_\sigma \mathbf{J}\cdot\mathbf{n}\,d\sigma.$$

By Stokes' theorem

$$\oint_C \mathbf{H}\cdot d\mathbf{r} = \iint_\sigma (\nabla\times\mathbf{H})\cdot\mathbf{n}\,d\sigma,$$

so we have

$$\iint_\sigma (\nabla\times\mathbf{H})\cdot\mathbf{n}\,d\sigma = \iint_\sigma \mathbf{J}\cdot\mathbf{n}\,d\sigma.$$

Since this is true for any σ, we have $\nabla\times\mathbf{H}=\mathbf{J}$, which is one of the Maxwell equations. Alternatively, we could start with the Maxwell equation and apply Stokes' theorem to get Ampère's law.

Conservative Fields We next want to state carefully, and use Stokes' theorem to prove, under what conditions a given field $\mathbf{F}$ is conservative (see Section 8). First, recall that in physical problems we are often interested only in a particular region of space, and our formulas (say for $\mathbf{F}$) may very well be correct *only* in that region. For example, the gravitational pull of the earth on an object is proportional to $1/r^2$ for $r \geq$ earths' radius R, but this is not a correct formula for $r < R$ (see Problem 8.21). The electric field in the region between the plates of a cylindrical capacitor is proportional to $1/r$ (problem 10.12), but only *in this region* is this formula correct. We must, then, consider the *kind of region* in which a given field $\mathbf{F}$ is defined. Consider the shaded regions in Figure 11.8. We say that a region is *simply connected* if any simple† closed curve in the region can be shrunk to a point without encountering any points not in the region. You can see in Figure 11.8c that the dotted curve surrounds the "hole" and so cannot be shrunk to a point in the region; this region is then not simply connected. The "hole" is sometimes only a single point, but this is enough to make the region *not* simply connected. In three dimensions the region between cylindrical capacitor plates (infinitely long) is not simply connected since a loop of string around the inner cylinder (see cross section, Figure 11.8c) cannot be drawn up to a knot. Similarly, the interior of an inner tube is not simply connected. The region between two concentric spheres *is* simply connected, however. You should see this by realizing that you could pull up into a

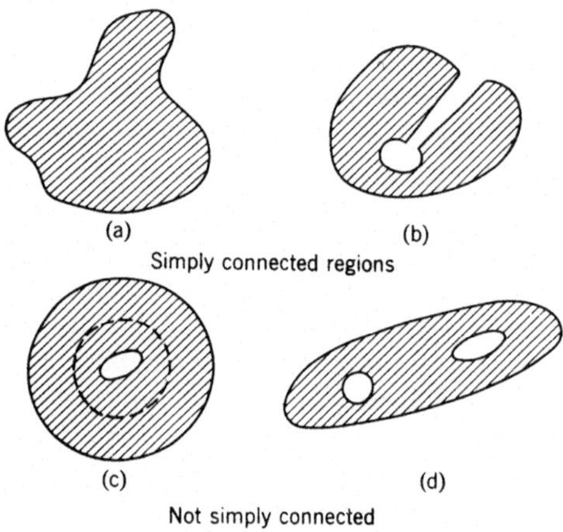

Figure 11.8

†A simple curve does not cross itself; for example, a figure eight is not a simple curve.

knot, a loop of string placed anywhere in this region. We shall now state and prove our theorem.

(11.10)

If the components of **F** and their first partial derivatives are continuous in a simply connected region, then any one of the following five conditions implies all the others.

(a) $\text{curl}\,\mathbf{F} = 0$ at every point of the region.

(b) $\oint \mathbf{F} \cdot d\mathbf{r} = 0$ around every simple closed curve in the region.

(c) **F** is conservative, that is $\int_A^B \mathbf{F} \cdot d\mathbf{r}$ is independent of the path of integration from A to B. (The path must, of course, lie entirely in the region.)

(d) $\mathbf{F} \cdot d\mathbf{r}$ is an exact differential of a single valued function.

(e) $\mathbf{F} = \text{grad}\, W$, W single-valued.

We shall show that each of these conditions implies the one following it. We can use Stokes' theorem to prove (b) assuming (a). First select any simple closed curve and let it be the bounding curve for the surface in Stokes' theorem. Since the region is simply connected we can think of shrinking the curve to a point in the region; as it shrinks it traces out a surface which we use as the Stokes' theorem surface. Assuming (a), we have $\text{curl}\,\mathbf{F} = 0$ at every point of the region and so also at every point of the surface. Thus the surface integral in Stokes' theorem is zero and therefore the line integral around the closed curve equals zero. This gives (b).

To show that (b) implies (c), consider any two paths I and II from A to B (Figure 11.9). From (b) we have

$$\int_{\substack{A \\ \text{path I}}}^{B} \mathbf{F} \cdot d\mathbf{r} + \int_{\substack{B \\ \text{path II}}}^{A} \mathbf{F} \cdot d\mathbf{r} = 0.$$

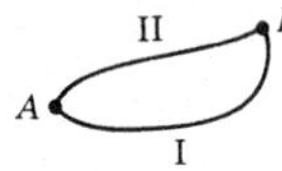

Figure 11.9

Since an integral from A to B is the negative of an integral from B to A, we have

$$\int_{\substack{A \\ \text{path I}}}^{B} \mathbf{F} \cdot d\mathbf{r} - \int_{\substack{A \\ \text{path II}}}^{B} \mathbf{F} \cdot d\mathbf{r} = 0.$$

which is (c).

To show that (c) implies (d), select some reference point O in the region and calculate $\int \mathbf{F} \cdot d\mathbf{r}$ from the reference point to every other point of the region. For each point P we find a single value of the integral no matter what path of integration we choose from O to P. Let this value be the value of the function W at the point P. We then have a single-valued function W such that

$$\int_{0 \text{ to } P} \mathbf{F} \cdot d\mathbf{r} = W(P).$$

Then (since $\mathbf{F}$ is continuous), $dW = \mathbf{F}\cdot d\mathbf{r}$, that is, $\mathbf{F}\cdot d\mathbf{r}$ is the differential of a single-valued function W. Since $dW = \nabla W\cdot d\mathbf{r} = \mathbf{F}\cdot d\mathbf{r}$ for arbitrary $d\mathbf{r}$, we have $\mathbf{F} = \nabla W$ which is (e).

Finally, (e) implies (a) as we proved in Section 8. (The continuity of the components of $\mathbf{F}$ and their partial derivatives makes the second-order mixed partial derivatives of W equal.) Thus we have shown that any one of the five conditions (a) to (e) implies the others under the conditions of the theorem. It is worth observing carefully the requirement that $\mathbf{F}$ and its partial derivatives must be continuous in a simply connected region. A simple example makes this clear. Look at Example 2 in Section 8; you can easily compute $\operatorname{curl}\mathbf{F}$ and find that it is zero everywhere except at the origin (where it is undefined). You might then be tempted to assume that $\oint \mathbf{F}\cdot d\mathbf{r} = 0$ around any closed path. But we found that $\mathbf{F}\cdot d\mathbf{r} = d\theta$, and the integral of $d\theta$ along a circle with center at the origin is 2π. What is wrong? The trouble is that $\mathbf{F}$ does not have continuous partial derivatives at the origin, and any simply connected region containing the circle of integration must contain the origin. Then $\operatorname{curl}\mathbf{F}$ is not zero at *every point* inside the integration curve. Notice also that $\mathbf{F}\cdot d\mathbf{r} = d\theta$ is an exact differential, but not of a single-valued function; θ increases by 2π every time we go around the origin.

A vector field $\mathbf{V}$ is called *irrotational* (or *conservative* or *lamellar*) if $\operatorname{curl}\mathbf{V} = 0$; in this case $\mathbf{V} = \operatorname{grad} W$, where W (or its negative) is called the *scalar potential.* If $\operatorname{div}\mathbf{V} = 0$, the vector field is called *solenoidal*; in this case $\mathbf{V} = \operatorname{curl}\mathbf{A}$, where A is a vector function called the *vector potential.* It is easy to prove (Problem 7.17d) that if $\mathbf{V} = \nabla\times\mathbf{A}$, then $\operatorname{div}\mathbf{V} = 0$. It is also possible to construct an $\mathbf{A}$ (actually an infinite number of $\mathbf{A}$'s) so that $\mathbf{V} = \operatorname{curl}\mathbf{A}$ if we know that $\nabla\cdot\mathbf{V} = 0$.

Example 2. Given $\mathbf{V} = \mathbf{i}(x^2 - yz) - \mathbf{j}2yz + \mathbf{k}(z^2 - 2zx)$, find $\mathbf{A}$ such that $\mathbf{V} = \nabla\times\mathbf{A}$. We find

$$\operatorname{div}\mathbf{V} = \frac{\partial}{\partial x}(x^2 - yz) + \frac{\partial}{\partial y}(-2yz) + \frac{\partial}{\partial z}(z^2 - 2zx)$$
$$= 2x - 2z + 2z - 2x = 0.$$

Thus $\mathbf{V}$ is solenoidal and we proceed to find $\mathbf{A}$. We are looking for an $\mathbf{A}$ such that

$$\mathbf{V} = \operatorname{curl}\mathbf{A} = \begin{vmatrix} \mathbf{i} & \mathbf{j} & \mathbf{k} \\ \dfrac{\partial}{\partial x} & \dfrac{\partial}{\partial y} & \dfrac{\partial}{\partial z} \\ A_x & A_y & A_z \end{vmatrix} = \mathbf{i}(x^2 - yz) - \mathbf{j}2yz + \mathbf{k}(z^2 - 2zx). \tag{11.11}$$

There are many $\mathbf{A}$'s satisfying this equation; we shall show first how to find one of them and then a general formula for all. It is possible to find an $\mathbf{A}$ with one zero component; let us take $A_x = 0$. Then the y and z components of $\operatorname{curl}\mathbf{A}$ each involve just one component of $\mathbf{A}$. From (11.11), the y and z components of $\operatorname{curl}\mathbf{A}$ are

$$-2yz = -\frac{\partial A_z}{\partial x}, \qquad z^2 - 2zx = \frac{\partial A_y}{\partial x}. \tag{11.12}$$

If we integrate (11.12) partially with respect to x (that is, with y and z constant), we find A_y and A_z except for possible functions of y and z which could be added without changing (11.12):

$$A_y = z^2x - zx^2 + f_1(y,z), \quad A_z = 2xyz + f_2(y,z). \tag{11.13}$$

Substituting (11.13) into the x component of (11.11), we get

$$x^2 - yz = \frac{\partial A_z}{\partial y} - \frac{\partial A_y}{\partial z} = 2xz + \frac{\partial f_2}{\partial y} - 2zx + x^2 - \frac{\partial f_1}{\partial z}. \tag{11.14}$$

We now select f_1 and f_2 to satisfy (11.14). There is much leeway here and this can easily be done by inspection. We could take $f_2 = 0$, $f_1 = \frac{1}{2}yz^2$, or $f_1 = 0$, $f_2 = -\frac{1}{2}y^2z$, and so forth. Using the second choice, we have

$$\mathbf{A} = \mathbf{j}(z^2x - zx^2) + \mathbf{k}(2xyz - \frac{1}{2}y^2z). \tag{11.15}$$

You may wonder why this process works and what $\operatorname{div}\mathbf{V} = 0$ has to do with it. We can answer both these questions by following the above process with a general $\mathbf{V}$ rather than a special example. Given that $\operatorname{div}\mathbf{V} = 0$, we want an $\mathbf{A}$ such that $\mathbf{V} = \operatorname{curl}\mathbf{A}$. We try to find one with $A_x = 0$. Then the y and z components of $\mathbf{V} = \operatorname{curl}\mathbf{A}$ are

$$V_y = -\frac{\partial A_z}{\partial x}, \qquad V_z = \frac{\partial A_y}{\partial x}. \tag{11.16}$$

Then we have

$$A_y = \int V_z\,dx + f(y,z), \qquad A_z = -\int V_y\,dx + g(y,z). \tag{11.17}$$

The x component of $\mathbf{V} = \operatorname{curl}\mathbf{A}$ is

$$V_x = \frac{\partial A_z}{\partial y} - \frac{\partial A_y}{\partial z} = -\int\left(\frac{\partial V_y}{\partial y} + \frac{\partial V_z}{\partial z}\right)dx + h(y,z). \tag{11.18}$$

Since $\operatorname{div}\mathbf{V} = 0$, we can put

$$-\left(\frac{\partial V_y}{\partial y} + \frac{\partial V_z}{\partial z}\right) = \frac{\partial V_x}{\partial x} \tag{11.19}$$

into (11.18), getting

$$V_x = \int \frac{\partial V_x}{\partial x}dx + h(y,z).$$

This is correct with proper choice of $h(y,z)$.

When we know one $\mathbf{A}$, for which a given $\mathbf{V}$ is equal to $\operatorname{curl}\mathbf{A}$, all others are of the form

$$\mathbf{A} + \nabla u, \tag{11.20}$$

where u is any scalar function. For (see Problem 7.17b), $\nabla \times \nabla u = 0$, so the addition of ∇u to $\mathbf{A}$ does not affect $\mathbf{V}$. Also we can show that all possible $\mathbf{A}$'s are of the form (11.20). For if $\mathbf{V} = \operatorname{curl}\mathbf{A}_1$ and $\mathbf{V} = \operatorname{curl}\mathbf{A}_2$, then $\operatorname{curl}(\mathbf{A}_1 - \mathbf{A}_2) = 0$, so $\mathbf{A}_1 - \mathbf{A}_2$ is the gradient of some scalar function.

A careful statement and proof that $\operatorname{div}\mathbf{V} = 0$ is a necessary and sufficient condition for $\mathbf{V} = \operatorname{curl}\mathbf{A}$ requires that $\mathbf{V}$ have continuous partial derivatives at every point of a region which is simply connected in the sense that every closed surface (rather than closed curve) can be shrunk to a point in the region (for example, the region between two concentric spheres is not simply connected in this sense).

PROBLEMS, SECTION 11

1. Do case (b) of Example 1 above.

2. Given the vector $\mathbf{A} = (x^2 - y^2)\mathbf{i} + 2xy\mathbf{j}$.

(a) Find $\nabla \times \mathbf{A}$.

(b) Evaluate $\iint (\nabla \times \mathbf{A}) \cdot d\boldsymbol{\sigma}$ over a rectangle in the (x, y) plane bounded by the lines $x = 0$, $x = a$, $y = 0$, $y = b$.

(c) Evaluate $\oint \mathbf{A} \cdot d\mathbf{r}$ around the boundary of the rectangle and thus verify Stokes' theorem for this case.

Use either Stokes' theorem or the divergence theorem to evaluate each of the following integrals in the easiest possible way.

3. $\iint_{\text{surface } \sigma} \operatorname{curl}(x^2\mathbf{i}+z^2\mathbf{j}-y^2\mathbf{k})\cdot\mathbf{n}\,d\sigma$, where σ is the part of the surface $z = 4-x^2-y^2$ above the (x, y) plane.

4. $\iint \operatorname{curl}(y\mathbf{i} + 2\mathbf{j}) \cdot \mathbf{n}\,d\sigma$, where σ is the surface in the first octant made up of part of the plane $2x + 3y + 4z = 12$, and triangles in the (x, z) and (y, z) planes, as indicated in the figure.

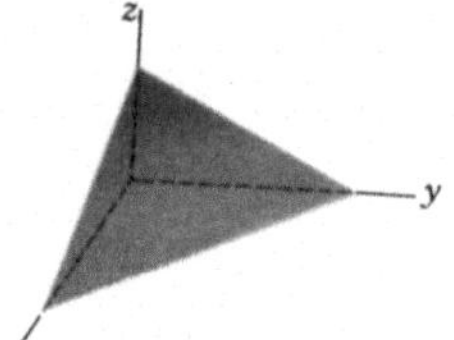

5. $\iint \mathbf{r} \cdot \mathbf{n}\,d\sigma$ over the surface in Problem 4, where $\mathbf{r} = \mathbf{i}x + \mathbf{j}y + \mathbf{k}z$. *Hint:* See Problem 10.9.

6. $\iint \mathbf{V} \cdot \mathbf{n}\,d\sigma$ over the closed surface of the tin can bounded by $x^2 + y^2 = 9$, $z = 0$, $z = 5$, if

$$\mathbf{V} = 2xy\mathbf{i} - y^2\mathbf{j} + (z + xy)\mathbf{k}.$$

7. $\iint (\operatorname{curl}\mathbf{V})\cdot\mathbf{n}\,d\sigma$ over any surface whose bounding curve is in the (x, y) plane, where

$$\mathbf{V} = (x - x^2z)\mathbf{i} + (yz^3 - y^2)\mathbf{j} + (x^2y - xz)\mathbf{k}.$$

8. $\iint \operatorname{curl}(x^2y\mathbf{i} - xz\mathbf{k}) \cdot \mathbf{n}\,d\sigma$ over the *closed* surface of the ellipsoid

$$\frac{x^2}{4} + \frac{y^2}{9} + \frac{z^2}{16} = 1.$$

Warning: Stokes' theorem applies only to an open surface. *Hints:* Could you cut the given surface into two halves? Also see (d) in the table of vector identities (page 339).

9. $\iint \mathbf{V} \cdot \mathbf{n}\,d\sigma$ over the entire surface of the volume in the first octant bounded by $x^2 + y^2 + z^2 = 16$ and the coordinate planes, where

$$\mathbf{V} = (x + x^2 - y^2)\mathbf{i} + (2xyz - 2xy)\mathbf{j} - xz^2\mathbf{k}.$$

10. $\iint (\operatorname{curl}\mathbf{V})\cdot\mathbf{n}\,d\sigma$ over the part of the surface $z = 9 - x^2 - 9y^2$ above the (x, y) plane, if $\mathbf{V} = 2xy\mathbf{i} + (x^2 - 2x)\mathbf{j} - x^2z^2\mathbf{k}$.

11. $\iint \mathbf{V} \cdot \mathbf{n}\, d\sigma$ over the entire surface of a cube in the first octant with edges of length 2 along the coordinate axes, where

$$\mathbf{V} = (x^2 - y^2)\mathbf{i} + 3y\mathbf{j} - 2xz\mathbf{k}.$$

12. $\oint \mathbf{V} \cdot d\mathbf{r}$ around the circle $(x-2)^2 + (y-3)^2 = 9$, $z = 0$, where

$$\mathbf{V} = (x^2 + yz^2)\mathbf{i} + (2x - y^3)\mathbf{j}.$$

13. $\iint (2x\mathbf{i} - 2y\mathbf{j} + 5\mathbf{k}) \cdot \mathbf{n}\, d\sigma$ over the surface of a sphere of radius 2 and center at the origin.

14. $\oint (y\mathbf{i} - x\mathbf{j} + z\mathbf{k}) \cdot d\mathbf{r}$ around the circumference of the circle of radius 2, center at the origin, in the (x, y) plane.

15. $\oint_C y\,dx + z\,dy + x\,dz$, where C is the curve of intersection of the surfaces whose equations are $x + y = 2$ and $x^2 + y^2 + z^2 = 2(x + y)$.

16. What is wrong with the following "proof" that there are no magnetic fields? By electromagnetic theory, $\nabla \cdot \mathbf{B} = 0$, and $\mathbf{B} = \nabla \times \mathbf{A}$. (The error is *not* in these equations.) Using them, we find

$$\iiint \nabla \cdot \mathbf{B}\, d\tau = 0 = \iint \mathbf{B} \cdot \mathbf{n}\, d\sigma \qquad \text{(by the divergence theorem)}$$

$$= \iint (\nabla \times \mathbf{A}) \cdot \mathbf{n}\, d\sigma = \int \mathbf{A} \cdot d\mathbf{r} \qquad \text{(by Stokes' theorem)}.$$

Since $\int \mathbf{A} \cdot d\mathbf{r} = 0$, $\mathbf{A}$ is conservative, or $\mathbf{A} = \nabla\psi$. Then $\mathbf{B} = \nabla \times \mathbf{A} = \nabla \times \nabla\psi = 0$, so $\mathbf{B} = 0$.

17. Derive the following vector integral theorems.

(a) $$\int_{\text{volume } \tau} \nabla\phi\, d\tau = \oint_{\substack{\text{surface} \\ \text{inclosing } \tau}} \phi\mathbf{n}\, d\sigma.$$

Hint: In the divergence theorem (10.17), substitute $\mathbf{V} = \phi\mathbf{C}$, where $\mathbf{C}$ is an arbitrary constant vector, to obtain $\mathbf{C} \cdot \int \nabla\phi\, d\tau = \mathbf{C} \cdot \oint \phi\mathbf{n}\, d\sigma$. Since $\mathbf{C}$ is arbitrary, let $\mathbf{C} = \mathbf{i}$ to show that the x components of the two integrals are equal; similarly, let $\mathbf{C} = \mathbf{j}$ and $\mathbf{C} = \mathbf{k}$ to show that the y components are equal and the z components are equal.

(b) $$\int_{\text{volume } \tau} \nabla \times \mathbf{V}\, d\tau = \oint_{\substack{\text{surface} \\ \text{inclosing } \tau}} \mathbf{n} \times \mathbf{V}\, d\sigma.$$

Hint: Replace $\mathbf{V}$ in the divergence theorem by $\mathbf{V} \times \mathbf{C}$, where $\mathbf{C}$ is an arbitrary constant vector. Follow the last part of the hint in (a).

(c) $$\int_{\substack{\text{curve} \\ \text{bounding } \sigma}} \phi\, d\mathbf{r} = \oint_{\text{surface } \sigma} (\mathbf{n} \times \nabla\phi) d\sigma.$$

(d) $$\oint_{\substack{\text{curve} \\ \text{bounding } \sigma}} d\mathbf{r} \times \mathbf{V} = \int_{\text{surface } \sigma} (\mathbf{n} \times \nabla) \times \mathbf{V}\, d\sigma.$$

Hints for (c) and (d): Use the substitutions suggested in (a) and (b) but in Stokes' theorem (11.9) instead of the divergence theorem.

(e) $$\int_{\text{volume } \tau} \phi\nabla \cdot \mathbf{V}\, d\tau = \oint_{\substack{\text{surface} \\ \text{inclosing } \tau}} \phi\mathbf{V} \cdot \mathbf{n}\, d\sigma - \int_{\text{volume } \tau} \mathbf{V} \cdot \nabla\phi\, d\tau.$$

Hint: Integrate (7.6) over volume τ and use the divergence theorem.

(f) $$\int_{\text{volume } \tau} \mathbf{V}\cdot(\nabla\times\mathbf{U})\,d\tau = \int_{\text{volume } \tau} \mathbf{U}\cdot(\nabla\times\mathbf{V})\,d\tau + \oint_{\substack{\text{surface}\\ \text{inclosing } \tau}} (\mathbf{U}\times\mathbf{V})\cdot\mathbf{n}\,d\sigma.$$

Hint: Integrate (h) in the Table of Vector Identities (page 339) and use the divergence theorem.

(g) $$\int_{\text{surface of } \sigma} \phi(\nabla\times\mathbf{V})\cdot\mathbf{n}\,d\sigma = \int_{\text{surface of } \sigma} (\nabla\times\nabla\phi)\cdot\mathbf{n}\,d\sigma + \oint_{\substack{\text{curve}\\ \text{bounding } \sigma}} \phi\mathbf{V}\cdot d\mathbf{r}.$$

Hint: Integrate (g) in the Table of Vector Identities (page 339) and use Stokes' Theorem.

Find vector fields $\mathbf{A}$ such that $\mathbf{V} = \operatorname{curl}\mathbf{A}$ for each given $\mathbf{V}$.

18. $\mathbf{V} = (x^2 - yz + y)\mathbf{i} + (x - 2yz)\mathbf{j} + (z^2 - 2zx + x + y)\mathbf{k}$

19. $\mathbf{V} = \mathbf{i}(x^2 - 2xz) + \mathbf{j}(y^2 - 2xy) + \mathbf{k}(z^2 - 2yz + xy)$

20. $\mathbf{V} = \mathbf{i}(ze^{zy} + x\sin zx) + \mathbf{j}x\cos xz - \mathbf{k}z\sin zx$

21. $\mathbf{V} = -\mathbf{k}$

22. $\mathbf{V} = (y+z)\mathbf{i} + (x-z)\mathbf{j} + (x^2+y^2)\mathbf{k}$

12. MISCELLANEOUS PROBLEMS

1. If $\mathbf{A}$ and $\mathbf{B}$ are unit vectors with an angle θ between them, and $\mathbf{C}$ is a unit vector perpendicular to both $\mathbf{A}$ and $\mathbf{B}$, evaluate $[(\mathbf{A}\times\mathbf{B})\times(\mathbf{B}\times\mathbf{C})]\times(\mathbf{C}\times\mathbf{A})$.

2. If $\mathbf{A}$ and $\mathbf{B}$ are the diagonals of a parallelogram, find a vector formula for the area of the parallelogram.

3. The force on a charge q moving with velocity $\mathbf{v} = d\mathbf{r}/dt$ in a magnetic field $\mathbf{B}$ is $\mathbf{F} = q(\mathbf{v}\times\mathbf{B})$. We can write $\mathbf{B}$ as $\mathbf{B} = \nabla\times\mathbf{A}$ where $\mathbf{A}$ (called the vector potential) is a vector function of x, y, z, t. If the position vector $\mathbf{r} = \mathbf{i}x + \mathbf{j}y + \mathbf{k}z$ of the charge q is a function of time t, show that

$$\frac{d\mathbf{A}}{dt} = \frac{\partial\mathbf{A}}{\partial t} + \mathbf{v}\cdot\nabla\mathbf{A}.$$

Thus show that

$$\mathbf{F} = q\mathbf{v}\times(\nabla\times\mathbf{A}) = q\left[\nabla(\mathbf{v}\cdot\mathbf{A}) - \frac{d\mathbf{A}}{dt} + \frac{\partial\mathbf{A}}{\partial t}\right].$$

4. Show that $\nabla\cdot(\mathbf{U}\times\mathbf{r}) = \mathbf{r}\cdot(\nabla\times\mathbf{U})$ where $\mathbf{U}$ is a vector function of x, y, z, and $\mathbf{r} = x\mathbf{i} + y\mathbf{j} + z\mathbf{k}$.

5. Use Green's theorem (Section 9) to do Problem 8.2.

6. Find the torque about the point $(1,-2,1)$ due to the force $\mathbf{F} = 2\mathbf{i} - \mathbf{j} + 3\mathbf{k}$ acting at the point $(1,1,-3)$.

7. Let $\mathbf{F} = 2\mathbf{i} - 3\mathbf{j} + \mathbf{k}$ act at the point $(5,1,3)$.

(a) Find the torque of $\mathbf{F}$ about the point $(4,1,0)$.

(b) Find the torque of $\mathbf{F}$ about the line $\mathbf{r} = 4\mathbf{i} + \mathbf{j} + (2\mathbf{i} + \mathbf{j} - 2\mathbf{k})t$.

8. The force $\mathbf{F} = \mathbf{i} - 2\mathbf{j} - 2\mathbf{k}$ acts at the point $(0,1,2)$. Find the torque of $\mathbf{F}$ about the line $\mathbf{r} = (2\mathbf{i} - \mathbf{j})t$.

9. Let $\mathbf{F} = \mathbf{i} - 5\mathbf{j} + 2\mathbf{k}$ act at the point $(2,1,0)$. Find the torque of $\mathbf{F}$ about the line $\mathbf{r} = (3\mathbf{j} + 4\mathbf{k}) - 2\mathbf{i}t$.

10. Given $u = xy + \sin z$, find

(a) the gradient of u at $(1, 2, \pi/2)$;

(b) how fast u is increasing, in the direction $4\mathbf{i} + 3\mathbf{j}$, at $(1, 2, \pi/2)$;

(c) the equation of the tangent plane to the surface $u = 3$ at $(1, 2, \pi/2)$.

11. Given $\phi = z^2 - 3xy$, find

(a) $\operatorname{grad} \phi$;

(b) the directional derivative of ϕ at the point $(1, 2, 3)$ in the direction $\mathbf{i} + \mathbf{j} + \mathbf{k}$;

(c) the equations of the tangent plane and of the normal line to $\phi = 3$ at the point $(1, 2, 3)$.

12. Given $u = xy + yz + z \sin x$, find

(a) ∇u at $(0, 1, 2)$;

(b) the directional derivative of u at $(0, 1, 2)$ in the direction $2\mathbf{i} + 2\mathbf{j} - \mathbf{k}$;

(c) the equations of the tangent plane and of the normal line to the level surface $u = 2$ at $(0, 1, 2)$;

(d) a unit vector in the direction of most rapid increase of u at $(0, 1, 2)$.

13. Given $\phi = x^2 - yz$ and the point $P(3, 4, 1)$, find

(a) $\nabla \phi$ at P;

(b) a unit vector normal to the surface $\phi = 5$ at P;

(c) a vector in the direction of most rapid increase of ϕ at P;

(d) the magnitude of the vector in (c);

(e) the derivative of ϕ at P in a direction parallel to the line $\mathbf{r} = \mathbf{i} - \mathbf{j} + 2\mathbf{k} + (6\mathbf{i} - \mathbf{j} - 4\mathbf{k})t$.

14. If the temperature is $T = x^2 - xy + z^2$, find

(a) the direction of heat flow at $(2, 1, -1)$;

(b) the rate of change of temperature in the direction $\mathbf{j} - \mathbf{k}$ at $(2, 1, -1)$.

15. Show that

$$\mathbf{F} = y^2 z \sinh(2xz)\mathbf{i} + 2y \cosh^2(xz)\mathbf{j} + y^2 x \sinh(2xz)\mathbf{k}$$

is conservative, and find a scalar potential ϕ such that $\mathbf{F} = -\nabla\phi$.

16. Given $\mathbf{F}_1 = 2xz\mathbf{i} + y\mathbf{j} + x^2\mathbf{k}$ and $\mathbf{F}_2 = y\mathbf{i} - x\mathbf{j}$:

(a) Which $\mathbf{F}$, if either, is conservative?

(b) If one of the given $\mathbf{F}$'s is conservative, find a function W so that $\mathbf{F} = \nabla \mathbf{W}$.

(c) If one of the $\mathbf{F}$'s is nonconservative, use it to evaluate $\int \mathbf{F} \cdot d\mathbf{r}$ along the straight line from $(0, 1)$ to $(1, 0)$.

(d) Do part (c) by applying Green's theorem to the triangle with vertices $(0, 0)$, $(0, 1)$, $(1, 0)$.

17. Find the value of $\int \mathbf{F} \cdot d\mathbf{r}$ along the circle $x^2 + y^2 = 2$ from $(1, 1)$ to $(1, -1)$ if

$$\mathbf{F} = (2x - 3y)\mathbf{i} - (3x - 2y)\mathbf{j}.$$

18. Is $\mathbf{F} = y\mathbf{i} + xz\mathbf{j} + z\mathbf{k}$ conservative? Evaluate $\int \mathbf{F} \cdot d\mathbf{r}$ from $(0, 0, 0)$ to $(1, 1, 1)$ along the paths

(a) broken line $(0,0,0)$ to $(1,0,0)$ to $(1,1,0)$ to $(1,1,1)$,

(b) straight line connecting the points.

19. Given $\mathbf{F}_1 = -2y\mathbf{i} + (z - 2x)\mathbf{j} + (y + z)\mathbf{k}$, $\mathbf{F}_2 = y\mathbf{i} + 2x\mathbf{j}$:

(a) Is $\mathbf{F}_1$ conservative? Is $\mathbf{F}_2$ conservative?

(b) Find the work done by $\mathbf{F}_2$ on a particle that moves around the ellipse $x = \cos\theta$, $y = 2\sin\theta$ from $\theta = 0$ to $\theta = 2\pi$.

(c) For any conservative force in this problem find a potential function V such that $\mathbf{F} = -\nabla V$.

(d) Find the work done by $\mathbf{F}_1$ on a particle that moves along the straight line from $(0,1,0)$ to $(0,2,5)$.

(e) Use Green's theorem and the result of Problem 9.7 to do Part (b) above.

In Problems 20 to 31, evaluate each integral in the simplest way possible.

20. $\iint \mathbf{P}\cdot\mathbf{n}\,d\sigma$ over the upper half of the sphere $r = 1$ if $\mathbf{P} = \text{curl}(\mathbf{j}x - \mathbf{k}z)$.

21. $\iint (\nabla\times\mathbf{V})\cdot\mathbf{n}\,d\sigma$ over the surface consisting of the four slanting faces of a pyramid whose base is the square in the (x,y) plane with corners at $(0,0)$, $(0,2)$, $(2,0)$, $(2,2)$ and whose top vertex is at $(1,1,2)$, where

$$\mathbf{V} = (x^2z - 2)\mathbf{i} + (x + y - z)\mathbf{j} - xyz\mathbf{k}.$$

22. $\iint \mathbf{V}\cdot\mathbf{n}\,d\sigma$ over the entire surface of the sphere $(x-2)^2 + (y+3)^2 + z^2 = 9$, if

$$\mathbf{V} = (3x - yz)\mathbf{i} + (z^2 - y^2)\mathbf{j} + (2yz + x^2)\mathbf{k}.$$

23. $\iint \mathbf{F}\cdot\mathbf{n}\,d\sigma$ where $\mathbf{F} = (y^2 - x^2)\mathbf{i} + (2xy - y)\mathbf{j} + 3z\mathbf{k}$ and σ is the entire surface of the tin can bounded by the cylinder $x^2 + y^2 = 16$, $z = 3$, $z = -3$.

24. $\iint \mathbf{r}\cdot\mathbf{n}\,d\sigma$ over the entire surface of the hemisphere $x^2 + y^2 + z^2 = 9$, $z \ge 0$, where $\mathbf{r} = x\mathbf{i} + y\mathbf{j} + z\mathbf{k}$.

25. $\iint \mathbf{V}\cdot\mathbf{n}\,d\sigma$ over the curved part of the hemisphere in Problem 24, if $\mathbf{V} = \text{curl}(y\mathbf{i} - x\mathbf{j})$.

26. $\iint (\text{curl}\,\mathbf{V})\cdot\mathbf{n}\,d\sigma$ over the entire surface of the cube in the first octant with three faces in the three coordinate planes and the other three faces intersecting at $(2,2,2)$, where

$$\mathbf{V} = (2 - y)\mathbf{i} + xz\mathbf{j} + xyz\mathbf{k}.$$

27. Problem 26, but integrate over the open surface obtained by leaving out the face of the cube in the (x,y) plane.

28. $\oint \mathbf{F}\cdot d\mathbf{r}$ around the circle $x^2 + y^2 + 2x = 0$, where $\mathbf{F} = y\mathbf{i} - x\mathbf{j}$.

29. $\oint \mathbf{V}\cdot d\mathbf{r}$ around the boundary of the square with vertices $(1,0)$, $(0,1)$, $(-1,0)$, $(0,-1)$, if $\mathbf{V} = x^2\mathbf{i} + 5x\mathbf{j}$.

30. $\int_C (x^2 - y)dx + (x + y^3)dy$, where C is the parallelogram with vertices at $(0,0)$, $(2,0)$, $(1,1)$, $(3,1)$.

31. $\int (y^2 - x^2)\,dx + (2xy + 3)dy$ along the x axis from $(0,0)$ to $(\sqrt{5},0)$ and then along a circular arc from $(\sqrt{5},0)$ to $(1,2)$. *Hint:* Use Green's theorem.

Table of Vector Identities Involving ∇

Note carefully that ϕ and ψ are scalar functions; $\mathbf{U}$ and $\mathbf{V}$ are vector functions. Formulas are given in rectangular coordinates; for other coordinate systems, see Chapter 10, Section 9.

(a) $\nabla \cdot \nabla\phi = \operatorname{div}\operatorname{grad}\phi = \nabla^2\phi = \text{Laplacian }\phi = \dfrac{\partial^2\phi}{\partial x^2} + \dfrac{\partial^2\phi}{\partial y^2} + \dfrac{\partial^2\phi}{\partial z^2}$

(b) $\nabla \times \nabla\phi = \operatorname{curl}\operatorname{grad}\phi = 0$

(c) $\nabla(\nabla \cdot \mathbf{V}) = \operatorname{grad}\operatorname{div}\mathbf{V}$

$$= \mathbf{i}\left(\frac{\partial^2 V_x}{\partial x^2} + \frac{\partial^2 V_y}{\partial x \partial y} + \frac{\partial^2 V_z}{\partial x \partial z}\right) + \mathbf{j}\left(\frac{\partial^2 V_x}{\partial x \partial y} + \frac{\partial^2 V_y}{\partial y^2} + \frac{\partial^2 V_z}{\partial y \partial z}\right) + \mathbf{k}\left(\frac{\partial^2 V_x}{\partial x \partial z} + \frac{\partial^2 V_y}{\partial y \partial z} + \frac{\partial^2 V_z}{\partial z^2}\right)$$

(d) $\nabla \cdot (\nabla \times \mathbf{V}) = \operatorname{div}\operatorname{curl}\mathbf{V} = 0$

(e) $\nabla \times (\nabla \times \mathbf{V}) = \operatorname{curl}\operatorname{curl}\mathbf{V} = \nabla(\nabla \cdot \mathbf{V}) - \nabla^2\mathbf{V} = \operatorname{grad}\operatorname{div}\mathbf{V} - \text{Laplacian }\mathbf{V}$

(f) $\nabla \cdot (\phi\mathbf{V}) = \phi(\nabla \cdot \mathbf{V}) + \mathbf{V} \cdot (\nabla\phi)$

(g) $\nabla \times (\phi\mathbf{V}) = \phi(\nabla \times \mathbf{V}) - \mathbf{V} \times (\nabla\phi)$

(h) $\nabla \cdot (\mathbf{U} \times \mathbf{V}) = \mathbf{V} \cdot (\nabla \times \mathbf{U}) - \mathbf{U} \cdot (\nabla \times \mathbf{V})$

(i) $\nabla \times (\mathbf{U} \times \mathbf{V}) = (\mathbf{V} \cdot \nabla)\mathbf{U} - (\mathbf{U} \cdot \nabla)\mathbf{V} - \mathbf{V}(\nabla \cdot \mathbf{U}) + \mathbf{U}(\nabla \cdot \mathbf{V})$

(j) $\nabla(\mathbf{U} \cdot \mathbf{V}) = \mathbf{U} \times (\nabla \times \mathbf{V}) + (\mathbf{U} \cdot \nabla)\mathbf{V} + \mathbf{V} \times (\nabla \times \mathbf{U}) + (\mathbf{V} \cdot \nabla)\mathbf{U}$

(k) $\nabla \cdot (\nabla\phi \times \nabla\psi) = 0$

CHAPTER 7

Fourier Series and Transforms

1. INTRODUCTION

Problems involving vibrations or oscillations occur frequently in physics and engineering. You can think of examples you have already met: a vibrating tuning fork, a pendulum, a weight attached to a spring, water waves, sound waves, alternating electric currents, etc. In addition, there are many more examples which you will meet as you continue to study physics. Some of them—for example, heat conduction, electric and magnetic fields, light—do not appear in elementary work to have anything oscillatory about them, but will turn out in your more advanced work to involve the sines and cosines which are used in describing simple harmonic motion and wave motion.

In Chapter 1 we discussed the use of power series to approximate complicated functions. In many problems, series called Fourier series, whose terms are sines and cosines, are more useful than power series. In this chapter we shall see how to find and use Fourier series. Then, in Chapter 13 (Sections 2 to 4), we shall consider several of the physics problems which Fourier was trying to solve when he invented Fourier series.

Since sines and cosines are periodic functions, Fourier series can represent only periodic functions. We will see in Section 12 how to represent a non-periodic function by a Fourier integral (Fourier transform).

2. SIMPLE HARMONIC MOTION AND WAVE MOTION; PERIODIC FUNCTIONS

We shall need much of the notation and terminology used in discussing simple harmonic motion and wave motion. Let's discuss these two topics briefly.

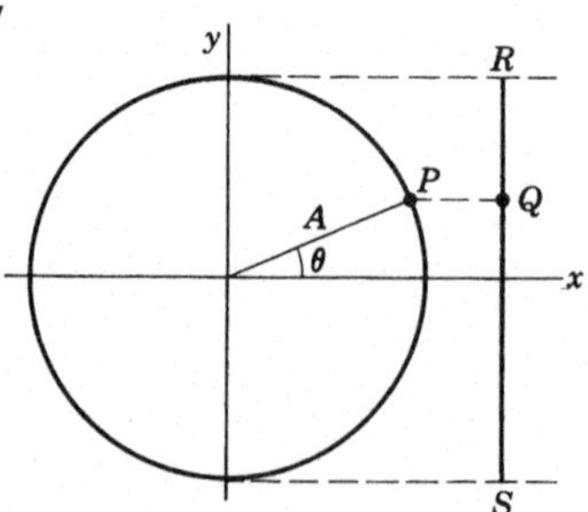

Figure 2.1

Let particle P (Figure 2.1) move at constant speed around a circle of radius A. At the same time, let particle Q move up and down along the straight line segment RS in such a way that the y coordinates of P and Q are always equal. If ω is the angular velocity of P in radians per second, and

(Figure 2.1) $\theta = 0$ when $t = 0$, then at a later time t

$$\theta = \omega t. \tag{2.1}$$

The y coordinate of Q (which is equal to the y coordinate of P) is

$$y = A \sin\theta = A \sin\omega t. \tag{2.2}$$

The back and forth motion of Q is called *simple harmonic motion.* By definition, an object is executing simple harmonic motion if its displacement from equilibrium can be written as $A\sin\omega t$ [or $A\cos\omega t$ or $A\sin(\omega t+\phi)$, but these two functions differ from $A\sin\omega t$ only in choice of origin; such functions are called *sinusoidal functions*]. You can think of many physical examples of this sort of simple vibration: a pendulum, a tuning fork, a weight bobbing up and down at the end of a spring.

The x and y coordinates of particle P in Figure 2.1 are

$$x = A\cos\omega t, \qquad y = A\sin\omega t. \tag{2.3}$$

If we think of P as the point $z = x+iy$ in the complex plane, we could replace (2.3) by a single equation to describe the motion of P:

$$\begin{aligned} z &= x + iy = A(\cos\omega t + i\sin\omega t) \\ &= Ae^{i\omega t}. \end{aligned} \tag{2.4}$$

It is often worth while to use this complex notation even to describe the motion of Q; we then understand that the actual position of Q is equal to the imaginary part of z (or with different starting conditions the real part of z). For example, the velocity of Q is the imaginary part of

$$\frac{dz}{dt} = \frac{d}{dt}(Ae^{i\omega t}) = Ai\omega e^{i\omega t} = Ai\omega(\cos\omega t + i\sin\omega t). \tag{2.5}$$

[The imaginary part of (2.5) is $A\omega\cos\omega t$, which is dy/dt from (2.2).]

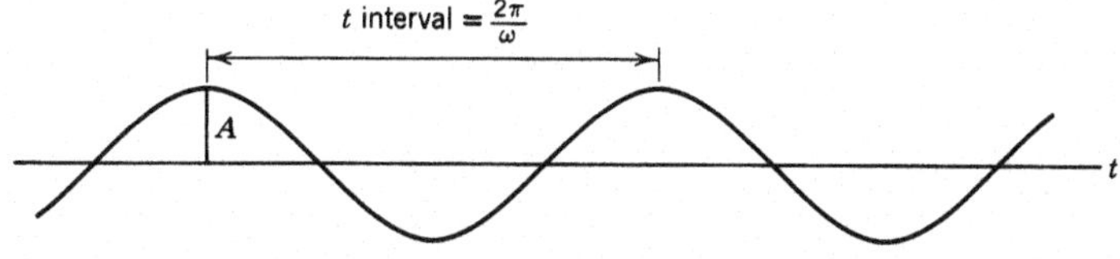

Figure 2.2

It is useful to draw a graph of x and y in (2.2) and (2.3) as a function of t. Figure 2.2 represents any of the functions $\sin\omega t$, $\cos\omega t$, $\sin(\omega t+\phi)$ if we choose the origin correctly. The number A is called the *amplitude of the vibration* or the *amplitude of the function.* Physically it is the maximum displacement of Q from its equilibrium position. The *period of the simple harmonic motion* or the *period of the function* is the time for one complete oscillation, that is, $2\pi/\omega$ (See Figure 2.2).

We could write the velocity of Q from (2.5) as

$$\frac{dy}{dt} = A\omega\cos\omega t = B\cos\omega t. \tag{2.6}$$

Here B is the maximum value of the velocity and is called the *velocity amplitude.* Note that the velocity has the same period as the displacement. If the mass of the particle Q is m, its kinetic energy is:

$$\text{Kinetic energy} = \frac{1}{2}m\left(\frac{dy}{dt}\right)^2 = \frac{1}{2}mB^2\cos^2\omega t. \tag{2.7}$$

We are considering an idealized harmonic oscillator which does not lose energy. Then the total energy (kinetic plus potential) must be equal to the largest value of the kinetic energy, that is, $\frac{1}{2}mB^2$. Thus we have:

$$\text{Total energy} = \frac{1}{2}mB^2. \tag{2.8}$$

Notice that the energy is proportional to the square of the (velocity) amplitude; we shall be interested in this result later when we discuss sound.

Waves are another important example of an oscillatory phenomenon. The mathematical ideas of wave motion are useful in many fields; for example, we talk about water waves, sound waves, and radio waves.

Example 1. Consider water waves in which the shape of the water surface is (unrealistically!) a sine curve. If we take a photograph (at the instant $t = 0$) of the water surface, the equation of this picture could be written (relative to appropriate axes)

$$y = A\sin\frac{2\pi x}{\lambda}, \tag{2.9}$$

where x represents horizontal distance and λ is the distance between wave crests. Usually λ is called the *wavelength*, but mathematically it is the same as the period of this function of x. Now suppose we take another photograph when the waves have moved forward a distance vt (v is the velocity of the waves and t is the time between photographs). Figure 2.3 shows the two photographs superimposed. Observe that the value of y at the point x on the graph labeled t, is just the same as the value of y at the point $(x-vt)$ on the graph labeled $t = 0$. If (2.9) is the equation representing the waves at $t = 0$, then

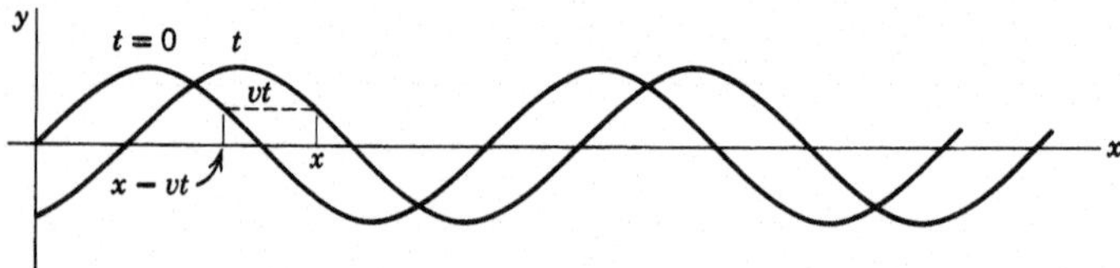

Figure 2.3

$$y = A\sin\frac{2\pi}{\lambda}(x - vt) \tag{2.10}$$

represents the waves at time t. We can interpret (2.10) in another way. Suppose you stand at one point in the water [fixed x in (2.10)] and observe the up and down motion of the water, that is, y in (2.10) as a function of t (for fixed x). This is a simple harmonic motion of amplitude A and period λ/v. You are doing something analogous to this when you stand still and listen to a sound (sound waves pass your ear and you observe their frequency) or when you listen to the radio (radio waves pass the receiver and it reacts to their frequency).

We see that y in (2.10) is a periodic function either of x (t fixed) or of t (x fixed); both interpretations are useful. It makes no difference in the basic mathematics, however, what letter we use for the independent variable. To simplify our notation we shall ordinarily use x as the variable, but if the physical problem calls for it, you can replace x by t.

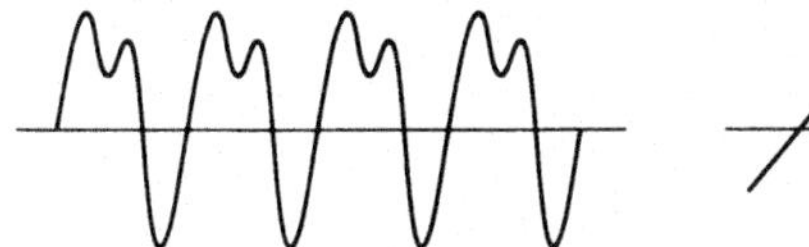

Figure 2.4

Sines and cosines are periodic functions; once you have drawn $\sin x$ from $x = 0$ to $x = 2\pi$, the rest of the graph from $x = -\infty$ to $x = +\infty$ is just a repetition over and over of the 0 to 2π graph. The number 2π is the period of $\sin x$. A periodic function need not be a simple sine or cosine, but may be any sort of complicated graph that repeats itself (Figure 2.4). The interval of repetition is the period.

Example 2. If we are describing the vibration of a seconds pendulum, the period is 2 sec (time for one complete back-and-forth oscillation). The reciprocal of the period is the *frequency*, the number of oscillations per second; for the seconds pendulum, the frequency is $\frac{1}{2}$ sec^{-1}. When radio announcers say, "operating on a frequency of 780 kilohertz," they mean that 780,000 radio waves reach you per second, or that the period of one wave is (1/780,000) sec.

By definition, the function $f(x)$ is periodic if $f(x + p) = f(x)$ for every x; the number p is the period. The period of $\sin x$ is 2π since $\sin(x+2\pi) = \sin x$; similarly, the period of $\sin 2\pi x$ is 1 since $\sin 2\pi(x + 1) = \sin(2\pi x + 2\pi) = \sin 2\pi x$, and the period of $\sin(\pi x/l)$ is $2l$ since $\sin(\pi/l)(x + 2l) = \sin(\pi x/l)$. In general, the period of $\sin 2\pi x/T$ is T.

PROBLEMS, SECTION 2

In Problems 1 to 6 find the amplitude, period, frequency, and velocity amplitude for the motion of a particle whose distance s from the origin is the given function.

1. $s = 3\cos 5t$

2. $s = 2\sin(4t - 1)$

3. $s = \frac{1}{2}\cos(\pi t - 8)$

4. $s = 5\sin(t - \pi)$

5. $s = 2\sin 3t\cos 3t$

6. $s = 3\sin(2t + \pi/8) + 3\sin(2t - \pi/8)$

In Problems 7 to 10 you are given a complex function $z = f(t)$. In each case, show that a particle whose coordinate is (a) $x = \operatorname{Re} z$, (b) $y = \operatorname{Im} z$ is undergoing simple harmonic motion, and find the amplitude, period, frequency, and velocity amplitude of the motion.

7. $z = 5e^{it}$

8. $z = 2e^{-it/2}$

9. $z = 2e^{i\pi t}$

10. $z = -4e^{i(2t+3\pi)}$

11. The charge q on a capacitor in a simple a-c circuit varies with time according to the equation $q = 3\sin(120\pi t + \pi/4)$. Find the amplitude, period, and frequency of this oscillation. By definition, the current flowing in the circuit at time t is $I = dq/dt$. Show that I is also a sinusoidal function of t, and find its amplitude, period, and frequency.

12. Repeat Problem 11: (a) if $q = \operatorname{Re} 4e^{30i\pi t}$; (b) if $q = \operatorname{Im} 4e^{30i\pi t}$.

13. A simple pendulum consists of a point mass m suspended by a (weightless) cord or rod of length l, as shown, and swinging in a vertical plane under the action of gravity. Show that for small oscillations (small θ), both θ and x are sinusoidal functions of time, that is, the motion is simple harmonic. *Hint:* Write the differential equation $\mathbf{F} = m\mathbf{a}$ for the particle m. Use the approximation $\sin\theta = \theta$ for small θ, and show that $\theta = A\sin\omega t$ is a solution of your equation. What are A and ω?

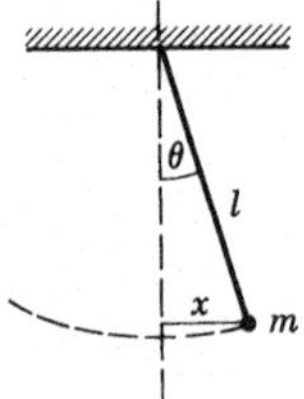

14. The displacements x of two simple pendulums (see Problem 13) are $4\sin(\pi t/3)$ and $3\sin(\pi t/4)$. They start together at $x = 0$. How long will it be before they are together again at $x = 0$? *Hint:* Sketch or computer plot the graphs.

15. As in Problem 14, the displacements x of two simple pendulums are $x = -2\cos(t/2)$ and $3\sin(t/3)$. They are *not* together at $t = 0$; plot graphs to see when they are first together.

16. As in Problem 14, let the displacements be $y_1 = 3\sin(t/\sqrt{2})$ and $y_2 = \sin t$. The pendulums start together at $t = 0$. Make computer plots to estimate when they will be together again and then, by computer, solve the equation $y_1 = y_2$ for the root near your estimate.

17. Show that equation (2.10) for a wave can be written in all these forms:

$$y = A\sin\frac{2\pi}{\lambda}(x - vt) = A\sin 2\pi\left(\frac{x}{\lambda} - \frac{t}{T}\right)$$
$$= A\sin\omega\left(\frac{x}{v} - t\right) = A\sin\left(\frac{2\pi x}{\lambda} - 2\pi f t\right) = A\sin\frac{2\pi}{T}\left(\frac{x}{v} - t\right).$$

Here λ is the wavelength, f is the frequency, v is the wave velocity, T is the period, and $\omega = 2\pi f$ is called the *angular frequency*. *Hint:* Show that $v = \lambda f$.

In Problems 18 to 20, find the amplitude, period, frequency, wave velocity, and wavelength of the given wave. By computer, plot on the same axes, y as a function of x for the given values of t, and label each graph with its value of t. Similarly, plot on the same axes, y as a function of t for the given values of x, and label each curve with its value of x.

18. $y = 2\sin\frac{2}{3}\pi(x - 3t)$; $t = 0, \frac{1}{4}, \frac{1}{2}, \frac{3}{4}$; $x = 0, 1, 2, 3$.

19. $y = \cos 2\pi(x - \frac{1}{4}t)$; $t = 0, 1, 2, 3$; $x = 0, \frac{1}{4}, \frac{1}{2}, \frac{3}{4}$.

20. $y = 3\sin\pi(x - \frac{1}{2}t)$; $t = 0, 1, 2, 3$; $x = 0, \frac{1}{2}, 1, \frac{3}{2}, 2$.

21. Write the equation for a sinusoidal wave of wavelength 4, amplitude 20, and velocity 6. (See Problem 17.) Make computer plots of y as a function of t for $x = 0, 1, 2, 3$, and of y as a function of x for $t = 0, \frac{1}{6}, \frac{1}{3}, \frac{1}{2}$. If this wave represents the shape of a long rope which is being shaken back and forth at one end, find the velocity $\partial y/\partial t$ of particles of the rope as a function of x and t. (Note that this velocity has nothing to do with the wave velocity v, which is the rate at which crests of the wave move forward.)

22. Do Problem 21 for a wave of amplitude 4, period 6, and wavelength 3. Make computer plots of y as a function of x when $t = 0, 1, 2, 3$, and of y as a function of t when $x = \frac{1}{2}, 1, \frac{3}{2}, 2$.

23. Write an equation for a sinusoidal sound wave of amplitude 1 and frequency 440 hertz (1 hertz means 1 cycle per second). (Take the velocity of sound to be 350 m/sec.)

24. The velocity of sound in sea water is about 1530 m/sec. Write an equation for a sinusoidal sound wave in the ocean, of amplitude 1 and frequency 1000 hertz.

25. Write an equation for a sinusoidal radio wave of amplitude 10 and frequency 600 kilohertz. *Hint:* The velocity of a radio wave is the velocity of light, $c = 3{\cdot}10^8$ m/sec.

3. APPLICATIONS OF FOURIER SERIES

We have said that the vibration of a tuning fork is an example of simple harmonic motion. When we hear the musical note produced, we say that a sound wave has passed through the air from the tuning fork to our ears. As the tuning fork vibrates it pushes against the air molecules, creating alternately regions of high and low

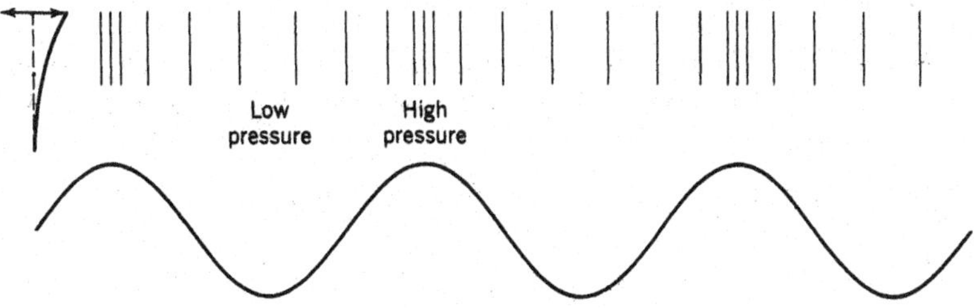

Figure 3.1

pressure (Figure 3.1). If we measure the pressure as a function of x and t from the tuning fork to us, we find that the pressure is of the form of (2.10); if we measure the pressure where we are as a function of t as the wave passes, we find that the pressure is a periodic function of t. The sound wave is a pure sine wave of a definite frequency (in the language of music, a pure tone). Now suppose that several pure tones are heard simultaneously. In the resultant sound wave, the pressure will not be a single sine function but a sum of several sine functions. If you strike a piano key you do not get a sound wave of just one frequency. Instead, you get a fundamental accompanied by a number of overtones (harmonics) of frequencies $2, 3, 4, \cdots$, times the frequency of the fundamental. Higher frequencies mean shorter periods. If $\sin\omega t$ and $\cos\omega t$ correspond to the fundamental frequency, then $\sin n\omega t$ and $\cos n\omega t$ correspond to the higher harmonics. The combination of the fundamental and the harmonics is a complicated periodic function with the period of the fundamental (Problem 5). Given the complicated function, we could ask how to write it as a sum of terms corresponding to the various harmonics. In general it might require all the harmonics, that is, an infinite series of terms. This is called a Fourier series. Expanding a function in a Fourier series then amounts to breaking it down into its various harmonics. In fact, this process is sometimes called harmonic analysis.

There are applications to other fields besides sound. Radio waves, visible light, and x rays are all examples of a kind of wave motion in which the "waves" correspond to varying electric and magnetic fields. Exactly the same mathematical equations apply as for water waves and sound waves. We could then ask what light frequencies (these correspond to the color) are in a given light beam and in what proportions. To find the answer, we would expand the given function describing the wave in a Fourier series.

You have probably seen a sine curve used to represent an alternating current (a-c) or voltage in electricity. This is a periodic function, but so are the functions shown in Figure 3.2. Any of these and many others might represent signals (voltages or currents) which are to be applied to an electric circuit. Then we could ask

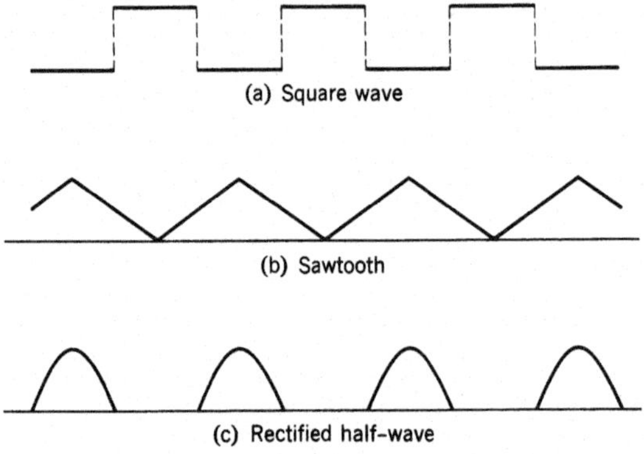

Figure 3.2

what a-c frequencies (harmonics) make up a given signal and in what proportions. When an electric signal is passed through a network (say a radio), some of the harmonics may be lost. If most of the important ones get through with their relative intensities preserved, we say that the radio possesses "high fidelity." To find out which harmonics are the important ones in a given signal, we expand it in a Fourier series. The terms of the series with large coefficients then represent the important harmonics (frequencies).

Since sines and cosines are themselves periodic, it seems rather natural to use series of them, rather than power series, to represent periodic functions. There is another important reason. The coefficients of a power series are obtained, you will recall (Chapter 1, Section 12), by finding successive derivatives of the function being expanded; consequently, only continuous functions with derivatives of all orders can be expanded in power series. Many periodic functions in practice are not continuous or not differentiable (Figure 3.2). Fortunately, Fourier series (unlike power series) can represent discontinuous functions or functions whose graphs have corners. On the other hand, Fourier series do not usually converge as rapidly as power series and much more care is needed in manipulating them. For example, a power series can be differentiated term by term (Chapter 1, Section 11), but differentiating a Fourier series term by term sometimes produces a series which doesn't converge. (See end of Section 9.)

Our problem then is to expand a given periodic function in a series of sines and cosines. We shall take this up in Section 5 after doing some preliminary work.

PROBLEMS, SECTION 3

For each of the following combinations of a fundamental musical tone and some of its overtones, make a computer plot of individual harmonics (all on the same axes) and then a plot of the sum. Note that the sum has the period of the fundamental (Problem 5).

1. $\sin t - \frac{1}{9}\sin 3t$

2. $2\cos t + \cos 2t$

3. $\sin \pi t + \sin 2\pi t + \frac{1}{3}\sin 3\pi t$

4. $\cos 2\pi t + \cos 4\pi t + \frac{1}{2}\cos 6\pi t$

5. Using the definition (end of Section 2) of a periodic function, show that a sum of terms corresponding to a fundamental musical tone and its overtones has the period of the fundamental.

In Problems 6 and 7, use a trigonometry formula to write the two terms as a single harmonic. Find the period and amplitude. Compare computer plots of your result and the given problem.

6. $\sin 2x + \sin 2(x + \pi/3)$

7. $\cos \pi x - \cos \pi(x - 1/2)$

8. A periodic modulated (AM) radio signal has the form

$$y = (A + B\sin 2\pi f t)\sin 2\pi f_c\left(t - \frac{x}{v}\right).$$

The factor $\sin 2\pi f_c(t - x/v)$ is called the carrier wave; it has a very high frequency (called radio frequency; f_c is of the order of 10^6 cycles per second). The amplitude of the carrier wave is $(A + B\sin 2\pi f t)$. This amplitude varies with time—hence the term "amplitude modulation"—with the much smaller frequency of the sound being transmitted (called audio frequency; f is of the order of 10^2 cycles per second). In order to see the general appearance of such a wave, use the following simple but unrealistic data to sketch a graph of y as a function of t for $x = 0$ over two periods of the *amplitude* function: $A = 3$, $B = 1$, $f = 1$, $f_c = 20$. Using trigonometric formulas, show that y can be written as a sum of three waves of frequencies f_c, $f_c + f$, and $f_c - f$; the first of these is the carrier wave and the other two are called side bands.

4. AVERAGE VALUE OF A FUNCTION

The concept of the average value of a function is often useful. You know how to find the average of a set of numbers: you add them and divide by the number of numbers.

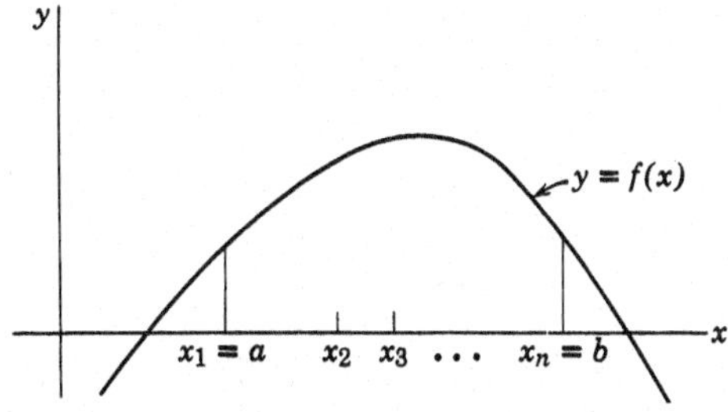

Figure 4.1

This process suggests that we ought to get an approximation to the average value of a function $f(x)$ on the interval (a, b) by averaging a number of values of $f(x)$ (Figure 4.1):

$$\text{(4.1)} \qquad \text{Average of } f(x) \text{ on } (a,b) \text{ is approximately equal to } \frac{f(x_1) + f(x_2) + \cdots + f(x_n)}{n}.$$

This should become a better approximation as n increases. Let the points $x_1, x_2, \cdots$ be Δx apart. Multiply the numerator and the denominator of the approximate average by Δx. Then (4.1) becomes:

$$\text{(4.2)} \qquad \text{Average of } f(x) \text{ on } (a,b) \text{ is approximately equal to } \frac{[f(x_1) + \cdots + f(x_n)]\Delta x}{n\,\Delta x}.$$

Now $n\,\Delta x = b - a$, the length of the interval over which we are averaging, no matter what n and Δx are. If we let $n \to \infty$ and $\Delta x \to 0$, the numerator approaches $\int_a^b f(x)\,dx$, and we have

$$\text{(4.3)} \qquad \text{Average of } f(x) \text{ on } (a,b) = \frac{\int_a^b f(x)\,dx}{b-a}.$$

In applications, it may happen that the average value of a given function is zero.

▶ **Example 1.** The average of $\sin x$ over any number of periods is zero. The average value of the velocity of a simple harmonic oscillator over any number of vibrations is zero. In such cases the average of the square of the function may be of interest.

▶ **Example 2.** If the alternating electric current flowing through a wire is described by a sine function, the square root of the average of the sine squared is known as the root-mean-square or effective value of the current, and is what you would measure with an a-c ammeter. In the example of the simple harmonic oscillator, the average kinetic energy (average of $\frac{1}{2}mv^2$) is $\frac{1}{2}m$ times the average of v^2.

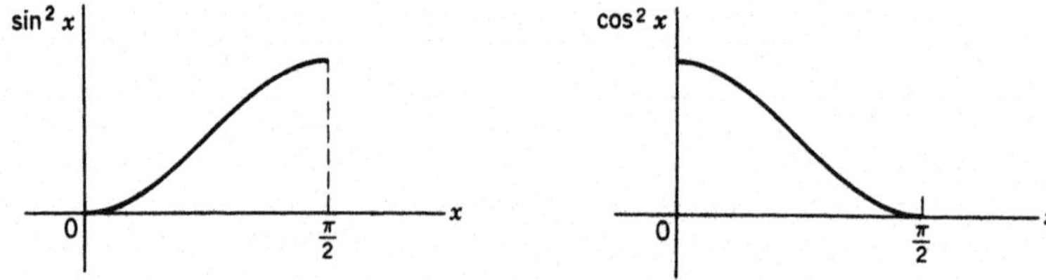

Figure 4.2

Now you can, of course, find the average value of $\sin^2 x$ over a period (say $-\pi$ to π) by evaluating the integral in (4.3). There is an easier way. Look at the graphs of $\cos^2 x$ and $\sin^2 x$ (Figure 4.2). You can probably convince yourself that the area under them is the same for any quarter-period from 0 to $\pi/2$, $\pi/2$ to π, etc. (Also see Problems 2 and 13.) Then

$$\int_{-\pi}^{\pi} \sin^2 x\,dx = \int_{-\pi}^{\pi} \cos^2 x\,dx. \tag{4.4}$$

Similarly (for integral $n \neq 0$),

$$\int_{-\pi}^{\pi} \sin^2 nx\,dx = \int_{-\pi}^{\pi} \cos^2 nx\,dx. \tag{4.5}$$

But since $\sin^2 nx + \cos^2 nx = 1$,

$$\int_{-\pi}^{\pi} (\sin^2 nx + \cos^2 nx)\,dx = \int_{-\pi}^{\pi} dx = 2\pi. \tag{4.6}$$

Using (4.5), we get

$$\int_{-\pi}^{\pi} \sin^2 nx\,dx = \int_{-\pi}^{\pi} \cos^2 nx\,dx = \pi. \tag{4.7}$$

Then using (4.3) we see that:

(4.8) The average value (over a period) of $\sin^2 nx$

$= $ the average value (over a period) of $\cos^2 nx$

$$= \frac{1}{2\pi}\int_{-\pi}^{\pi} \sin^2 nx\,dx = \frac{1}{2\pi}\int_{-\pi}^{\pi} \cos^2 nx\,dx = \frac{\pi}{2\pi} = \frac{1}{2}.$$

We can say all this more simply in words. By (4.5), the average value of $\sin^2 nx$ equals the average value of $\cos^2 nx$. The average value of $\sin^2 nx + \cos^2 nx = 1$ is 1. Therefore the average value of $\sin^2 nx$ or of $\cos^2 nx$ is $\frac{1}{2}$. (In each case the average value is taken over one or more periods.)

PROBLEMS, SECTION 4

1. Show that if $f(x)$ has period p, the average value of f is the same over any interval of length p. *Hint:* Write $\int_a^{a+p} f(x)\,dx$ as the sum of two integrals (a to p, and p to $a+p$) and make the change of variable $x = t + p$ in the second integral.

2. (a) Prove that $\int_0^{\pi/2} \sin^2 x\,dx = \int_0^{\pi/2} \cos^2 x\,dx$ by making the change of variable $x = \frac{1}{2}\pi - t$ in one of the integrals.

(b) Use the same method to prove that the averages of $\sin^2(n\pi x/l)$ and $\cos^2(n\pi x/l)$ are the same over a period.

In Problems 3 to 12, find the average value of the function on the given interval. Use equation (4.8) if it applies. If an average value is zero, you may be able to decide this from a quick sketch which shows you that the areas above and below the x axis are the same.

3. $\sin x + 2\sin 2x + 3\sin 3x$ on $(0, 2\pi)$ **4.** $1 - e^{-x}$ on $(0, 1)$

5. $\cos^2 \dfrac{x}{2}$ on $\left(0, \dfrac{\pi}{2}\right)$

6. $\sin x$ on $(0, \pi)$

7. $x - \cos^2 6x$ on $\left(0, \dfrac{\pi}{6}\right)$

8. $\sin 2x$ on $\left(\dfrac{\pi}{6}, \dfrac{7\pi}{6}\right)$

9. $\sin^2 3x$ on $(0, 4\pi)$

10. $\cos x$ on $(0, 3\pi)$

11. $\sin x + \sin^2 x$ on $(0, 2\pi)$

12. $\cos^2 \dfrac{7\pi x}{2}$ on $\left(0, \dfrac{8}{7}\right)$

13. Using (4.3) and equations similar to (4.5) to (4.7), show that

$$\int_a^b \sin^2 kx\,dx = \int_a^b \cos^2 kx\,dx = \frac{1}{2}(b-a)$$

if $k(b-a)$ is an integral multiple of π, or if kb and ka are both integral multiples of $\pi/2$.

Use the results of Problem 13 to evaluate the following integrals without calculation.

14. (a) $\displaystyle\int_0^{4\pi/3} \sin^2\left(\frac{3x}{2}\right) dx$ (b) $\displaystyle\int_{-\pi/2}^{3\pi/2} \cos^2\left(\frac{x}{2}\right) dx$

15. (a) $\displaystyle\int_{-1/4}^{11/4} \cos^2 \pi x\,dx$ (b) $\displaystyle\int_{-1}^{2} \sin^2\left(\frac{\pi x}{3}\right) dx$

16. (a) $\displaystyle\int_0^{2\pi/\omega} \sin^2 \omega t\,dt$ (b) $\displaystyle\int_0^{2} \cos^2 2\pi t\,dt$

5. FOURIER COEFFICIENTS

We want to expand a given periodic function in a series of sines and cosines. To simplify our formulas at first, we start with functions of period 2π; that is, we shall expand periodic functions of period 2π in terms of the functions $\sin nx$ and $\cos nx$. (Later we shall see how we can change the formulas to fit a different period—see Section 8.) The functions $\sin x$ and $\cos x$ have period 2π; so do $\sin nx$ and $\cos nx$ for any integral n since $\sin n(x + 2\pi) = \sin(nx + 2n\pi) = \sin nx$. (It is true that $\sin nx$ and $\cos nx$ also have shorter periods, namely $2\pi/n$, but the fact that they repeat every 2π is what we are interested in here, for this makes them reasonable functions to use in an expansion of a function of period 2π.) Then, given a function $f(x)$ of period 2π, we write

$$\text{(5.1)} \qquad f(x) = \frac{1}{2}a_0 + a_1 \cos x + a_2 \cos 2x + a_3 \cos 3x + \cdots$$
$$+ b_1 \sin x + b_2 \sin 2x + b_3 \sin 3x + \cdots,$$

and derive formulas for the coefficients a_n and b_n. (The reason for writing $\frac{1}{2}a_0$ as the constant term will be clear later—it makes the formulas for the coefficients simpler to remember—but you must not forget the $\frac{1}{2}$ in the series!)

In finding formulas for a_n and b_n in (5.1) we need the following integrals:

(5.2) The average value of $\sin mx \cos nx$ (over a period)

$$= \frac{1}{2\pi}\int_{-\pi}^{\pi} \sin mx \cos nx \, dx = 0.$$

The average value of $\sin mx \sin nx$ (over a period)

$$= \frac{1}{2\pi}\int_{-\pi}^{\pi} \sin mx \sin nx \, dx = \begin{cases} 0, & m \neq n, \\ \frac{1}{2}, & m = n \neq 0, \\ 0, & m = n = 0. \end{cases}$$

The average value of $\cos mx \cos nx$ (over a period)

$$= \frac{1}{2\pi}\int_{-\pi}^{\pi} \cos mx \cos nx \, dx = \begin{cases} 0, & m \neq n, \\ \frac{1}{2}, & m = n \neq 0, \\ 1, & m = n = 0. \end{cases}$$

We have already shown that the average values of $\sin^2 nx$ and $\cos^2 nx$ are $\frac{1}{2}$. The last integral in (5.2) is the average value of 1 which is 1. To show that the other average values in (5.2) are zero (unless $m = n \neq 0$), we could use the trigonometry formulas for products like $\sin\theta\cos\phi$ and then integrate. An easier way is to use the formulas for the sines and cosines in terms of complex exponentials. [See (7.1) or Chapter 2, Section 11.] We shall show this method for one integral

$$\int_{-\pi}^{\pi} \sin mx \cos nx \, dx = \int_{-\pi}^{\pi} \frac{e^{imx} - e^{-imx}}{2i} \cdot \frac{e^{inx} + e^{-inx}}{2} dx. \tag{5.3}$$

We can see the result without actually multiplying these out. All terms in the product are of the form e^{ikx}, where k is an integer $\neq 0$ (except for the cross-product terms when $n = m$, and these cancel). We can show that the integral of each such term is zero:

$$\int_{-\pi}^{\pi} e^{ikx} \, dx = \left.\frac{e^{ikx}}{ik}\right|_{-\pi}^{\pi} = \frac{e^{ik\pi} - e^{-ik\pi}}{ik} = 0 \tag{5.4}$$

because $e^{ik\pi} = e^{-ik\pi} = \cos k\pi$ (since $\sin k\pi = 0$). The other integrals in (5.2) may be evaluated similarly (Problem 12).

We now show how to find a_n and b_n in (5.1). To find a_0, we find the average value on $(-\pi, \pi)$ of each term of (5.1).

$$\begin{aligned} \frac{1}{2\pi}\int_{-\pi}^{\pi} f(x)\, dx &= \frac{a_0}{2}\frac{1}{2\pi}\int_{-\pi}^{\pi} dx + a_1 \frac{1}{2\pi}\int_{-\pi}^{\pi} \cos x \, dx \\ &\quad + a_2 \frac{1}{2\pi}\int_{-\pi}^{\pi} \cos 2x \, dx + \cdots + b_1 \frac{1}{2\pi}\int_{-\pi}^{\pi} \sin x \, dx + \cdots. \end{aligned} \tag{5.5}$$

By (5.2), all the integrals on the right-hand side of (5.5) are zero except the first, because they are integrals of $\sin mx \cos nx$ or of $\cos mx \cos nx$ with $n = 0$ and $m \neq 0$ (that is, $m \neq n$). Then we have

$$\frac{1}{2\pi}\int_{-\pi}^{\pi} f(x)\, dx = \frac{a_0}{2}\frac{1}{2\pi}\int_{-\pi}^{\pi} dx = \frac{a_0}{2},$$

$$a_0 = \frac{1}{\pi}\int_{-\pi}^{\pi} f(x)\, dx. \tag{5.6}$$

Given $f(x)$ to be expanded in a Fourier series, we can now evaluate a_0 by calculating the integral in (5.6).

To find a_1, multiply both sides of (5.1) by $\cos x$ and again find the average value of each term:

$$\begin{aligned}\frac{1}{2\pi}\int_{-\pi}^{\pi} f(x)\cos x\,dx = \frac{a_0}{2}\frac{1}{2\pi}\int_{-\pi}^{\pi}\cos x\,dx + a_1\frac{1}{2\pi}\int_{-\pi}^{\pi}\cos^2 x\,dx \\ + a_2\frac{1}{2\pi}\int_{-\pi}^{\pi}\cos 2x\cos x\,dx + \cdots \\ + b_1\frac{1}{2\pi}\int_{-\pi}^{\pi}\sin x\cos x\,dx + \cdots.\end{aligned} \tag{5.7}$$

This time, by (5.2), all terms on the right are zero except the a_1 term and we have

$$\frac{1}{2\pi}\int_{-\pi}^{\pi} f(x)\cos x\,dx = a_1\frac{1}{2\pi}\int_{-\pi}^{\pi}\cos^2 x\,dx = \frac{1}{2}a_1.$$

Solving for a_1, we have

$$a_1 = \frac{1}{\pi}\int_{-\pi}^{\pi} f(x)\cos x\,dx.$$

The method should be clear by now, so we shall next find a general formula for a_n. Multiply both sides of (5.1) by $\cos nx$ and find the average value of each term:

$$\begin{aligned}\frac{1}{2\pi}\int_{-\pi}^{\pi} f(x)\cos nx\,dx = \frac{a_0}{2}\frac{1}{2\pi}\int_{-\pi}^{\pi}\cos nx\,dx + a_1\frac{1}{2\pi}\int_{-\pi}^{\pi}\cos x\cos nx\,dx \\ + a_2\frac{1}{2\pi}\int_{-\pi}^{\pi}\cos 2x\cos nx\,dx + \cdots \\ + b_1\frac{1}{2\pi}\int_{-\pi}^{\pi}\sin x\cos nx\,dx + \cdots.\end{aligned} \tag{5.8}$$

By (5.2), all terms on the right are zero except the a_n term and we have

$$\frac{1}{2\pi}\int_{-\pi}^{\pi} f(x)\cos nx\,dx = a_n\frac{1}{2\pi}\int_{-\pi}^{\pi}\cos^2 nx\,dx = \frac{1}{2}a_n.$$

Solving for a_n, we have

$$a_n = \frac{1}{\pi}\int_{-\pi}^{\pi} f(x)\cos nx\,dx. \tag{5.9}$$

Notice that this includes the $n = 0$ formula, but only because we called the constant term $\frac{1}{2}a_0$.

To obtain a formula for b_n, we multiply both sides of (5.1) by $\sin nx$ and take average values just as we did in deriving (5.9). We find (Problem 13)

$$b_n = \frac{1}{\pi}\int_{-\pi}^{\pi} f(x)\sin nx\,dx. \tag{5.10}$$

The formulas (5.9) and (5.10) will be used repeatedly in problems and should be memorized.

Example 1. Expand in a Fourier series the function $f(x)$ sketched in Figure 5.1. This function might represent, for example, a periodic voltage pulse. The terms of our Fourier series would then correspond to the different a-c frequencies which are combined in this "square wave" voltage, and the magnitude of the Fourier coefficients would indicate the relative importance of the various frequencies.

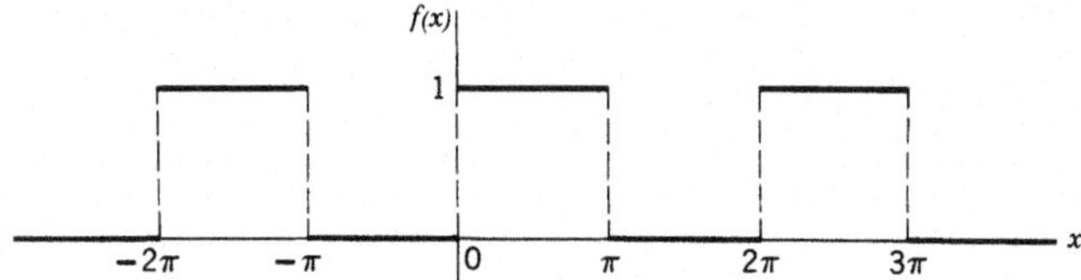

Figure 5.1

Note that $f(x)$ is a function of period 2π. Often in problems you will be given $f(x)$ for only one period; you should always sketch several periods so that you see clearly the periodic function you are expanding. For example, in this problem, instead of a sketch, you might have been given

$$f(x) = \begin{cases} 0, & -\pi < x < 0, \\ 1, & 0 < x < \pi. \end{cases} \tag{5.11}$$

It is then understood that $f(x)$ is to be continued periodically with period 2π outside the interval $(-\pi, \pi)$.

We use equations (5.9) and (5.10) to find a_n and b_n:

$$a_n = \frac{1}{\pi}\int_{-\pi}^{\pi} f(x)\cos nx\,dx = \frac{1}{\pi}\left[\int_{-\pi}^{0} 0\cdot\cos nx\,dx + \int_{0}^{\pi} 1\cdot\cos nx\,dx\right]$$

$$= \frac{1}{\pi}\int_0^{\pi}\cos nx\,dx = \begin{cases} \dfrac{1}{\pi}\cdot\dfrac{1}{n}\sin nx\Big|_0^{\pi} = 0 & \text{for } n \neq 0, \\ \dfrac{1}{\pi}\cdot\pi = 1 & \text{for } n = 0. \end{cases}$$

Thus $a_0 = 1$, and all other $a_n = 0$.

$$b_n = \frac{1}{\pi}\int_{-\pi}^{\pi} f(x)\sin nx\,dx = \frac{1}{\pi}\left[\int_{-\pi}^{0} 0\cdot\sin nx\,dx + \int_{0}^{\pi} 1\cdot\sin nx\,dx\right]$$

$$= \frac{1}{\pi}\int_0^{\pi}\sin nx\,dx = \frac{1}{\pi}\left[\frac{-\cos nx}{n}\right]_0^{\pi} = -\frac{1}{n\pi}[(-1)^n - 1]$$

$$= \begin{cases} 0 & \text{for even } n, \\ \dfrac{2}{n\pi} & \text{for odd } n. \end{cases}$$

Putting these values for the coefficients into (5.1), we have

$$f(x) = \frac{1}{2} + \frac{2}{\pi}\left(\frac{\sin x}{1} + \frac{\sin 3x}{3} + \frac{\sin 5x}{5} + \cdots\right). \tag{5.12}$$

Example 2. We can now find the Fourier series for some other functions without more evaluation of coefficients. For example, consider

$$g(x) = \begin{cases} -1, & -\pi < x < 0, \\ 1, & 0 < x < \pi. \end{cases} \tag{5.13}$$

Sketch this and verify that $g(x) = 2f(x) - 1$, where $f(x)$ is the function in Example 1. Then from (5.12), the Fourier series for $g(x)$ is

$$g(x) = \frac{4}{\pi}\left(\frac{\sin x}{1} + \frac{\sin 3x}{3} + \frac{\sin 5x}{5} + \cdots\right). \tag{5.14}$$

Similarly, verify that $h(x) = f(x + \pi/2)$ is Fig. 5.1 shifted $\pi/2$ to the left (sketch it), and its Fourier series is (replace x in (5.12) by $x + \pi/2$)

$$h(x) = \frac{1}{2} + \frac{2}{\pi}\left(\frac{\cos x}{1} - \frac{\cos 3x}{3} + \frac{\cos 5x}{5} + \cdots\right)$$

since $\sin(x + \pi/2) = \cos x$, $\sin(x + 3\pi/2) = -\cos 3x$, etc.

PROBLEMS, SECTION 5

In each of the following problems you are given a function on the interval $-\pi < x < \pi$. Sketch several periods of the corresponding periodic function of period 2π. Expand the periodic function in a sine-cosine Fourier series.

1. $f(x) = \begin{cases} 1, & -\pi < x < 0, \\ 0, & 0 < x < \pi. \end{cases}$

In this case the sketch is:

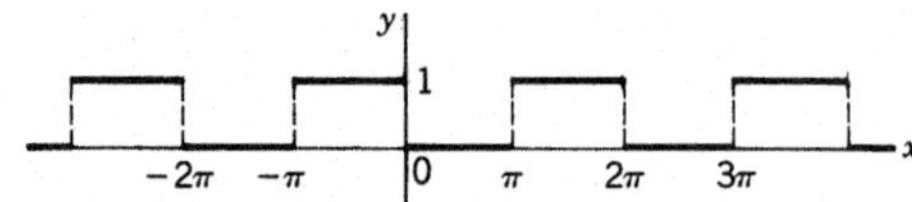

Your answer for the series is: $f(x) = \dfrac{1}{2} - \dfrac{2}{\pi}\left(\dfrac{\sin x}{1} + \dfrac{\sin 3x}{3} + \dfrac{\sin 5x}{5} \cdots\right)$.

Can you use the ideas of Example 2 to find this result without computation?

2. $f(x) = \begin{cases} 0, & -\pi < x < 0, \\ 1, & 0 < x < \dfrac{\pi}{2}, \\ 0, & \dfrac{\pi}{2} < x < \pi. \end{cases}$

Answer: $f(x) = \dfrac{1}{4} + \dfrac{1}{\pi}\left(\dfrac{\cos x}{1} - \dfrac{\cos 3x}{3} + \dfrac{\cos 5x}{5} \cdots\right)$

$+ \dfrac{1}{\pi}\left(\dfrac{\sin x}{1} + \dfrac{2\sin 2x}{2} + \dfrac{\sin 3x}{3} + \dfrac{\sin 5x}{5} \cdots\right)$.

3. $f(x) = \begin{cases} 0, & -\pi < x < \dfrac{\pi}{2}, \\ 1, & \dfrac{\pi}{2} < x < \pi. \end{cases}$

Answer: $f(x) = \dfrac{1}{4} - \dfrac{1}{\pi}\left(\dfrac{\cos x}{1} - \dfrac{\cos 3x}{3} + \dfrac{\cos 5x}{5} \cdots\right)$

$+ \dfrac{1}{\pi}\left(\dfrac{\sin x}{1} - \dfrac{2\sin 2x}{2} + \dfrac{\sin 3x}{3} + \dfrac{\sin 5x}{5} - \dfrac{2\sin 6x}{6} \cdots\right)$.

4. $f(x) = \begin{cases} -1, & -\pi < x < \dfrac{\pi}{2}, \\ 1, & \dfrac{\pi}{2} < x < \pi. \end{cases}$

Could you use Problem 3 to solve Problem 4 without computation?

5. $f(x) = \begin{cases} 0, & -\pi < x < 0, \\ -1, & 0 < x < \dfrac{\pi}{2}, \\ 1, & \dfrac{\pi}{2} < x < \pi. \end{cases}$

6. $f(x) = \begin{cases} 1, & -\pi < x < -\dfrac{\pi}{2}, \quad \text{and} \quad 0 < x < \dfrac{\pi}{2}; \\ 0, & -\dfrac{\pi}{2} < x < 0, \quad \text{and} \quad \dfrac{\pi}{2} < x < \pi. \end{cases}$

7. $f(x) = \begin{cases} 0, & -\pi < x < 0; \\ x, & 0 < x < \pi. \end{cases}$

Answer: $f(x) = \dfrac{\pi}{4} - \dfrac{2}{\pi}\left(\cos x + \dfrac{\cos 3x}{3^2} + \dfrac{\cos 5x}{5^2} + \cdots\right)$
$$+\left(\sin x - \frac{\sin 2x}{2} + \frac{\sin 3x}{3} - \cdots\right).$$

8. $f(x) = 1 + x, \quad -\pi < x < \pi.$

Answer: $f(x) = 1 + 2\left(\sin x - \dfrac{1}{2}\sin 2x + \dfrac{1}{3}\sin 3x - \dfrac{1}{4}\sin 4x + \cdots\right).$

9. $f(x) = \begin{cases} -x, & -\pi < x < 0, \\ x, & 0 < x < \pi. \end{cases}$

Answer: $f(x) = \dfrac{\pi}{2} - \dfrac{4}{\pi}\left(\cos x + \dfrac{1}{9}\cos 3x + \dfrac{1}{25}\cos 5x + \cdots\right).$

10. $f(x) = \begin{cases} \pi + x, & -\pi < x < 0, \\ \pi - x, & 0 < x < \pi. \end{cases}$

11. $f(x) = \begin{cases} 0, & -\pi < x < 0, \\ \sin x, & 0 < x < \pi. \end{cases}$

Answer: $f(x) = \dfrac{1}{\pi} + \dfrac{1}{2}\sin x - \dfrac{2}{\pi}\left(\dfrac{\cos 2x}{2^2 - 1} + \dfrac{\cos 4x}{4^2 - 1} + \dfrac{\cos 6x}{6^2 - 1} + \cdots\right).$

12. Show that in (5.2) the average values of $\sin mx \sin nx$ and of $\cos mx \cos nx$, $m \neq n$, are zero (over a period), by using the complex exponential forms for the sines and cosines as in (5.3).

13. Write out the details of the derivation of equation (5.10).

6. DIRICHLET CONDITIONS

Now we have a series, but there are still some questions that we ought to get answered. Does it converge, and if so, does it converge to the values of $f(x)$? You will find, if you try, that for most values of x the series in (5.12) does not respond to any of the tests for convergence that we discussed in Chapter 1. What is the sum of the series at $x = 0$ where $f(x)$ jumps from 0 to 1? You can see from the series (5.12) that the sum at $x = 0$ is $\frac{1}{2}$, but what does this have to do with $f(x)$?

These questions would not be easy for us to answer for ourselves, but they are answered for us for most practical purposes by the *theorem of Dirichlet*:

> If $f(x)$ is periodic of period 2π, and if between $-\pi$ and π it is single-valued, has a finite number of maximum and minimum values, and a finite number of discontinuities, and if $\int_{-\pi}^{\pi} |f(x)|\,dx$ is finite, then the Fourier series (5.1) [with coefficients given by (5.9) and (5.10)] converges to $f(x)$ at all the points where $f(x)$ is continuous; at jumps the Fourier series converges to the midpoint of the jump. (This includes jumps that occur at $\pm\pi$ for the periodic function.)

To see what all this means, we shall consider some special functions. We have already discussed what a periodic function means. A function $f(x)$ is single-valued if there is just one value of $f(x)$ for each x. For example, if $x^2 + y^2 = 1$, y is not a single-valued function of x, unless we select just $y = +\sqrt{1-x^2}$ or just $y = -\sqrt{1-x^2}$. An example of a function with an infinite number of maxima and minima is $\sin(1/x)$, which oscillates infinitely many times as $x \to 0$. If we imagine a function constructed from $\sin(1/x)$ by making $f(x) = 1$ for every x for which $\sin(1/x) > 0$, and $f(x) = -1$ for every x for which $\sin(1/x) < 0$, this function would have an infinite number of discontinuities. Now most functions in applied work do not behave like these, but will satisfy the Dirichlet conditions.

Finally, if $y = 1/x$, we find

$$\int_{-\pi}^{\pi} \left|\frac{1}{x}\right| dx = 2\int_0^{\pi} \frac{1}{x}\,dx = 2\ln x\Big|_0^1 = \infty,$$

so the function $1/x$ is ruled out by the Dirichlet conditions. On the other hand, if $f(x) = 1/\sqrt{|x|}$, then

$$\int_{-\pi}^{\pi} \frac{1}{\sqrt{|x|}}\,dx = 2\int_0^{\pi} \frac{dx}{\sqrt{x}} = 4\sqrt{x}\Big|_0^{\pi} = 4\sqrt{\pi},$$

so the periodic function which is $1/\sqrt{|x|}$ between $-\pi$ and π can be expanded in a Fourier series. In most problems it is not necessary to find the value of $\int_{-\pi}^{\pi} |f(x)|\,dx$; let us see why. If $f(x)$ is bounded (that is, all its values lie between $\pm M$ for some positive constant M), then

$$\int_{-\pi}^{\pi} |f(x)|dx \leq \int_{-\pi}^{\pi} M\,dx = M \cdot 2\pi$$

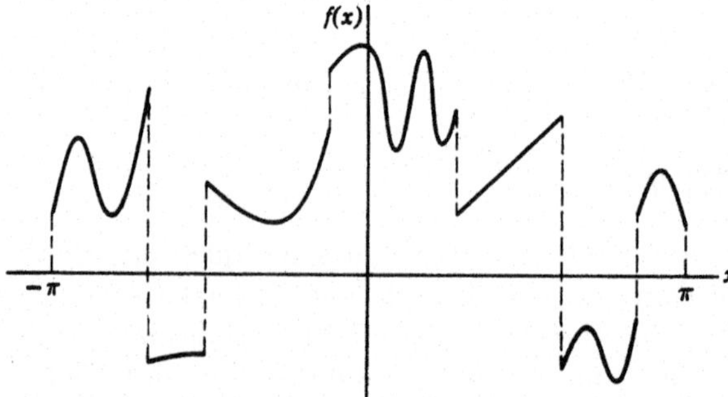

Figure 6.1

and so is finite. Thus you can simply verify that the function you are considering is bounded (if it is) instead of evaluating the integral. Figure 6.1 is an (exaggerated!) example of a function which satisfies the Dirichlet conditions on $(-\pi, \pi)$.

We see, then, that rather than testing Fourier series for convergence as we did power series, we instead check the given function; if it satisfies the Dirichlet conditions we are then sure that the Fourier series, when we get it, will converge to the function at points of continuity and to the midpoint of a jump. For example, consider the function $f(x)$ in Figure 5.1. Between $-\pi$ and π the given $f(x)$ is single-valued (one value for each x), bounded (between $+1$ and 0), has a finite number of maximum and minimum values (one of each), and a finite number of discontinuities (at $-\pi$, 0, and π), and therefore satisfies the Dirichlet conditions. Dirichlet's theorem then assures us that the series (5.12) actually converges to the function $f(x)$ in Figure 5.1 at all points except $x = n\pi$ where it converges to 1/2.

In Chapter 3, Sections 10 and 14, we defined a *basis* for ordinary 3-dimensional space as a set of linearly independent vectors (like **i**, **j**, **k**) in terms of which we could write every vector in the space. We then extended this idea to an n-dimensional space and to a space in which the basis vectors were functions. By analogy, we say here that the functions $\sin nx$, $\cos nx$ are a set of basis functions for the (infinite dimensional) space of all functions (satisfying Dirichlet conditions) defined on $(-\pi, \pi)$ or any 2π interval. (Also see "completeness relation" in Section 11. And for more examples of such sets of basis functions, see Chapters 12 and 13.)

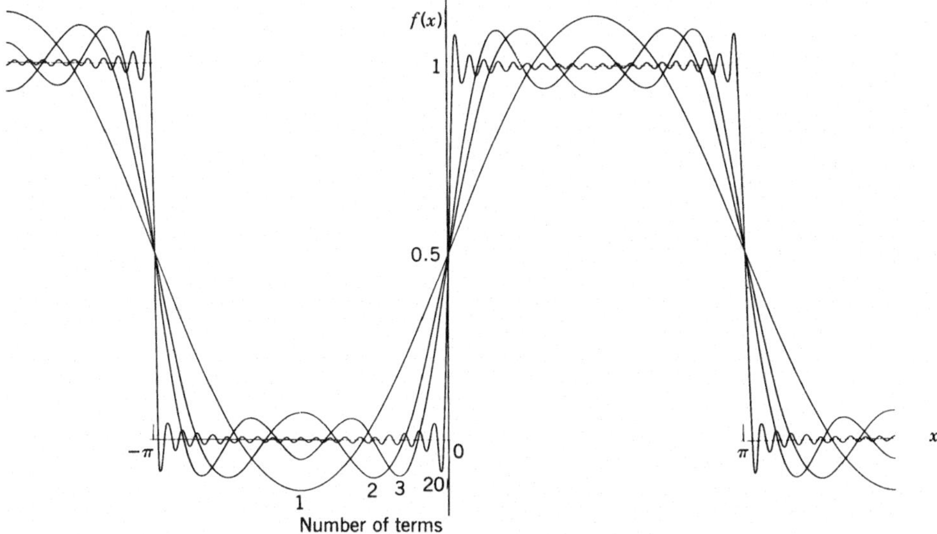

Figure 6.2

It is interesting to see a graph of the sum of a large number of terms of a Fourier series. Figure 6.2 shows several different partial sums of the series in (5.12) for the function in Figure 5.1. We can see that the sum of many terms of the series closely approximates the function away from the jumps and goes through the midpoint of the jump. The "overshoot" on either side of a jump bears comment. It does not disappear as we add more and more terms of the series. It simply becomes a narrower and narrower spike of height equal to about 9% of the jump. This fact is called the *Gibbs phenomenon.*

We ought to say here that the converse of Dirichlet's theorem is not true—if a function fails to satisfy the Dirichlet conditions, it still *may* be expandable in a Fourier series. The periodic function which is $\sin(1/x)$ on $(-\pi, \pi)$ is an example of such a function. However, such functions are rarely met with in practice.

Example. Fourier series can be useful in summing numerical series. Look at Problem 5.2 (sketch it). From Dirichlet's theorem, we see that the Fourier series converges to $1/2$ at $x = 0$. Let $x = 0$ in the Fourier series to get

$$\frac{1}{2} = \frac{1}{4} + \frac{1}{\pi}\left(1 - \frac{1}{3} + \frac{1}{5} - \frac{1}{7} + \cdots\right)$$

since $\sin 0 = 0$ and $\cos 0 = 1$. Thus

$$1 - \frac{1}{3} + \frac{1}{5} - \frac{1}{7} + \cdots = \frac{\pi}{4}.$$

PROBLEMS, SECTION 6

1 to **11**. For each of the periodic functions in Problems 5.1 to 5.11, use Dirichlet's theorem to find the value to which the Fourier series converges at $x = 0, \pm\pi/2, \pm\pi, \pm 2\pi$.

12. Use a computer to produce graphs like Fig. 6.2 showing Fourier series approximations to the functions in Problems 5.1 to 5.3, and 5.7 to 5.11. You might like to set up a computer animation showing the Gibbs phenomenon as the number of terms increases.

13. Repeat the example using the same Fourier series but at $x = \pi/2$.

14. Use Problem 5.7 to show that $\sum_{\text{odd } n} 1/n^2 = \pi^2/8$. Try $x = 0$, and $x = \pi$. What do you find at $x = \pi/2$?

15. Use Problem 5.11 to show that $\dfrac{1}{2^2-1} + \dfrac{1}{4^2-1} + \dfrac{1}{6^2-1} + \cdots = \dfrac{1}{2}$.

7. COMPLEX FORM OF FOURIER SERIES

Recall that real sines and cosines can be expressed in terms of complex exponentials by the formulas [Chapter 2, (11.3)]

$$\sin nx = \frac{e^{inx} - e^{-inx}}{2i}, \qquad \cos nx = \frac{e^{inx} + e^{-inx}}{2}. \tag{7.1}$$

If we substitute equations (7.1) into a Fourier series like (5.12), we get a series of terms of the forms e^{inx} and e^{-inx}. This is the complex form of a Fourier series. We can also find the complex form directly; this is often easier than finding the sine-cosine form. We can then, if we like, work back the other way and [using Euler's formula, Chapter 2, (9.3)] get the sine-cosine form from the exponential form.

We want to see how to find the coefficients in the complex form directly. We assume a series

$$\begin{aligned} f(x) &= c_0 + c_1 e^{ix} + c_{-1}e^{-ix} + c_2 e^{2ix} + c_{-2}e^{-2ix} + \cdots \\ &= \sum_{n=-\infty}^{n=+\infty} c_n e^{inx} \end{aligned} \tag{7.2}$$

and try to find the c_n's. From (5.4) we know that the average value of e^{ikx} on $(-\pi, \pi)$ is zero when k is an integer not equal to zero. To find c_0, we find the average values of the terms in (7.2):

$$\frac{1}{2\pi}\int_{-\pi}^{\pi} f(x)\,dx = c_0 \cdot \frac{1}{2\pi}\int_{-\pi}^{\pi} dx + \begin{cases}\text{average values of terms of the}\\ \text{form } e^{ikx} \text{ with } k \text{ an integer} \neq 0\end{cases}$$
$$= c_0 + 0, \tag{7.3}$$

$$c_0 = \frac{1}{2\pi}\int_{-\pi}^{\pi} f(x)\,dx. \tag{7.4}$$

To find c_n, we multiply (7.2) by e^{-inx} and again find the average value of each term. Note the minus sign in the exponent. In finding a_n, the coefficient of $\cos nx$ in equation (5.1), we multiplied by $\cos nx$; but here in finding the coefficient c_n of e^{inx}, we multiply by the complex conjugate e^{-inx}.

$$\frac{1}{2\pi}\int_{-\pi}^{\pi} f(x)e^{-inx}\,dx = c_0\frac{1}{2\pi}\int_{-\pi}^{\pi} e^{-inx}\,dx + c_1\frac{1}{2\pi}\int_{-\pi}^{\pi} e^{-inx}e^{ix}\,dx$$
$$+ c_{-1}\frac{1}{2\pi}\int_{-\pi}^{\pi} e^{-inx}e^{-ix}\,dx + \cdots. \tag{7.5}$$

The terms on the right are the average values of exponentials e^{ikx}, where the k values are integers. Therefore all these terms are zero except the one where $k = 0$; this is the term containing c_n. We then have

$$\frac{1}{2\pi}\int_{-\pi}^{\pi} f(x)e^{-inx}\,dx = c_n \cdot \frac{1}{2\pi}\int_{-\pi}^{\pi} e^{-inx}e^{inx}\,dx = c_n \cdot \frac{1}{2\pi}\int_{-\pi}^{\pi} dx = c_n,$$

$$c_n = \frac{1}{2\pi}\int_{-\pi}^{\pi} f(x)e^{-inx}\,dx. \tag{7.6}$$

Note that this formula contains the one for c_0 (no $\frac{1}{2}$ to worry about here!). Also, since (7.6) is valid for negative as well as positive n, you have only one formula to memorize here! You can easily show that for *real* $f(x)$, $c_{-n} = \bar{c}_n$ (Problem 12).

Example. Let us expand the same $f(x)$ we did before, namely (5.11). We have from (7.6)

$$c_n = \frac{1}{2\pi}\int_{-\pi}^{0} e^{-inx}\cdot 0\cdot dx + \frac{1}{2\pi}\int_{0}^{\pi} e^{-inx}\cdot 1\cdot dx$$
$$= \frac{1}{2\pi}\frac{e^{-inx}}{-in}\bigg|_0^{\pi} = \frac{1}{-2\pi in}(e^{-in\pi} - 1) = \begin{cases}\frac{1}{\pi in}, & n \text{ odd},\\ 0, & n \text{ even} \neq 0,\end{cases} \tag{7.7}$$
$$c_0 = \frac{1}{2\pi}\int_{0}^{\pi} dx = \frac{1}{2}.$$

Then

$$f(x) = \sum_{-\infty}^{\infty} c_n e^{inx} = \frac{1}{2} + \frac{1}{i\pi}\left(\frac{e^{ix}}{1} + \frac{e^{3ix}}{3} + \frac{e^{5ix}}{5} + \cdots\right) + \frac{1}{i\pi}\left(\frac{e^{-ix}}{-1} + \frac{e^{-3ix}}{-3} + \frac{e^{-5ix}}{-5} + \cdots\right). \tag{7.8}$$

It is interesting to verify that this is the same as the sine-cosine series we had before. We *could* use Euler's formula for each exponential, but it is easier to collect terms like this:

$$\begin{aligned} f(x) &= \frac{1}{2} + \frac{2}{\pi}\left(\frac{e^{ix} - e^{-ix}}{2i} + \frac{1}{3}\frac{e^{3ix} - e^{-3ix}}{2i} + \cdots\right) \\ &= \frac{1}{2} + \frac{2}{\pi}\left(\sin x + \frac{1}{3}\sin 3x + \cdots\right) \end{aligned} \tag{7.9}$$

which is the same as (5.12).

PROBLEMS, SECTION 7

1 to 11. Expand the same functions as in Problems 5.1 to 5.11 in Fourier series of complex exponentials e^{inx} on the interval $(-\pi, \pi)$ and verify in each case that the answer is equivalent to the one found in Section 5.

12. Show that if a real $f(x)$ is expanded in a complex exponential Fourier series $\sum_{-\infty}^{\infty} c_n e^{inx}$, then $c_{-n} = \bar{c}_n$, where $\bar{c}_n$ means the complex conjugate of c_n.

13. If $f(x) = \frac{1}{2}a_0 + \sum_1^\infty a_n \cos nx + \sum_1^\infty b_n \sin nx = \sum_{-\infty}^{\infty} c_n e^{inx}$, use Euler's formula to find a_n and b_n in terms of c_n and c_{-n}, and to find c_n and c_{-n} in terms of a_n and b_n.

8. OTHER INTERVALS

The functions $\sin nx$ and $\cos nx$ and e^{inx} have period 2π. We have been considering $(-\pi, \pi)$ as the basic interval of length 2π. Given $f(x)$ on $(-\pi, \pi)$, we have first sketched it for this interval, and then repeated our sketch for the intervals $(\pi, 3\pi)$, $(3\pi, 5\pi)$, $(-3\pi, -\pi)$, etc. There are (infinitely) many other intervals of length 2π, any one of which could serve as the basic interval. If we are given $f(x)$ on *any* interval of length 2π, we can sketch $f(x)$ for that given basic interval and then repeat it periodically with period 2π. We then want to expand the periodic function so obtained, in a Fourier series. Recall that in evaluating the Fourier coefficients, we used average values *over a period.* The formulas for the coefficients are then unchanged (except for the limits of integration) if we use other basic intervals of length 2π. In practice, the intervals $(-\pi, \pi)$ and $(0, 2\pi)$ are the ones most frequently used. For $f(x)$ defined on $(0, 2\pi)$ and then repeated periodically, (5.9), (5.10), and (7.6) would read

$$\begin{aligned} a_n &= \frac{1}{\pi}\int_0^{2\pi} f(x)\cos nx\,dx, \qquad b_n = \frac{1}{\pi}\int_0^{2\pi} f(x)\sin nx\,dx, \\ c_n &= \frac{1}{2\pi}\int_0^{2\pi} f(x)e^{-inx}\,dx, \end{aligned} \tag{8.1}$$

and (5.1) and (7.2) are unchanged.

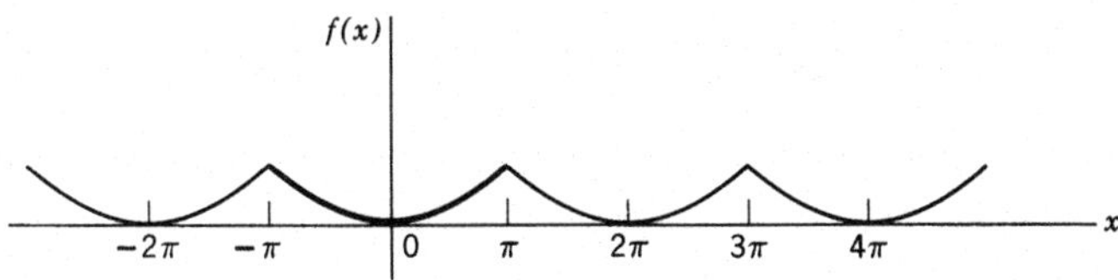

Figure 8.1

Notice how important it is to sketch a graph to see clearly what function you are talking about. For example, given $f(x) = x^2$ on $(-\pi, \pi)$, the extended function of period 2π is shown in Figure 8.1. But given $f(x) = x^2$ on $(0, 2\pi)$, the extended periodic function is different (see Figure 8.2). On the other hand, given $f(x)$ as in our example (5.11), or given $f(x) = 1$ on $(0, \pi)$, $f(x) = 0$ on $(\pi, 2\pi)$, you can easily verify by sketching that the graphs of the extended functions are identical. In this case you would get the same answer from either formulas (5.9), (5.10), and (7.6) or formulas (8.1).

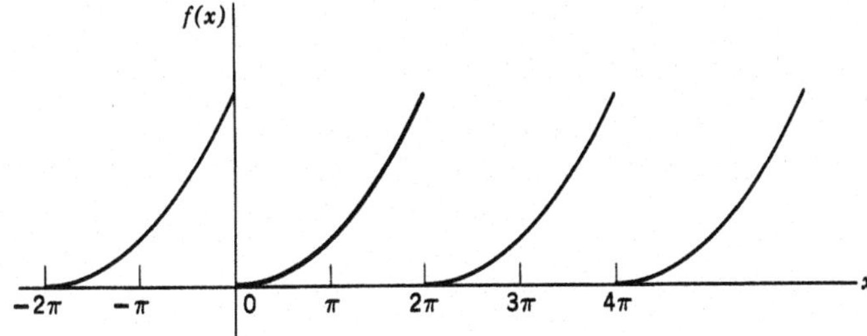

Figure 8.2

Physics problems do not always come to us with intervals of length 2π. Fortunately, it is easy now to change to other intervals. Consider intervals of length $2l$, say $(-l, l)$ or $(0, 2l)$. The function $\sin(n\pi x/l)$ has period $2l$, since

$$\sin \frac{n\pi}{l}(x + 2l) = \sin\left(\frac{n\pi x}{l} + 2n\pi\right) = \sin \frac{n\pi x}{l}.$$

Similarly, $\cos(n\pi x/l)$ and $e^{in\pi x/l}$ have period $2l$. Equations (5.1) and (7.2) are now replaced by

(8.2)
$$\begin{aligned} f(x) &= \frac{a_0}{2} + a_1 \cos\frac{\pi x}{l} + a_2 \cos\frac{2\pi x}{l} + \cdots \\ &\quad + b_1 \sin\frac{\pi x}{l} + b_2 \sin\frac{2\pi x}{l} + \cdots \\ &= \frac{a_0}{2} + \sum_1^\infty \left(a_n \cos\frac{n\pi x}{l} + b_n \sin\frac{n\pi x}{l}\right), \\ f(x) &= \sum_{-\infty}^{\infty} c_n e^{in\pi x/l}. \end{aligned}$$

We have already found the average values *over a period* of all the functions we need to use to find a_n, b_n, and c_n here. The period is now of length $2l$, say $-l$ to l, so in finding average values of the terms we replace

$$\frac{1}{2\pi}\int_{-\pi}^{\pi} \quad \text{by} \quad \frac{1}{2l}\int_{-l}^{l}.$$

Recall that the average of the square of either the sine or the cosine over a period is $\frac{1}{2}$ and the average of $e^{in\pi x/l}\cdot e^{-in\pi x/l}=1$ is 1. Then the formulas (5.9), (5.10), and (7.6) for the coefficients become

$$\begin{aligned}
a_n &= \frac{1}{l}\int_{-l}^{l} f(x)\cos\frac{n\pi x}{l}\,dx,\\
b_n &= \frac{1}{l}\int_{-l}^{l} f(x)\sin\frac{n\pi x}{l}\,dx, \qquad (8.3)\\
c_n &= \frac{1}{2l}\int_{-l}^{l} f(x)e^{-in\pi x/l}\,dx.
\end{aligned}$$

For the basic interval $(0, 2l)$ we need only change the integration limits to 0 to $2l$. The Dirichlet theorem just needs π replaced by l in order to apply here.

Example. Given $f(x)=\begin{cases}0, & 0<x<l,\\ 1, & l<x<2l.\end{cases}$

Expand $f(x)$ in an exponential Fourier series of period $2l$. [The function is given by the same formulas as (5.11) but on a different interval.]

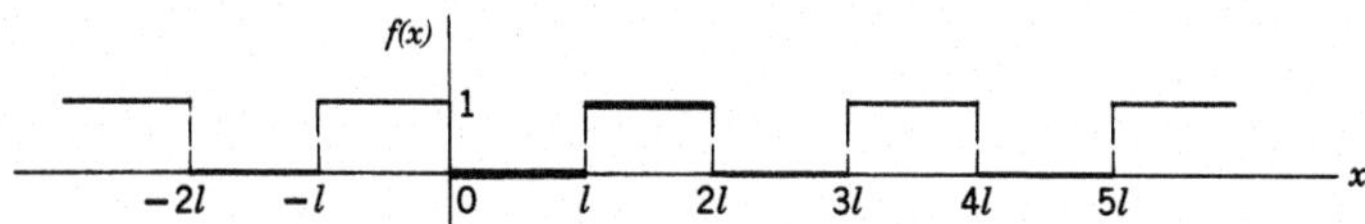

Figure 8.3

First we sketch a graph of $f(x)$ repeated with period $2l$ (Figure 8.3). By equations (8.3), we find

$$\begin{aligned}
c_n &= \frac{1}{2l}\int_0^l 0\cdot dx + \frac{1}{2l}\int_l^{2l} 1\cdot e^{-in\pi x/l}\,dx\\
&= \frac{1}{2l}\left.\frac{e^{-in\pi x/l}}{-in\pi/l}\right|_l^{2l} = \frac{1}{-2in\pi}(e^{-2in\pi}-e^{-in\pi})\\
&= \frac{1}{-2in\pi}(1-e^{in\pi}) = \begin{cases}0, & \text{even } n\neq 0,\\ -\dfrac{1}{in\pi}, & \text{odd } n,\end{cases} \qquad (8.4)\\
c_0 &= \frac{1}{2l}\int_l^{2l} dx = \frac{1}{2}.
\end{aligned}$$

Then,

$$f(x) = \frac{1}{2} - \frac{1}{i\pi}(e^{i\pi x/l} - e^{-i\pi x/l} + \frac{1}{3}e^{3i\pi x/l} - \frac{1}{3}e^{-3i\pi x/l} + \cdots)$$
$$= \frac{1}{2} - \frac{2}{\pi}\left(\sin\frac{\pi x}{l} + \frac{1}{3}\sin\frac{3\pi x}{l} + \cdots\right). \tag{8.5}$$

PROBLEMS, SECTION 8

1 to 9. In Problems 5.1 to 5.9, define each function by the formulas given but on the interval $(-l, l)$. [That is, replace $\pm\pi$ by $\pm l$ and $\pm\pi/2$ by $\pm l/2$.] Expand each function in a sine-cosine Fourier series and in a complex exponential Fourier series.

10. (a) Sketch several periods of the function $f(x)$ of period 2π which is equal to x on $-\pi < x < \pi$. Expand $f(x)$ in a sine-cosine Fourier series and in a complex exponential Fourier series.

Answer: $f(x) = 2(\sin x - \frac{1}{2}\sin 2x + \frac{1}{3}\sin 3x - \frac{1}{4}\sin 4x + \cdots)$.

(b) Sketch several periods of the function $f(x)$ of period 2π which is equal to x on $0 < x < 2\pi$. Expand $f(x)$ in a sine-cosine Fourier series and in a complex exponential Fourier series. Note that this is not the same function or the same series as (a).

Answer: $f(x) = \pi - 2\sum_{1}^{\infty}\frac{\sin nx}{n}$.

In Problems 11 to 14, parts (a) and (b), you are given in each case one period of a function. Sketch several periods of the function and expand it in a sine-cosine Fourier series, and in a complex exponential Fourier series.

11. (a) $f(x) = x^2$, $-\pi < x < \pi$; (b) $f(x) = x^2$, $0 < x < 2\pi$.

12. (a) $f(x) = e^x$, $-\pi < x < \pi$; (b) $f(x) = e^x$, $0 < x < 2\pi$.

13. (a) $f(x) = 2 - x$, $-2 < x < 2$; (b) $f(x) = 2 - x$, $0 < x < 4$.

14. (a) $f(x) = \sin\pi x$, $-\frac{1}{2} < x < \frac{1}{2}$; (b) $f(x) = \sin\pi x$, $0 < x < 1$.

15. Sketch (or computer plot) each of the following functions on the interval $(-1, 1)$ and expand it in a complex exponential series and in a sine-cosine series.

(a) $f(x) = x, \quad -1 < x < 1$.

Answer: $f(x) = \frac{2}{\pi}\sum_{1}^{\infty}(-1)^{n+1}\frac{\sin n\pi x}{n}$.

(b) $f(x) = \begin{cases} 1 + 2x, & -1 < x < 0, \\ 1 - 2x, & 0 < x < 1. \end{cases}$

Answer: $f(x) = \frac{8}{\pi^2}\sum_{\text{odd } n=1}^{\infty}\frac{\cos n\pi x}{n^2}$.

(c) $f(x) = \begin{cases} x + x^2, & -1 < x < 0, \\ x - x^2, & 0 < x < 1. \end{cases}$

Answer: $f(x) = \frac{8}{\pi^3}\sum_{\text{odd } n=1}^{\infty}\frac{\sin n\pi x}{n^3}$.

Each of the following functions is given over one period. Sketch several periods of the corresponding periodic function and expand it in an appropriate Fourier series.

16. $f(x) = x, \quad 0 < x < 2.$ *Answer:* $f(x) = 1 - \dfrac{2}{\pi}\displaystyle\sum_{1}^{\infty} \dfrac{\sin n\pi x}{n}.$

17. $f(x) = \begin{cases} 0, & -1 < x < 0, \\ 1, & 0 < x < 3. \end{cases}$

18. $f(x) = x^2, \quad 0 < x < 10.$

19. $f(x) = \begin{cases} 0, & -\frac{1}{2} < x < 0, \\ x, & 0 < x < \frac{1}{2}. \end{cases}$

20. $f(x) = \begin{cases} x/2, & 0 < x < 2, \\ 1, & 2 < x < 3. \end{cases}$

21. Write out the details of the derivation of the formulas (8.3).

9. EVEN AND ODD FUNCTIONS

An *even* function is one like x^2 or $\cos x$ (Figure 9.1) whose graph for negative x is just a reflection in the y axis of its graph for positive x. In formulas, the value of $f(x)$ is the same for a given x and its negative; that is

(9.1) $\qquad f(x) \quad \text{is even if} \quad f(-x) = f(x).$

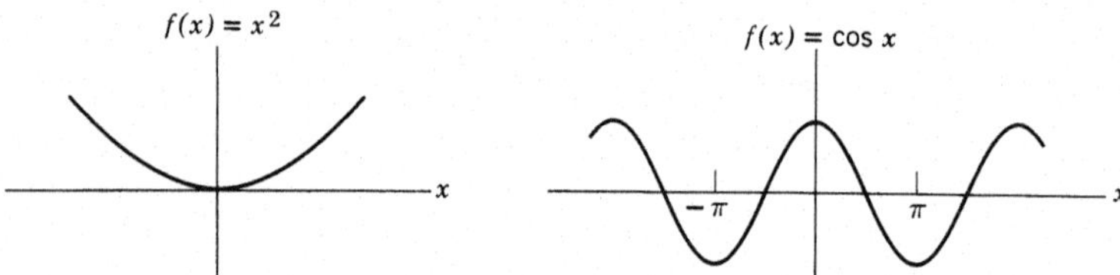

Figure 9.1

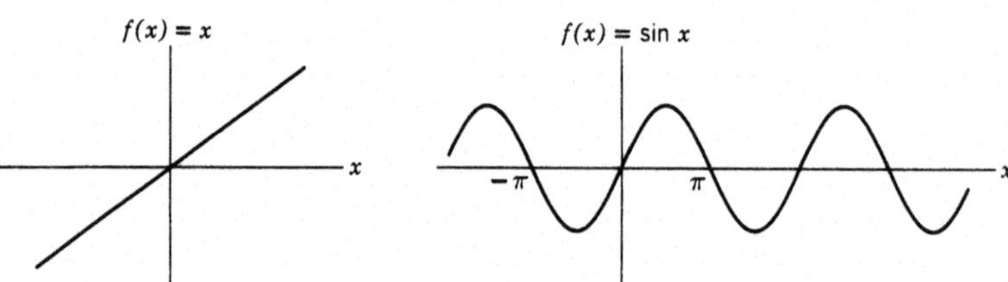

Figure 9.2

An odd function is one like x or $\sin x$ (Figure 9.2) for which the values of $f(x)$ and $f(-x)$ are negatives of each other. By definition

(9.2) $\qquad f(x) \quad \text{is odd if} \quad f(-x) = -f(x).$

Notice that even powers of x are even, and odd powers of x are odd; in fact, this is the reason for the names. You should verify (Problem 14) the following rules for the product of two functions: An even function times an even function, or an odd function times an odd function, gives an even function; an odd function times an even function gives an odd function. Some functions are even, some are odd, and some (for example, e^x) are neither. However, any function can be written as the sum of an even function and an odd function, like this:

$$f(x) = \frac{1}{2}[f(x) + f(-x)] + \frac{1}{2}[f(x) - f(-x)];$$

the first part is even and the second part is odd. For example,

$$e^x = \frac{1}{2}(e^x + e^{-x}) + \frac{1}{2}(e^x - e^{-x}) = \cosh x + \sinh x;$$

$\cosh x$ is even and $\sinh x$ is odd (look at the graphs).

Integrals of even functions or of odd functions, over symmetric intervals like $(-\pi, \pi)$ or $(-l, l)$, can be simplified. Look at the graph of $\sin x$ and think about $\int_{-\pi}^{\pi} \sin x\, dx$. The negative area from $-\pi$ to 0 cancels the positive area from 0 to π, so the integral is zero. This integral is still zero for any interval $(-l, l)$ which is symmetric about the origin, as you can see from the graph. The same is true for *any* odd $f(x)$; the areas to the left and to the right cancel. Next look at the cosine graph and the integral $\int_{-\pi/2}^{\pi/2} \cos x\, dx$. You see that the area from $-\pi/2$ to 0 is the same as the area from 0 to $\pi/2$. We could then just as well find the integral from 0 to $\pi/2$ and multiply it by 2. In general, if $f(x)$ is even, the integral of $f(x)$ from $-l$ to l is twice the integral from 0 to l. Then we have

$$\int_{-l}^{l} f(x)\, dx = \begin{cases} 0 & \text{if } f(x) \text{ is odd,} \\ 2\int_0^l f(x)\, dx & \text{if } f(x) \text{ is even.} \end{cases} \tag{9.3}$$

Suppose now that we are given a function on the interval $(0, l)$. If we want to represent it by a Fourier series of period $2l$, we must have $f(x)$ defined on $(-l, 0)$ too. There are several things we could do. We *could* define it to be zero (or, indeed, anything else) on $(-l, 0)$ and go ahead as we have done previously to find either an exponential or a sine-cosine series of period $2l$. However, it often happens in practice that we need (for physical reasons—see Chapter 13) to have an even function (or, in a different problem, an odd function). We first sketch the given function on $(0, l)$ (heavy lines in Figures 9.3 and 9.4). Then we extend the function on $(-l, 0)$ to be even or to be odd as required. To sketch more periods, just repeat the $(-l, l)$ sketch. (If the graph is complicated, it is helpful to trace it with a finger of one

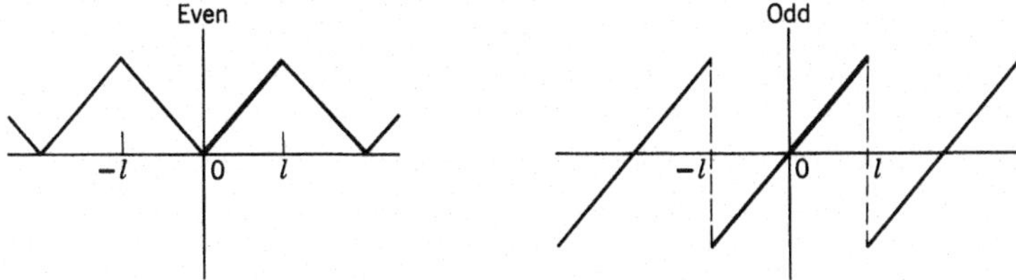

Figure 9.3

hand while you use the other hand to copy exactly what you are tracing. Turn the paper upside down to avoid crossing hands.)

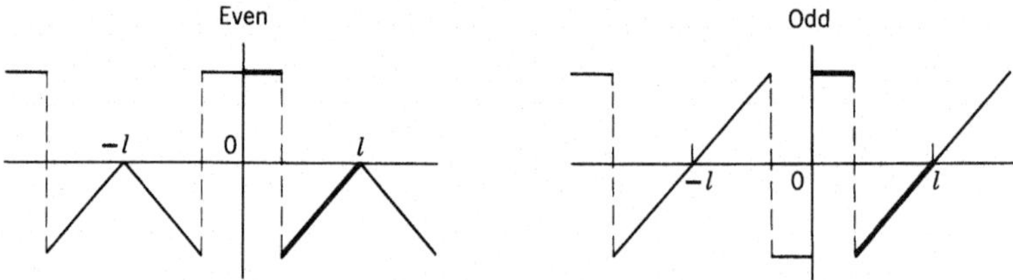

Figure 9.4

For even or odd functions, the coefficient formulas for a_n and b_n simplify. First suppose $f(x)$ is odd. Since sines are odd and cosines are even, $f(x)\sin(n\pi x/l)$ is even and $f(x)\cos(n\pi x/l)$ is odd. Then a_n is the integral, over a symmetric interval $(-l, l)$, of an odd function, namely $f(x)\cos(n\pi x/l)$; a_n is therefore zero. But b_n is the integral of an even function over a symmetric interval and is therefore twice the 0 to l integral. We have:

$$\text{If } f(x) \text{ is odd,} \quad \begin{cases} b_n = \dfrac{2}{l}\displaystyle\int_0^l f(x)\sin\frac{n\pi x}{l}\,dx, \\ a_n = 0. \end{cases} \tag{9.4}$$

We say that we have expanded $f(x)$ in a sine series ($a_n = 0$ so there are no cosine terms). Similarly, if $f(x)$ is even, all the b_n's are zero, and the a_n's are integrals of even functions. We have:

$$\text{If } f(x) \text{ is even,} \quad \begin{cases} a_n = \dfrac{2}{l}\displaystyle\int_0^l f(x)\cos\frac{n\pi x}{l}\,dx, \\ b_n = 0. \end{cases} \tag{9.5}$$

We say that $f(x)$ is expanded in a cosine series. (Remember that the constant term is $a_0/2$.)

You have now learned to find several different kinds of Fourier series that represent a given function $f(x)$ on, let us say, the interval $(0, 1)$. How do you know which to use in a given problem? You have to decide this from the physical problem when you are using Fourier series. There are two things to check: (1) the basic period involved in the physical problem; the functions in your series should have this period; and (2) the physical problem may require either an even function or an odd function for its solution; in these cases you must find the appropriate series. Now consider $f(x)$ defined on $(0, 1)$. We could find for it a sine-cosine or an exponential series of period 1 (that is, $l = \frac{1}{2}$):

$$f(x) = \sum_{-\infty}^{\infty} c_n e^{2in\pi x} \qquad \text{where} \qquad c_n = \int_0^1 f(x)e^{-2in\pi x}\,dx.$$

(The choice between sine-cosine and exponential series is just one of convenience in evaluating the coefficients—the series are really identical.) But we could also find two other Fourier series representing the same $f(x)$ on $(0,1)$. These series would have period 2 (that is, $l = 1$). One would be a cosine series

$$f(x) = \sum_{n=0}^{\infty} a_n \cos n\pi x, \qquad a_n = 2\int_0^1 f(x)\cos n\pi x\,dx, \qquad b_n = 0,$$

and represent an even function; the other would be a sine series and represent an odd function. In the problems, you may just be told to expand a function in a cosine series, say. You must then see for yourself what the period is when you have sketched an even function, and so choose the proper l in $\cos(n\pi x/l)$ and in the formula for a_n.

Example. Represent $f(x) = \begin{cases} 1, & 0 < x < \frac{1}{2}, \\ 0, & \frac{1}{2} < x < 1, \end{cases}$
by (a) a Fourier sine series, (b) a Fourier cosine series, (c) a Fourier series (the last ordinarily means a sine-cosine or exponential series whose period is the interval over which the function is given; in this case the period is 1).

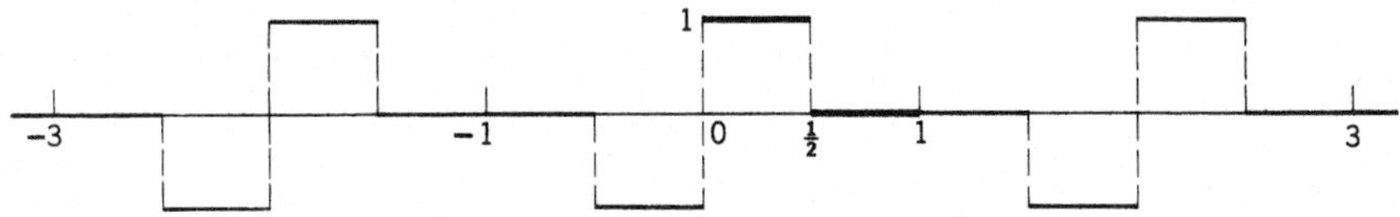

Figure 9.5

(a) Sketch the given function between 0 and 1. Extend it to the interval $(-1,0)$ making it odd. The period is now 2, that is, $l = 1$. Continue the function with period 2 (Figure 9.5). Since we now have an odd function, $a_n = 0$ and

$$\begin{aligned} b_n &= \frac{2}{1}\int_0^1 f(x)\sin n\pi x\,dx = 2\int_0^{1/2} \sin n\pi x\,dx \\ &= -\frac{2}{n\pi}\cos n\pi x\Big|_0^{1/2} = -\frac{2}{n\pi}\left(\cos\frac{n\pi}{2} - 1\right), \\ b_1 &= \frac{2}{\pi}, \quad b_2 = \frac{4}{2\pi}, \quad b_3 = \frac{2}{3\pi}, \quad b_4 = 0, \quad \cdots. \end{aligned}$$

Thus we obtain the *Fourier sine series* for $f(x)$:

$$f(x) = \frac{2}{\pi}\left(\sin \pi x + \frac{2\sin 2\pi x}{2} + \frac{\sin 3\pi x}{3} + \frac{\sin 5\pi x}{5} + \frac{2\sin 6\pi x}{6} + \cdots\right).$$

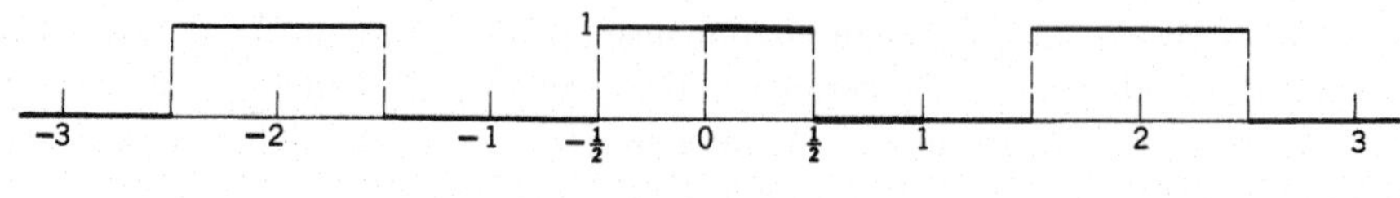

Figure 9.6

(b) Sketch an even function of period 2 (Figure 9.6).
Here $l = 1$, $b_n = 0$, and

$$a_0 = 2\int_0^1 f(x)\,dx = 2\int_0^{1/2} dx = 1,$$
$$a_n = 2\int_0^1 f(x)\cos n\pi x\,dx = \frac{2}{n\pi}\sin n\pi x\bigg|_0^{1/2} = \frac{2}{n\pi}\sin\frac{n\pi}{2}.$$

Then the *Fourier cosine series* for $f(x)$ is

$$f(x) = \frac{1}{2} + \frac{2}{\pi}\left(\frac{\cos\pi x}{1} - \frac{\cos 3\pi x}{3} + \frac{\cos 5\pi x}{5} - \cdots\right).$$

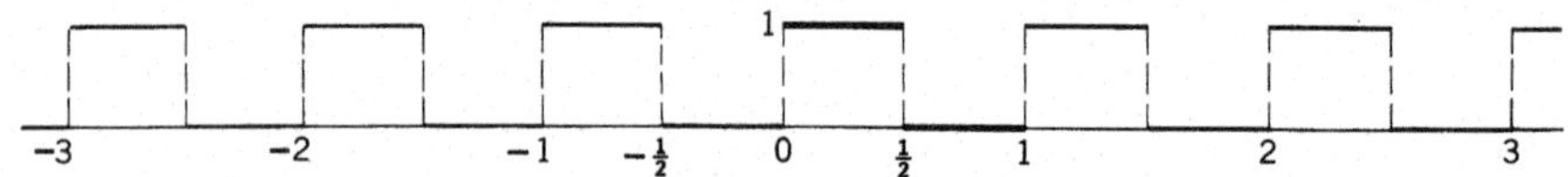

Figure 9.7

(c) Sketch the given function on $(0,1)$ and continue it with period 1 (Figure 9.7). Here $2l = 1$, and we find c_n as we did in the example of Section 8. As in that example, the exponential series here can then be put in sine-cosine form.

$$c_n = \int_0^1 f(x)e^{-2in\pi x}\,dx = \int_0^{1/2} e^{-2in\pi x}\,dx$$
$$= \frac{1-e^{-in\pi}}{2in\pi} = \frac{1-(-1)^n}{2in\pi} = \begin{cases}\dfrac{1}{in\pi}, & n \text{ odd},\\ 0, & n \text{ even} \neq 0.\end{cases}$$
$$c_0 = \int_0^{1/2} dx = \frac{1}{2}.$$
$$f(x) = \frac{1}{2} + \frac{1}{i\pi}\left(e^{2i\pi x} - e^{-2i\pi x} + \frac{1}{3}e^{6\pi ix} - \frac{1}{3}e^{-6\pi ix} + \cdots\right)$$
$$= \frac{1}{2} + \frac{2}{\pi}\left(\sin 2\pi x + \frac{\sin 6\pi x}{3} + \cdots\right).$$

Alternatively we can find both a_n and b_n directly.

$$a_0 = 2\int_0^1 f(x)\,dx = 2\int_0^{1/2} dx = 1.$$

$$a_n = 2\int_0^{1/2} \cos 2n\pi x\,dx = 0.$$

$$b_n = 2\int_0^{1/2} \sin 2n\pi x\,dx = \frac{1}{n\pi}(1-\cos n\pi) = \frac{1}{n\pi}[1-(-1)^n].$$

$$b_1 = \frac{2}{\pi}, \quad b_2 = 0, \quad b_3 = \frac{2}{3\pi}, \quad b_4 = 0, \quad \cdots.$$

There is one other very useful point to notice about even and odd functions. If you are given a function on $(-l, l)$ to expand in a sine-cosine series (of period $2l$) and happen to notice that it is an even function, you should realize that the b_n's are all going to be zero and you do not have to work them out. Also the a_n's can be written as twice an integral from 0 to l just as in (9.5). Similarly, if the given function is odd, you can use (9.4). Recognizing this may save you a good deal of algebra.

Differentiating Fourier Series Now that we have a supply of Fourier series for reference, let's discuss the question of differentiating a Fourier series term by term. First consider a Fourier series in which a_n and b_n are proportional to $1/n$. Since the derivative of $\frac{1}{n}\sin nx$ is $\cos nx$ (and a similar result for the cosine terms), we see that the differentiated series has no $1/n$ factors to make it converge. Now you might suspect (correctly) that if you can't differentiate the Fourier series, then the function $f(x)$ which it represents can't be differentiated either, at least not at all points. Turn back to examples and problems for which the Fourier series have coefficients proportional to $1/n$ and look at the graphs (or sketch them). Note in every case that $f(x)$ is discontinuous (that is, has jumps) at some points, and so can't be differentiated there. Next consider Fourier series with a_n and b_n proportional to $1/n^2$. If we differentiate such a series once, there are still $1/n$ factors left but we can't differentiate it twice. In that case we would (correctly) expect the function to be continuous with a discontinuous first derivative. (Look for examples.) Continuing, if a_n and b_n are proportional to $1/n^3$, we can find two derivatives, but the second derivative is discontinuous, and so on for Fourier coefficients proportional to higher powers of $1/n$. (See Problems 26 and 27.)

It is interesting to plot (by computer) a given function together with enough terms of its Fourier series to give a reasonable fit. In Section 5 we did this for discontinuous functions and it took many terms of the series. You will find (see Problems 26 and 27) that the more continuous derivatives a function has, the fewer terms of its Fourier series are required to approximate it. We can understand this: The higher order terms oscillate more rapidly (compare $\sin x$, $\sin 2x$, $\sin 10x$), and this rapid oscillation is what is needed to fit a curve which is changing rapidly (for example, a jump). But if $f(x)$ has several continuous derivatives, then it is quite "smooth" and doesn't require so much of the rapid oscillation of the higher order terms. This is reflected in the dependence of the Fourier coefficients on a power of $1/n$.

PROBLEMS, SECTION 9

The functions in Problems 1 to 3 are neither even nor odd. Write each of them as the sum of an even function and an odd function.

1. (a) e^{inx} (b) xe^x

2. (a) $\ln|1-x|$ (b) $(1+x)(\sin x+\cos x)$

3. (a) $x^5-x^4+x^3-1$ (b) $1+e^x$

4. Using what you know about even and odd functions, prove the first part of (5.2).

Each of the functions in Problems 5 to 12 is given over one period. For each function, sketch several periods and decide whether it is even or odd. Then use (9.4) or (9.5) to expand it in an appropriate Fourier series.

5. $f(x)=\begin{cases}-1, & -\pi<x<0,\\ 1, & 0<x<\pi.\end{cases}$

6. $f(x)=\begin{cases}-1, & -l<x<0,\\ 1, & 0<x<l.\end{cases}$

Answer: $f(x)=\dfrac{4}{\pi}\left(\sin\dfrac{\pi x}{l}+\dfrac{1}{3}\sin\dfrac{3\pi x}{l}+\dfrac{1}{5}\sin\dfrac{5\pi x}{l}+\cdots\right).$

7. $f(x)=\begin{cases}1, & -1<x<1,\\ 0, & -2<x<-1 \text{ and } 1<x<2.\end{cases}$

8. $f(x)=x,\quad -\dfrac{\pi}{2}<x<\dfrac{\pi}{2}.$

9. $f(x)=x^2,\quad -\frac{1}{2}<x<\frac{1}{2}.$

Answer: $f(x)=\dfrac{1}{12}-\dfrac{1}{\pi^2}\left(\cos 2\pi x-\dfrac{1}{2^2}\cos 4\pi x+\dfrac{1}{3^2}\cos 6\pi x-\cdots\right).$

10. $f(x)=|x|,\quad -\dfrac{\pi}{2}<x<\dfrac{\pi}{2}.$

11. $f(x)=\cosh x,\quad -\pi<x<\pi.$

Answer: $f(x)=\dfrac{2\sinh\pi}{\pi}\left(\dfrac{1}{2}-\dfrac{1}{2}\cos x+\dfrac{1}{5}\cos 2x-\dfrac{1}{10}\cos 3x+\dfrac{1}{17}\cos 4x-\cdots\right).$

12. $f(x)=\begin{cases}x+1, & -1<x<0,\\ x-1, & 0<x<1.\end{cases}$

13. Give algebraic proofs of (9.3). *Hint:* Write $\int_{-l}^{l}=\int_{-l}^{0}+\int_{0}^{l}$, make the change of variable $x=-t$ in $\int_{-l}^{0}$, and use the definition of even or odd function.

14. Give algebraic proofs that for even and odd functions:

(a) even times even = even; odd times odd = even; even times odd = odd;

(b) the derivative of an even function is odd; the derivative of an odd function is even.

15. Given $f(x)=x$ for $0<x<1$, sketch the even function f_c of period 2 and the odd function f_s of period 2, each of which equals $f(x)$ on $0<x<1$. Expand f_c in a cosine series and f_s in a sine series.

Answer: $f_c(x)=\dfrac{1}{2}-\dfrac{4}{\pi^2}\left(\cos\pi x+\dfrac{1}{3^2}\cos 3\pi x+\cdots\right),$

$f_s(x)=\dfrac{2}{\pi}\left(\sin\pi x-\dfrac{1}{2}\sin 2\pi x+\dfrac{1}{3}\sin 3\pi x-\cdots\right).$

16. Let $f(x) = \sin^2 x$, $0 < x < \pi$. Sketch (or computer plot) the even function f_c of period 2π, the odd function f_s of period 2π, and the function f_p of period π, each of which is equal to $f(x)$ on $(0, \pi)$. Expand each of these functions in an appropriate Fourier series.

In Problems 17 to 22 you are given $f(x)$ on an interval, say $0 < x < b$. Sketch several periods of the even function f_c of period $2b$, the odd function f_s of period $2b$, and the function f_p of period b, each of which equals $f(x)$ on $0 < x < b$. Expand each of the three functions in an appropriate Fourier series.

17. $f(x) = \begin{cases} 1, & 0 < x < \frac{1}{2}, \\ -1, & \frac{1}{2} < x < 1. \end{cases}$

18. $f(x) = \begin{cases} 1, & 0 < x < 1, \\ 0, & 1 < x < 3. \end{cases}$

19. $f(x) = |\cos x|$, $0 < x < \pi$.

20. $f(x) = x^2$, $0 < x < 1$.

21. $f(x) = \begin{cases} x, & 0 < x < 1, \\ 2 - x, & 1 < x < 2. \end{cases}$

22. $f(x) = \begin{cases} 10, & 0 < x < 10, \\ 20, & 10 < x < 20. \end{cases}$

23. If a violin string is plucked (pulled aside and let go), it is possible to find a formula $f(x, t)$ for the displacement at time t of any point x of the vibrating string from its equilibrium position. It turns out that in solving this problem we need to expand the function $f(x, 0)$, whose graph is the initial shape of the string, in a Fourier sine series. Find this series if a string of length l is pulled aside a small distance h at its center, as shown.

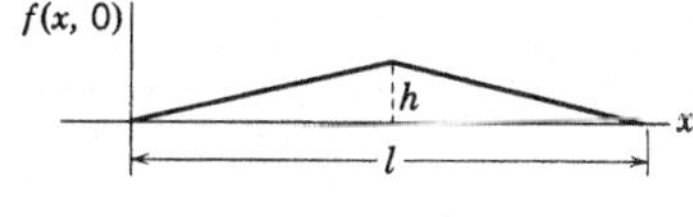

24. If, in Problem 23, the string is stopped at the center and half of it is plucked, then the function to be expanded in a sine series is shown here. Find the series. *Caution:* Note that $f(x, 0) = 0$ for $l/2 < x < l$.

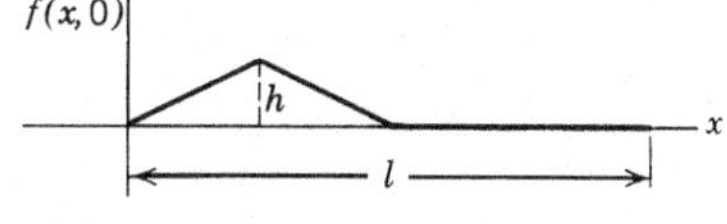

25. Suppose that $f(x)$ and its derivative $f'(x)$ are both expanded in Fourier series on $(-\pi, \pi)$. Call the coefficients in the $f(x)$ series a_n and b_n and the coefficients in the $f'(x)$ series a'_n and b'_n. Write the integral for a_n [equation (5.9)] and integrate it by parts to get an integral of $f'(x) \sin nx$. Recognize this integral in terms of b'_n [equation (5.10) for $f'(x)$] and so show that $b'_n = -na_n$. (In the integration by parts, the integrated term is zero because $f(\pi) = f(-\pi)$ since f is continuous—sketch several periods.). Find a similar relation for a'_n and b_n. Now show that this is the result you get by differentiating the $f(x)$ series term by term. Thus you have shown that the Fourier series for $f'(x)$ is correctly given by differentiating the $f(x)$ series term by term (assuming that $f'(x)$ is expandable in a Fourier series).

In Problems 26 and 27, find the indicated Fourier series. Then differentiate your result repeatedly (both the function and the series) until you get a discontinuous function. Use a computer to plot $f(x)$ and the derivative functions. For each graph, plot on the same axes one or more terms of the corresponding Fourier series. Note the number of terms needed for a good fit (see comment at the end of the section).

26. $f(x) = \begin{cases} 3x^2 + 2x^3, & -1 < x < 0, \\ 3x^2 - 2x^3, & 0 < x < 1. \end{cases}$

27. $f(x) = (x^2 - \pi^2)^2$, $-\pi < x < \pi$.

10. AN APPLICATION TO SOUND

We have said that when a sound wave passes through the air and we hear it, the air pressure where we are varies with time. Suppose the excess pressure above (and below) atmospheric pressure in a sound wave is given by the graph in Figure 10.1. (We shall not be concerned here with the units of p; however, reasonable units in Figure 10.1 would be p in 10^{-6} atmospheres.) Let us ask what frequencies we hear when we listen to this sound. To find out, we expand $p(t)$ in a Fourier series. The period of $p(t)$ is $\frac{1}{262}$; that is, the sound wave repeats itself 262 times per second. We have called the period $2l$ in our formulas, so here $l = \frac{1}{524}$. The functions we have called $\sin(n\pi x/l)$ here become $\sin 524n\pi t$. We can save some work by observing

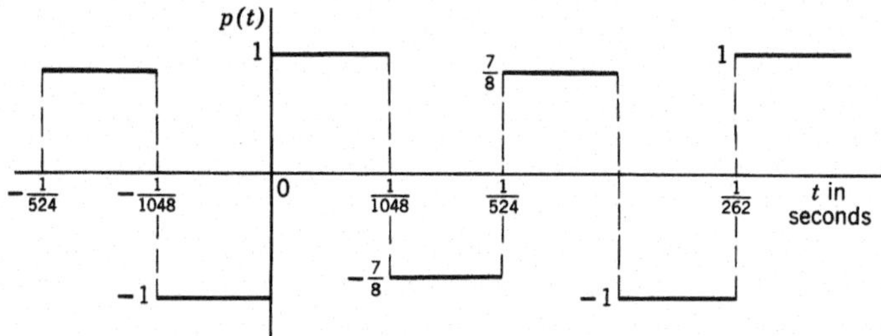

Figure 10.1

that $p(t)$ is an odd function; there are then only sine terms in its Fourier series and we need to compute only b_n. Using (9.4), we have

$$
\begin{aligned}
(10.1) \qquad b_n &= 2(524)\int_0^{1/524} p(t)\sin 524n\pi t\,dt \\
&= 1048\int_0^{1/1048} \sin 524n\pi t\,dt - \frac{7}{8}(1048)\int_{1/1048}^{1/524} \sin 524n\pi t\,dt \\
&= 1048\left(-\frac{\cos\frac{n\pi}{2}-1}{524n\pi} + \frac{7}{8}\frac{\cos n\pi - \cos\frac{n\pi}{2}}{524n\pi}\right) \\
&= \frac{2}{n\pi}\left(-\frac{15}{8}\cos\frac{n\pi}{2} + 1 + \frac{7}{8}\cos n\pi\right).
\end{aligned}
$$

From this we can compute the values of b_n for the first few values of n:

$$
(10.2) \qquad
\begin{aligned}
b_1 &= \frac{2}{\pi}\left(1-\frac{7}{8}\right) = \frac{2}{\pi}\left(\frac{1}{8}\right) = \frac{1}{\pi}\cdot\frac{1}{4} & b_5 &= \frac{1}{5\pi}\cdot\frac{1}{4} \\
b_2 &= \frac{2}{2\pi}\left(\frac{15}{8}+1+\frac{7}{8}\right) = \frac{1}{2\pi}\left(\frac{15}{2}\right) & b_6 &= \frac{1}{6\pi}\left(\frac{15}{2}\right) \\
b_3 &= \frac{2}{3\pi}\left(1-\frac{7}{8}\right) = \frac{1}{3\pi}\cdot\frac{1}{4} & b_7 &= \frac{1}{7\pi}\cdot\frac{1}{4} \\
b_4 &= \frac{2}{4\pi}\left(-\frac{15}{8}+1+\frac{7}{8}\right) = 0 & b_8 &= 0, \quad \text{etc.}
\end{aligned}
$$

Then we have

$$(10.3) \qquad p(t) = \frac{1}{4\pi}\left(\frac{\sin 524\pi t}{1} + \frac{30\sin(524\cdot 2\pi t)}{2} + \frac{\sin(524\cdot 3\pi t)}{3} + \frac{\sin(524\cdot 5\pi t)}{5} + \frac{30\sin(524\cdot 6\pi t)}{6} + \frac{\sin(524\cdot 7\pi t)}{7} + \cdots\right).$$

We can see just by looking at the coefficients that the most important term is the second one. The first term corresponds to the fundamental with frequency 262 vibrations per second (this is approximately middle C on a piano). But it is much weaker in this case than the first overtone (second harmonic) corresponding to the second term; this tone has frequency 524 vibrations per second (approximately high C). (You might like to use a computer to play one or several terms of the series.) The sixth harmonic (corresponding to $n = 6$) and also the harmonics for $n = 10$, 14, 18, 22, and 26 are all more prominent (that is, have larger coefficients) than the fundamental. We can be even more specific about the relative importance of the various frequencies. Recall that in discussing a simple harmonic oscillator, we showed that its average energy was proportional to the square of its velocity amplitude. It can be proved that the intensity of a sound wave (average energy striking unit area of your ear per second) is proportional to the average of the square of the excess pressure. Thus for a sinusoidal pressure variation $A\sin 2\pi ft$, the intensity is proportional to A^2. In the Fourier series for $p(t)$, the intensities of the various harmonics are then proportional to the squares of the corresponding Fourier coefficients. (The intensity corresponds roughly to the loudness of the tone—not exactly because the ear is not uniformly sensitive to all frequencies.) The relative intensities of the harmonics in our example are then:

n	=	1	2	3	4	5	6	7	8	9	10	$\cdots$
Relative intensity	=	1	225	$\frac{1}{9}$	0	$\frac{1}{25}$	25	$\frac{1}{49}$	0	$\frac{1}{81}$	9	$\cdots$

From this we see even more clearly that we would hear principally the second harmonic with frequency 524 (high C).

PROBLEMS, SECTION 10

In Problems 1 to 3, the graphs sketched represent one period of the excess pressure $p(t)$ in a sound wave. Find the important harmonics and their relative intensities. Use a computer to play individual terms or a sum of several terms of the series.

1.

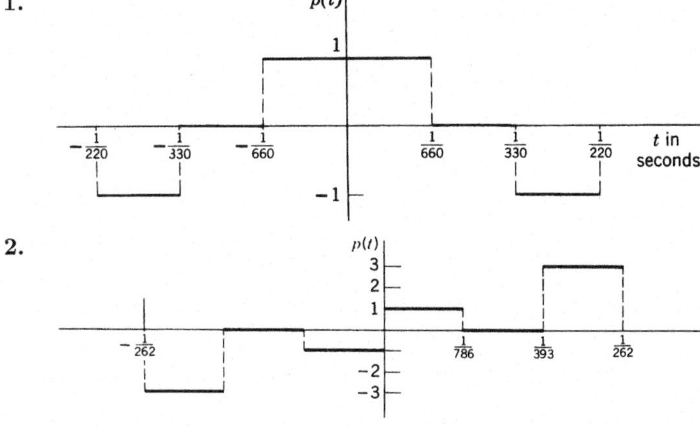

2.

3.

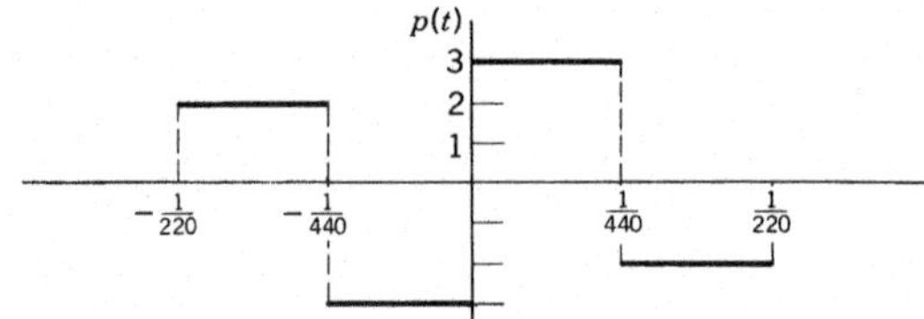

In Problems 4 to 10, the sketches show several practical examples of electrical signals (voltages or currents). In each case we want to know the harmonic content of the signal, that is, what frequencies it contains and in what proportions. To find this, expand each function in an appropriate Fourier series. Assume in each case that the part of the graph shown is repeated sixty times per second.

4. Output of a simple d-c generator; the shape of the curve is the absolute value of a sine function. Let the maximum voltage be 100 v.

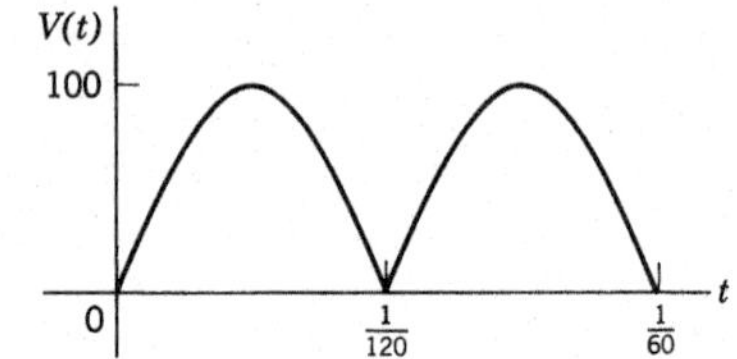

5. Rectified half-wave; the curve is a sine function for half the cycle and zero for the other half. Let the maximum current be 5 amp. *Hint:* Be careful! The value of l here is $1/60$, but $I(t) = \sin t$ only from $t = 0$ to $t = 1/120$.

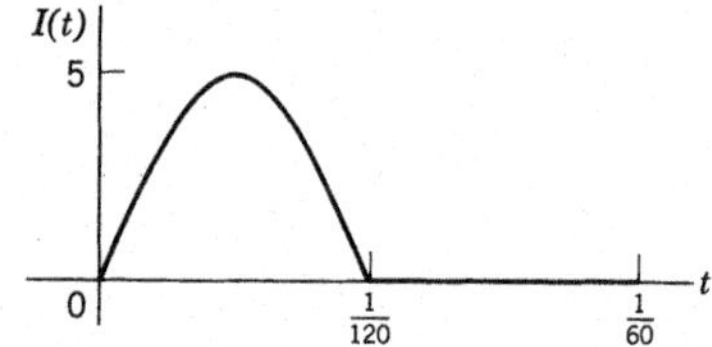

6. Triangular wave; the graph consists of two straight lines whose equations you must write! The maximum voltage of 100 v occurs at the middle of the cycle.

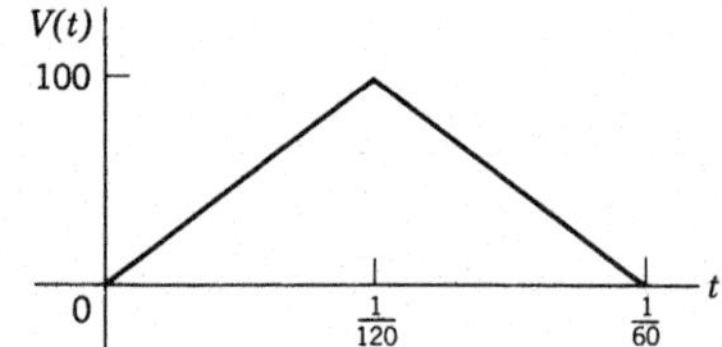

7. Sawtooth

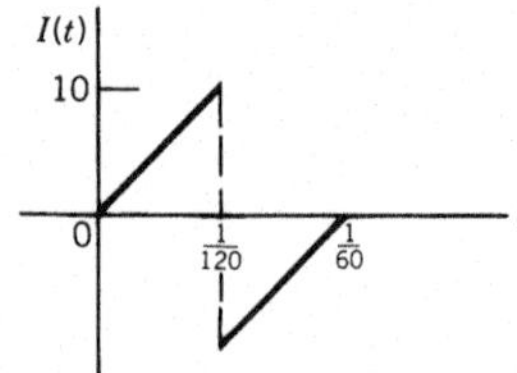

8. Rectified sawtooth

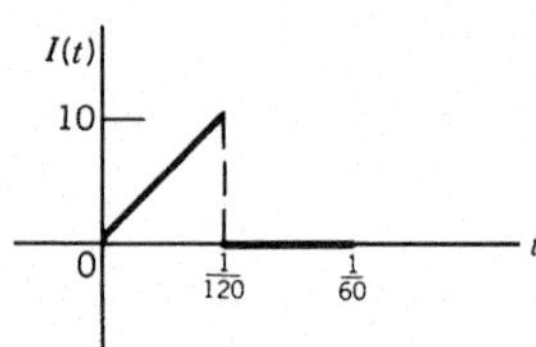

9. Square wave

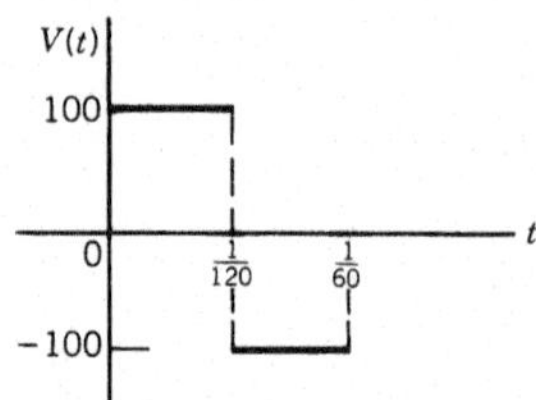

10. Periodic ramp function

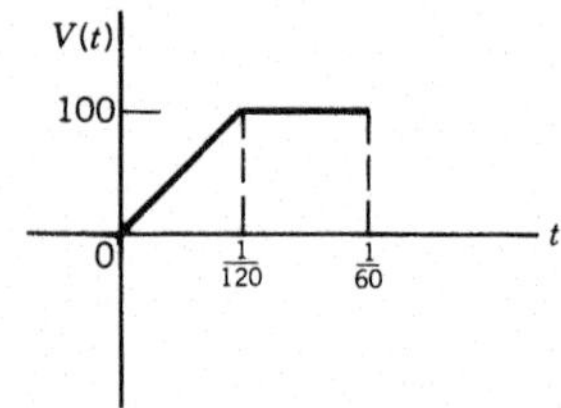

11. PARSEVAL'S THEOREM

We shall now find a relation between the average of the square (or absolute square) of $f(x)$ and the coefficients in the Fourier series for $f(x)$, assuming that $\int_{-\pi}^{\pi} |f(x)|^2\,dx$ is finite. The result is known as *Parseval's theorem* or the *completeness relation.* You should understand that the point of the theorem is *not* to get the average of the square of a given $f(x)$ by using its Fourier series. [Given $f(x)$, it is easy to get its average square just by doing the integration in (11.2) below.] The point of the theorem is to show the *relation* between the average of the square of $f(x)$ and the Fourier coefficients. We can derive a form of Parseval's theorem from any of the various Fourier expansions we have made; let us use (5.1).

$$f(x) = \frac{1}{2}a_0 + \sum_1^\infty a_n \cos nx + \sum_1^\infty b_n \sin nx. \tag{11.1}$$

We square $f(x)$ and then average the square over $(-\pi, \pi)$:

$$\text{The average of } [f(x)]^2 \text{ is } \frac{1}{2\pi}\int_{-\pi}^{\pi} [f(x)]^2 dx. \tag{11.2}$$

When we square the Fourier series in (11.1) we get many terms. To avoid writing out a large number of them, consider instead what types of terms there are and what the averages of the different kinds of terms are. First, there are the squares of the individual terms. Using the fact that the average of the square of a sine or cosine over a period is $\frac{1}{2}$, we have:

$$\begin{aligned} &\text{The average of } (\tfrac{1}{2}a_0)^2 && \text{is} \quad (\tfrac{1}{2}a_0)^2.\\ &\text{The average of } (a_n \cos nx)^2 && \text{is} \quad a_n^2 \cdot \tfrac{1}{2}.\\ &\text{The average of } (b_n \sin nx)^2 && \text{is} \quad b_n^2 \cdot \tfrac{1}{2}.\end{aligned} \tag{11.3}$$

Then there are cross-product terms of the forms $2 \cdot \frac{1}{2}a_0 a_n \cos nx$, $2 \cdot \frac{1}{2}a_0 b_n \sin nx$, and $2a_n b_m \cos nx \sin mx$ with $m \neq n$ (we write n in the cosine factor and m in the sine factor since every sine term must be multiplied times every cosine term). By (5.2), the average values of terms of all these types are zero. Then we have

$$\text{The average of } [f(x)]^2 \text{ (over a period)} = \left(\frac{1}{2}a_0\right)^2 + \frac{1}{2}\sum_1^\infty a_n^2 + \frac{1}{2}\sum_1^\infty b_n^2. \tag{11.4}$$

This is one form of Parseval's theorem. You can easily verify (Problem 1) that the theorem is unchanged if $f(x)$ has period $2l$ instead of 2π and its square is averaged over any period of length $2l$. You can also verify (Problem 3) that if $f(x)$ is written as a complex exponential Fourier series, and if in addition we include the possibility that $f(x)$ itself may be complex, then we find:

$$\text{The average of } |f(x)|^2 \text{ (over a period)} = \sum_{-\infty}^{\infty} |c_n|^2. \tag{11.5}$$

Parseval's theorem is also called the *completeness relation.* In the problem of representing a given sound wave as a sum of harmonics, suppose we had left one of the harmonics out of the series. It seems plausible physically, and it can be proved mathematically, that with one or more harmonics left out, we would not be able to represent sound waves containing the omitted harmonics. We say that the set of functions $\sin nx$, $\cos nx$ is a *complete set* of functions on any interval of length 2π; that is, any function (satisfying Dirichlet conditions) can be expanded in a Fourier series whose terms are constants times $\sin nx$ and $\cos nx$. If we left out some values of n, we would have an incomplete set of basis functions (see basis, page 357) and could not use it to expand some given functions. For example, suppose that you made a mistake in finding the period (that is, the value of l) of your given function and tried to use the set of functions $\sin 2nx$, $\cos 2nx$ in expanding a given function of period 2π. You would get a wrong answer because you used an incomplete set of basis functions (with the $\sin x$, $\cos x$, $\sin 3x$, $\cos 3x$, $\cdots$, terms missing). If your Fourier series is wrong because the set of basis functions you use is incomplete, then the results you get from Parseval's theorem (11.4) or (11.5) will be wrong too. In fact, if we use an incomplete basis set in, say, (11.5), then there are missing (non-negative) terms on the right-hand side, so the equation becomes the inequality: Average of $|f(x)|^2 \geq \sum_{-\infty}^{\infty} |c_n|^2$. This is known as Bessel's inequality. Conversely, if (11.4) and (11.5) are correct for *all* $f(x)$, then the set of basis functions used is a complete set. This is why Parseval's theorem is often called the completeness relation. (Also see page 377 and Chapter 12, Section 6.)

Let us look at some examples of the physical meaning and the use of Parseval's theorem.

▶ **Example 1.** In Section 10 we said that the intensity (energy per square centimeter per second) of a sound wave is proportional to the average value of the square of the excess pressure. If for simplicity we write (10.3) with letters instead of numerical values, we have

$$p(t) = \sum_1^\infty b_n \sin 2\pi n f t. \tag{11.6}$$

For this case, Parseval's theorem (11.4) says that:

$$\text{The average of } [p(t)]^2 = \sum_1^\infty b_n^2 \cdot \frac{1}{2} = \sum_1^\infty \text{ the average of } b_n^2 \sin^2 2n\pi f t\,. \tag{11.7}$$

Now the intensity or energy (per square centimeter per second) of the sound wave is proportional to the average of $[p(t)]^2$, and the energy associated with the nth harmonic is proportional to the average of $b_n^2 \sin^2 2n\pi ft$. Thus Parseval's theorem says that the total energy of the sound wave is equal to the sum of the energies associated with the various harmonics.

▶ **Example 2.** Let us use Parseval's theorem to find the sum of an infinite series. From Problem 8.15(a) written in complex exponential form we get:

The function $f(x)$ of period 2 which is equal to x on $(-1, 1)$

$$= -\frac{i}{\pi}\left(e^{i\pi x} - e^{-i\pi x} - \frac{1}{2}e^{2i\pi x} + \frac{1}{2}e^{-2i\pi x} + \frac{1}{3}e^{3i\pi x} - \frac{1}{3}e^{-3i\pi x} + \cdots\right).$$

Let us find the average of $[f(x)]^2$ on $(-1, 1)$.

$$\text{The average of } [f(x)]^2 = \frac{1}{2}\int_{-1}^{1} x^2\,dx = \frac{1}{2}\left[\frac{x^3}{3}\Big|_{-1}^{1}\right] = \frac{1}{3}.$$

By Parseval's theorem (11.5), this is equal to $\sum_{-\infty}^{\infty} |c_n|^2$, so we have

$$\frac{1}{3} = \sum_{-\infty}^{\infty} |c_n|^2 = \frac{1}{\pi^2}\left(1 + 1 + \frac{1}{4} + \frac{1}{4} + \frac{1}{9} + \frac{1}{9} + \cdots\right) = \frac{2}{\pi^2}\sum_{1}^{\infty}\frac{1}{n^2}.$$

Then we get the sum of the series

$$1 + \frac{1}{4} + \frac{1}{9} + \cdots = \sum_{1}^{\infty}\frac{1}{n^2} = \frac{\pi^2}{2}\cdot\frac{1}{3} = \frac{\pi^2}{6}.$$

We have seen that a function given on $(0, l)$ can be expanded in a sine series by defining it on $(-l, 0)$ to make it odd, or in a cosine series by defining it on $(-l, 0)$ to make it even. Here is another useful example of defining a function to suit our purposes. (We will need this in Chapter 13.) Suppose we want to expand a function defined on $(0, l)$ in terms of the basis functions $\sin(n + \frac{1}{2})\frac{\pi x}{l} = \sin\frac{(2n+1)\pi x}{2l}$. Can we do it, that is, do these functions make up a complete set for this problem? Note that our proposed basis functions have period $4l$, say $(-2l, 2l)$ (observe the $2l$ in the denominator where you are used to l). So given $f(x)$ on $(0, l)$, we can define it as we like on $(l, 2l)$ and on $(-2l, 0)$. We know (by the Dirichlet theorem) that the functions $\sin\frac{n\pi x}{2l}$ and $\cos\frac{n\pi x}{2l}$, *all* n, make up a complete set on $(-2l, 2l)$. We need to see how, on $(0, l)$ we can use just the sines (that's easy—make the function odd) and only the odd values of n. It turns out (see Problem 11) that if we define $f(x)$ on $(l, 2l)$ to make it symmetric around $x = l$, then all the b_n's for even n are equal to zero. So our desired basis set is indeed a complete set on $(0, l)$. Similarly we can show (Problem 11) that the functions $\cos\frac{(2n+1)\pi x}{2l}$ make up a complete set on $(0, l)$.

PROBLEMS, SECTION 11

1. Prove (11.4) for a function of period $2l$ expanded in a sine-cosine series.

2. Prove that if $f(x) = \sum_{-\infty}^{\infty} c_n e^{inx}$, then the average value of $[f(x)]^2$ is $\sum_{-\infty}^{\infty} c_n c_{-n}$. Show by Problem 7.12 that for real $f(x)$ this becomes (11.5).

3. If $f(x)$ is complex, we usually want the average of the square of the absolute value of $f(x)$. Recall that $|f(x)|^2 = f(x)\cdot\bar{f}(x)$, where $\bar{f}(x)$ means the complex conjugate of $f(x)$. Show that if a complex $f(x) = \sum_{-\infty}^{\infty} c_n e^{in\pi x/l}$, then (11.5) holds.

4. When a current I flows through a resistance R, the heat energy dissipated per second is the average value of RI^2. Let a periodic (not sinusoidal) current $I(t)$ be expanded in a Fourier series $I(t) = \sum_{-\infty}^{\infty} c_n e^{120in\pi t}$. Give a physical meaning to Parseval's theorem for this problem.

Use Parseval's theorem and the results of the indicated problems to find the sum of the series in Problems 5 to 9.

5. The series $1 + \frac{1}{3^2} + \frac{1}{5^2} + \cdots$, using Problem 9.6.

6. The series $\sum_{n=1}^{\infty}\frac{1}{n^4}$, using Problem 9.9.

7. The series $\displaystyle\sum_{n=1}^{\infty}\frac{1}{n^2}$, using Problem 5.8.

8. The series $\displaystyle\sum_{\text{odd } n}\frac{1}{n^4}$, using Problem 9.10.

9. The series $\dfrac{1}{3^2}+\dfrac{1}{15^2}+\dfrac{1}{35^2}+\cdots$, using Problem 5.11.

10. A general form of Parseval's theorem says that if two functions are expanded in Fourier series

$$f(x)=\frac{1}{2}a_0+\sum_1^{\infty}a_n\cos nx+\sum_1^{\infty}b_n\sin nx,$$

$$g(x)=\frac{1}{2}a_0'+\sum_1^{\infty}a_n'\cos nx+\sum_1^{\infty}b_n'\sin nx,$$

then the average value of $f(x)g(x)$ is $\frac{1}{4}a_0a_0'+\frac{1}{2}\sum_1^{\infty}a_na_n'+\frac{1}{2}\sum_1^{\infty}b_nb_n'$. Prove this.

11. (a) Let $f(x)$ on $(0,2l)$ satisfy $f(2l-x)=f(x)$, that is, $f(x)$ is symmetric about $x=l$. If you expand $f(x)$ on $(0,2l)$ in a sine series $\sum b_n\sin\frac{n\pi x}{2l}$, show that for even n, $b_n=0$. *Hint:* Note that the period of the sines is $4l$. Sketch an $f(x)$ which is symmetric about $x=l$, and on the same axes sketch a few sines to see that the even ones are antisymmetric about $x=l$. Alternatively, write the integral for b_n as an integral from 0 to l plus an integral from l to $2l$, and replace x by $2l-x$ in the second integral.

(b) Similarly, show that if we define $f(2l-x)=-f(x)$, the cosine series has $a_n=0$ for even n.

12. FOURIER TRANSFORMS

We have been expanding *periodic* functions in series of sines, cosines, and complex exponentials. Physically, we could think of the terms of these Fourier series as representing a set of harmonics. In music these would be an infinite set of frequencies nf, $n=1, 2, 3, \cdots$; notice that this set, although infinite, does not by any means include all possible frequencies. In electricity, a Fourier series could represent a periodic voltage; again we could think of this as made up of an infinite but discrete (that is, not continuous) set of a-c voltages of frequencies $n\omega$. Similarly, in discussing light, a Fourier series could represent light consisting of a discrete set of wavelengths λ/n, $n=1, 2, \cdots$, that is, a discrete set of colors. Two related questions might occur to us here. First, is it possible to represent a function which is *not* periodic by something analogous to a Fourier series? Second, can we somehow extend or modify Fourier series to cover the case of a continuous spectrum of wavelengths of light, or a sound wave containing a continuous set of frequencies?

If you recall that an integral is a limit of a sum, it may not surprise you very much to learn that the Fourier *series* (that is, a *sum* of terms) is replaced by a Fourier *integral* in the above cases. The Fourier integral can be used to represent nonperiodic functions, for example a single voltage pulse not repeated, or a flash of light, or a sound which is not repeated. The Fourier integral also represents a continuous set (spectrum) of frequencies, for example a whole range of musical tones or colors of light rather than a discrete set.

Recall from equations (8.2) and (8.3), these complex Fourier series formulas:

$$(12.1)\qquad \begin{aligned} f(x) &= \sum_{-\infty}^{\infty} c_n e^{in\pi x/l}, \\ c_n &= \frac{1}{2l}\int_{-l}^{l} f(x)e^{-in\pi x/l}\,dx. \end{aligned}$$

The period of $f(x)$ is $2l$ and the frequencies of the terms in the series are $n/(2l)$. We now want to consider the case of continuous frequencies.

Definition of Fourier Transforms We state without proof (see plausibility arguments below) the formulas corresponding to (12.1) for a continuous range of frequencies.

$$(12.2)\qquad \begin{aligned} f(x) &= \int_{-\infty}^{\infty} g(\alpha)e^{i\alpha x}\,d\alpha, \\ g(\alpha) &= \frac{1}{2\pi}\int_{-\infty}^{\infty} f(x)e^{-i\alpha x}\,dx. \end{aligned}$$

Compare (12.2) and (12.1); $g(\alpha)$ corresponds to c_n, α corresponds to n, and $\int_{-\infty}^{\infty}$ corresponds to $\sum_{-\infty}^{\infty}$. This agrees with our discussion of the physical meaning and use of Fourier integrals. The quantity α is a continuous analog of the integral-valued variable n, and so the set of coefficients c_n has become a function $g(\alpha)$; the sum over n has become an integral over α. The two functions $f(x)$ and $g(\alpha)$ are called a pair of *Fourier transforms.* Usually, $g(\alpha)$ is called the Fourier transform of $f(x)$, and $f(x)$ is called the inverse Fourier transform of $g(\alpha)$, but since the two integrals differ in form only in the sign in the exponent, it is rather common simply to call either a Fourier transform of the other. You should check the notation of any book or computer program you are using. Another point on which various references differ is the position of the factor $1/(2\pi)$ in (12.2); it is possible to have it multiply the $f(x)$ integral instead of the $g(\alpha)$ integral, or to have the factor $1/\sqrt{2\pi}$ multiply each of the integrals.

The *Fourier integral theorem* says that, if a function $f(x)$ satisfies the Dirichlet conditions (Section 6) on every finite interval, and if $\int_{-\infty}^{\infty} |f(x)|\,dx$ is finite, then (12.2) is correct. That is, if $g(\alpha)$ is computed and substituted into the integral for $f(x)$ [compare the procedure of computing the c_n's for a Fourier series and substituting them into the series for $f(x)$], then the integral gives the value of $f(x)$ anywhere that $f(x)$ is continuous; at jumps of $f(x)$, the integral gives the midpoint of the jump (again compare Fourier series, Section 6). The following discussion is not a mathematical proof of this theorem but is intended to help you see more clearly how Fourier integrals are related to Fourier series.

It might seem reasonable to think of trying to represent a function which is not periodic by letting the period $(-l, l)$ increase to $(-\infty, \infty)$. Let us try to do this, starting with (12.1). If we call $n\pi/l = \alpha_n$ and $\alpha_{n+1} - \alpha_n = \pi/l = \Delta\alpha$, then $1/(2l) = \Delta\alpha/(2\pi)$ and (12.1) can be rewritten as

$$f(x) = \sum_{-\infty}^{\infty} c_n e^{\alpha_n x}, \tag{12.3}$$

$$c_n = \frac{1}{2l}\int_{-l}^{l} f(x)e^{-i\alpha_n x}\,dx = \frac{\Delta\alpha}{2\pi}\int_{-l}^{l} f(u)e^{-i\alpha_n u}\,du. \tag{12.4}$$

(We have changed the dummy integration variable in c_n from x to u to avoid later confusion.) Substituting (12.4) into (12.3), we have

$$\begin{aligned} f(x) &= \sum_{-\infty}^{\infty}\left[\frac{\Delta\alpha}{2\pi}\int_{-l}^{l} f(u)e^{-i\alpha_n u}\,du\right]e^{i\alpha_n x} \\ &= \sum_{-\infty}^{\infty}\frac{\Delta\alpha}{2\pi}\int_{-l}^{l} f(u)e^{i\alpha_n(x-u)}\,du = \frac{1}{2\pi}\sum_{-\infty}^{\infty}F(\alpha_n)\,\Delta\alpha, \end{aligned} \tag{12.5}$$

where

$$F(\alpha_n) = \int_{-l}^{l} f(u)e^{i\alpha_n(x-u)}\,du. \tag{12.6}$$

Now $\sum_{-\infty}^{\infty} F(\alpha_n)\Delta\alpha$ looks rather like the formula in calculus for the sum whose limit, as $\Delta\alpha$ tends to zero, is an integral. If we let l tend to infinity [that is, let the period of $f(x)$ tend to infinity], then $\Delta\alpha = \pi/l \to 0$, and the sum $\sum_{-\infty}^{\infty} F(\alpha_n)\Delta\alpha$ goes over formally to $\int_{-\infty}^{\infty} F(\alpha)d\alpha$; we have dropped the subscript n on α now that it is a continuous variable. We also let l tend to infinity and $\alpha_n = \alpha$ in (12.6) to get

$$F(\alpha) = \int_{-\infty}^{\infty} f(u)e^{i\alpha(x-u)}\,du. \tag{12.7}$$

Replacing $\sum_{-\infty}^{\infty} F(\alpha_n)\Delta\alpha$ in (12.5) by $\int_{-\infty}^{\infty} F(\alpha)\,d\alpha$ and substituting from (12.7) for $F(\alpha)$ gives

$$\begin{aligned} f(x) &= \frac{1}{2\pi}\int_{-\infty}^{\infty} F(\alpha)\,d\alpha = \frac{1}{2\pi}\int_{-\infty}^{\infty}\int_{-\infty}^{\infty} f(u)e^{i\alpha(x-u)}\,du\,d\alpha \\ &= \frac{1}{2\pi}\int_{-\infty}^{\infty} e^{i\alpha x}\,d\alpha\int_{-\infty}^{\infty} f(u)e^{-i\alpha u}\,du. \end{aligned} \tag{12.8}$$

If we define $g(\alpha)$ by

$$g(\alpha) = \frac{1}{2\pi}\int_{-\infty}^{\infty} f(x)e^{-i\alpha x}\,dx = \frac{1}{2\pi}\int_{-\infty}^{\infty} f(u)e^{-i\alpha u}\,du, \tag{12.9}$$

then (12.8) gives

$$f(x) = \int_{-\infty}^{\infty} g(\alpha)e^{i\alpha x}\,d\alpha. \tag{12.10}$$

These equations are the same as (12.2). Notice that the actual requirement for the factor $1/(2\pi)$ is that the *product* of the constants multiplying the two integrals for $g(\alpha)$ and $f(x)$ should be $1/(2\pi)$; this accounts for the various notations we have discussed before.

Just as we have sine series representing odd functions and cosine series representing even functions (Section 9), so we have sine and cosine Fourier integrals which represent odd or even functions respectively. Let us prove that if $f(x)$ is odd, then $g(\alpha)$ is odd too, and show that in this case (12.2) reduces to a pair of sine transforms. The corresponding proof for even $f(x)$ is similar (Problem 1). We substitute

$$e^{-i\alpha x} = \cos\alpha x - i\sin\alpha x$$

into (12.9) to get

$$g(\alpha) = \frac{1}{2\pi}\int_{-\infty}^{\infty} f(x)(\cos\alpha x - i\sin\alpha x)\,dx. \tag{12.11}$$

Since $\cos\alpha x$ is even and we are assuming that $f(x)$ is odd, the product $f(x)\cos\alpha x$ is odd. Recall that the integral of an odd function over a symmetric interval about the origin (here, $-\infty$ to $+\infty$) is zero, so the term $\int_{-\infty}^{\infty} f(x)\cos\alpha x\,dx$ in (12.11) is zero. The product $f(x)\sin\alpha x$ is even (product of two odd functions); recall that the integral of an even function over a symmetric interval is twice the integral over positive x. Substituting these results into (12.11), we have

$$g(\alpha) = \frac{1}{2\pi}\int_{-\infty}^{\infty} f(x)(-i\sin\alpha x)\,dx = -\frac{i}{\pi}\int_0^{\infty} f(x)\sin\alpha x\,dx. \tag{12.12}$$

From (12.12), we can see that replacing α by $-\alpha$ changes the sign of $\sin\alpha x$ and so changes the sign of $g(\alpha)$. That is, $g(-\alpha) = -g(\alpha)$, so $g(\alpha)$ is an odd function as we claimed. Then expanding the exponential in (12.10) and arguing as we did to obtain (12.12), we find

$$f(x) = \int_{-\infty}^{\infty} g(\alpha)e^{i\alpha x}\,dx = 2i\int_0^{\infty} g(\alpha)\sin\alpha x\,d\alpha. \tag{12.13}$$

If we substitute $g(\alpha)$ from (12.12) into (12.13) to obtain an equation like (12.8), the numerical factor is $(-i/\pi)(2i) = 2/\pi$; thus the imaginary factors are not needed. The factor $2/\pi$ may multiply either of the two integrals or each integral may be multiplied by $\sqrt{2/\pi}$. Let us make the latter choice in giving the following definition.

Fourier Sine Transforms We define $f_s(x)$ and $g_s(\alpha)$, a pair of *Fourier sine transforms* representing *odd functions*, by the equations

$$\begin{aligned} f_s(x) &= \sqrt{\frac{2}{\pi}}\int_0^{\infty} g_s(\alpha)\sin\alpha x\,d\alpha, \\ g_s(x) &= \sqrt{\frac{2}{\pi}}\int_0^{\infty} f_s(x)\sin\alpha x\,dx. \end{aligned} \tag{12.14}$$

We discuss even functions in a similar way (Problem 1).

Fourier Cosine Transforms We define $f_c(x)$ and $g_c(\alpha)$, a pair of *Fourier cosine transforms* representing *even functions*, by the equations

$$(12.15)\qquad \begin{aligned} f_c(x) &= \sqrt{\frac{2}{\pi}}\int_0^\infty g_c(x)\cos\alpha x\,d\alpha,\\ g_c(\alpha) &= \sqrt{\frac{2}{\pi}}\int_0^\infty f_c(x)\cos\alpha x\,dx. \end{aligned}$$

▶ **Example 1.** Let us represent a nonperiodic function as a Fourier integral. The function

$$f(x) = \begin{cases} 1, & -1 < x < 1,\\ 0, & |x| > 1, \end{cases}$$

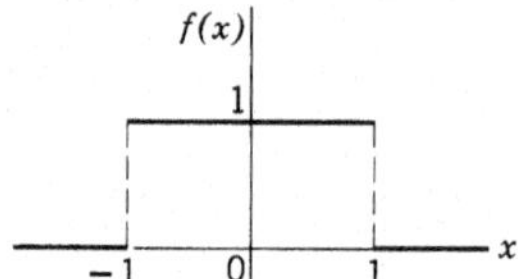

Figure 12.1

shown in Figure 12.1 might represent an impulse in mechanics (that is, a force applied only over a short time such as a bat hitting a baseball), or a sudden short surge of current in electricity, or a short pulse of sound or light which is not repeated. Since the given function is not periodic, it cannot be expanded in a Fourier *series*, since a Fourier series always represents a *periodic* function. Instead, we write $f(x)$ as a Fourier integral as follows. Using (12.9), we calculate $g(\alpha)$; this process is like finding the c_n's for a Fourier series. We find

$$(12.16)\qquad \begin{aligned} g(\alpha) &= \frac{1}{2\pi}\int_{-\infty}^\infty f(x)e^{-i\alpha x}\,dx = \frac{1}{2\pi}\int_{-1}^1 e^{-i\alpha x}\,dx\\ &= \frac{1}{2\pi}\frac{e^{-i\alpha x}}{-i\alpha}\bigg|_{-1}^1 = \frac{1}{\pi\alpha}\frac{e^{-i\alpha}-e^{i\alpha}}{-2i} = \frac{\sin\alpha}{\pi\alpha}. \end{aligned}$$

We substitute $g(\alpha)$ from (12.16) into the formula (12.10) for $f(x)$ (this is like substituting the evaluated coefficients into a Fourier series). We get

$$(12.17)\qquad \begin{aligned} f(x) &= \int_{-\infty}^\infty \frac{\sin\alpha}{\pi\alpha}e^{i\alpha x}\,dx\\ &= \frac{1}{\pi}\int_{-\infty}^\infty \frac{\sin\alpha(\cos\alpha x + i\sin\alpha x)}{\alpha}\,d\alpha = \frac{2}{\pi}\int_0^\infty \frac{\sin\alpha\cos\alpha x}{\alpha}\,d\alpha \end{aligned}$$

since $(\sin\alpha)/\alpha$ is an even function. We thus have an integral representing the function $f(x)$ shown in Figure 12.1.

▶ **Example 2.** We can use (12.17) to evaluate a definite integral. Using $f(x)$ in Figure 12.1, we find

$$(12.18)\qquad \int_0^\infty \frac{\sin\alpha\cos\alpha x}{\alpha}\,d\alpha = \frac{\pi}{2}f(x) = \begin{cases} \frac{\pi}{2} & \text{for } |x|<1, \quad \frac{\pi}{4} \text{ for } |x| = 1,\\ 0 & \text{for } |x|>1. \end{cases}$$

Notice that we have used the fact that the Fourier integral represents the midpoint of the jump in $f(x)$ at $|x| = 1$. If we let $x = 0$, we get

$$(12.19)\qquad \int_0^\infty \frac{\sin\alpha}{\alpha}\,d\alpha = \frac{\pi}{2}.$$

We could have done this problem by observing that $f(x)$ is an even function and so can be represented by a cosine transform. The final results (12.17) to (12.19) would be just the same (Problem 2).

In Section 9, we sometimes started with a function defined only for $x > 0$ and extended it to be even or odd so that we could represent it by a cosine series or by a sine series. Similarly, for Fourier transforms, we can represent a function defined for $x > 0$ by either a Fourier cosine integral (by defining it for $x < 0$ so that it is even), or by a Fourier sine integral (by defining it for $x < 0$ so that it is odd). (See Problem 2 and Problems 27 to 30.)

Parseval's Theorem for Fourier Integrals Recall (Section 11) that Parseval's theorem for a Fourier series $f(x) = \sum_{-\infty}^{\infty} c_n e^{in\pi x/l}$ relates $\int_{-l}^{l} |f|^2\, dx$ and $\sum_{-\infty}^{\infty} |c_n|^2$. In physical applications (see Section 11), Parseval's theorem says that the total energy (say in a sound wave, or in an electrical signal) is equal to the sum of the energies associated with the various harmonics. Remember that a Fourier integral represents a continuous spectrum of frequencies and that $g(\alpha)$ corresponds to c_n. Then we might expect that $\sum_{-\infty}^{\infty} |c_n|^2$ would be replaced by $\int_{-\infty}^{\infty} |g(\alpha)|^2 d\alpha$ (that is, a "sum" over a continuous rather than a discrete spectrum) and that Parseval's theorem would relate $\int_{-\infty}^{\infty} |f|^2 dx$ and $\int_{-\infty}^{\infty} |g|^2 d\alpha$. Let us try to find the relation.

We will first find a generalized form of Parseval's theorem involving two functions $f_1(x)$, $f_2(x)$ and their Fourier transforms $g_1(\alpha)$, $g_2(\alpha)$. Let $\bar{g}_1(\alpha)$ be the complex conjugate of $g_1(\alpha)$; from (12.1), we have

$$\bar{g}_1(\alpha) = \frac{1}{2\pi}\int_{-\infty}^{\infty} \bar{f}_1(x) e^{i\alpha x}\, dx. \tag{12.20}$$

We now multiply (12.20) by $g_2(\alpha)$ and integrate with respect to α:

$$\int_{-\infty}^{\infty} \bar{g}_1(\alpha) g_2(\alpha)\, d\alpha = \frac{1}{2\pi}\int_{-\infty}^{\infty}\left[\int_{-\infty}^{\infty} \bar{f}_1(x) e^{i\alpha x}\, dx\right] g_2(\alpha) d\alpha. \tag{12.21}$$

Let us rearrange (12.21) so that we integrate first with respect to α. [This is justified assuming that the absolute values of the functions f_1 and f_2 are integrable on $(-\infty, \infty)$.]

$$\frac{1}{2\pi}\int_{-\infty}^{\infty} \bar{f}_1(x)\, dx \left[\int_{-\infty}^{\infty} g_2(\alpha) e^{i\alpha x}\, d\alpha\right] = \frac{1}{2\pi}\int_{-\infty}^{\infty} \bar{f}_1(x) f_2(x)\, dx \tag{12.22}$$

by (12.2). Thus

$$\int_{-\infty}^{\infty} \bar{g}_1(\alpha) g_2(\alpha)\, d\alpha = \frac{1}{2\pi}\int_{-\infty}^{\infty} \bar{f}_1(x) f_2(x)\, dx. \tag{12.23}$$

(Compare this with the corresponding Fourier series theorem in Problem 11.10.) If we set $f_1 = f_2 = f$ and $g_1 = g_2 = g$, we get Parseval's theorem:

$$\int_{-\infty}^{\infty} |g(\alpha)|^2\, d\alpha = \frac{1}{2\pi}\int_{-\infty}^{\infty} |f(x)|^2\, dx. \tag{12.24}$$

PROBLEMS, SECTION 12

1. Following a method similar to that used in obtaining equations (12.11) to (12.14), show that if $f(x)$ is even, then $g(\alpha)$ is even too. Show that in this case $f(x)$ and $g(\alpha)$ can be written as Fourier cosine transforms and obtain (12.15).

2. Do Example 1 above by using a cosine transform (12.15). Obtain (12.17); for $x > 0$, the 0 to ∞ integral represents the function

$$f(x) = \begin{cases} 1, & 0 < x < 1, \\ 0, & x > 1. \end{cases}$$

Represent this function also by a Fourier sine integral (see the paragraph just before Parseval's theorem).

In Problems 3 to 12, find the exponential Fourier transform of the given $f(x)$ and write $f(x)$ as a Fourier integral [that is, find $g(\alpha)$ in equation (12.2) and substitute your result into the first integral in equation (12.2)].

3. $f(x) = \begin{cases} -1, & -\pi < x < 0 \\ 1, & 0 < x < \pi \\ 0, & |x| > \pi \end{cases}$

4. $f(x) = \begin{cases} 1, & \pi/2 < |x| < \pi \\ 0, & \text{otherwise} \end{cases}$

5. $f(x) = \begin{cases} 1, & 0 < x < 1 \\ 0, & \text{otherwise} \end{cases}$

6. $f(x) = \begin{cases} x, & |x| < 1 \\ 0, & |x| > 1 \end{cases}$

7. $f(x) = \begin{cases} |x|, & |x| < 1 \\ 0, & |x| > 1 \end{cases}$

8. $f(x) = \begin{cases} x, & 0 < x < 1 \\ 0, & \text{otherwise} \end{cases}$

9.

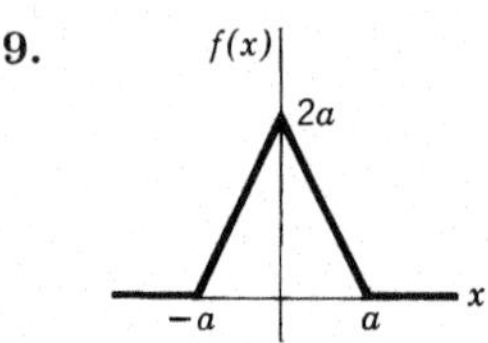

10.

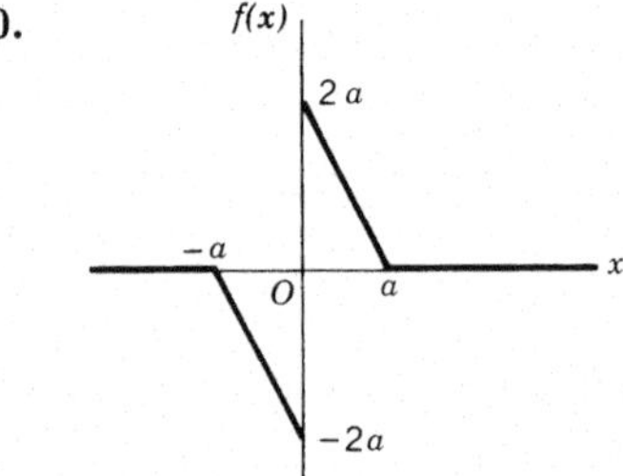

11. $f(x) = \begin{cases} \cos x, & -\pi/2 < x < \pi/2 \\ 0, & |x| > \pi/2 \end{cases}$

12. $f(x) = \begin{cases} \sin x, & |x| < \pi/2 \\ 0, & |x| > \pi/2 \end{cases}$

Hint: In Problems 11 and 12, use complex exponentials.

In Problems 13 to 16, find the Fourier cosine transform of the function in the indicated problem, and write $f(x)$ as a Fourier integral [use equation (12.15)]. Verify that the cosine integral for $f(x)$ is the same as the exponential integral found previously.

13. Problem 4.

14. Problem 7.

15. Problem 9.

16. Problem 11.

In Problems 17 to 20, find the Fourier sine transform of the function in the indicated problem, and write $f(x)$ as a Fourier integral [use equation (12.14)]. Verify that the sine integral for $f(x)$ is the same as the exponential integral found previously.

17. Problem 3. **18.** Problem 6.

19. Problem 10. **20.** Problem 12.

21. Find the Fourier transform of $f(x) = e^{-x^2/(2\sigma^2)}$. *Hint:* Complete the square in the x terms in the exponent and make the change of variable $y = x + \sigma^2 i\alpha$. Use tables or computer to evaluate the definite integral.

22. The function $j_1(\alpha) = (\alpha\cos\alpha - \sin\alpha)/\alpha$ is of interest in quantum mechanics. [It is called a spherical Bessel function; see Chapter 12, equation (17.4).] Using Problem 18, show that

$$\int_0^\infty j_1(\alpha)\sin\alpha x\,d\alpha = \begin{cases} \pi x/2, & -1 < x < 1, \\ 0, & |x| > 1. \end{cases}$$

23. Using Problem 17, show that

$$\int_0^\infty \frac{1-\cos\pi\alpha}{\alpha}\sin\alpha\,d\alpha = \frac{\pi}{2},$$
$$\int_0^\infty \frac{1-\cos\pi\alpha}{\alpha}\sin\pi\alpha\,d\alpha = \frac{\pi}{4}.$$

24. (a) Find the exponential Fourier transform of $f(x) = e^{-|x|}$ and write the inverse transform. You should find

$$\int_0^\infty \frac{\cos\alpha x}{\alpha^2+1}\,d\alpha = \frac{\pi}{2}e^{-|x|}.$$

(b) Obtain the result in (a) by using the Fourier cosine transform equations (12.15).

(c) Find the Fourier cosine transform of $f(x) = 1/(1+x^2)$. *Hint:* Write your result in (b) with x and α interchanged.

25. (a) Represent as an exponential Fourier transform the function

$$f(x) = \begin{cases} \sin x, & 0 < x < \pi, \\ 0, & \text{otherwise.} \end{cases}$$

Hint: Write $\sin x$ in complex exponential form.

(b) Show that your result can be written as

$$f(x) = \frac{1}{\pi}\int_0^\infty \frac{\cos\alpha x + \cos\alpha(x-\pi)}{1-\alpha^2}\,d\alpha.$$

26. Using Problem 15, show that

$$\int_0^\infty \frac{1-\cos\alpha}{\alpha^2}\,d\alpha = \frac{\pi}{2}.$$

Represent each of the following functions (a) by a Fourier cosine integral; (b) by a Fourier sine integral. *Hint:* See the discussion just before Parseval's theorem.

27. $f(x) = \begin{cases} 1, & 0 < x < \pi/2 \\ 0, & x > \pi/2 \end{cases}$

28. $f(x) = \begin{cases} 1, & 2 < x < 4 \\ 0, & 0 < x < 2,\ x > 4 \end{cases}$

29. $f(x) = \begin{cases} -1, & 0 < x < 2 \\ 1, & 2 < x < 3 \\ 0, & x > 3 \end{cases}$ **30.** $f(x) = \begin{cases} 1 - x/2, & 0 < x < 2 \\ 0, & x > 2 \end{cases}$

Verify Parseval's theorem (12.24) for the special cases in Problems 31 to 33.

31. $f(x)$ as in Figure 12.1. *Hint:* Integrate by parts and use (12.18) to evaluate

$$\int_{-\infty}^{\infty} |g(\alpha)|^2 \, d\alpha.$$

32. $f(x)$ and $g(\alpha)$ as in Problem 21.

33. $f(x)$ and $g(\alpha)$ as in Problem 24a.

34. Show that if (12.2) is written with the factor $1/\sqrt{2\pi}$ multiplying each integral, then the corresponding form of Parseval's theorem (12.24) is

$$\int_{-\infty}^{\infty} |f(x)|^2 \, dx = \int_{-\infty}^{\infty} |g(\alpha)|^2 \, d\alpha.$$

35. Starting with the symmetrized integrals as in Problem 34, make the substitutions $\alpha = 2\pi p/h$ (where p is the new variable, h is a constant), $f(x) = \psi(x)$, $g(\alpha) = \sqrt{h/2\pi}\,\phi(p)$; show that then

$$\psi(x) = \frac{1}{\sqrt{h}} \int_{-\infty}^{\infty} \phi(p) e^{2\pi i p x/h} \, dp,$$

$$\phi(p) = \frac{1}{\sqrt{h}} \int_{-\infty}^{\infty} \psi(x) e^{-2\pi i p x/h} \, dx,$$

$$\int_{-\infty}^{\infty} |\psi(x)|^2 \, dx = \int_{-\infty}^{\infty} |\phi(p)|^2 \, dp.$$

This notation is often used in quantum mechanics.

36. Normalize $f(x)$ in Problem 21; that is find the factor N so that $\int_{-\infty}^{\infty} |Nf(x)|^2 = 1$. Let $\psi(x) = Nf(x)$, and find $\phi(p)$ as given in Problem 35. Verify Parseval's theorem, that is, show that $\int_{-\infty}^{\infty} |\phi(p)|^2 \, dp = 1$.

13. MISCELLANEOUS PROBLEMS

1. The displacement (from equilibrium) of a particle executing simple harmonic motion may be either $y = A \sin \omega t$ or $y = A \sin(\omega t + \phi)$ depending on our choice of time origin. Show that the average of the kinetic energy of a particle of mass m (over a period of the motion) is the same for the two formulas (as it must be since both describe the same physical motion). Find the average value of the kinetic energy for the $\sin(\omega t + \phi)$ case in two ways:

(a) By selecting the integration limits (as you may by Problem 4.1) so that a change of variable reduces the integral to the $\sin \omega t$ case.

(b) By expanding $\sin(\omega t + \phi)$ by the trigonometric addition formulas and using (5.2) to write the average values.

2. The symbol $[x]$ means the greatest integer less than or equal to x (for example, $[3] = 3$, $[2.1] = 2$, $[-4.5] = -5$). Expand $x - [x] - \frac{1}{2}$ in an exponential Fourier series of period 1. *Hint:* Sketch the function.

Answer: $\dfrac{i}{2\pi}\left(\cdots - \dfrac{e^{-4\pi i x}}{2} - \dfrac{e^{-2\pi i x}}{1} + \dfrac{e^{2\pi i x}}{1} + \dfrac{e^{4\pi i x}}{2} + \cdots\right).$

3. We have said that Fourier series can represent discontinuous functions although power series cannot. It might occur to you to wonder why we could not substitute the power series for $\sin nx$ and $\cos nx$ (which converge for all x) into a Fourier series and collect terms to obtain a power series for a discontinuous function. As an example of what happens if we try this, consider the series in Problem 9.5. Show that the coefficients of x, if collected, form a divergent series; similarly, the coefficients of x^3 form a divergent series, and so on.

4. The diagram shows a "relaxation" oscillator. The charge q on the capacitor builds up until the neon tube fires and discharges the capacitor (we assume instantaneously). Then the cycle repeats itself over and over.

(a) The charge q on the capacitor satisfies the differential equation

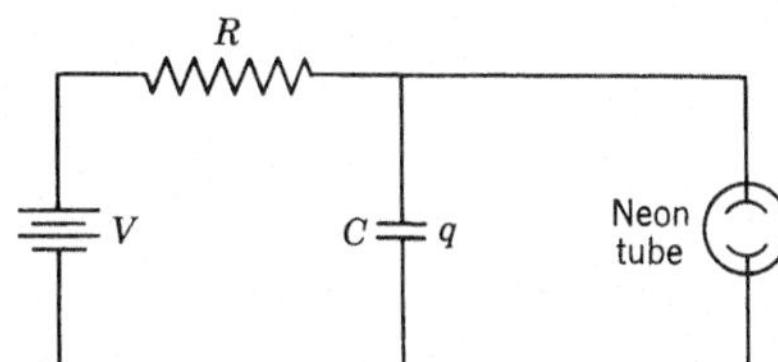

$$R\frac{dq}{dt} + \frac{q}{c} = V,$$

where R is the resistance, C is the capacitance, and V is the constant d-c voltage, as shown in the diagram. Show that if $q = 0$ when $t = 0$, then at any later time t (during one cycle, that is, before the neon tube fires)

$$q = CV(1 - e^{-t/RC}).$$

(b) Suppose the neon tube fires at $t = \frac{1}{2}RC$. Sketch q as a function of t for several cycles.

(c) Expand the periodic q in part (b) in an appropriate Fourier series.

5. Consider one arch of $f(x) = \sin x$. Show that the average value of $f(x)$ over the middle third of the arch is twice the average value over the end thirds.

6. Let $f(t) = e^{i\omega t}$ on $(-\pi, \pi)$. Expand $f(t)$ in a complex exponential Fourier series of period 2π. (Assume $\omega \neq$ integer.)

7. Given $f(x) = |x|$ on $(-\pi, \pi)$, expand $f(x)$ in an appropriate Fourier series of period 2π.

8. From facts you know, find in your head the average value of

(a) $x^3 - 3\sinh 2x + \sin^2 \pi x + \cos 3\pi x$ on $(-5, 5)$.

(b) $2\sin^2 3x - 4\cos x + 5x\cosh 2x - x\cos^2 x$ on $(-\pi, \pi)$.

9. Given $f(x) = \begin{cases} x, & 0 < x < 1, \\ -2, & 1 < x < 2. \end{cases}$

(a) Sketch at least three periods of the graph of the function represented by the sine series for $f(x)$. Without finding any series, answer the following questions:

(b) To what value does the sine series in (a) converge at $x = 1$? At $x = 2$? At $x = 0$? At $x = -1$?

(c) If the given function is continued with period 2 and then is represented by a complex exponential series $\sum_{n=-\infty}^{\infty} c_n e^{in\pi x}$, what is the value of $\sum_{n=-\infty}^{\infty} |c_n|^2$?

10. (a) Sketch at least three periods of the graph of the function represented by the cosine series for $f(x)$ in Problem 9.

(b) Sketch at least three periods of the graph of the exponential Fourier series of period 2 for $f(x)$ in Problem 9.

(c) To what value does the cosine series in (a) converge at $x = 0$? At $x = 1$? At $x = 2$? At $x = -2$?

(d) To what value does the exponential series in (b) converge at $x = 0$? At $x = 1$? At $x = \frac{3}{2}$? At $x = -2$.

11. Find the three Fourier series in Problems 9 and 10.

12. What would be the apparent frequency of a sound wave represented by

$$p(t) = \sum_{n=1}^{\infty} \frac{\cos 60n\pi t}{100(n-3)^2 + 1}?$$

13. (a) Given $f(x) = (\pi - x)/2$ on $(0, \pi)$, find the sine series of period 2π for $f(x)$.

(b) Use your result in (a) to evaluate $\sum 1/n^2$.

14. (a) Find the Fourier series of period 2 for $f(x) = (x-1)^2$ on $(0, 2)$.

(b) Use your result in (a) to evaluate $\sum 1/n^4$.

15. Given

$$f(x) = \begin{cases} 1, & -2 < x < 0, \\ -1, & 0 < x < 2, \end{cases}$$

find the exponential Fourier transform $g(\alpha)$ and the sine transform $g_s(\alpha)$. Write $f(x)$ as an integral and use your result to evaluate

$$\int_0^\infty \frac{(\cos 2\alpha - 1)\sin 2\alpha}{\alpha}\, d\alpha.$$

16. Given

$$f(x) = \begin{cases} x, & 0 \le x \le 1, \\ 2 - x, & 1 \le x \le 2, \\ 0, & x \ge 2, \end{cases}$$

find the cosine transform of $f(x)$ and use it to write $f(x)$ as an integral. Use your result to evaluate

$$\int_0^\infty \frac{\cos^2 \alpha \sin^2 \alpha/2}{\alpha^2}\, d\alpha.$$

17. Show that the Fourier sine transform of $x^{-1/2}$ is $\alpha^{-1/2}$. *Hint:* Make the change of variable $z = \alpha x$. The integral $\int_0^\infty z^{-1/2} \sin z\, dz$ can be found by computer or in tables.

18. Let $f(x)$ and $g(\alpha)$ be a pair of Fourier transforms. Show that df/dx and $i\alpha g(\alpha)$ are a pair of Fourier transforms. *Hint:* Differentiate the first integral in (12.2) under the integral sign with respect to x. Use (12.23) to show that

$$\int_{-\infty}^{\infty} \alpha |g(\alpha)|^2\, d\alpha = \frac{1}{2\pi i} \int_{-\infty}^{\infty} \bar{f}(x) \frac{d}{dx} f(x)\, dx.$$

Comment: This result is of interest in quantum mechanics where it would read, in the notation of Problem 12.35:

$$\int_{-\infty}^{\infty} p|\phi(p)|^2\, dp = \int_{-\infty}^{\infty} \psi^*(x) \left(\frac{-ih}{2\pi} \frac{d}{dx} \right) \psi(x)\, dx.$$

19. Find the form of Parseval's theorem (12.24) for sine transforms (12.14) and for cosine transforms (12.15).

20. Find the exponential Fourier transform of

$$f(x) = \begin{cases} 2a - |x|, & |x| < 2a, \\ 0, & |x| > 2a, \end{cases}$$

and use your result with Parseval's theorem to evaluate

$$\int_0^\infty \frac{\sin^4 a\alpha}{\alpha^4}\,d\alpha.$$

21. Define a function $h(x) = \sum_{k=-\infty}^{\infty} f(x + 2k\pi)$, assuming that the series converges to a function satisfying Dirichlet conditions (Section 6). Verify that $h(x)$ does have period 2π.

(a) Expand $h(x)$ in an exponential Fourier series $h(x) = \sum_{-\infty}^{\infty} c_n e^{inx}$; show that $c_n = g(n)$ where $g(\alpha)$ is the Fourier transform of $f(x)$. *Hint:* Write c_n as an integral from 0 to 2π and make the change of variable $u = x + 2k\pi$. Note that $e^{-2ink\pi} = 1$, and the sum on k gives a single integral from $-\infty$ to ∞.

(b) Let $x = 0$ in (a) to get *Poisson's summation formula* $\sum_{-\infty}^{\infty} f(2k\pi) = \sum_{-\infty}^{\infty} g(n)$. This result has many applications; for example: statistical mechanics, communication theory, theory of optical instruments, scattering of light in a liquid, and so on. (See Problem 22.)

22. Use Poisson's formula (Problem 21b) and Problem 20 to show that

$$\sum_{-\infty}^{\infty} \frac{\sin^2 n\theta}{n^2} = \pi\theta, \quad 0 < \theta < \pi.$$

(This sum is needed in the theory of scattering of light in a liquid.) *Hint:* Consider $f(x)$ and $g(\alpha)$ as in Problem 20. Note that $f(2k\pi) = 0$ except for $k = 0$ if $a < \pi$. Put $\alpha = n$, $a = \theta$.

23. Use Parseval's theorem and Problem 12.11 to evaluate

$$\int_0^\infty \frac{\cos^2(\alpha\pi/2)}{(1-\alpha^2)^2}\,d\alpha.$$

CHAPTER 8

Ordinary Differential Equations

1. INTRODUCTION

A great many applied problems involve rates, that is, derivatives. An equation containing derivatives is called a *differential equation.* If it contains partial derivatives, it is called a *partial differential equation*; otherwise it is called an *ordinary differential equation.* In this chapter we shall consider methods of solving ordinary differential equations which occur frequently in applications. Let us look at a few examples.

Newton's second law in vector form is $\mathbf{F} = m\mathbf{a}$. If we write the acceleration as $d\mathbf{v}/dt$, where $\mathbf{v}$ is the velocity, or as $d^2\mathbf{r}/dt^2$, where $\mathbf{r}$ is the displacement, we have a differential equation (or a set of differential equations, one for each component). Thus any mechanics problem in which we want to describe the motion of a body (automobile, electron, or satellite) under the action of a given force, involves the solution of a differential equation or a set of differential equations.

The rate at which heat Q escapes through a window or from a hot water pipe is proportional to the area A and to the rate of change of temperature with distance in the direction of flow of heat. Thus we have

$$\frac{dQ}{dt} = kA\frac{dT}{dx} \tag{1.1}$$

(k is called the thermal conductivity and depends on the material through which the heat is flowing). Here we have two different derivatives in the differential equation. In such a problem we might know either dT/dx or dQ/dt and solve the differential equation to find either T as a function of x, or Q as a function of t. (See Problems 2.23 to 2.25.)

Consider a simple series circuit (Figure 1.1) containing a resistance R, a capacitance C, an inductance L, and a source of emf V. If the current flowing around the circuit at time t is $I(t)$ and the charge on the capacitor is $q(t)$, then $I = dq/dt$. The voltage across R is RI, the voltage across C is q/C, and the voltage across L is

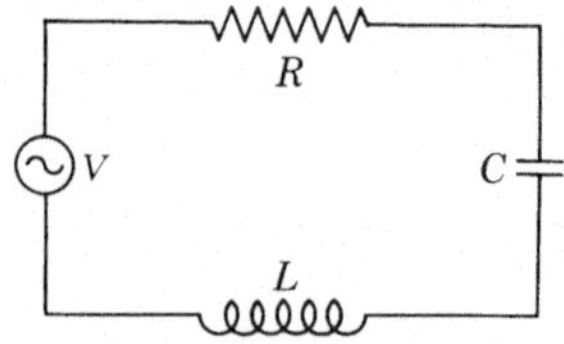

Figure 1.1

$L(dI/dt)$. Then at any time we must have

(1.2) $$L\frac{dI}{dt} + RI + \frac{q}{C} = V.$$

If we differentiate this equation with respect to t and substitute $dq/dt = I$, we have

(1.3) $$L\frac{d^2I}{dt^2} + R\frac{dI}{dt} + \frac{I}{C} = \frac{dV}{dt}$$

as the differential equation satisfied by the current I in a simple series circuit with given L, R, and C, and a given $V(t)$.

There are many more examples of physical problems leading to differential equations; we shall consider some of them later in the text and problems. You might find it interesting at this point to browse through the problems to see the wide range of topics giving rise to differential equations.

The *order* of a differential equation is the order of the highest derivative in the equation. Thus the equations

(1.4) $$\begin{aligned} y' + xy^2 &= 1, \\ xy' + y &= e^x, \\ \frac{dv}{dt} &= -g, \\ L\frac{dI}{dt} + RI &= V, \end{aligned}$$

are first-order equations, while (1.3) and

$$m\frac{d^2r}{dt^2} = -kr$$

are second-order equations. A *linear* differential equation (with x as independent and y as dependent variable) is one of the form

$$a_0y + a_1y' + a_2y'' + a_3y''' + \cdots = b,$$

where the a's and b are either constants or functions of x. The first equation in (1.4) is not linear because of the y^2 term; all the other equations we have mentioned so far are linear. Some other examples of nonlinear equations are:

$y' = \cot y$ (not linear because of the term $\cot y$);

$yy' = 1$ (not linear because of the product yy');

$y'^2 = xy$ (not linear because of the term y'^2).

Many of the differential equations which occur in applied problems are linear and of the first or second order; we shall be particularly interested in these.

> A *solution* of a differential equation (in the variables x and y) is a relation between x and y which, if substituted into the differential equation, gives an identity.

▶ **Example 1.** The relation

$$y = \sin x + C \tag{1.5}$$

is a solution of the differential equation

$$y' = \cos x \tag{1.6}$$

because if we substitute (1.5) into (1.6) we get the identity $\cos x = \cos x$.

▶ **Example 2.** The equation $y'' = y$ has solutions $y = e^x$ or $y = e^{-x}$ or $y = Ae^x + Be^{-x}$ as you can verify by substitution.

If we integrate $y' = f(x)$, the expression for y, namely $y = \int f(x)\,dx + C$, contains one arbitrary constant of integration. If we integrate $y'' = g(x)$ twice to get $y(x)$, then y contains two independent integration constants. We might expect that in general the solution of a differential equation of the nth order would contain n independent arbitrary constants. Note that in Example 1 above, the solution of the first-order equation $y' = \cos x$ contained one arbitrary constant C, and in Example 2 the solution $y = Ae^x + Be^{-x}$ of the second-order equation $y'' = y$ contained two arbitrary constants A and B.

> Any *linear* differential equation of order n has a solution containing n independent arbitrary constants, from which *all* solutions of the differential equation can be obtained by letting the constants have particular values. This solution is called the *general* solution of the linear differential equation.

(This may not be true for nonlinear equations; see Section 2.)

In applications, we usually want a *particular* solution, that is, one which satisfies the differential equation and some other requirements as well. Here are some examples of this.

▶ **Example 3.** Find the distance which an object falls under gravity in t seconds if it starts from rest.

Let x be the distance the object has fallen in time t. The acceleration of the object is g, the acceleration of gravity. Then we have

$$\frac{d^2x}{dt^2} = g. \tag{1.7}$$

Integrating, we get

$$\frac{dx}{dt} = gt + \text{const.} = gt + v_0, \tag{1.8}$$

$$x = \tfrac{1}{2}gt^2 + v_0 t + x_0, \tag{1.9}$$

where v_0 and x_0 are the values of v and x at $t = 0$. Now (1.9) is the *general* solution of (1.7) (because it is a solution of a second-order linear differential equation and contains two independent arbitrary constants). We want the *particular* solution for which $v_0 = 0$ (since the object starts from rest), and $x_0 = 0$ (since the distance the object has fallen is zero at $t = 0$). Then the desired particular solution is

$$x = \tfrac{1}{2}gt^2. \tag{1.10}$$

Example 4. Find the solution of $y'' = y$ which passes through the origin and through the point $(\ln 2, \frac{3}{4})$.

The general solution of the differential equation is

$$y = Ae^x + Be^{-x}$$

(see Example 2). If the given points satisfy the equation of the curve, we must have

$$0 = A + B \quad \text{or } A = -B,$$
$$\tfrac{3}{4} = Ae^{\ln 2} + Be^{-\ln 2} = A \cdot 2 + B \cdot \tfrac{1}{2} = 2A - \tfrac{1}{2}A = \tfrac{3}{2}A.$$

Thus we get

$$A = -B = \tfrac{1}{2},$$

and the desired particular solution is

$$y = \tfrac{1}{2}(e^x - e^{-x}) = \sinh x.$$

The given conditions which are to be satisfied by the particular solution are called *boundary conditions*, or when they are conditions at $t = 0$ they may be called *initial conditions*. For linear equations, the desired particular solution can be found from the general solution by determining the values of the constants as we did in Example 4. (For nonlinear equations, see Section 2.)

As you study methods of solving various types of differential equations in the following sections, you may wonder whether you can use computer solutions and not bother to learn these techniques. Just as for indefinite integrals (see Chapter 5, Section 1), there may be various forms for the solution of a differential equation, and your computer may not give the one you need. In order to make intelligent use of computer solutions, you need to know something about what to expect, and an effective way of gaining this knowledge is to solve some equations by hand. (See Example 1, Section 3.) By comparing your solutions with computer solutions, you will learn what you can (and cannot) expect from your computer.

The graphing capabilities of your computer are very useful in differential equations. Consider a first-order equation, say $y' = f(x, y)$. If the solution of this differential equation is $y = y(x)$, the differential equation gives the slope y' of the solution curve at each point (x, y). Suppose, for a large number of points, we draw a short line (or vector) centered on each point and with the slope y' at that point. (This would be a big job by hand, but your computer does it easily.) This plot is called a *slope field*, or a *direction field*, or a *vector field*. From such a diagram we can see the general trend of the solution curves even without solving the equation.

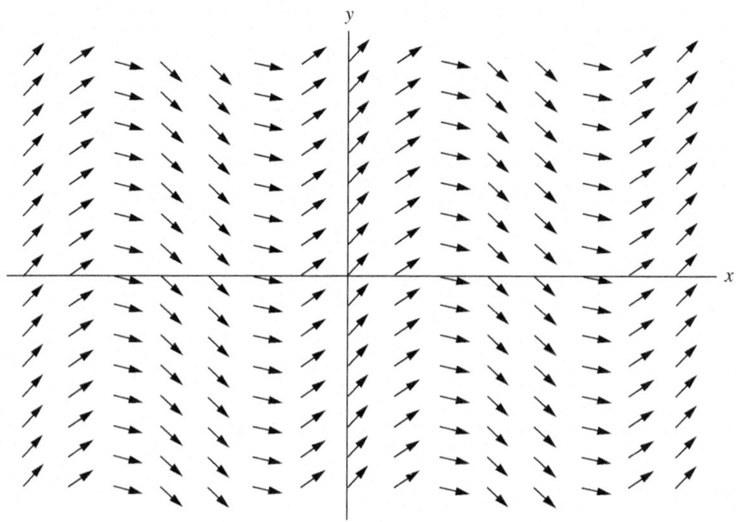

Figure 1.2

▶ **Example 5.** In Figure 1.2, we have plotted a "slope field" for the differential equation $y' = \cos x$. Note how you can trace the general shape of the solution curves, even without knowing from Example 1 that their equations are $y = \sin x + C$.

▶ PROBLEMS, SECTION 1

1. Verify the statement of Example 2. Also verify that $y = \cosh x$ and $y = \sinh x$ are solutions of $y'' = y$.

2. Solve Example 4 using the general solution $y = a \sinh x + b \cosh x$.

3. Verify that $y = \sin x$, $y = \cos x$, $y = e^{ix}$, and $y = e^{-ix}$ are all solutions of $y'' = -y$.

4. Find the distance which an object moves in time t if it starts from rest and has an acceleration $d^2x/dt^2 = ge^{-kt}$. Show that for small t the result is approximately (1.10), and for very large t, the speed dx/dt is approximately constant. The constant is called the terminal speed. (This problem corresponds roughly to the motion of a parachutist.)

5. Find the position x of a particle at time t if its acceleration is $d^2x/dt^2 = A \sin \omega t$.

6. A substance evaporates at a rate proportional to the exposed surface. If a spherical mothball of radius $\frac{1}{2}$ cm has radius 0.4 cm after 6 months, how long will it take:

(a) For the radius to be $\frac{1}{4}$ cm?

(b) For the volume of the mothball to be half of what it was originally?

7. The momentum p of an electron at speed v near the speed c of light increases according to the formula $p = mv/\sqrt{1 - v^2/c^2}$, where m is a constant (mass of the electron). If an electron is subject to a constant force F, Newton's second law describing its motion is

$$\frac{dp}{dt} = \frac{d}{dt}\frac{mv}{\sqrt{1 - v^2/c^2}} = F.$$

Find $v(t)$ and show that $v \to c$ as $t \to \infty$. Find the distance traveled by the electron in time t if it starts from rest.

2. SEPARABLE EQUATIONS

Every time you evaluate an integral

$$y = \int f(x)\,dx, \tag{2.1}$$

you are solving a differential equation, namely

$$y' = \frac{dy}{dx} = f(x). \tag{2.2}$$

This is a simple example of an equation which can be written with only y terms on one side of the equation and only x terms on the other:

$$dy = f(x)\,dx. \tag{2.3}$$

Whenever we can separate the variables this way, we call the equation *separable*, and we get the solution by just integrating each side of the equation.

Example 1. The rate at which a radioactive substance decays is proportional to the remaining number of atoms. If there are N_0 atoms at $t = 0$, find the number at time t.

The differential equation for this problem is

$$\frac{dN}{dt} = -\lambda N. \tag{2.4}$$

(The proportionality constant λ is called the decay constant.) This is a separable equation; we write it as $dN/N = -\lambda\,dt$. Then integrating both sides, we get $\ln N = -\lambda t + \text{const.}$ Since we are given $N = N_0$ at $t = 0$, we see that the constant is $\ln N_0$. Solving for N, we have

$$N = N_0 e^{-\lambda t}. \tag{2.5}$$

(For further discussion of radioactive decay problems, see Section 3, Example 2, and Problems 2.19b and 3.19 to 3.21.)

Example 2. Solve the differential equation

$$xy' = y + 1. \tag{2.6}$$

To separate variables, we divide both sides of (2.6) by $x(y+1)$; this gives

$$\frac{y'}{y+1} = \frac{1}{x} \quad \text{or} \quad \frac{dy}{y+1} = \frac{dx}{x}. \tag{2.7}$$

Integrating each side of (2.7), we have

$$\ln(y+1) = \ln x + \text{const.} = \ln x + \ln a = \ln(ax). \tag{2.8}$$

(We have called the constant of integration $\ln a$ for simplicity.) Then (2.8) gives the solution of (2.6), namely

$$y + 1 = ax. \tag{2.9}$$

This general solution represents a *family* of curves in the (x, y) plane, one curve for each value of the constant a. Or we may call the general solution (2.9) a *family of solutions* of the differential equation (2.6). Finding a particular solution means selecting one particular curve from the family.

Orthogonal Trajectories In Figure 2.1, the straight lines through $(0,-1)$ are the family of curves given by the solutions (2.9) of the differential equation (2.6). They might represent, for example, the lines of electric force due to an electric charge at $(0,-1)$. The circles in Figure 2.1 are then curves of constant electrostatic potential (called equipotentials—see Chapter 6, Sections 5 and 6). Note that the lines of force intersect the equipotential curves at right angles; each family of curves is called a set of *orthogonal trajectories* of the other family. It is often of interest to find the orthogonal trajectories of a given family of curves. Let us do this for the family (2.9). (In this case we know in advance that our answer will be the set of circles in Figure 2.1.)

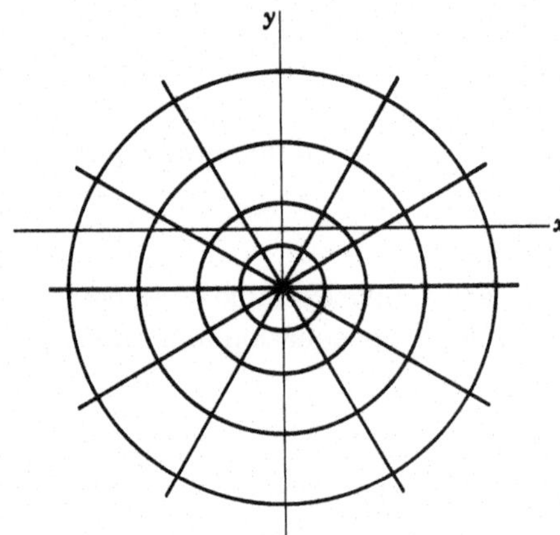

Figure 2.1

First we find the slope of a line of the family (2.9), namely,

$$y' = a. \tag{2.10}$$

For each a this gives the slope of *one* line. We want a formula (as a function of x and y) which gives the slope, at any point of the plane, of the line through that point. To obtain this, we eliminate a between (2.9) and (2.10) to get

$$y' = \frac{y+1}{x}. \tag{2.11}$$

[Or, given (2.6) rather than (2.9), we could simply solve for y'.] Now recall from analytic geometry that the slopes of two perpendicular lines are negative reciprocals. Then at each point we want the slope of the orthogonal trajectory curve to be the negative reciprocal of the slope of the line given by (2.11). Thus,

$$y' = -\frac{x}{y+1} \tag{2.12}$$

gives the slope of the orthogonal trajectories, and we solve (2.12) to obtain the equation of the orthogonal trajectory curves. Now (2.12) is separable; we obtain

$$\begin{aligned}
(y+1)\,dy &= -x\,dx, \\
\tfrac{1}{2}y^2 + y &= -\tfrac{1}{2}x^2 + C, \\
x^2 + y^2 + 2y &= 2C, \\
x^2 + (y+1)^2 &= 2C + 1.
\end{aligned}$$

This is, as we expected, a family of circles with centers at the point $(0,-1)$.

Nonlinear Differential Equations We have said that for linear differential equations of order n there is always a general solution containing n independent constants, and *all* solutions can be obtained by specializing the constants. You should be aware that this may not be true for some nonlinear equations, and routine methods of solution (including computer) may sometimes give partially incorrect or incomplete solutions. It is beyond our scope to discuss this in detail (see differential equations books), but here are some examples.

Example 3. Solve the differential equation $y' = \sqrt{1-y^2}$ and computer plot the slope field and a set of solution curves. Find particular solutions satisfying (a) $y = 0$ when $x = 0$, and (b) $y = 1$ when $x = 0$.

We separate variables and integrate to get

$$\frac{dy}{\sqrt{1-y^2}} = dx, \quad \arcsin y = x + \alpha, \quad y = \sin(x + \alpha).$$

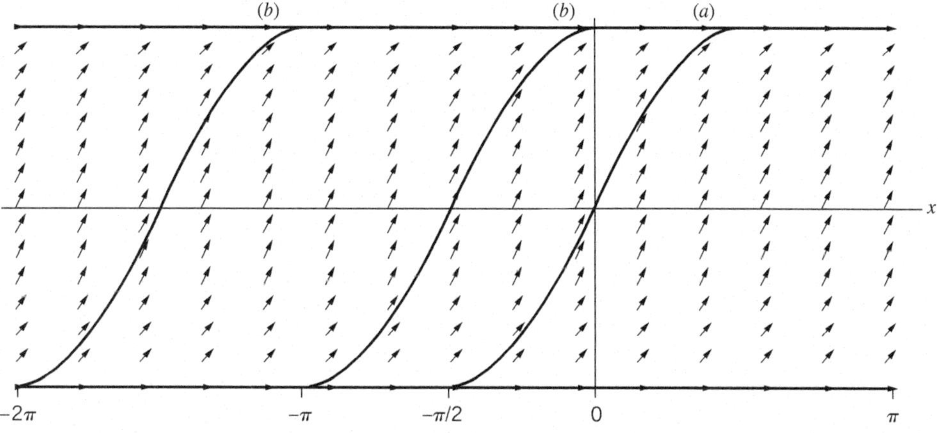

Figure 2.2

A computer gives the same answer. However if we look at either a computer plot of the slope field (Figure 2.2), or the differential equation itself, we see that the slope y' is always non-negative. Thus the solution of the given differential equation includes only the parts of the sine curves with non-negative slopes (Figure 2.2). A second difficulty is that part of the solution is missing. From either the slope field (Figure 2.2) or directly from the differential equation we can see that $y \equiv 1$ and $y \equiv -1$ are solutions not obtainable from the sine solution by any choice of α. (These are sometimes called *singular* solutions.) The fact that we did not find these solutions by separation of variables should not surprise us when we note that in separating variables we divided by $\sqrt{1-y^2}$ and this step is not valid if $y^2 = 1$.

Now for the particular solution (a) passing through $(0, 0)$, the sine solution gives either $y = \sin x$ or $y = \sin(x + \pi) = -\sin x$. But since we know that y' is non-negative, only the $y = \sin x$ solution is correct in the vicinity of $x = 0$. In fact (Figure 2.2), $y = \sin x$ is a correct particular solution from $x = -\pi/2$ to $x = \pi/2$. We could construct a continuous solution from $-\infty$ to ∞ by letting $y = -1$ from $-\infty$ to $-\pi/2$, $y = \sin x$ from $-\pi/2$ to $\pi/2$, and $y = 1$ from $\pi/2$ to ∞. Thus for case (a) [solution passing through the origin] we find just *one* particular solution.

For particular solution (b) [passing through $(0, 1)$], we find either $y = \sin(x + \frac{\pi}{2}) = \cos x$, or $y \equiv 1$; the $\cos x$ solution is valid from $x = -\pi$ to $x = 0$. As in (a), we can extend it by using parts of $y = -1$ and $y = 1$; this is one particular solution. But there are an infinite number of other particular solutions passing through $(0, 1)$ obtained by moving this one solution any distance to the left (Figure 2.2).

PROBLEMS, SECTION 2

For each of the following differential equations, separate variables and find a solution containing one arbitrary constant. Then find the value of the constant to give a particular solution satisfying the given boundary condition. Computer plot a slope field and some of the solution curves.

1. $xy' = y,$ $\qquad y = 3$ when $x = 2$.

2. $x\sqrt{1-y^2}\,dx + y\sqrt{1-x^2}\,dy = 0,$ $\qquad y = \frac{1}{2}$ when $x = \frac{1}{2}$.

3. $y' \sin x = y \ln y,$ $\qquad y = e$ when $x = \pi/3$.

4. $(1+y^2)\,dx + xy\,dy = 0,$ $\qquad y = 0$ when $x = 5$.

5. $xy' - xy = y,$ $\qquad y = 1$ when $x = 1$.

6. $y' = \dfrac{2xy^2 + x}{x^2y - y},$ $\qquad y = 0$ when $x = \sqrt{2}$.

7. $y\,dy + (xy^2 - 8x)\,dx = 0,$ $\qquad y = 3$ when $x = 1$.

8. $y' + 2xy^2 = 0,$ $\qquad y = 1$ when $x = 2$.

9. $(1+y)y' = y,$ $\qquad y = 1$ when $x = 1$.

10. $y' - xy = x,$ $\qquad y = 1$ when $x = 0$.

11. $2y' = 3(y-2)^{1/3},$ $\qquad y = 3$ when $x = 1$.

12. $(x + xy)y' + y = 0,$ $\qquad y = 1$ when $x = 1$.

In Problems 13 to 15, find a solution (or solutions) of the differential equation not obtainable by specializing the constant in your solution of the original problem. *Hint:* See Example 3.

13. Problem 2. **14.** Problem 8. **15.** Problem 11.

16. By separation of variables, find a solution of the equation $y' = \sqrt{y}$ containing one arbitrary constant. Find a particular solution satisfying $y = 0$ when $x = 0$. Show that $y \equiv 0$ is a solution of the differential equation which cannot be obtained by specializing the arbitrary constant in your solution above. Computer plot a slope field and some of the solution curves. Show that there are an infinite number of solution curves passing through any point on the x axis, but just one through any point for which $y > 0$. *Hint:* See Example 3. Problems 17 and 18 are physical problems leading to this differential equation.

17. The speed of a particle on the x axis, $x \geq 0$, is always numerically equal to the square root of its displacement x. If $x = 0$ when $t = 0$, find x as a function of t. Show that the given conditions are satisfied if the particle remains at the origin for any arbitrary length of time t_0 and then moves away; find x for $t > t_0$ for this case.

18. Let the rate of growth dN/dt of a colony of bacteria be proportional to the square root of the number present at any time. If there are no bacteria present at $t = 0$, how many are there at a later time? Observe here that the routine separation of variables solution gives an unreasonable answer, and the correct answer, $N \equiv 0$, is not obtainable from the routine solution. (You have to think, not just follow rules!)

19. (a) Consider a light beam traveling downward into the ocean. As the beam progresses, it is partially absorbed and its intensity decreases. The rate at which the intensity is decreasing with depth at any point is proportional to the intensity at that depth. The proportionality constant μ is called the *linear absorption coefficient.* Show that if the intensity at the surface is I_0, the intensity at a distance s below the surface is $I = I_0 e^{-\mu s}$. The linear absorption coefficient for water is of the order of 10^{-2} ft^{-1} (the exact value depending on the wavelength of the light and the impurities in the water). For this value of μ, find the intensity as a fraction of the surface intensity at a depth of 1 ft, 50 ft, 500 ft, 1 mile. When the intensity of a light beam has been reduced to half its surface intensity ($I = \frac{1}{2}I_0$), the distance the light has penetrated into the absorbing substance is called the *half-value thickness* of the substance. Find the half-value thickness in terms of μ. Find the half-value thickness for water for the value of μ given above.

(b) Note that the differential equation and its solution in this problem are mathematically the same as those in Example 1, although the physical problem and the terminology are different. In discussing radioactive decay, we call λ the *decay constant,* and we define the *half-life* T of a radioactive substance as the time when $N = \frac{1}{2}N_0$ (compare half-value thickness). Find the relation between λ and T.

20. Consider the following special cases of the simple series circuit [Figure 1.1 and equation (1.2)].

(a) RC circuit (that is, $L = 0$) with $V = 0$; find q as a function of t if q_0 is the charge on the capacitor at $t = 0$.

(b) RL circuit (that is, no capacitor; this means $1/C = 0$) with $V = 0$; find $I(t)$ given $I = I_0$ at $t = 0$.

(c) Again note that these are the same differential equations as in Problem 19 and Example 1. The terminology is again different; we define the time constant τ for a circuit as the time required for the charge (or current) to fall to $1/e$ times its initial value. Find the time constant for the circuits (a) and (b). If the same equation, say $y = y_0 e^{-at}$, represented either radioactive decay or light absorption or an RC or RL circuit, what would be the relations among the half-life, the half-value thickness, and the time constant?

21. Suppose the rate at which bacteria in a culture grow is proportional to the number present at any time. Write and solve the differential equation for the number N of bacteria as a function of time t if there are N_0 bacteria when $t = 0$. Again note that (except for a change of sign) this is the same differential equation and solution as in the preceding problems.

22. Solve the equation for the rate of growth of bacteria if the rate of increase is proportional to the number present but the population is being reduced at a constant rate by the removal of bacteria for experimental purposes.

23. Heat is escaping at a constant rate [dQ/dt in (1.1) is constant] through the walls of a long cylindrical pipe. Find the temperature T at a distance r from the axis of the cylinder if the inside wall has radius $r = 1$ and temperature $T = 100$ and the outside wall has $r = 2$ and $T = 0$.

24. Do Problem 23 for a spherical cavity containing a constant source of heat. Use the same radii and temperatures as in Problem 23.

25. Show that the thickness of the ice on a lake increases with the square root of the time in cold weather, making the following simplifying assumptions. Let the water temperature be a constant 10°C, the air temperature a constant $-10°$, and assume that at any given time the ice forms a slab of uniform thickness x. The rate of formation of ice is proportional to the rate at which heat is transferred from the water to the air. Let $t = 0$ when $x = 0$.

26. An object of mass m falls from rest under gravity subject to an air resistance proportional to its speed. Taking the y axis as positive down, show that the differential equation of motion is $m(dv/dt) = mg - kv$, where k is a positive constant. Find v as a function of t, and find the limiting value of v as t tends to infinity; this limit is called the *terminal speed*. Can you find the terminal speed directly from the differential equation without solving it? *Hint:* What is dv/dt after v has reached an essentially constant value?

Consider the following specific examples of this problem.

(a) A person drops from an airplane with a parachute. Find a reasonable value of k.

(b) In the Millikan oil drop experiment to measure the charge of an electron, tiny electrically charged drops of oil fall through air under gravity or rise under the combination of gravity and an electric field. Measurements can be made only after they have reached terminal speed. Find a formula for the time required for a drop starting at rest to reach 99% of its terminal speed.

27. According to Newton's law of cooling, the rate at which the temperature of an object changes is proportional to the difference between its temperature and that of its surroundings. A cup of coffee at 200° in a room of temperature 70° is stirred continually and reaches 100° after 10 min. At what time was it at 120°?

28. A glass of milk at 38° is removed from the refrigerator and left in a room at temperature 70°. If the temperature of the milk is 54° after 10 min, what will its temperature be in half an hour? (See Problem 27.)

29. A solution containing 90% by volume of alcohol (in water) runs at 1 gal/min into a 100-gal tank of pure water where it is continually mixed. The mixture is withdrawn at the rate of 1 gal/min. When will it start coming out 50% alcohol?

30. If P dollars are left in the bank at interest I percent per year compounded continuously, find the amount A at time t. *Hint:* Find dA, the interest on A dollars for time dt.

Find the orthogonal trajectories of each of the following families of curves. In each case, sketch or computer plot several of the given curves and several of their orthogonal trajectories. Be careful to eliminate the constant from y' for the original curves; this constant takes different values for different curves of the original family, and you want an expression for y' which is valid for all curves of the family crossed by the orthogonal trajectory you are trying to find. See equations (2.10) to (2.12).

31. $x^2 + y^2 = \text{const.}$

32. $y = kx^2$.

33. $y = kx^n$. (Assume that n is a given number; the different curves of the family have different values of k.)

34. $xy = k$.

35. $(y-1)^2 = x^2 + k$.

3. LINEAR FIRST-ORDER EQUATIONS

A first-order equation contains y' but no higher derivatives. A *linear* first-order equation means one which can be written in the form

$$y' + Py = Q, \tag{3.1}$$

where P and Q are functions of x. To see how to solve (3.1), let us first consider the simpler equation when $Q = 0$. The equation

$$y' + Py = 0 \qquad \text{or} \qquad \frac{dy}{dx} = -Py \tag{3.2}$$

is separable. As in Section 2, we obtain the solution as follows:

$$\begin{aligned} \frac{dy}{y} &= -P\,dx, \\ \ln y &= -\int P\,dx + C, \\ y &= e^{-\int P\,dx + C} = Ae^{-\int P\,dx} \end{aligned} \tag{3.3}$$

where $A = e^C$. Let us simplify the notation for future use; we write

$$I = \int P\,dx. \tag{3.4}$$

Then

$$\frac{dI}{dx} = P \tag{3.5}$$

and we can write (3.3) as $y = Ae^{-I}$ or

$$ye^I = A. \tag{3.6}$$

We can now see how to solve (3.1). If we differentiate (3.6) with respect to x and use (3.5), we get

$$\frac{d}{dx}(ye^I) = y'e^I + ye^I\frac{dI}{dx} = y'e^I + ye^I P = e^I(y' + Py), \tag{3.7}$$

which is the left-hand side of (3.1) multiplied by e^I. (We call e^I an integrating factor—see Section 4.) Thus, we can write (3.1) (times e^I) as

$$\frac{d}{dx}(ye^I) = e^I(y' + Py) = Qe^I. \tag{3.8}$$

Since Q and e^I are functions of x only, we can now integrate both sides of (3.8) with respect to x to get

$$\left.\begin{aligned} ye^I &= \int Qe^I\,dx + c, \qquad \text{or} \\ y &= e^{-I}\int Qe^I\,dx + ce^{-I}, \end{aligned}\right\} \quad \text{where} \quad I = \int P\,dx. \tag{3.9}$$

This is the general solution of (3.1). Note that it contains one arbitrary constant as expected for a first-order linear equation. The term ce^{-I} is a solution of equation (3.2); the first term in y is one particular solution of (3.1). Borrowing notation which we shall use in Section 6, let's call the term $ce^{-I} = y_c$ and the particular solution $= y_p$. Then $y_p + y_c$ is a solution of (3.1) for any value of c. Also note that $y_p e^I = \int Qe^I\,dx$ is an indefinite integral which, as we know (see Chapter 5, Section 1), has infinitely many answers differing from each other by constants of integration. Thus the particular solution obtained by you and by your computer may not be the same (see Example 1 and Problems).

▶ **Example 1.** Solve $(1+x^2)y' + 6xy = 2x$. In the form of (3.1), this is

$$y' + \frac{6x}{1+x^2}\,y = \frac{2x}{1+x^2}.$$

From (3.9), we get

$$\begin{aligned}
I &= \int \frac{6x}{1+x^2}\,dx = 3\ln(1+x^2)\\
e^I &= e^{3\ln(1+x^2)} = (1+x^2)^3\\
ye^I &= \int \frac{2x}{1+x^2}(1+x^2)^3\,dx = \int 2x(1+x^2)^2\,dx = \tfrac{1}{3}(1+x^2)^3 + c\\
y &= \tfrac{1}{3} + \frac{c}{(1+x^2)^3}.
\end{aligned}$$

A computer gives the answer

$$y = \frac{3x^2 + 3x^4 + x^6}{3(1+x^2)^3} + \frac{A}{(1+x^2)^3}.$$

Let us show that the answers agree (see comments just after (3.9)). If we put $A = c + 1/3$ in the computer solution above and combine terms, we get

$$y = \frac{3x^2 + 3x^4 + x^6 + 1}{3(1+x^2)^3} + \frac{c}{(1+x^2)^3} = \frac{(1+x^2)^3}{3(1+x^2)^3} + \frac{c}{(1+x^2)^3},$$

which, after cancelling, is our solution above. We see that the computer program chose a more complicated particular solution y_p which differed from our y_p by a multiple of $y_c = 1/(1+x^2)^3$. Always be aware of the possibility of simplifying a particular solution by adding a multiple of y_c.

▶ **Example 2.** Radium decays to radon which decays to polonium. If at $t = 0$, a sample is pure radium, how much radon does it contain at time t?

$$\begin{aligned}
\text{Let}\qquad N_0 &= \text{number of radium atoms at } t = 0,\\
N_1 &= \text{number of radium atoms at time } t,\\
N_2 &= \text{number of radon atoms at time } t,\\
\lambda_1 \text{ and } \lambda_2 &= \text{decay constants for Ra and Rn.}
\end{aligned}$$

As in Section 2, we have for radium

$$\frac{dN_1}{dt} = -\lambda_1 N_1, \qquad N_1 = N_0 e^{-\lambda_1 t}.$$

The rate at which radon is being created is the rate at which radium is decaying, namely $\lambda_1 N_1$ or $\lambda_1 N_0 e^{-\lambda_1 t}$. But the radon is also decaying at the rate $\lambda_2 N_2$. Hence, we have

$$\frac{dN_2}{dt} = \lambda_1 N_1 - \lambda_2 N_2, \qquad \text{or}$$
$$\frac{dN_2}{dt} + \lambda_2 N_2 = \lambda_1 N_1 = \lambda_1 N_0 e^{-\lambda_1 t}.$$

This equation is of the form (3.1), and we solve it as follows:

$$I = \int \lambda_2\, dt = \lambda_2 t,$$
$$(3.10) \qquad N_2 e^{\lambda_2 t} = \int \lambda_1 N_0 e^{-\lambda_1 t} e^{\lambda_2 t}\, dt + c$$
$$= \lambda_1 N_0 \int e^{(\lambda_2 - \lambda_1)t}\, dt + c = \frac{\lambda_1 N_0}{\lambda_2 - \lambda_1} e^{(\lambda_2 - \lambda_1)t} + c,$$

if $\lambda_1 \neq \lambda_2$. (For the case $\lambda_1 = \lambda_2$, see Problem 19.) Since $N_2 = 0$ at $t = 0$ (we assumed pure Ra at $t = 0$), we must have

$$0 = \frac{\lambda_1 N_0}{\lambda_2 - \lambda_1} + c \qquad \text{or} \qquad c = -\frac{\lambda_1 N_0}{\lambda_2 - \lambda_1}.$$

Substituting this value of c into (3.10) and solving for N_2, we get

$$N_2 = \frac{\lambda_1 N_0}{\lambda_2 - \lambda_1}\left(e^{-\lambda_1 t} - e^{-\lambda_2 t}\right).$$

PROBLEMS, SECTION 3

Using (3.9), find the general solution of each of the following differential equations. Compare a computer solution and, if necessary, reconcile it with yours. *Hint*: See comments just after (3.9), and Example 1.

1. $y' + y = e^x$

2. $x^2 y' + 3xy = 1$

3. $dy + (2xy - xe^{-x^2})\, dx = 0$

4. $2xy' + y = 2x^{5/2}$

5. $y' \cos x + y = \cos^2 x$

6. $y' + y/\sqrt{x^2+1} = 1/(x + \sqrt{x^2+1})$

7. $(1 + e^x)y' + 2e^x y = (1 + e^x)e^x$

8. $(x \ln x)y' + y = \ln x$

9. $(1 - x^2)y' = xy + 2x\sqrt{1 - x^2}$

10. $y' + y \tanh x = 2e^x$

11. $y' + y \cos x = \sin 2x$

12. $\dfrac{dx}{dy} = \cos y - x \tan y$

13. $dx + (x - e^y)\, dy = 0$

14. $\dfrac{dy}{dx} = \dfrac{3y}{3y^{2/3} - x}$

Hint: For Problems 12 to 14, solve for x in terms of y.

15. Water with a small salt content (5 lb in 1000 gal) is flowing into a very salty lake at the rate of $4 \cdot 10^5$ gal per hr. The salty water is flowing out at the rate of 10^5 gal per hr. If at some time (say $t = 0$) the volume of the lake is 10^9 gal, and its salt content is 10^7 lb, find the salt content at time t. Assume that the salt is mixed uniformly with the water in the lake at all times.

16. Find the general solution of (1.2) for an RL circuit ($1/C = 0$) with $V = V_0 \cos\omega t$ ($\omega =$ const.).

17. Find the general solution of (1.3) for an RC circuit ($L = 0$), with $V = V_0 \cos\omega t$.

18. Do Problems 16 and 17 using $V = V_0 e^{i\omega t}$, and find the solutions for 16 and 17 by taking real parts of the complex solutions.

19. If $\lambda_1 = \lambda_2 = \lambda$ in (3.10), then $\int e^{(\lambda_2-\lambda_1)t}\,dt = \int dt$. Find N_2 for this case.

20. Extend the radioactive decay problem (Example 2) one more stage, that is, let λ_3 be the decay constant of polonium and find how much polonium there is at time t.

21. Generalize Problem 20 to any number of stages.

22. Find the orthogonal trajectories of the family of curves $x = y + 1 + ce^y$. (See the instructions above Problem 2.31.)

23. Find the orthogonal trajectories of the family of curves $y = -e^{x^2}\operatorname{erf} x + Ce^{x^2}$. *Hint:* See Chapter 11, equation (9.1) for definition of $\operatorname{erf} x$, and Chapter 4, Section 12, for differentiation of an integral. Solve for x in terms of y.

4. OTHER METHODS FOR FIRST-ORDER EQUATIONS

Separable equations and linear equations are the two types of first-order equations you are most apt to meet in elementary applications. However, we shall also mention briefly a few other methods of solving special first-order equations. You will find more details in the problems and in most differential equations books.

The Bernoulli Equation The differential equation

$$y' + Py = Qy^n, \tag{4.1}$$

where P and Q are functions of x, is known as the Bernoulli equation. It is not linear but is easily reduced to a linear equation. We make the change of variable

$$z = y^{1-n}. \tag{4.2}$$

Then

$$z' = (1-n)y^{-n}y'. \tag{4.3}$$

Next multiply (4.1) by $(1-n)y^{-n}$ and make the substitutions (4.2) and (4.3) to get

$$\begin{aligned}(1-n)y^{-n}y' + (1-n)Py^{1-n} &= (1-n)Q,\\ z' + (1-n)Pz &= (1-n)Q.\end{aligned}$$

This is now a first-order linear equation which we can solve as we did the linear equations above. (See Section 7 for an example of a physical problem in which we need to solve a Bernoulli equation.)

Exact Equations; Integrating Factors Recall from Chapter 6, Section 8, that the expression $P(x,y)\,dx+Q(x,y)\,dy$ is an *exact differential* [that is, the differential of a function $F(x,y)$] if

$$\frac{\partial P}{\partial y}=\frac{\partial Q}{\partial x}. \tag{4.4}$$

If (4.4) holds, then there is a function $F(x,y)$ such that

$$P=\frac{\partial F}{\partial x},\quad Q=\frac{\partial F}{\partial y},\quad P\,dx+Q\,dy=dF. \tag{4.5}$$

In Chapter 6 we considered ways of finding F when (4.4) holds. The differential equation

$$P\,dx+Q\,dy=0 \qquad \text{or} \quad y'=-\frac{P}{Q} \tag{4.6}$$

is called *exact* if (4.4) holds. In this case

$$P\,dx+Q\,dy=dF=0,$$

and the solution of (4.6) is then

$$F(x,y)=\text{const.} \tag{4.7}$$

We find F as in Chapter 6, Section 8.

An equation which is not exact may often be made exact by multiplying it by an appropriate factor.

Example 1. The equation

$$x\,dy-y\,dx=0 \tag{4.8}$$

is not exact [by (4.4)]. But the equation

$$\frac{x\,dy-y\,dx}{x^2}=\frac{1}{x}\,dy-\frac{y}{x^2}\,dx=d\left(\frac{y}{x}\right)=0, \tag{4.9}$$

obtained by dividing (4.8) by x^2, *is* exact [use (4.4)], and its solution is

$$\frac{y}{x}=\text{const.} \tag{4.10}$$

We multiplied (4.8) by $1/x^2$ to make the equation exact; the factor $1/x^2$ is called an *integrating factor.* To see another example of an integrating factor, look back at Section 3. The expression e^I is an integrating factor for equations (3.1) and (3.2); as you can see in (3.8), multiplying (3.1) by e^I makes it an exact equation.

The method of finding an integrating factor and solving the resulting exact equation is useful mainly in simple cases when we can see the result by inspection. It is not usually worth while to spend much time searching for integrating factors.

Homogeneous Equations A *homogeneous function* of x and y of degree n means a function which can be written as $x^n f(y/x)$. For example, $x^3 - xy^2 = x^3[1-(y/x)^2]$ is a homogeneous function of degree 3. (Also see Problem 21.) An equation of the form

$$P(x,y)\,dx + Q(x,y)\,dy = 0, \tag{4.11}$$

where P and Q are homogeneous functions of *the same degree* is called *homogeneous.* (The term homogeneous is also used in another sense; see Section 5.) If we divide two homogeneous functions of the same degree, the x^n factors cancel and we have a function of y/x. Thus, from (4.11) we can write

$$y' = \frac{dy}{dx} = -\frac{P(x,y)}{Q(x,y)} = f\left(\frac{y}{x}\right), \tag{4.12}$$

and we can say that a differential equation is homogeneous if it can be written as $y' =$ a function of y/x. This suggests that we solve homogeneous equations by making the change of variables $v = y/x$, or

$$y = xv. \tag{4.13}$$

This substitution does, in fact, give us a separable equation in x and v (see Problem 22). We solve it to find a relation between v and x and then put back $v = y/x$ to find the solution of (4.11).

Also see Problem 23 for another way to solve homogeneous equations.

Change of Variables We have solved both Bernoulli equations and homogeneous equations by making changes of variables. Other equations may yield to this method also. If a differential equation contains some combination of the variables x, y (especially if this combination appears more than once), we try replacing this combination by a new variable. See Problems 11, 15, and 16 for examples.

PROBLEMS, SECTION 4

Use the methods of this section to solve the following differential equations. Compare computer solutions and reconcile differences.

1. $y' + y = xy^{2/3}$

2. $y' + \dfrac{1}{x}y = 2x^{3/2}y^{1/2}$

3. $3xy^2y' + 3y^3 = 1$

4. $(2xe^{3y} + e^x)\,dx + (3x^2e^{3y} - y^2)\,dy = 0$

5. $(x - y)\,dy + (y + x + 1)\,dx = 0$

6. $(\cos x \cos y + \sin^2 x)\,dx - (\sin x \sin y + \cos^2 y)\,dy = 0$

7. $x^2\,dy + (y^2 - xy)\,dx = 0$

8. $y\,dy = (-x + \sqrt{x^2 + y^2}\,)\,dx$

9. $xy\,dx + (y^2 - x^2)\,dy = 0$

10. $(y^2 - xy)\,dx + (x^2 + xy)\,dy = 0$

11. $y' = \cos(x + y)$ *Hint:* Let $u = x + y$; then $u' = 1 + y'$.

12. $y' = \dfrac{y}{x} - \tan\dfrac{y}{x}$

13. $yy' - 2y^2 \cot x = \sin x \cos x$

14. $(x-1)y' + y - x^{-2} + 2x^{-3} = 0$ **15.** $xy' + y = e^{xy}$ *Hint:* Let $u = xy$

16. Solve the differential equation $yy'^2 + 2xy' - y = 0$ by changing from variables y, x, to r, x, where $y^2 = r^2 - x^2$; then $yy' = rr' - x$.

17. If an incompressible fluid flows in a corner bounded by walls meeting at the origin at an angle of 60°, the streamlines of the flow satisfy the equation $2xy\,dx + (x^2 - y^2)\,dy = 0$. Find the streamlines.

18. Find the family of orthogonal trajectories of the circles $(x-h)^2 + y^2 = h^2$. (See the instructions above Problem 2.31.)

19. Find the family of curves satisfying the differential equation $(x+y)\,dy + (x-y)\,dx = 0$ and also find their orthogonal trajectories.

20. Find the shape of a mirror which has the property that rays from a point O on the axis are reflected into a parallel beam. *Hint:* Take the point O at the origin. Show from the figure that $\tan 2\theta = y/x$. Use the formula for $\tan 2\theta$ to express this in terms of $\tan\theta = dy/dx$ and solve the resulting differential equation. (*Hint:* See Problem 16.)

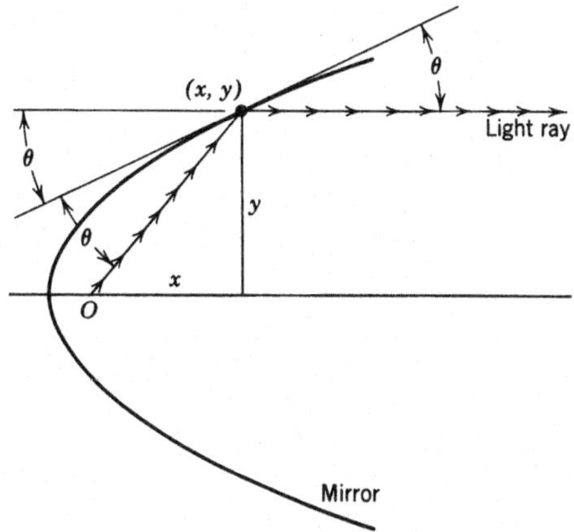

21. As in text just before (4.11), show that

(a) $x^2 - 5xy + y^3/x$ is a homogeneous function of degree 2;

(b) $x^{-1}(y^4 - x^3y) - xy^2\sin(x/y)$ is homogeneous of degree 3;

(c) $x^2y^3 + x^5\ln(y/x) - y^6/\sqrt{x^2+y^2}$ is homogeneous of degree 5;

(d) $x^2 + y$, $x + \cos y$, and $y + 1$ are not homogeneous.

See Chapter 4, Section 13, Problem 1 for a more general definition of a homogeneous function of any number of variables.

22. Show that the change of variables (4.13) in (4.11) or (4.12) gives a separable equation. *Hints:* Substitute $y = xv$ and $dy = x\,dv + v\,dx$ from (4.13) into (4.12) and rearrange terms to get the equation

(a) $$[f(v) - v]\,dx = x\,dv.$$

Alternatively, suppose P and Q are homogeneous of degree n; that is $P(x,y) = x^nP(1, y/x) = x^nP(1, v)$ and a similar equation for Q. Substitute these results and $dy = x\,dv + v\,dx$ into (4.11), divide by x^n, and rearrange terms to get

(b) $$[P(1,v) + Q(1,v)\,v]\,dx + Q(1,v)x\,dv = 0.$$

Write both (a) and (b) with variables separated.

23. Show that $(xP + yQ)^{-1}$ is an integrating factor for (4.11). *Hint:* You want to show that $(P\,dx + Q\,dy)/(xP + yQ)$ is an exact differential (see Chapter 6, Section 8). Remember that P and Q are homogeneous of the same degree. Divide numerator and denominator by Q and use $P/Q = -f(y/x)$ from (4.12). Now find the needed partial derivatives. *Comment:* If $(x\,P + y\,Q)$ turns out to be very simple, this may be an easier way to solve a homogeneous equation than the $v = y/x$ substitution (see Problem 24).

24. Solve Problems 9 and 10 by using an integrating factor as discussed in Problem 23.

25. An equation of the form $y' = f(x)y^2 + g(x)y + h(x)$ is called a *Riccati* equation. If we know one particular solution y_p, then the substitution $y = y_p + \frac{1}{z}$ gives a linear first-order equation for z. We can solve this for z and substitute back to find a solution of the y equation containing one arbitrary constant (see Problem 26). Following this method, check the given y_p, and then solve

(a) $y' = xy^2 - \dfrac{2}{x}y - \dfrac{1}{x^3}, \quad y_p = \dfrac{1}{x^2};$

(b) $y' = \dfrac{2}{x}y^2 + \dfrac{1}{x}y - 2x, \quad y_p = x;$

(c) $y' = e^{-x}y^2 + y - e^x, \quad y_p = e^x.$

26. Show that the substitution given in Problem 25 does in general give a solution of the Riccati equation. *Hints:* First show that the substitution $y = y_p + u$ yields the following equation for u: $u' - (g + 2fy_p)u = fu^2$. Note by text equation (4.1) that this is a Bernoulli equation with $n = 2$, so by equation (4.2) we let $z = u^{-1}$. Show that the z equation is the linear first-order equation $z' + (g + 2fy_p)z = -f$. Note that we could have obtained the z equation in one step by substituting $y = y_p + z^{-1}$ in the original equation as claimed in Problem 25.

5. SECOND-ORDER LINEAR EQUATIONS WITH CONSTANT COEFFICIENTS AND ZERO RIGHT-HAND SIDE

Because of their importance in applications, we are going to consider carefully the solution of differential equations of the form

$$a_2\frac{d^2y}{dx^2} + a_1\frac{dy}{dx} + a_0y = 0, \tag{5.1}$$

where a_2, a_1, a_0 are constants; also we shall consider (Section 6) the corresponding equation when the right-hand side of (5.1) is a function of x. Equations of the form (5.1) are called *homogeneous* because every term contains y or a derivative of y. Equations of the form (6.1) are called *inhomogeneous* because they contain a term which does not depend on y. (Note, however, that this use of the term homogeneous is completely unrelated to its use in Section 4.) Although we shall concentrate on second-order equations, which are the ones that occur most frequently in applications, most of our discussion can be extended immediately to linear equations of higher order with constant coefficients (see Problems 21 to 30).

These problems are pretty simple by hand; you may be able to write down answers faster than you can type the problem into a computer! Remember that a computer may not give an answer in the form you need. To use computer solutions effectively, you need to know what to expect, and you can learn this by studying the following methods and doing some problems by hand. Let us consider an equation of the form (5.1).

Example 1. Solve the equation

$$y'' + 5y' + 4y = 0. \tag{5.2}$$

It is convenient to let D stand for d/dx; then

$$Dy = \frac{dy}{dx} = y', \qquad D^2y = \frac{d}{dx}\left(\frac{dy}{dx}\right) = \frac{d^2y}{dx^2} = y''. \tag{5.3}$$

Expressions involving D, such as $D+1$ or D^2+5D+4, are called *differential operators.* (See Problem 31.) In this notation (5.2) becomes

$$D^2y + 5Dy + 4y = 0 \quad \text{or} \quad (D^2 + 5D + 4)y = 0. \tag{5.4}$$

The *algebraic* expression $D^2 + 5D + 4$ can be factored as $(D + 1)(D + 4)$ or $(D + 4)(D + 1)$. You should satisfy yourself that

$$(D+1)(D+4)y = (D+4)(D+1)y = (D^2 + 5D + 4)y \tag{5.5}$$

when $D = d/dx$, and, in fact, that a similar statement is true for $(D - a)(D - b)$ where a and b are any *constants.* (This is not necessarily true if a and b are functions of x; see Problem 31.) Then we can write (5.2) or (5.4) as

$$(D+1)(D+4)y = 0 \quad \text{or} \quad (D+4)(D+1)y = 0. \tag{5.6}$$

To solve (5.4) [or (5.6) which is the same equation rewritten], we shall first solve the simpler equations

$$(D+4)y = 0 \quad \text{and} \quad (D+1)y = 0. \tag{5.7}$$

These are separable equations (Section 2) with solutions

$$y = c_1 e^{-4x}, \quad y = c_2 e^{-x}. \tag{5.8}$$

Now if $(D + 4)y = 0$, then

$$(D+1)(D+4)y = (D+1) \cdot 0 = 0,$$

so any solution of $(D + 4)y = 0$ is a solution of the differential equation (5.6) or (5.4). Similarly, any solution of $(D + 1)y = 0$ is a solution of (5.6) or (5.4). Since the two solutions (5.8) are linearly independent [Problem 13; also see Chapter 3, equation (8.5)], a linear combination of them contains two arbitrary constants and so is the general solution. Thus

$$y = c_1 e^{-4x} + c_2 e^{-x} \tag{5.9}$$

is the general solution of (5.4). Note that we can think of the two solutions e^{-4x} and e^{-x} as basis vectors of a 2-dimensional linear vector space (see Chapter 3, Section 14). Then the general solution (5.9) gives all the vectors of that space. (See Problem 21.)

Now we must investigate whether we can solve all second-order linear equations with constant coefficients (and zero right-hand side) by this method. We first wrote the differential equation using D for d/dx, and then factored the D expression to get (5.5). In this last step, we treated D as if it were an algebraic letter instead of d/dx; this is justified by checking the result (5.5) when $D = d/dx$. Recall from algebra that saying that the algebraic expression D^2+5D+4 has the factors $(D+4)$ and $(D + 1)$ is equivalent to saying that the quadratic equation

$$D^2 + 5D + 4 = 0 \tag{5.10}$$

has roots -4 and -1. The equation (5.10) is called the *auxiliary* (or characteristic) equation for the given differential equation (5.2). From equations (5.6) to (5.9), we see that to solve a linear second-order equation with constant coefficients, we should first solve the auxiliary equation; if the roots of the auxiliary equation are a and b $(a \neq b)$, the general solution of the differential equation is a linear combination of e^{ax} and e^{bx}.

(5.11) $y = c_1 e^{ax} + c_2 e^{bx}$ is the general solution of $(D-a)(D-b)y = 0, \quad a \neq b.$

(If $a = b$, we get only one solution this way; we shall consider this case shortly.) Recall from algebra that the roots of a quadratic equation (with real coefficients; see Problem 19) can be real and unequal, real and equal, or a complex conjugate pair. The equation (5.2) which we have solved is an example in which the roots are real and unequal. Let us consider the other two cases.

Equal Roots of the Auxiliary Equation If the two roots of the auxiliary equation are equal, then the differential equation can be written

$$(D-a)(D-a)y = 0, \tag{5.12}$$

where a is the value of the two equal roots. From our previous discussion (5.5) to (5.11), we know that one solution of (5.12) is $y = c_1 e^{ax}$. But our previous second solution $y = c_2 e^{bx}$ in (5.11) is not a second solution here since $b = a$. To find the second solution for this case, we let

$$u = (D-a)y. \tag{5.13}$$

Then (5.12) becomes

$$(D-a)u = 0,$$

from which we get

$$u = Ae^{ax}. \tag{5.14}$$

We substitute (5.14) into (5.13) to get

$$(D-a)y = Ae^{ax} \quad \text{or} \quad y' - ay = Ae^{ax}.$$

This is a first-order linear equation which we solve as in Section 3:

$$ye^{-ax} = \int e^{-ax} Ae^{ax}\,dx = \int A\,dx = Ax + B.$$

Thus

(5.15) $y = (Ax + B)e^{ax}$ is the general solution of (5.12).

This is the general solution of (5.1) for the case of equal roots of the auxiliary equation. The solution e^{ax} we already know; what is new here is the fact that xe^{ax} is a second (linearly independent; see Problem 14) solution of the differential equation when a is a double root of the auxiliary equation. Equations (5.11) and (5.15) then give the general solution of (5.1) for both unequal and equal roots of the auxiliary equation.

Complex Conjugate Roots of the Auxiliary Equation Suppose the roots of the auxiliary equation are $\alpha \pm i\beta$. These are unequal roots, so by (5.11) the general solution of the differential equation is

$$y = Ae^{(\alpha+i\beta)x} + Be^{(\alpha-i\beta)x} = e^{\alpha x}(Ae^{i\beta x} + Be^{-i\beta x}). \tag{5.16}$$

There are two other very useful forms of (5.16). If we substitute $e^{\pm i\beta x} = \cos\beta x \pm i\sin\beta x$ [see Chapter 2, equation (9.3)] into (5.16), then the parenthesis becomes a linear combination of $\sin\beta x$ and $\cos\beta x$ and we can write (5.16) as

$$y = e^{\alpha x}(c_1\sin\beta x + c_2\cos\beta x), \tag{5.17}$$

where c_1 and c_2 are new arbitrary constants. We can also write (5.17) in the form

$$y = ce^{\alpha x}\sin(\beta x + \gamma), \tag{5.18}$$

where c and γ are now the arbitrary constants. An easy way to see that this is correct is to expand $\sin(\beta x + \gamma)$ by the trigonometric addition formula; this gives a linear combination of $\sin\beta x$ and $\cos\beta x$ as in (5.17). Although it is not hard to express any one of the sets of arbitrary constants [A, B in (5.16); c_1, c_2 in (5.17); and c, γ in (5.18)] in terms of either of the other sets, there is seldom any need to do this. In solving actual problems we simply write whichever one of the three forms seems best for the problem at hand and then determine the arbitrary constants in that form from the given data.

Example 2. Solve the differential equation

$$y'' - 6y' + 9y = 0. \tag{5.19}$$

We can write the equation as

$$(D^2 - 6D + 9)y = 0 \quad \text{or} \quad (D-3)(D-3)y = 0. \tag{5.20}$$

Since the roots of the auxiliary equation are equal, we know that the solution is of the form (5.15) and we simply write the result

$$y = (Ax + B)e^{3x}. \tag{5.21}$$

▶ **Example 3.** In Section 16, Chapter 2, we discussed the differential equation for the motion of a mass m oscillating at the end of a spring, and we solved it by guessing the solution. Now let's solve it by the methods of this chapter. The differential equation is [see Chapter 2, equation (16.21)]

$$m\frac{d^2y}{dt^2} = -ky \quad \text{or} \quad \frac{d^2y}{dt^2} = -\frac{k}{m}y = -\omega^2 y \quad \text{if} \quad \omega^2 = \frac{k}{m}. \tag{5.22}$$

We can write this differential equation as

$$D^2y + \omega^2 y = 0 \quad \text{or} \quad (D^2 + \omega^2)y = 0 \tag{5.23}$$

where $D = d/dt$. The roots of the auxiliary equation are $D = \pm i\omega$; the solution may be written in any of the three forms, (5.16), (5.17), or (5.18):

$$\begin{aligned} y &= Ae^{i\omega t} + Be^{-i\omega t} \\ &= c_1 \sin \omega t + c_2 \cos \omega t \\ &= c \sin(\omega t + \gamma). \end{aligned} \tag{5.24}$$

An object whose displacement from equilibrium satisfies (5.22) or (5.24) is said to be executing *simple harmonic motion.* (Recall Chapter 7, Section 2.)

Equations (5.24) are general solutions of (5.22), each containing two arbitrary constants. Let us find a particular solution corresponding to given initial conditions.

▶ **Example 4.** Suppose the mass is held at rest at a distance 10 cm below equilibrium and then suddenly let go. If we agree to call y positive when m is above the equilibrium position, then at $t = 0$, we have $y = -10$, and $dy/dt = 0$. Using the second solution in (5.24), we get

$$\frac{dy}{dt} = c_1\omega \cos \omega t - c_2\omega \sin \omega t,$$

so the initial conditions give

$$\begin{aligned} -10 &= c_1 \cdot 0 + c_2 \cdot 1, \\ 0 &= c_1\omega \cdot 1 - c_2\omega \cdot 0. \end{aligned}$$

Thus we find

$$c_1 = 0, \qquad c_2 = -10,$$

and the particular solution we wanted is

$$y = -10 \cos \omega t. \tag{5.25}$$

You can verify that either of the other solutions in (5.24) gives the same particular solution (5.25) for the same initial conditions (Problem 32).

This solution is pretty unrealistic from the practical viewpoint. Equations (5.24) and (5.25) imply that the mass m, once started, will simply oscillate up and down forever! This is certainly not true; what *will* happen is that the oscillations will gradually die down. The reason for the discrepancy between the physical facts and our mathematical answer is that we have neglected "friction" forces.

Example 5. A fairly reasonable assumption for this problem and many other similar ones is that there is a retarding force proportional to the velocity; let us call this force $-l(dy/dt)$ $(l > 0)$. Then (5.22), revised to include this force, becomes

$$m\frac{d^2y}{dt^2} = -ky - l\frac{dy}{dt} \qquad (l > 0) \tag{5.26}$$

or with the abbreviations

$$\omega^2 = \frac{k}{m}, \qquad 2b = \frac{l}{m} \qquad (b > 0)$$

it is

$$\frac{d^2y}{dt^2} + 2b\frac{dy}{dt} + \omega^2 y = 0. \tag{5.27}$$

To solve (5.27), we find the roots of the auxiliary equation

$$D^2 + 2bD + \omega^2 = 0, \tag{5.28}$$

which are

$$D = \frac{-2b \pm \sqrt{4b^2 - 4\omega^2}}{2} = -b \pm \sqrt{b^2 - \omega^2}. \tag{5.29}$$

There are three possible types of answer here depending on the relative size of b^2 and ω^2, and there are three special names given to the corresponding types of motion. We say that the motion is

overdamped if	$b^2 > \omega^2$,
critically damped if	$b^2 = \omega^2$,
underdamped or oscillatory if	$b^2 < \omega^2$.

Let us discuss the corresponding general solutions of the differential equation for the three cases.

Overdamped Motion Since $\sqrt{b^2 - \omega^2}$ is real and less than b, both roots of the auxiliary equation are negative, and the general solution is a linear combination of two negative exponentials:

$$y = Ae^{-\lambda t} + Be^{-\mu t}, \qquad \text{where} \quad \begin{cases} \lambda = b + \sqrt{b^2 - \omega^2}, \\ \mu = b - \sqrt{b^2 - \omega^2}. \end{cases} \tag{5.30}$$

Critically Damped Motion Since $b = \omega$, the auxiliary equation has equal roots and the general solution is

$$y = (A + Bt)e^{-bt}. \tag{5.31}$$

In both overdamped and critically damped motion, the mass m is subject to such a large retarding force that it slows down and returns to equilibrium rather than oscillating repeatedly.

Underdamped or Oscillatory Motion In this case $b^2 < \omega^2$ so $\sqrt{b^2 - \omega^2}$ is imaginary. Let $\beta = \sqrt{\omega^2 - b^2}$; then $\sqrt{b^2 - \omega^2} = i\beta$ and the roots (5.29) of the auxiliary equation are $-b \pm i\beta$. The general solution in the form (5.17) is then

$$y = e^{-bt}(A \sin \beta t + B \cos \beta t) \tag{5.32}$$

This result is more in accord with what we know actually happens to the mass m; because of the factor e^{-bt}, the oscillations in this case decrease in amplitude as time goes on. Also note that the frequency of the damped vibrations, namely $\beta = \sqrt{\omega^2 - b^2}$, is less than the frequency ω of the undamped vibrations.

Although we have stated a rather special physical problem, the mathematics we have just discussed applies to a great variety of problems. First, there are many kinds of mechanical vibrations besides a mass attached to a spring. Think of a tuning fork, a pendulum, the needle on the scale of a measuring device, and as more involved examples, the vibrations of complicated structures such as bridges or airplanes, and the vibrations of atoms in a crystal lattice. In such problems, we need to solve differential equations similar to the ones we have discussed. Differential equations of the same form arise in electricity. Consider equations (1.2) and (1.3) when $V = 0$. Remembering that $I = dq/dt$, we can write (1.2) as

$$L\frac{d^2q}{dt^2} + R\frac{dq}{dt} + \frac{1}{C}q = 0 \tag{5.33}$$

and (1.3) as

$$L\frac{d^2I}{dt^2} + R\frac{dI}{dt} + \frac{1}{C}I = 0. \tag{5.34}$$

Both these equations are of the form (5.27) which we have solved. Thus there is an analogy between a series circuit and the motion of a mass m described by (5.26); L corresponds to m, R to the "friction" constant l, and $1/C$ to the spring constant k.

PROBLEMS, SECTION 5

Solve the following differential equations by the methods discussed above and compare computer solutions.

1. $y'' + y' - 2y = 0$

2. $y'' - 4y' + 4y = 0$

3. $y'' + 9y = 0$

4. $y'' + 2y' + 2y = 0$

5. $(D^2 - 2D + 1)y = 0$

6. $(D^2 + 16)y = 0$

7. $(D^2 - 5D + 6)y = 0$

8. $D(D + 5)y = 0$

9. $(D^2 - 4D + 13)y = 0$

10. $y'' - 2y' = 0$

11. $4y'' + 12y' + 9 = 0$

12. $(2D^2 + D - 1)y = 0$

Recall from Chapter 3, equation (8.5), that a set of functions is linearly independent if their Wronskian is not identically zero. Calculate the Wronskian of each of the following sets to show that in each case they are linearly independent. For each set, write the differential equation of which they are solutions. Also note that each set of functions is a set of basis functions for a linear vector space (see Chapter 3, Section 14, Example 2) and that the general solution of the differential equation gives all vectors of the vector space.

13. e^{-x}, e^{-4x}

14. $e^{ax}, e^{bx}, a \neq b$ (a, b, real or complex)

15. e^{ax}, xe^{ax}

16. $\sin \beta x, \cos \beta x$

17. $1, x, x^2$

18. $e^{ax}, xe^{ax}, x^2 e^{ax}$

19. Solve the algebraic equation

$$D^2 + (1+2i)D + i - 1 = 0$$

(note the complex coefficients) and observe that the roots are complex but not complex conjugates. Show that the method of solution of (5.6) (case of unequal roots) is correct here, and so find the general solution of

$$y'' + (1+2i)y' + (i-1)y = 0.$$

20. As in Problem 19, solve $y'' + (1-i)y' - iy = 0$. *Hint:* See Chapter 2, Section 10, for a method of finding the square root of a complex number.

21. By the method used in solving (5.4) to get (5.9), show that the solution of the third-order equation

$$(D-a)(D-b)(D-c)y = 0$$

is

$$y = c_1 e^{ax} + c_2 e^{bx} + c_3 e^{cx}$$

if a, b, c are all different, and find the solutions if two or three of the roots of the auxiliary equation are equal. Generalize the result to higher-order equations. State your results in vector space language [see comment following equation (5.9)].

Use the results of Problem 21 to find the general solutions of the following equations and compare computer solutions.

22. $(D-1)(D+3)(D+5)y = 0$

23. $(D^2+1)(D^2-1)y = 0$ *Hint:* $D^2 + 1 = (D+i)(D-i)$.

24. $y''' + y = 0$

25. $(D^3 + D^2 - 6D)y = 0$

26. $y''' - 3y'' - 9y' - 5y = 0$

27. $D^2(D-1)^2(D+2)^3 y = 0$

28. $(D^4 + 4)y = 0$ *Hint:* Find the four 4th roots of -4 (see Chapter 2, Section 10).

29. $(D+1)^2(D^4 - 16)y = 0$

30. $(D^4 - 1)^2 y = 0$

31. Let D stand for d/dx, that is, $Dy = dy/dx$; then

$$D^2 y = D(Dy) = \frac{d}{dx}\left(\frac{dy}{dx}\right) = \frac{d^2y}{dx^2}, \quad D^3 y = \frac{d^3y}{dx^3}, \text{ etc.}$$

D (or an expression involving D) is called a differential operator. Two operators are equal if they give the same results when they operate on y. For example,

$$D(D+x)y = \frac{d}{dx}\left(\frac{dy}{dx} + xy\right) = \frac{d^2y}{dx^2} + x\frac{dy}{dx} + y = (D^2 + xD + 1)y$$

so we say that

$$D(D+x) = D^2 + xD + 1.$$

In a similar way show that:

(a) $(D-a)(D-b) = (D-b)(D-a) = D^2 - (b+a)D + ab$ for constant a and b.

(b) $D^3 + 1 = (D+1)(D^2 - D + 1)$.

(c) $Dx = xD + 1$. (Note that D and x do not commute, that is, $Dx \neq xD$.)

(d) $(D-x)(D+x) = D^2 - x^2 + 1$, but $(D+x)(D-x) = D^2 - x^2 - 1$.

Comment: The operator equations in (c) and (d) are useful in quantum mechanics; see Chapter 12, Section 22.

32. In Example 3, we used the second solution in (5.24), and obtained (5.25) as the particular solution satisfying the given initial conditions. Show that the first and third solutions in (5.24) also give the particular solution (5.25) satisfying the given initial conditions.

33. A particle moves along the x axis subject to a force toward the origin proportional to x (say $-kx$). Show that the particle executes simple harmonic motion (Example 3). Find the kinetic energy $\frac{1}{2}mv^2$ and the potential energy $\frac{1}{2}kx^2$ as functions of t and show that the total energy is constant. Find the time averages of the potential energy and the kinetic energy and show that these averages are each equal to one-half the total energy (see average values, Chapter 7, Section 4).

34. Find the equation of motion of a simple pendulum (see Chapter 7, Problem 2.13), that is, the differential equation for θ as a function of t. Show that, for small θ, this is approximately a simple harmonic motion equation, and find θ if $\theta = \theta_0$, $d\theta/dt = 0$ when $t = 0$.

35. The gravitational force on a particle of mass m inside the earth at a distance r from the center ($r <$ the radius of the earth R) is $F = -mgr/R$ (Chapter 6, Section 8, Problem 21). Show that a particle placed in an evacuated tube through the center of the earth would execute simple harmonic motion. Find the period of this motion.

36. Find (in terms of L and C) the frequency of electrical oscillations in a series circuit (Figure 1.1) if $R = 0$ and $V = 0$, but $I \neq 0$. (When you tune a radio, you are adjusting C and/or L to make this frequency equal to that of the radio station.)

37. A block of wood is floating in water; it is depressed slightly and then released to oscillate up and down. Assume that the top and bottom of the block are parallel planes which remain horizontal during the oscillations and that the sides of the block are vertical. Show that the period of the motion (neglecting friction) is $2\pi\sqrt{h/g}$, where h is the vertical height of the part of the block under water when it is floating at rest. *Hint:* Recall that the buoyant force is equal to the weight of displaced water.

38. Solve the RLC circuit equation [(5.33) or (5.34)] with $V = 0$ as we did (5.27), and write the conditions and solutions for overdamped, critically damped, and underdamped electrical oscillations in terms of the quantities R, L, and C.

39. (a) Find numerical values of the constants and computer plot together on the same axes graphs of (5.30), (5.31) and (5.32) in order to compare overdamped, critically damped, and oscillatory motion. *Suggested numbers:* Let $\omega = 1$, and $b = 13/5$, 1, 5/13 for the three kinds of motion. Let $y(0) = 1$ and $y'(0) = 0$.

(b) Repeat the problem with the same set of ω and b values and with $y(0) = 1$, but with $y'(0) = 1$.

(c) Again repeat, with $y'(0) = -1$.

40. The natural period of an undamped system is 3 sec, but with a damping force proportional to the velocity, the period becomes 5 sec. Find the differential equation of motion of the system and its solution.

6. SECOND-ORDER LINEAR EQUATIONS WITH CONSTANT COEFFICIENTS AND RIGHT-HAND SIDE NOT ZERO

So far we have considered second-order linear equations with constant coefficients and zero right-hand side (5.1). Such equations describe *free vibrations* or oscillations of mechanical or electrical systems. But often such systems are not free but are subject to an applied force or emf. The vibrations are then called *forced vibrations* and the differential equation describing the system is of the form

$$(6.1)\qquad \begin{aligned} &a_2\frac{d^2y}{dx^2} + a_1\frac{dy}{dx} + a_0y = f(x), \quad \text{or} \\ &\frac{d^2y}{dx^2} + \frac{a_1}{a_2}\frac{dy}{dx} + \frac{a_0}{a_2}y = F(x). \end{aligned}$$

The function $f(x)$ is often called the forcing function; it represents the applied force or emf. We want to find the general solution of equations of the form (6.1).

Example 1. Consider the equation

$$(6.2)\qquad (D^2 + 5D + 4)y = \cos 2x.$$

We already know (from Section 5, Example 1) the general solution of the corresponding equation (5.2) with the right-hand side equal to zero. This solution (5.9) is called the *complementary function*; it is not a solution of (6.2) but is related to it as we shall see. We shall denote the complementary function by y_c. Thus for equation (6.2) the complementary function is

$$(6.3)\qquad y_c = Ae^{-x} + Be^{-4x}.$$

Now suppose we know just any solution of (6.2); we call this solution a *particular solution* and denote it by y_p. You can easily verify that

$$(6.4)\qquad y_p = \tfrac{1}{10}\sin 2x$$

is a particular solution of (6.2), and we shall soon consider ways of finding such solutions. Then we have

$$(6.5)\qquad (D^2 + 5D + 4)y_p = \cos 2x$$

and from Section 5, Example 1,

$$(6.6)\qquad (D^2 + 5D + 4)y_c = 0.$$

Adding (6.5) and (6.6), we find

$$(D^2 + 5D + 4)(y_p + y_c) = \cos 2x + 0 = \cos 2x.$$

Thus

$$(6.7)\qquad y = y_c + y_p = Ae^{-x} + Be^{-4x} + \tfrac{1}{10}\sin 2x$$

is a solution of (6.2). In fact, it is the general solution of (6.2) since it contains two independent arbitrary constants (Problem 27).

Thus we see how to solve 6.1):

> The general solution of an equation of the form (6.1) is
>
> (6.8) $$y = y_c + y_p$$
>
> where the complementary function y_c is the general solution of the homogeneous equation (as in Section 5) and y_p is a particular solution of (6.1).

We shall now discuss some ways of finding particular solutions. It is worthwhile to know about this even if you are using a computer to find the solution. When you know what to expect, you are better able to judge whether a computer solution is in the best form for your purposes, and if not, to find a better form. (See problems.)

Inspection If there *is* a very simple particular solution, we may be able to guess and verify it.

Example 2. Consider the equation $y'' - 2y' + 3y = 5$.
It is easy to see that $y_p = \frac{5}{3}$ is a particular solution of this equation since if y is constant, y'' and y' are zero.

Example 3. As a less trivial problem, consider

(6.9) $$y'' - 6y' + 9y = 8e^x.$$

We might suspect that a multiple of e^x is a solution of this equation, and it is easy to verify that $y = 2e^x$ is a solution. But trying the same method for the equation

(6.10) $$y'' + y' - 2y = e^x,$$

we fail to find a particular solution since e^x satisfies

$$y'' + y' - 2y = 0.$$

The method of inspection is very good in simple cases where it gives us an answer quickly, but usually we need other methods.

Successive Integration of Two First-Order Equations This is a straightforward method which can always be used to solve equations of the form (6.1). In practice, however, it often involves more work than various special methods; we shall find it particularly useful in deriving the special methods.

Example 4. Let's solve (6.10) again. We can write this differential equation as

(6.11) $$(D - 1)(D + 2)y = e^x.$$

Let

(6.12) $$u = (D + 2)y.$$

Then the differential equation (6.11) becomes

$$(D-1)u = e^x \qquad \text{or} \quad u' - u = e^x. \tag{6.13}$$

This is a first-order linear differential equation which we solve as in Section 3.

$$\begin{aligned} I &= \int -dx = -x, \\ ue^{-x} &= \int e^{-x} e^x \, dx = x + c_1, \\ u &= xe^x + c_1 e^x. \end{aligned} \tag{6.14}$$

Then the differential equation for y becomes

$$(D+2)y = xe^x + c_1 e^x \qquad \text{or} \quad y' + 2y = xe^x + c_1 e^x.$$

This is again a linear first-order equation which we solve as follows:

$$\begin{aligned} I &= \int 2\,dx = 2x, \\ ye^{2x} &= \int e^{2x}(xe^x + c_1 e^x)\,dx = \tfrac{1}{3}xe^{3x} - \tfrac{1}{9}e^{3x} + \tfrac{1}{3}c_1 e^{3x} + c_2 \\ &= \frac{1}{3}xe^{3x} + c_1' e^{3x} + c_2, \\ y &= \tfrac{1}{3}xe^x + c_1' e^x + c_2 e^{-2x}. \end{aligned} \tag{6.15}$$

Notice that here we have obtained the general solution all in one process rather than finding the complementary function plus a particular solution in two separate processes. However, we could have obtained just the particular solution $xe^x/3$ by omitting the arbitrary constant at each integration (these led to the complementary function) and also dropping terms which are already in the complementary function ($-e^x/9$ in this example). Since it is easy to write the complementary function (by Section 5), it saves time to omit those terms when we are finding a particular solution. You may find that your computer gives a more complicated particular solution by including terms of the complementary function in the particular solution. Now that you know to watch for this, you can simplify a computer solution by removing those terms.

Exponential Right-Hand Side Let us consider how to find a particular solution when the right-hand side of (6.1) is $F(x) = ke^{cx}$ where k and c are given constants. Observe that c may be complex; we shall be especially interested in this case later. Let a and b be the roots of the auxiliary equation of (6.1); then (6.1) becomes

$$(D-a)(D-b)y = F(x) = ke^{cx}. \tag{6.16}$$

Let us first suppose that c is not equal to either a or b. Solving (6.16) by successive integration of two first-order equations as in the last paragraph is straightforward (Problem 28) and gives the result that the particular solution in this case is simply a multiple of e^{cx}. It is not necessary to remember the formula for the constant factor or to go through this process each time. Now that we know the form of the particular solution, we simply assume a solution of this form and solve for the constant.

Example 5. Solve the equation

$$(D-1)(D+5)y = 7e^{2x}. \tag{6.17}$$

We observe that $c = 2$ is not equal to either of the roots of the auxiliary equation. To find a particular solution we substitute $y_p = Ce^{2x}$ into (6.17) and get

$$y_p'' + 4y_p' - 5y_p = C(4e^{2x} + 8e^{2x} - 5e^{2x}) = 7e^{2x}.$$

Thus we must have $C = 1$, and the general solution of (6.17) is

$$y = Ae^x + Be^{-5x} + e^{2x}.$$

We have already seen in solving (6.11) that if c is equal to either a or b ($a \neq b$), the particular solution is of the form Cxe^{cx}. By the same method used for (6.11), you can easily discover that if $a = b = c$, the particular solution is of the form Cx^2e^{cx} (Problem 28c). In practice, then, we find a particular solution of (6.16) by assuming a solution of the form:

$$\begin{cases} Ce^{cx} & \text{if } c \text{ is not equal to either } a \text{ or } b; \\ Cxe^{cx} & \text{if } c \text{ equals } a \text{ or } b,\ a \neq b; \\ Cx^2e^{cx} & \text{if } c = a = b. \end{cases} \tag{6.18}$$

Now that we know this, we would solve (6.10) as follows. Substitute

$$y_p = Cxe^x, \quad y_p' = C(xe^x + e^x), \quad y_p'' = C(xe^x + 2e^x)$$

into (6.10) and get

$$y_p'' + y_p' - 2y_p = C(xe^x + 2e^x + xe^x + e^x - 2xe^x) = e^x.$$

Thus we find $C = \frac{1}{3}$ as in (6.15) (but with much less work).

Use of Complex Exponentials In applied problems, the function $F(x)$ on the right-hand side of (6.1) is very often a sine or a cosine representing alternating emf or a periodic force. We *could* find y_p for such a problem either by the method of integrating two successive first-order equations or by replacing the sine or cosine by its complex exponential form and using the method of the last paragraph. There is a still more efficient variation of the latter method which we shall now show.

Example 6. Solve

$$y'' + y' - 2y = 4\sin 2x. \tag{6.19}$$

Instead of tackling this problem directly, we are first going to solve the equation

$$Y'' + Y' - 2Y = 4e^{2ix}. \tag{6.20}$$

Since $e^{2ix} = \cos 2x + i\sin 2x$ is complex, the solution Y may be complex also. Then if $Y = Y_R + iY_I$, (6.20) is equivalent to two equations

$$\begin{aligned} Y_R'' + Y_R' - 2Y_R &= \operatorname{Re} 4e^{2ix} = 4\cos 2x, \\ Y_I'' + Y_I' - 2Y_I &= \operatorname{Im} 4e^{2ix} = 4\sin 2x. \end{aligned} \tag{6.21}$$

Since the second equation in (6.21) is the same as (6.19), we see that the solution of (6.19) is the imaginary part of Y. Thus to find y_p for (6.19), we find Y_p for (6.20) and take its imaginary part. We observe that $2i$ is not equal to either of the roots of the auxiliary equation in (6.20). Following the method of the last paragraph, we assume a solution of the form

$$Y_p = Ce^{2ix}$$

and substitute it into (6.20) to get

$$(-4 + 2i - 2)Ce^{2ix} = 4e^{2ix},$$
$$C = \frac{4}{2i - 6} = \frac{4(-2i - 6)}{40} = -\tfrac{1}{5}(i + 3),$$
$$Y_p = -\tfrac{1}{5}(i + 3)e^{2ix}.$$

Taking the imaginary part of Y_p, we find y_p for (6.19):

$$y_p = -\tfrac{1}{5}\cos 2x - \tfrac{3}{5}\sin 2x. \tag{6.22}$$

We summarize the method of complex exponentials:

(6.23) To find a particular solution of

$$(D - a)(D - b)y = \begin{cases} k\sin \alpha x, \\ k\cos \alpha x, \end{cases}$$

first solve

$$(D - a)(D - b)y = ke^{i\alpha x}$$

and then take the real or imaginary part.

Method of Undetermined Coefficients The method we have just discussed of assuming an exponential solution and determining the constant factor C is an example (and in practice the most important case) of the *method of undetermined coefficients.* In (6.18) we outlined the form of y_p to assume for equation (6.16), that is, when the right-hand side of (6.1) is an exponential. It is straightforward but tedious (Problems 29 and 32) to find the corresponding result (6.24) when the right-hand side is an exponential times a polynomial.

A particular solution y_p of $(D-a)(D-b)y = e^{cx}P_n(x)$ where $P_n(x)$ is a polynomial of degree n is

$$y_p = \begin{cases} e^{cx}Q_n(x) & \text{if } c \text{ is not equal to either } a \text{ or } b, \\ xe^{cx}Q_n(x) & \text{if } c \text{ equals } a \text{ or } b,\ a \neq b, \\ x^2e^{cx}Q_n(x) & \text{if } c = a = b, \end{cases} \tag{6.24}$$

where $Q_n(x)$ is a polynomial of the same degree as $P_n(x)$ with *undetermined coefficients* to be found to satisfy the given differential equation. Note that sines and cosines are included in e^{cx} by use of complex exponentials as in (6.19) to (6.23). (Also see Problem 29.)

Example 7. To illustrate using (6.24), let's find a particular solution of

$$(D-1)(D+2)y = y'' + y' - 2y = 18xe^x. \tag{6.25}$$

In the notation of (6.24) we have $a = 1$, $b = -2$, $c = 1$; also $P_n(x) = 18x = P_1(x)$ is a polynomial of degree 1. Then Q_1 is a polynomial of degree 1, namely $Ax + B$. Since $c = a \neq b$, we see by (6.24) that the form to assume for a particular solution of (6.25) is

$$y_p = xe^x(Ax + B) = e^x(Ax^2 + Bx).$$

We substitute this into (6.25) and find A and B so that we have an identity.

$$\begin{aligned} y_p{}' &= e^x(Ax^2 + Bx + 2Ax + B), \\ y_p{}'' &= e^x(Ax^2 + Bx + 4Ax + 2B + 2A) \\ y_p{}'' + y_p{}' - 2y_p &= e^x(6Ax + 3B + 2A) \equiv 18xe^x \end{aligned}$$

To make this an identity, we must have

$$6A = 18,\ 3B + 2A = 0, \quad \text{or} \quad A = 3,\ B = -2, \quad \text{so}$$

$$y_p = (3x^2 - 2x)e^x. \tag{6.26}$$

A computer solution may add to this a constant times e^x, but this is an unnecessary complication since e^x is a term in the complementary function.

If the right-hand side of a differential equation is a polynomial, then $c = 0$ in (6.24), and we assume for y_p a polynomial as indicated in (6.24).

Example 8. To solve

$$(D-1)(D+2)y = y'' + y' - 2y = x^2 - x \tag{6.27}$$

we assume $y_p = Ax^2 + Bx + C$, and find the particular solution

$$y_p = -\tfrac{1}{2}(x^2 + 1). \tag{6.28}$$

A computer solution gives the same result.

PROBLEMS, SECTION 6

Find the general solution of the following differential equations (complementary function + particular solution). Find the particular solution by inspection or by (6.18), (6.23), or (6.24). Also find a computer solution and reconcile differences if necessary, noticing especially whether the particular solution is in simplest form [see (6.26) and the discussion after (6.15)].

1. $y'' - 4y = 10$
2. $(D-2)^2 y = 16$
3. $y'' + y' - 2y = e^{2x}$
4. $(D+1)(D-3)y = 24e^{-3x}$
5. $(D^2 + 1)y = 2e^x$
6. $y'' + 6y' + 9y = 12e^{-x}$
7. $y'' - y' - 2y = 3e^{2x}$
8. $y'' - 16y = 40e^{4x}$

9. $(D^2 + 2D + 1)y = 2e^{-x}$ **10.** $(D - 3)^2 y = 6e^{3x}$

11. $y'' + 2y' + 10y = 100\cos 4x$ *Hint:* First solve $y'' + 2y' + 10y = 100e^{4ix}$.

12. $(D^2 + 4D + 12)y = 80\sin 2x$ **13.** $(D^2 - 2D + 1)y = 2\cos x$

14. $y'' + 8y' + 25y = 120\sin 5x$ **15.** $5y'' + 12y' + 20y = 120\sin 2x$

16. $(D^2 + 9)y = 30\sin 3x$ **17.** $y'' + 16y = 16\cos 4x$

18. $(D^2{+}2D{+}17)y = 60e^{-4x}\sin 5x$ *Hint:* First solve $(D^2{+}2D{+}17)y = 60e^{(-4+5i)x}$.

19. $(4D^2 + 4D + 5)y = 40e^{-3x/2}\sin 2x$ **20.** $y'' + 4y' + 8y = 30e^{-x/2}\cos 5x/2$

21. $5y'' + 6y' + 2y = x^2 + 6x$ **22.** $2y'' + y' = 2x$

23. $y'' + y = 2xe^x$ **24.** $y'' - 6y' + 9y = 12xe^{3x}$

25. $(D - 3)(D + 1)y = 16x^2e^{-x}$ **26.** $(D^2 + 1)y = 8x\sin x$

27. Verify that (6.4) is a particular solution of (6.2). Verify that another particular solution of (6.2) is

$$y_p = \tfrac{1}{10}\sin 2x - e^{-x}.$$

Observe that we obtain the same general solution (6.7) whichever particular solution we use [since $(A - 1)$ is just as good an arbitrary constant as A]. Show in general that the difference between two particular solutions of $(a_2D^2 + a_1D + a_0)y = f(x)$ is always a solution of the homogeneous equation $(a_2D^2 + a_1D + a_0)y = 0$, and thus show that the general solution is the same for all choices of a particular solution.

28. Solve (6.16) by the method used in solving (6.11), for the following three cases, to obtain the result (6.18).

(a) c is not equal to either a or b;

(b) $a \neq b$, $c = a$;

(c) $a = b = c$.

29. Consider the differential equation $(D - a)(D - b)y = P_n(x)$, where $P_n(x)$ is a polynomial of degree n. Show that a particular solution of this equation is given by (6.24) with $c = 0$; that is, y_p is

$$\begin{cases} \text{a polynomial } Q_n(x) \text{ of degree } n \text{ if } a \text{ and } b \text{ are both different from zero;} \\ xQ_n(x) \quad \text{if } a \neq 0, \text{ but } b = 0; \\ x^2Q_n(x) \quad \text{if } a = b = 0. \end{cases}$$

Hint: To show that $Q_n(x) = \sum a_nx^n$ is a solution of the differential equation for a given $P_n = \sum b_nx^n$, you have only to show that the coefficients a_n can be found so that $(D - a)(D - b)Q_n(x) \equiv P_n(x)$. Equate coefficients of x^n, x^{n-1}, $\cdots$, to see that this is always possible if $a \neq b$. For $b = 0$, the differential equation becomes $(D - a)Dy = P_n$; what is Dy if $y = xQ_n$? Similarly, consider D^2y if $y = x^2Q_n$.

30. (a) Show that

$$\begin{aligned} (D - a)e^{cx} &= (c - a)e^{cx}; \\ (D^2 + 5D - 3)e^{cx} &= (c^2 + 5c - 3)e^{cx}; \\ L(D)e^{cx} &= L(c)e^{cx}, \text{ where } L(D) \text{ is any polynomial in } D; \\ (D - c)xe^{cx} &= e^{cx}; \\ (D - c)^2x^2e^{cx} &= 2e^{cx}. \end{aligned}$$

(b) Define the expression $y = [1/L(D)]u(x)$ to mean a solution of the differential equation $L(D)y = u$. Using part (a), show that

$$\frac{1}{D-a}e^{cx} = \frac{e^{cx}}{c-a}, \quad c \neq a;$$
$$\frac{1}{D^2+5D-3}e^{cx} = \frac{e^{cx}}{c^2+5c-3};$$
$$\frac{1}{L(D)}e^{cx} = \frac{e^{cx}}{L(c)}, \quad L(c) \neq 0;$$
$$\frac{1}{D-c}e^{cx} = xe^{cx};$$
$$\frac{1}{(D-c)^2}e^{cx} = \tfrac{1}{2}x^2e^{cx}.$$

(c) The expressions $1/L(D)$ in (b) are called inverse operators. They can be used to find particular solutions of differential equations. As an example consider Problem 3. We write

$$(D^2+D-2)y = e^{2x},$$
$$y = \frac{1}{D^2+D-2}e^{2x} = \frac{e^{2x}}{2^2+2-2} = \frac{e^{2x}}{4}.$$

Using inverse operators, find particular solutions of Problems 4 to 20. Be careful to use parts 4 or 5 of (b) if c is a root of the auxiliary equation. For example,

$$\frac{1}{(D-a)(D-c)}e^{cx} = \frac{1}{D-c}\frac{1}{D-a}e^{cx} = \frac{1}{D-c}\frac{e^{cx}}{c-a} = \frac{xe^{cx}}{c-a}.$$

31. (a) Show that

$$D(e^{ax}y) = e^{ax}(D+a)y,$$
$$D^2(e^{ax}y) = e^{ax}(D+a)^2y,$$

and so on; that is, for any positive integral n,

$$D^n(e^{ax}y) = e^{ax}(D+a)^ny.$$

Thus show that if $L(D)$ is any polynomial in the operator D, then

$$L(D)(e^{ax}y) = e^{ax}L(D+a)y.$$

This is called the *exponential shift.*

(b) Use (a) to show that

$$(D-1)^3(e^xy) = e^xD^3y,$$
$$(D^2+D-6)(e^{-3x}y) = e^{-3x}(D^2-5D)y.$$

(c) Replace D by $D-a$, to obtain

$$e^{ax}P(D)y = P(D-a)e^{ax}y.$$

This is called the *inverse exponential shift.*

(d) Using (c), we can change a differential equation whose right-hand side is an exponential times a polynomial, to one whose right-hand side is just a polynomial. For example, consider $(D^2 - D - 6)y = 10xe^{3x}$; multiplying both sides by e^{-3x} and using (c), we get

$$e^{-3x}(D^2 - D - 6)y = [(D+3)^2 - (D+3) - 6]ye^{-3x}$$
$$= (D^2 + 5D)ye^{-3x} = 10x.$$

Show that a solution of $(D^2+5D)u = 10x$ is $u = x^2 - \frac{2}{5}x$; then $ye^{-3x} = x^2 - \frac{2}{5}x$ or $y = e^{3x}(x^2 - \frac{2}{5}x)$. Use this method to solve Problems 23 to 26.

32. Using Problems 29 and 31b, show that equation (6.24) is correct.

Several Terms on the Right-Hand Side: Principle of Superposition So far we have brushed over a question which may have occurred to you: What do we do if there are several terms on the right-hand side of the equation involving different exponentials?

Example 9. As an artificial problem to illustrate the ideas, consider the equation

$$y'' + y' - 2y = (D-1)(D+2)y = [e^x] + [4\sin 2x] + [x^2 - x]. \tag{6.29}$$

We have already solved differential equations with the same left-hand sides as (6.29) and with right-hand sides equal in turn to each of the three expressions in brackets in (6.29) [see (6.11) to (6.15), (6.19) to (6.22), (6.27), and (6.28)]. Thus we know that

$(D-1)(D+2)y = e^x$ has the particular solution $y_{p1} = \frac{1}{3}xe^x$;
$(D-1)(D+2)y = 4\sin 2x$ has the particular solution $y_{p2} = -\frac{1}{5}\cos 2x - \frac{3}{5}\sin 2x$;
$(D-1)(D+2)y = x^2 - x$ has the particular solution $y_{p3} = -\frac{1}{2}(x^2+1)$.

Adding these three solutions, we see that

$$y_p = y_{p1} + y_{p2} + y_{p3} = \tfrac{1}{3}xe^x - \tfrac{1}{5}\cos 2x - \tfrac{3}{5}\sin 2x - \tfrac{1}{2}(x^2+1) \tag{6.30}$$

is a particular solution of (6.29).

This is the easiest way of handling a complicated right-hand side: Solve a separate equation for *each different exponential* and add the solutions. The fact that this is correct for a linear equation is often called the *principle of superposition.* As we can see from (6.29) and (6.30), this amounts to a fancy name for the fact that the derivative (of any order) of a sum of terms is equal to the sum of the derivatives of the individual terms. Notice that the principle holds only for *linear* equations; for example, if the equation contained y'^2, the principle would not hold since $(y_1' + y_2')^2$ is not equal to $y_1'^2 + y_2'^2$. In fact, an operator (such as the D operators we have been using) which satisfies the principle of superposition is called a *linear operator.* [See Chapter 3, equation (7.4) and Problem 7.12.] Linear operators are of particular importance because they obey the principle of superposition; for example, $D^2(y_1 + y_2) = y_1'' + y_2'' = D^2y_1 + D^2y_2$, so D^2 is a linear operator. We shall make use of this principle shortly in our discussion of the use of Fourier series in finding particular solutions.

Forced Vibrations Let's return now to the physical problem we considered at the end of Section 5. There we set up and solved the differential equation which describes the free (zero right-hand side, no forcing function) vibrations of a damped oscillator. We commented that the same mathematics applies to a variety of mechanics problems and also to a simple RLC series electric circuit. As we know from experiment and as we can see from (5.30), (5.31), and (5.32), the free vibrations we considered in Section 5 die out as time passes. Such oscillations are referred to as *transients.* We next want to consider the vibrations obtained when a periodic force (or emf in the electric case) is applied. This means mathematically that we want to solve (5.27) with a function of t on the right-hand side. The solution will contain the appropriate one of (5.30), (5.31), (5.32); this is the complementary function and it is also the transient since it tends to zero as t tends to infinity. The solution will also contain a particular solution which does not tend to zero as t tends to infinity; this is the *steady-state solution* which we want to find.

▶ **Example 10.** Let us solve

$$\frac{d^2y}{dt^2}+2b\frac{dy}{dt}+\omega^2 y = F\sin\omega' t \qquad (F=\text{const.}). \tag{6.31}$$

By the method of complex exponentials, we solve first

$$\frac{d^2Y}{dt^2}+2b\frac{dY}{dt}+\omega^2 Y = Fe^{i\omega' t}. \tag{6.32}$$

Substitute

$$Y_p = Ce^{i\omega' t} \tag{6.33}$$

into (6.32) to get

$$\begin{gathered}(-\omega'^2+2bi\omega'+\omega^2)Ce^{i\omega' t} = Fe^{i\omega' t},\\ C=\frac{F}{(\omega^2-\omega'^2)+2bi\omega'}=\frac{[(\omega^2-\omega'^2)-2bi\omega']F}{\left(\omega^2-\omega'^2\right)^2+4b^2\omega'^2}.\end{gathered} \tag{6.34}$$

It is convenient to write the complex number C in the $re^{i\theta}$ form. We have

$$|C| = \frac{F}{\sqrt{\left(\omega^2-\omega'^2\right)^2+4b^2\omega'^2}}, \tag{6.35}$$

angle of $C=-\phi$, where ϕ is given by Figure 6.1.

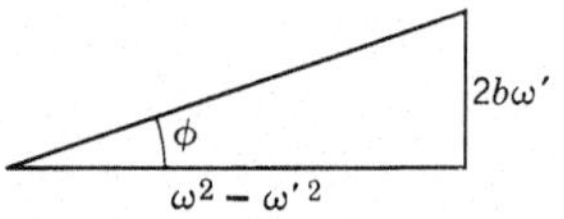

Figure 6.1

Thus

$$C=\frac{F}{\sqrt{\left(\omega^2-\omega'^2\right)^2+4b^2\omega'^2}}e^{-i\phi} \tag{6.36}$$

and from (6.33)

$$Y_p=\frac{F}{\sqrt{\left(\omega^2-\omega'^2\right)^2+4b^2\omega'^2}}e^{i(\omega' t-\phi)} \tag{6.37}$$

To find y_p we take the imaginary part of Y_p:

$$y_p = \frac{F}{\sqrt{\left(\omega^2 - \omega'^2\right)^2 + 4b^2\omega'^2}} \sin(\omega' t - \phi). \tag{6.38}$$

This is the steady-state solution, so-called because as t increases, the rest of the solution [given by (5.30), (5.31), or (5.32)] becomes negligible. For example, when you turn on an electric light, the current is given by (5.32) plus (6.38). The transient (5.32) tends to zero rapidly and the steady-state solution (6.38) becomes essentially the whole solution.

Resonance We note, by comparing (6.38) and the forcing function in (6.31), that the applied force (or emf) and the solution y (which represents displacement, current, etc.) are *out of phase*; that is, their maximum values do not occur at the same time because of the phase angle ϕ. We also see from (6.38) that for a given forcing frequency ω', the largest amplitude of y (also of dy/dt, Problem 40) occurs if the natural (undamped) frequency ω is equal to ω'. This situation is often called *resonance*. In the RLC series circuit problem, y represents the charge q on the capacitor if the forcing function is the emf, and y represents the current $I = dq/dt$ if the forcing function is the time derivative of the emf. For such a circuit, given the frequency ω' of the applied emf, the current (or charge) will have the largest amplitude when the natural (undamped) frequency ω is equal to ω'. This is almost always called the resonance condition for the electrical case. However, there is another question we could ask here which is of particular interest in mechanics. Given the natural (undamped) frequency ω of the system, what frequency of the forcing function will produce the largest amplitude of y? In (6.38), we want to maximize the coefficient of the sine; we can instead minimize the square of the denominator of the coefficient; that is, we want to find the value of ω' which minimizes $(\omega^2 - \omega'^2)^2 + 4b^2\omega'^2$ for given ω. Setting the derivative of this function (with respect to ω') equal to zero and solving for ω', we get

$$\begin{aligned} 2(\omega^2 - \omega'^2)(-2\omega') + 8b^2\omega' &= 0, \\ \omega'^2 &= \omega^2 - 2b^2. \end{aligned} \tag{6.39}$$

Note that this value of ω' is not equal to either the natural undamped frequency ω or the natural damped frequency β where $\beta^2 = \omega^2 - b^2$ [see (5.32)]. However, if we define *resonance* as the situation in which we get the maximum amplitude for y for a given value of ω, then the resonance condition is (6.39). (The maximum amplitude for the velocity—or current in the electrical case—is still obtained for $\omega' = \omega$; Problem 40.) The resonance condition (6.39) is of particular importance in mechanics where we are apt to be interested in the displacement y of a given system under the action of various forces. For example, consider a bridge; we would want to avoid periodic forces with an ω' given by (6.39) since such forces would produce large vibrations. In this case resonance is undesirable. It may in other cases be desirable; for example, when you tune your radio to the frequency of a given station, you are given ω' and you adjust the circuit in your radio to make its natural frequency ω equal to the given ω'.

Use of Fourier Series in Finding Particular Solutions In simple problems, the forcing function in either the electrical or mechanical case is just a sine or cosine and the problem can be solved as we have just done. In more complicated (and realistic) cases, however, the forcing function may very well be some more complicated function; it is often a periodic function, however, and we shall assume this. Suppose, for example, that the periodic emf applied to a circuit is given by one of the graphs in Figure 3.2 of Chapter 7. We learned in Chapter 7 how to expand such a function in a Fourier series. Let us suppose that this has been done, using for definiteness the complex exponential form of the Fourier series. Then we can write (6.1) as

$$a_2\frac{d^2y}{dx^2} + a_1\frac{dy}{dx} + a_0 y = f(x) = \sum_{n=-\infty}^{\infty} c_n e^{inx}. \tag{6.40}$$

We know how to solve the equation

$$a_2\frac{d^2y}{dx^2} + a_1\frac{dy}{dx} + a_0 y = c_n e^{inx} \tag{6.41}$$

with the right-hand side equal to any one term of the series. If we now add the solutions of all the equations (6.41) for all n, we have a solution of (6.40) (see *principle of superposition* above).

▶ **Example 11.** Solve

$$\frac{d^2y}{dt^2} + 2\frac{dy}{dt} + 10y = f(t), \tag{6.42}$$

where $f(t)$ is a function of period 2π and

$$f(t) = \begin{cases} 1, & 0 \le t < \pi, \\ 0, & \pi \le t < 2\pi. \end{cases}$$

The auxiliary equation is

$$D^2 + 2D + 10 = 0;$$

its roots are

$$D = -1 \pm 3i,$$

so the complementary function is

$$y_c = e^{-t}(A\cos 3t + B\sin 3t).$$

To find a particular solution, we first expand $f(t)$ in a Fourier series; from Chapter 7, equation (7.8), we have

$$f(t) = \frac{1}{2} + \frac{1}{i\pi}[e^{it} - e^{-it} + \tfrac{1}{3}(e^{3it} - e^{-3it}) + \cdots]. \tag{6.43}$$

We next write and solve a whole set of differential equations like (6.42) but each having just one term of the series (6.43) on the right-hand side. For the first term (namely $\frac{1}{2}$) we see by inspection that a particular solution of

$$\frac{d^2y}{dt^2} + 2\frac{dy}{dt} + 10y = \frac{1}{2}$$

is $y = \frac{1}{20}$. All the other terms of (6.43) are of the form $(1/ik\pi)e^{ikt}$, where k is a positive or negative odd integer. To solve

$$\frac{d^2y}{dt^2} + 2\frac{dy}{dt} + 10y = \frac{1}{ik\pi}e^{ikt}, \tag{6.44}$$

we substitute

$$y = Ce^{ikt} \tag{6.45}$$

into (6.44) and get

$$(-k^2 + 2ik + 10)Ce^{ikt} = \frac{1}{ik\pi}e^{ikt}.$$

Then we have

$$C = \frac{1}{ik\pi}\frac{1}{(10-k^2)+2ik} = \frac{1}{ik\pi}\frac{(10-k^2)-2ik}{(10-k^2)^2+4k^2}. \tag{6.46}$$

By letting $k = \pm1, \pm3, \cdots$, and substituting the values of C thus obtained into (6.45), we obtain the solutions of (6.44) for the various k values corresponding to the terms of the series (6.43). The sum of all the solutions corresponding to all the terms is the desired particular solution of (6.42). Thus

$$\begin{aligned} y_p &= \frac{1}{20} + \frac{1}{i\pi}\frac{9-2i}{85}e^{it} - \frac{1}{i\pi}\frac{9+2i}{85}e^{-it} \\ &\quad + \frac{1}{3i\pi}\frac{1-6i}{37}e^{3it} - \frac{1}{3i\pi}\frac{1+6i}{37}e^{-3it} + \cdots \\ &= \frac{1}{20} + \frac{2}{\pi}\frac{9}{85}\left(\frac{e^{it}-e^{-it}}{2i}\right) - \frac{2}{\pi}\frac{2}{85}\left(\frac{e^{it}+e^{-it}}{2}\right) \\ &\quad + \frac{2}{3\pi}\frac{1}{37}\left(\frac{e^{3it}-e^{-3it}}{2i}\right) - \frac{2}{3\pi}\frac{6}{37}\left(\frac{e^{3it}+e^{-3it}}{2}\right) + \cdots \\ &= \frac{1}{20} + \frac{2}{85\pi}(9\sin t - 2\cos t) + \frac{2}{111\pi}(\sin 3t - 6\cos 3t) + \cdots \end{aligned} \tag{6.47}$$

is a particular solution of (6.42).

PROBLEMS, SECTION 6

In Problem 33 to 38, solve the given differential equations by using the principle of superposition [see the solution of equation (6.29)]. For example, in Problem 33, solve three differential equations with right-hand sides equal to the three different brackets. Note that terms with the *same exponential factor* are kept together; thus a polynomial of any degree is kept together in one bracket.

33. $y'' + y = [x^3 - 1] + [2\cos x] + [(2-4x)e^x]$

34. $y'' - 5y' + 6y = 2e^x + 6x - 5$

35. $(D^2 - 1)y = \sinh x$

36. $(D^2 + 1)y = 2\sin x + 4x\cos x$

37. $(D-1)^2 y = 4e^x + (1-x)(e^{2x} - 1)$

38. $y'' - 2y' = 9xe^{-x} - 6x^2 + 4e^{2x}$

39. Find the solutions of (1.2) (put $I = dq/dt$) and (1.3), if $V = V_0 \sin \omega' t$ (ω' =const.).

40. In (6.38), show that for a given forcing frequency ω', the displacement y and the velocity dy/dt have their largest amplitude when $\omega = \omega'$.

For a given ω, we have shown in Section 6 that the maximum amplitude of y does not correspond to $\omega' = \omega$. Show, however, that the maximum amplitude of dy/dt for a given ω does correspond to $\omega' = \omega$.

State the corresponding results for an electric circuit in terms of L, R, C.

Solve Problems 41 and 42 by use of Fourier series. Assume in each case that the right-hand side is a periodic function whose values are stated for one period.

41. $y'' + 2y' + 2y = |x|, \qquad -\pi < x < \pi.$

42. $y'' + 9y = \begin{cases} x, & 0 < x < 1, \\ 0, & -1 < x < 0. \end{cases}$

43. Consider an equation for damped forced vibrations (mechanical or electrical) in which the right-hand side is a sum of several forces or emfs of different frequencies. For example, in (6.32) let the right-hand side be

$$F_1 e^{i\omega_1' t} + F_2 e^{i\omega_2' t} + F_3 e^{i\omega_3' t}.$$

Write the solution by the principle of superposition. Suppose, for given ω_1', ω_2', ω_3', that we adjust the system so that $\omega = \omega_1'$; show that the principal term in the solution is then the first one. Thus the system acts as a "filter" to select vibrations of one frequency from a given set (for example, a radio tuned to one station selects principally the vibrations of the frequency of that station).

7. OTHER SECOND-ORDER EQUATIONS

Although second-order linear equations with constant coefficients are the ones used most frequently in applications, there are a few other kinds of second-order equations and methods of solving them which are also important. We shall discuss several of these here, namely (a) equations with y missing; (b) equations with x missing; (c) equations of the form $y'' + f(y) = 0$; (d) Euler-Cauchy equations; (e) reduction of order. For still more methods, see Section 9 (Laplace transforms), Section 12 (Green functions), Problem 12.14b (variation of parameters), and Chapter 12 (special functions, series solutions, ladder operators). You can also find computer solutions but, as we have said, they may not always be in the simplest form or the form you need. Comparing hand solutions can show you what to expect and help you make more efficient use of computer solutions.

To solve either (a) or (b), we make the substitution

$$y' = p. \tag{7.1}$$

Case (a): Dependent variable y missing.

$$y' = p, \quad y'' = p'. \tag{7.2}$$

After these substitutions, an equation of the type (a) is of the first order with p

as the dependent variable and x as the independent variable. First, we solve it for p as a function of x; then we put back $p = y'$ and solve the resulting first-order equation for y.

Case (b): Independent variable x missing.

$$y' = p, \quad y'' = \frac{dp}{dx} = \frac{dp}{dy}\frac{dy}{dx} = p\frac{dp}{dy}. \tag{7.3}$$

What we are doing here is to change the independent variable from x to y. Observe that there is *one* independent variable in an ordinary differential equation. We were originally thinking of x as the independent variable with y and $p\,dy/dx$ as functions of x. Now we think of y as the independent variable with p a function of y; (7.3) is just the chain rule (Chapter 4, Section 5) for differentiating a function $p(y)$ with respect to x if y is a function of x. With the substitutions (7.3), a differential equation with x missing becomes a first-order equation with p as the dependent variable and y as the independent variable.

Example 1. In Section 5, we discussed the motion of a mass m subject to a restoring force $-ky$ and a damping force $l(dy/dt)$. Let us now consider a similar problem but with the damping force proportional to the square of the velocity. The differential equation of motion is then [compare (5.26)]

$$m\frac{d^2y}{dt^2} \pm l\left(\frac{dy}{dt}\right)^2 + ky = 0 \qquad (l > 0), \tag{7.4}$$

where the plus or minus sign must be chosen correctly at each stage of the motion so that the retarding force opposes the motion. Let us solve the following special case of this problem. Discuss the motion of a particle which is released from rest at the point $y = 1$ when $t = 0$, and obeys the equation of motion

$$4\frac{d^2y}{dt^2} \pm 2\left(\frac{dy}{dt}\right)^2 + y = 0. \tag{7.5}$$

This is an example of case (b) (for "x missing" read "t missing," that is, the independent variable missing). Using (7.3) (with x replaced by t), we have

$$\begin{aligned} \frac{dy}{dt} &= p, \\ \frac{d^2y}{dt^2} &= p\frac{dp}{dy}, \end{aligned} \tag{7.6}$$

so (7.5) becomes

$$4p\frac{dp}{dy} \pm 2p^2 + y = 0 \qquad \text{or} \qquad \frac{dp}{dy} \pm \tfrac{1}{2}p = -\tfrac{1}{4}yp^{-1}. \tag{7.7}$$

This is a Bernoulli equation [compare (4.1) with y replaced by p, and P and Q functions of y]. We have $n = -1$, and the substitution (4.2) is

$$z = p^2. \tag{7.8}$$

Then

$$\frac{dz}{dy} = 2p\frac{dp}{dy}$$

and (7.7) becomes

$$\frac{dz}{dy} \pm z = -\tfrac{1}{2}y. \tag{7.9}$$

This is a first-order linear equation; solving it (see Section 3), we get

$$\begin{aligned} ze^{\pm y} &= -\tfrac{1}{2}\int ye^{\pm y}\,dy = -\tfrac{1}{2}e^{\pm y}(\pm y - 1) + c, \\ z &= -\tfrac{1}{2}(\pm y - 1) + ce^{\mp y}. \end{aligned} \tag{7.10}$$

Since initially $dy/dt = 0$ and $y > 0$, we see from (7.5) that the initial acceleration is in the negative direction; since the particle starts from rest, its velocity for small t is also in the negative direction. Then the damping force must be in the positive direction so we must use the lower sign in (7.10) for the first part of the motion. Thus we have

$$z = \tfrac{1}{2}(y + 1) + ce^{y} \qquad \text{(for small } t\text{)}. \tag{7.11}$$

We determine c from the initial conditions $dy/dt = 0$, $y = 1$, at $t = 0$; we have $z = p^2 = (dy/dt)^2 = 0$ when $y = 1$; therefore from (7.11) we get

$$0 = 1 + ce, \quad c = -e^{-1}.$$

Then we have

$$z = \left(\frac{dy}{dt}\right)^2 = \tfrac{1}{2}(y + 1) - e^{y-1} \qquad \text{(for small } t\text{)}. \tag{7.12}$$

This is a valid solution as long as $dy/dt < 0$ (this is what small t means). Thus the particle initially moves in the negative direction for a while. To continue the problem we would need to find whether it stops and if so where. This means solving a transcendental equation which has to be done by some approximation method. It turns out that when it stops, y is negative; at this point the force $-y$ is in the positive direction and the particle, after stopping, moves in the positive direction. The solution for $(dy/dt)^2$ is then given by (7.10) with the upper sign. After another interval of time, the particle again reverses its motion and we again use the solution (7.11) (with a different c), and so on, the total motion appearing something like a damped vibration. We shall not continue the details further here since we have already accomplished our purpose of illustrating case (b) and the solution of a Bernoulli equation.

Case (c) appears to be very special and is obviously included by (b); however, it is very important to know the easy way to solve it because it so frequently arises in applications. The trick is simply to multiply the equation by y'; we can then integrate each term.

Case (c): To solve $y'' + f(y) = 0$, multiply by y'.

$$y'y'' + f(y)y' = 0, \qquad \text{or} \quad y'\,dy' + f(y)\,dy = 0.$$

Then integrate to get

$$\tfrac{1}{2}y'^2 + \int f(y)\,dy = \text{const.} \tag{7.13}$$

This equation is separable and so can be solved (except for possible difficulty in evaluating the integrals). We say that the problem is reduced to *quadratures* (indicated integrations); this means that we can write the answer in terms of integrals which may or may not be easy to evaluate!

Example 2. Consider a particle of mass m moving along the x axis under the action of a force $F(x)$. Then the equation of motion is

$$m\frac{d^2x}{dt^2} = F(x). \tag{7.14}$$

If we multiply this equation by $v = dx/dt$ and integrate with respect to t, we get

$$mv\frac{dv}{dt} = F(x)\frac{dx}{dt} \qquad \text{or} \quad mv\,dv = F(x)\,dx,$$

$$\tfrac{1}{2}mv^2 = \int F(x)\,dx + \text{const.} \tag{7.15}$$

Recall (Chapter 6, Section 8) that the potential energy of a particle is the negative of the work done by the force. Thus

$$\tfrac{1}{2}mv^2 - \int F(x)\,dx \tag{7.16}$$

is the kinetic energy plus the potential energy; equation (7.15) expresses the law of conservation of energy for this problem. This energy equation is often of more interest than the equation of motion (x as a function of t) and so it is useful to be able to find it directly, as we have done, without solving the differential equation for x. Equation (7.15) is known as a *first integral* of the differential equation since we have integrated a second-order equation *once* to get it.

Case (d): An equation of the form

$$a_2 x^2 \frac{d^2y}{dx^2} + a_1 x \frac{dy}{dx} + a_0 y = f(x) \tag{7.17}$$

(called an Euler or Cauchy equation) can be reduced to a linear equation with constant coefficients by changing the independent variable from x to z where

$$x = e^z. \tag{7.18}$$

For then we have (see Problem 14 and also Chapter 4, Section 11)

$$x\frac{dy}{dx} = \frac{dy}{dz} \quad \text{and} \quad x^2\frac{d^2y}{dx^2} = \frac{d^2y}{dz^2} - \frac{dy}{dz}. \tag{7.19}$$

Substituting (7.18) and (7.19) into (7.17) gives

$$a_2\frac{d^2y}{dz^2} + (a_1 - a_2)\frac{dy}{dz} + a_0 y = f(e^z). \tag{7.20}$$

This is a linear equation with constant coefficients which can be solved by the methods of Sections 5 and 6.

It is worth noting that the solutions of (7.17) when $f(x) = 0$ are often powers of x, so a way to solve this case is to assume $y = x^k$ and solve the resulting quadratic equation for k. However, if the values of k turn out to be complex, or equal, or if $f(x) \neq 0$, you may find it easier to use (7.18) which reduces the problem to a familiar one. (See Problems 15 to 23.)

Case (e): Reduction of order. To find a second solution of

$$y'' + f(x)y' + g(x)y = 0 \tag{7.21}$$

given one solution $u(x)$, substitute

$$y = u(x)v(x) \tag{7.22}$$

into (7.21) and solve for $v(x)$.

You can verify that when you substitute (7.22) into (7.21), the coefficient of $v(x)$ is $u'' + f(x)u' + g(x)u$. This expression is equal to zero because we assumed that $u(x)$ is a solution of (7.21). Then the equation for $v'(x)$ is a separable first-order equation (Problem 24).

▶ **Example 3.** Solve $x^3y'' + xy' - y = 0$, given that $u = x$ is a solution.

We let $y = uv = xv$. Then $y' = xv' + v$, $y'' = xv'' + 2v'$, and the differential equation becomes

$$x^3(xv'' + 2v') + x(xv' + v) - xv = 0 \qquad \text{or}$$
$$x^4v'' + (2x^3 + x^2)v' = 0.$$

Separating variables and integrating, we find

$$\frac{dv'}{v'} = -\left(\frac{2}{x} + \frac{1}{x^2}\right) dx, \quad \ln v' = -2\ln x + \frac{1}{x} + \ln K.$$

Solving for v', integrating again, and writing $y = uv$ gives

$$v' = \frac{K}{x^2} e^{1/x}, \quad v = -Ke^{1/x}, \quad y = -Kxe^{1/x}.$$

Thus the general solution of the given equation is $y = Ax + Bxe^{1/x}$.

PROBLEMS, SECTION 7

Solve the following differential equations by method (a) or (b) above.

1. $y''+yy' = 0$. Find a solution satisfying each of the following sets of initial conditions. If your computer says there is no such solution, don't believe it—do it by hand.

(a) $y(0) = 5, \quad y'(0) = 0$ (b) $y(0) = 2, \quad y'(0) = -2$
(c) $y(0) = 1, \quad y'(0) = -1$ (d) $y(0) = 0, \quad y'(0) = 2$

2. $y'' + 2xy' = 0$ *Hint:* The solution is $y = c_1 \operatorname{erf} x + c_2$; see Chapter 11, Section 9 for the definition of $\operatorname{erf} x$.

3. $2yy'' = y'^2$

4. $xy'' = y' + y'^3$

5. The differential equation of a hanging chain supported at its ends is

$$y''^2 = k^2\left(1 + y'^2\right).$$

Solve the equation to find the shape of the chain.

6. The curvature of a curve in the (x, y) plane is

$$K = y''\left(1 + y'^2\right)^{-3/2}.$$

With $K =$ const., solve this differential equation to show that curves of constant curvature are circles (or straight lines).

7. Solve $y'' + \omega^2 y = 0$ by method (c) above and compare with the solution as a linear equation with constant coefficients.

8. The force of gravitational attraction on a mass m at distance r from the center of the earth ($r >$ radius R of the earth) is mgR^2/r^2. Then the differential equation of motion of a mass m projected radially outward from the surface of the earth, with initial velocity v_0, is

$$md^2r/dt^2 = -mgR^2/r^2.$$

Use method (c) above to find v as a function of r if $v = v_0$ initially (that is, when $r = R$). Find the maximum value of r for a given v_0, that is, the value of r when $v = 0$. Find the *escape velocity*, that is, the smallest value of v_0 for which r can tend to infinity.

9. Show that (7.15) is a separable equation. [You may find it helpful to write $\int F(x)\,dx = f(x)$.] Thus solve (7.14) in terms of quadratures (that is, indicated integrations) as in Problem 2.

In Problems 10 and 11, solve (7.14) to find $v(x)$ and then $x(t)$ for the given $F(x)$ and initial conditions.

10. $F(x) = m/x^3$, $v = 0$, $x = 1$, at $t = 0$.

11. $F(x) = -2m/x^5$, $v = -1$, $x = 1$, at $t = 0$.

12. In Problem 11, find $v(x)$ if $v = 0$, $x = 1$, at $t = 0$. Then write an integral for $t(x)$.

13. The exact equation of motion of a simple pendulum is $d^2\theta/dt^2 = -\omega^2 \sin\theta$ where $\omega^2 = g/l$. By method (c) above, integrate this equation once to find $d\theta/dt$ if $d\theta/dt = 0$ when $\theta = 90°$. Write a formula for $t(\theta)$ as an integral. See Problem 5.34.

14. Verify (7.19) and (7.20). *Hint:* $dy/dz = (dy/dx)(dx/dz)$; write the first equation of (7.19) as $xD_x = D_z$, and find D_z^2.

15. If you solve (7.17) when $f(x) = 0$ by assuming a solution $y = x^k$, show that the quadratic equation for k is the same as the auxiliary equation for the z equation (7.20). Thus show (see Section 5) that if the two values of k are equal, the second solution is not a power of x but is $x^k \ln x$. Also show that if k is complex, say $k = a \pm bi$, the solutions are $x^a \cos(b \ln x)$ and $x^a \sin(b \ln x)$ or other equivalent forms [see (5.16) to (5.18)].

16. Solve the following equations either by method (d) above or by assuming $y = x^k$ (or try both methods to compare them). See Problem 15.

(a) $x^2y'' + 3xy' - 3y = 0$

(b) $x^2y'' + xy' - 4y = 0$

(c) $x^2y'' + 7xy' + 9y = 0$

(d) $x^2y'' - xy' + 6y = 0$

Solve the following equations using method (d) above.

17. $x^2y'' + xy' - 16y = 8x^4$

18. $x^2y'' + xy' - y = x - x^{-1}$

19. $x^2y'' - 5xy' + 9y = 2x^3$

20. $x^2y'' - 3xy' + 4y = 6x^2 \ln x$

21. $x^2y'' + y = 3x^2$

22. $x^2y'' + xy' + y = 2x$

23. Solve the two differential equations in Problem 5.11 of Chapter 13.

24. Substitute (7.22) into (7.21) to obtain the equation for $v'(x)$. Show that this equation is separable.

For the following problems, verify the given solution and then, by method (e) above, find a second solution of the given equation.

25. $x^2(2 - x)y'' + 2xy' - 2y = 0, \quad u = x$

26. $(x^2 + 1)y'' - 2xy' + 2y = 0, \quad u = x$

27. $xy'' - 2(x + 1)y' + (x + 2)y = 0, \quad u = e^x$

28. $3xy'' - 2(3x - 1)y' + (3x - 2)y = 0, \quad u = e^x$

29. $x^2y'' + (x + 1)y' - y = 0, \quad u = x + 1$

30. $x(x + 1)y'' - (x - 1)y' + y = 0, \quad u = x - 1$

8. THE LAPLACE TRANSFORM

As you will see in Section 9, Laplace transforms are useful in solving differential equations (for other uses see end of Section 9, page 442). Here we want to define the Laplace transform and obtain some needed formulas. We define $L(f)$, the Laplace transform of $f(t)$ [also written $F(p)$ since it is a function of p], by the equation

$$L(f) = \int_0^\infty f(t)e^{-pt}\,dt = F(p). \tag{8.1}$$

This is an example of an *integral transform* (also see Fourier transforms, Chapter 7, Section 12, and Hilbert transforms, Chapter 14, page 698). If we start with a function $f(t)$, multiply by a function of t and p, and find a definite integral with respect to t, we have a function $F(p)$ which is called an integral transform of $f(t)$. There are many named integral transforms which you may discover in tables and computer. Observe the notation for Laplace transforms in (8.1): we shall consistently use a small letter for the function of t, and the corresponding capital letter for the transform which is a function of p, for example $f(t)$ and $F(p)$, or $g(t)$ and $G(p)$, etc. Also note from (8.1) that since we integrate from 0 to ∞, $F(p)$ is the same no matter how $f(t)$ is defined for negative t. However, it is desirable to define $f(t) = 0$ for $t < 0$ (see footnote, page 447; also see Bromwich integral, page 696).

It is very convenient to have a table of corresponding $f(t)$ and $F(p)$ when we are using Laplace transforms to solve problems. Let us calculate some of the entries in the table of Laplace transforms at the end of the chapter (pages 469 to 471). Note that numbers preceded by L ($L1$, $L2$, $\cdots$, $L35$) refer to entries in the Laplace transform table.

Example 1. To obtain $L1$ in the table, we substitute $f(t) = 1$ into (8.1) and find

$$F(p) = \int_0^\infty 1 \cdot e^{-pt}\,dt = -\frac{1}{p}e^{-pt}\Big|_0^\infty = \frac{1}{p}, \qquad p > 0. \tag{8.2}$$

We have assumed $p > 0$ to make e^{-pt} zero at the upper limit; if p is complex, as it may be, then the real part of p must be positive ($\operatorname{Re} p > 0$), and this is the restriction we have stated in the table for $L1$.

Example 2. For $L2$, we have

$$\begin{aligned} f(t) &= e^{-at}, \\ F(p) &= \int_0^\infty e^{-(a+p)t}\,dt = \frac{1}{p+a}, \qquad \operatorname{Re}(p+a) > 0. \end{aligned} \tag{8.3}$$

We could continue in this way to obtain the function $F(p)$ corresponding to each $f(t)$ by using (8.1) and evaluating the integral. However, there are some easier methods which we now illustrate. First observe that the Laplace transform of a sum of two functions is the sum of their Laplace transforms; also the transform of

$cf(t)$ is $cL(f)$ when c is a constant:

$$L[f(t)+g(t)] = \int_0^\infty [f(t)+g(t)]e^{-pt}\,dt$$

(8.4)
$$= \int_0^\infty f(t)e^{-pt}\,dt + \int_0^\infty g(t)e^{-pt}\,dt = L(f)+L(g),$$

$$L[cf(t)] = \int_0^\infty cf(t)e^{-pt}\,dt = c\int_0^\infty f(t)e^{-pt}\,dt = cL(f).$$

In mathematical language, we say that the Laplace transform is *linear* (or is a *linear operator*—see Chapter 3, Section 7).

▶ **Example 3.** Now let us verify $L3$. In (8.3), replace the a by $-ia$; then we have

(8.5)
$$f(t) = e^{iat} = \cos at + i\sin at,$$
$$F(p) = \frac{1}{p-ia} = \frac{p+ia}{p^2+a^2}, \qquad \text{Re}(p-ia) > 0.$$

Remembering (8.4), we can write (8.5) as

(8.6)
$$L(\cos at + i\sin at) = L(\cos at) + iL(\sin at) = \frac{p}{p^2+a^2} + i\frac{a}{p^2+a^2}.$$

Similarly, replacing a by ia in (8.3), we get

(8.7)
$$L(\cos at - i\sin at) = \frac{p}{p^2+a^2} - i\frac{a}{p^2+a^2}, \qquad \text{Re}(p+ia) > 0.$$

Adding (8.6) and (8.7), we get $L4$; by subtracting, we get $L3$.

▶ **Example 4.** To verify $L11$, start with $L4$, namely

(8.8)
$$L(\cos at) = \int_0^\infty e^{-pt}\cos at\,dt = \frac{p}{p^2+a^2}.$$

Differentiate (8.8) with respect to the parameter a to get

$$\int_0^\infty e^{-pt}(-t\sin at)\,dt = \frac{p(-2a)}{(p^2+a^2)^2}$$

or

$$\int_0^\infty e^{-pt}t\sin at\,dt = \frac{2pa}{(p^2+a^2)^2}$$

which is $L11$. Ways of finding other entries in the table are outlined in the problems.

▶ PROBLEMS, SECTION 8

1. For integral k, verify $L5$ and $L6$ in the Laplace transform table. *Hint:* From $L2$, you can write: $\int_0^\infty e^{-pt}e^{-at}\,dt = 1/(p+a)$. Differentiate this equation repeatedly with respect to p. (See Chapter 4, Section 12, Example 4, page 235.) Also note $L32$. For the Γ function results in $L5$ and $L6$, see Chapter 11, Problem 5.7.

2. By using $L2$, verify $L7$ and $L8$ in the Laplace transform table.

3. Using either $L2$, or $L3$ and $L4$, verify $L9$ and $L10$.

4. By differentiating the appropriate formula with respect to a, verify $L12$.

5. By integrating the appropriate formula with respect to a, verify $L19$.

6. By replacing a in $L2$ by $a + ib$ and then by $a - ib$, and adding and subtracting the results [as in (8.6) and (8.7)], verify $L13$ and $L14$.

7. Verify $L15$ to $L18$, by combining appropriate preceding formulas using (8.4).

Find the inverse transforms of the functions $F(p)$ in Problems 8 to 13.

8. $\dfrac{1+p}{(p+2)^2}$ *Hint:* Use $L6$ and $L18$.

9. $\dfrac{5-2p}{p^2+p-2}$ *Hint:* Use $L7$ and $L8$.

10. $\dfrac{2p-1}{p^2-2p+10}$ *Hint:* You *can* use $L7$ and $L8$ with complex a and b, but $L13$ and $L14$ are more direct.

11. $\dfrac{3p+2}{3p^2+5p-2}$ **12.** $\dfrac{3p+10}{p^2-25}$ **13.** $\dfrac{6-p}{p^2+4p+20}$

14. Show that a combination of entries $L3$ to $L10$, $L13$, $L14$ and $L18$ in the table, will give the inverse transform of any function of the form

$$\frac{Ap+B}{Cp^2+Ep+F}, \quad \text{where } A, B, C, E, \text{and } F \text{ are constants.}$$

15. Prove $L32$ for $n = 1$. *Hint:* Differentiate equation (8.1) with respect to p.

16. Use $L32$ and $L3$ to obtain $L11$.

17. Use $L32$ and $L11$ to obtain $L(t^2 \sin at)$.

18. Use $L31$ to derive $L21$.

Table entries $L28$ and $L29$ are known as *translation* or *shifting* theorems. Do Problems 19 to 27 about them.

19. Prove the general formula $L29$ using (8.1).

20. Use $L29$ to verify $L6$, $L13$, $L14$, and $L18$.

21. Use $L29$ and $L11$ to obtain $L(te^{-at} \sin bt)$ which is not in the table.

22. Obtain $L(te^{-at} \cos bt)$ as in Problem 21.

23. Use the results which you have obtained in Problems 21 and 22 to find the inverse transform of $(p^2 + 2p - 1)/(p^2 + 4p + 5)^2$.

24. Sketch on the same axes graphs of $\sin t$, $\sin(t - \pi/2)$, and $\sin(t + \pi/2)$, and observe which way the graph shifts. *Hint:* You can, of course, have your calculator or computer plot these for you, but it's simpler and much more useful to do it in your head. *Hint:* What values of t make the sines equal to zero? For an even simpler example, sketch on the same axes $y = t$, $y = t - \pi/2$, $y = t + \pi/2$.

25. Use $L28$ to find the Laplace transform of

$$f(t) = \begin{cases} \sin(t - \pi/2), & t > \pi/2, \\ 0, & t < \pi/2. \end{cases}$$

26. Use $L28$ and $L4$ to find the inverse transform of $pe^{-p\pi}/(p^2 + 1)$.

27. Find the transform of

$$f(t) = \begin{cases} \sin(x - vt), & t > x/v, \\ 0, & t < x/v, \end{cases}$$

where x and v are constants.

9. SOLUTION OF DIFFERENTIAL EQUATIONS BY LAPLACE TRANSFORMS

We are going to discuss the solution of linear differential equations with constant coefficients (see Sections 5 and 6). Laplace transforms can reduce such an equation to an algebraic equation and so simplify solving it. Also, since Laplace transforms automatically use given values of initial conditions, we find immediately a desired particular solution without the extra step of determining constants to satisfy the initial conditions. Discontinuous forcing functions are messy to deal with by Section 6 methods; the Laplace transform method handles them easily.

We are going to take Laplace transforms of the terms in differential equations; to do this we need to know the transforms of derivatives $y' = dy/dt, y'' = d^2y/dt^2$, etc. To find $L(y')$, we use the definition (8.1) and integrate by parts, as follows

$$L(y') = \int_0^\infty y'(t)e^{-pt}\,dt = e^{-pt}y(t)\Big|_0^\infty - (-p)\int_0^\infty y(t)e^{-pt}\,dt = -y(0) + pL(y) = pY - y_0 \tag{9.1}$$

where for simplicity we have written $L(y) = Y$ and $y(0) = y_0$. To find $L(y'')$, we think of y'' as $(y')'$, and substitute y' for y in (9.1) to get

$$L(y'') = pL(y') - y'(0).$$

Using (9.1) again to eliminate $L(y')$, we finally have

$$L(y'') = p^2L(y) - py(0) - y'(0) = p^2Y - py_0 - y_0'. \tag{9.2}$$

Continuing this process, we obtain the transforms of the higher-order derivatives (Problem 1 and $L35$).

We are now ready to solve differential equations. We illustrate the method by some examples.

Example 1. Solve $y'' + 4y' + 4y = t^2e^{-2t}$ with initial conditions $y_0 = 0$, $y_0' = 0$.

We take the Laplace transform of each term in the equation, using $L35$ and $L6$ in the table of Laplace transforms. We get

$$p^2Y - py_0 - y_0' + 4pY - 4y_0 + 4Y = L(t^2e^{-2t}) = \frac{2}{(p+2)^3}.$$

But the initial conditions are $y_0 = y_0' = 0$. Thus we have

$$(p^2 + 4p + 4)Y = \frac{2}{(p+2)^3} \quad \text{or} \quad Y = \frac{2}{(p+2)^5}.$$

Now we want y, which is the inverse Laplace transform of Y. We look in the table for the inverse transform of $2/(p+2)^5$. By $L6$, we get

$$y = \frac{2t^4e^{-2t}}{4!} = \frac{t^4e^{-2t}}{12}.$$

This is much simpler than the general solution; we have obtained just the solution satisfying the given initial conditions.

Example 2. Solve $y'' + 4y = \sin 2t$, subject to the initial conditions $y_0 = 10$, $y_0' = 0$.
Using the table, take the Laplace transform of each term of the equation to get

$$p^2Y - py_0 - y_0' + 4Y = L(\sin 2t) = \frac{2}{p^2+4}.$$

Then we substitute the initial conditions and solve for Y as follows:

$$(p^2+4)Y - 10p = \frac{2}{p^2+4},$$
$$Y = \frac{10p}{p^2+4} + \frac{2}{(p^2+4)^2}.$$

Finally, taking the inverse transform using $L4$ and $L17$, we have the desired solution:

$$y = 10\cos 2t + \tfrac{1}{8}(\sin 2t - 2t\cos 2t) = 10\cos 2t + \tfrac{1}{8}\sin 2t - \tfrac{1}{4}t\cos 2t.$$

Example 3. Solve $y'' + 4y' + 13y = 20e^{-t}, \quad y_0 = 1, y_0' = 3$.
We take the transform of each term and solve for Y as follows:

$$p^2Y - p - 3 + 4pY - 4 + 13Y = \frac{20}{p+1},$$
$$Y = \frac{1}{p^2+4p+13}\left(\frac{20}{p+1} + p + 7\right) = \frac{p^2+8p+27}{(p+1)(p^2+4p+13)}.$$

Since this Y is not in our table, we can either use a larger table, or use partial fractions to split Y into fractions which are in our table (which you can do by computer) or find the inverse transform by computer. We find:

$$Y = \frac{2}{p+1} + \frac{-p+1}{p^2+4p+13} = \frac{2}{p+1} + \frac{3}{(p+2)^2+9} - \frac{p+2}{(p+2)^2+9}$$

and by $L2$, $L13$, and $L14$,

$$y = 2e^{-t} + e^{-2t}\sin 3t - e^{-2t}\cos 3t.$$

Sets of simultaneous differential equations can also be solved by using Laplace transforms. Here is an example.

Example 4. Solve the set of equations

$$y' - 2y + z = 0,$$
$$z' - y - 2z = 0,$$

subject to the initial conditions $y_0 = 1, z_0 = 0$.
We shall call $L(z) = Z$ and $L(y) = Y$ as before. We take the Laplace transform of each of the equations to get

$$pY - y_0 - 2Y + Z = 0,$$
$$pZ - z_0 - Y - 2Z = 0.$$

After substituting the initial conditions and collecting terms, we have

$$(p-2)Y + Z = 1,$$
$$Y - (p-2)Z = 0.$$

We solve this set of algebraic equations simultaneously for Y and Z (by any of the methods usually used for a pair of simultaneous equations—elimination, determinants, etc.). For example, we may multiply the first equation by $(p-2)$ and add the second to get

$$[(p-2)^2 + 1]Y = p - 2 \qquad \text{or} \quad Y = \frac{p-2}{(p-2)^2+1}.$$

We find y by looking up the inverse transform of Y using $L14$. We get

$$y = e^{2t}\cos t.$$

Similarly, solving for Z and looking up the inverse transform, we find

$$Z = \frac{1}{(p-2)^2+1}, \qquad z = e^{2t}\sin t$$

. Alternatively, we could find z from the first differential equation by substituting the y solution:

$$z = 2y - y' = 2e^{2t}\cos t + e^{2t}\sin t - 2e^{2t}\cos t = e^{2t}\sin t.$$

Solving linear differential equations with constant coefficients is not the only use of Laplace transforms. As you will see in Chapter 13, Section 10, we may solve some kinds of partial differential equations by Laplace transforms. Also a table of Laplace transforms can be used to evaluate definite integrals of the type $\int_0^\infty e^{-pt}f(t)dt$.

Example 5. By $L15$ with $a = 3$ and $p = 2$, we have

$$\int_0^\infty e^{-2t}(1-\cos 3t)\,dt = \frac{3^2}{2(2^2+3^2)} = \frac{9}{26}.$$

Actually, there is more to the subject than this. Although we are discussing in this chapter the use of Laplace transforms as a tool, they also can play a more theoretical role in applied problems. It is often possible to find desired information about a problem directly from the Laplace transform of the solution without ever finding the solution. Thus the use of Laplace transforms may lead to a better understanding of a problem or a simpler method of solution. (Compare the use of matrices, for example, or the use of Fourier transforms.)

PROBLEMS, SECTION 9

1. Continuing the method used in deriving (9.1) and (9.2), verify the Laplace transforms of higher-order derivatives of y given in the table ($L35$).

By using Laplace transforms, solve the following differential equations subject to the given initial conditions.

2. $y' - y = 2e^t$, $\qquad y_0 = 3$

3. $y'' + 4y' + 4y = e^{-2t}$, $y_0 = 0,\ y_0' = 4$

4. $y'' + y = \sin t$, $y_0 = 1,\ y_0' = 0$

5. $y'' + y = \sin t$, $y_0 = 0,\ y_0' = -\frac{1}{2}$

6. $y'' - 6y' + 9y = te^{3t}$, $y_0 = 0,\ y_0' = 5$

7. $y'' - 4y' + 4y = 4$, $y_0 = 0,\ y_0' = -2$

8. $y'' + 16y = 8\cos 4t$, $y_0 = y_0' = 0$

9. $y'' + 16y = 8\cos 4t$, $y_0 = 0,\ y_0' = 8$

10. $y'' - 4y' + 4y = 6e^{2t}$, $y_0 = y_0' = 0$

11. $y'' - 4y = 4e^{2t}$, $y_0 = 0,\ y_0' = 1$

12. $y'' - y = e^{-t} - 2te^{-t}$, $y_0 = 1,\ y_0' = 2$

13. $y'' + y = 5\sinh 2t$, $y_0 = 0,\ y_0' = 2$

14. $y'' - 4y' = -4te^{2t}$, $y_0 = 0,\ y_0' = 1$

15. $y'' + 9y = \cos 3t$, $y_0 = 0,\ y_0' = 6$

16. $y'' + 9y = \cos 3t$, $y_0 = 2,\ y_0' = 0$

17. $y'' + 5y' + 6y = 12$, $y_0 = 2,\ y_0' = 0$

18. $y'' - 4y = 3e^{-t}$, $y_0 = 1,\ y_0' = -3$

19. $y'' + y' - 5y = e^{2t}$, $y_0 = 1,\ y_0' = 2$

20. $y'' - 8y' + 16y = 32t$, $y_0 = 1,\ y_0' = 2$

21. $y'' + 4y' + 5y = 26e^{3t}$, $y_0 = 1,\ y_0' = 5$

22. $y''+2y'+5y = 10\cos t$, $y_0 = 2,\ y_0' = 1$

23. $y''+2y'+5y = 10\cos t$, $y_0 = 0,\ y_0' = 3$

24. $y'' - 2y' + y = 2\cos t$, $y_0 = 5,\ y_0' = -2$

25. $y'' + 4y' + 5y = 2e^{-2t}\cos t$, $y_0 = 0,\ y_0' = 3$

26. $y'' + 2y' + 10y = -6e^{-t}\sin 3t$, $y_0 = 0,\ y_0' = 1$

Solve the following sets of equations by the Laplace transform method.

27. $y' + z' - 3z = 0$ $\quad y_0 = y_0' = 0$
$y'' + z' = 0$ $\quad z_0 = \frac{4}{3}$

28. $y' + z = 2\cos t$ $\quad y_0 = -1$
$z' - y = 1$ $\quad z_0 = 1$

29. $y' + z' - 2y = 1$ $\quad y_0 = z_0 = 1$
$z - y' = t$

30. $y' + 2z = 1$ $\quad y_0 = 0$
$2y - z' = 2t$ $\quad z_0 = 1$

31. $y'' + z'' - z' = 0$ $\quad y_0 = 0,\quad y_0' = 1$
$y' + z' - 2z = 1 - e^t$ $\quad z_0 = 1,\quad z_0' = 1$

32. $z' + 2y = 0$ $\quad y_0 = z_0 = 0$
$y' - 2z = 2$

33. $y' - z' - y = \cos t$ $\qquad y_0 = -1$
$y' + y - 2z = 0$ $\qquad z_0 = 0$

Evaluate each of the following definite integrals by using the Laplace transform table.

34. $\int_0^\infty e^{-2t} \sin 3t \, dt = \frac{3}{13}$. *Hint:* In (8.1), let $p = 2$, $f(t) = \sin 3t$; use $L3$ with $a = 3$.

35. $\int_0^\infty te^{-t} \sin 5t \, dt$

36. $\int_0^\infty \frac{e^{-3t} \sin 2t}{t} \, dt$

37. $\int_0^\infty t^5 e^{-2t} \, dt$

38. $\int_0^\infty e^{-t}(1 - \cos 2t) \, dt$

39. $\int_0^\infty \frac{e^{-t} - e^{-2t}}{t} \, dt$

40. $\int_0^\infty \frac{e^{-2t} - e^{-2et}}{t} \, dt$

41. $\int_0^\infty \frac{1}{t} e^{-2t} \sin(t\sqrt{2}) \, dt$

42. $\int_0^\infty \frac{1}{t} e^{-t\sqrt{3}} \sin 2t \cos t \, dt$

10. CONVOLUTION

In solving differential equations by Laplace transforms in Section 9, we found Y and then found the inverse transform y either in a table or by computer. We had no way of writing a formula for y. We now want to consider another way of finding inverse transforms. (Also see Bromwich integral, Chapter 14, page 696.)

Let us first see why the method we are going to discuss in this section is useful. Consider differential equations of the kind discussed in Sections 5 and 6, namely linear second-order equations with constant coefficients. Recall that such equations describe the vibrations or oscillations of either a mechanical or an electrical system. If the right-hand side of the equation is a function of t, called the *forcing function*, then the differential equation describes forced vibrations.

Example 1. Let us solve the following representative equation by Laplace transforms, assuming that the system is initially at rest and that the force $f(t)$ starts being applied at $t = 0$.

$$Ay'' + By' + Cy = f(t), \qquad y_0 = y_0' = 0. \tag{10.1}$$

We take the Laplace transform of each term, substitute the initial conditions, and solve for Y as follows:

$$Ap^2Y + BpY + CY = L(f) = F(p), \qquad Y = \frac{1}{Ap^2 + Bp + C} F(p). \tag{10.2}$$

Note that Y is a product of two functions of p. We know the inverse transform of $F(p)$, namely $f(t)$. The factor $T(p) = (Ap^2 + Bp + C)^{-1}$ (called the *transfer function*) can always be written as

$$T(p) = \frac{1}{A(p+a)(p+b)}$$

by factoring the quadratic expression in the denominator. Hence by $L7$ (or $L6$ if $a = b$) we can find the inverse transform of $T(p)$ for any problem. Then y

[the inverse transform of Y in (10.2)] is the inverse transform of a product of two functions whose inverse transforms we know. We are going to show how to write y as an integral (that is, we are going to verify $L34$ in the table).

Let $G(p)$ and $H(p)$ be the transforms of $g(t)$ and $h(t)$. We want the inverse transform of the product $G(p)H(p)$. By the definition (8.1)

$$G(p)H(p) = \int_0^\infty e^{-pt}g(t)\,dt \cdot \int_0^\infty e^{-pt}h(t)\,dt. \tag{10.3}$$

Let us rewrite (10.3) replacing t by different dummy variables of integration so that we can write the product of the two integrals as a double integral. We then have

$$\begin{aligned} G(p)H(p) &= \int_0^\infty e^{-p\sigma}g(\sigma)\,d\sigma \cdot \int_0^\infty e^{-p\tau}h(\tau)\,d\tau \\ &= \int_0^\infty \int_0^\infty e^{-p(\sigma+\tau)}g(\sigma)h(\tau)\,d\sigma\,d\tau. \end{aligned} \tag{10.4}$$

Now we make a change of variables; in the σ integral (that is, with τ fixed), let $\sigma + \tau = t$. Then $\sigma = t - \tau$, $d\sigma = dt$, and the range of integration with respect to t is from $t = \tau$ (corresponding to $\sigma = 0$) to $t = \infty$ (corresponding to $\sigma = \infty$). Making these substitutions into (10.4), we get

$$G(p)H(p) = \int_{\tau=0}^\infty \int_{t=\tau}^\infty e^{-pt}g(t-\tau)h(\tau)\,dt\,d\tau. \tag{10.5}$$

Next we want to change the order of integration. From Figure 10.1, we see that the double integral in (10.5) is over the triangle in the first quadrant below the line $t = \tau$. The t integral ranges from the line $t = \tau$ to $t = \infty$ (indicated by a horizontal strip of width $d\tau$ from $t = \tau$ to ∞) and then the τ integral sums over the horizontal strips from $\tau = 0$ to $\tau = \infty$ covering the whole infinite triangle. Let us integrate with respect to τ first; τ then ranges from 0 to the line $\tau = t$ [indicated by a vertical strip in Figure 10.1] and then the t integral sums over the vertical strips from $t = 0$ to ∞. Making this change in (10.5), we get

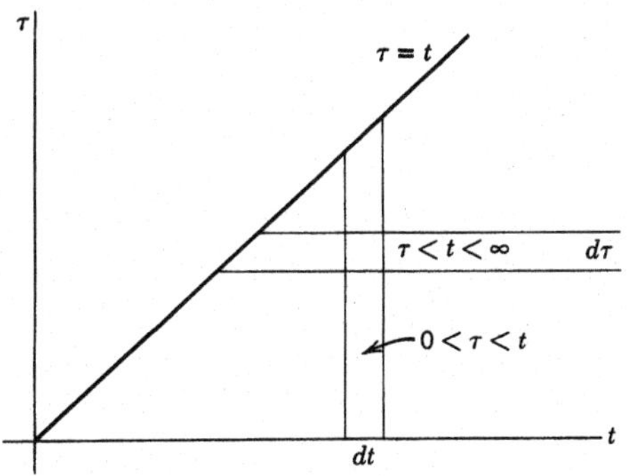

Figure 10.1

$$\begin{aligned} G(p)H(p) &= \int_{t=0}^\infty \int_{\tau=0}^t e^{-pt}g(t-\tau)h(\tau)\,d\tau\,dt \\ &= \int_0^\infty e^{-pt}\left[\int_0^t g(t-\tau)h(\tau)\,d\tau\right]dt \\ &= L\left[\int_0^t g(t-\tau)h(\tau)\,d\tau\right]. \qquad \text{(See } L34.) \end{aligned} \tag{10.6}$$

The last step follows from the definition (8.1) of a Laplace transform.

Definition of Convolution The integral

$$\int_0^t g(t-\tau)h(\tau)\,d\tau = g * h \tag{10.7}$$

is called the *convolution* of g and h (or the *resultant* or the *Faltung*). Note the abbreviation $g * h$ for the convolution integral, and do not confuse the symbol $*$, written *on* the line, with a star used as a superscript meaning complex conjugate. It is easy to show (Problem 1) that $g * h = h * g$; this result and (10.6) and (10.7) give $L34$ in the table.

Now let's see how to use (10.6) or $L34$ to solve the kind of problem indicated in (10.1) and (10.2).

▶ **Example 2.** Solve $y'' + 3y' + 2y = e^{-t}$, $\quad y_0 = y_0' = 0$.

Taking the Laplace transform of each term, substituting the initial conditions, and solving for Y, we get

$$p^2Y + 3pY + 2Y = L(e^{-t}),$$

$$Y = \frac{1}{p^2+3p+2}L(e^{-t}).$$

Since we are intending to use the convolution integral, we do not bother to look up the transform of e^{-t}. We do want, however, the inverse transform of $1/(p^2+3p+2)$; by $L7$, this is $e^{-t} - e^{-2t}$, so we have

$$Y = L(e^{-t} - e^{-2t})L(e^{-t}) = G(p)H(p),$$

with $g(t) = e^{-t} - e^{-2t}$ and $h(t) = e^{-t}$. We now use $L34$ to find y. Observe from $L34$ that we may use either $g(t-\tau)h(\tau)$ or $g(\tau)h(t-\tau)$ in the integral. It is well to choose whichever form is easier to integrate; usually it is best to put $(t-\tau)$ in the simpler function [here $h(t)$]. Then we have

$$\begin{aligned} y &= \int_0^t g(\tau)h(t-\tau)\,d\tau = \int_0^t (e^{-\tau} - e^{-2\tau})(e^{-(t-\tau)})\,d\tau \\ &= e^{-t}\int_0^t (1 - e^{-\tau})\,d\tau = e^{-t}(\tau + e^{-\tau})\Big|_0^t \\ &= e^{-t}(t + e^{-t} - 1) = te^{-t} + e^{-2t} - e^{-t}. \end{aligned}$$

It is not always as easy to evaluate the convolution integral as it was in this example. However, let us observe that, at the very worst, we can always write the solution to a forced vibrations problem [equation (10.1)] as an integral (which can, if necessary, be evaluated numerically). This is true because, as we showed just after (10.2), we can always find the inverse transform of the transfer function $T(p)$, and so have Y as a product of two functions whose inverse transforms we know. Then y is given by the convolution (10.7) of the forcing function $f(t)$ and the inverse transform of the the transfer function. Also note (Problem 16) that a combination of $L6$, $L7$, $L8$ and $L18$ will handle any terms arising in a problem with nonzero initial conditions.

Fourier Transform of a Convolution We have shown that the Laplace transform of the convolution of two functions is the product of their Laplace transforms. There is a similar theorem for Fourier transforms; let us see what it says. Let $g_1(\alpha)$ and $g_2(\alpha)$ be the Fourier transforms of $f_1(x)$ and $f_2(x)$. By analogy with equations (10.3), (10.4), (10.5), and (10.6), we might expect the product $g_1(\alpha)\cdot g_2(\alpha)$ to be the Fourier transform of something; let's investigate this idea. Assuming that $\int_{-\infty}^{\infty}|f_1(x)f_2(x)|dx$ is finite, then by the definition of a Fourier transform [Chapter 7, equation (12.2)], we have

$$(10.8)\qquad \begin{aligned} g_1(\alpha)\cdot g_2(\alpha) &= \frac{1}{2\pi}\int_{-\infty}^{\infty} f_1(v)e^{-i\alpha v}\,dv\cdot\frac{1}{2\pi}\int_{-\infty}^{\infty} f_2(u)e^{-i\alpha u}\,du \\ &= \left(\frac{1}{2\pi}\right)^2\int_{-\infty}^{\infty}\int_{-\infty}^{\infty} e^{-i\alpha(v+u)}f_1(v)f_2(u)\,dv\,du. \end{aligned}$$

[We have used different dummy integration variables as in (10.4).] Next we make the change of variables $x = v+u$, $dx = dv$, in the v integral, to get

$$(10.9)\qquad \begin{aligned} g_1(\alpha)g_2(\alpha) &= \left(\frac{1}{2\pi}\right)^2\int_{-\infty}^{\infty}\int_{-\infty}^{\infty} e^{-i\alpha x}f_1(x-u)f_2(u)\,dx\,du \\ &= \left(\frac{1}{2\pi}\right)^2\int_{-\infty}^{\infty} e^{-i\alpha x}\left[\int_{-\infty}^{\infty} f_1(x-u)f_2(u)\,du\right]dx. \end{aligned}$$

If we define the convolution of $f_1(x)$ and $f_2(x)$ by

$$(10.10)\qquad f_1 * f_2 = \int_{-\infty}^{\infty} f_1(x-u)f_2(u)\,du,^{\dagger}$$

then (10.9) becomes

$$(10.11)\quad g_1\cdot g_2 = \frac{1}{2\pi}\left[\frac{1}{2\pi}\int_{-\infty}^{\infty} f_1 * f_2\, e^{-i\alpha x}\,dx\right] = \frac{1}{2\pi}\cdot\text{Fourier transform of } f_1 * f_2.$$

In other words,

(10.12) $g_1\cdot g_2$ and $\dfrac{1}{2\pi}f_1 * f_2$ are a pair of Fourier transforms.

Because of the symmetry of the $f(x)$ and $g(\alpha)$ integrals, there is a similar result relating $f_1\cdot f_2$ and the convolution of g_1 and g_2. We find that (Problem 19)

(10.13) $g_1 * g_2$ and $f_1\cdot f_2$ are a pair of Fourier transforms.

As discussed in Chapter 7, after (12.2) and after (12.10), various references differ

†Note that (10.10) is really the same as (10.7) if we agree that, for Laplace transforms, $f(t) = 0$ when $t < 0$ (see the first paragraph of Section 8, page 437). For then in (10.7), $h(\tau) = 0$ for $\tau < 0$ and $g(t-\tau) = 0$ for $\tau > t$, so the integral would not really be different if written with infinite limits (in fact, it is sometimes written that way).

in the position of the factor $1/(2\pi)$. Some authors include factors of $1/(2\pi)$ or $1/\sqrt{2\pi}$ in the convolution definition (10.10); this definition as well as Chapter 7, equation (12.2), affects (10.12) and (10.13). Check the notation in any reference you are using.

PROBLEMS, SECTION 10

1. Show that $g * h = h * g$ as claimed in $L34$. *Hint:* Let $u = t - \tau$ in (10.7).

2. Use $L34$ and $L2$ to find the inverse transform of $G(p)H(p)$ when $G(p) = 1/(p+a)$ and $H(p) = 1/(p+b)$; your result should be $L7$.

Use the convolution integral to find the inverse transforms of:

3. $\dfrac{p}{(p^2-1)^2} = \dfrac{p}{p^2-1} \cdot \dfrac{1}{p^2-1}$

4. $\dfrac{1}{(p+a)(p+b)^2}$

5. $\dfrac{p}{(p+a)(p+b)^2}$

6. $\dfrac{1}{(p+a)(p^2-b^2)}$

7. $\dfrac{p}{(p+a)(p^2-b^2)}$

8. $\dfrac{1}{(p+a)(p+b)(p+c)}$

9. $\dfrac{2}{p^3(p+2)}$

10. $\dfrac{1}{p(p^2+a^2)^2}$

11. $\dfrac{p}{(p^2+a^2)(p^2+b^2)}$

12. $\dfrac{1}{p(p^2+a^2)(p^2+b^2)}$

Hint: In Problems 11 and 12 use $2\sin\theta\cos\phi = \sin(\theta+\phi) + \sin(\theta-\phi)$.

13. Use the Laplace transform table to find $f(t) = \int_0^t e^{-\tau}\sin(t-\tau)\,d\tau$. *Hint:* In $L34$, let $g(t) = e^{-t}$ and $h(t) = \sin t$, and find $G(p)H(p)$ which is the Laplace transform of the integral you want. Break the result into partial fractions and look up the inverse transforms.

Use the convolution integral (see Example 2) to solve the following differential equations.

14. $y'' + 5y' + 6y = e^{-2t}, \quad y_0 = y_0' = 0.$

15. $y'' + 3y' - 4y = e^{3t}, \quad y_0 = y_0' = 0.$

16. Consider solving an equation like (10.1) but with nonzero initial conditions.

(a) Write the corrected form of (10.2), writing the transfer function in factored form as indicated just after (10.2). Consider the extra terms in Y which arise from the initial conditions; show that the inverse transforms of such terms can always be found from $L6$, $L7$, $L8$, and $L18$.

(b) Find the explicit form of the inverse transform of the transfer function for $a \neq b$ (use $L7$), and so write the general solution of (10.2) with nonzero initial conditions as a convolution integral plus the terms which you found in (a).

17. Solve the differential equation $y'' - a^2 y = f(t)$, where

$$f(t) = \begin{cases} 0, & t < 0, \\ 1, & t > 0, \end{cases} \qquad \text{and } y_0 = y_0' = 0.$$

Hint: Use the convolution integral as in the example.

18. A mechanical or electrical system is described by the differential equation $y'' + \omega^2 y = f(t)$. Find y if

$$f(t) = \begin{cases} 1, & 0 < t < a, \\ 0, & \text{otherwise}, \end{cases} \qquad \text{and} \quad y_0 = y_0' = 0.$$

Hint: Use the convolution integral carefully. Consider $t < a$ and $t > a$ separately, remembering that $f(t) = 0$ for $t > a$. Show that

$$y = \begin{cases} \dfrac{1}{\omega^2}(1 - \cos \omega t), & t < a, \\ \dfrac{1}{\omega^2}[\cos \omega(t-a) - \cos \omega t], & t > a. \end{cases}$$

Sketch the motion if $a = \frac{1}{3}T$ where T is the period for free vibrations of the system; if $a = \frac{3}{2}T$; if $a = \frac{1}{10}T$.

19. Following the method of equations (10.8) to (10.12), show that $f_1 f_2$ and $g_1 * g_2$ are a pair of Fourier transforms.

11. THE DIRAC DELTA FUNCTION

In mechanics we consider the idea of an impulsive force such as a hammer blow which lasts for a very short time. We usually do not know the exact shape of the force function $f(t)$, and so we proceed as follows. Let the impulsive force $f(t)$ lasting from $t = t_0$ till $t = t_1$ be applied to a mass m; then by Newton's second law we have

$$\int_{t_0}^{t_1} f(t)\,dt = \int_{t_0}^{t_1} m\frac{dv}{dt}\,dt = \int_{v_0}^{v_1} m\,dv = m(v_1 - v_0). \tag{11.1}$$

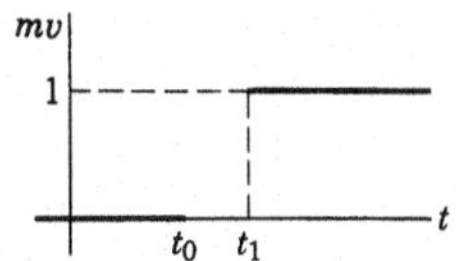

Figure 11.1

This says that the integral of $f(t)$ [called the impulse of $f(t)$] is equal to the change in the momentum of m, and we note that the result is independent of the shape of $f(t)$ but depends only on the area under the $f(t)$ curve. If this area is 1, we call the impulse a *unit impulse*. If $t_1 - t_0$ is very small, we may simply ignore the motion of m during this small time, and say only that the momentum jumped from mv_0 to mv_1 during the time $t_1 - t_0$. If $v_0 = 0$, the graph of the momentum as a function of time would be as in Figure 11.1, where we have simply omitted the (unknown) part of the graph between t_0 and t_1. We note that if $t_1 - t_0$ is very small, the graph in Figure 11.1 is almost the unit step function ($L24$). Let us imagine making $t_1 - t_0$ smaller and smaller while keeping the jump in mv always 1.

In Figures 11.2, 11.3 and 11.4 we have sketched some possible sequences of functions $f_n(t)$ which would do this. We could draw many other similar sets of graphs; the essential requirement is that $f(t)$ should become taller and narrower (that is, that the force should become more intense but act over a shorter time) in such a way that the impulse [area under the $f(t)$ curve] remains 1. We might then consider the limiting case in which Figure 11.1 has a jump of 1 at t_0; the force $f(t)$ required to produce this result would have to be infinite and act instantaneously. Also from equation (11.1), we see that the function $f(t)$ is the slope of the mv graph; thus we are asking for $f(t)$ to be the derivative of a step function at the jump. We see immediately that no ordinary function has these properties. However, we also note

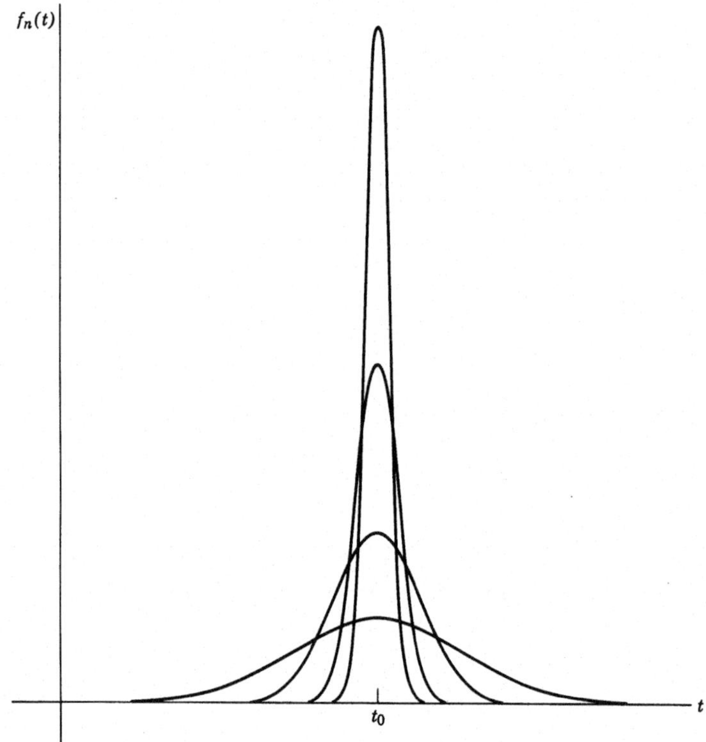

Figure 11.2

that we are not so much interested in $f(t)$ as in the results it produces. Figure 11.1 with a jump at t_0 makes perfectly good sense; for any $t > t_0$ we could choose a sufficiently tall and narrow $f_n(t)$ so that mv would already have its final value. We shall see that it is convenient to introduce a symbol $\delta(t - t_0)$ to represent the force which produces a jump of 1 in mv at t_0; $\delta(t - t_0)$ is called the *Dirac delta function* although it is not an ordinary function as we have seen. (It may properly be called a *generalized function* or a *distribution*, and is one of a whole class of such functions.) Introducing and using this symbol is much like introducing and using the symbol ∞. It is convenient to write equations like $1/\infty = 0$, but we must not write $\infty/\infty = 1$; that is, such symbolic equations must be abbreviations for correct limiting processes. Let us investigate, then, how we can use the δ function correctly.

▶ **Example 1.** Consider the differential equation

$$y'' + \omega^2 y = f(t), \qquad y_0 = y_0' = 0. \tag{11.2}$$

This equation might describe the oscillations of a mass suspended by a spring, or a simple series electric circuit with negligible resistance. Let us assume that the system is initially at rest ($y_0 = y_0' = 0$); then suppose that, at $t = t_0$ the mass is struck a sharp blow, or a sudden short surge of current is sent through the electric circuit. The function $f(t)$ may be one of those shown in Figures 11.2 to 11.4 or another similar function. Let us solve (11.2) with $f(t)$ equal to one of the functions in Figure 11.3, that is, $f(t) = ne^{-n(t-t_0)}, t > t_0$. Using Laplace transforms $L28$

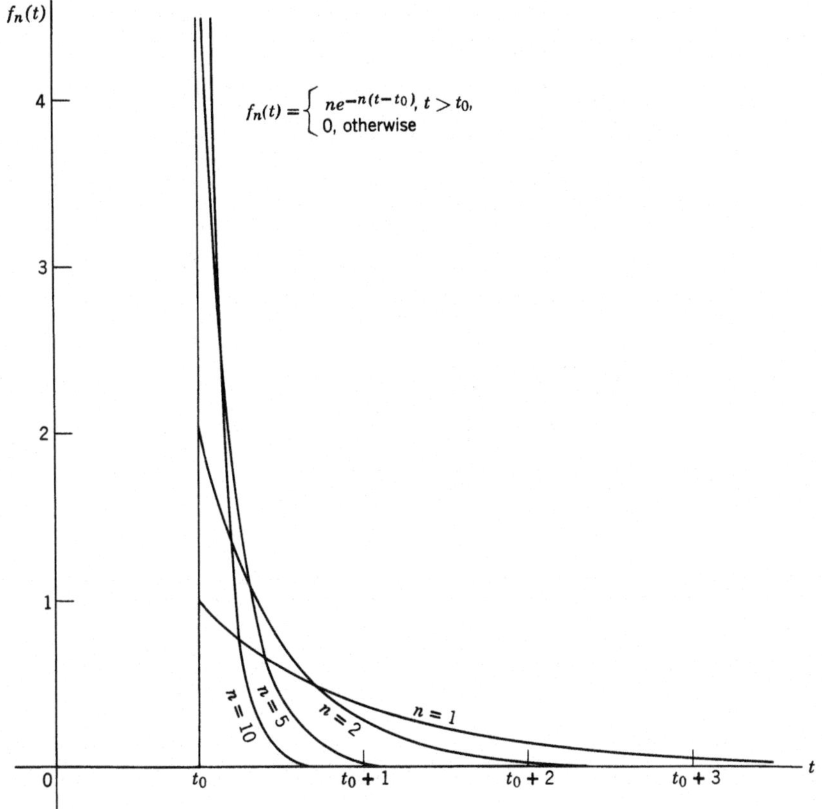

Figure 11.3

and $L2$ we find

$$(11.3)\qquad \begin{aligned} (p^2+\omega^2)Y &= L(ne^{-n(t-t_0)}) = n\cdot\frac{e^{-pt_0}}{p+n}, \\ Y = n\cdot\frac{e^{-pt_0}}{(p+n)(p^2+\omega^2)} &= \frac{ne^{-pt_0}}{(n^2+\omega^2)}\left[\frac{1}{p+n}+\frac{n}{p^2+\omega^2}-\frac{p}{p^2+\omega^2}\right]. \end{aligned}$$

(You can easily verify the partial fractions expansion in the last step.) Then by $L28$ with $a = t_0$, $L2$ with $a = n$, and $L3$, $L4$, with $a = \omega$, we find:

$$(11.4)\qquad y = n\left(\frac{e^{-n(t-t_0)}}{n^2+\omega^2}+\frac{n\sin\omega(t-t_0)}{(n^2+\omega^2)\omega}-\frac{\cos\omega(t-t_0)}{n^2+\omega^2}\right),\qquad t > t_0.$$

(Of course, $y = 0$ for $t < t_0$.) By making $f(t)$ sufficiently narrow and peaked (that is, by making n large enough), we can make the first and third terms in y negligible, and the coefficient of $\sin\omega(t-t_0)$ approximately equal to $1/\omega$. Thus the solution is approximately

$$(11.5)\qquad y = \frac{1}{\omega}\sin\omega(t-t_0),\qquad t > t_0,$$

for a unit impulse of very short duration at $t = t_0$. (We have shown this only for the functions of Figure 11.3; however, the same result would be found for other sets of functions, such as those in Figure 11.4, for example—see Problem 5.)

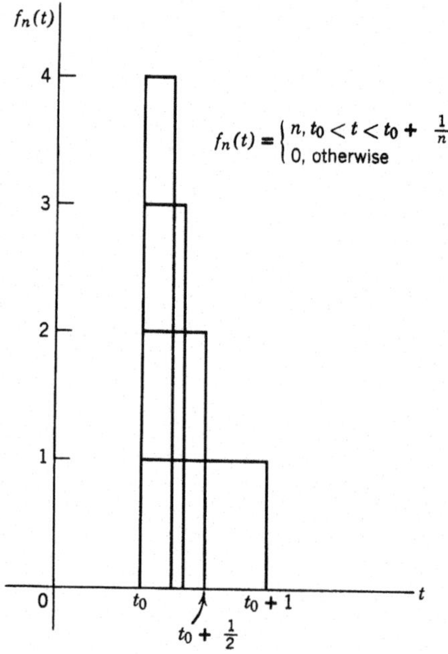

Figure 11.4

Now we would like to be able to find (11.5) without finding (11.4), in fact, without choosing a specific set of functions $f_n(t)$. Our discussion above suggests that we try using the symbol $\delta(t - t_0)$ for $f(t)$ on the right-hand side of (11.2). In solving the equation, we would then like to take the Laplace transform of $\delta(t - t_0)$.

Laplace Transform of a δ Function Let us investigate whether we can make sense out of the Laplace transform of $\delta(t - t_0)$. More generally, let us try to attach meaning to the integral $\int \phi(t)\delta(t - t_0)\,dt$, where $\phi(t)$ is any continuous function and $\delta(t - t_0)$ is the symbol indicating an impulse at t_0. We consider the integrals $\int \phi(t) f_n(t - t_0)\,dt$, where the functions $f_n(t - t_0)$ are more and more strongly peaked at t_0 as n increases (Figure 11.5), but the area under each graph is 1. When $f_n(t - t_0)$ is so narrow that $\phi(t)$ is essentially constant [equal to $\phi(t_0)$] over the width of $f_n(t - t_0)$, the integral becomes nearly $\phi(t_0) \int f_n(t - t_0)\,dt = \phi(t_0) \cdot 1 = \phi(t_0)$; that is, the sequence of integrals $\int \phi(t) f_n(t - t_0)\,dt$ tends to $\phi(t_0)$ as n tends to infinity. It then seems reasonable to say that

$$\text{(11.6)} \qquad \int_a^b \phi(t)\delta(t - t_0)\,dt = \begin{cases} \phi(t_0), & a < t_0 < b, \\ 0, & \text{otherwise.} \end{cases}$$

Equation (11.6) is the defining equation for the δ function; when we operate with δ functions, we use them in integrals and (11.6) tells us the value of the integral. The integral in (11.6) is not a Riemann integral; it is just a very useful symbol indicating that we have found the limit of $\int \phi(t) f_n(t - t_0)\,dt$ as $n \to \infty$. You may then ask how we can carry out familiar operations like integration by parts. When you treat an integral containing a δ function as an ordinary integral, you can, if you like, think

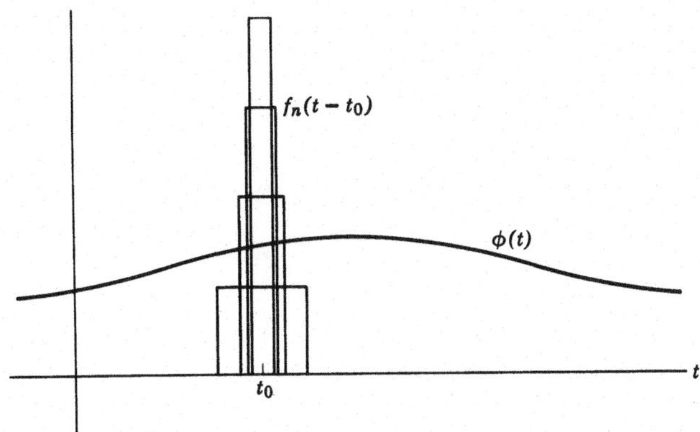

Figure 11.5

that you are really working with the functions $f_n(t - t_0)$ and then taking the limit at the end. Of course, all this needs mathematical justification which exists but is beyond our scope. (For two different mathematical developments of generalized functions, see Lighthill, and Chapter 9 of Folland.) Our purpose is just to make understandable the δ function formulas which are so useful in applications.

Example 2. We can now easily find the Laplace transform of $\delta(t - t_0)$. In the notation used in $L27$ (which we are about to derive), we have, using (8.1),

$$L[\delta(t-a)] = \int_0^\infty \delta(t-a)e^{-pt}\,dt = e^{-pa}, \qquad a > 0, \tag{11.7}$$

since, by 11.6, the integral of the product of $\delta(t - a)$ and a function "picks out" the value of the function at $t = a$. Now let us use our results to obtain (11.5) more easily.

Example 3. Solve

$$y'' + \omega^2 y = \delta(t - t_0), \qquad y_0 = y_0' = 0. \tag{11.8}$$

Taking Laplace transforms and using (11.7), we get

$$(p^2 + \omega^2)Y = L[\delta(t - t_0)] = e^{-pt_0}. \tag{11.9}$$

Then

$$Y = \frac{e^{-pt_0}}{p^2 + \omega^2} \tag{11.10}$$

and, by $L3$ and $L28$,

$$y = \frac{1}{\omega}\sin\omega(t - t_0), \qquad t > t_0, \tag{11.11}$$

as in (11.5).

Fourier Transform of a δ Function Using (11.6) and the definition of a Fourier transform [Chapter 7, equation (12.2)], we may write

$$g(\alpha) = \frac{1}{2\pi}\int_{-\infty}^{\infty} \delta(x-a)e^{-i\alpha x}\,dx = \frac{1}{2\pi}e^{-i\alpha a}. \tag{11.12}$$

Formally, then (12.2) of Chapter 7 would give for the inverse transform

$$\delta(x-a) = \frac{1}{2\pi}\int_{-\infty}^{\infty} e^{i\alpha(x-a)}\,d\alpha. \tag{11.13}$$

We say "formally" because the integral in (11.13) does not converge. However, if we replace the limits $-\infty, \infty$ by $-n, n$, we obtain a set of functions (Problem 12) which, like the functions $f_n(t)$ in Figures 11.2 to 11.4, are increasingly peaked around $x = a$ as n increases, but all have area 1. In this sense, then, (11.13) is a representation of the δ function. Equations (11.12) and (11.13) are useful in quantum mechanics.

Another Physical Application of δ functions What is the density (mass per unit length) of a point mass on the x axis? Compare the concept of a point mass with our discussion of δ functions. We could think of a point mass as corresponding to the limiting case of a density function like those in Figures 11.2 to 11.4. A point mass at $x = a$ requires that the density be zero everywhere except at $x = a$ but the integral of the density function across $x = a$ should be the mass m. Thus we can write the density function for a point mass m at $x = a$ as $m\delta(x-a)$. Similarly, we can represent the charge density for point electrical charges using δ functions.

▶ **Example 4.** The charge density for a charge of 2 at $x = 3$, a charge of -5 at $x = 7$ and a charge of 3 at $x = -4$ would be $2\delta(x-3) - 5\delta(x-7) + 3\delta(x+4)$.

Derivatives of the δ Function To see that we can attach a meaning to the derivative of $\delta(x-a)$, we write $\int_{-\infty}^{\infty} \phi(x)\delta'(x-a)\,dx$ and integrate by parts to get

$$\int_{-\infty}^{\infty} \phi(x)\delta'(x-a)\,dx = \phi(x)\delta(x-a)\Big|_{-\infty}^{\infty} - \int_{-\infty}^{\infty} \phi'(x)\delta(x-a)\,dx = -\phi'(a). \tag{11.14}$$

The integrated term is zero at $\pm\infty$, and we evaluated the integral using equation (11.6). Thus, just as $\delta(x-a)$ "picks out" the value of $\phi(x)$ at $x = a$ [see equation (11.6)], so $\delta'(x-a)$ picks out the negative of $\phi'(x)$ at $x = a$. Integrating by parts twice (Problem 14), we find

$$\int_{-\infty}^{\infty} \phi(x)\delta''(x-a)\,dx = \phi''(a). \tag{11.15}$$

Repeated integrations by parts gives the formula for the derivative of any order of the δ function (Problem 14):

$$\int_{-\infty}^{\infty} \phi(x)\delta^{(n)}(x-a)\,dx = (-1)^n\phi^{(n)}(a). \tag{11.16}$$

We have written the integrals in (11.14) to (11.16) with limits $-\infty$ to ∞, but all that is necessary is that the range of integration include $x = a$; otherwise, as for (11.6), the integrals are zero.

Some Formulas Involving δ Functions Our discussion at the beginning of this section (see Figure 11.1 and $L24$ in the Laplace transform table) implied that the derivative at $x = a$ of the unit step function ought to be $\delta(x-a)$.

$$\text{(11.17)} \quad \begin{aligned} &(a) \quad u(x-a) = \begin{cases} 1, & x > a \\ 0, & x < a \end{cases} \\ &(b) \quad u'(x-a) = \delta(x-a). \end{aligned}$$

What does the $u' = \delta$ equation mean? By definition, two generalized functions (distributions) are equal, say $G_1(x) = G_2(x)$, if $\int \phi(x)G_1\,dx = \int \phi(x)G_2\,dx$ for any test function $\phi(x)$. Test functions are assumed to be very well behaved functions; let's assume that they are continuous with continuous derivatives of all orders and that they are identically zero outside some finite interval so that the integrated term in an integration by parts is always zero. You can think of generalized functions as being operators; given a test function $\phi(x)$, they "operate" on it to produce a value such as $\phi(0)$. Compare the differential operators in Problem 5.31. We wrote $Dx = xD + 1$ where $D = d/dx$. This would be nonsense as an elementary calculus formula, but as an operator equation to be applied to $y(x)$, it means $D(xy) = (xD+1)y = xy' + y$ which is correct. In a similar way, two generalized functions are equal if they give the same results when they operate on any test function. Let's try this for (11.17b). We multiply u' by $\phi(x)$, integrate by parts (noting that the integrated term is zero because we require test functions to be zero for large $|x|$), substitute the values of $u(x-a)$, and integrate again to get

$$\int_{-\infty}^{\infty} \phi(x)u'(x-a)\,dx = -\int_{-\infty}^{\infty} \phi'(x)u(x-a)\,dx = -\int_{a}^{\infty} \phi'(x)\,dx = \phi(a).$$

This is indeed the value of the integral of $\phi(x)\delta(x-a)$, so $u'(x-a) = \delta(x-a)$ is a valid operator equation (generalized function equation).

Since we think of $\delta(x)$ and its derivatives as being zero except at the origin, and x is zero at the origin, it might seem plausible that $x\delta, x^2\delta, x\delta'$, etc. would be identically zero. It turns out that some of these are zero and some are not; to find out, we multiply by an arbitrary test function $\phi(x)$ and integrate. We state a few results; also see Problems 17 and 18.

$$\text{(11.18)} \quad \begin{aligned} &(a) \quad x\delta(x) = 0 \\ &(b) \quad x\delta'(x) = -\delta(x) \\ &(c) \quad x^2\delta''(x) = 2\delta(x) \end{aligned}$$

To check (b), we multiply by $\phi(x)$ and integrate using (11.14) with $\phi(x)$ replaced by $x\phi(x)$.

$$\int_{-\infty}^{\infty} x\delta'(x)\phi(x)\,dx = -(x\phi)'\Big|_{x=0} = -(x\phi' + \phi)\Big|_{x=0} = -\phi(0) = -\int_{-\infty}^{\infty} \delta(x)\phi(x)\,dx.$$

Here is another way to produce valid generalized function identities like those in (11.18). Suppose $G_1(x) = G_2(x)$; then we can show (Problem 19a) that $\frac{d}{dx}G_1(x) = \frac{d}{dx}G_2(x)$ and $x^n G_1(x) = x^n G_2(x)$. For example, if we differentiate (11.18a) we get $x\delta'(x) + \delta(x) = 0$, or $x\delta'(x) = -\delta(x)$ which is (11.18b).

We list a few more operator equations (see Problems 20 and 21).

$$(11.19)\quad \begin{aligned} &(a)\quad \delta(-x)=\delta(x) \text{ and } \delta(x-a)=\delta(a-x);\\ &(b)\quad \delta'(-x)=-\delta'(x) \text{ and } \delta'(x-a)=-\delta'(a-x);\\ &(c)\quad \delta(ax)=\frac{1}{|a|}\delta(x),\ a\neq 0;\\ &(d)\quad \delta[(x-a)(x-b)]=\frac{1}{|a-b|}[\delta(x-a)+\delta(x-b)],\ a\neq b;\\ &(e)\quad \delta[f(x)]=\sum_i \frac{\delta(x-x_i)}{|f'(x_i)|} \quad \text{if } f(x_i)=0 \text{ and } f'(x_i)\neq 0. \end{aligned}$$

We first prove (c) when a is negative, say $a=-b$, $b>0$. Let $u=-bx$, then $du=-b\,dx$, and the limits $x=-\infty,\infty$, become $u=\infty,-\infty$.

$$\begin{aligned} \int_{-\infty}^{\infty}\phi(x)\delta(-bx)\,dx &= \int_{\infty}^{-\infty}\phi\left(\frac{u}{-b}\right)\delta(u)\left(\frac{du}{-b}\right)=\frac{1}{b}\int_{-\infty}^{\infty}\phi\left(\frac{u}{-b}\right)\delta(u)\,du\\ &=\frac{1}{b}\phi(0)=\frac{1}{|a|}\phi(0)=\frac{1}{|a|}\int_{-\infty}^{\infty}\phi(x)\delta(x)\,dx. \end{aligned}$$

From the second integral to the third, we have reversed the order of integration (one minus sign) and also changed $\frac{du}{-b}$ to $\frac{du}{b}$ (another minus sign, which cancels the first). Now, if we repeat the calculation using $a>0$ instead of $-b$, neither of these sign reversals occurs, and so we get the result $\frac{1}{a}\phi(0)$ instead of $\frac{1}{b}\phi(0)$. But when $a>0$, a and $|a|$ are the same. Thus we get the result stated in (11.19c).

δ functions in 2 or 3 dimensions It is now straightforward to write the defining equations in rectangular coordinates for δ functions in 2 or 3 dimensions. We have

$$(11.20)\qquad \int_{-\infty}^{\infty}\int_{-\infty}^{\infty}\phi(x,y)\delta(x-x_0)\delta(y-y_0)\,dx\,dy=\phi(x_0,y_0).$$

$$(11.21)\qquad \begin{aligned}\int_{-\infty}^{\infty}\int_{-\infty}^{\infty}\int_{-\infty}^{\infty}\phi(x,y,z)\delta(x-x_0)\delta(y-y_0)\delta(z-z_0)\,dx\,dy\,dz\\ =\phi(x_0,y_0,z_0).\end{aligned}$$

As in one dimension, the delta function "picks out" the value of the test function ϕ at the "peak" of the δ function. The integrals need not be over all space, just over a region containing the point $\mathbf{r}_0$; otherwise the integral is zero. The abbreviations $\delta(\mathbf{r})$ or $\delta^3(\mathbf{r})$ are often used for $\delta(x)\delta(y)\delta(z)$, but note carefully that they do not mean functions of the vector $\mathbf{r}$, but rather functions of the components x,y,z of $\mathbf{r}$. Similarly you may see $\delta(\mathbf{r}-\mathbf{r}_0)$ or $\delta^3(\mathbf{r}-\mathbf{r}_0)$ meaning the δ function in (11.21).

In spherical coordinates, let's use f instead of ϕ to mean a test function (since ϕ is a spherical coordinate angle). By the definition of δ functions, we want $\iiint f(r,\theta,\phi)\delta(r-r_0)\delta(\theta-\theta_0)\delta(\phi-\phi_0)\,dr\,d\theta\,d\phi=f(r_0,\theta_0,\phi_0)$. But since we would like to use the volume element $d\tau=r^2\sin\theta\,dr\,d\theta\,d\phi=r^2\,dr\,d\Omega$, we need to write

(also see Problem 22)

$$\delta(\mathbf{r}-\mathbf{r}_0) = \frac{\delta(r-r_0)\delta(\theta-\theta_0)\delta(\phi-\phi_0)}{r^2\sin\theta}; \qquad \text{then}$$

(11.22)

$$\iiint f(r,\theta,\phi)\delta(\mathbf{r}-\mathbf{r}_0)\,d\tau = f(r_0,\theta_0,\phi_0).$$

Similarly in cylindrical coordinates, with $d\tau = r\,dr\,d\theta\,dz$,

$$\delta(\mathbf{r}-\mathbf{r}_0) = \frac{\delta(r-r_0)\delta(\theta-\theta_0)\delta(z-z_0)}{r}; \qquad \text{then}$$

(11.23)

$$\iiint f(r,\theta,z)\delta(\mathbf{r}-\mathbf{r}_0)\,d\tau = f(r_0,\theta_0,z_0).$$

Note that we can use these formulas to write mass density or charge density functions in the various coordinate systems.

Example 5. Suppose there is a unit charge or unit mass at the point $(x,y,z) = (-1,\sqrt{3},-2)$; then in rectangular coordinates, the density is

$$\rho = \delta(x+1)\delta(y-\sqrt{3})\delta(z+2).$$

In cylindrical coordinates the point is $(r,\theta,z) = (2, 2\pi/3, -2)$ so in cylindrical coordinates the density is

$$\rho = \delta(r-2)\delta(\theta-2\pi/3)\delta(z+2)/r.$$

In spherical coordinates, the point is $(r,\theta,\phi) = (2\sqrt{2}, 3\pi/4, 2\pi/3)$, so in spherical coordinates the density is

$$\rho = \delta(r-2\sqrt{2})\delta(\theta-3\pi/4)\delta(\phi-2\pi/3)/(r\sin\theta).$$

Finally, let's verify two useful operator equations for δ functions in 3 dimensions.

(11.24) $$\boldsymbol{\nabla}\cdot\frac{\mathbf{e}_r}{r^2} = 4\pi\delta(\mathbf{r});$$

(11.25) $$\nabla^2\frac{1}{r} = -4\pi\delta(\mathbf{r}).$$

You can easily show (Problem 24a) that $\boldsymbol{\nabla}\cdot(\mathbf{e}_r/r^2)$ is zero for any $r \neq 0$ (and undefined for $r=0$). Also, by the divergence theorem [Chapter 6, equation (10.17)] in spherical coordinates, we find

$$\iiint\limits_{\text{volume }\tau} \boldsymbol{\nabla}\cdot\frac{\mathbf{e}_r}{r^2}\,d\tau = \iint\limits_{\substack{\text{surface}\\ \text{inclosing }\tau}} \frac{\mathbf{e}_r}{r^2}\cdot\mathbf{e}_r\,d\sigma = \int_{\phi=0}^{2\pi}\int_{\theta=0}^{\pi}\frac{1}{r^2}r^2\sin\theta\,d\theta\,d\phi = 4\pi.$$

Thus $\boldsymbol{\nabla}\cdot(\mathbf{e}_r/r^2)$ has the properties that it is zero for all $r > 0$ but its integral over any volume including the origin $= 4\pi$; this suggests that it is equal to $4\pi\delta(\mathbf{r})$. Let's verify that this is correct. (Compare Problem 25.) Since $\boldsymbol{\nabla}\cdot(\mathbf{e}_r/r^2)$ depends only on r (Problem 24a), we use a test function $f(r)$. We want to show that

$\iiint f(r)\nabla\cdot(\mathbf{e}_r/r^2)\,d\tau$, over any volume containing the origin, is equal to $4\pi f(0)$. For convenience we integrate over the volume inside the sphere $r=a$. (Since the integrand is zero for $r>0$, the answer is the same for any volume containing the origin.) By Problem 11.17(e) of Chapter 6 with $\phi=f$, $\mathbf{V}=\mathbf{e}_r/r^2$, and $\mathbf{n}=\mathbf{e}_r$, we find

$$\iiint\limits_{\text{volume } r<a} f(r)\nabla\cdot\frac{\mathbf{e}_r}{r^2}\,d\tau=\iint\limits_{\text{surface } r=a} f(r)\frac{\mathbf{e}_r}{r^2}\cdot\mathbf{e}_r\,d\sigma-\iiint\limits_{\text{volume } r<a}\nabla f(r)\cdot\frac{\mathbf{e}_r}{r^2}\,d\tau.$$

On the surface $r=a$, the integrand of the surface integral is $f(a)\frac{1}{a^2}a^2\,d\Omega$, so the surface integral is $4\pi f(a)$. In the volume integral, $\nabla f(r)\cdot\mathbf{e}_r$ is the r component of $\nabla f(r)$; in spherical coordinates this is just $\partial f/\partial r$ (Chapter 6, equation 6.8). Thus the volume integral on the right-hand side is

$$\iiint\limits_{\text{volume } r<a}\frac{\partial f}{\partial r}\frac{1}{r^2}r^2\,dr\,d\Omega=4\pi f(r)\Big|_0^a=4\pi[f(a)-f(0)]$$

and we have

$$\iiint\limits_{\text{volume } r<a} f(r)\nabla\cdot\frac{\mathbf{e}_r}{r^2}\,d\tau=4\pi f(a)-4\pi[f(a)-f(0)]=4\pi f(0)$$

as we expected. Thus (11.24) is a valid operator equation. You can show (Problem 24b) that $\nabla(1/r)=-\mathbf{e}_r/r^2$). Since $\nabla\cdot\nabla=\nabla^2$ (that is, div grad = Laplacian), we have $\nabla^2(1/r)=\nabla\cdot\nabla(1/r)=-\nabla\cdot(\mathbf{e}_r/r^2)$. Thus (11.25) is also valid.

We can write (11.24) and (11.25) with the peak of the δ function shifted from the origin to $\mathbf{r}_0$. The unit vector from $\mathbf{r}_0$ to $\mathbf{r}$ can be written as $(\mathbf{r}-\mathbf{r}_0)/|\mathbf{r}-\mathbf{r}_0|$. Then we have

$$\nabla^2\frac{1}{|\mathbf{r}-\mathbf{r}_0|}=-\nabla\cdot\frac{\mathbf{r}-\mathbf{r}_0}{|\mathbf{r}-\mathbf{r}_0|^3}=-4\pi\delta(\mathbf{r}-\mathbf{r}_0). \tag{11.26}$$

It is now interesting to note that we have seen this before without recognizing that we were dealing with a δ function. Look back to Chapter 6, Equations (6.10.19) to (6.10.25). With $\mathbf{D}$ given by (6.10.19), we have from (8.11.24) above, $\nabla\cdot\mathbf{D}=\frac{q}{4\pi}4\pi\delta(\mathbf{r})=q\delta(\mathbf{r})$ which we recognize as the charge density for a charge q at the origin. Thus, although we wrote (6.10.25) with ρ as the density of a charge distribution, we could now write it with $\rho=$ charge density of a point charge $=q\delta(\mathbf{r})$, and then (6.10.25) becomes (6.10.22).

PROBLEMS, SECTION 11

1. Find the inverse Laplace transform of e^{-2p}/p^2 in the following ways:
   (a) using $L5$ and $L27$ and the convolution integral of Section 10;
   (b) using $L28$.
2. Verify $L24$ in the table by using $L1$, $L27$, and the convolution integral.
3. Verify $L28$ in the table by using $L27$ and the convolution integral.
4. Show that $\int_{-\infty}^{\infty} f_n(t)\,dt=1$ for the functions $f_n(t)$ in Figures 11.3 and 11.4.

5. Solve the differential equation $y'' + \omega^2 y = f(t)$, $y_0 = y_0' = 0$, with $f(t)$ given by the functions in Figure 11.4, by the following methods.

(a) Use the convolution integral, being careful to consider separately the three intervals 0 to t_0, t_0 to $t_0 + 1/n$, and $t_0 + 1/n$ to ∞.

(b) Write $f_n(t)$ as a difference of unit step functions as in $L25$, and use $L25$ to find $L(f_n)$. Expand $\dfrac{1}{p(p^2+\omega^2)}$ by partial fractions and use $L28$ to find $y_n(t)$. Your result should agree with (a).

(c) Let $n \to \infty$ and show that your solution in (a) and (b) tends to the same solution (11.5) obtained using the functions of Figure 11.3; that is, either set of functions gives, in the limit, the same solution (11.11) obtained by using the δ function. Note that, when you let $n \to \infty$, you do not need to consider the interval t_0 to $t_0 + 1/n$ since, if $t > t_0$, then for sufficiently large n, $t > t_0 + 1/n$.

6. (a) Let a mechanical or electrical system be described by the differential equation $Ay'' + By' + Cy = f(t)$, $y_0 = y_0' = 0$. As in Problem 10.16b, write the solution as a convolution (assume $a \neq b$). Let $f(t)$ be one of the functions in Figure 11.4 and Problem 5. Find y and then let $n \to \infty$.

(b) Solve (a) with $f(t) = \delta(t - t_0)$; your result should be the same as in (a).

(c) The solution y as found in (a) and (b) is called the *response* of the system to a unit impulse. Show that the response of a system to a unit impulse at $t_0 = 0$ is the inverse Laplace transform of the transfer function.

Using the δ function method, find the response (see Problem 6c) of each of the following systems to a unit impulse.

7. $y'' + 2y' + y = \delta(t - t_0)$

8. $y'' + 4y' + 5y = \delta(t - t_0)$

9. $y'' + 2y' + 10y = \delta(t - t_0)$

10. $y'' - 9y = \delta(t - t_0)$

11. $\dfrac{d^4y}{dt^4} - y = \delta(t - t_0)$

12. Evaluate the functions $f_n(x-a)$ defined by the integral in (11.13) with limits $-n$, n. Show that $\int_{-\infty}^{\infty} f_n(x-a)\,dx = 1$ for all n. Sketch or computer plot graphs of several f_n's to show that as n increases the functions $f_n(x)$ are increasingly peaked around $x = a$, and that as $|x - a|$ increases, they oscillate with decreasing amplitude.

13. Using δ functions, write the following mass or charge density functions.

(a) Mass 5 at $x = 2$, and mass 3 at $x = -7$.

(b) Charge 3 at $x = -5$ and charge -4 at $x = 10$.

14. Integrate by parts as we did for (11.14) to obtain (11.15) and (11.16).

15. Use (11.6) and (11.14) to (11.16) to evaluate the following integrals. *Warning hint:* See comments just after (11.6) and (11.16) about the range of integration.

(a) $\displaystyle\int_0^{\pi} \sin x\ \delta\left(x - \frac{\pi}{2}\right) dx$ (b) $\displaystyle\int_0^{\pi} \sin x\ \delta\left(x + \frac{\pi}{2}\right) dx$

(c) $\displaystyle\int_{-1}^{1} e^{3x}\delta'(x)\,dx$ (d) $\displaystyle\int_0^{\pi} \cosh x\ \delta''(x - 1)\,dx$

16. Verify the operator equation $\frac{d}{dx}\text{sgn } x = 2\delta(x)$ where the function signum x, meaning "sign of x," and abbreviated sgn x, is defined by

$$\text{sgn } x = \begin{cases} 1, & x > 0, \\ -1, & x < 0. \end{cases}$$

17. Verify (11.18a) and (11.18c) by multiplying by a test function and integrating.

18. Use equation (11.16) to generalize the operator equations (11.18) as follows:

(a) Show that $x^m\delta^{(n)}(x) = 0$ if $m > n$; compare equation (11.18a).

(b) Show that $x^n\delta^{(n)}(x) = (-1)^n n!\,\delta(x)$; compare (11.18b) and (11.18c).

(c) Show that $x^m\delta^{(n)}(x) = (-1)^m\frac{n!}{(n-m)!}\,\delta^{(n-m)}(x)$, $m \le n$.

(d) Use the results in (a) and (b) to show that

$$(x^2 + y^2 + z^2)\nabla^2[\delta(x)\delta(y)\delta(z)] = 6\delta(x)\delta(y)\delta(z).$$

19. (a) Show that you can differentiate a generalized function equation or multiply it by a power of x. This means to show that if $\int \phi(x)G_1(x)\,dx = \int \phi(x)G_2(x)\,dx$ for all test functions ϕ, then

$$\int \phi(x)G_1'(x)\,dx = \int \phi(x)G_2'(x)\,dx \qquad \text{and}$$

$$\int \phi(x)x^n G_1(x)\,dx = \int \phi(x)x^n G_2(x)\,dx.$$

Hints: For the differentiation proof, integrate by parts. For the multiplication by x^n proof, consider whether $x^n\phi(x)$ is a test function if ϕ is. See comment just after equation (11.17).

(b) Multiply (11.18b) by x and use (11.18a). Differentiate the result and simplify to get (11.18c).

(c) Multiply (11.18c) by x, use (11.18a), differentiate and simplify to find $x^3\delta'''(x)$ in terms of $\delta(x)$. Check your result by Problem (18b).

(d) Try a few more examples as in (b) and (c) and check your results by Problem 18.

20. Verify the operator equations in (11.19) not done in text.
Hints for (a) and (b): Follow the text method of proof of (c), making the change of variable $u = -x$ or $u = a - x$. *Hints* for (c) and (d): Split the integral into a sum of integrals each including just one x_i. In (d), what is the value of $(x - a)$ when x is in the vicinity of b? Use part (c).

21. Make use of the operator equations (11.19) and previous equations to evaluate the following integrals.

(a) $\displaystyle\int_0^3 (5x-2)\delta(2-x)\,dx$ (b) $\displaystyle\int_0^\infty \phi(x)\delta(x^2-a^2)\,dx$

(c) $\displaystyle\int_{-1}^1 \cos x\,\delta(-2x)\,dx$ (d) $\displaystyle\int_{-\pi/2}^{\pi/2} \cos x\,\delta(\sin x)\,dx$

22. You may find the spherical coordinate δ function written as

$$\delta(\mathbf{r} - \mathbf{r}_0) = \delta(r - r_0)\delta(\cos\theta - \cos\theta_0)\delta(\phi - \phi_0)/r^2.$$

Show that this equation is equivalent to (11.22). *Hints:* You want to show that $\delta(\cos\theta - \cos\theta_0) = \delta(\theta - \theta_0)/\sin\theta$. See (11.19e). Also note that it doesn't really matter whether we write $r^2\sin\theta$ or $r_0^2\sin\theta_0$ in the denominator of (11.22) since the δ functions are zero unless $r = r_0$ and $\theta = \theta_0$.

23. Write a formula in rectangular coordinates, in cylindrical coordinates, and in spherical coordinates for the density of a unit point charge or mass at the point with the given rectangular coordinates:

(a) $(-5, 5, 0)$ (b) $(0, -1, -1)$

(c) $(-2, 0, 2\sqrt{3})$ (d) $(3, -3, -\sqrt{6})$

24. (a) Show that $\nabla \cdot (\mathbf{e}_r/r^2) = 0$ for $r > 0$. *Hint:* You can do this in rectangular coordinates, but it is easier in spherical coordinates. See Chapter 6, equation (7.9). Show that $\nabla \cdot [\mathbf{e}_r F(r)]$, for any $F(r)$, is a function of r only.

(b) Show that $\nabla(1/r) = -\mathbf{e}_r/r^2$. See Chapter 6, equation (6.8).

25. Let

$$F(x) = \begin{cases} x - 2, & x > 0, \\ 0, & x < 0. \end{cases}$$

Show that $F''(x) = 0$ for all $x \neq 0$, and $\int_{-\infty}^{\infty} F''(x)\,dx = 1$, which leads you to think that $F''(x)$ might $= \delta(x)$. Show in two ways, as outlined in (a) and (b), that this is not true.

(a) Show that $\int_{-\infty}^{\infty} \phi(x)F''(x)\,dx = \phi(0) + 2\phi'(0)$, where ϕ is any test function. Then by (11.6) and (11.14), what is $F''(x)$?

(b) Show that $F(x) = (x - 2)u(x)$ where $u(x)$ is the unit step function in (11.17). Differentiate this equation twice and simplify using (11.17) and (11.18). Compare your result in (a).

(c) As in (a) and (b), find $G''(x)$ in terms of δ and δ' if

$$G(x) = \begin{cases} 3x + 1, & x > 0, \\ 2x - 4, & x < 0. \end{cases}$$

12. A BRIEF INTRODUCTION TO GREEN FUNCTIONS

Let's do some examples to see what a Green function is and how we can use it to solve ordinary differential equations. Also see Chapter 13, Section 8, for an application to partial differential equations. (You might find it interesting to read "The Green of Green Functions", Physics Today, December 2003, 41–46.)

Example 1. We reconsider the differential equation (11.2), namely

$$y'' + \omega^2 y = f(t), \quad y_0 = y_0' = 0 \tag{12.1}$$

where $f(t)$ is some given forcing function. Using (11.6), we can write

$$f(t) = \int_0^{\infty} f(t')\delta(t' - t)\,dt', \tag{12.2}$$

that is, we can think of the force $f(t)$ as (a limiting case of) a whole sequence of impulses. (You might reflect that, on the molecular level, air pressure is the force per unit area due to a tremendous number of impacts of individual molecules.) Now suppose that we have solved (12.1) with $f(t)$ replaced by $\delta(t' - t)$, that is, we find the response of the system to a unit impulse at t'. Let us call this response $G(t, t')$, that is, $G(t, t')$ is the solution of

$$\frac{d^2}{dt^2}G(t, t') + \omega^2 G(t, t') = \delta(t' - t). \tag{12.3}$$

Then, given some forcing function $f(t)$, we try to find a solution of (12.1) by "adding up" the responses of many such impulses. We shall show that this solution is

$$y(t) = \int_0^\infty G(t,t')f(t')\,dt'. \tag{12.4}$$

Substituting (12.4) into (12.1) and using (12.3) and (12.2), we find

$$\begin{aligned} y'' + \omega^2 y &= \left(\frac{d^2}{dt^2} + \omega^2\right) y = \left(\frac{d^2}{dt^2} + \omega^2\right) \int_0^\infty G(t,t')f(t')\,dt' \\ &= \int_0^\infty \left(\frac{d^2}{dt^2} + \omega^2\right) G(t,t')f(t')\,dt' = \int_0^\infty \delta(t'-t)f(t')\,dt' = f(t). \end{aligned}$$

Thus (12.4) is a solution of (12.1).

The function $G(t,t')$ is called a *Green function* (or Green's function). The Green function is the response of the system to a unit impulse at $t = t'$. Solving (12.3) with initial conditions $G = 0$ and $dG/dt = 0$ at $t = 0$, we find (Problem 1)

$$G(t,t') = \begin{cases} 0, & 0 < t < t', \\ \frac{1}{\omega}\sin\omega(t-t'), & 0 < t' < t. \end{cases} \tag{12.5}$$

Then (12.4) gives the solution of (12.1) with $y_0 = y_0' = 0$, namely

$$y(t) = \int_0^t \frac{1}{\omega}\sin\omega(t-t')f(t')\,dt'. \tag{12.6}$$

(The upper limit is $t' = t$ since $G = 0$ for $t' > t$.) Thus, given a forcing function $f(t)$, we can find the response $y(t)$ of the system (12.1) by integrating (12.6) (see Problems 2 to 5). Similarly for other differential equations we can find the solution in terms of an appropriate Green function (see Problems 6 to 8).

Example 2. As we will see later (Chapter 13, Section 8), in using Green functions in three-dimensional problems, we usually want a solution which is zero on the boundary of some region. In order to have a similar problem here, let us ask for a solution of

$$y'' + y = f(x) \tag{12.7}$$

such that $y = 0$ at $x = 0$ and at $x = \pi/2$. A physical interpretation of this problem may be useful. If a string is stretched along the x axis from $x = 0$ to $x = \pi/2$, and then caused to vibrate by a force proportional to $-f(x)\sin\omega t$, then $|y(x)|$ in (12.7) gives the amplitude of small vibrations.

We first find a solution of [compare (12.3)]

$$\frac{d^2}{dx^2}G(x,x') + G(x,x') = \delta(x'-x) \tag{12.8}$$

satisfying $G(0,x') = G(\pi/2,x') = 0$; this solution is the Green function for our problem. Then [compare (12.4)]

$$y(x) = \int_0^{\pi/2} G(x,x')f(x')\,dx' \tag{12.9}$$

gives a solution of (12.7) satisfying the conditions $y(0) = y(\pi/2) = 0$ (Problem 9).

To construct the desired Green function, we first note that for any $x \neq x'$, the equation (12.8) becomes

$$\frac{d^2}{dx^2}G(x,x') + G(x,x') = 0, \qquad x \neq x'. \tag{12.10}$$

The solutions of (12.10) are $\sin x$ and $\cos x$; we observe that $\sin x = 0$ at $x = 0$ and $\cos x = 0$ at $x = \pi/2$. Thus we try to find a Green function of the form

$$G(x,x') = \begin{cases} A(x')\sin x, & 0 < x < x' < \pi/2, \\ B(x')\cos x, & 0 < x' < x < \pi/2. \end{cases} \tag{12.11}$$

The next step may be clarified by thinking about the string problem. If the string is oscillated by a concentrated force at x' [see (12.8)], then the amplitude of the vibration given by (12.11) is shown in Figure 12.1. At $x = x'$, $G(x, x')$ is continuous, that is, from (12.11)

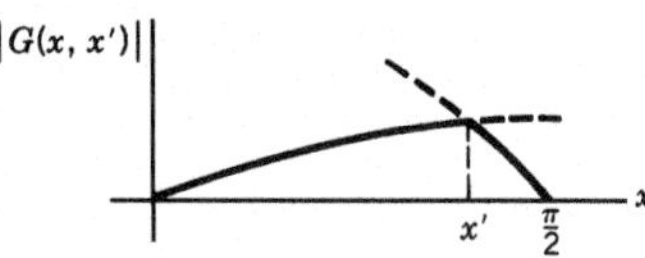

Figure 12.1

$$A(x')\sin x' = B(x')\cos x'. \tag{12.12}$$

However (see Figure 12.1), the slope changes abruptly at x'. From (12.11), we find

$$\frac{d}{dx}G(x,x') = \begin{cases} A(x')\cos x, & x < x', \\ -B(x')\sin x, & x > x'. \end{cases}$$

$$\text{Change in } \frac{dG}{dx} \text{ at } x' \text{ is } -B(x')\sin x' - A(x')\cos x'. \tag{12.13}$$

We can evaluate this change in dG/dx by integrating (12.8) from $x = x' - \epsilon$ to $x = x' + \epsilon$ and letting $\epsilon \to 0$. Since $\int d^2G/dx^2 = dG/dx$, we find

$$\left.\frac{dG}{dx}\right|_{x'-\epsilon}^{x'+\epsilon} + \int_{x'-\epsilon}^{x'+\epsilon} G(x,x')\,dx = \int_{x'-\epsilon}^{x'+\epsilon} \delta(x'-x)\,dx = 1,$$

or, letting $\epsilon \to 0$:

$$\left(\text{Change in slope } \frac{dG}{dx} \text{ at } x'\right) \text{ is } 1.$$

Then from (12.13)

$$-B(x')\sin x' - A(x')\cos x' = 1. \tag{12.14}$$

We solve (12.12) and (12.14) for $A(x')$ and $B(x')$ (Problem 10) and get

$$A(x') = -\cos x', \qquad B(x') = -\sin x'. \tag{12.15}$$

Thus we have

$$G(x,x') = \begin{cases} -\cos x' \sin x, & 0 < x < x' < \pi/2, \\ -\sin x' \cos x, & 0 < x' < x < \pi/2. \end{cases} \tag{12.16}$$

Then from (12.9), the solution of (12.7) with $y(0) = y(\pi/2) = 0$ is

$$y(x) = -\cos x \int_0^x (\sin x')f(x')\,dx' - \sin x \int_x^{\pi/2} (\cos x')f(x')\,dx'. \tag{12.17}$$

Example 3. If $f(x) = \csc x$, we find from (12.17):

$$y(x) = -\cos x \int_0^x \sin x' \csc x'\,dx' - \sin x \int_x^{\pi/2} \cos x' \csc x'\,dx'$$
$$= (-\cos x)(x) - (\sin x)(\ln \sin x')\Big|_x^{\pi/2} = -x\cos x + (\sin x)(\ln \sin x).$$

It is interesting to note that we can use the Green function method to obtain a particular solution of a nonhomogeneous differential equation (nonzero right-hand side) when we know the solutions of the corresponding homogeneous equation (zero right-hand side). (See Problems 14 to 18.) In (12.17) each integral gives a function of x minus a constant (from the constant limits); these constants times $\sin x$ and $\cos x$ give a solution of the homogeneous equation. Thus the remaining terms give a particular solution of the nonhomogeneous equation. We can write this particular solution in a simple form by changing $\int_x^{\pi/2}$ to $-\int_{\pi/2}^x$, dropping the constant limits and writing indefinite integrals. Then a particular solution $y_p(x)$ of (12.7) is given by

$$y_p(x) = -\cos x \int (\sin x) f(x)\,dx + \sin x \int (\cos x) f(x)\,dx. \tag{12.18}$$

Example 4. By the same methods used above, you can verify (Problem 14) that a solution of the differential equation

$$y'' + p(x)y' + q(x)y = f(x) \tag{12.19}$$

with $y(a) = y(b) = 0$ is given by

$$y(x) = y_2(x) \int_a^x \frac{y_1(x')f(x')}{W(x')}\,dx' + y_1(x) \int_x^b \frac{y_2(x')f(x')}{W(x')}\,dx', \tag{12.20}$$

where $y_1(x)$ and $y_2(x)$ are solutions of the homogeneous equation with $y_1(a) = 0$, $y_2(b) = 0$, and W is the Wronskian of $y_1(x)$ and $y_2(x)$ [See Chapter 3, equation (8.5)]. Just as in (12.18), we find that a particular solution y_p of (12.19) is

$$y_p(x) = y_2(x) \int \frac{y_1(x)f(x)}{W(x)}\,dx - y_1(x) \int \frac{y_2(x)f(x)}{W(x)}\,dx. \tag{12.21}$$

The particular solution (12.18) and (12.21) are exactly the same as those obtained by the method of *variation of parameters* (see Problem 14b) but the Green function method may seem less arbitrary.

PROBLEMS, SECTION 12

1. Solve (12.3) if $G = 0$ and $dG/dt = 0$ at $t = 0$ to obtain (12.5). *Hint:* Use *L*28 and *L*3 to find the inverse transform.

In Problems 2 and 3, use (12.6) to solve (12.1) when $f(t)$ is as given.

2. $f(t) = \sin \omega t$

3. $f(t) = e^{-t}$

4. Use equation (12.6) to solve Problem 10.18.

5. Obtain (12.6) by using the convolution integral to solve (12.1).

6. For Problem 10.17, show (as in Problem 1) that the Green function is

$$G(t,t') = \begin{cases} 0, & 0 < t < t', \\ (1/a)\sinh a(t-t'), & 0 < t' < t. \end{cases}$$

Thus write the solution of Problem 10.17 as an integral [similar to (12.6)] and evaluate it.

7. Use the Green function of Problem 6 to solve

$$y'' - a^2 y = e^{-t}, \qquad y_0 = y_0' = 0.$$

8. Solve the differential equation $y'' + 2y' + y = f(t)$, $y_0 = y_0' = 0$, where

$$f(t) = \begin{cases} 1, & 0 < t < a, \\ 0, & t > a. \end{cases}$$

As in Problems 6 and 7, find the Green function for the problem and use it in equation (12.4). Consider the cases $t < a$ and $t > a$ separately.

9. Following the proof of (12.4), show that (12.9) gives a solution of (12.7).

10. Solve (12.12) and (12.14) to get (12.15). *Hint:* Use Cramer's rule (Chapter 3, Section 3); note that the denominator determinant is the Wronskian [Chapter 3, equation (8.5)] of the functions $\sin x$ and $\cos x$.

In Problems 11 to 13, use (12.17) to find the solution of (12.7) with $y(0) = y(\pi/2) = 0$ when the forcing function is given $f(x)$.

11. $f(x) = \sin 2x$

12. $f(x) = \sec x$

13. $f(x) = \begin{cases} x, & 0 < x < \pi/4 \\ \pi/2 - x, & \pi/4 < x < \pi/2. \end{cases}$

Hint: Write separate formulas for $y(x)$ for $x < \pi/4$ and $x > \pi/4$.

14. (a) Given that $y_1(x)$ and $y_2(x)$ are solutions of (12.19) with $f(x) = 0$, and that $y_1(a) = 0$, $y_2(b) = 0$, find the Green function [as in (12.11) to (12.16)] and so obtain the solution (12.20). Then find the particular solution (12.21) as discussed for (12.18) and (12.21).

(b) The method of *variation of parameters* is an elementary way of finding a particular solution of (12.19) when you know the solutions of the homogeneous equation. Show as follows that this method leads to the same result (12.21) as the Green function method. Start with the known solution of the homogeneous equation, say $y = c_1y_1 + c_2y_2$ and allow the "constants" to be functions of x to be determined so that y satisfies (12.19). (The c's are the "parameters" which are to be "varied" in the expression "variation of parameters".) You want to find y' and y'' to substitute into (12.19). First find y' and set the sum of the terms involving derivatives of the c's equal to zero. Differentiate the rest of y' again to get y''. Now substitute y, y' and y'' into (12.19) and use the fact that y_1 and y_2 both satisfy the homogeneous equation [that is, (12.19) with $f(x) = 0$]. You should have the two equations:

$$\begin{aligned} c_1'y_1 + c_2'y_2 &= 0, \\ c_1'y_1' + c_2'y_2' &= f(x). \end{aligned}$$

Solve this pair of equations for c_1' and c_2' [say by determinants, and note that the denominator determinant is the Wronskian as in (12.20) and (12.21)]. Write the indefinite integrals for c_1 and c_2, and write $y = c_1y_1 + c_2y_2$ to get (12.21).

In Problems 15 to 18, use the given solutions of the homogeneous equation to find a particular solution of the given equation. You can do this either by the Green function formulas in the text or by the method of variation of parameters in Problem 14b.

15. $y'' - y = \operatorname{sech} x; \quad \sinh x, \cosh x$

16. $x^2y'' - 2xy' + 2y = x \ln x; \quad x, x^2$

17. $y'' - 2(\csc^2 x)y = \sin^2 x; \quad \cot x, 1 - x \cot x$

18. $(x^2 + 1)y'' - 2xy' + 2y = (x^2 + 1)^2; \quad x, 1 - x^2$

13. MISCELLANEOUS PROBLEMS

Identify each of the differential equations in Problems 1 to 24 as to type (for example, separable, linear first order, linear second order, etc.), and then solve it.

1. $x^2y' - xy = 1/x$

2. $x(\ln y)y' - y \ln x = 0$

3. $y''' + 2y'' + 2y' = 0$

4. $\dfrac{d^2r}{dt^2} - 6\dfrac{dr}{dt} + 9r = 0$

5. $(2x - y \sin 2x)\, dx = (\sin^2 x - 2y)\, dy$

6. $y'' + 2y' + 2y = 10e^x + 6e^{-x} \cos x$

7. $3x^3y^2y' - x^2y^3 = 1$

8. $x^2y'' - xy' + y = x$

9. $dy - (2y + y^2e^{3x})\, dx = 0$

10. $u(1 - v)\, dv + v^2(1 - u)\, du = 0$

11. $(y + 2x)\, dx - x\, dy = 0$

12. $xy'' + y' = 4x$

13. $y'' + 4y' + 5y = 26e^{3x}$

14. $y'' + 4y' + 5y = 2e^{-2x} \cos x$

15. $y'' - 4y' + 4y = 6e^{2x}$

16. $y'' - 5y' + 6y = e^{2x}$

17. $(2x + y)\, dy - (x - 2y)\, dx = 0$

18. $(x \cos y - e^{-\sin y})\, dy + dx = 0$

19. $\sin^2 x\, dy + [\sin^2 x + (x + y) \sin 2x]\, dx = 0$

20. $y'' - 2y' + 5y = 5x + 4e^x(1 + \sin 2x)$

21. $y' + xy = x/y$

22. $(D - 2)^2(D^2 + 9)y = 0$

23. $\sin\theta \cos\theta\, dr - \sin^2\theta\, d\theta = r \cos^2\theta\, d\theta$

24. $x(yy'' + y'^2) = yy'$ *Hint:* Let $u = yy'$.

In Problems 25 to 28, find a particular solution satisfying the given conditions.

25. $3x^2y\, dx + x^3\, dy = 0, \quad y = 2$ when $x = 1$.

26. $xy' - y = x^2, \quad y = 6$ when $x = 2$

27. $y'' + y' - 6y = 6, \quad y = 1, y' = 4$ when $x = 0$

28. $yy'' + y'^2 + 4 = 0 \quad y = 3, y' = 0$ when $x = 1$

29. If 10 kg of rock salt is placed in water, it dissolves at a rate proportional to the amount of salt still undissolved. If 2kg dissolve during the first 10 minutes, how long will it be until only 2kg remain undissolved?

30. A mass m falls under gravity (force mg) through a liquid whose viscosity is decreasing so that the retarding force is $-2mv/(1+t)$, where v is the speed of m. If the mass starts from rest, find its speed, its acceleration, and how far it has fallen (in terms of g) when $t = 1$.

31. The acceleration of an electron in the electric field of a positively charged sphere is inversely proportional to the square of the distance between the electron and the center of the sphere. Let an electron fall from rest at infinity to the sphere. What is the electron's velocity when it reaches the surface of the sphere?

32. Suppose that the rate at which you work on a hot day is inversely proportional to the excess temperature above 75°. One day the temperature is rising steadily, and you start studying at 2 p.m. You cover 20 pages the first hour and 10 pages the second hour. At what time was the temperature 75°?

33. Compare the temperatures of your cup of coffee at time t

(a) if you add cream and let the mixture cool;

(b) if you let the coffee and cream sit on the table and mix them at time t.

Hints: Assume Newton's law of cooling (Problem 2.27) for both coffee and cream (where it is a law of heating). Combine n' units of cream initially at temperature T_0' with n units of coffee initially at temperature T_0, and find the temperature at time t in (a) and in (b) assuming that the air temperature remains a constant T_a, and that the proportionality constant in the law of cooling is the same for both coffee and cream.

34. A flexible chain of length l is hung over a peg with one end of the chain slightly longer than the other. Assuming that the chain slides off with no friction, write and solve the differential equation of motion to show that $y = y_0 \cosh t\sqrt{2g/l}$, $0 < y < l/2$, where $2y$ is the difference in length of the two ends, and $y = y_0$ when $t = 0$.

35. A raindrop falls through a cloud, increasing in size as it picks up moisture. Assume that its shape always remains spherical. Also assume that the rate of increase of its volume with respect to distance fallen is proportional to the cross-sectional area of the drop at any time (that is, the mass increase $dm = \rho dV$ is proportional to the volume $\pi r^2\,dy$ swept out by the drop as it falls a distance dy). Show that the radius r of the drop is proportional to the distance y the drop has fallen if $r = 0$ when $y = 0$. Recall that when m is not constant, Newton's second law is properly stated as $(d/dt)(mv) = F$. Use this equation to find the distance y which the drop falls in time t under the force of gravity, if $y = \dot{y} = 0$ at $t = 0$. Show that the acceleration of the drop is $g/7$ where g is the acceleration of gravity.

36. (a) A rocket of (variable) mass m is propelled by steadily ejecting part of its mass at velocity u (constant with respect to the rocket). Neglecting gravity, the differential equation of the rocket is $m(dv/dm) = -u$ as long as $v \ll c$, $c =$ speed of light. Find v as a function of m if $m = m_0$ when $v = 0$.

(b) In the relativistic region (v/c not negligible), the rocket equation is

$$m\frac{dv}{dm} = -u\left(1 - \frac{v^2}{c^2}\right).$$

Solve this differential equation to find v as a function of m. Show that $v/c = (1-x)/(1+x)$, where $x = (m/m_0)^{2u/c}$.

37. The differential equation for the path of a planet around the sun (or any object in an inverse square force field) is, in polar coordinates,

$$\frac{1}{r^2}\frac{d}{d\theta}\left(\frac{1}{r^2}\frac{dr}{d\theta}\right) - \frac{1}{r^3} = -\frac{k}{r^2}.$$

Make the substitution $u = 1/r$ and solve the equation to show that the path is a conic section.

38. Use $L15$ and $L31$ to find the Laplace transform of $(1 - \cos at)/t$.

39. Use $L32$ and $L9$ to find the Laplace transform of $t \sinh at$. Verify your result by finding its inverse transform using the convolution integral.

Use the Laplace transform table to evaluate:

40. $\int_0^\infty t^3 e^{-4t} \sinh 2t\, dt$ **41.** $\sum_{n=0}^{\infty} (-1)^n \int_n^{n+1} te^{-2t}\, dt$

Find the inverse Laplace transform of:

42. $\frac{p}{(p+a)^3}$ **43.** $\frac{p^2}{(p^2+a^2)^2}$ **44.** $\frac{1}{(p^2+a^2)^3}$

45. Prove the following shifting or translation theorems for Fourier transforms. If $g(\alpha)$ is the Fourier transform of $f(x)$, then

(a) the Fourier transform of $f(x-a)$ is $e^{-i\alpha a} g(\alpha)$;

(b) the Fourier transform of $e^{i\beta x} f(x)$ is $g(\alpha - \beta)$.

Compare Problems 8.19 to 8.27.

46. Use the table of Laplace transforms to find the sine and cosine Fourier transforms of e^{-x}; of xe^{-x}.

Solve Problems 47 and 48 either by Laplace transforms and the convolution integral or by Green functions.

47. $y'' + y = \sec^2 t$ **48.** $y'' + y = t \sin t$

Table of Laplace Transforms

	$y = f(t),\ t > 0$ $[y = f(t) = 0,\ t < 0]$	$Y = L(y) = F(p) = \int_0^\infty e^{-pt} f(t)\,dt$	
$L1$	1	$\frac{1}{p}$	$\text{Re } p > 0$
$L2$	e^{-at}	$\frac{1}{p+a}$	$\text{Re } (p+a) > 0$
$L3$	$\sin at$	$\frac{a}{p^2+a^2}$	$\text{Re } p > \lvert\text{Im } a\rvert$
$L4$	$\cos at$	$\frac{p}{p^2+a^2}$	$\text{Re } p > \lvert\text{Im } a\rvert$
$L5$	$t^k,\ k > -1$	$\frac{k!}{p^{k+1}}$ or $\frac{\Gamma(k+1)}{p^{k+1}}$	$\text{Re } p > 0$
$L6$	$t^k e^{-at},\ k > -1$	$\frac{k!}{(p+a)^{k+1}}$ or $\frac{\Gamma(k+1)}{(p+a)^{k+1}}$	$\text{Re } (p+a) > 0$
$L7$	$\frac{e^{-at} - e^{-bt}}{b-a}$	$\frac{1}{(p+a)(p+b)}$	$\text{Re } (p+a) > 0$ $\text{Re } (p+b) > 0$
$L8$	$\frac{ae^{-at} - be^{-bt}}{a-b}$	$\frac{p}{(p+a)(p+b)}$	$\text{Re } (p+a) > 0$ $\text{Re } (p+b) > 0$
$L9$	$\sinh at$	$\frac{a}{p^2-a^2}$	$\text{Re } p > \lvert\text{Re } a\rvert$
$L10$	$\cosh at$	$\frac{p}{p^2-a^2}$	$\text{Re } p > \lvert\text{Re } a\rvert$
$L11$	$t \sin at$	$\frac{2ap}{(p^2+a^2)^2}$	$\text{Re } p > \lvert\text{Im } a\rvert$
$L12$	$t \cos at$	$\frac{p^2-a^2}{(p^2+a^2)^2}$	$\text{Re } p > \lvert\text{Im } a\rvert$
$L13$	$e^{-at} \sin bt$	$\frac{b}{(p+a)^2+b^2}$	$\text{Re } (p+a) > \lvert\text{Im } b\rvert$
$L14$	$e^{-at} \cos bt$	$\frac{p+a}{(p+a)^2+b^2}$	$\text{Re } (p+a) > \lvert\text{Im } b\rvert$
$L15$	$1 - \cos at$	$\frac{a^2}{p(p^2+a^2)}$	$\text{Re } p > \lvert\text{Im } a\rvert$
$L16$	$at - \sin at$	$\frac{a^3}{p^2(p^2+a^2)}$	$\text{Re } p > \lvert\text{Im } a\rvert$
$L17$	$\sin at - at \cos at$	$\frac{2a^3}{(p^2+a^2)^2}$	$\text{Re } p > \lvert\text{Im } a\rvert$

Table of Laplace Transforms (continued)

	$y = f(t),\ t > 0$ $[y = f(t) = 0,\ t < 0]$	$Y = L(y) = F(p) = \int_0^\infty e^{-pt} f(t)\, dt$	
*L*18	$e^{-at}(1 - at)$	$\dfrac{p}{(p+a)^2}$	Re $(p+a) > 0$
*L*19	$\dfrac{\sin at}{t}$	$\arctan \dfrac{a}{p}$	Re $p > \lvert \text{Im } a \rvert$
*L*20	$\dfrac{1}{t} \sin at \cos bt,$ $a > 0,\ b > 0$	$\dfrac{1}{2}\left(\arctan \dfrac{a+b}{p} + \arctan \dfrac{a-b}{p}\right)$	Re $p > 0$
*L*21	$\dfrac{e^{-at} - e^{-bt}}{t}$	$\ln \dfrac{p+b}{p+a}$	Re $(p+a) > 0$ Re $(p+b) > 0$
*L*22	$1 - \text{erf}\left(\dfrac{a}{2\sqrt{t}}\right), \quad a > 0$ (See Chapter 11, Section 9)	$\dfrac{1}{p} e^{-a\sqrt{p}}$	Re $p > 0$
*L*23	$J_0(at)$ (See Chapter 12, Section 12)	$(p^2 + a^2)^{-1/2}$	Re $p > \lvert \text{Im}\, a \rvert$; or Re $p \geq 0$ for real $a \neq 0$
*L*24	$u(t-a) = \begin{cases} 1, & t > a > 0 \\ 0, & t < a \end{cases}$ (unit step, or Heaviside function)	$\dfrac{1}{p} e^{-pa}$	Re $p > 0$
*L*25	$f(t) = u(t-a) - u(t-b)$ 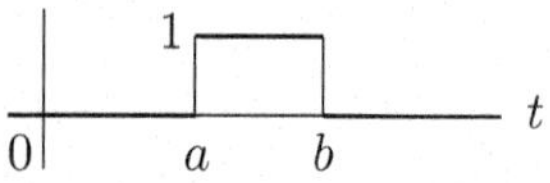	$\dfrac{e^{-ap} - e^{-bp}}{p}$	All p
*L*26	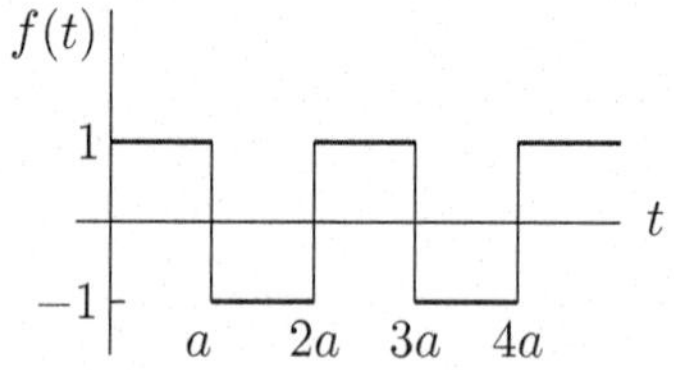	$\dfrac{1}{p} \tanh\left(\tfrac{1}{2}ap\right)$	Re $p > 0$
*L*27	$\delta(t-a),\ a \geq 0$ (See Section 11)	e^{-pa}	
*L*28	$f(t) = \begin{cases} g(t-a), & t > a > 0 \\ 0, & t < a \end{cases}$ $= g(t-a)u(t-a)$	$e^{-pa}G(p)$ $[G(p)$ means $L(g).]$	
*L*29	$e^{-at}g(t)$	$G(p+a)$	

Table of Laplace Transforms (continued)

	$y = f(t),\ t > 0$ $[y = f(t) = 0,\ t < 0]$	$Y = L(y) = F(p) = \int_0^\infty e^{-pt} f(t)\,dt$
*L*30	$g(at),\ a > 0$	$\frac{1}{a} G\left(\frac{p}{a}\right)$
*L*31	$\frac{g(t)}{t}$ (if integrable)	$\int_p^\infty G(u)\,du$
*L*32	$t^n g(t)$	$(-1)^n \frac{d^n G(p)}{dp^n}$
*L*33	$\int_0^t g(\tau)\,d\tau$	$\frac{1}{p} G(p)$
*L*34	$\int_0^t g(t-\tau)h(\tau)\,d\tau = \int_0^t g(\tau)h(t-\tau)\,d\tau$ (convolution of g and h, often written as $g * h$; see Section 10)	$G(p)H(p)$
*L*35	Transforms of derivatives of y (see Section 9): $L(y') = pY - y_0$ $L(y'') = p^2Y - py_0 - y_0'$ $L(y''') = p^3Y - p^2y_0 - py_0' - y_0''$, etc. $L(y^{(n)}) = p^nY - p^{n-1}y_0 - p^{n-2}y_0' - \cdots - y_0^{(n-1)}$	

CHAPTER 9

Calculus of Variations

1. INTRODUCTION

What is the shortest distance between two points? You probably laugh at such a simple question because you know the answer so well. Can you prove it? We shall see how to prove it shortly. Meanwhile we ask the same question about a sphere, for example, the earth. What is the shortest distance between two points on the surface of the earth, measured along the surface? Again you probably know that the answer is the distance measured along a great circle. But suppose you were asked the same question about some other surface, say an ellipsoid or a cylinder or a cone. The curve along a surface which marks the shortest distance between two neighboring points is called a *geodesic* of the surface. Finding geodesics is one of the problems which we can solve using the calculus of variations.

There are many others. To understand what the basic problem is, think about finding maximum and minimum values of $f(x)$ in ordinary calculus. You find $f'(x)$ and set it equal to zero. The values of x you find may correspond to maximum points ⌒, minimum points ⌣, or points of inflection with a horizontal tangent ⌐. Suppose that in solving a given physical problem you want the minimum values of a function $f(x)$. The equation $f'(x) = 0$ is a necessary (but not a sufficient) condition for an interior minimum point. To find the desired minimum, you would find all the values of x such that $f'(x) = 0$, and then rely on the physics or on further mathematical tests to sort out the minimum points. We use the general term *stationary point* to mean simply that $f'(x) = 0$ there; that is, stationary points include maximum points, minimum points, and points of inflection with horizontal tangent. In the calculus of variations, we often state problems by saying that a certain quantity is to be minimized. However, what we actually always do is something similar to putting $f'(x) = 0$, above; that is, we make the quantity stationary. The question of whether we have a maximum, a minimum, or neither, is, in general, a difficult mathematical problem (see calculus of variations texts) so we shall rely on the physics or geometry. Fortunately, in many applications, "stationary" is all that is required (Fermat's principle, Problems 1 to 3; Lagrange's equations, Section 5).

Now what is the quantity which we want to make stationary? It is an integral

$$I = \int_{x_1}^{x_2} F(x, y, y')\,dx \qquad \left(\text{where } y' = \frac{dy}{dx}\right), \tag{1.1}$$

and our problem is this: Given the points (x_1, y_1) and (x_2, y_2) and the form of the function F of x, y, and y', find the curve $y = y(x)$ (passing through the given points) which makes the integral I have the smallest possible value (or stationary value). Before we try to do this, let us look at several examples.

Example 1. Geodesics: Find the equation $y = y(x)$ of a curve joining two points (x_1, y_1) and (x_2, y_2) in the plane so that the distance between the points measured along the curve (arc length) is a minimum. Thus we want to minimize

$$I = \int_{x_1}^{x_2} \sqrt{1 + y'^2}\,dx; \tag{1.2}$$

this is equation (1.1) with $F(x, y, y') = \sqrt{1 + y'^2}$. See Section 2 and Example 3.4.

Example 2. The famous brachistochrone problem (from the Greek: *brachistos* = shortest, *chronos* = time, as in chronometer): Find the shape of a wire joining two given points so that a bead will slide down under gravity from one point to the other (without friction) in the shortest time. Here we must minimize $\int dt$. If ds is an element of arc length, then the velocity of the particle is $v = ds/dt$. Then we have

$$dt = \frac{1}{v}\,ds = \frac{1}{v}\sqrt{1 + y'^2}\,dx.$$

We shall see later that (using the law of conservation of energy) we can find v as a function of x and y. Then the integral which we want to minimize, namely

$$\int dt = \int \frac{1}{v}\sqrt{1 + y'^2}\,dx,$$

is of the form (1.1). See Section 4.

Example 3. Soap film problem: Suppose a soap film is suspended between two circular wire hoops as shown in Figure 1.1; what is the shape of the surface? It is clear from symmetry that it is a surface of revolution (neglecting gravity), and it is known that the soap film will adjust itself so that the surface area is a minimum. The surface area can be written as an integral and again our problem is to minimize an integral. See Section 4.

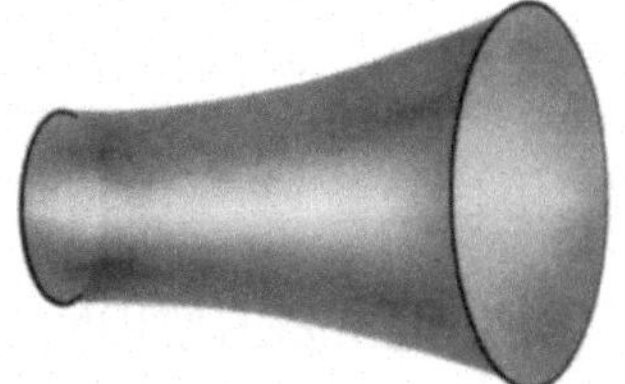

Figure 1.1

There are many other examples from physics. A chain suspended between two points hangs so that its center of gravity is as low as possible; the z coordinate of the center of gravity is given by an integral. Fermat's principle in optics says that light traveling between two given points follows the path requiring the least time. (This is a simple, but inaccurate, statement; we *should* say that $t = \int dt$ is stationary—there are examples where it is a maximum! See Problem 3). Various other basic principles in physics are stated in the form that certain integrals have stationary values.

PROBLEMS, SECTION 1

The speed of light in a medium of index of refraction n is $v = ds/dt = c/n$. Then the time of transit from A to B is $t = \int_A^B dt = c^{-1} \int_A^B n\, ds$. By Fermat's principle above, t is stationary. If the path consists of two straight line segments with n constant over each segment, then

$$\int_A^B n\, ds = n_1 d_1 + n_2 d_2,$$

and the problem can be done by ordinary calculus. Thus solve the following problems:

1. Derive the optical law of reflection. *Hint:* Let light go from the point $A = (x_1, y_1)$ to $B = (x_2, y_2)$ via an arbitrary point $P = (x, 0)$ on a mirror along the x axis. Set $dt/dx = (n/c)dD/dx = 0$, where $D =$ distance APB, and show that then $\theta = \phi$.

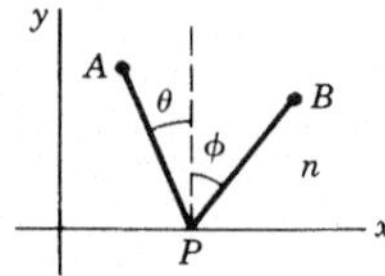

2. Derive Snell's law of refraction: $n_1 \sin\theta_1 = n_2 \sin\theta_2$ (see figure).

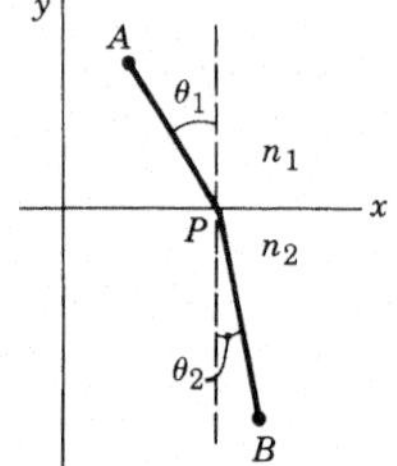

3. Show that the actual path is not necessarily one of minimum time. *Hint:* In the diagram, A is a source of light; CD is a cross section of a reflecting surface, and B is a point to which a light ray is to be reflected. APB is to be the actual path and $AP'B$, $AP''B$ represent varied paths. Then show that the varied paths:

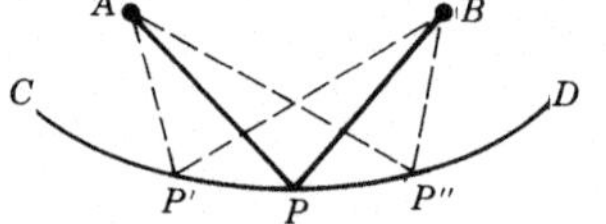

   (a) Are the same length as the actual path if CD is an ellipse with A and B as foci.

   (b) Are longer than the actual path if CD is a line tangent at P to the ellipse in (a).

   (c) Are shorter than the actual path if CD is an arc of a curve tangent to the ellipse at P and lying inside it. Note that in this case the time is a *maximum*!

   (d) Are longer on one side and shorter on the other if CD crosses the ellipse at P but is tangent to it (that is, CD has a point of inflection at P).

2. THE EULER EQUATION

Before we do the general problem, let us first do the problem of a geodesic on a plane; we shall show that a straight line gives the shortest distance between two points. (The reason for doing this is to clarify the theory; you will not do problems this way.) Our problem is to find $y = y(x)$ which will make

$$I = \int_{x_1}^{x_2} \sqrt{1 + y'^2}\, dx$$

as small as possible. The $y(x)$ which does this is called an *extremal.* Now we want some way to represent algebraically all the curves passing through the given end-points, but differing from the (as yet unknown) extremal by small amounts. (We assume that all the curves have continuous second derivatives so that we can carry out needed differentiations later.) These curves are called varied curves; there are infinitely many of them as close as we like to the extremal. We construct a function representing these varied curves in the following way (Figure 2.1). Let $\eta(x)$ represent a function of x which is zero at x_1 and x_2, and has a continuous second derivative in the interval x_1 to x_2, but is otherwise completely arbitrary. We define the function $Y(x)$ by the equation

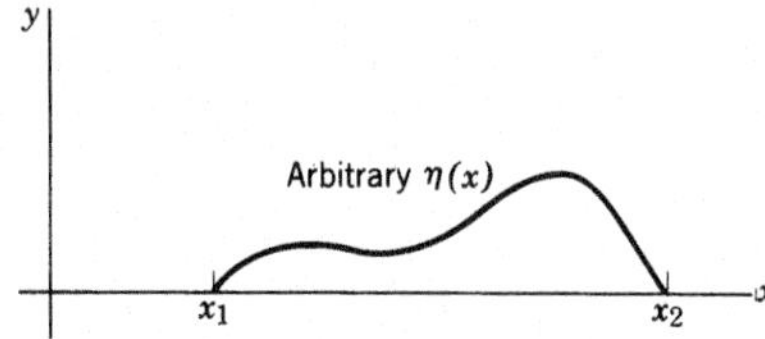

Figure 2.1

$$Y(x) = y(x) + \epsilon\eta(x), \tag{2.1}$$

where $y(x)$ is the desired extremal and ϵ is a parameter. Because of the arbitrariness of $\eta(x)$, $Y(x)$ represents any (single-valued) curve (with continuous second derivative) you want to draw through (x_1, y_1) and (x_2, y_2). Out of all these curves $Y(x)$ we want to pick the one curve that makes

$$I = \int_{x_1}^{x_2} \sqrt{1 + Y'^2}\, dx \tag{2.2}$$

a minimum. Now I is a function of the parameter ϵ; when $\epsilon = 0$, $Y = y(x)$, the desired extremal. Our problem then is to make $I(\epsilon)$ take its minimum value when $\epsilon = 0$. In other words, we want

$$\frac{dI}{d\epsilon} = 0 \quad \text{when } \epsilon = 0. \tag{2.3}$$

Differentiating (2.2) under the integral sign with respect to the parameter ϵ, we get

$$\frac{dI}{d\epsilon} = \int_{x_1}^{x_2} \frac{1}{2} \frac{1}{\sqrt{1 + Y'^2}} 2Y' \left(\frac{dY'}{d\epsilon}\right) dx. \tag{2.4}$$

Differentiating (2.1) with respect to x, we get

$$Y'(x) = y'(x) + \epsilon\eta'(x). \tag{2.5}$$

Then from (2.5) we have

$$\frac{dY'}{d\epsilon} = \eta'(x). \tag{2.6}$$

We see from (2.1) that putting $\epsilon = 0$ means putting $Y(x) = y(x)$. Then substituting (2.6) into (2.4) and putting $dI/d\epsilon$ equal to zero when $\epsilon = 0$, we get

$$\left(\frac{dI}{d\epsilon}\right)_{\epsilon=0} = \int_{x_1}^{x_2} \frac{y'(x)\eta'(x)}{\sqrt{1 + y'^2}}\, dx = 0. \tag{2.7}$$

We can integrate this by parts (since we assumed that η and y have continuous second derivatives). Let

$$u = y'/\sqrt{1 + y'^2}, \quad dv = \eta'(x)dx.$$

Then

$$du = \frac{d}{dx}\left(\frac{y'}{\sqrt{1+y'^2}}\right) dx, \quad v = \eta(x),$$

and

$$\left(\frac{dI}{d\epsilon}\right)_{\epsilon=0} = \frac{y'}{\sqrt{1+y'^2}}\eta(x)\Bigg|_{x_1}^{x_2} - \int_{x_1}^{x_2} \eta(x)\frac{d}{dx}\left(\frac{y'}{\sqrt{1+y'^2}}\right) dx.$$

The first term is zero because $\eta(x) = 0$ at the endpoints. In the second term, recall that $\eta(x)$ is an arbitrary function. This means that

$$\frac{d}{dx}\left(\frac{y'}{\sqrt{1+y'^2}}\right) = 0, \tag{2.8}$$

for otherwise we could select some function $\eta(x)$ so that the integral would not be zero. Notice carefully here that we are *not* saying that when an integral is zero, the integrand is also zero; this is not true (as, for example $\int_0^{2\pi} \sin x\, dx = 0$ shows). What we *are* saying is that the only way $\int_{x_1}^{x_2} f(x)\eta(x)\, dx$ can *always* be zero for *every* $\eta(x)$ is for $f(x)$ to be zero. You can prove this by contradiction in the following way. If $f(x)$ is not zero, then, since $\eta(x)$ is arbitrary, choose η to be positive where f is positive and negative where f is negative. Then $f\eta$ is positive, so its integral is not zero, in contradiction to the statement that $\int f\eta\, dx = 0$ for every η.

Integrating (2.8) with respect to x, we get

$$\frac{y'}{\sqrt{1+y'^2}} = \text{const.}$$

or $y' = \text{const.}$ Thus the slope of $y(x)$ is constant, so $y(x)$ is a straight line as we expected.

Now we *could* go through this process with every calculus of variations problem. It is much simpler to do the general problem once for all and find a differential equation which we can use to solve later problems. The problem is to find the y which will make stationary the integral

$$I = \int_{x_1}^{x_2} F(x, y, y')\, dx, \tag{2.9}$$

where F is a given function. The $y(x)$ which makes I stationary is called an extremal whether I is a maximum or minimum or neither. The method is the one we have just used with the straight line. We consider a set of varied curves

$$Y(x) = y(x) + \epsilon\eta(x)$$

just as before. Then we have

$$I(\epsilon) = \int_{x_1}^{x_2} F(x,\ Y,\ Y')\, dx, \tag{2.10}$$

and we want $(d/d\epsilon)I(\epsilon) = 0$ when $\epsilon = 0$. Remembering that Y and Y' are functions of ϵ, and differentiating under the integral sign with respect to ϵ, we get

$$\frac{dI}{d\epsilon} = \int_{x_1}^{x_2} \left(\frac{\partial F}{\partial Y}\frac{dY}{d\epsilon} + \frac{\partial F}{\partial Y'}\frac{dY'}{d\epsilon}\right) dx. \tag{2.11}$$

Substituting (2.1) and (2.5) into (2.11), we have

$$\frac{dI}{d\epsilon} = \int_{x_1}^{x_2} \left[\frac{\partial F}{\partial Y} \eta(x) + \frac{\partial F}{\partial Y'} \eta'(x) \right] dx. \tag{2.12}$$

We want $dI/d\epsilon = 0$ at $\epsilon = 0$; recall that $\epsilon = 0$ means $Y = y$. Then (2.12) gives

$$\left(\frac{dI}{d\epsilon}\right)_{\epsilon=0} = \int_{x_1}^{x_2} \left[\frac{\partial F}{\partial y} \eta(x) + \frac{\partial F}{\partial y'} \eta'(x) \right] dx = 0. \tag{2.13}$$

Assuming that y'' is continuous, we can integrate the second term by parts just as in the straight-line problem:

$$\int_{x_1}^{x_2} \frac{\partial F}{\partial y'} \eta'(x)\, dx = \frac{\partial F}{\partial y'} \eta(x)\bigg|_{x_1}^{x_2} - \int_{x_1}^{x_2} \frac{d}{dx}\left(\frac{\partial F}{\partial y'}\right) \eta(x)\, dx. \tag{2.14}$$

The integrated term is zero as before because $\eta(x)$ is zero at x_1 and x_2. Then we have

$$\left(\frac{dI}{d\epsilon}\right)_{\epsilon=0} = \int_{x_1}^{x_2} \left[\frac{\partial F}{\partial y} - \frac{d}{dx}\frac{\partial F}{\partial y'} \right] \eta(x)\, dx = 0. \tag{2.15}$$

As before, since $\eta(x)$ is arbitrary, we must have

$$\frac{d}{dx}\frac{\partial F}{\partial y'} - \frac{\partial F}{\partial y} = 0. \qquad \text{Euler equation} \tag{2.16}$$

This is the Euler (or Euler-Lagrange) equation.

Any problem in the calculus of variations, then, is solved by setting up the integral which is to be stationary, writing what the function F is, substituting it into the Euler equation, and solving the resulting differential equation.

Example. Let's find the geodesics in a plane again, this time using the Euler equation as you will do in problems.

We are to minimize

$$\int_{x_1}^{x_2} \sqrt{1 + y'^2}\, dx,$$

so we have $F = \sqrt{1 + y'^2}$. Then

$$\frac{\partial F}{\partial y'} = \frac{y'}{\sqrt{1 + y'^2}}, \qquad \frac{\partial F}{\partial y} = 0,$$

and the Euler equation gives

$$\frac{d}{dx}\left(\frac{y'}{\sqrt{1 + y'^2}}\right) = 0,$$

as we had in (2.8).

PROBLEMS, SECTION 2

Write and solve the Euler equations to make the following integrals stationary. In solving the Euler equations, the integrals in Chapter 5, Section 1, may be useful.

1. $\displaystyle\int_{x_1}^{x_2} \sqrt{x}\sqrt{1+y'^2}\,dx$

2. $\displaystyle\int_{x_1}^{x_2} \frac{ds}{x}$

3. $\displaystyle\int_{x_1}^{x_2} x\sqrt{1-y'^2}\,dx$

4. $\displaystyle\int_{x_1}^{x_2} x\,ds$

5. $\displaystyle\int_{x_1}^{x_2} (y'^2+y^2)\,dx$

6. $\displaystyle\int_{x_1}^{x_2} (y'^2+\sqrt{y}\,)\,dx$

7. $\int_{x_1}^{x_2} e^x\sqrt{1+y'^2}\,dx$ *Hint:* In the last integration, let $u=e^x$ and see Chapter 5, Problem 1.6.

8. $\displaystyle\int_{x_1}^{x_2} x\sqrt{y'^2+x^2}\,dx$

9. $\displaystyle\int_{x_1}^{x_2} (1+yy')^2\,dx$

10. $\displaystyle\int_{x_1}^{x_2} \frac{x^2\,dx}{xy'+1}$

3. USING THE EULER EQUATION

Other Variables We have used x and y as our variables. But the mathematics is just the same if we use some other letters, for example polar coordinates r and θ. To minimize (make stationary) the integral

$$\int F(r,\theta,\theta')\,dr \qquad \text{where} \qquad \theta'=d\theta/dr,$$

we solve the Euler equation

$$\frac{d}{dr}\left(\frac{\partial F}{\partial \theta'}\right)-\frac{\partial F}{\partial \theta}=0. \tag{3.1}$$

To minimize $\int F(t,x,\dot{x})\,dt$ where $\dot{x}=dx/dt$, we solve

$$\frac{d}{dt}\frac{\partial F}{\partial \dot{x}}-\frac{\partial F}{\partial x}=0. \tag{3.2}$$

Notice that the first derivative in the Euler equation [d/dx in (2.16), d/dr in (3.1), d/dt in (3.2)] is with respect to the integration variable in the integral. The partial derivatives are with respect to the other variable and its derivative [y and y' in (2.16), θ and θ' in (3.1), x and $\dot{x}$ in (3.2)].

▶ **Example 1.** Find the path followed by a light ray if the index of refraction (in polar coordinates) is proportional to r^{-2}. We want to make stationary

$$\int n\,ds \quad \text{or} \quad \int r^{-2}\,ds=\int r^{-2}\sqrt{dr^2+r^2\,d\theta^2}=\int r^{-2}\sqrt{1+r^2\theta'^2}\,dr.$$

The Euler equation is then (3.1) with $F = r^{-2}\sqrt{1+r^2\theta'^2}$. Since $\partial F/\partial\theta = 0$, we have

$$\frac{d}{dr}\left(\frac{r^{-2}r^2\theta'}{\sqrt{1+r^2\theta'^2}}\right) = 0 \qquad \text{or} \qquad \frac{\theta'}{\sqrt{1+r^2\theta'^2}} = \text{const.} = K.$$

Solve for θ' and integrate (see Chapter 5, Problem 1.5):

$$\begin{aligned}\theta'^2 &= K^2(1+r^2\theta'^2) \quad \text{so} \quad \theta'^2(1-K^2r^2) = K^2,\\ \theta' &= \frac{d\theta}{dr} = \frac{K}{\sqrt{1-K^2r^2}},\\ \theta &= \arcsin Kr + \text{const.}\end{aligned}$$

First Integrals of the Euler Equation In some problems the integrand F in I [see equation (1.1)] does not contain y (that is, F does not contain the dependent variable). Then $\partial F/\partial y = 0$ and the Euler equation becomes

$$\frac{d}{dx}\frac{\partial F}{\partial y'} = 0, \qquad \frac{\partial F}{\partial y'} = \text{const.}$$

This happened in the example and most of your problems in Section 2. Because $\partial F/\partial y$ was zero, we were able to integrate the Euler equation *once*; the equation $\partial F/\partial y' = \text{const.}$ is for this reason called a *first integral* of the Euler equation.

There is another less obvious case in which we can easily find a first integral of the Euler equation. Let us show this by an example (the soap film problem mentioned in Section 1).

Example 2. Our problem is this: Given two points P_1 and P_2 (not too far apart), we are going to draw a curve joining P_1 and P_2 and revolve it about the x axis to form a surface of revolution. We want the equation of the curve so that the surface area will be a minimum. That is, we want to minimize $I = \int 2\pi y\,ds$. We usually write $ds = \sqrt{1+y'^2}\,dx$. Instead, let us write $ds = \sqrt{1+x'^2}\,dy$, where $x' = dx/dy$. Then $I = \int 2\pi y\sqrt{1+x'^2}\,dy$. Recall from (3.1) and (3.2) and the discussion following them how to write the Euler equation in various sets of variables. Here y is the variable of integration, $F = y\sqrt{1+x'^2}$, and the Euler equation is

(3.3)
$$\frac{d}{dy}\frac{\partial F}{\partial x'} - \frac{\partial F}{\partial x} = 0.$$

Since $\partial F/\partial x = 0$, (3.3) becomes

$$\frac{d}{dy}\left(\frac{yx'}{\sqrt{1+x'^2}}\right) = 0.$$

This is the simplified equation we wanted. We integrate once, solve for x' and integrate again (see Chapter 5, Problem 1.3):

$$\frac{yx'}{\sqrt{1+x'^2}} = c_1,$$

$$x' = \frac{dx}{dy} = \frac{c_1}{\sqrt{y^2 - c_1{}^2}},$$

$$x = c_1 \cosh^{-1}\frac{y}{c_1} + c_2,$$

$$y = c_1 \cosh\frac{x - c_2}{c_1}.$$

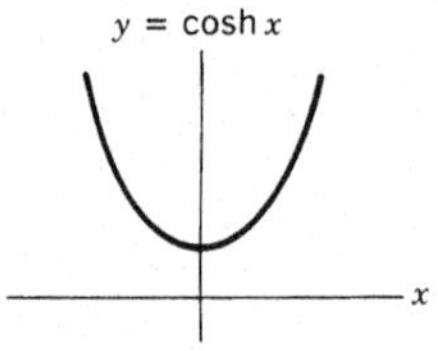

Figure 3.1

The graph of this equation is called a *catenary*; it is shown in Figure 3.1 for the special case $c_1 = 1$, $c_2 = 0$, $y = \cosh x = \frac{1}{2}(e^x + e^{-x})$. The catenary does not always give the solution to the soap film problem. If the given points (or the hoops in Figure 1.1) are too far apart, the soap film may break into two parts (circular films on the hoops). For further discussion see Courant and Robbins, Chapter 7, Section 11, and Arfken and Weber, Chapter 17. For another problem involving a catenary, see Problem 6.4.

Observe that the method used in this example will simplify any problem in which $I = \int F(y, y')\,dx$ does not have the independent variable x in the integrand. We change to y as the integration variable making the substitutions

$$(3.4) \qquad x' = \frac{dx}{dy} = \left(\frac{dy}{dx}\right)^{-1}, \quad y' = \frac{1}{x'}, \quad dx = \frac{dx}{dy}\,dy = x'\,dy$$

in I. Then the integrand is a function of y and x', so the Euler equation [now (3.3)] simplifies since $\partial F/\partial x = 0$. (See also Problem 8.1.)

Example 3. Find a first integral of the Euler equation to make stationary the integral

$$(3.5) \qquad I = \int \frac{\sqrt{1+y'^2}}{\sqrt{y}}\,dx.$$

Since x is missing in the integrand, we change to y as the integration variable; then by (3.4)

$$\sqrt{1+y'^2}\,dx = \sqrt{1+y'^2}\,x'\,dy = \sqrt{x'^2+1}\,dy,$$

$$I = \int \frac{\sqrt{x'^2+1}}{\sqrt{y}}\,dy = \int F(y, x')\,dy.$$

We see that $\partial F/\partial x = 0$; from (3.3) the Euler equation is

$$\frac{d}{dy}\left(\frac{\partial F}{\partial x'}\right) = \frac{d}{dy}\left(\frac{x'}{\sqrt{y}\sqrt{x'^2+1}}\right) = 0.$$

The first integral of the Euler equation is, then,

$$\frac{x'}{\sqrt{y}\sqrt{x'^2+1}} = \text{const.} \tag{3.6}$$

Example 4. Find the geodesics on the cone $z^2 = 8(x^2+y^2)$. Using cylindrical coordinates, we have $z^2 = 8r^2$, $z = r\sqrt{8}$, $dz = dr\sqrt{8}$, so

$$ds^2 = dr^2 + r^2 d\theta^2 + dz^2 = dr^2 + r^2 d\theta^2 + 8\,dr^2 = 9\,dr^2 + r^2 d\theta^2.$$

We want to minimize

$$I = \int ds = \int \sqrt{9\,dr^2 + r^2 d\theta^2} = \int \sqrt{9 + r^2\theta'^2}\,dr.$$

Note that we use r as the integration variable since the integrand contains r but not θ. Then $\partial F/\partial\theta = 0$, and we can immediately write a first integral of the Euler equation:

$$\frac{d}{dr}\left(\frac{\partial F}{\partial\theta'}\right) = 0, \qquad \frac{\partial F}{\partial\theta'} = \frac{r^2\theta'}{\sqrt{9+r^2\theta'^2}} = \text{const.} = K.$$

We solve for θ' and integrate again.

$$r^4\theta'^2 = K^2(9 + r^2\theta'^2),$$
$$\theta'^2(r^4 - K^2r^2) = 9K^2,$$
$$\int d\theta = \int \frac{3K\,dr}{r\sqrt{r^2-K^2}}.$$

From computer or tables (or see Chapter 5, Problem 1.6):

$$\theta + \alpha = 3\arccos\frac{K}{r} \qquad (\alpha = \text{const. of integration})$$

$$\cos\left(\frac{\theta+\alpha}{3}\right) = \frac{K}{r} \quad \text{or} \quad r\cos\left(\frac{\theta+\alpha}{3}\right) = K.$$

PROBLEMS, SECTION 3

Change the independent variable to simplify the Euler equation, and then find a first integral of it.

1. $\displaystyle\int_{x_1}^{x_2} y^{3/2}\,ds$

2. $\displaystyle\int_{x_1}^{x_2} \frac{\sqrt{1+y'^2}}{y^2}\,dx$

3. $\displaystyle\int_{y_1}^{y_2} \frac{x'^2}{\sqrt{x'^2+x^2}}\,dy$

4. $\displaystyle\int_{x_1}^{x_2} y\sqrt{y'^2+y^2}\,dx$

Write and solve the Euler equations to make the following integrals stationary. Change the independent variable, if needed, to make the Euler equation simpler.

5. $\displaystyle\int_{x_1}^{x_2} \sqrt{1+y^2y'^2}\,dx$

6. $\displaystyle\int_{x_1}^{x_2} \frac{yy'^2}{1+yy'}\,dx$

7. $\displaystyle\int_{x_1}^{x_2} (y'^2 + y^2)\,dx$

8. $\displaystyle\int_{\theta_1}^{\theta_2} \sqrt{r'^2 + r^2}\,d\theta,\quad r' = dr/d\theta$

9. $\displaystyle\int_{\phi_1}^{\phi_2} \sqrt{\theta'^2 + \sin^2\theta}\,d\phi,\quad \theta' = d\theta/d\phi$

10. $\displaystyle\int_{t_1}^{t_2} s^{-1}\sqrt{s^2 + s'^2}\,dt,\quad s' = ds/dt$

Use Fermat's principle to find the path followed by a light ray if the index of refraction is proportional to the given function.

11. $x + 1$

12. y^{-1}

13. $\sqrt{y}$

14. r^{-1}

15. Find the geodesics on a plane using polar coordinates.

16. Show that the geodesics on a circular cylinder (with elements parallel to the z axis) are helices $az + b\theta = c$, where a, b, c are constants depending on the given endpoints. (*Hint:* Use cylindrical coordinates.) Note that the equation $az + b\theta = c$ includes the circles $z =$ const. (for $b = 0$), straight lines $\theta =$ const. (for $a = 0$), and the special helices $az + b\theta = 0$.

17. Find the geodesics on the cone $x^2 + y^2 = z^2$. *Hint:* Use cylindrical coordinates.

18. Find the geodesics on a sphere. *Hints:* Use spherical coordinates with constant $r = a$. Choose your integration variable so that you can write a first integral of the Euler equation. For the second integration, make the change of variable $w = \cot\theta$. To recognize your result as a great circle, find, in terms of spherical coordinates θ and ϕ, the equation of intersection of the sphere with a plane through the origin.

4. THE BRACHISTOCHRONE PROBLEM; CYCLOIDS

We have already mentioned this problem in Section 1. We are given the points (x_1, y_1) and (x_2, y_2); we choose axes through the point 1 with the y axis positive downward as shown in Figure 4.1. Our problem is to find the curve joining the two points, down which a bead will slide (from rest) in the least time; that is, we want to minimize $\int dt$. Let $v = 0$ initially, and let $y = 0$ be our reference level for potential energy. Then at the point (x, y) we have

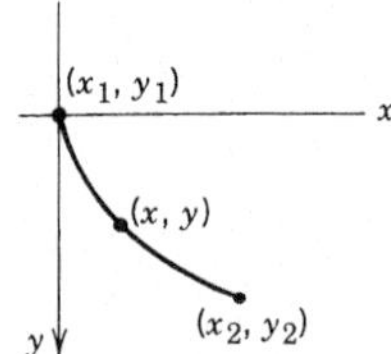

Figure 4.1

$$\text{kinetic energy} = \frac{1}{2}mv^2 = \frac{1}{2}m\left(\frac{ds}{dt}\right)^2,$$

$$\text{potential energy} = -mgy.$$

The sum of the two energies is zero initially and therefore zero at any time since the total energy is constant when there is no friction. Hence we have

$$\frac{1}{2}mv^2 - mgy = 0 \quad\text{or}\quad v = \sqrt{2gy}.$$

Then the integral which we want to minimize is

$$\int dt = \int \frac{ds}{v} = \int \frac{ds}{\sqrt{2gy}} = \frac{1}{\sqrt{2g}}\int_{x_1}^{x_2} \frac{\sqrt{1 + y'^2}}{\sqrt{y}}\,dx.$$

This is the integral (3.5) in Example 3, Section 3. Then the first integral of the Euler equation is given by (3.6):

$$\frac{x'}{\sqrt{y}\sqrt{x'^2+1}} = \sqrt{c}.$$

Solving for x', we get

$$x' = \frac{dx}{dy} = \sqrt{\frac{cy}{1-cy}}. \tag{4.1}$$

This simplifies if we let $cy = \sin^2\frac{\theta}{2} = \frac{1}{2}(1-\cos\theta)$. We find (Problem 1)

$$\begin{aligned} dx &= \frac{1}{c}\sin^2\frac{\theta}{2}\,d\theta = \frac{1}{2c}(1-\cos\theta)\,d\theta, \\ x &= \frac{1}{2c}(\theta-\sin\theta)+c'. \end{aligned} \tag{4.2}$$

The equations for x and y as functions of θ are parametric equations of the curve along which the particle slides in minimum time. Since we have chosen axes to make the curve pass through the origin, $x = y = 0$ must satisfy the equations of the curve, so $c' = 0$, and we have

$$\begin{aligned} x &= \frac{1}{2c}(\theta-\sin\theta), \\ y &= \frac{1}{2c}(1-\cos\theta). \end{aligned} \tag{4.3}$$

We shall now show that these are the parametric equations of a cycloid. Imagine a circle of radius a (say a wheel) in the (x, y) plane rolling along the x axis. Let it start tangent to the x axis at the origin O in Figure 4.2. Place a mark on the *circle* at O. As the circle rolls, the mark traces out a cycloid as shown in Figure 4.3. Let point P in Figure 4.2 be the position of the mark when the circle is tangent to the x axis at A; let (x, y) be the coordinates of P. Since the circle rolled, $OA = PA = a\theta$ with θ in radians. Then from Figure 4.2 we have

$$\begin{aligned} x &= OA - PB = a\theta - a\sin\theta = a(\theta-\sin\theta), \\ y &= AB = AC - BC = a - a\cos\theta = a(1-\cos\theta). \end{aligned} \tag{4.4}$$

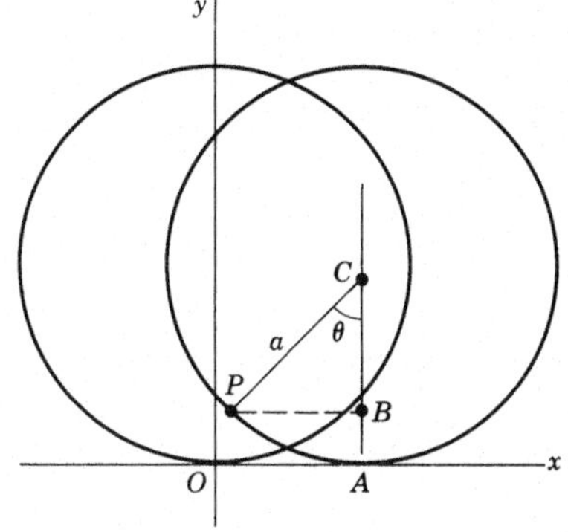

Figure 4.2

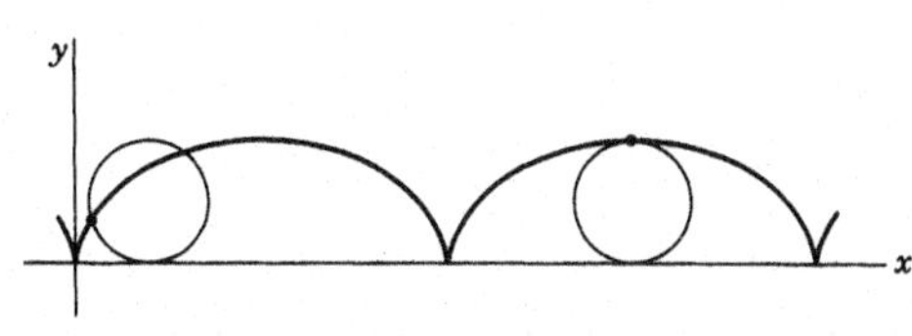

Figure 4.3

Equations (4.4) are the parametric equations of a cycloid. Comparing (4.3), we see that the brachistochrone is a cycloid as we claimed. Note that, since we have taken the y axis positive down (Figures 4.1 and 4.4), the circle which generates the brachistochrone rolls along the under side of the x axis.

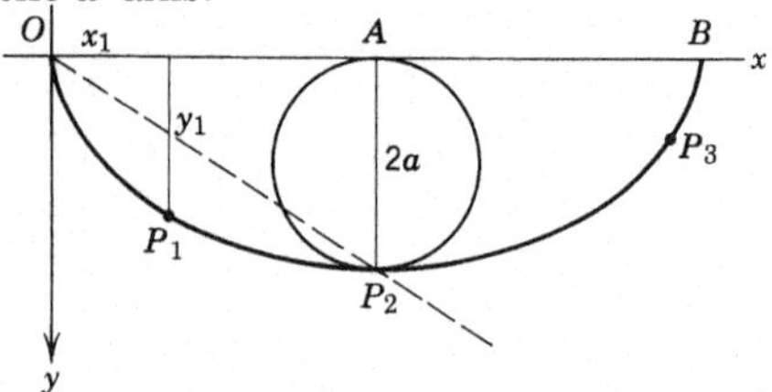

Figure 4.4

From either (4.3) or (4.4), we see that all cycloids are similar; that is they differ from each other only in size (determined by a or c) and not in shape. Figure 4.4 is a sketch of a cycloid for arbitrary a. If the given endpoints for the wire along which the bead slides are O and P_3, we see that the particle slides down to P_2 and back up to P_3 in minimum time! At point P_2 the circle has rolled halfway around so $OA = \frac{1}{2} \cdot 2\pi a = \pi a$. For any point P_1 on arc OP_2, P_1 is below the line OP_2, and the coordinates (x_1, y_1) of P_1 have

$$\frac{y_1}{x_1} > \frac{P_2 A}{AO} = \frac{2a}{\pi a} = \frac{2}{\pi}$$

or $x_1/y_1 < \pi/2$. For points like P_3 on P_2B, $x_3/y_3 > \pi/2$, whereas *at* P_2, we have $x_2/y_2 = \pi/2$ (Problem 2). Then if the right-hand endpoint is (x, y) and the origin is the left-hand endpoint, we can say that the bead just slides down, or slides down and back up, depending on whether x/y is less than or greater than $\pi/2$ (Problem 2).

▶ PROBLEMS, SECTION 4

1. Verify equations (4.2).

2. Show, in Figure 4.4, that for a point like P_3, $x_3/y_3 > \pi/2$ and for P_2, $x_2/y_2 = \pi/2$.

3. In the brachistochrone problem, show that if the particle is given an initial velocity $v_0 \neq 0$, the path of minimum time is still a cycloid.

4. Consider a rapid transit system consisting of frictionless tunnels bored through the earth between points A and B on the earth's surface (see figure). The unpowered passenger trains would move under gravity. Using polar coordinates, set up $\int dt$ to be minimized to find the path through the earth requiring the least time. See Chapter 6, Problem 8.21, for the potential inside the earth. Find a first integral of the Euler equation. Evaluate the constant of integration using $dr/d\theta = 0$ when $r = r_0$ (where r_0 is the deepest point of the tunnel—see figure). Now solve for $\theta' = d\theta/dr$ as a function of r. Substitute this into the integral for t and evaluate the integral to show that the transit time is

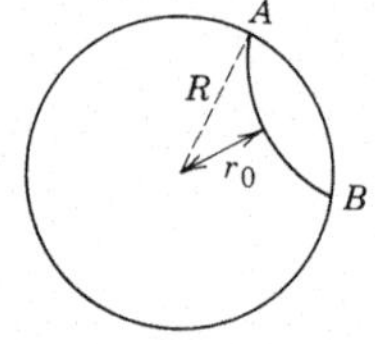

$$T = \pi\sqrt{\frac{R^2 - r_0{}^2}{gR}}. \quad \text{Hint: Find } 2\int_{r=r_0}^{R} dt.$$

Evaluate T for $r_0 = 0$ (path through the center of the earth—see Chapter 8, Problem 5.35); for $r_0 = 0.99R$. [For more detail, see Am. J. Phys. **34** 701–704 (1966).]

In Problems 5 to 7, use Fermat's principle to find the path followed by a light ray if the index of refraction is proportional to the given function.

5. $x^{-1/2}$ **6.** $(y - 1)^{-1/2}$ **7.** $(2x + 5)^{-1/2}$

5. SEVERAL DEPENDENT VARIABLES; LAGRANGE'S EQUATIONS

It is not necessary to restrict ourselves to problems with one dependent variable y. Recall that in ordinary calculus problems the necessary condition for a minimum point on $z = z(x)$ is $dz/dx = 0$; for a function of two variables $z = z(x, y)$, we have the two conditions $\partial z/\partial x = 0$ and $\partial z/\partial y = 0$. We have a somewhat analogous situation in the calculus of variations. Suppose that we are given an F which is a function of y, z, dy/dx, dz/dx, and x, and we want to find *two* curves $y = y(x)$ and $z = z(x)$ which make $I = \int F\,dx$ stationary. Then the value of the integral I depends on *both* $y(x)$ and $z(x)$ and you might very well guess that in this case we would have two Euler equations, one for y and one for z, namely

$$\begin{aligned} &\frac{d}{dx}\left(\frac{\partial F}{\partial y'}\right) - \frac{\partial F}{\partial y} = 0, \\ &\frac{d}{dx}\left(\frac{\partial F}{\partial z'}\right) - \frac{\partial F}{\partial z} = 0. \end{aligned} \tag{5.1}$$

By carrying through calculations similar to those we used in deriving the single Euler equation for the one dependent variable case you can show (Problem 1(a)) that this guess is correct. If there are still more dependent variables (but one independent variable), then we write an Euler equation for each dependent variable. It is also possible to consider a problem with more than one independent variable (see Problem 1(b)) or with F depending on y'' as well as x, y, y' (see Problem 1(c)).

There is a very important application of equations like (5.1) to mechanics. In elementary physics, Newton's second law $\mathbf{F} = m\mathbf{a}$ is a fundamental equation. In more advanced mechanics, it is often useful to start from a different assumption (which can be proved equivalent to Newton's law; see mechanics text books.) This assumption is called *Hamilton's principle.* It says that any particle or system of particles always moves in such a way that $I = \int_{t_1}^{t_2} L\,dt$ is stationary, where $L = T - V$ is called the *Lagrangian*; T is the kinetic energy, and V is the potential energy of the particle or system.

Example 1. Use Hamilton's principle to find the equations of motion of a single particle of mass m moving (near the earth) under gravity.

We first write the formulas for the kinetic energy T and the potential energy V of the particle. (It is convenient to use a dot to mean a derivative with respect to t just as we use a prime to indicate a derivative with respect to x; thus $dx/dt = \dot{x}$, $dy/dt = \dot{y}$, $dy/dx = y'$, $d^2x/dt^2 = \ddot{x}$, etc.) The equations for T, V, and $L = T - V$, are:

$$\begin{aligned} T &= \frac{1}{2}mv^2 = \frac{1}{2}m(\dot{x}^2 + \dot{y}^2 + \dot{z}^2), \\ V &= mgz, \\ L &= T - V = \frac{1}{2}m(\dot{x}^2 + \dot{y}^2 + \dot{z}^2) - mgz. \end{aligned} \tag{5.2}$$

Here t is the independent variable; x, y, and z are the dependent variables, and L corresponds to what we have called F previously. Then to make $I = \int_{t_1}^{t_2} L\,dt$ stationary, we write the corresponding Euler equations. There are three Euler equations, one for x, one for y, and one for z. The Euler equations are called *Lagrange's equations* in mechanics [see (5.3) next page].

$$\frac{d}{dt}\frac{\partial L}{\partial \dot{x}} - \frac{\partial L}{\partial x} = 0,$$
$$\frac{d}{dt}\frac{\partial L}{\partial \dot{y}} - \frac{\partial L}{\partial y} = 0, \qquad \text{Lagrange's equations} \tag{5.3}$$
$$\frac{d}{dt}\frac{\partial L}{\partial \dot{z}} - \frac{\partial L}{\partial z} = 0.$$

Substituting L in (5.2) into Lagrange's equations (5.3), we get

$$\begin{cases} \frac{d}{dt}(m\dot{x}) = 0 \\ \frac{d}{dt}(m\dot{y}) = 0 \\ \frac{d}{dt}(m\dot{z}) + mg = 0 \end{cases} \quad \text{or} \quad \begin{cases} \dot{x} = \text{const.}, \\ \dot{y} = \text{const.}, \\ \ddot{z} = -g. \end{cases} \tag{5.4}$$

These are just the familiar equations obtained from Newton's law; they say that in the gravitational field near the surface of the earth, the horizontal velocity is constant and the vertical acceleration is $-g$. In this problem you may say that it would have been simpler just to write the equations from Newton's law in the first place! This is true in simple cases, but in more complicated problems it may be much simpler to find one scalar function (that is, L) than to find six functions (that is, the components of the two vectors, force and acceleration). For example, the acceleration components in spherical coordinates are quite complicated to derive by elementary methods (see mechanics text books), but you should have no trouble deriving the equations of motion in polar, cylindrical or spherical coordinates using the Lagrangian. Let's do some examples.

▶ **Example 2.** Use Lagrange's equations to find the equations of motion of a particle in terms of the polar coordinate variables r and θ.

The element of arc length in polar coordinates is ds where

$$ds^2 = dr^2 + r^2\, d\theta^2. \tag{5.5}$$

The velocity of a moving particle is ds/dt; from (5.5) we get

$$v^2 = \left(\frac{ds}{dt}\right)^2 = \left(\frac{dr}{dt}\right)^2 + r^2\left(\frac{d\theta}{dt}\right)^2 = \dot{r}^2 + r^2\dot{\theta}^2. \tag{5.6}$$

The kinetic energy is $\frac{1}{2}mv^2$, so we have

$$T = \frac{1}{2}m(\dot{r}^2 + r^2\dot{\theta}^2),$$
$$L = T - V = \frac{1}{2}m(\dot{r}^2 + r^2\dot{\theta}^2) - V(r,\theta), \tag{5.7}$$

where $V(r, \theta)$ is the potential energy of the particle. Lagrange's equations in the variables r, θ are:

$$\begin{aligned} &\frac{d}{dt}\frac{\partial L}{\partial \dot{r}} - \frac{\partial L}{\partial r} = 0, \\ &\frac{d}{dt}\frac{\partial L}{\partial \dot{\theta}} - \frac{\partial L}{\partial \theta} = 0. \end{aligned} \tag{5.8}$$

Substituting L from (5.7) into (5.8), we get

$$\begin{aligned} \frac{d}{dt}(m\dot{r}) - mr\dot{\theta}^2 + \frac{\partial V}{\partial r} &= 0, \\ \frac{d}{dt}(mr^2\dot{\theta}) + \frac{\partial V}{\partial \theta} &= 0. \end{aligned} \tag{5.9}$$

The r equation of motion is, then,

$$m(\ddot{r} - r\dot{\theta}^2) = -\frac{\partial V}{\partial r}. \tag{5.10}$$

The θ equation is

$$m(r^2\ddot{\theta} + 2r\dot{r}\dot{\theta}) = -\frac{\partial V}{\partial \theta},$$

or, dividing by r,

$$m(r\ddot{\theta} + 2\dot{r}\dot{\theta}) = -\frac{1}{r}\frac{\partial V}{\partial \theta}. \tag{5.11}$$

Now the quantities $-\partial V/\partial r$ and $-(1/r)(\partial V/\partial \theta)$ are the components of the force ($\mathbf{F} = -\nabla V$) on the particle in the r and θ directions. (See Chapter 6.) Then equations (5.10) and (5.11) are just the components of $m\mathbf{a} = \mathbf{F}$; the acceleration components are then

$$\begin{aligned} a_r &= \ddot{r} - r\dot{\theta}^2, \\ a_\theta &= r\ddot{\theta} + 2\dot{r}\dot{\theta}. \end{aligned}$$

The second term in a_r is a familiar one; it is just the centripetal acceleration v^2/r when $v = r\dot{\theta}$ (the minus sign indicates that it is toward the origin). The second term in a_θ is called the Coriolis acceleration.

We show by an example another important point about Lagrange's equations.

Example 3. A mass m_1 moves without friction on the surface of the cone shown (Figure 5.1). Mass m_2 is joined to m_1 by a string of constant length; m_2 can move only vertically up and down. Find the Lagrange equations of motion of the system.

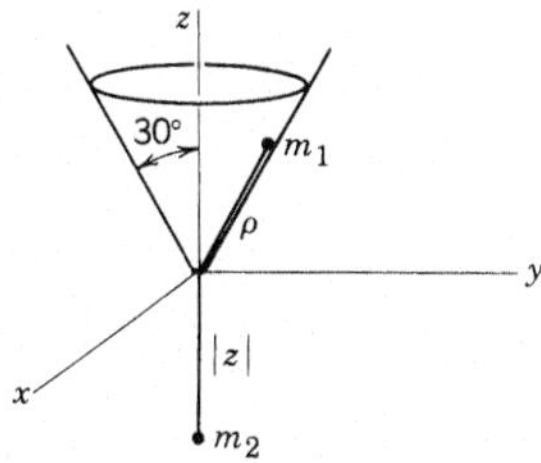

Figure 5.1

Let's use spherical coordinates ρ, θ, ϕ for m_1, and coordinate z for m_2. Then for m_1, $v^2 = (ds/dt)^2 = \dot{\rho}^2 + \rho^2\dot{\theta}^2 + \rho^2 \sin^2\theta\, \dot{\phi}^2$ [Chapter 5, equation (4.20)], and for m_2, $v^2 = \dot{z}^2$. The potential energy mgh of m_1 is $m_1 g\rho\cos\theta$ and of m_2 is $m_2 gz$. Note that we have used four variables: ρ, θ, ϕ, z; however, there are *not* four Lagrange equations We must use the equation of the cone ($\theta = 30°$) and the equation $\rho + |z| = l$ (string of constant length) to eliminate θ and either ρ or z. The Lagrangian L must always be written using the smallest possible number of variables (we say that we eliminate the constraint equations). Then, with $\theta = 30°$, $\sin\theta = \frac{1}{2}$, $\cos\theta = \frac{1}{2}\sqrt{3}$, $\dot{\theta} = 0$, and $z = -|z| = -(l - \rho)$, we find L in terms of ρ and ϕ:

$$L = \frac{1}{2}m_1(\dot{\rho}^2 + \rho^2\dot{\phi}^2/4) + \frac{1}{2}m_2\dot{\rho}^2 - \frac{1}{2}m_1 g\rho\sqrt{3} + m_2 g(l - \rho).$$

Thus the Lagrange equations are

$$\frac{d}{dt}(m_1\dot{\rho} + m_2\dot{\rho}) - m_1\rho\dot{\phi}^2/4 + \frac{1}{2}m_1 g\sqrt{3} + m_2 g = 0,$$

$$\frac{d}{dt}(m_1\rho^2\dot{\phi}/4) = 0 \qquad \text{or} \qquad \rho^2\dot{\phi} = \text{const.}$$

PROBLEMS, SECTION 5

1. (a) Consider the case of two dependent variables. Show that if $F = F(x, y, z, y', z')$ and we want to find $y(x)$ and $z(x)$ to make $I = \int_{x_1}^{x_2} F\,dx$ stationary, then y and z should each satisfy an Euler equation as in (5.1). *Hint:* Construct a formula for a varied path Y for y as in Section 2 [$Y = y + \epsilon\eta(x)$ with $\eta(x)$ arbitrary] and construct a similar formula for z [let $Z = z + \epsilon\zeta(x)$, where $\zeta(x)$ is *another* arbitrary function]. Carry through the details of differentiating with respect to ϵ, putting $\epsilon = 0$, and integrating by parts as in Section 2; then use the fact that *both* $\eta(x)$ and $\zeta(x)$ are arbitrary to get (5.1).

(b) Consider the case of two independent variables. You want to find the function $u(x, y)$ which makes stationary the double integral

$$\int_{y_1}^{y_2}\int_{x_1}^{x_2} F(u, x, y, u_x, u_y)\,dx\,dy.$$

Hint: Let the varied $U(x, y) = u(x, y) + \epsilon\eta(x, y)$ where $\eta(x, y) = 0$ at $x = x_1$, $x = x_2$, $y = y_1$, $y = y_2$, but is otherwise arbitrary. As in Section 2, differentiate with respect to ϵ, set $\epsilon = 0$, integrate by parts, and use the fact that η is arbitrary. Show that the Euler equation is then

$$\frac{\partial}{\partial x}\frac{\partial F}{\partial u_x} + \frac{\partial}{\partial y}\frac{\partial F}{\partial u_y} - \frac{\partial F}{\partial u} = 0.$$

(c) Consider the case in which F depends on x, y, y', and y''. Assuming zero values of the variation $\eta(x)$ and its derivative at the endpoints x_1 and x_2, show that then the Euler equation becomes

$$\frac{d^2}{dx^2}\frac{\partial F}{\partial y''} - \frac{d}{dx}\frac{\partial F}{\partial y'} + \frac{\partial F}{\partial y} = 0.$$

2. Set up Lagrange's equations in cylindrical coordinates for a particle of mass m in a potential field $V(r, \theta, z)$. *Hint:* $v = ds/dt$; write ds in cylindrical coordinates.

3. Do Problem 2 in spherical coordinates.

4. Use Lagrange's equations to find the equation of motion of a simple pendulum. (See Chapter 7, Problem 2.13.)

5. Find the equation of motion of a particle moving along the x axis if the potential energy is $V = \frac{1}{2}kx^2$. (This is a simple harmonic oscillator.)

6. A particle moves on the surface of a sphere of radius a under the action of the earth's gravitational field. Find the θ, ϕ equations of motion. (*Comment:* This is called a spherical pendulum. It is like a simple pendulum suspended from the center of the sphere, except that the motion is not restricted to a plane.)

7. Prove that a particle constrained to stay on a surface $f(x, y, z) = 0$, but subject to no other forces, moves along a geodesic of the surface. *Hint:* The potential energy V is constant, since constraint forces are normal to the surface and so do no work on the particle. Use Hamilton's principle and show that the problem of finding a geodesic and the problem of finding the path of the particle are identical mathematics problems.

8. Two particles each of mass m are connected by an (inextensible) string of length l. One particle moves on a horizontal table (assume no friction), The string passes through a hole in the table and the particle at the lower end moves up and down along a vertical line. Find the Lagrange equations of motion of the particles. *Hint:* Let the coordinates of the particle on the table be r and θ, and let the coordinate of the other particle be z. Eliminate one variable from L (using $r + |z| = l$) and write two Lagrange equations.

9. A mass m moves without friction on the surface of the cone $r = z$ under gravity acting in the negative z direction. Here r is the cylindrical coordinate $r = \sqrt{x^2 + y^2}$. Find the Lagrangian and Lagrange's equations in terms of r and θ (that is, eliminate z).

10. Do Example 3 above, using cylindrical coordinates for m_1. *Hint:* Use z_1 and z_2 for the z coordinates of m_1 and m_2. What is the equation of the cone in terms of r and z_1? Note that $r \neq \rho$, and θ in cylindrical coordinates is not the same as in spherical coordinates (see Chapter 5, Figures 4.4 and 4.5).

11. A yo-yo (as shown) falls under gravity. Assume that it falls straight down, unwinding as it goes. Find the Lagrange equation of motion. *Hints:* The kinetic energy is the sum of the translational energy $\frac{1}{2}m\dot{z}^2$ and the rotational energy $\frac{1}{2}I\dot{\theta}^2$ where I is the moment of inertia. What is the relation between $\dot{z}$ and $\dot{\theta}$? Assume the yo-yo is a solid cylinder with inner radius a and outer radius b.

12. Find the Lagrangian and Lagrange's equations for a simple pendulum (Problem 4) if the cord is replaced by a spring with spring constant k. *Hint:* If the unstretched spring length is r_0, and the polar coordinates of the mass m are (r, θ), the potential energy of the spring is $\frac{1}{2}k(r - r_0)^2$.

13. A particle moves without friction under gravity on the surface of the paraboloid $z = x^2 + y^2$. Find the Lagrangian and the Lagrange equations of motion. Show that motion in a horizontal circle is possible and find the angular velocity of this motion. Use cylindrical coordinates.

14. A hoop of mass M and radius a rolls without slipping down an inclined plane of angle α. Find the Lagrangian and the Lagrange equation of motion. *Hint:* The kinetic energy of a body which is both translating and rotating is a sum of two terms: the translational kinetic energy $\frac{1}{2}Mv^2$ where v is the velocity of the center of mass, and the rotational kinetic energy $\frac{1}{2}I\omega^2$ where ω is the angular velocity and I is the moment of inertia around the rotation axis through the center of mass.

15. Generalize Problem 14 to any mass M of circular cross section and moment of inertia I. Consider a hoop, a disk, a spherical shell, a solid spherical ball; order them as to which would first reach the bottom of the inclined plane. (For moments of inertia, see Chapter 5, Section 4.)

16. Find the Lagrangian and the Lagrange equation for the pendulum shown. The vertical circle is fixed. The string winds up or unwinds as the mass m swings back and forth. Assume that the unwound part of the string at any time is in a straight line tangent to the circle. Let l be the length of unwound string when the pendulum hangs straight down.

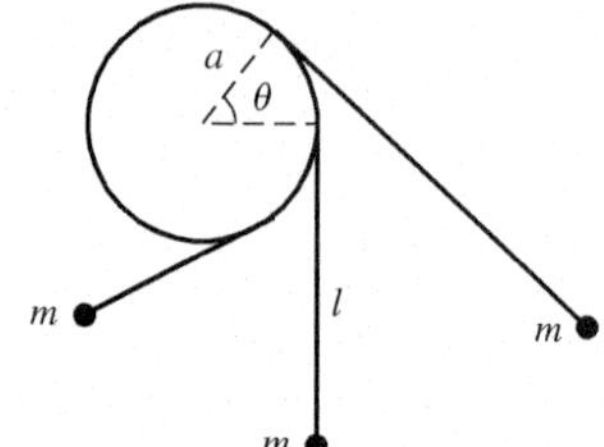

17. A simple pendulum (Problem 4) is suspended from a mass M which is free to move without friction along the x axis. The pendulum swings in the xz plane and gravity acts in the negative z direction. Find the Lagrangian and Lagrange's equations for the system.

18. A hoop of mass m in a vertical plane rests on a frictionless table. A thread is wound many times around the circumference of the hoop. The free end of the thread extends from the bottom of the hoop along the table, passes over a pulley (assumed weightless), and then hangs straight down with a mass m (equal to the mass of the hoop) attached to the end of the thread. Let x be the length of thread between the bottom of the hoop and the pulley, let y be the length of thread between the pulley and the hanging mass m, and let θ be the angle of rotation of the hoop about its center if the thread unwinds. What is the relation between x, y, and θ? Find the Lagrangian and Lagrange's equations for the system. If the system starts from rest, how does the hoop move?

For the following problems, use the Lagrangian to find the equations of motion and then refer to Chapter 3, Section 12.

19. For small vibrations, find the characteristic frequencies and the characteristic modes of vibration of the coupled pendulums shown. All motion takes place in a single vertical plane. Assume the spring unstretched when both pendulums hang vertically, and take the spring constant as $k = mg/l$ to simplify the algebra. *Hints*: Write the kinetic and potential energies in terms of the rectangular coordinates of the masses relative to their positions hanging at rest. Don't forget the gravitational potential energies. Then write the rectangular coordinates x and y in terms of θ and ϕ, and for small vibrations approximate $\sin\theta = \theta$, $\cos\theta = 1 - \theta^2/2$, and similar equations for ϕ.

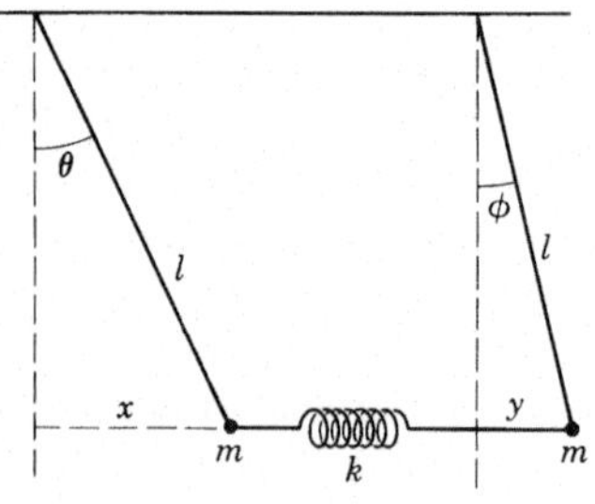

20. Do Problem 19 if the spring constant is $k = 3mg/l$.

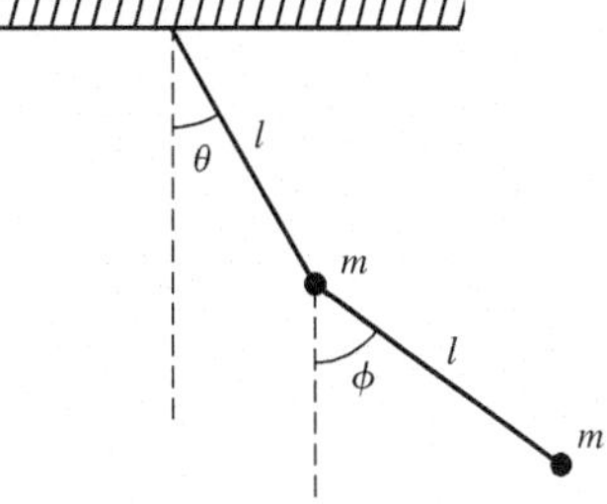

21. Find the Lagrangian and Lagrange's equations for the double pendulum shown. All motion takes place in a single vertical plane. *Hint*: See the hint in Problem 19.

22. Do Problem 21 if the two masses are different. Let m be the lower mass and let M be the sum of the two masses.

23. For small oscillations of the double pendulum in Problem 22, let $M = 4m$ and find the characteristic frequencies and characteristic modes of vibration.

24. Do Problem 23 if $M/m = 9/4$

25. Do Problem 23 in general, that is, in terms of the ratio M/m. *Hint*: You may find it helpful to use a single letter to represent $\sqrt{m/M}$, say $\alpha^2 = m/M$.

6. ISOPERIMETRIC PROBLEMS

Recall that in ordinary calculus we sometimes want to maximize a quantity subject to a condition (for example, find the volume of the largest box you can make with given surface area). Also recall that the method of Lagrange multipliers was useful in such problems (see Chapter 4, Section 9). There are similar problems in the calculus of variations. The original question which gave this class of problems its name was this: Of all the closed plane curves of given perimeter (isoperimetric = same perimeter), which one incloses the largest area? To solve this problem, we must maximize the area, $\int y\,dx$, subject to the condition that the arc, $\int ds$, is the given length l. In other words, we want to maximize an integral subject to the condition that another integral has a given (constant) value; any such problem is called an isoperimetric problem. Let

$$I = \int_{x_1}^{x_2} F(x, y, y')\,dx$$

be the integral we want to make stationary; at the same time,

$$J = \int_{x_1}^{x_2} G(x, y, y')\,dx,$$

with the same integration variable and the same limits, is to have a given constant value. (This means that the allowed varied paths must be paths for which J has the given value.) By using the Lagrange multiplier method, it can be shown that the desired condition is that

$$\int_{x_1}^{x_2} (F + \lambda G)\,dx$$

should be stationary, that is, that $F + \lambda G$ should satisfy the Euler equation. The Lagrange multiplier λ is a constant. It will appear in the solution $y(x)$ of the Euler equation; having found $y(x)$, we can substitute it into $\int_{x_1}^{x_2} G(x, y, y')\,dx = \text{const.}$ and so find λ if we like. However, for many purposes we do not need to find λ.

Example 1. Given two points x_1 and x_2 on the x axis, and an arc length $l > x_2 - x_1$, find the shape of the curve of length l joining the given points which, with the x axis, incloses the largest area.

We want to maximize $I = \int_{x_1}^{x_2} y\,dx$ subject to the condition $J = \int_{x_1}^{x_2} ds = l$. Here $F = y$ and $G = \sqrt{1+y'^2}$ so

$$F + \lambda G = y + \lambda\sqrt{1+y'^2}. \tag{6.1}$$

We want the Euler equation for $F + \lambda G$. Since

$$\frac{\partial}{\partial y'}(F+\lambda G) = \frac{\lambda y'}{\sqrt{1+y'^2}} \quad \text{and} \quad \frac{\partial}{\partial y}(F+\lambda G) = 1,$$

the Euler equation is

$$\frac{d}{dx}\left(\frac{\lambda y'}{\sqrt{1+y'^2}}\right) - 1 = 0. \tag{6.2}$$

The solution of (6.2) is (Problem 7):

$$(x+c)^2 + (y+c')^2 = \lambda^2 \tag{6.3}$$

We see that the answer to our problem is an arc of a circle passing through the two given points, and the Lagrange multiplier λ is the radius of the circle. The center and radius of the circle are determined by the given points x_1 and x_2, and the given arc length l (Problem 7).

PROBLEMS, SECTION 6

In Problems 1 and 2, given the length l of a curve joining two given points, find the equation of the curve so that:

1. The surface of revolution formed by rotating the curve about the x axis has minimum area.

2. The plane area between the curve and a straight line joining the points is a maximum.

3. Given 10 cc of lead, find how to form it into a solid of revolution of height 1 cm and minimum moment of inertia about its axis.

4. A uniform flexible chain of given length is suspended at given points (x_1, y_1) and (x_2, y_2). Find the curve in which it hangs. *Hint:* It will hang so that its center of gravity is as low as possible.

5. A curve $y = y(x)$, joining two points x_1 and x_2 on the x axis, is revolved around the x axis to produce a surface and a volume of revolution. Given the surface area, find the shape of the curve $y = y(x)$ to maximize the volume. *Hint*: You should find a first integral of the Euler equation of the form $yf(y, x', \lambda) = C$. Since $y = 0$ at the endpoints, $C = 0$. Then either $y = 0$ for all x, or $f = 0$. But $y \equiv 0$ gives zero volume of the solid of revolution, so for maximum volume you want to solve $f = 0$.

6. In Problem 5, given the volume, find the shape of the curve $y = y(x)$ to minimize the surface area. *Hint:* See the hint in Problem 5.

7. Integrate (6.2), simplify the result and integrate again to get (6.3) where c and c' are constants of integration. If $x_1 = -\sqrt{3}$, $x_2 = \sqrt{3}$, and $l = 4\pi/3$, show that the center and radius of the circle are $(0, -1)$ and $\lambda =$ radius $= 2$.

7. VARIATIONAL NOTATION

The symbol δ was used in the early days of the development of the calculus of variations to indicate what we have called differentiation with respect to the parameter ϵ. It is just like the symbol d in a differential except that it warns you that ϵ and not x is the differentiation variable. The δ notation is not used much any more in mathematics, but you will find it in applications and so should understand its meaning. The quantity δI is just the differential

$$\delta I = \frac{dI}{d\epsilon} d\epsilon,$$

where $dI/d\epsilon$ is evaluated for $\epsilon = 0$. The symbol δ (read "the variation of") is also treated as a differential operator acting on F, y, and y'; we shall define δy, $\delta y'$, and δF in terms of our previous notation. We had in Section 2:

$$\begin{aligned} Y(x,\epsilon) &= y(x) = \epsilon\eta(x), \\ Y'(x,\epsilon) &= y'(x) + \epsilon\eta'(x). \end{aligned} \tag{7.1}$$

Then the meaning of δy is

$$\delta y = \left(\frac{\partial Y}{\partial \epsilon}\right)_{\epsilon=0} d\epsilon = \eta(x) d\epsilon; \tag{7.2}$$

this is just like a differential dY if ϵ is the variable. The meaning of $\delta y'$ is

$$\delta y' = \left(\frac{\partial Y'}{\partial \epsilon}\right)_{\epsilon=0} d\epsilon = \eta'(x) d\epsilon. \tag{7.3}$$

This is identical with

$$\frac{d}{dx}(\delta y) = \frac{d}{dx}[\eta(x) d\epsilon] = \eta'(x) d\epsilon \tag{7.4}$$

since x and ϵ are independent variables; in other words, d and δ commute. The meaning of δF is

$$\delta F = \frac{\partial F}{\partial y}\delta y + \frac{\partial F}{\partial y'}\delta y'; \tag{7.5}$$

this is just a total differential $dF = (\partial F/\partial \epsilon)_{\epsilon=0} d\epsilon$ of the function $F[x, Y(x,\epsilon), Y'(x,\epsilon)]$ at $\epsilon = 0$ with ϵ considered the only variable. Then the variation in I is

$$\begin{aligned} \delta I &= \delta \int_{x_1}^{x_2} F\,dx = \int_{x_1}^{x_2} \delta F\,dx \\ &= \int_{x_1}^{x_2} \left(\frac{\partial F}{\partial y}\delta y + \frac{\partial F}{\partial y'}\delta y'\right) dx \\ &= \int_{x_1}^{x_2} \left[\frac{\partial F}{\partial y}\eta(x) d\epsilon + \frac{\partial F}{\partial y'}\eta'(x) d\epsilon\right] dx. \end{aligned} \tag{7.6}$$

If you compare (7.6) with (2.13), you find that the following two statements about $I = \int F(x, y, y') dx$ mean the same thing:

(a) I is stationary; that is, $dI/d\epsilon = 0$ at $\epsilon = 0$ as in (2.13).

(b) The variation of I is zero; that is, $\delta I = 0$ as in (7.6).

8. MISCELLANEOUS PROBLEMS

1. (a) In Section 3, we showed how to obtain a first integral of the Euler equation when $F = F(y, y')$. There is an alternative method of handling this case. You can show that if $F = F(y, y')$, then $F - y'\partial F/\partial y' = \text{const}$. To prove this, differentiate the left-hand side with respect to x, and show that the result is zero if F satisfies the Euler equation. Note that what you have is a first integral of the Euler equation.

(b) Use the method of (a) to do the problems at the end of Section 3.

(c) Consider the motion of a particle along the x axis; then $L = T - V = \frac{1}{2}m\dot{x}^2 - V(x)$. Note that L does not contain the independent variable t; this corresponds to the case $F = F(y, y')$ in (a). Show that the first integral found in (a) is just the equation of conservation of energy for the mechanics problem.

Find a first integral of the Euler equation to make stationary the integrals in Problems 2 to 4.

2. $\displaystyle\int_a^b \frac{x^2\,dy}{\sqrt{1+x'^2}}$ **3.** $\displaystyle\int_a^b \frac{yy'^2\,dx}{\sqrt{1+y'^2}}$ **4.** $\displaystyle\int_\alpha^\beta \sqrt{r^2r'^2+r^4}\,d\theta$

Write and solve the Euler equations to make stationary the integrals in Problems 5 to 7.

5. $\displaystyle\int_a^b \sqrt{\frac{y'^2}{y^2}+1}\,dx$ **6.** $\displaystyle\int_a^b \frac{\sqrt{1+y'^2}}{1+y}\,dx$ **7.** $\displaystyle\int_a^b \sqrt{1+x}\sqrt{1+x'^2}\,dy$

8. Find the geodesics on the cylinder $r = 1 + \cos\theta$.

9. Find the geodesics on the cone $z = r\cot\alpha$, where $r^2 = x^2 + y^2$.

10. Find the geodesics on the parabolic cylinder $y = x^2$.

In Problems 11 to 18, use Fermat's principle to find the path of a light ray through a medium of index of refraction proportional to the given function.

11. $r^{-1/2}$ **12.** e^y **13.** $(2x+3)^{-1}$ **14.** $(y+2)^{1/2}$

15. $x^{1/3}$ *Hint:* In the last integration, let $x = u^3$.

16. r *Hint:* In the last integration, let $u = r^2$.

17. $r^{-1}\ln r$ *Hint:* In the last integration, let $u = \ln r$.

18. $(x+y)^{1/2}$ *Hint:* Make the change of variables (45° rotation)

$$X = \frac{1}{\sqrt{2}}(x+y), \quad Y = \frac{1}{\sqrt{2}}(x-y); \quad \text{what is} \quad dX^2 + dY^2?$$

19. Find Lagrange's equations in polar coordinates for a particle moving in a plane if the potential energy is $V = \frac{1}{2}kr^2$.

20. Repeat Problem 19 if $V = -K/r$.

21. Write Lagrange's equations in cylindrical coordinates for a particle moving in the gravitational field $V = mgz$.

22. In spherical coordinates, find the θ Lagrange equation for a particle moving in the potential field $V = V(r, \theta, \phi)$. What is the θ component of the acceleration? *Hint:* The θ Lagrange equation is the θ component of $m\mathbf{a} = \mathbf{F} = -\nabla V$; for components of ∇V, see Chapter 6, end of Section 6, or Chapter 10, Section 9.

23. A particle slides without friction around a vertical circle under the force of gravity. Set up the Lagrange equation of motion.

24. Write and simplify the Euler equation to make stationary the integral

$$\int_a^b [P(x,y) + Q(x,y)y'] \, dx.$$

Show that if the Euler equation is satisfied, the integral has the same value for *all* paths joining a and b. (See Problem 1.3. Also see Chapter 6, Section 9, Example 2.)

25. Find the shape of a curve of minimum length which will inclose a given area A lying in the (x, y) plane. Find the length in terms of A.

26. A wire carrying a uniform distribution of positive charge lies in the (x, y) plane and joins two given points. Find its shape to minimize the electrostatic potential at the origin.

27. Find a first integral of the Euler equation for Problem 26 if the length of the wire is given.

28. Write the θ Lagrange equation for a particle moving in a plane if $V = V(r)$ (that is, a central force). Use the θ equation to show that:

(a) The angular momentum $\mathbf{r} \times m\mathbf{v}$ is constant.

(b) The vector $\mathbf{r}$ sweeps out equal areas in equal times (Kepler's second law).

CHAPTER 10

Tensor Analysis

1. INTRODUCTION

You already know something about tensors although you may not have used the term tensor. Tensors of *rank* (or *order*) zero are just scalars and tensors of rank one are just vectors; you are already familiar with these. In 3-dimensional space a scalar has $3^0 = 1$ components and a vector has $3^1 = 3$ components; a second-rank tensor has $3^2 = 9$ components; and in general a tensor of rank n has 3^n components. After scalars and vectors, second-rank tensors are the ones you are most likely to find in applications, so let's consider an example of such a tensor.

Example 1. Think of a beam carrying a load; there are stresses and strains in the material of the beam. If we imagine cutting the beam in two by a plane perpendicular to the x direction, we realize that there is a force per unit area exerted *by* the material on one side of our imaginary cut *on* the other side. This is a vector, so it has three components P_{xx}, P_{xy}, P_{xz}, where the first subscript x is to emphasize that this is a force across a plane perpendicular to the x direction. Similarly, if we consider a plane perpendicular to the y direction, there is a force per unit area across this plane with components P_{yx}, P_{yy}, P_{yz}; and finally across a plane perpendicular to the z direction there is a force per unit area with components P_{zx}, P_{zy}, P_{zz}. At a point in the material, then, we have a set of nine quantities which could be displayed as a matrix:

$$\begin{pmatrix} P_{xx} & P_{xy} & P_{xz} \\ P_{yx} & P_{yy} & P_{yz} \\ P_{zx} & P_{zy} & P_{zz} \end{pmatrix} \tag{1.1}$$

This is a second-rank tensor known as the stress tensor. The forces (per unit area) P_{xx}, P_{yy}, P_{zz} are pressures or tensions; the others are shear forces (per unit area). For example P_{zy} is a force per unit area in the y direction acting across a plane perpendicular to the z direction; this force tends to shear the beam.

So far, we have simply indicated the number of components that tensors of the various ranks have. This is not the whole story. To see what else is required, let us talk about first-rank tensors, that is vectors, which are already familiar to you. In elementary work a vector is usually defined either as a magnitude and a direction, or as a set of three components. To see that we need to give a more careful definition, consider this example.

Example 2. We can draw an arrow to represent a given rotation of a rigid body in the following way. Draw the arrow along the axis of rotation, make its length equal to the rotation angle in radians, and let its sense be given by the right-hand rule. Then, apparently, a rotation is a vector according to the magnitude and direction definition. But this is not so! Take a book and rotate it 90° about the x axis, then 90° about the y axis. (See Chapter 3, Problem 7.31.) Repeat, rotating this time first about the y axis and then about the x axis. The final positions of the book are different. But the sum of the two vectors does not depend on the order in which they are added (in mathematical language, vector addition is commutative). The arrows associated with rotations are not vectors.

Example 3. Now let us consider the idea of a vector as a set of three components. In order to talk about components, we must have a coordinate system. There are infinitely many coordinate systems—even for rectangular axes (x, y, z) there are infinitely many sets of rotated axes. Thus we must say that a vector consists of a set of three components *in each coordinate system.* If the components of a vector relative to one set of axes are given, we know from elementary vector analysis that the component of the vector in any direction, or its components relative to any rotated set of axes, can be found by taking projections. Then the new components are definite combinations of the old components. This fact allows us to decide whether a physical quantity is really a vector or not. There is a similar requirement for tensors, for example the second-rank stress tensor we have described. We could imagine cutting the beam by a plane oriented in any given direction and ask for the force per unit area acting across this plane. It can be shown (see Section 7) that each component of this force is a certain combination of the nine components of the stress tensor (1.1). Thus the components of the stress tensor in any other coordinate system are definite combinations of the nine components of the tensor relative to the (x, y, z) axes. In other words, tensors of all ranks, like vectors, have a physical meaning which is independent of the reference coordinate system and there are definite mathematical laws which relate their components in two systems.

You may wonder why we cannot make just any set of components (3 for a vector, 9 for a second-rank tensor, etc.), given in *one* coordinate system, a tensor by *defining* its components in other systems by the correct transformation laws. Mathematically, we could! But for a physical entity, we are not free to define its components in various coordinate systems; they are determined by physical fact. We merely give a mathematical description of the entity and identify it as a scalar, a vector, a second-rank tensor, etc. (or perhaps none of these). We can see again now why an arrow associated with a rotation is not a vector. If we treat the arrow as a vector and take components of it, these component vectors do not represent rotations which can be combined to give the original rotation. Thus a vector which looks superficially like the arrow we have defined is not a correct mathematical representation of the physical entity (a rotation) we are trying to describe.

What is the relationship between the vectors we are going to define here and the vectors of a linear vector space (Chapter 3, Sections 10 and 14)? The ideas of an abstract vector space grew out of the geometry of three-dimensional displacement vectors. A change of coordinate system (for example, rotation of axes) corresponds to a change of basis in a vector space. Because the definitions of a vector space are set up to parallel the geometry, displacement vectors are vectors in the vector space sense. Whether other physical entities (force, temperature, stress, etc.) are properly modeled as vectors then depends on whether they transform under a change of coordinate system (that is, change of basis) in the same way that displacement vectors do. Here we want the word "vector" to refer to all physical quantities which transform properly. Thus we shall find the transformation law for a displacement vector, and then define a vector as any entity which obeys the same law.

A tensor which transforms properly under a rotation of rectangular (x, y, z) axes is called a Cartesian tensor; we will study these in some detail. Things become a little more complicated when we consider transformations to other coordinate systems such as spherical coordinates; we will consider this at the end of the chapter. But for Sections 1 to 7, the term tensor will mean Cartesian tensor.

2. CARTESIAN TENSORS

In Chapter 3, Section 7, we considered the effect of rotations on vectors, and emphasized active rotations (vector rotated, axes fixed). Now we want to consider passive rotations (vector fixed, axes rotated), in order to find how the components of a displacement vector in one coordinate system are related to its components in a rotated system. Let (x, y, z) be a set of rectangular axes and (x', y', z') another set obtained by rotating the axes in any manner keeping the origin fixed (Figure 2.1). In the table (2.1), we list the cosines of the nine angles between the (x, y, z) axes and the (x', y', z') axes.

(2.1)

	x	y	z
x'	l_1	m_1	n_1
y'	l_2	m_2	n_2
z'	l_3	m_3	n_3

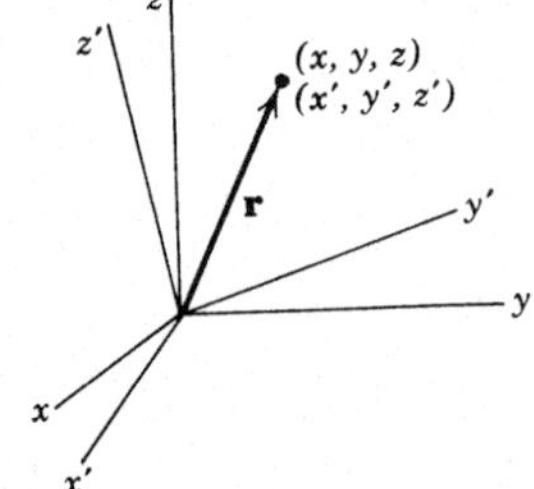

Figure 2.1

In the table, l_2 means the cosine of the angle between the x axis and the y' axis, etc. A vector $\mathbf{r}$ (Figure 2.1) has components x, y, z or x', y', z' relative to the two coordinate systems; we want to find the relations between the two sets of components.

Example 1. Let $\mathbf{i}$, $\mathbf{j}$, $\mathbf{k}$ be unit basis vectors along the (x, y, z) axes and $\mathbf{i}'$, $\mathbf{j}'$, $\mathbf{k}'$ be unit basis vectors along the (x', y', z') axes. Then the vector $\mathbf{r}$ can be written in terms of either set of components and basis vectors as follows:

$$\mathbf{r} = \mathbf{i}x + \mathbf{j}y + \mathbf{k}z = \mathbf{i}'x' + \mathbf{j}'y' + \mathbf{k}'z'. \tag{2.2}$$

Taking the dot product of this equation with $\mathbf{i}'$, we get

$$\mathbf{r} \cdot \mathbf{i}' = \mathbf{i} \cdot \mathbf{i}'x + \mathbf{j} \cdot \mathbf{i}'y + \mathbf{k} \cdot \mathbf{i}'z = x' \tag{2.3}$$

(since $\mathbf{i}' \cdot \mathbf{i}' = 1$, and $\mathbf{i}' \cdot \mathbf{j}' = \mathbf{i}' \cdot \mathbf{k}' = 0$). Now $\mathbf{i} \cdot \mathbf{i}'$ is the cosine of the angle between $\mathbf{i}$ and $\mathbf{i}'$, that is, between the x and x' axes, since $\mathbf{i}$ and $\mathbf{i}'$ are unit vectors; thus $\mathbf{i} \cdot \mathbf{i}' = l_1$ from the table (2.1). Similarly, $\mathbf{j} \cdot \mathbf{i}' = m_1$ and $\mathbf{k} \cdot \mathbf{i}' = n_1$ and (2.3) becomes

$$x' = l_1 x + m_1 y + n_1 z. \tag{2.4}$$

Similarly, dotting $\mathbf{r}$ into $\mathbf{j}'$ nd $\mathbf{k}'$, and using (2.1) we get

$$\begin{aligned} y' &= l_2 x + m_2 y + n_2 z, \\ z' &= l_3 x + m_3 y + n_3 z. \end{aligned} \tag{2.5}$$

The equations (2.4) and (2.5) are called the transformation equations from the coordinate system (x, y, z) to (x', y', z').

Example 2. In the same way, dotting $\mathbf{r}$ with $\mathbf{i}$, $\mathbf{j}$, $\mathbf{k}$ in turn, we get equations for x, y, z in terms of x', y', z':

$$\begin{aligned} x &= l_1 x' + l_2 y' + l_3 z', \\ y &= m_1 x' + m_2 y' + m_3 z', \\ z &= n_1 x' + n_2 y' + n_3 z'. \end{aligned} \tag{2.6}$$

These transformation equations may be written more concisely in matrix notation. Equations (2.4) and (2.5) become the matrix equation:

$$\begin{pmatrix} x' \\ y' \\ z' \end{pmatrix} = \begin{pmatrix} l_1 & m_1 & n_1 z \\ l_2 & m_2 & n_2 z \\ l_3 & m_3 & n_3 z \end{pmatrix} \begin{pmatrix} x \\ y \\ z \end{pmatrix} \quad \text{or} \quad \mathrm{r}' = \mathrm{A}\mathrm{r}, \tag{2.7}$$

where r', A, and r stand for the matrices in (2.7). [Compare Chapter 3, equation (7.13) for the two-dimensional case.] Similarly, (2.6) becomes

$$\mathrm{r} = \mathrm{A}^{\mathrm{T}}\mathrm{r}' \tag{2.8}$$

where A^{T} is the transpose of A. Recall from Chapter 3, Sections 7 and 9, that a rotation matrix is an orthogonal matrix, and for an orthogonal matrix, $\mathrm{A}^{\mathrm{T}} = \mathrm{A}^{-1}$. Also see Problems 3 and 4.

Equations (2.7) or (2.8) tell us how displacement vectors in a rectangular coordinate system transform under a rotation of axes. We now use this result to define *Cartesian* vectors, that is, vectors which transform in the same way that displacement vectors do under rotations of rectangular (Cartesian) axes. We will then generalize this to define Cartesian tensors of other ranks.

Definition of Cartesian Vectors A Cartesian vector $\mathbf{V}$ consists of a set of three numbers (components) in *every* rectangular coordinate system; if V_x, V_y, V_z are the components in one system and V_x', V_y', V_z' are the components in a rotated system, these two sets of components are related by an equation similar to (2.7), namely,

$$\begin{pmatrix} V_x' \\ V_y' \\ V_z' \end{pmatrix} = \mathrm{A} \begin{pmatrix} V_x \\ V_y \\ V_z \end{pmatrix} \quad \text{or} \quad \mathrm{V}' = \mathrm{A}\mathrm{V}, \tag{2.9}$$

where A is the rotation matrix in (2.7). Alternatively, we could use (2.8) and require that $\mathrm{V} = \mathrm{A}^{\mathrm{T}}\mathrm{V}'$.

We can simplify our notation by making the following changes.

$$
\begin{array}{lll}
 & \text{Replace } x, y, z & \text{by } x_1, x_2, x_3 \\
 & \text{Replace } x', y', z' & \text{by } x_1', x_2', x_3' \\
 & \text{Replace } V_x, V_y, V_z & \text{by } V_1, V_2, V_3 \\
(2.10) & \text{Replace } V_x', V_y', V_z' & \text{by } V_1', V_2', V_3' \\
 & \text{Replace A in (2.7)} & \text{by } \begin{pmatrix} a_{11} & a_{12} & a_{13} \\ a_{21} & a_{22} & a_{23} \\ a_{31} & a_{32} & a_{33} \end{pmatrix}
\end{array}
$$

In this notation (2.7) and (2.9) become (2.11) and (2.12):

(2.11)
$$x_i' = \sum_{j=1}^{3} a_{ij} x_j, \qquad i = 1, 2, 3,$$

(2.12)
$$V_i' = \sum_{j=1}^{3} a_{ij} V_j, \qquad i = 1, 2, 3.$$

Alternatively, we could solve (2.11) for the x coordinates in terms of the x' coordinates as in (2.8), to get, in the summation form: $x_i = \sum_{j=1}^{3} a_{ji} x_j'$, and a similar companion formula to (2.12), namely

(2.13)
$$V_i = \sum_{j=1}^{3} a_{ji} V_j'.$$

Since we will occasionally want the transformation formula for a Cartesian vector solved for the unprimed components as in (2.13), you should be sure you understand (2.13). Compare carefully the indices in (2.12) and (2.13). In matrix form (2.12) is $\mathrm{V}' = \mathrm{AV}$ and (2.13) is $\mathrm{V} = \mathrm{A}^{\mathrm{T}}\mathrm{V}'$ [see equations (2.7) and (2.8)]. Now element i, j of A^{T} is the same as element j, i of A, so the coefficients in (2.13) are a_{ji} instead of a_{ij} as they were in (2.12). It is now straightforward to define tensors.

Definition of Cartesian Tensors A tensor of rank zero has one component which is unchanged by a rotation of axes; it is called an invariant or a scalar. Simple examples are the length of a vector, or the dot product of two vectors. A first rank tensor is just a vector. A tensor of second rank has nine components (in three dimensions) in every rectangular coordinate system. If we call the components in one system T_{ij}, the components T_{kl}' in a rotated coordinate system are given by (2.14), where the a's are the direction cosines in the rotation matrix A.

(2.14)
$$T_{kl}' = \sum_{i=1}^{3} \sum_{j=1}^{3} a_{ki} a_{lj} T_{ij}, \qquad k, l = 1, 2, 3.$$

Direct Product We can give a very simple example of a second-rank tensor.

Example 3. Let **U** and **V** be vectors; we form the following array (in each coordinate system) from the components U_1, U_2, U_3 and V_1, V_2, V_3 of **U** and **V** (in that coordinate system):

$$\begin{matrix} U_1V_1 & U_1V_2 & U_1V_3 \\ U_2V_1 & U_2V_2 & U_2V_3 \\ U_3V_1 & U_3V_2 & U_3V_3 \end{matrix} \tag{2.15}$$

We can show that these nine quantities are the components of a second-rank tensor which we shall denote by **UV**. Note that this is not a dot product or a cross product; it is called the *direct product* of **U** and **V** (or *outer product* or *tensor product*). Since **U** and **V** are vectors, their components in a rotated coordinate system are, by (2.12):

$$U'_k = \sum_{i=1}^{3} a_{ki}U_i, \qquad V'_l = \sum_{j=1}^{3} a_{lj}V_j. \tag{2.16}$$

Hence the components of the second-rank tensor **UV** are

$$U'_kV'_l = \sum_{i=1}^{3} a_{ki}U_i \sum_{j=1}^{3} a_{lj}V_j = \sum_{i,j=1}^{3} a_{ki}a_{lj}U_iV_j, \tag{2.17}$$

which is just (2.14) with $T_{ij} = U_iV_j$ and $T'_{kl} = U'_kV'_l$.

Equation (2.14) generalizes immediately. For example, a 4th-rank Cartesian tensor is defined as a set of 3^4 or 81 components T_{ijkl}, in every rectangular coordinate system, which transform to a rotated coordinate system by the equations

$$T'_{\alpha\beta\gamma\delta} = \sum_{i,j,k,l} a_{\alpha i}a_{\beta j}a_{\gamma k}a_{\delta l}T_{ijkl}, \tag{2.18}$$

where i, j, k, l take the values 1, 2, 3. Note that a 4th-rank tensor has 4 indices and requires four a's in its definition. Similarly, an nth-rank tensor has n indices and requires n a's in its definition. Also we can generalize (2.17) to show, for example, that the direct product of a vector and a 3rd-rank tensor produces a 4th-rank tensor, and, in general, the direct product of tensors of ranks n and m is a tensor of rank $m + n$ (see Problem 7).

PROBLEMS, SECTION 2

1. Verify equations (2.6).
2. Show that the sum of the squares of the direction cosines of a line through the origin is equal to 1 *Hint:* Let (a, b, c) be a point on the line at distance 1 from the origin. Write the direction cosines in terms of (a, b, c).
3. Consider the matrix A in (2.7) or (2.10). Think of the elements in each row (or column) as the components of a vector. Show that the row vectors form an orthonormal triad (that is each is of unit length and they are all mutually orthogonal), and the column vectors form an orthonormal triad.

4. Any rotation of axes in three dimensions can be described by giving the nine direction cosines of the angle between the (x, y, z) axes and the (x', y', z') axes. Show that the matrix A of these direction cosines in (2.7) or (2.10) is an orthogonal matrix. *Hint:* See Chapter 3, Section 9. Find AA^T and use Problem 3.

5. Write equations (2.12) out in detail and solve the three simultaneous equations (say by determinants) for x_1, x_2, x_3 in terms of x_1', x_2', x_3' to verify equations (2.13). Use your results in Problem 4.

6. Write the transformation equation for a 3rd-rank tensor; for a 5th-rank tensor.

7. Following what we did in equations (2.14) to (2.17), show that the direct product of a vector and a 3rd-rank tensor is a 4th-rank tensor. Also show that the direct product of two 2nd-rank tensors is a 4th-rank tensor. Generalize this to show that the direct product of two tensors of ranks m and n is a tensor of rank $m + n$.

8. Write the equations in (2.16) and so in (2.17) solved for the unprimed components in terms of the primed components.

3. TENSOR NOTATION AND OPERATIONS

Summation Convention As you may have noticed in the last section, tensor equations use a lot of summation signs—it would be a simplification if we could get along without them. Using the *summation convention* (or Einstein summation convention), we omit the summation signs in equations like (2.11) to (2.14), and (2.16) to (2.18), and simply understand a summation over any index which appears exactly twice in one term. Here are some examples using summation convention (in three dimensions).

Examples.

(3.1)
$$\begin{aligned}
a_{ii} \text{ or } a_{jj} \text{ or } a_{\beta\beta}, \text{ etc.} \quad &\text{means} \quad a_{11} + a_{22} + a_{33};\\
x_i x_i \text{ or } x_\alpha x_\alpha, \text{ etc.} \quad &\text{means} \quad x_1^2 + x_2^2 + x_3^2;\\
a_{ij} b_{jk} \quad &\text{means} \quad a_{i1} b_{1k} + a_{i2} b_{2k} + a_{i3} b_{3k};\\
\frac{\partial u}{\partial x_j}\frac{\partial x_j}{\partial x_i'} \quad &\text{means} \quad \frac{\partial u}{\partial x_1}\frac{\partial x_1}{\partial x_i'} + \frac{\partial u}{\partial x_2}\frac{\partial x_2}{\partial x_i'} + \frac{\partial u}{\partial x_3}\frac{\partial x_3}{\partial x_i'};\\
T_{ijkl} S_{ij} V_k U_l \quad &\text{means} \quad \sum_i \sum_j \sum_k \sum_l T_{ijkl} S_{ij} V_k U_l;
\end{aligned}$$

and so on. The repeated index (which is summed over) is called a *dummy* index; like an integration variable in a definite integral, it does not matter what letter is used for it. An index which is not repeated is called a *free* index.

When summation convention is being used, we are not warned by a summation sign what letters to sum over; we just have to inspect the indices and see which ones appear twice. In writing terms using the summation convention, we must be careful not to re-use an index. For example, if we already have two i subscripts indicating a sum over i, and we want another sum in the same term, we must use a different dummy index, say j or m or α, etc. In the following discussion we will use summation convention; watch carefully for the repeated dummy indices.

Contraction The transformation equations for a 4th-rank tensor are [see (2.18)]

(3.2)
$$T'_{\alpha\beta\gamma\delta} = a_{\alpha i} a_{\beta j} a_{\gamma k} a_{\delta l} T_{ijkl}.$$

(Note the sums over i, j, k, and l).

Example 1. Now suppose we put $\delta = \beta$ which, by summation convention, means a further sum over β. Then we have

$$T'_{\alpha\beta\gamma\beta} = a_{\alpha i}a_{\beta j}a_{\gamma k}a_{\beta l}T_{ijkl}. \tag{3.3}$$

Now $a_{\beta j}a_{\beta l}$ (summed over β) is the dot product of columns j and l of the rotation matrix A [see Problem (2.3)]. This dot product is 1 if $j = l$, and 0 otherwise. In other words $a_{\beta j}a_{\beta l} = \delta_{jl}$ [see Chapter 3, equation (9.4)]. Then $\delta_{jl}T_{ijkl}$ becomes T_{ijkj} since δ_{jl} is zero unless j and l are equal. (The repeated dummy index could be either j or l or anything else except the dummy indices i and k which are already used, and the free indices α and γ). Thus we have

$$T'_{\alpha\beta\gamma\beta} = a_{\alpha i}a_{\gamma k}\delta_{jl}T_{ijkl} = a_{\alpha i}a_{\gamma k}T_{ijkj} \tag{3.4}$$

Now (3.4) says that T_{ijkj} are the components of a 2nd-rank tensor since there are two free indices and two a factors are required [compare equation (2.14)]. This process of setting two indices of a tensor equal to each other and then summing is called *contraction*. Contraction reduces the rank of a tensor by 2. Note that in (3.2) we started with a 4th-rank tensor and after contracting we have a tensor of rank 2 in (3.4).

It is interesting to observe that the dot (or scalar or inner) product of two vectors in elementary vector analysis is an example of contraction. In Section 2 we showed that the direct product of two vectors [see (2.17)] is a 2nd-rank tensor. If we contract U_iV_j to get U_iV_i we have the dot product of vectors $\mathbf{U}$ and $\mathbf{V}$, which is a scalar. Again note that contraction has reduced the rank of a tensor by 2 (a scalar is a tensor of rank zero).

Tensors and Matrices The components of first or second rank tensors can be displayed as matrices and this is often useful. We have frequently (see Chapter 3) written the components of a vector (1st-rank tensor) as a column or row matrix. The components T_{ij} of a 2nd-rank tensor can be written as the elements of a square matrix (see inertia matrix, Section 4). Then note that in the tensor equation, $U_i = T_{ij}V_j$, the contraction (sum on j) corresponds exactly to row times column multiplication for matrices.

Symmetric and Antisymmetric Tensors A 2nd-rank tensor T_{ij} is called *symmetric* if $T_{ij} = T_{ji}$, and *antisymmetric* (or *skew symmetric*) if $T_{ij} = -T_{ji}$. Note that these agree with the corresponding definitions for matrices [Chapter 3, (9.2)]. Any 2nd-rank tensor can be written as a sum of a symmetric tensor and an antisymmetric tensor as in (3.5) (Problem 13).

$$T_{ij} = \frac{1}{2}(T_{ij} + T_{ji}) + \frac{1}{2}(T_{ij} - T_{ji}). \tag{3.5}$$

For tensors of higher rank, similar terminology is used. If an exchange of two indices leaves the tensor component unchanged, we say that the tensor is symmetric with respect to those two indices. If an exchange of two indices changes the tensor component to its negative, we say that the tensor is antisymmetric with respect to those two indices.

Combining tensors The sum or difference (in fact linear combination) of two tensors of rank n is a tensor of rank n (Problems 6 and 7). For example, $T_{ij}+R_{ijk}V_k$ is a tensor of rank 2. Note the summation convention and the contraction which makes $R_{ijk}V_k$ also a tensor of rank 2 so that we can add it to T_{ij}. (Addition is not defined for tensors of different ranks.)

Quotient Rule Let us suppose we know that, for every vector V_j, the quantities $U_i = T_{ij}V_j$ are the components of a non-zero vector and that this holds true in all rotated coordinate systems. Then we can prove that the quantities T_{ij} are the components of a 2nd-rank tensor. This is an example of the *quotient rule.*

▶ **Example 2.** To prove this, we need the following equations:

(3.6)
$$\begin{aligned} T'_{\alpha\beta}V'_\beta &= U'_\alpha, && \text{given equation in rotated system;} \\ U'_\alpha &= a_{\alpha i}U_i, && \mathbf{U} \text{ is a vector;} \\ U_i &= T_{ij}V_j, && \text{given equation;} \\ V_j &= a_{\beta j}V'_\beta, && \mathbf{V} \text{ is a vector; see equation (2.13).} \end{aligned}$$

Now, putting this all together we have

$$T'_{\alpha\beta}V'_\beta = U'_\alpha = a_{\alpha i}U_i = a_{\alpha i}T_{ij}V_j = a_{\alpha i}T_{ij}a_{\beta j}V'_\beta. \tag{3.7}$$

Factoring out V'_β from the first and last steps, we have

$$(T'_{\alpha\beta} - a_{\alpha i}a_{\beta j}T_{ij})V'_\beta = 0 \quad \text{for } \textit{all} \text{ vectors } \mathbf{V}'. \tag{3.8}$$

Since $\mathbf{V}'$ is arbitrary, the parenthesis in (3.8) is equal to zero (Problem 8). Thus we have

$$T'_{\alpha\beta} = a_{\alpha i}a_{\beta j}T_{ij}. \tag{3.9}$$

Now (3.9) is the transformation equation for a 2nd-rank tensor [compare (2.14)], so, as claimed, the quantities T_{ij} are the components of a 2nd-rank tensor.

The quotient rule is useful in determining whether some given quantities are the components of a tensor. [As an example of this, see (4.1).] Suppose **X** is a set of 3^n components (the right number for a tensor of rank n in 3 dimensions). The quotient rule says that if the product of **X** and an arbitrary tensor is a non-zero tensor, then **X** is a tensor. The product may be either a direct product or a direct product combined with one or more contractions. We have proved the quotient rule for one case but the proof of any case follows this same pattern. Given **XA** = **B**, where **A** is an arbitrary tensor and **B** is a non-zero tensor, we use the transformation equations for **A** and **B**, and the fact that **A** is arbitrary, to find the transformation equations for **X** (see Problems 9 to 12).

PROBLEMS, SECTION 3

1. Write equations (2.11), (2.12), (2.13), (2.14), (2.16), (2.17), and (2.18) using summation convention.

2. Show that the fourth expression in (3.1) is equal to $\partial u/\partial x_i'$. By equations (2.6) and (2.10), show that $\partial x_j/\partial x_i' = a_{ij}$, so

$$\frac{\partial u}{\partial x_i'} = \frac{\partial u}{\partial x_j}\frac{\partial x_j}{\partial x_i'} = a_{ij}\frac{\partial u}{\partial x_j}.$$

Compare this with equation (2.12) to show that ∇u is a Cartesian vector. *Hint:* Watch the summation indices carefully and if it helps, put back the summation signs or write sums out in detail as in (3.1) until you get used to summation convention.

3. As we did in (3.3), show that the contracted tensor T_{iij} is a first-rank tensor, that is, a vector.

4. Show that the contracted tensor $T_{ijk}V_k$ is a 2nd-rank tensor.

5. Show that $T_{ijklm}S_{lm}$ is a tensor and find its rank (assuming that **T** and **S** are tensors of the rank indicated by the indices).

6. Show that the sum of two 3rd-rank tensors is a 3rd-rank tensor. *Hint:* Write the transformation law for each tensor and then add your two equations. Divide out the a factors to leave the result $T'_{\alpha\beta\gamma} + S'_{\alpha\beta\gamma} = a_{\alpha i}a_{\beta j}a_{\gamma k}(T_{ijk} + S_{ijk})$ using summation convention.

7. As in problem 6, show that the sum of two 2nd-rank tensors is a 2nd-rank tensor; that the sum of two 4th-rank tensors is a 4th-rank tensor.

8. Show that (3.9) follows from (3.8). *Hint:* Give a proof by contradiction. Let $S_{\alpha\beta}$ be the parenthesis in (3.8); you may find it useful to think of the components written as a matrix. You want to prove that all 9 components of $S_{\alpha\beta}$ are zero. Suppose it is claimed that S_{12} is not zero. Since V'_β is an arbitrary vector, take it to be the vector $(0, 1, 0)$, and observe that $S_{\alpha\beta}V'_\beta$ is then not zero in contradiction to (3.8). Similarly show that all components of $S_{\alpha\beta}$ are zero as (3.9) claims.

Prove the quotient rule in each of the following problems, that is, given **XA** = **B** where **A** is any arbitrary tensor and **B** is a non-zero tensor, show that **X** is a tensor. *Hints:* Follow the general method in (3.6) to (3.9). See the last sentence of the section.

9. $X_iA_{ij} = B_j$

10. $X_iA_j = B_{ij}$

11. $X_{ij}A_k = B_{ijk}$

12. $X_{ijkl}A_{kl} = B_{ij}$

13. Show that the first parenthesis in (3.5) is a symmetric tensor and the second parenthesis is antisymmetric.

4. INERTIA TENSOR

Inertia tensor If a rigid body is rotating about a fixed axis, then from elementary mechanics we know that $\boldsymbol{\tau} = d\mathbf{L}/dt$ where $\boldsymbol{\tau}$ is the torque and $\mathbf{L}$ is the angular momentum about the rotation axis. The angular velocity $\boldsymbol{\omega}$ and the angular momentum $\mathbf{L}$ are related by the equation $\mathbf{L} = I\boldsymbol{\omega}$ where I is the moment of inertia of the body about the rotation axis. For rotation about a fixed axis, $\mathbf{L}$ and $\boldsymbol{\omega}$ are parallel vectors, and I is a scalar. But if the rotation axis is not fixed, the angular velocity and the angular momentum may not be parallel.

Example 1. Try the following experiment. Take a small book bound by a rubber band, hold it by one corner and toss it upward giving it a spin. As it falls observe that it tumbles, that is, the angular velocity $\boldsymbol{\omega}$ about the center of mass is not fixed in direction. However, by definition of the center of mass, the gravitational torque $\boldsymbol{\tau}$ about the center of mass is zero so $\boldsymbol{\tau} = d\mathbf{L}/dt = 0$. (We are neglecting air resistance.) Thus $\mathbf{L}$ is a constant vector, and a constant $\mathbf{L}$ and a changing $\boldsymbol{\omega}$ are not parallel. Then if the equation $\mathbf{L} = I\boldsymbol{\omega}$ is to be true, I cannot be a scalar.

We have seen this situation before; look at the discussion of the quotient rule in Section 3 and the proof of the case we have here in (3.5) to (3.8). Since $\mathbf{L}$ and $\boldsymbol{\omega}$ are vectors, we see by the quotient rule that (when $\mathbf{L}$ and $\boldsymbol{\omega}$ are not parallel) the scalar I must be replaced by a 2nd-rank tensor with components I_{jk}. Then in component form we have

$$L_j = I_{jk}\omega_k \tag{4.1}$$

Example 2. Next we want to find the components of the inertia tensor. For simplicity, first consider a point mass m at the tip of a vector $\mathbf{r}$ with tail at the origin O. From Chapter 6, end of Section 3, the angular momentum of m about the origin is $\mathbf{L} = m\mathbf{r} \times (\boldsymbol{\omega} \times \mathbf{r})$ where $\boldsymbol{\omega}$ is the angular velocity of the mass m about O. (See Chapter 6, Figures 2.6 and 3.8.) We can expand the triple vector product [see Chapter 6, equation (3.8)] to get

$$\mathbf{L} = m\mathbf{r} \times (\boldsymbol{\omega} \times \mathbf{r}) = m[r^2\boldsymbol{\omega} - (\mathbf{r}\cdot\boldsymbol{\omega})\mathbf{r}] = m[r^2\boldsymbol{\omega} - (x\omega_x + y\omega_y + z\omega_z)\mathbf{r}]. \tag{4.2}$$

Next we write the components of $\mathbf{L}$ in terms of the components of $\boldsymbol{\omega}$. For example, taking the x component of (4.2), we find

$$L_x = m[r^2\omega_x - (x\omega_x + y\omega_y + z\omega_z)x] = m[(r^2 - x^2)\omega_x - xy\omega_y - xz\omega_z]. \tag{4.3}$$

Thus three components of the inertia tensor are

$$I_{xx} = m(r^2 - x^2) = m(y^2 + z^2), \quad I_{xy} = -mxy, \quad I_{xz} = -mxz. \tag{4.4}$$

The other 6 components can be found similarly by taking the y and z components of (4.2) (Problem 1).

Example 3. If, instead of a single mass, we have a set of masses or an extended body, then the expressions for the components of the inertia tensor become sums or integrals.

$$\begin{aligned} I_{xx} &= \sum_i m_i(y_i^2 + z_i^2) \quad \text{or} \quad \int (y^2 + z^2)\, dm, \\ I_{xy} &= -\sum_i m_i x_i y_i \quad \text{or} \quad -\int xy\, dm, \text{ etc. (Problem 1.)} \end{aligned} \tag{4.5}$$

It is useful to write (4.1) as a matrix equation (see discussion in Section 3 about contraction). Then the inertia tensor components form a square matrix. This matrix is symmetric and so we know from Chapter 3, Section 11, that it can be diagonalized by an orthogonal similarity transformation. The new axes are called the *principal axes of inertia* and the three eigenvalues are called the *principal moments of inertia*. We see that the equations of motion are simpler relative to the principal axes.

Example 4. Find the inertia tensor about the origin for the mass distribution consisting of a mass 1 at $(0,1,1)$ and a mass 2 at $(1,-1,0)$. Find the principal moments of inertia and the principal axes.

Substituting $(x_1, y_1, z_1) = (0,1,1)$, $m_1 = 1$, and $(x_2, y_2, z_2) = (1,-1,0)$, $m_2 = 2$ into (4.5), we find $I_{xx} = (1^2 + 1^2) + 2(-1)^2 = 4$, $I_{xy} = I_{yx} = -0 - 2(-1) = 2$. Continuing in the same way, we can find the rest of the components (Problem 2) and write them as an inertia matrix

$$\mathrm{I} = \begin{pmatrix} 4 & 2 & 0 \\ 2 & 3 & -1 \\ 0 & -1 & 5 \end{pmatrix}$$

Either by hand or by computer we find that the eigenvalues of the matrix I are 6 and $3 \pm \sqrt{3}$; these are the principal moments of inertia. The corresponding eigenvectors are $(1,1,-1)$, $(-1-\sqrt{3}, 2+\sqrt{3}, 1)$, $(-1+\sqrt{3}, 2-\sqrt{3}, 1)$; these are vectors along the principal axes of inertia.

Example 5. Find the inertia tensor about the origin for a mass of uniform density $= 1$, inside the part of the unit sphere in the first octant, that is, $x > 0$, $y > 0$, $z > 0$.

We will write the integrals for the components of the inertia tensor first in rectangular coordinates and then switch to spherical coordinates [see Chapter 5, equation (4.5)] to evaluate them since the limits are then simpler. Satisfy yourself that in order to cover the required volume, the limits are: r from 0 to 1, θ from 0 to $\pi/2$, and ϕ from 0 to $\pi/2$. Then

$$I_{xx} = \iiint (r^2 - x^2)\,dV =$$

$$\int_{r=0}^{1}\int_{\theta=0}^{\pi/2}\int_{\phi=0}^{\pi/2} (r^2 - r^2\sin^2\theta\cos^2\phi)r^2\sin\theta\,dr\,d\theta\,d\phi = \frac{\pi}{15}.$$

$$I_{xy} = \iiint (-xy)\,dV =$$

$$-\int_{r=0}^{1}\int_{\theta=0}^{\pi/2}\int_{\phi=0}^{\pi/2} (r^2\sin^2\theta\cos\phi\sin\phi)r^2\sin\theta\,dr\,d\theta\,d\phi = -\frac{1}{15}.$$

Similarly, the other integrals can be written and evaluated (Problem 3). Alternatively, it may be clear that by symmetry the three diagonal components are all the same, and all the off-diagonal components are the same. Then the inertia matrix is

$$\mathrm{I} = \frac{1}{15}\begin{pmatrix} \pi & -1 & -1 \\ -1 & \pi & -1 \\ -1 & -1 & \pi \end{pmatrix}.$$

As in Example 4, we find (Problem 3):

Principal moments of inertia: $\dfrac{(\pi-2, \pi+1, \pi+1)}{15}$

Principal axes of inertia: $(1,1,1)$, and any two orthogonal vectors in the plane $x + y + z = 0$, for example, $(1,-1,0)$ and $(1,1,-2)$.

PROBLEMS, SECTION 4

1. As in (4.3) and (4.4), find the y and z components of (4.2) and the other 6 components of the inertia tensor. Write the corresponding components of the inertia tensor for a set of masses or an extended body as in (4.5).

2. Complete Example 4 to verify the rest of the components of the inertia tensor and the principal moments of inertia and principal axes. Verify that the three principal axes form an orthogonal triad.

3. As in Problem 2, complete Example 5.

4. Find the inertia tensor about the origin for a mass of uniform density $= 1$, inside the part of the unit sphere where $x > 0$, $y > 0$, and find the principal moments of inertia and the principal axes. Note that this is similar to Example 5 but the mass is both above and below the (x, y) plane. *Warning hint:* This time don't make the assumptions about symmetry that we did in Example 5.

For the mass distributions in Problems 5 to 7, find the inertia tensor about the origin, and find the principal moments of inertia and the principal axes.

5. Point masses 1 at $(1, 1, 1)$ and at $(-1, 1, 1)$.

6. Point masses 1 at $(1, 1, -2)$ and 2 at $(1, 1, 1)$.

7. Mass of uniform density $= 1$, bounded by the coordinate planes and the plane $x + y + z = 1$.

8. For the point mass m we considered in (4.2) to (4.4), the velocity is $\mathbf{v} = \boldsymbol{\omega} \times \mathbf{r}$ so the kinetic energy is $T = \frac{1}{2}mv^2 = \frac{1}{2}m(\boldsymbol{\omega} \times \mathbf{r}) \cdot (\boldsymbol{\omega} \times \mathbf{r})$. Show that T can be written in matrix notation as $T = \frac{1}{2}\omega^{\mathrm{T}}\mathrm{I}\omega$ where I is the inertia matrix, ω is a column matrix, and ω^{T} is a row matrix with elements equal to the components of $\boldsymbol{\omega}$.

5. KRONECKER DELTA AND LEVI-CIVITA SYMBOL

The Kronecker δ is defined in Chapter 3, equation (9.3) but let's repeat it here for convenience.

$$\delta_{ij} = \begin{cases} 1 & \text{if } \ i = j, \\ 0 & \text{otherwise.} \end{cases} \tag{5.1}$$

The definition of the Levi-Civita symbol (or permutation symbol) is

$$\epsilon_{ijk} = \begin{cases} 1 & \text{if } i, j, k = 1, 2, 3 \text{ or } 2, 3, 1 \text{ or } 3, 1, 2; \\ -1 & \text{if } i, j, k = 3, 2, 1 \text{ or } 2, 1, 3 \text{ or } 1, 3, 2; \\ 0 & \text{if any indices are repeated.} \end{cases} \tag{5.2}$$

Note in (5.2) that if you read the indices i, j, k, cyclically (as if they were written around a circle so you can start anywhere), then if the indices read in the direction $1, 2, 3, 1, 2, 3, 1, \cdots$, the result is $+1$; if the indices read in the opposite direction the result is -1.

We can say (5.2) in another way which is sometimes useful. Start with the fact that $\epsilon_{123} = +1$. Now if we exchange any two indices, we change the sign; for example $\epsilon_{321} = -1$ (we exchanged 1 and 3). If we now continue this process and exchange 1 and 2 in $\epsilon_{321} = -1$, we have $\epsilon_{312} = +1$. [Try a few more and compare with (5.2).] The result of an even number of exchanges in 123 is called an even permutation of 123 and the result of an odd number of exchanges is called an odd permutation of 123. Thus we could replace (5.2) by the definition

$$\epsilon_{ijk} = \begin{cases} 1 & \text{if } i, j, k = \text{an even permutation of } 1, 2, 3; \\ -1 & \text{if } i, j, k = \text{an odd permutation of } 1, 2, 3; \\ 0 & \text{if any indices are repeated.} \end{cases} \tag{5.3}$$

We say that ϵ_{ijk} is *totally antisymmetric* (see Section 3), that is, it is antisymmetric with respect to every pair of indices, since each exchange of indices produces a change in sign.

Isotropic Tensors A Cartesian *isotropic* tensor means a tensor which has the same components in all rotated coordinate systems. The definitions (5.1) to (5.3) are general, independent of any reference system. Thus to show that δ_{ij} and ϵ_{ijk} are isotropic *tensors*, we need to show that a tensor transformation simply reproduces the tensor we start with, that is, $\delta' = \delta$ and $\epsilon' = \epsilon$. In this section we shall show this and develop some useful formulas.

Kronecker delta To show that δ_{ij} is an isotropic 2nd-rank tensor, we write the tensor transformation to a rotated system and show that it gives $\delta' = \delta$.

$$\delta'_{mn} = a_{mi}a_{nj}\delta_{ij} = a_{mj}a_{nj} = \delta_{mn}. \tag{5.4}$$

Remember summation convention and follow carefully the sums in (5.4). Note in the second step that a_{mi} becomes a_{mj} because δ_{ij} is zero unless $i = j$. (We could just as well change a_{nj} to a_{ni} and sum on i.) In the last step, $a_{mj}a_{nj}$ (or $a_{mi}a_{ni}$) is the dot product of rows m and n of the rotation matrix and this is δ_{mn} (see Problem 2.3). Thus the Kronecker δ is a 2nd-rank isotropic Cartesian tensor.

Determinants We can write a useful formula for the value of a 3-by-3 determinant using the Levi-Civita symbol:

$$\det \mathrm{A} = a_{1i}a_{2j}a_{3k}\epsilon_{ijk}. \tag{5.5}$$

It is straightforward to show (Problem 1) that (5.5) is equivalent to a Laplace development. Another useful formula is

$$\epsilon_{\alpha\beta\gamma} \det \mathrm{A} = a_{\alpha i}a_{\beta j}a_{\gamma k}\epsilon_{ijk}. \tag{5.6}$$

Again it is straightforward (although lengthy) to show that this is equivalent to a Laplace development (Problem 2).

Levi-Civita Symbol To show that ϵ_{ijk} is an isotropic tensor, we write the transformation equation to a rotated system. We find

$$\epsilon'_{\alpha\beta\gamma} = a_{\alpha i}a_{\beta j}a_{\gamma k}\epsilon_{ijk} = \epsilon_{\alpha\beta\gamma}. \tag{5.7}$$

In the last step we used (5.6) with $\det \mathrm{A} = 1$ (recall from Chapter 3, Section 7, that if A is a rotation matrix, $\det \mathrm{A} = 1$). Thus $\epsilon_{\alpha\beta\gamma}$ is a 3rd-rank isotropic Cartesian tensor (assuming only rotated coordinate systems; for reflections see Section 6).

Products of Isotropic Tensors We can find other isotropic tensors from direct products of the two we have, or from direct products followed by contraction. Recall from Sections 2 and 3 that the direct product of two tensors of ranks n and m is a tensor of rank $n+m$ and that each contraction produces another tensor of rank smaller by 2. If the tensors you multiply are isotropic, the products are also isotropic (Problems 3 and 4).

To simplify products of two Levi-Civita tensors, the following formula is useful.

$$\epsilon_{ijk}\epsilon_{imn} = \delta_{jm}\delta_{kn} - \delta_{jn}\delta_{km} \tag{5.8}$$

Both sides of (5.8) are 4th-rank tensors (contracted 6th-rank on the left) with free indices j, k, m, n. We want to see that (5.8) is true for any choice of these four indices. Most choices will just give $0 = 0$; let's consider what is required for the product of ϵ's to be different from zero.

▶ **Example 1.** Remember that an ϵ is zero unless its three indices are all different. Since the first index is the same in ϵ_{ijk} and ϵ_{imn}, the product is different from zero only if the other two indices (j, k and m, n) are the same pair in both ϵ's. (For example, if $i = 1$, then j, k and m, n must be 2, 3 or 3, 2.) This means that either (1) $j = m$ and $k = n$, or (2) $j = n$ and $k = m$. In case (1), the two ϵ's are the same (both $= +1$ or both $= -1$) so the product is $+1$; this is the same as $\delta_{jm}\delta_{kn}$ on the right side of (5.8). In case (2), the indices on the two ϵ's are ijk and ikj so one of them is an even permutation of 1, 2, 3 and the other is an odd permutation. Thus the product of the two ϵ's is -1, and this is the same as $-\delta_{jn}\delta_{kn}$ on the right side of (5.8). Note that, given j, k, m, n satisfying either $j,k = m,n$ or $j,k = n,m$, only one term in the sum over i is different from zero, that is, the term with i different from either j or k. Also see Problem 5.

Now that we have (5.8), it is easy to write a similar formula with the contraction (sum) over a different pair of indices. Suppose we want $\epsilon_{abc}\epsilon_{pqb}$. Recall that an ϵ is not changed by cyclic permutation of its indices [see discussion after (5.2)]. Thus $\epsilon_{abc} = \epsilon_{bca}$ and $\epsilon_{pqb} = \epsilon_{bpq}$ [we have cyclically permuted the indices so that the summation index b appears as the first index for each ϵ as it does in (5.8)]. Now, this is the same pattern as in (5.8), with the sum over the first index of each ϵ [i in (5.8), b in the second step in (5.9)], so we have

$$\epsilon_{abc}\epsilon_{pqb} = \epsilon_{bca}\epsilon_{bpq} = \delta_{cp}\delta_{aq} - \delta_{cq}\delta_{ap} \tag{5.9}$$

It may be helpful in writing this to repeat what we said in getting (5.8). In (5.9), the product $\epsilon_{bca}\epsilon_{bpq}$ is zero unless either $c,a = p,q$, or $c,a = q,p$, as indicated by the right side of (5.9). For practice, do Problem 7.

We can further contract (5.8) to get (Problem 8).

$$(5.10) \qquad \begin{cases} \epsilon_{ijk}\epsilon_{ijn} = 2\delta_{kn}, \\ \epsilon_{ijk}\epsilon_{ijk} = 6. \end{cases}$$

Vector Identities The familiar formulas in vector analysis can be written in tensor form using δ_{ij} and ϵ_{ijk}. [See Am. J. Phys. **34**, 503–507 (1966).] We have already commented (Section 3) that the dot product $\mathbf{A} \cdot \mathbf{B}$ is the contracted direct product, $A_i B_i$. Now let's show that the components of the cross product of two vectors can be written as

$$(5.11) \qquad (\mathbf{B} \times \mathbf{C})_i = \epsilon_{ijk} B_j C_k.$$

To see that this is correct we look at one component at a time and compare the result with Chapter 3, equation (4.19), replacing x, y, z by 1, 2, 3. To find the first component of $\mathbf{B} \times \mathbf{C}$ in (5.11), we let $i = 1$. Then the only nonzero terms on the right side of (5.11) are the two with $j, k = 2, 3$ or $3, 2$, so we find that the first component of $\mathbf{B} \times \mathbf{C}$ is $(B_2 C_3 - B_3 C_2)$, in agreement with Chapter 3. Similarly the other components agree with the vector analysis definition of a cross product (Problem 9a).

Example 2. Now let's use (5.11) to write a triple vector product in tensor form, and then use (5.8) or (5.9) to simplify it (Problem 9b).

$$(5.12) \qquad \begin{aligned} [\mathbf{A} \times (\mathbf{B} \times \mathbf{C})]_n &= \epsilon_{nip} A_i (\mathbf{B} \times \mathbf{C})_p = \epsilon_{nip} A_i [\epsilon_{pjk} B_j C_k] \\ &= \epsilon_{nip}\epsilon_{pjk} A_i B_j C_k = \epsilon_{pni}\epsilon_{pjk} A_i B_j C_k \\ &= (\delta_{nj}\delta_{ik} - \delta_{nk}\delta_{ij}) A_i B_j C_k = B_n(A_i C_i) - C_n(A_i B_i) \\ &= \text{components of } \mathbf{B}(\mathbf{A} \cdot \mathbf{C}) - \mathbf{C}(\mathbf{A} \cdot \mathbf{B}). \end{aligned}$$

We recognize the final step as the formula [Chapter 6, (3.8)] for the triple vector product; we have just derived it in tensor form. Similarly we can prove other vector formulas (see Problems 10 to 13).

Recall from Chapter 6 that we treated $\boldsymbol{\nabla}$ as if it were "almost" a vector. Here we can similarly treat it as a first rank tensor, always remembering that it is also a differential operator. The components of $\boldsymbol{\nabla}$ are $\partial/\partial x_i$, so as in (5.11) we write

$$(5.13) \qquad (\boldsymbol{\nabla} \times \mathbf{V})_i = \epsilon_{ijk} \frac{\partial}{\partial x_j} V_k.$$

Then following the method of (5.12), we next find the components of curl curl $\mathbf{V}$ in tensor form [compare part (e) in the table at the end of Chapter 6].

$$(5.14)\qquad \begin{aligned}[\nabla\times(\nabla\times\mathbf{V})]_n &= \epsilon_{nip}\frac{\partial}{\partial x_i}(\nabla\times\mathbf{V})_p\\ &= \epsilon_{nip}\frac{\partial}{\partial x_i}[\epsilon_{pjk}\frac{\partial}{\partial x_j}V_k]\\ &= \epsilon_{pni}\epsilon_{pjk}\frac{\partial}{\partial x_i}\frac{\partial}{\partial x_j}V_k\\ &= (\delta_{nj}\delta_{ik}-\delta_{nk}\delta_{ij})\frac{\partial}{\partial x_i}\frac{\partial}{\partial x_j}V_k\\ &= \frac{\partial}{\partial x_n}\left(\frac{\partial}{\partial x_i}V_i\right)-\frac{\partial}{\partial x_i}\frac{\partial}{\partial x_i}V_n\\ &= \text{components of } \nabla(\nabla\cdot\mathbf{V})-\nabla^2\mathbf{V}.\end{aligned}$$

Dual tensors Let T_{ij} be an antisymmetric 2nd-rank tensor, that is, $T_{ij} = -T_{ji}$. If we display the components T_{ij} as elements of a matrix, it looks like this (see Problem 14).

$$(5.15)\qquad \mathrm{T} = \begin{pmatrix} 0 & T_{12} & -T_{31} \\ -T_{12} & 0 & T_{23} \\ T_{31} & -T_{23} & 0 \end{pmatrix}$$

Observe that there are just 3 independent nonzero components, just enough to be the components of a vector. (Note that this happens only in 3 dimensions—see Problem 15.) If we define

$$(5.16)\qquad V_i = \frac{1}{2}\epsilon_{ijk}T_{jk}$$

then we find (Problem 16)

$$(5.17)\qquad V_1 = T_{23},\quad V_2 = T_{31},\quad V_3 = T_{12}.$$

Since ϵ_{ijk} and T_{jk} are tensors and V_i is a contracted direct product of them, we are assured that V_i is a first rank tensor, that is, a vector (but see Section 6). Thus the three quantities in (5.17) can be considered as the three independent components of an antisymmetric 2nd-rank tensor T_{ij}, or as the three components of a vector V_k called the *dual* of T_{ij}. We can also start with a vector V_k and define T_{ij} in terms of it (Problem 16).

$$(5.18)\qquad T_{ij} = \epsilon_{ijk}V_k$$

Now suppose A_j and B_k are vectors. Then $T_{jk} = A_jB_k - A_kB_j$ is a 2nd-rank antisymmetric tensor, and the three independent components of T_{jk} are just the components of $\mathbf{A}\times\mathbf{B}$ (Problem 17). Thus we see that the vector product can be considered as either a vector or a 2nd-rank antisymmetric tensor.

▶ PROBLEMS, SECTION 5

1. Verify that (5.5) agrees with a Laplace development, say on the first row (Chapter 3, Section 3). *Hints:* You will find 6 terms corresponding to the 6 non-zero values of ϵ_{ijk}. First let $i = 1$; then j, k can be $2, 3$ or $3, 2$. These two terms give you a_{11} times its cofactor. Next let $i = 2$ with $j, k = 1, 3$ and $3, 1$, and show that you get a_{12} times its cofactor. Finally let $i = 3$. Watch all the signs carefully.

2. Verify for a few representative cases that (5.6) gives the same results as a Laplace development. First note that if $\alpha, \beta, \gamma = 1, 2, 3$, then (5.6) is just (5.5). Then try letting $\alpha, \beta, \gamma =$ an even permutation of $1, 2, 3$, and then try an odd permutation, to see that the signs work out correctly. Finally try a case when $\epsilon_{\alpha\beta\gamma} = 0$ (that is when two of the indices are equal) to see that the right hand side of (5.6) is zero because you are evaluating a determinant which has two identical rows.

3. Show that $\delta_{ij}\epsilon_{klm}$ is an isotropic tensor of rank 5. *Hint:* Combine equations (5.4) and (5.7).

4. Generalize Problem 3 to see that the direct product of any two isotropic tensors (or a direct product contracted) is an isotropic tensor. For example show that $\epsilon_{ijk}\epsilon_{lmn}$ is an isotropic tensor (what is its rank?) and $\epsilon_{ijk}\epsilon_{lmn}\delta_{jn}$ is an isotropic tensor (what is its rank?).

5. Let T_{jkmn} be the tensor in (5.8). This is a 4th-rank tensor and so has $3^4 = 81$ components. Most of the components are zero. Find the nonzero components and their values. *Hint:* See discussion after (5.8).

6. Evaluate:

(a) $\delta_{ij}\delta_{jk}\delta_{km}\delta_{im}$
(b) $\epsilon_{ijk}\delta_{jk}$
(c) $\epsilon_{jk2}\epsilon_{k2j}$
(d) $\epsilon_{3jk}\epsilon_{kj3}$
(e) $\epsilon_{23i}\epsilon_{2i3}$
(f) $\epsilon_{k31}\epsilon_{3k1}$

7. Write in terms of δ's as in (5.8) and (5.9):

(a) $\epsilon_{ijk}\epsilon_{pjq}$
(b) $\epsilon_{abc}\epsilon_{pqc}$

8. Show that the equations (5.10) are correct. *Hints:* You can do these by further contracting (5.8). You can also do them by direct argument as follows: In the first equation, why must $k = n$? If $k = n$, then how many choices are there for i and j? In the second equation, in how many ways can you arrange the three numbers 1, 2, 3, and for each arrangement, what is the product of the ϵ's?

9. (a) Finish the work of showing that the cross product components are correctly given by (5.11). *Hints:* Follow the text discussion just after (5.11). For the second component, let $i = 2$; etc.

(b) Go through the sums in (5.12) carefully to verify each step. *Hints:* Use (5.11) twice being careful about repeated indices, and look at the discussion after equation (5.4).

(c) Similarly check (5.14).

10. (a) Write the triple scalar product $\mathbf{A} \cdot (\mathbf{B} \times \mathbf{C})$ in tensor form and show that it is equal to the determinant in Chapter 6, equation (3.2). *Hint:* See (5.5).

(b) Write equation (3.2) of Chapter 6 in tensor form to show the equivalence of the various expressions for the triple scalar product. *Hint:* Change the dummy indices as needed.

11. Using problem 10, write $\mathbf{A} \cdot (\mathbf{B} \times \mathbf{A})$ in tensor notation and show that it is $= 0$.

12. Write and prove in tensor notation:

(a) Chapter 6, Problem 3.13.

(b) Chapter 6, Problem 3.14.

(c) Lagrange's identity: $(\mathbf{A}\times\mathbf{B})\cdot(\mathbf{C}\times\mathbf{D}) = (\mathbf{A}\cdot\mathbf{C})(\mathbf{B}\cdot\mathbf{D}) - (\mathbf{A}\cdot\mathbf{D})(\mathbf{B}\cdot\mathbf{C})$.

(d) $(\mathbf{A}\times\mathbf{B})\times(\mathbf{C}\times\mathbf{D}) = (\mathbf{ABD})\mathbf{C} - (\mathbf{ABC})\mathbf{D}$, where the symbol $(\mathbf{XYZ})$ means the triple scalar product of the three vectors.

13. Write in tensor notation and prove the following vector operator identities in the table at the end of Chapter 6: parts (b), (d), (f), (g), (h), (k).

14. Show that the diagonal elements of an antisymmetric tensor are zero and that (5.15) is a correct display of the components of an antisymmetric 2nd-rank tensor in 3 dimensions.

15. Write a 4-by-4 antisymmetric matrix to show that there are 6 different components, not the 4 components of a vector in 4 dimensions.

16. Verify that (5.16) gives (5.17). Also verify that (5.18) gives (5.17).

17. Write out the components of $T_{jk} = A_j B_k - A_k B_j$ to show that T_{jk} is a 2nd-rank antisymmetric tensor with elements which are the components of $\mathbf{A}\times\mathbf{B}$.

6. PSEUDOVECTORS AND PSEUDOTENSORS

So far we have considered only rotations of rectangular coordinate systems in our definitions of tensors. Recall that an orthogonal transformation includes both rotations and reflections (Chapter 3, Sections 7 and 11). Now we want to consider how the entities we have called tensors behave under reflections. Remember that the determinant of an orthogonal matrix is $+1$ for a rotation (sometimes called a "proper" rotation) and the determinant is -1 if a reflection is involved (sometimes called an "improper" rotation).

When $\det \mathrm{A} = -1$, at least one eigenvalue of matrix A is -1 (see Chapter 3, Section 11). The -1 eigenvalue corresponds to the reversal of one principal axis, that is, a reflection through the plane perpendicular to the axis [for example a reflection through the (x, y) plane which reverses the z axis]. The other two eigenvalues correspond to a rotation [see Chapter 3, equation (7.19)]; this includes the case of a 180° rotation which is equivalent to reversal of the other two axes (see Problems 1 and 2). So in thinking about reflections, we can think of reversing all three axes (called an inversion) or reversing just one, since a rotation doesn't affect the sign of $\det \mathrm{A}$. It is important to realize that reversing either one or all three axes changes the coordinate system from a right-handed to a left-handed coordinate system.

Example 1. Let's look at a simple example of something we usually think of as a vector (namely a cross product) which doesn't obey the vector transformation laws under reflections. Let $\mathbf{U}$ and $\mathbf{V}$ be displacement vectors. Recall (Section 2) that, by definition, a vector transforms the way displacement vectors do. Also remember that we are considering passive transformations: vectors remain fixed in space while the axes are changed (rotated or reflected). Now if the z axis is reversed [reflected through the (x, y) plane], then the z components of the displacement vectors $\mathbf{U}$ and $\mathbf{V}$ change signs; this is then a requirement for all vectors. But the z component of $\mathbf{U}\times\mathbf{V}$ (which is $U_xV_y - U_yV_x$) does not change sign (Problems 3 and 4). Thus $\mathbf{U}\times\mathbf{V}$ is not a vector under reflections. We call $\mathbf{U}\times\mathbf{V}$ a *pseudovector*. We will discover other pseudovectors as we continue.

Levi-Civita symbols We want to use (5.6) when the matrix A is the matrix of an orthogonal transformation. Remember (Chapter 3, Section 7) that if A is orthogonal, $\det A = \pm 1$ so $(\det A)^2 = 1$. Multiply (5.6) by $\det A$ to get the equation $\epsilon_{\alpha\beta\gamma} = (\det A) a_{\alpha i} a_{\beta j} a_{\gamma k} \epsilon_{ijk}$. Then the transformation which gives $\epsilon' = \epsilon$ (see isotropic tensors in Section 5) is

$$\epsilon'_{\alpha\beta\gamma} = (\det A) a_{\alpha i} a_{\beta j} a_{\gamma k} \epsilon_{ijk} = \epsilon_{\alpha\beta\gamma}. \tag{6.1}$$

Now this is not the right transformation equation for a 3rd-rank tensor—the factor $\det A$ would not be there for, say, the direct product of three displacement vectors. Of course, we got away with calling ϵ_{ijk} a 3rd-rank tensor in Section 5 because we were discussing just rotations and $\det A = 1$ if A is a rotation matrix. But now we are dealing with general orthogonal transformations, and when $\det A = -1$ (reflection) there is an extra factor -1 in the transformation equation. We call ϵ_{ijk} a 3rd-rank *pseudotensor*. A *pseudovector* or *pseudotensor* obeys the tensor transformation equations under rotations (that is, $\det A = 1$), but if the transformation includes a reflection (that is, $\det A = -1$), then the transformation equation contains an extra factor of -1. If we have a direct product of two pseudotensors (or such a product contracted), this will be a tensor because the product of the two $\det A$ factors is $(\det A)^2 = 1$. (Problem 5).

Polar and Axial Vectors If a vector (under rotations) also satisfies the vector transformation equations (that is, behaves like a displacement vector) under reflections, it is called a *polar vector* (or true vector or just a vector). If there is a change in sign when $\det A = -1$, it is called an *axial vector* (or pseudovector). In Example 1, **U** and **V** were polar vectors and $\mathbf{U} \times \mathbf{V}$ was an axial vector.

In order to understand pseudotensors we need to discuss left-handed coordinate systems. These are relatively unfamiliar in elementary work and for good reason. When we define a cross product or specify a vector to represent a rotation, the right hand rule is a part of our definition. It would be confusing to deal with this in a left-handed system so you are always warned to use right-handed systems. But we are now considering the general case of orthogonal transformations which includes reflections and so produces left-handed reference systems which we must learn to cope with.

Let's consider the physics and geometry of this by comparing linear velocity and angular velocity, both vectors under rotations. Is there a difference when we consider reflections and so have a left-handed coordinate system? The linear velocity vector indicates a path along which something moves; it has a direct physical meaning, and under passive transformations, it stays fixed in space. In the case of angular velocity, the physical motion is taking place in the plane perpendicular to the angular velocity vector, say a wheel rotating, or a mass or charge moving in a circle. The angular velocity "vector" is something *we choose via the right hand rule* to represent the motion. We might guess (correctly) that linear velocity is a vector (polar vector) and angular velocity is a pseudovector (axial vector). Remember that in Example 1 we found that the cross product (defined using the right hand rule) is a pseudovector. As we continue, watch for this; when the right hand rule is used in the *definition* of a vector, you suspect that it is a pseudovector.

Cross Product In Example 1, we found that the cross product of two displacement vectors does not satisfy the vector transformation equations under reflections. Now we want to write a formula to show exactly how a cross product transforms under a general orthogonal transformation. By (5.11), we write

$$(\mathbf{U}\times\mathbf{V})_i = \epsilon_{ijk}U_jV_k. \tag{6.2}$$

Then using (6.1), (6.2), and the vector transformation equations for the displacement vectors **U** and **V**, we find

$$\begin{aligned}(\mathbf{U}'\times\mathbf{V}')_\alpha &= \epsilon'_{\alpha\beta\gamma}U'_\beta V'_\gamma \\ &= (\det \mathrm{A})a_{\alpha i}a_{\beta j}a_{\gamma k}\epsilon_{ijk}a_{\beta m}U_m a_{\gamma p}V_p \\ &= (\det \mathrm{A})a_{\alpha i}\delta_{jm}\delta_{kp}\epsilon_{ijk}U_mV_p \\ &= (\det \mathrm{A})a_{\alpha i}(\epsilon_{ijk}U_jV_k) = (\det \mathrm{A})a_{\alpha i}(\mathbf{U}\times\mathbf{V})_i.\end{aligned} \tag{6.3}$$

If $\det \mathrm{A} = 1$ (no reflection, just a rotation), then (6.3) is the transformation equation for a vector. If $\det \mathrm{A} = -1$ (reflection) then the transformation has an extra -1 factor. Thus the vector product of two polar vectors is a pseudovector, as we have seen before and as we guessed from the fact that the right hand rule is used in defining cross product.

▶ **Example 2.** Find the triple scalar product of 3 polar vectors.

Here we have one $\det \mathrm{A}$ factor (from the cross product), so the triple scalar product of 3 polar vectors is a pseudoscalar (Problem 7).

▶ **Example 3.** What is the tensor character of $\mathbf{W}\times\mathbf{S}$ if $\mathbf{W}$ is a polar vector and $\mathbf{S}$ is a pseudovector?

In the transformation equation for $\mathbf{W}\times\mathbf{S}$, there is one factor of $\det \mathrm{A}$ for $\mathbf{S}$, and another $\det \mathrm{A}$ for the cross product as in (6.3). The two minus signs cancel, so $\mathbf{W}\times\mathbf{S}$ is a polar vector (Problem 8).

▶ **Example 4.** Show that acceleration $\mathbf{a}$ and force $\mathbf{F}$ are polar vectors.

By definition, the displacement $\mathbf{r}$ is a polar vector (we define vectors as quantities which transform the way displacements do). Then the velocity $\mathbf{v} = d\mathbf{r}/dt$ and the acceleration $\mathbf{a} = d^2\mathbf{r}/dt^2$ are vectors (since time t is a scalar) and $\mathbf{F} = m\mathbf{a}$ is a vector since m is a scalar.

▶ **Example 5.** Find the tensor character of each symbol in $\mathbf{v} = \boldsymbol{\omega}\times\mathbf{r}$.

By Example 4, $\mathbf{v}$ is a vector so $\boldsymbol{\omega}\times\mathbf{r}$ must be a vector (both sides of a tensor equation must have the same tensor character). Then $\boldsymbol{\omega}$ must be a pseudovector so that there are two $\det \mathrm{A}$ factors, one from the cross product and one from $\boldsymbol{\omega}$. Recall that we predicted this because the right hand rule is used in defining angular velocity.

PROBLEMS, SECTION 6

1. Show that in 2 dimensions (say the x, y plane), an inversion through the origin (that is, $x' = -x, y' = -y$) is equivalent to a 180° rotation of the (x, y) plane about the z axis. *Hint:* Compare Chapter 3, equation (7.13) with the negative unit matrix.

2. In Chapter 3, we said that any 3-by-3 orthogonal matrix with determinant $= -1$ can be written in the form (7.19). Use this and Problem 1 to show that in 3 dimensions, an inversion (that is a reflection through the origin so that all three axes are reversed) is equivalent to a reflection through a plane combined with a rotation about the line perpendicular to the plane [say a reflection through the (x, y) plane—that is, a reversal of the z axis—and a rotation of the (x, y) plane about the z axis]. *Hint:* Consider the matrix B in Chapter 3, (7.19).

3. For Example 1, write out the components of $\mathbf{U}$, $\mathbf{V}$, and $\mathbf{U} \times \mathbf{V}$ in the original right-handed coordinate system S and in the left-handed coordinate system S' with the z axis reflected. Show that each component of $\mathbf{U} \times \mathbf{V}$ in S' has the "wrong" sign to obey the vector transformation laws.

4. Do Example 1 and Problem 3 if the transformation to a left-handed system is an inversion (see Problem 2).

5. Write the tensor transformation equations for $\epsilon_{ijk}\epsilon_{mnp}$ to show that this is a (rank 6) tensor (*not* a pseudotensor). *Hint:* Write (6.1) for each ϵ and multiply them, being careful not to re-use a pair of summation indices.

6. Write the transformation equations to show that $\nabla \times \mathbf{V}$ is a pseudovector if $\mathbf{V}$ is a vector. *Hint:* See equations (5.13), (6.2) and (6.3).

7. Write the transformation equations for the triple scalar product $\mathbf{W} \cdot (\mathbf{U} \times \mathbf{V})$ remembering that now det A $= -1$ if the transformation involves a reflection. Thus show that the triple scalar product of three polar vectors is a pseudoscalar as claimed in Example 2. *Hint:* Use the result in (6.3).

8. Write the transformation equations for $\mathbf{W} \times \mathbf{S}$ to verify the results of Example 3.

In the physics formulas of Problems 9 to 14, identify each symbol as a vector (polar vector) or a pseudovector (axial vector). Use results from the text and the fact that both sides of an equation must have the same tensor character. The definition of the symbols used is: $\mathbf{r}$ = displacement, t = time, m = mass, q = electric charge, $\mathbf{v}$ = velocity, $\mathbf{F}$ = force, $\boldsymbol{\omega}$ = angular velocity, $\boldsymbol{\tau}$ = torque, $\mathbf{L}$ = angular momentum, T = kinetic energy, $\mathbf{E}$ = electric field, $\mathbf{B}$ = magnetic field. Assume that t, m, and q are scalars. Note that we are working in 3 dimensional physical space and assuming classical (that is nonrelativistic) physics.

9. $\mathbf{E} = \dfrac{\mathbf{F}}{q}$

10. $\mathbf{L} = m\mathbf{r} \times \mathbf{v} = m\mathbf{r} \times (\boldsymbol{\omega} \times \mathbf{r})$

11. $\boldsymbol{\tau} = \mathbf{r} \times \mathbf{F}$

12. $\mathbf{F} = q(\mathbf{E} + \mathbf{v} \times \mathbf{B})$

13. $\dfrac{\partial \mathbf{B}}{\partial t} = -\nabla \times \mathbf{E}$

14. $T = \frac{1}{2}m(\boldsymbol{\omega} \times \mathbf{r}) \cdot (\boldsymbol{\omega} \times \mathbf{r})$

15. In equation (5.12), find whether $\mathbf{A} \times (\mathbf{B} \times \mathbf{C})$ is a vector or a pseudovector assuming

(a) $\mathbf{A}$, $\mathbf{B}$, $\mathbf{C}$ are all vectors;

(b) $\mathbf{A}$, $\mathbf{B}$, $\mathbf{C}$ are all pseudovectors;

(c) $\mathbf{A}$ is a vector and $\mathbf{B}$ and $\mathbf{C}$ are pseudovectors.

Hint: Count up the number of det A factors from pseudovectors and cross products.

16. In equation (5.14), is $\nabla \times (\nabla \times \mathbf{V})$ a vector or a pseudovector?

17. In equation (5.16), show that if T_{jk} is a tensor (that is, not a pseudotensor), then V_i is a pseudovector (axial vector). Also show that if T_{jk} is a pseudotensor, then V_i is a vector (true or polar vector). You know that if V_i is a cross product of polar vectors, then it is a pseudovector. Is its dual T_{jk} a tensor or a pseudotensor?

7. MORE ABOUT APPLICATIONS

Stress Tensor We started our discussion of tensors with a description of the stress tensor (you may want to review this in Section 1). Now let's show that the nine quantities P_{ij} displayed in the matrix (1.1) really are the components of a 2nd-rank tensor. For simplicity in notation (and to use summation convention), we make the replacements indicated in (2.10); we also replace $\mathbf{i}$, $\mathbf{j}$, $\mathbf{k}$ by $\mathbf{e}_1$, $\mathbf{e}_2$, $\mathbf{e}_3$ and $\mathbf{i}'$, $\mathbf{j}'$, $\mathbf{k}'$ by $\mathbf{e}'_1$, $\mathbf{e}'_2$, $\mathbf{e}'_3$. Our problem is to write the components $P'_{\alpha\beta}$ relative to a rotated coordinate system in terms of the components P_{ij} to show that $P'_{\alpha\beta} = a_{\alpha i}a_{\beta j}P_{ij}$ as in (2.14) or (3.9).

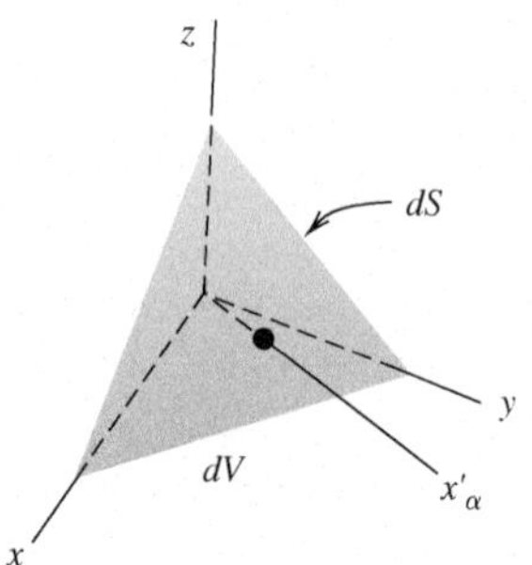

Figure 7.1

Figure 7.1 shows the unprimed axes and one of the rotated axes. (With $\alpha = 1, 2, 3$, the x'_α axis represents any one of the rotated axes.) We draw a slanted plane, as shown, perpendicular to the x'_α axis, and consider the forces on the small volume element dV bounded by the unprimed coordinate planes and the slanted plane. Recall (Section 1) that pressure is force per unit area, so the force acting across a face is the pressure times the area of the face. Let the area of the slanted face (call it face α) be dS. Then the area of the face perpendicular to the x_i axis (call it face i) is $a_{\alpha i}\,dS$ where $a_{\alpha i}$ [see (2.10)] is the cosine of the angle between the x'_α and x_i axes (Problem 1).

$$\text{Area of face } i \text{ is equal to } a_{\alpha i}\,dS. \tag{7.1}$$

The pressure across face i is $P_{ij}\mathbf{e}_j$ (note the sum on j and see Problem 2). Multiplying this by (7.1) (force = pressure times area of face) and summing on i, we find that the total force acting on the material in the volume element dV, across the three faces in the unprimed coordinate planes is

$$(P_{ij}\mathbf{e}_j)a_{\alpha i}\,dS. \tag{7.2}$$

For equilibrium, the sum of these three forces must be equal to the force acting across face α on the neighboring material. This force is

$$P'_{\alpha\beta}\mathbf{e}'_\beta\,dS \tag{7.3}$$

Setting (7.2) and (7.3) equal, taking the dot product of both sides with $\mathbf{e}'_\beta$, and canceling dS, we have (Problem 3)

$$P'_{\alpha\beta} = a_{\alpha i}a_{\beta j}P_{ij} \tag{7.4}$$

Thus we see that the stress P_{ij} is, as claimed, a 2nd-rank tensor.

Example 1. Suppose the following matrix is a display of the elements of a stress tensor.

$$\mathrm{P} = \begin{pmatrix} 1 & 3 & 0 \\ 3 & -2 & -1 \\ 0 & -1 & 1 \end{pmatrix}$$

We note that P is symmetric (this is true of stress tensors) so we can diagonalize P by an orthogonal transformation. In Chapter 3, Section 12, Example 2, we found that the eigenvalues of this matrix are 1, −4, 3. Thus a rotation of axes (matrix C in the Chapter 3 example) produces a stress tensor P′ with stress components only along the principal axes. The positive eigenvalues are tensions and the negative are compressions. Relative to the principal axes there are no shear forces.

Strain and Stress; Hooke's Law The strain tensor specifies the deformation of a solid body under stress. For a simple case such as a wire supporting a weight, strain (change in length per unit length) and stress (force per unit cross sectional area) are proportional (Hooke's Law). But for a 3 dimensional problem, stress is a 2nd-rank tensor P_{ij} (as we have seen above), and strain is also a 2nd-rank tensor S_{ij}. If the components of **P** are linear combinations of the components of **S**, then we can write

$$P_{ij} = C_{ijkm} S_{km} \tag{7.5}$$

By the quotient rule, C_{ijkm} is a 4th-rank tensor (Problem 5). The components of C_{ijkm} depend on the kind of material under stress and are called the elastic constants of the material (see Problem 6).

Inertia Tensor Revisited In Section 4 we considered the inertia tensor using vector notation. Now let's look at it using the tensor form for vector identities that we discussed in Section 5.

Example 2. In (4.2) we had $\mathbf{L} = m\mathbf{r} \times (\boldsymbol{\omega} \times \mathbf{r})$. Using (5.12) with $\mathbf{A} = \mathbf{C} = \mathbf{r}$ and $\mathbf{B} = \boldsymbol{\omega}$, we find

$$L_n = m[\mathbf{r} \times (\boldsymbol{\omega} \times \mathbf{r})]_n = m(\delta_{nj}\delta_{ik} - \delta_{nk}\delta_{ij})x_i\omega_j x_k. \tag{7.6}$$

Now sum over i and k to get $\delta_{nj}\delta_{ik}x_i x_k = \delta_{nj}x_k x_k = \delta_{nj}r^2$ and $\delta_{nk}\delta_{ij}x_i x_k = x_j x_n$. Thus we have [compare (4.2)]

$$L_n = m(\delta_{nj}r^2 - x_n x_j)\omega_j. \tag{7.7}$$

The coefficient of ω_j is then the component I_{nj} of the inertia tensor.

$$I_{nj} = m(\delta_{nj}\, r^2 - x_n x_j). \tag{7.8}$$

We can easily verify that these components are the same as we found in Section 4. For example [compare (4.4)]:

$$I_{11} = m(r^2 - x_1^2), \quad I_{12} = -m x_1 x_2, \quad I_{13} = -m x_1 x_3, \tag{7.9}$$

and similarly for the other components (Problem 7).

Other Applications In your study of electric fields in matter, you will find the equation $\mathbf{P} = \chi\mathbf{E}$; this relates the electric field $\mathbf{E}$ applied to a dielectric and the resulting polarization $\mathbf{P}$ of the dielectric. For some materials it may be true that $\mathbf{P}$ and $\mathbf{E}$ are parallel vectors with $\chi =$ scalar, but for other materials $\mathbf{P}$ and $\mathbf{E}$ are not parallel. Now this should remind you of our work in Section 4 with the equation $\mathbf{L} = I\boldsymbol{\omega}$ when we realized that $\mathbf{L}$ and $\boldsymbol{\omega}$ are not always parallel. Just as we replaced the scalar I by a 2nd-rank tensor, so we replace χ by a 2nd-rank tensor. In the equation $P_i = \chi_{ij}E_j$, the quotient rule (see Section 3) tells us that χ_{ij} is a 2nd-rank tensor. You will find other equations of this sort in various applications.

Tensor Fields Recall from Chapter 6 that a scalar field (temperature, for example) means a single number at each point, that is, a single function $f(x, y, z)$. A vector field (such as the electric field) means a set of three numbers at each point, that is, a set of three functions $V_i(x, y, z)$. Similarly, a 2nd-rank tensor field means a set of 9 numbers at each point, that is, a set of 9 functions $T_{ij}(x, y, z)$. Think of our discussion of stress and strain. At every point in the material under stress, we can think of three vectors giving the force per unit area across the three perpendicular planes through the point, that is, a set of 9 functions. The 4th-rank tensor C_{ijkm} in (7.5) is then a set of $3^4 = 81$ functions, and so on. (Of course, in order to be tensors these sets must transform properly under rotations as discussed in this chapter.)

PROBLEMS, SECTION 7

1. Verify (7.1). *Hints:* In Figure 7.1, consider the projection of the slanted face of area dS onto the three unprimed coordinate planes. In each case, show that the projection angle is equal to an angle between the x'_α axis and one of the unprimed axes. Find the cosine of the angle from the matrix A in (2.10).

2. Write out the sums $P_{ij}\mathbf{e}_j$ for each value of i and compare the discussion of (1.1). *Hint:* For example, if $i = 2$ [or y in (1.1)], then the pressure across the face perpendicular to the x_2 axis is $P_{21}\mathbf{e}_1 + P_{22}\mathbf{e}_{22} + P_{23}\mathbf{e}_3$, or, in the notation of (1.1), $P_{yx}\mathbf{i} + P_{yy}\mathbf{j} + P_{yz}\mathbf{k}$.

3. Carry through the details of getting (7.4) from (7.2) and (7.3). *Hint:* You need the dot product of $\mathbf{e}'_\beta$ and $\mathbf{e}_j$. This is the cosine of an angle between two axes since each $\mathbf{e}$ is a unit vector. Identify the result from matrix A in (2.10).

4. Interpret the elements of the matrices in Chapter 3, Problems 11.18 to 11.21, as components of stress tensors. In each case diagonalize the matrix and so find the principal axes of the stress (along which the stress is pure tension or compression). Describe the stress relative to these axes. (See Example 1.)

5. Show by the quotient rule (Section 3) that C_{ijkm} in (7.5) is a 4th-rank tensor.

6. If P and S are 2nd-rank tensors, show that $9^2 = 81$ coefficients are needed to write each component of P as a linear combination of the components of S. Show that $81 = 3^4$ is the number of components in a 4th-rank tensor. If the components of the 4th-rank tensor are C_{ijkm}, then equation (7.5) gives the components of P in terms of the components of S. If P and S are both symmetric, show that we need only 36 different non-zero components in C_{ijkm}. *Hint:* Consider the number of different components in P and S when they are symmetric. *Comment:* The stress and strain tensors can both be shown to be symmetric. Further symmetry reduces the 36 components of C in (7.5) to 21 or less.

7. In (7.9) we have written the first row of elements in the inertia matrix. Write the formulas for the other 6 elements and compare with Section 4.

8. Do Problem 4.8 in tensor notation and compare the result with your solution of 4.8.

8. CURVILINEAR COORDINATES

Before we discuss non-Cartesian tensors we need to talk about some properties of curvilinear coordinate systems such as spherical or cylindrical coordinates. To make the discussion concrete, we shall illustrate the ideas involved by using two familiar coordinate systems—rectangular coordinates (x, y, z) and cylindrical coordinates (r, θ, z). The elements of arc length in these two systems are given by

$$(8.1)\qquad \begin{aligned} ds^2 &= dx^2 + dy^2 + dz^2 \qquad \text{(rectangular coordinates)} \\ ds^2 &= dr^2 + r^2\,d\theta^2 + dz^2 \qquad \text{(cylindrical coordinates)} \end{aligned}$$

These expressions for ds are called *line elements*; they have much greater significance than just their use in computing arc lengths. First consider how we can find ds^2 for a given coordinate system. In the case of a well-known coordinate system, the answer may be obvious from the geometry. For example in polar coordinates in the plane we have (from Figure 8.1 and the Pythagorean theorem)

$$(8.2)\qquad ds^2 = dr^2 + r^2\,d\theta^2.$$

For an unfamiliar or complicated change of variables, however, we need a systematic method of finding ds; we illustrate the method by finding the value of ds^2 for cylindrical coordinates as given in (8.1). From the equations

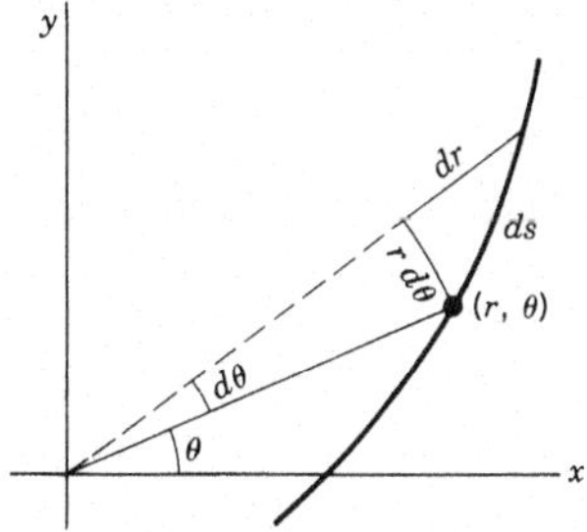

Figure 8.1

$$(8.3)\qquad \begin{aligned} x &= r\cos\theta, \\ y &= r\sin\theta, \\ z &= z, \end{aligned}$$

we get

$$(8.4)\qquad \begin{aligned} dx &= \cos\theta\,dr - r\sin\theta\,d\theta, \\ dy &= \sin\theta\,dr + r\cos\theta\,d\theta, \\ dz &= dz. \end{aligned}$$

Squaring each equation in (8.4) and adding the results, we find

$$(8.5)\qquad ds^2 = dx^2 + dy^2 + dz^2 = dr^2 + r^2\,d\theta^2 + dz^2.$$

Notice particularly here that all the cross products ($dr\,d\theta$, etc.) canceled out. This will not always happen, but it often does; when it does we call the coordinate system *orthogonal*. Such coordinate systems have some particularly simple and useful properties. Geometrically, an orthogonal system means that the *coordinate surfaces* are mutually perpendicular. For the cylindrical system (Figure 8.2), the coordinate surfaces are $r =$ const. (set of concentric cylinders), $\theta =$ const. (set of half-planes), and $z =$ const. (set of planes). The three coordinate surfaces through a given point intersect at right angles. The three curves of intersection of the coordinate surfaces in pairs intersect at right

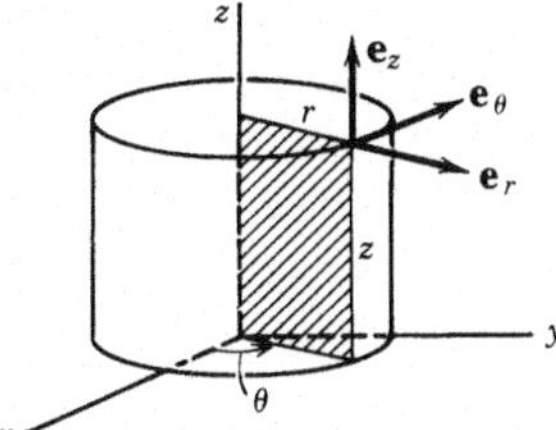

Figure 8.2

angles; these curves are called the *coordinate "lines"* or directions. We draw unit basis vectors tangent to the coordinate directions; for the cylindrical system (Figure 8.2) we might call them $\mathbf{e}_r$, $\mathbf{e}_\theta$, $\mathbf{e}_z$ ($\mathbf{e}_z$ is identical to $\mathbf{k}$). These unit vectors form an orthogonal triad like $\mathbf{i}$, $\mathbf{j}$, $\mathbf{k}$. We refer to such coordinate systems as *curvilinear coordinate systems* when the coordinate surfaces (or some of them) are not planes and the coordinate lines (or some of them) are curves rather than straight lines. We shall be principally interested in orthogonal curvilinear coordinate systems.

Scale Factors and Basis Vectors In the rectangular system, if x, y, z, are the coordinates of a particle, and x changes by dx with y and z constant, then the distance the particle moves is $ds = dx$. However, in the cylindrical system, if θ changes by $d\theta$ with r and z constant, the distance the particle moves is *not* $d\theta$, but $ds = r\,d\theta$. Factors like the r in $r\,d\theta$ which must multiply the differentials of the coordinates to get distances are known as *scale factors* and are very important as we shall see. A straightforward way to get them is to calculate ds^2 as we did in (8.5); if the transformation is orthogonal, then the scale factors can be read off from ds^2. (Note that the coefficients in ds^2 are the squares of the scale factors.) From (8.5), we see that the scale factors for cylindrical coordinates are 1, r, 1.

It is also useful to consider a vector $d\mathbf{s}$ which (in cylindrical coordinates) has components dr, $r\,d\theta$, dz in the coordinate directions $\mathbf{e}_r$, $\mathbf{e}_\theta$, $\mathbf{e}_z$:

$$d\mathbf{s} = \mathbf{e}_r\,dr + \mathbf{e}_\theta\,r\,d\theta + \mathbf{e}_z\,dz. \tag{8.6}$$

Then $ds^2 = d\mathbf{s} \cdot d\mathbf{s}$ which gives (8.1), since the $\mathbf{e}$ vectors are orthonormal.

We can write the unit basis vectors of a curvilinear coordinate system ($\mathbf{e}_r$, $\mathbf{e}_\theta$, $\mathbf{e}_z$ in cylindrical coordinates) in terms of $\mathbf{i}$, $\mathbf{j}$, $\mathbf{k}$. This is useful when we want to differentiate a vector which is expressed in terms of the curvilinear coordinate basis vectors. The unit vectors $\mathbf{i}$, $\mathbf{j}$, $\mathbf{k}$ are constant in magnitude *and direction*, but $\mathbf{e}_r$ and $\mathbf{e}_\theta$ are not fixed in direction, so their derivatives are not zero. We illustrate an algebraic method of finding the relation between two sets of basis vectors by finding them for the cylindrical system. (Compare the geometrical method shown in Chapter 6, Section 4.)

▶ **Example 1.** Using (8.4) and collecting coefficients of dr, $d\theta$, and dz, we find

$$\begin{aligned} d\mathbf{s} &= \mathbf{i}\,dx + \mathbf{j}\,dy + \mathbf{k}\,dz \\ &= \mathbf{i}(\cos\theta\,dr - r\sin\theta\,d\theta) + \mathbf{j}\,(\sin\theta\,dr + r\cos\theta\,d\theta) + \mathbf{k}\,dz \\ &= (\mathbf{i}\cos\theta + \mathbf{j}\sin\theta)\,dr + (-\mathbf{i}r\sin\theta + \mathbf{j}\,r\cos\theta)\,d\theta + \mathbf{k}\,dz. \end{aligned} \tag{8.7}$$

Comparing (8.7) with (8.6), we have

$$\begin{aligned} \mathbf{e}_r &= \mathbf{i}\cos\theta + \mathbf{j}\sin\theta \\ r\mathbf{e}_\theta &= -\mathbf{i}r\sin\theta + \mathbf{j}\,r\cos\theta \\ \mathbf{e}_z &= \mathbf{k}. \end{aligned} \tag{8.8}$$

Notice that $\mathbf{e}_r$ is already a unit vector since $\sin^2\theta + \cos^2\theta = 1$, but $r\mathbf{e}_\theta$ must be divided by the scale factor r to get the unit vector $\mathbf{e}_\theta$. It is often convenient to use basis vectors which we shall call $\mathbf{a}_r$ and $\mathbf{a}_\theta$ (which are not necessarily of unit length), given by the coefficients of dr and $d\theta$ in (8.7). Then we just have to divide each **a** vector by its magnitude to get the corresponding **e** vector. Thus from (8.7)

$$\begin{aligned} &\mathbf{a}_r = \mathbf{e}_r \text{ is already a unit vector,} \\ &\mathbf{a}_\theta = -\mathbf{i} r\sin\theta + \mathbf{j}\, r\cos\theta \text{ has magnitude } r \text{, so} \\ &\mathbf{e}_\theta = \frac{1}{r}\mathbf{a}_\theta = -\mathbf{i}\sin\theta + \mathbf{j}\cos\theta \end{aligned} \tag{8.9}$$

We can use these formulas to find the velocity and acceleration of a particle in cylindrical coordinates, and similar formulas for any coordinate system. The displacement of a particle from the origin at time t is, in cylindrical coordinates (Figure 8.3),

$$\mathbf{s} = r\mathbf{e}_r + z\mathbf{e}_z.$$

Then

$$\frac{d\mathbf{s}}{dt} = \frac{dr}{dt}\mathbf{e}_r + r\frac{d}{dt}(\mathbf{e}_r) + \frac{dz}{dt}\mathbf{e}_z.$$

By (8.8),

$$\frac{d}{dt}(\mathbf{e}_r) = -\mathbf{i}\sin\theta\frac{d\theta}{dt} + \mathbf{j}\cos\theta\frac{d\theta}{dt} = \mathbf{e}_\theta\frac{d\theta}{dt},$$

so

$$\frac{d\mathbf{s}}{dt} = \dot{r}\mathbf{e}_r + r\dot{\theta}\mathbf{e}_\theta + \dot{z}\mathbf{e}_z. \tag{8.10}$$

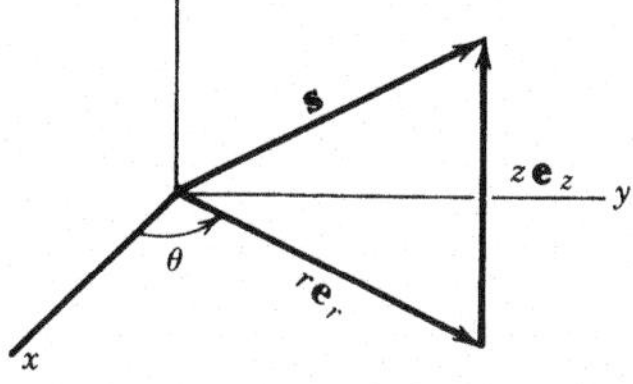

Figure 8.3

By differentiating again with respect to t and using (8.8) to find $(d/dt)(\mathbf{e}_\theta)$, we can find the acceleration $d^2\mathbf{s}/dt^2$ in cylindrical coordinates (Problem 2).

General Curvilinear Coordinates In general, let x_1, x_2, x_3 be the set of variables or coordinates we are considering (for example, in cylindrical coordinates, $x_1 = r$, $x_2 = \theta$, $x_3 = z$). Then the three sets of coordinate surfaces are $x_1 =$ const., $x_2 =$ const., $x_3 =$ const. The three coordinate surfaces through a given point intersect in three coordinate lines.

Example 2. Given x, y, z as functions of x_1, x_2, x_3, we can find $d\mathbf{s}$ and the **a** vectors as we did for cylindrical coordinates in (8.7) and (8.9).

$$\begin{aligned} d\mathbf{s} &= \mathbf{i}\,dx + \mathbf{j}\,dy + \mathbf{k}\,dz \\ &= \mathbf{i}\frac{\partial x}{\partial x_n}\,dx_n + \mathbf{j}\frac{\partial y}{\partial x_n}\,dx_n + \mathbf{k}\frac{\partial z}{\partial x_n}\,dx_n \\ &= \mathbf{a}_1\,dx_1 + \mathbf{a}_2\,dx_2 + \mathbf{a}_3\,dx_3 = \mathbf{a}_n\,dx_n, \end{aligned} \tag{8.11}$$

where

$$\mathbf{a}_n = \frac{\partial}{\partial x_n}\mathbf{s} = \mathbf{i}\frac{\partial x}{\partial x_n} + \mathbf{j}\frac{\partial y}{\partial x_n} + \mathbf{k}\frac{\partial z}{\partial x_n}. \tag{8.12}$$

Now defining $g_{ij} = \mathbf{a}_i \cdot \mathbf{a}_j$, we can write $ds^2 = d\mathbf{s} \cdot d\mathbf{s}$ in matrix form as follows:

$$ds^2 = \begin{pmatrix} dx_1 & dx_2 & dx_3 \end{pmatrix} \begin{pmatrix} g_{11} & g_{12} & g_{13} \\ g_{21} & g_{22} & g_{23} \\ g_{31} & g_{32} & g_{33} \end{pmatrix} \begin{pmatrix} dx_1 \\ dx_2 \\ dx_3 \end{pmatrix}, \tag{8.13}$$

Note that g_{ij} is symmetric since the dot product of two vectors is the same in either order. In simpler form using summation convention (8.13) becomes

$$ds^2 = g_{ij}\, dx_i\, dx_j. \tag{8.14}$$

We will see later (Section 10) that the g_{ij} are the components of a tensor known as the *metric tensor.*

If the coordinate system is orthogonal, that is, if the basis vectors (**e** or **a**) form an orthogonal triad, then $d\mathbf{s}$ and ds^2 can be written in terms of the scale factors as follows:

$$d\mathbf{s} = \mathbf{e}_1 h_1\, dx_1 + \mathbf{e}_2 h_2\, dx_2 + \mathbf{e}_3 h_3\, dx_3, \tag{8.15}$$

$$ds^2 = \begin{pmatrix} dx_1 & dx_2 & dx_3 \end{pmatrix} \begin{pmatrix} h_1^2 & 0 & 0 \\ 0 & h_2^2 & 0 \\ 0 & 0 & h_3^2 \end{pmatrix} \begin{pmatrix} dx_1 \\ dx_2 \\ dx_3 \end{pmatrix}. \tag{8.16}$$

Also note that the volume element in an orthogonal system is $h_1 h_2 h_3\, dx_1\, dx_2\, dx_3$ (volume of a small rectangular parallelepiped with edges $h_1\, dx_1$, $h_2\, dx_2$, $h_3\, dx_3$). For example, in cylindrical coordinates, the volume element is $dr \cdot r\, d\theta \cdot dz = r\, dr\, d\theta\, dz$.

PROBLEMS, SECTION 8

1. Find ds^2 in spherical coordinates by the method used to obtain (8.5) for cylindrical coordinates. Use your result to find for spherical coordinates, the scale factors, the vector $d\mathbf{s}$, the volume element, the basis vectors $\mathbf{a}_r$, $\mathbf{a}_\theta$, $\mathbf{a}_\phi$ and the corresponding unit basis vectors $\mathbf{e}_r$, $\mathbf{e}_\theta$, $\mathbf{e}_\phi$. Write the g_{ij} matrix.

2. Observe that a simpler way to find the velocity $d\mathbf{s}/dt$ in (8.10) is to divide the vector $d\mathbf{s}$ in (8.6) by dt. Complete the problem to find the acceleration in cylindrical coordinates.

3. Use the results of Problem 1 to find the velocity and acceleration components in spherical coordinates. Find the velocity in two ways: starting with $d\mathbf{s}$ and starting with $\mathbf{s} = r\mathbf{e}_r$.

4. In the text and problems so far, we have found the **e** vectors for various coordinate systems in terms of **i** and **j** (or **i**, **j**, **k** in three dimensions). We can solve these equations to find **i** and **j** in terms of the **e** vectors, and so express a vector given in rectangular form in terms of the basis vectors of another coordinate system. Carry out this process to express in cylindrical coordinates the vector $\mathbf{V} = y\mathbf{i} - x\mathbf{j} + \mathbf{k}$. *Hint:* Use matrices (as in Chapter 3) to solve the set of equations for **i** and **j**.

5. Using the results of Problem 1, express the vector in Problem 4 in spherical coordinates.

As in Problem 1, find ds^2, the scale factors, the vector $d\mathbf{s}$, the volume (or area) element, the **a** vectors, and the **e** vectors for each of the following coordinate systems.

6. Parabolic cylinder coordinates u, v, z:

$$x = \frac{1}{2}(u^2 - v^2),$$
$$y = uv,$$
$$z = z.$$

7. Elliptic cylinder coordinates u, v, z:

$$x = a \cosh u \cos v,$$
$$y = a \sinh u \sin v,$$
$$z = z.$$

8. Parabolic coordinates u, v, ϕ:

$$x = uv\cos\phi,$$
$$y = uv\sin\phi,$$
$$z = \frac{1}{2}(u^2 - v^2).$$

9. Bipolar coordinates u, v:

$$x = \frac{a\sinh u}{\cosh u + \cos v},$$
$$y = \frac{a\sin v}{\cosh u + \cos v}.$$

10. Sketch or computer plot the coordinate surfaces in Problems 6 to 9.

Using the expression you have found for ds, and for the $\mathbf{e}$ vectors, find the velocity and acceleration components in the coordinate systems indicated.

11. Parabolic cylinder

12. Elliptic cylinder

13. Parabolic

14. Bipolar

15. Let $x = u + v$, $y = v$. Find ds, the $\mathbf{a}$ vectors, and ds^2 for the u, v coordinate system and show that it is not an orthogonal system. *Hint:* Show that the $\mathbf{a}$ vectors are not orthogonal, and that ds^2 contains $du\,dv$ terms. Write the g_{ij} matrix and observe that it is symmetric but not diagonal. Sketch the lines $u =$ const. and $v =$ const. and observe that they are not perpendicular to each other.

9. VECTOR OPERATORS IN ORTHOGONAL CURVILINEAR COORDINATES

We have previously (Chapter 6, Sections 6 and 7) defined the gradient (∇u), the divergence ($\nabla \cdot \mathbf{V}$), the curl ($\nabla \times \mathbf{V}$), and the Laplacian ($\nabla^2 u$) in rectangular coordinates x, y, z. Since in many practical problems it is better to use some other coordinate system (cylindrical or spherical, for example), we need to see how to express the vector operators in terms of general orthogonal coordinates x_1, x_2, x_3. (We consider only orthogonal coordinate systems here; see Section 10 for the more general case.) We shall outline proofs of the formulas; some of the details of the proofs are left to the problems.

Gradient, ∇u. In Chapter 6, Section 6, we showed that the directional derivative du/ds in a given direction is the component of ∇u in that direction.

Example 1. In cylindrical coordinates, if we go in the r direction (θ and z constant), then by (8.5) $ds = dr$. Thus the r component of ∇u is du/ds when $ds = dr$, that is, $\partial u/\partial r$. Similarly, the θ component of ∇u is du/ds when $ds = r\,d\theta$, that is, $(1/r)(\partial u/\partial\theta)$. Thus ∇u in cylindrical coordinates is

$$\nabla u = \mathbf{e}_r\frac{\partial u}{\partial r} + \mathbf{e}_\theta\frac{1}{r}\frac{\partial u}{\partial\theta} + \mathbf{e}_z\frac{\partial u}{\partial z}. \tag{9.1}$$

In general orthogonal coordinates x_1, x_2, x_3, the component of ∇u in the x_1 direction (x_2 and x_3 constant) is du/ds if $ds = h_1\,dx_1$ [from (8.11)]; that is, the component of ∇u in the direction $\mathbf{e}_1$ is $(1/h_1)(\partial u/\partial x_1)$. Similar formulas hold for the other components and we have

$$\begin{aligned}\nabla u &= \mathbf{e}_1\frac{1}{h_1}\frac{\partial u}{\partial x_1} + \mathbf{e}_2\frac{1}{h_2}\frac{\partial u}{\partial x_2} + \mathbf{e}_3\frac{1}{h_3}\frac{\partial u}{\partial x_3}\\ &= \sum_{i=1}^{3}\mathbf{e}_i\frac{1}{h_i}\frac{\partial u}{\partial x_i}.\end{aligned} \tag{9.2}$$

Divergence, $\nabla \cdot \mathbf{V}$ Let

$$\mathbf{V} = \mathbf{e}_1 V_1 + \mathbf{e}_2 V_2 + \mathbf{e}_3 V_3 \tag{9.3}$$

be a vector with components V_1, V_2, V_3 in an orthogonal system. We can prove (Problem 1) that

$$\nabla \cdot \left(\frac{\mathbf{e}_3}{h_1 h_2}\right) = 0, \qquad \nabla \cdot \left(\frac{\mathbf{e}_2}{h_1 h_3}\right) = 0, \qquad \nabla \cdot \left(\frac{\mathbf{e}_1}{h_2 h_3}\right) = 0. \tag{9.4}$$

Let us write (9.3) as

$$\mathbf{V} = \frac{\mathbf{e}_1}{h_2 h_3}(h_2 h_3 V_1) + \frac{\mathbf{e}_2}{h_1 h_3}(h_1 h_3 V_2) + \frac{\mathbf{e}_3}{h_1 h_2}(h_1 h_2 V_3). \tag{9.5}$$

We find $\nabla \cdot \mathbf{V}$ by taking the divergence of each term on the right side of (9.5). Using (7.6) of Chapter 6, namely

$$\nabla \cdot (\phi \mathbf{v}) = \mathbf{v} \cdot (\nabla \phi) + \phi \nabla \cdot \mathbf{v}, \tag{9.6}$$

with $\phi = h_2 h_3 V_1$ and $\mathbf{v} = \mathbf{e}_1 / h_2 h_3$, we find that the divergence of the first term on the right side of (9.5) is

$$\nabla \cdot \left(h_2 h_3 V_1 \frac{\mathbf{e}_1}{h_2 h_3}\right) = \frac{\mathbf{e}_1}{h_2 h_3} \cdot \nabla (h_2 h_3 V_1) + h_2 h_3 V_1 \nabla \cdot \left(\frac{\mathbf{e}_1}{h_2 h_3}\right). \tag{9.7}$$

By (9.4), the last term in (9.7) is zero. In the first term on the right side of (9.7), the dot product of $\mathbf{e}_1$ with $\nabla(h_2 h_3 V_1)$ is the first component of $\nabla(h_2 h_3 V_1)$. By (9.2), this is

$$\frac{1}{h_1}\frac{\partial}{\partial x_1}(h_2 h_3 V_1).$$

Calculating the divergence of the other terms of (9.5) in a similar way, we get

$$\nabla \cdot \mathbf{V} = \frac{1}{h_2 h_3}\frac{1}{h_1}\frac{\partial}{\partial x_1}(h_2 h_3 V_1) + \frac{1}{h_1 h_3}\frac{1}{h_2}\frac{\partial}{\partial x_2}(h_1 h_3 V_2) + \frac{1}{h_1 h_2}\frac{1}{h_3}\frac{\partial}{\partial x_3}(h_1 h_2 V_3)$$

or

$$\nabla \cdot \mathbf{V} = \frac{1}{h_1 h_2 h_3}\left[\frac{\partial}{\partial x_1}(h_2 h_3 V_1) + \frac{\partial}{\partial x_2}(h_1 h_3 V_2) + \frac{\partial}{\partial x_3}(h_1 h_2 V_3)\right]. \tag{9.8}$$

▶ **Example 2.** In cylindrical coordinates, $h_1 = 1$, $h_2 = r$, $h_3 = 1$. By (9.8), the divergence in cylindrical coordinates is

$$\begin{aligned} \nabla \cdot \mathbf{V} &= \frac{1}{r}\left[\frac{\partial}{\partial r}(r V_r) + \frac{\partial}{\partial \theta}(V_\theta) + \frac{\partial}{\partial z}(r V_z)\right] \\ &= \frac{1}{r}\frac{\partial}{\partial r}(r V_r) + \frac{1}{r}\frac{\partial V_\theta}{\partial \theta} + \frac{\partial V_z}{\partial z}. \end{aligned} \tag{9.9}$$

Laplacian, $\nabla^2 u$. Since $\nabla^2 u = \nabla \cdot \nabla u$ we can find $\nabla^2 u$ by combining (9.2) and (9.8) with $\mathbf{V} = \nabla u$. We get

(9.10)
$$\nabla^2 u = \frac{1}{h_1 h_2 h_3}\left[\frac{\partial}{\partial x_1}\left(\frac{h_2 h_3}{h_1}\frac{\partial u}{\partial x_1}\right) + \frac{\partial}{\partial x_2}\left(\frac{h_1 h_3}{h_2}\frac{\partial u}{\partial x_2}\right) + \frac{\partial}{\partial x_3}\left(\frac{h_1 h_2}{h_3}\frac{\partial u}{\partial x_3}\right)\right].$$

Example 3. In cylindrical coordinates, the Laplacian is then

$$\begin{aligned}\nabla^2 u &= \frac{1}{r}\left[\frac{\partial}{\partial r}\left(r\frac{\partial u}{\partial r}\right) + \frac{\partial}{\partial \theta}\left(\frac{1}{r}\frac{\partial u}{\partial \theta}\right) + \frac{\partial}{\partial z}\left(r\frac{\partial u}{\partial z}\right)\right]\\ &= \frac{1}{r}\frac{\partial}{\partial r}\left(r\frac{\partial u}{\partial r}\right) + \frac{1}{r^2}\frac{\partial^2 u}{\partial \theta^2} + \frac{\partial^2 u}{\partial z^2}.\end{aligned}$$

Curl, $\nabla \times \mathbf{V}$. By methods similar to those used in finding $\nabla \cdot \mathbf{V}$ we can find $\nabla \times \mathbf{V}$ (Problem 2). The result is

(9.11)
$$\begin{aligned}\nabla \times \mathbf{V} &= \frac{1}{h_1 h_2 h_3}\begin{vmatrix} h_1\mathbf{e}_1 & h_2\mathbf{e}_2 & h_3\mathbf{e}_3 \\ \dfrac{\partial}{\partial x_1} & \dfrac{\partial}{\partial x_2} & \dfrac{\partial}{\partial x_3} \\ h_1 V_1 & h_2 V_2 & h_3 V_3 \end{vmatrix}\\ &= \frac{\mathbf{e}_1}{h_2 h_3}\left[\frac{\partial}{\partial x_2}(h_3 V_3) - \frac{\partial}{\partial x_3}(h_2 V_2)\right]\\ &\quad + \frac{\mathbf{e}_2}{h_1 h_3}\left[\frac{\partial}{\partial x_3}(h_1 V_1) - \frac{\partial}{\partial x_1}(h_3 V_3)\right]\\ &\quad + \frac{\mathbf{e}_3}{h_1 h_2}\left[\frac{\partial}{\partial x_1}(h_2 V_2) - \frac{\partial}{\partial x_2}(h_1 V_1)\right]\end{aligned}$$

Example 4. In cylindrical coordinates, we find

$$\begin{aligned}\nabla \times \mathbf{V} &= \frac{1}{r}\begin{vmatrix} \mathbf{e}_r & r\mathbf{e}_\theta & \mathbf{e}_z \\ \dfrac{\partial}{\partial r} & \dfrac{\partial}{\partial \theta} & \dfrac{\partial}{\partial z} \\ V_r & rV_\theta & V_z \end{vmatrix}\\ &= \mathbf{e}_r\left(\frac{1}{r}\frac{\partial V_z}{\partial \theta} - \frac{\partial V_\theta}{\partial z}\right) + \mathbf{e}_\theta\left(\frac{\partial V_r}{\partial z} - \frac{\partial V_z}{\partial r}\right) + \frac{1}{r}\mathbf{e}_z\left(\frac{\partial}{\partial r}(rV_\theta) - \frac{\partial V_r}{\partial \theta}\right).\end{aligned}$$

PROBLEMS, SECTION 9

1. Prove (9.4) in the following way. Using (9.2) with $u = x_1$, show that $\nabla x_1 = \mathbf{e}_1/h_1$. Similarly, show that $\nabla x_2 = \mathbf{e}_2/h_2$ and $\nabla x_3 = \mathbf{e}_3/h_3$. Let $\mathbf{e}_1$, $\mathbf{e}_2$, $\mathbf{e}_3$ in that order form a right-handed triad (so that $\mathbf{e}_1 \times \mathbf{e}_2 = \mathbf{e}_3$, etc.) and show that $\nabla x_1 \times \nabla x_2 = \mathbf{e}_3/(h_1 h_2)$. Take the divergence of this equation and, using the vector identities (h) and (b) in the table at the end of Chapter 6, show that $\nabla \cdot (\mathbf{e}_3/h_1 h_2) = 0$. The other parts of (9.4) are proved similarly.

2. Derive the expression (9.11) for curl $\mathbf{V}$ in the following way. Show that $\nabla x_1 = \mathbf{e}_1/h_1$ and $\nabla \times (\nabla x_1) = \nabla \times (e_1/h_1) = 0$. Write $\mathbf{V}$ in the form

$$\mathbf{V} = \frac{\mathbf{e}_1}{h_1}(h_1 V_1) + \frac{\mathbf{e}_2}{h_2}(h_2 V_2) + \frac{\mathbf{e}_3}{h_3}(h_3 V_3)$$

and use vector identities from Chapter 6 to complete the derivation.

3. Using cylindrical coordinates write the Lagrange equations for the motion of a particle acted on by a force $\mathbf{F} = -\nabla V$, where V is the potential energy. Divide each Lagrange equation by the corresponding scale factor so that the components of $\mathbf{F}$ (that is, of $-\nabla V$) appear in the equations. Thus write the equations as the component equations of $\mathbf{F} = m\mathbf{a}$, and so find the components of the acceleration $\mathbf{a}$. Compare the results with Problem 8.2.

4. Do Problem 3 in spherical coordinates; compare the results with Problem 8.3.

5. Write out ∇U, $\nabla \cdot \mathbf{V}$, $\nabla^2 U$, and $\nabla \times \mathbf{V}$ in spherical coordinates.

Do Problem 3 for the coordinate systems indicated in Problems 6 to 9. Compare the results with Problems 8.11 to 8.14.

6. Parabolic cylinder

7. Elliptic cylinder

8. Parabolic

9. Bipolar

Do Problem 5 for the coordinate systems indicated in Problems 10 to 13.

10. Parabolic cylinder

11. Elliptic cylinder

12. Parabolic

13. Bipolar

In each of the following coordinate systems, find the scale factors h_u and h_v; the basis vectors $\mathbf{e}_u$ and $\mathbf{e}_v$; the u and v Lagrange equations, and from them the acceleration components (see Problem 3).

14. $\begin{cases} x = u - v, \\ y = 2\sqrt{uv}. \end{cases}$

15. $\begin{cases} x = uv, \\ y = u\sqrt{1-v^2}. \end{cases}$

Use equations (9.2), (9.8), and (9.11) to evaluate the following expressions

16. In cylindrical coordinates, $\nabla \cdot \mathbf{e}_r$, $\nabla \cdot \mathbf{e}_\theta$, $\nabla \times \mathbf{e}_r$, $\nabla \times \mathbf{e}_\theta$.

17. In spherical coordinates, $\nabla \cdot \mathbf{e}_r$, $\nabla \cdot \mathbf{e}_\theta$, $\nabla \times \mathbf{e}_\theta$, $\nabla \times \mathbf{e}_\phi$.

18. In cylindrical coordinates, $\nabla \times \mathbf{k} \ln r$, $\nabla \ln r$, $\nabla \cdot (r\mathbf{e}_r + z\mathbf{e}_z)$.

19. In spherical coordinates, $\nabla \times (r\mathbf{e}_\theta)$, $\nabla (r \cos\theta)$, $\nabla \cdot \mathbf{r}$.

20. In cylindrical coordinates, $\nabla^2 r$, $\nabla^2 (1/r)$, $\nabla^2 \ln r$.

21. In spherical coordinates, $\nabla^2 r$, $\nabla^2 (r^2)$, $\nabla^2 (1/r^2)$, $\nabla^2 e^{ikr\cos\theta}$.

10. NON-CARTESIAN TENSORS

So far we have considered only the behavior of the rectangular components of tensors under orthogonal transformations. Now let's generalize this to include any change of variables.

Example 1. In spherical coordinates r, θ, ϕ,

$$x = r\sin\theta\cos\phi,\quad y = r\sin\theta\sin\phi,\quad z = r\cos\theta. \tag{10.1}$$

This is not a linear transformation, and we cannot write equations like (2.4) to (2.9) for the relations between the *variables*. However, we *can* write such equations for the relations between the *differentials* of the variables. From (10.1), we find the differentials dx, dy, dz, in terms of dr, $d\theta$, $d\phi$:

$$\begin{pmatrix} dx \\ dy \\ dz \end{pmatrix} = \begin{pmatrix} \sin\theta\cos\phi & r\cos\theta\cos\phi & -r\sin\theta\sin\phi \\ \sin\theta\sin\phi & r\cos\theta\sin\phi & r\sin\theta\cos\phi \\ \cos\theta & -r\sin\theta & 0 \end{pmatrix} \begin{pmatrix} dr \\ d\theta \\ d\phi \end{pmatrix}. \tag{10.2}$$

Example 2. For general coordinates x_1, x_2, x_3, and x'_1, x'_2, x'_3, if we are given the relations [like (10.1)] between the two sets of variables, we can write the relations between the two sets of differentials as follows:

$$\begin{pmatrix} dx'_1 \\ dx'_2 \\ dx'_3 \end{pmatrix} = \begin{pmatrix} \dfrac{\partial x'_1}{\partial x_1} & \dfrac{\partial x'_1}{\partial x_2} & \dfrac{\partial x'_1}{\partial x_3} \\ \dfrac{\partial x'_2}{\partial x_1} & \dfrac{\partial x'_2}{\partial x_2} & \dfrac{\partial x'_2}{\partial x_3} \\ \dfrac{\partial x'_3}{\partial x_1} & \dfrac{\partial x'_3}{\partial x_2} & \dfrac{\partial x'_3}{\partial x_3} \end{pmatrix} \begin{pmatrix} dx_1 \\ dx_2 \\ dx_3 \end{pmatrix}. \tag{10.3}$$

More simply, using index notation and summation convention, (10.3) becomes

$$dx'_i = \frac{\partial x'_i}{\partial x_j}\,dx_j. \tag{10.4}$$

Compare this with the transformation for the partial derivatives of a function u,

$$\frac{\partial u}{\partial x'_i} = \frac{\partial u}{\partial x_j}\frac{\partial x_j}{\partial x'_i} = \frac{\partial x_j}{\partial x'_i}\frac{\partial u}{\partial x_j}, \tag{10.5}$$

and compare both (10.4) and (10.5) with the transformation for a Cartesian vector

$$V'_i = a_{ij}V_j, \qquad \text{(Cartesian)}. \tag{10.6}$$

For Cartesian vectors you can easily verify that

$$\frac{\partial x'_i}{\partial x_j} = a_{ij} = \frac{\partial x_j}{\partial x'_i}, \qquad \text{(Cartesian)}, \tag{10.7}$$

since both the partial derivatives in (10.7) equal the cosine of the angle between the x'_i and the x_j axes (Problem 1). This is not true for general coordinate systems; for example, in (10.1), $\partial x/\partial\theta \neq \partial\theta/\partial x$ (see Problem 2). Thus in general we have two possible definitions of a vector, which become identical for Cartesian vectors.

Contravariant and Covariant Vectors By definition, **V** is a *contravariant vector* if its components transform like this:

$$V_i' = \frac{\partial x_i'}{\partial x_j} V_j, \qquad \text{(contravariant vector)}, \tag{10.8}$$

and **V** is a covariant vector if its components transform like this:

$$V_i' = \frac{\partial x_j}{\partial x_i'} V_j, \qquad \text{(covariant vector)}. \tag{10.9}$$

By comparing (10.4) and (10.8), we see that the differentials of the coordinates are the components of a contravariant vector. Similarly, by comparing (10.5) and (10.9), we see that the partial derivatives of a function are the components of a covariant vector.

Notation Before we define tensors in general, we need to discuss a few things about notation. It is customary to write the indices of contravariant vectors and tensors as superscripts rather than subscripts. Be careful not to confuse them with exponents! (You may find the mnemonic "low-co" useful; *lower* indices are *co*variant indices, so, of course, upper indices are contravariant indices.) In this notation, equation (10.8) for a contravariant vector becomes

$$V'^i = \frac{\partial x_i'}{\partial x_j} V^j, \qquad \text{(contravariant vector)}. \tag{10.10}$$

(In fact, to be strictly consistent, since the differentials are contravariant, we should write $\partial x'^i/\partial x^j$. For our purposes this seems unnecessary so we will leave the partial derivative notation as it is.) Also note that the summation convention now applies to a pair of indices, one upper and one lower. (An index in the denominator counts as a lower index and an index in the numerator counts as an upper index.) Note that this new rule about summation convention applies in (10.9) and (10.10) and watch for it in future formulas.

Components and basis vectors You may be wondering how the vectors you studied in vector analysis (Section 9 and Chapter 6) are related to covariant and contravariant vectors. Actually we should speak of covariant and contravariant components, but the former terminology is customary. Any vector has various sets of components relative to various sets of basis vectors. Let's discuss this for orthogonal coordinate systems where it is especially simple. Recall that in vector analysis, we use the unit basis vectors such as **i**, **j**, **k** or $\mathbf{e}_r$, $\mathbf{e}_\theta$, $\mathbf{e}_\phi$; for example, the vectors $\mathbf{e}_i$ in Section 9 are all unit vectors. Then the components of a vector **V** in vector analysis are the projections $\mathbf{e}_i \cdot \mathbf{V}$ of the vector on the coordinate directions. To be able to refer to these components, let's call them the physical components (they have the right physical dimensions—see Problem 6). We would like to see the relation between the physical components and the covariant and contravariant components of a vector, and the relation between the unit basis vectors and the contravariant and covariant basis vectors.

Example 3. You have learned that, in polar coordinates, the (physical) components of ds are dr and $r\,d\theta$. Now (10.4) and (10.10) tell us that the contravariant components of ds are just dr and $d\theta$ (**not** $r\,d\theta$). Thus we may guess (correctly) that the contravariant components of a vector are the physical components divided by the scale factors. By considering the components of the gradient (Problem 4), you can show that the covariant components of a vector are the physical components multiplied by the scale factors.

Example 4. In polar coordinates we can write [see equation (8.9)]

$$ds = \mathbf{e}_r\,dr + \mathbf{e}_\theta\,r\,d\theta = \mathbf{a}_r\,dr + \mathbf{a}_\theta\,d\theta. \tag{10.11}$$

We have written ds in terms of its physical components and the unit $\mathbf{e}_i$ vectors, and in terms of its contravariant components and the covariant $\mathbf{a}_i$ basis vectors. From (10.11) and from Section 8 we can see that the $\mathbf{a}_i$ basis vectors are the $\mathbf{e}_i$ unit vectors multiplied by the scale factors. Note that the components and the basis vectors used with them vary in opposite ways so that the scale factors cancel. Similarly we can write a vector in terms of its covariant components and the contravariant basis vectors $\mathbf{a}^i$ which are the unit vectors divided by the scale factors (Problem 5). Note carefully that what we have just said applies only to orthogonal coordinate systems. If a coordinate system is not orthogonal, then $\mathbf{a}_i$ and $\mathbf{a}^i$ are not in general parallel; see the discussion just after (10.19).

Definition of Tensors Tensors may be covariant of any rank, contravariant of any rank, or mixed. Here are some sample tensor definitions; you should be able to write the corresponding definitions for tensors of any rank or kind in a similar way (Problem 7).

$$\begin{aligned}
T'_{ij} &= \frac{\partial x_k}{\partial x'_i}\frac{\partial x_l}{\partial x'_j}T_{kl} && \text{(2}^{\text{nd}}\text{-rank covariant tensor)},\\
T'^{ijk} &= \frac{\partial x'_i}{\partial x_l}\frac{\partial x'_j}{\partial x_m}\frac{\partial x'_k}{\partial x_n}T^{lmn} && \text{(3}^{\text{rd}}\text{-rank contravariant tensor)},\\
T'^{ij}_{\ \ k} &= \frac{\partial x'_i}{\partial x_l}\frac{\partial x'_j}{\partial x_m}\frac{\partial x_n}{\partial x'_k}T^{lm}_{\ \ n} && \text{(3}^{\text{rd}}\text{-rank mixed tensor, one covariant and two contravariant indices)}.
\end{aligned} \tag{10.12}$$

Kronecker delta We showed in Section 5 that δ_{ij} is a 2nd-rank isotropic Cartesian tensor. In a general coordinate system, the 2nd-rank tensor which is equal to 1 if $i = j$ and 0 otherwise in all coordinate systems, is a mixed tensor so we write it as δ^i_j. To show that this is correct we write the tensor transformation equation for δ^k_l to see that we get $\delta'^i_{\ j}$.

$$\frac{\partial x'_i}{\partial x_k}\frac{\partial x_l}{\partial x'_j}\delta^k_l = \frac{\partial x'_i}{\partial x_k}\frac{\partial x_k}{\partial x'_j} = \frac{\partial x'_i}{\partial x'_j} = \delta'^i_{\ j}. \tag{10.13}$$

Thus we see that δ^i_j is an isotropic 2nd-rank tensor in general coordinate systems.

Quotient Rule In Section 3, we discussed the quotient rule for Cartesian tensors. A similar rule applies in general. To give proofs, we must replace the a_{ij} by the appropriate partial derivatives, noting carefully that summation convention now applies to a sum over one lower and one upper index.

▶ **Example 5.** If we are given $T_{ij}V^j = U_i$ where $\mathbf{V}$ is an arbitrary contravariant vector and $\mathbf{U}$ is a non-zero covariant vector, we want to show that T_{ij} is a 2nd-rank covariant tensor. We write [compare equations (3.6) to (3.9)]

$$T'_{\alpha\beta}V'^{\beta} = U'_{\alpha} = \frac{\partial x_i}{\partial x'_{\alpha}}U_i = \frac{\partial x_i}{\partial x'_{\alpha}}T_{ij}V^j = \frac{\partial x_i}{\partial x'_{\alpha}}T_{ij}\frac{\partial x_j}{\partial x'_{\beta}}V'^{\beta}. \tag{10.14}$$

Set the first and last steps equal; then since V'^{β} is arbitrary, its coefficient $= 0$ and we have

$$T'_{\alpha\beta} = \frac{\partial x_i}{\partial x'_{\alpha}}T_{ij}\frac{\partial x_j}{\partial x'_{\beta}}$$

which is the transformation equation for a 2nd-rank covariant tensor.

Metric Tensor; Raising and Lowering Indices

▶ **Example 6.** From (8.14) we have (with the contravariant dx indices now written as superscripts)

$$ds^2 = g_{ij}\, dx^i\, dx^j. \tag{10.15}$$

Since ds^2 is a scalar, and each dx is a contravariant vector, it follows by the quotient rule (Problem 8) that g_{ij} is a 2nd-rank covariant tensor. It is known as the *metric tensor.* If the elements of g_{ij} are written as a matrix [see (8.13)], then we define g^{ij} as the elements of the inverse matrix. We can interpret $g_{ij}g^{jk}$ as either the contracted direct product of two tensors, or as the row times column product of two matrices which are inverses of each other, that is, a unit matrix. Thus we can write

$$g_{ij}g^{jk} = \delta_i^k. \tag{10.16}$$

Then by (10.13) and the quotient rule, g^{ij} is a 2nd-rank contravariant tensor.

▶ **Example 7.** If V^i is a contravariant vector then $V_i = g_{ij}V^j$ is a covariant vector (Problem 10). We can also show that $g^{ij}V_j$ gives back the V^i we started with:

$$g^{ij}V_j = g^{ij}g_{jk}V^k = \delta_k^i V^k = V^i. \tag{10.17}$$

This process of finding the contracted product of a vector (or tensor) with g^{ij} or g_{ij} is called *raising* or *lowering indices.* The vectors V^i and V_i are called the contravariant and covariant components of the vector $\mathbf{V}$.

In equation (8.12), we defined the covariant basis vectors $\mathbf{a}_i$ which we use with contravariant components to write a vector [see (8.11) for example, remembering that the differentials are the contravariant components of $d\mathbf{s}$]. The contravariant basis vectors to use with covariant components are given by $\mathbf{a}^i = g^{ij}\mathbf{a}_j$. We can then write a vector in two ways (Problem 11):

$$\mathbf{V} = \mathbf{a}_i V^i = \mathbf{a}^i V_i, \quad \text{where} \quad \begin{cases} V_i = g_{ij}V^j, & V^i = g^{ij}V_j, \\ \mathbf{a}^i = g^{ij}\mathbf{a}_j, & \mathbf{a}_i = g_{ij}\mathbf{a}^j. \end{cases} \tag{10.18}$$

It is interesting to consider the directions of the vectors $\mathbf{a}_i$ and $\mathbf{a}^i$. We have defined $\mathbf{a}^i = g^{ij}\mathbf{a}_j$ but you can show (Problem 12) that $\mathbf{a}^i = \nabla x_i$. Thus we have

$$\begin{aligned} \mathbf{a}_i &= \frac{\partial}{\partial x_i}\mathbf{s} = \mathbf{i}\frac{\partial x}{\partial x_i} + \mathbf{j}\frac{\partial y}{\partial x_i} + \mathbf{k}\frac{\partial z}{\partial x_i}, \\ \mathbf{a}^i &= g^{ij}\mathbf{a}_j = \nabla x_i = \mathbf{i}\frac{\partial x_i}{\partial x} + \mathbf{j}\frac{\partial x_i}{\partial y} + \mathbf{k}\frac{\partial x_i}{\partial z}. \end{aligned} \tag{10.19}$$

We see from the displacement vector $d\mathbf{s} = \mathbf{a}_i\, dx^i$ that the basis vectors $\mathbf{a}_i$ are tangent to the coordinate lines. The vectors $\mathbf{a}^i = \nabla x_i$ are orthogonal to the coordinate surfaces $x_i =$ const. (Recall that grad u is orthogonal to $u =$ const.) For orthogonal coordinates, $\mathbf{a}_i$ and $\mathbf{a}^i$ are in the same direction. (For example, in spherical coordinates, $\mathbf{a}_r$ points in the radial direction, and $\mathbf{a}^r$ is orthogonal to the sphere $r =$ const.; these are the same direction.) Thus for orthogonal coordinates, if we normalize each $\mathbf{a}^i$, we get the same set of unit basis vectors that we get if we normalize each $\mathbf{a}_i$. However, if the coordinate system is not orthogonal, then at each point we have two different sets of basis vectors $\mathbf{a}_i$ and $\mathbf{a}^i$ (see Problems 16 and 17).

Just as we did for vectors, any tensor, say T^i_{jk}, can be written in various different forms by raising and lowering indices to get T_{ijk}, T^{ijk}, T^{ij}_k. These tensors are called *associated tensors*. They really all represent the same tensor $\mathbf{T}$, with components relative to various bases.

Orthogonal coordinate systems For orthogonal coordinate systems, formulas involving g_{ij} can be written in terms of the scale factors h_1, h_2, h_3 [compare (8.13) and (8.16)]. Remember that the g^{ij} matrix is the inverse of the g_{ij} matrix [see equation (10.16)]. Also let g represent the determinant of the g_{ij} matrix. Then you can show (Problem 13).

$$\begin{aligned} g_{ij} &= \begin{cases} 0, & i \neq j, \\ h_i^2, & i = j, \end{cases} \qquad g^{ij} = \begin{cases} 0, & i \neq j, \\ \dfrac{1}{h_i^2}, & i = j, \end{cases} \\ g &= h_1^2 h_2^2 h_3^2, \qquad \sqrt{g} = h_1 h_2 h_3 \end{aligned} \tag{10.20}$$

Vector Operators in Tensor Notation We state without proof the following tensor expressions for ∇u, $\nabla \cdot \mathbf{V}$, and $\nabla^2 u$. They are correct for any coordinate system, orthogonal or not. Using (10.20), you can specialize them to orthogonal coordinate systems and so obtain the expressions given in Section 9. (Problems 14 and 15).

(10.21) The covariant components of ∇u are $\dfrac{\partial u}{\partial x_i}$.

(10.22) $\nabla \cdot \mathbf{V} = \dfrac{1}{\sqrt{g}}\dfrac{\partial}{\partial x_i}(\sqrt{g}\, V^i)$, where V^i are contravariant components of $\mathbf{V}$.

$$\nabla^2 u = \frac{1}{\sqrt{g}}\frac{\partial}{\partial x_i}\left(\sqrt{g}\, g^{ij}\frac{\partial u}{\partial x_j}\right). \tag{10.23}$$

PROBLEMS, SECTION 10

1. Verify equation (10.7). *Hint:* Use equations (2.4) to (2.6) and (2.10). For example, $\partial y'/\partial z = \partial z/\partial y' = n_2 = a_{23}$.

2. From (10.1) find $\partial\theta/\partial x = (1/r)\cos\theta\cos\phi$ and show that $\partial x/\partial\theta \neq \partial\theta/\partial x$. Note carefully that $\partial x/\partial\theta$ means that r and ϕ are constant, but $\partial\theta/\partial x$ means that y and z are constant. (See Chapter 4, Example 7.6 for further discussion.)

3. Divide equation (10.4) by dt to show that the velocity $\mathbf{v} = d\mathbf{s}/dt$ is a contravariant vector. Note that the contravariant components of the velocity in polar coordinates are $\dot{r}$ and $\dot{\theta}$ (*not* $\dot{r}$ and $r\dot{\theta}$ which are physical components). As we did in (10.11), write the velocity $\mathbf{v}$ in polar coordinates in terms of the unit $\mathbf{e}$ vectors and in terms of the covariant $\mathbf{a}$ vectors. Repeat the problem in spherical coordinates.

4. What are the physical components of the gradient in polar coordinates? [See (9.1)]. The partial derivatives in (10.5) are the covariant components of ∇u. What relation do you deduce between physical and covariant components? Answer the same questions for spherical coordinates, and for an orthogonal coordinate system with scale factors h_1, h_2, h_3.

5. Write ∇u in polar coordinates in terms of its physical components and the unit basis vectors $\mathbf{e}_i$, and in terms of its covariant components and the contravariant basis vectors $\mathbf{a}^i$. What is the relation between the contravariant basis vectors and the unit basis vectors? *Hint:* Compare equation (10.11) and our discussion of it.

6. Show that, in polar coordinates, the θ contravariant component of $d\mathbf{s}$ is $d\theta$ which is unitless, the θ physical component of $d\mathbf{s}$ is $r\,d\theta$ which has units of length, and the θ covariant component of $d\mathbf{s}$ is $r^2\,d\theta$ which has units (length)^2.

7. As in (10.12), write the transformation equations for the following tensors: 2nd-rank contravariant, 3rd-rank covariant, 4th-rank mixed with 2 covariant and 2 contravariant indices.

8. Using (10.15) show that g_{ij} is a 2nd-rank covariant tensor. *Hint:* Write the transformation equation for each dx, and set the scalar $ds'^2 = ds^2$ to find the transformation equation for g_{ij}.

9. If U^i is a contravariant vector and V_j is a covariant vector, show that U^iV_j is a 2nd-rank mixed tensor. *Hint:* Write the transformation equations for $\mathbf{U}$ and $\mathbf{V}$ and multiply them.

10. Show that if V^i is a contravariant vector then $V_i = g_{ij}V^j$ is a covariant vector, and that if V_i is a covariant vector, then $V^i = g^{ij}V_j$ is a contravariant vector.

11. In (10.18), show by raising and lowering indices that $\mathbf{a}_iV^i = \mathbf{a}^iV_i$. Also write (10.18) for an orthogonal coordinate system with g_{ij} and g^{ij} written in terms of the scale factors.

12. Show that in a general coordinate system with variables x_1, x_2, x_3, the contravariant basis vectors are given by

$$\mathbf{a}^i = \nabla x_i = \mathbf{i}\frac{\partial x_i}{\partial x} + \mathbf{j}\frac{\partial x_i}{\partial y} + \mathbf{k}\frac{\partial x_i}{\partial z}.$$

Hint: Write the gradient in terms of its covariant components and the $\mathbf{a}^i$ basis vectors to get $\nabla u = \mathbf{a}^j\partial u/\partial x_j$ and let $u = x_i$.

13. Verify (10.20).

14. Using equations (10.20) to (10.23), write the gradient, divergence, and Laplacian in cylindrical coordinates and in spherical coordinates. Change covariant or contravariant components to physical components and compare with the formulas stated in Chapter 6, Sections 6 and 7.

15. Do Problem 14 for an orthogonal coordinate system with scale factors h_1, h_2, h_3, and compare with the Section 9 formulas.

16. Continue Problem 8.15 to find the g^{ij} matrix and the contravariant basis vectors. Check your result by solving the given equations for u and v in terms of x and y, and finding the contravariant basis vectors using Problem 12. On your Problem 8.15 sketches of the lines $u = $ const. and $v = $ const., also sketch the covariant and contravariant basis vectors. Observe that the covariant basis vectors lie along the lines $u = $ const. and $v = $ const. and the contravariant basis vectors lie along the normals to these lines.

17. Repeat Problems 8.15 and 10.16 above for the (u, v) coordinate system if $x = 2u - v$, $y = u - 2v$.

18. Using (10.19), show that $\mathbf{a}^i \cdot \mathbf{a}_i = \delta^i_j$.

11. MISCELLANEOUS PROBLEMS

1. Show that the transformation equation for a 2nd-rank Cartesian tensor is equivalent to a similarity transformation. *Warning hint:* Note that the matrix C in Chapter 3, Section 11, is the inverse of the matrix A we are using in Chapter 10 (compare $\mathbf{r}' = \mathrm{A}\mathbf{r}$ and $\mathbf{r} = \mathrm{C}\mathbf{r}'$). Thus a similarity transformation of the matrix T with tensor components T_{ij} is $\mathrm{T}' = \mathrm{ATA}^{-1}$. Also see "Tensors and Matrices" in Section 3 and remember that A is orthogonal.

2. Let $\mathbf{e}_1$, $\mathbf{e}_2$, $\mathbf{e}_3$ be a set of orthogonal unit vectors forming a right-handed system if taken in cyclic order. Show that the triple scalar product $\mathbf{e}_i \cdot (\mathbf{e}_j \times \mathbf{e}_k) = \epsilon_{ijk}$.

3. In Chapter 3, Problem 6.6, you are asked to prove some identities among the Pauli spin matrices (called A, B, C, in that problem). Call the Pauli spin matrices σ_1, σ_2, σ_3; then show that the identities can be written in the following summation forms:

$$\sigma_k\sigma_m = i\epsilon_{kmn}\sigma_n + \delta_{km};$$

$$\sigma_k\sigma_m\epsilon_{kmn} = 2i\sigma_n.$$

4. If $\mathbf{E}$ = electric field and $\mathbf{B}$ = magnetic field, is $\mathbf{E} \times \mathbf{B}$ a vector or a pseudovector? *Comment*: $\mathbf{E} \times \mathbf{B}/\mu_0$ is called the Poynting vector; it points in the direction of transfer of energy. Does that tell you from the physics whether it is a vector or a pseudovector?

Do Problems 5 to 8 for the (u, v) coordinate system if $x = u(1 - v)$, $y = u\sqrt{2v - v^2}$.

5. Find ds^2, the scale factors, the area element, the vector $d\mathbf{s}$, the unit basis vectors, and the covariant and contravariant basis vectors.

6. Use Lagrange's equations to find the u and v acceleration components.

7. Write ∇U, $\nabla \cdot \mathbf{V}$, and $\nabla^2 U$.

8. Evaluate $\nabla \cdot \mathbf{e}_u$, $\nabla \times \mathbf{e}_v$, $\nabla^2 \ln u$.

9. If $\mathbf{u}$ is a vector specifying the displacement under stress of each point of a deformable medium, then $\nabla\mathbf{u}$ is a 2nd-rank Cartesian tensor (see Problem 11) which describes the strain at each point. Display the components of $\nabla\mathbf{u}$ as a matrix. Write $\nabla\mathbf{u}$ as the sum of a symmetric tensor and an antisymmetric tensor [see (3.5)]. *Comment*: The symmetric part of $\nabla\mathbf{u}$ is called the *stress tensor* and the antisymmetric part the *rotation tensor*.

10. Show that elements R_{ij} of a rotation matrix are the elements of a Cartesian tensor. *Hints:* Could you use the quotient rule? Could you use Problem 1?

11. Show that the nine quantities $T_{ij} = \partial V_i/\partial x_j$ (which are the Cartesian components of $\nabla\mathbf{V}$ where $\mathbf{V}$ is a vector) satisfy the transformation equations (2.14) for a Cartesian 2nd-rank tensor. Show that they do not satisfy the general tensor transformation equations as in (10.12). *Hint:* Differentiate (10.9) or (10.10) partially with respect to, say, x'_k. You should get the expected terms [as in (10.12)] plus some extra terms; these extraneous terms show that $\partial V_i/\partial x_j$ is not a tensor under general transformations. *Comment:* It is possible to express the components of $\nabla\mathbf{V}$ correctly in general coordinate systems by taking into account the variation of the basis vectors in length and direction.

12. The square matrix in equation (10.3) is called the Jacobian matrix J; the determinant of this matrix is the Jacobian $J = \det \mathrm{J}$ which we used in Chapter 5, Section 4 to find volume elements in multiple integrals. (Note that as in Chapter 3, J represents a matrix; J in italics is its determinant.) For the transformation to spherical coordinates in (10.1) and (10.2) show that $J = \det \mathrm{J} = r^2 \sin\theta$. Recall that the spherical coordinate volume element is $r^2 \sin\theta\, dr\, d\theta\, d\phi$. *Hint:* Find $\mathrm{J}^{\mathrm{T}}\mathrm{J}$ and note that $\det(\mathrm{J}^{\mathrm{T}}\mathrm{J}) = (\det \mathrm{J})^2$.

13. In equation (10.13), let the x' variables be rectangular coordinates x, y, z, and let x_1, x_2, x_3, be general curvilinear coordinates, orthogonal or not (see end of Section 8). Show that $\mathrm{J}^{\mathrm{T}}\mathrm{J}$ is the g_{ij} matrix in (8.13) [or in (8.16) for an orthogonal system]. Thus show that the volume element in a general coordinate system is $dV = \sqrt{g}\, dx_1\, dx_2\, dx_3$ where $g = \det(g_{ij})$, and that for an orthogonal system, this becomes [by (8.16) or (10.19)], $dV = h_1 h_2 h_3\, dx_1\, dx_2\, dx_3$. *Hint:* To evaluate the products of partial derivatives in $\mathrm{J}^{\mathrm{T}}\mathrm{J}$, observe that the same expressions arise as in finding ds^2. In fact, from (8.11) and (8.12), you can show that row i times column j in $\mathrm{J}^{\mathrm{T}}\mathrm{J}$ is just $\mathbf{a}_i \cdot \mathbf{a}_j = g_{ij}$ in equations (8.11) to (8.14).

CHAPTER 11

Special Functions

1. INTRODUCTION

The integrals and series and functions of this chapter arise in a variety of physical problems. Just as you learn about trigonometric functions, logarithms, etc., and use them in applied problems, so you should learn something about these special functions so that you can use them and understand their use as they come up in your more advanced work. An enormous amount of detail is known about these functions, and numerous formulas involving them exist and can be looked up in books or found in your computer program. Our purpose is not to study them intensively, but to give definitions and some of the simpler relations and show their use. This should develop your ability and confidence to cope with more complicated formulas and many other similar functions and relations that may crop up occasionally in texts or computer results.

Now you may be thinking that your computer will give you the answers for definite integrals and functions so you really don't need to bother with this chapter. If all you want is a numerical approximation, this may be true. However, in theoretical work, you often need an exact expression (say in terms of π or $\sqrt{3}$ or $\ln 2$) and your computer may not give you the form you need.

Example 1. Suppose you want $\int_0^{\pi/2} d\theta/\sqrt{\cos\theta}$. One computer program gives you the result $\sqrt{2}\,K(1/\sqrt{2})$ and another gives you $2\sqrt{\pi}\,\Gamma(5/4)/\Gamma(3/4)$. In books you find answers $\frac{1}{2}B(1/4, 1/2)$ and $[\Gamma(1/4)]^2/\sqrt{8\pi}$. What's going on here and which is right? They all are! And when you have studied the formulas in this chapter, you will be able to show this (Problem 12.21) just as you now recognize that $\sin^2\theta = 1 - \cos^2\theta$.

Also in some problems you may want an algebraic approximation for a complicated expression rather than a numerical answer.

Example 2. You will find in thermal physics the approximation $\ln N! \cong N\ln N - N$; we will discuss this approximation and its accuracy. (See Problem 11.3).

2. THE FACTORIAL FUNCTION

Let us calculate the values of some integrals. For $\alpha > 0$,

$$\int_0^\infty e^{-\alpha x} dx = -\frac{1}{\alpha}e^{-\alpha x}\Big|_0^\infty = \frac{1}{\alpha}. \tag{2.1}$$

Next differentiate both sides of this equation repeatedly with respect to α (see Chapter 4, Section 12):

$$\int_0^\infty -xe^{-\alpha x}\,dx = -\frac{1}{\alpha^2} \quad \text{or} \quad \int_0^\infty xe^{-\alpha x}\,dx = \frac{1}{\alpha^2},$$
$$\int_0^\infty x^2e^{-\alpha x}\,dx = \frac{2}{\alpha^3},$$
$$\int_0^\infty x^3e^{-\alpha x}\,dx = \frac{3!}{\alpha^4}.$$

or in general

$$\int_0^\infty x^n e^{-\alpha x}\,dx = \frac{n!}{\alpha^{n+1}}. \tag{2.2}$$

Putting $\alpha = 1$, we get

$$\int_0^\infty x^n e^{-x}\,dx = n!, \quad n = 1, 2, 3, \cdots. \tag{2.3}$$

Thus we have a definite integral whose value is $n!$ for positive integral n. We can use (2.3) to give a meaning to $0!$. Putting $n = 0$ in (2.3), we get

$$0! = \int_0^\infty e^{-x}\,dx = -e^{-x}\Big|_0^\infty = 1. \tag{2.4}$$

(This agrees with our previous definition of $0!$ in Chapter 1.)

PROBLEMS, SECTION 2

In Chapter 4, Section 12, do Problems 14 to 17.

3. DEFINITION OF THE GAMMA FUNCTION; RECURSION RELATION

So far n has been a nonnegative integer; it is natural to *define* the factorial function for nonintegral n by the definite integral (2.3). There is no real objection to the notation $n!$ for nonintegral n (and we shall occasionally use it), but it is customary to reserve the factorial notation for integral n and to call the corresponding function for nonintegral n the gamma (Γ) function. It is also rather common practice to replace n by the letter p when we do not necessarily mean an integer. Following these conventions, we define, for *any* $p > 0$

$$\Gamma(p) = \int_0^\infty x^{p-1}e^{-x}\,dx, \quad p > 0. \tag{3.1}$$

For $0 < p < 1$, this is an improper integral because x^{p-1} becomes infinite at the lower limit. However, it is a convergent integral for $p > 0$ (Problem 1). For $p \leq 0$, the integral diverges and so cannot be used to define $\Gamma(p)$; we shall see in Section 4 how to define $\Gamma(p)$ when $p \leq 0$. Then from (3.1) and (2.3) we have

$$\Gamma(n) = \int_0^\infty x^{n-1} e^{-x}\,dx = (n-1)!,$$
$$\Gamma(n+1) = \int_0^\infty x^n e^{-x}\,dx = n!. \tag{3.2}$$

Thus

$$\Gamma(1) = 0! = 1,\quad \Gamma(2) = 1! = 1,\quad \Gamma(3) = 2! = 2,\quad \Gamma(4) = 3! = 6,\quad \cdots,$$

with the usual meaning of factorial for positive integral n. The fact that $\Gamma(n) = (n-1)!$ and not $n!$ is unfortunate but that's the notation which is used, so watch out for it. Replacing p by $p+1$ in (3.1), we have

$$\Gamma(p+1) = \int_0^\infty x^p e^{-x}\,dx = p!, \quad p > -1. \tag{3.3}$$

Some authors use the factorial notation $p! = \Gamma(p+1)$ even though p is not an integer; this avoids the nuisance of the $p+1$.

Let us integrate (3.3) by parts, calling $x^p = u$, $e^{-x}dx = dv$; then we get

$$du = p\,x^{p-1}\,dx, \quad v = -e^{-x},$$
$$\Gamma(p+1) = -x^p e^{-x}\Big|_0^\infty - \int_0^\infty (-e^{-x})p\,x^{p-1}\,dx$$
$$= p\int_0^\infty x^{p-1}e^{-x}\,dx = p\Gamma(p).$$

This equation

$$\Gamma(p+1) = p\Gamma(p) \tag{3.4}$$

is called the *recursion relation* for the Γ function. It can be used to simplify expressions involving Γ functions or to write them in a different form (much as you use trigonometric identities).

Example. By (3.4) we find $\Gamma(9/4) = (5/4)\Gamma(5/4) = (5/4)(1/4)\Gamma(1/4)$; then $\Gamma(1/4) \div \Gamma(9/4) = 16/5$.

PROBLEMS, SECTION 3

1. The integral in (3.1) is improper because of the infinite upper limit and it is also improper for $0 < p < 1$ because x^{p-1} becomes infinite at the lower limit. However, the integral is convergent for any $p > 0$. Prove this.

Use the recursion relation (3.4), and if needed, equation (3.2) to simplify:

2. $\Gamma(2/3)/\Gamma(5/3)$ **3.** $\Gamma(2/3)/\Gamma(8/3)$ **4.** $\Gamma(2/5)/\Gamma(12/5)$

5. $\Gamma(1/2)\Gamma(4)/\Gamma(9/2)$ **6.** $\Gamma(10)/\Gamma(8)$ **7.** $\Gamma(4)\Gamma(3/4)/\Gamma(7/4)$

Express each of the following integrals as a Γ function. By computer, evaluate numerically both the Γ function and the original integral.

8. $\displaystyle\int_0^\infty x^{2/3}e^{-x}\,dx$

9. $\displaystyle\int_0^\infty e^{-x^4}dx$ *Hint:* Put $x^4 = u$.

10. $\displaystyle\int_0^\infty x^{-2/5}e^{-x}\,dx$

11. $\displaystyle\int_0^\infty x^5e^{-x^2}dx$ *Hint:* Put $x^2 = u$.

12. $\displaystyle\int_0^\infty xe^{-x^3}dx$

13. $\displaystyle\int_0^1 x^2\left(\ln\frac{1}{x}\right)^3 dx$ *Hint:* Put $x = e^{-u}$.

14. $\displaystyle\int_0^1 \sqrt[3]{\ln x}\,dx$

15. $\displaystyle\int_0^\infty x^{-1/3}e^{-8x}\,dx$

16. A particle starting from rest at $x = 1$ moves along the x axis toward the origin. Its potential energy is $V = \frac{1}{2}m\ln x$. Write the Lagrange equation and integrate it to find the time required for the particle to reach the origin. *Caution:* $dx/dt < 0$. *Answer:* $\Gamma(\frac{1}{2})$.

17. Express as a Γ function

$$\int_0^1\left[\ln\left(\frac{1}{x}\right)\right]^{p-1}dx. \qquad \textit{Hint}\text{: See Problem 13.}$$

4. THE GAMMA FUNCTION OF NEGATIVE NUMBERS

For $p \le 0$, $\Gamma(p)$ has not so far been defined. We shall now define it by the recursion relation (3.4) solved for $\Gamma(p)$.

$$\Gamma(p) = \frac{1}{p}\Gamma(p+1) \tag{4.1}$$

defines $\Gamma(p)$ for $p < 0$.

Example.

$$\Gamma(-0.3) = \frac{1}{-0.3}\Gamma(0.7), \quad \Gamma(-1.3) = \frac{1}{(-1.3)(-0.3)}\Gamma(0.7),$$

and so on. Since $\Gamma(1) = 1$, we see that

$$\Gamma(p) = \frac{\Gamma(p+1)}{p} \to \infty \quad \text{as } p \to 0.$$

From this and successive use of (4.1) it follows that $\Gamma(p)$ becomes infinite not only at zero but also at all the negative integers. In the intervals between the negative integers, it is alternately positive and negative, negative from 0 to -1, positive from -1 to -2, and so on, as you can see from computations like those for $\Gamma(-0.3)$ and $\Gamma(-1.3)$ above. See Problems 5.1 and 5.2.

5. SOME IMPORTANT FORMULAS INVOLVING GAMMA FUNCTIONS

First we evaluate $\Gamma\left(\frac{1}{2}\right)$. By definition

$$\Gamma(\tfrac{1}{2}) = \int_0^\infty \frac{1}{\sqrt{t}} e^{-t}\, dt. \tag{5.1}$$

(Note that it does not matter what letter we use for the dummy variable of integration in a definite integral.) Put $t = y^2$ in (5.1); then $dt = 2y\, dy$, and (5.1) becomes

$$\Gamma(\tfrac{1}{2}) = \int_0^\infty \frac{1}{y} e^{-y^2} 2y\, dy = 2\int_0^\infty e^{-y^2}\, dy$$

or, with x as the dummy integration variable,

$$\Gamma(\tfrac{1}{2}) = 2\int_0^\infty e^{-x^2}\, dx. \tag{5.2}$$

Multiply these two integrals for $\Gamma(\frac{1}{2})$ together and write the result as a double integral:

$$\left[\Gamma(\tfrac{1}{2})\right]^2 = 4\int_0^\infty \int_0^\infty e^{-(x^2+y^2)} dx\, dy.$$

This is an integral over the first quadrant; it can be more easily evaluated in polar coordinates:

$$\left[\Gamma(\tfrac{1}{2})\right]^2 = 4\int_0^{\pi/2} \int_0^\infty e^{-r^2} r\, dr\, d\theta = 4 \cdot \frac{\pi}{2} \cdot \left.\frac{e^{-r^2}}{-2}\right|_0^\infty = \pi.$$

Therefore

$$\Gamma(\tfrac{1}{2}) = \sqrt{\pi}. \tag{5.3}$$

We state here another important formula involving Γ functions (for proof, see Chapter 14, Section 7, Example 5):

$$\Gamma(p)\Gamma(1-p) = \frac{\pi}{\sin \pi p}. \tag{5.4}$$

Notice that (5.4) also gives $\Gamma\left(\frac{1}{2}\right) = \sqrt{\pi}$ if we put $p = \frac{1}{2}$.

PROBLEMS, SECTION 5

1. Using (5.3) with (3.4) and (4.1), find $\Gamma(3/2)$, $\Gamma(-1/2)$, and $\Gamma(-3/2)$ in terms of $\sqrt{\pi}$.

2. Without computer or tables, but just using facts you know, sketch a quick rough graph of the Γ function from -2 to 3. *Hint:* This is easy; don't make a big job of it. From Section 3, you know the values of the Γ function at the positive integers in terms of factorials. From Problem 1, you can easily find and plot the Γ function at $\pm 1/2$, $\pm 3/2$. (Approximate $\sqrt{\pi}$ as a little less than 2.) From (4.1) and the discussion following it, you know that the Γ function tends to plus or minus infinity at 0 and the negative integers, and you know the intervals where it is positive or negative. After sketching your graph, make a computer plot of the Γ function from -5 to 5 and compare your sketch.

3. In Chapter 1, equations (13.5) and (13.6), we defined the binomial coefficients $\binom{p}{n}$ where n is a non-negative integer but p may be negative or fractional. Show that $\binom{p}{n}$ can be written in terms of Γ functions as

$$\binom{p}{n} = \frac{\Gamma(p+1)}{n!\,\Gamma(p-n+1)}.$$

4. Prove that, for positive integral n:

$$\Gamma(n+\tfrac{1}{2}) = \frac{1\cdot 3\cdot 5\cdots(2n-1)}{2^n}\sqrt{\pi} = \frac{(2n)!}{4^n n!}\sqrt{\pi}.$$

5. Use (5.4) to show that

(a) $\Gamma(\frac{1}{2}-n)\Gamma(\frac{1}{2}+n) = (-1)^n\pi$ if $n =$ a positive integer;

(b) $(z!)(-z)! = \pi z/\sin\pi z$, where z is not necessarily an integer; see comment after equation (3.3).

6. Prove that

$$\frac{d}{dp}\Gamma(p) = \int_0^\infty x^{p-1}e^{-x}\ln x\,dx,$$
$$\frac{d^n}{dp^n}\Gamma(p) = \int_0^\infty x^{p-1}e^{-x}(\ln x)^n\,dx.$$

7. In the Table of Laplace Transforms (end of Chapter 8, page 469), verify the Γ function results for $L5$ and $L6$. Also show that $L(1/\sqrt{t}) = \sqrt{\pi/p}$.

6. BETA FUNCTIONS

The *beta function* is also defined by a definite integral:

$$B(p,q) = \int_0^1 x^{p-1}(1-x)^{q-1}\,dx, \quad p>0, \quad q>0. \tag{6.1}$$

There are a number of simple transformations of (6.1) which are useful to know [see (6.3), (6.4), (6.5)]. It is easy to show that (Problem 1)

$$B(p,q) = B(q,p). \tag{6.2}$$

The range of integration in (6.1) can be changed by putting $x = y/a$; then $x = 1$ corresponds to $y = a$, and (6.1) becomes

$$(6.3)\quad B(p,q) = \int_0^a \left(\frac{y}{a}\right)^{p-1}\left(1-\frac{y}{a}\right)^{q-1}\frac{dy}{a} = \frac{1}{a^{p+q-1}}\int_0^a y^{p-1}(a-y)^{q-1}\,dy.$$

To obtain the trigonometric form of the beta function, let $x = \sin^2\theta$; then

$$dx = 2\sin\theta\cos\theta\,d\theta, \quad (1-x) = 1-\sin^2\theta = \cos^2\theta,$$
$$x = 1 \text{ corresponds to } \theta = \pi/2.$$

With these substitutions, (6.1) becomes

$$B(p,q) = \int_0^{\pi/2} (\sin^2\theta)^{p-1}(\cos^2\theta)^{q-1} 2\sin\theta\cos\theta\,d\theta \qquad \text{or}$$

$$(6.4)\qquad B(p,q) = 2\int_0^{\pi/2} (\sin\theta)^{2p-1}(\cos\theta)^{2q-1}\,d\theta.$$

Finally, let $x = y/(1+y)$ in (6.1); then we get (Problem 2):

$$(6.5)\qquad B(p,q) = \int_0^\infty \frac{y^{p-1}dy}{(1+y)^{p+q}}.$$

PROBLEMS, SECTION 6

1. Prove that $B(p,q) = B(q,p)$. *Hint:* Put $x = 1-y$ in Equation (6.1).

2. Prove equation (6.5).

3. Show that for integral n, m,

$$1/B(n,m) = m\binom{n+m-1}{n-1} = n\binom{n+m-1}{m-1}.$$

Hint: See Chapter 1, Section 13C, Problem 13.3.

7. BETA FUNCTIONS IN TERMS OF GAMMA FUNCTIONS

Beta functions are easily expressed in terms of Γ functions. We shall show that

$$(7.1)\qquad B(p,q) = \frac{\Gamma(p)\Gamma(q)}{\Gamma(p+q)}.$$

Thus we can evaluate a B function in terms of Γ functions (see example below).

To prove (7.1), we start with

$$\Gamma(p) = \int_0^\infty t^{p-1} e^{-t}\, dt$$

and put $t = y^2$. Then we have

$$\Gamma(p) = 2\int_0^\infty y^{2p-1} e^{-y^2}\, dy. \tag{7.2}$$

Similarly (remember that the dummy integration variable can be any letter),

$$\Gamma(q) = 2\int_0^\infty x^{2q-1} e^{-x^2}\, dx.$$

Next we multiply these two equations together and change to polar coordinates:

$$\begin{aligned}\Gamma(p)\Gamma(q) &= 4\int_0^\infty \int_0^\infty x^{2q-1} y^{2p-1} e^{-(x^2+y^2)} dx\, dy \\ &= 4\int_0^\infty \int_0^{\pi/2} (r\cos\theta)^{2q-1}(r\sin\theta)^{2p-1} e^{-r^2}\, r\, dr\, d\theta \\ &= 4\int_0^\infty r^{2p+2q-1} e^{-r^2}\, dr \int_0^{\pi/2} (\cos\theta)^{2q-1}(\sin\theta)^{2p-1} d\theta.\end{aligned} \tag{7.3}$$

The r integral in (7.3) is $\frac{1}{2}\Gamma(p+q)$ by (7.2). The θ integral in (7.3) is $\frac{1}{2}B(p,q)$ by (6.4). Then $\Gamma(p)\Gamma(q) = 4 \cdot \frac{1}{2}\Gamma(p+q) \cdot \frac{1}{2}B(p,q)$ and (7.1) follows.

▶ **Example.** Find

$$I = \int_0^\infty \frac{x^3\, dx}{(1+x)^5}.$$

This is (6.5) with $(p+q) = 5$, $p - 1 = 3$ or $p = 4$, $q = 1$. Then $I = B(4,1)$. By (7.1), this is

$$\frac{\Gamma(4)\Gamma(1)}{\Gamma(5)} = \frac{3!}{4!} = \frac{1}{4}.$$

▶ PROBLEMS, SECTION 7

Express the following integrals as B functions, and then, by (7.1), in terms of Γ functions. When possible, use Γ function formulas to write an exact answer in terms of π, $\sqrt{2}$, etc. Compare your answers with computer results and reconcile any discrepancies.

1. $\displaystyle\int_0^1 \frac{x^4\, dx}{\sqrt{1-x^2}}$

2. $\displaystyle\int_0^{\pi/2} \sqrt{\sin^3 x \cos x}\, dx$

3. $\displaystyle\int_0^1 \frac{dx}{\sqrt{1-x^3}}$

4. $\displaystyle\int_0^1 x^2(1-x^2)^{3/2}\, dx$

5. $\displaystyle\int_0^\infty \frac{y^2\, dy}{(1+y)^6}$

6. $\displaystyle\int_0^\infty \frac{y\, dy}{(1+y^3)^2}$

7. $\displaystyle\int_0^{\pi/2} \frac{d\theta}{\sqrt{\sin\theta}}$

8. $\displaystyle\int_0^2 \frac{x^2\, dx}{\sqrt{2-x}}$

9. Prove $B(n, n) = B(n, \frac{1}{2})/2^{2n-1}$. *Hint:* In (6.4), use the identity $2\sin\theta\cos\theta = \sin 2\theta$ and put $2\theta = \phi$. Use this result and (5.3) to derive the *duplication formula* for Γ functions:

$$\Gamma(2n) = \frac{1}{\sqrt{\pi}} 2^{2n-1}\Gamma(n)\Gamma(n + \tfrac{1}{2}).$$

Check this formula for the case $n = \frac{1}{4}$ by using (5.4).

Computer plot the graph of $x^3 + y^3 = 8$. Write the integrals for the following quantities (see Chapter 5 if needed) and evaluate them as B functions.

10. The first quadrant area bounded by the curve.

11. The centroid of this area.

12. The volume generated when the area is revolved about the y axis.

13. The moment of inertia of this volume about its axis.

8. THE SIMPLE PENDULUM

A simple pendulum means a mass m suspended by a string (or weightless rod) of length l so that it can swing in a plane, as shown in Figure 8.1. The kinetic energy of m is then

$$T = \frac{1}{2}mv^2 = \frac{1}{2}m(l\dot{\theta})^2. \tag{8.1}$$

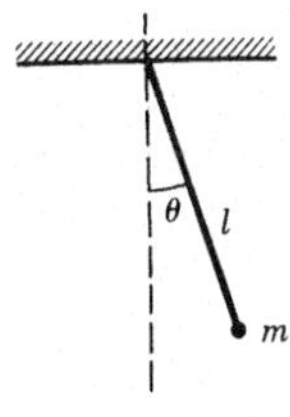

Figure 8.1

If the potential energy is zero when the string is horizontal, then at angle θ it is

$$V = -mgl\cos\theta.$$

Then the Lagrangian is (see Chapter 9, Section 5)

$$L = T - V = \frac{1}{2}ml^2\dot{\theta}^2 + mgl\cos\theta,$$

and the Lagrange equation of motion is

$$\frac{d}{dt}(ml^2\dot{\theta}) + mgl\sin\theta = 0$$

or

$$\ddot{\theta} = -\frac{g}{l}\sin\theta. \tag{8.2}$$

Example 1. Suppose the pendulum executes such small vibrations that $\sin\theta$ can be approximated by θ. Then (8.2) becomes the usual equation for the simple harmonic motion of a pendulum executing small vibrations, namely

$$\ddot{\theta} = -\frac{g}{l}\theta. \tag{8.3}$$

The solutions of (8.3) are $\sin\omega t$ and $\cos\omega t$ where $\omega = 2\pi\nu = \sqrt{g/l}$; the period of the motion is then (see Chapter 7, Problem 2.13, and Chapter 8, Problem 5.34)

$$T = \frac{1}{\nu} = 2\pi\sqrt{l/g}. \tag{8.4}$$

We now want to replace this approximate solution by one which is exact even for large θ.

Example 2. Going back to the differential equation of motion (8.2), we multiply both sides of it by $\dot{\theta}$ and integrate, thus obtaining

$$\dot{\theta}\ddot{\theta} = -\frac{g}{l}\sin\theta\,\dot{\theta} \qquad \text{or} \qquad \dot{\theta}\,d\dot{\theta} = -\frac{g}{l}\sin\theta\,d\theta;$$

$$\frac{1}{2}\dot{\theta}^2 = \frac{g}{l}\cos\theta + \text{const.} \tag{8.5}$$

We shall come back to the general solution of this equation when we discuss elliptic integrals; for now let us find the period for 180° swings (back and forth from −90° to +90°). For this case, $\dot{\theta} = 0$ when $\theta = 90°$, so the constant in (8.5) is zero, and we have (compare Chapter 8, Problem 7.13)

$$\frac{1}{2}\dot{\theta}^2 = \frac{g}{l}\cos\theta, \qquad \frac{d\theta}{dt} = \sqrt{\frac{2g}{l}}\sqrt{\cos\theta}, \qquad \frac{d\theta}{\sqrt{\cos\theta}} = \sqrt{\frac{2g}{l}}\,dt.$$

From $\theta = 0$ to $\theta = 90°$ is one-quarter of a period; hence the period for 180° swings is given by T in the equation

$$\int_0^{\pi/2}\frac{d\theta}{\sqrt{\cos\theta}} = \sqrt{\frac{2g}{l}}\int_0^{T/4}dt = \sqrt{\frac{2g}{l}}\cdot\frac{T}{4}.$$

Then the period is

$$T = 4\sqrt{\frac{l}{2g}}\int_0^{\pi/2}\frac{d\theta}{\sqrt{\cos\theta}}. \tag{8.6}$$

We can see by comparing (8.6) with (6.4) that this is a B function. By computer or tables we find that $T \cong 7.42\sqrt{l/g}$ (see Problem 1 and Problem 12.21). We can find the period for only this one special case (180° swings) by B functions; the general case gives an elliptic integral (Section 12).

PROBLEMS, SECTION 8

1. Complete the pendulum problem to find the period for 180° swings as a multiple of $\sqrt{l/g}$ [that is, evaluate the integral in (8.6)].

2. Suppose that a car with a door open at right angles ($\theta = 90°$) starts up and accelerates at a constant rate $a = 1$ mph/sec. The differential equation for $\theta(t)$ is $\ddot{\theta} = -A\sin\theta$ where $A = 3a/2w$ for a uniform door of width w. If $w = 3.5$ ft, find how long it takes for the door to close.

3. The figure is part of a cycloid with parametric equations

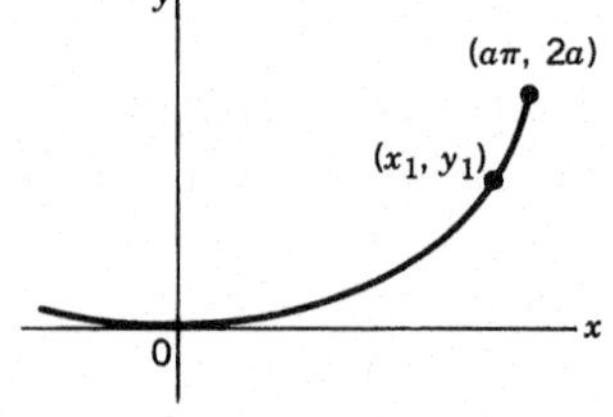

$$x = a(\theta + \sin\theta), \quad y = a(1 - \cos\theta).$$

(The graph shown is like Figure 4.4 of Chapter 9 with the origin shifted to P_2.) Show that the time for a particle to slide without friction along the curve from (x_1, y_1) to the origin is given by

$$t = \sqrt{\frac{a}{g}} \int_0^{y_1} \frac{dy}{\sqrt{y(y_1 - y)}}.$$

Hint: Show that the arc length element is $ds = \sqrt{2a/y}\,dy$. Evaluate the integral to show that the time is independent of the starting height y_1.

9. THE ERROR FUNCTION

You will meet this function in probability theory (Chapter 15, Section 8), and consequently in statistical mechanics and other applications of probability theory. You have probably heard of "grading on a curve." The "curve" means the bell-shaped graph of $y = e^{-x^2}$ (see Problem 1); the error function is the area under part of this curve. We define the error function as

$$\text{erf}(x) = \frac{2}{\sqrt{\pi}} \int_0^x e^{-t^2}\,dt. \tag{9.1}$$

Although this is the usual definition of erf(x), there are other closely related integrals which are used and sometimes referred to as the error function. Consequently, you need to look carefully at the definition in the reference you are using (text, tables, computer). Here are some integrals you may find and their relation to (9.1) (see Problem 2).

The standard normal or Gaussian cumulative distribution function $\Phi(x)$ [see Chapter 15, equation (8.5)]:

$$\Phi(x) = \frac{1}{\sqrt{2\pi}} \int_{-\infty}^x e^{-t^2/2}\,dt = \tfrac{1}{2} + \tfrac{1}{2}\,\text{erf}(x/\sqrt{2}\,), \tag{9.2a}$$

$$\Phi(x) - \tfrac{1}{2} = \frac{1}{\sqrt{2\pi}} \int_0^x e^{-t^2/2}\,dt = \tfrac{1}{2}\,\text{erf}(x/\sqrt{2}\,). \tag{9.2b}$$

The complementary error function:

$$\text{erfc}(x) = \frac{2}{\sqrt{\pi}} \int_x^\infty e^{-t^2}\,dt = 1 - \text{erf}(x), \tag{9.3a}$$

$$\text{erfc}\left(\frac{x}{\sqrt{2}}\right) = \sqrt{\frac{2}{\pi}} \int_x^\infty e^{-t^2/2}\,dt. \tag{9.3b}$$

We can also use (9.2) to write erf(x) in terms of the standard normal cumulative distribution function [Chapter 15, equation (8.5)].

$$\text{erf}(x) = 2\Phi(x\sqrt{2}\,) - 1. \tag{9.4}$$

We next consider several useful facts about the error function. You can easily prove that the error function is odd; that is, $\operatorname{erf}(-x) = -\operatorname{erf}(x)$ (Problem 3). We show that $\operatorname{erf}(\infty) = 1$ as follows:

$$(9.5) \qquad \operatorname{erf}(\infty) = \frac{2}{\sqrt{\pi}} \int_0^\infty e^{-t^2} dt = \frac{2}{\sqrt{\pi}} \tfrac{1}{2}\Gamma(\tfrac{1}{2}) = \frac{2}{\sqrt{\pi}} \tfrac{1}{2}\sqrt{\pi} = 1$$

by (5.2) and (5.3). For very small values of x, $\operatorname{erf}(x)$ can be approximated by expanding e^{-t^2} in a power series and integrating term by term. We get

$$(9.6) \qquad \begin{aligned} \operatorname{erf}(x) &= \frac{2}{\sqrt{\pi}} \int_0^x e^{-t^2} dt = \frac{2}{\sqrt{\pi}} \int_0^x \left(1 - t^2 + \frac{t^4}{2!} - \cdots\right) dt \\ &= \frac{2}{\sqrt{\pi}} \left(x - \frac{x^3}{3} + \frac{x^5}{5 \cdot 2!} - \cdots\right). \end{aligned}$$

[Use this when $|x| \ll 1$. Compare (10.4).]

For large x, say $x > 3$, $\operatorname{erf}(x)$ differs from $\operatorname{erf}(\infty) = 1$ [see (9.5)] by less than 10^{-4} (and of course even less for larger x). We are then usually interested in $1 - \operatorname{erf}(x) = \operatorname{erfc}(x)$. This is best approximated by an asymptotic series; we shall discuss such expansions in Section 10.

The function $\operatorname{erfi}(x)$, called the imaginary error function, is similar to the error function but with a positive exponential. We define

$$(9.7) \qquad \operatorname{erfi}(x) = \frac{2}{\sqrt{\pi}} \int_0^x e^{t^2}\, dt.$$

You can show (Problem 5) that $\operatorname{erf}(ix) = i \operatorname{erfi}(x)$.

The Fresnel integrals (Chapter 1, Section 15) are related to the error function (Problem 6). Also see Section 10, Problem 3 for other relations involving error functions.

PROBLEMS, SECTION 9

1. Sketch or computer plot a graph of the function $y = e^{-x^2}$.

2. Verify equations (9.2), (9.3), and (9.4). *Hint:* In (9.2a), you want to write $\Phi(x)$ in terms of an error function. Make the change of variable $t = u\sqrt{2}$ in the $\Phi(x)$ integral. *Warning:* Don't forget to adjust the limits; when $t = x$, $u = x/\sqrt{2}$.

3. Prove that $\operatorname{erf}(x)$ is an odd function of x. *Hint:* Put $t = -s$ in (9.1).

4. Show that $\displaystyle\int_{-\infty}^{\infty} e^{-y^2/2}\, dy = \sqrt{2\pi}$

   (a) by using (9.5) and (9.2a);

   (b) by reducing it to a Γ function and using (5.3).

5. Replace x by ix in (9.1) and let $t = iu$ to show that $\operatorname{erf}(ix) = i \operatorname{erfi}(x)$, where $\operatorname{erfi}(x)$ is defined in (9.7).

6. Assuming that x is real, show the following relation between the error function and the Fresnel integrals.

$$\operatorname{erf}\left(\frac{1-i}{\sqrt{2}}x\right) = (1-i)\sqrt{\frac{2}{\pi}}\int_0^x (\cos u^2 + i\sin u^2)\,du.$$

Hint: In (9.1), make the change of variables $t = \dfrac{1-i}{\sqrt{2}}u$.

10. ASYMPTOTIC SERIES

Since you have spent some time learning to test series for convergence, it may surprise you to learn that there are divergent series which can be of practical use. We can show this best by an example.

Example 1. From (9.3a)

$$\operatorname{erfc}(x) = 1 - \operatorname{erf}(x) = \frac{2}{\sqrt{\pi}}\int_x^\infty e^{-t^2}\,dt. \tag{10.1}$$

We are going to expand the integral in (10.1) in a series of inverse powers of x. To do this we write

$$e^{-t^2} = \frac{1}{t}te^{-t^2} = \frac{1}{t}\frac{d}{dt}\left(-\frac{1}{2}e^{-t^2}\right) \tag{10.2}$$

and integrate by parts as follows:

$$\begin{aligned}\int_x^\infty e^{-t^2}\,dt &= \int_x^\infty \frac{1}{t}\frac{d}{dt}\left(-\frac{1}{2}e^{-t^2}\right)dt\\ &= \frac{1}{t}\left(-\frac{1}{2}e^{-t^2}\right)\Big|_x^\infty - \int_x^\infty\left(-\frac{1}{2}e^{-t^2}\right)\left(-\frac{1}{t^2}\right)dt\\ &= \frac{1}{2x}e^{-x^2} - \frac{1}{2}\int_x^\infty \frac{1}{t^2}e^{-t^2}\,dt.\end{aligned} \tag{10.3}$$

Now in the last integral in (10.3), write $(1/t^2)e^{-t^2} = (1/t^3)(d/dt)(-\frac{1}{2}e^{-t^2})$, and again integrate by parts:

$$\begin{aligned}\int_x^\infty \frac{1}{t^2}e^{-t^2}\,dt &= \int_x^\infty \frac{d}{dt}\left(-\frac{1}{2}e^{-t^2}\right)dt\\ &= \frac{1}{t^3}\left(-\frac{1}{2}e^{-t^2}\right)\Big|_x^\infty - \int_x^\infty\left(-\frac{1}{2}e^{-t^2}\right)\left(-\frac{3}{t^4}\right)dt\\ &= \frac{1}{2x^3}e^{-x^2} - \frac{3}{2}\int_x^\infty \frac{1}{t^4}e^{-t^2}\,dt.\end{aligned}$$

Continue this process, and substitute (10.3) and the following steps into (10.1) to get (Problem 1)

$$\operatorname{erfc}(x) = 1 - \operatorname{erf}(x) \sim \frac{e^{-x^2}}{x\sqrt{\pi}}\left(1 - \frac{1}{2x^2} + \frac{1\cdot 3}{(2x^2)^2} - \frac{1\cdot 3\cdot 5}{(2x^2)^3} + \cdots\right). \tag{10.4}$$

[Use this when $|x| \gg 1$. Compare (9.6).]

(We shall explain the exact meaning of the symbol $\sim$ shortly.) This series diverges for every x because of the factors in the numerator. However, suppose we stop after a few terms and keep the integral at the end so that we have an exact equation. If we stop after the second term, we have

$$\text{(10.5)} \qquad \operatorname{erfc}(x) = \frac{e^{-x^2}}{x\sqrt{\pi}}\left(1 - \frac{1}{2x^2}\right) + \frac{3}{2\sqrt{\pi}}\int_x^\infty t^{-4}e^{-t^2}\,dt.$$

There is no approximation here. This is not an infinite series so there is no question of convergence. However, we shall show that the integral at the end is negligible for large enough x; this will then make it possible for us to use the rest of (10.5) [that is, the first two terms of (10.4)] as a good approximation for $\operatorname{erfc}(x)$ for large x. This is the meaning of an asymptotic series. As an infinite series it may diverge, but we do not use the infinite series. Instead, using an exact equation [like (10.5) for this example], we show that the first few terms which we *do* use give a good approximation if x is large.

▶ **Example 2.** Now let's look at the integral in (10.5); we want to estimate its size for large x. The t in the integrand takes values from x to ∞; therefore $t \geq x$ or $1/x \geq 1/t$ for all values of t from x to ∞. Let us write the integral as

$$\int_x^\infty t^{-4}e^{-t^2}\,dt = \int_x^\infty \frac{1}{t^5}\left(te^{-t^2}\right)\,dt.$$

We *increase* the value of this integral if we replace $1/t^5$ by $1/x^5$ since $1/x \geq 1/t$. Thus

$$\int_x^\infty t^{-4}e^{-t^2}\,dt < \int_x^\infty \frac{1}{x^5}\left(te^{-t^2}\right)\,dt = \frac{1}{x^5}\int_x^\infty te^{-t^2}\,dt$$
$$= \frac{1}{x^5}\left(-\frac{1}{2}e^{-t^2}\right)\Big|_x^\infty = \frac{e^{-x^2}}{2x^5}.$$

When we stop in (10.5) with the term in e^{-x^2}/x^3, the error is of the order of e^{-x^2}/x^5, which becomes much smaller than e^{-x^2}/x^3 as x increases. Thus we have shown that two terms of (10.4) give a good approximation for $\operatorname{erfc}(x)$ when $x \gg 1$. A similar result can be shown for an approximation using any number of terms of the asymptotic series (10.4) with the error depending on the "left-over" integral and the value of x.

We can make the above discussion more precise. For (10.4), we have seen that if we stop after the term in $x^{-3}e^{-x^2}$, the error is of the order of $x^{-5}e^{-x^2}$. Then the ratio of the error to the last term kept (namely $x^{-5}e^{-x^2} \div x^{-3}e^{-x^2} = x^{-2}$) tends to zero as x tends to infinity, that is, the approximation becomes increasingly good for larger x as we have said. The "error" in an asymptotic expansion means in general the difference between the function being expanded and a partial sum (first N terms) of the series. A series is called an asymptotic expansion (about ∞) of a function $f(x)$ if, for each fixed N, the ratio of the error to the last (nonzero) term kept, tends to zero as $x \to \infty$. In symbols

$$f(x) \sim \sum_{n=0}^{\infty} \phi_n(x)$$

(10.6) $$\left(\text{read } \sum_{n=0}^{\infty} \phi_n(x) \text{ is an asymptotic expansion of } f(x)\right)$$

if for each fixed N

$$\left| f(x) - \sum_{n=0}^{N} \phi_n(x) \right| \div \phi_N(x) \to 0 \quad \text{as } x \to \infty.$$

Frequently, the terms of an asymptotic series (about ∞) are inverse powers of x. [We could write (10.4) this way by multiplying through by e^{x^2}.] Then (10.6) becomes

$$f(x) \sim \sum_{n=0}^{\infty} \frac{a_n}{x^n}$$

(10.7) if for each fixed N

$$\left| f(x) - \sum_{n=0}^{N} \frac{a_n}{x^n} \right| \cdot x^N \to 0 \quad \text{as } x \to \infty.$$

We can also have asymptotic series about the origin (or any point—compare Taylor series). We say that

$$f(x) \sim \sum_{n=0}^{\infty} a_n x^n$$

(10.8) if for each fixed N

$$\left| f(x) - \sum_{n=0}^{N} a_n x^n \right| \div x^N \to 0 \quad \text{as } x \to 0.$$

Although we have discussed the particularly interesting case of *divergent* asymptotic series, it is not necessary for such series to diverge. Note that to test a series for convergence, we fix x and let n tend to infinity; to see if a series is asymptotic, we fix n and let x tend to a limit. A given series may meet both tests, or only one or the other (or neither).

PROBLEMS, SECTION 10

1. Carry through the algebra to get equation (10.4).

2. The integral $\int_x^\infty t^{p-1}e^{-t}\,dt = \Gamma(p,x)$ is called an *incomplete* Γ function. [Note that if $x = 0$, this integral is $\Gamma(p)$.] By repeated integration by parts, find several terms of the asymptotic series for $\Gamma(p,x)$.

3. Express the complementary error function $\operatorname{erfc}(x)$ as an incomplete Γ function (see Problem 2) and use your result in Problem 2 to obtain (again) the asymptotic expansion of $\operatorname{erfc}(x)$ as in (10.4).

4. $E_n(x) = \int_1^\infty \frac{e^{-xt}}{t^n}\,dt$, $n = 0, 1, 2, \cdots$, and $\mathrm{Ei}(x) = \int_{-\infty}^{x} \frac{e^t}{t}\,dt$, and other similar integrals are called *exponential integrals*. By making appropriate changes of variable, show that

(a) $E_1(x) = \int_x^\infty \frac{e^{-t}}{t}\,dt$ (b) $\mathrm{Ei}(x) = -\int_{-x}^\infty \frac{e^{-t}}{t}\,dt$

(c) $E_1(x) = -\mathrm{Ei}(-x)$ (d) $\int_0^x \frac{e^{1/t}}{t}\,dt = E_1(-1/x)$

(*Caution:* Various notations are used; check carefully the notation in references you are using.)

5. (a) Express $E_1(x)$ as an incomplete Γ function.
(b) Find the asymptotic series for $E_1(x)$.

6. The *logarithmic integral* is $\mathrm{li}(x) = \int_0^x \frac{dt}{\ln t}$. Express as exponential integrals

(a) $\mathrm{li}(x)$ (b) $\mathrm{li}(e^x)$ (c) $\int_0^x \frac{dt}{\ln(1/t)}$

7. Computer plot graphs of

(a) $E_n(x)$ for $n = 0$ to 10 and $x = 0$ to 2;
(b) $E_1(x)$ and $\mathrm{Ei}(x)$ for $x = 0$ to 2;
(c) the *sine integral* $\mathrm{Si}(x) = \int_0^x \frac{\sin t}{t}\,dt$ and the *cosine integral* $\mathrm{Ci}(x) = -\int_x^\infty \frac{\cos t}{t}\,dt$ for $x = 0$ to 4π.

11. STIRLING'S FORMULA

Formulas involving $n!$ or $\Gamma(p)$ are not convenient to simplify algebraically or to differentiate. Here is an approximate formula for the factorial or Γ function known as Stirling's formula which can be used to simplify formulas involving factorials:

$$(11.1)\quad n! \sim n^n e^{-n}\sqrt{2\pi n} \quad \text{or} \quad \Gamma(p+1) \sim p^p e^{-p}\sqrt{2\pi p}. \qquad \text{Stirling's formula}$$

The sign $\sim$ (read "is asymptotic to") means that the ratio of the two sides

$$\frac{n!}{n^n e^{-n}\sqrt{2\pi n}}$$

tends to 1 as $n \to \infty$. Thus we get better approximations to $n!$ as n becomes large. Actually the absolute error (difference between the Stirling approximation and the correct value) *increases*, but the relative error (ratio of the error to the value of $n!$) tends to zero as n increases. To get some idea of how this formula arises, we outline what could, with a little more detail, be a derivation of it. (For more detail, consult advanced calculus books.) Start with

$$(11.2)\qquad \Gamma(p+1) = p! = \int_0^\infty x^p e^{-x}\,dx = \int_0^\infty e^{p\ln x - x}\,dx.$$

Substitute a new variable y such that

$$x = p + y\sqrt{p}.$$

Then

$$dx = \sqrt{p}\,dy,$$
$$x = 0 \text{ corresponds to } y = -\sqrt{p},$$

and (11.2) becomes

$$p! = \int_{-\sqrt{p}}^{\infty} e^{p\ln(p+y\sqrt{p})-p-y\sqrt{p}}\sqrt{p}\,dy. \tag{11.3}$$

For large p, the logarithm can be expanded in the following power series:

$$\ln(p+y\sqrt{p}) = \ln p + \ln\left(1+\frac{y}{\sqrt{p}}\right) = \ln p + \frac{y}{\sqrt{p}} - \frac{y^2}{2p} + \cdots. \tag{11.4}$$

Substituting (11.4) into (11.3), we get

$$\begin{aligned} p! &\sim \int_{-\sqrt{p}}^{\infty} e^{p\ln p + y\sqrt{p} - (y^2/2) - p - y\sqrt{p}}\sqrt{p}\,dy \\ &= e^{p\ln p - p}\sqrt{p}\int_{-\sqrt{p}}^{\infty} e^{-y^2/2}\,dy \\ &= p^p e^{-p}\sqrt{p}\left[\int_{-\infty}^{\infty} e^{-y^2/2}\,dy - \int_{-\infty}^{-\sqrt{p}} e^{-y^2/2}\,dy\right]. \end{aligned}$$

The first integral is easily shown to be $\sqrt{2\pi}$ (Problem 9.4). The second integral tends to zero as $p \to \infty$, and we have

$$p! \sim p^p e^{-p}\sqrt{2\pi p}$$

which is (11.1). With more work, it is possible to find an asymptotic expansion for $\Gamma(p+1)$:

$$\Gamma(p+1) = p! = p^p e^{-p}\sqrt{2\pi p}\left(1 + \frac{1}{12p} + \frac{1}{288p^2} + \cdots\right). \tag{11.5}$$

This is another example of an asymptotic series which is divergent as an infinite series; however, the first term alone (Stirling's formula) is a good approximation when p is large, and the second term can be used to estimate the relative error (Problem 1).

PROBLEMS, SECTION 11

1. Use the term $1/(12p)$ in (11.5) to show that the error in Stirling's formula (11.1) is $< 10\%$ for $p > 1$; $< 1\%$ for $p > 10$; $< 0.1\%$ for $p > 100$; $< 0.01\%$ for $p > 1000$.

2. (a) To see the results in Problem 1 graphically, computer plot the percentage error in Stirling's formula as a function of p for values of p from 1 to 1000. Make separate plots, say for $p = 1$ to 10, 10 to 100, 100 to 1000, to make it easier to read values from your plots.

(b) Repeat part (a) for the percentage error in (11.5) using two terms of the asymptotic series, that is, Stirling's formula times $[1 + 1/(12p)]$.

3. In statistical mechanics, we frequently use the approximation $\ln N! = N \ln N - N$, where N is of the order of Avogadro's number. Write out $\ln N!$ using Stirling's formula, compute the approximate value of each term for $N = 10^{23}$, and so justify this commonly used approximation.

4. Use Stirling's formula to evaluate $\lim_{n\to\infty} \dfrac{(2n)!\,\sqrt{n}}{2^{2n}(n!)^2}$.

5. Use Stirling's formula to evaluate $\lim_{n\to\infty} \dfrac{\Gamma(n+\frac{3}{2})}{\sqrt{n}\,\Gamma(n+1)}$.

6. Use equations (3.4) and (11.5) to show that $\Gamma(p) \sim p^p e^{-p}\sqrt{2\pi/p}\left(1 + \frac{1}{12p} + \cdots\right)$.

7. The function $\psi(p) = \frac{d}{dp}\ln\Gamma(p)$ is called the digamma function, and the polygamma functions are defined by $\psi^n(p) = \frac{d^n}{dp^n}\psi(p)$. [*Warning:* Some authors define $\psi(p)$ as $\frac{d}{dp}\ln p! = \frac{d}{dp}\ln\Gamma(p+1)$.]

(a) Show that $\psi(p+1) = \psi(p) + \frac{1}{p}$. *Hint:* See (3.4).

(b) Use Problem 6 to obtain $\psi(p) \sim \ln p - \frac{1}{2p} - \frac{1}{12p^2}\cdots$.

8. Sketch or computer plot a graph of $y = \ln x$ for $x > 0$. Show that $\ln n!$ is between the values of the integrals $\int_2^{n+1} \ln x\, dx$ and $\int_1^n \ln x\, dx$. (*Hint:* $\ln n! = \ln 1 + \ln 2 + \ln 3 + \cdots$ is the sum of the areas of rectangles of width 1 and height up to the $\ln x$ curve at $x = 1, 2, 3, \cdots$.) By considering the values of the two integrals for very large n as in Problem 3, show that $\ln n! = n\ln n - n$ approximately for large n.

9. The following expression occurs in statistical mechanics:

$$P = \frac{n!}{(np+u)!\,(nq-u)!}p^{np+u}q^{nq-u}.$$

Use Stirling's formula to show that

$$\frac{1}{P} \sim x^{npx}y^{nqy}\sqrt{2\pi npqxy},$$

where $x = 1 + \dfrac{u}{np}$, $y = 1 - \dfrac{u}{nq}$, and $p + q = 1$. *Hint:* Show that

$$(np)^{np+u}(nq)^{nq-u} = n^n p^{np+u}q^{nq-u}$$

and divide numerator and denominator of P by this expression.

10. Use Stirling's formula to find $\lim_{n\to\infty}(n!)^{1/n}/n$.

12. ELLIPTIC INTEGRALS AND FUNCTIONS

This is another collection of integrals and related functions which may arise in applied problems and as computer answers (see problems). We shall merely summarize the basic definitions and properties—there are whole books on the subject—and you may find useful formulas and information in your computer program and in reference books and tables.

Legendre Forms The *Legendre* forms of the elliptic integrals of the first and second kinds are:

$$(12.1)\qquad \begin{aligned} F(\phi, k) &= \int_0^\phi \frac{d\theta}{\sqrt{1-k^2\sin^2\theta}}, \qquad 0 \le k \le 1,\\ E(\phi, k) &= \int_0^\phi \sqrt{1-k^2\sin^2\theta}\, d\theta, \quad 0 \le k \le 1. \end{aligned}$$

There is also an elliptic integral of the third kind which occurs less frequently. In (12.1), ϕ is called the *amplitude* and k is called the *modulus* of the elliptic integral.

Jacobi Forms If we put $t = \sin\theta$, $x = \sin\phi$ in the *Legendre* forms (12.1), we obtain the *Jacobi* forms of the elliptic integrals of the first and second kind:

$$t = \sin\theta,$$

$$dt = \cos\theta\, d\theta \quad \text{or} \quad d\theta = \frac{dt}{\cos\theta} = \frac{dt}{\sqrt{1-t^2}}.$$

The limits $\theta = 0$ to ϕ become $t = 0$ to x.

Then

$$(12.2)\qquad \begin{aligned} F(\phi, k) &= \int_0^\phi \frac{d\theta}{\sqrt{1-k^2\sin^2\theta}} = \int_0^x \frac{dt}{\sqrt{1-t^2}\,\sqrt{1-k^2t^2}}\\ E(\phi, k) &= \int_0^\phi \sqrt{1-k^2\sin^2\theta}\, d\theta = \int_0^x \frac{\sqrt{1-k^2t^2}}{\sqrt{1-t^2}}\, dt. \end{aligned}$$

Complete Elliptic Integrals The *complete* elliptic integrals of the first and second kind are the values of F and E when $\phi = \pi/2$ or $x = \sin\phi = 1$:

$$(12.3)\qquad \begin{aligned} K \text{ or } K(k) &= F\left(\frac{\pi}{2}, k\right) = \int_0^{\pi/2} \frac{d\theta}{\sqrt{1-k^2\sin^2\theta}} = \int_0^1 \frac{dt}{\sqrt{1-t^2}\,\sqrt{1-k^2t^2}},\\ E \text{ or } E(k) &= E\left(\frac{\pi}{2}, k\right) = \int_0^{\pi/2} \sqrt{1-k^2\sin^2\theta}\, d\theta = \int_0^1 \frac{\sqrt{1-k^2t^2}}{\sqrt{1-t^2}}\, dt. \end{aligned}$$

Warning: The notation used for elliptic integrals is not uniform. Most references use F and E, but you may find ϕ replaced by $x = \sin\phi$, and instead of k you may find $m = k^2$, or $\sin^{-1} k$. Also (ϕ, k) may be written as (k, ϕ), and other variations exist. So check carefully the notation of any book or computer program you are using and reconcile the results with the notation used here.

Example 1. $\int_0^{\pi/3} \sqrt{1-(1/2)\sin^2\theta}\, d\theta = E(\phi, k) = E(\pi/3, 1/\sqrt{2})$ in our notation. Other books or computer programs might give: $E(\phi, m) = E(\pi/3, 1/2)$, or $E(x, k) = E(\sqrt{3}/2, 1/\sqrt{2})$ or $E(\phi, \sin^{-1} k) = E(\pi/3, \pi/4)$, etc. Of course, all of them will give the same numerical approximation 0.964951.

Many integrals can be written in the form of one of the integrals in (12.2).

Example 2. $\int_0^{\pi/3} \sqrt{16 - 8\sin^2\theta}\, d\theta$ becomes 4 times the integral in Example 1 if we divide out a factor of 4 to get $4\int_0^{\pi/3} \sqrt{1 - (1/2)\sin^2\theta}\, d\theta$.

Example 3. $\int_0^{2/5} \frac{dt}{\sqrt{1-t^2}\sqrt{1-4t^2}} = F(\phi, k) = F(\sin^{-1}\frac{2}{5}, 2)$ in the notation of (12.2), except that we have previously required $k < 1$, and here $k = 2$. However, we can put this integral in the standard form with $k < 1$ by making the change of variable $4t^2 = r^2$, or $r = 2t$. Substituting this into the given integral gives

$$\int_0^{4/5} \frac{dr/2}{\sqrt{1 - r^2/4}\sqrt{1-r^2}}$$

which, by (12.2), is $\frac{1}{2}F(\phi, k) = \frac{1}{2}F(\sin^{-1}\frac{4}{5}, \frac{1}{2})$. (See Problem 24.)

It is sometimes useful to note that the integrands in elliptic integrals are all functions of $\sin^2\theta$ and so are even functions of θ. Thus an elliptic integral from $-\phi_1$ to ϕ_2 (ϕ_1 and ϕ_2 both positive) is equal to the integral from 0 to ϕ_1 plus the integral from 0 to ϕ_2 and we have

$$\int_{-\phi_1}^{\phi_2} \sqrt{1 - k^2\sin^2\theta}\, d\theta = E(\phi_1, k) + E(\phi_2, k)$$

and a similar formula for $F(\phi, k)$. Also we may note that a function of $\sin^2\theta$ has period π and is symmetric about $\theta = n\pi + \pi/2$ (look at a graph of $\sin^2\theta$). Thus, using the complete elliptic integrals in (12.3), we can write (Problem 2)

$$\begin{aligned} F(n\pi \pm \phi, k) &= 2nK \pm F(\phi, k), \\ E(n\pi \pm \phi, k) &= 2nE \pm E(\phi, k). \end{aligned} \tag{12.4}$$

Since $k^2\sin^2\theta < 1$ (for $k^2 < 1$), we get convergent infinite series for elliptic integrals by expanding their integrands using the binomial theorem, and then integrating term by term (Problem 1). For small k these series converge rapidly and provide a good method for approximating elliptic integrals when $k \ll 1$.

Here are some examples where elliptic integrals occur.

Example 4. Find the arc length of an ellipse. This is the problem that gave elliptic integrals their name. We write the equation of the ellipse in the parametric form

$$\begin{aligned} x &= a\sin\theta, \\ y &= b\cos\theta, \end{aligned}$$

for the case $a > b$. (If $b > a$, use the form $x = a\cos\theta$, $y = b\sin\theta$; see Problem 15.) Then for $a > b$, we have

$$ds^2 = dx^2 + dy^2 = (a^2\cos^2\theta + b^2\sin^2\theta)\, d\theta^2.$$

Since $a^2 - b^2 > 0$, we can write

$$\int ds = \int \sqrt{a^2 - (a^2 - b^2)\sin^2\theta}\, d\theta = a\int \sqrt{1 - \frac{a^2 - b^2}{a^2}\sin^2\theta}\, d\theta.$$

This is an elliptic integral of the second kind where $k^2 = (a^2 - b^2)/a^2 = e^2$ (e is the eccentricity of the ellipse in analytic geometry). If we want the complete circumference, θ goes from 0 to 2π, and the answer is $4aE(\pi/2, k) = 4aE(k)$. For a smaller arc, we use the appropriate limits ϕ_1 and ϕ_2 and obtain $E(\phi_2, k) - E(\phi_1, k)$. For any given ellipse (that is, given a and b), we can find the numerical value of the desired arc length from computer or tables.

Example 5. Let a pendulum swing through large angles. We had in Section 8

$$\dot{\theta}^2 = \frac{2g}{l}\cos\theta + \text{ const.,} \tag{12.5}$$

and we considered 180° swings, that is of amplitude 90°. Now we want to consider swings of any amplitude, say α; then $\dot{\theta} = 0$ when $\theta = \alpha$, and (12.5) becomes

$$\dot{\theta}^2 = \frac{2g}{l}(\cos\theta - \cos\alpha). \tag{12.6}$$

Integrating (12.6), we get

$$\int_0^\alpha \frac{d\theta}{\sqrt{\cos\theta - \cos\alpha}} = \sqrt{\frac{2g}{l}}\frac{T_\alpha}{4}, \tag{12.7}$$

where T_α is the period for swings from $-\alpha$ to $+\alpha$ and back. This integral can be written as an elliptic integral; its value (Problem 17) is

$$\sqrt{2}\,K\left(\sin\frac{\alpha}{2}\right). \tag{12.8}$$

Then (12.7) gives for the period

$$T_\alpha = 4\sqrt{\frac{l}{2g}}\sqrt{2}\,K\left(\sin\frac{\alpha}{2}\right) = 4\sqrt{\frac{l}{g}}\,K\left(\sin\frac{\alpha}{2}\right).$$

For α not too large (say $\alpha < 90°$, $\frac{1}{2}\alpha < 45°$, so that $\sin^2(\alpha/2) < \frac{1}{2}$), we can get a good approximation to T_α by series (Problem 1):

$$T_\alpha = 4\sqrt{\frac{l}{g}}\frac{\pi}{2}\left(1 + \left(\frac{1}{2}\right)^2\sin^2\frac{\alpha}{2} + \left(\frac{1\cdot 3}{2\cdot 4}\right)^2\sin^4\frac{\alpha}{2} + \cdots\right). \tag{12.9}$$

For α small enough so that $\sin\alpha/2$ can be approximated by $\alpha/2$, we can write

$$T_\alpha = 2\pi\sqrt{\frac{l}{g}}\left(1 + \frac{\alpha^2}{16} + \cdots\right). \tag{12.10}$$

For very small α, we get the familiar formula for simple harmonic motion, $T = 2\pi\sqrt{l/g}$ independent of α. For somewhat larger α, say $\alpha = \frac{1}{2}$ radian (about 30°), we get

$$T_{\alpha=1/2} = 2\pi\sqrt{\frac{l}{g}}\left(1 + \frac{1}{64} + \cdots\right). \tag{12.11}$$

This would mean that a pendulum started at 30° would get exactly out of phase with one of very small amplitude in about 32 periods.

For another physics problem giving rise to an elliptic integral, see Am. J. Phys. **55**, 763 (1987).

Elliptic Functions Recall that

$$u = \int_0^x \frac{dt}{\sqrt{1-t^2}} = \sin^{-1} x$$

defines u as a function of x, or x as a function of u; in fact $x = \sin u$. In a similar way, $u = F(\phi, k)$ in (12.2) defines u as a function of ϕ (or of $x = \sin\phi$) or it defines x or ϕ as functions of u (we are assuming k fixed). We write

$$u = \int_0^x \frac{dt}{\sqrt{1-t^2}\,\sqrt{1-k^2t^2}} = \operatorname{sn}^{-1} x. \tag{12.12}$$

or $x = \operatorname{sn} u$. The function $\operatorname{sn} u$ (read ess-en of u) is an elliptic function. Since $x = \sin\phi$, we have

$$x = \operatorname{sn} u = \sin\phi. \tag{12.13}$$

There are other elliptic functions, related to $\operatorname{sn} u$; you will notice [in (12.14)] that they have some resemblance to the trigonometric functions. We define

$$\begin{aligned} \operatorname{cn} u &= \cos\phi = \sqrt{1-\sin^2\phi} = \sqrt{1-\operatorname{sn}^2 u} = \sqrt{1-x^2}, \\ \operatorname{dn} u &= \frac{d\phi}{du} = \frac{1}{du/d\phi} = \sqrt{1-k^2\sin^2\phi} = \sqrt{1-k^2\operatorname{sn}^2 u} = \sqrt{1-k^2x^2}. \end{aligned} \tag{12.14}$$

[The value of $du/d\phi$ is found from $u = F(\phi, k)$ in (12.2).] There are many formulas relating these functions—for example, addition formulas, integrals, derivatives, etc. These can be looked up or, in some cases, easily worked out. For example, since $\operatorname{sn} u = \sin\phi$, we have

$$\frac{d}{du}(\operatorname{sn} u) = \frac{d}{du}(\sin\phi) = \cos\phi\frac{d\phi}{du} = \operatorname{cn} u \operatorname{dn} u.$$

For a physical problem using elliptic functions, see Am. J. Phys. **68**, 888–895 (2000).

PROBLEMS, SECTION 12

1. Expand the integrands of K and E [see (12.3)] in power series in $k^2\sin^2\theta$ (assuming small k), and integrate term by term to find power series approximations for the complete elliptic integrals K and E.

2. Use a graph of $\sin^2\theta$ and the text discussion just before (12.4) to verify the equations (12.4). Note that the area under the $\sin^2\theta$ graph from 0 to $\pi/2$ and the area from $\pi/2$ to π are mirror images of each other, and this will be true also for any function of $\sin^2\theta$.

3. Computer plot graphs of $K(k)$ and $E(k)$ in (12.3) for k from 0 to 1. Also plot 3D graphs of $F(\phi, k)$ and $E(\phi, k)$ in (12.1) for k from 0 to 1 and ϕ from 0 to $\pi/2$ and also from 0 to 2π. *Warning:* Be sure you understand the notation used by your computer program; see text discussion just after (12.3) and Example 1.

In Problems 4 to 13, identify each of the integrals as an elliptic integral (see Examples 1 and 2). Learn the notation of your computer program (see Problem 3) and then evaluate the integral by computer.

4. $\displaystyle\int_0^1 \frac{dt}{\sqrt{1-t^2}\sqrt{1-t^2/4}}$

5. $\displaystyle\int_0^{\pi/2} \sqrt{1-\frac{1}{9}\sin^2\theta}\,d\theta$

6. $\displaystyle\int_0^{\pi/3} \frac{d\theta}{\sqrt{9-\sin^2\theta}}$

7. $\displaystyle\int_0^{5\pi/4} \sqrt{25-\sin^2\theta}\,d\theta$

8. $\displaystyle\int_0^{\sqrt{3}/2} \frac{\sqrt{49-4t^2}}{\sqrt{1-t^2}}\,dt$

9. $\displaystyle\int_{-1/2}^{1/2} \frac{dt}{\sqrt{1-t^2}\sqrt{4-3t^2}}$

10. $\displaystyle\int_0^{\pi/4} \frac{d\theta}{\sqrt{4-\sin^2\theta}}$

11. $\displaystyle\int_{-\pi/2}^{3\pi/8} \frac{d\theta}{\sqrt{1-\frac{9}{10}\sin^2\theta}}$

12. $\displaystyle\int_0^{1/2} \frac{\sqrt{100-t^2}}{\sqrt{1-t^2}}\,dt$

13. $\displaystyle\int_{-1/2}^{3/4} \frac{\sqrt{9-4t^2}}{\sqrt{1-t^2}}\,dt$

14. Find the circumference of the ellipse $4x^2 + 9y^2 = 36$.

15. Find the length of arc of the ellipse $x^2 + (y^2/4) = 1$ between $(0, 2)$ and $(\frac{1}{2}, \sqrt{3})$. (Note that here $b > a$; see Example 4.)

16. Find the arc length of one arch of $y = \sin x$.

17. Write the integral in equation (12.7) as an elliptic integral and show that (12.8) gives its value. *Hints:* Write $\cos\theta = 1 - 2\sin^2(\theta/2)$ and a similar equation for $\cos\alpha$. Then make the change of variable $x = \sin(\theta/2)/\sin(\alpha/2)$.

18. Computer plot graphs of $\operatorname{sn} u$, $\operatorname{cn} u$, and $\operatorname{dn} u$, for several values of k, say, for example, $k = 1/4,\ 1/2,\ 3/4,\ 0.9,\ 0.99$. Also plot 3D graphs of sn, cn, and dn as functions of u and k.

19. If $u = \ln(\sec\phi + \tan\phi)$, then ϕ is a function of u called the *Gudermannian* of u, $\phi = \operatorname{gd} u$. Prove that:

$$u = \ln\tan\left(\frac{\pi}{4} + \frac{\phi}{2}\right), \quad \tan\operatorname{gd} u = \sinh u, \quad \sin\operatorname{gd} u = \tanh u, \quad \frac{d}{du}\operatorname{gd} u = \operatorname{sech} u.$$

20. Show that for $k = 0$:

$$u = F(\phi, 0) = \phi, \quad \operatorname{sn} u = \sin u, \quad \operatorname{cn} u = \cos u, \quad \operatorname{dn} u = 1;$$

and for $k = 1$:

$$u = F(\phi, 1) = \ln(\sec\phi + \tan\phi) \quad \text{or} \quad \phi = \operatorname{gd} u \quad \text{(Problem 19)},$$
$$\operatorname{sn} u = \tanh u, \quad \operatorname{cn} u = \operatorname{dn} u = \operatorname{sech} u.$$

21. Show that the four answers given in Section 1 for $\int_0^{\pi/2} d\theta/\sqrt{\cos\theta}$ are all correct. *Hints:* For the beta function result, use (6.4). Then get the gamma function results by using (7.1) and the various Γ function formulas. For the elliptic integral, use the hint of Problem 17 with $\alpha = \pi/2$.

22. In the pendulum problem, $\theta = \alpha \sin \sqrt{g/l}\, t$ is an approximate solution when the amplitude α is small enough for the motion to be considered simple harmonic. Show that the corresponding exact solution when α is not small is

$$\sin\frac{\theta}{2} = \sin\frac{\alpha}{2}\,\mathrm{sn}\sqrt{\frac{g}{l}}\,t$$

where $k = \sin(\alpha/2)$ is the modulus of the elliptic function. Show that this reduces to the simple harmonic motion solution for small amplitude α.

23. A uniform solid sphere of density $\frac{1}{2}$ is floating in water. (Compare Chapter 8, Problem 5.37.) It is pushed down just under water and released. Write the differential equation of motion (neglecting friction) and solve it to obtain the period in terms of $K(5^{-1/2})$. Show that this period is approximately 1.16 times the period for small oscillations.

24. Sometimes you may find the notation $F(\phi, k)$ in (12.2) used when $k > 1$. Allowing this notation, show that $\frac{1}{3}F(\sin^{-1}\frac{3}{5}, \frac{4}{3}) = \frac{1}{4}F(\sin^{-1}\frac{4}{5}, \frac{3}{4})$. *Hints:* Using the Jacobi form of F in (12.2), write the integral which is equal to $\frac{1}{3}F(\sin^{-1}\frac{3}{5}, \frac{4}{3})$. Follow Example 3 to make a change of variable, write the corresponding integral, and verify that it is equal to $\frac{1}{4}F(\sin^{-1}\frac{4}{5}, \frac{3}{4})$.

25. As in Problem 24, show that $\frac{1}{2}F(\sin^{-1}\frac{4}{15}, \frac{5}{2}) = \frac{1}{5}F(\sin^{-1}\frac{2}{3}, \frac{2}{5})$.

13. MISCELLANEOUS PROBLEMS

1. Show that

$$\int_0^\infty \frac{y^m\,dy}{(1+y)^{n+1}} = \frac{1}{(n-m)C(n,m)}$$

for positive integral m and n, $n > m$, where $C(n,m) = \binom{n}{m}$.

2. Show that $B(m,n)B(m+n,k) = B(n,k)B(n+k,m)$.

3. Use Stirling's formula to show that

$$\lim_{n\to\infty} n^x B(x,n) = \Gamma(x).$$

4. Verify the asymptotic series

$$\int_0^\infty \frac{e^{-t}\,dt}{(1+xt)} \sim \sum(-1)^n n!\,x^n$$

[see equation (10.8)]. *Hint:* Integrate by parts repeatedly, integrating $e^{-t}\,dt$ and differentiating the powers of $(1+xt)^{-1}$.

5. Use gamma and beta function formulas to show that $\displaystyle\int_0^\infty \frac{dx}{(1+x)\sqrt{x}} = \pi.$

6. Generalize Problem 5 to show that $\displaystyle\int_0^\infty \frac{dx}{(1+x)x^p} = \frac{\pi}{\sin \pi p}$, $0 < p < 1$.

Identify each of the following integrals or expressions as one of the functions of this chapter. Check your work by evaluating both your answer and the original problem by computer. Be sure you understand your computer program's notation.

7. $\displaystyle\int_0^\infty x^3 e^{-x}\,dx$ **8.** $\displaystyle\int_0^1 e^{-x^2}\,dx$ **9.** $\displaystyle\int_0^1 \sqrt{\frac{4-3x^2}{1-x^2}}\,dx$

10. $\displaystyle\int_{-\pi/4}^{3\pi/4} \frac{d\phi}{\sqrt{1+\cos^2\phi}}$ **11.** $\displaystyle\int_0^{3/5} \frac{dt}{\sqrt{1-t^2}\,\sqrt{16-25t^2}}$

12. $\displaystyle\int_0^{\pi/2} \frac{dx}{\sqrt{2-\sin^2 x}}$ **13.** $\displaystyle\frac{d}{du}(\operatorname{cn} u)$ **14.** $\displaystyle\int_1^{\infty} e^{-x^2/2}\,dx$

15. $\displaystyle\int_0^{\infty} x^{5/2}e^{-x}\,dx$ **16.** $\displaystyle\int_{-\infty}^{\infty} e^{-x^2}\,dx$ **17.** $\displaystyle\int_0^{\pi/2} \sqrt{\sin^3\theta\cos^5\theta}\,d\theta$

18. $\displaystyle\int_0^{\infty} \frac{e^{-x}\,dx}{x^{1/4}}$ **19.** $\displaystyle\int_5^{\infty} e^{-x^2}\,dx$ **20.** $\displaystyle\int_0^{\pi/2} (\cos x)^{5/2}\,dx$

21. $\displaystyle\int_0^{5} x^{-1/3}(5-x)^{10/3}\,dx$ **22.** $\displaystyle\int_0^{7\pi/8} \sqrt{4-\sin^2 x}\,dx$

23. Find an expression for the exact value of $\Gamma(55.5)$ in terms of double factorials (!!), powers of 2 and $\sqrt{\pi}$. For !!, see Chapter 1, Section 13C, Example 2.

24. Using your result in Problem 23 and equation (5.4), find an expression for the exact value of $\Gamma(-54.5)$.

25. As in problems 23 and 24, find expressions for the exact values of $\Gamma(28.5)$ and $\Gamma(-27.5)$.

CHAPTER

12

Series Solutions of Differential Equations; Legendre, Bessel, Hermite, and Laguerre Functions

1. INTRODUCTION

By now you are well aware that physical problems in many fields lead to differential equations to be solved. In Chapter 13, we will discuss a variety of physical problems which lead to partial differential equations. To solve them, we will need the solutions of some ordinary differential equations which cannot be solved in terms of elementary functions. So in this chapter we will learn about these equations and their solutions. However, if you would prefer to see some of the physics before you study the math, and if you've studied Chapters 7 and 8, you could first do Sections 1 to 4 of Chapter 13, and then come back to Chapter 12 to learn the material needed for the rest of Chapter 13. (See the Preface.)

Now you may be thinking that your computer will give you the solutions of these differential equations so you don't need to study this. What your computer may give you is the *name* of a function. What you need to know is something about the function: graphs; formulas for derivatives and integrals; formulas that correspond to trigonometric identities for sine and cosine functions; and other useful information so that you can work with these named functions which occur often in applications. This is what we will discuss in this chapter.

The differential equations we are going to solve are linear, like the equations of Chapter 8, Section 5, but with coefficients which are functions of x instead of constants, that is, of the form $y'' + f(x)y' + g(x)y = 0$. A method of solving such equations which we will find useful is to assume an infinite series solution.

Example 1. We illustrate the method of series solution by solving the following simple equation (which you can easily solve by elementary methods also!):

$$y' = 2xy. \tag{1.1}$$

We assume a solution of this differential equation in the form of a power series, namely

$$(1.2)\qquad \begin{aligned} y &= a_0 + a_1x + a_2x^2 + a_3x^3 + \cdots + a_nx^n + \cdots \\ &= \sum_{n=0}^{\infty} a_nx^n, \end{aligned}$$

where the a's are to be found. Differentiating (1.2) term by term, we get

$$(1.3)\qquad \begin{aligned} y' &= a_1 + 2a_2x + 3a_3x^2 + \cdots + na_nx^{n-1} + \cdots \\ &= \sum_{n=1}^{\infty} na_nx^{n-1}. \end{aligned}$$

We substitute (1.2) and (1.3) into the differential equation (1.1); we then have two power series equal to each other. Now the original differential equation is to be satisfied for all values of x, that is, y' and $2xy$ are to be the same function of x. Since a given function has only one series expansion in powers of x (see Chapter 1, Section 11), the two series must be identical, that is, the coefficients of corresponding powers of x must be equal. We get the following set of equations for the a's:

$$(1.4)\qquad a_1 = 0, \qquad a_2 = a_0, \qquad a_3 = \tfrac{2}{3}a_1 = 0, \qquad a_4 = \tfrac{1}{2}a_0,$$

or in general:

$$(1.5)\qquad na_n = 2a_{n-2}, \qquad a_n = \begin{cases} 0, & \text{odd } n, \\ \frac{2}{n}a_{n-2}, & \text{even } n. \end{cases}$$

Putting $n = 2m$ (since only even terms appear in this series), we get

$$(1.6)\qquad a_{2m} = \frac{2}{2m}a_{2m-2} = \frac{1}{m}a_{2m-2} = \frac{1}{m}\frac{1}{m-1}a_{2m-4} = \cdots = \frac{1}{m!}a_0.$$

Substituting these values of the coefficients into the assumed solution (1.2) gives the solution

$$(1.7)\qquad y = a_0 + a_0x^2 + \frac{1}{2!}a_0x^4 + \cdots + \frac{1}{m!}a_0x^{2m} + \cdots \quad = a_0\sum_{m=0}^{\infty}\frac{x^{2m}}{m!}.$$

Example 2. Compare this with the solution by an elementary method (in this case, separation of variables):

$$\frac{dy}{y} = 2x\,dx, \quad \ln y = x^2 + \ln c, \quad y = ce^{x^2}.$$

Expanding this in a series of powers of x^2, we get:

$$y = c\left(1 + x^2 + \frac{x^4}{2!} + \cdots\right) = c\sum_{n=0}^{\infty}\frac{x^{2n}}{n!}$$

which, with $c = a_0$, is the same as the series solution (1.7).

You cannot always expect to find the closed form of a power series solution (that is, an elementary function for which your series solution is the power series expansion), but in simple cases you may recognize it. Of course, in that case, the problem could have been done without series; the real need for series is in problems for which there is no closed form in terms of elementary functions. Also you should realize that not all solutions have series expansions in powers of x, for example, $\ln x$ or $1/x^2$. All we can say is that if there is a solution which can be represented by a convergent power series this method will find it. We shall discuss later (Section 21) some theorems which tell us when we can expect to find such a solution.

In the following sections we consider some differential equations which occur frequently in applied problems and which are usually solved by series methods.

PROBLEMS, SECTION 1

Solve the following differential equations by series and also by an elementary method and verify that your solutions agree. Note that the goal of these problems is not to get the answer (that's easy by computer or by hand) but to become familiar with the method of series solutions which we will be using later. Check your results by computer.

1. $xy' = xy + y$

2. $y' = 3x^2y$

3. $xy' = y$

4. $y'' = -4y$

5. $y'' = y$

6. $y'' - 2y' + y = 0$

7. $x^2y'' - 3xy' + 3y = 0$

8. $(x^2 + 2x)y'' - 2(x + 1)y' + 2y = 0$

9. $(x^2 + 1)y'' - 2xy' + 2y = 0$

10. $y'' - 4xy' + (4x^2 - 2)y = 0$

2. LEGENDRE'S EQUATION

The Legendre differential equation is

$$(1 - x^2)y'' - 2xy' + l(l + 1)y = 0, \tag{2.1}$$

where l is a constant. This equation arises in the solution of partial differential equations in spherical coordinates (see Problem 10.2 and Chapter 13, Section 7) and so in problems in mechanics, quantum mechanics, electromagnetic theory, heat, etc., with spherical symmetry. Also see an application in Section 5.

Although the most useful solutions of this equation are polynomials (called the *Legendre polynomials*), one way to find them is to assume a series solution of the differential equation, and show that the series terminates after a finite number of terms. [There are other ways of finding the Legendre polynomials; see Sections 4 and 5, and Chapter 3, Section 14, Example 6.] We assume the series solution (1.2) for y and differentiate it term by term twice to get y' and y'':

$$\begin{cases} y = a_0 + a_1x + a_2x^2 + a_3x^3 + a_4x^4 + \cdots + a_nx^n + \cdots, \\ y' = a_1 + 2a_2x + 3a_3x^2 + 4a_4x^3 + \cdots + na_nx^{n-1} + \cdots, \\ y'' = 2a_2 + 6a_3x + 12a_4x^2 + 20a_5x^3 + \cdots + n(n-1)a_nx^{n-2} + \cdots. \end{cases} \tag{2.2}$$

We substitute (2.2) into (2.1) and collect the coefficients of the various powers of x; it is convenient to tabulate them as follows:

	const.	x	x^2	x^3	$\cdots\ x^n\ \cdots$
y''	$2a_2$	$6a_3$	$12a_4$	$20a_5$	$(n+2)(n+1)a_{n+2}$
$-x^2y''$			$-2a_2$	$-6a_3$	$-n(n-1)a_n$
$-2xy'$		$-2a_1$	$-4a_2$	$-6a_3$	$-2na_n$
$l(l+1)y$	$l(l+1)a_0$	$l(l+1)a_1$	$l(l+1)a_2$	$l(l+1)a_3$	$l(l+1)a_n$

Next we set the total coefficient of each power of x equal to zero [because, as discussed in Section 1, y must satisfy (2.1) identically]. For the first few powers of x we get

$$\begin{aligned} 2a_2 + l(l+1)a_0 &= 0 \qquad \text{or} \quad a_2 = -\frac{l(l+1)}{2}a_0; \\ 6a_3 + (l^2+l-2)a_1 &= 0 \qquad \text{or} \quad a_3 = -\frac{(l-1)(l+2)}{6}a_1; \\ 12a_4 + (l^2+l-6)a_2 &= 0 \qquad \text{or} \quad a_4 = -\frac{(l-2)(l+3)}{12}a_2 \\ &\qquad\qquad\qquad\quad = \frac{l(l+1)(l-2)(l+3)}{4!}a_0; \end{aligned} \tag{2.3}$$

and from the x^n coefficient we get

$$(n+2)(n+1)a_{n+2} + (l^2+l-n^2-n)a_n = 0. \tag{2.4}$$

The coefficient of a_n in (2.4) can be factored to give

$$l^2 - n^2 + l - n = (l+n)(l-n) + (l-n) = (l-n)(l+n+1). \tag{2.5}$$

Then we can write a general formula for a_{n+2} in terms of a_n. This formula (2.6) includes the formulas (2.3) for a_2, a_3, and a_4, and makes it possible for us to find any even coefficient as a multiple of a_0, and any odd coefficient as a multiple of a_1. Solving (2.4) for a_{n+2} and using (2.5), we have

$$a_{n+2} = -\frac{(l-n)(l+n+1)}{(n+2)(n+1)}a_n. \tag{2.6}$$

The general solution of (2.1) is then a sum of two series containing (as the solution of a second-order differential equation should) two constants a_0 and a_1 to be determined by the given initial conditions:

$$\begin{aligned} y = a_0 &\left[1 - \frac{l(l+1)}{2!}x^2 + \frac{l(l+1)(l-2)(l+3)}{4!}x^4 - \cdots\right] \\ + a_1 &\left[x - \frac{(l-1)(l+2)}{3!}x^3 + \frac{(l-1)(l+2)(l-3)(l+4)}{5!}x^5 - \cdots\right]. \end{aligned} \tag{2.7}$$

From equation (2.6) you can see by the ratio test that these series converge for $x^2 < 1$. It can be shown that, in general, they do not converge for $x^2 = 1$.

Example. Consider the a_1 series for $l = 0$. If $x^2 = 1$, this series is $1 + \frac{1}{3} + \frac{1}{5} + \cdots$, which is divergent by the integral test (Chapter 1, Section 6B). Now in many applications x is the cosine of an angle θ, and l is a (nonnegative) integer. We want a solution which converges for all θ, that is, a solution which converges at $x = \pm 1$ as well as for $|x| < 1$. We can always find one (but not two) such solutions when l is an integer; let us see how.

Legendre Polynomials We have seen that for $l = 0$ the a_1 series in (2.7) diverges. But look at the a_0 series; it gives just $y = a_0$ for $l = 0$ since all the rest of the terms contain the factor l. If $l = 1$, the a_0 series is divergent at $x^2 = 1$, but the a_1 series stops with $y = a_1 x$ [since all the rest of the terms in the a_1 series contain the factor $(l - 1)$]. For any integral l, one series terminates giving a polynomial solution; the other series is divergent at $x^2 = 1$. (Negative integral values of l would simply give solutions already obtained for positive l's; for example, $l = -2$ gives the polynomial solution $y = a_1 x$ which is the same as the $l = 1$ solution. Consequently, it is customary to restrict l to nonnegative values.) Thus we obtain a set of polynomial solutions of the Legendre equation, one for each nonnegative integral l. Each solution contains an arbitrary constant factor (a_0 or a_1); for $l = 0$, $y = a_0$; for $l = 1$, $y = a_1 x$, and so on. If the value of a_0 or a_1 in each polynomial is selected so that $y = 1$ when $x = 1$, the resulting polynomials are called *Legendre Polynomials*, written $P_l(x)$. From (2.6) and (2.7) and the requirement $P_l(1) = 1$, we find the following expressions for the first few Legendre polynomials:

$$P_0(x) = 1, \qquad P_1(x) = x, \qquad P_2(x) = \tfrac{1}{2}(3x^2 - 1). \tag{2.8}$$

Finding a few more Legendre polynomials by this method and other methods will be left to the problems. Although $P_l(x)$ for any integral l may be found by this method, simpler ways of obtaining the Legendre polynomials for larger l will be outlined in Sections 4 and 5. Of course, if you just want the formula for a particular P_l, you can find it by computer or in reference books.

Eigenvalue Problems In finding the Legendre polynomials as solutions of Legendre's equation (2.1), we have solved an *eigenvalue problem.* (See Chapter 3, Sections 11 and 12.) Recall that in an eigenvalue problem we are given an equation or a set of equations containing a parameter, and we want solutions that satisfy some special requirement; in order to obtain such solutions we must choose particular values (called eigenvalues) for the parameter in the problem. In finding the Legendre polynomials, we asked for series solutions of Legendre's equation (2.1) which converged at $x = \pm 1$. We saw that we could obtain such solutions if the parameter took on any integral value. The values of l, namely 0, 1, 2, $\cdots$, are called *eigenvalues* (or *characteristic values*); the corresponding solutions $P_l(x)$ are called *eigenfunctions* (or *characteristic functions*).

Note the parallel between the eigenvalue-eigenvector problems of Chapter 3 and the eigenvalue-eigenfunction problems of this chapter. Recall that in Chapter 3, we wrote an eigenvalue equation as $\mathrm{Mr} = \lambda \mathrm{r}$ where M was a matrix operator which operated on the eigenvector r to produce a multiple of r. The Legendre equation is of the form $f(D)y(x) = l(l+1)y(x)$ where $f(D)$ is a differential operator which operates on the eigenfunction $y(x)$ to produce a multiple of $y(x)$. See Section 22 and Chapter 13 for further examples of differential equations whose solutions are eigenfunctions.

The Legendre polynomials are also called Legendre functions of the first kind. The second solution for each l, which is an infinite series (convergent for $x^2 < 1$), is called a Legendre function of the second kind and is denoted by $Q_l(x)$ (See Problem 4.) The functions $Q_l(x)$ are not used as frequently as the polynomials $P_l(x)$. For fractional l both solutions are infinite series; these again occur less frequently in applications.

PROBLEMS, SECTION 2

1. Using (2.6) and (2.7) and the requirement that $P_l(l) = 1$, find $P_2(x)$, $P_3(x)$, and $P_4(x)$. Check your results by computer.
2. Show that $P_l(-1) = (-1)^l$. *Hint:* When is $P_l(x)$ an even function and when is it an odd function?
3. Computer plot graphs of $P_l(x)$ for $l = 0, 1, 2, 3, 4$, and x from -1 to 1.
4. Use the method of reduction of order [Chapter 8, Section 7(e)] and the known solution $P_l(x)$ of Legendre's equation to find the second solution $Q_l(x)$ (in terms of an integral). Evaluate the integral for the cases $l = 0$ and $l = 1$ to find Q_0 and Q_1. Note the divergence of the logarithms at $x = \pm 1$. Expand the logarithms in Q_0 to get the divergent series mentioned above [a_1 series in (2.7) with $l = 0$, $x^2 = 1$].

3. LEIBNIZ' RULE FOR DIFFERENTIATING PRODUCTS

Let us digress for a moment to discuss a very useful formula called *Leibniz rule* for finding a high order derivative of a product. We shall first illustrate this by a numerical example. We could, of course, do a numerical problem by computer, but our purpose is to understand the general formula which we will need in derivations. Also, when you know Leibniz rule, you may find in simple numerical cases that you can write down the answer for a high order derivative of a product faster than you can type the problem into the computer (see Problems 2 to 5).

Example. Find $(d^9/dx^9)(x \sin x)$.

Leibniz rule says that the answer is

$$x\frac{d^9}{dx^9}(\sin x) + 9\frac{d}{dx}(x)\frac{d^8}{dx^8}(\sin x) + \frac{9\cdot 8}{2!}\frac{d^2}{dx^2}(x)\frac{d^7}{dx^7}(\sin x) + \cdots. \tag{3.1}$$

This should remind you of a binomial expansion

$$(a+b)^9 = a^0b^9 + 9ab^8 + \frac{9\cdot 8}{2!}a^2b^7 + \cdots.$$

The coefficients in (3.1) are, in fact, binomial coefficients, and the sum of the orders of the two derivatives in each term is 9. (You may find the second hint in Problem 6 useful in understanding and remembering this.) Now if it happens, as here, that the derivatives of one factor become zero after the first few, the rule saves much work. In (3.1), $(d^2/dx^2)(x) = 0$ and all higher derivatives of x are zero so we get

$$\frac{d^9}{dx^9}(x\sin x) = x\frac{d^9}{dx^9}(\sin x) + 9\frac{d^8}{dx^8}(\sin x) = x\cos x + 9\sin x.$$

PROBLEMS, SECTION 3

1. By Leibniz' rule, write the formula for $(d^n/dx^n)(uv)$.

Use Problem 1 to find the following derivatives.

2. $(d^{10}/dx^{10})(xe^x)$

3. $(d^6/dx^6)(x^2 \sin x)$

4. $d^{25}/dx^{25})(x \cos x)$

5. $d^{100}/dx^{100})(x^2 e^{-x})$

6. Verify Problem 1. *Hints:* One method is to use mathematical induction. Another method is to write

$$\frac{d}{dx}(uv) = D(uv) = (D_u + D_v)(uv),$$

where D_u acts only on u and D_v acts only on v, that is, $D_u(uv)$ means $v(du/dx)$, etc. Then

$$\frac{d^n}{dx^n}(uv) = (D_u + D_v)^n(uv).$$

Expand $(D_u + D_v)^n$ by the binomial theorem and interpret the terms to get Leibniz' rule.

4. RODRIGUES' FORMULA

We have obtained the Legendre polynomials as solutions of Legendre's equation when l is an integer; there are other ways of obtaining them. We shall prove that Rodrigues' formula

$$P_l(x) = \frac{1}{2^l l!}\frac{d^l}{dx^l}(x^2-1)^l \tag{4.1}$$

gives correctly the Legendre polynomials $P_l(x)$. There are two parts to the proof. First we show that if

$$v = (x^2-1)^l, \tag{4.2}$$

then $d^l v/dx^l$ is a solution of Legendre's equation; then we show that $P_l(1) = 1$ in (4.1). To prove the first part, find dv/dx in (4.2) and multiply it by $x^2 - 1$:

$$(x^2-1)\frac{dv}{dx} = (x^2-1)l(x^2-1)^{l-1}\cdot 2x = 2lxv. \tag{4.3}$$

Differentiate (4.3) $l+1$ times by Leibniz' rule:

$$\begin{aligned}(x^2-1)\frac{d^{l+2}v}{dx^{l+2}} + (l+1)(2x)\frac{d^{l+1}v}{dx^{l+1}} + \frac{(l+1)l}{2!}\cdot 2\cdot\frac{d^l v}{dx^l}&\\ = 2lx\frac{d^{l+1}v}{dx^{l+1}} + 2l(l+1)\frac{d^l v}{dx^l}&.\end{aligned} \tag{4.4}$$

Simplifying (4.4), we get (Problem 1)

$$(1-x^2)\left(\frac{d^l v}{dx^l}\right)'' - 2x\left(\frac{d^l v}{dx^l}\right)' + l(l+1)\frac{d^l v}{dx^l} = 0. \tag{4.5}$$

This is just Legendre's equation (2.1) with $y = d^l v/dx^l$; thus we see that $d^l v/dx^l = (d^l/dx^l)(x^2-1)^l$ is a solution of Legendre's equation as we claimed. It *is* a polynomial of degree l, and since we have previously called the polynomial solution of degree l the Legendre polynomial $P_l(x)$, this must be it with the possible exception of the numerical factor which must give $P_l(1) = 1$. A simple method of showing that $P_l(1) = 1$ for the functions $P_l(x)$ in (4.1) is outlined in Problem 2.

PROBLEMS, SECTION 4

1. Verify equations (4.4) and (4.5).

2. Show that $P_l(1) = 1$, with $P_l(x)$ given by (4.1), in the following way. Factor $(x^2-1)^l$ into $(x+1)^l(x-1)^1$ and differentiate the product l times by Leibniz' rule. Without writing out very many terms you should see that every term but one contains the factor $x-1$ and so becomes zero when $x = 1$. Use this to evaluate $P_l(x)$ in (4.1) when $x = 1$ to get $P_l(1) = 1$.

3. Find $P_0(x)$, $P_1(x)$, $P_2(x)$, $P_3(x)$, and $P_4(x)$ from Rodrigues' formula (4.1). Check your results by computer.

4. Show that $\int_{-1}^{1} x^m P_l(x)\,dx = 0$ if $m < l$. *Hint:* Use Rodrigues' formula (4.1) and integrate repeatedly by parts, differentiating the power of x and integrating the derivative each time.

5. GENERATING FUNCTION FOR LEGENDRE POLYNOMIALS

The expression

$$\Phi(x,h) = (1-2xh+h^2)^{-1/2}, \qquad |h| < 1, \tag{5.1}$$

is called the generating function for Legendre polynomials. We shall show that

$$\Phi(x,h) = P_0(x) + hP_1(x) + h^2P_2(x) + \cdots = \sum_{l=0}^{\infty} h^l P_l(x), \tag{5.2}$$

where the functions $P_l(x)$ are the Legendre polynomials. (For discussion of convergence of the series, see Chapter 14, Problem 2.43.) Let us first verify a few terms of (5.2). For simplicity put $2xh - h^2 = y$ into (5.1), expand $(1-y)^{-1/2}$ in powers of y, then substitute back $y = 2xh - h^2$ and collect powers of h to get

$$\begin{aligned}
\Phi &= (1-y)^{-1/2} = 1 + \tfrac{1}{2}y + \frac{\frac{1}{2}\cdot\frac{3}{2}}{2!}y^2 + \cdots \\
&= 1 + \tfrac{1}{2}(2xh-h^2) + \tfrac{3}{8}(2xh-h^2)^2 + \cdots \\
&= 1 + xh - \tfrac{1}{2}h^2 + \tfrac{3}{8}(4x^2h^2 - 4xh^3 + h^4) + \cdots \\
&= 1 + xh + h^2(\tfrac{3}{2}x^2 - \tfrac{1}{2}) + \cdots \\
&= P_0(x) + hP_1(x) + h^2P_2(x) + \cdots .
\end{aligned} \tag{5.3}$$

This is not a proof that the functions called $P_l(x)$ in (5.2) are really Legendre polynomials, but merely a verification of the first few terms. To prove in general that the polynomials called $P_l(x)$ in (5.2) are Legendre polynomials we must show that they satisfy Legendre's equation and that they have the property $P_l(1) = 1$. The latter is easy to prove; putting $x = 1$ in (5.1) and (5.2), we get

$$\begin{aligned}\Phi(1,h) &= (1-2h+h^2)^{-1/2} = \frac{1}{1-h} = 1 + h + h^2 + \cdots \\ &\equiv P_0(1) + P_1(1)h + P_2(1)h^2 + \cdots .\end{aligned} \tag{5.4}$$

Since this is an identity in h, the functions $P_l(x)$ in (5.2) have the property $P_l(1) = 1$. To show that they satisfy Legendre's equation, we shall use the following identity which can be verified from (5.1) by straightforward differentiation and some algebra (Problem 2):

$$(1-x^2)\frac{\partial^2\Phi}{\partial x^2} - 2x\frac{\partial\Phi}{\partial x} + h\frac{\partial^2}{\partial h^2}(h\Phi) = 0. \tag{5.5}$$

Substituting the series (5.2) for Φ into (5.5), we get

$$(1-x^2)\sum_{l=0}^{\infty} h^l P_l''(x) - 2x\sum_{l=0}^{\infty} h^l P_l'(x) + \sum_{l=0}^{\infty} l(l+1)h^l P_l(x) = 0. \tag{5.6}$$

This is an identity in h, so the coefficient of each power of h must be zero. Setting the coefficient of h^l equal to zero, we get

$$(1-x^2)P_l''(x) - 2xP_l'(x) + l(l+1)P_l(x) = 0. \tag{5.7}$$

This is Legendre's equation, so we have proved that the functions $P_l(x)$ in (5.2) satisfy it as claimed.

Recursion Relations The generating function is useful in deriving the *recursion relations* (also called *recurrence relations*) for Legendre polynomials. These recursion relations are identities in x and are used (as trigonometric identities are) to simplify work and to help in proofs and derivations. Some examples of recursion relations are:

$$\begin{aligned}
&\text{(a)}\quad lP_l(x) = (2l-1)xP_{l-1}(x) - (l-1)P_{l-2}(x),\\
&\text{(b)}\quad xP_l'(x) - P_{l-1}'(x) = lP_l(x),\\
&\text{(c)}\quad P_l'(x) - xP_{l-1}'(x) = lP_{l-1}(x),\\
&\text{(d)}\quad (1-x^2)P_l'(x) = lP_{l-1}(x) - lxP_l(x),\\
&\text{(e)}\quad (2l+1)P_l(x) = P_{l+1}'(x) - P_{l-1}'(x),\\
&\text{(f)}\quad (1-x^2)P_{l-1}'(x) = lxP_{l-1}(x) - lP_l(x).
\end{aligned} \tag{5.8}$$

We shall now derive (5.8a); the problems outline derivations of the other equations.

From (5.1) we get

$$\begin{aligned}\frac{\partial\Phi}{\partial h} &= -\tfrac{1}{2}(1-2xh+h^2)^{-3/2}(-2x+2h);\\ (1-2xh+h^2)\frac{\partial\Phi}{\partial h} &= (x-h)\Phi.\end{aligned} \tag{5.9}$$

Substituting the series (5.2) and its derivative with respect to h into (5.9), we get

$$(1-2xh+h^2)\sum_{l=1}^{\infty} lh^{l-1}P_l(x) = (x-h)\sum_{l=0}^{\infty} h^l P_l(x).$$

This is an identity in h so we equate coefficients of h^{l-1}. Carefully adjusting indices so that we select the term in h^{l-1} each time, we find

$$lP_l(x) - 2x(l-1)P_{l-1}(x) + (l-2)P_{l-2}(x) = xP_{l-1}(x) - P_{l-2}(x) \tag{5.10}$$

which simplifies to (5.8a). The recursion relation (5.8a) gives the simplest way of finding any Legendre polynomial when we know the Legendre polynomials for smaller l (Problem 3).

Expansion of a Potential The generating function is useful in problems involving the potential associated with any inverse square force. Recall that the gravitational force between two point masses separated by a distance d is proportional to $1/d^2$ and the associated potential energy is proportional to $1/d$. Similarly, the electrostatic force between two electric charges a distance d apart is proportional to $1/d^2$ and the associated electrostatic potential energy is proportional to $1/d$.

Example 1. In either case we can write the potential as

$$V = \frac{K}{d}, \tag{5.11}$$

where K is an appropriate constant. In Figure 5.1, let the two masses (or charges) be at the heads of vectors $\mathbf{r}$ and $\mathbf{R}$. Then, by the law of cosines, the distance between them is

$$\begin{aligned} d &= |\mathbf{R}-\mathbf{r}| \\ &= \sqrt{R^2 - 2Rr\cos\theta + r^2} \\ &= R\sqrt{1 - 2\frac{r}{R}\cos\theta + \left(\frac{r}{R}\right)^2} \end{aligned} \tag{5.12}$$

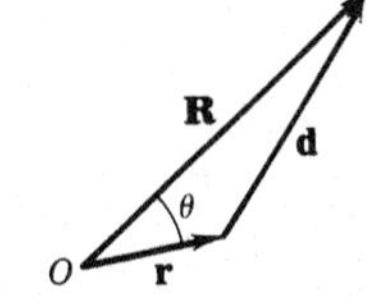

Figure 5.1

and the gravitational or electric potential is

$$V = \frac{K}{R}\left[1 - \frac{2r}{R}\cos\theta + \left(\frac{r}{R}\right)^2\right]^{-1/2}. \tag{5.13}$$

For $|\mathbf{r}| < |\mathbf{R}|$, we make the change of variables

$$\begin{aligned} h &= \frac{r}{R}, \\ x &= \cos\theta. \end{aligned} \tag{5.14}$$

(*Note:* x is *not* a coordinate but just a new variable standing for $\cos\theta$.) Then in terms of the generating function Φ in (5.1) we have

$$\begin{aligned} d &= R\sqrt{1-2hx+h^2} \\ V &= \frac{K}{R}(1-2hx+h^2)^{-1/2} = \frac{K}{R}\Phi. \end{aligned} \tag{5.15}$$

Using (5.2), we can write the potential V as an infinite series

$$V = \frac{K}{R}\sum_{l=0}^{\infty} h^l P_l(x) \tag{5.16}$$

or in terms of r and θ [using (5.14)]

$$V = \frac{K}{R}\sum_{l=0}^{\infty} \frac{r^l P_l(\cos\theta)}{R^l} = K\sum_{l=0}^{\infty} \frac{r^l P_l(\cos\theta)}{R^{l+1}}. \tag{5.17}$$

In many applications the distance $|\mathbf{R}|$ is much larger than $|\mathbf{r}|$. Then the terms of the series (5.17) decrease rapidly in magnitude because of the factor $(r/R)^l$, and the potential can be approximated by using only a few terms in the series.

We can make (5.17) more general and useful by considering the following problem. (We shall discuss the electrical case for definiteness—the gravitational case could be discussed in parallel fashion.)

Example 2. Suppose there are a large number of charges q_i at points $\mathbf{r}_i$. The electrostatic potential V_i at the point $\mathbf{R}$ due to the charge q_i at $\mathbf{r}_i$ means the electrostatic potential energy of a pair of charges, namely, a *unit* charge at $\mathbf{R}$ and the charge q_i at $\mathbf{r}_i$; this is given by (5.11) and (5.12), or by (5.17), with $r = r_i$, $\theta = \theta_i$, and $K = q_i \cdot 1 \cdot K'$, where K' is a numerical constant depending on the choice of units:

$$V_i = K' q_i \sum_{l=0}^{\infty} \frac{r_i^l P_l(\cos\theta_i)}{R^{l+1}}. \tag{5.18}$$

The total potential V at $\mathbf{R}$ due to all the charges q_i is then a sum over i of all the series (5.18), namely

$$V = \sum_i V_i = K' \sum_i q_i \sum_{l=0}^{\infty} \frac{r_i^l P_l(\cos\theta_i)}{R^{l+1}} = K' \sum_{l=0}^{\infty} \frac{\sum_i q_i r_i^l P_l(\cos\theta_i)}{R^{l+1}}. \tag{5.19}$$

Example 3. If, instead of a set of discrete charges, we have a continuous charge distribution, then the sum over i becomes an integral, namely

$$\int r^l P_l(\cos\theta)\,dq \qquad \text{or} \qquad \iiint r^l P_l(\cos\theta)\rho\,d\tau, \tag{5.20}$$

where ρ is the charge density, and the integral is over the space occupied by the charge distribution. Then (5.19) becomes

$$V = K' \sum_l \frac{1}{R^{l+1}} \iiint r^l P_l(\cos\theta)\rho\,d\tau. \tag{5.21}$$

The terms of the series (5.21) can be interpreted physically. The $l = 0$ term is

$$\frac{1}{R}\iiint \rho\,d\tau = \frac{1}{R}\cdot(\text{total charge}). \tag{5.22}$$

Thus if R is large enough compared to all the r_i or all the values of r at points of the charge distribution, we can approximate the potential of the distribution as that of a single charge at the origin of magnitude equal to the total charge of the distribution. The $l = 1$ term of the series (5.21) is

$$\frac{1}{R^2}\iiint r\cos\theta\,\rho\,d\tau. \tag{5.23}$$

To interpret this recall that the *electric dipole moment* of a pair of charges $+q$ and $-q$ a distance d apart (as in Figure 5.2) is defined as the vector $q\mathbf{d}$, where $\mathbf{d}$ is the vector from $-q$ to $+q$. Since the vector $q\mathbf{d}$ is equal to $q(\mathbf{r}_1 - \mathbf{r}_2) = q\mathbf{r}_1 - q\mathbf{r}_2$, we often call $q\mathbf{r}_1$ and $-q\mathbf{r}_2$ the dipole moments of $+q$ and $-q$ about O; then the total dipole moment due to the two charges is just the sum of the two moments. Suppose we calculate the dipole moment about O of all the charges q_i; this is the vector sum $\sum_i q_i r_i \cos\theta_i$, since θ_i is the angle between $\mathbf{R}$ and $\mathbf{r}_i$. In the case of a continuous charge distribution this sum becomes

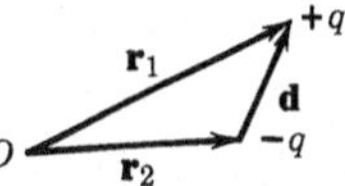

Figure 5.2

$$\iiint r\cos\theta\,\rho\,d\tau. \tag{5.24}$$

Thus we see from (5.23) and (5.24) that the second term of the series (5.21) is $1/R^2$ times the component in the $\mathbf{R}$ direction of the dipole moment of the charge distribution. If you consider the fact that the first term of (5.21) involves the total charge (a scalar, that is, a tensor of rank zero) and the second term involves the dipole moment (a vector, that is, a tensor of rank one), it may not surprise you to learn that the third term involves a 2nd-rank tensor known as the quadrupole moment of the charge distribution, the fourth term involves a 3rd-rank tensor known as the octopole moment, etc. (See Problem 15 for more detail.)

Example 4. Given a charge or mass distribution, the moments of various ranks and the terms in (5.21) can be computed. The opposite process is often of great interest in applied problems. Consider a satellite circling the earth; it is moving in the gravitational field of the earth's mass. If the mass distribution of the earth were spherically symmetric, then only the first term would appear in the series for the gravitational potential [this series would be (5.21) with ρ a mass density instead of a charge density]. But since the earth is not a perfect sphere (equatorial bulge, etc.), other terms are present in (5.21) and the corresponding forces affect the motion of satellites. From accurate measurements of the satellite orbits, it is now possible to calculate many terms of the series (5.21). Similarly, in the electrical case, experimental measurements give us information about the distribution of electric charge inside atoms and nuclei; our discussion here and equation (5.21) provide the basis for the interpretation of such measurements, and the terminology used in discussing them.

PROBLEMS, SECTION 5

1. Find $P_3(x)$ by getting one more term in the generating function expansion (5.3).

2. Verify (5.5) using (5.1).

3. Use the recursion relation (5.8a) and the values of $P_0(x)$ and $P_1(x)$ to find $P_2(x)$, $P_3(x)$, $P_4(x)$, $P_5(x)$, and $P_6(x)$. [After you have found $P_3(x)$, use it to find $P_4(x)$, and so on for the higher order polynomials.]

4. Show from (5.1) that

$$(x-h)\frac{\partial \Phi}{\partial x} = h\frac{\partial \Phi}{\partial h}.$$

Substitute the series (5.2) for Φ, and so prove the recursion relation (5.8b).

5. Differentiate the recursion relation (5.8a) and use the recursion relation (5.8b) with l replaced by $l-1$ to prove the recursion relation (5.8c).

6. From (5.8b) and (5.8c), obtain (5.8d) and (5.8f). Then differentiate (5.8d) with respect to x and eliminate $P'_{l-1}(x)$ using (5.8b). Your result should be the Legendre equation. The derivation of Problems 4 to 6 constitutes an alternative proof [to that of equations (5.5) to (5.7)] that the functions $P_l(x)$ in (5.2) are Legendre polynomials.

7. Write (5.8c) with l replaced by $l+1$ and use it to eliminate the $xP'_l(x)$ term in (5.8b). You should get (5.8e).

Express each of the following polynomials as linear combinations of Legendre polynomials. *Hint:* Start with the highest power of x and work down in finding the correct combination.

8. $5-2x$ **9.** $3x^2+x-1$ **10.** x^4

11. $x-x^3$ **12.** $7x^4-3x+1$ **13.** x^5

14. Show that any polynomial of degree n can be written as a linear combination of Legendre polynomials with $l \le n$.

15. Expand the potential $V = K/d$ in (5.11) in the following way in order to see how the terms depend on the tensors mentioned above. In Figure 5.1 let $\mathbf{R}$ have the coordinates X, Y, Z and $\mathbf{r}$ have coordinates x, y, z. [*Note:* The coordinate x here is *not* the x in (5.14).] Then

$$V = \frac{K}{d} = K[(X-x)^2 + (Y-y)^2 + (Z-z)^2]^{-1/2}.$$

Consider X, Y, Z as constants and expand $V(x, y, z)$ in a three variable power series about the origin. (See Chapter 4, Section 2, for discussion of two-variable power series and generalize the method.) You should find

$$\begin{aligned} V =& \frac{K}{R} + \frac{K}{R^2}\left(\frac{X}{R}x + \cdots\right) \\ &+ \frac{K}{R^3}\left[\left(\frac{3}{2}\frac{X^2}{R^2} - \frac{1}{2}\right)x^2 + \cdots + \frac{3}{2}\frac{X}{R}\frac{Y}{R}2xy + \cdots\right] + \cdots \end{aligned}$$

and similar terms in y, z, y^2, xz, and so on. Now letting $\mathbf{r} = \mathbf{r}_i$ and $K = K'q_i$ for a charge distribution as in (5.18), and summing (or integrating) over the charge distribution, show that: the first term is just $(K'/R)\cdot$ total charge; the next group of terms (in x, y, z) involve the three components of the electric dipole moment; the sum of these terms is $(K'/R^2)\cdot$component of the dipole moment in the $\mathbf{R}$ direction; the next group (quadratic terms) involve six quantities of the form

$$\iiint x^2 \rho\, d\tau \qquad \text{and similar } y,\ z \text{ integrals,}$$

$$\iiint 2xy\rho\, d\tau \qquad \text{and similar } xz,\ yz \text{ integrals.}$$

If we split the $2xy$ term into xy and yx (and similarly for the $2xz$ and $2yz$ terms), we have the nine components of a 2nd-rank tensor called the quadrupole moment. Use the "direct product" method of Chapter 10, Section 2 to show that it *is* a 2nd-rank tensor. (Remember from Chapter 10 that, by definition, x, y, z are the components of a vector, that is, a 1st-rank tensor.) Just as two charges $+q$ and $-q$ form an electric dipole, so four charges like this $\substack{+\bullet\bullet-\\-\bullet\bullet+}$ form an electric quadrupole and the quadratic terms in the V series give the potential of such a charge configuration. Again using Chapter 10, Section 2, show that the third-order terms in x, y, z form a 3rd-rank tensor; this is known as the octopole moment. It can be represented physically by two quadrupoles side by side just as the quadrupole above was formed by two dipoles side by side.

6. COMPLETE SETS OF ORTHOGONAL FUNCTIONS

Orthogonal functions Two vectors **A** and **B** are orthogonal (perpendicular) if their scalar product is zero, that is, if

$$\sum_i A_i B_i = 0. \tag{6.1}$$

[See Chapter 3, equations (4.12) and (10.3).] Recall from Chapter 3, Section 14, that we can think of functions as elements of a vector space. Then by analogy with (6.1) we say that two functions $A(x)$ and $B(x)$ are orthogonal on (a, b) if

$$\int_a^b A(x)B(x)\,dx = 0. \tag{6.2}$$

If the functions $A(x)$ and $B(x)$ are complex, the definition of orthogonality is [see Chapter 3, equation (14.3)]

> $A(x)$ and $B(x)$ are orthogonal on (a, b) if
>
> $$\int_a^b A^*(x)B(x)\,dx = 0, \tag{6.3}$$
>
> where $A^*(x)$ is the complex conjugate of $A(x)$ (see Problem 1).

Since (6.3) is identical with (6.2) if $A(x)$ and $B(x)$ are real, we can take (6.3) as the general definition of orthogonality of $A(x)$ and $B(x)$ on (a, b).

If we have a whole set of functions $A_n(x)$ where $n = 1, 2, 3, \cdots$, and

$$\int_a^b A_n^*(x)A_m(x)\,dx = \begin{cases} 0 & \text{if } m \neq n, \\ \text{const.} \neq 0 & \text{if } m = n, \end{cases} \tag{6.4}$$

we call the functions $A_n(x)$ a *set of orthogonal functions.* We have already used such sets of functions in Fourier series. Recall that [Chapter 7, equation (5.2)]

$$\int_{-\pi}^{\pi} \sin nx \sin mx\,dx = \begin{cases} 0 & \text{if } m \neq n, \\ \pi & \text{if } m = n \neq 0. \end{cases} \tag{6.5}$$

Thus $\sin nx$ is a set of orthogonal functions on $(-\pi, \pi)$, or in fact on any other interval of length 2π. Similarly, the functions $\cos nx$ are orthogonal on $(-\pi, \pi)$.

Also the whole set consisting of $\sin nx$ *and* $\cos nx$ is a set of orthogonal functions on $(-\pi, \pi)$ since

$$\int_{-\pi}^{\pi} \sin nx \cos mx \, dx = 0 \qquad \text{for } any\ n \text{ and } m.$$

We have used complex functions also, namely the set e^{inx}. For this set the orthogonality property is given by (6.4), namely

$$(6.6) \qquad \int_{-\pi}^{\pi} (e^{inx})^* e^{imx} \, dx = \int_{-\pi}^{\pi} e^{-inx} e^{imx} \, dx = \begin{cases} 0 & \text{if } m \neq n, \\ 2\pi & \text{if } m = n. \end{cases}$$

Recall that $\sin nx$ and $\cos nx$ (or e^{inx}) were the functions used in a Fourier series expansion on $(-\pi, \pi)$. You should now realize that it was the orthogonality property that we used in getting the coefficients. When we multiplied the equation $f(x) = \sum_{m=-\infty}^{\infty} c_m e^{imx}$ by e^{-inx} and integrated, the integrals of all the terms in the series except the c_n term were zero by the orthogonality property (6.6). There are many other sets of orthogonal functions besides the trigonometric or exponential ones. Just as we used the sine-cosine or exponential set to expand a function in a Fourier series, so we can expand a function in a series using other sets of orthogonal functions. We shall show this for the functions $P_l(x)$ after we prove that they are orthogonal.

Complete sets There is another important point to consider when we want to expand a function in terms of a set of orthogonal functions. Again let us consider the vector analogy. We write vectors in terms of their components and the basis vectors **i**, **j**, **k**. In two dimensions we need only two basis vectors, say **i** and **j**. But if we tried to write three-dimensional vectors in terms of just **i** and **j**, there would be some vectors we could not represent; we say that (in three dimensions) **i** and **j** are not a *complete* set of basis vectors. A simple way of expressing this (which generalizes to n dimensions) is to say that there is another vector (namely **k**) which is orthogonal to both **i** and **j**. Thus we define a set of orthogonal basis vectors as complete if there is no other vector orthogonal to all of them (in the space of the number of dimensions we are considering). By analogy, we define a set of orthogonal functions as *complete* on a given interval if there is no other function orthogonal to all of them on that interval. Now it is easy to see that there are some vectors in three dimensions which cannot be represented using only **i** and **j**. Similarly, there are functions which cannot be represented by a series using an incomplete set of orthogonal functions. We have discussed one example of this in Fourier series (Chapter 7, Section 11). If we are trying to represent a sound wave by a Fourier series, we must not leave out any of the harmonics; that is, the set of functions $\sin nx$, $\cos nx$ on $(-\pi, \pi)$ would not be complete if we left out some of the values of n. As another example, the set of functions $\sin nx$ is an orthogonal set on $(-\pi, \pi)$. However, it is not complete; to have a complete set we must include also the functions $\cos nx$, and you should recall that this is what we did in Fourier series. On the other hand, $\sin nx$ is a complete set on $(0, \pi)$; we used this fact when we started with a function given on $(0, \pi)$, defined it on $(-\pi, 0)$ to make it odd, and then expanded it in a sine series. Similarly, $\cos nx$ is a complete set on $(0, \pi)$. In this chapter, we are particularly interested in the fact (which we state without proof) that the Legendre polynomials are a complete set on $(-1, 1)$.

PROBLEMS, SECTION 6

1. Show that if $\int_a^b A^*(x)B(x)\,dx = 0$ [see (6.3)], then $\int_a^b A(x)B^*(x)\,dx = 0$, and vice versa.

2. Show that the functions $e^{in\pi x/l}$, $n = 0, \pm 1, \pm 2, \cdots$, are a set of orthogonal functions on $(-l, l)$.

3. Show that the functions x^2 and $\sin x$ are orthogonal on $(-1, 1)$. *Hint:* See Chapter 7, Section 9.

4. Show that the functions $f(x)$ and $g(x)$ are orthogonal on $(-a, a)$ if $f(x)$ is even and $g(x)$ is odd. (See Problem 3.)

5. Evaluate $\int_{-1}^{1} P_0(x)P_2(x)\,dx$ to show that these functions are orthogonal on $(-1, 1)$.

6. Show in two ways that $P_l(x)$ and $P_l'(x)$ are orthogonal on $(-1, 1)$. *Hint:* See Problem 4 and Problem 4.4.

7. Show that the set of functions $\sin nx$ is not a complete set on $(-\pi, \pi)$ by trying to expand the function $f(x) = 1$ on $(-\pi, \pi)$ in terms of them.

8. Show that the functions $\cos(n + \frac{1}{2})x$, $n = 0, 1, 2, \cdots$, are orthogonal on $(0, \pi)$. Expand the function $f(x) = 1$ on $(0, \pi)$ in terms of them. (Is it a complete set? See Chapter 7, end of Section 11.)

9. Show in two ways that $\int_{-1}^{1} P_{2n+1}(x)\,dx = 0$.

7. ORTHOGONALITY OF THE LEGENDRE POLYNOMIALS

We are going to show that the Legendre polynomials are a set of orthogonal functions on $(-1, 1)$, that is, that

$$\int_{-1}^{1} P_l(x)P_m(x)\,dx = 0 \qquad \text{unless } l = m. \tag{7.1}$$

To prove this we rewrite the Legendre differential equation (2.1) in the form

$$\frac{d}{dx}[(1 - x^2)P_l'(x)] + l(l + 1)P_l(x) = 0. \tag{7.2}$$

Write (7.2) for $P_l(x)$ and for $P_m(x)$; multiply the $P_l(x)$ equation by $P_m(x)$, and the $P_m(x)$ equation by $P_l(x)$ and subtract to get

$$\begin{aligned} P_m(x)\frac{d}{dx}[(1 - x^2)P_l'(x)] - P_l(x)\frac{d}{dx}[(1 - x^2)P_m'(x)]& \\ + [l(l + 1) - m(m + 1)]P_m(x)P_l(x) &= 0. \end{aligned} \tag{7.3}$$

The first two terms of (7.3) can be written as

$$\frac{d}{dx}[(1 - x^2)(P_m P_l' - P_l P_m')] \tag{7.4}$$

where, for simplicity, we have used $P_l = P_l(x)$, and so forth. Integrating (7.3) between -1 and 1 and using (7.4), we get

$$(1 - x^2)(P_m P_l' - P_l P_m')\Big|_{-1}^{1} + [l(l + 1) - m(m + 1)]\int_{-1}^{1} P_m(x)P_l(x)\,dx = 0. \tag{7.5}$$

The integrated term is zero because $(1 - x^2) = 0$ at $x = \pm 1$, and $P_m(x)$ and $P_l(x)$ are finite. The bracket in front of the integral is not zero unless $m = l$. Therefore the integral must be zero for $l \neq m$ and we have (7.1).

The method we have used here is a standard one which can be used for many other sets of orthogonal functions to prove the orthogonality property by using the differential equation satisfied by the functions. (See Problems 1 and 2; also see Section 19 and Problems 10.3, 22.7, 22.16, 22.24, 23.24b, and 23.25.)

Recall (Section 5, Problems 8 to 14) that we can write any polynomial of degree n as a linear combination of Legendre polynomials of degree $\leq n$. Thus, by (7.1), any polynomial of degree $< l$ is orthogonal to $P_l(x)$:

$$\int_{-1}^{1} P_l(x) \cdot (\text{any polynomial of degree } < l)\, dx = 0. \tag{7.6}$$

PROBLEMS, SECTION 7

1. By a method similar to that we used to show that the P_l's are an orthogonal set of functions on $(-1, 1)$, show that the solutions of $y_n'' = -n^2 y_n$ are an orthogonal set on $(-\pi, \pi)$. *Hint:* You should know what functions the solutions y_n are; do not use the functions themselves, but you may use their values and the values of their derivatives at $-\pi$ and π to evaluate the integrated part of your equation.

2. Following the method in (7.2) to (7.5), show that the solutions of the differential equation
$$(1 - x^2)y'' - 2xy' + [l(l+1) - (1 - x^2)^{-1}]y = 0$$
are a set of orthogonal functions on $(-1, 1)$.

3. Use Problem 4.4 to show that $\int_{-1}^{1} P_m(x) P_l(x)\, dx = 0$ if $m < l$. *Comment:* This amounts to a different proof of orthogonality—via Rodrigues' formula instead of the differential equation.

4. Use equation (7.6) to show that $\int_{-1}^{1} P_l(x) P_{l-1}'(x)\, dx = 0$. *Hint:* What is the degree of $P_{l-1}'(x)$? Also show that $\int_{-1}^{1} P_l'(x) P_{l+1}(x)\, dx = 0$.

5. Show that $\int_{-1}^{1} P_l(x)\, dx = 0$, $l > 0$. *Hint:* Consider $\int_{-1}^{1} P_l(x) P_0(x)\, dx$.

6. Show that $P_1(x)$ is orthogonal to $[P_l(x)]^2$ on $(-1, 1)$. *Hint:* See Problem 6.4.

8. NORMALIZATION OF THE LEGENDRE POLYNOMIALS

If we take the scalar product of a vector with itself, $\mathbf{A} \cdot \mathbf{A} = A^2$, we get the square of the length (or norm) of the vector. If we divide $\mathbf{A}$ by its length, we get a unit vector. In Chapter 3, Section 14 we showed that we can think of functions as the vectors of a vector space and we defined the *norm* N of a function $A(x)$ on (a, b) by [see Chapter 3, equation (14.2)]

$$\int_a^b A^*(x) A(x)\, dx = \int_a^b |A(x)|^2\, dx = N^2.$$

We also say that the function $N^{-1}A(x)$ is *normalized*; like a unit vector, a normalized function has norm $= 1$. The factor N^{-1} is called the *normalization factor*. For example, $\int_0^\pi \sin^2 nx\,dx = \pi/2$. Then the norm of $\sin nx$ on $(0,\pi)$ is $\sqrt{\pi/2}$, and the functions $\sqrt{2/\pi}\sin nx$ have norm 1 on $(0,\pi)$, that is, they are normalized. A set of normalized orthogonal functions is called *orthonormal*. For example, $\sqrt{2/\pi}\sin nx$ is an orthonormal set on $(0,\pi)$.

Such a set of orthonormal functions may remind us of $\mathbf{i},\mathbf{j},\mathbf{k}$; like these unit vectors, the functions are orthogonal and have norm $= 1$. If the elements of a vector space are functions, we can then use a (complete) orthonormal subset of the functions as the basis vectors of the space. We think of expanding other functions in terms of them (by analogy with writing a three-dimensional vector in terms of $\mathbf{i},\mathbf{j},\mathbf{k}$). For example, suppose we have expanded a given function $f(x)$ on $(0,\pi)$ in a Fourier sine series:

$$f(x) = \sum B_n \sqrt{\frac{2}{\pi}}\sin nx.$$

We call $f(x)$ a vector with components B_n in terms of the basis vectors $\sqrt{2/\pi}\sin nx$. Thus, in quantum mechanics, we often refer to a function which describes the state of a physical system as either a state function or a state vector. Just as we can write a three-dimensional vector in terms of $\mathbf{i},\mathbf{j},\mathbf{k}$, or in terms of another basis, say $\mathbf{e}_r,\mathbf{e}_\theta,\mathbf{e}_\phi$, so we can expand a given $f(x)$ in terms of another orthonormal set of functions and find its components relative to this new basis. In Section 9, we shall see how to expand functions in Legendre series.

Just as we needed the norm of $\sin nx$ in Fourier series, so we shall need the norm of $P_l(x)$ in expanding functions in Legendre series. We shall prove that

$$\int_{-1}^{1} [P_l(x)]^2\,dx = \frac{2}{2l+1}. \tag{8.1}$$

Then the functions $\sqrt{(2l+1)/2}\,P_l(x)$ are an orthonormal set of functions on $(-1,1)$.

To prove (8.1), we use the recursion relation (5.8b), namely,

$$lP_l(x) = xP_l'(x) - P_{l-1}'(x). \tag{8.2}$$

Multiply (8.2) by $P_l(x)$ and integrate to get

$$l\int_{-1}^{1} [P_l(x)]^2\,dx = \int_{-1}^{1} xP_l(x)P_l'(x)\,dx - \int_{-1}^{1} P_l(x)P_{l-1}'(x)\,dx. \tag{8.3}$$

The last integral is zero by Problem 7.4. To evaluate the middle integral in (8.3), we integrate by parts:

$$\begin{aligned}\int_{-1}^{1} xP_l(x)P_l'(x)\,dx &= \frac{x}{2}[P_l(x)]^2\Big|_{-1}^{1} - \frac{1}{2}\int_{-1}^{1}[P_l(x)]^2\,dx\\ &= 1 - \frac{1}{2}\int_{-1}^{1}[P_l(x)]^2\,dx\end{aligned}$$

(see Problem 2.2). Then (8.3) gives

$$l\int_{-1}^{1}[P_l(x)]^2\,dx = 1 - \frac{1}{2}\int_{-1}^{1}[P_l(x)]^2\,dx$$

which simplifies to (8.1). We can combine (7.1) and (8.1) to write

$$\int_{-1}^{1} P_l(x)P_m(x)\,dx = \frac{2}{2l+1}\,\delta_{lm}. \tag{8.4}$$

PROBLEMS, SECTION 8

Find the norm of each of the following functions on the given interval and state the normalized function.

1. $\cos nx$ on $(0, \pi)$

2. $P_2(x)$ on $(-1, 1)$

3. $xe^{-x/2}$ on $(0, \infty)$

4. $e^{-x^2/2}$ on $(-\infty, \infty)$

5. $xe^{-x^2/2}$ on $(0, \infty)$ *Hint:* See Chapter 4, Section 12.

6. Give another proof of (8.1) as follows. Multiply (5.8e) by $P_l(x)$ and integrate from -1 to 1. To evaluate the middle term, integrate by parts. Then use Problem 7.4.

7. Using (8.1), write the first four normalized Legendre polynomials and compare with the answers we found by a different method in Chapter 3, Section 14, Example 6.

9. LEGENDRE SERIES

Since the Legendre polynomials form a complete orthogonal set on $(-1, 1)$, we can expand functions in Legendre series just as we expanded functions in Fourier series.

Example 1. Expand in a Legendre series the function $f(x)$ given by

$$f(x) = \begin{cases} 0, & -1 < x < 0, \\ 1, & 0 < x < 1, \end{cases} \tag{9.1}$$

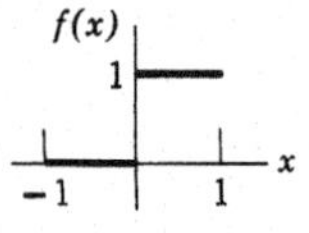

(see Figure 9.1). We put

Figure 9.1

$$f(x) = \sum_{l=0}^{\infty} c_l P_l(x). \tag{9.2}$$

Our problem is to find the coefficients c_l. We do this by a method parallel to the one we used in finding the formulas for the coefficients in a Fourier series. We multiply both sides of (9.2) by $P_m(x)$ and integrate from -1 to 1. Because the Legendre polynomials are orthogonal, all the integrals on the right are zero except the one containing c_m, and we can evaluate it by (8.1). Thus we get

$$\int_{-1}^{1} f(x)P_m(x)\,dx = \sum_{l=0}^{\infty} c_l \int_{-1}^{1} P_l(x)P_m(x)\,dx = c_m \cdot \frac{2}{2m+1}. \tag{9.3}$$

Using this result in our example (9.1), we find

$$\int_{-1}^{1} f(x)P_0(x)\,dx = c_0\int_{-1}^{1}[P_0(x)]^2\,dx \qquad \text{or} \qquad \int_0^1 dx = c_0\cdot 2, \quad c_0=\tfrac{1}{2};$$

$$\int_{-1}^{1} f(x)P_1(x)\,dx = c_1\int_{-1}^{1}[P_1(x)]^2\,dx \qquad \text{or} \qquad \int_0^1 x\,dx = c_1\cdot \tfrac{2}{3}, \quad c_1=\tfrac{3}{4};$$

$$\int_{-1}^{1} f(x)P_2(x)\,dx = c_2\int_{-1}^{1}[P_2(x)]^2\,dx \qquad \text{or} \qquad \int_0^1 (\tfrac{3}{2}x^2-\tfrac{1}{2})\,dx = c_2\cdot \tfrac{2}{5}, \quad c_2=0.$$

Continuing in this way we find for the function given in (9.1)

$$f(x)=\tfrac{1}{2}P_0(x)+\tfrac{3}{4}P_1(x)-\tfrac{7}{16}P_3(x)+\tfrac{11}{32}P_5(x)+\cdots. \tag{9.4}$$

It is unnecessary for $f(x)$ to be continuous as it must be for expansion in a Maclaurin series. Just as for Fourier series, the Dirichlet conditions (see Chapter 7, Section 6) are a convenient set of sufficient conditions for a function $f(x)$ to be expandable in a Legendre series. If $f(x)$ satisfies the Dirichlet conditions on $(-1,1)$, then at points inside $(-1,1)$ (not necessarily at the endpoints), the Legendre series converges to $f(x)$ anywhere $f(x)$ is continuous and converges to the midpoint of the jump at discontinuities.

Example 2. Here is an interesting fact about Legendre series. Sometimes we want to fit a given curve as closely as possible by a polynomial of a given degree, say a cubic. The criterion of "Least Squares" is often used to determine the best fit. This means that if, say, we want to fit a given $f(x)$ on $(-1,1)$ by a cubic, we find the coefficients a,b,c,d so that

$$\int_{-1}^{1}[f(x)-(ax^3+bx^2+cx+d)]^2\,dx \tag{9.5}$$

is as small as possible. Then

$$f(x)\cong ax^3+bx^2+cx+d \tag{9.6}$$

is called the best approximation (by a cubic) in the least squares sense. It can be proved that an expansion (as far as the desired degree of the polynomial approximation) in Legendre polynomials gives this best least squares approximation (Problem 16).

PROBLEMS, SECTION 9

Expand the following functions in Legendre series.

1. $f(x)=\begin{cases}-1, & -1<x<0\\ 1, & 0<x<1\end{cases}$

2. $f(x)=\begin{cases}0, & -1<x<0\\ x, & 0<x<1\end{cases}$

3. $f(x)=P_3'(x)$

4. $f(x)=\arcsin x$

5.

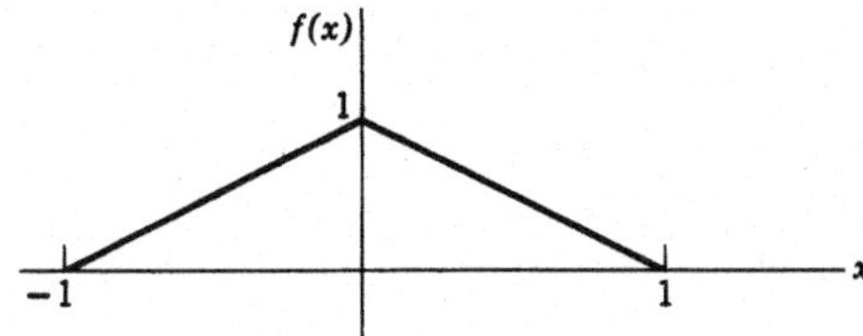

6. $f(x) = \begin{cases} 0 & \text{on } (-1,0) \\ \left(\ln \frac{1}{x}\right)^2 & \text{on } (0,1) \end{cases}$ *Hint:* See Chapter 11, Section 3, Problem 13.

7. $f(x) = \begin{cases} 0 & \text{on } (-1,0) \\ \sqrt{1-x} & \text{on } (0,1) \end{cases}$ *Hint:* See Chapter 11, Sections 6 and 7.

8. *Hint:* Solve the recursion relation (5.8e) for $P_l(x)$ and show that

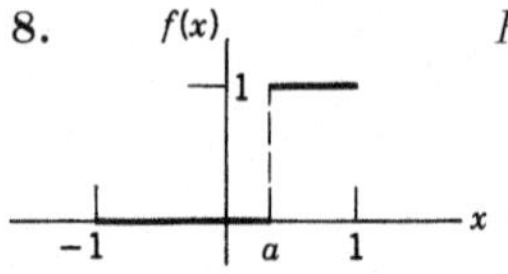

$$\int_a^1 P_l(x)\,dx = \frac{1}{2l+1}[P_{l-1}(a) - P_{l+1}(a)].$$

9. $f(x) = P_n'(x)$. *Hint:* For $l \geq n$, $\int_{-1}^1 P_n'(x)P_l(x)\,dx = 0$ (Why?); for $l < n$, integrate by parts.

Expand each of the following polynomials in a Legendre series. You should get the same results that you got by a different method in the corresponding problems in Section 5.

10. $3x^2 + x - 1$ **11.** $7x^4 - 3x + 1$ **12.** $x - x^3$

Find the best (in the least squares sense) second-degree polynomial approximation to each of the given functions over the interval $-1 < x < 1$. (See Problem 16.)

13. x^4 **14.** $|x|$ **15.** $\cos \pi x$

16. Prove the least squares approximation property of Legendre polynomials [see (9.5) and (9.6)] as follows. Let $f(x)$ be the given function to be approximated. Let the functions $p_l(x)$ be the normalized Legendre polynomials, that is,

$$p_l(x) = \sqrt{\frac{2l+1}{2}}P_l(x) \qquad \text{so that} \qquad \int_{-1}^1 [p_l(x)]^2\,dx = 1.$$

Show that the Legendre series for $f(x)$ as far as the $p_2(x)$ term is

$$f(x) = c_0p_0(x) + c_1p_1(x) + c_2p_2(x) \qquad \text{with} \quad c_l = \int_{-1}^1 f(x)p_l(x)\,dx.$$

Write the quadratic polynomial satisfying the least squares condition as $b_0p_0(x) + b_1p_1(x) + b_2p_2(x)$ (by Problem 5.14 any quadratic polynomial can be written in this form). The problem is to find b_0, b_1, b_2 so that

$$I = \int_{-1}^1 [f(x) - (b_0p_0(x) + b_1p_1(x) + b_2p_2(x))]^2\,dx$$

is a minimum. Square the bracket and write I as a sum of integrals of the individual terms. Show that some of the integrals are zero by orthogonality, some are 1 because the p_l's are normalized, and others are equal to the coefficients c_l. Add and subtract $c_0^2 + c_1^2 + c_2^2$ and show that

$$I = \int_{-1}^1 [f^2(x) + (b_0 - c_0)^2 + (b_1 - c_1)^2 + (b_2 - c_2)^2 - c_0^2 - c_1^2 - c_2^2]\,dx.$$

Now determine the values of the b's to make I as small as possible. (*Hint:* The smallest value the square of a real number can have is zero.) Generalize the proof to polynomials of degree n.

10. THE ASSOCIATED LEGENDRE FUNCTIONS

A differential equation closely related to the Legendre equation is

$$(10.1) \qquad (1-x^2)y'' - 2xy' + \left[l(l+1) - \frac{m^2}{1-x^2}\right] y = 0$$

with $m^2 \le l^2$. We *could* solve this equation by series; however, it is more useful to know how the solutions are related to Legendre polynomials, so we shall simply verify the known solution. First we substitute

$$(10.2) \qquad y = (1-x^2)^{m/2}u$$

into (10.1) and obtain (Problem 1)

$$(10.3) \qquad (1-x^2)u'' - 2(m+1)xu' + [l(l+1) - m(m+1)]u = 0.$$

For $m = 0$, this is Legendre's equation with solutions $P_l(x)$. Differentiate (10.3), obtaining (Problem 1)

$$(10.4) \quad (1-x^2)(u')'' - 2[(m+1)+1]x(u')' + [l(l+1) - (m+1)(m+2)]u' = 0.$$

But this is just (10.3) with u' in place of u, and $(m+1)$ in place of m. In other words, if $P_l(x)$ is a solution of (10.3) with $m = 0$, $P_l'(x)$ is a solution of (10.3) with $m = 1$, $P_l''(x)$ is a solution with $m = 2$, and in general for integral m, $0 \le m \le l$, $(d^m/dx^m)P_l(x)$ is a solution of (10.3). Then

$$(10.5) \qquad y = (1-x^2)^{m/2}\frac{d^m}{dx^m}P_l(x)$$

is a solution of (10.1). The functions in (10.5) are called *associated Legendre functions* are are denoted by

$$(10.6) \qquad P_l^m(x) = (1-x^2)^{m/2}\frac{d^m}{dx^m}P_l(x) \qquad \text{Associated Legendre functions}$$

[Some authors include a factor $(-1)^m$ in the definition of $P_l^m(x)$.]

A negative value for m in (10.1) does not change m^2, so a solution of (10.1) for positive m is also a solution for the corresponding negative m. Thus many references define $P_l^m(x)$ for $-l \le m \le l$ as equal to $P_l^{|m|}(x)$. Alternatively, we may use Rodrigues' formula (4.1) for $P_l(x)$ in (10.6) to get

$$(10.7) \qquad P_l^m(x) = \frac{1}{2^l l!}(1-x^2)^{m/2}\frac{d^{l+m}}{dx^{l+m}}(x^2-1)^l.$$

It can be shown that (10.7) is a solution of (10.1) for either positive or negative m; however $P_l^{-m}(x)$ and $P_l^m(x)$ are then proportional rather than equal (see Problem 8).

For each m, the functions $P_l^m(x)$ are a set of orthogonal functions on $(-1,1)$ (Problem 3). The normalization constants can be evaluated; for the definition (10.7) we find (Problem 10)

$$\int_{-1}^{1} [P_l^m(x)]^2\,dx = \frac{2}{2l+1}\frac{(l+m)!}{(l-m)!}. \tag{10.8}$$

The associated Legendre functions arise in many of the same problems in which Legendre polynomials appear (see the first paragraph of Section 2); in fact, the Legendre polynomials are just the special case of the functions $P_l^m(x)$ when $m = 0$.

PROBLEMS, SECTION 10

1. Verify equations (10.3) and (10.4).

2. The equation for the associated Legendre functions (and for Legendre functions when $m = 0$) usually arises in the form (see, for example, Chapter 13, Section 7)

$$\frac{1}{\sin\theta}\frac{d}{d\theta}\left(\sin\theta\frac{dy}{d\theta}\right) + \left[l(l+1) - \frac{m^2}{\sin^2\theta}\right]y = 0.$$

Make the change of variable $x = \cos\theta$, and obtain (10.1).

3. Show that the functions $P_l^m(x)$ for each m are a set of orthogonal functions on (-1,1), that is, show that

$$\int_{-1}^{1} P_l^m(x)P_n^m(x)\,dx = 0, \qquad l \neq n.$$

Hint: Use the differential equations (10.1) and follow the method of Section 7.

Substitute the $P_l(x)$ you found in Problems 4.3 or 5.3 into equation (10.6) to find $P_l^m(x)$; then let $x = \cos\theta$ to evaluate:

4. $P_1^1(\cos\theta)$ **5.** $P_4^1(\cos\theta)$ **6.** $P_3^2(\cos\theta)$

7. Show that

$$\frac{d^{l-m}}{dx^{l-m}}(x^2-1)^l = \frac{(l-m)!}{(l+m)!}(x^2-1)^m\frac{d^{l+m}}{dx^{l+m}}(x^2-1)^l.$$

Hint: Write $(x^2-1)^l = (x-1)^l(x+1)^l$ and find the derivatives by Leibniz' rule.

8. Write (10.7) with m replaced by $-m$; then use Problem 7 to show that

$$P_l^{-m}(x) = (-1)^m\frac{(l-m)!}{(l+m)!}P_l^m(x).$$

Comment: This shows that (10.7) is a solution of (10.1) when m is negative.

9. Use Problem 7 to show that

$$P_l^m(x) = (-1)^m\frac{(l+m)!}{(l-m)!}\frac{(1-x^2)^{-m/2}}{2^l l!}\frac{d^{l-m}}{dx^{l-m}}(x^2-1)^l.$$

10. Derive (10.8) as follows: Multiply together the two formulas for $P_l^m(x)$ given in (10.7) and Problem 9. Then integrate by parts repeatedly lowering the $l+m$ derivative and raising the $l-m$ derivative until both are l derivatives. Then use (8.1).

11. GENERALIZED POWER SERIES OR THE METHOD OF FROBENIUS

It may happen that the solution of a differential equation is not a power series $\sum_{n=0}^{\infty} a_n x^n$ but may either

(a) contain some negative powers of x, for example,

$$y = \frac{\cos x}{x^2} = \frac{1}{x^2} - \frac{1}{2!} + \frac{x^2}{4!} - \cdots$$

or

(b) have a fractional power of x as a factor, for example,

$$y = \sqrt{x}\sin x = x^{1/2}\left(x - \frac{x^3}{3!} + \cdots\right).$$

Both these cases (and others—see Section 21) are covered by a series of the form

$$y = x^s \sum_{n=0}^{\infty} a_n x^n = \sum_{n=0}^{\infty} a_n x^{n+s}, \tag{11.1}$$

where s is a number to be found to fit the problem; it may be either positive or negative and it may be a fraction. (In fact, it may even be complex, but we shall not consider this case.) Since $a_0 x^s$ is to be the first term of the series, we assume that a_0 is not zero. The series (11.1) is called a *generalized power series*. We shall consider some differential equations which can be solved by assuming a series of the form (11.1); this way of solving differential equations is called the *method of Frobenius*.

Example 1. As an illustration of this method we solve the equation

$$x^2 y'' + 4xy' + (x^2 + 2)y = 0. \tag{11.2}$$

From (11.1) we have

$$\begin{aligned}
y &= a_0 x^s + a_1 x^{s+1} + a_2 x^{s+2} + \cdots = \sum_{n=0}^{\infty} a_n x^{n+s},\\
y' &= s a_0 x^{s-1} + (s+1)a_1 x^s + (s+2)a_2 x^{s+1} + \cdots\\
&= \sum_{n=0}^{\infty} (n+s) a_n x^{n+s-1}, \qquad (11.3)\\
y'' &= s(s-1)a_0 x^{s-2} + (s+1)s a_1 x^{s-1} + (s+2)(s+1)a_2 x^s + \cdots\\
&= \sum_{n=0}^{\infty} (n+s)(n+s-1) a_n x^{n+s-2}.
\end{aligned}$$

We substitute (11.3) into (11.2) and set up a table of powers of x as we did for the Legendre equation:

	x^s	x^{s+1}	x^{s+2}	$\cdots$	x^{n+s}
x^2y''	$s(s-1)a_0$	$(s+1)sa_1$	$(s+2)(s+1)a_2$		$(n+s)(n+s-1)a_n$
$4xy'$	$4sa_0$	$4(s+1)a_1$	$2(s+2)a_2$		$4(n+s)a_n$
x^2y			a_0		a_{n-2}
$2y$	$2a_0$	$2a_1$	$2a_2$		$2a_n$

The total coefficient of each power of x must be zero. From the coefficient of x^s we get $(s^2 + 3s + 2)a_0 = 0$, or since $a_0 \neq 0$ by hypothesis,

$$s^2 + 3s + 2 = 0. \tag{11.4}$$

This equation for s is called the *indicial* equation. We solve it and find

$$s = -2, \qquad s = -1.$$

From here on we solve two separate problems, one when $s = -2$, and another when $s = -1$; a linear combination of the two solutions so obtained is then the general solution just as $A \sin x + B \cos x$ is the general solution of $y'' + y = 0$.

Example 2. For $s = -1$, the coefficient of x^{s+1} in the table gives $a_1 = 0$. From the x^{s+2} column on, we can use the general formula given by the last column. Notice, however, that the first two columns in the table do not contain the a_{n-2} term, so you must be careful about using the general term at first (Problems 13 and 14). From the general column with $s = -1$, we have

$$a_n[(n-1)(n+2) + 2] = -a_{n-2}$$

or

$$a_n = \frac{-a_{n-2}}{n(n+1)} \qquad \text{for } n \geq 2.$$

Since $a_1 = 0$, this gives all odd a's equal to zero. For even a's:

$$a_2 = -\frac{a_0}{3!}, \qquad a_4 = \frac{a_0}{5!}, \qquad a_6 = -\frac{a_0}{7!}, \qquad \cdots. \tag{11.5}$$

Then one solution of (11.2) is

$$\begin{aligned} y &= a_0 x^{-1} - \frac{a_0}{3!}x + \frac{a_0}{5!}x^3 - \cdots \\ &= a_0 x^{-2}\left(x - \frac{x^3}{3!} + \frac{x^5}{5!} - \cdots\right) = \frac{a_0 \sin x}{x^2}. \end{aligned} \tag{11.6}$$

The other solution, when $s = -2$, will be left to Problem 1.

PROBLEMS, SECTION 11

1. Finish the solution of equation (11.2) when $s = -2$. Write your solution in closed form as in (11.6). To avoid confusion with the a_n values we found when $s = -1$, you may want to call the coefficients in your series a_n' or b_n; however, this is not essential as long as you realize that there are two separate problems, one when $s = -1$ and one when $s = -2$, and each series has its own coefficients.

Solve the following differential equations by the method of Frobenius (generalized power series). Remember that the point of doing these problems is to learn about the method (which we will use later), not just to find a solution. You may recognize some series [as we did in (11.6)] or you can check your series by expanding a computer answer.

2. $x^2y'' + xy' - 9y = 0$
3. $x^2y'' + 2xy' - 6y = 0$
4. $x^2y'' - 6y = 0$
5. $2xy'' + y' + 2y = 0$
6. $3xy'' + (3x+1)y' + y = 0$
7. $x^2y'' - (x^2 + 2)y = 0$
8. $x^2y'' + 2x^2y' - 2y = 0$
9. $xy'' - y' + 9x^5y = 0$
10. $2xy'' - y' + 2y = 0$
11. $36x^2y'' + (5 - 9x^2)y = 0$

12. $3xy'' - 2(3x-1)y' + (3x-2)y = 0$

Consider each of the following problems as illustrations showing that, in a power series solution, we must be cautious about using the general recursion relation between the coefficients for the first few terms of the series.

13. Solve $y'' + y'/x^2 = 0$ by power series to find the relation

$$a_{n+1} = -\frac{n(n-1)}{n+1}a_n.$$

If, without thinking carefully, we test the series $\sum_{n=0}^{\infty} a_n x^n$ for convergence by the ratio test, we find

$$\lim_{n\to\infty} \frac{|a_{n+1}x^{n+1}|}{|a_n x^n|} = \infty. \qquad \text{(Show this.)}$$

Thus we might conclude that the series diverges and that there is no power series solution of this equation. Show why this is wrong, and that the power series solution is $y = \text{const}$.

14. Solve $y'' = -y$ by the Frobenius method. You should find that the roots of the indicial equation are $s = 0$ and $s = 1$. The value $s = 0$ leads to the solutions $\cos x$ and $\sin x$ as you would expect. For $s = 1$, call the series $y = \sum_{n=0}^{\infty} b_n x^{n+1}$, and find the relation

$$b_{n+2} = -\frac{b_n}{(n+3)(n+2)}.$$

Show that the b_0 series obtained from this relation is just $\sin x$, but that the b_1 series is *not* a solution of the differential equation. What is wrong?

12. BESSEL'S EQUATION

Like Legendre's equation, Bessel's equation is another of the "named" equations which have been studied extensively. There are whole books on Bessel functions, and you will find numerous formulas, graphs, and numerical values available in your computer program and in reference tables. You can think of Bessel functions as being something like damped sines and cosines. In fact, if you had first learned about $\sin nx$ and $\cos nx$ as power series solutions of $y'' = -n^2 y$ instead of in elementary trigonometry, you would not feel that Bessel functions were appreciably more difficult or strange than trigonometric functions. Like sines and cosines, Bessel functions are solutions of a differential equation; they can be represented by power series, their graphs can be drawn, and many formulas involving them (compare trigonometric identities) are known. Of special interest to science students is the fact that they occur in many applications. The following list of some of the problems in which they arise will give you an idea of the great range of topics which may involve Bessel functions: problems in electricity, heat, hydrodynamics, elasticity, wave motion, quantum mechanics, etc., involving cylindrical symmetry (for this reason Bessel functions are sometimes called cylinder functions); the motion of a pendulum whose length increases steadily; the small oscillations of a flexible chain; railway transition curves; the stability of a vertical wire or beam; Fresnel integrals in optics; the current distribution in a conductor; Fourier series for the arc of a circle. We shall discuss some of these applications later (see Section 18, and Chapter 13, Sections 5 and 6).

Bessel's equation in the usual standard form is

$$x^2y'' + xy' + (x^2 - p^2)y = 0, \tag{12.1}$$

where p is a constant (not necessarily an integer) called the *order* of the Bessel function y which is the solution of (12.1). You can easily verify that $x(xy')' = x^2y''+xy'$, so we can write (12.1) in the simpler form

$$x(xy')' + (x^2 - p^2)y = 0. \tag{12.2}$$

We find a generalized power series for (12.2) in the same way that we solved (11.2). [In fact, (11.2) is a form of Bessel's equation! See Problems 16.1 and 17.1.] Writing only the general terms in the series for y and the derivatives we need in (12.2), we have

$$\begin{aligned} y &= \sum_{n=0}^{\infty} a_n x^{n+s} \\ y' &= \sum_{n=0}^{\infty} a_n(n+s)x^{n+s-1} \\ xy' &= \sum_{n=0}^{\infty} a_n(n+s)x^{n+s} \\ (xy')' &= \sum_{n=0}^{\infty} a_n(n+s)^2x^{n+s-1} \\ x(xy')' &= \sum_{n=0}^{\infty} a_n(n+s)^2x^{n+s} \end{aligned} \tag{12.3}$$

We substitute (12.3) into (12.2) and tabulate the coefficients of powers of x:

	x^s	x^{s+1}	x^{s+2}	$\cdots$	x^{s+n}
$x(xy')'$	s^2a_0	$(1+s)^2a_1$	$(2+s)^2a_2$		$(n+s)^2a_n$
x^2y			a_0		a_{n-2}
$-p^2y$	$-p^2a_0$	$-p^2a_1$	$-p^2a_2$		$-p^2a_n$

The coefficient of x^s gives the indicial equation and the values of s:

$$s^2 - p^2 = 0, \qquad s = \pm p.$$

The coefficient of x^{s+1} gives $a_1 = 0$. The coefficient of x^{s+2} gives a_2 in terms of a_0, etc., but we may as well write the general formula from the last column at this point. We get

$$[(n+s)^2 - p^2]a_n + a_{n-2} = 0$$

or

$$a_n = -\frac{a_{n-2}}{(n+s)^2 - p^2}. \tag{12.4}$$

First we shall find the coefficients for the case $s = p$. From (12.4) we have

$$a_n = -\frac{a_{n-2}}{(n+p)^2 - p^2} = -\frac{a_{n-2}}{n^2 + 2np} = -\frac{a_{n-2}}{n(n+2p)}. \tag{12.5}$$

Since $a_1 = 0$, all odd a's are zero. For even a's it is convenient to replace n by $2n$; then from (12.5) we have

$$a_{2n} = -\frac{a_{2n-2}}{2n(2n+2p)} = -\frac{a_{2n-2}}{2^2 n(n+p)}. \tag{12.6}$$

The formulas for the coefficients can be simplified by the use of the Γ function notation (Chapter 11, Sections 2 to 5) as you can see by examining (12.7) below. Recall that $\Gamma(p+1) = p\Gamma(p)$ for any p, so,

$$\Gamma(p+2) = (p+1)\Gamma(p+1),$$
$$\Gamma(p+3) = (p+2)\Gamma(p+2) = (p+2)(p+1)\Gamma(p+1),$$

and so on. Then from (12.6) we find

$$\begin{aligned}
a_2 &= -\frac{a_0}{2^2(1+p)} = -\frac{a_0\Gamma(1+p)}{2^2\Gamma(2+p)},\\
a_4 &= -\frac{a_2}{2^3(2+p)} = \frac{a_0}{2!2^4(1+p)(2+p)} = \frac{a_0\Gamma(1+p)}{2!2^4\Gamma(3+p)},\\
a_6 &= -\frac{a_4}{3!2(3+p)} = -\frac{a_0}{3!2^6(1+p)(2+p)(3+p)}\\
&= -\frac{a_0\Gamma(1+p)}{3!2^6\Gamma(4+p)},
\end{aligned} \tag{12.7}$$

and so on. Then the series solution (for the $s = p$ case) is

$$\begin{aligned}
y &= a_0 x^p \Gamma(1+p)\left[\frac{1}{\Gamma(1+p)} - \frac{1}{\Gamma(2+p)}\left(\frac{x}{2}\right)^2\right.\\
&\quad \left. + \frac{1}{2!\Gamma(3+p)}\left(\frac{x}{2}\right)^4 - \frac{1}{3!\Gamma(4+p)}\left(\frac{x}{2}\right)^6 + \cdots\right]\\
&= a_0 2^p \left(\frac{x}{2}\right)^p \Gamma(1+p)\left[\frac{1}{\Gamma(1)\Gamma(1+p)} - \frac{1}{\Gamma(2)\Gamma(2+p)}\left(\frac{x}{2}\right)^2\right.\\
&\quad \left. + \frac{1}{\Gamma(3)\Gamma(3+p)}\left(\frac{x}{2}\right)^4 - \frac{1}{\Gamma(4)\Gamma(4+p)}\left(\frac{x}{2}\right)^6 + \cdots\right].
\end{aligned} \tag{12.8}$$

We have inserted $\Gamma(1)$ and $\Gamma(2)$ (which are both equal to 1) in the first two terms and written $x^p = 2^p(x/2)^p$ to make the series appear more systematic. If we take

$$a_0 = \frac{1}{2^p\Gamma(1+p)} \quad \text{or} \quad \frac{1}{2^p p!},$$

then y is called the Bessel function of the first kind of order p, and written $J_p(x)$.

$$
\begin{aligned}
J_p(x) &= \frac{1}{\Gamma(1)\Gamma(1+p)}\left(\frac{x}{2}\right)^{p} - \frac{1}{\Gamma(2)\Gamma(2+p)}\left(\frac{x}{2}\right)^{2+p} \\
&\quad + \frac{1}{\Gamma(3)\Gamma(3+p)}\left(\frac{x}{2}\right)^{4+p} - \frac{1}{\Gamma(4)\Gamma(4+p)}\left(\frac{x}{2}\right)^{6+p} + \cdots \\
&= \sum_{n=0}^{\infty} \frac{(-1)^n}{\Gamma(n+1)\Gamma(n+1+p)}\left(\frac{x}{2}\right)^{2n+p}.
\end{aligned}
\tag{12.9}
$$

PROBLEMS, SECTION 12

1. Show by the ratio test that the infinite series (12.9) for $J_p(x)$ converges for all x.

Use (12.9) to show that:

2. $J_2(x) = (2/x)J_1(x) - J_0(x)$

3. $J_1(x) + J_3(x) = (4/x)J_2(x)$

4. $(d/dx)J_0(x) = -J_1(x)$

5. $(d/dx)[xJ_1(x)] = xJ_0(x)$

6. $J_0(x) - J_2(x) = 2(d/dx)J_1(x)$

7. $\lim_{x\to 0} J_1(x)/x = \frac{1}{2}$

8. $\lim_{x\to 0} x^{-3/2}J_{3/2}(x) = 3^{-1}\sqrt{2/\pi}$ *Hint:* See Chapter 11, equations (3.4) and (5.3).

9. $\sqrt{\pi x/2}\, J_{1/2}(x) = \sin x$

13. THE SECOND SOLUTION OF BESSEL'S EQUATION

We have found just one of the two solutions of Bessel's equation, that is, the one when $s = p$; we must next find the solution when $s = -p$. It is unnecessary to go through the details again; we can just replace p by $-p$ in (12.9). In fact, the solution when $s = -p$ is usually written J_{-p}. From (12.9) we have

$$
J_{-p}(x) = \sum_{n=0}^{\infty} \frac{(-1)^n}{\Gamma(n+1)\Gamma(n-p+1)}\left(\frac{x}{2}\right)^{2n-p}. \tag{13.1}
$$

If p is not an integer, $J_p(x)$ is a series starting with x^p and J_{-p} is a series starting with x^{-p}. Then $J_p(x)$ and $J_{-p}(x)$ are two independent solutions and a linear combination of them is a general solution. But if p is an integer, then the first few terms in J_{-p} are zero because $\Gamma(n-p+1)$ in the denominator is Γ of a negative integer, which is infinite. You can show (Problem 2) that $J_{-p}(x)$ starts with the term x^p (for integral p) just as $J_p(x)$ does, and that

$$
J_{-p}(x) = (-1)^p J_p(x) \qquad \text{for integral } p; \tag{13.2}
$$

thus $J_{-p}(x)$ is not an independent solution when p is an integer. The second solution in this case is not a Frobenius series (11.1) but contains a logarithm. $J_p(x)$ is finite at the origin, but the second solution is infinite and so is useful only in applications involving regions not containing the origin.

Although $J_{-p}(x)$ is a satisfactory second solution when p is not an integer, it is customary to use a linear combination of $J_p(x)$ and $J_{-p}(x)$ as the second solution.

This is much as if $\sin x$ and $(2\sin x - 3\cos x)$ were used as the two solutions of $y'' + y = 0$ instead of $\sin x$ and $\cos x$. Remember that the general solution of this differential equation is a linear combination of $\sin x$ and $\cos x$ with arbitrary coefficients. But $A\sin x + B(2\sin x - 3\cos x)$ is just as good a linear combination as $c_1 \sin x + c_2 \cos x$. Similarly, any combination of $J_p(x)$ and $J_{-p}(x)$ is a satisfactory second solution of Bessel's equation. The combination which is used is called either the Neumann or the Weber function and is denoted by either N_p or Y_p:

$$N_p(x) = Y_p(x) = \frac{\cos(\pi p)J_p(x) - J_{-p}(x)}{\sin \pi p}. \tag{13.3}$$

For integral p this expression is an indeterminate form $0/0$. However, for any $x \neq 0$ it has a limit (as p tends to an integral value) which gives a second solution. This is why the special form (13.3) is used; it is valid for any p. N_p or Y_p are called Bessel functions of the second kind. The general solution of Bessel's equation (12.1) or (12.2) may then be written as

$$y = AJ_p(x) + BN_p(x), \tag{13.4}$$

where A and B are arbitrary constants.

PROBLEMS, SECTION 13

1. Using equations (12.9) and (13.1), write out the first few terms of $J_0(x)$, $J_1(x)$, $J_{-1}(x)$, $J_2(x)$, $J_{-2}(x)$. Show that $J_{-1}(x) = -J_1(x)$ and $J_{-2}(x) = J_2(x)$.

2. Show that, in general for integral n, $J_{-n}(x) = (-1)^n J_n(x)$, and $J_n(-x) = (-1)^n J_n(x)$.

Use equations (12.9) and (13.1) to show that:

3. $\sqrt{\pi x/2}\, J_{-1/2}(x) = \cos x$

4. $J_{3/2}(x) = x^{-1} J_{1/2}(x) - J_{-1/2}(x)$.

5. Using equation (13.3), show that $N_{1/2}(x) = -J_{-1/2}(x)$; that $N_{3/2}(x) = J_{-3/2}(x)$.

6. Show from (13.3) that $N_{(2n+1)/2}(x) = (-1)^{n+1} J_{-(2n+1)/2}(x)$.

14. GRAPHS AND ZEROS OF BESSEL FUNCTIONS

You can find the values of Bessel functions both from your computer program and in reference books, and you can use your computer to plot graphs of Bessel functions (see problems). Except for $J_0(x)$, all the J_p's start at the origin behaving like x^p and then oscillate something like $\sin x$ but with decreasing amplitude. $J_0(x)$ is equal to 1 at $x = 0$ and so looks something like a damped cosine. All the N's are $\pm\infty$ at the origin, but away from it they also oscillate with decreasing amplitude.

The values of x for which $\sin x = 0$ (called the zeros for $\sin x$) do not need to be computed because they are just $x = n\pi$ for $n = 0, 1, 2, \cdots$. The zeros of the Bessel functions, however, do not occur at regular intervals; they have to be computed numerically. You can find their values by computer or in tables. It is worth noticing that the difference between two successive zeros becomes approximately π (as it is for $\sin x$ and $\cos x$) when x is large. You can see this from graphs of the functions or from tables of the zeros, or from the approximate formulas for the Bessel functions when x is large. (See Section 20).

PROBLEMS, SECTION 14

1. By computer, plot graphs of $J_p(x)$ for $p = 0, 1, 2, 3$, and x from 0 to 15.

2. From the graphs in Problem 1, read approximate values of the first three zeros of each of the functions. Then, by computer, find more accurate values of the zeros.

3. By computer, plot $N_0(x)$ for x from 0 to 15, and $N_p(x)$ for $p = 1, 2, 3$, and x from 1 to 15.

4. From the graphs in Problem 3, read approximate values of the first three zeros of each of the functions, and then find more accurate values by computer.

5. By computer, plot $\sqrt{x}\, J_{1/2}(x)$ for x from 0 to 4π. Do you recognize the curve? See Problem 12.9.

6. By computer, find 30 zeros of J_0 and note that the spacing between consecutive zeros is tending to π.

15. RECURSION RELATIONS

The following useful relations hold among Bessel functions and their derivatives. Although we state them and outline proofs for $J_p(x)$, they also hold for $N_p(x)$.

$$\frac{d}{dx}[x^p J_p(x)] = x^p J_{p-1}(x), \tag{15.1}$$

$$\frac{d}{dx}[x^{-p} J_p(x)] = -x^{-p} J_{p+1}(x), \tag{15.2}$$

$$J_{p-1}(x) + J_{p+1}(x) = \frac{2p}{x} J_p(x), \tag{15.3}$$

$$J_{p-1}(x) - J_{p+1}(x) = 2J_p'(x), \tag{15.4}$$

$$J_p'(x) = -\frac{p}{x} J_p(x) + J_{p-1}(x) = \frac{p}{x} J_p(x) - J_{p+1}(x). \tag{15.5}$$

To prove (15.1), first multiply (12.9) by x^p and differentiate to get

$$\begin{aligned} \frac{d}{dx}[x^p J_p(x)] &= \frac{d}{dx} \sum_{n=0}^{\infty} \frac{(-1)^n}{\Gamma(n+1)\Gamma(n+1+p)} \frac{x^{2n+2p}}{2^{2n+p}} \\ &= \sum_{n=0}^{\infty} \frac{(-1)^n (2n+2p)}{\Gamma(n+1)\Gamma(n+1+p)} \frac{x^{2n+2p-1}}{2^{2n+p}}. \end{aligned}$$

Use the fact that $\Gamma(n+1+p) = (n+p)\Gamma(n+p)$, and cancel the factors 2 and $(n+p)$ to get

$$\frac{d}{dx}[x^p J_p(x)] = \sum_{n=0}^{\infty} \frac{(-1)^n}{\Gamma(n+1)\Gamma(n+p)} \frac{x^{2x+2p-1}}{2^{2n+p-1}}.$$

Divide by x^p and compare with (12.9); this gives

$$\frac{1}{x^p}\frac{d}{dx}[x^p J_p(x)] = \sum_{n=0}^{\infty} \frac{(-1)^n}{\Gamma(n+1)\Gamma(n+p)} \left(\frac{x}{2}\right)^{2n+p-1} = J_{p-1}(x),$$

since this series is just (12.9) with p replaced by $p-1$. Proofs of the other relations are outlined in Problems 1 to 3.

PROBLEMS, SECTION 15

1. Prove equation (15.2) by a method similar to the one used above to prove (15.1).

2. Solve equations (15.1) and (15.2) for $J_{p+1}(x)$ and $J_{p-1}(x)$. Add and subtract these two equations to get (15.3) and (15.4).

3. Carry out the differentiation in equations (15.1) and (15.2) to get (15.5).

4. Use equations (15.1) to (15.5) to do Problems 12.2 to 12.6.

5. Using equations (15.4) and (15.5), show that $J_0(x) = J_2(x)$ at every maximum or minimum of $J_1(x)$, and $J_0(x) = -J_2(x) = J_1'(x)$ at every positive zero of $J_1(x)$. Computer plot $J_0(x)$, $J_1(x)$, and $J_2(x)$ on the same axes, and verify that these results are true.

6. As in Problem 5, show that $J_{p-1}(x) = J_{p+1}(x)$ at every maximum or minimum of $J_p(x)$, and $J_{p-1}(x) = -J_{p+1}(x) = J_p'(x)$ at every positive zero of $J_p(x)$. Computer plot, say, J_2, J_3, and J_4 on the same axes, (or any other set of three consecutive J's or three consecutive N's) and check to see that the results are true.

7. (a) Using (15.2), show that

$$\int_0^\infty J_1(x)\,dx = -J_0(x)\Big|_0^\infty = 1.$$

(b) Use L23 of the Laplace Transform Table (page 469) to show that $\int_0^\infty J_0(t)\,dt = 1$. (Also see Problem 23.29.)

8. From equation (15.4), show that

$$\int_0^\infty J_1(x)\,dx = \int_0^\infty J_3(x)\,dx = \cdots = \int_0^\infty J_{2n+1}(x)\,dx,$$

and

$$\int_0^\infty J_0(x)\,dx = \int_0^\infty J_2(x)\,dx = \cdots = \int_0^\infty J_{2n}(x)\,dx.$$

Then, by Problem 7, show that

$$\int_0^\infty J_n(x)\,dx = 1 \qquad \text{for all integral } n.$$

9. Use L23 and L32 of the Laplace Transform Table (page 469) to evaluate $\int_0^\infty tJ_0(2t)e^{-t}\,dt$.

16. DIFFERENTIAL EQUATIONS WITH BESSEL FUNCTION SOLUTIONS

Many differential equations occur in practice that are not of the standard form (12.1) but whose solutions can be written in terms of Bessel functions. It can be shown (Problem 13) that the differential equation

(16.1)
$$y'' + \frac{1-2a}{x}y' + \left[(bcx^{c-1})^2 + \frac{a^2 - p^2c^2}{x^2}\right]y = 0$$

has the solution

(16.2)
$$y = x^a Z_p(bx^c),$$

where Z stands for J or N or any linear combination of them, and a, b, c, p are constants. To see how to use this, let us "solve" the differential equation

$$y'' + 9xy = 0. \tag{16.3}$$

If (16.3) is of the type (16.1), then we must have

$$1 - 2a = 0, \qquad (bc)^2 = 9, \qquad 2(c-1) = 1, \qquad a^2 - p^2c^2 = 0.$$

From these equations we find

$$a = \tfrac{1}{2}, \qquad c = \tfrac{3}{2}, \qquad b = 2, \qquad p = \frac{a}{c} = \frac{1}{3}.$$

Then the solution of (16.3) is

$$y = x^{1/2} Z_{1/3}(2x^{3/2}). \tag{16.4}$$

This means that the general solution of (16.3) is

$$y = x^{1/2}[AJ_{1/3}(2x^{3/2}) + BN_{1/3}(2x^{3/2})],$$

where A and B are arbitrary constants.

It is useful to write the differential equation whose solutions are $J_p(Kx)$ and $N_p(Kx)$, where K is a constant. We substitute Kx for x in (12.2). Then $x(dy/dx)$ becomes $Kx[dy/d(Kx)] = x(dy/dx)$ and similarly, $x(xy')'$ is unchanged. Thus the only change in (12.2) is to replace $x^2 - p^2$ by $K^2x^2 - p^2$ and we have:

(16.5) $\quad x(xy')' + (K^2x^2 - p^2)y = 0 \quad$ has solutions $J_p(Kx)$ and $N_p(Kx)$.

PROBLEMS, SECTION 16

Find the solutions of the following differential equations in terms of Bessel functions by using equations (16.1) and (16.2).

1. Equation (11.2)

2. $y'' + 4x^2y = 0$

3. $xy'' + 2y' + 4y = 0$

4. $3xy'' + 2y' + 12y = 0$

5. $y'' - \dfrac{1}{x}y' + \left(4 + \dfrac{1}{x^2}\right)y = 0$

6. $4xy'' + y = 0$

7. $xy'' + 3y' + x^3y = 0$

8. $y'' + xy = 0$

9. $3xy'' + y' + 12y = 0$

10. $xy'' - y' + 9x^2y = 0$

11. $xy'' + 5y' + xy = 0$

12. $4xy'' + 2y' + y = 0$

13. Verify by direct substitution that the text solution of equation (16.3) and your solutions in the problems above are correct. Also prove in general that the solution (16.2) given for (16.1) is correct. *Hint:* These are exercises in partial differentiation. To verify the solution (16.4) of (16.3), we would change variables from x, y to say z, u where

$$y = x^{1/2}u, \qquad u = J_{1/3}(z), \qquad z = 2x^{3/2},$$

and show that if x, y satisfy (16.3), then u, z satisfy (12.1), that is,

$$z^2\frac{d^2u}{dz^2} + z\frac{du}{dz} + (z^2 - \tfrac{1}{9})u = 0.$$

Use (16.5) to write the solutions of the following problems. Remember that $x(xy')' = x^2y'' + xy'$.

14. $x^2y'' + xy' + (4x^2 - 9)y = 0$

15. $x(xy')' + (25x^2 - 4)y = 0$

16. $x^2y'' + xy' + (16x^2 - 1)y = 0$

17. $xy'' + y' + 9xy = 0$

17. OTHER KINDS OF BESSEL FUNCTIONS

We have discussed $J_p(x)$ and $N_p(x)$ which are called *Bessel functions of the first and second kinds*, respectively. Since Bessel's equation is of second order, there are, of course, only two independent solutions. However, there are a number of related functions which are also called Bessel functions. Here again there is a close analogy to sines and cosines. We may think of $\cos x$ and $\sin x$ as the solutions of $y'' + y = 0$. But $\cos x \pm i \sin x$ are also solutions which we usually write as $e^{\pm ix}$. If we replace x by ix, we get the functions e^x, e^{-x}, $\cosh x$, $\sinh x$, which are solutions of $y'' - y = 0$. We list a number of Bessel functions which are frequently used and their trigonometric analogues:

Hankel Functions or Bessel Functions of the Third Kind

$$H_p^{(1)}(x) = J_p(x) + iN_p(x),$$
$$H_p^{(2)}(x) = J_p(x) - iN_p(x). \tag{17.1}$$

(Compare $e^{\pm ix} = \cos x \pm i \sin x$.)

Modified or Hyperbolic Bessel Functions The solutions of

$$x^2y'' + xy' - (x^2 + p^2)y = 0 \tag{17.2}$$

are, by (16.1) $Z_p(ix)$. (Compare this with the standard Bessel equation and by analogy consider the relation between $y'' + y = 0$ and $y'' - y = 0$.) The two independent solutions of (17.2) which are ordinarily used are

$$I_p(x) = i^{-p}J_p(ix),$$
$$K_p(x) = \frac{\pi}{2}i^{p+1}H_p^{(1)}(ix). \tag{17.3}$$

These should be compared with $\sinh x = -i \sin(ix)$ and $\cosh x = \cos(ix)$; because of the analogy, I and K are called hyperbolic Bessel functions. The i factors are adjusted to make I and K real for real x.

Spherical Bessel Functions If $p = (2n+1)/2 = n + \frac{1}{2}$, n an integer, then $J_p(x)$ and $N_p(x)$ are called Bessel functions of half-odd integral order; they can be expressed in terms of $\sin x$, $\cos x$, and powers of x. The spherical Bessel functions are closely related to them as you can see from the formulas (17.4) below. Spherical Bessel functions arise in a variety of vibration problems especially when spherical coordinates are used. We define the spherical Bessel functions $j_n(x)$, $y_n(x)$, $h_n^{(1)}(x)$, $h_n^{(2)}(x)$, for $n = 0, 1, 2, \cdots$, and state their values in terms of elementary functions (see Problems 2 and 3). For the use of these functions, see Chapter 13, Problems 7.15, 7.16, 7.19, and 10.20.

(17.4)
$$\begin{aligned}
j_n(x) &= \sqrt{\frac{\pi}{2x}}\,J_{(2n+1)/2}(x) = x^n\left(-\frac{1}{x}\frac{d}{dx}\right)^n\left(\frac{\sin x}{x}\right),\\
y_n(x) &= \sqrt{\frac{\pi}{2x}}\,Y_{(2n+1)/2}(x) = -x^n\left(-\frac{1}{x}\frac{d}{dx}\right)^n\left(\frac{\cos x}{x}\right),\\
h_n^{(1)} &= j_n(x) + iy_n(x),\\
h_n^{(2)} &= j_n(x) - iy_n(x).
\end{aligned}$$

The Kelvin Functions A standard method of solving vibration problems is to assume a solution involving $e^{i\omega t}$; the resulting equation may contain imaginary terms. As an example, the following equation arises in the problem of the distribution of alternating current in wires (skin effect) (Relton, p. 177):

(17.5)
$$y'' + \frac{1}{x}y' - iy = 0.$$

The solution of this equation is (Problem 8a)

(17.6)
$$y = Z_0(i^{3/2}x).$$

This is complex, and it is customary to separate it into its real and imaginary parts, called (for $Z = J$) ber and bei; these stand for Bessel-real and Bessel-imaginary. We define the ber, bei, ker, kei functions by

(17.7)
$$\begin{aligned}
J_0(i^{3/2}x) &= \operatorname{ber} x + i\operatorname{bei} x,\\
K_0(i^{1/2}x) &= \operatorname{ker} x + i\operatorname{kei} x.
\end{aligned}$$

There are also similar functions for $n \neq 0$. These functions occur in problems in heat flow and in the theory of viscous fluids, as well in electrical engineering.

The Airy Functions The Airy differential equation is

(17.8)
$$y'' - xy = 0.$$

By Section 16, the solutions are (Problem 8b)

(17.9)
$$\sqrt{x}\,Z_{1/3}(\tfrac{2}{3}ix^{3/2}),$$

so by (17.3) they can be written in terms of $I_{1/3}$ and $K_{1/3}$. The Airy functions are defined as

$$(17.10)\qquad \begin{aligned} \mathrm{Ai}(x) &= \frac{1}{\pi}\sqrt{\frac{x}{3}}\,K_{1/3}\left(\tfrac{2}{3}x^{3/2}\right), \\ \mathrm{Bi}(x) &= \sqrt{\frac{x}{3}}\left[I_{-1/3}\left(\tfrac{2}{3}x^{3/2}\right) + I_{1/3}\left(\tfrac{2}{3}x^{3/2}\right)\right]. \end{aligned}$$

For negative x, Ai and Bi can be expressed in terms of $J_{1/3}$ and $N_{1/3}$, or the Hankel functions (17.1) of order 1/3. Airy functions are of use in electrodynamics and quantum mechanics.

PROBLEMS, SECTION 17

1. Write the solutions of Problem 16.1 as spherical Bessel functions using the definitions (17.4) of $j_n(x)$ and $y_n(x)$ in terms of $J_{(2n+1)/2}(x)$ and $Y_{(2n+1)/2}(x)$. Then, using (17.4), obtain the solutions in terms of $\sin x$ and $\cos x$. Compare with the answers in equation (11.6) and Problem 11.1.

2. From Problem 12.9, $J_{1/2}(x) = \sqrt{2/\pi x}\,\sin x$. Use (15.2) to obtain $J_{3/2}(x)$ and $J_{5/2}(x)$. Substitute your results for the J's into (17.4) to verify the formulas stated for j_0, j_1, and j_2 in terms of $\sin x$ and $\cos x$.

3. From Problems 13.3 and 13.5, $Y_{1/2}(x) = -\sqrt{2/\pi x}\,\cos x$. As in Problem 2, obtain $Y_{3/2}$ and $Y_{5/2}$ and verify the formulas (17.4) for y_0, y_1, and y_2 in terms of $\sin x$ and $\cos x$.

4. Using (17.3) and the results stated in Problems 2 and 3 for $J_{1/2}$ and $Y_{1/2}$ $(= N_{1/2})$, show that

$$I_{1/2}(x) = \sqrt{\frac{2}{\pi x}}\,\sinh x, \qquad \text{and} \quad K_{1/2}(x) = \sqrt{\frac{\pi}{2x}}\,e^{-x}.$$

5. Show from (17.4) that $h_n^{(1)}(x) = -ix^n\left(-\dfrac{1}{x}\dfrac{d}{dx}\right)^n\left(\dfrac{e^{ix}}{x}\right)$.

6. Using (16.1) and (17.4) show that the spherical Bessel functions satisfy the differential equation

$$x^2y'' + 2xy' + [x^2 - n(n+1)]y = 0.$$

7. (a) Solve the differential equation $xy'' = y$ using (16.1), and then express the answer in terms of a function I_p by (17.3).

(b) As in (a), find a solution of of $y'' - x^4y = 0$.

8. Using (16.1) and (16.2), verify that

(a) the solution of (17.5) is (17.6);

(b) the solution of (17.8) is (17.9).

9. Using (17.3) and (15.1) to (15.5), find the recursion relations for $I_p(x)$. In particular, show that $I_0' = I_1$.

10. Computer plot

(a) $I_0(x)$, $I_1(x)$, $I_2(x)$, from $x = 0$ to 2.

(b) $K_0(x)$, $K_1(x)$, $K_2(x)$, from $x = 0.1$ to 2.

(c) $\mathrm{Ai}(x)$ from $x = -10$ to 10.

(d) $\mathrm{Bi}(x)$ from $x = -10$ to 1.

11. From (17.4), show that $h_0^{(1)}(ix) = -e^{-x}/x$.

Use the Section 15 recursion relations and (17.4) to obtain the following recursion relations for spherical Bessel functions. We have written them for j_n, but they are valid for y_n and for the h_n's.

12. $j_{n-1}(x) + j_{n+1}(x) = (2n+1)j_n(x)/x$ **13.** $(d/dx)j_n(x) = nj_n(x)/x - j_{n+1}(x)$

14. $(d/dx)j_n(x) = j_{n-1}(x) - (n+1)j_n(x)/x$

15. $(d/dx)[x^{n+1}j_n(x)] = x^{n+1}j_{n-1}(x)$ **16.** $(d/dx)[x^{-n}j_n(x)] = -x^{-n}j_{n+1}(x)$

18. THE LENGTHENING PENDULUM

As an example of the use of Bessel functions we consider the following problem. Suppose that a simple pendulum (see Chapter 11, Section 8) has the length l of its string increased at a steady rate (for example, a weight swaying as it is lowered by a crane). (This problem was considered as early as 1707; see L. LeCornu, *Acta Mathematica* 19 (1895), 201–249. Also see Relton, and Problem 8.) Find the equation of motion and the solution for small oscillations.

From Chapter 11, Section 8, we have the equation of motion

$$\frac{d}{dt}(ml^2\dot{\theta}) + mgl\sin\theta = 0. \tag{18.1}$$

Let the length of the string at time t be

$$l = l_0 + vt, \tag{18.2}$$

and change from t to l as the independent variable. For small oscillations, we may replace $\sin\theta$ by θ. Then (18.1) becomes (Problem 1):

$$l\frac{d^2\theta}{dl^2} + 2\frac{d\theta}{dl} + \frac{g}{v^2}\theta = 0. \tag{18.3}$$

(This equation could also describe the damped vibration of a variable mass, or an RLC circuit with variable L.)

We solve (18.3) by comparing it with the standard equation (16.1) to get (Problem 2)

$$\theta = l^{-1/2}Z_1(bl^{1/2}) \qquad \text{where } b = 2g^{1/2}/v. \tag{18.4}$$

To simplify the notation, let

$$u = b\,l^{1/2} = (2g^{1/2}/v)l^{1/2}. \tag{18.5}$$

The general solution of (18.3) is then

$$\theta = Au^{-1}J_1(u) + Bu^{-1}N_1(u). \tag{18.6}$$

We can find $d\theta/du$ from (18.6) using (15.2):

$$\frac{d\theta}{du} = -[Au^{-1}J_2(u) + Bu^{-1}N_2(u)]. \tag{18.7}$$

The constants A and B must be found from the starting conditions just as they are for the ordinary simple pendulum with constant l. For example, in the ordinary case, if $\theta = \theta_0$ and $\dot{\theta} = 0$ at $t = 0$, then the general solution $\theta = A\cos\omega t + B\sin\omega t$ becomes just $\theta = \theta_0 \cos\omega t$. For the lengthening pendulum, let's take the same simple initial conditions, namely $\theta = \theta_0$ and $\dot{\theta} = 0$ at $t = 0$. For these initial conditions, we find (after some calculations—see Problems 3 to 6)

$$A = -\frac{\pi u_0^2}{2}\theta_0 N_2(u_0), \qquad B = \frac{\pi u_0^2}{2}\theta_0 J_2(u_0). \tag{18.8}$$

The solution has a particularly simple form if we adjust the constants v and l_0 so that

$$u_0 = 2(gl_0)^{1/2}/v \qquad \text{is a zero of } J_2(u). \tag{18.9}$$

Then $B = 0$ and the second term of (18.6) is zero, so we have

$$\theta = Au^{-1}J_1(u) = Cl^{-1/2}J_1(b\,l^{1/2}), \tag{18.10}$$

where (Problem 7)

$$b = \frac{2g^{1/2}}{v} = \frac{u_0}{l_0^{1/2}}, \qquad C = \frac{\theta_0 l_0^{1/2}}{J_1(u_0)}. \tag{18.11}$$

For this simple case, $\dot{\theta}$ is a multiple of $J_2(u)$ (Problem 8); thus $\theta = 0$ corresponds to zeros of $J_1(u)$ and $\dot{\theta} = 0$ corresponds to zeros of $J_2(u)$. A "quarter" period corresponds to the time from $\theta = 0$ to $\dot{\theta} = 0$, or $\dot{\theta} = 0$ to $\theta = 0$. These quarter periods can be found from the zeros of $J_1(u)$ and $J_2(u)$ (Problem 8).

PROBLEMS, SECTION 18

1. Verify equation (18.3). *Hint:* From equation (18.2), $dl = v\,dt$, so

$$\frac{d}{dt} = v\frac{d}{dl}.$$

2. Solve equation (18.3) to get equation (18.4).

3. Prove

$$J_p(x)J'_{-p}(x) - J_{-p}(x)J'_p(x) = -\frac{2}{\pi x}\sin p\pi$$

as follows: Write Bessel's equation (12.1) with $y = J_p$ and with $y = J_{-p}$; multiply the J_p equation by J_{-p} and the J_{-p} equation by J_p and subtract to get

$$\frac{d}{dx}[x(J_pJ'_{-p} - J_{-p}J'_p)] = 0.$$

Then $J_pJ'_{-p} - J_{-p}J'_p = c/x$. To find c, use equation (12.9) for each of the four functions and pick out the $1/x$ terms in the products. Then use equation (5.4) of Chapter 11.

4. Using equation (13.3) and Problem 3, show that

$$J_p(x)N'_p(x) - J'_p(x)N_p(x) = \frac{J'_p(x)J_{-p}(x) - J_p(x)J'_{-p}(x)}{\sin p\pi} = \frac{2}{\pi x}.$$

5. Use the recursion relations of Section 15 (for N's as well as for J's) and Problem 4 to show that

$$J_n(x)N_{n+1}(x) - J_{n+1}(x)N_n(x) = -\frac{2}{\pi x}.$$

Hint: Do it first for $n = 0$; then use the result in proving the $n = 1$ case, and so on.

6. For the initial conditions $\theta = \theta_0$, $\dot{\theta} = 0$, show that the constants A and B in equations (18.6) and (18.7) are as given in (18.8). *Hints:* Show that $d\theta/du = 0$ if $\dot{\theta} = 0$. In equations (18.6) and (18.7), set $\theta = \theta_0$ and $d\theta/du = 0$ when $u = u_0$ and solve for A and B. Then use the formula in Problem 5 to simplify your results to get equation (18.8).

7. Verify the values of b and C given in equation (18.11). Note that C can be found in two ways: (1) in equation (18.10), $u = bl^{1/2}$, so $Au^{-1} = (A/b)l^{-1/2}$, $C = A/b$. Use Problem 5 to simplify this. (2) Set $\theta = \theta_0$, $u = u_0$, $l = l_0$ in equation (18.10) and solve for C.

8. Find

$$\dot{\theta} = \frac{d\theta}{dt} = \frac{d\theta}{du}\frac{du}{dl}\frac{dl}{dt}$$

either from equations (18.10) and (15.2) or from equation (18.7) with $B = 0$. Thus show that $\theta = 0$ when $J_1(u) = 0$ and $\dot{\theta} = 0$ when $J_2(u) = 0$. Show that the successive (variable) quarter periods of the lengthening pendulum are $(v/4g)(r_2^2 - r_1^2)$ or $(v/4g)(r_1^2 - r_2^2)$, where r_1 and r_2 are successive zeros of J_1 and J_2. Use a computer or tables to find the needed zeros and calculate several quarter periods (as multiples of $v/(4g)$). Observe that an inward swing takes longer than either the preceding or the following outward swing. [This result is proved by Ll. G. Chambers, *Proceedings of the Edinburgh Mathematical Society* (2) 12, 17–18 (1960).]

9. Consider the "shortening pendulum" problem. Follow the method in the text but with $l = l_0 - vt$. Does the θ amplitude of the vibration increase or decrease as the pendulum shortens? Restate the result of Problem 8 about quarter periods for this case.

10. The differential equation for transverse vibrations of a string whose density increases linearly from one end to the other is $y'' + (Ax + B)y = 0$, where A and B are constants. Find the general solution of this equation in terms of Bessel functions. *Hint:* Make the change of variable $Ax + B = Au$.

11. A straight wire clamped vertically at its lower end stands vertically if it is short, but bends under its own weight if it is long. It can be shown that the greatest length for vertical equilibrium is l, where $kl^{3/2}$ is the first zero of $J_{-1/3}$ and

$$k = \frac{4}{3r^2}\sqrt{\frac{\rho g}{\pi Y}},$$

r = radius of the wire, ρ = linear density, g = acceleration of gravity, Y = Young's modulus. Find l for a steel wire of radius 1 mm; for a lead wire of the same radius.

19. ORTHOGONALITY OF BESSEL FUNCTIONS

You may expect here that we are going to prove that two J_p's for different p values are orthogonal. However, this is *not* what we are going to do—as a matter of fact it isn't true! To see what we *are* going to prove, look at the following comparison between Bessel functions and sines and cosines.

	Two functions: $\sin x$ and $\cos x$.	Two functions for each p: $J_p(x)$ and $N_p(x)$.
	Consider just $\sin x$.	Consider just $J_p(x)$ for one value of p.
(19.1)	At the zeros of $\sin x$, namely, $x = n\pi$, $\sin x = 0$.	At the zeros of $J_p(x)$, say $x = \alpha, \beta, \cdots$, $J_p(x) = 0$.
	At $x = 1$, $\sin n\pi x = 0$.	At $x = 1$, $J_p(\alpha x) = 0$, $J_p(\beta x) = 0, \cdots$.
	The differential equation satisfied by $y = \sin n\pi x$ is $y'' + (n\pi)^2 y = 0$.	The differential equation satisfied by $y = J_p(\alpha x)$ is [see (16.5)] $x(xy')' + (\alpha^2 x^2 - p^2)y = 0$.

(In comparing the differential equations remember that p is a fixed constant. The correspondence is between the zeros of $\sin x$, namely $n\pi$, and the zeros of $J_p(x)$, namely, α, β, etc.)

We have proved (Chapter 7):	We shall prove:
$\int_0^1 \sin n\pi x \sin m\pi x \, dx = 0$ for $n \neq m$.	$\int_0^1 x J_p(\alpha x) J_p(\beta x) \, dx = 0$ for $\alpha \neq \beta$.

By (16.5), the differential equation satisfied by $J_p(\alpha x)$ is

$$x(xy')' + (\alpha^2 x^2 - p^2)y = 0 \tag{19.2}$$

and the differential equation satisfied by $J_p(\beta x)$ is

$$x(xy')' + (\beta^2 x^2 - p^2)y = 0. \tag{19.3}$$

Let us for simplicity call $J_p(\alpha x) = u$ and $J_p(\beta x) = v$; then (19.2) and (19.3) become

$$\begin{aligned} x(xu')' + (\alpha^2 x^2 - p^2)u &= 0, \\ x(xv')' + (\beta^2 x^2 - p^2)v &= 0. \end{aligned} \tag{19.4}$$

We are going to use equations (19.4) to prove the last equation in (19.1) by a method parallel to that used in proving the orthogonality of Legendre polynomials (Section 7). Multiply the first equation of (19.4) by v, the second by u, subtract the two equations and cancel an x to get

$$v(xu')' - u(xv')' + (\alpha^2 - \beta^2)xuv = 0. \tag{19.5}$$

The first two terms of (19.5) are equal to

$$\frac{d}{dx}(vxu' - uxv'). \tag{19.6}$$

Using (19.6) and integrating (19.5), we get

$$(19.7) \qquad (vxu' - uxv')\Big|_0^1 + (\alpha^2 - \beta^2)\int_0^1 xuv\,dx = 0.$$

At the lower limit the integrated term is zero because $x = 0$ and u, v, u', v' are finite. To evaluate the integrated term at the upper limit, recall that $u = J_p(\alpha x)$, $v = J_p(\beta x)$; then at $x = 1$, $u = J_p(\alpha) = 0$, $v = J_p(\beta) = 0$ since α and β are zeros of J_p. The integrated term is therefore zero at the upper limit also. Thus (19.7) becomes

$$(19.8) \qquad (\alpha^2 - \beta^2)\int_0^1 xuv\,dx = 0$$

or

$$(19.9) \qquad (\alpha^2 - \beta^2)\int_0^1 xJ_p(\alpha x)J_p(\beta x)\,dx = 0.$$

If $\alpha \neq \beta$, that is, if α and β are different zeros of J_p, the integral must be zero. If $\alpha = \beta$, the integral is not zero; it can be evaluated, but we shall just state the answer (see Problem 1):

$$(19.10) \qquad \int_0^1 xJ_p(\alpha x)J_p(\beta x)\,dx = \begin{cases} 0 & \text{if } \alpha \neq \beta, \\ \frac{1}{2}J_{p+1}^2(\alpha) = \frac{1}{2}J_{p-1}^2(\alpha) = \frac{1}{2}J_p'^2(\alpha) & \text{if } \alpha = \beta, \end{cases}$$

where α and β are zeros of $J_p(x)$.

[You can see that the three answers for the case $\alpha = \beta$ are equal by equations (15.3) to (15.5), remembering that α is a zero of J_p.]

We can state (19.10) in words in two different ways; if α_n, $n = 1, 2, 3, \cdots$, are the zeros of $J_p(x)$, then we say either that

(a) the functions $\sqrt{x}J_p(\alpha_n x)$ are orthogonal on $(0, 1)$;

or that

(b) the functions $J_p(\alpha_n x)$ are orthogonal on $(0, 1)$ with respect to the *weight function* x.

You may meet other sets of functions which are orthogonal with respect to a weight function. (See, for example, Section 22.) In general, we say that $y_n(x)$ is a set of orthogonal functions on (x_1, x_2) with respect to the weight function $w(x)$ if

$$\int_{x_1}^{x_2} y_n(x)y_m(x)w(x)\,dx = 0 \qquad \text{for} \quad n \neq m.$$

The fact that the Bessel functions $J_p(\alpha_n x)$ obey (19.10) makes it possible to expand a given function in a series of Bessel functions much as we expand functions in Fourier series and Legendre series. We shall do this later (Chapter 13) when we need it in a physical example.

Just as we generalized Fourier series to an interval $(0, l)$, here we can generalize (19.10) to an interval $(0, a)$. In (19.10), let $x = r/a$. Then the limits are $x = r/a = 0$ to 1, that is, $r = 0$ to a. The integral in (19.10) becomes

$$\int_0^a (r/a)J_p(\alpha r/a)J_p(\beta r/a)\,d(r/a) = \frac{1}{a^2}\int_0^a rJ_p(\alpha r/a)J_p(\beta r/a)\,dr.$$

Thus we have

$$(19.11)\quad \int_0^a rJ_p(\alpha r/a)J_p(\beta r/a)\,dr$$

$$= \begin{cases} 0 & \text{if } \alpha \neq \beta, \\ \frac{a^2}{2}J_{p+1}^2(\alpha) = \frac{a^2}{2}J_{p-1}^2(\alpha) = \frac{a^2}{2}J_p'^2(\alpha) & \text{if } \alpha = \beta. \end{cases}$$

PROBLEMS, SECTION 19

1. Prove equation (19.10) in the following way. First note that (19.2) and (19.3) and therefore (19.7) hold whether α and β are zeros of $J_p(x)$ or not. Let α be a zero, but let β be just any number. From (19.7) show that then

$$\int_0^1 xuv\,dx = \frac{J_p(\beta)\alpha J_p'(\alpha)}{\beta^2 - \alpha^2}.$$

Now let $\beta \to \alpha$ and evaluate the indeterminate form by L'Hôpital's rule (that is, differentiate numerator and denominator with respect to β and let $\beta \to \alpha$). Hence find

$$\int_0^1 xuv\,dx = \tfrac{1}{2}J_p'^2(\alpha)$$

for $\alpha = \beta$, that is, for $u = v = J_p(\alpha x)$ as in (19.10). Use equations (15.3) to (15.5) to show that the other two expressions given in (19.10) are equivalent.

2. Given that

$$J_{3/2}(x) = \sqrt{\frac{2}{\pi x}}\left(\frac{\sin x}{x} - \cos x\right),$$

use (19.10) to evaluate

$$\int_0^1 \left(\frac{\sin \alpha x}{\alpha x} - \cos \alpha x\right)^2 dx$$

where α is a root of the equation $\tan x = x$.

3. Use (17.4) and (19.10) to write the orthogonality condition and the normalization integral for the spherical Bessel functions $j_n(x)$.

4. Define $J_p(z)$ for complex z by the power series (12.9) with x replaced by z. (By Problem 12.1, the series converges for all z.) Show by (19.10) that all the zeros of $J_p(z)$ are real. *Hint:* Suppose α and β in (19.10) were a complex conjugate pair; show that then the integrand would be positive so the integral could not be zero.

5. We obtained (19.10) for $J_p(x)$, $p \geq 0$. It is, however, valid for $p > -1$, that is for $N_p(x)$, $0 \leq p < 1$. The difficulty in the proof occurs just after (19.7); we said that u, v, u', v' are finite at $x = 0$ which is not true for $N_p(x)$. However, the negative powers of x cancel if $p < 1$. Show this for $p = \frac{1}{2}$ by using two terms of the power series (12.9) or (13.1) for the function $N_{1/2}(x) = -J_{-1/2}(x)$ [see (13.3)].

6. By Problem 5, $\int_0^1 xN_{1/2}(\alpha x)N_{1/2}(\beta x)\,dx = 0$ if α and β are different zeros of $N_{1/2}(x)$. Using (17.4), find $N_{1/2}(x)$ in terms of $\cos x$ and so find the zeros of $N_{1/2}(x)$. Show that the functions $\cos(n + \frac{1}{2})\pi x$ are an orthogonal set on $(0, 1)$. Use (19.10) to find the normalization constant. (Compare Problem 6.8.)

20. APPROXIMATE FORMULAS FOR BESSEL FUNCTIONS

There are often cases in which it is useful to have an approximate formula giving the behavior of a Bessel function when x is near zero or when x is very large. We list some of these formulas here for reference. The symbol $O(x^n)$ is read "terms of the order of x^n or less," and means that the error in the given approximation is less than a constant times x^n; thus $O(1)$ means bounded terms. Note that $p \geq 0$.

Function	*Small* x	*Large* x *(asymptotic formulas)*
$J_p(x)$	$\dfrac{1}{\Gamma(p+1)}\left(\dfrac{x}{2}\right)^p + O(x^{p+2})$	$\sqrt{\dfrac{2}{\pi x}}\cos\left(x - \dfrac{2p+1}{4}\pi\right) + O(x^{-3/2})$
$N_p(x)$	$p=0$: $\dfrac{2}{\pi}\ln x + O(1)$; $p>0$: $-\dfrac{\Gamma(p)}{\pi}\left(\dfrac{2}{x}\right)^p + \begin{cases} O(x^p), & p<1 \\ O(x\ln\frac{1}{x}), & p=1 \\ O(x^{2-p}), & p>1 \end{cases}$	$\sqrt{\dfrac{2}{\pi x}}\sin\left(x - \dfrac{2p+1}{4}\pi\right) + O(x^{-3/2})$
$H_p^{(1) \text{ or } (2)}(x)$	Like $\pm iN_p(x)$	$\sqrt{\dfrac{2}{\pi x}}\, e^{\pm i[x-(2p+1)\pi/4]} + O(x^{-3/2})$
$I_p(x)$	Like $J_p(x)$	$\dfrac{1}{\sqrt{2\pi x}}e^x + O\left(\dfrac{e^x}{x}\right)$
$K_p(x)$	Like $-\dfrac{\pi}{2}N_p(x)$	$\sqrt{\dfrac{\pi}{2x}}\, e^{-x} + O\left(\dfrac{e^{-x}}{x}\right)$
$j_n(x)$	$\dfrac{x^n}{(2n+1)!!} + O(x^{n+2})$	$\dfrac{1}{x}\sin\left(x - \dfrac{n\pi}{2}\right) + O(x^{-2})$
$y_n(x)$	$\dfrac{-(2n-1)!!}{x^{n+1}} + O(x^{1-n})$	$-\dfrac{1}{x}\cos\left(x - \dfrac{n\pi}{2}\right) + O(x^{-2})$

Note: $(2n+1)!!$ means $1\cdot3\cdot5\cdot7\cdots(2n+1) = \dfrac{(2n+1)!}{2^n n!}$. See Chapter 1, Section 13C.

PROBLEMS, SECTION 20

Use the table above to evaluate the following limits:

1. $\lim_{x\to0} J_4(x)/[J_2(x)]^2$ **2.** $\lim_{x\to\infty} I_3(x)/I_5(x)$ **3.** $\lim_{x\to0} N_0(x^2)/\ln(x)$

4. $\lim_{x\to0} J_p(x)/N_p(x)$ **5.** $\lim_{x\to\infty} xI_p(x)K_p(x)$ **6.** $\lim_{x\to0} xj_n(x)y_n(x)$

Use the table above and the definitions in Section 17 to find approximate formulas for large x for:

7. $h_n^{(1)}(x)$ **8.** $h_n^{(2)}(x)$

9. $h_n^{(1)}(ix)$ **10.** $h_n^{(2)}(ix)$

To study the approximations in the table, computer plot on the same axes the given function together with its small x approximation and its asymptotic approximation. Use an interval large enough to show the asymptotic approximation agreeing with the function for large x. If the small x approximation is not clear, plot it alone with the function over a small interval.

11. $J_1(x)$	**12.** $J_2(x)$	**13.** $J_3(x)$	**14.** $N_2(x)$
15. $N_3(x)$	**16.** $j_1(x)$	**17.** $j_2(x)$	**18.** $y_2(x)$

19. Computer plot on the same axes several $I_p(x)$ functions together with their common asymptotic approximation. Then computer plot each function with its small x approximation.

20. As in Problem 19, study the $K_p(x)$ functions. It is interesting to note (see Problem 17.4) that $K_{1/2}(x)$ is equal to the asymptotic approximation.

21. SERIES SOLUTIONS; FUCHS'S THEOREM

We have discussed two examples of differential equations solvable by the Frobenius method (Legendre and Bessel equations). There are many other "named" equations and the corresponding "named" functions which are their solutions. (See a few more examples in Section 22.) All of them have much in common with our two examples and you should not hesitate to look them up and use them without a formal introduction when you run into them on your computer or in a text or reference book. You may discover any or all of the following things about such a new (to you) set of functions: that they are the set of solutions of a differential equation with one or more parameters (like the p in Bessel's equation); that the values of the functions, their derivatives, their zeros, and many formulas involving them are available in references (tables and computer); that they have orthogonality properties, perhaps with respect to a weight function, and consequently (suitably restricted) functions can be expanded in series of them; that there is a generating function for the set of functions; that there are physical problems whose solutions involve the functions, often in the solution of a partial differential equation; etc.

Now you may wonder whether all differential equations can be solved by the Frobenius method. A general theorem due to Fuchs tells when this method will work; we shall state it for second-order differential equations which are the most important ones in applications. Write the differential equation as

$$y'' + f(x)y' + g(x)y = 0. \tag{21.1}$$

If $xf(x)$ and $x^2g(x)$ are expandable in convergent power series $\sum_{n=0}^{\infty} a_n x^n$, we say that the differential equation (21.1) is regular (or has a nonessential singularity) at the origin. Let us call these the Fuchsian conditions. Fuchs's theorem says that these conditions are necessary and sufficient for the general solution of (21.1) to consist of either

(1) two Frobenius series, or

(2) one solution $S_1(x)$ which is a Frobenius series, and a second solution which is $S_1(x)\ln x + S_2(x)$ where $S_2(x)$ is another Frobenius series.

Case (2) occurs only when the roots of the indicial equation are equal or differ by an integer, and not always then. [See, for example, equation (11.2) and Problems 11.1 to 11.4, and 11.7 to 11.9.] Note the *necessary* condition: If the Fuchsian conditions are not met, we cannot find the general solution by the method of generalized power series (see Problems 11 to 13). However, the equations most commonly found in applications do meet these conditions.

If the first Frobenius series $S_1(x)$ happens to break off, or you can easily write its sum in closed form, then the method of "reduction of order" [Chapter 8, Section 7(e)] gives a way of finding the second solution without using infinite series (see Problems 1 to 4). However, note that our main interest in series solutions is not to solve differential equations this way in general, but to study sets of functions (like Legendre polynomials and Bessel functions) which are solutions of differential equations that occur in applications. So the purpose in using series to solve a few simple differential equations (for which there are easier methods) is to learn how and when the series method works—to watch Fuchs's theorem in action (see problems).

PROBLEMS, SECTION 21

For Problems 1 to 4, find one (simple) solution of each differential equation by series, and then find the second solution by the "reduction of order" method, Chapter 8, Section 7(e).

1. $(x^2+1)y''-xy'+y=0$

2. $x^2y''+(x+1)y'-y=0$

3. $x^2y''+x^2y'-2y=0$

4. $(x-1)y''-xy'+y=0$

Solve the differential equations in Problems 5 to 10 by the Frobenius method; observe that you get only one solution. (Note, also, that the two values of s are equal or differ by an integer, and in the latter case the larger s gives the one solution.) Show that the conditions of Fuchs's theorem are satisfied. Knowing that the second solution is $\ln x$ times the solution you have, plus another Frobenius series, find the second solution.

5. $x(x+1)y''-(x-1)y'+y=0$

6. $4x^2(x+1)y''-4x^2y'+(3x+1)y=0$

7. $x(x-1)^2y''-2y=0$

8. $xy''+xy'-2y=0$

9. $x^2y''+(x^2-3x)y'+(4-2x)y=0$

10. $x^2(x-1)y''-x(5x-4)y'+(9x-6)y=0$

11. For the differential equation in Problem 2, verify that it does not satisfy the Fuchsian conditions, and that your second solution cannot be expanded in a Frobenius series.

12. Verify that the differential equation $x^4y''+y=0$ is not Fuchsian; that it has the two independent solutions $x\sin(1/x)$ and $x\cos(1/x)$; and that these solutions are not expandable in Frobenius series.

13. Verify that the the differential equation in Problem 11.13 is not Fuchsian. Solve it by separation of variables to find the obvious solution $y=$ const. and a second solution in the form of an integral. Show that the second solution is not expandable in a Frobenius series.

22. HERMITE FUNCTIONS; LAGUERRE FUNCTIONS; LADDER OPERATORS

In this section, we shall outline some of the important formulas for two more sets of named functions. Both Hermite and Laguerre functions are of interest in quantum mechanics where they arise as solutions of eigenvalue problems (see Problem 27, and Chapter 13, Problems 7.20 to 7.22). We shall also consider an operator method which is a useful alternative to series solution for some differential equations.

Hermite Functions The differential equation for Hermite functions is

$$y_n'' - x^2 y_n = -(2n+1)y_n, \qquad n = 0, 1, 2, \cdots. \tag{22.1}$$

This equation can be solved by power series (Problem 5), but here we shall consider an operator method which is particularly efficient for this equation. Let's use the operator D to mean d/dx; then (see Problem 5.31 of Chapter 8)

$$\begin{aligned}&(D-x)(D+x)y = \left(\frac{d}{dx} - x\right)(y' + xy) = y'' - x^2 y + y, \quad \text{and similarly}\\ &(D+x)(D-x)y = y'' - x^2 y - y.\end{aligned} \tag{22.2}$$

Using (22.2), we can write (22.1) in two ways:

$$(D-x)(D+x)y_n = -2ny_n \quad \text{or} \tag{22.3}$$

$$(D+x)(D-x)y_n = -2(n+1)y_n. \tag{22.4}$$

Now let us operate on (22.3) with $(D+x)$ and on (22.4) with $(D-x)$, and change n to m for later convenience:

$$(D+x)(D-x)[(D+x)y_m] = -2m[(D+x)y_m], \tag{22.5}$$

$$(D-x)(D+x)[(D-x)y_m] = -2(m+1)[(D-x)y_m]. \tag{22.6}$$

(The brackets have been inserted to clarify our next step.)

Now compare (22.3) and (22.6); if $y_n = [(D-x)y_m]$ and $n = m+1$, the equations are identical. We write

$$y_{m+1} = (D-x)y_m \tag{22.7}$$

and we see that, given a solution y_m of (22.1) for one value of n, namely $n = m$, we can find a solution when $n = m+1$ by applying the "raising operator" $(D-x)$ to y_m. Similarly, from (22.4) and (22.5), we find that (Problem 1)

$$y_{m-1} = (D+x)y_m. \tag{22.8}$$

We may call $(D+x)$ a "lowering operator"; these operators are called *creation* and *annihilation* operators in quantum theory. Operators of this kind (see Problems 29, 30, and 23.27 for other examples) are called *ladder operators* since, like the rungs of a ladder, they enable us to go up or down in a set of functions.

Now if $n = 0$, we find a solution of (22.3) [and therefore of (22.1)] by requiring

$$(D+x)y_0 = 0. \tag{22.9}$$

We solve this equation (Problem 2) to get

$$y_0 = e^{-x^2/2}. \tag{22.10}$$

Then, by (22.7), $y_n = (D - x)^n e^{-x^2/2}$. These are the Hermite functions; they can be written in the simpler form $y_n = e^{x^2/2}(d^n/dx^n)e^{-x^2}$ (Problem 3):

(22.11) $$y_n = (D - x)^n e^{-x^2/2} \quad \text{or} \quad y_n = e^{x^2/2}(d^n/dx^n)e^{-x^2}.$$ Hermite functions

If we multiply (22.11) by $(-1)^n e^{x^2/2}$, we obtain the *Hermite polynomials*; the following equation may be called a Rodrigues formula for them:

(22.12) $$H_n(x) = (-1)^n e^{x^2} \frac{d^n}{dx^n} e^{-x^2}.$$ Hermite polynomials

We find (Problems 4 and 5):

(22.13) $$H_0(x) = 1, \qquad H_1(x) = 2x, \qquad H_2(x) = 4x^2 - 2.$$

The Hermite polynomials satisfy the differential equation (Problem 6):

(22.14) $$y'' - 2xy' + 2ny = 0.$$ Hermite equation

Using the differential equation, we can prove (Problem 7) that the Hermite polynomials are orthogonal on $(-\infty, \infty)$ with respect to the weight function e^{-x^2}. The normalization integral can be evaluated (Problem 10). Thus we have:

(22.15) $$\int_{-\infty}^{\infty} e^{-x^2} H_n(x) H_m(x)\, dx = \begin{cases} 0, & n \neq m, \\ \sqrt{\pi}\, 2^n n! & n = m. \end{cases}$$

The generating function for the Hermite polynomials is (Problem 8):

$$\Phi(x,h) = e^{2xh-h^2} = \sum_{n=0}^{\infty} H_n(x)\frac{h^n}{n!}. \tag{22.16}$$

The generating function can be used to derive recursion relations for the Hermite polynomials. Two useful relations are (Problem 9):

$$\begin{aligned} &\text{(a)} \quad H_n'(x) = 2nH_{n-1}(x),\\ &\text{(b)} \quad H_{n+1}(x) = 2xH_n(x) - 2nH_{n-1}(x). \end{aligned} \tag{22.17}$$

Laguerre functions The *Laguerre polynomials* may be defined by a Rodrigues formula:

$$L_n(x) = \frac{1}{n!}e^x \frac{d^n}{dx^n}(x^n e^{-x}). \tag{22.18}$$

Carrying out the differentiation (Problem 12), we find:

$$\begin{aligned} L_n(x) &= 1 - nx + \frac{n(n-1)}{2!}\frac{x^2}{2!} - \frac{n(n-1)(n-2)}{3!}\frac{x^3}{3!} + \cdots + \frac{(-1)^n x^n}{n!} \\ &= \sum_{m=0}^{n}(-1)^m \binom{n}{m}\frac{x^m}{m!}. \qquad \text{Laguerre polynomials} \end{aligned} \tag{22.19}$$

The symbol $\binom{n}{m}$ is a binomial coefficient (see Chapter 1, Section 13C). Some authors omit the $1/n!$ in (22.18); then the series in (22.19) is multiplied by $n!$. It is convenient to note that the series in (22.19) is like the binomial expansion of $(1-x)^n$ except that each power of x, say x^m, is divided by an extra $m!$. We find (Problem 13):

$$L_0(x) = 1, \qquad L_1(x) = 1 - x, \qquad L_2(x) = 1 - 2x + x^2/2. \tag{22.20}$$

The Laguerre polynomials are solutions of the differential equation (Problems 14 and 15):

$$xy'' + (1-x)y' + ny = 0, \qquad y = L_n(x). \tag{22.21}$$

Using the differential equation, we can prove (Problem 16) that the Laguerre polynomials are orthogonal on $(0, \infty)$ with respect to the weight function e^{-x}. In fact, we find (Problem 19) that, with the definition (22.18), the functions $e^{-x/2}L_n(x)$ are an orthonormal set on $(0, \infty)$.

$$\int_0^\infty e^{-x} L_n(x) L_k(x)\, dx = \delta_{nk} = \begin{cases} 0, & n \neq k, \\ 1, & n = k. \end{cases} \tag{22.22}$$

The generating function for the Laguerre polynomials is (Problem 17):

$$\Phi(x,h) = \frac{e^{-xh/(1-h)}}{1-h} = \sum_{n=0}^{\infty} L_n(x) h^n. \tag{22.23}$$

Using it, we can derive recursion relations; some examples are (Problem 18):

$$\begin{aligned} &\text{(a)}\quad L'_{n+1}(x) - L'_n(x) + L_n(x) = 0, \\ &\text{(b)}\quad (n+1)L_{n+1}(x) - (2n+1-x)L_n(x) + nL_{n-1}(x) = 0, \\ &\text{(c)}\quad xL'_n(x) - nL_n(x) + nL_{n-1}(x) = 0. \end{aligned} \tag{22.24}$$

Warning: These formulas will be different if the factor $1/n!$ is omitted in the definition (22.18), so check the notation of any reference you are using (computer, text, tables).

Derivatives of the Laguerre polynomials are called associated Laguerre polynomials; they may be found by differentiating (22.18), (22.19), or (22.20) (Problem 20). We define:

$$L_n^k(x) = (-1)^k \frac{d^k}{dx^k} L_{n+k}(x). \qquad \text{Associated Laguerre polynomials} \tag{22.25}$$

Warning: The notation in various references may be confusing; some authors define $L_n^k(x)$ as $(d^k/dx^k)L_n(x)$ [compare our definition in (22.25)], so read carefully the definition in the reference you are using. For example, associated Laguerre polynomials are used in the theory of the hydrogen atom in quantum mechanics. In various references you will find them denoted by $L_{n-l-1}^{2l+1}(x)$ and by $L_{n+l}^{2l+1}(x)$; both these notations mean (except for sign) $(d^{2l+1}/dx^{2l+1})L_{n+l}(x)$. (See Problems 26 to 28.)

By differentiating the Laguerre equation (22.21), we find the differential equation satisfied by the polynomials $L_n^k(x)$ (Problem 21):

$$xy'' + (k+1-x)y' + ny = 0, \qquad y = L_n^k(x). \tag{22.26}$$

The polynomials $L_n^k(x)$ may also be found from the Rodrigues formula (Problem 22):

$$L_n^k(x) = \frac{x^{-k}e^x}{n!}\frac{d^n}{dx^n}(x^{n+k}e^{-x}). \tag{22.27}$$

Note that in this form k does not have to be an integer; in fact, (22.27) is used to define $L_n^k(x)$ for any $k > -1$.

Recursion relations for the polynomials $L_n^k(x)$ may be found by differentiating recursion relations for the Laguerre polynomials. Some examples are (Problem 23):

(22.28)

$$\text{(a)}\quad (n+1)L_{n+1}^k(x) - (2n+k+1-x)L_n^k(x) + (n+k)L_{n-1}^k(x) = 0,$$

$$\text{(b)}\quad x\frac{d}{dx}L_n^k(x) - nL_n^k(x) + (n+k)L_{n-1}^k(x) = 0.$$

Using the differential equation (22.26), we can show (Problem 24) that the functions $L_n^k(x)$ are orthogonal on $(0,\infty)$ with respect to the weight function $x^k e^{-x}$. We find (Problem 25):

$$\int_0^\infty x^k e^{-x} L_n^k(x) L_m^k(x)\,dx = \begin{cases} 0, & m \neq n, \\ \frac{(n+k)!}{n!}, & m = n. \end{cases} \tag{22.29}$$

The normalization integral needed in the theory of the hydrogen atom is not (22.29), but instead has the factor x^{k+1}. We find (see Problems 25 to 27):

$$\int_0^\infty x^{k+1} e^{-x} [L_n^k(x)]^2\,dx = (2n+k+1)\frac{(n+k)!}{n!}. \tag{22.30}$$

Again warning: The formulas (22.28), (22.29), and (22.30) will be different in references which omit the $1/n!$ in (22.18) and/or use a different definition of $L_n^k(x)$ in (22.25).

PROBLEMS, SECTION 22

1. Verify equations (22.2), (22.3), (22.4), and (22.8).

2. Solve (22.9) to get (22.10). If needed, see Chapter 8, Section 2.

3. Show that $e^{x^2/2}D[e^{-x^2/2}f(x)] = (D-x)f(x)$. Now set

$$f(x) = (D-x)g(x) = e^{x^2/2}D[e^{-x^2/2}g(x)]$$

to get

$$(D-x)^2 g(x) = e^{x^2/2}D^2[e^{-x^2/2}g(x)].$$

Continue this process to show that

$$(D-x)^n F(x) = e^{x^2/2}D^n[e^{-x^2/2}F(x)]$$

for any $F(x)$. Then let $F(x) = e^{-x^2/2}$ to get (22.11).

4. Using (22.12) find the Hermite polynomials given in (22.13). Then use (22.17b) to find $H_3(x)$ and $H_4(x)$.

5. By power series, solve the Hermite differential equation

$$y'' - 2xy' + 2py = 0$$

You should find an a_0 series and an a_1 series as for the Legendre equation in Section 2. Show that the a_0 series terminates when p is an even integer, and the a_1 series terminates when p is an odd integer. Thus for each integer n, the differential equation (22.14) has one polynomial solution of degree n. These polynomials with a_0 or a_1 chosen so that the highest order term is $(2x)^n$ are the Hermite polynomials. Find $H_0(x)$, $H_1(x)$, and $H_2(x)$. Observe that you have solved an eigenvalue problem (see end of Section 2), namely to find values of p for which the given differential equation has polynomial solutions, and then to find the corresponding solutions (eigenfunctions).

6. Substitute $y_n = e^{-x^2/2}H_n(x)$ into (22.1) to show that the differential equation satisfied by $H_n(x)$ is (22.14).

7. Prove that the functions $H_n(x)$ are orthogonal on $(-\infty, \infty)$ with respect to the weight function e^{-x^2}. *Hint:* Write the differential equation (22.14) as

$$e^{x^2}\frac{d}{dx}(e^{-x^2}y') + 2ny = 0,$$

and see Sections 7 and 19.

8. In the generating function (22.16), expand the exponential in a power series and collect powers of h to obtain the first few Hermite polynomials. Verify the identity

$$\frac{\partial^2\Phi}{\partial x^2} - 2x\frac{\partial\Phi}{\partial x} + 2h\frac{\partial\Phi}{\partial h} = 0.$$

Substitute the series in (22.16) into this identity to prove that the functions $H_n(x)$ in (22.16) satisfy equation (22.14). Verify that the highest term in $H_n(x)$ in (22.16) is $(2x)^n$. [You have then proved that the functions called $H_n(x)$ are really the Hermite polynomials since, by Problem 5, (22.14) has just one polynomial solution of degree n.]

9. Use the generating function to prove the recursion relations in (22.17). *Hint for* (a): Differentiate (22.16) with respect to x and equate coefficients of h^n. *Hint for* (b): Differentiate (22.16) with respect to h and equate coefficients of h^n.

10. Evaluate the normalization integral in (22.15). *Hint:* Use (22.12) for one of the $H_n(x)$ factors, integrate by parts, and use (22.17a); then use your result repeatedly.

11. Show that we have solved the following eigenvalue problem (see Problem 5 and end of Section 2): Given the differential equation $y'' + (\epsilon - x^2)y = 0$ [compare equation (22.1)]. find the possible values of ϵ (eigenvalues) such that the solutions $y(x)$ of the given differential equation tend to zero as $x \to \pm\infty$; for these values of ϵ, find the eigenfunctions $y(x)$. What is ϵ, and what are the eigenfunctions?

12. Using Leibniz' rule (Section 3), carry out the differentiation in (22.18) to obtain (22.19).

13. Using (22.19), verify (22.20) and also find $L_3(x)$ and $L_4(x)$.

14. Show that $y = L_n(x)$ given in (22.18) satisfies (22.21). *Hint:* Follow a method similar to that used in Section 4. Let $v = x^n e^{-x}$ and show that $xv' = (n-x)v$. Differentiate this last equation $(n+1)$ times by Leibniz' rule, and use $d^n v/dx^n = n!\,e^{-x}L_n(x)$ from (22.18).

15. Solve the Laguerre differential equation

$$xy'' + (1-x)y' + py = 0$$

by power series. Show that the a_0 series terminates if p is an integer. Thus for each integer n, the differential equation (22.21) has one solution which is a polynomial of degree n. These polynomials with $a_0 = 1$ are the Laguerre polynomials $L_n(x)$. Find $L_0(x)$, $L_1(x)$, $L_2(x)$, and $L_3(x)$. (This is an eigenvalue problem—compare Problem 5 and Section 2.)

16. Prove that the functions $L_n(x)$ are orthogonal on $(0, \infty)$ with respect to the weight function e^{-x}. *Hint:* Write the differential equation (22.21) as

$$e^x \frac{d}{dx}(xe^{-x}y') + ny = 0,$$

and see Sections 7 and 19.

17. In (22.23), write the series for the exponential and collect powers of h to verify the first few terms of the series. Verify the identity

$$x\frac{\partial^2\Phi}{\partial x^2} + (1-x)\frac{\partial\Phi}{\partial x} + h\frac{\partial\Phi}{\partial h} = 0.$$

Substitute the series in (22.23) into this identity to show that the functions $L_n(x)$ in (22.23) satisfy Laguerre's equation (22.21). Verify that the constant term is 1 by putting $x = 0$ in the generating function. [You have then proved that the functions called $L_n(x)$ in (22.23) are really Laguerre polynomials since, by Problem 15, (22.21) has just one polynomial solution of degree n.]

18. Verify the recursion relations (22.24) as follows:

(a) Differentiate (22.23) with respect to x to get $h\Phi = (h-1)(\partial\Phi/\partial x)$; equate coefficients of h^{n+1}.

(b) Differentiate (22.23) with respect to h to get $(1-h)^2(\partial\Phi/\partial h) = (1-h-x)\Phi$; equate coefficients of h^n.

(c) Combine (a) and (b) to get $x(\partial\Phi/\partial x) + h\Phi - h(1-h)\partial\Phi/\partial h = 0$. Substitute the series for Φ and equate coefficients of h^n.

19. Evaluate the normalization integral in (22.22). *Hint:* Use (22.18) for one of the $L_n(x)$ factors; integrate by parts n times. Use (22.19) to find $(d^n/dx^n)L_n(x)$ and Chapter 11, Section 3, to evaluate $\int_0^\infty x^n e^{-x}\,dx$.

20. Using (22.25), (22.20), and Problem 13, find $L_n^k(x)$ for $n = 0, 1, 2$, and $k = 1, 2$.

21. Verify that the polynomials $L_n^k(x)$ in (22.25) satisfy (22.26). *Hint:* Write (22.21) with n replaced by $n + k$ and differentiate k times by Leibniz' rule.

22. Verify that the polynomials given by (22.27) are the same as the $L_n^k(x)$ defined in (22.25). *Hints:* Show that the functions in (22.27) satisfy (22.26) as follows. Let $v = e^{-x}x^{n+k}$ and show that $xv' = (n+k-x)v$. (Compare Problem 14.) Differentiate this equation $n + 1$ times by Leibniz' rule, and use $d^n v/dx^n = n!\,e^{-x}x^k L_n^k(x)$ from (22.27). Also show that the coefficient of x^n in both (22.25) and (22.27) is $(-1)^n/n!$ [Thus, assuming that (22.26) for one k has only one polynomial solution of degree n (which can be shown by series solution), (22.27) gives the same polynomials as (22.25) for integral k.]

23. Verify the recursion relation relations (22.28) as follows:

(a) In (22.24b), replace n by $n+k$ and differentiate k times by Leibniz' rule; in (22.24a), replace k by $n+k$ and differentiate $k-1$ times. Subtract k times the second result from the first.

(b) In (22.24c), replace n by $n+k$ and differentiate k times.

24. Show that the functions $L_n^k(x)$ are orthogonal on $(0,\infty)$ with respect to the weight function $x^k e^{-x}$. *Hint:* Write the differential equation (22.26) as

$$x^{-k}e^{x}\frac{d}{dx}(x^{k+1}e^{-x}y')+ny=0$$

and see Sections 7 and 19.

25. Evaluate the normalization integrals (22.29) and (22.30). *Hints:* Use (22.27) for one of the $L_n^k(x)$ factors in (22.29); integrate by parts n times. Use (22.25) and then (22.19) to evaluate $d^n/dx^n L_n^k(x)$. Compare Problem 19. To evaluate (22.30), multiply (22.28a) by $x^k e^{-x}$ and integrate; use (22.29) both for $m\neq n$ and $m=n$.

26. Solve the following eigenvalue problem (see end of Section 2 and Problem 11): Given the differential equation

$$y''+\left(\frac{\lambda}{x}-\frac{1}{4}-\frac{l(l+1)}{x^2}\right)y=0$$

where l is an integer ≥ 0, find values of λ such that $y\to 0$ as $x\to\infty$, and find the corresponding eigenfunctions. *Hint:* let $y=x^{l+1}e^{-x/2}v(x)$, and show that $v(x)$ satisfies the differential equation

$$xv''+(2l+2-x)v'+(\lambda-l-1)v=0.$$

Compare (22.26) to show that if λ is an integer $>l$, there is a polynomial solution $v(x)=L_{\lambda-l-1}^{2l+1}(x)$.

27. The functions which are of interest in the theory of the hydrogen atom are

$$f_n(x)=x^{l+1}e^{-x/2n}L_{n-l-1}^{2l+1}\left(\frac{x}{n}\right)$$

where n and l are integers with $0\leq l\leq n-1$. (Note that here $k=2l+1$, and we have replaced n by $n-l-1$; in this problem L_2^3 , say, means $l=1, n=4$.) For $l=1$, show that

$$f_2(x)=x^2e^{-x/4},\quad f_3(x)=x^2e^{-x/6}\left(4-\frac{x}{3}\right),\quad f_4(x)=x^2e^{-x/8}\left(10-\frac{5x}{4}+\frac{x^2}{32}\right).$$

Hint: Find the polynomials L_0^3, L_1^3, L_2^3 as in Problem 20 (with $k=3$) and then replace x by x/n. The functions $f_n(x)$ are very different from those in (22.29) since x/n changes from one function to the next. However, it can be shown (Problem 23.25) that for one fixed l, the set of functions $f_n(x), n\geq l+1$, is an orthogonal set on $(0,\infty)$. Verify this for these three functions. *Hint:* The integrals are Γ functions—see Chapter 11, Section 3.

28. Repeat Problem 27 for $l=0$, $n=1,2,3$.

29. Show that $R_p=\frac{p}{x}-D$ and $L_p=\frac{p}{x}+D$ where $D=d/dx$, are raising and lowering operators for Bessel functions, that is, show that $R_pJ_p(x)=J_{p+1}(x)$ and $L_pJ_p(x)=J_{p-1}(x)$. *Hint:* Use equations (15.5). Note that these operators depend on p as well as x, so they are not as simple as the Hermite function raising and lowering operators (22.7) and (22.8). If you want to operate, say, on J_{p+1}, you must change p in R or L to $p+1$, etc. Making this adjustment, show that the equations $LRJ_p=J_p$ and $RLJ_p=J_p$ both give Bessel's equation.

30. Find raising and lowering operators (see Problem 29) for spherical Bessel functions. *Hint:* See problems 17.15 and 17.16.

23. MISCELLANEOUS PROBLEMS

1. Use the generating function (5.1) to find the normalizing factor for Legendre polynomials. *Hint:* Square equation (5.2) with Φ as in (5.1) and integrate from -1 to 1. Expand the integral of Φ^2 (after integrating) in powers of h and equate coefficients.

2. Use the generating function to show that

$$P_{2n+1}(0) = 0 \qquad \text{and} \quad P_{2n}(0) = \binom{-1/2}{n} = \frac{(-1)^n(2n-1)!!}{2^n n!};$$

Hints: Expand (5.1) for $x = 0$ in powers of h and equate coefficients of powers of h in (5.2). See Chapter 1, Section 13C.

3. Use (5.8e) to show that $\int_0^1 P_l(x)\,dx = [P_{l-1}(0) - P_{l+1}(0)]/(2l+1)$. Then use the result of Problem 2 and Chapter 1, Section 13C to show that

$$\int_0^1 P_{2n}(x)\,dx = 0, n > 0, \qquad \text{and} \qquad \int_0^1 P_{2n+1}(x)\,dx = \frac{(-1)^n(2n-1)!!}{2^{n+1}(n+1)!} = \binom{1/2}{n+1}.$$

4. Obtain the binomial coefficient result in Problem 3 directly by integrating the generating function from 0 to 1 and expanding the result in powers of h. Equate the coefficients of h^l in the identity obtained by integrating (5.2) from 0 to 1, and use Chapter 1, Section 13C.

5. Show that $\sum_0^n (2l+1)P_l(x) = P'_n(x) + P'_{n+1}(x)$. *Hint:* Use mathematical induction as follows:

(a) Verify the formula for $n = 0$.

(b) Assuming that the formula is true for $l = n - 1$, show [using (5.8e)] that it is true for $l = n$.

6. Using (10.6), (5.8), and Problem 2, evaluate $P^1_{2n+1}(0)$.

7. Show that, for $l > 0$, $\int_a^b P_l(x)\,dx = 0$ if a and b are any two maximum or minimum points of $P_l(x)$, or ± 1. *Hint:* Integrate (7.2).

8. Show that $(2l+1)(x^2-1)P'_l(x) = l(l+1)[P_{l+1}(x) - P_{l-1}(x)]$. *Hint:* Integrate (5.8e) and (7.2) and combine the results. Thus show that $P_{l+1}(x) = P_{l-1}(x)$ at maximum and minimum points of $P_l(x)$ and at ± 1.

9. Evaluate $\int_{-1}^1 xP_l(x)P_n(x)\,dx$, $n \le l$. *Hint:* Write (5.8a) with l replaced by $l+1$, multiply by $P_n(x)$ and integrate.

Use the recursion relations of Section 15 (and, as needed, Sections 12, 13, 17, and 20) to verify the formulas in Problems 10 to 14.

10. $$\int_0^\infty x^{-p} J_{p+1}(x)\,dx = \frac{1}{2^p\Gamma(1+p)}.$$

11. $$\int_0^\infty x^{-n} j_{n+1}(x)\,dx = \frac{1}{(2n+1)!!}.$$

12. $$\frac{d}{dx}K_p(x) = -\tfrac{1}{2}[K_{p-1}(x) + K_{p+1}(x)].$$

13. $$\frac{d}{dx}j_n(x) = [nj_{n-1}(x) - (n+1)j_{n+1}(x)]/(2n+1).$$

14. $$\int x^3 J_0(x)\,dx = x^3 J_1(x) - 2x^2 J_2(x).$$

15. Use the result of Problem 18.4 and equations (17.4) to show that

$$j_n(x)y_n'(x) - y_n(x)j_n'(x) = \frac{1}{x^2}.$$

Then use Problem 17.14 (for y's as well as j's) to show that

$$j_n(x)y_{n-1}(x) - y_n(x)j_{n-1}(x) = \frac{1}{x^2}.$$

16. Use (15.2) repeatedly to show that

$$J_1(x) = x\left(-\frac{1}{x}\frac{d}{dx}\right)J_0(x), \qquad J_2(x) = x^2\left(-\frac{1}{x}\frac{d}{dx}\right)^2 J_0(x),$$

and, in general,

$$J_n(x) = x^n\left(-\frac{1}{x}\frac{d}{dx}\right)^n J_0(x).$$

17. Let α be the first positive zero of $J_1(x)$ and let β_n be the zeros of $J_0(x)$. In terms of α and β_n, find the values of x at the maximum and minimum points of the function $y = xJ_1(\alpha x)$. By computer or tables, find the needed zeros and compute the coordinates of the maximum and minimum points on the graph of $y(x)$ for x between 0 and 5. Computer plot y from $x = 0$ to 5 and compare your computed maximum and minimum points with what the plot shows.

18. (a) Make the change of variables $z = e^x$ in the differential equation $y'' + e^{2x}y = 0$, and so find a solution of the differential equation in terms of Bessel functions.

(b) Make the change of variables $z = e^{x^2/2}$ in the differential equation $xy'' - y' + x^3(e^{x^2} - p^2)y = 0$, and solve the equation in terms of Bessel functions.

19. (a) The generating function for Bessel functions of integral order $p = n$ is

$$\Phi(x,h) = e^{(1/2)x(h-h^{-1})} = \sum_{n=-\infty}^{\infty} h^n J_n(x).$$

By expanding the exponential in powers of $x(h - h^{-1})$ show that the $n = 0$ term is $J_0(x)$ as claimed.

(b) Show that

$$x^2\frac{\partial^2\Phi}{\partial x^2} + x\frac{\partial\Phi}{\partial x} + x^2\Phi - \left(h\frac{\partial}{\partial h}\right)^2\Phi = 0.$$

Use this result and $\Phi(x,h) = \sum_{n=-\infty}^{\infty} h^n J_n(x)$ to show that the functions $J_n(x)$ satisfy Bessel's equation. By considering the terms in h^n in the expansion of $e^{(1/2)x(h-h^{-1})}$ in part (a), show that the coefficient of h^n is a series starting with the term $(1/n!)(x/2)^n$. (You have then proved that the functions called $J_n(x)$ in the expansion of $\Phi(x,h)$ are indeed the Bessel functions of integral order previously defined by (12.9) and (13.1) with $p = n$.)

20. In the generating function equation of Problem 19, put $h = e^{i\theta}$ and separate real and imaginary parts to derive the equations

$$\cos(x\sin\theta) = J_0(x) + 2J_2(x)\cos 2\theta + 2J_4(x)\cos 4\theta + \cdots$$
$$= J_0(x) + 2\sum_{n=1}^{\infty} J_{2n}(x)\cos 2n\theta,$$
$$\sin(x\sin\theta) = 2[J_1(x)\sin\theta + J_3(x)\sin 3\theta + \cdots]$$
$$= 2\sum_{n=0}^{\infty} J_{2n+1}(x)\sin(2n+1)\theta.$$

These are Fourier series with Bessel functions as coefficients. (In fact the J_n's for integral n are often called Bessel coefficients because they occur in many series like these.) Use the formulas for the coefficients in a Fourier series to find integrals representing J_n for even n and for odd n. Show that these results can be combined to give

$$J_n(x) = \frac{1}{\pi}\int_0^{\pi} \cos(n\theta - x\sin\theta)\,d\theta$$

for *all* integral n. These series and integrals are of interest in astronomy and in the theory of frequency modulated waves.

21. In the generating function equation, Problem 19, put $x = iy$ and $h = -ik$ and show that

$$e^{(1/2)y(k+k^{-1})} = \sum_{n=-\infty}^{\infty} k^n I_n(y).$$

22. In the $\cos(x\sin\theta)$ series of Problem 20, let $\theta = 0$, and then let $\theta = \pi/2$, and add the results to show that (recall Problem 13.2)

$$\sum_{n=-\infty}^{\infty} J_{4n}(x) = \tfrac{1}{2}(1 + \cos x).$$

23. Solve by power series $(1 - x^2)y'' - xy' + n^2 y = 0$. The polynomial solutions of this equation with coefficients determined to make $y(1) = 1$ are called Chebyshev polynomials $T_n(x)$. Find T_0, T_1, and T_2.

24. (a) The following differential equation is often called a Sturm-Liouville equation:

$$\frac{d}{dx}[A(x)y'] + [\lambda B(x) + C(x)]y = 0$$

(λ is a constant parameter). This equation includes many of the differential equations of mathematical physics as special cases. Show that the following equations can be written in the Sturm-Liouville form: the Legendre equation (7.2); Bessel's equation (19.2) for a *fixed* p, that is, with the parameter λ corresponding to α^2; the simple harmonic motion equation $y'' = -n^2 y$; the Hermite equation (22.14); the Laguerre equations (22.21) and (22.26).

(b) By following the methods of the orthogonality proofs in Sections 7 and 19, show that if y_1 and y_2 are two solutions of the Sturm-Liouville equation (corresponding to the two values λ_1 and λ_2 of the parameter λ), then y_1 and y_2 are orthogonal on (a, b) with respect to the weight function $B(x)$ if $A(x)(y_1'y_2 - y_2'y_1)|_a^b = 0$.

25. In Problem 22.26, replace x by x/n in the y differential equation and set $\lambda = n$ to show that the differential equation satisfied by the functions $f_n(x)$ in Problem 22.27 is

$$y'' + \left(\frac{1}{x} - \frac{1}{4n^2} - \frac{l(l+1)}{x^2}\right)y = 0.$$

Hence show by Problem 24 that the functions $f_n(x)$ are orthogonal on $(0, \infty)$.

26. Verify *Bauer's formula* $e^{ixw} = \sum_0^\infty (2l+1)i^l j_l(x)P_l(w)$ as follows. Write the integral for the coefficients c_l in the Legendre series for $e^{ixw} = \sum c_l P_l(w)$. You want to show that $c_l(x) = (2l+1)i^l j_l(x)$. First show that $y = c_l(x)$ satisfies the differential equation (Problem 17.6) for spherical Bessel functions. *Hints:* Differentiate with respect to x under the integral sign to find y' and y''; substitute into the left side of the differential equation. Now integrate by parts with respect to w to show that the integrand is zero because $P_l(w)$ satisfies Legendre's equation. Thus $c_l(x)$ must be a linear combination of $j_l(x)$ and $n_l(x)$. Now consider the $c_l(x)$ integral for small x; expand e^{iwx} in series and evaluate the lowest term (which is x^l since $\int_{-1}^1 w^n P_l(w)\,dw = 0$ for $n < l$). Compare with the approximate formulas for $j_1(x)$ and $n_l(x)$ in Section 20.

27. Show that $R = lx - (1-x^2)D$ and $L = lx + (1-x^2)D$, where $D = d/dx$, are raising and lowering operators for Legendre polynomials [compare Hermite functions, (22.1) to (22.11) and Bessel functions, Problems 22.29 and 22.30]. More precisely, show that $RP_{l-1}(x) = lP_l(x)$ and $LP_l(x) = lP_{l-1}(x)$. *Hint:* Use equations (5.8d) and (5.8f). Note that, unlike the raising and lowering operators for Hermite functions, here R and L depend on l as well as x, so you must be careful about indices. The L operator operates on P_l, but the R operator as given operates on P_{l-1} to produce lP_l. [If you prefer, you could replace l by $l+1$ to rewrite R as $(l+1)x - (1-x^2)D$; then it operates on P_l to produce $(l+1)P_{l+1}$.] Assuming that all $P_l(1) = 1$, solve $LP_0(x) = 0$ to find $P_0(x) = 1$, and then use raising operators to find $P_1(x)$ and $P_2(x)$.

28. Show that the functions $J_0(t)$ and $J_0(\pi - t)$ are orthogonal on $(0, \pi)$. *Hints:* See the Laplace transform table (page 469), L23 and L24 with $g = h = J_0$. What is the inverse transform of $(p^2 + a^2)^{-1}$?

29. Show that the Fourier cosine transform (Chapter 7, Section 12) of $J_0(x)$ is

$$\begin{cases} \sqrt{\dfrac{2}{\pi}}\dfrac{1}{\sqrt{1-\alpha^2}}, & 0 \le \alpha < 1, \\ 0, & \alpha > 1. \end{cases}$$

Hence show that $\int_0^\infty J_0(x)\,dx = 1$. *Hints:* Show that the integral in Problem 20 gives $J_0(x) = (2/\pi)\int_0^{\pi/2} \cos(x \sin\theta)\,d\theta$. (Replace θ by $\pi - \theta$ in the $\pi/2$ to π integral.) Let $\sin\theta = \alpha$ to find J_0 as a cosine transform; write the inverse transform. Now let $\alpha = 0$.

30. Use the results of Chapter 7, Problems 12.18 and 13.19 to evaluate $\int_0^\infty [j_1(\alpha)]^2\,d\alpha$.

CHAPTER

13

Partial Differential Equations

1. INTRODUCTION

Many of the problems of mathematical physics involve the solution of partial differential equations. The same partial differential equation may apply to a variety of physical problems; thus the mathematical methods which you will learn in this chapter apply to many more problems than those we shall discuss in the illustrative examples. Let us outline the partial differential equations we shall consider, and the kinds of physical problems which lead to each of them.

(1.1) **Laplace's equation** $\nabla^2 u = 0$

The function u may be the gravitational potential in a region containing no mass, the electrostatic potential in a charge-free region, the steady-state temperature (that is, temperature not changing with time) in a region containing no sources of heat, or the velocity potential for an incompressible fluid with no vortices and no sources or sinks.

(1.2) **Poisson's equation** $\nabla^2 u = f(x, y, z)$

The function u may represent the same physical quantities listed for Laplace's equation, but in a region containing mass, electric charge, or sources of heat or fluid, respectively, for the various cases. The function $f(x, y, z)$ is called the source density; for example, in electricity it is proportional to the density of the electric charge.

(1.3) **The diffusion or heat flow equation** $\nabla^2 u = \dfrac{1}{\alpha^2}\dfrac{\partial u}{\partial t}$

Here u may be the non-steady-state temperature (that is, temperature varying with time) in a region with no heat sources; or it may be the concentration of a diffusing substance (for example, a chemical, or particles such as neutrons). The quantity α^2 is a constant known as the diffusivity.

(1.4) **Wave equation** $\nabla^2 u = \dfrac{1}{v^2}\dfrac{\partial^2 u}{\partial t^2}$

Here u may represent the displacement from equilibrium of a vibrating string or membrane or (in acoustics) of the vibrating medium (gas, liquid, or solid); in electricity u may be the current or potential along a transmission line; or u may be a component of $\mathbf{E}$ or $\mathbf{B}$ in an electromagnetic wave (light, radio waves, etc.). The quantity v is the speed of propagation of the waves; for example, for light in a vacuum it is c, the speed of light, and for sound waves it is the speed at which sound travels in the medium under consideration. The operator $\nabla^2 - \frac{1}{c^2}\frac{\partial^2}{\partial t^2}$ is called the d'Alembertian.

(1.5) **Helmholtz equation** $\nabla^2 F + k^2 F = 0$

As you will see later, the function F here represents the space part (that is, the time-independent part) of the solution of either the diffusion or the wave equation.

(1.6) **Schrödinger equation** $-\dfrac{\hbar^2}{2m}\nabla^2\Psi + V\Psi = i\hbar\dfrac{\partial}{\partial t}\Psi$

This is the wave equation of quantum mechanics. In this equation, $\hbar$ is Planck's constant divided by 2π, m is the mass of a particle, $i = \sqrt{-1}$, and V is the potential energy of the particle. The wave function Ψ is complex, and its absolute square is proportional to the position probability of the particle.

We shall be principally concerned with the solution of these equations rather than their derivation. If you like, you could say that it is true experimentally that the physical quantities mentioned above satisfy the given equations. However, it is also true that the equations can be derived from somewhat simpler experimental assumptions. Let us indicate briefly an example of how this can be done. In Chapter 6, Sections 10 and 11, we considered the flow of a fluid. We showed (Chapter 6, Problem 10.15) that $\nabla \cdot \mathbf{v} = 0$ for an incompressible fluid in a region containing no sources or sinks. If it is also true that there are no vortices (that is, the flow is irrotational), then $\operatorname{curl}\mathbf{v} = 0$, and $\mathbf{v}$ can be written as the gradient of a scalar function: $\mathbf{v} = \nabla u$. Combining these two equations, we have $\nabla \cdot \nabla u = \nabla^2 u = 0$. The function u is called the velocity potential and we see that (under the given conditions) it satisfies Laplace's equation as we claimed. A few more examples of such derivations are outlined in the problems.

In the following sections, we shall consider a number of physical problems to illustrate the very useful method of solving partial differential equations known as *separation of variables* (no relation to the same term used in ordinary differential equations, Chapter 8). In Sections 2 to 4, we consider problems in rectangular coordinates leading to Fourier series solutions—problems similar to those solved by Fourier. In later sections, we consider use of other coordinate systems (cylindrical, spherical) leading to solutions using Legendre or Bessel series.

PROBLEMS, SECTION 1

1. Assume from electrostatics the equations $\nabla \cdot \mathbf{E} = \rho/\epsilon_0$ and $\mathbf{E} = -\nabla\phi$ ($\mathbf{E}$ = electric field, ρ = charge density, ϵ_0 = constant, ϕ = electrostatic potential). Show that the electrostatic potential satisfies Laplace's equation (1.1) in a charge-free region and satisfies Poisson's equation (1.2) in a region of charge density ρ.

2. (a) Show that the expression $u = \sin(x - vt)$ describing a sinusoidal wave (see Chapter 7, Figure 2.3), satisfies the wave equation (1.4). Show that, in general,

$u = f(x - vt)$ and $u = f(x + vt)$ satisfy the wave equation, where f is any function with a second derivative. This is the d'Alembert solution of the wave equation. (See Chapter 4, Section 11, Example 1.) The function $f(x - vt)$ represents a wave moving in the positive x direction and $f(x + vt)$ represents a wave moving in the opposite direction.

(b) Show that $u(r,t) = (1/r)f(r - vt)$ and $u(r,t) = (1/r)f(r + vt)$ satisfy the wave equation in spherical coordinates. [Use the first term of (7.1) for $\nabla^2 u$ since here u is independent of θ and ϕ.] These functions represent spherical waves spreading out from the origin or converging on the origin.

3. Assume from electrodynamics the following equations which are valid in free space. (They are called Maxwell's equations.)

$$\nabla \cdot \mathbf{E} = 0 \qquad \nabla \cdot \mathbf{B} = 0$$

$$\nabla \times \mathbf{E} = -\frac{\partial \mathbf{B}}{\partial t} \qquad \nabla \times \mathbf{B} = \frac{1}{c^2}\frac{\partial \mathbf{E}}{\partial t}$$

where $\mathbf{E}$ and $\mathbf{B}$ are the electric and magnetic fields, and c is the speed of light in a vacuum. From them show that any component of $\mathbf{E}$ or $\mathbf{B}$ satisfies the wave equation (1.4) with $v = c$.

4. Obtain the heat flow equation (1.3) as follows: The quantity of heat Q flowing across a surface is proportional to the normal component of the (negative) temperature gradient, $(-\nabla T)\cdot\mathbf{n}$. Compare Chapter 6, equation (10.4), and apply the discussion of flow of water given there to the flow of heat. Thus show that the rate of gain of heat per unit volume per unit time is proportional to $\nabla \cdot \nabla T$. But $\partial T/\partial t$ is proportional to this gain in heat; thus show that T satisfies (1.3).

2. LAPLACE'S EQUATION; STEADY-STATE TEMPERATURE IN A RECTANGULAR PLATE

We want to solve the following problem: A long rectangular metal plate has its two long sides and the far end at 0° and the base at 100° (Figure 2.1). The width of the plate is 10 cm. Find the steady-state temperature distribution inside the plate. (This problem is mathematically identical to the problem of finding the electrostatic potential in the region $0 < x < 10, y > 0$, if the given temperatures are replaced by potentials—see, for example, Jackson, 3rd edition, p.73)

To simplify the problem, we shall assume at first that the plate is so long compared to its width that we may make the mathematical approximation that it extends to infinity in the y direction. It is then called a semi-infinite plate. This is a good approximation if we are interested in temperatures not too near the far end.

The temperature T satisfies Laplace's equation inside the plate where there are no sources of heat, that is,

$$\nabla^2 T = 0 \quad \text{or} \quad \frac{\partial^2 T}{\partial x^2} + \frac{\partial^2 T}{\partial y^2} = 0 \tag{2.1}$$

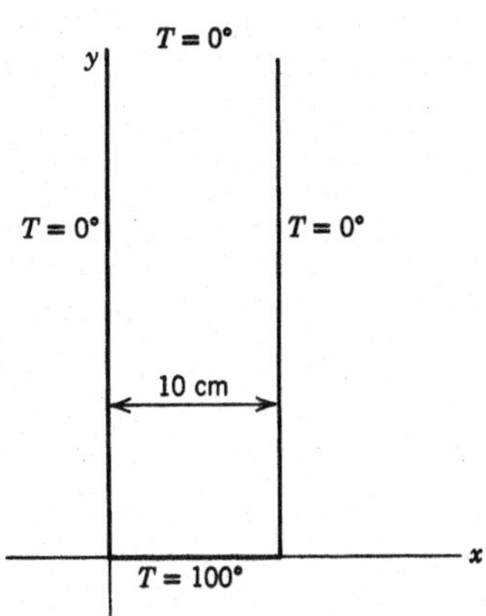

Figure 2.1

We have written ∇^2 in rectangular coordinates because the boundary of the plate is rectangular and we have omitted the z term because the plate is in two dimensions. To solve this equation, we are going to *try* a solution of the form

$$T(x, y) = X(x)Y(y), \tag{2.2}$$

where, as indicated, X is a function only of x, and Y is a function only of y. Immediately you may raise the question: But how do we know that the solution is of this form? The answer is that it is not! However, as you will see, once we have solutions of the form (2.2) we can combine them to get the solution we want. [Note that a sum of solutions of (2.1) is a solution of (2.1).] Substituting (2.2) into (2.1), we have

$$Y\frac{d^2X}{dx^2} + X\frac{d^2Y}{dy^2} = 0. \tag{2.3}$$

(Ordinary instead of partial derivatives are now correct since X depends only on x, and y depends only on y.) Divide (2.3) by XY to get

$$\frac{1}{X}\frac{d^2X}{dx^2} + \frac{1}{Y}\frac{d^2Y}{dy^2} = 0. \tag{2.4}$$

The next step is really the key to the process of *separation of variables.* We are going to say that each of the terms in (2.4) is a constant because the first term is a function of x alone and the second term is a function of y alone. Why is this correct? Recall that when we say $u = \sin t$ is a *solution* of $\ddot{u} = -u$, we mean that if we substitute $u = \sin t$ into the differential equation, we get an identity ($\ddot{u} = -u$ becomes $-\sin t = -\sin t$), which is true for all values of t. Although we speak of an *equation*, when we substitute the solution into a differential equation, we have an *identity* in the independent variable. (We made use of this fact in series solutions of differential equations in Chapter 12, Sections 1 and 2.) In (2.1) to (2.4) we have two independent variables, x and y. Saying that (2.2) is a solution of (2.1) means that (2.4) is an identity in the two independent variables x and y [recall that (2.4) was obtained by substituting (2.2) into (2.1)]. In other words, if (2.2) is a solution of (2.1), then (2.4) must be true for any and all values of the two independent variables x and y. Since X is a function only of x and Y of y, the first term of (2.4) is a function only of x and the second term is a function only of y. Suppose we substitute a particular x into the first term; that term is then some numerical constant. To have (2.4) satisfied, the second term must be minus the same constant. While x remains fixed, let y vary (remember that x and y are independent). We have said that (2.4) is an identity; it is then true for our fixed x and *any* y. Thus the second term remains constant as y varies. Similarly, if we fix y and let x vary, we see that the first term of (2.4) is a constant. To say this more concisely, the equation $f(x) = g(y)$, with x and y independent variables, is an identity only if both functions are the same constant; this is the basis of the process of separation of variables. From (2.4) we then write

$$\begin{gathered}\frac{1}{X}\frac{d^2X}{dx^2} = -\frac{1}{Y}\frac{d^2Y}{dy^2} = \text{const.} = -k^2, \qquad k \geq 0, \qquad \text{or} \\ X'' = -k^2X \qquad \text{and} \qquad Y'' = k^2Y.\end{gathered} \tag{2.5}$$

The constant k^2 is called the *separation constant.* The solutions of (2.5) are

$$X = \begin{cases}\sin kx, \\ \cos kx,\end{cases} \qquad Y = \begin{cases}e^{ky}, \\ e^{-ky},\end{cases} \tag{2.6}$$

and the solutions of (2.1) [of the form (2.2)] are

$$T = XY = \left\{ \begin{matrix} e^{ky} \\ e^{-ky} \end{matrix} \right\} \left\{ \begin{matrix} \sin kx \\ \cos kx \end{matrix} \right\}. \tag{2.7}$$

None of the four solutions in (2.7) satisfies the given boundary temperatures. What we must do now is to take a combination of the solutions (2.7), with the constant k properly selected, which *will* satisfy the given boundary conditions. [Any linear combination of solutions of (2.1) is a solution of (2.1) because the differential equation (2.1) is *linear*; see Chapter 3, Section 7, and Chapter 8, Sections 1 and 6.] We first discard the solutions containing e^{ky} since we are given $T \to 0$ as $y \to \infty$. (We are assuming $k > 0$; see Problem 5.) Next we discard solutions containing $\cos kx$ since $T = 0$ when $x = 0$. This leaves us just $e^{-ky} \sin kx$, but the value of k is still to be determined. When $x = 10$, we are to have $T = 0$; this will be true if $\sin(10k) = 0$, that is, if $k = n\pi/10$ for $n = 1, 2, \cdots$. Thus for any integral n, the solution

$$T = e^{-n\pi y/10} \sin \frac{n\pi x}{10} \tag{2.8}$$

satisfies the given boundary conditions on the three $T = 0$ sides.

Finally, we must have $T = 100$ when $y = 0$; this condition is not satisfied by (2.8) for any n. But a linear combination of solutions like (2.8) is a solution of (2.1); let us try to find such a combination which does satisfy $T = 100$ when $y = 0$. In order to allow all possible n's we write an infinite series for T, namely

$$T = \sum_{n=1}^{\infty} b_n e^{-n\pi y/10} \sin \frac{n\pi x}{10}. \tag{2.9}$$

For $y = 0$, we must have $T = 100$; from (2.9) with $y = 0$ we get

$$T_{y=0} = \sum_{n=1}^{\infty} b_n \sin \frac{n\pi x}{10} = 100. \tag{2.10}$$

But this is just the Fourier sine series (Chapter 7, Section 9) for $f(x) = 100$ with $l = 10$. We can find the coefficients b_n, as we did in Chapter 7; we get

$$b_n = \frac{2}{l} \int_0^l f(x) \sin \frac{n\pi x}{l}\, dx = \frac{2}{10} \int_0^{10} 100 \sin \frac{n\pi x}{10}\, dx = \begin{cases} \frac{400}{n\pi}, & \text{odd } n, \\ 0, & \text{even } n. \end{cases} \tag{2.11}$$

Then (2.9) becomes

$$T = \frac{400}{\pi} \left(e^{-\pi y/10} \sin \frac{\pi x}{10} + \frac{1}{3} e^{-3\pi y/10} \sin \frac{3\pi x}{10} + \cdots \right). \tag{2.12}$$

Equation (2.12) can be used for computation if $\pi y/10$ is not too small since then the series converges rapidly. (See also Problem 6.) For example, at $x = 5$ (central line of the plate) and $y = 5$, we find

$$T = \frac{400}{\pi} \left(e^{-\pi/2} \sin \frac{\pi}{2} + \frac{1}{3} e^{-3\pi/2} \sin \frac{3\pi}{2} + \cdots \right) \simeq 26.1^\circ. \tag{2.13}$$

To see how the temperature varies with x and y over a rectangle, you can computer plot a 3-dimensional graph of several terms of $T(x, y)$ in (2.12). Or you can make a 2-dimensional contour plot which shows the isothermals (curves of constant T). If the temperature on the bottom edge is any function $f(x)$ instead of $100°$ (with the other three sides at $0°$ as before), we can do the problem by the same method. We have only to expand the given $f(x)$ in a Fourier sine series and substitute the coefficients into (2.9).

Next, let us consider a finite plate of height 30 cm with the top edge at $T = 0°$, and other dimensions and temperatures as in Figure 2.1. We no longer have any reason to discard the e^{ky} solution since y does not become infinite. We now replace e^{-ky} by a linear combination $ae^{-ky} + be^{ky}$ which is zero when $y = 30$. The most convenient way to do this is to use the combination

$$\tfrac{1}{2}e^{k(30-y)} - \tfrac{1}{2}e^{-k(30-y)} \tag{2.14}$$

(that is, let $a = \frac{1}{2}e^{30k}$ and $b = -\frac{1}{2}e^{-30k}$). Then, when $y = 30$, (2.14) gives $e^0 - e^0 = 0$ as we wanted. Now (2.14) is just $\sinh k(30 - y)$ (see Chapter 2, Section 12), so for the finite plate, we can write the solution as [compare (2.9)]

$$T = \sum_{n=1}^{\infty} B_n \sinh \frac{n\pi}{10}(30 - y) \sin \frac{n\pi x}{10}. \tag{2.15}$$

Each term of this series is zero on the three $T = 0$ sides of the plate. When $y = 0$, we want $T = 100$:

$$T_{y=0} = 100 = \sum_{n=1}^{\infty} B_n \sinh(3n\pi) \sin \frac{n\pi x}{10} = \sum_{n=1}^{\infty} b_n \sin \frac{n\pi x}{10} \tag{2.16}$$

where $b_n = B_n \sinh 3n\pi$ or $B_n = b_n / \sinh 3n\pi$. We find b_n, solve for B_n and substitute into (2.15) to get the temperature distribution in the finite plate:

$$T = \sum_{\text{odd } n} \frac{400}{n\pi \sinh 3n\pi} \sinh \frac{n\pi}{10}(30 - y) \sin \frac{n\pi x}{10}. \tag{2.17}$$

In (2.12) and (2.17) we have found functions $T(x, y)$ satisfying both (2.1) and all the given boundary conditions. For a bounded region with given boundary temperatures, it is an experimental fact (and it can also be shown mathematically—see Problem 16 and Chapter 14, Problem 11.38) that there is only one $T(x, y)$ satisfying Laplace's equation and the given boundary conditions. Thus (2.17) is the desired solution for the rectangular plate. It can also be shown that there is only one solution for the semi-infinite plate provided $T \to 0$ at ∞; thus (2.12) is the solution for that case.

It may have occurred to you to wonder why we took the constant in (2.5) to be $-k^2$ and what would happen if we took $+k^2$ instead. As far as getting solutions of the differential equation is concerned it would be perfectly correct to use $+k^2$; we would get instead of (2.7):

$$T = XY = \begin{Bmatrix} e^{kx} \\ e^{-kx} \end{Bmatrix} \begin{Bmatrix} \sin ky \\ \cos ky \end{Bmatrix}. \tag{2.18}$$

[We are assuming that k is real; an imaginary k in (2.18) would simply give combinations of the solutions (2.7) over again. Also see Problem 5.] The solutions (2.18) would not be of any use for the semi-infinite plate problem since none of them tends to zero as $y \to \infty$, and a linear combination of e^{kx} and e^{-kx} cannot be zero both at $x = 0$ and at $x = 10$. However, if we had considered a semi-infinite plate with its long sides parallel to the x axis instead of the y axis, and $T = 100°$ along the short end on the y axis, the solutions (2.18) would have been the ones needed. Or, for the finite plate, if the 100° side were along the y axis, then we would want (2.18).

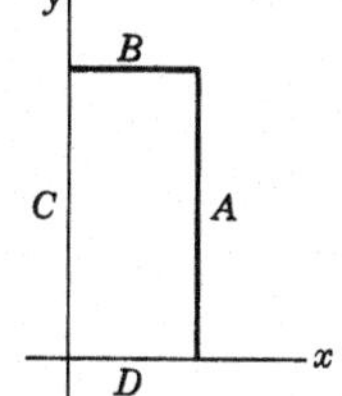

Figure 2.2

Finally, let us see how to find the temperature distribution in a plate if two adjacent sides are held at 100° and the other two at 0° (or, in general, if any values are given for the four sides). We can find the solution to this problem by a combination of the results we have already obtained. Let us call the sides of the rectangular plate A, B, C, D (Figure 2.2). If sides A, B, and C are held at 0°, and D at 100°, we can find the temperature distribution by the same method we used in finding (2.17) if we take the x axis along D. Next suppose that for the same plate (Figure 2.2) sides A, B, and D are held at 0° and C at 100°. This is the same kind of problem over again, but this time we want to use the solutions (2.18). [Or to shorten the work, we could write the solution like (2.17) with the x axis taken along C and then interchange x and y in the result to agree with Figure 2.2.] Having obtained the two solutions (one for C at 100° and one for D at 100°), let us add these two answers. The result is a solution of the differential equation (2.1) (linearity: the sum of any two solutions is a solution). The temperatures on the boundary (as well as inside) are the sums of the temperatures in the two solutions we added, that is, 0° on A, 0° on B, 0° + 100° on C, and 100° + 0° on D. These are the given boundary conditions we wanted to satisfy. Thus the sum of the solutions of two simple problems gives the answer to the more complicated one (see Problems 11 to 13).

Before solving more problems, let us stop for a moment to summarize this process of separation of variables which is basically the same for all the partial differential equations we shall discuss. We first assume a solution which is a product of functions of the independent variables [like (2.2)], and separate the partial differential equation into several ordinary differential equations [like (2.5)]. We solve these ordinary differential equations; the solutions may be exponential functions, trigonometric functions, powers (positive or negative), Bessel functions, Legendre polynomials, etc. Any linear combination of these solutions, with any values of the separation constants, is a solution of the partial differential equation. The problem is to determine both the values of the separation constants and the correct linear combination to fit the given boundary or initial conditions.

The problem of finding the solution of a given differential equation subject to given boundary conditions is called a *boundary value problem.* Such problems often lead to *eigenvalue problems.* Recall (Chapter 3, Section 11, and Chapter 12, end of Section 2) that in an eigenvalue (or characteristic value) problem, there is a parameter whose values are to be selected so that the solutions of the problem meet some given requirements. The separation constants we have been using are just such parameters; their values are determined by demanding that the solutions satisfy some of the boundary conditions. [For example, we found $k = n\pi/10$ just before (2.8) by requiring that $T = 0$ when $x = 10$.] The resulting values of the separation constants are called *eigenvalues* and the solutions of the differential equation [for example (2.8)] corresponding to the eigenvalues are called *eigenfunctions.* It may also happen that in addition to the separation constants there is a parameter in the original partial differential equation [for example, E in the Schrödinger equation (1.6)]. Again the possible values of this parameter (for which the equation has solutions meeting specified requirements) are called eigenvalues, and the corresponding solutions are called eigenfunctions.

Having found the eigenfunctions, the next step is to expand the given function (boundary or initial conditions) in terms of them. [See, for example (2.10) and (2.16) and many examples in later sections.] As we have discussed (see Chapter 7, Section 8, and Chapter 12, Section 6), the eigenfunctions are a set of basis functions for this expansion. Thus we select the functions [for example $e^{-ky}\sin kx$ in (2.7)] and values of the separation constants (eigenvalues) to fit the given boundary (or initial) conditions; this determines the basis functions for a problem.

PROBLEMS, SECTION 2

After you find the series solution of a problem, make computer plots of your results as discussed just after equation (2.13).

1. Find the steady-state temperature distribution for the semi-infinite plate problem if the temperature of the bottom edge is $T = f(x) = x$ (in degrees; that is, the temperature at x cm is x degrees), the temperature of the other sides is 0°, and the width of the plate is 10 cm.

Answer: $T = \dfrac{20}{\pi}\displaystyle\sum_{n=1}^{\infty}\frac{(-1)^{n+1}}{n}e^{-n\pi y/10}\sin(n\pi x/10).$

2. Solve the semi-infinite plate problem if the bottom edge of width 20 is held at

$$T = \begin{cases} 0°, & 0 < x < 10, \\ 100°, & 10 < x < 20, \end{cases}$$

and the other sides are at 0°.

3. Solve the semi-infinite plate problem if the bottom edge of width π is held at $T = \cos x$ and the other sides are at 0°.

Answer: $T = \dfrac{4}{\pi}\displaystyle\sum_{\text{even } n}\frac{n}{n^2-1}e^{-ny}\sin nx.$

4. Solve the semi-infinite plate problem if the bottom edge of width 30 is held at

$$T = \begin{cases} x, & 0 < x < 15, \\ 30 - x, & 15 < x < 30, \end{cases}$$

and the other sides are at 0°.

5. Show that the solutions of (2.5) can also be written as

$$X = \begin{cases} e^{ikx}, \\ e^{-ikx}, \end{cases} \qquad Y = \begin{cases} \sinh ky, \\ \cosh ky. \end{cases}$$

Also show that these solutions are equivalent to (2.7) if k is real and equivalent to (2.18) if k is pure imaginary. (See Chapter 2, Section 12.) Also show that $X = \sin k(x - a), Y = \sinh k(y - b)$ are solutions of (2.5).

6. Show that the series in (2.12) can be summed to get

$$T = \frac{200}{\pi}\arctan\left(\frac{\sin(\pi x/10)}{\sinh(\pi y/10)}\right)$$

(with the arc tangent in radians). Use this formula to check the value $T = 26.1°$ at $x = y = 5$. *Hints for summing the series:* Use $\sin(n\pi x/10) = \operatorname{Im} e^{in\pi x/10}$ to write the series as $\operatorname{Im}\sum_{\text{odd } n} z^n/n$. (What is z?) Compare this with the series for $\ln[(1+z)/(1-z)]$ (see Chapter 1, Problem 13.17). Then use (13.5) of Chapter 2.

7. Solve Problem 3 if the plate is cut off at height 1 and the temperature at $y = 1$ is held at $0°$.

Answer: $T = \dfrac{4}{\pi}\displaystyle\sum_{\text{even } n} \frac{n}{(n^2-1)\sinh n}\sinh n(1-y)\sin nx.$

8. Find the steady-state temperature distribution in a rectangular plate 30 cm by 40 cm given that the temperature is $0°$ along the two long sides and along one short end; the other short end along the x axis has temperature

$$T = \begin{cases} 100°, & 0 < x < 10, \\ 0°, & 10 < x < 30. \end{cases}$$

9. Solve Problem 2 if the plate is cut off at height 10 and the temperature of the top edge is $0°$.

10. Find the steady-state temperature distribution in a metal plate 10 cm square if one side is held at $100°$ and the other three sides at $0°$. Find the temperature at the center of the plate.

Answer: $T = \displaystyle\sum_{\text{odd } n} \frac{400}{n\pi\sinh n\pi}\sinh\frac{n\pi}{10}(10-y)\sin\frac{n\pi x}{10},$

$T(5,5) \simeq 25°$.

11. Find the steady-state temperature distribution in the plate of Problem 10 if two adjacent sides are at $100°$ and the other two at $0°$. *Hint:* Use your solution of Problem 10. You should not have to do any calculation—just write down the answer!

12. Find the temperature distribution in a rectangular plate 10 cm by 30 cm if two adjacent sides are held at $100°$ and the other two sides at $0°$.

13. Find the steady-state temperature distribution in a rectangular plate covering the area $0 < x < 10$, $0 < y < 20$, if the two adjacent sides along the axes are held at temperatures $T = x$ and $T = y$ and the other two sides at $0°$.

14. In the rectangular plate problem, we have so far had the temperature specified all around the boundary. We could, instead, have some edges insulated. The heat flow across an edge is proportional to $\partial T/\partial n$, where n is a variable in the direction normal to the edge (see normal derivatives, Chapter 6, Section 6). For example, the heat flow across an edge lying along the x axis is proportional to $\partial T/\partial y$. Since the heat flow across an insulated edge is zero, we must have not T, but a partial derivative of T, equal to zero on an insulated boundary. Use this fact to find the steady-state temperature distribution in a semi-infinite plate of width 10 cm if the two long sides are insulated, the far end (at ∞ as in Figure 2.1) is at $0°$, and the bottom edge is at $T = f(x) = x - 5$.

Note that you used $T \to 0$ as $y \to \infty$ only to discard the solutions e^{+ky}; it would be just as satisfactory to say that T does not become infinite as $y \to \infty$. Actually, the temperature (assumed finite) as $y \to \infty$ in this problem is determined by the given temperature at $y = 0$. Let $T = f(x) = x$ at $y = 0$, repeat your calculations above to find the temperature distribution, and find the value of T for large y. Don't forget the $k = 0$ term in the series!

15. Consider a finite plate, 10 cm by 30 cm, with two insulated sides, one end at 0° and the other at a given temperature $T = f(x)$. Try $f(x) = 100°$; $f(x) = x$. You should convince yourself that this problem cannot be done using just the solutions (2.7). To see what is wrong, go back to the differential equations (2.5) and solve them if $k = 0$. You should find solutions x, y, xy and constant [the constant is already contained in (2.7) for $k = 0$, but the other three solutions are not]. Now go back over each of the problems we have done so far and see why we could ignore these $k = 0$ solutions; then including the $k = 0$ solutions, finish the problem of the finite plate with insulated sides.

For the case $f(x) = x$, the answer is:

$$T = \frac{1}{6}(30 - y) - \frac{40}{\pi^2} \sum_{\text{odd } n} \frac{1}{n^2 \sinh 3n\pi} \sinh \frac{n\pi}{10}(30 - y) \cos \frac{n\pi x}{10}.$$

16. Show that there is only one function u which takes given values on the (closed) boundary of a region and satisfies Laplace's equation $\nabla^2 u = 0$ in the interior of the region. *Hints:* Suppose u_1 and u_2 are both solutions with the same boundary conditions so that $U = u_1 - u_2 = 0$ on the boundary. In Green's first identity (Chapter 6, Problem 10.16), let $\phi = \Psi = U$ to show that $\boldsymbol{\nabla} U \equiv 0$. Thus show $U \equiv 0$ everywhere inside the region.

3. THE DIFFUSION OR HEAT FLOW EQUATION; THE SCHRÖDINGER EQUATION

The heat flow equation is

$$\nabla^2 u = \frac{1}{\alpha^2} \frac{\partial u}{\partial t}, \tag{3.1}$$

where u is the temperature and α^2 is a constant characteristic of the material through which heat is flowing. It is worthwhile to do first a partial separation of (3.1) into a space equation and a time equation; the space equation in more than one dimension then must be further separated into ordinary differential equations in x and y, or x, y, and z, or r, θ, ϕ, etc. We assume a solution of (3.1) of the form

$$u = F(x, y, z)T(t). \tag{3.2}$$

(Note the change in meaning of T; we have previously used it for temperature; here u is temperature and T is the time-dependent factor in u.) Substitute (3.2) into (3.1); we get

$$T\nabla^2 F = \frac{1}{\alpha^2} F \frac{dT}{dt}. \tag{3.3}$$

Next divide (3.3) by FT to get

$$\frac{1}{F}\nabla^2 F = \frac{1}{\alpha^2} \frac{1}{T} \frac{dT}{dt}. \tag{3.4}$$

The left side of this identity is a function only of the space variables x, y, z, and the right side is a function only of time. Therefore both sides are the same constant and we can write

$$(3.5)\qquad \begin{aligned} &\frac{1}{F}\nabla^2 F = -k^2 \qquad \text{or} \qquad \nabla^2 F + k^2 F = 0 \qquad \text{and}\\ &\frac{1}{\alpha^2}\frac{1}{T}\frac{dT}{dt} = -k^2 \qquad \text{or} \qquad \frac{dT}{dt} = -k^2\alpha^2 T. \end{aligned}$$

The time equation can be integrated to give

$$(3.6)\qquad T = e^{-k^2\alpha^2 t}.$$

We can see a physical reason here for choosing the separation constant $(-k^2)$ to be negative. As t increases, the temperature of a body might decrease to zero as in (3.6), but it could not increase to infinity as it would if we had used $+k^2$ in (3.5) and (3.6). The space equation in (3.5) is the Helmholtz equation (1.5) as promised. You will find (Problem 10) that the space part of the wave equation is also the Helmholtz equation.

Example 1. Let us now consider the flow of heat through a slab of thickness l (for example, the wall of a refrigerator). We shall assume that the faces of the slab are so large that we may neglect any end effects and assume that heat flows only in the x direction (Figure 3.1). This problem is then identical to the problem of heat flow in a bar of length l with insulated sides, because in both cases the heat flow is just in the x direction. Suppose the slab has initially a steady-state temperature distribution with the $x = 0$ wall at 0° and the $x = l$ wall at 100°. From $t = 0$ on, let the $x = l$ wall (as well as the $x = 0$ wall) be held at 0°. We want to find the temperature at any x (in the slab) at any later time.

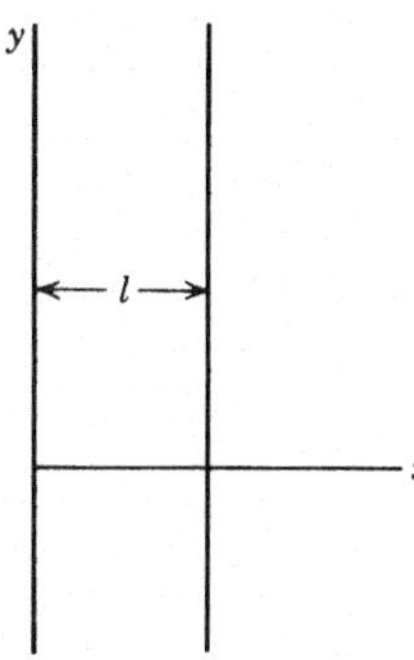

Figure 3.1

First, we find the initial steady-state temperature distribution. You probably already know that this is linear, but it is interesting to see this from our equations. The initial steady-state temperature u_0 satisfies Laplace's equation, which in this one-dimensional case is $d^2u_0/dx^2 = 0$. The solution of this equation is $u_0 = ax + b$, where a and b are constants which must be found to fit the given conditions. Since $u_0 = 0$ at $x = 0$ and $u_0 = 100$ at $x = l$, we have

$$(3.7)\qquad u_0 = \frac{100}{l}x.$$

From $t = 0$ on, u satisfies the heat flow equation (3.1). We have already separated this; the solutions are (3.2) where $T(t)$ is given by (3.6) and $F(x)$ satisfies the first of equations (3.5), namely

$$(3.8)\qquad \nabla^2 F + k^2 F = 0 \qquad \text{or} \qquad \frac{d^2F}{dx^2} + k^2 F = 0.$$

(For this one-dimensional problem, F is a function only of x.) The solutions of (3.8) are

$$(3.9)\qquad F(x) = \begin{cases} \sin kx, \\ \cos kx, \end{cases}$$

and the solutions (3.2) are

$$u = \begin{cases} e^{-k^2\alpha^2 t}\sin kx \\ e^{-k^2\alpha^2 t}\cos kx \end{cases} \tag{3.10}$$

We discard the $\cos kx$ solution for this problem because we are given $u = 0$ at $x = 0$. Also we want $u = 0$ at $x = l$; this will be true if $\sin kl = 0$, that is, $kl = n\pi$, or $k = n\pi/l$ (eigenvalues). Our basis functions (eigenfunctions) are then

$$u = e^{-(n\pi\alpha/l)^2 t}\sin\frac{n\pi x}{l} \tag{3.11}$$

and the solution of our problem will be the series

$$u = \sum_{n=1}^{\infty} b_n e^{-(n\pi\alpha/l)^2 t}\sin\frac{n\pi x}{l}. \tag{3.12}$$

At $t = 0$, we want $u = u_0$ as in (3.7), that is,

$$u = \sum_{n=1}^{\infty} b_n \sin\frac{n\pi x}{l} = u_0 = \frac{100}{l}x. \tag{3.13}$$

This means finding the Fourier sine series for $(100/l)x$ on $(0, l)$; the result (from Problem 1) for the coefficients is

$$b_n = \frac{100}{l}\frac{2l}{\pi}\frac{1}{n}(-1)^{n-1} = \frac{200}{\pi}\frac{(-1)^{n-1}}{n}. \tag{3.14}$$

Then we get the final solution by substituting (3.14) into (3.12); this gives

$$u = \frac{200}{\pi}\sum_{n=1}^{\infty}\frac{(-1)^{n-1}}{n}e^{-(n\pi\alpha/l)^2 t}\sin\frac{n\pi x}{l}. \tag{3.15}$$

▶ **Example 2.** We can now do some variations of this problem. Suppose the final temperatures of the faces are given as two different constant values different from zero. Then, as for the initial steady state, the final steady state is a linear function of distance. The series (3.12) tends to a final steady state of zero; to obtain a solution tending to some other final steady state, we add to (3.12) the linear function u_f representing the correct final steady state. Thus we write instead of of (3.12)

$$u = \sum_{n=1}^{\infty} b_n e^{-(n\pi\alpha/l)^2 t}\sin\frac{n\pi x}{l} + u_f. \tag{3.16}$$

Then for $t = 0$, the equation corresponding to (3.13) is

$$u_0 = \sum_{n=1}^{\infty} b_n \sin\frac{n\pi x}{l} + u_f \tag{3.17}$$

or

$$u_0 - u_f = \sum_{n=1}^{\infty} b_n \sin\frac{n\pi x}{l}. \tag{3.18}$$

Thus when $u_f \neq 0$, it is $u_0 - u_f$ rather than u_0 which must be expanded in a Fourier series.

Insulated boundaries So far we have had the boundary temperatures given. We could, instead, have the faces insulated; then no heat flows in or out of the body. This will be true if the normal derivative $\partial u/\partial n$ (see Problem 2.14) of the temperature is zero at the boundary. (When the boundary values of u are given, the problem is called a *Dirichlet problem;* when the boundary values of the normal derivative $\partial u/\partial n$ are given, the problem is called a *Neumann problem.*) For the one-dimensional case we have considered, we replace the condition $u = 0$ at $x = 0$ and l by the condition $\partial u/\partial x = 0$ at $x = 0$ and l if the faces are insulated. This means that the useful solution in (3.10) is now the one containing $\cos kx$; note carefully that we must include the constant term (corresponding to $k = 0$). See Problem 7.

The Schrödinger Equation Compare equations (1.3) and (1.6). If $V = 0$ in (1.6), the two equations have the same form (a ∇^2 term and a first partial with respect to t). For future reference (see problems, Section 7), let's first separate variables in the general equation (1.6). We assume [compare (3.2)]

$$\Psi = \psi(x, y, z)T(t). \tag{3.19}$$

Substitute (3.19) into (1.6) and divide by Ψ to get

$$-\frac{\hbar^2}{2m}\frac{1}{\psi}\nabla^2\psi + V = i\hbar\frac{1}{T}\frac{dT}{dt} = E \tag{3.20}$$

where E is the separation constant [compare (3.5)]. (In quantum mechanics, E has the meaning of energy of the particle.) Then integrating the time equation gives (compare 3.6)

$$T = e^{-iEt/\hbar} \tag{3.21}$$

and the space equation (called the *time-independent Schrödinger equation*) is

$$-\frac{\hbar^2}{2m}\nabla^2\psi + V\psi = E\psi. \qquad \text{Time-independent Schrödinger equation} \tag{3.22}$$

For the one-dimensional problems that we consider in this section, and with $V = 0$, we have

$$-\frac{\hbar^2}{2m}\frac{d^2\psi}{dx^2} = E\psi \qquad \text{or} \qquad \frac{d^2\psi}{dx^2} + \frac{2mE}{\hbar^2}\psi = 0 \tag{3.23}$$

which is (3.8) with $k^2 = \frac{2mE}{\hbar^2}$. Thus the solutions of (3.23) are the same as in (3.9) and the corresponding Ψ solutions are

$$\Psi = \psi(x)T(t) = \begin{Bmatrix} \sin kx \\ \cos kx \end{Bmatrix} e^{-iEt/\hbar}. \tag{3.24}$$

Example 3. The "particle in a box" problem in quantum mechanics requires the solution of the Schrödinger equation with $V = 0$ on $(0, l)$ and $\Psi = 0$ at the endpoints $x = 0$ and $x = l$ for all t. (The wave function Ψ then describes a particle trapped between 0 and l.) As in the heat flow problem, $\Psi = 0$ at $x = 0$ requires the sine solutions in (3.24) and $\Psi = 0$ at $x = l$ requires $k = n\pi/l$. Since $k^2 = 2mE/\hbar^2$, we find $E = \frac{\hbar^2}{2m}\frac{n^2\pi^2}{l^2}$ which we will call E_n. (The meaning of this equation in quantum mechanics is that the energy of a particle trapped between 0 and l can have only a discrete set of values called eigenvalues. We say that the energy is quantized.) The basis functions for this problem are then the eigenfunctions

$$\Psi_n = \sin\frac{n\pi x}{l} e^{-iE_n t/\hbar}, \tag{3.25}$$

and we write $\Psi(x, t)$ as a linear combination of them.

$$\Psi(x,t) = \sum_{n=1}^{\infty} b_n \sin\frac{n\pi x}{l} e^{-iE_n t/\hbar}, \tag{3.26}$$

(compare (3.12) for the heat flow problem). If the initial state $\Psi(x, 0)$ is the same function as in (3.7), the b_n coefficients are the same as in (3.14), so we have

$$\Psi(x,t) = \frac{200}{\pi}\sum_{n=1}^{\infty}\frac{(-1)^{n-1}}{n}\sin\frac{n\pi x}{l} e^{-iE_n t/\hbar}. \tag{3.27}$$

See Problems 11 and 12; also see Problems 6.6 to 6.8, and 7.17 to 7.22.

PROBLEMS, SECTION 3

As in Section 2, make computer plots of your results.

1. Verify the coefficients in equation (3.14).

2. A bar 10 cm long with insulated sides is initially at 100°. Starting at $t = 0$, the ends are held at 0°. Find the temperature distribution in the bar at time t.

   Answer: $u = \frac{400}{\pi}\sum_{\text{odd } n}\frac{1}{n}e^{-(n\pi\alpha/10)^2 t}\sin\frac{n\pi x}{10}.$

3. In the initial steady state of an infinite slab of thickness l, the face $x = 0$ is at 0° and the face $x = l$ is at 100°. From $t = 0$ on, the $x = 0$ face is held at 100° and the $x = l$ face at 0°. Find the temperature distribution at time t.

   Answer: $u = 100 - \frac{100x}{l} - \frac{400}{\pi}\sum_{\text{even } n}\frac{1}{n}e^{-(n\pi\alpha/l)^2 t}\sin\frac{n\pi x}{l}.$

4. At $t = 0$, two flat slabs each 5 cm thick, one at 0° and one at 20°, are stacked together, and then the surfaces are kept at 0°. Find the temperature as a function of x and t for $t > 0$.

5. Two slabs, each 1 inch thick, each have one surface at 0° and the other surface at 100°. At $t = 0$, they are stacked with their 100° faces together and then the outside surfaces are held at 100°. Find $u(x, t)$ for $t > 0$.

6. Show that the following problem is easily solved using (3.15): The ends of a bar are initially at 20° and 150°; at $t = 0$ the 150° end is changed to 50°. Find the time-dependent temperature distribution.

7. A bar of length l with insulated sides has its ends also insulated from time $t = 0$ on. Initially the temperature is $u = x$, where x is the distance from one end. Determine the temperature distribution inside the bar at time t. *Hints and comments:* See the discussion above and also Problem 2.14. Show that the $k = 0$ solutions are x and constant (time independent). Note that here (unlike Problem 2.15) you do not need the extra solution (namely x) for $k = 0$ since the final steady state is a constant and this is included in the solutions (3.10). Also note that we *did* need the $k = 0$ solutions in the discussion following (3.15) but were able to simplify the work by observing that these linear solutions simply give the final steady state.

Answer: $u = \dfrac{l}{2} - \dfrac{4l}{\pi^2} \sum_{\text{odd } n} \dfrac{1}{n^2} \cos \dfrac{n\pi x}{l} e^{-(n\pi\alpha/l)^2 t}.$

8. A bar of length 2 is initially at 0°. From $t = 0$ on, the $x = 0$ end is held at 0° and the $x = 2$ end at 100°. Find the time-dependent temperature distribution.

9. Solve Problem 8 if, for $t > 0$, the $x = 0$ end of the bar is insulated and the $x = 2$ end is held at 100°. See Problem 7 above, and Chapter 7, end of Section 11.

10. Separate the wave equation (1.4) into a space equation and a time equation as we did the heat flow equation, and show that the space equation is the Helmholtz equation for this case also.

11. Solve the "particle in a box" problem to find $\Psi(x,t)$ if $\Psi(x,0) = 1$ on $(0,\pi)$. What is E_n? The function of interest here which you should plot is $|\Psi(x,t)|^2$.

12. Do Problem 11 if $\Psi(x,0) = \sin^2 \pi x$ on $(0,1)$.

4. THE WAVE EQUATION; THE VIBRATING STRING

Let a string (for example, a piano or violin string) be stretched tightly and its ends fastened to supports at $x = 0$ and $x = l$. When the string is vibrating, its vertical displacement y from its equilibrium position along the x axis depends on x and t. We assume that the displacement y is always very small and that the slope $\partial y/\partial x$ of the string at any point at any time is small. In other words, we assume that the string never gets very far away from its stretched equilibrium position; in fact, we do not distinguish between the length of the string and the distance between the supports, although it is clear that the string must stretch a little as it vibrates out of its equilibrium position. Under these assumptions, the displacement $y(x,t)$ satisfies the (one-dimensional) wave equation

$$\frac{\partial^2 y}{\partial x^2} = \frac{1}{v^2}\frac{\partial^2 y}{\partial t^2}. \tag{4.1}$$

The constant v depends on the tension and the linear density of the string; it is called the wave velocity because it is the velocity with which a disturbance at one point of the string would travel along the string. To separate the variables, we substitute

$$y = X(x)T(t) \tag{4.2}$$

into (4.1) and get (Problem 3.10)

$$\frac{1}{X}\frac{d^2 X}{dx^2} = \frac{1}{v^2}\frac{1}{T}\frac{d^2 T}{dt^2} = -k^2,$$

or

$$X'' + k^2 X = 0, \qquad (4.3)$$
$$\ddot{T} + k^2 v^2 T = 0.$$

We can see from the physical problem why we use a negative separation constant here; the solutions are to describe vibrations which are represented by sines and cosines, not by real exponentials. Of course, if we tried using $+k^2$ with k real, we would also discover mathematically that we could not satisfy the boundary conditions.

Recall the following notation used in discussing wave phenomena (see Chapter 7, Problem 2.17):

$\nu =$ frequency (sec^{-1})	$\omega = 2\pi\nu =$ angular frequency (radians)
$\lambda =$ wavelength	$k = \dfrac{2\pi}{\lambda} = \dfrac{2\pi\nu}{v} = \dfrac{\omega}{v} =$ wave number
$v = \lambda\nu$	

The solutions of the two equations in (4.3) are

$$X = \begin{cases} \sin kx, \\ \cos kx, \end{cases} \qquad T = \begin{cases} \sin kvt = \sin \omega t, \\ \cos kvt = \cos \omega t, \end{cases} \qquad (4.4)$$

and so the solutions (4.2) for y are are

$$y = \begin{Bmatrix} \sin kx \\ \cos kx \end{Bmatrix} \begin{Bmatrix} \sin \omega t \\ \cos \omega t \end{Bmatrix} \quad \text{where } \omega = kv. \qquad (4.5)$$

Since the string is fastened at $x = 0$ and $x = l$, we must have $y = 0$ for these values of x and all t. This means that we want only the $\sin kx$ factors in (4.5), and also we select k so that $\sin kl = 0$ or $k = n\pi/l$. The solutions then become

$$y = \begin{cases} \sin \dfrac{n\pi x}{l} \sin \dfrac{n\pi vt}{l}, \\ \sin \dfrac{n\pi x}{l} \cos \dfrac{n\pi vt}{l}. \end{cases} \qquad (4.6)$$

The particular combination of solutions (4.6) that we should take to solve a given problem depends on the initial conditions. For example, suppose the string is started vibrating by plucking (that is, pulling it aside a small distance h at the center and letting go). Then we are given the shape of the string at $t = 0$, namely $y_0 = f(x)$ as in Figure 4.1, and also the fact that the velocity $\partial y/\partial t$ of points on the string is zero at $t = 0$. (Do not confuse $\partial y/\partial t$ with the wave velocity v; there is no relation between

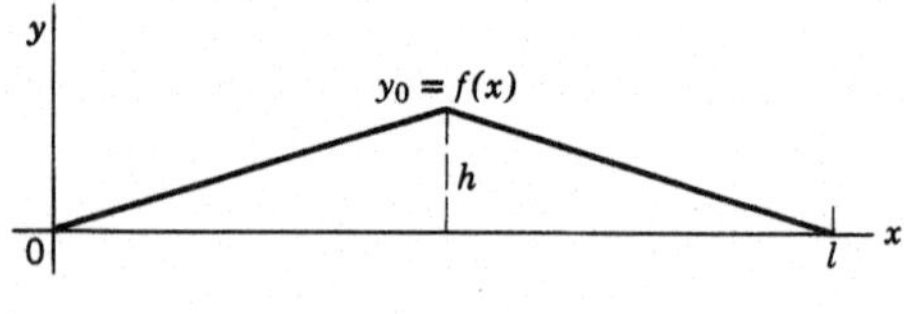

Figure 4.1

them.) In (4.6) we must then discard the term containing $\sin(n\pi vt/l)$ since its time derivative is not zero when $t = 0$. Thus the basis functions for this problem are $\sin(n\pi x/l)\cos(n\pi vt/l)$ and we write the solution in the form

$$y = \sum_{n=1}^{\infty} b_n \sin\frac{n\pi x}{l}\cos\frac{n\pi vt}{l}. \tag{4.7}$$

The coefficients b_n are to be determined so that at $t = 0$ we have $y_0 = f(x)$, that is,

$$y_0 = \sum_{n=1}^{\infty} b_n \sin\frac{n\pi x}{l} = f(x). \tag{4.8}$$

As in previous problems, we find the coefficients in the Fourier sine series for the given $f(x)$ and substitute them into (4.7). The result is (Problem 1)

$$y = \frac{8h}{\pi^2}\left(\sin\frac{\pi x}{l}\cos\frac{\pi vt}{l} - \frac{1}{9}\sin\frac{3\pi x}{l}\cos\frac{3\pi vt}{l} + \cdots\right). \tag{4.9}$$

Another way to start the string vibrating is to hit it (a piano string, for example). In this case the initial conditions would be $y = 0$ at $t = 0$, with the velocity $\partial y/\partial t$ at $t = 0$ given as a function of x (that is, the velocity of each point of the string is given at $t = 0$). This time we discard in (4.6) the term containing $\cos(n\pi vt/l)$ because it is not zero at $t = 0$. Then, for this problem, the basis functions are $\sin(n\pi x/l)\sin(n\pi vt/l)$ and the solution is of the form

$$y = \sum_{n=1}^{\infty} B_n \sin\frac{n\pi x}{l}\sin\frac{n\pi vt}{l}. \tag{4.10}$$

Here the coefficients must be determined so that

$$\left(\frac{\partial y}{\partial t}\right)_{t=0} = \sum_{n_1}^{\infty} B_n\frac{n\pi v}{l}\sin\frac{n\pi x}{l} = \sum_{n=1}^{\infty} b_n \sin\frac{n\pi x}{l} = V(x), \tag{4.11}$$

that is, $V(x)$, the given initial velocity, must be expanded in a Fourier sine series (see Problems 5 to 8).

Suppose the string is vibrating in such a way that, instead of an infinite series for y, we have just one of the solutions (4.6), say

$$y = \sin\frac{n\pi x}{l}\sin\frac{n\pi vt}{l} \tag{4.12}$$

for some one value of n. The largest value of $\sin(n\pi vt/l)$, for any t, is 1, and the shape of the string then is

$$y = \sin\frac{n\pi x}{l}. \tag{4.13}$$

Graphs of (4.13) are sketched in Figure 4.2 for $n = 1, 2, 3, 4$. (The graphs are exaggerated! Remember that the displacements are actually very small.) Consider

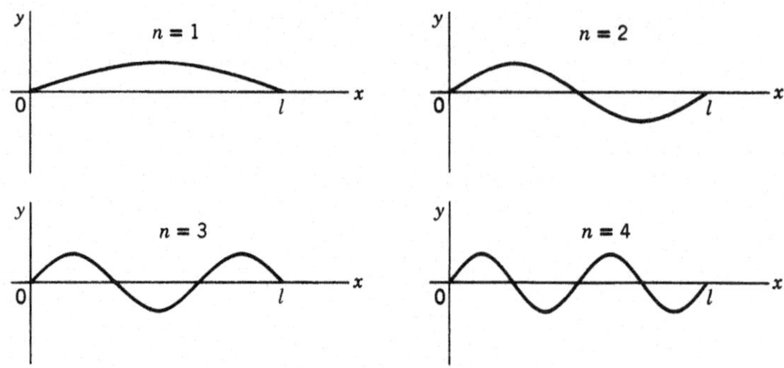

Figure 4.2

a point x on the string; for this point $\sin(n\pi x/l)$ is some number, say A. Then the displacement of this point at time t is [from (4.12)]

$$y = A \sin \frac{n\pi vt}{l}. \tag{4.14}$$

As time passes, this point of the string oscillates up and down with frequency ν_n given by $\omega_n = n\pi v/l = 2\pi\nu_n$ or $\nu_n = nv/(2l)$; the amplitude of the oscillation at this point is $A = \sin(n\pi x/l)$ (see Figure 4.2). Other points of the string oscillate with different amplitudes but the *same* frequency. This is the frequency of the musical note which the string is producing. (See Chapter 7, Section 10.) If $n = 1$ (see Figure 4.2), the frequency is $v/(2l)$; in music this tone is called the fundamental or first harmonic. If $n = 2$, the frequency is just twice that of the fundamental; this tone is called the first overtone or the second harmonic; etc. All the frequencies which this string can produce are multiples of the fundamental. These frequencies are called the *characteristic frequencies* of the string. (They are proportional to the *characteristic values* or *eigenvalues*, $k = n\pi/l$.) The corresponding ways in which the string may vibrate producing a pure tone of just one frequency [that is, with y given by (4.12) for one value of n] are called the *normal modes of vibration*. The first four normal modes are indicated in Figure 4.2. Any vibration is a combination of these normal modes [for example, (4.9) or (4.10)]. The solution (4.12) (for *one* n) describing one normal mode, is a *characteristic function* or *eigenfunction*.

The waves in Figure 4.2 are called standing waves. The d'Alembert solution of the wave equation (see Problem 1.2) represents traveling waves. Suppose we combine two traveling waves moving in opposite directions as follows:

$$\cos k(x - vt) - \cos k(x + vt) = 2 \sin kx \sin kvt \tag{4.15}$$

(by a trigonometry formula). This is one of the solutions (4.5) so we see that this combination of two traveling waves produces a standing wave. Suppose these two traveling waves are moving along a string which is fastened at $x = 0$ and at $x = l$. First consider the wave $\cos k(x + vt)$ which is moving in the negative x direction toward $x = 0$. When it reaches $x = 0$, it will be reflected, and the combination of the incident and reflected waves must equal zero at $x = 0$ for all t. We see that this is true in (4.15), so the wave $\cos k(x - vt)$ is the reflection of $-\cos k(x + vt)$. Now consider $\cos k(x - vt)$ traveling toward $x = l$. When it reaches $x = l$ and is reflected, we can verify (Problem 10) that, if $k = n\pi/l$, then the reflection at $x = l$ is $-\cos \frac{n\pi}{l}(x + vt)$. We can think of a wave traveling back and forth between $x = 0$

and $x = l$, being reflected at each end. The net result as we see from (4.15) is a standing wave.

So far we have been considering problems in which a string is pinned at both ends. We could, instead, have a "free" end; this means free to move up and down along $x = 0$ or $x = l$, say by allowing the end to slide along a frictionless track. The mathematical condition for this is $\partial y/\partial x = 0$ at the free end (compare the condition for an insulated face in Section 3). If the $x = 0$ end is free, we choose the solution containing $\cos kx$ (since $\frac{\partial}{\partial x}\cos kx = -k\sin kx = 0$ at $x = 0$). Then, if the string is pinned at $x = l$, we want $\cos kl = 0$, so $kl = (n + \frac{1}{2})\pi$. Thus the basis functions when the $x = 0$ end is free, the $x = l$ end is pinned, and the initial string velocity is zero, are

$$y = \cos\frac{\left(n+\frac{1}{2}\right)\pi x}{l}\cos\frac{\left(n+\frac{1}{2}\right)\pi vt}{l}. \tag{4.16}$$

For a discussion of these functions, see Chapter 7, Section 11 and Problem 11.11.

PROBLEMS, SECTION 4

As in Sections 2 and 3, use a computer to plot your answers.

1. Complete the plucked string problem to get equation (4.9).

2. A string of length l has a zero initial velocity and a displacement $y_0(x)$ as shown. (This initial displacement might be caused by stopping the string at the center and plucking half of it.) Find the displacement as a function of x and t.

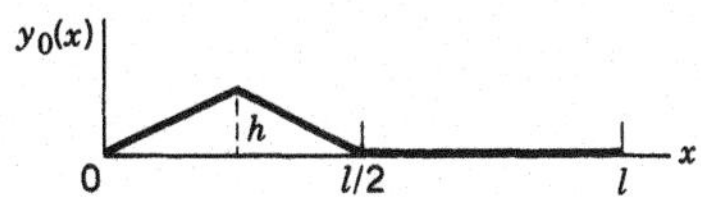

Answer: $y = \dfrac{8h}{\pi^2}\displaystyle\sum_{n=1}^{\infty} B_n \sin\frac{n\pi x}{l}\cos\frac{n\pi vt}{l}$, where $B_n = (2\sin n\pi/4 - \sin n\pi/2)/n^2$.

3. Solve Problem 2 if the initial displacement is:

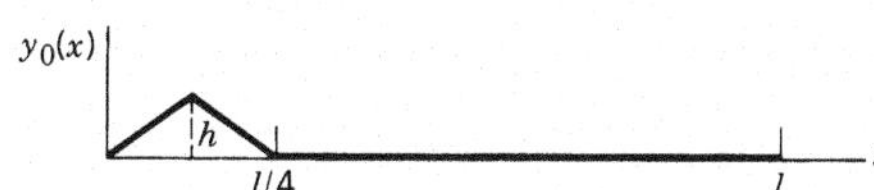

4. Solve Problem 2 if the initial displacement is:

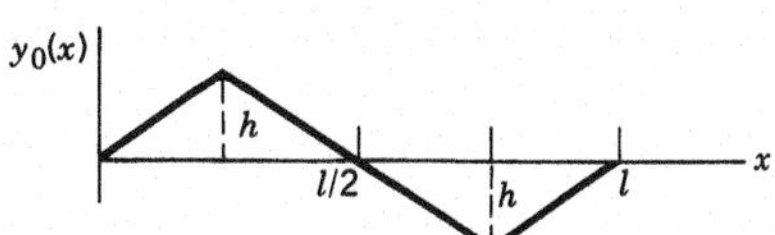

5. A string of length l is initially stretched straight; its ends are fixed for all t. At time $t = 0$, its points are given the *velocity* $V(x) = (\partial y/\partial t)_{t=0}$ as indicated in the diagram (for example, by hitting the string). Determine the shape of the string at time t, that is, find the displacement y as a function of x and t in the form of a series similar to (4.9). *Warning:* What basis functions do you need here?

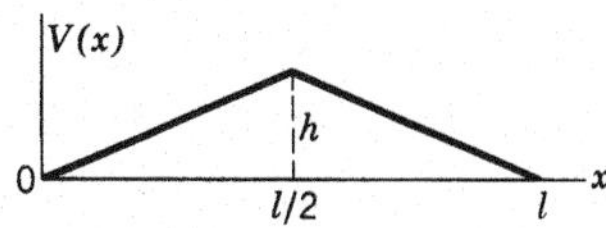

Answer: $y = \dfrac{8hl}{\pi^3 v}\left(\sin\dfrac{\pi x}{l}\sin\dfrac{\pi vt}{l} - \dfrac{1}{3^3}\sin\dfrac{3\pi x}{l}\sin\dfrac{3\pi vt}{l} + \dfrac{1}{5^3}\sin\dfrac{5\pi x}{l}\sin\dfrac{5\pi vt}{l} - \cdots\right)$.

6. Do Problem 5 if the initial velocity $V(x) = (\partial y/\partial t)_{t=0}$ is as shown.

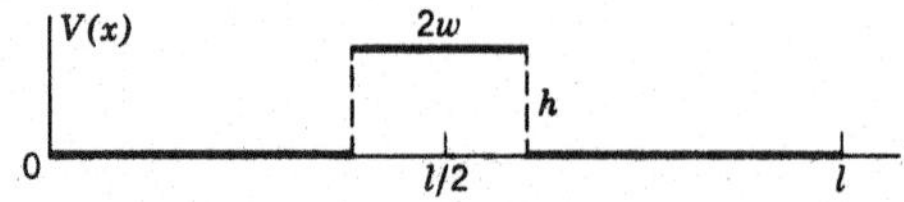

Answer: $y = \dfrac{4hl}{\pi^2 v}\left(\sin\dfrac{\pi w}{l}\sin\dfrac{\pi x}{l}\sin\dfrac{\pi vt}{l} - \dfrac{1}{9}\sin\dfrac{3\pi w}{l}\sin\dfrac{3\pi x}{l}\sin\dfrac{3\pi vt}{l} + \cdots\right).$

7. Solve Problem 5 if the initial velocity is:

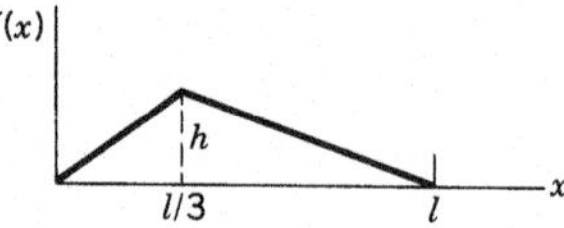

8. Solve Problem 5 if the initial velocity is

$$V(x) = \begin{cases} \sin 2\pi x/l, & 0 < x < l/2, \\ 0, & l/2 < x < l. \end{cases}$$

9. In each of the Problems 1 to 8, find the frequency of the most important harmonic.

10. Verify that, if $k = \frac{n\pi}{l}$, then the sum of the two traveling waves in equation (4.15) is zero at $x = l$, for all t.

11. Verify (4.16) and find a similar formula for a string pinned at $x = 0$ and free at $x = l$. Solve Problems 2, 3, and 4, for a string with a free end (a) at $x = 0$; (b) at $x = l$.

12. In Sections 2, 3, 4, we have solved a number of physics problems which led to the expansion of a given $f(x)$ in a Fourier sine series. Look at (2.9) and (2.25), temperature in a plate; (3.12), heat flow; (3.26), wave function for a particle in a box; (4.7) and (4.10), displacement of a vibrating string plucked or struck. If we have expanded a given $f(x)$ in a Fourier sine series on $(0, l)$, we can immediately write the corresponding solutions for these six different physics problems on the same interval. Do this for $f(x) = x - x^2$ on $(0, 1)$, that is with $l = 1$. Make computer plots of your results.

13. Do Problem 12 for $f(x) = 1 - \cos 2x$ on $(0, \pi)$.

14. Do Problem 12 for $f(x) = x - x^3$ on $(0, 1)$.

5. STEADY-STATE TEMPERATURE IN A CYLINDER

Consider the following problem. Find the steady-state temperature distribution u in a semi-infinite solid cylinder (Figure 5.1) of radius a if the base is held at 100° and the curved sides at 0°. This sounds very much like the problem of the temperature distribution in a semi-infinite plate. However, it is not convenient here to use the solutions in rectangular coordinates, because the boundary condition $u = 0$ is given for $r = a$ rather than for constant values of x or y. The natural variables for this problem are the cylindrical coordinates r, θ, z. The temperature u inside the cylinder satisfies Laplace's equation since there are no sources of heat there.

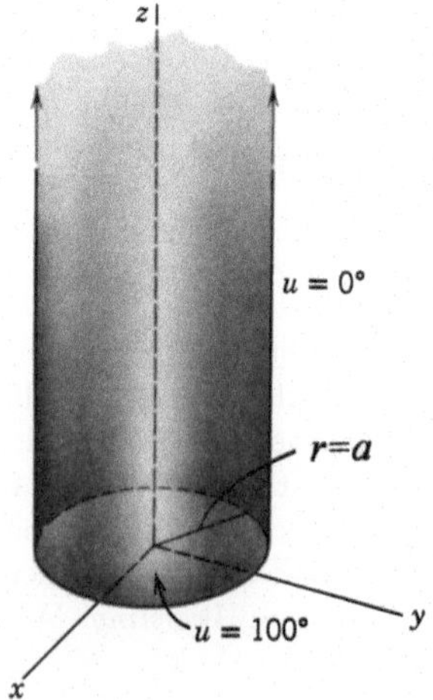

Figure 5.1

Laplace's equation in cylindrical coordinates is (see Chapter 10, Section 9)

$$\nabla^2 u = \frac{1}{r}\frac{\partial}{\partial r}\left(r\frac{\partial u}{\partial r}\right) + \frac{1}{r^2}\frac{\partial^2 u}{\partial \theta^2} + \frac{\partial^2 u}{\partial z^2} = 0. \tag{5.1}$$

To separate the variables, we assume a solution of the form

$$u = R(r)\Theta(\theta)Z(z). \tag{5.2}$$

Substitute (5.2) into (5.1) and divide by $R\Theta Z$ to get

$$\frac{1}{R}\frac{1}{r}\frac{d}{dr}\left(r\frac{dR}{dr}\right) + \frac{1}{\Theta}\frac{1}{r^2}\frac{d^2\Theta}{d\theta^2} + \frac{1}{Z}\frac{d^2 Z}{dz^2} = 0. \tag{5.3}$$

The last term is a function only of z, while the other two terms do not contain z. Therefore the last term is a constant and the *sum* of the first two terms is minus the same constant. Notice that neither of the first two terms is constant alone since both contain r.

> In order to say that a term is constant, we must be sure that:
> (a) it is a function of only one variable, and
> (b) that variable does not appear elsewhere in the equation.

Thus we have

$$\frac{1}{Z}\frac{d^2 Z}{dz^2} = K^2, \qquad Z = \begin{cases} e^{Kz}, \\ e^{-Kz}. \end{cases} \tag{5.4}$$

Since we want the temperature u to tend to zero as z tends to infinity, we call the separation constant $+K^2$ ($K > 0$) and then use only the e^{-Kz} solution. Next write (5.3) with the last term replaced by K^2—see (5.4).

$$\frac{1}{R}\frac{1}{r}\frac{d}{dr}\left(r\frac{dR}{dr}\right) + \frac{1}{\Theta}\frac{1}{r^2}\frac{d^2\Theta}{d\theta^2} + K^2 = 0.$$

We can separate the variables by multiplying by r^2.

$$\frac{r}{R}\frac{d}{dr}\left(r\frac{dR}{dr}\right) + \frac{1}{\Theta}\frac{d^2\Theta}{d\theta^2} + K^2 r^2 = 0. \tag{5.5}$$

In (5.5) the second term is a function of θ only, and the other terms are independent of θ. Thus we have

$$\frac{1}{\Theta}\frac{d^2\Theta}{d\theta^2} = -n^2, \qquad \Theta = \begin{cases} \sin n\theta, \\ \cos n\theta. \end{cases} \tag{5.6}$$

Here we must use $-n^2$ as the separation constant and then require n to be an integer for the following reason. When we locate a point using polar coordinates, we can choose the angle as θ or as $\theta + 2m\pi$ where m is any integer. But regardless of the value of m, there is *one* physical point and *one* temperature there. The mathematical formula for the temperature at the point must give the same value at

θ as at $\theta+2m\pi$, that is, the temperature must be a periodic function of θ with period 2π. This is true only if the Θ solutions are sines and cosines instead of exponentials (hence the negative separation constant) and the constant n is an integer (to give period 2π). The solutions of (5.6) when $n = 0$ are θ and constant. Since θ is not periodic, we can use only the constant solution which is already contained in the $\cos n\theta$ solution when $n = 0$.

Finally, the r equation is

$$\frac{r}{R}\frac{d}{dr}\left(r\frac{dR}{dr}\right) - n^2 + K^2r^2 = 0$$

or

$$r\frac{d}{dr}\left(r\frac{dR}{dr}\right) + \left(K^2r^2 - n^2\right)R = 0. \tag{5.7}$$

This is a Bessel equation with solutions $J_n(Kr)$ and $N_n(Kr)$ [see Chapter 12, equation (16.5)]. Since the base of the cylinder contains the origin, we can use only the J_n and not the N_n solutions since N_n becomes infinite at the origin. Hence we have

$$R(r) = J_n(Kr). \tag{5.8}$$

We can find the possible values of K from the condition that the temperature is zero on the curved surface of the cylinder. Thus $u = 0$ when $r = a$ (for all θ and z) or $R(r) = 0$ when $r = a$. So from (5.8) we see that $J_n(Ka) = 0$, that is, the possible values of Ka are the zeros of J_n. If we define $k = Ka$, or $K = k/a$, then

$$R(r) = J_n(kr/a) \qquad \text{and} \qquad Z(z) = e^{-kz/a}. \tag{5.9}$$

Thus the solutions for u are

$$u = \begin{cases} J_n(kr/a)\,\sin n\theta\; e^{-kz/a}, \\ J_n(kr/a)\,\cos n\theta\; e^{-kz/a}, \end{cases} \tag{5.10}$$

where k is a zero of J_n.

For our problem, the base of the cylinder is held at a constant temperature of 100°. If we turn the cylinder through any angle the boundary conditions are not changed; thus the solution does not depend on the angle θ. This means that we use $\cos n\theta$ with $n = 0$ in (5.10). The possible values of k are the zeros of J_0; call these zeros k_m, where $m = 1, 2, 3, \cdots$. Thus we have the basis functions for the problem and write the solution in terms of them:

$$u = \sum_{m=1}^{\infty} c_m J_0(k_m r/a) e^{-k_m z/a}. \tag{5.11}$$

When $z = 0$, we want $u = 100$, that is,

$$u_{z=0} = \sum_{m=1}^{\infty} c_m J_0(k_m r/a) = 100. \tag{5.12}$$

This should remind you of a Fourier series; here we want to expand 100 in a series of Bessel functions instead of a series of sines or cosines. We proved [see Chapter 12, equation (19.11)] that the functions $J_0(k_m r/a)$ are orthogonal on $(0, a)$ with respect to the weight function r. We can then find the coefficients c_m in (5.12) by the same method used in finding the coefficients in a Fourier sine or cosine series. (In fact, series like (5.12) are often called Fourier-Bessel series.) Multiply (5.12) by $rJ_0(k_\mu r/a), \mu = 1, 2, 3, \cdots$, and integrate term by term from $r = 0$ to $r = a$. Because of the orthogonality [see Chapter 12, equation (19.11)], all terms of the series drop out except the term with $m = \mu$, and we have

$$c_\mu \int_0^a r\left[J_0(k_\mu r/a)\right]^2 dr = \int_0^a 100 r J_0(k_\mu r/a)\, dr. \tag{5.13}$$

For each value of $\mu = 1, 2, 3, \cdots$, equation (5.13) gives one of the coefficients in (5.11) and (5.12); thus any c_m in (5.11) is given by (5.13) with μ replaced by m.

We need to evaluate the integrals in (5.13). Equation (19.11) of Chapter 12 gives (for $p = 0, \alpha = \beta = k_m$)

$$\int_0^a r\left[J_0(k_m r/a)\right]^2 dr = \frac{a^2}{2} J_1^2(k_m). \tag{5.14}$$

By equation (15.1) of Chapter 12

$$\frac{d}{dx}\left[xJ_1(x)\right] = xJ_0(x).$$

If we put $x = k_m r/a$ in this formula, we get

$$\frac{a}{k_m}\frac{d}{dr}\left[(k_m r/a) J_1(k_m r/a)\right] = (k_m r/a) J_0(k_m r/a).$$

Cancelling one k_m/a factor and integrating from 0 to a, we have

$$\int_0^a rJ_0(k_m r/a)\, dr = \frac{a}{k_m} rJ_1(k_m r/a)\Big|_0^a = \frac{a^2}{k_m} J_1(k_m). \tag{5.15}$$

Now we write (5.13) for c_m, substitute the values of the integrals from (5.14) and (5.15), and solve for c_m. The result is

$$c_m = \frac{100 a^2 J_1(k_m)}{k_m} \cdot \frac{2}{a^2 J_1^2(k_m)} = \frac{200}{k_m J_1(k_m)}. \tag{5.16}$$

The solution of our problem is now (5.11) with the values of c_m given by (5.16). The numerical value of the temperature at any point can be found by computing a few terms of the series (Problem 1). The values of the zeros and of the Bessel functions can be found either from your computer or from tables. *Warning:* Remember that k_m is a zero of J_0, not of J_1.

Suppose the given temperature of the base of the cylinder is more complicated than just a constant value, say $f(r, \theta)$, some function of r and θ. Down to (5.10) we proceed as before. But now the series solution is more complicated than (5.11) since we must include all J_n's instead of just J_0. We need a double subscript on the numbers k which are the zeros of the Bessel functions; by k_{mn} we shall mean the

mth positive zero of J_n, where $n = 0, 1, 2, \cdots$ and $m = 1, 2, 3, \cdots$. The temperature u is a double infinite series, summed over the indices m, n of all zeros of all the J_n's:

$$u = \sum_{m=1}^{\infty}\sum_{n=0}^{\infty} J_n(k_{mn}r/a)(A_{mn}\cos n\theta + B_{mn}\sin n\theta)e^{-k_{mn}z/a}. \tag{5.17}$$

At $z = 0$, we want $u = f(r, \theta)$. Thus we write

$$u_{z=0} = \sum_{m=1}^{\infty}\sum_{n=0}^{\infty} J_n(k_{mn}r/a)(A_{mn}\cos n\theta + B_{mn}\sin n\theta) = f(r, \theta). \tag{5.18}$$

To determine the coefficients A_{mn}, multiply this equation by $J_\nu(k_{\mu\nu}r/a)\cos\nu\theta$ and integrate over the whole base of the cylinder, (0 to 2π for θ, 0 to a for r). Because of the orthogonality of the functions $\sin n\theta$ and $\cos n\theta$ on $(0, 2\pi)$, all the B_{mn} terms drop out, and only the A_{mn} terms for $n = \nu$ remain. Because of the orthogonality of the functions $J_n(k_{mn}r/a)$ (one n, all m), only the one term $A_{\mu\nu}$ remains. Thus we have

$$\begin{aligned}\int_0^a\int_0^{2\pi} & f(r,\theta)J_\nu(k_{\mu\nu}r/a)\cos\nu\theta\, r\,dr\,d\theta \\ &= A_{\mu\nu}\int_0^a\int_0^{2\pi} J_\nu^2(k_{\mu\nu}r/a)\cos^2\nu\theta\, r\,dr\,d\theta = A_{\mu\nu}\cdot\frac{a^2}{2}J_{\nu+1}^2(k_{\mu\nu})\cdot\pi.\end{aligned} \tag{5.19}$$

[The r integral is given by (19.11) of Chapter 12, and the θ integral by Chapter 7, Section 4]. Notice how the weight function r in the Bessel function integral arises here as part of the polar coordinate area element. Similarly, we can find

$$B_{\mu\nu} = \frac{2}{\pi a^2 J_{\nu+1}^2(k_{\mu\nu})}\int_0^a\int_0^{2\pi} f(r,\theta)J_\nu(k_{\mu\nu}r/a)\sin\nu\theta\, r\,dr\,d\theta. \tag{5.20}$$

By substituting the values of the A and B coefficients from (5.19) and (5.20) into (5.17), we find the solution to the problem.

PROBLEMS, SECTION 5

1. (a) Compute numerically the coefficients (5.16) of the first three terms of the series (5.11) for the steady-state temperature in a solid semi-infinite cylinder when $u = 0$ at $r = 1$, and $u = 100$ at $z = 0$. Find u at $r = \frac{1}{2}, z = 1$.

(b) In part (a), if $u = 0$ at $r = 10$ and $u = 100$ at $z = 0$, find u at $r = 5, z = 10$. What is the relation between parts (a) and (b)? *Hint:* Suppose in part (a) that the length units for r and z are centimeters. Consider the identical physics problem but with distances measured in millimeters, and compare part (b). Note that in equation (5.10), r/a and z/a are just measurements as multiples of the radius a.

2. (a) Find the steady-state temperature distribution in a solid semi-infinite cylinder if the boundary temperatures are $u = 0$ at $r = 1$ and $u = y = r\sin\theta$ at $z = 0$. *Hints:* In (5.10) you want the solution containing $\sin\theta$; therefore you want the functions J_1. You will need to integrate r^2J_1; follow the text method of integrating rJ_0 just before (5.15).

(b) Do part (a) if the cylinder radius is $r = a$.

$$\text{Answer:} \quad u = \sum_{m=1}^{\infty} \frac{2a}{k_m J_2(k_m)} J_1(k_m r/a) e^{-k_m z/a} \sin\theta, \qquad k_m = \text{zeros of } J_1.$$

If $a = 2$, find u when $r = 1, z = 1, \theta = \pi/2$.

3. (a) Find the steady-state temperature distribution in a solid cylinder of height 10 and radius 1 if the top and curved surface are held at 0° and the base at 100°. *Hint*: See Section 2.

(b) Generalize part (a) to a cylinder of height H and radius a.

4. A flat circular plate of radius a is initially at temperature 100°. From time $t = 0$ on, the circumference of the plate is held at 0°. Find the time-dependent temperature distribution $u(r, \theta, t)$. *Hint:* Separate variables in equation (3.1) in polar coordinates.

5. Do Problem 4 if the initial temperature distribution is $u(r, \theta, t = 0) = 100r \sin\theta$.

6. Consider Problem 4 if the initial temperature distribution is given as some function $f(r, \theta)$. The solution is, in general, a double infinite series similar to (5.17). Find formulas for the coefficients in the series.

7. Find the steady-state temperature distribution in a solid cylinder of height 20 and radius 3 if the flat ends are held at 0° and the curved surface at 100°. *Hints:* Use $-K^2$ in (5.4). Also see Chapter 12, Sections 17 and 20.

8. Water at 100° is flowing through a long pipe of radius 1 rapidly enough so that we may assume that the temperature is 100° at all points. At $t = 0$, the water is turned off and the surface of the pipe is maintained at 40° from then on (neglect the wall thickness of the pipe). Find the temperature distribution in the water as a function of r and t. Note that you need only consider a cross section of the pipe.

$$\text{Answer:} \quad u = 40 + \sum_{m=1}^{\infty} \frac{120}{k_m J_1(k_m)} J_0(k_m r) e^{-(\alpha k_m)^2 t}, \qquad \text{where } J_0(k_m) = 0.$$

9. Find the steady-state distribution of temperature in a cube of side 10 if the temperature is 100° on the face $z = 0$ and 0° on the other five faces. *Hint:* Separate Laplace's equation in three dimensions in rectangular coordinates, and follow the methods of Section 2. You will want to expand 100 in the double Fourier series

$$\sum_{n=1}^{\infty} \sum_{m=1}^{\infty} a_{nm} \sin\frac{n\pi x}{l} \sin\frac{m\pi y}{l}.$$

The coefficients a_{nm} are determined by using the orthogonality of the functions $\sin(n\pi x/l)\sin(m\pi y/l)$ over the square, that is,

$$\int_0^l \int_0^l \sin\frac{n\pi x}{l} \sin\frac{m\pi y}{l} \sin\frac{p\pi x}{l} \sin\frac{q\pi y}{l}\, dx\, dy = 0 \qquad \text{unless} \quad \begin{cases} n = p, \\ m = q. \end{cases}$$

10. A cube is originally at 100°. From $t = 0$ on, the faces are held at 0°. Find the time-dependent temperature distribution. *Hint:* This problem leads to a triple Fourier series; see the double Fourier series in Problem 9 and generalize it to three dimensions.

11. The following two $R(r)$ equations arise in various separation of variables problems in polar, cylindrical, or spherical coordinates:

$$r\frac{d}{dr}\left(r\frac{dR}{dr}\right) = n^2R,$$
$$\frac{d}{dr}\left(r^2\frac{dR}{dr}\right) = l(l+1)R.$$

There are various ways of solving them: They are a standard kind of equation (often called Euler or Cauchy equations—see Chapter 8, Section 7d); you could use power series methods; given the fact that the solutions are just powers of r, it is easy to find the powers. Choose any method you like, and solve the two equations for future reference. Consider the case $n = 0$ separately. Is this necessary for $l = 0$?

12. Separate Laplace's equation in two dimensions in polar coordinates [equation (5.1) without the z term] and solve the r and θ equations. (See Problem 11.) Remember that for the θ equation, only periodic solutions are of interest. Use your results to solve the problem of the steady-state temperature in a circular plate if the upper semicircular boundary is held at 100° and the lower at 0°.

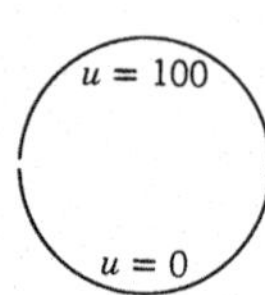

Comment: Another physical problem whose mathematical solution is identical with this temperature problem is this: Find the electrostatic potential inside a capacitor formed by two half-cylinders, insulated from each other and maintained at potentials 0 and 100.

Answer: $u = 50 + \frac{200}{\pi}\sum_{\text{odd } n}\left(\frac{r}{a}\right)^n \frac{\sin n\theta}{n}.$

13. Find the steady-state distribution of temperature in the sector of a circular plate of radius 10 and angle $\pi/4$ if the temperature is maintained at 0° along the radii and at 100° along the curved edge. *Hint:* See Problem 12.

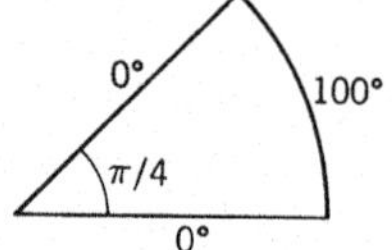

14. Find the steady state temperature distribution in a circular annulus (shaded area) of inner radius 1 and outer radius 2 if the inner circle is held at 0° and the outer circle has half its circumference at 0° and half at 100°. *Hint:* Don't forget the r solutions corresponding to $k = 0$.

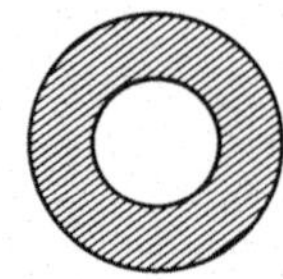

15. Solve Problem 14 if the temperatures of the two circles are interchanged.

6. VIBRATION OF A CIRCULAR MEMBRANE

A circular membrane (for example, a drumhead) is attached to a rigid support along its circumference. Find the characteristic vibration frequencies and the corresponding normal modes of vibration.

Take the (x, y) plane to be the plane of the circular support and take the origin at its center. Let $z(x, y, t)$ be the displacement of the membrane from the (x, y) plane. Then z satisfies the wave equation

$$\nabla^2 z = \frac{1}{v^2}\frac{\partial^2 z}{\partial t^2}. \tag{6.1}$$

Putting

$$z = F(x, y)T(t), \tag{6.2}$$

we separate (6.1) into a space equation (Helmholtz) and a time equation (see Problem 3.10 and Section 3). We get the two equations

$$\nabla^2 F + K^2 F = 0 \quad \text{and} \quad \ddot{T} + K^2 v^2 T = 0. \tag{6.3}$$

Because the membrane is circular we write ∇^2 in polar coordinates (see Chapter 10, Section 9); then the F equation is

$$\frac{1}{r}\frac{\partial}{\partial r}\left(r\frac{\partial F}{\partial r}\right) + \frac{1}{r^2}\frac{\partial^2 F}{\partial \theta^2} + K^2 F = 0. \tag{6.4}$$

When we put

$$F = R(r)\Theta(\theta), \tag{6.5}$$

(6.4) becomes (5.5), and the separated equations and their solutions are just (5.6), (5.7), and (5.8). The solutions of the time equation (6.3) are $\sin Kvt$ and $\cos Kvt$. Thus the solutions for z are $z = R(r)\Theta(\theta)T(t)$, where $R(r) = J_n(Kr), \Theta(\theta) = \{\sin n\theta, \cos n\theta\}$ and $T(t) = \{\sin Kvt, \cos Kvt\}$. Just as in Section 5, n must be an integer. To find possible values of K, we use the fact that the membrane is attached to a rigid frame at $r = a$, so we must have $z = 0$ at $r = a$ for all values of θ and t. Thus $J_n(Ka) = 0$ so the possible values of Ka are the zeros of J_n. As in Section 5, let $k = Ka$, that is, $K = k/a$. Then the possible values of k for each J_n are k_{mn}, the zeros of J_n. We can now write the solutions for z as

$$z = J_n(kr/a)\begin{Bmatrix} \sin n\theta \\ \cos n\theta \end{Bmatrix}\begin{Bmatrix} \sin kvt/a \\ coskvt/a \end{Bmatrix}. \tag{6.6}$$

For a given initial displacement or velocity of the membrane, we could find z as a double series as we found (5.17) in the cylinder temperature problem. However, here we shall do something different, namely investigate the separate normal modes of vibration and their frequencies. Recall that for the vibrating string (Section 4), each n gives a different frequency and a corresponding normal mode of vibration (Figure 4.2). The frequencies of the string are $\nu = nv/(2l)$; all frequencies are integral multiples of the frequency $\nu_1 = v/(2l)$ of the fundamental. For the circular membrane, the frequencies are [from (6.6)]

$$\nu = \frac{\omega}{2\pi} = \frac{kv}{2\pi a}.$$

The possible values of k are the zeros k_{mn} of the Bessel functions. Each value of k_{mn} gives a frequency $\nu_{mn} = k_{mn}v/(2\pi a)$, so we have a doubly infinite set of characteristic frequencies and the corresponding normal modes of vibration. All these frequencies are different, and they are not integral multiples of the fundamental as is true for the string. This is why a drum is less musical than a violin. From your computer or tables you can find several k_{mn} values (Problem 2) and find the frequencies as (nonintegral) multiples of the fundamental (which corresponds to k_{10}, the first zero of J_0). Let us sketch a few graphs (Figure 6.1) of the normal vibration modes corresponding to those in Figure 4.2 for the string, and write the corresponding formulas (eigenfunctions) for the displacement z given in (6.6). (For simplicity, we have used just the $\cos n\theta \cos kvt/a$ solutions in Figure 6.1.) In the fundamental mode of vibration corresponding to k_{10}, the membrane vibrates as a whole. In the

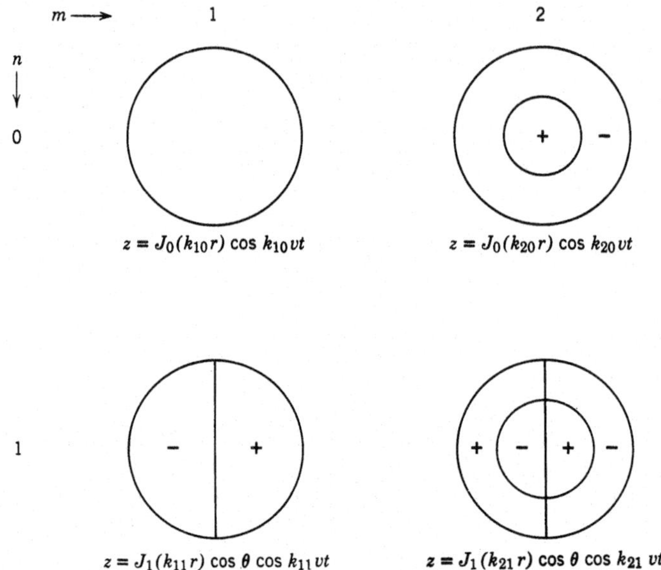

Figure 6.1

k_{20} mode, it vibrates in two parts as shown, the $+$ part vibrating up while the $-$ part vibrates down, and vice versa, with the circle between them at rest. We can show that there is such a circle (called a nodal line) and find its radius. Since $k_{20} > k_{10}$, the circle $r = ak_{10}/k_{20}$ is a circle of radius less than a. Hence it is a circle on the membrane. For this value of r, $J_0(k_{20}r/a) = J_0(k_{20}k_{10}/k_{20})) = J_0(k_{10}) = 0$, so points on this circle are at rest. For the k_{11} mode, $\cos\theta = 0$ when $\theta = \pm\pi/2$ and is positive or negative as shown. Continuing in this way you can sketch any normal mode (Problem 1).

It is difficult experimentally to obtain pure normal modes of a vibrating object. However, a complicated vibration will have nodal lines of some kind and it is easy to observe these. Fine sand sprinkled on the vibrating object will collect along the nodal lines (where there is no vibration) so that you can see them clearly—but see Am. J. Phys. **72**, 1345–1346, (2004). [For experimental work on the vibrating circular membrane, see Am. J. Phys. **35**, 1029–1031, (1967); Am. J. Phys. **40**, 186–188, (1972); Am. J. Phys. **59**, 376–377, (1991). Also see Problem 1(b).]

▶ PROBLEMS, SECTION 6

1. (a) Continue Figure 6.1 to show the fundamental modes of vibration of a circular membrane for $n = 0, 1, 2$, and $m = 1, 2, 3$. As in Figure 6.1, write the formula for the displacement z under each sketch.

(b) Use a computer to set up animations of the various modes of vibration of a circular membrane. [This has been discussed in a number of places. See, for example, Am. J. Phys. **67**, 534–537, (1999).]

2. Find, from computer or tables, the first three zeros k_{mn} of each of the Bessel functions J_0, J_1, J_2, and J_3. Find the first six frequencies of a vibrating circular membrane as (non-integral) multiples of the fundamental frequency.

3. Separate the wave equation in two-dimensional rectangular coordinates x, y. Consider a rectangular membrane as shown, rigidly attached to supports along its sides.

Show that its characteristic frequencies are

$$\nu_{nm} = (v/2)\sqrt{(n/a)^2 + (m/b)^2},$$

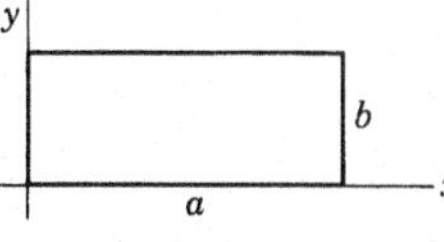

where n and m are positive integers, and sketch the normal modes of vibration corresponding to the first few frequencies. That is, indicate the nodal lines as we did for the circular membrane in Figure 6.1 and Problem 1.

Next suppose the membrane is square. Show that in this case there may be two or more normal modes of vibration corresponding to a single frequency. (*Hint for one example:* $7^2 + 1^1 = 1^2 + 7^2 = 5^2 + 5^2$.) This is an example of what is called *degeneracy*; we say that there is degeneracy when several different solutions of the wave equation (eigenfunctions) correspond to the same frequency (eigenvalue). Sketch several normal modes giving rise to the same frequency. *Comment:* Compare Chapter 3, Section 11, where an eigenvalue of a matrix is called degenerate if several eigenvectors correspond to it.

4. Find the characteristic frequencies for sound vibration in a rectangular box (say a room) of sides a, b, c. *Hint:* Separate the wave equation in three dimensions in rectangular coordinates. This problem is like Problem 3 but for three dimensions instead of two. Discuss degeneracy (see Problem 3).

5. A square membrane of side l is distorted into the shape

$$f(x, y) = xy(l - x)(l - y)$$

and released. Express its shape at subsequent times as an infinite series. *Hint:* Use a double Fourier series as in Problem 5.9.

6. Let $V = 0$ in the Schrödinger equation (3.22) and separate variables in 2-dimensional rectangular coordinates. Solve the problem of a particle in a 2-dimensional square box, $0 < x < l, 0 < y < l$. This means to find solutions of the Schrödinger equation which are 0 for $x = 0, x = l, y = 0, y = l$, that is, on the boundary of the box, and to find the corresponding energy eigenvalues. *Comments:* If we extend the idea of a "particle in a box" (see Section 3, Example 3) to two or three dimensions, the box in 2D might be a square (as in this problem) or a circle (Problem 8); in 3D it might be a cube (Problem 7.17) or a sphere (Problem 7.19). In all cases, the mathematical problem is to find solutions of the Schrödinger equation with $V = 0$ inside the box and $\Psi = 0$ on the boundary of the box, and to find the corresponding energy eigenvalues. In quantum mechanics, Ψ describes a particle trapped inside the box and the energy eigenvalues are the possible values of the energy of the particle.

7. In your Problem 6 solutions, find some examples of degeneracy. (See Problem 3. Degeneracy means that several eigenfunctions correspond to the same energy eigenvalue.)

8. Do Problem 6 in polar coordinates to find the eigenfunctions and energy eigenvalues of a particle in a circular box $r < a$. You want $\Psi = 0$ when $r = a$.

7. STEADY-STATE TEMPERATURE IN A SPHERE

Find the steady-state temperature inside a sphere of radius a when the surface of the upper half is held at 100° and the surface of the lower half at 0°.

Inside the sphere, the temperature u satisfies Laplace's equation. In spherical coordinates this is (see Chapter 10, Section 9)

$$\nabla^2 u = \frac{1}{r^2}\frac{\partial}{\partial r}\left(r^2\frac{\partial u}{\partial r}\right) + \frac{1}{r^2\sin\theta}\frac{\partial}{\partial\theta}\left(\sin\theta\frac{\partial u}{\partial\theta}\right) + \frac{1}{r^2\sin^2\theta}\frac{\partial^2 u}{\partial\phi^2} = 0. \tag{7.1}$$

We separate this equation following our standard procedure. Substitute

$$u = R(r)\Theta(\theta)\Phi(\phi) \tag{7.2}$$

into (7.1) and multiply by $r^2/R\Theta\Phi$ to get

$$\frac{1}{R}\frac{d}{dr}\left(r^2\frac{dR}{dr}\right)+\frac{1}{\Theta}\frac{1}{\sin\theta}\frac{d}{d\theta}\left(\sin\theta\frac{d\Theta}{d\theta}\right)+\frac{1}{\Phi}\frac{1}{\sin^2\theta}\frac{d^2\Phi}{d\phi^2}=0. \tag{7.3}$$

If we multiply (7.3) by $\sin^2\theta$, the last term becomes a function of ϕ only and the other terms do not contain ϕ. Thus we obtain the ϕ equation and its solutions:

$$\frac{1}{\Phi}\frac{d^2\Phi}{d\phi^2}=-m^2, \qquad \Phi=\begin{cases}\sin m\phi,\\ \cos m\phi.\end{cases} \tag{7.4}$$

The separation constant must be negative and m an integer to make Φ a periodic function of ϕ [see the discussion after (5.6)].

Equation (7.3) can now be written as

$$\frac{1}{R}\frac{d}{dr}\left(r^2\frac{dR}{dr}\right)+\frac{1}{\Theta}\frac{1}{\sin\theta}\frac{d}{d\theta}\left(\sin\theta\frac{d\Theta}{d\theta}\right)-\frac{m^2}{\sin^2\theta}=0. \tag{7.5}$$

The first term is a function of r and the last two terms are functions of θ, so we have two equations

$$\frac{1}{R}\frac{d}{dr}\left(r^2\frac{dR}{dr}\right)=k, \tag{7.6}$$

$$\frac{1}{\sin\theta}\frac{d}{d\theta}\left(\sin\theta\frac{d\Theta}{d\theta}\right)-\frac{m^2}{\sin^2\theta}\Theta+k\Theta=0. \tag{7.7}$$

If you compare (7.7) with the equation of Problem 10.2 in Chapter 12, you will see that (7.7) is the equation for the associated Legendre functions if $k = l(l+1)$. Recall that l must be an integer in order for the solution of Legendre's equation to be finite at $x = \cos\theta = \pm 1$, that is, at $\theta = 0$ or π; the same statement is true for the equation for the associated Legendre functions. The corresponding result for (7.7) is that k must be a product of two successive integers; it is then convenient to replace k by $l(l+1)$, where l is an integer. The solutions of (7.7) are then the associated Legendre functions (see Problem 10.2, Chapter 12)

$$\Theta = P_l^m(\cos\theta). \tag{7.8}$$

In (7.6), we put $k = l(l+1)$; you can then easily verify (Problem 5.11) that the solutions of (7.6) are

$$R=\begin{cases}r^l,\\ r^{-l-1}.\end{cases} \tag{7.9}$$

Since we are interested in the interior of the sphere, we discard the solutions r^{-l-1} because they become infinite at the origin. If we were discussing a problem (say about water flow or electrostatic potential) outside the sphere, we would use the r^{-l-1} solutions and discard the solutions r^l because they become infinite at infinity.

The basis functions for our problem are then

$$u = r^l P_l^m(\cos\theta)\begin{cases}\sin m\phi,\\ \cos m\phi.\end{cases} \tag{7.10}$$

[The functions $P_l^m(\cos\theta)\sin m\phi$ and $P_l^m(\cos\theta)\cos m\phi$ are called *spherical harmonics*and are often denoted by $Y_l^m(\theta,\phi)$; also see Problem 16.] If the surface temperature at $r = a$ were given as a function of θ and ϕ, we would have a double series (summed on l and m). For the given surface temperatures in our problem (100° on the top hemisphere and 0° on the lower hemisphere), the temperature is independent of ϕ; thus in (7.10) we must have $m = 0, \cos m\phi = 1$. The solutions (7.10) then reduce to $r^l P_l(\cos\theta)$. We write the solution of the problem as a series of these basis functions:

$$u = \sum_{l=0}^{\infty} c_l r^l P_l(\cos\theta). \tag{7.11}$$

We determine the coefficients c_l by using the given temperatures when $r = a$; that is, we must have

$$\begin{aligned} u_{r=a} &= \sum_{l=0}^{\infty} c_l a^l P_l(\cos\theta) \\ &= \begin{cases} 100, & 0 < \theta < \frac{\pi}{2}, \quad \text{that is,} \quad 0 < \cos\theta < 1, \\ 0, & \frac{\pi}{2} < \theta < \pi, \quad \text{that is,} \quad -1 < \cos\theta < 0, \end{cases} \end{aligned} \tag{7.12}$$

or, with $x = \cos\theta$,

$$u_{r=a} = \sum_{l=0}^{\infty} c_l a^l P_l(x) = 100 f(x) \tag{7.13}$$

where

$$f(x) = \begin{cases} 0, & -1 < x < 0, \\ 1, & 0 < x < 1. \end{cases}$$

(Note that here x just stands for $\cos\theta$ and is not the coordinate x.) In Section 9 of Chapter 12, we expanded this $f(x)$ in a series of Legendre polynomials and obtained:

$$f(x) = \frac{1}{2}P_0(x) + \frac{3}{4}P_1(x) - \frac{7}{16}P_3(x) + \frac{11}{32}P_5(x) + \cdots. \tag{7.14}$$

The coefficients c_l in (7.13) are just these coefficients times $100/a^l$. Substituting the c's into (7.11), we get the final solution:

$$\begin{aligned} u = 100\Big[&\tfrac{1}{2}P_0(\cos\theta) + \tfrac{3}{4}\tfrac{r}{a}P_1(\cos\theta) - \tfrac{7}{16}\left(\tfrac{r}{a}\right)^3 P_3(\cos\theta) \\ &+ \tfrac{11}{32}\left(\tfrac{r}{a}\right)^5 P_5(\cos\theta) + \cdots\Big]. \end{aligned} \tag{7.15}$$

We can do variations of this problem. Notice that we have not even mentioned so far what temperature scale we are using (Celsius, Fahrenheit, absolute, etc.). This is a very easy adjustment to make once we have a solution in any one scale. To see why, observe that if u is a solution of Laplace's equation $\nabla^2 u = 0$ or of the heat flow equation $\nabla^2 u = (1/\alpha^2)(\partial u/\partial t)$, then $u + C$ and Cu are also solutions for any constant C. If we add, say, 50° to the solution (7.15), we have the temperature distribution inside a sphere with the top half of the surface at 150° and the lower half at 50°. If we multiply the solution (7.15) by 2, we find the temperature distribution with given surface temperatures of 200° and 0°, and so on.

The temperature of the equatorial plane $\theta = \pi/2$ or $\cos\theta = 0$ as given by equations (7.11) to (7.15) is halfway between the top and bottom surface temperatures, because Legendre series, like Fourier series, converge to the midpoint of a jump in the function which was expanded to get the series. To solve the problem of the temperature in a hemisphere given the temperatures of the curved surface and of the equatorial plane, we need only imagine the lower hemisphere in place and at the proper temperature to give the desired average on the equatorial plane. When the temperature of the equatorial plane is 0°, this amounts to defining the function $f(x)$ in (7.13) on $(-1, 0)$ to make it an odd function.

PROBLEMS, SECTION 7

Find the steady-state temperature distribution inside a sphere of radius 1 when the surface temperatures are as given in Problems 1 to 10.

1. $35\cos^4\theta$

2. $\cos\theta - \cos^3\theta$

3. $\cos\theta - 3\sin^2\theta$

4. $5\cos^3\theta - 3\sin^2\theta$

5. $|\cos\theta|$

6. $\pi/2 - \theta$. See Chapter 12, Problem 9.4.

7. $\begin{cases} \cos\theta, & 0 < \theta < \pi/2, \quad \text{that is, upper hemisphere,} \\ 0, & \pi/2 < \theta < \pi, \quad \text{that is, lower hemisphere.} \end{cases}$

8. $\begin{cases} 100°, & 0 < \theta < \pi/3, \\ 0°, & \text{otherwise.} \end{cases}$ *Hint:* See Problem 9.8 of Chapter 12.

9. $3\sin\theta\cos\theta\sin\phi$. *Hint:* See equation (7.10) and Chapter 12, equation (10.6).

10. $\sin^2\theta\cos\theta\cos 2\phi - \cos\theta$. (See Problem 9.)

11. Find the steady-state temperature distribution inside a hemisphere if the spherical surface is held at 100° and the equatorial plane at 0°. *Hint:* See the last paragraph of this section above.

12. Do Problem 11 if the curved surface is held at $\cos^2\theta$ and the equatorial plane at zero. *Careful:* The answer does *not* involve P_2; read the last sentence of this section.

13. Find the electrostatic potential outside a conducting sphere of radius a placed in an originally uniform electric field, and maintained at zero potential. *Hint:* Let the original field $\mathbf{E}$ be in the negative z direction so that $\mathbf{E} = -E_0\mathbf{k}$. Then since $\mathbf{E} = -\nabla\Phi$, where Φ is the potential, we have $\Phi = E_0 z = E_0 r\cos\theta$ (Verify this!) for the original potential. You then want a solution of Laplace's equation $\nabla^2 u = 0$ which is zero at $r = a$ and becomes $u \sim \Phi$ for large r (that is, far away from the sphere). Select the solutions of Laplace's equation in spherical coordinates which have the right θ and ϕ dependence (there are just two such solutions) and find the combination which reduces to zero for $r = a$.

14. Find the steady-state temperature distribution in a spherical shell of inner radius 1 and outer radius 2 if the inner surface is held at 0° and the outer surface has its upper half at 100° and its lower half at 0°. *Hint:* $r = 0$ is not in the region of interest, so the solutions r^{-l-1} in (7.9) should be included. Replace $c_l r^l$ in (7.11) by $(c_l r^l + b_l r^{-l-1})$.

15. A sphere initially at 0° has its surface kept at 100° from $t = 0$ on (for example, a frozen potato in boiling water!). Find the time-dependent temperature distribution. *Hint:* Subtract 100° from all temperatures and solve the problem; then add the 100° to the answer. Can you justify this procedure? Show that the Legendre function required for this problem is P_0 and the r solution is $(1/\sqrt{r})J_{1/2}$ or j_0 [see (17.4) in Chapter 12]. Since spherical Bessel functions can be expressed in terms of elementary functions, the series in this problem can be thought of as either a Bessel series or a Fourier series. Show that the results are identical.

16. Separate the wave equation in spherical coordinates, and show that the θ, ϕ solutions are the spherical harmonics $Y_l^m(\theta, \phi) = P_l^m(\cos\theta)e^{\pm im\phi}$ and the r solutions are the spherical Bessel functions $j_l(kr)$ and $y_l(kr)$ [Chapter 12, equations (17.4)].

17. Do Problem 6.6 in 3 dimensional rectangular coordinates. That is, solve the "particle in a box" problem for a cube.

18. Separate the time-independent Schrödinger equation (3.22) in spherical coordinates assuming that $V = V(r)$ is independent of θ and ϕ. (If V depends only on r, then we are dealing with central forces, for example, electrostatic or gravitational forces.) *Hints:* You may find it helpful to replace the mass m in the Schrödinger equation by M when you are working in spherical coordinates to avoid confusion with the letter m in the spherical harmonics (7.10). Follow the separation of (7.1) but with the extra term $[V(r) - E]\Psi$. Show that the θ, ϕ solutions are spherical harmonics as in (7.10) and Problem 16. Show that the r equation with $k = l(l+1)$ is [compare (7.6)]

$$\frac{1}{R}\frac{d}{dr}\left(r^2\frac{dR}{dr}\right) - \frac{2Mr^2}{\hbar^2}[V(r) - E] = l(l+1).$$

19. Find the eigenfunctions and energy eigenvalues for a "particle in a spherical box" $r < a$. *Hints:* See Problem 6.6. Write the R equation from Problem 18 with $V = 0$, and compare Chapter 12, Problem 17.6, with $y = R, x = \beta r$ where $\beta = \sqrt{2ME/\hbar^2}$, and $n = l$.

20. Write the Schrödinger equation (3.22) if ψ is a function of x, and $V = \frac{1}{2}m\omega^2x^2$ (this is a one-dimensional harmonic oscillator). Find the solutions $\psi_n(x)$ and the energy eigenvalues E_n. *Hints:* In Chapter 12, equation (22.1) and the first equation in (22.11), replace x by αx where $\alpha = \sqrt{m\omega/\hbar}$. (Don't forget appropriate factors of α for the x's in the denominators of $D = d/dx$ and $\psi'' = d^2\psi/dx^2$.) Compare your results for equation (22.1) with the Schrödinger equation you wrote above to see that they are identical if $E_n = (n + \frac{1}{2})\hbar\omega$. Write the solutions $\psi_n(x)$ of the Schrödinger equation using Chapter 12, equations (22.11) and (22.12).

21. Separate the Schrödinger equation (3.22) in rectangular coordinates in 3 dimensions assuming that $V = \frac{1}{2}m\omega^2(x^2+y^2+z^2)$. (This is a 3-dimensional harmonic oscillator). Observe that each of the separated equations is of the form of the one-dimensional oscillator equation in Problem 20. Thus write the solutions $\psi_n(x, y, z)$ for the 3-dimensional problem, where $n = n_x + n_y + n_z$. Find the energy eigenvalues E_n and their degree of degeneracy (see Problem 6.7 and Chapter 15, Problem 4.21).

22. Find the energy eigenvalues and eigenfunctions for the hydrogen atom. The potential energy is $V(r) = -e^2/r$ in Gaussian units, where e is the charge of the electron and r is in spherical coordinates. Since V is a function of r only, you know from Problem 18 that the eigenfunctions are $R(r)$ times the spherical harmonics $Y_l^m(\theta, \phi)$, so you only have to find $R(r)$. Substitute $V(r)$ into the R equation in Problem 18 and make the following simplifications: Let $x = 2r/\alpha$, $y = rR$; show that then

$$r = \alpha x/2, \quad R(r) = \frac{2}{\alpha x}y(x), \quad \frac{d}{dr} = \frac{2}{\alpha}\frac{d}{dx}, \quad \frac{d}{dr}\left(r^2\frac{dR}{dr}\right) = \frac{2}{\alpha}xy''.$$

Let $\alpha^2 = -2ME/\hbar^2$ (note that for a bound state, E is negative, so α^2 is positive) and $\lambda = Me^2\alpha/\hbar^2$, to get the first equation in Problem 22.26 of Chapter 12. Do this problem to find $y(x)$, and the result that λ is an integer, say n. [*Caution:* **not** the same n as in equation (22.26)]. Hence find the possible values of α (these are the radii of the Bohr orbits), and the energy eigenvalues. You should have found α proportional to n; let $\alpha = na$, where a is the value of α when $n = 1$, that is, the radius of the first Bohr orbit. Write the solutions $R(r)$ by substituting back $y = rR$, and $x = 2r/(na)$, and find E_n from α.

8. POISSON'S EQUATION

We are going to derive Poisson's equation (1.2) for a simple problem whose answer we know in advance. Using our known solution, we shall be able to see a method of solving more difficult problems.

Recall from Chapter 6, Section 8, that the gravitational field is conservative, that is, $\operatorname{curl}\mathbf{F} = 0$, and there is a potential function V such that $\mathbf{F} = -\boldsymbol{\nabla}V$. If we consider the gravitational field at a point P due to a point mass m a distance r away, we have

$$V = -\frac{Gm}{r} \quad \text{and} \quad \mathbf{F} = -\frac{Gm}{r^2}\mathbf{u} \tag{8.1}$$

where $\mathbf{u}$ is a unit vector along r toward P. It is straightforward to show that $\operatorname{div}\mathbf{F} = 0$ and V satisfies Laplace's equation (Problem 1), that is,

$$\boldsymbol{\nabla}\cdot\mathbf{F} = -\boldsymbol{\nabla}\cdot\boldsymbol{\nabla}V = -\nabla^2 V = 0. \tag{8.2}$$

Now suppose there are many masses m_i at distances r_i from P. The total potential at P is the sum of the potentials due to the individual m_i, that is,

$$V = \sum_i V_i = -\sum_i \frac{Gm_i}{r_i}$$

and the total gravitational field at P is the vector sum of the fields $\mathbf{F}_i$, that is,

$$\mathbf{F} = -\sum_i \boldsymbol{\nabla}V_i = -\boldsymbol{\nabla}V.$$

Note that we are taking it for granted that none of the masses m_i are *at* P, that is, that no r_i is zero. Since

$$\boldsymbol{\nabla}\cdot\mathbf{F}_i = -\nabla^2 V_i = 0,$$

we have also

$$\nabla \cdot \mathbf{F} = -\nabla^2 V = 0.$$

Instead of a number of masses m_i, we can consider a continuous distribution of mass inside a volume τ (Figure 8.1). Let ρ be the mass density of the distribution; then the mass in an element $d\tau$ is $\rho\, d\tau$. The gravitational potential at P due to this mass $\rho d\tau$ is $-(G\rho/r)\, d\tau$. Then the total gravitational potential at P due to the whole mass distribution is the triple integral over the volume τ:

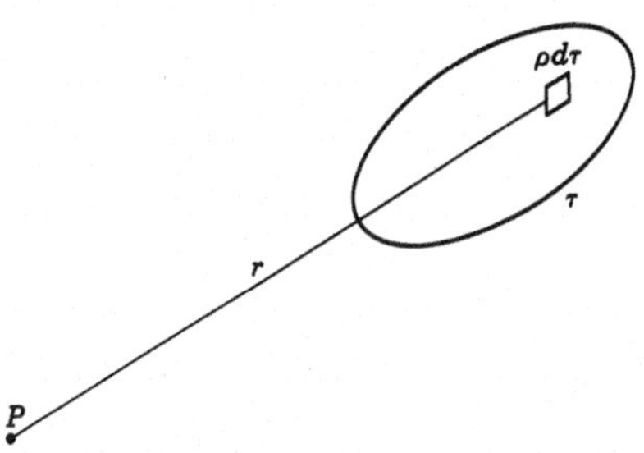

Figure 8.1

$$V = -\iiint\limits_{\text{volume } \tau} \frac{G\rho\, d\tau}{r}. \tag{8.3}$$

As before, the contribution to V at P due to each bit of mass satisfies Laplace's equation and therefore V satisfies Laplace's equation. Also the total field $\mathbf{F}$ at P is the vector sum of the fields due to the elements of mass, and as before we have

$$\nabla \cdot \mathbf{F} = -\nabla^2 V = 0.$$

Again note that we are implicitly assuming that none of the mass distribution coincides with P, that is, that $r \neq 0$, which means that point P is not a point of the region τ.

Now let us investigate what happens if P *is* a point of τ. Can we find V from (8.3) and does V satisfy Laplace's equation? Let S be a small sphere of radius a about P; imagine all the mass removed from inside S (Figure 8.2). Then our previous discussion holds at points inside S since these points are not in the mass distribution. If $\mathbf{F}'$ and V' are the new field and potential (with the matter inside S removed), then $\nabla \cdot \mathbf{F}' = -\nabla^2 V' = 0$ at points inside S. Now restore the mass inside S; let $\mathbf{F}$ and V represent the field and potential due to the whole distribution and let $\mathbf{F}_S$ and V_S represent the field and potential due to just the mass inside S. Then $\mathbf{F} = \mathbf{F}' + \mathbf{F}_S$ and at points inside S

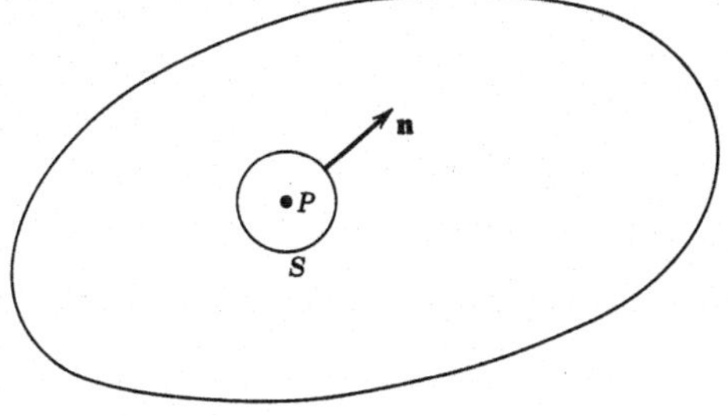

Figure 8.2

$$\nabla \cdot \mathbf{F} = \nabla \cdot \mathbf{F}' + \nabla \cdot \mathbf{F}_S = \nabla \cdot \mathbf{F}_S \tag{8.4}$$

since $\nabla \cdot \mathbf{F}' = 0$ inside S.

By the divergence theorem (see Figure 8.2 and Chapter 6, Section 10)

$$\iiint\limits_{\text{volume of } S} \nabla \cdot \mathbf{F}_S\, d\tau = \iint\limits_{\text{surface of } S} \mathbf{F}_S \cdot \mathbf{n}\, d\sigma. \tag{8.5}$$

If we let the radius a of S tend to zero, the density ρ of matter inside S tends to its value at P; thus for small a, S contains a total mass M approximately equal to

$\frac{4}{3}\pi a^3\rho$, where ρ is evaluated at P. The gravitational field at the surface of S due to this mass is of magnitude

$$F_s = \frac{GM}{a^2} = G\frac{4}{3}\pi a\rho$$

directed *toward* P. Thus in (8.5), $\mathbf{F}_S \cdot \mathbf{n} = -\frac{4}{3}G\pi a\rho$ because $\mathbf{F}_S$ and $\mathbf{n}$ are antiparallel. Since F_S is constant over the surface S, the right-hand side of (8.5) is $\mathbf{F}_S \cdot \mathbf{n}$ times the area of the sphere. The left-hand side is, for small a, approximately the value of $\boldsymbol{\nabla} \cdot \mathbf{F}_S$ at P times the volume of S. Then we have

$$(\boldsymbol{\nabla} \cdot \mathbf{F}_S)(\frac{4}{3}\pi a^3) = (-\frac{4}{3}G\pi a\rho)(4\pi a^2)$$

or

$$\boldsymbol{\nabla} \cdot \mathbf{F}_S = -4\pi G\rho \qquad \text{at } P. \tag{8.6}$$

Since

$$\boldsymbol{\nabla} \cdot \mathbf{F}_S = \boldsymbol{\nabla} \cdot \mathbf{F} = -\boldsymbol{\nabla} \cdot \boldsymbol{\nabla} V = -\nabla^2 V,$$

we have

$$\nabla^2 V = 4\pi G\rho. \tag{8.7}$$

This is Poisson's equation; we see that the gravitational potential in a region containing matter satisfies Poisson's equation as claimed in (1.2). Note that if $\rho = 0$, (8.7) becomes (8.2) as it should.

Next we must consider whether our formula (8.3) for V is valid when P is a point of the mass distribution. The integral appears to diverge at $r = 0$, but this is not really so as we see most easily by using spherical coordinates. Then (8.3) becomes

$$V = -\iiint\limits_{\text{volume } \tau} \frac{G\rho}{r} r^2 \sin\theta \, dr \, d\theta \, d\phi$$

and we see that there is no trouble when $r = 0$. Thus (8.3) is valid in general and gives a solution for (8.7).

Using the notation of (1.2) for Poisson's equation [that is, replacing $4\pi G\rho$ by f and V by u in (8.7) and (8.3)] we can write

$$u = -\frac{1}{4\pi}\iiint \frac{f \, d\tau}{r} \qquad \text{is a solution of} \quad \nabla^2 u = f. \tag{8.8}$$

In the more detailed notation needed when we use this solution in a problem, (8.8) becomes (see Figure 8.3):

$$u(x, y, z) = -\frac{1}{4\pi}\iiint \frac{f(x', y', z')}{\sqrt{(x - x')^2 + (y - y')^2 + (z - z')^2}} \, dx' \, dy' \, dz' \tag{8.9}$$

is a solution of

$$\nabla^2 u(x, y, z) = f(x, y, z)$$

In (8.9) and Figure 8.3, the point (x, y, z) is the point at which we are calculating the potential u; the point (x', y', z') is a point in the mass distribution over which we integrate; r in (8.8) is the distance between these two points and is written out in full in (8.9).

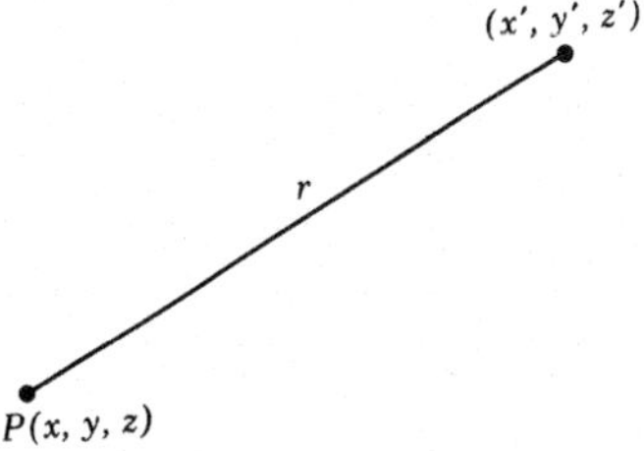

Figure 8.3

Equations (8.8) or (8.9) actually give a very special solution of Poisson's equation. Recall that it is customary to take the zero point for gravitational (and electrostatic) potential energy at infinity, and this is what we have done. Thus (8.8) or (8.9) gives a solution of Poisson's equation which tends to zero at infinity. In another problem this may not be what we want. For example, suppose we have an electrostatic charge distribution near a grounded plane. The electrostatic potential satisfies Poisson's equation, but here we want a solution which is zero on the grounded plane rather than at infinity. To see how we might find such a solution, observe that if u is a solution of Poisson's equation, and w is any solution of Laplace's equation ($\nabla^2 w = 0$), then

$$\nabla^2(u + w) = \nabla^2 u + \nabla^2 w = \nabla^2 u = f; \tag{8.10}$$

thus $u + w$ is a solution of Poisson's equation. Then we can add to the solution (8.9) any solution of Laplace's equation; the combination must be adjusted to fit the given boundary conditions just as we have done in the problems in previous paragraphs.

Example 1. Let us do the following simple problem to illustrate this process. In Figure 8.4, a point charge q at $(0, 0, a)$ is outside a grounded sphere of radius R and center

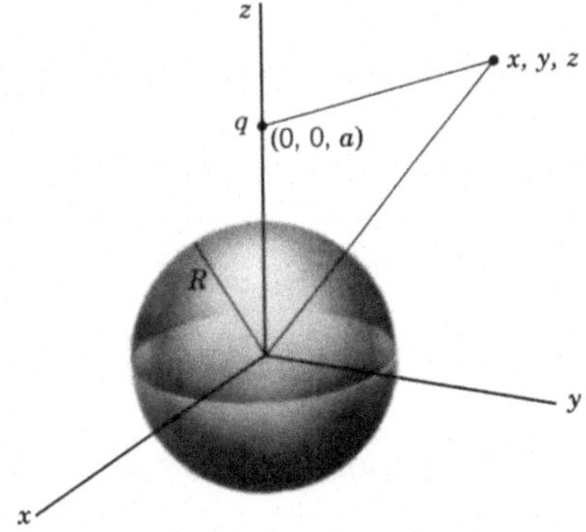

FIgure 8.4

at the origin. Our problem is to find the electrostatic potential V at points outside the sphere. The potential V and the charge density ρ are related by Poisson's equation

$$\nabla^2 V = -4\pi\rho \qquad \text{(in Gaussian units).} \tag{8.11}$$

The potential at (x, y, z) due to a given charge distribution ρ is given by (8.8) or

(8.9) with $f = -4\pi\rho$:

$$V(x, y, z) = -\frac{1}{4\pi}\iiint \frac{-4\pi\rho(x', y', z')}{\sqrt{(x-x')^2 + (y-y')^2 + (z-z')^2}}\,dx'\,dy'\,dz'. \tag{8.12}$$

For a given space-charge distribution, we would next evaluate this integral. For the single point charge q, we have $(x', y', z') = (0, 0, a)$ and we replace $\iiint \rho\,dx'\,dy'\,dz'$ (which is simply the total charge) by q to obtain

$$V = \frac{q}{\sqrt{x^2 + y^2 + (z-a)^2}}. \tag{8.13}$$

[We could, of course, simply have written down (8.13) without using (8.8); (8.13) is just the electrostatic formula corresponding to the gravitational formula (8.1) with which we started.]

Now we want to add to (8.13) a solution of Laplace's equation such that the combination is zero on the given sphere (Figure 8.4). It will be convenient to change to spherical coordinates and to use solutions of Laplace's equation in spherical coordinates. [Note a change in the meaning of r from now on. We have been using r to mean the distance from q at (x', y', z') to (x, y, z); from now on we want to use it to mean the distance from $(0, 0, 0)$ to (x, y, z). See, for example, Figures 8.3 and 8.4.] Writing V_q for V in (8.13) (to distinguish it from our final answer which will be a sum of V_q and a solution of Laplace's equation) and changing to spherical coordinates, we get

$$V_q = \frac{q}{\sqrt{r^2 - 2ar\cos\theta + a^2}}. \tag{8.14}$$

The solutions of Laplace's equation in spherical coordinates are (Section 7):

$$\left\{\begin{matrix} r^l \\ r^{-l-1} \end{matrix}\right\} P_l^m(\cos\theta) \left\{\begin{matrix} \sin m\phi \\ \cos m\phi \end{matrix}\right\}. \tag{8.15}$$

Since we are interested in the region outside the sphere, we want r solutions which do not become infinite at infinity; thus we use r^{-l-1} and discard the r^l solutions. Because the physical problem is symmetric about the z axis, we look for solutions independent of ϕ; that is, we choose $m = 0, \cos m\phi = 1$. Then the basis functions for our problem are $r^{-l-1}P_l(\cos\theta)$ and we try to find a solution of the form

$$V = V_q + \sum_l c_l r^{-l-1} P_l(\cos\theta). \tag{8.16}$$

We must satisfy the boundary condition $V = 0$ when $r = R$. This gives

$$V_{r=R} = \frac{q}{\sqrt{R^2 - 2aR\cos\theta + a^2}} + \sum_l c_l R^{-l-1} P_l(\cos\theta) = 0. \tag{8.17}$$

Thus we want to expand V_q in a Legendre series. Since V_q is essentially the generating function for Legendre polynomials, this is very easy. Comparing (8.17) and the formulas of Chapter 12, Section 5 [(5.1) and (5.2), or more simply, (5.12) and (5.17)], we find

$$\frac{q}{\sqrt{R^2 - 2aR\cos\theta + a^2}} = q\sum_l \frac{R^l P_l(\cos\theta)}{a^{l+1}}. \tag{8.18}$$

Thus the coefficients c_l in (8.17) are given by

$$c_l R^{-l-1} = -\frac{qR^l}{a^{l+1}} \quad \text{or} \quad c_l = -\frac{qR^{2l+1}}{a^{l+1}}. \tag{8.19}$$

Substituting (8.19) into (8.16), we obtain the final solution for V:

$$V = \frac{q}{\sqrt{r^2 - 2ar\cos\theta + a^2}} - q\sum_l \frac{R^{2l+1}r^{-l-1}P_l(\cos\theta)}{a^{l+1}}. \tag{8.20}$$

Since the second term in (8.20) is of the same general form as (8.18), we can simplify (8.20) by summing the series to get (Problem 2)

$$V = \frac{q}{\sqrt{r^2 - 2ar\cos\theta + a^2}} - \frac{(R/a)q}{\sqrt{r^2 + (R^2/a)^2 - 2r(R^2/a)\cos\theta}}. \tag{8.21}$$

Formula (8.21) has a very interesting physical interpretation. The second term is the potential of a charge $-(R/a)q$ at the point $(0, 0, R^2/a)$; thus we could replace the grounded sphere by this charge and have the same potential for $r > R$. This result can be shown also by elementary analytic geometry and is known as the "method of images." For problems with simple geometry (involving planes, spheres, circular cylinders), it may offer a simpler method of solution than the one we have discussed; however, our purpose was to illustrate the more general method.

Use of Green Functions In Chapter 8, Section 12, we used Green functions to solve ordinary differential equations with a nonzero right-hand side. Here we consider the use of Green functions to solve a corresponding partial differential equation in three dimensions, namely Poisson's equation

$$\nabla^2 u = f(\mathbf{r}) = f(x, y, z). \tag{8.22}$$

Suppose that we have a solution of Poisson's equation when the right hand side is a 3-dimensional δ function (see Chapter 8, Sections 11 and 12):

$$\nabla^2 G(\mathbf{r}, \mathbf{r}') = \delta(\mathbf{r} - \mathbf{r}') = \delta(x - x')\delta(y - y')\delta(z - z'). \tag{8.23}$$

The three-dimensional δ function has the property that

$$\iiint f(x', y', z')\delta(\mathbf{r} - \mathbf{r}')\, d\tau' = f(x, y, z) \tag{8.24}$$

if the volume of integration includes the point (x, y, z) (and the integral is zero otherwise). Recall that the right-hand side of Poisson's equation is proportional to the mass density or the charge density. The volume integral of the density gives the total mass or total charge. Since $\iiint \delta(\mathbf{r} - \mathbf{r}')\, d\tau' = 1$, the right-hand side of (8.23) corresponds to a point mass or point charge. That is, the Green function in (8.23) is the potential due to a point source. Just as we showed in Chapter 8, Section 12, that (12.4) is a solution of (12.1), we find here that a solution of (8.22) is given by (see Problem 6)

$$u(\mathbf{r}) = \iiint G(\mathbf{r}, \mathbf{r}')f(\mathbf{r}')\, d\tau'. \tag{8.25}$$

In equation (8.9) we found that a solution of (8.22) is

$$u(\mathbf{r}) = -\frac{1}{4\pi}\iiint \frac{f(\mathbf{r}')}{|\mathbf{r}-\mathbf{r}'|}\,d\tau'. \tag{8.26}$$

Comparing (8.25) and (8.26), we conclude that a solution of (8.23) is

$$G(\mathbf{r},\mathbf{r}') = -\frac{1}{4\pi|\mathbf{r}-\mathbf{r}'|}. \tag{8.27}$$

Now (8.26) and (8.27) give solutions which are zero at infinity; usually we want solutions which are zero on some given surface (for example, zero electrostatic potential on a grounded sphere or plane). In order to obtain such a solution, we add to (8.27) a solution $F(\mathbf{r},\mathbf{r}')$ of Laplace's equation chosen so that the new Green function

$$G(\mathbf{r},\mathbf{r}') = -\frac{1}{4\pi|\mathbf{r}-\mathbf{r}'|} + F(\mathbf{r},\mathbf{r}') \tag{8.28}$$

satisfies the desired zero boundary conditions. Then (8.25) with $G(\mathbf{r},\mathbf{r}')$ as in (8.28) gives a solution of (8.22) which is zero of the boundary. For example, in equation (8.21), V is the potential outside the grounded sphere $r = R$ due to a point charge at $r = a > R$. Rewriting that result in our present notation gives the Green function (8.28) which satisfies (8.23) and is zero on the sphere $r = R$, namely (Problem 7)

$$G(\mathbf{r},\mathbf{r}') = -\frac{1}{4\pi|\mathbf{r}-\mathbf{r}'|} + \frac{R/r'}{4\pi|\mathbf{r}-R^2\mathbf{r}'/r'^2|}. \tag{8.29}$$

(Also see Problems 8 and 9.)

PROBLEMS, SECTION 8

1. Show that the gravitational potential $V = -Gm/r$ satisfies Laplace's equation, that is, show that $\nabla^2(1/r) = 0$ where $r^2 = x^2 + y^2 + z^2, r \neq 0$.
2. Using the formulas of Chapter 12, Section 5, sum the series in (8.20) to get (8.21).
3. Do the problem in Example 1 for the case of a charge q inside a grounded sphere to obtain the potential V inside the sphere. Sum the series solution and state the image method of solving this problem.
4. Do the two-dimensional analogue of the problem in Example 1. A "point charge" in a plane means physically a uniform charge along an infinite line perpendicular to the plane; a "circle" means an infinitely long circular cylinder perpendicular to the plane. However, since all cross sections of the parallel line and cylinder are the same, the problem is a two-dimensional one. *Hint:* The potential must satisfy Laplace's equation in charge-free regions. What are the solutions of the two-dimensional Laplace equation?
5. Find the method of images for problem 4.
6. Substitute (8.25) into (8.22) and use (8.23) and (8.24) to show that (8.25) is a solution of (8.22).
7. Verify that the Green function in (8.29) is zero when $r = R$. Also verify that the point at which the second term becomes infinite is *inside* the sphere, so outside the sphere this term satisfies Laplace's equation as required. Thus write a triple integral for the solution of (8.22) for $r > R$ which is zero on the sphere $r = R$.

8. Show that the Green function (8.28) which is zero on the plane $z = 0$ is

$$G(\mathbf{r}, \mathbf{r}') = -\frac{1}{4\pi}\left[(x-x')^2 + (y-y')^2 + (z-z')^2\right]^{-1/2} + \frac{1}{4\pi}\left[(x-x')^2 + (y-y')^2 + (z+z')^2\right]^{-1/2}.$$

Hence write a triple integral for the solution of (8.22) for $z > 0$ which is zero for $z = 0$.

9. Show that our results can be extended to find the following solution of (8.22) which satisfies given nonzero boundary conditions:

$$u(\mathbf{r}) = \iiint G(\mathbf{r}, \mathbf{r}')f(\mathbf{r}')\,d\tau' + \iint u(\mathbf{r}')\frac{\partial G(\mathbf{r}, \mathbf{r}')}{\partial n'}\,d\sigma'$$

where $G(r, r')$ is the Green function (8.28) which is zero on the surface σ, and $\partial G/\partial n' = \boldsymbol{\nabla} G \cdot \mathbf{n}'$ is the normal derivative of G (see Chapter 6, Section 6). *Hints:* In Green's second identity (Chapter 6, Problem 10.16) let $\phi = u(\mathbf{r})$ and $\psi = G(\mathbf{r}, \mathbf{r}')$, and use (8.22) and (8.23) to find $\nabla^2\phi$ and $\nabla^2\psi$. *Comment:* Although we derived the divergence theorem and so Green's identities only for bounded regions in Chapter 6, they are valid for unbounded regions if the functions involved tend to zero sufficiently rapidly.

9. INTEGRAL TRANSFORM SOLUTIONS OF PARTIAL DIFFERENTIAL EQUATIONS

Laplace Transform Solutions We have seen (Chapter 8, Section 9) that taking the Laplace transform of an ordinary differential equation converts it into an algebraic equation. Taking the Laplace transform of a partial differential equation reduces the number of independent variables by one, and so converts a two-variable partial differential equation into an ordinary differential equation. To illustrate this, we solve the following problem.

Example 1. A semi-infinite bar (extending from $x = 0$ to $x = \infty$), with insulated sides, is initially at the uniform temperature $u = 0°$. At $t = 0$, the end at $x = 0$ is brought to $u = 100°$ and held there. Find the temperature distribution in the bar as a function of x and t.

The differential equation satisfied by u is

$$\frac{\partial^2 u}{\partial x^2} = \frac{1}{\alpha^2}\frac{\partial u}{\partial t}. \tag{9.1}$$

We are going to take the t Laplace transform of (9.1); the variable x will just be a parameter in this process. Let U be the Laplace transform of u, that is,

$$U(x, p) = \int_0^\infty u(x, t)e^{-pt}\,dt. \tag{9.2}$$

By Chapter 8, equation (9.1) we have

$$L\left(\frac{\partial u}{\partial t}\right) = pU - u_{t=0} = pU$$

since $u = 0$ when $t = 0$. Also

$$L\left(\frac{\partial^2 u}{\partial x^2}\right) = \frac{\partial^2}{\partial x^2}L(u) = \frac{\partial^2 U}{\partial x^2}$$

(remember that x is just a parameter here; we are taking a t Laplace transform). The transform of (9.1) is then

$$\frac{\partial^2 U}{\partial x^2} = \frac{1}{\alpha^2} pU. \tag{9.3}$$

Now if we think of p as a constant and x as the variable, this is an ordinary differential equation for U as a function of x. Its solutions are

$$U = \begin{cases} e^{(\sqrt{p}/\alpha)x}, \\ e^{-(\sqrt{p}/\alpha)x}. \end{cases} \tag{9.4}$$

To find the correct combination of these solutions to fit our problem, we need the Laplace transforms of the boundary conditions on u since these give the conditions on U. Using $L1$ (see Laplace Transform Table, page 469) to find the transforms, we have

$$\begin{aligned} &u = 100 \quad \text{at} \quad x = 0, \qquad U = L(100) = \tfrac{100}{p} \quad \text{at} \quad x = 0; \\ &u \to 0 \quad \text{as} \quad x \to \infty, \quad U \to L(0) = 0 \quad \text{as} \quad x \to \infty. \end{aligned} \tag{9.5}$$

Since $U \to 0$ as $x \to \infty$, we see that we must use the solution $e^{-(\sqrt{p}/\alpha)x}$ from (9.4) and discard the positive exponential solution. We determine the constant multiple of this solution which fits our problem from the condition that $U = 100/p$ at $x = 0$. Thus we find that the U solution satisfying the given boundary conditions is

$$U = \frac{100}{p} e^{-(\sqrt{p}/\alpha)x}. \tag{9.6}$$

We find u by looking up the inverse transform of (9.6); it is, by $L22$

$$u = 100 \left[1 - \operatorname{erf} \frac{x}{2\alpha\sqrt{t}}\right] \tag{9.7}$$

and this is the solution of the problem.

Fourier Transform Solutions In the examples in Sections 2, 3 and 4, we expanded a given function in a Fourier series. This was possible because the function was to be represented by a series over a finite interval. We could then take that interval as the period for the Fourier series. If we are dealing with a function which is given over an infinite interval (and not periodic), then instead of representing it by a Fourier series we represent it by a Fourier integral (Chapter 7, Section 12). Let us do this for a specific problem.

▶ **Example 2.** An infinite metal plate (Figure 9.1) covering the first quadrant has the edge along the y axis held at 0°, and the edge along the x axis held at

$$u(x, 0) = \begin{cases} 100^\circ, & 0 < x < 1, \\ 0^\circ, & x > 1. \end{cases} \tag{9.8}$$

Find the steady-state temperature distribution as a function of x and y.

The differential equation and its solutions are the same as in the semi-infinite plate problem discussed in Section 2, equations (2.1), (2.6), and (2.7). As in that problem, we assume $u \to 0$ as $y \to \infty$, and use only the e^{-ky} terms. Since $u = 0$ when $x = 0$, we use only the sine solutions. The basis functions we want are then $u = e^{-ky} \sin kx$. We do not have any requirement here which determines k as we did in Section 2. We must then allow all k's and try to find a solution in the form of an integral over k. Instead of coefficients b_n in a series, we have a coefficient function $B(k)$ to determine. Remember that $k > 0$ since e^{-ky} must tend to zero as $y \to \infty$. Thus we try to find a solution of the form

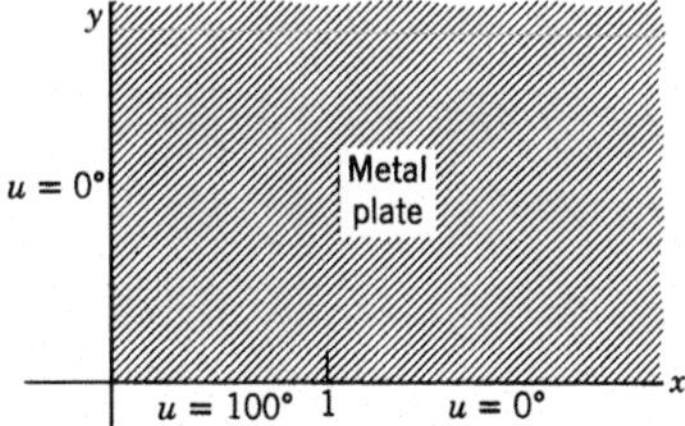

Figure 9.1

$$u(x,y) = \int_0^\infty B(k) e^{-ky} \sin kx \, dk. \tag{9.9}$$

When $y = 0$, we have

$$u(x,0) = \int_0^\infty B(k) \sin kx \, dk. \tag{9.10}$$

This is the first of equations (12.14) in Chapter 7, if we identify k with α, $u(x,0)$ with $f_s(x)$, and $B(k)$ with $\sqrt{2/\pi}\, g_s(\alpha)$. Thus the given temperature on the x axis is a Fourier sine transform of the desired coefficient function, so $B(k)$ can be found as the inverse transform. Using the second of equations (12.14) in Chapter 7, we get

$$B(k) = \sqrt{\frac{2}{\pi}} g_s(k) = \frac{2}{\pi} \int_0^\infty f_s(x) \sin kx \, dx = \frac{2}{\pi} \int_0^\infty u(x,0) \sin kx \, dx. \tag{9.11}$$

For the given $u(x,0)$ in (9.8), we find

$$B(k) = \frac{2}{\pi} \int_0^1 100 \sin kx \, dx = -\frac{200}{\pi} \frac{\cos kx}{k} \bigg|_0^1 = \frac{200}{\pi k}(1 - \cos k). \tag{9.12}$$

Finding $B(k)$ corresponds to evaluating the coefficients in a Fourier series. Substituting (9.12) into (9.9), we get the solution to our problem in the form of an integral instead of a series:

$$u(x,y) = \frac{200}{\pi} \int_0^\infty \frac{1 - \cos k}{k} e^{-ky} \sin kx \, dk. \tag{9.13}$$

An integral can, of course, be evaluated numerically just as a convergent series can be approximated by calculating a few terms. However, (9.13) can be integrated; a convenient way to do it is to recognize that it is a Laplace transform of $f(k) = [(1 - \cos k) \sin kx]/k$, where x is just a parameter and y corresponds to p and k to t. From $L19$ and $L20$

$$u(x,y) = \frac{200}{\pi} \left[\arctan \frac{x}{y} - \frac{1}{2} \arctan \frac{x+1}{y} - \frac{1}{2} \arctan \frac{x-1}{y} \right]. \tag{9.14}$$

This can also be written in polar coordinates as (Problem 1)

$$u = \frac{100}{\pi} \left(\frac{\pi}{2} - \arctan \frac{r^2 - \cos 2\theta}{\sin 2\theta} \right). \tag{9.15}$$

PROBLEMS, SECTION 9

1. Verify that (9.15) follows from (9.14). *Hint:* Use the formulas for $\tan(\alpha \pm \beta)$, $\tan 2\alpha$, etc., to condense (9.14) and then change to polar coordinates. You may find

$$u = \frac{100}{\pi} \arctan \frac{\sin 2\theta}{r^2 - \cos 2\theta}.$$

Show that if you use principal values of the arc tangent, this formula does not give the correct boundary conditions on the x axis, whereas (9.15) does.

2. A metal plate covering the first quadrant has the edge which is along the y axis insulated and the edge which is along the x axis held at

$$u(x,0) = \begin{cases} 100(2-x), & \text{for } 0 < x < 2, \\ 0, & \text{for } x > 2. \end{cases}$$

Find the steady-state temperature distribution as a function of x and y. *Hint:* Follow the procedure of Example 2, but use a cosine transform (because $\partial u/\partial x = 0$ for $x = 0$). Leave your answer as an integral like (9.13).

3. Consider the heat flow problem of Section 3. Solve this by Laplace transforms (with respect to t) by starting as in Example 1. You should get

$$\frac{\partial^2 U}{\partial x^2} - \frac{p}{\alpha^2} U = -\frac{100}{\alpha^2 l} x \qquad \text{and } U(0,p) = U(l,p) = 0.$$

Solve this differential equation to get

$$U(x,p) = -\frac{100 \sinh (p^{1/2}/\alpha)x}{p \sinh (p^{1/2}/\alpha)l} + \frac{100}{pl} x.$$

Assume the following expansion, and find u by looking up the inverse Laplace transforms of the individual terms of U:

$$\frac{\sinh (p^{1/2}/\alpha)x}{p \sinh (p^{1/2}/\alpha)l} = \frac{x}{pl} - \frac{2}{\pi}\left[\frac{\sin(\pi x/l)}{p + (\pi^2\alpha^2/l^2)} - \frac{\sin(2\pi x/l)}{2[p + (4\pi^2\alpha^2/l^2)]} + \frac{\sin(3\pi x/l)}{3[p + (9\pi^2\alpha^2/l^2)]} \cdots\right].$$

Your answer should be (3.15).

4. A semi-infinite bar is initially at temperature 100° for $0 < x < 1$, and 0° for $x > 1$. Starting at $t = 0$, the end $x = 0$ is maintained at 0° and the sides are insulated. Find the temperature in the bar at time t, as follows. Separate variables in the heat flow equation and get elementary solutions $e^{-\alpha^2 k^2 t} \sin kx$ and $e^{-\alpha^2 k^2 t} \cos kx$. Discard the cosines since $u = 0$ at $x = 0$. Look for a solution

$$u(x,t) = \int_0^\infty B(k) e^{-\alpha^2 k^2 t} \sin kx \, dk.$$

and proceed as in Example 2. Leave your answer as an integral.

5. A long wire occupying the x axis is initially at rest. The end $x = 0$ is oscillated up and down so that

$$y(0,t) = 2 \sin 3t, \qquad t > 0.$$

Find the displacement $y(x,t)$. The initial and boundary conditions are $y(0,t) = 2\sin 3t$, $y(x,0) = 0$, $\partial y/\partial t|_{t=0} = 0$. Take Laplace transforms of these conditions and of the wave equation with respect to t as in Example 1. Solve the resulting differential equation to get

$$Y(x,p) = \frac{6e^{-(p/v)x}}{p^2 + 9}.$$

Use $L3$ and $L28$ to find

$$y(x,t) = \begin{cases} 2\sin 3\left(t - \frac{x}{v}\right), & x < vt, \\ 0, & x > vt. \end{cases}$$

6. Continue the problem of Example 2 in the following way: Instead of using the explicit form of $B(k)$ from (9.12), leave it as an integral and write (9.13) in the form

$$u(x,y) = \frac{200}{\pi}\int_0^\infty e^{-ky}\sin kx\,dk\int_0^1 \sin kt\,dt.$$

Change the order of integration and evaluate the integral with respect to k first. (*Hint:* Write the product of sines as a difference of cosines.) Now do the t integration and get (9.14).

7. Continue with Problem 4 as in Problem 6.

10. MISCELLANEOUS PROBLEMS

1. Find the steady-state temperature distribution in a rectangular plate covering the area $0 < x < 1$, $0 < y < 2$, if $T = 0$ for $x = 0$, $x = 1$, $y = 2$, and $T = 1 - x$ for $y = 0$.

2. Solve Problem 1 if $T = 0$ for $x = 0$, $x = 1$, $y = 0$, and $T = 1 - x$ for $y = 2$. *Hint:* Use $\sinh ky$ as the y solution; then $T = 0$ when $y = 0$ as required.

3. Solve Problem 1 if the sides $x = 0$ and $x = 1$ are insulated (see Problems 2.14 and 2.15), and $T = 0$ for $y = 2$, $T = 1 - x$ for $y = 0$.

4. Find the steady-state temperature distribution in a plate with the boundary temperatures $T = 30°$ for $x = 0$ and $y = 3$; $T = 20°$ for $y = 0$ and $x = 5$. *Hint:* Subtract $20°$ from all temperatures and solve the problem; then add $20°$. (Also see Problem 2.)

5. A bar of length l is initially at $0°$. From $t = 0$ on, the ends are held at $20°$. Find $u(x,t)$ for $t > 0$.

6. Do Problem 5 if the $x = 0$ end is insulated and the $x = l$ end held at $20°$ for $t > 0$. (See Problem 3.9.)

7. Solve Problem 2 if the sides $x = 0$ and $x = 1$ are insulated.

8. A slab of thickness 10 cm has its two faces at $10°$ and $20°$. At $t = 0$, the face temperatures are interchanged. Find $u(x,t)$ for $t > 0$.

9. A string of length l has initial displacement $y_0 = x(l - x)$. Find the displacement as a function of x and t.

10. Solve Problem 5.7 if half the curved surface of the cylinder is held at $100°$ and the other half at $-100°$ with the ends at $0°$.

11. The series in Problem 5.12 can be summed (see Problem 2.6). Show that

$$u = 50 + \frac{100}{\pi}\arctan\frac{2ar\sin\theta}{a^2 - r^2}.$$

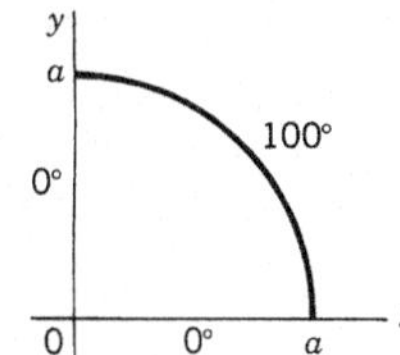

12. A plate in the shape of a quarter circle has boundary temperatures as shown. Find the interior steady-state temperature $u(r,\theta)$. (See Problem 5.12.)

13. Sum the series in Problem 12 to get

$$u = \frac{200}{\pi} \arctan \frac{2a^2 r^2 \sin 2\theta}{a^4 - r^4}.$$

Hint: See Problem 2.6.

14. A long cylinder has been cut into quarter cylinders which are insulated from each other; alternate quarter cylinders are held at potentials +100 and −100. Find the electrostatic potential inside the cylinder. *Hints:* Do you see a relation to Problem 12 above? Also see Problem 5.12.

15. Repeat Problems 12 and 13 for a plate in the shape of a circular sector of angle 30° and radius 10 if the boundary temperatures are 0° on the straight sides and 100° on the circular arc. Can you then state and solve a problem like 14?

16. Consider the normal modes of vibration for a square membrane of side π (see Problem 6.3). Sketch the 2, 1 and 1, 2 modes. Show that the line $y = x$ is a nodal line for the combination $\sin x \sin 2y - \sin 2x \sin y$ of these two modes. Thus find a vibration frequency of a membrane in the shape of a 45° right triangle.

17. Sketch some of the normal modes of vibration for a semicircular drumhead and find the characteristic vibration frequencies as multiples of the fundamental for the corresponding circular drumhead.

18. Repeat Problem 17 for a membrane in the shape of a circular sector of angle 60°.

19. A long conducting cylinder is placed parallel to the z axis in an originally uniform electric field in the negative x direction. The cylinder is held at zero potential. Find the potential in the region outside the cylinder. *Hints:* See Problem 7.13. You want solutions of Laplace's equation in polar coordinates (Problem 5.12).

20. Use Problem 7.16 to find the characteristic vibration frequencies of sound in a spherical cavity.

21. The surface temperature of a sphere of radius 1 is held at $u = \sin^2\theta + \cos^3\theta$. Find the interior temperature $u(r, \theta, \phi)$.

22. Find the interior temperature in a hemisphere if the curved surface is held at $u = \cos\theta$ and the equatorial plane at $u = 1$.

23. Find the steady-state temperature in the region between two spheres $r = 1$ and $r = 2$ if the surface of the outer sphere has its upper half held at 100° and its lower half at −100° and these temperatures are reversed for the inner sphere. *Hint:* See Problem 7.14. Here you will need to find two Legendre series (when $r = 1$ and when $r = 2$) and solve for a_l and b_l.

24. Find the general solution for the steady-state temperature in Figure 2.2 if the boundary temperatures are the constants $T = A$, $T = B$, etc., on the four sides, and the rectangle covers the area $0 < x < a$, $0 < y < b$. *Hints:* You can subtract, say, A from all four temperatures, solve the problem, and then add A back again. Thus a solution with one side at $T = 0$ and the other three at given temperatures solves the general problem. You have previously solved problems (Section 2) with temperatures C and D given. For B, see Problem 2.

25. The Klein-Gordon equation is $\nabla^2 u = (1/v^2)\partial^2 u/\partial t^2 + \lambda^2 u$. This equation is of interest in quantum mechanics, but it also has a simpler application. It describes, for example, the vibration of a stretched string which is embedded in an elastic medium. Separate the one-dimensional Klein-Gordon equation and find the characteristic frequencies of such a string.

Answer: $\nu_n = \dfrac{v}{2}\sqrt{(n/l)^2 + (\lambda/\pi)^2}$.

26. Find the characteristic frequencies of a circular membrane which satisfies the Klein-Gordon equation (Problem 25). *Hint:* Separate the equation in two dimensions in polar coordinates.

27. Do Problem 26 for a rectangular membrane.

28. Find the steady-state temperature in a semi-infinite plate covering the region $x > 0$, $0 \le y \le 1$, if the edges along the x axis and y axis are insulated (see Problem 2.14) and the top edge is held at

$$u(x,1) = \begin{cases} 100^\circ, & 0 < x < 1, \\ 0^\circ, & x > 1. \end{cases}$$

Hint: Look for a solution as a Fourier integral. Leave your answer as an integral (just as we usually give answers as series.)

CHAPTER

14

Functions of a Complex Variable

1. INTRODUCTION

In Chapter 2 we discussed plotting complex numbers $z = x + iy$ in the complex plane (see Figure 1.1) and finding values of the elementary functions of z such as roots, trigonometric functions, logarithms, etc. Now we want to discuss the calculus of functions of z, differentiation, integration, power series, etc. As you know from such topics as differential equations, Fourier series and integrals, mechanics, electricity, etc., it is often very convenient to use complex expressions. The basic facts and theorems about functions of a complex variable not only simplify many calculations but often lead to a better understanding of a problem and consequently to a more efficient method of solution. We are going to state some of the basic definitions and theorems of the subject (omitting the longer proofs), and show some of their uses.

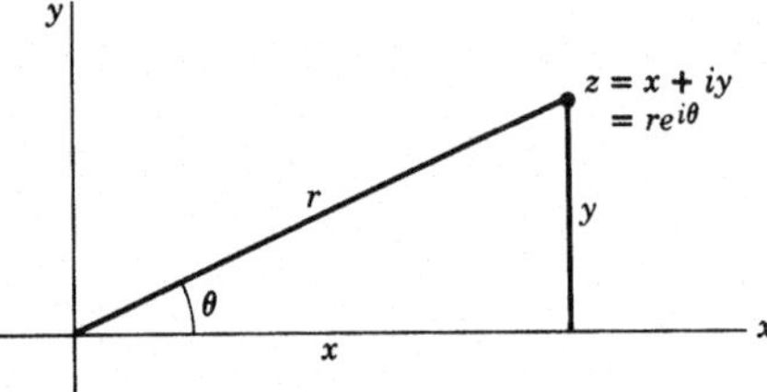

Figure 1.1

As in Chapter 2, the value of a function of z for a given z is a complex number.

Example. Consider a simple function of z, namely $f(z) = z^2$. We may write

$$f(z) = z^2 = (x + iy)^2 = x^2 - y^2 + 2ixy = u(x, y) + iv(x, y),$$

where $u(x, y) = x^2 - y^2$ and $v(x, y) = 2xy$.

In Chapter 2, we observed that a complex number $z = x + iy$ is equivalent to a pair of real numbers x, y. Here we see that a function of z is equivalent to a pair of real functions, $u(x, y)$ and $v(x, y)$, of the real variables x and y. In general, we write

$$f(z) = f(x + iy) = u(x, y) + iv(x, y), \tag{1.1}$$

where it is understood that u and v are real functions of the real variables x and y.

Recall that functions are customarily *single-valued*, that is, $f(z)$ has just one (complex) value for each z. Does this mean that we cannot define a function by a formula such as $\ln z$ or $\arctan z$? By Chapter 2, we have

$$\ln z = \ln|z| + i(\theta + 2n\pi),$$

where $\tan\theta = y/x$. For each z, $\ln z$ has an infinite set of values. But if θ is allowed a range of only 2π , then $\ln z$ has one value for each z and this single-valued function is called a *branch* of $\ln z$. Thus in using formulas such as $\sqrt{z}$, $\ln z$, $\arctan z$, to define functions, we always discuss a single branch at a time so that we have a single-valued function. (As a matter of terminology, however, you should know that the whole collection of branches is sometimes called a "multiple-valued function.")

PROBLEMS, SECTION 1

Find the real and imaginary parts $u(x, y)$ and $v(x, y)$ of the following functions.

1. z^3 **2.** z **3.** $\bar{z}$

4. $|z|$ **5.** $\text{Re}\, z$ **6.** e^z

7. $\cosh z$ **8.** $\sin z$ **9.** $\dfrac{1}{z}$

10. $\dfrac{2z+3}{z+2}$ **11.** $\dfrac{2z-i}{iz+2}$ **12.** $\dfrac{z}{z^2+1}$

13. $\ln|z|$ **14.** $z^2\bar{z}$ **15.** $\overline{e^z}$

16. $z^2 - \bar{z}^2$ **17.** $\cos\bar{z}$ **18.** $\sqrt{\bar{z}}$

19. $\ln z$ (Use $0 < \theta < 2\pi$.) **20.** $(1+2i)z^2 + (i-1)z + 3$

21. e^{iz} (*Careful:* $\cos z$ and $\sin z$ are *not* u and v.)

2. ANALYTIC FUNCTIONS

Definition The derivative of $f(z)$ is defined (just as it is for a function of a real variable) by the equation

$$f'(z) = \frac{df}{dz} = \lim_{\Delta z \to 0} \frac{\Delta f}{\Delta z}, \tag{2.1}$$

where $\Delta f = f(z + \Delta z) - f(z)$ and $\Delta z = \Delta x + i\Delta y$.

Definition: A function $f(z)$ is *analytic* (or *regular* or *holomorphic* or *monogenic*) in a region* of the complex plane if it has a (unique) derivative at every point of the region. The statement "$f(z)$ is analytic *at a point* $z = a$" means that $f(z)$ has a derivative at every point inside some small circle about $z = a$.

*Isolated points and curves are not regions; a region must be two-dimensional.

Let us consider what it means for $f(z)$ to have a derivative. First think about a function $f(x)$ of a real variable x; it is possible for the limit of $\Delta f/\Delta x$ to have two values at a point x_0, as shown in Figure 2.1—one value when we approach x_0 from the left and a different value when we approach x_0 from the right. When we say that $f(x)$ has a derivative at $x = x_0$, we mean that these two values are equal. However, for a function $f(z)$ of a complex variable z, there are an infinite number of ways we can approach a point z_0; a few ways are shown in Figure 2.2. When we say that $f(z)$ has a derivative at $z = z_0$, we mean that $f'(z)$ [as defined by (2.1)] has the same value no matter how we approach z_0. This is an amazingly stringent requirement and we might well wonder whether there *are* any analytic functions. On the other hand, it is hard to imagine making any progress in calculus unless we can find derivatives!

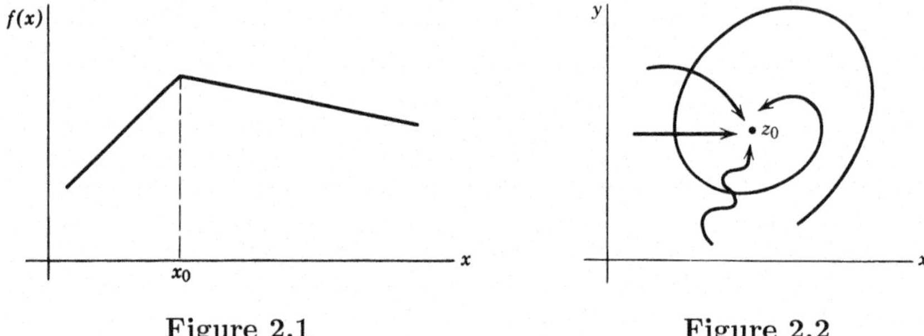

Figure 2.1 **Figure 2.2**

Let us immediately reassure ourselves that there *are* analytic functions by using the definition (2.1) to find the derivatives of some simple functions.

▶ **Example 1.** Show that $(d/dz)(z^2) = 2z$. By (2.1) we have

$$\begin{aligned}\frac{d}{dz}(z^2) &= \lim_{\Delta z\to 0}\frac{(z+\Delta z)^2 - z^2}{\Delta z} = \lim_{\Delta z\to 0}\frac{z^2 + 2z\Delta z + (\Delta z)^2 - z^2}{\Delta z}\\ &= \lim_{\Delta z\to 0}(2z + \Delta z) = 2z.\end{aligned}$$

We see that the result is independent of *how* Δz tends to zero; thus z^2 is an analytic function. By the same method it follows that $(d/dz)(z^n) = nz^{n-1}$ if n is a positive integer (Problem 30).

Observe that the definition (2.1) of a derivative is of exactly the same form as the corresponding definition for a function of a real variable. Because of this similarity, many familiar formulas can be proved by the same methods used in the real case, as we have just discovered in differentiating z^2. You can easily show (Problems 25 to 28) that derivatives of sums, products, and quotients follow the familiar rules and that the chain rule holds [if $f = f(g)$ and $g = g(z)$, then $df/dz = (df/dg)(dg/dz)$]. Then derivatives of rational functions of z follow the familiar real-variable formulas. If we assume the definitions and theorems of Chapters 1 and 2, we can see that the derivatives of the other elementary functions also follow the familiar formulas; for example, $(d/dz)(\sin z) = \cos z$, etc. (Problems 29 to 33).

Now you may be wondering what is new here since all our results so far seem to be just the same as for functions of a real variable. The reason for this is that we have been discussing only functions $f(z)$ that *have* derivatives. Comparing Figures 2.1 and 2.2, we pointed out the essential difference between finding $(d/dx)f(x)$

and finding $(d/dz)f(z)$, namely that there are an infinite number of ways we can approach z_0 in Figure 2.2.

Example 2. Find $(d/dz)(|z|^2)$. Note that $|x|^2 = x^2$, and its derivative is $2x$. If $|z|^2$ has a derivative, it is given by (2.1), that is, by

$$\lim_{\Delta z\to 0} = \lim_{\Delta z\to 0} \frac{|z+\Delta z|^2 - |z|^2}{\Delta z}.$$

The numerator of this fraction is always real (because absolute values are real—recall $|z| = \sqrt{x^2+y^2} = r$.) Consider the denominator $\Delta z = \Delta x + i\Delta y$. As we approach z_0 in Figure 2.2 (that is, let $\Delta z \to 0$), Δz has different values depending on our method of approach. For example, if we come in along a horizontal line, then $\Delta y = 0$ and $\Delta z = \Delta x$; along a vertical line $\Delta x = 0$ so $\Delta z = i\Delta y$, and along other directions Δz is some complex number; in general, Δz is neither real nor pure imaginary. Since the numerator of $\Delta f/\Delta z$ is real and the denominator may be real or imaginary (in general, complex), we see that $\lim_{\Delta z\to 0} \Delta f/\Delta z$ has different values for different directions of approach to z_0, that is, $|z|^2$ is not analytic.

Now we have seen examples of both analytic and nonanalytic functions, but we still do not know how to tell whether a function has a derivative [except to appeal to (2.1)]. The following theorems answer this question.

Theorem I (which we shall prove). If $f(z) = u(x,y) + iv(x,y)$ is analytic in a region, then in that region

$$\frac{\partial u}{\partial x} = \frac{\partial v}{\partial y}, \qquad \frac{\partial v}{\partial x} = -\frac{\partial u}{\partial y}. \tag{2.2}$$

These equations are called the *Cauchy-Riemann conditions.*

Proof. Remembering that $f = f(z)$, where $z = x + iy$, we find by the rules of partial differentiation (see Problem 28 and also Chapter 4)

$$\begin{aligned} \frac{\partial f}{\partial x} &= \frac{df}{dz}\frac{\partial z}{\partial x} = \frac{df}{dz}\cdot 1, \\ \frac{\partial f}{\partial y} &= \frac{df}{dz}\frac{\partial z}{\partial y} = \frac{df}{dz}\cdot i. \end{aligned} \tag{2.3}$$

Since $f = u(x,y) + iv(x,y)$ by (1.1), we also have

$$\frac{\partial f}{\partial x} = \frac{\partial u}{\partial x} + i\frac{\partial v}{\partial x} \quad \text{and} \quad \frac{\partial f}{\partial y} = \frac{\partial u}{\partial y} + i\frac{\partial v}{\partial y}. \tag{2.4}$$

Notice that if f has a derivative with respect to z, then it also has partial derivatives with respect to x and y by (2.3). Since a complex function has a derivative with respect to a real variable if and only if its real and imaginary parts do [see (1.1)], then by (2.4) u and v also have partial derivatives with respect to x and y. Combining (2.3) and (2.4) we have

$$\frac{df}{dz} = \frac{\partial f}{\partial x} = \frac{\partial u}{\partial x} + i\frac{\partial v}{\partial x} \quad \text{and} \quad \frac{df}{dz} = \frac{1}{i}\frac{\partial f}{\partial y} = \frac{1}{i}\left(\frac{\partial u}{\partial y} + i\frac{\partial v}{\partial y}\right) = \frac{\partial v}{\partial y} - i\frac{\partial u}{\partial y}.$$

Since we assumed that df/dz exists and is unique (this is what analytic means), these two expressions for df/dz must be equal. Taking real and imaginary parts, we get the Cauchy-Riemann equations (2.2).

Theorem II (which we state without proof). If $u(x, y)$ and $v(x, y)$ and their partial derivatives with respect to x and y are continuous and satisfy the Cauchy-Riemann conditions in a region, then $f(z)$ is analytic at all points inside the region (not necessarily on the boundary).

Although we shall not prove this (see texts on complex variables), we can make it plausible by showing that it is true when we approach z_0 along any straight line.

Example 3. Find df/dz assuming that we approach z_0 along a straight line of slope m, and show that df/dz does not depend on m if u and v satisfy (2.2). The equation of the straight line of slope m through the point $z_0 = x_0 + iy_0$ is

$$y - y_0 = m(x - x_0)$$

and along this line we have $dy/dx = m$. Then we find

$$\begin{aligned}\frac{df}{dz} = \frac{du + i\,dv}{dx + i\,dy} &= \frac{\dfrac{\partial u}{\partial x}\,dx + \dfrac{\partial u}{\partial y}\,dy + i\left(\dfrac{\partial v}{\partial x}\,dx + \dfrac{\partial v}{\partial y}\,dy\right)}{dx + i\,dy} \\ &= \frac{\dfrac{\partial u}{\partial x} + \dfrac{\partial u}{\partial y}m + i\left(\dfrac{\partial v}{\partial x} + \dfrac{\partial v}{\partial y}m\right)}{1 + im}.\end{aligned}$$

Using the Cauchy-Riemann equations (2.2), we get

$$\begin{aligned}\frac{df}{dz} &= \frac{\dfrac{\partial u}{\partial x} - \dfrac{\partial v}{\partial x}m + i\left(\dfrac{\partial v}{\partial x} + \dfrac{\partial u}{\partial x}m\right)}{1 + im} \\ &= \frac{\dfrac{\partial u}{\partial x}(1 + im) + i\dfrac{\partial v}{\partial x}(1 + im)}{1 + im} = \frac{\partial u}{\partial x} + i\frac{\partial v}{\partial x}.\end{aligned}$$

Thus df/dz has the same value for approach along *any* straight line. The theorem states that it also has the same value for approach along *any curve.*

Some definitions:

A *regular point* of $f(z)$ is a point at which $f(z)$ is analytic.

A *singular point* or *singularity* of $f(z)$ is a point at which $f(z)$ is not analytic. It is called an *isolated* singular point if $f(z)$ is analytic everywhere else inside some small circle about the singular point.

Theorem III (which we state without proof). If $f(z)$ is analytic in a region (R in Figure 2.3), then it has derivatives of all orders at points inside the region and can be expanded in a Taylor series about any point z_0 inside the region. The power series converges *inside* the circle about z_0 that extends to the nearest singular point (C in Figure 2.3).

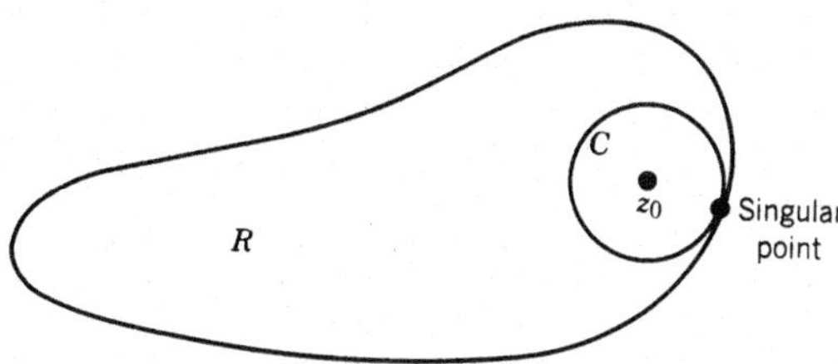

Figure 2.3

Notice again what a strong condition it is on $f(z)$ to say that it has a derivative. It is quite possible for a function of a real variable $f(x)$ to have a first derivative but not higher derivatives. But if $f(z)$ has a first derivative with respect to z, then it has derivatives of all orders, and all these derivatives are analytic functions.

This theorem also explains a fact about power series which may have puzzled you. The function $f(x) = 1/(1+x^2)$ does not have anything peculiar about its behavior at $x = \pm 1$. Yet if we expand it in a power series

$$\frac{1}{1+x^2} = 1 - x^2 + x^4 - x^6 + \cdots \tag{2.5}$$

we see that the series converges only for $|x| < 1$. We can see why this happens if we consider instead

$$f(z) = \frac{1}{1+z^2} = 1 - z^2 + z^4 - z^6 + \cdots . \tag{2.6}$$

When $z = \pm i$, $f(z)$ and its derivatives become infinite; that is, $f(z)$ is not analytic in any region containing $z = \pm i$. The point z_0 of the theorem is the origin and the circle C (bounding the disk of convergence of the series) passes through the nearest singular points $\pm i$ (Figure 2.4). Since a power series in z always converges inside its disk of convergence and diverges outside (Chapter 2, Problem 6.14), we see that (2.5) [which is (2.6) for $y = 0$] converges for $|x| < 1$ and diverges for $|x| > 1$. This simple example shows an important reason for studying functions of a complex variable; our study of $f(z)$ gives us insights about the corresponding $f(x)$. Formulas involving not only the elementary functions but also Γ functions, Bessel functions, and many others are more easily derived and understood by considering them as functions of z.

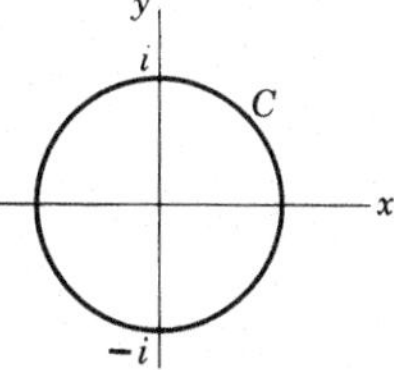

Figure 2.4

A function $\phi(x, y)$ which satisfies Laplace's equation in two dimensions, namely, $\nabla^2\phi = \partial^2\phi/dx^2 + \partial^2\phi/\partial y^2$, is called a *harmonic* function. A great many physical problems lead to Laplace's equation, and consequently we are very much interested in finding solutions of it. (See Section 10 and Chapter 13.) The following theorem should then give you a clue as to one reason why the theory of functions of a complex variable is important in applications.

Theorem IV. Part 1 (to be proved in Problem 44). If $f(z) = u + iv$ is analytic in a region, then u and v satisfy Laplace's equation in the region (that is, u and v are harmonic functions).

Part 2 (which we state without proof). Any function u (or v) satisfying Laplace's equation in a simply-connected region, is the real or imaginary part of an analytic function $f(z)$.

Thus we can find solutions of Laplace's equation simply by taking the real or imaginary parts of an analytic function of z. It is also often possible, starting with a simple function which satisfies Laplace's equation, to find the explicit function $f(z)$ of which it is, say, the real part.

Example 4. Consider the function $u(x, y) = x^2 - y^2$. We find that

$$\nabla^2 u = \frac{\partial^2 u}{\partial x^2} + \frac{\partial^2 u}{\partial y^2} = 2 - 2 = 0,$$

that is, u satisfies Laplace's equation (or u is a harmonic function). Let us find the function $v(x, y)$ such that $u+iv$ is an analytic function of z. By the Cauchy-Riemann equations

$$\frac{\partial v}{\partial y} = \frac{\partial u}{\partial x} = 2x.$$

Integrating partially with respect to y, we get

$$v(x, y) = 2xy + g(x),$$

where $g(x)$ is a function of x to be found. Differentiating partially with respect to x and again using the Cauchy-Riemann equations, we have

$$\frac{\partial v}{\partial x} = 2y + g'(x) = -\frac{\partial u}{\partial y} = 2y.$$

Thus we find

$$g'(x) = 0, \qquad \text{or} \qquad g = \text{const.}$$

Then

$$f(z) = u + iv = x^2 - y^2 + 2ixy + \text{const.} = z^2 + \text{const.}$$

The pair of functions u, v are called *conjugate harmonic functions*. (Also see Problem 64.)

PROBLEMS, SECTION 2

1 to **21**. Use the Cauchy-Riemann conditions to find out whether the functions in Problems 1.1 to 1.21 are analytic. Similarly, find out whether the following functions are analytic.

22. $y + ix$ **23.** $\dfrac{x - iy}{x^2 + y^2}$ **24.** $\dfrac{y - ix}{x^2 + y^2}$

Using the definition (2.1) of $(d/dz)f(z)$, show that the following familiar formulas hold. *Hint:* Use the same methods as for functions of a real variable.

25. $\dfrac{d}{dz}[Af(z) + Bg(z)] = A\dfrac{df}{dz} + B\dfrac{dg}{dz}$. **26.** $\dfrac{d}{dz}[f(z)g(z)] = f(z)\dfrac{dg}{dz} + g(z)\dfrac{df}{dz}$.

27. $\dfrac{d}{dz}\left(\dfrac{f(z)}{g(z)}\right) = \dfrac{gf' - fg'}{g^2}$, $g(z) \neq 0$. **28.** $\dfrac{d}{dz}f[g(z)] = \dfrac{df}{dg}\dfrac{dg}{dz}$. (See hint below.)

Problem 28 is the chain rule for the derivative of a function of a function. *Hint:* Assume that df/dg and dg/dz exist, and write equations like (3.5) of Chapter 4 for Δf and Δg; substitute Δg into Δf, divide by Δz, and take limits.

29. $\dfrac{d}{dz}(z^3) = 3z^2$. **30.** $\dfrac{d}{dz}(z^n) = nz^{n-1}$.

31. $\dfrac{d}{dz}\ln z = \dfrac{1}{z}$, $z \neq 0$. *Hint:* Expand $\ln\left(1 + \dfrac{\Delta z}{z}\right)$ in series.

32. Using the definition of e^z by its power series [(8.1) of Chapter 2], and the theorem (Chapters 1 and 2) that power series may be differentiated term by term (within the disk of convergence), and the result of Problem 30, show that $(d/dz)(e^z) = e^z$.

33. Using the definitions of $\sin z$ and $\cos z$ [Chapter 2, equation (11.4)], find their derivatives. Then using Problem 27, find $(d/dz)(\cot z)$, $z \neq n\pi$.

Using series you know from Chapter 1, write the power series (about the origin) of the following functions. Use Theorem III to find the disk of convergence of each series. What you are looking for is the point (anywhere in the complex plane) nearest the origin, at which the function does not have a derivative. Then the disk of convergence has center at the origin and extends to that point. The series converges *inside* the disk.

34. $\ln(1 - z)$ **35.** $\cos z$ **36.** $\sqrt{1 + z^2}$

37. $\tanh z$ **38.** $\dfrac{1}{2i + z}$ **39.** $\dfrac{z}{z^2 + 9}$

40. $(1 - z)^{-1}$ **41.** e^{iz} **42.** $\sinh z$

43. In Chapter 12, equations (5.1) and (5.2), we expanded the function $\phi(x, h)$ in a series of powers of h. Use Theorem III (see instructions for Problems 34 to 42 above) to show that the series for $\phi(x, h)$ converges for $|h| < 1$ and $-1 \leq x \leq 1$. Here h is the variable and x is a parameter; you should find the (complex) value of h which makes Φ infinite, and show that the absolute value of this complex number is 1 (independent of x when $x^2 \leq 1$). This proves that the series for real h converges for $|h| < 1$.

44. Prove Theorem IV, Part 1. *Hint:* Recall the equality of the second cross partial derivatives; see Chapter 4, end of Section1.

45. Let $f(z) = u + iv$ be an analytic function, and let $\mathbf{F}$ be the vector $\mathbf{F} = v\mathbf{i} + u\mathbf{j}$. Show that the equations $\operatorname{div}\mathbf{F} = 0$ and $\operatorname{curl}\mathbf{F} = 0$ are equivalent to the Cauchy-Riemann equations.

46. Find the Cauchy-Riemann equations in polar coordinates. *Hint*: Write $z = re^{i\theta}$ and $f(z) = u(r, \theta) + iv(r, \theta)$. Follow the method of equations (2.3) and (2.4).

47. Using your results in Problem 46 and the method of Problem 44, show that u and v satisfy Laplace's equation in polar coordinates (see Chapter 10, Section 9) if $f(z) = u + iv$ is analytic.

Using polar coordinates (Problem 46), find out whether the following functions satisfy the Cauchy-Riemann equations.

48. $\sqrt{z}$ **49.** $|z|$ **50.** $\ln z$

51. z^n **52.** $|z|^2$ **53.** $|z|^{1/2}e^{i\theta/2}$

Show that the following functions are harmonic, that is, that they satisfy Laplace's equation, and find for each a function $f(z)$ of which the given function is the real part. Show that the function $v(x, y)$ (which you find) also satisfies Laplace's equation.

54. y **55.** $3x^2y - y^3$ **56.** xy **57.** $x + y$

58. $\cosh y \cos x$ **59.** $e^x \cos y$ **60.** $\ln(x^2 + y^2)$

61. $\dfrac{x}{x^2 + y^2}$ **62.** $e^{-y} \sin x$ **63.** $\dfrac{y}{(1 - x)^2 + y^2}$

64. It can be shown that, if $u(x, y)$ is a harmonic function which is defined at $z_0 = x_0 + iy_0$, then an analytic function of which $u(x, y)$ is the real part is given by

$$f(z) = 2u\left(\frac{z + \bar{z}_0}{2}, \frac{z - \bar{z}_0}{2i}\right) + \text{const.}$$

[See Struble, Quart. Appl. Math., 37 (1979), 79-81.] Use this formula to find $f(z)$ in Problems 54 to 63. *Hint:* If $u(0, 0)$ is defined, take $z_0 = 0$.

3. CONTOUR INTEGRALS

Theorem V Cauchy's theorem (see discussion below). Let C be a simple[†] closed curve with a continuously turning tangent except possibly at a finite number of points (that is, we allow a finite number of corners, but otherwise the curve must be "smooth"). If $f(z)$ is analytic on and inside C, then

$$\oint_{\text{around } C} f(z)\, dz = 0. \tag{3.1}$$

(This is a line integral as in vector analysis; it is called a *contour integral* in the theory of complex variables.)

Proof. We shall prove Cauchy's theorem assuming that $f'(z)$ is continuous. (With more effort it is possible to prove it without this assumption, and then show that if $f'(z)$ exists in a region, it is, in fact, continuous there. See also Theorem III which we stated without proof; it is usually proved using the results of Cauchy's theorem.)

$$\begin{aligned} \oint_C f(z)\, dz &= \oint_C (u + iv)(dx + i\, dy) \\ &= \oint_C (u\, dx - v\, dy) + i \oint_C (v\, dx + u\, dy). \end{aligned} \tag{3.2}$$

[†]A simple curve is one which does not cross itself.

Green's theorem in the plane (Chapter 6, Section 9) says that if $P(x, y)$, $Q(x, y)$, and their partial derivatives are continuous in a simply-connected region R, then

$$\oint_C P\,dx + Q\,dy = \iint_{\substack{\text{area}\\ \text{inside } C}} \left(\frac{\partial Q}{\partial x} - \frac{\partial P}{\partial y}\right) dx\,dy, \tag{3.3}$$

where C is a simple closed curve lying entirely in R. The curve C is traversed in a direction so that the area inclosed is always to the left; the area integral is over the area inside C. Applying (3.3) to the first integral in (3.2), we get

$$\oint_C (u\,dx - v\,dy) = \iint_{\substack{\text{area}\\ \text{inside } C}} \left(-\frac{\partial v}{\partial x} - \frac{\partial u}{\partial y}\right) dx\,dy. \tag{3.4}$$

Since we are assuming that $f'(z)$ is continuous, then u and v and their derivatives are continuous; by the Cauchy-Riemann equations the integrand on the right of (3.4) is zero at every point of the area of integration, so the integral is equal to zero. In the same way the second integral in (3.2) is zero; thus (3.1) is proved.

Theorem VI Cauchy's integral formula (which we shall prove). If $f(z)$ is analytic on and inside a simple closed curve C, the value of $f(z)$ at a point $z = a$ inside C is given by the following contour integral along C:

$$f(a) = \frac{1}{2\pi i}\oint \frac{f(z)}{z-a}\,dz.$$

Proof. Let a be a fixed point inside the simple closed curve C and consider the function

$$\phi(z) = \frac{f(z)}{z-a}, \tag{3.5}$$

where $f(z)$ is analytic on and inside C. Let C' be a small circle (inside C) with center at a and radius ρ. Make a cut between C and C' along AB (Figure 3.1); two cuts are shown to make the picture clear, but later we shall make them coincide. We are now going to integrate along the path shown in Figure 3.1 (in the direction shown by the arrows) from A, around C, to B, around C', and back to A. Notice that the area between the curves C and C' is always to the left of the path of integration and is inclosed by it. In this area between C and C', the function $\phi(z)$ is analytic; we have cut out a small disk about the point $z = a$ at which $\phi(z)$ is not analytic. Cauchy's theorem then applies to the integral along the combined path consisting of C counterclockwise, C' clockwise, and the two cuts. The two integrals, in opposite directions along the cuts, cancel when the cuts are made to coincide. Thus we have

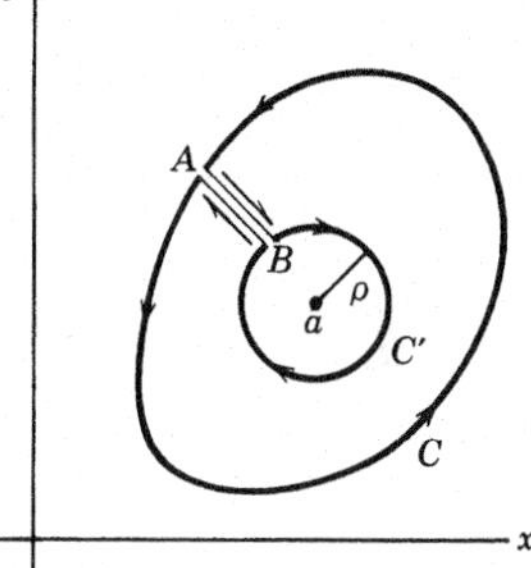

Figure 3.1

$$\oint_{\substack{C \text{ counter-}\\ \text{clockwise}}} \phi(z)\,dz + \oint_{C' \text{ clockwise}} \phi(z)\,dz = 0 \quad \text{or}$$

(3.6)

$$\oint_C \phi(z)\,dz = \oint_{C'} \phi(z)\,dz \qquad \text{where both are counterclockwise.}$$

Along the circle C', $z = a + \rho e^{i\theta}$, $dz = \rho i e^{i\theta}\,d\theta$, and (3.6) becomes

$$\begin{aligned}\oint_C \phi(z)\,dz &= \oint_{C'} \phi(z)\,dz = \oint_{C'} \frac{f(z)}{z-a}\,dz \\ &= \int_0^{2\pi} \frac{f(z)}{\rho e^{i\theta}} \rho i e^{i\theta}\,d\theta = \int_0^{2\pi} f(z) i\,d\theta.\end{aligned} \tag{3.7}$$

Since our calculation is valid for any (sufficiently small) value of ρ, we shall let $\rho \to 0$ (that is, $z \to a$) to simplify the formula. Because $f(z)$ is continuous at $z = a$ (it is analytic inside C), $\lim_{z\to a} f(z) = f(a)$. Then (3.7) becomes

$$\oint_C \phi(z)\,dz = \oint_C \frac{f(z)}{z-a}\,dz = \int_0^{2\pi} f(z) i\,d\theta = \int_0^{2\pi} f(a) i\,d\theta = 2\pi i f(a) \tag{3.8}$$

or

$$f(a) = \frac{1}{2\pi i} \oint_C \frac{f(z)}{z-a}\,dz, \qquad a \text{ inside } C. \tag{3.9}$$

This is Cauchy's integral formula. Note carefully that the point a is inside C; if a were outside C, then $\phi(z)$ would be analytic everywhere inside C and the integral would be zero by Cauchy's theorem. A useful way to look at (3.9) is this: If the values of $f(z)$ are given on the boundary of a region (curve C), then (3.9) gives the value of $f(z)$ at any point a inside C. With this interpretation you will find Cauchy's integral formula written with a replaced by z, and z replaced by some different dummy integration variable, say w:

$$f(z) = \frac{1}{2\pi i} \oint_C \frac{f(w)}{w-z}\,dw, \qquad z \text{ inside } C. \tag{3.10}$$

For some important uses of this theorem, see Problems 11.3 and 11.36 to 11.38.

▶ PROBLEMS, SECTION 3

Evaluate the following line integrals in the complex plane by direct integration, that is, as in Chapter 6, Section 8, *not* using theorems from this chapter. (If you see that a theorem applies, use it to check your result.)

1. $\int_i^{i+1} z\,dz$ along a straight line parallel to the x axis.

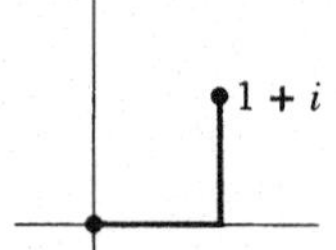

2. $\int_0^{1+i} (z^2 - z)\,dz$

(a) along the line $y = x$;

(b) along the indicated broken line.

3. $\oint_C z^2\,dz$ along the indicated paths:

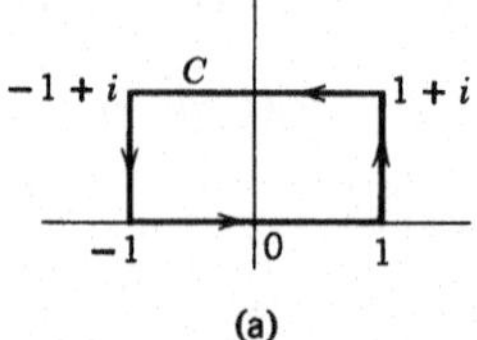

(a)

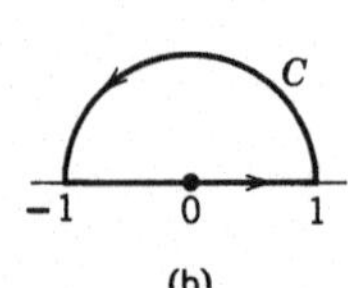

(b)

4. $\int dz/(1-z^2)$ along the whole positive imaginary axis, that is, the y axis; this is frequently written as $\int_0^{i\infty} dz/(1-z^2)$.

5. $\int e^{-z}$ along the positive part of the line $y = \pi$; this is frequently written as $\int_{i\pi}^{\infty+i\pi} e^{-z}\,dz$.

6. $\int_1^i z\,dz$ along the indicated paths:

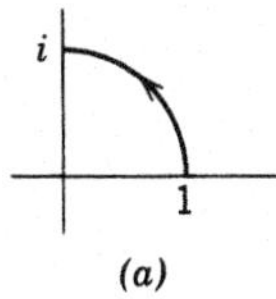

(a)

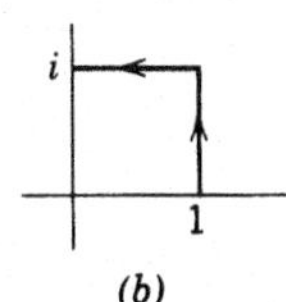

(b)

7. $\displaystyle\int \frac{dz}{8i+z^2}$ along the line $y = x$ from 0 to ∞.

8. $\displaystyle\int_{2\pi}^{2\pi+i\infty} e^{2iz}\,dz$ **9.** $\displaystyle\int_{1+2i}^{\infty+2i} \frac{dz}{(x-2i)^2}$ **10.** $\displaystyle\int_2^{2+i\infty} ze^{iz}\,dz$

11. Evaluate $\oint_C (\bar{z}-3)\,dz$ where C is the indicated closed curve along the first quadrant part of the circle $|z| = 2$, and the indicated parts of the x and y axes. *Hint:* Don't try to use Cauchy's theorem! (Why not? *Further hint:* See Problem 2.3.)

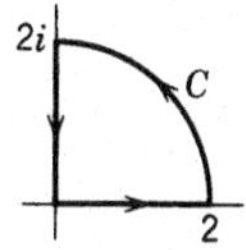

12. $\int_0^{1+2i} |z|^2\,dz$ along the indicated paths:

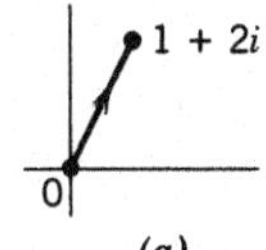

(a)

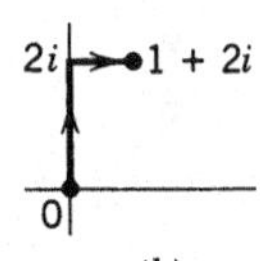

(b)

13. In Chapter 6, Section 11, we showed that a necessary condition for $\int_a^b \mathbf{F}\cdot d\mathbf{r}$ to be independent of the path of integration, that is, for $\oint_C \mathbf{F}\cdot d\mathbf{r}$ around a simple closed curve C to be zero, was $\operatorname{curl}\mathbf{F} = 0$, or in two dimensions, $\partial F_y/\partial x = \partial F_x/\partial y$. By considering (3.2), show that the corresponding condition for $\oint_C f(z)\,dz$ to be zero is that the Cauchy-Riemann conditions hold.

14. In finding complex Fourier series in Chapter 7, we showed that

$$\int_0^{2\pi} e^{inx}e^{-imx}\,dx = 0, \qquad n \neq m.$$

Show this by applying Cauchy's theorem to

$$\oint_C z^{n-m-1}\,dz, \qquad n > m,$$

where C is the circle $|z| = 1$. (Note that although we take $n > m$ to make z^{n-m-1} analytic at $z = 0$, an identical proof using z^{m-n-1} with $n < m$ completes the proof for all $n \neq m$.)

15. If $f(z)$ is analytic on and inside the circle $|z| = 1$, show that $\int_0^{2\pi} e^{i\theta} f(e^{i\theta})\,d\theta = 0$.

16. If $f(z)$ is analytic in the disk $|z| \le 2$, evaluate $\int_0^{2\pi} e^{2i\theta} f(e^{i\theta})\,d\theta$.

Use Cauchy's theorem or integral formula to evaluate the integrals in Problems 17 to 20.

17. $\displaystyle\oint_C \frac{\sin z\,dz}{2z-\pi}$ where C is the circle (a) $|z| = 1$, (b) $|z| = 2$.

18. $\displaystyle\oint_C \frac{\sin 2z\,dz}{6z-\pi}$ where C is the circle $|z| = 3$.

19. $\oint \frac{e^{3z}\,dz}{z-\ln 2}$ if C is the square with vertices $\pm 1 \pm i$.

20. $\oint_C \frac{\cosh z\,dz}{2\ln 2 - z}$ if C is the circle (a) $|z| = 1$, (b) $|z| = 2$.

21. Differentiate Cauchy's formula (3.9) or (3.10) to get

$$f'(z) = \frac{1}{2\pi i}\oint_C \frac{f(w)\,dw}{(w-z)^2} \qquad \text{or} \qquad f'(a) = \frac{1}{2\pi i}\oint_C \frac{f(z)\,dz}{(z-a)^2}.$$

By differentiating n times, obtain

$$f^{(n)}(z) = \frac{n!}{2\pi i}\oint_C \frac{f(w)\,dw}{(w-z)^{n+1}} \qquad \text{or} \qquad f^{(n)}(a) = \frac{n!}{2\pi i}\oint_C \frac{f(z)\,dz}{(z-a)^{n+1}}.$$

Use Problem 21 to evaluate the following integrals.

22. $\oint_C \frac{\sin 2z\,dz}{(6z-\pi)^3}$ where C is the circle $|z| = 3$.

23. $\oint_C \frac{e^{3z}\,dz}{(z-\ln 2)^4}$ where C is the square in Problem 19.

24. $\oint_C \frac{\cosh z\,dz}{(2\ln 2 - z)^5}$ where C is the circle $|z| = 2$.

4. LAURENT SERIES

Theorem VII Laurent's theorem [equation (4.1)] (which we shall state without proof). Let C_1 and C_2 be two circles with center at z_0. Let $f(z)$ be analytic in the region R between the circles. Then $f(z)$ can be expanded in a series of the form

$$f(z) = a_0 + a_1(z-z_0) + a_2(z-z_0)^2 + \cdots + \frac{b_1}{z-z_0} + \frac{b_2}{(z-z_0)^2} + \cdots \tag{4.1}$$

convergent in R. Such a series is called a *Laurent series*. The "b" series in (4.1) is called the *principal part* of the Laurent series.

Example 1. Consider the Laurent series

$$f(z) = 1 + \frac{z}{2} + \frac{z^2}{4} + \frac{z^3}{8} + \cdots + \left(\frac{z}{2}\right)^n + \cdots + \frac{2}{z} + 4\left(\frac{1}{z^2} - \frac{1}{z^3} + \cdots + \frac{(-1)^n}{z^n} + \cdots\right). \tag{4.2}$$

Let us see where this series converges. First consider the series of positive powers; by the ratio test (see Chapters 1 and 2), this series converges for $|z/2| < 1$, that is, for $|z| < 2$. Similarly, the series of negative powers converges for $|1/z| < 1$, that is, $|z| > 1$. Then both series converge (and so the Laurent series converges) for $|z|$ between 1 and 2, that is, in a ring between two circles of radii 1 and 2.

We expect this result in general. The "a" series is a power series, and a power series converges *inside* some circle (say C_2 in Figure 4.1). The "b" series is a series of inverse powers of z, and so converges for $|1/z| <$ some constant; thus the "b" series converges *outside* some circle (say C_1 in Figure 4.1). Then a Laurent series converges between two circles (if it converges at all). (Note that the inner circle may be a point and the outer circle may have infinite radius).

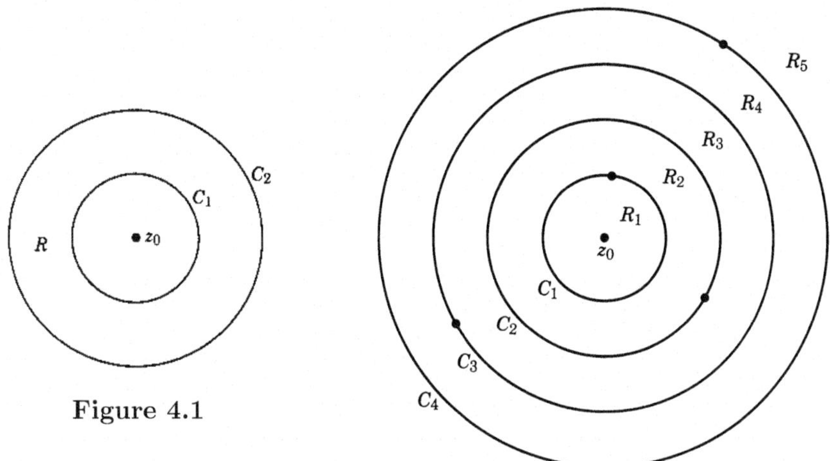

Figure 4.1

Figure 4.2

The formulas for the coefficients in (4.1) are (Problem 5.2)

$$a_n = \frac{1}{2\pi i}\oint_C \frac{f(z)\,dz}{(z-z_0)^{n+1}}, \qquad b_n = \frac{1}{2\pi i}\oint_C \frac{f(z)\,dz}{(z-z_0)^{-n+1}}, \tag{4.3}$$

where C is any simple closed curve surrounding z_0 and lying in R. However, this is not usually the easiest way to find a Laurent series. Like power series about a point, the Laurent series (about z_0) for a function in a given annular ring (about z_0) where the function is analytic, is unique, and we can find it by any method we choose. (See examples below.) *Warning:* If $f(z)$ has several isolated singularities (Figure 4.2), there are several annular rings, R_1, R_2, $\cdots$, in which $f(z)$ is analytic; then there are several different Laurent series for $f(z)$, one for each ring. The Laurent series which we usually want is the one that converges near z_0. If you have any doubt about the ring of convergence of a Laurent series, you can find out by testing the "a" series and the "b" series separately.

Example 2. The function from which we obtained (4.2) was

$$f(z) = \frac{12}{z(2-z)(1+z)}. \tag{4.4}$$

This function has three singular points, at $z = 0$, $z = 2$, and $z = -1$. Thus there are two circles C_1 and C_2 about $z_0 = 0$ in Figure 4.2, and three Laurent series about $z_0 = 0$, one series valid in each of the three regions R_1 $(0 < |z| < 1)$, $R_2(1 < |z| < 2)$, and $R_3(|z| > 2)$. To find these series we first write $f(z)$ in the following form using partial fractions (Problem 2):

$$f(z) = \frac{4}{z}\left(\frac{1}{1+z} + \frac{1}{2-z}\right). \tag{4.5}$$

Now, for $0 < |z| < 1$, we expand each of the fractions in the parenthesis in (4.5) in powers of z. This gives (Problem 2):

$$f(z) = -3 + 9z/2 - 15z^2/4 + 33z^3/8 + \cdots + 6/z. \tag{4.6}$$

This is the Laurent series for $f(z)$ which is valid in the region $0 < |z| < 1$. To obtain the series valid in the region $|z| > 2$, we write the fractions in (4.5) as

$$\frac{1}{1+z} = \frac{1}{z}\frac{1}{1+1/z}, \qquad \frac{1}{2-z} = -\frac{1}{z}\frac{1}{1-2/z} \tag{4.7}$$

and expand each fraction in powers of $1/z$. This gives the Laurent series valid for $|z| > 2$ (problem 2):

$$f(z) = -(12/z^3)(1 + 1/z + 3/z^2 + 5/z^3 + 11/z^4 + \cdots). \tag{4.8}$$

Finally, to obtain (4.2), we expand the fraction $1/(2-z)$ in powers of z, and the fraction $1/(1+z)$ in powers of $1/z$; this gives a Laurent series which converges for $1 < |z| < 2$. Thus the Laurent series (4.6), (4.2) and (4.8) all represent $f(z)$ in (4.4), but in three different regions.

Let z_0 in Figure 4.2 be either a regular point or an isolated singular point and assume that there are no other singular points inside C_1. Let $f(z)$ be expanded in the Laurent series about z_0 which converges inside C_1 (except possibly at z_0); we say that we have expanded $f(z)$ in the Laurent series which converges near z_0. Then we have the following definitions.

Definitions:

If all the b's are zero, $f(z)$ is analytic at $z = z_0$, and we call z_0 a *regular point.* (See Problem 4.1)

If $b_n \neq 0$, but all the b's after b_n are zero, $f(z)$ is said to have a *pole of order n* at $z = z_0$. If $n = 1$, we say that $f(z)$ has a *simple pole.*

If there are an infinite number of b's different from zero, $f(z)$ has an *essential singularity* at $z = z_0$.

The coefficient b_1 of $1/(z - z_0)$ is called the *residue* of $f(z)$ at $z = z_0$.

Example 3.

(a)
$$e^z = 1 + z + \frac{z^2}{2!} + \frac{z^3}{3!} + \cdots$$

is analytic at $z = 0$; the residue of e^z at $z = 0$ is 0.

(b)
$$\frac{e^z}{z^3} = \frac{1}{z^3} + \frac{1}{z^2} + \frac{1}{2!z} + \frac{1}{3!} + \cdots$$

has a pole of order 3 at $z = 0$; the residue of $\frac{e^z}{z^3}$ at $z = 0$ is $\frac{1}{2!}$.

(c)
$$e^{1/z} = 1 + \frac{1}{z} + \frac{1}{2!z^2} + \cdots$$

has an essential singularity at $z = 0$; the residue of $e^{1/z}$ at $z = 0$ is 1.

Most of the functions we shall consider will be analytic except for poles—such functions are called *meromorphic* functions. If $f(z)$ has a pole at $z = z_0$, then $|f(z)| \to \infty$ as $z \to z_0$. A three-dimensional graph with $|f(z)|$ plotted vertically over a horizontal complex plane would look like a tapered pole near $z = z_0$. We can often see that a function has a pole and find the order of the pole without finding the Laurent series.

Example 4.

(a)
$$\frac{z+3}{z^2(z-1)^3(z+1)}$$

has a pole of order 2 at $z = 0$, a pole of order 3 at $z = 1$, and a simple pole at $z = -1$.

(b) $\dfrac{\sin^2 z}{z^3}$ has a simple pole at $z = 0$.

To see that these results are correct, consider finding the Laurent series for $f(z) = g(z)/(z - z_0)^n$. We write $g(z) = a_0 + a_1(z - z_0) + \cdots$; then the Laurent series for $f(z)$ starts with the term $(z - z_0)^{-n}$ unless $a_0 = 0$, that is unless $g(z_0) = 0$. Then the order of the pole of $f(z)$ is n unless some factors cancel. In Example 4b, the $\sin z$ series starts with z, so $\sin^2 z$ has a factor z^2; thus $(\sin^2 z)/z^3$ has a simple pole at $z = 0$.

PROBLEMS, SECTION 4

1. Show that the sum of a power series which converges inside a circle C is an analytic function inside C. *Hint:* See Chapter 2, Section 7, and Chapter 1, Section 11, and the definition of an analytic function.

2. Show that equation (4.4) can be written as (4.5). Then expand each of the fractions in the parenthesis in (4.5) in powers of z and in powers of $1/z$ [see equation (4.7)] and combine the series to obtain (4.6), (4.8), and (4.2).

For each of the following functions find the first few terms of each of the Laurent series about the origin, that is, one series for each annular ring between singular points. Find the residue of each function at the origin. (*Warning:* To find the residue, you must use the Laurent series which converges near the origin.) *Hints:* See Problem 2. Use partial fractions as in equations (4.5) and (4.7). Expand a term $1/(z - a)$ in powers of z to get a series convergent for $|z| < a$, and in powers of $1/z$ to get a series convergent for $|z| > a$.

3. $\dfrac{1}{z(z-1)(z-2)}$ **4.** $\dfrac{1}{z(z-1)(z-2)^2}$ **5.** $\dfrac{z-1}{z^3(z-2)}$

6. $\dfrac{1}{z^2(1+z)^2}$ **7.** $\dfrac{2-z}{1-z^2}$ **8.** $\dfrac{30}{(1+z)(z-2)(3+z)}$

For each of the following functions, say whether the indicated point is regular, an essential singularity, or a pole, and if a pole of what order it is.

9. (a) $\dfrac{\sin z}{z}$, $z = 0$ (b) $\dfrac{\cos z}{z^3}$, $z = 0$

(c) $\dfrac{z^3 - 1}{(z-1)^3}$, $z = 1$ (d) $\dfrac{e^z}{z-1}$, $z = 1$

10. (a) $\dfrac{e^z - 1}{z^2 + 4}$, $z = 2i$ (b) $\tan^2 z$, $z = \pi/2$

(c) $\dfrac{1 - \cos z}{z^4}$, $z = 0$ (d) $\cos\left(\dfrac{\pi}{z - \pi}\right)$, $z = \pi$

11. (a) $\dfrac{e^z - 1 - z}{z^2}$, $z = 0$ (b) $\dfrac{\sin z}{z^3}$, $z = 0$

(c) $\dfrac{z^2 - 1}{(z-1)^2}$, $z = 1$ (d) $\dfrac{\cos z}{(z - \pi/2)^4}$, $z = \pi/2$

12. (a) $\dfrac{\sin z - z}{z^6}$, $z = 0$ (b) $\dfrac{z^2 - 1}{(z^2 + 1)^2}$, $z = i$

(c) $ze^{1/z}$, $z = 0$ (d) $\Gamma(z)$, $z = 0$ [See Chapter 11, equation (4.1)]

5. THE RESIDUE THEOREM

Let z_0 be an isolated singular point of $f(z)$. We are going to find the value of $\oint_C f(z)\,dz$ around a simple closed curve C surrounding z_0 but inclosing no other singularities. Let $f(z)$ be expanded in the Laurent series (4.1) about $z = z_0$ that converges near $z = z_0$. By Cauchy's theorem (V), the integral of the "a" series is zero since this part is analytic. To evaluate the integrals of the terms in the "b" series in (4.1), we replace the integrals around C by integrals around a circle C' with center at z_0 and radius ρ as in (3.6), (3.7), and Figure 3.1. Along C', $z = z_0 + \rho e^{i\theta}$; calculating the integral of the b_1 term in (4.1), we find

$$\oint_C \frac{b_1 dz}{(z - z_0)} = b_1 \int_0^{2\pi} \frac{\rho i e^{i\theta}\,d\theta}{\rho e^{i\theta}} = 2\pi i b_1. \tag{5.1}$$

It is straightforward to show (Problem 1) that the integrals of all the other b_n terms are zero. Then $\oint_C f(z)\,dz = 2\pi i b_1$, or since b_1 is called the residue of $f(z)$ at $z = z_0$, we can say

$$\oint_C f(z)\,dz = 2\pi i \cdot \text{residue of } f(z) \text{ at the singular point inside } C.$$

The only term of the Laurent series which has survived the integration process is the b_1 term; you can see the reason for the term "residue." If there are several isolated singularities inside C, say at z_0, z_1, z_2, $\cdots$, we draw small circles about each as shown in Figure 5.1 so that $f(z)$ is analytic in the region between C and the circles. Then, introducing cuts as in Figure 3.1, we find that the integral around C counterclockwise, plus the integrals around the circles clockwise, is zero (since the

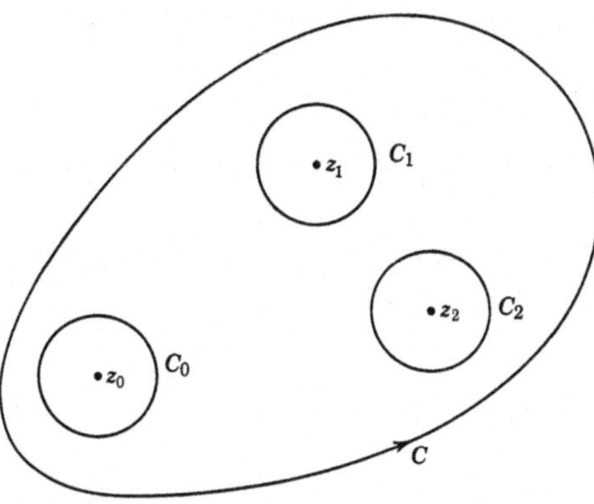

Figure 5.1

integrals along the cuts cancel), or the integral along C is the sum of the integrals around the circles (all counterclockwise). But by (5.1), the integral around each circle is $2\pi i$ times the residue of $f(z)$ at the singular point inside. Thus we have the *residue theorem*:

(5.2) $$\oint_C f(z)\,dz = 2\pi i \cdot \text{sum of the residues of } f(z) \text{ inside } C,$$

where the integral around C is in the counterclockwise direction.

The residue theorem is useful in evaluating many definite integrals; we shall consider this in Section 7. But first, in Section 6, we need to develop some techniques for finding residues.

PROBLEMS, SECTION 5

1. If C is a circle of radius ρ about z_0, show that

$$\oint_C \frac{dz}{(z-z_0)^n} = 2\pi i \qquad \text{if } n = 1,$$

but for any other integral value of n, positive or negative, the integral is zero. *Hint:* Use the fact that $z = z_0 + \rho e^{i\theta}$ on C.

2. Verify the formulas (4.3) for the coefficients in a Laurent series. *Hint:* To get a_n, divide equation (4.1) by $(z-z_0)^{n+1}$ and use the results of Problem 1 to evaluate the integrals of the terms of the series. Use a similar method to find b_n.

3. Obtain Cauchy's integral formula (3.9) from the residue theorem (5.2).

6. METHODS OF FINDING RESIDUES

A. Laurent Series If it is easy to write down the Laurent series for $f(z)$ about $z = z_0$ that is valid near z_0, then the residue is just the coefficient b_1 of the term $1/(z-z_0)$. *Caution*: Be sure you have the expansion about $z = z_0$; the series you have memorized for e^z, $\sin z$, etc., are expansions about $z = 0$ and so can be used only for finding residues at the origin (see Section 4, Example 3). Here is another example: Given $f(z) = e^z/(z-1)$, find the residue, $R(1)$, of $f(z)$ at $z = 1$. We want to expand e^z in powers of $z - 1$; we write

$$\frac{e^z}{z-1} = \frac{e \cdot e^{z-1}}{z-1} = \frac{e}{z-1}\left[1 + (z-1) + \frac{(z-1)^2}{2!} + \cdots\right] = \frac{e}{z-1} + e + \cdots.$$

Then the residue is the coefficient of $1/(z-1)$, that is, $R(1) = e$.

B. Simple Pole If $f(z)$ has a simple pole at $z = z_0$, we find the residue by multiplying $f(z)$ by $(z - z_0)$ and evaluating the result at $z = z_0$ (Problem 10).

▶ **Example 1.** Find $R(-\frac{1}{2})$ and $R(5)$ for

$$f(z) = \frac{z}{(2z+1)(5-z)}.$$

Multiply $f(z)$ by $(z + \frac{1}{2})$, [*Caution*: not by $(2z + 1)$], and evaluate the result at $z = -\frac{1}{2}$. We find

$$(z+\tfrac{1}{2})f(z) = (z+\tfrac{1}{2})\frac{z}{(2z+1)(5-z)} = \frac{z}{2(5-z)},$$

$$R(-\tfrac{1}{2}) = \frac{-\frac{1}{2}}{2(5+\frac{1}{2})} = -\frac{1}{22}.$$

Similarly,

$$(z-5)f(z) = (z-5)\frac{z}{(2z+1)(5-z)} = -\frac{z}{2z+1},$$

$$R(5) = -\frac{5}{11}.$$

▶ **Example 2.** Find $R(0)$ for $f(z) = (\cos z)/z$. Since $zf(z) = \cos z$, we have

$$R(0) = (\cos z)_{z=0} = \cos 0 = 1.$$

To use this method, we may in some problems have to evaluate an indeterminate form, so in general we write

(6.1) $$R(z_0) = \lim_{z \to z_0} (z - z_0) f(z) \quad \text{when } z_0 \text{ is a simple pole.}$$

▶ **Example 3.** Find the residue of $\cot z$ at $z = 0$. By (6.1)

$$R(0) = \lim_{z \to 0} \frac{z \cos z}{\sin z} = \cos 0 \cdot \lim_{z \to 0} \frac{z}{\sin z} = 1 \cdot 1 = 1.$$

If, as often happens, $f(z)$ can be written as $g(z)/h(z)$, where $g(z)$ is analytic and not zero at z_0 and $h(z_0) = 0$, then (6.1) becomes

$$R(z_0) = \lim_{z \to z_0} \frac{(z-z_0)g(z)}{h(z)} = g(z_0) \lim_{z \to z_0} \frac{z-z_0}{h(z)} = g(z_0) \lim_{z \to z_0} \frac{1}{h'(z)} = \frac{g(z_0)}{h'(z_0)}$$

by L'Hôpital's rule or the definition of $h'(z)$ (Problem 11).

Thus we have

$$(6.2) \qquad R(z_0) = \frac{g(z_0)}{h'(z_0)} \quad \text{if} \begin{cases} f(z) = g(z)/h(z), \text{ and} \\ g(z_0) = \text{finite const.} \neq 0, \text{ and} \\ h(z_0) = 0, \; h'(z_0) \neq 0. \end{cases}$$

Often (6.2) gives the most convenient way of finding the residue at a simple pole.

Example 4. Find the residue of $(\sin z)/(1 - z^4)$ at $z = i$. By (6.2) we have

$$R(i) = \left.\frac{\sin z}{-4z^3}\right|_{z=i} = \frac{\sin i}{-4i^3} = \frac{e^{-1} - e}{(2i)(4i)} = \frac{1}{8}(e - e^{-1}) = \frac{1}{4}\sinh 1.$$

Now you may ask how you know, without finding the Laurent series, that a function has a simple pole. Perhaps the simplest answer is that if the limit obtained using (6.1) is some constant (not 0 or ∞), then $f(z)$ *does* have a simple pole and the constant is the residue. [If the limit $= 0$, the function is analytic and the residue $= 0$; if the limit is infinite, the pole is of higher order.] However, you can often recognize the order of a pole in advance. [See end of Section 4 for the simple case in which $(z - z_0)^n$ is a factor of the denominator.] Suppose $f(z)$ is written in the form $g(z)/h(z)$, where $g(z)$ and $h(z)$ are analytic. Then you can think of $g(z)$ and $h(z)$ as power series in $(z - z_0)$. If the denominator has the factor $(z - z_0)$ to *one* higher power than the numerator, then $f(z)$ has a simple pole at z_0. For example,

$$z\cot^2 z = \frac{z\cos^2 z}{\sin^2 z} = \frac{z(1 - z^2/2 + \cdots)^2}{(z - z^3/3! + \cdots)^2} = \frac{z(1 + \cdots)}{z^2(1 + \cdots)}$$

has a simple pole at $z = 0$. By the same method we can see whether a function has a pole of any order

C. Multiple Poles When $f(z)$ has a pole of order n, we can use the following method of finding residues.

Multiply $f(z)$ by $(z - z_0)^m$, where m is an integer greater than or equal to the order n of the pole, differentiate the result $m - 1$ times, divide by $(m - 1)!$, and evaluate the resulting expression at $z = z_0$.

It is easy to prove that this rule is correct (Problem 12) by using the Laurent series (4.1) for $f(z)$ and showing that the result of the outlined process is b_1.

Example 5. Find the residue of $f(z) = (z \sin z)/(z-\pi)^3$ at $z = \pi$.

We take $m = 3$ to eliminate the denominator before differentiating; this is an allowed choice for m because the order of the pole of $f(z)$ at π is not greater than 3 since $z \sin z$ is finite at π. (The pole is actually of order 2, but we do not need this fact.) Then following the rule stated, we get

$$R(\pi) = \frac{1}{2!}\frac{d^2}{dz^2}(z\sin z)\bigg|_{z=\pi} = \frac{1}{2}[-z\sin z + 2\cos z]_{z=\pi} = -1.$$

(To compute the derivative quickly, use Leibniz' rule for differentiating a product; see Chapter 12, Section 3.)

Much of this work can be done by computer. However, remember that the point of doing these problems is to gain skill in using the ideas and techniques of complex variable theory. So a good study method is to do the problems as outlined above and then check your results by computer.

PROBLEMS, SECTION 6

Find the Laurent series for the following functions about the indicated points; hence find the residue of the function at the point. (Be sure you have the Laurent series which converges near the point.)

1. $\dfrac{1}{z(z+1)}$, $z = 0$

2. $\dfrac{1}{z(z-1)}$, $z = 1$

3. $\dfrac{\sin z}{z^4}$, $z = 0$

4. $\dfrac{\cosh z}{z^2}$, $z = 0$

5. $\dfrac{e^z}{z^2-1}$, $z = 1$

6. $\sin\dfrac{1}{z}$, $z = 0$

7. $\dfrac{\sin \pi z}{4z^2-1}$, $z = \dfrac{1}{2}$

8. $\dfrac{1+\cos z}{(z-\pi)^2}$, $z = \pi$

9. $\dfrac{1}{z^2-5z+6}$, $z = 2$

10. Show that rule B is correct by applying it to (4.1).

11. Derive (6.2) by using the limit definition of the derivative $h'(z_0)$ instead of using L'Hôpital's rule. Remember that $h(z_0) = 0$ because we are assuming that $f(z)$ has a simple pole at z_0.

12. Prove rule C for finding the residue at a multiple pole, by applying it to (4.1). Note that the rule is valid for $n = 1$ (simple pole) although we seldom use it for that case.

13. Prove rule C by using (3.9). *Hints:* If $f(z)$ has a pole of order n at $z = a$, then $f(z) = g(z)/(z-a)^n$ with $g(z)$ analytic at $z = a$. By (3.9)

$$\int_C \frac{g(z)}{(z-a)}\,dz = 2\pi i g(a)$$

with C a contour inclosing a but no other singularities. Differentiate this equation $(n-1)$ times with respect to a. (Or, use Problem 3.21.)

Find the residues of the following functions at the indicated points. Try to select the easiest of the methods outlined above. Check your results by computer.

14. $\dfrac{1}{(3z+2)(2-z)}$ at $z = -\frac{2}{3}$; at $z = 2$

15. $\dfrac{1}{(1-2z)(5z-4)}$ at $z = \frac{1}{2}$; at $z = \frac{4}{5}$

16. $\dfrac{z-2}{z(1-z)}$ at $z = 0$; at $z = 1$

17. $\dfrac{z+2}{4z^2-1}$ at $z = \dfrac{1}{2}$; at $z = -\dfrac{1}{2}$

18. $\dfrac{z+2}{z^2+9}$ at $z = 3i$

19. $\dfrac{\sin^2 z}{2z-\pi}$ at $z = \pi/2$

20. $\dfrac{z}{1-z^4}$ at $z = i$

21. $\dfrac{z^2}{z^4+16}$ at $z = \sqrt{2}(1+i)$

22. $\dfrac{e^{2z}}{1+e^z}$ at $z = i\pi$

23. $\dfrac{e^{iz}}{9z^2+4}$ at $z = \dfrac{2i}{3}$

24. $\dfrac{1-\cos 2z}{z^3}$ at $z = 0$

25. $\dfrac{e^{2z}-1}{z^2}$ at $z = 0$

26. $\dfrac{e^{2\pi i z}}{1-z^3}$ at $z = e^{2\pi i/3}$

27. $\dfrac{\cos z}{1-2\sin z}$ at $z = \pi/6$

28. $\dfrac{z+2}{(z^2+9)(z^2+1)}$ at $z = 3i$

29. $\dfrac{e^{2z}}{4\cosh z - 5}$ at $z = \ln 2$

30. $\dfrac{\cosh z - 1}{z^7}$ at $z = 0$

31. $\dfrac{e^{3z}-3z-1}{z^4}$ at $z = 0$

32. $\dfrac{e^{iz}}{(z^2+4)^2}$ at $z = 2i$

33. $\dfrac{1+\cos z}{(\pi-z)^3}$ at $z = \pi$

34. $\dfrac{z-2}{z^2(1-2z)^2}$ at $z = 0$ and at $z = \dfrac{1}{2}$

35. $\dfrac{z}{(z^2+1)^2}$ at $z = i$

14′ to 35′ Use the residue theorem to evaluate the contour integrals of each of the functions in Problems 14 to 35 around a circle of radius $\frac{3}{2}$ and center at the origin. Check carefully to see which singular points are inside the circle. You may use your results in the previous problems as far as they go, but you may have to compute some more residues.

36. For complex z, $J_p(z)$ can be defined by the series (12.9) in Chapter 12. Use this definition to find the Laurent series about $z = 0$ for $z^{-3}J_0(z)$. Find the residue of the function at $z = 0$.

37. The gamma function $\Gamma(z)$ is analytic except for poles at $z = x = 0, -1, -2, -3 \cdots$ (all the negative integers). Find the residues at these poles. *Hints:* See Example 1 above and Chapter 11, Equation (4.1).

7. EVALUATION OF DEFINITE INTEGRALS BY USE OF THE RESIDUE THEOREM

We are going to use (5.2) and the techniques of Section 6 to evaluate several different types of definite integrals. The methods are best shown by examples.

Example 1. Find $I = \displaystyle\int_0^{2\pi} \frac{d\theta}{5+4\cos\theta}$.

If we make the change of variable $z = e^{i\theta}$, then as θ goes from 0 to 2π, z traverses the unit circle $|z| = 1$ (Figure 7.1) in the counterclockwise direction, and we have a contour integral. We shall evaluate this integral by the residue theorem. If $z = e^{i\theta}$, we have

$$dz = ie^{i\theta}\,d\theta = iz\,d\theta \quad \text{or} \quad d\theta = \frac{1}{iz}\,dz,$$

$$\cos\theta = \frac{e^{i\theta}+e^{-i\theta}}{2} = \frac{z+\frac{1}{z}}{2}.$$

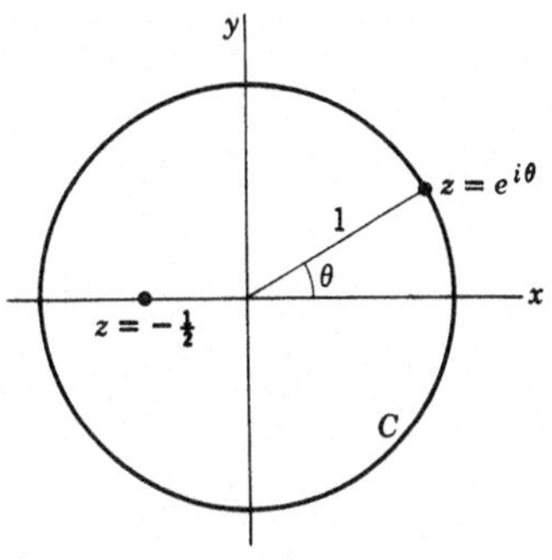

Figure 7.1

Making these substitutions in I, we get

$$I = \oint_C \frac{\frac{1}{iz}\,dz}{5+2(z+1/z)} = \frac{1}{i}\oint_C \frac{dz}{5z+2z^2+2}$$
$$= \frac{1}{i}\oint_C \frac{dz}{(2z+1)(z+2)},$$

where C is the unit circle. The integrand has poles at $z=-\frac{1}{2}$ and $z=-2$; only $z=-\frac{1}{2}$ is inside the contour C. The residue of $1/[(2z+1)(z+2)]$ at $z=-\frac{1}{2}$ is

$$R(-\tfrac{1}{2}) = \lim_{z\to -1/2}(z+\tfrac{1}{2})\cdot\frac{1}{(2z+1)(z+2)} = \left.\frac{1}{2(z+2)}\right|_{z=-1/2} = \tfrac{1}{3}.$$

Then by the residue theorem

$$I = \frac{1}{i}\,2\pi i R(-\tfrac{1}{2}) = 2\pi\cdot\tfrac{1}{3} = \frac{2\pi}{3}.$$

This method can be used to evaluate the integral of any rational function of $\sin\theta$ and $\cos\theta$ between 0 and 2π, provided the denominator is never zero for any value of θ. You can also find an integral from 0 to π if the integrand is even, since the integral from 0 to 2π of an even periodic function is twice the integral from 0 to π of the same function. (See Chapter 7, Section 9 for discussion of even and odd functions.)

▶ **Example 2.** Evaluate $I = \displaystyle\int_{-\infty}^{\infty}\frac{dx}{1+x^2}$.

Here we could easily find the indefinite integral and so evaluate I by elementary methods. However, we shall do this simple problem by contour integration to illustrate a method which is useful for more complicated problems.

This time we are not going to make a change of variable in I. We are going to start with a different integral and show how to find I from it. We consider

$$\oint_C \frac{dz}{1+z^2},$$

where C is the closed boundary of the semicircle shown in Figure 7.2. For any $\rho>1$, the semicircle incloses the singular point $z=i$ and no others; the residue of the integrand at $z=i$ is

$$R(i) = \lim_{z\to i}(z-i)\frac{1}{(z-i)(z+i)} = \frac{1}{2i}.$$

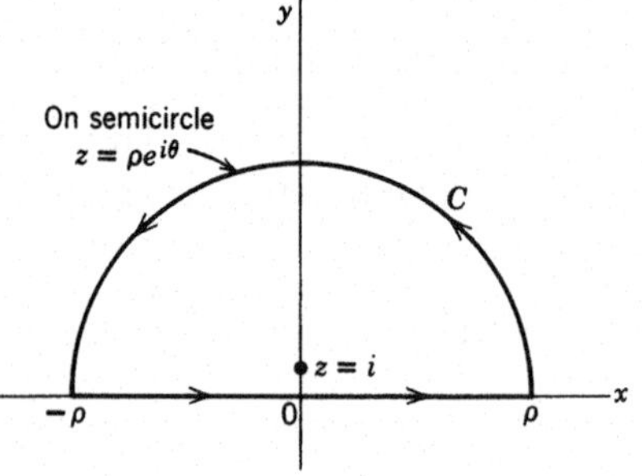

Figure 7.2

Then the value of the contour integral is $2\pi i(1/2i)=\pi$. Let us write the integral in two parts: (1) an integral along the x axis from $-\rho$ to ρ; for this part $z=x$; (2) an integral along the semicircle, where $z=\rho e^{i\theta}$. Then we have

$$\int_C \frac{dz}{1+z^2} = \int_{-\rho}^{\rho}\frac{dx}{1+x^2} + \int_0^{\pi}\frac{\rho i e^{i\theta}\,d\theta}{1+\rho^2 e^{2i\theta}}. \tag{7.1}$$

We know that the value of the contour integral is π no matter how large ρ becomes since there are no other singular points besides $z = i$ in the upper half-plane. Let $\rho \to \infty$; then the second integral on the right in (7.1) tends to zero since the numerator contains ρ and the denominator ρ^2. Thus the first term on the right tends to π (the value of the contour integral) as $\rho \to \infty$, and we have

$$I = \int_{-\infty}^{\infty} \frac{dx}{1+x^2} = \pi.$$

This method can be used to evaluate any integral of the form

$$\int_{-\infty}^{\infty} \frac{P(x)}{Q(x)}\,dx$$

if $P(x)$ and $Q(x)$ are polynomials with the degree of Q at least two greater than the degree of P, and if $Q(z)$ has no real zeros (that is, zeros on the x axis). If the integrand $P(x)/Q(x)$ is an even function, then we can also find the integral from 0 to ∞.

Example 3. Evaluate $I = \displaystyle\int_0^{\infty} \frac{\cos x\,dx}{1+x^2}$.

We consider the contour integral

$$\oint_C \frac{e^{iz}\,dz}{1+z^2},$$

where C is the same semicircular contour as in Example 2. The singular point inclosed is again $z = i$, and the residue there is

$$\lim_{z\to i}(z-i)\frac{e^{iz}}{(z-i)(z+i)} = \frac{e^{-1}}{2i} = \frac{1}{2ie}.$$

The value of the contour integral is $2\pi i(1/2ie) = \pi/e$. As in Example 2 we write the contour integral as a sum of two integrals:

$$(7.2) \qquad \oint_C \frac{e^{iz}\,dz}{1+z^2} = \int_{-\rho}^{\rho} \frac{e^{ix}\,dx}{1+x^2} + \int_{\substack{\text{along upper half}\\ \text{of } z=\rho e^{i\theta}}} \frac{e^{iz}\,dz}{1+z^2}.$$

As before, we want to show that the second integral on the right of (7.2) tends to zero as $\rho \to \infty$. This integral is the same as the corresponding integral in (7.1) except for the e^{iz} factor. Now

$$|e^{iz}| = |e^{ix-y}| = |e^{ix}||e^{-y}| = e^{-y} \le 1$$

since $y \ge 0$ on the contour we are considering. Since $|e^{iz}| \le 1$, this factor does not change the proof given in Example 2 that the integral along the semicircle tends to zero as the radius $\rho \to \infty$. We have then

$$\int_{-\infty}^{\infty} \frac{e^{ix}}{1+x^2}\,dx = \frac{\pi}{e},$$

or taking the real part of both sides of this equation,

$$\int_{-\infty}^{\infty} \frac{\cos x\, dx}{1+x^2} = \frac{\pi}{e}.$$

Since the integrand $(\cos x)/(1+x^2)$ is an even function, the integral from 0 to ∞ is half the integral from $-\infty$ to ∞. Hence we have

$$I = \int_0^{\infty} \frac{\cos x\, dx}{1+x^2} = \frac{\pi}{2e}.$$

Observe that the same proof would work if we replaced e^{iz} by e^{imz} $(m > 0)$ in the above integrals. At the point where we said $e^{-y} \le 1$ (since $y \ge 0$) we would then want $e^{-my} \le 1$ for $y \ge 0$, which is true if $m > 0$. [For $m < 0$, we *could* use a semicircle in the lower half-plane $(y < 0)$; then we would have $e^{my} \le 1$ for $y \le 0$. This is an unnecessary complication, however, in evaluating integrals containing $\sin mx$ or $\cos mx$ since we *can* then choose m to be positive.] Although we have assumed here that (as in Example 2) $Q(x)$ is of degree at least 2 higher than $P(x)$, a more detailed proof (see books on complex variables) shows that degree at least one higher is enough to make the integral

$$\int \frac{P(z)}{Q(z)} e^{imz}\, dz$$

around the semicircle tend to zero as $\rho \to \infty$. Thus

$$\int_{-\infty}^{\infty} \frac{P(x)}{Q(x)} e^{imx}\, dx = 2\pi i \cdot \text{sum of the residues of } \frac{P(z)}{Q(z)} e^{imz}$$

in the upper half-plane if all the following requirements are met:

1. $P(x)$ and $Q(x)$ are polynomials, and
2. $Q(x)$ has no real zeros, and
3. the degree of $Q(x)$ is at least 1 greater than the degree of $P(x)$, and $m > 0$.

By taking real and imaginary parts, we then find the integrals

$$\int_{-\infty}^{\infty} \frac{P(x)}{Q(x)} \cos mx\, dx, \qquad \int_{-\infty}^{\infty} \frac{P(x)}{Q(x)} \sin mx\, dx.$$

▶ **Example 4.** Evaluate $\displaystyle\int_{-\infty}^{\infty} \frac{\sin x}{x}\, dx.$

Here we remove the restriction of Examples 2 and 3 that $Q(x)$ has no real zeros. As in Example 3, we consider

$$\int \frac{e^{iz}}{z}\, dz.$$

To avoid the singular point at $z = 0$, we integrate around the contour shown in Figure 7.3. We then let the radius r shrink to zero so that in effect we are integrating straight through the simple pole at the origin. We are going to show (later in this section and Problem 21) that the net result of integrating in the counterclockwise direction around a closed contour which passes straight[‡] through one or more simple poles is $2\pi i \cdot$ (sum of the residues at interior points plus one-half the sum of the residues at the simple poles on the boundary). (*Warning:* this rule does not hold in general for a multiple pole on a boundary.) You might expect this result. If a pole is inside a contour, it contributes $2\pi i\cdot$ residue, to the integral; if it is outside, it contributes nothing; if it is *on* the straight line boundary, its contribution is just halfway between zero and $2\pi i\cdot$ residue. [See Am. J. Phys. **52**, 276 (1984).] Using this fact, and observing that, as in Example 3, the integral along the large semicircle tends to zero as R tends to infinity, we have

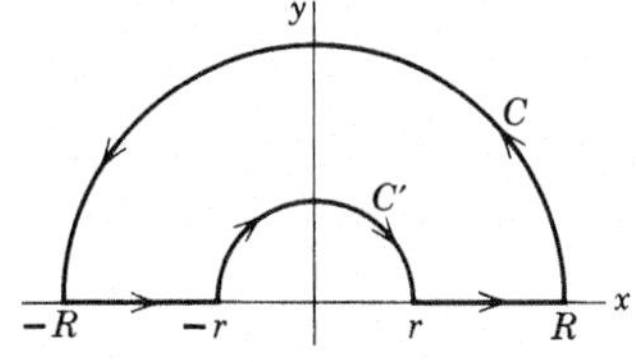

Figure 7.3

$$\int_{-\infty}^{\infty} \frac{e^{ix}}{x}\,dx = 2\pi i \cdot \frac{1}{2}\left(\text{residue of } \frac{e^{iz}}{z} \text{ at } z = 0\right) = 2\pi i \cdot \frac{1}{2} \cdot 1 = i\pi.$$

Taking the imaginary parts of both sides, we get

$$\int_{-\infty}^{\infty} \frac{\sin x}{x}\,dx = \pi.$$

To show more carefully that our result is correct, let us return to the contour of Figure 7.3. Since e^{iz}/z is analytic inside this contour, the integral around the whole contour is zero. As we have said, the integral along C tends to zero as $R \to \infty$ by the theorem at the end of Example 3. Along the small semicircle C', we have

$$z = re^{i\theta}, \quad dz = re^{i\theta} i d\theta, \quad \frac{dz}{z} = i d\theta,$$

$$\int_{C'} \frac{e^{iz}\,dz}{z} = \int_{C'} e^{iz} i d\theta.$$

As $r \to 0$, $z \to 0$, $e^{iz} \to 1$, and the integral (along C' in the direction indicated in Figure 7.3) tends to

$$\int_{\pi}^{0} i\,d\theta = -i\pi.$$

Then we have as $R \to \infty$, and $r \to 0$,

$$\int_{-\infty}^{\infty} \frac{e^{ix}}{x}\,dx - i\pi = 0$$

or

$$\int_{-\infty}^{\infty} \frac{e^{ix}}{x}\,dx = i\pi$$

as before. Taking real and imaginary parts of this equation (and using Euler's formula $e^{ix} = \cos x + i \sin x$), we get

$$\int_{-\infty}^{\infty} \frac{\cos x}{x}\,dx = 0, \qquad \int_{-\infty}^{\infty} \frac{\sin x}{x}\,dx = \pi.$$

[‡]By "straight" we mean that the contour curve has a tangent at the pole, that is, it does not turn a corner there.

Since $(\sin x)/x$ is an even function, we have

$$\int_0^\infty \frac{\sin x}{x}\,dx = \frac{1}{2}\int_{-\infty}^\infty \frac{\sin x}{x}\,dx = \frac{\pi}{2}.$$

[For another way of evaluating this integral, see Chapter 7, equation (12.19).]

Principal value Now consider the cosine integral.

$$\int_0^\infty \frac{\cos x}{x}\,dx$$

is a divergent integral since the integrand $(\cos x)/x$ is approximately $1/x$ near $x = 0$. The value zero which we found for $I = \int_{-\infty}^\infty (\cos x)/x\,dx$ is called the *principal value* (or Cauchy principal value) of I. To see what this means, consider a simpler integral, namely

$$\int_0^5 \frac{dx}{x-3}.$$

The integrand becomes infinite at $x = 3$, and both $\int_0^3 dx/(x-3)$ and $\int_3^5 dx/(x-3)$ are divergent. Suppose we cut out a small symmetric interval about $x = 3$, and integrate from 0 to $3 - r$ and from $3 + r$ to 5. We find

$$\int_0^{3-r} \frac{dx}{x-3} = \ln|x-3|\Big|_0^{3-r} = \ln r - \ln 3,$$

$$\int_{3+r}^5 \frac{dx}{x-3} = \ln 2 - \ln r.$$

The sum of these two integrals is

$$\ln 2 - \ln 3 = \ln\frac{2}{3};$$

this sum is independent of r. Thus, if we let $r \to 0$, we get the result $\ln\frac{2}{3}$ which is called the principal value of

$$\int_0^5 \frac{dx}{x-3} \qquad \left(\text{often written } PV\int_0^5 \frac{dx}{x-3} = \ln\frac{2}{3}\right).$$

The terms $\ln r$ and $-\ln r$ have been allowed to cancel each other; graphically an infinite area above the x axis and a corresponding infinite area below the x axis have been canceled. In computing the contour integral we integrated along the x axis from $-\infty$ up to $-r$, and from $+r$ to $+\infty$, and then let $r \to 0$; this is just the process we described for finding principal values, so the result we found for the improper integral $\int_{-\infty}^\infty (\cos x)/x\,dx$, namely zero, was the principal value of this integral.

▶ **Example 5.** Evaluate

$$\int_0^\infty \frac{r^{p-1}}{1+r}\,dr, \qquad 0 < p < 1,$$

and use the result to prove (5.4) of Chapter 11.

We first find

$$(7.3) \qquad \oint \frac{z^{p-1}}{1+z}\,dz, \qquad 0 < p < 1, \qquad \text{around } C \text{ in Figure 7.4.}$$

Before we can evaluate this integral, we must ask what z^{p-1} means, since for each z there may be more than one value of z^{p-1}. (See discussion of branches at the end of Section 1.) For example, consider the case $p = \frac{1}{2}$; then $z^{p-1} = z^{-1/2}$. Recall from Chapter 2, Section 10, that there are two square roots of any complex number. At a point where $\theta = \pi/4$, say, we have

$$z = re^{i\pi/4}, \qquad z^{-1/2} = r^{-1/2}e^{-i\pi/8}.$$

But if θ increases by 2π (we think of following a circle around the origin and back to our starting point), we have

$$z = re^{i(\pi/4+2\pi)}, \qquad z^{-1/2} = r^{-1/2}e^{-i(\pi/8+\pi)} = -r^{-1/2}e^{-i\pi/8}.$$

Similarly, for any starting point (with $r \neq 0$), we find that $z^{-1/2}$ or z^{p-1} comes back to a different value (different branch) when θ increases by 2π and we return to our starting point. If we want to use the formula z^{p-1} to define a (single-valued) function, we must decide on some interval of length 2π for θ (that is, we must select one branch of z^{p-1}). Let us agree to restrict θ to the values of 0 to 2π in evaluating the contour integral (7.3). We may imagine an artificial barrier or cut (which we agree not to cross) along the positive x axis; this is called a *branch cut.* (See Example 3, Section 9.) A point which we cannot encircle (on an arbitrarily small circle) without crossing a branch cut (thus changing to another branch) is called a *branch point*; the origin is a branch point here.

In Figure 7.4, then, $\theta = 0$ along AB (upper side of the positive x axis); when we follow C around to DE, θ increases by 2π, so $\theta = 2\pi$ on the lower side of the positive x axis. Note that the contour in Figure 7.4 never takes us outside the 0 to 2π interval, so the factor z^{p-1} in (7.3) is a single-valued function. The integrand in (7.3), namely $z^{p-1}/(1+z)$, is now an analytic function inside the closed curve C in Figure 7.4 except for the pole at $z = -1 = e^{i\pi}$. The residue there is $(e^{i\pi})^{p-1} = -e^{i\pi p}$. Then we have

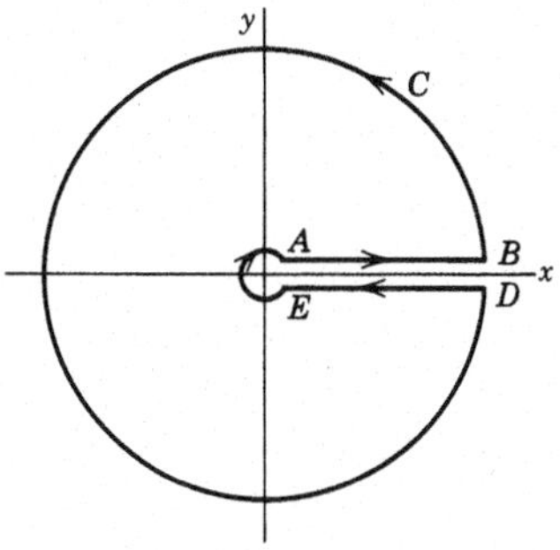

$$(7.4) \qquad \oint_C \frac{z^{p-1}}{1+z}\,dz = -2\pi i e^{i\pi p}, \qquad 0 < p < 1.$$

Figure 7.4

Along either of the two circles in Figure 7.4 we have $z = re^{i\theta}$ and the integral is

$$\int \frac{r^{p-1}e^{i(p-1)\theta}}{1+re^{i\theta}} rie^{i\theta}\,d\theta = i\int \frac{r^p e^{ip\theta}}{1+re^{i\theta}}\,d\theta.$$

This integral tends to zero if $r \to 0$ or if $r \to \infty$. (Verify this; note that the denominator is approximately 1 for small r, and approximately $re^{i\theta}$ for large r.) Thus the integrals along the circular parts of the contour tend to zero as the little circle shrinks to a point and the large circle expands indefinitely. We are left with the two integrals along the positive x axis with AB now extending from 0 to ∞ and DE from ∞ to 0. Along AB we agreed to have $\theta = 0$, so $z = re^{i\cdot 0} = r$, and this integral is

$$\int_{r=0}^{\infty} \frac{r^{p-1}}{1+r}\,dr.$$

Along DE, we have $\theta = 2\pi$, so $z = re^{2\pi i}$ and this integral is

$$\int_{r=\infty}^{0} \frac{(re^{2\pi i})^{p-1}}{1+re^{2\pi i}} e^{2\pi i}\,dr = -\int_{0}^{\infty} \frac{r^{p-1}e^{2\pi ip}}{1+r}\,dr.$$

Adding the AB and DE integrals, we get

$$(1-e^{2\pi ip})\int_0^\infty \frac{r^{p-1}}{1+r}\,dr = -2\pi i e^{i\pi p}$$

by (7.4). Then the desired integral is

$$\int_0^\infty \frac{r^{p-1}}{1+r}\,dr = \frac{-2\pi i e^{i\pi p}}{1-e^{2\pi ip}} = \frac{\pi\cdot 2i}{e^{i\pi p}-e^{-i\pi p}} = \frac{\pi}{\sin \pi p}. \tag{7.5}$$

Let us use (7.5) to obtain (5.4) of Chapter 11. Putting $q = 1-p$ in (6.5) and (7.1) of Chapter 11, we have

$$\begin{aligned} B(p,1-p) &= \int_0^\infty \frac{y^{p-1}}{1+y}\,dy \qquad \text{and} \\ B(p,1-p) &= \Gamma(p)\Gamma(1-p) \qquad \Gamma(1)=1. \end{aligned} \tag{7.6}$$

Combining (7.5) and (7.6) gives (5.4) of Chapter 11, namely

$$\Gamma(p)\Gamma(1-p) = B(p,1-p) = \int_0^\infty \frac{y^{p-1}}{1+y}\,dy = \frac{\pi}{\sin \pi p}.$$

Argument Principle Since $w = f(z)$ is a complex number for each z, we can write $w = Re^{i\Theta}$ (just as we write $z = re^{i\theta}$) where $R = |w|$ and Θ is the angle of w [or we could call it the angle of $f(z)$]. As z changes, $w = f(z)$ also changes and so R and Θ vary as we go from point to point in the complex (x, y) plane. We want to show that

(a) if $f(z)$ is analytic on and inside a simple closed curve C and $f(z) \neq 0$ on C, then the number of zeros of $f(z)$ inside C is equal to $(1/2\pi)\cdot$ (change in the angle of $f(z)$ as we traverse the curve C);

(b) if $f(z)$ has a finite number of poles inside C, but otherwise meets the requirements stated,[§] then the change in the angle of $f(z)$ around C is equal to $(2\pi)\cdot$ (the number of zeros minus the number of poles).

(Just as we say that a quadratic equation with equal roots has *two* equal roots, so here we mean that a zero of order n counts as n zeros and a pole of order n counts as n poles.)

To show (a) and (b) we consider

$$\oint_C \frac{f'(z)}{f(z)}\,dz.$$

[§] A function which is analytic except for poles is called *meromorphic*.

By the residue theorem, the integral is equal to $2\pi i\cdot$(sum of the residues at singularities inside C). It is straightforward to show (Problem 42) that the residue of $F(z) = f'(z)/f(z)$ at a zero of $f(z)$ of order n is n, and the residue of $F(z)$ at a pole of $f(z)$ of order p is $-p$. Then if N is the number of zeros and P the number of poles of $f(z)$ inside C, the integral is $2\pi i(N-P)$. Now by direct integration, we have

$$\oint_C \frac{f'(z)}{f(z)}\,dz = \ln f(z)\big|_C = \ln Re^{i\Theta}\big|_C = \operatorname{Ln} R\big|_C + i\Theta\big|_C, \tag{7.7}$$

where $R = |f(z)|$ and Θ is the angle of $f(z)$. Recall from Chapter 2, Section 13, that $\operatorname{Ln} R$ means the ordinary real logarithm (to the base e) of the positive number R, and is single-valued; $\ln f(z)$ is multiple-valued because Θ is multiple-valued. Then if we integrate from a point A on C all the way around the curve and back to A, $\operatorname{Ln} R$ has the same value at A both at the beginning and at the end, so the term $\operatorname{Ln} R|_C$ is $\operatorname{Ln} R$ at A minus $\operatorname{Ln} R$ at A; this is zero. The same result may not be true for Θ; that is, the angle may have changed as we go from point A all the way around C and back to A. (Think, for example, of the angle of z as we go from $z=1$ around the unit circle and back to $z=1$; the angle of z has increased from 0 to 2π.) Collecting our results, we have

$$\begin{aligned} N - P &= \frac{1}{2\pi i}\oint_C \frac{f'(z)}{f(z)}\,dz = \frac{1}{2\pi i} i\Theta_C \\ &= \frac{1}{2\pi}\cdot(\text{change in the angle of } f(z) \text{ around } C), \end{aligned} \tag{7.8}$$

where N is the number of zeros and P the number of poles of $f(z)$ inside C, with poles of order n counted as n poles and similarly for zeros of order n. Equation (7.8) is known as the *argument principle* (recall from Chapter 2 that *argument* means *angle*).

This principle is often used to find out how many zeros (or poles) a given function has in a given region. (Locating the zeros of a function has important applications to determining the stability of linear systems such as electric circuits and servomechanisms.

Example 6. Let us show that $f(z) = z^3 + 4z + 1$ has exactly one zero in the first quadrant. The closed curve C in (7.8) is, for this problem, the contour OPQ in Figure 7.5, where PQ is a large quarter circle. We first observe that on the x axis, $x^3 + 4x + 1 > 0$ for $x > 0$, and on the y axis, $(iy)^3 + 4iy + 1 \neq 0$ for any y (since its real part, namely 1, $\neq 0$). Then $f(z) \neq 0$ on OP or OQ. Also $f(z) \neq 0$ on PQ if we choose a circle large enough to inclose all zeros. We now want to find the change in the angle Θ of $f(z) = Re^{i\Theta}$ as we go around C. Along OP, $z = x$; then $f(z) = f(x)$ is real and so $\Theta = 0$. Along PQ, $z = re^{i\theta}$, with r constant and very large. For very large r, the z^3 term in $f(z)$ far outweighs the other terms, and we have $f(z) \cong z^3 = r^3 e^{3i\theta}$. As θ goes from 0 to $\pi/2$ along PQ,

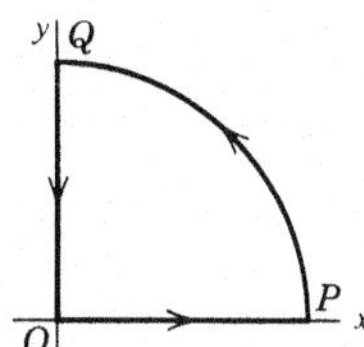

Figure 7.5

$\Theta = 3\theta$ goes from 0 to $3\pi/2$. On QO, $z = iy$, $f(z) = -iy^3 + 4iy + 1$; then

$$\tan\Theta = \frac{\text{imaginary part of } f(z)}{\text{real part of } f(z)} = \frac{4y - y^3}{1}.$$

For very large y (that is, at Q), we had $\Theta \cong 3\pi/2$ (for $y = \infty$, we would have $\tan\theta = -\infty$, and Θ would be exactly $3\pi/2$). Now as y decreases along QO, the value of $\tan\Theta = 4y - y^3$ decreases in magnitude but remains negative until it becomes 0 at $y = 2$. This means that Θ changes from $3\pi/2$ to 2π. Between $y = 2$ and $y = 0$, the tangent becomes positive, but then decreases to zero again without becoming infinite. This means that the angle Θ increases beyond 2π but not as far as $2\pi + \pi/2$, and then decreases again to 2π. Thus the total change in Θ around C is 2π, and by (7.8), the number of zeros of $f(z)$ in the first quadrant is $(1/2\pi)\cdot 2\pi = 1$. If we realize that (for a polynomial with real coefficients) the zeros off the real axis always occur in conjugate pairs, we see that there must also be one zero for z in the fourth quadrant, and the third zero must be on the negative x axis.

Bromwich integral (Inverse Laplace Transform) In Chapter 8 (Section 8ff), we found inverse Laplace transforms from a table (pages 469–471), or by computer, but we had no general formula for the inverse transform. By analogy with Fourier transforms (Chapter 7, Section 12), where we have similar integrals for the direct and inverse transforms, we might reasonably wonder whether an inverse Laplace transform could be given by an integral. To discuss this, we repeat here for convenience the definitions of Laplace and Fourier transforms.

$$L(f) = \int_0^\infty f(t)e^{-pt}dt = F(p) \tag{7.9}$$

$$\begin{aligned} f(x) &= \int_{-\infty}^{\infty} g(\alpha)e^{i\alpha x}\,d\alpha \\ g(\alpha) &= \frac{1}{2\pi}\int_{-\infty}^{\infty} f(x)e^{-i\alpha x}\,dx \end{aligned} \tag{7.10}$$

If we compare the Laplace transform (7.9) with the Fourier transform [$g(\alpha)$ in (7.10)], we observe that if p were imaginary, the integrals would be almost the same. This suggests that we should consider complex p, and that the integral we want for the inverse Laplace transform might be an integral in the complex p plane (that is, a contour integral). Let's investigate this idea.

In the definition (7.9) of the Laplace transform of $f(t)$, let p be complex, say $p = z = x + iy$. (Note that this possibility has already been considered in Chapter 8.) Then (7.9) becomes

$$\begin{aligned} F(p) = F(z) = F(x+iy) &= \int_0^\infty e^{-pt} f(t)\,dt \\ &= \int_0^\infty e^{-(x+iy)t} f(t)\,dt = \int_0^\infty e^{-xt} f(t) e^{-iyt}\,dt, \quad x = \operatorname{Re} p > k. \end{aligned} \tag{7.11}$$

[Recall (Chapter 8, Section 8) that we must have some restriction on $\operatorname{Re} p$ to make the integral converge at infinity. The restriction depends on what the function $f(t)$ is, but is always of the form $\operatorname{Re} p > k$, for *some* real k, as you can see in the table of Laplace transforms, pages 469–471.] Now (7.11) is of the form of a Fourier transform. To see this, compare (7.11) with (7.10) making the following correspondences: $e^{-iyt}\,dt$ corresponds to $e^{-i\alpha x}\,dx$, that is, y corresponds to α and t to x [the x in (7.11) is just a constant parameter in this discussion]; the function

$$\phi(t) = \begin{cases} e^{-xt}f(t), & t > 0, \\ 0, & t < 0, \end{cases} \tag{7.12}$$

corresponds to $f(x)$ in (7.10) and $F(p) = F(x+iy)$ corresponds to $g(\alpha)$; and finally, we recall that the $1/(2\pi)$ factor may be in either integral in (7.10). Then assuming that $\phi(t)$ satisfies the required conditions for a function to have a Fourier transform [see Chapter 7, Section 12: Dirichlet conditions, and $\int_{-\infty}^{\infty} |\phi(t)|\,dt$ finite], we can write the inverse transform to get

$$\phi(t) = \frac{1}{2\pi}\int_{-\infty}^{\infty} F(x+iy)e^{iyt}\,dy. \tag{7.13}$$

Using the definition (7.12) of $\phi(t)$, we find

$$f(t) = e^{xt}\cdot\frac{1}{2\pi}\int_{-\infty}^{\infty} F(x+iy)e^{iyt}\,dy = \frac{1}{2\pi}\int_{-\infty}^{\infty} F(x+iy)e^{(x+iy)t}\,dy \tag{7.14}$$

for $t > 0$. Since x is constant, say $x = c$, we have $dz = d(x+iy) = i\,dy$, and we can write (7.14) as

$$f(t) = \frac{1}{2\pi i}\int_{c-i\infty}^{c+i\infty} F(z)e^{zt}\,dz, \qquad t > 0, \tag{7.15}$$

where the notation means (see Problem 3.4) that we integrate along a vertical line $x = c$ in the z plane. [This can be *any* vertical line on which $x = c > k$ as required by the restriction on $\operatorname{Re} p$ in (7.11).] The integral (7.15) for the inverse Laplace transform is known as the *Bromwich integral.*

We would like to use contour integration and the residue theorem to evaluate $f(t)$ in (7.15) for a given $F(p)$ [which we call $F(z)$ since we consider complex p]. In Examples 2 and 3, we have evaluated integrals along a straight line (the x axis) by considering the contour made up of the x axis and a large semicircle inclosing the upper half plane. If we rotate this contour 90°, we have a contour consisting of a vertical straight line and a semicircle inclosing a left half-plane (that is, the area to the left of $x = c$). Let's use this contour to evaluate (7.15). We restrict $F(z)$ to be of the form $P(z)/Q(z)$ with $P(z)$ and $Q(z)$ polynomials, and $Q(z)$ of degree at least one higher than $P(z)$ (compare the conditions in Example 3). Then it can be shown that, as in Examples 2 and 3, the integral along the semicircle tends to zero as the radius tends to infinity. Thus the integral along the straight line is equal to $2\pi i$ times the sum of the residues of $F(z)e^{zt}$ at its poles, or, cancelling the factor $2\pi i$ in (7.15),

$$f(t) = \text{sum of residues of } F(z)e^{zt} \text{ at all poles.} \tag{7.16}$$

We must include *all* poles in (7.16); to see this we can argue as follows. We know that (7.15) is true for any value of $c > k$. Suppose we use a value of c which is large enough so that all poles lie to the left of $x = c$; then we know that our answer is correct. Turning the argument around, we can say that since we would get a different answer if we did not take $x = c$ to the right of all poles, we *must* integrate along a vertical line such that all poles of $F(z)e^{zt}$ are included in the contour to the left of the line.

▶ **Example 7.** Find the inverse transform of $F(p) = \dfrac{5}{(p+2)(p^2+1)}$.
We first find the poles of $F(z)e^{zt}$ and factor the denominator to get

$$F(z)e^{zt} = \frac{5e^{zt}}{(z+2)(z+i)(z-i)}.$$

Evaluating the residues at the three simple poles (Section 6, method B), we find

residue at $z = -2$	is	$\dfrac{5e^{-2t}}{5} = e^{-2t}$
residue at $z = i$	is	$\dfrac{5e^{it}}{(2+i)(2i)}$
residue at $z = -i$	is	$\dfrac{5e^{-it}}{(2-i)(-2i)}$

Then by (7.16) we have

$$f(t) = e^{-2t} + \frac{5e^{it}(2-i) - 5e^{-it}(2+i)}{(2+i)(2-i)(2i)} = e^{-2t} + 2\sin t - \cos t.$$

Dispersion relations Consider $\displaystyle\int \frac{f(z)}{z-a}\,dz$ around the upper half plane as in Problem 21. Let a be real. Let $f(z)$ be analytic for $y \geq 0$, and $\to 0$ rapidly enough at ∞ so that the integral around the semicircle in the upper half plane $\to 0$ as the radius of the semicircle $\to \infty$. Then by Example 4 and Problem (21b) we get

$$PV \int_{-\infty}^{\infty} \frac{f(x)}{x-a}\,dx = i\pi f(a). \tag{7.17}$$

Now we write $f(x) = u(x) + iv(x)$, and take real and imaginary parts of (7.17):

$$PV \int_{-\infty}^{\infty} \frac{u(x)}{x-a}\,dx = -\pi v(a), \qquad PV \int_{-\infty}^{\infty} \frac{v(x)}{x-a}\,dx = \pi u(a). \tag{7.18}$$

These (and similar integrals relating the real and imaginary parts of a function satisfying the given conditions) are called *dispersion relations.* From them, you can find the *Kramers-Kronig relations* (see Problem 66) named for the two people who developed similar relations involving the complex index of refraction for light. (Light traveling through a material medium is both refracted and absorbed. The real part of the complex index of refraction is related to refraction and the imaginary part to absorption.) These formulas have widespread applications, to optics, electricity, solid state, elementary particle theory, quantum mechanics, etc.

The integrals in (7.18) are called *Hilbert transforms,* and (7.18) may be stated in the form: $u(x)$ and $v(x)$ are Hilbert transforms of each other. Compare Fourier transforms (Chapter 7, Section 12), or a Laplace transform and the corresponding Bromwich integral. In each case two functions have the property that each is given by an integral involving the other. This is what an integral transform means, and there are other integral transforms which you may discover in tables or computer.

PROBLEMS, SECTION 7

The values of the following integrals are known and can be found in integral tables or by computer. Your goal in evaluating them is to learn about contour integration by applying the methods discussed in the examples above. Then check your answers by computer.

1. $\displaystyle\int_0^{2\pi} \frac{d\theta}{13+5\sin\theta}$

2. $\displaystyle\int_0^{2\pi} \frac{d\theta}{5-3\cos\theta}$

3. $\displaystyle\int_0^{2\pi} \frac{d\theta}{5-4\sin\theta}$

4. $\displaystyle\int_0^{2\pi} \frac{\sin^2\theta\, d\theta}{5+3\cos\theta}$

5. $\displaystyle\int_0^{\pi} \frac{d\theta}{1-2r\cos\theta+r^2} \quad (0\le r<1)$

6. $\displaystyle\int_0^{\pi} \frac{d\theta}{(2+\cos\theta)^2}$

7. $\displaystyle\int_0^{2\pi} \frac{\cos 2\theta\, d\theta}{5+4\cos\theta}$

8. $\displaystyle\int_0^{\pi} \frac{\sin^2\theta\, d\theta}{13-12\cos\theta}$

9. $\displaystyle\int_0^{2\pi} \frac{d\theta}{1+\sin\theta\cos\alpha} \quad (\alpha - \text{const.})$

10. $\displaystyle\int_{-\infty}^{\infty} \frac{dx}{x^2+4x+5}$

11. $\displaystyle\int_0^{\infty} \frac{dx}{(4x^2+1)^3}$

12. $\displaystyle\int_0^{\infty} \frac{x^2\, dx}{x^4+16}$

13. $\displaystyle\int_0^{\infty} \frac{x^2\, dx}{(x^2+4)(x^2+9)}$

14. $\displaystyle\int_{-\infty}^{\infty} \frac{\sin x\, dx}{x^2+4x+5}$

15. $\displaystyle\int_0^{\infty} \frac{\cos 2x\, dx}{9x^2+4}$

16. $\displaystyle\int_0^{\infty} \frac{x\sin x\, dx}{9x^2+4}$

17. $\displaystyle\int_{-\infty}^{\infty} \frac{x\sin x\, dx}{x^2+4x+5}$

18. $\displaystyle\int_0^{\infty} \frac{\cos \pi x\, dx}{1+x^2+x^4}$

19. $\displaystyle\int_0^{\infty} \frac{\cos 2x\, dx}{(4x^2+9)^2}$

20. $\displaystyle\int_0^{\infty} \frac{\cos x\, dx}{(1+9x^2)^2}$

21. In Example 4 we stated a rule for evaluating a contour integral when the contour passes through simple poles. We proved that the result was correct for

$$PV\int_\Gamma \frac{e^{iz}}{z}\,dz$$

around the contour Γ shown here.

(a) By following the same method (integrating around C' of Figure 7.3 and letting $r\to 0$) show that the result is correct if we replace e^{iz} by any $f(z)$ which is analytic at $z=0$.

(b) Repeat the proof in (a) for

$$PV\int_\Gamma \frac{f(z)}{(z-a)}\,dz, \qquad a \text{ real}$$

(that is, a pole on the x axis), with $f(z)$ analytic at $z=a$.

Using the rule of Example 4 (also see problem 21), evaluate the following integrals. Find principal values if necessary.

22. $\displaystyle\int_{-\infty}^{\infty} \frac{dx}{(x-1)(x^2+1)}$

23. $\displaystyle\int_{-\infty}^{\infty} \frac{dx}{(x^2+4)(2-x)}$

24. $\displaystyle\int_{-\infty}^{\infty} \frac{x\sin\pi x}{1-x^2}\,dx$

25. $\displaystyle\int_{0}^{\infty} \frac{x\sin x}{9x^2-\pi^2}\,dx$

26. $\displaystyle\int_{-\infty}^{\infty} \frac{x\,dx}{(x-1)^4-1}$

27. $\displaystyle\int_{0}^{\infty} \frac{\cos\pi x}{1-4x^2}\,dx$

28. $\displaystyle\int_{0}^{\infty} \frac{dx}{1-x^4}$

29. $\displaystyle\int_{0}^{\infty} \frac{\sin ax}{x}\,dx$

30. (a) By the method of Example 2 evaluate $\displaystyle\int_{0}^{\infty} \frac{dx}{1+x^4}$.

(b) Evaluate the same integral by using tables or computer to get the indefinite integral; unless you are very careful you may get zero. Explain why.

(c) Make the change of variables $u = x^4$ in the integral in (a) and evaluate the u integral using (7.5).

31. Use the method of Problem 30(c) to evaluate $\displaystyle\int_{0}^{\infty} \frac{dx}{1+x^6}$.

32. Use the method of Problem 30(c) and the contour and method of Example 5 to evaluate $\displaystyle\int_{0}^{\infty} \frac{dx}{(1+x^4)^2}$.

Evaluate the following integrals by the method of Example 5.

33. $\displaystyle\int_{0}^{\infty} \frac{\sqrt{x}\,dx}{1+x^2}$

34. $\displaystyle\int_{0}^{\infty} \frac{\sqrt{x}\,dx}{(1+x)^2}$

35. $\displaystyle\int_{0}^{\infty} \frac{x^{1/3}\,dx}{(1+x)(2+x)}$

36. $\displaystyle\int_{0}^{\infty} \frac{\ln x}{x^{3/4}(1+x)}\,dx$

37.

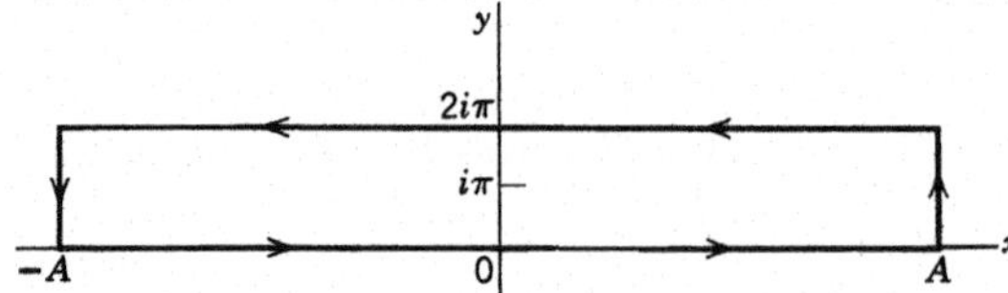

(a) Show that

$$\int_{-\infty}^{\infty} \frac{e^{px}}{1+e^x}\,dx = \frac{\pi}{\sin\pi p}$$

for $0 < p < 1$. *Hint:* Find $\int e^{pz}\,dz/(1+e^z)$ around the rectangular contour shown. Show that the integrals along the vertical sides tend to zero as $A \to \infty$. Note that the integral along the upper side is a multiple of the integral along the x axis.

(b) Make the change of variable $y = e^x$ in the x integral of part (a), and using (6.5) of Chapter 11, show that this integral is the beta function, $B(p, 1-p)$. Then using (7.1) of Chapter 11, show that $\Gamma(p)\Gamma(1-p) = \pi/\sin\pi p$.

38. Using the same contour and method as in Problem 37a evaluate

$$\int_{-\infty}^{\infty} \frac{e^{px}}{1-e^x}\,dx, \qquad 0<p<1.$$

Hint: The only difference between this problem and Problem 37a is that you now have two simple poles on the contour instead of a pole inside. Use the rule of Example 4.

39. Evaluate

$$\int_{-\infty}^{\infty} \frac{e^{2\pi x/3}}{\cosh \pi x}\,dx.$$

Hint: Use a rectangle as in Problem 37a but of height 1 instead of 2π. Note that there is a pole at $i/2$.

40. Evaluate

$$\int_0^{\infty} \frac{x\,dx}{\sinh x}.$$

Hint: First find the $-\infty$ to ∞ integral. Use a rectangle of height π and note the simple pole at $i\pi$ on the contour.

41. The Fresnel integrals, $\int_0^u \sin u^2\,du$ and $\int_0^u \cos u^2\,du$, are important in optics. For the case of infinite upper limits, evaluate these integrals as follows: Make the change of variable $x = u^2$; to evaluate the resulting integrals, find $\oint z^{-1/2} e^{iz}\,dz$ around the contour shown. Let $r \to 0$ and $R \to \infty$ and show that the integrals along these quarter-circles tend to zero. Recognize the integral along the y axis as a Γ function and so evaluate it. Hence evaluate the integral along the x axis; the real and imaginary parts of this integral are the integrals you are trying to find.

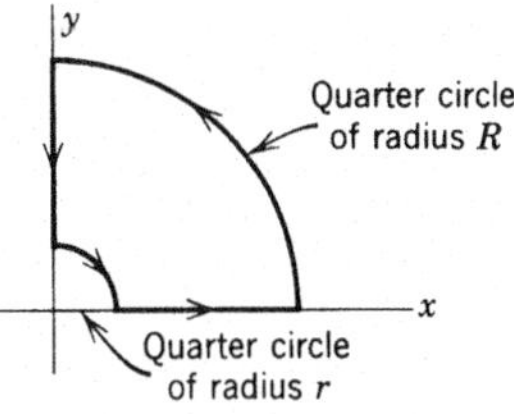

42. If $F(z) = f'(z)/f(z)$,

(a) show that the residue of $F(z)$ at an nth order zero of $f(z)$, is n. *Hint:* If $f(z)$ has a zero of order n at $z=a$, then

$$f(z) = a_n(z-a)^n + a_{n+1}(z-a)^{n+1} + \cdots.$$

(b) Also show that the residue of $F(z)$ at a pole of order p of $f(z)$, is $-p$. *Hint:* See the definition of a pole of order p at the end of Section 4.

43. By using theorem (7.8), show that $z^3+z^2+9=0$ has exactly one root in the first quadrant. Hence show that it has one root in the fourth quadrant and one on the negative real axis. *Hint*: See Example 6.

44. The *fundamental theorem of algebra* says that every equation of the form $f(z) = a_n z^n + a_{n-1}z^{n-1} + \cdots + a_0 = 0$, $a_n \neq 0$, $n \geq 1$, has at least one root. From this it follows that an nth degree equation has n roots. Prove this by using the argument principle. *Hint:* Follow the increase in the angle of $f(z)$ around a very large circle $z = re^{i\theta}$; for sufficiently large r, all roots are inclosed, and $f(z)$ is approximately $a_n z^n$.

As in Problem 43 find out in which quadrants the roots of the following equations lie:

45. $z^3+z^2+z+4=0$

46. $z^3+3z^2+4z+2=0$

47. $z^3+4z^2+12=0$

48. $z^4-z^3+6z^2-3z+5=0$

49. $z^4 - 4z^3 + 11z^2 - 14z + 10 = 0$

50. $z^4 + z^3 + 4z^2 + 2z + 3 = 0$

51. Use (7.8) to evaluate

$$\oint_C \frac{f'(z)}{f(z)}\,dz, \qquad \text{where} \quad f(z) = \frac{z^3(z+1)^2 \sin z}{(z^2+1)^2(z-3)},$$

around the circle $|z| = 2$; around $|z| = \frac{1}{2}$.

52. Use (7.8) to evaluate $\oint \dfrac{z^3\,dz}{1+2z^4}$ around $|z| = 1$.

53. Use (7.8) to evaluate $\oint \dfrac{z^3+4z}{z^4+8z^2+16}\,dz$ around the circle $|z-2i| = 2$.

54. Use (7.8) to evaluate

$$\oint_C \frac{\sec^2(z/4)\,dz}{1-\tan(z/4)},$$

where C is the rectangle formed by the lines $y = \pm 1$, $x = \pm\frac{5}{2}\pi$.

Find the inverse Laplace transform of the following functions by using (7.16).

55. $\dfrac{p^3}{p^4+4}$ *Hint:* Use (6.2).

56. $\dfrac{1}{p^4-1}$

57. $\dfrac{p+1}{p(p^2+1)}$

58. $\dfrac{p^3}{p^4-16}$

59. $\dfrac{3p^2}{p^3+8}$

60. $\dfrac{1}{p^2(p+1)}$

61. $\dfrac{p^5}{p^6-64}$

62. $\dfrac{(p-1)^2}{p(p+1)^2}$

63. $\dfrac{p}{p^4-1}$

64. $\dfrac{p^2}{(p^2-1)(p^2-4)}$

65. $\dfrac{p}{(p+1)(p^2+4)}$

66. In equation (7.18), let $u(x)$ be an even function and $v(x)$ be an odd function.

(a) If $f(x) = u(x)+iv(x)$, show that these conditions are equivalent to the equation $f^*(x) = f(-x)$.

(b) Show that

$$\pi u(a) = PV\int_0^\infty \frac{2xv(x)}{x^2-a^2}\,dx, \qquad \pi v(a) = -PV\int_0^\infty \frac{2au(x)}{x^2-a^2}\,dx.$$

These are the Kramers-Kronig relations. *Hint:* To find $u(a)$, write the integral for $u(a)$ in (7.18) as an integral from $-\infty$ to 0 plus an integral from 0 to ∞. Then in the $-\infty$ to 0 integral, replace x by $-x$ to get an integral from 0 to ∞, and use $v(-x) = -v(x)$. Add the two 0 to ∞ integrals and simplify. Similarly find $v(a)$.

8. THE POINT AT INFINITY; RESIDUES AT INFINITY

It is often useful to think of the complex plane as corresponding to the surface of a sphere in the following way. In Figure 8.1, the sphere is tangent to the plane at the origin O. Let O be the south pole of the sphere, and N be the north pole of the sphere. If a line through N intersects the sphere at P and the plane at Q, we say that the point P on the sphere and the point Q on the plane are corresponding points. Then we have a one-to-one correspondence between points on the sphere (except N) and points of the plane (at finite distances from O). Imagine point Q moving farther and farther out away from O; then P moves nearer and nearer to N. If $z = x + iy$ is the complex coordinate of Q, then as Q moves out farther and farther from O, we would say $z \to \infty$. It is customary to say that the point N corresponds to the *point at infinity* in the complex plane. Observe that straight lines through the origin in the plane correspond to meridians of the sphere. The meridians all pass through both the north pole and the south pole. Corresponding to this, straight lines through the origin in the complex plane pass through the point at infinity. Circles in the complex plane with center at O correspond to parallels of latitude on the sphere. This mapping of the complex plane onto a sphere (or the mapping of the sphere onto a tangent plane) is called a *stereographic projection.*

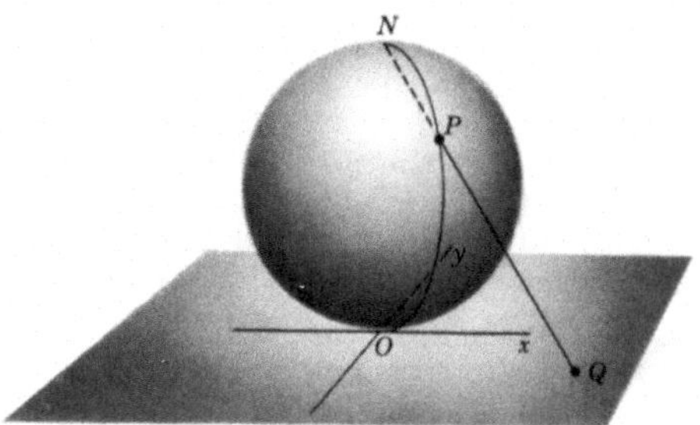

Figure 8.1

To investigate the behavior of a function at infinity, we replace z by $1/z$ and consider how the new function behaves at the origin. We then say that infinity is a regular point, a pole, etc., of the original function, depending on what the new function does at the origin. For example, consider z^2 at infinity; $1/z^2$ has a pole of order 2 at the origin, so z^2 has a pole of order 2 at infinity. Or consider $e^{1/z}$; since e^z is analytic at $z = 0$, $e^{1/z}$ is analytic at ∞.

Next we want to see how to find the residue of a function at ∞. To do this, we want to replace z by $1/z$ and work around the origin. In order to keep our notation straight, let us use two variables, namely Z which takes on values near ∞, and $z = 1/Z$ which takes on values near 0. The residue of a function at ∞ is defined so that the residue theorem holds, that is

$$\oint_C f(Z)\, dZ = 2\pi i \cdot (\text{residue of } f(Z) \text{ at } Z = \infty) \tag{8.1}$$

if C is a closed path around the point at ∞ but inclosing no other singular points. Now what does it mean to integrate "around ∞"? Recall that we have agreed to traverse contours so that the area inclosed always lies to our left. The area we wish to "inclose" is the area "around ∞"; if C is a circle, this area would lie *outside* the circle in our usual terminology. Figure 8.1 may clarify this. Imagine a small circle about the north pole; the area inside this circle (that is, the area including N) corresponds to points in the plane which are outside a large circle C. We must go around C in the clockwise direction in order to have the area "around ∞" to our left. Note that if $Z = Re^{i\Theta}$, then in going clockwise around C, we are going in the direction of *decreasing* Θ. Let us make the following change of variable in the integral (8.1):

$$Z = \frac{1}{z}, \qquad dZ = -\frac{1}{z^2}\, dz.$$

If $Z = Re^{i\Theta}$ traverses a circle C of radius R in the direction of decreasing Θ, then $z = 1/Z = (1/R)e^{-i\Theta} = re^{i\theta}$ traverses a circle C' of radius $r = 1/R$ in the counterclockwise direction (that is, $\theta = -\Theta$ increases as Θ decreases). Thus (8.1) becomes

$$\oint_{C'} -\frac{1}{z^2} f\left(\frac{1}{z}\right) dz = 2\pi i \cdot \text{residue of } f(Z) \text{ at } Z = \infty. \tag{8.2}$$

The integral in (8.2) is an integral about the origin and so can be evaluated by calculating the residue of $(-1/z^2)f(1/z)$ at the origin. (There are no other singular points of $f(1/z)$ inside C' because we assumed that there were no singular points of $f(Z)$ outside C except perhaps ∞.) Thus we have

$$(\text{residue of } f(Z) \text{ at } Z = \infty) = -\left(\text{residue of } \frac{1}{z^2} f\left(\frac{1}{z}\right) \text{ at } z = 0\right) \tag{8.3}$$

and we can use the methods we already know for computing residues at the origin.

Note that a function may be analytic at ∞ and still have a residue there.

Example. $f(Z) = 1/Z$ is analytic at ∞ because z is analytic at the origin. But the residue of $f(Z) = 1/Z$ at $Z = \infty$ is

$$-\left(\text{residue of } \frac{1}{z^2} \cdot z \text{ at } z = 0\right) = -1.$$

PROBLEMS, SECTION 8

1. Let $f(z)$ be expanded in the Laurent series that is valid for all z *outside* some circle, that is, $|z| > M$ (see Section 4). This series is called the Laurent series "about infinity." Show that the result of integrating the Laurent series term by term around a very large circle (of radius $> M$) in the positive direction, is $2\pi i b_1$ (just as in the original proof of the residue theorem in Section 5). Remember that the integral "around ∞ " is taken in the negative direction, and is equal to $2\pi i$·(residue at ∞). Conclude that $R(\infty) = -b_1$. *Caution*: In using this method of computing $R(\infty)$, be sure you have the Laurent series that converges for all sufficiently large z.

2. (a) Show that if $f(z)$ tends to a finite limit as z tends to infinity, then the residue of $f(z)$ at infinity is $\lim_{z\to\infty} z^2 f'(z)$.

   (b) Also show that if $f(z)$ tends to zero as z tends to infinity, then the residue of $f(z)$ at infinity is $-\lim_{z\to\infty} z f(z)$.

Find out whether infinity is a regular point, an essential singularity, or a pole (and if a pole, of what order) for each of the following functions. Using Problem 1, or Problem 2, or (8.3), find the residue of each function at infinity. Check your results by computer.

3. $\dfrac{z}{z^2+1}$
4. $\dfrac{2z+3}{(z+2)^2}$
5. $\sin\dfrac{1}{z}$
6. $\dfrac{z^2+5}{z}$
7. $\dfrac{4z^3+2z+3}{z^2}$
8. $\dfrac{z^2+2}{3z^2}$
9. $\dfrac{z^2-1}{z^2+1}$
10. $\dfrac{1+z}{1-z}$
11. $\tan\dfrac{1}{z}$
12. $\ln\dfrac{z+1}{z-1}$

13. Give another proof of the fundamental theorem of algebra (see Problem 7.44) as follows. Let $I = \oint f'(z)/f(z)\,dz$ about infinity, that is, in the negative direction around a very large circle C. Use the argument principle (7.8), and also evaluate I by finding the residue of $f'(z)/f(z)$ at infinity; thus show that $f(z)$ has n zeros inside C.

Evaluate the following integrals by computing residues at infinity. Check your answers by computing residues at all the finite poles. (It is understood that $\oint$ means in the positive direction.)

14. $\displaystyle\oint \frac{1-z^2}{1+z^2}\frac{dz}{z}$ around $|z|=2$.

15. $\displaystyle\oint \frac{z^2\,dz}{(2z+1)(z^2+9)}$ around $|z|=5$.

16. Observe that in Problems 14 and 15 the sum of the residues at finite points plus the residue at infinity is zero. Prove that this is always true for a function which has a finite number of singularities.

9. MAPPING

We often find it useful to sketch a graph of a given function $y = f(x)$ of a real variable x. Imagine trying to make a similar sketch for a function $w = f(z)$ of a complex variable z. We need a plane to plot values of z and another plane to plot values of $w = f(z)$, that is, we need a four-dimensional space. Lacking this, we must resort to a different method. Imagine trying to "graph" $y = f(x)$ using only two straight lines, but not a plane. A "graph" of $y = x^2$ might look like Figure 9.1. Given a point on the x axis, we can locate a corresponding point $y = f(x)$ on the y axis and label the two points with the same letter to indicate this correspondence. (Note that to finish our "graph," we really need a second positive y axis to hold the y points corresponding to negative values of x.)

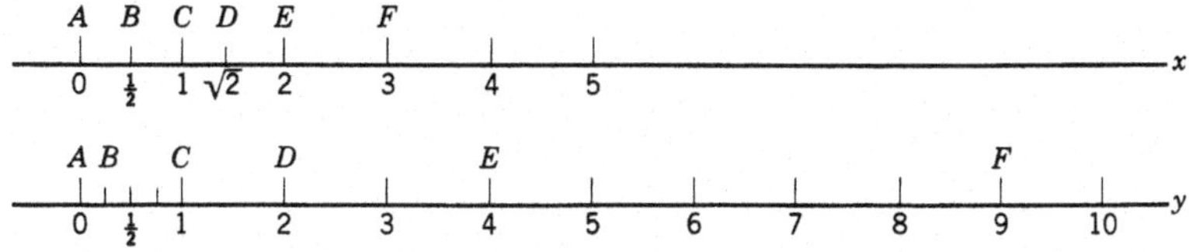

Figure 9.1

Now consider a similar method of representing a function of a complex variable $w = f(z)$. We use a z plane and a w plane; a given point in the z plane (that is, a value of z) determines a corresponding value of w, that is, a point in the w plane. The pair of points, one z and one w, are called *images* of each other. Although we *could* label pairs of corresponding z and w points (as we did corresponding x and y points in Figure 9.1), it is usually more interesting to sketch corresponding curves or regions in the two planes. The correspondence between a point (or curve or region) in the z plane, and the image point (or curve or region) in the w plane, is called a *mapping* or a *transformation*.

Example 1. Consider the function $w = i + ze^{i\pi/4}$, and let us map the grid of coordinate lines $x =$ const., $y =$ const. (z plane in Figure 9.2) into the w plane. You may be able to see at once that this transformation amounts to a rotation of the grid through an angle of $\pi/4$ (since $ze^{i\pi/4} = re^{i(\theta+\pi/4)}$) plus a translation i (the image of $z = 0$ is $w = i$), giving the result shown in the w plane in Figure 9.2. Alternatively,

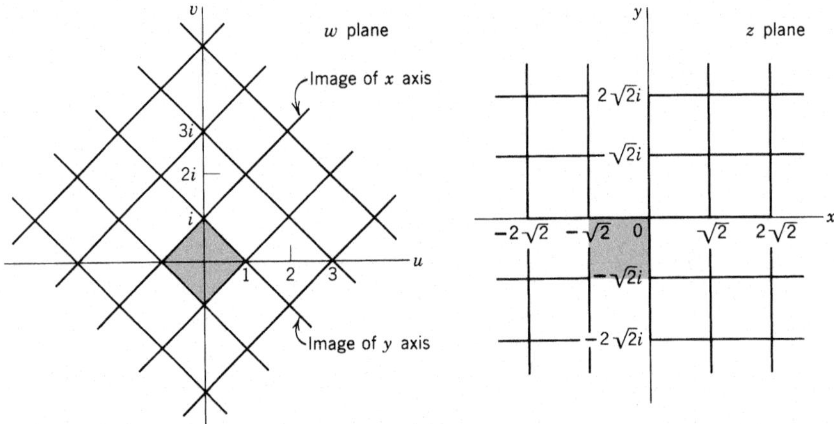

Figure 9.2

we can compute u and v as follows:

$$\begin{aligned} w = i + ze^{i\pi/4} &= i + (x+iy)\left(\cos\frac{\pi}{4} + i\sin\frac{\pi}{4}\right) \\ &= i + (x+iy)\left(\frac{1+i}{\sqrt{2}}\right) = \frac{x-y}{\sqrt{2}} + i\left(1 + \frac{x+y}{\sqrt{2}}\right). \end{aligned}$$

Since $w = u + iv$, we have

$$u = \frac{x-y}{\sqrt{2}}, \qquad v = 1 + \frac{x+y}{\sqrt{2}}. \tag{9.1}$$

Then (eliminating x and y in turn), we have

$$u - v = -1 - y\sqrt{2}, \qquad u + v = 1 + x\sqrt{2}. \tag{9.2}$$

The image of the x axis ($y = 0$) is, from the first equation in (9.2), $u - v = -1$; the image of the y axis ($x = 0$) is, from the second equation in (9.2), $u+v = 1$. Plotting these lines in the w plane, and also plotting the images of $x = \pm\sqrt{2}$, $x = \pm 2\sqrt{2}$, $y = \pm\sqrt{2}$, $y = \pm 2\sqrt{2}$ [using the equations (9.2)], we get Figure 9.2. (Verify that the shaded squares are images of each other.)

If the elimination [to get (9.2)] is not easy, we can use equations (9.1) directly. Suppose that we want the image of $y = 0$. With $y = 0$, equations (9.1) become $u = x/\sqrt{2}$, $v = 1 + x/\sqrt{2}$; these are a pair of parametric equations for a curve in the (u, v) plane, with x as the parameter. Similarly, to find the image of $x =$ const., we substitute the value of x into (9.1); we then have a pair of parametric equations with y as the parameter.

Note that we could just as easily have found the images in the z plane of the lines $u =$ const., and $v =$ const. For example, letting $u = 0$ in (9.1), we get $x - y = 0$; the image of the v axis ($u = 0$) is the 45° line in the (x, y) plane. (We might have guessed that going back to the z plane would involve a rotation through $-45°$.) In any given problem, we may start with simple curves (or regions) in either the z plane or the w plane, and find their images in the other plane.

Example 2. Let us map the coordinate grid $u =$ const., $v =$ const., into the z plane by the function $w = z^2$. We have

$$w = z^2 = (x + iy)^2 = x^2 - y^2 + 2ixy, \tag{9.3}$$
$$u = x^2 - y^2, \qquad v = 2xy.$$

Then the images of $u =$ const. are hyperbolas $x^2 - y^2 =$ const., and the images of $v =$ const. are also hyperbolas $xy =$ const. (Figure 9.3). Alternatively, we could map the lines $x =$ const., $y =$ const. into the w plane (Problem 1); this gives two sets of parabolas in the (u, v) plane. Accurate graphs can be obtained by computer.

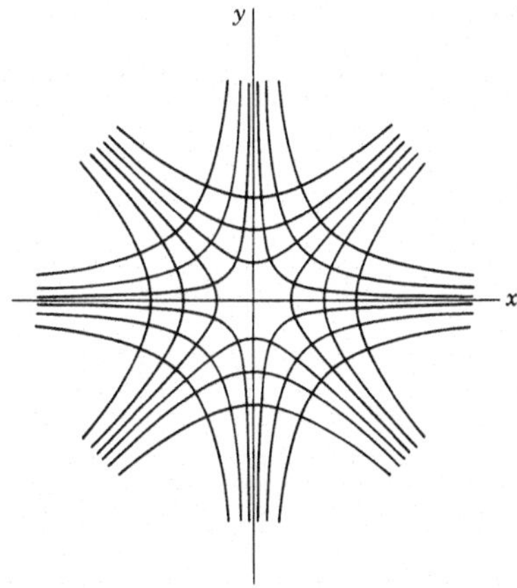

Figure 9.3

Example 3. Let us consider still another useful way of discussing the mapping by $w = z^2$. Using polar coordinates, we have

$$z = re^{i\theta}, \qquad w = z^2 = r^2e^{2i\theta}. \tag{9.4}$$

Consider the region inside the circle $r = 1$ in the (x, y) plane. If $r = 1$ in (9.4), we have $z = e^{i\theta}$, $w = e^{2i\theta}$. The angle of w is twice the angle of z; thus the first-quadrant part of the area inside the circle $r = 1$ in the z plane maps into a semicircular area in the w plane as indicated by the shading in Figure 9.4. The second quadrant of the z plane disk (θ between $\pi/2$ and π) maps into the lower half of the disk in the w plane (angle of w between π and 2π) as indicated. We have now used up the whole area of the disk in the w plane and only half of the z plane disk. (Compare Figure 9.1 and the comment about a second y axis.) In order to have a one-to-one correspondence between points in the z plane and their images in the w plane, we draw a second w plane (w plane II in Figure 9.4) to contain the images of points in the lower half of the z plane. (Convince yourself that the two lower quarter-disks in the z plane and their images in w plane II are correctly indicated by the shading.) We agree that as we reach the angle 2π in w plane I, we go over to w plane II, and as we reach the angle 4π in w plane II, we go back to w plane I. The two w planes joined in this way are called a *Riemann surface*; each plane is called a *sheet* of the Riemann surface. Note that the line along which the sheets of the Riemann surface are joined (positive real axis here) is a branch cut, and the origin is a branch point (see Example 5, Section 7). Here the branch cut and Riemann surface are in

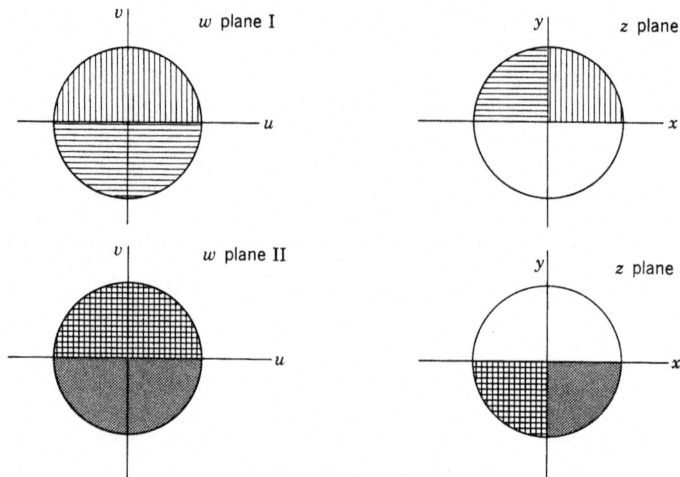

Figure 9.4

the w plane because $z = \sqrt{w}$ has two branches; for $w = \sqrt{z}$, the Riemann surface would be in the z plane (as in Section 7). It is not necessary to take the branch cut along the positive x axis; we can select any 2π interval for one branch of $\sqrt{w}$, for example from $-\pi$ to π instead of 0 to 2π. A Riemann surface may have many sheets; for example for $w = z^5$, there are 5 sheets, and for $w = \ln z$, there are an infinite number. For more detail, see complex variables texts.

Conformal Mapping We have been discussing mappings or transformations. We used the term *transformation* in Chapters 3 and 10, meaning a change of variables or a change of coordinate system or a change of basis; let us see the connection between the two discussions. In Chapter 10 we used only one plane [the (x, y) plane]; we located a point in the (x, y) plane by giving its rectangular coordinates (x, y), or its polar coordinates (r, θ), or some other coordinates (u, v). In polar coordinates, the circles $r =$ const., and the rays $\theta =$ const., were sketched in the (x, y) plane. Similarly, for any coordinate system (u, v) (in Chapter 10, see Section 8 and the Section 8 problems), we sketched the curves $u =$ const., $v =$ const., in the (x, y) plane. In the complex variable language we are now using, this amounts to mapping the w plane lines $u =$ const., $v =$ const., into the z plane. In Chapter 10 we were particularly interested in transformations to orthogonal curvilinear coordinates. Let us see that any analytic function $w = f(z) = u + iv$ gives us a transformation to an orthogonal coordinate system (u, v). We have

$$\begin{aligned} dz &= dx + i\,dy, & dw &= du + i\,dv, \\ |dz|^2 &= dx^2 + dy^2, & |dw|^2 &= du^2 + dv^2. \end{aligned} \tag{9.5}$$

Then the square of the arc length element in the (x, y) plane is

$$ds^2 = dx^2 + dy^2 = |dz|^2 = \left|\frac{dz}{dw}\right|^2 |dw|^2 = \left|\frac{dz}{dw}\right|^2 (du^2 + dv^2). \tag{9.6}$$

Since there is no $du\,dv$ term in ds^2, the (u, v) coordinate system is orthogonal (Chapter 10, Section 8). By this we mean that if we obtain $u(x, y)$ and $v(x, y)$ from

$f(z) = u + iv$ and plot the curves $u(x, y) = $ const., $v(x, y) = $ const. in the (x, y) plane, we have two sets of mutually orthogonal curves. These are the coordinate curves for the (u, v) coordinate systems as in Chapter 10. If we solve the equations $u = u(x, y)$, $v = v(x, y)$ for x and y in terms of u and v, we have the transformation equations from the variables x, y to the variables u, v as in Problems 8.6 to 8.9 of Chapter 10, and by (9.6) we know that the coordinate system (u, v) is an orthogonal system [if $f(z)$ is analytic]. We see an example of this in Figure 9.3 (two orthogonal sets of hyperbolas). Note from (9.6) that the two scale factors in a (u, v) coordinate system obtained this way are equal.

Although we used only one plane in Chapter 10, for complex variables we find it useful to consider both the z plane [that is the (x, y) plane] and the w plane [that is, the (u, v) plane]. In the (x, y) plane, the arc length element ds is given by $ds^2 = dx^2 + dy^2$. Similarly, in the (u, v) plane, the arc length element (which we shall call dS) is given by $dS^2 = du^2 + dv^2$. From (9.5) we see that $ds = |dz|$ and $dS = |dw|$. Then the ratio of dS to ds is $|dw/dz|$. Consider a point z (and its image w) at which $w(z)$ is analytic and dw/dz is not zero. If we stay near z, the value of dw/dz is almost constant, and the ratio dS/ds is nearly constant. This says that if we consider a small area in the z plane ($ABCD$ in Figure 9.5) and its image ($A'B'C'D'$ in Figure 9.5) in the w plane, then

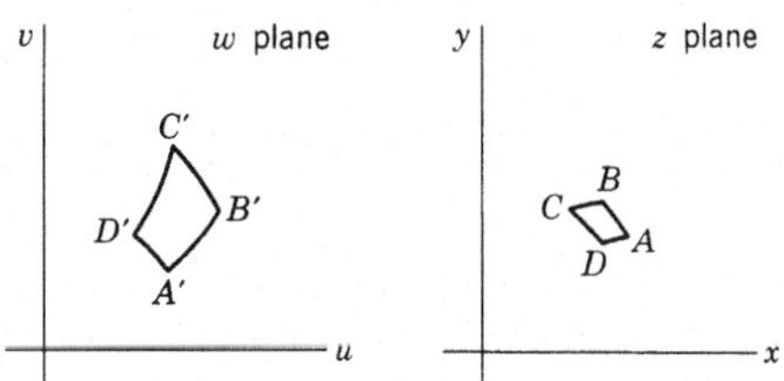

Figure 9.5

$$(9.7) \qquad \frac{A'B'}{AB} = \frac{B'C'}{BC} = \frac{C'D'}{CD} = \frac{D'A'}{DA} = \frac{dS}{ds} = \left|\frac{dw}{dz}\right|,$$

that is, the two small areas are similar figures (since corresponding sides are proportional). Because of this property of any mapping by an analytic function, we call the mapping or transformation *conformal* (same form or shape). Corresponding angles are equal ($A = A'$, etc.) and the net result of the transformation is to magnify (or minify) and rotate each infinitesimal area. Note that the conformal property is a local one; since the value of dw/dz changes from point to point, each tiny bit of a figure is magnified and rotated by a different amount, and so a large figure will not have the same shape after mapping. Also note that we do not have conformality in the neighborhood of a point where $dw/dz = 0$; for example, in Figure 9.4 a tiny quarter-circle about the origin in the z plane maps into a tiny semicircle in the w plane.

PROBLEMS, SECTION 9

In these problems you should be able to make rough sketches by hand, but for accurate graphs use a computer.

1. Solve equations (9.3) for x and y in terms of u and v. Use your equations to sketch the images in the w plane of the z plane lines $x = $ const. (for several values of x) and similarly of $y = $ const.

For each of the following functions $w = f(z) = u + iv$, find u and v as functions of x and y. Sketch the graphs in the (x, y) plane of the images of $u =$ const. and $v =$ const. for several values of u and several values of v as was done for $w = z^2$ in Figure 9.3. The curves $u =$ const. should be orthogonal to the curves $v =$ const.

2. $w = \dfrac{z+1}{2i}$ **3.** $w = \dfrac{1}{z}$ **4.** $w = e^z$ **5.** $w = \dfrac{z-i}{z+i}$

6. $w = \sqrt{z}$. *Hint:* This is equivalent to $w^2 = z$; find x and y in terms of u and v and then solve the pair of equations for u and v in terms of x and y. Note that this is really the same problem as Problem 1 with the z and w planes interchanged.

7. $w = \sin z$ **8.** $w = \cosh z$

Describe the Riemann surface for

9. $w = z^3$ **10.** $w = \sqrt{z}$ **11.** $w = \ln z$

12. If $w = f(z) = u(x, y) + iv(x, y)$, $f(z)$ analytic, defines a transformation from the variables x, y to the variables u, v, show that the Jacobian of the transformation (Chapter 5, Section 4) is $\partial(u, v)/\partial(x, y) = |f'(z)|^2$. *Hint:* To simplify the determinant, use the Cauchy-Riemann equations and the equations (Section 2) used in obtaining them.

13. Verify the matrix equation

$$\begin{pmatrix} du \\ dv \end{pmatrix} = \mathrm{J} \begin{pmatrix} dx \\ dy \end{pmatrix},$$

where J is a matrix whose determinant is the Jacobian in Problem 12. Multiply the matrix equation by its transpose and use Problem 12 to obtain $dS/ds = |dw/dz|$ as in (9.7).

14. We have discussed the fact that a conformal transformation magnifies and rotates an infinitesimal geometrical figure. We showed that $|dw/dz|$ is the magnification factor. Show that the angle of dw/dz is the rotation angle. *Hint:* Consider the rotation and magnification of an arc $dz = dx + idy$ (of length ds and angle arc tan dy/dx) which is required to obtain the image of dz, namely dw.

15. Compare the directional derivative $d\phi/ds$ (Chapter 6, Section 6) at a point and in the direction given by dz in the z plane, and the directional derivative $d\phi/dS$ in the direction in the w plane given by the image dw of dz. Hence show that the rate of change of T in a given direction in the z plane is proportional to the corresponding rate of change of T in the image direction in the w plane. (See Section 10, Example 2.) Show that the proportionality constant is $|dw/dz|$. *Hint:* See equations (9.6) and (9.7).

10. SOME APPLICATIONS OF CONFORMAL MAPPING

Many different physical problems require solution of Laplace's equation. We are going to show how to solve a few such problems by conformal mapping. Much of this work can be done by computer. But before you can use the computer, you need to know the basic theory behind the use of conformal mapping. Our purpose in this section is to learn this background. First we consider a very simple problem for which we know the answer from elementary physics.

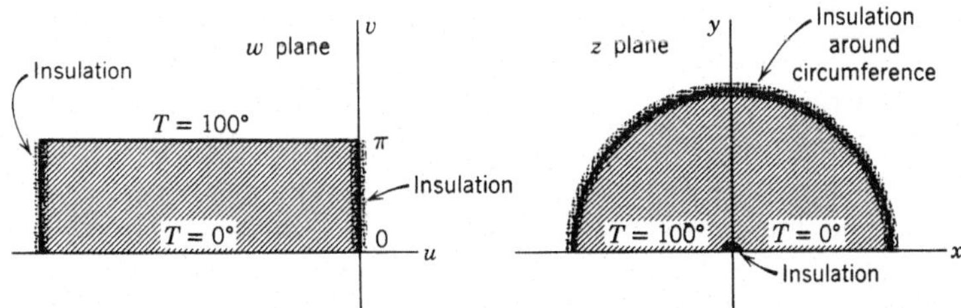

Figure 10.1

Example 1. In Figure 10.1, the shaded area in the (u, v) plane represents a rectangular plate. The ends and faces of the plate are insulated, the bottom edge is held at temperature $T = 0°$, and the top edge at $T = 100°$. Then we know from elementary physics that the temperature increases linearly from the bottom edge $(v = 0)$ to the top edge $(v = \pi)$, that is, $T = (100/\pi)v$ at any point of the plate. Let us also derive this answer by a more advanced method. It is known from the theory of heat that the temperature T of a body satisfies Laplace's equation in regions where there is no source of heat. (See Chapter 13, Section 2.) In our problem we want a solution of Laplace's equation which satisfies the *boundary conditions*, that is, $T = 100°$ when $v = \pi$, $T = 0°$ when $v = 0$, and $\partial T/\partial u = 0$ on the ends (see Chapter 13, Problem 2.14). You should verify that $T = 100v/\pi$ satisfies $\partial^2 T/\partial u^2 + \partial^2 T/\partial v^2 = 0$, and satisfies all the boundary conditions. Also note that an easy way to know that v satisfies Laplace's equation is to observe that it is the imaginary part of $w = u + iv$, and use Theorem IV of Section 2 which says that the real and imaginary parts of an analytic function of a complex variable satisfy Laplace's equation.

Now let us use our results to solve a harder problem.

Example 2. Consider the mapping of the rectangle in the w plane into the z plane by the function $w = \ln z$ (Figure 10.1, z plane). We have

$$\begin{aligned} &w = \ln z = \ln(re^{i\theta}) = \ln r + i\theta = u + iv, \\ &u = \ln r, \qquad v = \theta. \end{aligned} \tag{10.1}$$

Then $v = 0$ maps into $\theta = 0$, that is, the positive x axis; $v = \pi$ maps into $\theta = \pi$, that is, the negative x axis (z plane, Figure 10.1). The insulated end of the rectangle at $u = 0$ maps into $\ln r = 0$ or $r = 1$; the left-hand end of the rectangle maps into a small semicircle about the origin which we can think of as a bit of insulation at the origin separating the 0° and 100° parts of the x axis. (If the left-hand end of the rectangle is at $u = -\infty$, we have $\ln r = -\infty$, $r = 0$, and the image is just the origin; for finite negative u, the image is a semicircle with $r < 1$.) We can now solve the problem indicated by the picture in the z plane of Figure 10.1. A semicircular plate has its faces and its curved boundary insulated, and has half its flat boundary at 0° and the other half at 100° (with a bit of insulation at the center). Find the temperature T at any point of the plate. To solve this problem we need only transform our solution in the (u, v) plane to the variables x, y by using (10.1). Thus we find

$$T = \frac{100}{\pi}V = \frac{100}{\pi}\theta = \frac{100}{\pi}\arctan\frac{y}{x}, \qquad 0 \le \theta \le \pi. \tag{10.2}$$

It is not hard to justify our method; we need to show that our solution satisfies Laplace's equation and that it satisfies the boundary conditions. It is straightforward to show (Problem 1) that if a function $\phi(u, v)$ satisfies Laplace's equation $\partial^2\phi/\partial u^2 + \partial^2\phi/\partial v^2 = 0$, then the function of x and y obtained by substituting $u = u(x, y)$, $v = v(x, y)$ in ϕ satisfies Laplace's equation in x and y, where u and v are the real and imaginary parts of an analytic function $w = f(z)$. Thus we know that (10.2) satisfies Laplace's equation (or in this case you can easily verify the fact directly). We must also know that the transformed T satisfies the boundary conditions; this is where conformal mapping is so useful. Observe in Figure 10.1 that we had a transformation which took the boundaries of a simple region (a rectangle) for which we knew the solution of the temperature problem, into the boundaries of a more complicated region for which we wanted the solution. This is the basic method of conformal mapping—to transform from a simple region where you know the answer to a given problem, to the region in which you want the solution. The temperature at any (x, y) point is the same as the temperature at the (u, v) image point, since we obtain the temperature as a function of x and y by the same substitution $u = u(x, y)$, $v = v(x, y)$ that we use to obtain image points. Thus the temperatures on the boundaries of the transformed region are the same as the temperatures on the corresponding boundaries of the simpler (u, v) region. Similarly, isothermals (curves of constant temperature) transform into isothermals; in this problem the (u, v) isothermals are the lines $v =$ const., and so the (x, y) isothermals are $\theta =$ const. You can show that the rate of change of T in a direction perpendicular to a boundary in the (u, v) plane is proportional to the corresponding rate of change of T in a direction perpendicular to the image boundary in the (x, y) plane (Problem 9.15). Thus insulated boundaries (across which the rate of change of T is zero) map into insulated boundaries. The lines (or curves) perpendicular to the isothermals give the direction of flow of heat; in Figure 10.1 heat flows along the lines $u =$ const. in the w plane, and along the circles $r =$ const. (which are the images of $u =$ const.) in the z plane.

Using the same mapping function $w = \ln z$, we can solve a number of other physics problems. Observe first that if we think of Figure 10.1 as representing a cross section of a three-dimensional problem (with all parallel cross sections identical), then (10.2) gives the solution of the three-dimensional problem also. In Figure 10.1 the (u, v) diagram would be the cross section of a slab with faces at $T = 100°$ and $T = 0°$ and all other surfaces insulated (or extending to infinity); the (x, y) diagram would similarly represent half a cylinder. Now let us do a three-dimensional problem in electrostatics.

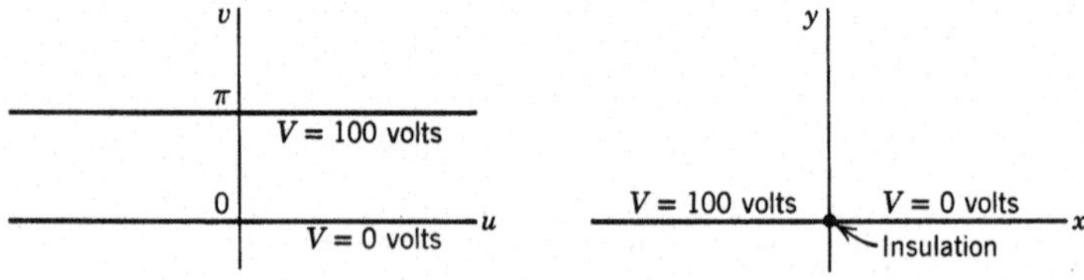

Figure 10.2

▶ **Example 3.** In Figure 10.2 the (u, v) diagram represents (the cross section of) two infinite parallel plates, one at potential $V = 0$ volts and one at potential $V = 100$ volts. The (x, y) diagram represents (the cross section of) one plane with its right-hand half at potential $V = 0$ volts and its left-hand half at $V = 100$ volts. From electricity we know that the electrostatic potential V satisfies Laplace's equation in regions where there is no free charge (see Chapter 13, Section 1). You should convince yourself that the mapping by (10.1) gives the result shown in Figure 10.2, and that the potential is given by

$$V = \frac{100}{\pi} v = \frac{100}{\pi} \theta = \frac{100}{\pi} \arctan \frac{y}{x}, \qquad 0 \le \theta \le \pi$$

as in (10.2). The equipotentials ($V =$ const.) in the (x, y) plane are the lines $\theta =$ const. Recall that the electric field is given by $\mathbf{E} = -\boldsymbol{\nabla} V$, and that the gradient of V is perpendicular to $V =$ const. (Chapter 6, Section 6). Then the direction of the electric field at any point is perpendicular to the equipotential through that point. Thus if we sketch the curves $r =$ const. which are perpendicular to the equipotentials $\theta =$ const., then the tangent to a circle at a point gives the direction of the electric field $\mathbf{E}$ at that point. Note the correspondence between the isothermals of the temperature problem and the equipotentials here, and between the lines of electric flux (curves tangent to $\mathbf{E}$) and the lines or curves along which heat flows.

We can also solve problems in hydrodynamics (see Chapter 6, Section 10) by conformal mapping. We consider a two-dimensional flow of water by which we mean either that we think of the water as flowing in a thin sheet over the (x, y) [or (u, v)] plane, or if it has depth, the flow is the same in all planes parallel to the (x, y) [or (u, v)] plane. Although it is convenient to talk about water, what we actually require is an irrotational flow (see Chapter 6, Section 11) of a nonviscous incompressible fluid. For then (see Problem 2) the velocity $\mathbf{V}$ of the liquid is given by $\mathbf{V} = \boldsymbol{\nabla}\Phi$, where Φ (called the *velocity potential*) satisfies Laplace's equation. Water approximately meets these requirements.

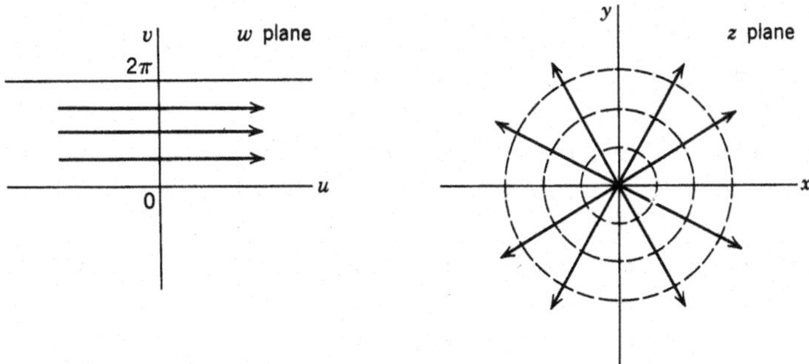

Figure 10.3

Example 4. Figure 10.3 shows two flow patterns related by the same transformation we have used in the heat problem and the electrostatics problem, namely $w = \ln z$. In the w plane of Figure 10.3, we picture water flowing in the u direction at constant speed V_0 down a channel between $v = 0$ and $v = 2\pi$. (Note that v is the imaginary part of $w = u + iv$ and has nothing to do with velocity.) The velocity potential is $\Phi = V_0 u$; for then the velocity, $\mathbf{V} = \boldsymbol{\nabla}\Phi$, has components $\partial\Phi/\partial u = V_0$ in the u direction and $\partial\Phi/\partial v = 0$ in the v direction as we have assumed. The function $\Phi + i\Psi = V_0 w = V_0(u+iv)$ is called the *complex potential*; the function Ψ (conjugate to Φ; see Section 2) is called the *stream function.* The lines $\Psi =$ const. (that is,

$v =$ const. in the w plane) are the lines along which the water flows and are called *streamlines*. Observe that the lines $\Phi =$ const. and the lines $\Psi =$ const. are mutually perpendicular sets of lines. The water flows across lines of constant Φ and along streamlines (constant Ψ); boundaries of the channel ($v = 0$ and $v = 2\pi$) must then be streamlines. The water comes from the left (Figure 10.3, w plane) and goes off to the right; we say that there is a *source* at the left and a *sink* at the right.

Now consider the mapping of the w plane flow of Figure 10.3 into the z plane by the function $w = \ln z$. The complex potential is

$$\Phi + i\Psi = V_0 w = V_0 \ln z = V_0(\ln r + i\theta).$$

The streamlines are $\Psi =$ const., or $\theta =$ const., that is, radial lines; the curves $\Phi =$ const. are circles $r =$ const. and are perpendicular to the streamlines. The velocity is given by

$$\mathbf{V} = \boldsymbol{\nabla}\Phi = V_0 \boldsymbol{\nabla}(\ln r) = V_0 \left(\mathbf{e}_r \frac{\partial}{\partial r} + \mathbf{e}_\theta \frac{1}{r}\frac{\partial}{\partial \theta} \right) \ln r = \mathbf{e}_r \frac{V_0}{r}.$$

What we are describing, then, is the flow of water from a source at the origin out along radial lines. Since the same amount of water crosses a small circle (about the origin) or a large one, the velocity of the water decreases with r as we have found ($|\mathbf{V}| = V_0/r$).

We can obtain another flow pattern from any given one by interchanging the equipotentials and the streamlines. In Figure 10.3, z plane, this new flow would have the circles $r =$ const. as streamlines and would correspond to a whirlpool motion of the water about the origin (called a *vortex*). There are still other applications of this diagram. The circles $r =$ const. give the direction of the magnetic field about a long current-carrying wire perpendicular to the (x, y) plane and passing through the origin. The radial lines $\theta =$ const. give the direction of the electric field about a similar long wire with a static charge on it. The radial lines give the direction of heat flow from a small hot object at the origin, and the circles $r =$ const. are then the isothermals. By starting with problems like these to which we know the answers and using various conformal transformations, we can solve many other physics problems involving fluid flow, electricity, heat, and so on. Some examples are outlined in the problems and you will find many more in books on complex variables.

▶ **Example 5.** Let us consider one somewhat more complicated example of the use of conformal mapping. We shall be able to solve two interesting physics problems in this example: (1) to find the flow pattern for water flowing out of the end of a straight channel into the open, and (2) to find the edge effect (fringing) at the ends of a parallel-plate capacitor.

We consider the mapping function

$$\begin{aligned} &z = w + e^w = u + iv + e^u e^{iv} = u + iv + e^u(\cos v + i \sin v), \\ &x = u + e^u \cos v, \qquad y = v + e^u \sin v. \end{aligned} \tag{10.3}$$

In Figure 10.4, w plane, we picture a parallel flow of water at constant velocity in the region between the lines DEF and GHI; this is just like the flow of Figure 10.3,

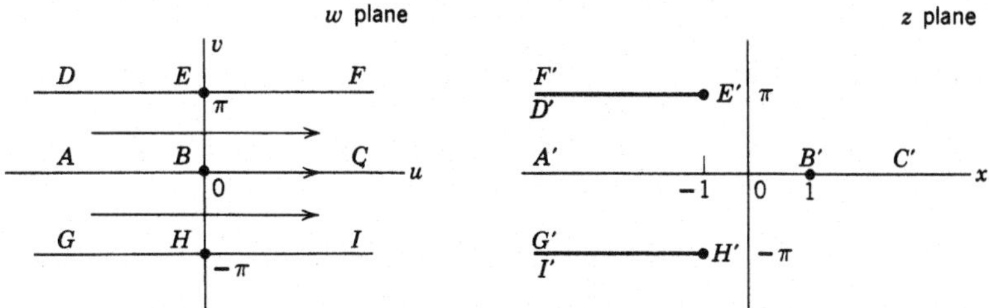

Figure 10.4

w plane. Now let us map the w plane streamlines into the z plane using (10.3). On the u axis, $v = 0$; putting $v = 0$ in (10.3), we find $y = 0$, and $x = u + e^u$. Thus the u axis maps into the x axis ($y = 0$) with $u = -\infty$ corresponding to $x = -\infty$, $u = 0$ corresponding to $x = 1$, and $u = +\infty$ corresponding to $x = +\infty$ as shown in Figure 10.4 (line ABC maps into $A'B'C'$). Now on DEF, $v = \pi$; substituting $v = \pi$ into (10.3), we find $y = \pi$, $x = u + e^u \cos\pi = u - e^u$. However, the image of $v = \pi$ is not the entire line $y = \pi$. To see this consider $x = u - e^u$. We find the maximum value of x for $dx/du = 1 - e^u = 0$, $d^2x/du^2 = -e^u < 0$. These equations are satisfied for $u = 0$, $x = -1$. The point E ($u = 0$, $v = \pi$) maps into the point E' ($x = -1$, $y = \pi$). Thus DE in the w plane maps into the part of the line $y = \pi$ in the z plane up to $x = -1$ with $u = -\infty$ corresponding to $x = -\infty$ and $u = 0$ corresponding to $x = -1$. To see how to map EF, we realize that x has its largest value at $u = 0$ and so decreases as u increases; for very large positive u, $x = u - e^u$ is negative and of large absolute value since $e^u \gg u$. Thus the positive part of $v = \pi$ (EF) maps into the same line segment ($y = \pi$, $x \le -1$) that we obtained for the mapping of the negative part (DE), but this time the line segment ($E'F'$, z plane) is traversed backward. It is as if the line $y = \pi$ were broken at $x = -1$ and bent back upon itself through an angle of 180°. By a parallel discussion of the line GHI, we find that it maps as shown in Figure 10.4 into $G'H'I'$. Other streamlines in the w plane are given by $v =$ const. for any v between $-\pi$ and π. If we substitute $v =$ const. into the x and y equations in (10.3), we have parametric equations (with u as the parameter) for the streamlines in the z plane. For any value of v, these streamlines can be plotted in the z plane: some of them are shown by the solid curves in Figure 10.5. Think of $D'E'$ and $G'H'$ as boundaries of a channel (in the z plane) down which water flows coming from $x = -\infty$. The boundaries stop at $x = -1$ and the water flows out of the channel spreading over the whole plane, including spreading back along the outsides ($E'F'$ and $H'I'$) of the channel boundaries. This is correct according to our mapping, for the boundary streamline DEF mapped into the broken-and-folded-back line $D'E'F'$, and similarly for GHI to $G'H'I'$.

For the electrical application, let DEF and GHI represent (the cross section of) a large parallel plate capacitor. Then the lines $v =$ const. are the equipotentials and the lines $u =$ const. give the direction of the electric field $\mathbf{E}$. The image in the z plane represents (a cross section of) the end of a parallel plate capacitor. The images of the equipotentials $v =$ const. are the equipotentials in the z plane (same as the streamlines, shown as solid curves in Figure 10.5). The images of the lines $u =$ const. (shown as dotted curves in Figure 10.5) give the direction of the electric field at the end of a parallel-plate capacitor. Well inside the plates the $\mathbf{E}$ lines are vertical, but at the end they bulge out; this effect is known as *fringing*.

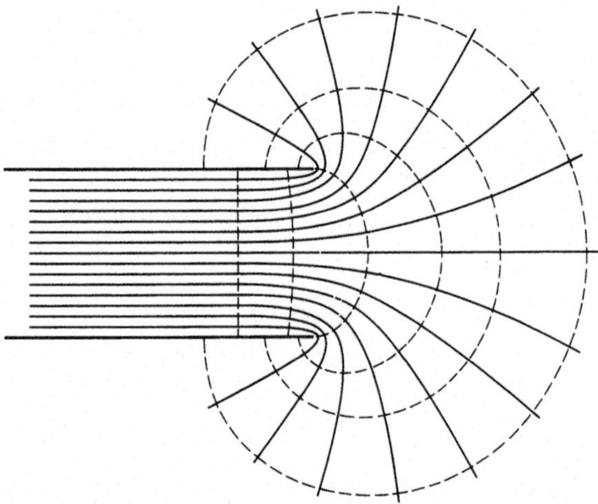

Figure 10.5

PROBLEMS, SECTION 10

1. Prove the theorem stated just after (10.2) as follows. Let $\phi(u, v)$ be a harmonic function (that is, ϕ satisfies $\partial^2\phi/\partial u^2 + \partial^2\phi/\partial v^2 = 0$). Show that there is then an analytic function $g(w) = \phi(u, v) + i\psi(u, v)$ (see Section 2). Let $w = f(z) = u + iv$ be another analytic function (this is the mapping function). Show that the function $h(z) = g(f(z))$ is analytic. *Hint:* Show that $h(z)$ has a derivative. (How do you find the derivative of a function of a function, for example, $\ln \sin z$?) Then (by Section 2) the real part of $h(z)$ is harmonic. Show that this real part is $\phi(u(x, y), v(x, y))$.

2. A fluid flow is called irrotational if $\nabla \times \mathbf{V} = 0$ where $\mathbf{V}$ = velocity of fluid (Chapter 6, Section 11); then $\mathbf{V} = \nabla\Phi$. Use Problem 10.15 of Chapter 6 to show that if the fluid is incompressible, the Φ satisfies Laplace's equation. (*Caution:* In Chapter 6, we used $\mathbf{V} = \mathbf{v}\rho$, with $\mathbf{v}$ = velocity; here $\mathbf{V}$ = velocity.)

3. Assuming from electricity the equations $\nabla \cdot \mathbf{D} = \rho$, $\mathbf{E} = -\nabla V$, $\mathbf{D} = \epsilon\mathbf{E}$ (ϵ = const.), show that in regions where the free charge density ρ is zero, V satisfies Laplace's equation.

4. Let a flat plate in the shape of a quarter-circle, as shown, have its faces and curved boundary insulated, and its two straight edges held at 0° and 100°. Find the temperature distribution $T(x, y)$ in the plate, and the equations of the isothermals. *Hint:* Use the mapping function $w = \ln z$ as in Figure 10.1; what w plane line maps into the y axis?

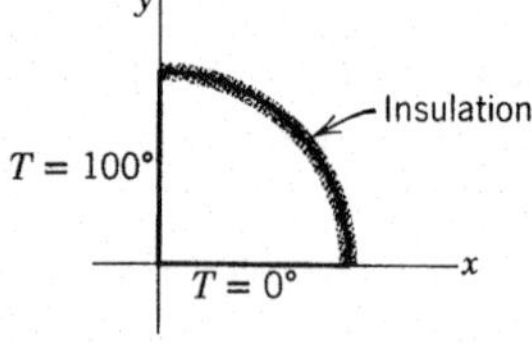

5. Consider a capacitor made of two very large perpendicular plates. (Let the positive x and y axes in the diagram of Problem 4 represent a cross section of the capacitor.) Let one plate (x axis) be held at potential $V = 0$, and the other plate (y axis) be held at potential $V = 100$ volts. Find the potential $V(x, y)$ for $x > 0$, $y > 0$, and the equations of the equipotentials. *Hint:* This problem is mathematically the same as Problem 4.

6. Let the figure represent (the cross section of) a hot cylinder (say $T = 100°$) lying on a cold plane (say $T = 0°$). (Separate the two by a bit of insulation.) Find the temperature in the shaded region. Alternatively, let the cylinder and the plane be held at two different electric potentials (with insulation between), and find the electric potential in the shaded region. Find and sketch some of the isothermals (equipotentials) and some of the curves (perpendicular to the isothermals) along which heat flows (lines of flux for the electric case). *Hint:* Use the mapping function $w = 1/z$, and consider the image of the w plane region between $v = 0$ and $v = -1$.

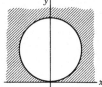

7. Use the mapping function $w = z^2$ to find the streamlines for the flow of water around the inside of a right-angle boundary. Find the velocity potential Φ, the stream function Ψ, and the velocity $\mathbf{V} = \boldsymbol{\nabla}\Phi$.

8. Observe that the magnitude of the velocity in Problem 7 can be obtained from $V = V_0|dw/dz|$. Show that this result holds in general as follows. Let $w = f(z)$ be an analytic mapping function such that the lines $v =$ const. map into the streamlines of the flow you want to consider in the z plane. Then

$$V_0 w = V_0(u + iv) = \Phi(x, y) + i\Psi(x, y).$$

Show that

$$V_0 \frac{dw}{dz} = \frac{\partial \Phi}{\partial x} - i\frac{\partial \Phi}{\partial y} = V_x - iV_y$$

(this expression is called the *complex velocity*). Hence show that $V = V_0|dw/dz|$.

9. Find and sketch the streamlines for the flow of water over a semicircular hump (say a half-buried log at the bottom of a stream) as shown. *Hint:* Use the mapping function $w = z + z^{-1}$. Show that the u axis maps into the contour $ABCDE$ with the correspondence shown.

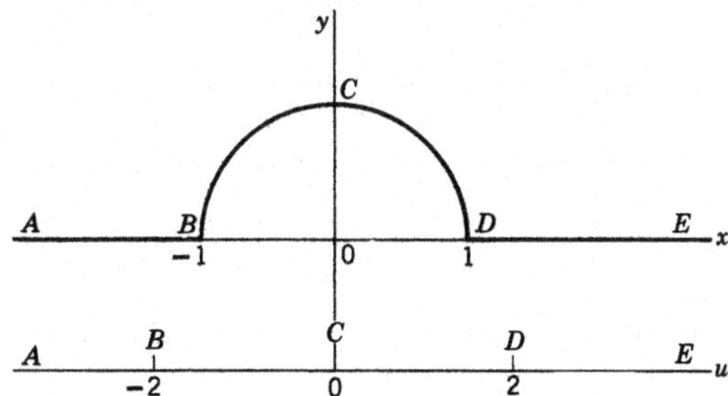

10. Find and sketch the streamlines for the indicated flow of water inside a rectangular boundary. *Hint:* Consider $w = \sin z$; map the u axis into the boundary of the rectangle.

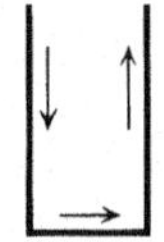

11. For $w = \ln[(z + 1)/(z - 1)]$, show that the images of $u =$ const. and $v =$ const. are two orthogonal sets of circles. Find centers and radii of five or six circles of each set and sketch them. Include the circle with center at the origin.

Use the results of Problem 11 to solve the following physics problems.

12. The figure represents the cross section of a long cylinder (assume it infinitely long) cut in half, with the top half and the bottom half insulated

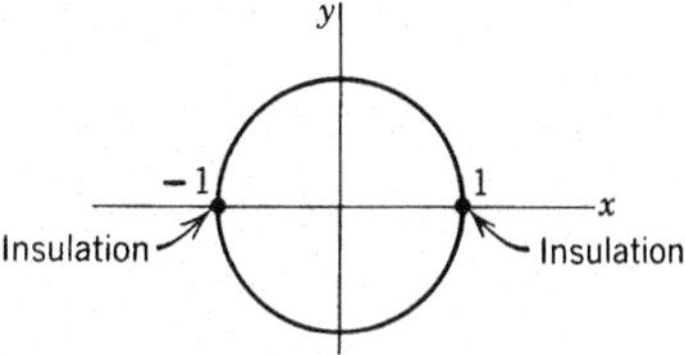

from each other. Let the surface of the top half be held at temperature $T = 30°$ and the surface of the bottom half at $T = 10°$. Find the temperature $T(x, y)$ inside the cylinder. *Hint:* Show that the line $v = \pi/2$ maps into the lower half of the circle $|z| = 1$, and the line $v = 3\pi/2$ maps into the upper half of the circle.

13. Let the figure in Problem 12 represent (the cross section of) a capacitor with the lower half at potential V_1 and the upper half at potential V_2. Find the potential $V(x, y)$ between the plates (that is, inside the circle). *Hint:* This is almost like Problem 12. Observe that in the text and in Problem 12, the w-plane temperature is of the form Av, with A constant; here you need the potential of the form $Av + B$, A and B constants.

14. In the figure in Problem 12, let $z = -1$ be a source and $z = +1$ a sink, and let the water flow inside the circular boundary. Find $\mathbf{\Phi}$, $\mathbf{\Psi}$, and $\mathbf{V}$. Sketch the streamlines.

15. In Problem 14, the streamlines were the images of $v =$ const. Consider the flow (over the whole plane, that is, with no boundaries) with streamlines $u =$ const. This flow may be described as two vortices rotating in opposite directions. Sketch a number of streamlines indicating the direction of the velocity with arrows. Since a boundary is a streamline, a flow is not disturbed by inserting a boundary along a streamline. Insert two circular boundaries corresponding to $u = a$ and $u = -a$. Show that the velocity through the narrow neck (say at $z = 0$) is greater than the velocity elsewhere (say at $z = i$). You can simplify your calculation of the velocity by showing that the result in Problem 8 holds here also.

16. Two long parallel cylinders form a capacitor. (Let their cross sections be the images of $u = a$ and $u = -a$.) If they are held at potentials V_0 and $-V_0$, find the potential $V(x, y)$ at points between them. given that the charge (per unit length) on a cylinder is $q = V_0/(2a)$, show that the capacitance (per unit length), that is, $q/(2V_0)$, is given by $1/(4\cosh^{-1} d/(2r))$, where d is the distance between the centers of the two cylinders, and their radii are r.

17. Other problems to consider using the mapping function of Problem 11: (a) a capacitor consisting of two long cylinders one inside the other, but not concentric; (b) the magnetic field in a plane perpendicular to two long parallel wires carrying equal but opposite currents; (c) the electric field in a plane perpendicular to two long parallel wires, one charged positive and the other negative; (d) other flow problems obtained by inserting boundaries along streamlines.

11. MISCELLANEOUS PROBLEMS

In Problems 1 and 2, verify that the given function is harmonic, and find a function $f(z)$ of which it is the real part. *Hint:* Use Problem 2.64. For Problem 2, see Chapter 2, Section 17, Problem 19.

1. $\ln\sqrt{(1+x)^2+y^2}$

2. $\arctan\dfrac{y}{x+1}$

3. Liouville's theorem: Suppose $f(z)$ is analytic for all z (except ∞), and bounded [that is, $|f(z)| \le M$ for all z and some M]. Prove that $f(z)$ is a constant. *Hints:* If $f'(z) = 0$, then $f(z) =$ const. To show this, write $f'(z)$ as in Problem 3.21 where C is a circle of radius R and center z, that is, $w = z + Re^{i\theta}$. Show that $|f'(z)| \le M/R$, and let $R \to \infty$.

4. Use Liouville's theorem (Problem 3) to prove the fundamental theorem of algebra (see Problem 7.44). *Hint:* Let $P(z)$ be a polynomial of degree ≥ 1; then $f(z) = 1/P(z)$ is a bounded analytic function in a region not containing any zeros of $P(z)$. Disprove the assumption that $P(z)$ has no zeros anywhere.

In Problems 5 to 8, find the residues of the given function at all poles. Take $z = re^{i\theta}$, $0 \le \theta < 2\pi$.

5. $\dfrac{z^{1/3}}{1+z^2}$ **6.** $\dfrac{\sqrt{z}}{1+8z^3}$ **7.** $\dfrac{\ln z}{1+z^2}$ **8.** $\dfrac{\ln z}{(2z-1)^2}$

In Problems 9 to 10, use Laurent series to find the residues of the given functions at the origin.

9. $\dfrac{\sin z^2}{z^7}$ **10.** $\dfrac{\ln(1-z)}{\sin^2 z}$

11. Find the Laurent series of $f(z) = e^z/(1-z)$ for $|z| < 1$ and $|z| > 1$. *Hints:* For $|z| < 1$, multiply two power series; you should find $f(z) = \sum_{n=0}^{\infty} a_n z^n$ with $a_n = \sum_{k=0}^{n} 1/k!$. For $|z| > 1$, use (4.3) where C is a circle $|z| = a$ with $a > 1$. Evaluate the integrals by finding residues at 1 and 0. You should find $f(z) = \sum_{n=0}^{\infty} a_n z^n + \sum_{n=1}^{\infty} b_n z^{-n}$ where all $b_n = -e$ and $a_n = -e + \sum_{k=0}^{n} 1/k!$.

12. Let $f(z)$ be the branch of $\sqrt{z^2-1}$ which is positive for large positive real values of z. Expand the square root in powers of $1/z$ to obtain the Laurent series of $f(z)$ about ∞. Thus by Problem 8.1 find the residue of $f(z)$ at ∞. Check your result by using equation (8.2).

In Problems 13 and 14, find the residues at the given points.

13. (a) $\dfrac{\cos z}{(2z-\pi)^4}$ at $\dfrac{\pi}{2}$ (b) $\dfrac{2z^2+3z}{z-1}$ at ∞

(c) $\dfrac{z^3}{1+32z^5}$ at $z = -\frac{1}{2}$ (d) $\csc(2z-3)$ at $z = \frac{3}{2}$

14. (a) $\dfrac{\ln(1+2z)}{z^2}$ at 0 (b) $\dfrac{1}{z}\sin(2z+5)$ at ∞

(c) $\dfrac{z^3}{4z^4+1}$ at $\frac{1}{2}(1+i)$ (d) $\dfrac{z\sin 2z}{(z+\pi)^2}$ at $-\pi$

In Problem 15 to 20, evaluate the integrals by contour integration.

15. $\displaystyle\int_0^{\pi} \frac{\cos\theta\, d\theta}{5-4\cos\theta}$ **16.** $\displaystyle\int_0^{2\pi} \frac{\sin\theta\, d\theta}{5+3\sin\theta}$

17. $\displaystyle\int_0^{\infty} \frac{\cos x\, dx}{(4x^2+1)(x^2+9)}$ **18.** $\displaystyle\int_0^{\infty} \frac{x\sin(\pi x/2)}{x^4+4}\, dx$

19. $\displaystyle PV\int_{-\infty}^{\infty} \frac{\sin x\, dx}{(3x-\pi)(x^2+\pi^2)}$ **20.** $\displaystyle PV\int_{-\infty}^{\infty} \frac{\cos x\, dx}{x(1-x)(x^2+1)}$

Verify the formulas in Problem 21 to 27 by contour integration or as indicated. Assume $a > 0$, $m > 0$.

21. $\displaystyle\int_0^{2\pi} \frac{d\theta}{a+b\sin\theta} = \int_0^{2\pi} \frac{d\theta}{a+b\cos\theta} = \frac{2\pi}{\sqrt{a^2-b^2}},\qquad |b| < a$

22. $\displaystyle\int_0^{2\pi} \frac{d\theta}{(a+b\sin\theta)^2} = \int_0^{2\pi} \frac{d\theta}{(a+b\cos\theta)^2} = \frac{2\pi a}{(a^2-b^2)^{3/2}},\qquad |b| < a$

Hint: You can do this directly by contour integration, but it is easier to differentiate Problem 21 with respect to a.

23. $\displaystyle\int_0^{2\pi} \frac{\sin\theta\, d\theta}{a+b\sin\theta} = \int_0^{2\pi} \frac{\cos\theta\, d\theta}{a+b\cos\theta} = \frac{2\pi}{b}\left(1-\frac{a}{\sqrt{a^2-b^2}}\right),\qquad |b| < a$

24. $\displaystyle\int_0^\infty \frac{\cos mx\,dx}{x^2+a^2} = \frac{\pi}{2a}e^{-ma}$

25. $\displaystyle PV\int_0^\infty \frac{\cos mx\,dx}{x^2-a^2} = -\frac{\pi}{2a}\sin ma$

26. $\displaystyle\int_0^\infty \frac{x\sin mx\,dx}{x^2+a^2} = \frac{\pi}{2}e^{-ma}$

27. $\displaystyle PV\int_0^\infty \frac{x\sin mx\,dx}{x^2-a^2} = \frac{\pi}{2}\cos ma$

Hint for Problems 26 and 27: Differentiate Problems 24 and 25 with respect to m.

28. Evaluate $\displaystyle\int_0^\infty \frac{\sqrt{x}\,\ln x\,dx}{(1+x)^2}$ by using the contour of Figure 7.4.
Hint: Along DE, $z = re^{2\pi i}$ so $\ln z = \ln r + 2\pi i$.

29. Evaluate $\displaystyle\int_0^\infty \frac{(\ln x)^2}{1+x^2}\,dx$ by using the contour of Figure 7.3. *Comment:* Note that your work also shows that $\displaystyle\int_0^\infty \frac{\ln x}{1+x^2}\,dx = 0$.

30. Show that

$$PV\int_0^\infty \frac{\cos(\ln x)}{x^2+1}\,dx = \frac{\pi}{2\cosh(\pi/2)}$$

by integrating $e^{i\ln z}/(z^2-1)$ around a contour like Figure 7.3 but rotated 90° clockwise so the straight side is along the y axis.

As in Section 7, find out how many roots the equations in Problem 31 to 34 have in each quadrant.

31. $z^4 + 3z + 5 = 0$

32. $z^3 + 2z^2 + 5z + 6 = 0$

33. $z^6 + z^3 + 9z + 64 = 0$
(no real roots)

34. $z^8 + 5z^3 + 3z + 4 = 0$
(2 negative real roots)

35. Show that the Cauchy-Riemann equations [see (2.2) and Problem 2.46] in a general orthogonal curvilinear coordinate system [see Chapter 10, Sections 8 and 9] are

$$\frac{1}{h_1}\frac{\partial u}{\partial x_1} = \frac{1}{h_2}\frac{\partial v}{\partial x_2}, \qquad \frac{1}{h_1}\frac{\partial v}{\partial x_1} = -\frac{1}{h_2}\frac{\partial u}{\partial x_2}$$

where, as in Chapter 10, the variables are x_1, x_2 and the scale factors are h_1, h_2. *Hint:* Consider the directional derivatives (Chapter 6, Section 6) in two perpendicular directions. (Compare Problem 2.46.) Also show that u and v satisfy Laplace's equation, Chapter 10, equation (9.10) (drop the x_3 term and set $h_3 = 1$).

36. Show that a harmonic function $u(x, y)$ is equal at every point a to its average value on any circle centered at a [and lying in the region where $f(z) = u(x,y) + iv(x,y)$ is analytic]. *Hint:* In (3.9), let $z = a + re^{i\theta}$ (that is, C is a circle with center at a), and show that the average value of $f(z)$ on the circle is $f(a)$ (see Chapter 7, Section 4 for discussion of the average of a function). Take real and imaginary parts of $f(a) = [u(x,y) + iv(x,y)]_{z=a}$.

37. A (nonconstant) harmonic function takes its maximum value and its minimum value on the boundary of any region (not at an interior point). Thus, for example, the electrostatic potential V in a region containing no free charge takes on its largest and smallest values on the boundary of the region; similarly, the temperature T of a body containing no sources of heat takes its largest and smallest values on the surface of the body. Prove this fact (for two-dimensional regions) as follows: Suppose that it is claimed that $u(x, y)$ takes its maximum value at some interior point a; this means that, at all points of some small disk about a, the values of $u(x, y)$ are no larger than at a. Show by Problem 36 that such a claim leads to a contradiction (unless u = const.). Similarly prove that $u(x, y)$ cannot take its minimum value at an interior point.

38. Show that a Dirichlet problem (see Chapter 13, Section 3) for Laplace's equation in a finite region has a unique solution; that is, two solutions u_1 and u_2 with the same boundary values are identical. *Hint:* Consider $u_2 - u_1$ and use Problem 37. [Also see Chapter 13, discussion following equation (2.17).]

39. Use the following sequence of mappings to find the steady state temperature $T(x, y)$ in the semi-infinite strip $y \geq 0$, $0 \leq x \leq \pi$ if $T(x, 0) = 100°$, $T(0, y) = T(\pi, y) = 0$, and $T(x, y) \to 0$ as $y \to \infty$. (See Chapter 13, Section 2 and Problem 2.6.)

(a) Use $w = (z' - 1)/(z' + 1)$ to map the half plane $v \geq 0$ on the upper half plane $y' > 0$, with the positive u axis corresponding to the two rays $x' > 1$ and $x' < -1$, and the negative u axis corresponding to the interval $-1 \leq x \leq 1$ of the x' axis.

(b) Use $z' = -\cos z$ to map the half-strip $0 < x < \pi$, $y > 0$ on the z' half plane described in (a). The interval $-1 \leq x' < 1$, $y' = 0$ corresponds to the base $0 < x < \pi$, $y = 0$ of the strip.

Comments: The temperature problem in the (u, v) plane is like the problems shown in the z plane of Figures 10.1 and 10.2, and so is given by $T = (100/\pi) \arctan(v/u)$. In the z plane you will find

$$T(x, y) = \frac{100}{\pi} \arctan \frac{2 \sin x \sinh y}{\sinh^2 y - \sin^2 x}.$$

Put $\tan \alpha = \dfrac{\sin x}{\sinh y}$ and use the formula for $\tan 2\alpha$ to get

$$T(x, y) = \frac{200}{\pi} \arctan \frac{\sin x}{\sinh y}.$$

Note that this is the same answer as in Chapter 13, Problem 2.6, if we replace 10 by π.

40. Use $L13$ of the Laplace transform table to find the Laplace transform of $\sin at \sinh at$. Verify your result by finding its inverse transform using the Bromwich integral.

41. Evaluate by contour integration $\displaystyle\int_0^\infty \frac{\cos^2(\alpha\pi/2)}{(1 - \alpha^2)^2}\, d\alpha$.

Hint: $\cos^2(\alpha\pi/2) = (1 + \cos \alpha\pi)/2$. Evaluate $\displaystyle\oint \frac{1 + e^{i\pi z}}{(z - 1)^2(z + 1)^2}\, dz$ around the upper half plane; note that the poles are actually simple poles (see Section 7, Example 4).

CHAPTER 15

Probability and Statistics

1. INTRODUCTION

The theory of probability has many applications in the physical sciences. It is of basic importance in quantum mechanics where results may be expressed in terms of probabilities (see Chapter 13, Schrödinger equation). It is needed whenever we are dealing with large numbers of particles or variables where it is impossible or impractical to have complete information about each one, such as in kinetic theory and statistical mechanics and a great variety of engineering problems. Statistics is the part of probability theory which deals with the interpretation of sets of data. You need statistical terms and methods every time you make a set of laboratory measurements. In this chapter, we shall discuss some of the basic ideas of probability and statistics which are most useful in applications.

The word "probably" is frequently used in everyday life. We say "The test will probably be hard," "It will probably snow today," "We will probably win this game," and so on. Such statements always imply a state of partial ignorance about the outcome of some event; we do not say "probably" about something whose outcome we know. The theory of probability tries to express more precisely just what our state of ignorance is. We say that the probability of getting a head in one toss of a coin is $\frac{1}{2}$, and similarly for a tail. We mean by this that there are two possible outcomes of the experiment (if we do not consider the possibility of the coin's standing on edge) and that we have no reason to expect one outcome more than the other; therefore we assign equal probabilities to the two possible outcomes. (See end of Section 2 for further discussion of this.)

Consider the following problem. You and I each toss a coin and look at our own coins but not each other's. The question is "What is the probability that both coins show heads?" Suppose you see that your coin shows tails; you say that the probability that both coins are heads is zero because you *know* that yours is tails. On the other hand, suppose I see that my coin is heads; then I say that the probability of both heads is $\frac{1}{2}$ because I don't know whether your coin shows heads or tails. Now suppose neither of us looks at either coin, but a third person looks at both coins and gives us the information that at least one is heads. Without this information, there are four possibilities, namely

$$hh \qquad tt \qquad th \qquad ht \tag{1.1}$$

to each of which we would ordinarily assign the probability $\frac{1}{4}$ (see end of Section 2, and Section 3). The information "at least one head" rules out tt, but gives no new information about the other three cases. Since hh, th, ht were equally likely before, we still consider them equally likely and say that the probability of hh is $\frac{1}{3}$.

Notice in the above discussion that the answer to a probability problem depends on the state of knowledge (or ignorance) of the person giving the answer. Notice also that in order to find the probability of an event, we consider all the different equally likely outcomes which are possible according to our information. We say that these are mutually exclusive (for example, if a coin is heads it cannot be tails), collectively exhaustive (we must consider *all* possibilities), and equally likely (we have no information which makes us expect one result more than another so we assume the same probability for each one of the set of outcomes). Let us now formalize this notion of probability as a definition (also see Section 2).

(1.2) If there are several equally likely, mutually exclusive, and collectively exhaustive outcomes of an experiment, the probability of an event E is

$$p = \frac{\text{number of outcomes favorable to } E}{\text{total number of outcomes}}.$$

Example 1. Find the probability that a single card drawn from a shuffled deck of cards will be either a diamond or a king (or both).

There are 52 different possible outcomes of the drawing; since the deck is shuffled, we assume all cards equally likely. Of the 52 cards, 16 are favorable (13 diamonds and the 3 other kings); therefore by (1.2) the desired probability is $\frac{16}{52} = \frac{4}{13}$.

Example 2. A three-digit number (that is, a number from 100–999) is selected "at random." ("At random" means that we assume all numbers to have the same probability of being selected.) What is the probability that all three digits are the same?

There are 900 three-digit numbers; 9 of them (namely 111, 222, $\cdots$, 999) have all three digits the same. Hence the desired probability is $\frac{9}{900} = \frac{1}{100}$.

PROBLEMS, SECTION 1

1. If you select a three-digit number at random, what is the probability that the units digit is 7? What is the probability that the hundreds digit is 7?

2. Three coins are tossed; what is the probability that two are heads and one tails? That the first two are heads and the third tails? If at least two are heads, what is the probability that all are heads?

3. In a box there are 2 white, 3 black, and 4 red balls. If a ball is drawn at random, what is the probability that it is black? That it is *not* red?

4. A single card is drawn at random from a shuffled deck. What is the probability that it is red? That it is the ace of hearts? That it is either a three or a five? That it is either an ace or red or both?

5. Given a family of two children (assume boys and girls equally likely, that is, probability 1/2 for each), what is the probability that both are boys? That at least one is a girl? Given that at least one is a girl, what is the probability that both are girls? Given that the first two are girls, what is the probability that an expected third child will be a boy?

6. A trick deck of cards is printed with the hearts and diamonds black, and the spades and clubs red. A card is chosen at random from this deck (after it is shuffled). Find the probability that it is either a red card or the queen of hearts. That it is either a red face card or a club. That it is either a red ace or a diamond.

7. A letter is selected at random from the alphabet. What is the probability that it is one of the letters in the word "probability?" What is the probability that it occurs in the first half of the alphabet? What is the probability that it is a letter after x?

8. An integer N is chosen at random with $1 \leq N \leq 100$. What is the probability that N is divisible by 11? That $N > 90$? That $N \leq 3$? That N is a perfect square?

9. You are trying to find instrument A in a laboratory. Unfortunately, someone has put both instruments A and another kind (which we shall call B) away in identical unmarked boxes mixed at random on a shelf. You know that the laboratory has 3 A's and 7 B's. If you take down one box, what is the probability that you get an A? If it is a B and you put it on the table and take down another box, what is the probability that you get an A this time?

10. A shopping mall has four entrances, one on the North, one on the South, and two on the East. If you enter at random, shop and then exit at random, what is the probability that you enter and exit on the same side of the mall?

2. SAMPLE SPACE

It is frequently convenient to make a list of the possible outcomes of an experiment [as we did in (1.1)]. Such a set of all possible mutually exclusive outcomes is called a *sample space*; each individual outcome is called a *point* of the sample space. There are many different sample spaces for any given problem. For example, instead of (1.1), we could say that a set of all mutually exclusive outcomes of two tosses of a coin is

$$\text{2 heads,} \qquad \text{1 head,} \qquad \text{no heads.} \tag{2.1}$$

Still another sample space for the same problem is

$$\text{no heads,} \qquad \text{at least 1 head.} \tag{2.2}$$

(Can you list some more examples?) On the other hand, the set of outcomes

$$\text{2 heads,} \qquad \text{at least 1 head,} \qquad \text{exactly 1 tail.}$$

cannot be used as a sample space, because these outcomes are not mutually exclusive. "At least 1 head" includes "2 heads" and also includes "exactly 1 tail" (which means also "exactly 1 head").

In order to use a sample space to solve problems, we need to have the probabilities corresponding to the different points in the sample space. We usually assign probability 1/4 to each of the outcomes listed in (1.1). (See end of Section 2 and Section 3.) We call such a list of equally likely outcomes a *uniform* sample space. Now suppose the outcomes are not equally likely. Satisfy yourself that the probabilities associated with the points of (2.1) and (2.2) are as follows.

$$\text{For (2.1):}\quad \begin{matrix} 2h & 1h & \text{no } h \\ \frac{1}{4} & \frac{1}{2} & \frac{1}{4} \end{matrix} \qquad \text{and for (2.2):}\quad \begin{matrix} \text{no } h & \text{at least } 1\ h \\ \frac{1}{4} & \frac{3}{4} \end{matrix}$$

The sample spaces (2.1) and (2.2) with different probabilities associated with different points are called *nonuniform* sample spaces. For some problems, there may be both uniform and nonuniform sample spaces; for example, (1.1) is a uniform sample space, and (2.1) and (2.2) are nonuniform sample spaces for a toss of two coins. But sometimes there *is* no uniform sample space; for example, consider a weighted coin which has a probability $\frac{1}{3}$ for heads and $\frac{2}{3}$ for tails. In such cases, we cannot use the definition (1.2) of probability, and we need the following more general definition.

Definition of Probability. Given any sample space (uniform or not) and the probabilities associated with the points, we find the probability of an event by adding the probabilities associated with all the sample points favorable to the event.

For a given nonuniform sample space, we must use this definition since (1.2) does not apply. If the given sample space is uniform, or if there is an underlying uniform sample space [for example, (1.1) is the uniform space underlying (2.1) and (2.2)], then this definition is consistent with the definition (1.2) by equally likely cases (Problems 15 and 16), and we may use either definition. As an example, let us find from (2.1) the probability of at least one head; this is the probability of one head plus the probability of two heads or $\frac{1}{2} + \frac{1}{4} = \frac{3}{4}$. We get the same result from the uniform sample space (1.1) using either (1.2) or the definition above.

If we can easily construct several sample spaces for a given problem, we must choose an appropriate one for the question we want to answer. Suppose we ask the question: In two tosses of a coin, what is the probability that both are heads? From either (1.1) or (2.1) we find the answer $\frac{1}{4}$; (2.2) is not an appropriate sample space to use in answering this question. (Why not?) To find the probability of both tails, we could use any of the three listed sample spaces, and to find the probability that the first toss gave a head and the second a tail, we could use only (1.1) since the other sample spaces do not give enough information. Let us now consider some less trivial examples.

Example 1. A coin is tossed three times. A uniform sample space for this problem contains eight points,

$$\text{(2.3)}\qquad \begin{matrix} hhh & hth & ttt & tht \\ hht & thh & tth & htt \end{matrix}$$

and we attach probability $\frac{1}{8}$ to each. Now let us use this sample space to answer some questions.

What is the probability of at least two tails in succession? By actual count, we see that there are three such cases, so the probability is $\frac{3}{8}$.

What is the probability that two consecutive coins fall the same? Again by actual count, this is true in six cases, so the probability is $\frac{6}{8}$ or $\frac{3}{4}$.

If we know that there was at least one tail, what is the probability of all tails? The point hhh is now ruled out; we have a new sample space consisting of seven points. Since the new information (at least one tail) tells us nothing new about these seven outcomes, we consider them equally probable, each with probability $\frac{1}{7}$. Thus the probability of all tails when all heads is ruled out is $\frac{1}{7}$.

(See problems 11 and 12 for further discussion of this example.)

▶ **Example 2.** Let two dice be thrown; the first die can show any number from 1 to 6 and similarly for the second die. Then there are 36 possible outcomes or points in a uniform sample space for this problem; with each point we associate the probability $\frac{1}{36}$. We can indicate a 3 on the first die and a 2 on the second die by the symbol 3,2. Then the sample space is as shown in (2.4). (Ignore the circling of some points and the letters a and b right now; they are for use in the problems below.)

(2.4)

	1,1	1,2	1,3	1,4	1,5	1,6	
	2,1	2,2	2,3	2,4	2,5	2,6	
	3,1	3,2	3,3	3,4	3,5	3,6	
a	4,1	4,2	4,3	4,4	4,5	4,6	b
	5,1	5,2	5,3	5,4	5,5	5,6	
	6,1	6,2	6,3	6,4	6,5	6,6	

Let us now ask some questions and use the sample space (2.4) to answer them.

(a) What is the probability that the sum of the numbers on the dice will be 5? The sample space points circled and marked a in (2.4) give all the cases for which the sum is 5. There are four of these sample points; therefore the probability that the sum is 5 is $\frac{4}{36}$ or $\frac{1}{9}$.

(b) What is the probability that the sum on the dice is divisible by 5? This means a sum of 5 or 10; the four points circled and marked a in (2.4) correspond to a sum of 5, and the three points circled and marked b correspond to a sum of 10. Thus there are seven points in the sample space corresponding to a sum divisible by 5, so the probability of a sum divisible by 5 is $\frac{7}{36}$ (7 favorable cases out of 36 possible cases, or 7 times the probability $\frac{1}{36}$ of each of the favorable sample points).

(c) Set up a sample space in which the points correspond to the possible sums of the two numbers on the dice, and find the probabilities associated with the points of this nonuniform sample space. The possible sums range from 2 (that is, $1+1$) to 12 (that is, $6+6$). From (2.4) we see that the points corresponding to any given sum lie on a diagonal (parallel to the diagonal elements marked a or b). There is one point corresponding to the sum 2; there are two points giving the sum 3, three points for sum 4, etc. Thus we have:

(2.5)

Sample Space	2	3	4	5	6	7	8	9	10	11	12
Associated probabilities	$\frac{1}{36}$	$\frac{2}{36}$	$\frac{3}{36}$	$\frac{4}{36}$	$\frac{5}{36}$	$\frac{6}{36}$	$\frac{5}{36}$	$\frac{4}{36}$	$\frac{3}{36}$	$\frac{2}{36}$	$\frac{1}{36}$

(d) What is the most probable sum in a toss of two dice? Although we can answer this from the sample space (2.4) (Try it!), it is easier from (2.5). We see that the sum 7 has the largest probability, namely $\frac{6}{36} = \frac{1}{6}$.

(e) What is the probability that the sum on the dice is greater than or equal to 9? Using (2.5), we add the probabilities associated with the sums 9, 10, 11, and 12. Thus the desired probability is

$$\frac{4}{36} + \frac{3}{36} + \frac{2}{36} + \frac{1}{36} = \frac{10}{36} = \frac{5}{18}.$$

So far we have been talking as if it were perfectly obvious and unquestionable that heads and tails are equally likely in the toss of a coin. If you have felt skeptical about this, you are perfectly right. It is *not* obvious; it is not even necessarily true, as a bent or weighted coin would show. We must distinguish here between the mathematical theory of probability and its application to a problem about the physical world. Mathematical probability (like all of mathematics) starts with a set of assumptions and shows that *if* the assumptions are true, *then* various results follow. The basic assumptions in a mathematical probability problem are the probabilities associated with the points of the sample space. Thus in a coin tossing problem, we *assume* that for each toss the probability of heads and the probability of tails are both $\frac{1}{2}$, and then we show that the probability of both heads in two tosses is $\frac{1}{4}$. (See Section 3.) The question of whether the assumptions are correct is not a mathematical one. Here we must ask what physical problem we are trying to solve. If we are dealing with a weighted coin, and if we know or can somehow estimate experimentally the probability p of heads (and so $1 - p$ of tails), then the mathematical theory starts with these values instead of $\frac{1}{2}, \frac{1}{2}$. In the absence of any information as to whether heads or tails is more likely, we often make the "natural" or "intuitive" assumption that the probabilities are both $\frac{1}{2}$. The only possible answer to the question of whether this is correct or not lies in experiment. If the results predicted on the basis of our assumptions agree with experiment, then the assumptions are good; otherwise we must revise the assumptions. (See Section 4, Example 5.)

In this chapter we shall consider mainly the mathematical methods of calculating the probabilities of complicated happenings if we are given the probabilities associated with the points of the sample space. For simplicity, we shall often assume these probabilities to be the "natural" ones; the mathematical theory we develop applies, however, if we replace these "natural" probabilities ($\frac{1}{2}$, $\frac{1}{2}$ in the coin toss problem, etc.) by any set of non-negative fractions whose sum is 1.

PROBLEMS, SECTION 2

1 to 10. Set up an appropriate sample space for each of Problems 1.1 to 1.10 and use it to solve the problem. Use either a uniform or nonuniform sample space or try both.

11. Set up several nonuniform sample spaces for the problem of three tosses of a coin (Example 1, above).

12. Use the sample space of Example 1 above, or one or more of your sample spaces in Problem 11, to answer the following questions.

(a) If there were more heads than tails, what is the probability of one tail?

(b) If two heads did not appear in succession, what is the probability of all tails?

(c) If the coins did not all fall alike, what is the probability that two in succession were alike?

(d) If N_t = number of tails and N_h = number of heads, what is the probability that $|N_h - N_t| = 1$?

(e) If there was at least one head, what is the probability of exactly two heads?

13. A student claims in Problem 1.5 that if one child is a girl, the probability that both are girls is $\frac{1}{2}$. Use appropriate sample spaces to show what is wrong with the following argument: It doesn't matter whether the girl is the older child or the younger; in either case the probability is $\frac{1}{2}$ that the other child is a girl.

14. Two dice are thrown. Use the sample space (2.4) to answer the following questions.

(a) What is the probability of being able to form a two-digit number greater than 33 with the two numbers on the dice? (Note that the sample point 1, 4 yields the two-digit number 41 which is greater than 33, etc.)

(b) Repeat part (a) for the probability of being able to form a two-digit number greater than or equal to 42.

(c) Can you find a two-digit number (or numbers) such that the probability of being able to form a larger number is the same as the probability of being able to form a smaller number? [See note, part (a).]

15. Use both the sample space (2.4) and the sample space (2.5) to answer the following questions about a toss of two dice.

(a) What is the probability that the sum is ≥ 4?

(b) What is the probability that the sum is even?

(c) What is the probability that the sum is divisible by 3?

(d) If the sum is odd, what is the probability that it is equal to 7?

(e) What is the probability that the product of the numbers on the two dice is 12?

16. Given an nonuniform sample space and the probabilities associated with the points, we defined the probability of an event A as the sum of the probabilities associated with the sample points favorable to A. [You used this definition in Problem 15 with the sample space (2.5).] Show that this definition is consistent with the definition by equally likely cases if there is also a uniform sample space for the problem (as there was in Problem 15). *Hint:* Let the uniform sample space have N points each with the probability N^{-1}. Let the nonuniform sample space have $n < N$ points, the first point corresponding to N_1 points of the uniform space, the second to N_2 points, etc. What is

$$N_1 + N_2 + \cdots + N_n?$$

What are $p_1, p_2, \ldots$, the probabilities associated with the first, second, etc., points of the nonuniform space? What is $p_1 + p_2 + \cdots + p_n$? Now consider an event for which several points, say i, j, k, of the nonuniform sample space are favorable. Then using the nonuniform sample space, we have, by definition of the probability p of the event, $p = p_i + p_j + p_k$. Write this in terms of the N's and show that the result is the same as that obtained by equally likely cases using the uniform space. Refer to Problem 15 as a specific example if you need to.

17. Two dice are thrown. Given the information that the number on the first die is even, and the number on the second is < 4, set up an appropriate sample space and answer the following questions.

(a) What are the possible sums and their probabilities?

(b) What is the most probable sum?

(c) What is the probability that the sum is even?

18. Are the following correct nonuniform sample spaces for a throw of two dice? If so, find the probabilities of the given sample points. If not show what is wrong. *Suggestion:* Copy sample space (2.4) and circle on it the regions corresponding to the points of the proposed nonuniform spaces.

(a) First die shows an even number.
First die shows an odd number.

(b) Sum of two numbers on dice is even.
First die is even and second odd.
First die is odd and second even.

(c) First die shows a number ≤ 3.
At least one die shows a number > 3.

19. Consider the set of all permutations of the numbers 1, 2, 3. If you select a permutation at random, what is the probability that the number 2 is in the middle position? In the first position? Do your answers suggest a simple way of answering the same questions for the set of all permutations of the numbers 1 to 7?

3. PROBABILITY THEOREMS

It is not always easy to make direct use of our definitions to calculate probabilities. Definition (1.2) asks us to find a uniform sample space for a problem, that is, a set of all possible *equally likely*, mutually exclusive outcomes of an experiment, and then determine how many of these are favorable to a given event. The definition in Section 2 similarly requires a sample space, that is, a list of the possible outcomes and their probabilities. Such lists may be prohibitively long; we want to consider some theorems which will shorten our work.

Suppose there are 5 black balls and 10 white balls in a box; we draw one ball "at random" (this means we are assuming that each ball has probability $\frac{1}{15}$ of being drawn), and then without replacing the first ball, we draw another. Let us ask for the probability that the first ball is white and the second one is black. The probability of drawing a white ball the first time is $\frac{10}{15}$ (10 of the 15 balls are white). The probability of *then* drawing a black ball is $\frac{5}{14}$ since there are 14 balls left and 5 of them are black. We are going to show that the probability of drawing first a white ball and then (without replacement) a black is the product $\frac{10}{15} \cdot \frac{5}{14}$. We reason in the following way, using a uniform sample space. Imagine that the balls are numbered 1 to 15. The symbol 5,3 will mean that ball 5 was drawn the first time and ball 3 the second time. In such pairs of two (different) numbers representing a drawing of two balls in succession, there are 15 choices for the first number and 14 for the second (the first ball was not replaced). Thus the uniform sample space representing all possible drawings consists of a rectangular array of symbols (like 5,3) with 15 columns (for the 15 different choices for the first number) and 14 rows (for the 14 choices for the second number). Thus there are $15 \cdot 14$ points in the sample space. [See also (4.1)]. How many of these sample points correspond to drawing first a white ball and then a black ball? Ten numbers correspond to white balls and the other five to black balls. Thus to obtain a sample point corresponding to drawing first a white and then a black ball, we can choose the first number in 10 ways and then the second number in 5 ways, and so choose the sample point in $10 \cdot 5$ ways; that is, there are $10 \cdot 5$ sample points favorable to the desired drawing. Then by the definition (1.2), the desired probability is $(10 \cdot 5)/(15 \cdot 14)$ as claimed.

Let us state in general the theorem we have just illustrated. We are interested in two successive events A and B. Let $P(A)$ be the probability that A will happen, $P(AB)$ be the probability that both A and B will happen, and $P_A(B)$ be the probability that B will happen if know that A has happened. Then

$$P(AB) = P(A) \cdot P_A(B) \tag{3.1}$$

or in words, the probability of the compound event "A *and* B" is the product of the probability that A will happen times the probability that B will happen if A does. Using the idea of a uniform sample space, we can prove (3.1) by following the method in the ball drawing problem. Let N be the total number of sample points in a uniform sample space, $N(A)$ and $N(B)$ be the numbers of sample points corresponding to the events A and B respectively, and $N(AB)$ be the number of sample points corresponding to the compound event A *and* B. It is useful to picture the sample space geometrically (Figure 3.1) as an array of N points [compare with sample space (2.4)]. We can then circle all points which correspond to A's happening and mark this region A; it contains $N(A)$ points. Similarly, we can circle the $N(B)$ points which correspond to B's happening and call this region B. The overlapping region we call AB; it is part of both A and B and contains $N(AB)$ points which correspond to the compound event A *and* B. Then by the definition (1.2):

$$\begin{aligned} P(AB) &= \frac{N(AB)}{N}, \\ P(A) &= \frac{N(A)}{N}, \\ P_A(B) &= \frac{N(AB)}{N(A)}. \end{aligned} \tag{3.2}$$

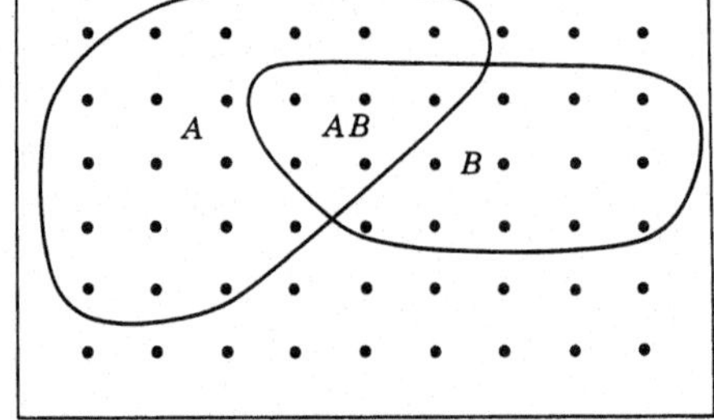

Figure 3.1

Perhaps this last formula for $P_A(B)$ needs some discussion. Recall from Section 2, Example 1, the uniform sample space (2.3) for three tosses of a coin. To find the probability of all tails given that there was at least one tail, we reduced our sample space to seven points (eliminating hhh). We then assumed that the seven points of the new sample space had the same relative probability as before the deletion of the point hhh; thus each of the seven points had probability $\frac{1}{7}$. (This is no more and no less "obvious" than the original assumption that the eight points had equal probability; it is an additional assumption which we make in the absence of any information to the contrary; see end of Section 2.) Now let us look at the third equation of (3.2). $N(A)$ is the number of sample points corresponding to event A; the N points in the original sample apace all had the same probability so we now assume that when we cross off all the points corresponding to A's *not* happening, the remaining $N(A)$ points also have equal probability. Thus we have a new uniform sample space consisting of $N(A)$ points. $N(AB)$ of these $N(A)$ points correspond to the event B (assuming A). Thus by (1.2), the probability of "B if A" is $N(AB)/N(A)$. From the three equations (3.2), we then have (3.1). In a similar way we can show that

$$P(BA) = P(B) \cdot P_B(A) = P(AB) \tag{3.3}$$

(see Problem 1). [We have proved (3.1) assuming a uniform sample space. This assumption is not necessary; (3.1) is true whether or not we can construct a uniform sample space; see Problem 2.]

Suppose, now, in our example of 5 black and 10 white balls in a box, we draw a ball and replace it and then draw a second ball. The probability of a black ball on the second drawing is then $\frac{5}{15} = \frac{1}{3}$; this is exactly the same result we would get if we had not drawn and replaced the first ball. In the notation of the last paragraph

$$P(B) = P_A(B), \qquad A \text{ and } B \text{ independent.} \tag{3.4}$$

When (3.4) is true, we say that the event B is *independent* of event A and (3.1) becomes

$$P(AB) = P(A) \cdot P(B), \qquad A \text{ and } B \text{ independent.} \tag{3.5}$$

Because of the symmetry of (3.5), we may simply say that A and B are independent if (3.5) is true. (Also see Problem 7.)

Example 1. (a) In three tosses of a coin, what is the probability that all three are heads? We found $p = \frac{1}{8}$ for this problem in Section 2 by seeing that one sample point out of eight corresponds to all heads. Now we can do the problem more simply by saying that the probability of heads on each toss is $\frac{1}{2}$, the tosses are independent, and therefore

$$p = \frac{1}{2} \cdot \frac{1}{2} \cdot \frac{1}{2} = \frac{1}{8}.$$

(b) If we should want the probability of all heads when a coin is tossed ten times, the sample space would be unwieldy; instead of using the sample space, we can say that since the tosses are independent, the desired probability is $p = (\frac{1}{2})^{10}$.

(c) To find the probability of at least one tail in ten tosses, we see that this event corresponds to all the rest of the sample space except the "all heads" point. Since the sum of the probabilities of all the sample points is 1, the desired probability is $1 - (\frac{1}{2})^{10}$.

In Figure 3.1 or Figure 3.2 the region AB corresponds to the happening of *both* A and B. The whole region consisting of points in A or B or both corresponds to the happening of *either* A *or* B *or* both. We write $P(AB)$ for the probability that both A and B occur. We shall write $P(A + B)$ for the probability that either or both occur. Then we can prove that

Figure 3.2 Figure 3.3

$$(3.6)\qquad P(A+B) = P(A) + P(B) - P(AB).$$

To see why this is true, consider Figure 3.2. To find $P(A+B)$ we add the probabilities of all the sample points in the region consisting of A or B or both. But if we add $P(A)$ and $P(B)$, we have included the probabilities of all the sample points in AB twice [once in $P(A)$ and once in $P(B)$]. Thus we must subtract $P(AB)$, which is the sum of the probabilities of all the sample points in AB. This is just what (3.6) says.

If the sample space diagram is like the one in Figure 3.3, so that $P(AB) = 0$, we say that A and B are mutually exclusive. Then (3.6) becomes

$$(3.7)\qquad P(A+B) = P(A) + P(B), \qquad A \text{ and } B \text{ mutually exclusive.}$$

▶ **Example 2.** Two students are working separately on the same problem. If the first student has probability $\frac{1}{2}$ of solving it and the second student has probability $\frac{3}{4}$ of solving it, what is the probability that at least one of them solves it?

Let A be the event "first student succeeds," and B be the event "second student succeeds." Then $P(AB) = \frac{1}{2} \cdot \frac{3}{4} = \frac{3}{8}$ (assume A and B independent since the students work separately). Then by (3.6) the probability that one or the other or both students solve the problem is

$$P(A+B) = \frac{1}{2} + \frac{3}{4} - \frac{3}{8} = \frac{7}{8}.$$

Conditional Probability; Bayes' Formula If we are asked for the probability of event B assuming that event A occurs [that is, $P_A(B)$], it is often useful to find it from (3.1):

$$(3.8)\qquad P_A(B) = \frac{P(AB)}{P(A)}.$$

Equation (3.8) is called Bayes' formula. In any conditional probability problem to which the answer is not immediately obvious, you should consider whether you can easily find $P(A)$ and $P(AB)$; if so, the conditional probability $P_A(B)$ is given by (3.8).

Example 3. A preliminary test is customarily given to the students at the beginning of a certain course. The following data are accumulated after several years:

(a) 95% of the students pass the course, 5% fail.

(b) 96% of the students who pass the course also passed the preliminary test.

(c) 25% of the students who fail the course passed the preliminary test.

What is the probability that a student who has failed the preliminary test will pass the course?

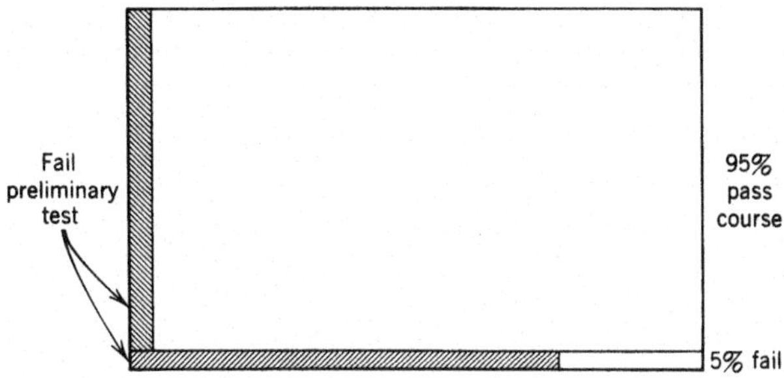

Figure 3.4

Let A be the event "fails preliminary test" and B be the event "Passes course." The probability we want is then $P_A(B)$ in (3.8), so we need $P(AB)$ and $P(A)$. $P(AB)$ is the probability that the student both fails the preliminary test and passes the course; this is $P(AB) = (0.95)(0.04) = 0.038$. (See Figure 3.4; 95% of the students passed the course and of these 4% had failed the preliminary test.) We also want $P(A)$, the probability that a students fails the preliminary test; this event corresponds to the shaded area in Figure 3.4. Thus $P(A)$ is the sum of the probabilities of the two events "passes course after failing test," "fails course after failing test." Then

$$P(A) = (0.095)(0.04) + (0.05)(0.75) = 0.0755$$

(See Figure 3.4; of the 95% of students who passed the course, 4% failed the preliminary test; of the 5% of the students who failed the course, 75% failed the preliminary test since we are given that 25% passed.) By (3.8) we have

$$P_A(B) = \frac{P(AB)}{P(A)} = \frac{0.038}{0.0755} = 50\%,$$

that is, half of the students who fail the preliminary test succeed in passing the course.

Note that in Figure 3.4, the shaded area corresponds to event A (fails preliminary test). We are interested in event B (passes course) given event A. Thus instead of the original sample space (whole rectangle in Figure 3.4) we consider a smaller sample space (shaded area in Figure 3.4). We then want to know what part of this sample space corresponds to event B (passes course). This fraction is $P(AB)/P(A)$ which we computed.

PROBLEMS, SECTION 3

1. (a) Set up a sample space for the 5 black and 10 white balls in a box discussed above assuming the first ball is not replaced. *Suggestions:* Number the balls, say 1 to 5 for black and 6 to 15 for white. Then the sample points form an array something like (2.4), but the point 3,3 for example is not allowed. (Why? What other points are not allowed?) You might find it helpful to write the numbers for black balls and the numbers for white balls in different colors.

(b) Let A be the event "first ball is white" and B be the event "second ball is black." Circle the region of your sample space containing points favorable to A and mark this region A. Similarly, circle and mark region B. Count the number of sample points in A and in B; these are $N(A)$ and $N(B)$. The region AB is the region inside both A and B; the number of points in this region is $N(AB)$. Use the numbers you have found to verify (3.2) and (3.1). Also find $P(B)$ and $P_B(A)$ and verify (3.3) numerically.

(c) Use Figure 3.1 and the ideas of part (b) to prove (3.3) in general.

2. Prove (3.1) for a nonuniform sample space. *Hints:* Remember that the probability of an event is the sum of the probabilities of the sample points favorable to it. Using Figure 3.1, let the points in A but not in AB have probabilities $p_1, p_2, \ldots, p_n$, the points in AB have probabilities $p_{n+1}, p_{n+2}, \ldots, p_{n+k}$, and the points in B but not in AB have probabilities $p_{n+k+1}, p_{n+k+2}, \ldots, p_{n+k+l}$. Find each of the probabilities in (3.1) in terms of the p's and show that you then have an identity.

3. What is the probability of getting the sequence *hhhttt* in six tosses of a coin? If you know the first three are heads, what is the probability that the last three are tails?

4. (a) A weighted coin has probability of $\frac{2}{3}$ of showing heads and $\frac{1}{3}$ of showing tails. Find the probabilities of *hh*, *ht*, *th* and *tt* in two tosses of the coin. Set up the sample space and the associated probabilities. Do the probabilities add to 1 as they should? What is the probability of at least one head? What is the probability of two heads if you know there was at least one head?

(b) For the coin in (a), set up the sample space for three tosses, find the associated probabilities, and use it to answer the questions in Problem 2.12.

5. What is the probability that a number n, $1 \le n \le 99$, is divisible by *both* 6 and 10? By *either* 6 *or* 10 or both?

6. A card is selected from a shuffled deck. What is the probability that it is either a king or a club? That it is both a king and a club?

7. (a) Note that (3.4) assumes $P(A) \neq 0$ since $P_A(B)$ is meaningless if $P(A) = 0$. Assuming both $P(A) \neq 0$ and $P(B) \neq 0$, show that if (3.4) is true, then $P(A) = P_B(A)$; that is if B is independent of A, then A is independent of B. If either $P(A)$ or $P(B)$ is zero, then we use (3.5) to define independence.

(b) When is an event E independent of itself? When is E independent of "not E"?

8. Show that

$$P(A+B+C) = P(A) + P(B) + P(C) - P(AB) - P(AC) - P(BC) + P(ABC).$$

Hint: Start with Figure 3.2 and sketch in a region C overlapping some of the points of each of the regions A, B, and AB.

9. Two cards are drawn at random from a shuffled deck and laid aside without being examined. Then a third card is drawn. Show that the probability that the third card is a spade is $\frac{1}{4}$ just as it was for the first card. *Hint:* Consider all the (mutually exclusive) possibilities (two discarded cards spades, third card spade or not spade, etc.).

10. (a) Three typed letters and their envelopes are piled on a desk. If someone puts the letters into the envelopes at random (one letter in each), what is the probability that each letter gets into its own envelope? Call the envelopes A, B, C, and the corresponding letters a, b, c, and set up the sample space. Note that "a in C, b in B, c in A" is *one* point in the sample space.

(b) What is the probability that at least one letter gets into its own envelope? *Hint:* What is the probability that no letter gets into its own envelope?

(c) Let A mean that a got into envelope A, and so on. Find the probability $P(A)$ that a got into A. Find $P(B)$ and $P(C)$. Find the probability $P(A+B)$ that either a or b or both got into their correct envelopes, and the probability $P(AB)$ that both got into their correct envelopes. Verify equation (3.6).

11. In paying a bill by mail, you want to put your check and the bill (with a return address printed on it) into a window envelope so that the address shows right side up and is not blocked by the check. If you put check and bill at random into the envelope, what is the probability that the address shows correctly?

12. (a) A loaded die has probabilities $\frac{1}{21}, \frac{2}{21}, \frac{3}{21}, \frac{4}{21}, \frac{5}{21}, \frac{6}{21}$, of showing 1, 2, 3, 4, 5, 6. What is the probability of throwing two 3's in succession?

(b) What is the probability of throwing a 4 the first time and not a 4 the second time with a die loaded as in (a)?

(c) If two dice loaded as in (a) are thrown, and we know that the sum of the numbers on the faces is greater than or equal to 10, what is the probability that both are 5's?

(d) How many times must we throw a die loaded as in (a) to have probability greater than $\frac{1}{2}$ of getting an ace?

(e) A die, loaded as in (a), is thrown twice. What is the probability that the number on the die is even the first time > 4 the second time?

13. (a) A candy vending machine is out of order. The probability that you get a candy bar (with or without return of your money) is $\frac{1}{2}$, the probability that you get your money back (with or without candy) is $\frac{1}{3}$, and the probability that you get both the candy and your money back is $\frac{1}{12}$. What is the probability that you get nothing at all? *Suggestion:* Sketch a geometric diagram similar to Figure 3.1, indicate regions representing the various possibilities and their probabilities; then set up a four-point sample space and the associated probabilities of the points.

(b) Suppose you try again to get a candy bar as in part (a). Set up the 16-point sample space corresponding to the possible results of your two attempts to buy a candy bar, and find the probability that you get two candy bars (and no money back); that you get no candy and lose your money both times; that you just get your money back both times.

14. A basketball player succeeds in making a basket 3 tries out of 4. How many tries are necessary in order to have probability > 0.99 of at least one basket?

15. Use Bayes' formula (3.8) to repeat these simple problems previously done by using a reduced sample space.

(a) In a family of two children, what is the probability that both are girls if at least one is a girl?

(b) What is the probability of all heads in three tosses of a coin if you know that at least one is a head?

16. Suppose you have 3 nickels and 4 dimes in your right pocket and 2 nickels and a quarter in your left pocket. You pick a pocket at random and from it select a coin at random. If it is a nickel, what is the probability that it came from your right pocket?

17. (a) There are 3 red and 5 black balls in one box and 6 red and 4 white balls in another. If you pick a box at random, and then pick a ball from it at random, what is the probability that it is red? Black? White? That it is either red or white?

(b) Suppose the first ball selected is red and is not replaced before a second ball is drawn. What is the probability that the second ball is red also?

(c) If both balls are red, what is the probability that they both came from the same box?

18. Two cards are drawn at random from a shuffled deck.

(a) What is the probability that at least one is a heart?

(b) If you know that at least one is a heart, what is the probability that both are hearts?

19. Suppose it is known that 1% of the population have a certain kind of cancer. It is also known that a test for this kind of cancer is positive in 99% of the people who have it but is also positive in 2% of the people who do not have it. What is the probability that a person who tests positive has cancer of this type?

20. Some transistors of two different kinds (call them N and P) are stored in two boxes. You know that there are 6 N's in one box and that 2 N's and 3 P's got mixed in the other box, but you don't know which box is which. You select a box and a transistor from it at random and find that it is an N; what is the probability that it came from the box with the 6 N's? From the other box? If another transistor is picked from the same box as the first, what is the probability that it is also an N?

21. Two people are taking turns tossing a pair of coins; the first person to toss two alike wins. What are the probabilities of winning for the first player and for the second player? *Hint:* Although there are an infinite number of possibilities here (win on first turn, second turn, third turn, etc.), the sum of the probabilities is a geometric series which can be summed; see Chapter 1 if necessary.

22. Repeat Problem 21 if the players toss a pair of dice trying to get a double (that is, both dice showing the same number).

23. A thick coin has probability $\frac{3}{7}$ of falling heads, $\frac{3}{7}$ of falling tails, and $\frac{1}{7}$ of standing on edge. Show that if it is tossed repeatedly it has probability 1 of eventually standing on edge.

4. METHODS OF COUNTING

Let us digress for a bit to review some ideas and formulas we need in computing probabilities in more complicated problems.

Let us ask how many two-digit numbers have either 5 or 7 for the tens digit and either 3, 4, or 6 for the units digit. The answer becomes obvious if we arrange the possible numbers in a rectangle

$$\begin{matrix} 53 & 54 & 56 \\ 73 & 74 & 76 \end{matrix}$$

with two rows corresponding to the two choices of the tens digit and three columns corresponding to the three choices of the units digit. This is an example of the *fundamental principle of counting*:

(4.1) If one thing can be done N_1 ways, and after that a second thing can be done in N_2 ways, the two things can be done in succession in that order in $N_1 \cdot N_2$ ways. This can be extended to doing any number of things one after the other, the first N_1 ways, the second N_2 ways, the third N_3 ways, etc. Then the total number of ways to perform the succession of acts is the product $N_1 N_2 N_3 \cdots$.

Now consider a set of n things lined up in a row; we ask how many ways we can arrange (permute) them. This result is called the number of *permutations* of n things n at a time, and is denoted by ${}_nP_n$ or $P(n,n)$ or P_n^n. To find this number, we think of seating n people in a row of n chairs. We can place anyone in the first chair, that is, we have n possible ways of filling the first chair. Once we have selected someone for the first chair, there are $(n-1)$ choices left for the second chair, then $(n-2)$ choices for the third chair, and so on. Thus by the fundamental principle, there are $n(n-1)(n-2)\cdots 2\cdot 1 = n!$ ways of arranging the n people in the row of n chairs. The number of permutations of n things n at a time is

$$P(n,n) = n!. \tag{4.2}$$

Next suppose there are n people but only $r < n$ chairs and we ask how many ways we can select groups of r people and seat them in the r chairs. The result is called the number of permutations of n things r at a time and is denoted by ${}_nP_r$ or $P(n,r)$ or P_r^n. Arguing as before, we find that there are n ways to fill the first chair, $(n-1)$ ways to fill the second chair, $(n-2)$ ways for the third [note that we could write $(n-2)$ as $(n-3+1)$], etc., and finally $(n-r+1)$ ways of filling chair r. Thus we have for the number of permutations of n things r at a time

$$P(n,r) = n(n-1)(n-2)\cdots(n-r+1).$$

By multiplying and dividing by $(n-r)!$ we can write this as

$$P(n,r) = n(n-1)(n-2)\cdots(n-r+1)\frac{(n-r)!}{(n-r)!} = \frac{n!}{(n-r)!}. \tag{4.3}$$

So far we have been talking about arranging things in a definite order. Suppose, instead that we ask how many committees of r people can be chosen from a group of n people $(n \geq r)$. Here the order of the people in the committee is not considered; the committee made up of people A, B, C, is the same as the committee made up of people B, A, C. We call the number of such committees of r people which we can select from n people, the number of *combinations* or *selections* of n things r at a time, and denote this number by ${}_nC_r$ or $C(n,r)$ or $\binom{n}{r}$. To find $C(n,r)$, we go back to the problem of selecting r people from a group of n and seating them in r chairs; we found that the number of ways of doing this is $P(n,r)$ as given in

(4.3). We can perform this job by first selecting r people from the total n and then arranging the r people in r chairs. The selection of r people can be done in $C(n,r)$ ways (this is the number we are trying to find), and after r people are selected, they can be arranged in r chairs in $P(r,r)$ ways by (4.2). By the fundamental principle (4.1), the total number of ways $P(n,r)$ of selecting and seating r people out of n is the product $C(n,r)\cdot P(r,r)$. Thus we have

$$P(n,r) = C(n,r)\cdot P(r,r). \tag{4.4}$$

We can solve this equation to find the value $C(n,r)$ which we wanted. Substituting the values of $P(n,r)$ and $P(r,r)$ from (4.3) and (4.2) into (4.4) and solving for $C(n,r)$, we find for the number of combinations of n things r at a time

$$C(n,r) = \frac{P(n,r)}{P(r,r)} = \frac{n!}{(n-r)!r!} = \binom{n}{r}. \tag{4.5}$$

Each time we select r people to be seated, we leave $n-r$ people without chairs. Then there are exactly the same number of combinations of n things $n-r$ at a time as there are combinations of n things r at a time. Hence we write

$$C(n,n-r) = C(n,r) = \frac{n!}{(n-r)!r!}. \tag{4.6}$$

We can also obtain (4.6) from (4.5) by replacing r by $(n-r)$.

▶ **Example 1.** A club consists of 50 members. In how many ways can a president, vice-president, secretary, and treasurer be chosen? In how many ways can a committee of 4 members be chosen?

In the selection of officers, we must not only select 4 people, but decide which one is president, etc.; we could think of seating the 4 people in chairs labeled president, vice-president, etc. Thus the number of ways of selecting the officers is

$$P(50,4) = \frac{50!}{(50-4)!} = \frac{50!}{46!} = 50\cdot 49\cdot 48\cdot 47.$$

The committee members, however, are all equivalent (we are neglecting the possibility that one is named chairman), so the number of ways of selecting committees of 4 people is

$$C(50,4) = \frac{50!}{46!4!} = \frac{50\cdot 49\cdot 48\cdot 47}{24}.$$

▶ **Example 2.** Find the coefficient of x^8 in the binomial expansion of $(1+x)^{15}$.

Think of multiplying out

$$(1+x)(1+x)(1+x)\cdots(1+x), \qquad \text{(with 15 factors).}$$

We obtain a term in x^8 each time we multiply 1's from seven of the parentheses by x's from eight of the parentheses. The number of ways of selecting 8 parentheses out of 15 is

$$C(15,8) = \frac{15!}{8!7!}.$$

This is the desired coefficient of x^8.

Generalizing this example, we see that in the expansion of $(a+b)^n$, the coefficient of $a^{n-r}b^r$ is $C(n,r)$, usually written $\binom{n}{r}$ when used in connection with a binomial expansion (see Chapter 1, Section 13C). Thus the expressions $C(n,r)$ are just the binomial coefficients, and we can write

$$(a+b)^n = \sum_{r=0}^{n} \binom{n}{r} a^{n-r}b^r. \tag{4.7}$$

Example 3. A basic problem in statistical mechanics is this: Given N balls, and n boxes, in how many ways can the balls be put into the boxes so that there will be given numbers of balls in the boxes, say N_1 balls in the first box, N_2 balls in the second box, N_3 in the third, $\cdots$, N_n in the nth, and what is the probability that this given distribution will occur when the balls are put into the boxes? In statistical mechanics the "balls" may be molecules, electrons, photons, etc., and each "box" corresponds to a small range of values of position and momentum of a particle. We can state many other problems in this same language of putting balls into boxes. For example, in tossing a coin, we can equate heads with box 1, and tails with box 2; in tossing a die, there are six "boxes." In putting letters into envelopes, the letters are the balls, and the envelopes are the boxes. In dealing cards, the cards are the balls and the players who receive them are the boxes. In an alpha scattering experiment, the alpha particles are the balls, and the boxes are elements of area on the detecting screen which the particles hit after they are scattered. (Also see Problems 14 and 21 and Feller, pp. 10–11.)

Let us do a special case of this problem in which we have 15 balls and 6 boxes, and the numbers of balls we are to put into the various boxes are:

Number of balls:	3	1	4	2	3	2
In box number:	1	2	3	4	5	6

We first ask how many ways we can select 3 balls to go in the first box from the 15 balls; this is $C(15,3)$. (Note that the order of the balls in the boxes is not considered; this is like the committee problem in Example 1.) Now we have 12 balls left, of which we are to select 1 for box 2; we can do this in $C(12,1)$ ways. We can then select the 4 balls for box 3 from the remaining 11 balls in $C(11,4)$ ways, the 2 balls for box 4 in $C(7,2)$ ways, the 3 balls for box 5 in $C(5,3)$ ways, and finally the balls for box 6 in $C(2,2)$ ways (verify that this is 1). By the fundamental principle, the total number of ways of putting the required numbers of balls into the boxes is

$$\begin{aligned}
&C(15,3)\cdot C(12,1)\cdot C(11,4)\cdot C(7,2)\cdot C(5,3)\cdot C(2,2)\\
&\quad = \frac{15!}{3!\cdot 12!}\cdot\frac{12!}{1!\cdot 11!}\cdot\frac{11!}{4!\cdot 7!}\cdot\frac{7!}{2!\cdot 5!}\cdot\frac{5!}{3!\cdot 2!}\cdot\frac{2!}{2!\cdot 0!}\\
&\quad = \frac{15!}{3!\cdot 1!\cdot 4!\cdot 2!\cdot 3!\cdot 2!}.
\end{aligned}$$

(Remember from Chapters 1 and 11 that $0! = 1$.)

Next we want the probability of this particular distribution. Let us assume that the balls are distributed "at random" into the boxes; by this we mean that a ball has the same probability (namely $\frac{1}{6}$) of being put into any one box as into any other box. We can put the first ball into any one of the 6 boxes, the second ball into any one of the 6 boxes, and so on. Thus by the fundamental principle, the total number of ways of distributing the 15 balls into the 6 boxes is $6 \cdot 6 \cdot 6 \cdot 6 \cdots 6 = 6^{15}$ and we are assuming that these distributions are equally probable. Then the probability that, when 15 balls are distributed "at random" into 6 boxes, there will be 3 balls in box 1, 1 in box 2, etc., as given, is, by (1.2) (favorable cases ÷ total)

$$\frac{15!}{3! \cdot 1! \cdot 4! \cdot 2! \cdot 3! \cdot 2!} \div 6^{15}.$$

▶ **Example 4.** In Example 3, we assumed that the 6^{15} possible distributions of 15 balls into 6 boxes were equally likely. This seems very reasonable if we think of putting the balls into the boxes by tossing a die for each ball; if the die shows 1 we put the ball into box 1, etc. However, we can think of situations to which this method and result do not apply. For example, suppose we are putting letters into envelopes or seating people in chairs; then we may reasonably require only one letter per envelope, not more than one person per chair, that is, one ball (or none) per box. Consider the problem of seating 4 people in 6 chairs, that is of putting 4 balls into 6 boxes. If we number the chairs from 1 to 6 and let each person choose a chair by tossing a die, we may have two or more people choosing the same chair. The result 6^4 (which the method of Example 3 gives for the problem of 4 balls in 6 boxes) then does not apply to this problem. However, let us consider the uniform sample space of 6^4 points and select from it the points corresponding to our restriction (one ball or none per box). The new sample space contains $C(6,4) \cdot 4!$ points (number of ways of selecting the 4 chairs to be occupied times the number of ways of then arranging 4 people in 4 chairs). Since these points were equally probable in the original (uniform) sample space, we still consider them equally probable. Now let us ask for the probability that the first two chairs are vacant when the 4 people are seated. The number of sample points corresponding to this event is 4! (the number of ways of arranging the 4 people in the last 4 chairs). Thus the desired probability is

$$\frac{4!}{C(6,4) \cdot 4!} = \frac{1}{C(6,4)}.$$

We can now see an easier way of doing problems of this kind. The factor 4!, which canceled in the probability calculation, was the number of rearrangements of the 4 people among the 4 occupied chairs. Since this is the same for any given set of 4 chairs, we can lump together all the sample points corresponding to each given set of 4 chairs, and have a smaller (still uniform) sample space of $C(6,4)$ points. Each point now corresponds to a given set of 4 occupied chairs; the quantity $C(6,4)$ is just the number of ways of picking 4 occupied chairs out of 6. The probability that the first two chairs are vacant when 4 people are seated is $1/C(6,4)$ since there is only one way to select 4 occupied chairs leaving the first two chairs vacant.

Another useful way of looking at this problem is to consider a set of 4 *identical* balls to be put into 6 boxes. Since the balls are identical, the 4! arrangements of the 4 balls in 4 given boxes all look alike. We can say that there are $C(6,4)$ *distinguishable* arrangements of the 4 identical balls in 6 boxes (one ball or none per box). Since all these arrangements are equally probable, the probability of any one arrangement (say the first two boxes empty) is $1/C(6,4)$ as we found previously.

Example 5. In Example 4 we found the same answer for the probability that two particular boxes were empty whether or not we considered the balls distinguishable. This was true because the allowed distinguishable arrangements were equally probable. Without the restriction of one ball or none per box, all distinguishable arrangements are not equally probable according to the methods of Examples 3 and 4. For example, the probability of all balls in box 1 is $1/6^4$; compare this with the probability of no balls in the first 2 boxes and one ball in each of the other 4 boxes, which is $4! \div 6^4 = \frac{1}{54}$. We see that the concentrated arrangements (all or several balls in one box) are less probable than the more uniform arrangements.

Now we are going to try to imagine a situation in which *all* distinguishable arrangements *are* equally probable. Suppose the 6 boxes are benches in a waiting room and the 4 balls are people who are going to come in and sit on the benches. Then if the people are friends, there will be a certain tendency for them to sit together and the probabilities we have been calculating will not apply—the probabilities of the concentrated arrangements will increase. Consider the following mathematical model. (This is a modification of Pólya's urn model.) We have 6 boxes labeled 1 to 6, and 4 balls. From 6 cards labeled 1 to 6 we draw one at random and place a ball in the box numbered the same as the card drawn. We then replace the card and also add another card of the same number so that there are now 7 cards, two with the number first drawn. We now select a card at random from these 7, put a ball in the corresponding box and again replace the card adding a duplicate to make 8 cards. We repeat this process two more times (until all balls are distributed). Then the probability that all balls are in box 1 is $\frac{1}{6} \cdot \frac{2}{7} \cdot \frac{3}{8} \cdot \frac{4}{9}$. The probability of one ball in each of the first 4 boxes is $\frac{1}{6} \cdot \frac{1}{7} \cdot \frac{1}{8} \cdot \frac{1}{9} \cdot 4!$ (here $\frac{1}{6} \cdot \frac{1}{7} \cdot \frac{1}{8} \cdot \frac{1}{9}$ is the probability that the first ball is in box 1, the second in box 2, etc.; we must add to this the probability that the first ball is in box 3, the second in box 1, etc.; there are 4! such possibilities all giving one ball in each of the first 4 boxes). We see that the distributions "all balls in box 1" and "one ball in each of the first 4 boxes" are equally probable. Further calculation (Problem 20) shows that all distinguishable arrangements are equally probable.

To find the number of distinguishable arrangements, consider the following picture of the 4 balls in the 6 boxes.

	| o |	|	| oo |	|	| o |	|
Box number:	1	2	3	4	5	6
Number of balls:	1	0	2	0	1	0

The lines mean the sides of the boxes and the circles are the balls; note that it requires 7 lines to picture the 6 boxes. This picture shows one of many possible arrangements of the 4 balls in 6 boxes. In any such picture there must be a line at the beginning and at the end, but the rest of the lines (5 of them) and the 4 circles can be arranged in any order. You should convince yourself that every arrangement of the balls in the boxes can be so pictured. Then the number of such distinguishable arrangements is just the number of ways we can select 4 positions for the 4 circles out of 9 positions for the 5 lines and 4 circles. Thus there are $C(9, 4)$ equally likely arrangements in this problem.

We see then that putting balls in boxes is not quite as simple as we thought; we must say *how* we propose to distribute them and even before that we must think what practical problem we are trying to solve; this is what determines the sample space and the probabilities to be associated with the sample points. Unfortunately, it may not always be clear what the sample space probabilities should be; then the best we can do is to try various assumptions. In statistical mechanics it is found that certain particles (for example, the molecules of a gas) are correctly described if we assume that they behave like the balls of Example 3 (all 6^{15} arrangements equally likely); we then say that they obey Maxwell-Boltzmann statistics. Other particles (for example, electrons) behave like the people to be seated in Example 4 (one particle or none per box); we say that such particles obey Fermi-Dirac statistics. Finally some particles (for example, photons) act something like the friends who want to sit near each other (all distinguishable arrangements of identical particles are equally likely); we say that these particles obey Bose-Einstein statistics. For the problem of 4 particles in 6 boxes, there are then 6^4 equally likely arrangements for Maxwell-Boltzmann particles, $C(6,4)$ for Fermi-Dirac particles, and $C(9,4)$ for Bose-Einstein particles. (See Problems 15 to 20.)

PROBLEMS, SECTION 4

1. (a) There are 10 chairs in a row and 8 people to be seated. In how many ways can this be done?

(b) There are 10 questions on a test and you are to do 8 of them. In how many ways can you choose them?

(c) In part (a) what is the probability that the first two chairs in the row are vacant?

(d) In part (b), what is the probability that you omit the first two problems in the test?

(e) Explain why the answer to parts (a) and (b) are different, but the answers to (c) and (d) are the same.

2. In the expansion of $(a+b)^n$ (see Example 2), let $a = b = 1$, and interpret the terms of the expansion to show that the total number of combinations of n things taken $1, 2, 3, \cdots, n$ at a time, is $2^n - 1$.

3. A bank allows one person to have only one savings account insured to \$100,000. However, a larger family may have accounts for each individual, and also accounts in the names of any 2 people, any 3 and so on. How many accounts are possible for a family of 2? Of 3? Of 5? Of n? *Hint:* See Problem 2.

4. Five cards are dealt from a shuffled deck. What is the probability that they are all of the same suit? That they are all diamond? That they are all face cards? That the five cards are a sequence in the same suit (for example, 3, 4, 5, 6, 7 of hearts)?

5. A bit (meaning binary digit) is 0 or 1. An ordered array of eight bits (such as 01101001) is a byte. How many different bytes are there? If you select a byte at random, what is the probability that you select 11000010? What is the probability that you select a byte containing three 1's and five 0's?

6. A so-called 7-way lamp has three 60-watt bulbs which may be turned on one or two or all three at a time, and a large bulb which may be turned to 100 watts, 200 watts or 300 watts. How many different light intensities can the lamp be set to give if the completely off position is not included? (The answer is *not* 7.)

7. What is the probability that the 2 and 3 of clubs are next to each other in a shuffled deck? *Hint:* Imagine the two cards accidentally stuck together and shuffled as one card.

8. Two cards are drawn from a shuffled deck. What is the probability that both are aces? If you know that at least one is an ace, what is the probability that both are aces? If you know that one is the ace of spades, what is the probability that both are aces?

9. Two cards are drawn from a shuffled deck. What is the probability that both are red? If at least one is red, what is the probability that both are red? If at least one is a red ace, what is the probability that both are red? If exactly one is a red ace, what is the probability that both are red?

10. What is the probability that you and a friend have different birthdays? (For simplicity, let a year have 365 days.) What is the probability that three people have three different birthdays? Show that the probability that n people have n different birthdays is

$$p = \left(1 - \frac{1}{365}\right)\left(1 - \frac{2}{365}\right)\left(1 - \frac{3}{365}\right)\cdots\left(1 - \frac{n-1}{365}\right).$$

Estimate this for $n \ll 365$ by calculating $\ln p$ [recall that $\ln(1+x)$ is approximately x for $x \ll 1$]. Find the smallest (integral) n for which $p < \frac{1}{2}$. Hence, show that for a group of 23 people or more, the probability is greater than $\frac{1}{2}$ that two of them have the same birthday. (Try it with a group of friends or a list of people such as the presidents of the United States.)

11. The following game was being played on a busy street: Observe the last two digits on each license plate. What is the probability of observing at least two cars with the same last two digits among the first 5 cars? 10 cars? 15 cars? How many cars must you observe in order for the probability to be greater than $\frac{1}{2}$ of observing two with the same last two digits?

12. Consider Problem 10 for different months of birth. What is the smallest number of people for which the probability is greater than $\frac{1}{2}$ that two of them were born in the same month?

13. Generalize Example 3 to show that the number of ways of putting N balls in n boxes with N_1 in box 1, N_2 in box 2, etc., is

$$\left(\frac{N!}{N_1! \cdot N_2! \cdot N_3! \cdots N_n!}\right).$$

14. (a) Find the probability that in two tosses of a coin, one is heads and one tails. That in six tosses of a die, all six of the faces show up. That in 12 tosses of a 12-sided die, all 12 faces show up. That in n tosses of an n-sided die, all n faces show up.

(b) The last problem in part (a) is equivalent to finding the probability that, when n balls are distributed at random into n boxes, each box contains exactly one ball. Show that for large n, this is approximately $e^{-n}\sqrt{2\pi n}$.

15. Set up the uniform sample spaces for the problem of putting 2 particles in 3 boxes: for Maxwell-Boltzmann particles, for Fermi-Dirac particles, and for Bose-Einstein particles. See Example 5. (You should find 9 sample points for MB, 3 for FD, and 6 for BE.)

16. Do Problem 15 for 2 particles in 2 boxes. Using the model discussed in Example 5, find the probability of each of the three sample points in the Bose-Einstein case. (You should find that each has probability $\frac{1}{3}$, that is, they are equally probable.)

17. Find the number of ways of putting 2 particles in 4 boxes according to the three kinds of statistics.

18. Find the number of ways of putting 3 particles in 5 boxes according to the three kinds of statistics.

19. (a) Following the methods of Examples 3, 4, and 5, show that the number of equally likely ways of putting N particles in n boxes, $n > N$, is n^N for Maxwell-Boltzmann particles, $C(n, N)$ for Fermi-Dirac particles, and $C(n-1+N, N)$ for Bose-Einstein particles.

(b) Show that if n is much larger than N (think , for example, of $n = 10^6, N = 10$), then both the Bose-Einstein and the Fermi-Dirac results in part (a) contain products of N numbers, each number approximately equal to n. Thus show that for $n \gg N$, both the BE and the FD results are approximately equal to $n^N/N!$, which is $1/N!$ times the MB result.

20. (a) In Example 5, a mathematical model is discussed which claims to give a distribution of identical balls into boxes in such a way that all distinguishable arrangements are equally probable (Bose-Einstein statistics). Prove this by showing that the probability of a distribution of N balls into n boxes (according to this model) with N_1 balls in the first box, N_2 in the second, $\cdots, N_n$ in the nth, is $1/C(n-1+N, N)$ for any set of numbers N_i such that $\sum_{i=1}^{n} N_i = N$.

(b) Show that the model in (a) leads to Maxwell-Boltzmann statistics if the drawn card is replaced (but no extra card added) and to Fermi-Dirac statistics if the drawn card is not replaced. *Hint:* Calculate in each case the number of possible arrangements of the balls in the boxes. First do the problem of 4 particles in 6 boxes as in the example, and then do N particles in n boxes ($n > N$) to get the results in Problem 19.

21. The following problem arises in quantum mechanics (see Chapter 13, Problem 7.21). Find the number of ordered triples of nonnegative integers a, b, c whose sum $a+b+c$ is a given positive integer n. (For example, if $n = 2$, we could have $(a, b, c) = (2, 0, 0)$ or (0, 2, 0) or (0, 0, 2) or (0, 1, 1) or (1, 0, 1) or (1, 1, 0).) *Hint:* Show that this is the same as the number of distinguishable distributions of n identical balls in 3 boxes, and follow the method of the diagram in Example 5.

22. Suppose 13 people want to schedule a regular meeting one evening a week. What is the probability that there is an evening when everyone is free if each person is already busy one evening a week?

23. Do Problem 22 if one person is busy 3 evenings, one is busy 2 evenings, two are each busy one evening, and the rest are free every evening.

5. RANDOM VARIABLES

In the problem of tossing two dice (Example 2, Section 2), we may be more interested in the value of the sum of the numbers on the two dice than we are in the individual numbers. Let us call this sum x; then for each point of the sample space in (2.4), x has a value. For example, for the point 2,1, we have $x = 2+1 = 3$; for the point 6,2, we have $x = 8$, etc. Such a variable, x, which has a definite value for each sample point, is called a *random variable.* We can easily construct many more examples of random variables for the sample space (2.4); here are a few (Can you construct some more?):

$$x = \text{number on first die minus number on second;}$$
$$x = \text{number on second die;}$$
$$x = \text{probability } p \text{ associated with the sample point;}$$
$$x = \begin{cases} 1 \text{ if the sum is 7 or 11,} \\ 0 \text{ otherwise.} \end{cases}$$

For each of these random variables x, we could set up a table listing all the sample points in (2.4) and, next to each sample point, the corresponding value of x. This table may remind you of the tables of values we could use in plotting the graph of a function. In analytical geometry or in a physics problem, knowing x as a function of t means that for any given t we can find the corresponding value of x. In probability the sample point corresponds to the independent variable t; given the sample point, we can find the corresponding value of the random variable x if we are given a description of x (for example, $x =$ the sum of numbers on dice). The "description" corresponds to the formula $x(t)$ that we use in plotting a graph in analytic geometry. Thus we may say that a *random variable x is a function defined on a sample space.*

Probability Functions Let us consider further the random variable $x =$ "sum of numbers on dice" for a toss of two dice [sample space (2.4)]. We note that there are several sample points for which $x = 5$, namely the points marked a in (2.4). Similarly, there are several sample points for most of the other values of x. It is then convenient to lump together all the sample points corresponding to a given value of x, and consider a new sample space in which each point corresponds to one value of x; this is the sample space (2.5). The probability associated with each point of the new sample space is obtained as in Section 2, by adding the probabilities associated with all the points in the original sample space corresponding to the particular value of x. Each value of x, say x_i, has a probability p_i of occurrence; we may write $p_i = f(x_i) =$ probability that $x = x_i$, and call the function $f(x)$ the *probability function* for the random variable x. In (2.5) we have listed on the first line the values of x and on the second line the values of $f(x)$. [In this problem, x and $f(x)$ take on only a finite number of discrete values; in some later problems they will take on a continuous set of values.] We could also exhibit these values graphically (Figure 5.1).

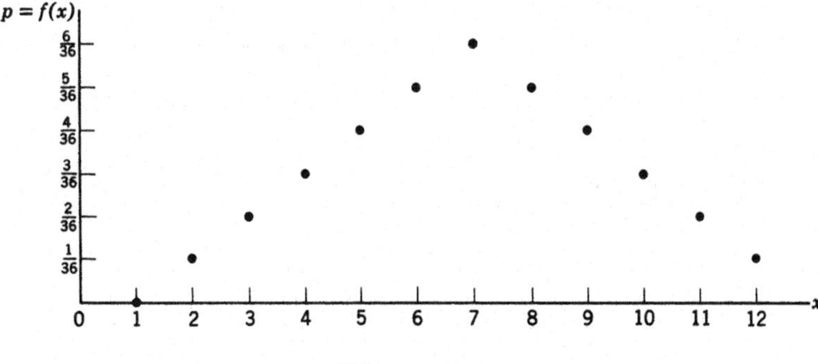

Figure 5.1

Now that we have the table of values (2.5) or the graph (Figure 5.1) to describe the random variable x and its probability function $f(x)$, we can dispense with the original sample space (2.4). But since we used (2.4) in defining what is meant by a random variable, let us now give another definition using (2.5) or Figure 5.1. We can say that *x is a random variable if it takes various values x_i with probabilities* $p_i = f(x_i)$. This definition may explain the name random variable; x is called a variable since it takes various values. A random (or stochastic) process is one whose outcome is not known in advance. The way the two dice fall is such an unknown outcome, so the value of x is unknown in advance, and we call x a *random* variable.

You may note that at first we thought of x as a dependent variable or function with the sample point as the independent variable. Although we didn't say much about it, there was also a value of the probability p attached to each sample point, that is p and x were both functions of the sample point. In the last paragraph, we have thought of x as an independent variable with p as a function of x. This is quite analogous to having both x and p given as functions of t and eliminating t to obtain p as a function of x. We have here eliminated the sample point from the forefront of our discussion in order to consider directly the probability function $p = f(x)$.

Example 1. Let x = number of heads when three coins are tossed. The uniform sample space is (2.3) and we could write the value of x for each sample point in (2.3). Instead, let us go immediately to a table of x and $p = f(x)$. [Can you verify this table by using (2.3), or otherwise?]

(5.1)

x	0	1	2	3
$p = f(x)$	$\frac{1}{8}$	$\frac{3}{8}$	$\frac{3}{8}$	$\frac{1}{8}$

Other terms used for the probability function $p = f(x)$ are: *probability density function*, *frequency function*, or *probability distribution* (*caution*: **not** distribution function, which means the *cumulative distribution* as we will discuss later; see Figure 5.2). The origins of these terms will become clearer as we go on (Sections 6 and 7) but we can get some idea of the terms frequency and distribution from (5.1). Suppose we toss three coins repeatedly; we might reasonably expect to get three heads in about $\frac{1}{8}$ of the tosses, two heads in about $\frac{3}{8}$ of the tosses, etc. That is, each value of $p = f(x)$ is proportional to the *frequency* of occurrence of that value of x—hence the term *frequency function* (see also Section 7). Again in (5.1), imagine four boxes labeled $x = 0, 1, 2, 3$, and put a marble into the appropriate box for each toss of three coins. Then $p = f(x)$ indicates approximately how the marbles are distributed into the boxes after many tosses—hence the term *distribution.*

Mean Value; Standard Deviation The probability function $f(x)$ of a random variable x gives us detailed information about it, but for many purposes we want a simpler description. Suppose, for example, that x represents experimental measurements of the length of a rod, and that we have a large number N of measurements x_i. We might reasonably take $p_i = f(x_i)$ proportional to the number of times N_i we obtained the value x_i, that is $p_i = N_i/N$. We are especially interested in two numbers, namely a mean or average value of all our measurements, and some number which indicates how widely the original set of values spreads out about that average. Let us define two such quantities which are customarily used to describe a random variable. To calculate the average of a set of N numbers, we add them and divide by N. Instead of adding the large number of measurements, we can multiply each measurement by the number of times it occurs and add the results. This gives for the average of the measurements, the value

$$\frac{1}{N}\cdot\sum_i N_i x_i = \sum_i p_i x_i.$$

By analogy with this calculation, we now define the *average* or *mean value* μ *of a random variable* x whose probability function is $f(x)$ by the equation

$$\mu = \text{average of } x = \sum_i x_i p_i = \sum_i x_i f(x_i). \tag{5.2}$$

To obtain a measure of the spread or dispersion of our measurements, we might first list how much each measurement differs from the average. Some of these deviations are positive and some are negative; if we average them, we get zero (Problem 10). Instead, let us square each deviation and average the squares. We define the *variance* of a random variable x by the equation

$$\operatorname{Var}(x) = \sum_i (x_i - \mu)^2 f(x_i). \tag{5.3}$$

(The variance is sometimes called the dispersion.) If nearly all the measurements x_i are very close to μ, then $\operatorname{Var}(x)$ is small; if the measurements are widely spread, $\operatorname{Var}(x)$ is large. Thus we have a number which indicates the spread of the measurements; this is what we wanted. The square root of $\operatorname{Var}(x)$, called the *standard deviation* of x, is often used instead of $\operatorname{Var}(x)$:

$$\sigma_x = \text{standard deviation of } x = \sqrt{\operatorname{Var}(x)}. \tag{5.4}$$

Example 2. For the data in (5.1) we can compute:

By (5.2), $\mu = \text{average of } x = 0\cdot\frac{1}{8} + 1\cdot\frac{3}{8} + 2\cdot\frac{3}{8} + 3\cdot\frac{1}{8} = \frac{12}{8} = \frac{3}{2}.$

By (5.3), $\operatorname{Var}(x) = \left(0-\frac{3}{2}\right)^2\cdot\frac{1}{8} + \left(1-\frac{3}{2}\right)^2\cdot\frac{3}{8} + \left(2-\frac{3}{2}\right)^2\cdot\frac{3}{8} + \left(3-\frac{3}{2}\right)^2\cdot\frac{1}{8}$

$= \frac{9}{4}\cdot\frac{1}{8} + \frac{1}{4}\cdot\frac{3}{8} + \frac{1}{4}\cdot\frac{3}{8} + \frac{9}{4}\cdot\frac{1}{8} = \frac{3}{4}.$

By (5.4), $\sigma_x = \text{standard deviation of } x = \sqrt{\operatorname{Var}(x)} = \frac{1}{2}\sqrt{3}.$

The mean or average value of a random variable x is also called its *expectation* or its *expected value* or (especially in quantum mechanics) its *expectation value.* Instead of μ, the symbols $\overline{x}$ or $E(x)$ or $\langle x\rangle$ may be used to denote the mean value of x.

$$\overline{x} = E(x) = \langle x \rangle = \mu = \sum_i x_i f(x_i). \tag{5.5}$$

The term expectation comes from games of chance.

Example 3. Suppose you will be paid \$5 if a die shows a 5, \$2 if it shows a 2 or a 3, and nothing otherwise. Let x represent your gain in playing the game. Then the possible values of x and the corresponding probabilities are $x = 5$ with $p = \frac{1}{6}$, $x = 2$ with $p = \frac{1}{3}$, and $x = 0$ with $p = \frac{1}{2}$. We find for the average or expectation of x:

$$E(x) = \sum x_i p_i = \$5 \cdot \frac{1}{6} + \$2 \cdot \frac{1}{3} + \$0 \cdot \frac{1}{2} = \$1.50.$$

If you play the game many times, this is a reasonable estimate of your average gain per game; this is what your expectation means. It is also a reasonable amount to pay as a fee for each game you play. The term *expected value* (which means the same as *expectation* or *average*) may be somewhat confusing and misleading if you try to interpret "expected" in an everyday sense. Note that the expected value (\$1.50) of x is not one of the possible values of x, so you cannot ever "expect" to have $x = \$1.50$. If you think of expected value as a technical term meaning the same as average, then there is no difficulty. Of course, in some cases, it makes reasonable sense with its everyday meaning; for example, if a coin is tossed n times, the expected number of heads is $n/2$ (Problem 11) and it is true that we may reasonably "expect" a fair approximation to this result (see Section 7).

Cumulative Distribution Functions So far we have been using the probability function $f(x)$ which gives the probability $p_i = f(x_i)$ that x is exactly x_i. In some problems we may be more interested in the probability that x is less than some particular value. For example, in an election we would like to know the probability that less than half the votes would be cast for the opposing candidate, that is, that our candidate would win. In an experiment on radioactivity, we would like to know the probability that the background radiation always remains below a certain level. Given the probability function $f(x)$, we can obtain the probability that x is less than or equal to a certain value x_i by adding all the probabilities of values of x less than or equal to x_i. For example, consider the sum of the numbers on two dice; the probability function $p = f(x)$ is plotted in Figure 5.1. The probability that x is, say, less than or equal to 4 is the sum of the probabilities that x is 2 or 3 or 4, that is, $\frac{1}{36} + \frac{2}{36} + \frac{3}{36} = \frac{1}{6}$. Similarly, we could find the probability that x is less than or equal to any given number. The resulting function of x is plotted in Figure 5.2. Such a function $F(x)$ is called a *cumulative distribution function*; we can write

$$F(x_i) = (\text{probability that } x \le x_i) = \sum_{x_j \le x_i} f(x_j). \tag{5.6}$$

Note carefully that, although the probability function $f(x)$ may be referred to as a *probability distribution*, the term *distribution function* means the *cumulative distribution* $F(x)$.

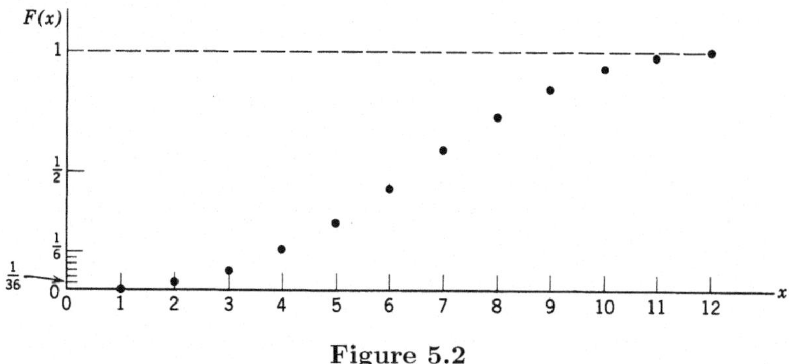

Figure 5.2

PROBLEMS, SECTION 5

Set up sample spaces for Problems 1 to 7 and list next to each sample point the value of the indicated random variable x, and the probability associated with the sample point. Make a table of the different values x_i of x and the corresponding probabilities $p_i = f(x_i)$. Compute the mean, the variance, and the standard deviation for x. Find and plot the cumulative distribution function $F(x)$.

1. Three coins are tossed; $x =$ number of heads minus number of tails.

2. Two dice are thrown; $x =$ sum of the numbers on the dice.

3. A coin is tossed repeatedly; $x =$ number of the toss at which a head first appears.

4. Suppose that Martian dice are 4-sided (tetrahedra) with points labeled 1 to 4. When a pair of these dice is tossed, let x be the product of the two numbers at the tops of the dice if the product is odd; otherwise $x = 0$.

5. A random variable x takes the values 0, 1, 2, 3, with probabilities $\frac{5}{12}, \frac{1}{3}, \frac{1}{12}, \frac{1}{6}$.

6. A card is drawn from a shuffled deck. Let $x = 10$ if it is an ace or a face card; $x = -1$ if it is a 2; and $x = 0$ otherwise.

7. A weighted coin with probability p of coming down heads is tossed three times; $x =$ number of heads minus number of tails.

8. Would you pay \$10 per throw of two dice if you were to receive a number of dollars equal to the product of the numbers on the dice? *Hint:* What is your expectation? If it is more than \$10, then the game would be favorable for you.

9. Show that the expectation of the sum of two random variables defined over the same sample space is the sum of the expectations. *Hint:* Let $p_1, p_2, \cdots, p_n$ be the probabilities associated with the n sample points; let $x_1, x_2, \cdots, x_n$, and $y_1, y_2, \cdots, y_n$, be the values of the random variables x and y for the n sample points. Write out $E(x)$, $E(y)$, and $E(x + y)$.

10. Let μ be the average of the random variable x. Then the quantities $(x_i - \mu)$ are the deviations of x from its average. Show that the average of these deviations is zero. *Hint:* Remember that the sum of all the p_i must equal 1.

11. Show that the expected number of heads in a single toss of a coin is $\frac{1}{2}$. Show in two ways that the expected number of heads in two tosses of a coin is 1:

(a) Let $x =$ number of heads in two tosses and find $\bar{x}$.

(b) Let $x =$ number of heads in toss 1 and $y =$ number of heads in toss 2; find the average of $x + y$ by Problem 9. Use this method to show that the expected number of heads in n tosses of a coin is $\frac{1}{2}n$.

12. Use Problem 9 to find the expected value of the sum of the numbers on the dice in Problem 2.

13. Show that adding a constant K to a random variable increases the average by K but does not change the variance. Show that multiplying a random variable by K multiplies both the average and the standard deviation by K.

14. As in Problem 11, show that the expected number of 5's in n tosses of a die is $n/6$.

15. Use Problem 9 to find $\overline{x}$ in Problem 7.

16. Show that $\sigma^2 = E(x^2) - \mu^2$. *Hint:* Write the definition of σ^2 from (5.3) and (5.4) and use Problems 9 and 13.

17. Use Problem 16 to find σ in Problems 2, 6, and 7.

6. CONTINUOUS DISTRIBUTIONS

In Section 5, we discussed random variables x which took a discrete set of values x_i. It is not hard to think of cases in which a random variable takes a continuous set of values.

Example 1. Consider a particle moving back and forth along the x axis from $x = 0$ to $x = l$, rebounding elastically at the turning points so that its speed is constant. (This could be a simple-minded model of an alpha particle in a radioactive nucleus, or of a gas molecule bouncing back and forth between the walls of a container.) Let the position x of the particle be the random variable; then x takes a continuous set of values from $x = 0$ to $x = l$. Now suppose that, following Section 5, we ask for the probability that the particle is *at* a particular point x; this probability must be the same, say k, for all points (because the speed is constant). In Section 5, with a finite number of points, we would say $k = 1/N$. In the continuous case, there are an infinite number of points so we would find $k = 0$, that is, the probability that the particle is *at* a given point) must be zero. But this is not a very useful result. Let us instead divide $(0, l)$ into small intervals dx; since the particle has constant speed, the time it spends in each dx is proportional to the length of dx. In fact, since the particle spends the fraction $(dx)/l$ of its time in a given interval dx, the probability of finding it in dx is just $(dx)/l$.

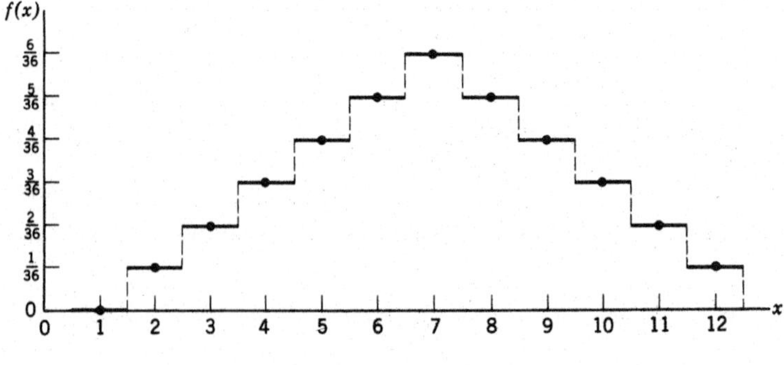

Figure 6.1

Comparison of Discrete and Continuous Probability Functions To see how to define a probability function for the continuous case and to correlate this discussion with the discrete case, let us return for a moment to Figure 5.1. There we plotted a vertical *distance* to represent the probability $p = f(x)$ of each value of x. Instead of a dot (as in Figure 5.1) to indicate p for each x, let us now draw a horizontal line segment of length 1 centered on each dot, as in Figure 6.1. Then the *area* under the horizontal line segment at a particular x_i is $f(x_i) \cdot 1 = f(x_i) = p_i$ (since the length of each horizontal line segment is 1), and we could use this *area* instead of the ordinate as a measure of the probability. Such a graph is called a *histogram*.

Example 2. Now let us apply this area idea to Example 1. Consider Figure 6.2. We have plotted the function

$$f(x) = \begin{cases} 1/l, & 0 < x < l, \\ 0, & x < 0 \quad \text{and} \quad x > l. \end{cases}$$

If we consider any interval x to $x + dx$ on $(0, l)$, the area under the curve $f(x) = 1/l$ for this interval is $(1/l)\,dx$ or $f(x)\,dx$, and this is just the probability that the particle is in this interval. The probability that the particle is in some longer subinterval of $(0, l)$, say (a, b), is $(b - a)/l$ or $\int_a^b f(x)\,dx$, that is, the area under the curve from a to b. If the interval (a, b) is outside $(0, l)$, then $\int_a^b f(x)\,dx = 0$ since $f(x)$ is zero, and again this is the correct value of the probability of finding the particle on the given interval.

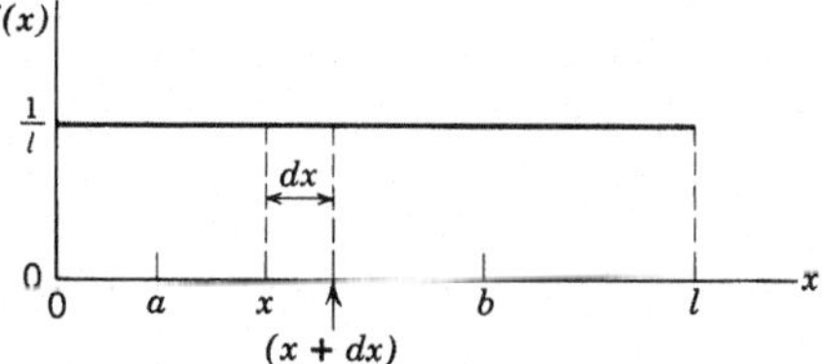

Figure 6.2

When $f(x)$ is constant over an interval (as in Figure 6.2), we say that x is *uniformly* distributed on that interval. Let us consider an example in which $f(x)$ is not constant.

Example 3. This time suppose the particle of Example 1 is sliding up and down an inclined plane (no friction) rebounding elastically (no energy loss) against a spring at the bottom and reaching zero speed at height $y = h$ (Figure 6.3). The total energy, namely $\frac{1}{2}mv^2 + mgy$ is constant and equal to mgh since $v = 0$ at $y = h$. Thus we have

$$v^2 = \frac{2}{m}(mgh - mgy) = 2g(h - y). \tag{6.1}$$

The probability of finding the particle within an interval dy at a given height y is proportional to the time dt spent in that interval. From $v = ds/dt$, we have $dt = (ds)/v$; from Figure 6.3, we find $ds = (dy)\csc\alpha$. Combining these with (6.1) we have

$$dt = \frac{ds}{v} = \frac{(dy)\csc\alpha}{\sqrt{2g}\sqrt{h - y}}$$

Since the probability $f(y)\,dy$ of finding the particle in the interval dy at height y is proportional to dt, we can drop the constant factor $(\csc\alpha)/\sqrt{2g}$, and say that

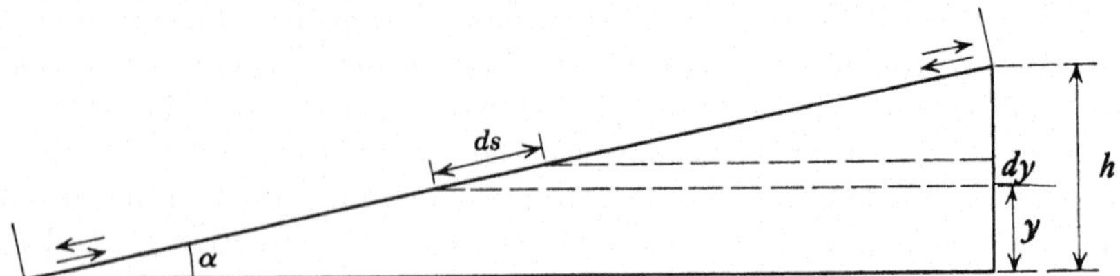

Figure 6.3

$f(y)\,dy$ is proportional to $dy/\sqrt{h-y}$. In order to find $f(y)$, we must multiply by a constant factor which makes the total probability $\int_0^h f(y)\,dy$ equal to 1 since this is the probability that the particle is *somewhere*. You can easily verify that

$$f(y)\,dy = \frac{1}{2\sqrt{h}}\frac{dy}{\sqrt{h-y}} \quad \text{or} \quad f(y) = \frac{1}{2\sqrt{h(h-y)}}$$

A graph of $f(y)$ is plotted in Figure 6.4. Note that although $f(y)$ becomes infinite at $y = h$, the area under the $f(y)$ curve for any interval is finite; this area represents the probability that the particle is in that height interval.

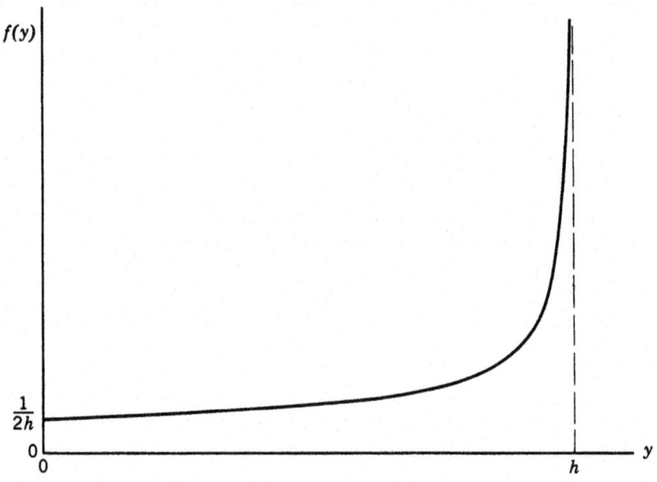

Figure 6.4

We can now extend the definitions of mean (expectation), variance, standard deviation, and cumulative distribution function to the continuous case. Let $f(x)$ be a probability density function; remember that $\int_{-\infty}^{\infty} f(x)\,dx = 1$ just as $\sum_{i=1}^{n} p_i = 1$. The average of a random variable x with probability density function $f(x)$ is

$$\mu = \bar{x} = E(x) = \langle x \rangle = \int_{-\infty}^{\infty} x f(x)\,dx. \tag{6.2}$$

(In writing the limits $-\infty, \infty$ here, we assume that $f(x)$ is defined to be zero on intervals where the probability is zero.) Note that (6.2) is a natural extension of the sum in (5.5). Having found the mean of x, we now define the variance as in Section 5 as the average of $(x-\mu)^2$, that is,

$$\text{(6.3)} \qquad \operatorname{Var}(x) = \int_{-\infty}^{\infty} (x-\mu)^2 f(x)\,dx = \sigma_x^2.$$

As before, the standard deviation σ_x is the square root of the variance. Finally, the cumulative distribution function $F(x)$ gives for each x the probability that the random variable is less than or equal to that x. But this probability is just the area under the $f(x)$ curve from $-\infty$ up to the point x. Also, of course, the integral of $f(x)$ from $-\infty$ to ∞ must $= 1$ since that is the total probability for all values of x. Thus we have

$$\text{(6.4)} \qquad F(x) = \int_{-\infty}^{x} f(u)\,du, \qquad \int_{-\infty}^{\infty} f(x)\,dx = F(\infty) = 1.$$

Example 4. For the problem in Example 3, we find:

By (6.2), $\mu_y = \displaystyle\int_0^h y f(y)\,dy = \frac{1}{2\sqrt{h}}\int_0^h y\frac{1}{\sqrt{h-y}}\,dy = \frac{2}{3}h.$

By (6.3), $\operatorname{Var}(y) = \displaystyle\int_0^h (y-\mu_y)^2 f(y)\,dy = \int_0^h \left(y - \frac{2}{3}h\right)^2 \frac{1}{\sqrt{h-y}}\,dy = \frac{4h^2}{45},$

so standard deviation $\sigma_y = \sqrt{\operatorname{Var}(y)} = 2h/\sqrt{45}$.

By (6.4), cumulative distribution function $F(y) = \displaystyle\int_0^y f(u)\,du$

$$= \frac{1}{2\sqrt{h}}\int_0^y \frac{du}{\sqrt{h-u}}.$$

Why "density function"? In Section 5, we mentioned that the probability function $f(x)$ is often called the *probability density*. We can now explain why. Consider (6.2). If $f(x)$ represents the density (mass per unit length) of a thin rod, then the center of mass of the rod is given by [see Chapter 5, (3.3)]

$$\text{(6.5)} \qquad \bar{x} = \int x f(x)\,dx \Big/ \int f(x)\,dx,$$

where the integrals are over the length of the rod, or from $-\infty$ to ∞ as in (6.2) with $f(x) = 0$ outside the rod. But in (6.2), $\int f(x)\,dx$ is the total probability that x has *some* value, and so this integral is equal to 1. Then (6.5) and (6.2) are really the same; we see that it is reasonable to call $f(x)$ a density, and also that the mean of x corresponds to the center of mass of a linear mass distribution of density $f(x)$. In a similar way, we can interpret (6.3) as giving the moment of inertia of the mass distribution about the center of mass (see Chapter 5, Section 3).

Joint Distributions We can easily generalize the ideas and formulas above to two (or more) dimensions. Suppose we have two random variables x and y; we define their joint probability density function $f(x,y)$ so that $f(x_i,y_j)\,dx\,dy$ is the probability that the point (x,y) is in an element of area $dx\,dy$ at $x=x_i$, $y=y_j$. Then the probability that the point (x,y) is in a given region of the (x,y) plane, is the integral of $f(x,y)$ over that area. The average or expected values of x and y, the variances and standard deviations of x and y, and the covariance of x, y (see Problems 13 to 16) are given by

$$
\begin{aligned}
\overline{x} &= \int_{-\infty}^{\infty}\int_{-\infty}^{\infty} xf(x,y)\,dx\,dy,\\
\overline{y} &= \int_{-\infty}^{\infty}\int_{-\infty}^{\infty} yf(x,y)\,dx\,dy,\\
\mathrm{Var}(x) &= \int_{-\infty}^{\infty}\int_{-\infty}^{\infty} (x-\overline{x})^2 f(x,y)\,dx\,dy = \sigma_x^2,\\
\mathrm{Var}(y) &= \int_{-\infty}^{\infty}\int_{-\infty}^{\infty} (y-\overline{y})^2 f(x,y)\,dx\,dy = \sigma_y^2,\\
\mathrm{Cov}(x,y) &= \int_{-\infty}^{\infty}\int_{-\infty}^{\infty} (x-\overline{x})(y-\overline{y}) f(x,y)\,dx\,dy.
\end{aligned}
\tag{6.6}
$$

You should see that these are generalizations of (6.2) and (6.3); that (6.6) can be interpreted as giving the coordinates of the center of mass and the moments of inertia of a two-dimensional mass distribution; and that similar formulas can be written for three (or more) random variables (that is, in three or more dimensions). Also note that the formulas in (6.6) could be written in terms of polar coordinates (see Problems 6 to 9).

We have discussed a number of probability distributions both discrete and continuous, and you will find others in the problems. We will discuss three very important named distributions (binomial, normal, and Poisson) in the following sections. Learning about these and related graphs, formulas, and terminology should make it possible for you to cope with any of the many other named distributions you find in texts, reference books, and computer programs.

PROBLEMS, SECTION 6

1. (a) Find the probability density function $f(x)$ for the position x of a particle which is executing simple harmonic motion on $(-a,a)$ along the x axis. (See Chapter 7, Section 2, for a discussion of simple harmonic motion.) *Hint:* The value of x at time t is $x=a\cos\omega t$. Find the velocity dx/dt; then the probability of finding the particle in a given dx is proportional to the time it spends there which is inversely proportional to its speed there. Don't forget that the total probability of finding the particle *somewhere* must be 1.

   (b) Sketch the probability density function $f(x)$ found in part (a) and also the cumulative distribution function $F(x)$ [see equation (6.4)].

   (c) Find the average and the standard deviation of x in part (a).

2. It is shown in the kinetic theory of gases that the probability for the distance a molecule travels between collisions to be between x and $x+dx$, is proportional to $e^{-x/\lambda}\,dx$, where λ is a constant. Show that the average distance between collisions (called the "mean free path") is λ. Find the probability of a free path of length $\geq 2\lambda$.

3. A ball is thrown straight up and falls straight back down. Find the probability density function $f(h)$ so that $f(h)\,dh$ is the probability of finding it between height h and $h + dh$. *Hint:* Look at Example 3.

4. In Problem 1 we found the probability density function for a classical harmonic oscillator. In quantum mechanics, the probability density function for a harmonic oscillator (in the ground state) is proportional to $e^{-\alpha^2 x^2}$, where α is a constant and x takes values from $-\infty$ to ∞. Find $f(x)$ and the average and standard deviation of x. (In quantum mechanics, the standard deviation of x is called the uncertainty in position and is written Δx.)

5. The probability for a radioactive particle to decay between time t and time $t + dt$ is proportional to $e^{-\lambda t}$. Find the density function $f(t)$ and the cumulative distribution function $F(t)$. Find the expected lifetime (called the mean life) of the radioactive particle. Compare the mean life and the so-called "half life" which is defined as the value of t when $e^{-\lambda t} = 1/2$.

6. A circular garden bed of radius 1 m is to be planted so that N seeds are uniformly distributed over the circular area. Then we can talk about the number n of seeds in some particular area A, or we can call n/N the probability for any one particular seed to be in the area A. Find the probability $F(r)$ that a seed (that is, some particular seed) is within r of the center. (*Hint:* What is F(1)?) Find $f(r)\,dr$, the probability for a seed to be between r and $r + dr$ from the center. Find $\bar{r}$ and σ.

7. (a) Repeat Problem 6 where the "circular" area is now on the curved surface of the earth, say all points at distance s from Chicago (measured along a great circle on the earth's surface) with $s \le \pi R/3$ where R = radius of the earth. The seeds could be replaced by, say, radioactive fallout particles (assuming these to be uniformly distributed over the surface of the earth). Find $F(s)$ and $f(s)$.

(b) Also find $F(s)$ and $f(s)$ if $s \le 1 \ll R$ (say $s \le 1$ mile where $R = 4000$ miles). Do your answers then reduce to those in Problem 6?

8. Given that a particle is inside a sphere of radius 1, and that it has equal probabilities of being found in any two volume elements of the same size, find the cumulative distribution function $F(r)$ for the spherical coordinate r, and from it find the density function $f(r)$. *Hint:* $F(r)$ is the probability that the particle is inside a sphere of radius r. Find $\bar{r}$ and σ.

9. A hydrogen atom consists of a proton and an electron. According to the Bohr theory, the electron revolves about the proton in a circle of radius a ($a = 5 \cdot 10^{-9}$cm for the ground state). According to quantum mechanics, the electron may be at any distance r (from 0 to ∞) from the proton; for the ground state, the probability that the electron is in a volume element dV, at a distance r to $r + dr$ from the proton, is proportional to $e^{-2r/a}dV$, where a is the Bohr radius. Write dV in spherical coordinates (see Chapter 5, Section 4) and find the density function $f(r)$ so that $f(r)\,dr$ is the probability that the electron is at a distance between r and $r + dr$ from the proton. (Remember that the probability for the electron to be somewhere must be 1.) Computer plot $f(r)$ and show that its maximum value is at $r = a$; we then say that the most probable value of r is a. Also show that the average value of r^{-1} is a^{-1}.

10. Do Problem 5.10 for a continuous distribution.

11. Do Problem 5.13 for a continuous distribution.

12. Do Problem 5.16 for a continuous distribution.

13. Given a joint distribution function $f(x, y)$ as in (6.6), show that $E(x + y) = E(x) + E(y)$ and $\text{Var}(x + y) = \text{Var}(x) + \text{Var}(y) + 2\,\text{Cov}(x, y)$.

14. Recall that two events A and B are called independent if $p(AB) = p(A)p(B)$. Similarly two random variables x and y are called independent if the joint probability function $f(x, y) = g(x)h(y)$. Show that if x and y are independent, then the expectation or average of xy is $E(xy) = E(x)E(y) = \mu_x \mu_y$.

15. Show that the covariance of two independent (see Problem 14) random variables is zero, and so by Problem 13, the variance of the sum of two independent random variables is equal to the sum of their variances.

16. By Problem 15, if x and y are independent, then $\text{Cov}(x, y) = 0$. The converse is not always true, that is, if $\text{Cov}(x, y) = 0$, it is not necessarily true that the joint distribution function is of the form $f(x, y) = g(x)h(y)$. For example, suppose $f(x, y) = (3y^2 + \cos x)/4$ on the rectangle $-\pi/2 < x < \pi/2, -1 < y < 1$, and $f(x, y) = 0$ elsewhere. Show that $\text{Cov}(x, y) = 0$, but x and y are not independent, that is, $f(x, y)$ is not of the form $g(x)h(y)$. Can you construct some more examples?

7. BINOMIAL DISTRIBUTION

Example 1. Let a coin be tossed 5 times; what is the probability of exactly 3 heads out of the 5 tosses? We can represent any sequence of 5 tosses by a symbol such as *thhth*. The probability of this particular sequence (or any other particular sequence) is $(\frac{1}{2})^5$ since the tosses are independent (see Example 1 of Section 3). The number of such sequences containing 3 heads and 2 tails is the number of ways we can select 3 positions out of 5 for heads (or 2 for tails), namely $C(5, 3)$. Hence, the probability of exactly 3 heads in 5 tosses of a coin is $C(5, 3)(\frac{1}{2})^5$. Suppose a coin is tossed repeatedly, say n times; let x be the number of heads in the n tosses. We want to find the probability density function $p = f(x)$ which gives the probability of exactly x heads in n tosses. By generalizing the case of 3 heads in 5 tosses, we see that

$$f(x) = C(n, x)(\tfrac{1}{2})^n \tag{7.1}$$

Example 2. Let us do a similar problem with a die, asking this time for the probability of exactly 3 aces in 5 tosses of the die. If A means ace and N not ace, the probability of a particular sequence such as $ANNAA$ is $\frac{1}{6} \cdot \frac{5}{6} \cdot \frac{5}{6} \cdot \frac{1}{6} \cdot \frac{1}{6}$ since the probability of A is $\frac{1}{6}$, the probability of N is $\frac{5}{6}$, and the tosses are independent. The number of such sequences containing 3 A's and 2 N's is $C(5, 3)$; thus the probability of exactly 3 aces in 5 tosses of a die is $C(5, 3)(\frac{1}{6})^3(\frac{5}{6})^2$. Generalizing this, we find that the probability of exactly x aces in n tosses of a die is

$$f(x) = C(n, x)(\tfrac{1}{6})^x(\tfrac{5}{6})^{n-x}. \tag{7.2}$$

Bernoulli Trials In the two examples we have just done, we have been concerned with repeated independent trials, each trial having two possible outcomes (h or t, A or N) of given probability. There are many examples of such problems; let's consider a few. A manufactured item is good or defective; given the probability of a defect we want the probability of x defectives out of n items. An archer has probability p of hitting a target; we ask for the probability of x hits out of n tries. Each atom of a radioactive substance has probability p of emitting an alpha particle during the next minute; we are to find the probability that x alpha particles will be emitted in the next minute from the n atoms in the sample. A particle moves back and forth along the x axis in unit jumps; it has, at each step, equal probabilities of

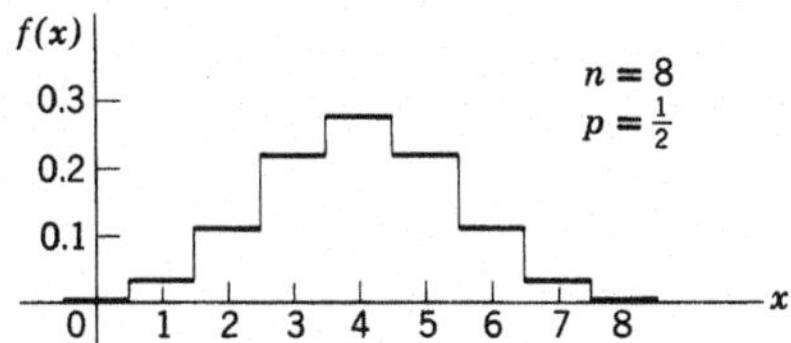

Figure 7.1

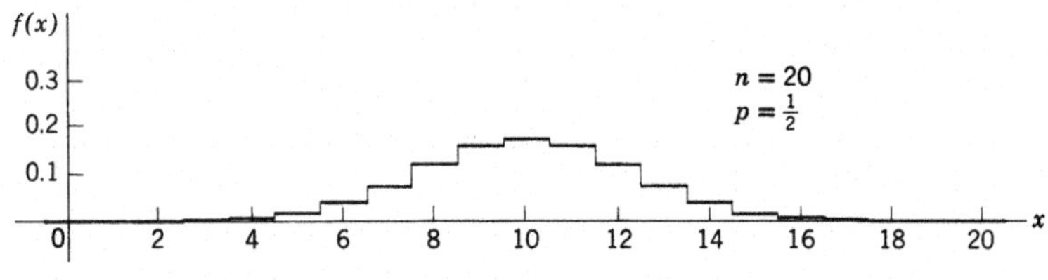

Figure 7.2

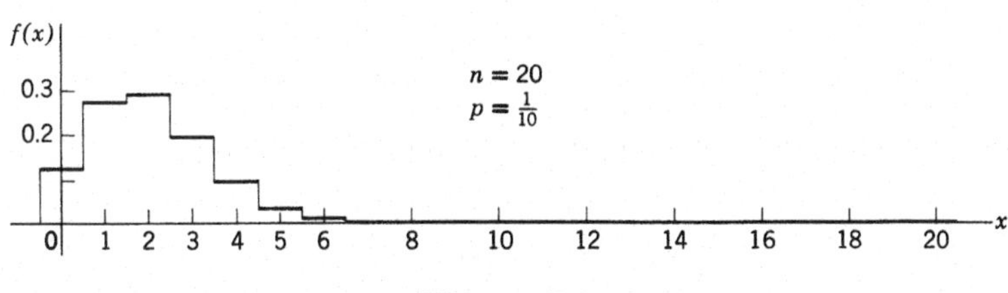

Figure 7.3

jumping forward or backward. (This motion is called a *random walk*; it can be used as a model of a diffusion process.) We want to know the probability that, after n jumps, the particle is at a distance

$$d = \text{number } x \text{ of positive jumps} - \text{number } (n - x) \text{ of negative jumps},$$

from its starting point; this probability is the probability of x positive jumps out of a total of n jumps.

In all these problems, something is tried repeatedly. At each trial there are two possible outcomes of probabilities p (usually called the probability of "success") and

Binomial distribution graphs of $nf(x)$ plotted against x/n

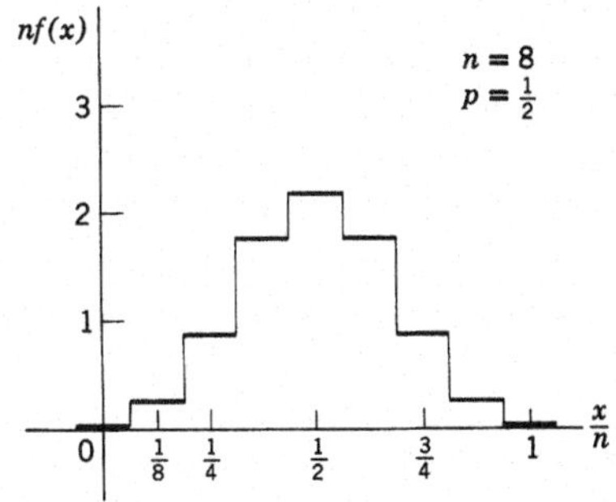

Figure 7.4

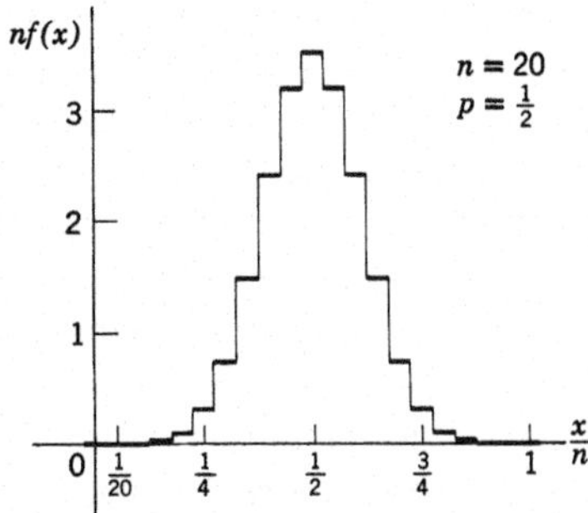

Figure 7.5

$q = 1 - p$ (where q = probability of "failure"). Such repeated independent trials with constant probabilities p and q are called *Bernoulli trials.*

Binomial Probability Functions Let us generalize (7.1) and (7.2) to obtain a formula which applies to any similar problem, namely the probability $f(x)$ of exactly x successes in n Bernoulli trials. Reasoning as we did to obtain (7.1) and (7.2), we find that

$$f(x) = C(n, x)p^x q^{n-x}. \tag{7.3}$$

We might also ask for the probability of *not more than* x successes in n trials. This is the sum of the probabilities of $0, 1, 2, \cdots, x$ successes, that is, it is the cumulative distribution function $F(x)$ for the random variable x whose probability density function is (7.3) [see (5.6)]. We can write

$$\begin{aligned} F(x) &= f(0) + f(1) + \cdots + f(x) \\ &= C(n, 0)p^0 q^n + C(n, 1)p^1 q^{n-1} + \cdots + C(n, x)p^x q^{n-x} \\ &= \sum_{u=0}^{x} C(n, u)p^u q^{n-u} = \sum_{u=0}^{x} \binom{n}{u} p^u q^{n-u}. \end{aligned} \tag{7.4}$$

Observe that (7.3) is one term of the binomial expansion of $(p + q)^n$ and (7.4) is a sum of several terms of this expansion (see Section 4, Example 2). For this reason, the functions $f(x)$ in (7.1), (7.2), or (7.3) are called *binomial probability* (or *density*) *functions* or *binomial distributions,* and the function $F(x)$ in (7.4) is called a *binomial cumulative distribution function.*

We shall find it very useful to computer plot graphs of the binomial density function $f(x)$ for various values of p and n. (See Figures 7.1 to 7.5 and Problems 1 to 8.) Instead of a point at $y = f(x)$ for each x, we plot a horizontal line segment of length 1 centered on each x as in Figure 6.1; the probabilities are then represented by *areas* under the broken line, rather than by ordinates. From Figures 7.1 to 7.3 and similar graphs, we can draw a number of conclusions. The most probable value of x [corresponding to the largest value of $f(x)$] is approximately $x = np$ (Problems 10 and 11); for example for $p = \frac{1}{2}$, the most probable value of x is $\frac{1}{2}n$ for even n; for odd n, there are two consecutive values of x, namely $\frac{1}{2}(n \pm 1)$, for which the probability is largest. The graphs for $p = \frac{1}{2}$ are symmetric about $x = \frac{1}{2}n$. For $p \neq \frac{1}{2}$, the curve is asymmetric, favoring small x values for small p and large x values for large p. As n increases, the graph of $f(x)$ becomes wider and flatter (the total area under the graph must remain 1). The probability of the most probable value of x decreases with n. For example, the most probable number of heads in 8 tosses of a coin is 4 with probability 0.27; the most probable number of heads in 20 tosses is 10 with probability 0.17; for 10^6 tosses, the probability of exactly 500,000 heads is less than 10^{-3}.

Let us redraw Figures 7.1 and 7.2 plotting $nf(x)$ against the relative number of successes x/n (Figures 7.4 and 7.5). Since this change of scale (ordinate times n, abscissa divided by n) leaves the area unchanged, we can still use the area to represent probability. Note that now the curves become narrower and taller as n

increases. This means that values of the ratio x/n tend to cluster about their most probable value, namely $np/n = p$. For example, if we toss a coin repeatedly, the difference "number of heads $-\frac{1}{2}$ number of tosses" is apt to be large and to increase with n (Figures 7.1 and 7.2), but the ratio "number of heads $\div$ number of tosses" is apt to be closer and closer to $\frac{1}{2}$ as n increases (Figures 7.4 and 7.5). It is for this reason that we can use experimentally determined values of x/n as a reasonable estimate of p.

Chebyshev's Inequality This is a simple but very general result which we will find useful. We consider a random variable x with probability function $f(x)$, and let μ be the mean value and σ the standard deviation of x. We are going to prove that if we select any number t, the probability that x differs from its mean value μ by more than t, is less than σ^2/t^2. This means that x is unlikely to differ from μ by more than a few standard deviations; for example, if t is twice the standard deviation σ, we find that the probability for x to differ from μ by more than 2σ is less than $\sigma^2/t^2 = \sigma^2/(2\sigma)^2 = \frac{1}{4}$. The proof is simple. By definition of σ, we have

$$\sigma^2 = \sum (x-\mu)^2 f(x)$$

where the sum is over all x. Then if we sum just over the values of x for which $|x-\mu| \geq t$, we get less than σ^2:

$$\sigma^2 > \sum_{|x-\mu|\geq t} (x-\mu)^2 f(x). \tag{7.5}$$

If we replace each $x-\mu$ by the number t in (7.5), the sum is decreased, so we have

$$\sigma^2 > \sum_{|x-\mu|\geq t} t^2 f(x) = t^2 \sum_{|x-\mu|\geq t} f(x) \qquad \text{or} \qquad \sum_{|x-\mu|\geq t} f(x) < \frac{\sigma^2}{t^2}. \tag{7.6}$$

But $\sum_{|x-\mu|\geq t} f(x)$ is just the sum of all probabilities of x values which differ from μ by more than t, and (7.6) says that this probability is less than σ^2/t^2, as we claimed.

Laws of Large Numbers Statements and proofs which make more precise our general comments about the effect of large n are known as *laws of large numbers.* Let us state and prove one such law. We apply Chebyshev's inequality to a random variable whose probability function is the binomial distribution (7.3). From Problems 9 and 13 we have $\mu = np$ and $\sigma = \sqrt{npq}$. Then by Chebyshev's inequality,

$$(\text{probability of } |x-np| \geq t) \qquad \text{is less than} \quad npq/t^2. \tag{7.7}$$

Let us choose the arbitrary value of t in (7.7) proportional to n, that is, $t = n\epsilon$ where ϵ is now arbitrary. Then (7.7) becomes

$$(\text{probability of } |x-np| \geq n\epsilon) \qquad \text{is less than} \quad npq/n^2\epsilon^2, \tag{7.8}$$

or, when we divide the first inequality by n,

$$\left(\text{probability of } \left|\frac{x}{n}-p\right| \geq \epsilon\right) \qquad \text{is less than} \quad \frac{pq}{n\epsilon^2}. \tag{7.9}$$

Recall that x/n is the relative number of successes; we intuitively expect x/n to be near p for large n. Now (7.9) says that, if ϵ is any small number, the probability is less than $pq/(n\epsilon^2)$ for x/n to differ from p by ϵ; that is, as n tends to infinity, this probability tends to zero. (Note, however, that x/n need not tend to p.) This is one form of the law of large numbers and it justifies our intuitive ideas.

PROBLEMS, SECTION 7

For the values of n indicated in Problems 1 to 4:

(a) Write the probability density function $f(x)$ for the probability of x heads in n tosses of a coin and computer plot a graph of $f(x)$ as in Figures 7.1 and 7.2. Also computer plot a graph of the corresponding cumulative distribution function $F(x)$.

(b) Computer plot a graph of $nf(x)$ as a function of x/n as in Figures 7.4 and 7.5.

(c) Use your graphs and other calculations if necessary to answer these questions: What is the probability of exactly 7 heads? Of at most 7 heads? [Hint: Consider $F(x)$.] Of at least 7 heads? What is the most probable number of heads? The expected number of heads?

1. $n = 7$ **2.** $n = 12$ **3.** $n = 15$ **4.** $n = 18$

5. Write the formula for the binomial density function $f(x)$ for the case $n = 6, p = 1/6$, representing the probability of, say, x aces in 6 throws of a die. Computer plot $f(x)$ as in Figure (7.3). Also plot the cumulative distribution function $F(x)$. What is the probability of at least 2 aces out of 6 tosses of a die? *Hint:* Can you read the probability of at most one ace from one of your graphs?

For the given values of n and p in Problems 6 to 8, computer plot graphs of the binomial density function for the probability of x successes in n Bernoulli trials with probability p of success.

6. $n = 6$, $p = 5/6$ (Compare Problem 5)

7. $n = 50$, $p = 1/5$ **8.** $n = 50$, $p = 4/5$

9. Use the second method of Problem 5.11 to show that the expected number of successes in n Bernoulli trials with probability p of success is $\overline{x} = np$. *Hint:* What is the expected number of successes in one trial?

10. Show that the most probable number of heads in n tosses of a coin is $\frac{1}{2}n$ for even n [that is, $f(x)$ in (7.1) has its largest value for $x = n/2$] and that for odd n, there are two equal "largest" values of $f(x)$, namely for $x = \frac{1}{2}(n+1)$ and $x = \frac{1}{2}(n-1)$. *Hint:* Simplify the fraction $f(x+1)/f(x)$, and then find the values of x for which it is greater than 1 [that is, $f(x+1) > f(x)$], and less than or equal to 1 [that is, $f(x+1) \le f(x)$]. Remember that x must be an integer.

11. Use the method of Problem 10 to show that for the binomial distribution (7.3), the most probable value of x is approximately np (actually within 1 of this value).

12. Let x = number of heads in one toss of a coin. What are the possible values of x and their probabilities? What is μ_x? Hence show that $\text{Var}(x) =$ [average of $(x - \mu_x)^2$] $= \frac{1}{4}$, so the standard deviation is $\frac{1}{2}$. Now use the result from Problem 6.15 "variance of a sum of independent random variables = sum of their variances" to show that if x = number of heads in n tosses of a coin, $\text{Var}(x) = \frac{1}{4}n$ and the standard deviation $\sigma_x = \frac{1}{2}\sqrt{n}$.

13. Generalize Problem 12 to show that for the general binomial distribution (7.3), $\text{Var}(x) = npq$, and $\sigma = \sqrt{npq}$.

8. THE NORMAL OR GAUSSIAN DISTRIBUTION

The graph of the *normal* or *Gaussian distribution* is the bell-shaped curve you may know as the normal error curve (Figure 8.1). The normal distribution is used a great deal because, as we shall see, it is not only of interest in itself (see Problems 2 and 3), but also other distributions become almost normal when n (the number of trials or measurements) becomes large (see Figures 8.2 and 8.3).

The probability density function $f(x)$ and the cumulative distribution function $F(x)$ for the normal or Gaussian distribution are given by

$$(8.1)\qquad \begin{aligned} f(x) &= \frac{1}{\sigma\sqrt{2\pi}} e^{-(x-\mu)^2/(2\sigma^2)}, \\ F(x) &= \frac{1}{\sigma\sqrt{2\pi}} \int_{-\infty}^{x} e^{-(t-\mu)^2/(2\sigma^2)}\, dt. \end{aligned} \qquad \text{Normal distribution}$$

It is straightforward to show (Problem 1) that if x is a random variable with probability density $f(x)$ in (8.1), then the mean of x is μ and the standard deviation is σ. Also we can show that the integral of $f(x)$ from $-\infty$ to ∞ is equal to 1 as it must be for a probability function. Then the probability that a normally distributed random variable x lies between x_1 and x_2 is the area under the $f(x)$ curve between x_1 and x_2 which is

$$(8.2)\qquad F(x_2) - F(x_1) = \text{probability that } x_1 \le x \le x_2.$$

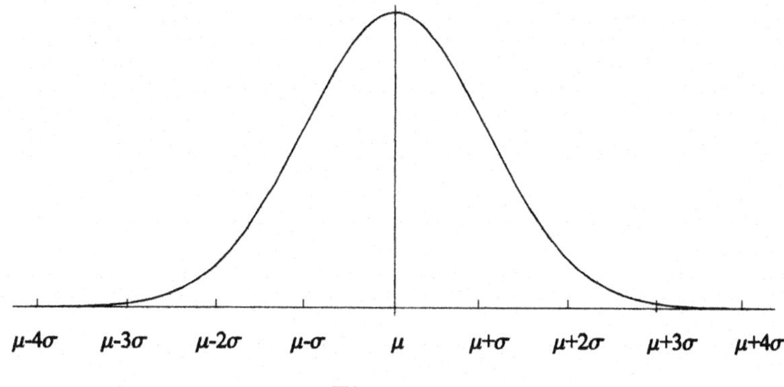

Figure 8.1

A normal density function graph (Figure 8.1) has its peak at $x = \mu$ and is symmetric with respect to the line $x = \mu$. Since the area from $-\infty$ to ∞ is 1, the area from $-\infty$ to μ is $\frac{1}{2}$ (that is, $F(\mu) = \frac{1}{2}$), and similarly the area from μ to ∞ is $\frac{1}{2}$. A change in μ merely translates the graph with no change in shape. An increase in σ widens and flattens the graph so that the area remains 1, and similarly a decrease in σ makes the graph taller and narrower. (Problems 4 to 6). The area from $\mu - \sigma$ to $\mu + \sigma$ is 0.6827, that is, the probability that x differs from its mean value by 1 standard deviation or less, is just over 68%. The probability that $|x - \mu| \le 2\sigma$

is over 95% and the probability that $|x - \mu| \le 3\sigma$ is over 99.7%. Note that these probabilities are independent of the values of μ and σ (Problem 7).

Normal Approximation to the Binomial Distribution As an example of approximating another distribution by a normal distribution, let's consider the binomial distribution (7.3). For large n and large np, we can use Stirling's formula (Chapter 11, Section 11) to approximate the factorials in $C(n, x)$ in (7.3) and make other approximations to find

$$f(x) = C(n,x)p^x q^{n-x} \sim \frac{1}{\sqrt{2\pi npq}} e^{-(x-np)^2/(2npq)}. \tag{8.3}$$

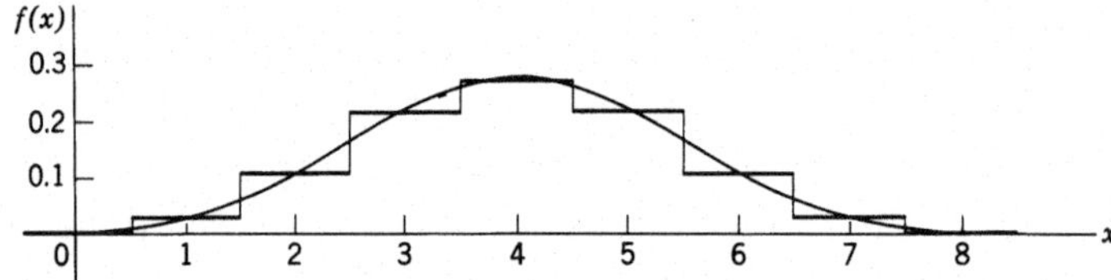

Figure 8.2 Binomial distribution for $n = 8$, $p = \frac{1}{2}$, and the normal approximation.

The sign $\sim$ means (as in Chapter 11, Section 11) that the ratio of the exact binomial distribution (7.3) and the right-hand side of (8.3) tends to 1 as $n \to \infty$. An outline of a derivation of (8.3) is given in Problem 8, but you may be more impressed by doing some computer plotting of graphs like Figures 8.2 and 8.3 (Problems 9 and 10). Although we have said that equation (8.3) gives an approximation valid for large n, the agreement is quite good even for fairly small values of n. Figure 8.2 shows this for the case $n = 8$. The binomial distribution $f(x)$ is defined only for integral x; you should compare the values of $f(x)$ with the values of the approximating normal curve at integral values of x. When n is very large (Figure 8.3), a graph of the exact binomial distribution is very close to the normal approximation (Problem 9).

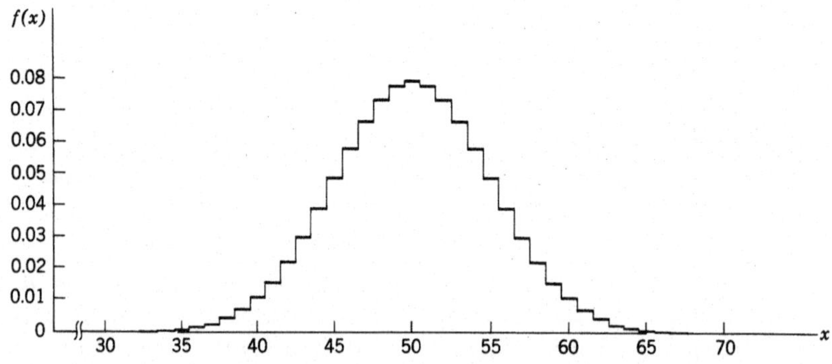

Figure 8.3 Binomial distribution for $n = 100$, $p = \frac{1}{2}$.

In (8.3), the left-hand side is the exact binomial distribution and the right-hand side is a normal distribution with $\mu = np$ and $\sigma = \sqrt{npq}$ as we see by comparing (8.3) and (8.1). Recall from Problems 7.9 and 7.13 that the mean value

μ and standard deviation σ for a random variable whose probability function is the binomial distribution (7.3) are also $\mu = np$ and $\sigma = \sqrt{npq}$.

(8.4) For the binomial distribution and its normal approximation,
$$\mu = np, \qquad \sigma = \sqrt{npq}\,.$$

We can expect this in general; whatever the μ and σ are for a given distribution, the normal approximation will have the same μ and σ.

Example 1. Find the probability of exactly 52 heads in 100 tosses of a coin using the binomial distribution and using the normal approximation.

See Figure (8.3) which is a plot of the binomial probability density function with $n = 100, p = \frac{1}{2}$. We find by computer for $x = 52$, binomial $f(52) = 0.07353$, which you could also read approximately from Figure (8.3).

For the normal approximation, we find from (8.4), $\mu = np = 100 \cdot \frac{1}{2} = 50$, $\sigma = \sqrt{npq} = \sqrt{100 \cdot \frac{1}{2} \cdot \frac{1}{2}} = 5$. Then for the normal approximation with $\mu = 50$, $\sigma = 5$, we find by computer for $x = 52$, normal $f(52) = 0.07365$.

Example 2. Find the probability $P(45, 55)$ of between 45 and 55 heads in 100 tosses of a coin, that is $45 \leq x \leq 55$.

As in Example 1, for the binomial distribution we have $n = 100, p = \frac{1}{2}$. The cumulative binomial distribution function $F(x)$ in (7.4) gives $P(45, 55)$ as a sum of terms; we want the sum of the 11 terms with $x = 45, 46, \cdots 55$. By computer, we can find $F(55)$, the binomial cumulative distribution function with $x = 55$, which is the probability of 55 heads or less, and then find and subtract $F(44)$, the probability of 44 heads or less. Thus we find $P(45, 55) =$ binomial $F(55)$ – binomial $F(44) =$ 0.72875.

For the normal approximation, we find by computer from (8.2), $P(45, 55) =$ normal $F(55)$ – normal $F(45) = 0.68269$. We can get a better approximation by integrating from 44.5 to 55.5; this corresponds more closely to the appropriate area under the exact binomial graph in Figure 8.3 by including the whole steps at $x = 45$ and $x = 55$. This gives $P(44.5, 55.5) =$ normal $F(55.5)$ – normal $F(44.5) = 0.72867$.

Standard Normal Distribution This is just the normal distribution in (8.1) for the special case $\mu = 0$ and $\sigma = 1$. The density function is often denoted by $\phi(z)$, and the corresponding cumulative distribution function by $\Phi(z)$:

(8.5) Standard normal distribution
$$\phi(z) = \frac{1}{\sqrt{2\pi}} e^{-z^2/2},$$
$$\Phi(z) = \frac{1}{\sqrt{2\pi}} \int_{-\infty}^{z} e^{-u^2/2}\, du.$$

The cumulative distribution function $\Phi(z)$ is related to the error function (see Chapter 11, Section 9).

It is sometimes convenient to write the functions in (8.1) in terms of $\phi(z)$ and $\Phi(z)$. We can do this by making the change of variables $z = (x-\mu)/\sigma$. The result is (Problem 21)

$$(8.6)\qquad \begin{aligned} f(x) &= \frac{1}{\sigma}\phi(z), \\ F(x) &= \Phi(z), \end{aligned} \qquad \text{where } z = \frac{(x-\mu)}{\sigma}.$$

The functions $\phi(z)$ and $\Phi(z)$ [or sometimes $\Phi(z) - \frac{1}{2}$] are tabulated so you can use either tables or computer to do problems.

▶ **Example 3.** Find the number r such that the area under the normal distribution curve $y = f(x)$ from $\mu - r$ to $\mu + r$ is equal to 1/2.

Look at Figure 8.1 and recall that the area from $-\infty$ to ∞ is 1 and that the graph is symmetric about $x = \mu$. Then the integral from $-\infty$ to $\mu - r$ and the integral from $\mu + r$ to ∞ are equal to each other and so each is equal to 1/4. Thus the integral from $-\infty$ to $\mu + r$ must be 3/4, that is $F(\mu + r) = 3/4$. By (8.6) this is $\Phi(z) = 3/4$ where $z = (\mu + r - \mu)/\sigma = r/\sigma$. By computer or tables we find that if $\Phi(z) = 3/4$, then $z = 0.6745$. Thus $r = 0.6745\sigma$.

▶ **Example 4.** You have taken a test (academic like the SAT, or medical like a bone density test) and a report gives your z-score as 1.14. What percent of your peers scored higher than you?

If we call the actual test scores x, and their average is μ and standard deviation σ, then the term z-score means the value of $z = (x-\mu)/\sigma$ as in (8.6). (In words, the z-score is the difference between x and its average, measured in units of the standard deviation.) Now we want the area $1 - F(x) = 1 - \Phi(z)$ by (8.6). By computer (or tables) we find $\Phi(1.14) = 0.87$; then $1 - 0.87 = 0.13$, so 13% of your peers scored higher than you. If your z-score is negative, then you are below average—bad if it's a physics test, good if it's your cholesterol! For example, if $z = -0.25$, then $\Phi(z) = 0.40$, so 60% of your peers scored higher than you.

▶ **Example 5.** Suppose that boxes of a certain kind of cereal have an average weight of 16 ounces and it is known that 70% of the boxes weigh within 1 ounce of the average. What is the probability that the box you buy weighs less than 14 ounces?

If x represents the weight of a box, then we are given that the probability of $15 < x < 17$ is 0.7. Assuming a normal distribution, the area under the $f(x)$ curve up to $x = \mu = 16$ is $\frac{1}{2}$ and the area from $x = 16$ to $x = 17$ is half of 0.7 (by symmetry; see Figure 8.1). Thus $F(17) = 0.5 + 0.35 = 0.85$. We want to find the probability that $x < 14$; this is $F(14)$. Using (8.6), $x = 17$ gives $z = (17-16)/\sigma = 1/\sigma$, and similarly $x = 14$ gives $z = -2/\sigma$. So we are given $\Phi(1/\sigma) = 0.85$, and we want to find $\Phi(-2/\sigma)$. By computer (or tables) we find that if $\Phi(1/\sigma) = 0.85$, then $1/\sigma = 1.0364$, so $2/\sigma = 2.0728$, and $\Phi(-2/\sigma) = 0.019$. So there is almost a 2% chance that we would get a box weighing less than 14 ounces.

Note that in Examples 4 and 5 we assumed a normal distribution with no obvious justification. It is a very interesting and useful fact that such an assumption is reasonable if the number of measurements is very large. We will discuss this further at the end of Section 10.

PROBLEMS, SECTION 8

1. Verify that for a random variable x with normal density function $f(x)$ as in (8.1), the mean value of x is μ , the standard deviation is σ, and the integral of $f(x)$ from $-\infty$ to ∞ is 1 as it must be for a probability function. *Hint:* Write and evaluate the integrals $\int_{-\infty}^{\infty} f(x)\,dx$, $\int_{-\infty}^{\infty} xf(x)\,dx$, $\int_{-\infty}^{\infty}(x-\mu)^2 f(x)\,dx$. See equations (6.2), (6.3), and (6.4).

2. Do Problem 6.4 by comparing e^{-ax^2} with $f(x)$ in (8.1).

3. The probability density function for the x component of the velocity of a molecule of an ideal gas is proportional to $e^{-mv^2/(2kT)}$ where v is the x component of the velocity, m is the mass of the molecule, T is the temperature of the gas and k is the Boltzmann constant. By comparing this with (8.1), find the mean and standard deviation of v, and write the probability density function $f(v)$.

4. Computer plot on the same axes the normal probability density functions with $\mu = 0$, $\sigma = 1$, and with $\mu = 3$, $\sigma = 1$ to note that they are identical except for a translation.

5. Computer plot on the same axes the normal density functions with $\mu = 0$ and $\sigma = 1$, 2, and 5. Label each curve with its σ.

6. Do Problem 5 for $\sigma = \frac{1}{6}, \frac{1}{3}, 1$.

7. By computer find the value of the normal cumulative distribution function at $\mu+\sigma$, $\mu + 2\sigma$, $\mu + 3\sigma$, and satisfy yourself that these are independent of your choices for μ and σ. Find the probabilities that x is within 1, 2, or 3 standard deviations of its mean value μ to verify the results stated in the paragraph following (8.2). *Hint*: See Figure (8.1). The probability that x is within 1 standard deviation of its mean value is the area from $\mu-\sigma$ to $\mu+\sigma$; this is twice the area from μ to $\mu+\sigma$. Subtract $\frac{1}{2}$ (that is the area from $-\infty$ to μ) from your value of $F(\mu+\sigma)$ and then double the result.

8. Carry through the following details of a derivation of (8.3). Start with (7.3); we want an approximation to (7.3) for large n. First approximate the factorials in $C(n, x)$ by Stirling's formula (Chapter 11, Section 11) and simplify to get

$$f(x) \sim \left(\frac{np}{x}\right)^x \left(\frac{nq}{n-x}\right)^{n-x} \sqrt{\frac{n}{2\pi x(n-x)}}.$$

Show that if $\delta = x - np$, then $x = np + \delta$ and $n - x = nq - \delta$. Make these substitutions for x and $n - x$ in the approximate $f(x)$. To evaluate the first two factors in $f(x)$ (ignore the square root for now): Take the logarithm of the first two factors; show that

$$\ln\frac{np}{x} = -\ln\left(1+\frac{\delta}{np}\right)$$

and a similar formula for $\ln[nq/(n-x)]$; expand the logarithms in a series of powers of $\delta/(np)$, collect terms and simplify to get

$$\ln\left(\frac{np}{x}\right)^x \left(\frac{nq}{n-x}\right)^{n-x} \sim -\frac{\delta^2}{2npq}\left(1 + \text{powers of } \frac{\delta}{n}\right).$$

Hence

$$\left(\frac{np}{x}\right)^x \left(\frac{nq}{n-x}\right)^{n-x} \sim e^{-\delta^2/(2npq)}$$

for large n. [We really want δ/n small, that is, x near enough to its average value np so that $\delta/n = (x - np)/n$ is small. This means that our approximation is valid for the central part of the graph (see Figures 7.1 to 7.3) around $x = np$ where $f(x)$ is large. Since $f(x)$ is negligibly small anyway for x far from np, we ignore the fact that our approximation may not be good there. For more detail on this point, see Feller, p. 192]. Returning to the square root factor in $f(x)$, approximate x by np and $n - x$ by nq (assuming $\delta \ll np$ or nq) and obtain (8.3).

9. Computer plot a graph like Figure 8.3 of the binomial distribution with $n = 1000$, $p = \frac{1}{2}$, and observe that you have practically the corresponding normal approximation.

10. Computer plot graphs like Figure 8.2 but with $p \neq \frac{1}{2}$ to see that as n increases, the normal approximation becomes good (at least in the region around $x = \mu$ where the probabilities are large) even though the binomial graph is not symmetric (see Figure 7.3).

As in Examples 1 and 2, use (a) the binomial distribution; (b) the corresponding normal approximation, to find the probabilities of each of the following:

11. Exactly 50 heads in 100 tosses of a coin.

12. Exactly 120 aces in 720 tosses of a die.

13. Between 100 and 140 aces in 720 tosses of a die.

14. Between 499,000 and 501,000 heads in 10^6 tosses of a coin.

15. Exactly 195 tails in 400 tosses of a coin.

16. Between 195 and 205 tails in 400 tosses of a coin.

17. Exactly 31 4's in 180 tosses of a die.

18. Between 29 and 33 4's in 180 tosses of a die.

19. Exactly 21 successes in 100 Bernoulli trials with probability $\frac{1}{5}$ of success.

20. Between 17 and 21 successes in 100 Bernoulli trials with probability $\frac{1}{5}$ of success.

21. Verify equations (8.6). *Hints:* In $F(x)$, let $u = (t-\mu)/\sigma$; note that $dt = \sigma du$. What is u when $t = -\infty$? When $t = x$? Remember that by definition $z = (x - \mu)/\sigma$.

22. Using (8.6), do Problem 7.

23. Using (8.6), find h such that 90% of the area under a normal $f(x)$ lies between $\mu - h$ and $\mu + h$. Repeat for 95%. *Hint:* See Example 3.

24. Write out a proof of Chebyshev's inequality (see end of Section 7) for the case of a continuous probability function $f(x)$.

25. An instructor who grades "on the curve" computes the mean and standard deviation of the grades, and then, assuming a normal distribution with this μ and σ, sets the border lines between the grades at: C from $\mu - \frac{1}{2}\sigma$ to $\mu + \frac{1}{2}\sigma$, B from $\mu + \frac{1}{2}\sigma$ to $\mu + \frac{3}{2}\sigma$, A from $\mu + \frac{3}{2}\sigma$ up, etc. Find the percentages of the students receiving the various grades. Where should the border lines be set to give the percentages: A and F, 10%; B and D, 20%; C, 40%?

9. THE POISSON DISTRIBUTION

The Poisson distribution is useful in a variety of problems in which the probability of some occurrence is small and constant. (See Example 1 and Problems 3 to 9.) It is also a good approximation to the binomial distribution when p is so small that np is small even though n is large (see Example 2).

Let's derive the Poisson distribution by considering the following experiment. Suppose we observe and count the number of particles emitted per unit time by a radioactive substance. We assume that our period of observation is much less than the half-life of the substance, so that the average counting rate does not decrease during the experiment. Then the probability that one particle is emitted during a small time interval Δt is $\mu\Delta t$, μ =const., if Δt is short enough so that the probability of two particles during Δt is negligible. We want to find the probability $P_n(t)$ of observing exactly n counts during a time interval t. The probability $P_n(t+\Delta t)$ is the probability of observing n counts in the time interval $t+\Delta t$. For $n>0$, this is the sum of the probabilities of the two mutually exclusive events, "n particles in t, none in Δt" and "$(n-1)$ particles in t, one in Δt"; in symbols,

$$P_n(t+\Delta t) = P_n(t)P_0(\Delta t) + P_{n-1}(t)P_1(\Delta t). \tag{9.1}$$

Now $P_1(\Delta t)$ is the probability of one particle in Δt; this, by assumption, is $\mu\Delta t$. Then the probability of no particles in Δt is $1 - P_1(\Delta t) = 1 - \mu\Delta t$. Substituting these values into (9.1), we get

$$P_n(t+\Delta t) = P_n(t)(1-\mu\Delta t) + P_{n-1}(t)\mu\Delta t, \tag{9.2}$$

or,

$$\frac{P_n(t+\Delta t) - P_n(t)}{\Delta t} = \mu P_{n-1}(t) - \mu P_n(t). \tag{9.3}$$

Letting $\Delta t \to 0$, we have

$$\frac{dP_n(t)}{dt} = \mu P_{n-1}(t) - \mu P_n(t). \tag{9.4}$$

For $n=0$, (9.1) simplifies since the only possible event is "no particles in t, no particles in Δt," and (9.4) becomes, for $n=0$,

$$\frac{dP_0(t)}{dt} = -\mu P_0(t). \tag{9.5}$$

Then, since $P_0(0) =$ "probability that no particle is emitted during a zero time interval" $= 1$, integration of (9.5) gives

$$P_0 = e^{-\mu t}. \tag{9.6}$$

Substituting (9.6) into (9.4) with $n=1$ gives a differential equation for $P_1(t)$; its solution (Problem 1) is $P_1(t) = \mu t e^{-\mu t}$. Solving (9.4) successively (Problem 1) for $P_2, P_3, \cdots, P_n$, we obtain

$$P_n(t) = \frac{(\mu t)^n}{n!}e^{-\mu t}. \tag{9.7}$$

Putting $t = 1$, we get for the probability of exactly n counts per unit time

$$(9.8) \qquad P_n = \frac{\mu^n}{n!} e^{-\mu}. \qquad \text{Poisson distribution}$$

The probability density function (9.8) is called the *Poisson distribution* or the *Poisson probability density function.* You can show (Problem 2) that for the random variable n, the mean (that is the average number of counts per unit time) is μ, and the variance is also μ so the standard deviation is $\sqrt{\mu}$.

Example 1. The number of particles emitted each minute by a radioactive source is recorded for a period of 10 hours; a total of 1800 counts are registered. During how many 1-minute intervals should we expect to observe no particles; exactly one; etc.?

The average number of counts per minute is $1800/(10\cdot 60) = 3$ counts per minute; this is the value of μ. Then by (9.8), the probability of n counts per minute is

$$P_n = \frac{3^n}{n!} e^{-3}.$$

A graph of this probability function is shown in Figure 9.1. For $n = 0$, we find $P_0 = e^{-3} = 0.05$; then we should expect to observe no particles in about 5% of the 600 1-minute intervals, that is, during 30 1-minute intervals. Similarly we could compute the expected number of 1-minute intervals during which 1, 2, $\cdots$, particles would be observed.

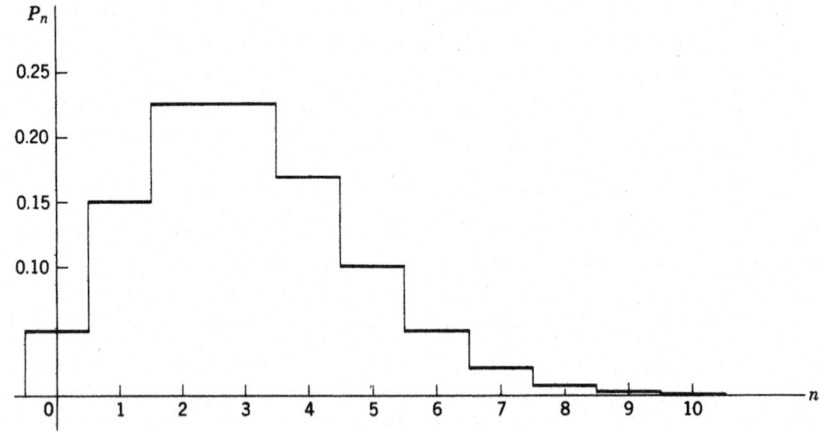

Figure 9.1 Poisson distribution $\mu = 3$.

Poisson Approximation of the Binomial Distribution In Section 8, we discussed the fact that the binomial distribution can be approximated by the normal distribution for large n and large np. If p is very small so that np is very much less than n (say, for example, $p = 10^{-3}, n = 2000, np = 2$), the normal approximation is not good. In this case you can show (Problem 10) that the Poisson distribution gives a good approximation to the binomial distribution (7.3), that is, that

$$(9.9) \qquad C(n,x)p^x q^{n-x} \sim \frac{(np)^x e^{-np}}{x!}, \qquad \text{Large } n, \text{ small } p.$$

[The exact meaning of (9.9) is that, for any fixed x, the ratio of the two sides approaches 1 as $n \to \infty$ and $p \to 0$ with np remaining constant.]

Example 2. If 1500 people each select a number at random between 1 and 500, what is the probability that 2 people selected the number 29?

The answer is given by the binomial distribution (7.3) with $n = 1500$, $p = 1/500$, $x = 2$. This is

$$C(n,x)p^x q^{n-x} = \frac{1500!}{2!1498!}\left(\frac{1}{500}\right)^2\left(\frac{499}{500}\right)^{998} = 0.2241.$$

(Or from your computer: the binomial probability density function with $n = 1500$, $p = 1/500$, $x = 2$, is 0.2241 to four decimal places.). A simpler formula from (9.9) is the Poisson approximation with $\mu = np = 3$, $x = 2$, namely $\mu^x e^{-x}/x! = 3^2 e^{-2}/2! = 0.2240$. (Or from your computer, the Poisson probability density function with $\mu = 3$, $x = 2$, is 0.2240 to four decimal places.) It is interesting to computer plot on the same axes the binomial distribution with $n = 1500$, $p = 1/500$, and the Poisson distribution with $\mu = 3$ as in Figure 9.1 to discover that they are almost identical (Problem 12).

Approximations by the Normal Distribution We have commented that many distributions can be approximated by the normal distribution when n and $\mu = np$ are both large, and have shown this for the binomial distribution in (8.1). The Poisson distribution when μ is large is also fairly well approximated by the normal distribution as in (9.10).

$$(9.10) \qquad \frac{\mu^x e^{-\mu}}{x!} \cong \frac{1}{\sqrt{2\pi\mu}} e^{-(x-\mu)^2/(2\mu)}, \qquad \mu \text{ large.}$$

Note that the normal distribution in (9.10) has the same mean and variance as the Poisson distribution it is approximating (see Problem 2 for the Poisson mean and variance). It is useful to computer plot on the same axes graphs of the Poisson distribution and their normal approximations (Problem 13).

PROBLEMS, SECTION 9

1. Solve the sequence of differential equations (9.4) for successive n values [as started in (9.5) and (9.6)] to obtain (9.7).

2. Show that the average value of a random variable n whose probability function is the Poisson distribution (9.8) is the number μ in (9.8). Also show that the standard deviation of the random variable is $\sqrt{\mu}$. *Hint:* Write the infinite series for e^x, differentiate it and multiply by x to get $xe^x = \sum(nx^n/n!)$; put $x = \mu$. To find σ^2 differentiate the xe^x series again, etc.

3. In an alpha-particle counting experiment the number of alpha particles is recorded each minute for 50 hours. The total number of particles is 6000. In how many 1-minute intervals would you expect no particles? Exactly n particles, for $n = 1$, 2, 3, 4, 5? Plot the Poisson distribution.

4. Suppose you receive an average of 4 phone calls per day. What is the probability that on a given day you receive no phone calls? Just one call? Exactly 4 calls?

5. Suppose that you have 5 exams during the 5 days of exam week. Find the probability that on a given day you have no exams; just 1 exam; 2 exams; 3 exams.

6. If you receive, on the average, 5 email messages per day, in how many days out of a 365-day year would you expect to receive exactly 5 messages? Fewer than 5? Exactly 10? More than 10? Just 1? None at all?

7. In a club with 500 members, what is the probability that exactly two people have birthdays on July 4?

8. If there are 100 misprints in a magazine of 40 pages, on how many pages would you expect to find no misprints? Two misprints? Five misprints?

9. If there are, on the average, 7 defects in a new car, what is the probability that your new car has only 2 defects? That it has 6 or 7? That it has more than 10?

10. Derive equation (9.9) as follows: In $C(n,x)$, show that $n!/(n-x)! \sim n^x$ for fixed x and large n [write $n!/(n-x)!$ as a product of x factors, divide by n^x, and show that the limit is 1 as $n \to \infty$]. Then write $q^{n-x} = (1-p)^{n-x}$ as $(1-p)^n(1-p)^{-x} = (1-np/n)^n(1-p)^{-x}$; evaluate the limit of the first factor as $n \to \infty$, np fixed; the limit of the second factor as $p \to 0$ is 1. Collect your results to obtain equation (9.9).

11. Suppose 520 people each have a shuffled deck of cards and draw one card from the deck. What is the probability that exactly 13 of the 520 cards will be aces of spades? Write the binomial formula and approximate it. Which is best, the normal or the Poisson approximation? Although you only need values at one x to answer the question, you might like to computer plot on the same axes graphs of the three distributions for the given n and p.

12. Computer plot on the same axes graphs of the binomial distribution in Example 2 and the Poisson and normal approximations.

13. Computer plot on the same axes a graph of the Poisson distribution and the corresponding normal approximation for the cases $\mu = 1$, 5, 10, 20, 30.

10. STATISTICS AND EXPERIMENTAL MEASUREMENTS

Statistics uses probability theory to consider sets of data and draw reasonable conclusions from them. So far in this chapter, we have been discussing problems for which we could write down a density function formula (normal, Poisson, etc.). Suppose that, instead, we have only a table of data, say a set of laboratory measurements of some physical quantity. Presumably, if we spent more time, we could enlarge this table of data as much as we liked. We can then imagine an infinite set of measurements of which we have only a sample. The infinite set is called the *parent population* or *universe*. What we would really like to know is the probability function for the parent population, or at least the average value μ (often thought of as the "true" value of the quantity being measured) and the standard deviation σ of the parent population. We must content ourselves with the best estimates we can make of these quantities using our available sample, that is, the set of measurements which we have made.

Estimate of Population Average As a quick estimate of μ we might take the median of our measurements x_i (a value such that there are equal numbers of larger and smaller measurements), or the mode (the measurement we obtained the most times, that is the most probable measurement). The most frequently used estimate of μ is, however, the arithmetic mean (or average) of the measurements, that is the sample mean $\overline{x} = (1/n)\sum_{i=1}^{n} x_i$. Thus we have

(10.1) Estimate of population mean is $\mu \simeq \overline{x} = (1/n)\sum_{i=1}^{n} x_i.$

For a large set of measurements we can justify this choice as follows (also see Problem 1). Assuming that the parent population for our measurements has probability density function $f(x)$ with expected value μ and standard deviation σ, it is easy to show (Problem 2) that the expected value of $\overline{x}$ is μ and the standard deviation of $\overline{x}$ is $\sigma/\sqrt{n}$. Now Chebyshev's inequality (end of Section 7) says that a random variable is unlikely to differ from its expected value by more than a few standard deviations. For our problem this says that $\overline{x}$ is unlikely to differ from μ by more than a few multiples of $\sigma/\sqrt{n}$, which becomes small as n increases. Thus $\overline{x}$ becomes an increasingly good estimate of μ as we increase the number n of measurements. Note that this just says mathematically what you would assume from experience, that the average of a large number of measurements is more likely to be accurate than the average of a small number. For example, two measurements might both be too large, but it's unlikely that 20 would all be too large.

Estimate of Population Variance Our first guess for an estimate of σ^2 might be $s^2 = (1/n)\sum_{i=1}^{n}(x_i - \overline{x})^2$, but we would be wrong. To see what is reasonable, we find the expected value of s^2 assuming that our measurements are from a population with mean μ and variance σ^2. The result is (Problem 3), $E(s^2) = [(n-1)/n]\sigma^2$. We conclude that a reasonable estimate of σ^2 is $\frac{n}{n-1}s^2$.

(10.2) Estimate of population variance is $\sigma^2 \simeq \dfrac{1}{n-1}\sum_{i=1}^{n}(x_i - \overline{x})^2.$

(*Caution:* The term "sample variance" is used in various references—texts, reference books, computer programs—to mean either our s^2 or our estimate of σ^2, so check the definition carefully in any reference you use. We shall avoid using the term.)

The quantity σ which we have just estimated is the standard deviation for the parent population whose probability function we call $f(x)$. Consider just a single measurement x. The function $f(x)$ (if we knew it) would give us the probabilities of the different possible values of x, the population mean μ would tell us approximately the value we are apt to find for x, and the standard deviation σ would tell us roughly the spread of x values about μ. Since σ tells us something about a single measurement, it is often called the *standard deviation of a single measurement.*

Standard Deviation of the Mean; Standard Error Instead of a single measurement, let us consider $\overline{x}$, the average (mean) of a set of n measurements. (The mean, $\overline{x}$, will be what we will use or report as the result of an experiment.) Just as we originally imagined obtaining the probability function $f(x)$ by making a large number of single measurements, so we can imagine obtaining a probability function $g(\overline{x})$ by making a large number of *sets* of n measurements with each set giving us a value of $\overline{x}$. The function $g(\overline{x})$ (if we knew it) would give us the probability of different values of $\overline{x}$. We have seen (Problem 2) that $\text{Var}(\overline{x}) = \sigma^2/n$, so the *standard deviation of the mean* (that is, of $\overline{x}$) is

$$(10.3) \qquad \sigma_m = \sqrt{\text{Var}(\overline{x})} = \frac{\sigma}{\sqrt{n}}.$$

The quantity σ_m is also called the *standard error*; it gives us an estimate of the spread of values of $\overline{x}$ about μ. We see that the new probability function $g(\overline{x})$ must be much more peaked than $f(x)$ about the value μ because the standard deviation $\sigma/\sqrt{n}$ is much smaller than σ. Collecting formulas (10.2) and (10.3), we have

$$(10.4) \qquad \sigma_m \cong \sqrt{\frac{\sum_{i=1}^{n}(x_i - \overline{x})^2}{n(n-1)}}$$

▶ **Example 1.** To illustrate our discussion, let's consider the following set of measurements: {7.2, 7.1, 6.7, 7.0, 6.8, 7.0, 6.9, 7.4, 7.0, 6.9}. [Note that, to show methods but minimize computation, we consider unrealistically small sets of measurements.]

From (10.1) we find $\mu \simeq \overline{x} = \frac{1}{10}\sum_{i=1}^{10} x_i = \frac{70}{10} = 7.0.$

From (10.2) we find $\sigma^2 \simeq \frac{1}{9}\sum_{i=1}^{10}(x_i - 7)^2 = \frac{0.36}{9} = 0.04, \sigma \simeq 0.2.$

From (10.4), the standard error is $\sigma_m \simeq \sqrt{\frac{0.36}{10 \cdot 9}} = 0.0632.$

Combination of Measurements We have discussed how we can use a set of measurements x_i to estimate μ (the population average) by $\overline{x}$ (the sample average) and to estimate the standard error $\sigma_{mx} = \sqrt{\text{Var}(\overline{x})}$ [equation (10.4)]. Now suppose we have done this for two quantities, x and y, and we want to use a known formula $w = w(x, y)$ to estimate a value for w and the standard error in w. First we consider the simple example $w = x + y$. Then, by Problem 6.13,

$$(10.5) \qquad E(w) = E(x) + E(y) = \mu_x + \mu_y$$

where μ_x and μ_y are population averages. As discussed above, we estimate μ_x and μ_y by $\overline{x}$ and $\overline{y}$ and conclude that a reasonable estimate of w is

$$(10.6) \qquad \overline{w} = \overline{x} + \overline{y}.$$

Now let us assume that x and y are independently measured quantities. Then by Problem 6.15,

$$\text{(10.7)}\qquad \begin{aligned}\operatorname{Var}(\overline{w}) &= \operatorname{Var}(\overline{x}) + \operatorname{Var}(\overline{y}) = \sigma_{mx}^2 + \sigma_{my}^2,\\ \sigma_{mw} &= \sqrt{\sigma_{mx}^2 + \sigma_{my}^2}.\end{aligned}$$

Next consider the case $w = 4 - 2x + 3y$. As in equations (10.5) and (10.6), we find $\overline{w} = 4 - 2\overline{x} + 3\overline{y}$. Now by Problem 5.13, we have $\operatorname{Var}(x+K) = \operatorname{Var}(x)$, and $\operatorname{Var}(Kx) = K^2 \operatorname{Var}(x)$, where K is a constant. Thus,

$$\text{(10.8)}\qquad \begin{aligned}\operatorname{Var}(\overline{w}) &= \operatorname{Var}(4 - 2\overline{x} + 3\overline{y}) = \operatorname{Var}(-2\overline{x} + 3\overline{y})\\ &= (-2)^2 \operatorname{Var}(\overline{x}) + (3)^2 \operatorname{Var}(\overline{y}) = 4\sigma_{mx}^2 + 9\sigma_{my}^2,\end{aligned}$$

$$\text{(10.9)}\qquad \sigma_{mw} = \sqrt{4\sigma_{mx}^2 + 9\sigma_{my}^2}.$$

We can now see how to find $\overline{w}$ and σ_{mw} for any function $w(x,y)$ which can be approximated by the linear terms of its Taylor series about the point (μ_x, μ_y), namely (see Chapter 4, Section 2)

$$\text{(10.10)}\qquad w(x,y) \cong w(\mu_x, \mu_y) + \left(\frac{\partial w}{\partial x}\right)(x - \mu_x) + \left(\frac{\partial w}{\partial y}\right)(y - \mu_y)$$

where the partial derivatives are evaluated at $x = \mu_x$, $y = \mu_y$, and so are constants. [Practically speaking, this means that the first partial derivatives should not be near zero—we can't expect good results near a maximum or minimum of w—and the higher derivatives should not be large, that is, w should be "smooth" near the point (μ_x, μ_y).] Assuming (10.10), and remembering that $w(\mu_x, \mu_y)$ and the partial derivatives are constants, we find

$$\text{(10.11)}\qquad \begin{aligned}E[w(x,y)] &\cong w(\mu_x, \mu_y) + \left(\frac{\partial w}{\partial x}\right)[E(x) - \mu_x] + \left(\frac{\partial w}{\partial y}\right)[E(y) - \mu_y]\\ &= w(\mu_x, \mu_y).\end{aligned}$$

Since we have agreed to estimate μ_x and μ_y by $\overline{x}$ and $\overline{y}$, we conclude that a reasonable estimate of w is

$$\text{(10.12)}\qquad \overline{w} = w(\overline{x}, \overline{y}).$$

(This may look obvious, but see Problem 7.)

Then, putting $x = \overline{x}, y = \overline{y}$ in (10.10) and remembering the comment just before (10.11), we find as in (10.8)

$$\begin{aligned}\operatorname{Var}(\overline{w}) &= \operatorname{Var}[w(\overline{x}, \overline{y})]\\ &= \operatorname{Var}\left[w(\mu_x, \mu_y) + \left(\frac{\partial w}{\partial x}\right)(\overline{x} - \mu_x) + \left(\frac{\partial w}{\partial y}\right)(\overline{y} - \mu_y)\right]\\ &= \left(\frac{\partial w}{\partial x}\right)^2 \sigma_{mx}^2 + \left(\frac{\partial w}{\partial y}\right)^2 \sigma_{my}^2,\end{aligned}$$

$$\text{(10.13)}\qquad \sigma_{mw} = \sqrt{\left(\frac{\partial w}{\partial x}\right)^2 \sigma_{mx}^2 + \left(\frac{\partial w}{\partial y}\right)^2 \sigma_{my}^2}.$$

We can use (10.12) and (10.13) to estimate the value of a given function w of two measured quantities x and y and to find the standard error in w.

Example 2. From Example 1 we have $\overline{x} = 7$ and $\sigma_{mx} = 0.0632$. Suppose we have also found from measurements that $\overline{y} = 5$ and $\sigma_{my} = 0.0591$. If $w = x/y$, find $\overline{w}$ and σ_{mw}. From (10.12) we have $\overline{w} = \overline{x}/\overline{y} = 7/5 = 1.4$. From (10.13) we find

$$\sigma_{mw} = \sqrt{\left(\frac{1}{\overline{y}}\right)^2 \sigma_{mx}^2 + \left(\frac{-\overline{x}}{\overline{y}^2}\right)^2 \sigma_{my}^2} = \sqrt{\left(\frac{1}{5}\right)^2 (0.0632)^2 + \left(\frac{-7}{25}\right)^2 (0.0591)^2}$$
$$= 0.0208.$$

Central Limit Theorem So far we have not assumed any special form (such as normal, etc.) for the density function $f(x)$ of the parent population, so that our results for computation of approximate values of μ, σ, and σ_m from a set of measurements apply whether or not the parent distribution is normal. (And, in fact, it may not be; for example, Poisson distributions are quite common.) You will find, however, that most discussions of experimental errors are based on an assumed normal distribution. Let us discuss the justification for this. We have seen above that we can think of the sample average $\overline{x}$ as a random variable with average μ and standard deviation $\sigma/\sqrt{n}$. We have said that we might think of a density function $g(\overline{x})$ for $\overline{x}$ and that it would be more strongly peaked about μ than the density function $f(x)$ for a single measurement, but we have not said anything so far about the form of $g(\overline{x})$. There is a basic theorem in probability (which we shall quote without proof) which gives us some information about the probability function for $\overline{x}$. The *central limit theorem* says that no matter what the parent probability function $f(x)$ is (provided μ and σ exist), the probability function for $\overline{x}$ is approximately the normal distribution with standard deviation $\sigma/\sqrt{n}$ if n is large.

Confidence Intervals, Probable Error If we assume that the probability function for $\overline{x}$ is normal (a reasonable assumption if n is large), then we can give a more specific meaning to σ_m (standard deviation of the mean) than our vague statement that it gives us an estimate of the spread of $\overline{x}$ values about μ. Since the probability for a normally distributed random variable to have values between $\mu - \sigma$ and $\mu + \sigma$ is 0.6827 (see Section 8 and Problem 8.7), we can say that the probability is about 68% for a measurement of $\overline{x}$ to lie between $\mu - \sigma_m$ and $\mu + \sigma_m$. This interval is called the 68% *confidence interval.* Similarly we can find an interval $\mu \pm r$ such that the probability is $\frac{1}{2}$ that a new measurement would fall in this interval (and so also the probability is $\frac{1}{2}$ that it would fall outside!), that is, a 50% confidence interval. From Section 8, Example 3, this is $r = 0.6745\sigma_m$. The number r is called the *probable error.* When we have found σ_m as in Examples 1 and 2, we just have to multiply it by 0.6745 to find the corresponding probable error. Similarly we can find the error corresponding to other choices of confidence interval (see Problem 4).

PROBLEMS, SECTION 10

1. Let $m_1, m_2, \cdots, m_n$ be a set of measurements, and define the values of x_i by $x_1 = m_1 - a, x_2 = m_2 - a, \cdots, x_n = m_n - a$, where a is some number (as yet unspecified, but the same for all x_i). Show that in order to minimize $\sum_{i=1}^n x_i^2$, we should choose $a = (1/n)\sum_{i=1}^n m_i$. *Hint:* Differentiate $\sum_{i=1}^n x_i^2$ with respect to a. You have shown that the arithmetic mean is the "best" average in the least squares sense, that is, that if the sum of the squares of the deviations of the measurements from their

"average" is a minimum, the "average" is the arithmetic mean (rather than, say, the median or mode).

2. Let $x_1, x_2, \cdots, x_n$ be independent random variables, each with density function $f(x)$, expected value μ, and variance σ^2. Define the sample mean by $\overline{x} = \sum_{i=1}^{n} x_i$. Show that $E(\overline{x}) = \mu$, and $\mathrm{Var}(\overline{x}) = \sigma^2/n$. (See Problems 5.9, 5.13, and 6.15.)

3. Define s by the equation $s^2 = (1/n)\sum_{i=1}^{n}(x_i - \overline{x})^2$. Show that the expected value of s^2 is $[(n-1)/n]\sigma^2$. *Hints:* Write

$$\begin{aligned}(x_i - \overline{x})^2 &= [(x_i - \mu) - (\overline{x} - \mu)]^2 \\ &= (x_i - \mu)^2 - 2(x_i - \mu)(\overline{x} - \mu) + (\overline{x} - \mu)^2.\end{aligned}$$

Find the average value of the first term from the definition of σ^2 and the average value of the third term from Problem 2. To find the average value of the middle term write

$$(\overline{x} - \mu) = \left(\frac{x_1 + x_2 + \cdots + x_n}{n} - \mu\right) = \frac{1}{n}[(x_1 - \mu) + (x_2 - \mu) + \cdots + (x_n - \mu)].$$

Show by Problem 6.14 that

$$E[(x_i - \mu)(x_j - \mu)] = E(x_i - \mu)E(x_j - \mu) = 0 \qquad \text{for } i \neq j,$$

and evaluate $E[(x_i - \mu)^2]$ (same as the first term). Collect terms to find

$$E(s^2) = \frac{n-1}{n}\sigma^2.$$

4. Assuming a normal distribution, find the limits $\mu \pm h$ for a 90% confidence interval; for a 95% confidence interval; for a 99% confidence interval. What percent confidence interval is $\mu \pm 1.3\sigma$? *Hints:* See Section 8, Example 3, and Problems 8.7, 8.22, and 8.23.

5. Show that if $w = xy$ or $w = x/y$, then (10.14) gives the convenient formula for relative error

$$\frac{r_w}{w} = \sqrt{\left(\frac{r_x}{x}\right)^2 + \left(\frac{r_y}{y}\right)^2}.$$

6. By expanding $w(x, y, z)$ in a three-variable power series similar to (10.10), show that

$$r_w = \sqrt{\left(\frac{\partial w}{\partial x}\right)^2 r_x^2 + \left(\frac{\partial w}{\partial y}\right)^2 r_y^2 + \left(\frac{\partial w}{\partial z}\right)^2 r_z^2}.$$

7. Equation (10.12) is only an approximation (but usually satisfactory). Show, however, that if you keep the second order terms in (10.10), then

$$\overline{w} = w(\overline{x}, \overline{y}) + \frac{1}{2}\left(\frac{\partial^2 w}{\partial x^2}\right)\sigma_x^2 + \frac{1}{2}\left(\frac{\partial^2 w}{\partial y^2}\right)\sigma_y^2.$$

8. The following measurements of x and y have been made.

$$x : 5.1, 4.9, 5.0, 5.2, 4.9, 5.0, 4.8, 5.1$$

$$y : 1.03, 1.05, 0.96, 1.00, 1.02, 0.95, 0.99, 1.01, 1.00, 0.99$$

Find the mean value and the probable error of x, y, $x + y$, xy, $x^3 \sin y$, and $\ln x$. *Hint*: See Examples 1 and 2 and the last paragraph of this section.

9. Given the measurements

$$x: 98, 101, 102, 100, 99$$
$$y: 21.2, 20.8, 18.1, 20.3, 19.6, 20.4, 19.5, 20.1$$

find the mean value and probable error of $x - y$, x/y, x^2y^3, and $y \ln x$.

10. Given the measurements

$$x: 5.8, 6.1, 6.4, 5.9, 5.7, 6.2, 5.9$$
$$y: 2.7, 3.0, 2.9, 3.3, 3.1$$

find the mean value and probable error of $2x - y$, $y^2 - x$, e^y, and x/y^2.

11. MISCELLANEOUS PROBLEMS

1. (a) Suppose you have two quarters and a dime in your left pocket and two dimes and three quarters in your right pocket. You select a pocket at random and from it a coin at random. What is the probability that it is a dime?

(b) Let x be the amount of money you select. Find $E(x)$.

(c) Suppose you selected a dime in (a). What is the probability that it came from your right pocket?

(d) Suppose you do not replace the dime, but select another coin which is also a dime. What is the probability that this second coin came from your right pocket?

2. (a) Suppose that Martian dice are regular tetrahedra with vertices labeled 1 to 4. Two such dice are tossed and the sum of the numbers showing is even. Let x be this sum. Set up the sample space for x and the associated probabilities.

(b) Find $E(x)$ and σ_x.

(c) Find the probability of exactly fifteen 2's in 48 tosses of a Martian die using the binomial distribution.

(d) Approximate (c) using the normal distribution.

(e) Approximate (c) using the Poisson distribution.

3. There are 3 red and 2 white balls in one box and 4 red and 5 white in the second box. You select a box at random and from it pick a ball at random. If the ball is red, what is the probability that it came from the second box?

4. If 4 letters are put at random into 4 envelopes, what is the probability that at least one letter gets into the correct envelope?

5. Two decks of cards are "matched," that is, the order of the cards in the decks is compared by turning the cards over one by one from the two decks simultaneously; a "match" means that the two cards are identical. Show that the probability of at least one match is nearly $1 - 1/e$.

6. Find the number of ways of putting 2 particles in 5 boxes according to the different kinds of statistics.

7. Suppose a coin is tossed three times. Let x be a random variable whose value is 1 if the number of heads is divisible by 3, and 0 otherwise. Set up the sample space for x and the associated probabilities. Find $\overline{x}$ and σ.

8. (a) A weighted coin has probability $\frac{2}{3}$ of coming up heads and probability $\frac{1}{3}$ of coming up tails. The coin is tossed twice. Let $x =$ number of heads. Set up the sample space for x and the associated probabilities.

(b) Find $\bar{x}$ and σ.

(c) If in (a) you know that there was at least one tail, what is the probability that both were tails?

9. (a) One box contains one die and another box contains two dice. You select a box at random and take out and toss whatever is in it (that is, toss *both* dice if you have picked box 2). Let $x =$ number of 3's showing. Set up the sample space and associated probabilities for x.

(b) What is the probability of at least one 3?

(c) If at least one 3 turns up, what is the probability that you picked the first box?

(d) Find $\bar{x}$ and σ.

Do Problems 10 to 12 using both the binomial distribution and the normal approximation.

10. A true coin is tossed 10^4 times.

(a) Find the probability of getting exactly 5000 heads.

(b) Find the probability of between 4900 and 5075 heads.

11. A die is thrown 720 times.

(a) Find the probability that 3 comes up exactly 125 times.

(b) Find the probability that 3 comes up between 115 and 130 times.

12. Consider a biased coin with probability 1/3 of heads and 2/3 of tails and suppose it is tossed 450 times.

(a) Find the probability of getting exactly 320 tails.

(b) Find the probability of getting between 300 and 320 tails.

13. A radioactive source emits 1800 α particles during an observation lasting 10 hours. In how many one minute intervals do you expect no α's? 5α's?

14. Suppose a 200-page book has, on the average, one misprint every 10 pages. On about how many pages would you expect to find 2 misprints?

In Problems 15 and 16, find the binomial probability for the given problem, and then compare the normal and the Poisson approximations.

15. Out of 1095 people, what is the probability that exactly 2 were born on Jan. 1? Assume 365 days in a year.

16. Find the probability of x successes in 100 Bernoulli trials with probability $p = 1/5$ of success (a) if $x = 25$; (b) if $x = 21$.

17. Given the measurements

$$x : 2.3, 2.1, 1.8, 1.7, 2.1$$

$$y : 1.0, 1.1, 0.9$$

find the mean value and the probable error for $x - y$, xy, and x/y^3.

18. Given the measurements

$$x : 5.7, 4.5, 4.8, 5.1, 4.9$$

$$y : 61.5, 60.1, 59.7, 60.3, 58.4$$

find the mean value and the probable error for $x + y$, y/x, and x^2.

References

This list includes the details of references cited in the text, plus a few other books you might find useful.

Abramowitz, Milton, and Irene A. Stegun, editors, *Handbook of Mathematical Functions With Formulas, Graphs, and Mathematical Tables*, National Bureau of Standards, Applied Mathematics Series, 55, U. S. Government Printing Office, Washington, D. C., 1964.

Arfken, George B., and Hans J. Weber, *Mathematical Methods for Physicists*, Academic Press, fifth edition, 2001.

Boyce, William E., and Richard C. DiPrima, *Introduction to Differential Equations*, Wiley, 1970.

Butkov, Eugene, *Mathematical Physics*, Addison-Wesley, 1968.

Callen, Herbert B., *Thermodynamics and an Introduction to Thermostatistics*, Wiley, second edition, 1985.

Cantrell, C. D., *Modern Mathematical Methods for Physicists and Engineers*, Cambridge University Press, 2000.

Chow, Tai L., *Mathematical Methods for Physicists: A Concise Introduction*, Cambridge University Press, 2000.

Courant, Richard, and Herbert Robbins, *What Is Mathematics?*, Oxford University Press, second edition revised by Ian Stewart, 1996.

CRC Standard Mathematical Tables, CRC Press, any recent edition.

Feller, William, *An Introduction to Probability Theory and Its Applications*, Wiley, second edition, 1966.

Folland, G. B., *Fourier analysis and its applications*, Brooks/Cole, 1992.

Goldstein, Herbert, Charles P. Poole, and John L. Safko, *Classical Mechanics*, Addison Wesley, third edition, 2002.

Griffiths, David J., *Introduction to Electrodynamics*, Prentice Hall, third edition, 1999.

Griffiths, David J., *Introduction to Quantum Mechanics*, Prentice Hall, second edition, 2004.

Hassani, Sadri, *Mathematical Methods: For Students of Physics and Related Fields*, Springer, 2000.

Jackson, John David, *Classical Electrodynamics*, Wiley, third edition, 1999.

Jahnke, E., and F. Emde, *Tables of Higher Functions*, McGraw-Hill, sixth edition revised by Friedrich Lösch, 1960.

Jeffreys, Harold, *Cartesian Tensors*, Cambridge University Press, 1965 reprint.

Jordan, D. W., and Peter Smith, *Mathematical Techniques: An Introduction for the Engineering, Physical, and Mathematical Sciences*, Oxford University Press, third edition, 2002.

Kittel, Charles, *Elementary Statistical Physics*, Dover edition, 2004.

Kreyszig, Erwin, *Advanced Engineering Mathematics*, Wiley, eighth edition, 1999.

Lighthill, M. J., *Introduction to Fourier Analysis and Generalised Functions*, Cambridge University Press, 1958.

Lyons, Louis, *All You Wanted To Know About Mathematics but Were Afraid To Ask: Mathematics for Science Students*, two volumes, Cambridge University Press, 1995–1998.

Mathews, Jon, and R. L. Walker, *Mathematical Methods of Physics*, Benjamin, second edition, 1970.

McQuarrie, Donald A., *Mathematical Methods for Scientists and Engineers*, University Science Books, 2003.

Morse, Philip M., and Herman Feshbach, *Methods of Theoretical Physics*, McGraw-Hill, 1953.

NBS Tables. See Abramowitz and Stegun.

Parratt, Lyman G., *Probability and Experimental Errors in Science*, Dover edition, 1971.

Relton, F. E., *Applied Bessel Functions*, Dover edition, 1965.

Riley, K. F., M. P. Hobson, and S. J. Bence, *Mathematical Methods for Physics and Engineering: A Comprehensive Guide*, Cambridge University Press, second edition, 2002.

Schey, H. M., *Div, Grad, Curl, and All That: An Informal Text on Vector Calculus*, Norton, fourth edition, 2004.

Snieder, Roel, *A Guided Tour of Mathematical Methods for the Physical Sciences*, Cambridge University Press, second edition, 2004.

Strang, Gilbert, *Linear Algebra and Its Applications*, Harcourt, Brace, Jovanovich, third edition, 1988.

Weinstock, Robert, *Calculus of Variations, with Applications to Physics and Engineering*, Dover edition, 1974.

Weisstein, Eric W., *CRC Concise Encyclopedia of Mathematics*, Chapman & Hall /CRC, second edition, 2003.

Woan, Graham, *Cambridge Handbook of Physics Formulas*, Cambridge University Press, reprinted 2003 with corrections.

Young, Hugh D., *Statistical Treatment of Experimental Data*, McGraw-Hill, 1962.

Answers to Selected Problems

Chapter 1

1.1 0.0173 yd; 0.104 yd (compared to a total of 5 yd)

1.3	$\frac{5}{9}$	1.5	$\frac{7}{12}$	1.9	$\frac{6}{7}$	1.11	$\frac{19}{28}$	1.15	1
2.1	1	2.4	∞	2.7	e^2	2.9	1		

4.2 $a_n = \frac{1}{5^{n-1}} \to 0$; $S_n = \frac{5}{4}\left(1 - \frac{1}{5^n}\right) \to \frac{5}{4}$; $R_n = \frac{1}{4 \cdot 5^{n-1}} \to 0$

4.4 $a_n = \frac{1}{3^n} \to 0$; $S_n = \frac{1}{2}\left(1 - \frac{1}{3^n}\right) \to \frac{1}{2}$; $R_n = \frac{1}{2 \cdot 3^n} \to 0$

4.6 $a_n = \frac{1}{n(n+1)} \to 0$; $S_n = 1 - \frac{1}{n+1} \to 1$; $R_n = \frac{1}{n+1} \to 0$

5.2	Test further	5.4	D	5.5	D
5.6	Test further	5.8	Test further	5.9	D

6.5 b	D	6.7	D	6.9	C	6.10	C	6.14	D
6.18	D	6.20	C	6.22	C	6.23	D	6.24	D
6.26	C	6.29	D	6.31	D	6.32	D	6.35	C
6.36	D								
7.1	C	7.2	D	7.4	C	7.6	D	7.8	C
9.2	D	9.3	C	9.7	D	9.8	C	9.9	D
9.10	D	9.12	C	9.13	C	9.15	D	9.16	C
9.20	C	9.21	C	9.22	(b) D				

10.1	$\lvert x\rvert < 1$	10.3	$\lvert x\rvert \le 1$	10.4	$\lvert x\rvert \le \sqrt{2}$
10.5	All x	10.9	$\lvert x\rvert < 1$	10.10	$\lvert x\rvert \le 1$
10.11	$-5 \le x < 5$	10.13	$-1 < x \le 1$	10.15	$-1 < x < 5$
10.17	$-2 < x \le 0$	10.18	$-\frac{3}{4} \le x \le -\frac{1}{4}$	10.20	All x
10.21	$0 \le x \le 1$	10.22	No x	10.24	$\lvert x\rvert < \frac{1}{2}\sqrt{5}$

10.25 $n\pi - \frac{\pi}{6} < x < n\pi + \frac{\pi}{6}$

13.4 $\binom{-1/2}{0} = 1$; $\binom{-1/2}{n} = \dfrac{(-1)^n(2n-1)!!}{(2n)!!}$

13.6 $\sum_0^\infty \binom{1/2}{n} x^{n+1}$ (see Example 2)

13.8 $\sum_0^\infty \binom{-1/2}{n}(-x^2)^n$ (see Problem 13.4)

13.11 $\displaystyle\sum_0^\infty \frac{(-1)^n x^n}{(2n+1)!}$ 13.14 $\displaystyle\sum_0^\infty \frac{x^{2n+1}}{2n+1}$

13.15 $\displaystyle\sum_0^\infty \binom{-1/2}{n}(-1)^n \frac{x^{2n+1}}{2n+1}$ 13.17 $\displaystyle 2\sum_{\text{odd } n} \frac{x^n}{n}$

13.21 $x^2 + 2x^4/3 + 17x^6/45 \cdots$

13.22 $1 + 2x + 5x^2/2 + 8x^3/3 + 65x^4/24 \cdots$

13.25 $1 - x + x^2/3 - x^4/45 \cdots$

13.27 $1 + x + x^2/2 - x^4/8 - x^5/15 \cdots$

13.28 $x - x^2/2 + x^3/6 - x^5/12 \cdots$

13.29 $1 + x/2 - 3x^2/8 + 17x^3/48 \cdots$

13.34 $x - x^2 + x^3 - 13x^4/12 + 5x^5/4 \cdots$

13.35 $1 + x^2/3! + 7x^4/(3 \cdot 5!) + 31x^6/(3 \cdot 7!) \cdots$

13.41 $e^3[1 + (x-3) + (x-3)^2/2! + (x-3)^3/3! \cdots]$

13.44 $5 + (x-25)/10 - (x-25)^2/10^3 + (x-25)^3/(5 \times 10^4) \cdots$

14.8 For $x < 0$, error < 0.001; for $x > 0$, error < 0.002.

15.1 $-x^4/24 - x^5/30 \cdots \cong -3.376 \times 10^{-16}$

15.3 $x^5/15 - 2x^7/45 \cdots \cong 6.667 \times 10^{-17}$

15.6 12 15.8 1/2 15.10 -1 15.12 1/3

15.14 $t - \frac{t^3}{3}$, error $< 10^{-6}$ 15.17 $\cos(\pi/2) = 0$

15.19 $\sqrt{2}$ 15.20 (b) $5e$ 15.21 (b) 0.937548

15.22 (b) 1.202057 15.23 (a) 1/2 (c) 1/3

15.24 (a) $-\pi$ (d) 0 (f) 0

15.27 (a) $1 - \frac{v}{c} = 1.3 \times 10^{-5}$, or $v = 0.999987c$

(d) $1 - \frac{v}{c} = 1.3 \times 10^{-11}$

15.28 $mc^2 + \frac{1}{2}mv^2$

15.29 (b) $\dfrac{F}{W} = \dfrac{x}{l} + \dfrac{x^3}{2l^3} + \dfrac{3x^5}{8l^5} \cdots$

15.30 (b) $T = \dfrac{1}{2}\dfrac{F}{\theta}\left(1 + \dfrac{\theta^2}{6} + 7\dfrac{\theta^4}{360} \cdots\right)$

15.31 (a) finite (b) infinite

16.6 C 16.7 D 16.9 $-1 \le x < 1$

16.10 $-4 < x < 4$ 16.13 $-5 < x \le 1$

16.15 $-x^2/6 - x^4/180 - x^6/2835 \cdots$

16.16 $1 - x/2 + 3x^2/8 - 11x^3/48 + 19x^4/128 \cdots$

16.19 $-(x-\pi) + (x-\pi)^3/3! - (x-\pi)^5/5! \cdots$

16.20 $2 + \dfrac{x-8}{12} - \dfrac{(x-8)^2}{2^5 \cdot 3^2} + \dfrac{5(x-8)^3}{2^8 \cdot 3^4} \cdots$

16.26 $-1/3$ 16.28 1

16.31 (b) 2.66×10^{86} terms. For $N = 15$, $1.6905 < S < 1.6952$

Chapter 2

	x	y	r	θ	
4.1	1	1	$\sqrt{2}$	$\pi/4$	See Fig. 5.1
4.2	-1	1	$\sqrt{2}$	$3\pi/4$	See Fig. 9.6
4.3	1	$-\sqrt{3}$	2	$-\pi/3$	
4.5	0	2	2	$\pi/2$	See Fig. 5.2
4.7	-1	0	1	π	See Fig. 9.2
4.9	-2	2	$2\sqrt{2}$	$3\pi/4$	
4.11	$\sqrt{3}$	1	2	$\pi/6$	See Fig. 9.1
4.14	$\sqrt{2}$	$\sqrt{2}$	2	$\pi/4$	
4.15	-1	0	1	$-\pi$ or π	See Fig. 9.2
4.17	1	-1	$\sqrt{2}$	$-\pi/4$	See Fig. 9.5
4.20	-2.39	-6.58	7	-110° $= -1.92$ radians	
5.2	$-1/2$	$-1/2$	$1/\sqrt{2}$	$-3\pi/4$ or $5\pi/4$	
5.4	0	2	2	$\pi/2$	
5.6	-1	0	1	π	
5.8	1.6	-2.7	3.14	-59.3°	
5.10	$-25/17$	$19/17$	$\sqrt{58/17}$	142.8°	
5.12	2.65	1.41	3	28°	
5.14	1.27	-2.5	2.8	-1.1 radians $= -63^\circ$	
5.16	1.53	-1.29	2	-40°	
5.17	-7.35	-10.9	13.1	-124°	
5.18	-0.94	-0.36	1	201° or -159°	

5.19 $(2+3i)/13$; $(x-yi)/(x^2+y^2)$
5.21 $(1+i)/6$; $(x+1-yi)/[(x+1)^2+y^2]$
5.23 $(-6-3i)/5$; $(1-x^2-y^2+2yi)/[(1-x)^2+y^2]$
5.26 1 5.30 $3/2$ 5.31 1
5.32 169 5.34 1 5.35 $x=-4,\ y=3$
5.36 $x=-1/2,\ y=3$ 5.39 $x=y=$ any real number
5.42 $x=-1/7,\ y=-10/7$ 5.43 $(x,y)=(0,0)$, or $(1,1)$, or $(-1,1)$
5.45 $x=0$, any real y; or $y=0$, any real x
5.46 $y=-x$ 5.48 $x=36/13,\ y=2/13$
5.49 $y=0,\ x=1/2$
5.53 Circle (Find center and radius)
5.55 Straight line (What is its equation?)
5.56 Part of a straight line (Describe it.)
5.57 Hyperbola (What is its equation?)
5.60 Circle (Find center and radius)
5.62 Ellipse (Find its equation; where are the foci?)
5.63 Two straight lines (What lines?)
5.68 $v=2,\ a=4$

6.2 D 6.3 C 6.4 D
6.5 D 6.10 C 6.12 C

7.1 All z 7.3 All z 7.6 $|z| < 1/3$
7.7 All z 7.10 $|z| < 1$ 7.12 $|z| < 4$
7.14 $|z - 2i| < 1$ 7.16 $|z + (i - 3)| < 1/\sqrt{2}$

8.3 See Problem 17.30

9.3 $-9i$ 9.4 $-e(1 + i\sqrt{3})/2$ 9.6 1
9.7 $3e^2$ 9.8 $-\sqrt{3} + i$ 9.10 -2
9.11 $-1 - i$ 9.13 $-4 + 4i$ 9.14 64
9.17 $-(1 + i)/4$ 9.19 16 9.20 i
9.21 1 9.24 $4i$ 9.26 $(1 + i\sqrt{3})/2$
9.29 1 9.32 $3e^2$ 9.34 $4/e$
9.35 21 9.38 $1/\sqrt{2}$

10.3 $\pm 1, \ \pm i$ 10.4 $\pm 2, \ \pm 2i$
10.7 $\pm\sqrt{2}, \ \pm i\sqrt{2}, \ \pm 1 \pm i$
10.9 $1, \ 0.309 \pm 0.951i, \ -0.809 \pm 0.588i$
10.16 $\pm i, \ (\pm\sqrt{3} \pm i)/2$
10.17 $-1, \ 0.809 \pm 0.588i, \ -0.309 \pm 0.951i$
10.18 $\pm(1 + i)/\sqrt{2}$ 10.21 $\pm(\sqrt{3} + i)$
10.22 $r = \sqrt{2}$, $\theta = 45° + 120°n$: $1 + i, \ -1.366 + 0.366i, \ 0.366 - 1.366i$
10.24 $\pm(\sqrt{3} + i)/2, \ \pm(1 - i\sqrt{3})/2, \ \pm(0.259 + 0966i), \ \pm(0.966 - 0.259i)$
10.25 $0.758(1 + i), -0.487 + 0.955i, -1.059 - 0.168i, -0.168 - 1.059i,$
$0.955 - 0.487i$

11.3 $3(1 - i)/\sqrt{2}$ 11.5 $1 + i$ 11.8 $-41/9$ 11.9 $4i/3$

12.25 $\sin x \cosh y - i \cos x \sinh y, \ \sqrt{\sin^2 x + \sinh^2 y}$
12.26 $\cosh 2 \cos 3 - i \sinh 2 \sin 3 = -3.72 - 0.51i, \ 3.76$
12.28 $\tanh 1 = 0.762$
12.30 $-i$ 12.32 $-4i/3$
12.33 $i \tanh 1 = 0.762i$ 12.35 $-\cosh 2 = -3.76$

14.2 $-i\pi/2$ or $3i\pi/2$ 14.3 $\text{Ln}\, 2 + i\pi/6$
14.5 $\text{Ln}\, 2 + 5i\pi/4$ 14.6 $-i\pi/4$ or $7i\pi/4$
14.8 $-1, \ (1 \pm i\sqrt{3})/2$ 14.10 $e^{-\pi^2/4}$
14.11 $\cos(\text{Ln}\, 2) + i \sin(\text{Ln}\, 2) = 0.769 + 0.639i$
14.14 $0.3198i - 0.2657$ 14.15 $e^{-\pi \sinh 1} = 0.0249$
14.18 -1 14.20 1
14.23 $e^{\pi/2} = 4.81$

15.2 $\pi/2 + n\pi + (i\, \text{Ln}\, 3)/2$ 15.3 $i(\pm\pi/3 + 2n\pi)$
15.4 $i(2n\pi + \pi/6), \ i(2n\pi + 5\pi/6)$ 15.5 $\pm[\pi/2 + 2n\pi - i\, \text{Ln}(3 + \sqrt{8})]$
15.8 $\pi/2 + 2n\pi \pm i\, \text{Ln}\, 3$ 15.9 $i(\pi/3 + n\pi)$
15.12 $i(2n\pi \pm \pi/6)$ 15.14 $2n\pi + i\, \text{Ln}\, 2, \ (2n + 1)\pi - i\, \text{Ln}\, 2$
15.15 $n\pi + 3\pi/8 + (i/4)\, \text{Ln}\, 2$

16.3 $|z| = \sqrt{2}$; motion around a circle of radius $\sqrt{2}$, at constant speed $v = \sqrt{2}$, constant acceleration $a = \sqrt{2}$.

16.5 $v = |z_1 - z_2|$; $a = 0$
16.6 (a) Series: $3 - 2i$; parallel: $5 + i$
(b) Series: $2(1 + i\sqrt{3})$; parallel: $i\sqrt{3}$
16.8 $[R - i(\omega CR^2 + \omega^3 L^2 C - \omega L)]/[(\omega CR)^2 + (\omega^2 LC - 1)^2]$; this simplifies to $L/(RC)$ at resonance.
16.9 (b) $\omega = 1/\sqrt{LC}$
16.12 $(1 + r^4 - 2r^2 \cos\theta)^{-1}$

17.2 $(\sqrt{3} + i)/2$
17.4 $i \cosh 1 = 1.54i$
17.6 $-e^{-\pi^2} = -5.17 \times 10^{-5}$
17.7 $e^{\pi/2} = 4.81$
17.9 $\pi/2 \pm 2n\pi$
17.11 i
17.13 $x = 0,\ y = 4$
17.15 $|z| < 1/e$
17.26 1
17.27 (c) $e^{-2(x-t)^2}$
17.28 $1 + (a^2 + b^2)^2 (2ab)^{-2} \sinh^2 b$
17.30 $e^x \cos x = \sum_{n=0}^{\infty} \left(2^{n/2} x^n / n!\right) \cos n\pi/4$
$e^x \sin x = \sum_{n=0}^{\infty} \left(2^{n/2} x^n / n!\right) \sin n\pi/4$

Chapter 3

2.4 $\begin{pmatrix} 1 & 0 & -\frac{1}{2} & \frac{1}{2} \\ 0 & 1 & 0 & 1 \end{pmatrix}$, $x = \frac{1}{2}(z+1)$, $y = 1$

2.8 $\begin{pmatrix} 1 & -1 & 0 & -11 \\ 0 & 0 & 1 & 7 \\ 0 & 0 & 0 & 0 \end{pmatrix}$, $x = y - 11$, $z = 7$

2.9 $\begin{pmatrix} 1 & 0 & 1 & 0 \\ 0 & 1 & -1 & 0 \\ 0 & 0 & 0 & 1 \end{pmatrix}$, inconsistent, no solution

2.12 $\begin{pmatrix} 1 & 0 & 0 & -2 \\ 0 & 1 & 0 & 1 \\ 0 & 0 & 1 & 1 \end{pmatrix}$, $x = -2$, $y = 1$, $z = 1$

2.17 $R = 2$

3.1 -11
3.5 -544
3.12 16
3.16 $A = -(K + ik)/(K - ik)$, $|A| = 1$

4.12 $\arccos(-1/\sqrt{2}) = 3\pi/4$
4.14 (a) $\arccos(1/3) = 70.5°$
4.14 (c) $\arccos\sqrt{2/3} = 35.3°$
4.15 (b) $8\mathbf{i} - 4\mathbf{j} + 8\mathbf{k}$
4.18 $2\mathbf{i} - 8\mathbf{j} - 3\mathbf{k}$
4.19 $\mathbf{i} + \mathbf{j} + \mathbf{k}$
4.22 Law of cosines
4.24 $A^2 B^2$

5.1 $\mathbf{r} = (2\mathbf{i} - 3\mathbf{j}) + (4\mathbf{i} + 3\mathbf{j})t$ [Note that $2\mathbf{i} - 3\mathbf{j}$ may be replaced by *any* point on the line; $4\mathbf{i} + 3\mathbf{j}$ may be replaced by *any* vector along the line. Thus, for example, $\mathbf{r} = 6\mathbf{i} - (8\mathbf{i} + 6\mathbf{j})t$ is just as good an answer, and similarly for all such problems.]
5.4 $\mathbf{r} = \mathbf{i} + (2\mathbf{i} + \mathbf{j})t$
5.6 $(x-1)/1 = (y+1)/(-2) = (z+5)/2$, or $\mathbf{r} = \mathbf{i} - \mathbf{j} - 5\mathbf{k} + (\mathbf{i} - 2\mathbf{j} + 2\mathbf{k})t$
5.8 $x/3 = (z-4)/(-5)$, $y = -2$; or $\mathbf{r} = -2\mathbf{j} + 4\mathbf{k} + (3\mathbf{i} - 5\mathbf{k})t$
5.9 $x = -1$, $z = 7$; or $\mathbf{r} = -\mathbf{i} + 7\mathbf{k} + \mathbf{j}t$

5.11 $(x-4)/1 = (z-3)/(-2)$, $y = -1$; or $\mathbf{r} = 4\mathbf{i} - \mathbf{j} + 3\mathbf{k} + (\mathbf{i} - 2\mathbf{k})t$

5.12 $(x-5)/5 = (y+4)/(-2) = (z-2)/1$; or $\mathbf{r} = 5\mathbf{i} - 4\mathbf{j} + 2\mathbf{k} + (5\mathbf{i} - 2\mathbf{j} + \mathbf{k})t$

5.14 $36x - 3y - 22z = 23$ 5.16 $5x - 2y + z = 35$

5.18 $x + 6y + 7z + 5 = 0$ 5.20 $x - 4y - z + 5 = 0$

5.21 $\cos\theta = 25/(7\sqrt{30}) = 0.652$, $\theta = 49.3°$

5.22 $\cos\theta = 2/\sqrt{6}$, $\theta = 35.3°$

5.24 $\mathbf{r} = 2\mathbf{i} + \mathbf{j} + (\mathbf{j} + 2\mathbf{k})t$, $d = 2\sqrt{6/5}$

5.25 $\mathbf{r} = \mathbf{i} - 2\mathbf{j} + (4\mathbf{i} + 9\mathbf{j} - \mathbf{k})t$, $d = (3\sqrt{3})/7$

5.29 $2/\sqrt{6}$ 5.31 $5/7$ 5.33 $\sqrt{43/15}$

5.34 $\sqrt{11/10}$ 5.36 3 5.38 $\arccos\sqrt{21/22} = 12.3°$

5.39 Intersect at $(3, 2, 0)$; $\cos\theta = 5/\sqrt{60}$, $\theta = 49.8°$

5.42 $1/\sqrt{5}$ 5.43 $20/\sqrt{21}$ 5.45 $d = \sqrt{2}$, $t = -1$

6.2 $\text{AB} = \begin{pmatrix} -2 & -2 \\ 1 & 2 \end{pmatrix}$, $\text{BA} = \begin{pmatrix} -6 & 17 \\ -2 & 6 \end{pmatrix}$, $\text{A} + \text{B} = \begin{pmatrix} 1 & -1 \\ -1 & 5 \end{pmatrix}$, $\text{A} - \text{B} = \begin{pmatrix} 3 & -9 \\ -1 & 1 \end{pmatrix}$, $\text{A}^2 = \begin{pmatrix} 9 & -25 \\ -5 & 14 \end{pmatrix}$, $\text{B}^2 = \begin{pmatrix} 1 & 4 \\ 0 & 4 \end{pmatrix}$, $5\text{A} = \begin{pmatrix} 10 & -25 \\ -5 & 15 \end{pmatrix}$, $3\text{B} = \begin{pmatrix} -3 & 12 \\ 0 & 6 \end{pmatrix}$, $\det(5\text{A}) = 5^2 \det \text{A}$ for a 2×2 matrix

6.4 You should have found BA, C^2, CB, C^3, C^2B, and CBA; all others are meaningless. $\text{C}^2\text{B} = \begin{pmatrix} 32 & 12 \\ 53 & 7 \\ -13 & -9 \end{pmatrix}$, $\text{CBA} = \begin{pmatrix} 36 & 46 & 14 & -36 \\ 40 & 22 & 1 & 91 \\ -8 & -2 & 1 & -29 \end{pmatrix}$

6.13 $\begin{pmatrix} 5/3 & -3 \\ -1 & 2 \end{pmatrix}$ 6.15 $-\dfrac{1}{2}\begin{pmatrix} 4 & 5 & 8 \\ -2 & -2 & -2 \\ 2 & 3 & 4 \end{pmatrix}$

6.19 $\text{A}^{-1} = \dfrac{1}{7}\begin{pmatrix} 1 & 2 \\ -3 & 1 \end{pmatrix}$, $(x, y) = (5, 0)$

6.22 $\text{A}^{-1} = \dfrac{1}{12}\begin{pmatrix} 4 & 4 & 0 \\ -7 & -1 & 3 \\ 1 & -5 & 3 \end{pmatrix}$, $(x, y, z) = (1, -1, 2)$

6.30 $\sin k\text{A} = \text{A} \sin k = \begin{pmatrix} 0 & \sin k \\ \sin k & 0 \end{pmatrix}$, $\cos k\text{A} = \text{I} \cos k = \begin{pmatrix} \cos k & 0 \\ 0 & \cos k \end{pmatrix}$, $e^{k\text{A}} = \begin{pmatrix} \cosh k & \sinh k \\ \sinh k & \cosh k \end{pmatrix}$, $e^{ik\text{A}} = \begin{pmatrix} \cos k & i \sin k \\ i \sin k & \cos k \end{pmatrix}$

7.1 Not linear 7.4 Linear 7.6 Not linear

7.8 Not linear 7.11 Not linear 7.12 Linear

7.14 Not linear 7.15 Linear

7.22 $D = 1$, rotation $\theta = -45°$ 7.24 $D = -1$, reflection line $x + y = 0$

7.26 $D = -1$, reflection line $x = 2y$

7.30 $\text{R} = \begin{pmatrix} 0 & -1 & 0 \\ 1 & 0 & 0 \\ 0 & 0 & 1 \end{pmatrix}$, $\text{S} = \begin{pmatrix} 1 & 0 & 0 \\ 0 & 0 & -1 \\ 0 & 1 & 0 \end{pmatrix}$,
R is a 90° rotation about the z axis, S is a 90° rotation about the x axis.

7.32 180° rotation about $\mathbf{i} - \mathbf{k}$

7.35 Reflection through the (x, y) plane and 90° rotation about the z axis.

8.1 In terms of basis $\mathbf{u} = \frac{1}{9}(9,0,7)$, $\mathbf{v} = \frac{1}{9}(0,-9,13)$, the vectors are: $\mathbf{u}-4\mathbf{v}$, $5\mathbf{u}-2\mathbf{v}$, $2\mathbf{u}+\mathbf{v}$, $3\mathbf{u}+6\mathbf{v}$.

8.3 Basis $\mathbf{i}, \mathbf{j}, \mathbf{k}$

8.6 $\mathbf{V} = 3\mathbf{A} - \mathbf{B}$

8.19 $x = y = z = w = 0$

8.20 $x = -z,\ y = z$

8.23 For $\lambda = 3$, $x = 2y$; for $\lambda = 8$, $x = -2y$

8.25 For $\lambda = 2$: $x = 0$, $y = -3z$; for $\lambda = -3$: $x = -5y$, $z = 3y$; for $\lambda = 4$: $z = 3y$, $x = 2y$

8.26 $\mathbf{r} = (3,1,0) + (-1,1,1)z$

9.4 $$\mathrm{A}^\dagger = \begin{pmatrix} 0 & i & 3 \\ -2i & 2 & 0 \\ -1 & 0 & 0 \end{pmatrix}, \quad \mathrm{A}^{-1} = \frac{1}{6}\begin{pmatrix} 0 & 0 & 2 \\ 0 & 3 & i \\ -6 & 6i & -2 \end{pmatrix}$$

9.14 $\mathrm{C^TBA^T}$, $\mathrm{C^{-1}M^{-1}C}$, H

10.1 (b) $d = 8$

10.2 The number of basis vectors given is the dimension of the space. We list one possible basis; other bases consist of the same number of independent linear combinations of the vectors given.
(b) $(1,0,0,5,0,1)$, $(0,1,0,0,6,4)$, $(0,0,1,0,-3,0)$

10.3 (a) Label the vectors $\mathbf{A}, \mathbf{B}, \mathbf{C}, \mathbf{D}$. Then $\cos(\mathbf{A},\mathbf{B}) = 1/\sqrt{15}$, $\cos(\mathbf{A},\mathbf{C}) = \sqrt{2}/3$, $\cos(\mathbf{B},\mathbf{D}) = \sqrt{17/690}$.

10.4 (b) $\mathbf{e}_1 = (0,0,0,1)$, $\mathbf{e}_2 = (1,0,0,0)$, $\mathbf{e}_3 = (0,1,1,0)/\sqrt{2}$

10.5 (b) $\|\mathbf{A}\| = 7$, $\|\mathbf{B}\| = \sqrt{60}$, $|\text{Inner product of } \mathbf{A} \text{ and } \mathbf{B}| = \sqrt{5}$

11.5 $\theta = 1.1 = 63.4°$

11.11 $$\begin{pmatrix} x \\ y \end{pmatrix} = \frac{1}{5}\begin{pmatrix} 1 & 3 \\ -1 & 2 \end{pmatrix}\begin{pmatrix} x' \\ y' \end{pmatrix},$$ not orthogonal

In the following answers, for each eigenvalue, the components of a corresponding eigenvector are listed in parentheses.

Problem	Eigenvalue	Eigenvector
11.12	4	$(1,1)$
	-1	$(3,-2)$
11.15	1	$(0,0,1)$
	-1	$(1,-1,0)$
	5	$(1,1,0)$
11.18	4	$(2,1,3)$
	2	$(0,-3,1)$
	-3	$(5,-1,-3)$
11.20	3	$(0,-1,2)$
	4	$(1,2,1)$
	-2	$(-5,2,1)$
11.22	-4	$(-4,1,1)$
	5	$(1,2,2)$
	-2	$(0,-1,1)$

Problem	Eigenvalue	Eigenvector	
11.23	18	$(2,2,-1)$	The two eigenvectors corresponding to the eigenvalue 9
	9	$(1,-1,0)$	may be any two vectors orthogonal to $(2,2,-1)$ and or-
	9	$(1,1,4)$	thogonal to each other.
11.26	4	$(1,1,1)$	
	1	$(1,-1,0)$	
	1	$(1,1,-2)$	

11.27 $\mathrm{D} = \begin{pmatrix} 3 & 0 \\ 0 & 1 \end{pmatrix}$, $\mathrm{C} = \frac{1}{\sqrt{2}}\begin{pmatrix} 1 & 1 \\ -1 & 1 \end{pmatrix}$

11.29 $\mathrm{D} = \begin{pmatrix} 11 & 0 \\ 0 & 1 \end{pmatrix}$, $\mathrm{C} = \frac{1}{\sqrt{5}}\begin{pmatrix} 1 & -2 \\ 2 & 1 \end{pmatrix}$

11.31 $\mathrm{D} = \begin{pmatrix} 5 & 0 \\ 0 & 1 \end{pmatrix}$, $\mathrm{C} = \frac{1}{\sqrt{2}}\begin{pmatrix} 1 & -1 \\ 1 & 1 \end{pmatrix}$

11.41 $\lambda = 1,\ 3$; $\mathrm{U} = \frac{1}{\sqrt{2}}\begin{pmatrix} 1 & i \\ i & 1 \end{pmatrix}$

11.44 $\lambda = 3,\ -7;\quad U = \dfrac{1}{5\sqrt{2}}\begin{pmatrix} 5 & -3-4i \\ 3-4i & 5 \end{pmatrix}$

11.52 60° rotation about $-\mathbf{i}\sqrt{2}+\mathbf{k}$ and reflection through the plane $z = x\sqrt{2}$

11.53 180° rotation about $\mathbf{i}+\mathbf{j}+\mathbf{k}$

11.56 45° rotation about $\mathbf{j}-\mathbf{k}$

11.58 $M^{10} = \dfrac{1}{5}\begin{pmatrix} 1+4\cdot 6^{10} & 2-2\cdot 6^{10} \\ 2-2\cdot 6^{10} & 4+6^{10} \end{pmatrix}$

11.59 $e^{M} = e^{3}\begin{pmatrix} \cosh 1 & -\sinh 1 \\ -\sinh 1 & \cosh 1 \end{pmatrix}$

12.2 $3x'^2 - 2y'^2 = 24$ 12.3 $10x'^2 = 35$

12.6 $3x'^2 + \sqrt{3}\,y'^2 - \sqrt{3}\,z'^2 = 12$

12.15 $y = 2x$ with $\omega = \sqrt{3k/m}$; $x = -2y$ with $\omega = \sqrt{8k/m}$

12.17 $x = -2y$ with $\omega = \sqrt{2k/m}$; $3x = 2y$ with $\omega = \sqrt{2k/(3m)}$

12.19 $y = -x$ with $\omega = \sqrt{3k/m}$; $y = 2x$ with $\omega = \sqrt{3k/(2m)}$

12.22 $y = -x$ with $\omega = \sqrt{2k/m}$; $y = 3x$ with $\omega = \sqrt{2k/(3m)}$

13.6 The cyclic group

13.11 The four matrices of the symmetry group of the rectangle are:

$$I = \begin{pmatrix} 1 & 0 \\ 0 & 1 \end{pmatrix},\ P = \begin{pmatrix} -1 & 0 \\ 0 & 1 \end{pmatrix},\ \begin{pmatrix} 1 & 0 \\ 0 & -1 \end{pmatrix} = -P,\ \begin{pmatrix} -1 & 0 \\ 0 & -1 \end{pmatrix} = -I$$

This group is isomorphic to the 4's group.

13.21 SO(2) is Abelian; SO(3) is not Abelian.

14.3 $x,\ \cos x,\ x\cos x,\ e^{x}\cos x$ 14.5 $1,\ x+x^3,\ x^2,\ x^4,\ x^5$

14.6 Not a vector space 14.8 $1,\ x^2,\ x^4,\ x^6$

15.3 (a) $(x-4)/1 = (y+1)/(-2) = (z-2)/(-2)$; or $\mathbf{r} = (4,-1,2) + (1,-2,-2)t$
(b) $x - 5y + 3z = 0$ (c) $5/7$
(d) $5\sqrt{2}/3 = 2.36$ (e) $\arcsin 19/21 = 64.8°$

15.5 (a) $y = 7,\ (x-2)/3 = (z+1)/4$; or $\mathbf{r} = (2,7,-1) + (3,0,4)t$
(b) $x - 4y - 9z = 0$ (c) $\arcsin(\frac{33}{70}\sqrt{2}) = 41.8°$
(d) $12/\sqrt{98} = 1.21$ (e) $\sqrt{29}/5 = 1.08$

15.7 You should have found all except A^TB^T, BA^T, ABC, AB^TC, $B^{-1}C$, and CB^T, which are meaningless.

$$B^TAC = \begin{pmatrix} 2 & 2 \\ 1-3i & 1 \\ -1-5i & -1 \end{pmatrix},\qquad C^{-1}A = \begin{pmatrix} 0 & -i \\ 1 & -1 \end{pmatrix}$$

15.9 $\dfrac{1}{f} = (n-1)\left[\dfrac{1}{R_1} - \dfrac{1}{R_2} + \dfrac{(n-1)d}{nR_1R_2}\right]$

15.13 Area $= \frac{1}{2}\left|\overrightarrow{PQ} \times \overrightarrow{PR}\right| = 7/2$

15.14 $x'' = -x,\ y'' = -y,$ 180° rotation

15.15 $x'' = -y,\ y'' = x$; 90° rotation of vectors or −90° rotation of axes

15.18 1 (1, 1)
−2 (0, 1)

15.20 1 (1, 1)
9 (1, −1)

15.22 1 (1, 0, 1)
4 (0, 1, 0)
5 (1, 0, −1)

15.24 2 (0, 4, 3)
7 (5, −3, 4)
−3 (5, 3, −4)

15.27 $3x'^2 - y'^2 - 5z'^2 = 15,\ d = \sqrt{5}$ 15.29 $3x'^2 + 6y'^2 - 4z'^2 = 54,\ d = 3$

Chapter 4

1.1 $\partial u/\partial x = 2xy^2/(x^2+y^2)^2$, $\partial u/\partial y = -2x^2y/(x^2+y^2)^2$
1.3 $\partial z/\partial u = u/(u^2+v^2+w^2)$
1.4 At $(0,0)$, both $= 0$; at $(-2/3,\ 2/3)$, both $= -4$

1.7 $2x$ 1.9 $2x(1+2\tan^2\theta)$ 1.11 $2y$
1.13 $4r^2\tan\theta$ 1.15 $r^2\sin 2\theta$ 1.17 $4r$
1.19 0 1.21 $-4x\csc^2\theta$ 1.23 $2r\sin 2\theta$
1.8′ $-2r^4/x^3$ 1.10′ $2y+4y^3/x^2$ 1.12′ $2y\sec^2\theta$
1.14′ $2y^2\sec^2\theta\tan\theta$ 1.16′ $2r\tan^2\theta$ 1.18′ $-2ry^4/(r^2-y^2)^2$
1.20′ $4x(\tan\theta\sec^2\theta)(\tan^2\theta+\sec^2\theta)$
1.22′ $-8r^3/x^3$ 1.24′ $-8y^3/x^3$

2.1 $y+y^3/6-x^2y/2+x^4y/24-x^2y^3/12+y^5/120\cdots$
2.3 $x-x^2/2-xy+x^3/3+x^2y/2+xy^2\cdots$
2.5 $1+xy/2-x^2y^2/8+x^3y^3/16-5x^4y^4/128\cdots$
2.8 $e^x\cos y = 1+x+(x^2-y^2)/2+(x^3-3xy^2)/6\cdots$

4.2 2.5×10^{-13} 4.4 12.2 4.6 9%
4.8 5% 4.10 4.28 nt 4.11 3.95
4.15 8×10^{23}

5.1 $e^{-y}\sinh t + z\sin t$ 5.3 $2r(q^2-p^2)$
5.7 $(1-2b-e^{2a})\cos(a-b)$

6.2 $y'=1$, $y''=0$ 6.3 $y'=4(\ln 2-1)/(2\ln 2-1)$
6.5 $2x+11y-24=0$ 6.6 $1800/11^3$
6.10 $x+y=0$ 6.11 $y''=4$

7.1 $dx/dy = z-y+\tan(y+z)$, $d^2x/dy^2 = \frac{1}{2}\sec^3(y+z)+\frac{1}{2}\sec(y+z)-2$
7.4 $\partial w/\partial u = -2(rv+s)w$, $\partial w/\partial v = -2(ru+2s)w$
7.7 $(\partial y/\partial\theta)_r = x$, $(\partial y/\partial\theta)_x = r^2/x$, $(\partial\theta/\partial y)_x = x/r^2$
7.8 $\partial x/\partial s = -19/13$, $\partial x/\partial t = -21/13$, $\partial y/\partial s = 24/13$, $\partial y/\partial t = 6/13$
7.10 $\partial x/\partial s = 1/6$, $\partial x/\partial t = 13/6$, $\partial y/\partial s = 7/6$, $\partial y/\partial t = -11/6$
7.13 $(\partial p/\partial q)_m = -p/q$, $(\partial p/\partial q)_a = 1/(a\cos p-1)$,
$(\partial p/\partial q)_b = 1-b\sin q$, $(\partial b/\partial a)_p = (\sin p)(b\sin q-1)/\cos q$,
$(\partial a/\partial q)_m = [q+p(a\cos p-1)]/(q\sin p)$
7.15 $(\partial x/\partial u)_v = (2yv^2-x^2)/(2yv+2xu)$, $(\partial x/\partial u)_y = (x^2u+y^2v)/(y^2-2xu^2)$
7.17 $(\partial p/\partial s)_t = -9/7$, $(\partial p/\partial s)_q = 3/2$
7.19 $(\partial x/\partial z)_s = 7/2$, $(\partial x/\partial z)_r = 4$, $(\partial x/\partial z)_y = 3$

8.3 $(-1,2)$ is a minimum point 8.4 $(-1,-2)$ is a saddle point
8.8 $\theta=\pi/3$; bend up 8 cm on each side
8.9 $l=w=2h$ 8.11 $\theta=30^\circ$, $x=y\sqrt{3}=z/2$
8.13 $(4/3,5/3)$ 8.16 $m=5/2$, $b=1/3$

9.2 $r:l:s=\sqrt{5}:(1+\sqrt{5}):3$ 9.4 $4/\sqrt{3}$ by $6/\sqrt{3}$ by $10/\sqrt{3}$
9.6 $V=1/3$ 9.8 $(8/13,12/13)$
9.12 Let legs of right triangle be a and b, height of prism $=h$; then $a=b$, $h=(2-\sqrt{2})a$.

10.2 4, 2
10.4 $d = 1$
10.6 $d = 2$
10.7 $\frac{1}{2}\sqrt{11}$
10.10 (a) $\max T = \frac{1}{2}$, $\min T = -\frac{1}{2}$
(b) $\max T = 1$, $\min T = -\frac{1}{2}$
(c) $\max T = 1$, $\min T = -\frac{1}{2}$
10.12 Largest sum $= 180°$
Smallest sum $= 3 \operatorname{arc\,cos}(1/\sqrt{3})$
$= 164.2°$
10.13 Largest sum $= 3 \operatorname{arc\,sin}(1/\sqrt{3}) = 105.8°$, smallest sum $= 90°$

11.1 $z = f(y + 2x) + g(y + 3x)$
11.6 $d^2y/dz^2 + dy/dz - 5y = 0$
11.11 $H = p\dot{q} - L$

12.1 $\frac{1}{2}x^{-1/2} \sin x$
12.3 $dz/dx = -\sin(\cos x)\tan x - \sin(\sin x)\cot x$
12.4 $\frac{1}{2}\sin 2$
12.7 $(\partial u/\partial x)_y = -e^4$, $(\partial u/\partial y)_x = e^4/\ln 2$, $(\partial y/\partial x)_u = \ln 2$
12.10 $dy/dx = (e^x - 1)/x$
12.12 $(2x + 1)/\ln(x + x^2) - 2/\ln(2x)$
12.14 $\pi/(4y^3)$

13.2 (a) and (b) $d = 4/\sqrt{13}$
13.4 $-\csc\theta\cot\theta$
13.5 $-6x$, $2x^2\tan\theta\sec^2\theta$, $4x\tan\theta\sec^2\theta$
13.9 $dz/dt = 1 + (t/z)(2 - x - y)$, $z \neq 0$
13.10 $[x\ln x - (y^2/x)]x^y$ where $x = r\cos\theta$, $y = r\sin\theta$
13.13 -1
13.14 $(\partial w/\partial x)_y = (\partial f/\partial x)_{s,t} + 2(\partial f/\partial s)_{x,t} + 2(\partial f/\partial t)_{x,s} = f_1 + 2f_2 + 2f_3$
13.18 $\sqrt{26/3}$
13.21 $T(2) = 4$, $T(5) = -5$
13.23 $t\cot t$
13.25 $-e^x/x$
13.29 $dt = 3.9$

Chapter 5

2.1	3	2.3	4	2.5	$\frac{1}{4}e^2 - \frac{5}{12}$	2.7	5/3	2.9	6
2.11	36	2.13	7/4	2.15	3/2	2.17	$\frac{1}{2}\ln 2$	2.19	32
2.21	131/6	2.23	9/8	2.25	3/2	2.27	32/5	2.29	2
2.31	6	2.33	16/3	2.36	1/6	2.37	7/6	2.39	70
2.41	5	2.43	9/2	2.45	$46k/15$	2.47	16/3	2.49	1/3

3.2 (b) $Ml^2/12$ (c) $Ml^2/3$
3.3 (a) $M = 140$ (b) $\bar{x} = 130/21$ (c) $I_m = 6.92M$
(d) $I = 150M/7$
3.5 (a) $Ma^2/3$ (b) $Ma^2/12$ (c) $2Ma^2/3$
3.7 (a) $M = 9$ (b) $(\bar{x}, \bar{y}) = (2, 4/3)$
(c) $I_x = 2M$, $I_y = 9M/2$ (d) $I_m = 13M/18$
3.9 (a) 1/6 (b) $(1/4, 1/4, 1/4)$ (c) $M = 1/24$, $\bar{z} = 2/5$
3.11 (a) $M = (5\sqrt{5} - 1)/6 = 1.7$
(b) $\bar{x} = 0$, $\bar{y} = (313 + 15\sqrt{5})/620 = 0.56$
3.14 $V = 2\pi^2 a^2 b$, $A = 4\pi^2 ab$, where a = radius of revolving circle, and b = distance to axis from center of this circle.
3.15 For area, $(\bar{x}, \bar{y}) = (0, \frac{4}{3}r/\pi)$, for arc, $(\bar{x}, \bar{y}) = (0, 2r/\pi)$

3.18 $s = [3\sqrt{2} + \ln(1+\sqrt{2})]/2$
3.20 $13\pi/3$
3.21 $s\bar{x} = [51\sqrt{2} - \ln(1+\sqrt{2})]/32$, $s\bar{y} = 13/6$, s as in Problem 3.18
3.23 $(149/130, 0, 0)$
3.25 I/M has the same numerical value as $\bar{x}$ in Problem 3.21
3.26 $2M/3$ 3.27 $149M/130$ 3.29 2 3.30 $32/5$

4.1 (b) $\bar{x} = \bar{y} = \frac{4}{3}a/\pi$ (c) $I = Ma^2/4$ (e) $\bar{x} = \bar{y} = 2a/\pi$
4.2 (c) $\bar{y} = \frac{4}{3}a/\pi$
(d) $I_x = Ma^2/4$, $I_y = 5Ma^2/4$, $I_z = 3Ma^2/2$
(e) $\bar{y} = 2a/\pi$
(f) $\bar{x} = 6a/5$, $I_x = 48Ma^2/175$, $I_y = 288Ma^2/175$, $I_z = 48Ma^2/25$
(g) $A = (\frac{2}{3}\pi - \frac{1}{2}\sqrt{3})a^2$
4.4 (b) $(0, 0, a/2)$ (c) $2Ma^2/3$ (e) $(0, 0, 3a/8)$
4.5 $7\pi/3$
4.11 12π
4.12 (c) $M = (16\rho/9)(3\pi - 4) = 9.64\rho$
$I = (128\rho/15^2)(15\pi - 26) = 12.02\rho = 1.25M$
4.14 $\pi(1 - e^{-1})/4$ 4.16 $u^2 + v^2$ 4.19 $\pi/4$
4.22 $12(1 + 36\pi^2)^{1/2}$ 4.24 $\rho G\pi a/2$ 4.26 (a) $\frac{7}{5}Ma^2$
4.27 $2\pi ah$ (where h = distance between parallel planes)

5.1 $\frac{9}{5}\pi\sqrt{30}$ 5.3 $\pi(37^{3/2} - 1)/6$
5.5 8π for each nappe 5.6 4
5.8 $\frac{3}{16}\sqrt{6} + \frac{9}{16}\ln(\sqrt{2} + \sqrt{3})$ 5.9 $\pi\sqrt{2}$
5.12 $M = \frac{1}{6}\sqrt{3}$, $(\bar{x}, \bar{y}, \bar{z}) = (\frac{1}{2}, \frac{1}{4}, \frac{1}{4})$ 5.14 $M = \frac{1}{2}\pi - \frac{4}{3}$
5.16 $\bar{x} = 0$, $\bar{y} = 1$, $\bar{z} = [32/(9\pi)]\sqrt{2/5} = 0.716$

6.2 $45(2 + \sqrt{2})/112$ 6.3 $15\pi/8$
6.4 (a) $\frac{1}{2}MR^2$ (b) $\frac{3}{2}MR^2$ 6.6 (a) $(4\pi - 3\sqrt{3})/6$
6.7 $(8\pi - 3\sqrt{3})(4\pi - 3\sqrt{3})^{-1}M$ 6.8 (b) $27/20$
6.10 (a) $(\bar{x}, \bar{y}) = (\pi/2, \pi/8)$ 6.10 (c) $3M/8$
6.12 $(abc)^2/6$ 6.14 $16a^3/3$
6.15 $I_x = \frac{8}{15}Ma^2$, $I_y = \frac{7}{15}Ma^2$ 6.16 $\bar{x} = \bar{y} = 2a/5$
6.18 $(0, 0, 5h/6)$
6.19 $I_x = I_y = 20Mh^2/21$, $I_z = 10Mh^2/21$, $I_m = 65Mh^2/252$
6.21 $\pi G\rho h(2 - \sqrt{2})$ 6.24 $(0, 0, 2c/3)$
6.26 $\frac{1}{2}\sinh 1$ 6.27 $e^2 - e - 1$

Chapter 6

3.1 $(\mathbf{A}\cdot\mathbf{B})\mathbf{C} = 6\mathbf{C}$, $(\mathbf{A}\times\mathbf{B})\cdot\mathbf{C} = \mathbf{A}\cdot(\mathbf{B}\times\mathbf{C}) = -8$,
$\mathbf{A}\times(\mathbf{B}\times\mathbf{C}) = -4(\mathbf{i} + 2\mathbf{k})$
3.3 -5
3.6 $\mathbf{v} = (2/\sqrt{6})(\mathbf{A}\times\mathbf{B}) = (2/\sqrt{6})(\mathbf{i} - 7\mathbf{j} - 3\mathbf{k})$,
$\mathbf{r}\times\mathbf{F} = (\mathbf{A} - \mathbf{C})\times\mathbf{B} = 3\mathbf{i} + 3\mathbf{j} - \mathbf{k}$,
$\mathbf{n}\cdot(\mathbf{r}\times\mathbf{F}) = [(\mathbf{A} - \mathbf{C})\times\mathbf{B}]\cdot\mathbf{C}/|\mathbf{C}| = 8/\sqrt{26}$
3.7 (a) $11\mathbf{i} + 3\mathbf{j} - 13\mathbf{k}$, (b) 3, (c) 17

3.9 $-9\mathbf{i} - 23\mathbf{j} + \mathbf{k}$, $1/\sqrt{21}$
3.15 $\mathbf{u}_1 \cdot \mathbf{u} = -\mathbf{u}_3 \cdot \mathbf{u}$, $n_1\mathbf{u}_1 \times \mathbf{u} = n_2\mathbf{u}_2 \times \mathbf{u}$
3.17 $\mathbf{a} = (\boldsymbol{\omega} \cdot \mathbf{r})\boldsymbol{\omega} - \omega^2\mathbf{r}$; for $\mathbf{r} \perp \boldsymbol{\omega}$, $\mathbf{a} = -\omega^2\mathbf{r}$, $|\mathbf{a}| = v^2/r$.
3.19 (a) $16\mathbf{i} - 2\mathbf{j} - 5\mathbf{k}$ (b) $8/\sqrt{6}$
3.20 (b) 12

4.2 (a) $t = 2$
(b) $\mathbf{v} = 4\mathbf{i} - 2\mathbf{j} + 6\mathbf{k}$, $|\mathbf{v}| = 2\sqrt{14}$
(c) $(x-4)/4 = (y+4)/(-2) = (z-8)/6$, $2x - y + 3z = 36$
4.5 $|d\mathbf{r}/dt| = \sqrt{2}$; $|d^2\mathbf{r}/dt^2| = 1$; path is a helix.
4.8 $d\mathbf{r}/dt = \mathbf{e}_r(dr/dt) + \mathbf{e}_\theta(r\,d\theta/dt)$;
$d^2\mathbf{r}/dt^2 = \mathbf{e}_r[d^2r/dt^2 - r(d\theta/dt)^2] + \mathbf{e}_\theta[r\,d^2\theta/dt^2 + 2(dr/dt)(d\theta/dt)]$

6.2 $-\mathbf{i}$
6.4 $\pi e/(3\sqrt{5})$
6.6 $6x + 8y - z = 25$, $(x-3)/6 = (y-4)/8 = (z-25)/(-1)$
6.9 (a) $2\mathbf{i} - 2\mathbf{j} - \mathbf{k}$ (b) $5/\sqrt{6}$
(c) $\mathbf{r} = (1,1,1) + (2,-2,-1)t$
6.12 (a) $2\sqrt{5}$, $-2\mathbf{i} + \mathbf{j}$ (b) $3\mathbf{i} + 2\mathbf{j}$ (c) $\sqrt{10}$
6.14 (b) Down, at the rate $11\sqrt{2}$
6.17 $\mathbf{e}_r$ 6.19 $\mathbf{j}$

7.1 $\boldsymbol{\nabla} \cdot \mathbf{r} = 3$, $\boldsymbol{\nabla} \times \mathbf{r} = 0$
7.2 $\boldsymbol{\nabla} \cdot \mathbf{r} = 2$, $\boldsymbol{\nabla} \times \mathbf{r} = 0$
7.4 $\boldsymbol{\nabla} \cdot \mathbf{V} = 0$, $\boldsymbol{\nabla} \times \mathbf{V} = -(\mathbf{i} + \mathbf{j} + \mathbf{k})$
7.6 $\boldsymbol{\nabla} \cdot \mathbf{V} = 5xy$, $\boldsymbol{\nabla} \times \mathbf{V} = \mathbf{i}xz - \mathbf{j}yz + \mathbf{k}(y^2 - x^2)$
7.7 $\boldsymbol{\nabla} \cdot \mathbf{V} = 0$, $\boldsymbol{\nabla} \times \mathbf{V} = \mathbf{i}x - \mathbf{j}y - \mathbf{k}x\cos y$
7.10 0 7.11 $-(x^2 + y^2)/(x^2 - y^2)^{3/2}$
7.13 $2xy$ 7.14 0
7.16 $2(x^2 + y^2 + z^2)^{-1}$ 7.19 $2/r$

8.1 $-11/3$ 8.2 (a) -4π (b) -16 (c) -8
8.3 (a) 5/3 (b) 1 (c) 2/3 8.4 (a) 3 (b) 8/3
8.7 (b) 0 (d) 2π 8.8 $yz - x$
8.9 $3xy - x^3yz - z^2$ 8.11 $-y\sin^2 x$
8.14 $-\arcsin xy$ 8.18 (a) $\pi + \pi^2/2$ (b) $\pi^2/2$

9.2 40 9.4 $-3/2$ 9.7 πab
9.8 24π 9.10 -20 9.11 2

10.2 3 10.4 36π 10.5 $4\pi \cdot 5^2$ 10.7 48π 10.9 16π

10.12 $\phi = \begin{cases} 0, & r \le R_1; \\ (k/2\pi\epsilon_0)\ln(R_1/r), & R_1 \le r \le R_2; \\ (k/2\pi\epsilon_0)\ln(R_1/R_2), & r \ge R_2. \end{cases}$

11.2 $2ab^2$ 11.3 0 11.4 -12
11.5 36 11.6 45π 11.7 0
11.10 -6π 11.12 18π 11.15 $-2\pi\sqrt{2}$
11.18 $\mathbf{A} = (xz - yz^2 - y^2/2)\mathbf{i} + (x^2/2 - x^2z + yz^2/2 - yz)\mathbf{j} + \boldsymbol{\nabla}u$, any u
11.20 $\mathbf{A} = \mathbf{i}\sin zx + \mathbf{j}\cos zx + \mathbf{k}e^{zy} + \boldsymbol{\nabla}u$, any u

12.1 $(\sin\theta\cos\theta)\mathbf{C}$
12.7 (a) $9\mathbf{i}+5\mathbf{j}-3\mathbf{k}$ (b) $29/3$ 12.9 24
12.11 (a) $\text{grad}\,\phi = -3y\mathbf{i}-3x\mathbf{j}+2z\mathbf{k}$ (b) $-\sqrt{3}$
(c) $2x+y-2z+2=0$, $\mathbf{r}=(1,2,3)+(2,1,-2)t$
12.13 (a) $6\mathbf{i}-\mathbf{j}-4\mathbf{k}$ (b) $53^{-1/2}(6\mathbf{i}-\mathbf{j}-4\mathbf{k})$ (c) same as (a)
(d) $53^{1/2}$ (e) $53^{1/2}$
12.18 Not conservative (a) 1/2 (b) 4/3
12.21 4 12.23 192π 12.25 -18π
12.27 4 12.29 10 12.31 29/3

Chapter 7

		Amplitude	Period	Frequency	Velocity Amplitude
2.2		2	$\pi/2$	$2/\pi$	8
2.3		1/2	2	1/2	$\pi/2$
2.6	$s=6\cos(\pi/8)\sin(2t)$	$6\cos(\pi/8)$ $=5.54$	π	$1/\pi$	$12\cos(\pi/8)$ $=11.1$
2.8		2	4π	$1/(4\pi)$	1
2.10		4	π	$1/\pi$	8
2.11	q	3	1/60	60	
	I	360π	1/60	60	

2.13 $A=$ maximum value of θ, $\omega=\sqrt{g/l}$
2.16 $t\cong 4.91\cong 281^\circ$
2.19 $A=1$, $T=4$, $f=1/4$, $v=1/4$, $\lambda=1$
2.21 $y=20\sin\frac{1}{2}\pi(x-6t)$, $\partial y/\partial t=-60\pi\cos\frac{1}{2}\pi(x-6t)$
2.23 $y=\sin 880\pi((x/350)-t)$
2.25 $y=10\sin[\pi(x-3\cdot 10^8 t)/250]$

3.6 $\sin(2x+\frac{1}{3}\pi)$

4.5 $\pi^{-1}+\frac{1}{2}$ 4.6 $2/\pi$ 4.8 0
4.11 1/2 4.14 (a) $2\pi/3$ (b) π 4.15 (a) 3/2

$x\to$	-2π	$-\pi$	$-\pi/2$	0	$\pi/2$	π	2π
6.2	1/2	0	0	1/2	1/2	0	1/2
6.4	-1	0	-1	-1	0	0	-1
6.6	1/2	1/2	1/2	1/2	1/2	1/2	1/2
6.8	1	1	$1-\frac{1}{2}\pi$	1	$1+\frac{1}{2}\pi$	1	1
6.10	π	0	$\pi/2$	π	$\pi/2$	0	π

7.1 $f(x) = \frac{1}{2} + \frac{i}{\pi} \sum_{\substack{-\infty \\ \text{odd } n}}^{\infty} \frac{1}{n} e^{inx}$

7.2 $f(x) = \frac{1}{4} + \frac{1}{2\pi}\Big[(1-i)e^{ix} + (1+i)e^{-ix} - i(e^{2ix} - e^{-2ix}) - \frac{1+i}{3}e^{3ix} - \frac{1-i}{3}e^{-3ix} + \frac{1-i}{5}e^{5ix} + \frac{1+i}{5}e^{-5ix}\cdots\Big]$

7.7 $f(x) = \frac{\pi}{4} - \sum_{\substack{-\infty \\ \text{odd } n}}^{\infty} \left(\frac{1}{n^2\pi} + \frac{i}{2n}\right) e^{inx} + \sum_{\substack{-\infty \\ \text{even } n\neq 0}}^{\infty} \frac{i}{2n} e^{inx}$

7.11 $f(x) = \frac{1}{\pi} + \frac{e^{ix} - e^{-ix}}{4i} - \frac{1}{\pi} \sum_{\substack{-\infty \\ \text{even } n\neq 0}}^{\infty} \frac{e^{inx}}{n^2 - 1}$

8.2 $f(x) = \frac{1}{4} + \frac{1}{\pi}\left(\cos\frac{\pi x}{l} - \frac{1}{3}\cos\frac{3\pi x}{l} + \frac{1}{5}\cos\frac{5\pi x}{l}\cdots\right) + \frac{1}{\pi}\left(\sin\frac{\pi x}{l} + \frac{2}{2}\sin\frac{2\pi x}{l} + \frac{1}{3}\sin\frac{3\pi x}{l} + \frac{1}{5}\sin\frac{5\pi x}{l} + \frac{2}{6}\sin\frac{6\pi x}{l}\cdots\right)$

8.6 $f(x) = \frac{1}{2} + \frac{4}{\pi}\sum \frac{1}{n}\sin\frac{n\pi x}{l} \quad (n = 2,\ 6,\ 10,\ \cdots)$

8.11 (a) $f(x) = \frac{\pi^2}{3} + 4\sum_{1}^{\infty} \frac{(-1)^n}{n^2}\cos nx$

(b) $f(x) = \frac{4\pi^2}{3} + 2\sum_{-\infty}^{\infty}\left(\frac{1}{n^2} + \frac{i\pi}{n}\right)e^{inx}, \quad n \neq 0$

8.14 (a) $f(x) = \frac{8}{\pi}\sum_{1}^{\infty} \frac{n(-1)^{n+1}}{4n^2-1}\sin 2n\pi x$

(b) $f(x) = \frac{2}{\pi} - \frac{4}{\pi}\sum_{1}^{\infty}\frac{\cos 2n\pi x}{4n^2-1} = -\frac{2}{\pi}\sum_{-\infty}^{\infty}\frac{1}{4n^2-1}e^{2in\pi x}$

8.19 $f(x) = \frac{1}{8} - \frac{1}{\pi^2}\sum_{\text{odd } n=1}^{\infty}\frac{1}{n^2}\cos 2n\pi x + \frac{1}{2\pi}\sum_{1}^{\infty}\frac{(-1)^{n+1}}{n}\sin 2n\pi x$

8.20 $f(x) = \frac{2}{3} - \frac{9}{8\pi^2}\left[\cos\frac{2\pi x}{3} + \frac{1}{2^2}\cos\frac{4\pi x}{3} + \frac{1}{4^2}\cos\frac{8\pi x}{3} + \cdots\right] - \left(\frac{3\sqrt{3}}{8\pi^2} + \frac{1}{\pi}\right)\sin\frac{2\pi x}{3} + \left(\frac{3\sqrt{3}}{32\pi^2} - \frac{1}{2\pi}\right)\sin\frac{4\pi x}{3} - \frac{1}{3\pi}\sin\frac{6\pi x}{3} - \left(\frac{3\sqrt{3}}{128\pi^2} + \frac{1}{4\pi}\right)\sin\frac{8\pi x}{3}\cdots$

9.2 (a) $\frac{1}{2}\ln|1-x^2| + \frac{1}{2}\ln|(1-x)/(1+x)|$

9.5 $f(x) = \frac{4}{\pi}\sum_{\text{odd } n=1}^{\infty}\frac{1}{n}\sin nx$

9.19 $f_c(x) = f_p(x) = \frac{2}{\pi} - \frac{4}{\pi}\sum_1^\infty \frac{(-1)^n \cos 2nx}{4n^2-1}$

$f_s(x) = \frac{2}{\pi}\left(\sin x + \sin 3x + \frac{1}{3}\sin 5x + \frac{1}{3}\sin 7x + \frac{1}{5}\sin 9x + \frac{1}{5}\sin 11x \cdots\right)$

9.20 $f_c(x) = \frac{1}{3} + \frac{4}{\pi^2}\sum_1^\infty \frac{(-1)^n}{n^2}\cos n\pi x$

$f_s(x) = \frac{2}{\pi}\sum_1^\infty \frac{(-1)^{n+1}}{n}\sin n\pi x - \frac{8}{\pi^3}\sum_{\text{odd } n=1}^\infty \frac{1}{n^3}\sin n\pi x$

$f_p(x) = \frac{1}{3} + \frac{1}{\pi^2}\sum_1^\infty \frac{1}{n^2}\cos 2n\pi x - \frac{1}{\pi}\sum_1^\infty \frac{1}{n}\sin 2n\pi x$

9.22 $f_c(x) = 15 - \frac{20}{\pi}\left(\cos\frac{\pi x}{20} - \frac{1}{3}\cos\frac{3\pi x}{20} + \frac{1}{5}\cos\frac{5\pi x}{20}\cdots\right)$

$f_s(x) = \frac{20}{\pi}\left(3\sin\frac{\pi x}{20} - \frac{2}{2}\sin\frac{2\pi x}{20} + \frac{3}{3}\sin\frac{3\pi x}{20} + \frac{3}{5}\sin\frac{5\pi x}{20} - \frac{2}{6}\sin\frac{6\pi x}{20}\cdots\right)$

$f_p(x) = 15 - \frac{20}{\pi}\sum_{\text{odd } n=1}^\infty \frac{1}{n}\sin\frac{n\pi x}{10}$

9.23 $f(x,0) = \frac{8h}{\pi^2}\left(\sin\frac{\pi x}{l} - \frac{1}{3^2}\sin\frac{3\pi x}{l} + \frac{1}{5^2}\sin\frac{5\pi x}{l}\cdots\right)$

10.1 Relative intensities $= 1:0:0:0:\frac{1}{25}:0:\frac{1}{49}:0:0:0$

10.3 Relative intensities $= 1:25:\frac{1}{9}:0:\frac{1}{25}:\frac{25}{9}:\frac{1}{49}:0:\frac{1}{81}:1$

10.5 $I(t) = \frac{5}{\pi}\left[1 - 2\sum_{\text{even } n=2}^\infty \frac{1}{n^2-1}\cos 120n\pi t\right] + \frac{5}{2}\sin 120\pi t$

10.6 $V(t) = 50 - \frac{400}{\pi^2}\sum_{\text{odd } n=1}^\infty \frac{1}{n^2}\cos 120n\pi t$

10.7 $I(t) = -\frac{20}{\pi}\sum_1^\infty \frac{(-1)^n}{n}\sin 120n\pi t$

10.10 $V(t) = 75 - \frac{200}{\pi^2}\sum_{\text{odd } n=1}^\infty \frac{1}{n^2}\cos 120n\pi t - \frac{100}{\pi}\sum_1^\infty \frac{1}{n}\sin 120n\pi t$

Relative intensities $= 1.4 : 0.25 : 0.12 : 0.06 : 0.04$

11.5 $\pi^2/8$ 11.7 $\pi^2/6$ 11.9 $\frac{\pi^2}{16} - \frac{1}{2}$

12.2 $f_s(x) = \frac{2}{\pi}\int_0^\infty \frac{1-\cos\alpha}{\alpha}\sin\alpha x\, d\alpha$

12.4 $f(x) = \int_{-\infty}^\infty \frac{\sin\alpha\pi - \sin(\alpha\pi/2)}{\alpha\pi}e^{i\alpha x}\, d\alpha$

12.6 $f(x) = \int_{-\infty}^\infty \frac{\sin\alpha - \alpha\cos\alpha}{i\pi\alpha^2}e^{i\alpha x}\, d\alpha$

12.8 $f(x) = \int_{-\infty}^\infty \frac{(i\alpha+1)e^{-i\alpha} - 1}{2\pi\alpha^2}e^{i\alpha x}\, d\alpha$

12.10 $f(x) = 2\int_{-\infty}^{\infty} \frac{\alpha a - \sin\alpha a}{i\pi\alpha^2} e^{i\alpha x}\, d\alpha$

12.11 $f(x) = \frac{1}{\pi}\int_{-\infty}^{\infty} \frac{\cos(\alpha\pi/2)}{1-\alpha^2} e^{i\alpha x}\, d\alpha$

12.13 $f_c(x) = \frac{2}{\pi}\int_0^{\infty} \frac{\sin\alpha\pi - \sin(\alpha\pi/2)}{\alpha} \cos\alpha x\, d\alpha$

12.16 $f_c(x) = \frac{2}{\pi}\int_0^{\infty} \frac{\cos(\alpha\pi/2)}{1-\alpha^2} \cos\alpha x\, d\alpha$

12.18 $f_s(x) = \frac{2}{\pi}\int_0^{\infty} \frac{\sin\alpha - \alpha\cos\alpha}{\alpha^2} \sin\alpha x\, d\alpha$

12.19 $f_s(x) = \frac{4}{\pi}\int_0^{\infty} \frac{\alpha a - \sin\alpha a}{\alpha^2} \sin\alpha x\, d\alpha$

12.21 $g(\alpha) = \sigma(2\pi)^{-1/2} e^{-\alpha^2\sigma^2/2}$

12.25 (a) $f(x) = \frac{1}{2\pi}\int_{-\infty}^{\infty} \frac{1+e^{-i\alpha\pi}}{1-\alpha^2} e^{i\alpha x}\, d\alpha$

12.28 (a) $f_c(x) = \frac{4}{\pi}\int_0^{\infty} \frac{\cos 3\alpha \sin\alpha}{\alpha} \cos\alpha x\, d\alpha$

(b) $f_s(x) = \frac{4}{\pi}\int_0^{\infty} \frac{\sin 3\alpha \sin\alpha}{\alpha} \sin\alpha x\, d\alpha$

12.30 (a) $f_c(x) = \frac{1}{\pi}\int_0^{\infty} \frac{1-\cos 2\alpha}{\alpha^2} \cos\alpha x\, d\alpha$

(b) $f_s(x) = \frac{1}{\pi}\int_0^{\infty} \frac{2\alpha - \sin 2\alpha}{\alpha^2} \sin\alpha x\, d\alpha$

13.7 $f(x) = \frac{\pi}{2} - \frac{4}{\pi}\sum_{\substack{1 \\ \text{odd } n}}^{\infty} \frac{1}{n^2}\cos nx$

13.8 (b) 1

13.10 (d) -1, $-1/2$, -2, -1

13.14 (a) $f(x) = \frac{1}{3} + \frac{4}{\pi^2}\sum_1^{\infty} \frac{\cos n\pi x}{n^2}$

(b) $\pi^4/90$

13.15 $-\pi/4$

13.23 $\pi^2/8$

Chapter 8

1.5 $x = -A\omega^{-2}\sin\omega t + v_0 t + x_0$

1.7 $x = (c/F)[(m^2c^2 + F^2t^2)^{1/2} - mc]$

2.2 $(1-x^2)^{1/2} + (1-y^2)^{1/2} = C$, $C = \sqrt{3}$

2.3 $\ln y = A(\csc x - \cot x)$, $A = \sqrt{3}$

2.6 $2y^2 + 1 = A(x^2-1)^2$, $A = 1$

2.7 $y^2 = 8 + e^{K-x^2}$, $K = 1$

2.9 $ye^y = ae^x$, $a = 1$

2.13 $y \equiv 1$, $y \equiv -1$, $x \equiv 1$, $x \equiv -1$

2.19 (a) $I/I_0 = e^{-0.5} = 0.6$ for $s = 50$ ft

Half value thickness $= (\ln 2)/\mu = 69.3$ ft

(b) Half life $T = (\ln 2)/\lambda$

2.20 (c) $\tau = RC$, $\tau = L/R$. Corresponding quantities are a, $\lambda = (\ln 2)/T$, μ, $1/\tau$.

2.22 $N = N_0 e^{Kt} - (R/K)(e^{Kt} - 1)$ where N_0 = number of bacteria at $t = 0$, KN = rate of increase, R = removal rate.

2.23 $T = 100[1 - (\ln r)/(\ln 2)]$

2.26 (a) k = weight divided by terminal speed
(b) $t = g^{-1}\cdot$ (terminal speed) $\cdot$ (ln 100); typical terminal speeds are 0.02 to 0.1 cm/sec, so t is of the order of 10^{-4} sec.

2.27 $t = 10(\ln\frac{5}{13})/(\ln\frac{3}{13}) = 6.6$ min

2.29 $t = 100\ln\frac{9}{4} = 81.1$ min

2.31 $ay = bx$

2.33 $x^2 + ny^2 = C$

2.35 $x(y-1) = C$

3.1 $y = \frac{1}{2}e^x + Ce^{-x}$

3.3 $y = (\frac{1}{2}x^2 + C)e^{-x^2}$

3.6 $y = (x+C)/(x+\sqrt{x^2+1})$

3.8 $y = \frac{1}{2}\ln x + C/\ln x$

3.9 $y(1-x^2)^{1/2} = x^2 + C$

3.11 $y = 2(\sin x - 1) + Ce^{-\sin x}$

3.13 $x = \frac{1}{2}e^y + Ce^{-y}$

3.14 $x = y^{2/3} + Cy^{-1/3}$

3.15 $S = (10^7/2)[(1+3t/10^4) + (1+3t/10^4)^{-1/3}]$, where S = number of pounds of salt, and t is in hours.

3.17 $I = Ae^{-t/(RC)} - V_0\omega C(\sin\omega t - \omega RC\cos\omega t)/(1+\omega^2R^2C^2)$

3.21 $N_n = c_1e^{-\lambda_1 t} + c_2e^{-\lambda_2 t} + \cdots$ where

$$c_1 = \frac{\lambda_1\lambda_2\cdots\lambda_{n-1}N_0}{(\lambda_2-\lambda_1)(\lambda_3-\lambda_1)\ldots(\lambda_n-\lambda_1)},\quad c_2 = \frac{\lambda_1\lambda_2\cdots\lambda_{n-1}N_0}{(\lambda_1-\lambda_2)(\lambda_3-\lambda_2)\cdots(\lambda_n-\lambda_2)},$$

etc. (all λ's different)

3.22 $y = x + 1 + Ke^x$

4.1 $y^{1/3} = x - 3 + Ce^{-x/3}$

4.4 $x^2e^{3y} + e^x - \frac{1}{3}y^3 = C$

4.5 $x^2 - y^2 + 2x(y+1) = C$

4.7 $x = y(\ln x + C)$

4.9 $y^2 = Ce^{-x^2/y^2}$

4.11 $\tan\frac{1}{2}(x+y) = x + C$

4.13 $y^2 = -\sin^2 x + C\sin^4 x$

4.16 $y^2 = C(C \pm 2x)$

4.18 $x^2 + (y-k)^2 = k^2$

4.19 $r = Ae^{-\theta}$, $r = Be^{\theta}$

5.1 $y = Ae^x + Be^{-2x}$

5.3 $y = Ae^{3ix} + Be^{-3ix}$ or other forms as in (5.24)

5.5 $y = (Ax+B)e^x$

5.7 $y = Ae^{3x} + Be^{2x}$

5.9 $y = Ae^{2x}\sin(3x+\gamma)$

5.11 $y = (A+Bx)e^{-3x/2}$

5.20 $y = Ae^{-x} + Be^{ix}$

5.22 $y = Ae^x + Be^{-3x} + Ce^{-5x}$

5.24 $y = Ae^{-x} + Be^{x/2}\sin(\frac{1}{2}x\sqrt{3}+\gamma)$

5.26 $y = Ae^{5x} + (Bx+C)e^{-x}$

5.28 $y = e^x(A\sin x + B\cos x) + e^{-x}(C\sin x + D\cos x)$

5.29 $y = (A+Bx)e^{-x} + Ce^{2x} + De^{-2x} + E\sin(2x+\gamma)$

5.35 $T = 2\pi\sqrt{R/g} \cong 85$ min.

6.1 $y = Ae^{2x} + Be^{-2x} - \frac{5}{2}$

6.3 $y = Ae^x + Be^{-2x} + \frac{1}{4}e^{2x}$

6.5 $y = Ae^{ix} + Be^{-ix} + e^x$

6.7 $y = Ae^{-x} + Be^{2x} + xe^{2x}$

6.9 $y = (Ax + B + x^2)e^{-x}$

6.11 $y = e^{-x}(A\sin 3x + B\cos 3x) + 8\sin 4x - 6\cos 4x$

6.13 $y = (Ax+B)e^x - \sin x$

6.15 $y = e^{-6x/5}[A\sin(8x/5) + B\cos(8x/5)] - 5\cos 2x$

6.17 $y = A\sin 4x + B\cos 4x + 2x\sin 4x$

6.18 $y = e^{-x}(A\sin 4x + B\cos 4x) + 2e^{-4x}\cos 5x$

6.20 $y = Ae^{-2x}\sin(2x+\gamma) + 4e^{-x/2}\sin(5x/2)$

6.22 $y = A + Be^{-x/2} + x^2 - 4x$

6.24 $y = (A + Bx + 2x^3)e^{3x}$

6.26 $y = A\sin x + B\cos x - 2x^2\cos x + 2x\sin x$

6.33 $y = A\sin(x+\gamma) + x^3 - 6x - 1 + x\sin x + (3-2x)e^x$

6.34 $y = Ae^{3x} + Be^{2x} + e^x + x$

6.37 $y = (A + Bx)e^x + 2x^2e^x + (3 - x)e^{2x} + x + 1$

6.41 $y = e^{-x}(A\cos x + B\sin x) + \frac{1}{4}\pi$
$$+ \sum_{\text{odd } n=1}^{\infty} [4(n^2 - 2)\cos nx - 8n\sin nx]/[\pi n^2(n^2 + 4)]$$

7.1 (a) $y \equiv 5$ (b) $y = 2/(x + 1)$
(c) $y = \tan(\frac{\pi}{4} - \frac{x}{2}) = \sec x - \tan x$ (d) $y = 2\tanh x$

7.4 $x^2 + (y - b)^2 = a^2$, or $y = C$

7.11 $x = (1 - 3t)^{1/3}$

7.12 $t = \int_1^x u^2(1 - u^4)^{-1/2}\,du$

7.16 (a) $y = Ax + Bx^{-3}$

7.16 (c) $y = (A + B\ln x)/x^3$

7.18 $y = Ax + Bx^{-1} + \frac{1}{2}\left(x + x^{-1}\right)\ln x$

7.20 $y = x^2(A + B\ln x) + x^2(\ln x)^3$

7.22 $y = A\cos\ln x + B\sin\ln x + x$

7.25 $x^{-1} - 1$

7.27 x^3e^x

7.29 $xe^{1/x}$

8.8 $e^{-2t} - te^{-2t}$

8.10 $\frac{1}{3}e^t\sin 3t + 2e^t\cos 3t$

8.12 $3\cosh 5t + 2\sinh 5t$

8.21 $2b(p + a)/[(p + a)^2 + b^2]^2$

8.23 $y = te^{-2t}(\cos t - \sin t)$

8.25 $e^{-p\pi/2}/(p^2 + 1)$

9.3 $y = e^{-2t}(4t + \frac{1}{2}t^2)$

9.4 $y = \cos t + \frac{1}{2}(\sin t - t\cos t)$

9.7 $y = 1 - e^{2t}$

9.9 $y = (t + 2)\sin 4t$

9.11 $y = te^{2t}$

9.12 $y = \frac{1}{2}(t^2e^{-t} + 3e^t - e^{-t})$

9.13 $y = \sinh 2t$

9.17 $y = 2$

9.19 $y = e^{2t}$

9.21 $y = e^{3t} + 2e^{-2t}\sin t$

9.23 $y = \sin t + 2\cos t - 2e^{-t}\cos 2t$

9.25 $y = (3 + t)e^{-2t}\sin t$

9.27 $\begin{cases} y = t + \frac{1}{4}(1 - e^{4t}) \\ z = \frac{1}{3} + e^{4t} \end{cases}$

9.28 $\begin{cases} y = t\cos t - 1 \\ z = \cos t + t\sin t \end{cases}$

9.30 $\begin{cases} y = t - \sin 2t \\ z = \cos 2t \end{cases}$

9.32 $\begin{cases} y = \sin 2t \\ z = \cos 2t - 1 \end{cases}$

9.36 $\arctan(2/3)$

9.38 $4/5$

9.40 1

9.42 $\pi/4$

10.3 $\frac{1}{2}t\sinh t$

10.5 $[b(b - a)te^{-bt} + a(e^{-bt} - e^{-at})]/(b - a)^2$

10.7 $(a\cosh bt - b\sinh bt - ae^{-at})/(a^2 - b^2)$

10.9 $(2t^2 - 2t + 1 - e^{-2t})/4$

10.12 $(b^2 - a^2)^{-1}(b^{-2}\cos bt - a^{-2}\cos at) + a^{-2}b^{-2}$

10.13 $\frac{1}{2}(e^{-t} + \sin t - \cos t)$

10.15 $\frac{1}{14}e^{3t} + \frac{1}{35}e^{-4t} - \frac{1}{10}e^t$

10.17 $y = \begin{cases} (\cosh at - 1)/a^2, & t > 0 \\ 0, & t < 0 \end{cases}$

11.7 $y = \begin{cases} (t - t_0)e^{-(t-t_0)}, & t > t_0 \\ 0, & t < t_0 \end{cases}$

11.9 $y = \begin{cases} \frac{1}{3}e^{-(t-t_0)} \sin 3(t-t_0), & t > t_0 \\ 0, & t < t_0 \end{cases}$

11.11 $y = \begin{cases} \frac{1}{2}[\sinh(t-t_0) - \sin(t-t_0)], & t > t_0 \\ 0, & t < t_0 \end{cases}$

11.13 (b) $3\delta(x+5) - 4\delta(x-10)$

11.15 (b) 0 (d) $\cosh 1$

11.21 (b) $\phi(|a|)/(2|a|)$ (c) $1/2$

11.23 (a) $\delta(x+5)\delta(y-5)\delta(z)$, $\delta(r-5\sqrt{2}\,)\delta(\theta-\frac{3\pi}{4})\delta(z)/r$, $\delta(r-5\sqrt{2}\,)\delta(\theta-\frac{\pi}{2})\delta(\phi-\frac{3\pi}{4})/(r\sin\theta)$
(c) $\delta(x+2)\delta(y)\delta(z-2\sqrt{3}\,)$, $\delta(r-2)\delta(\theta-\pi)\delta(z-2\sqrt{3}\,)/r$, $\delta(r-4)\delta(\theta-\frac{\pi}{6})\delta(\phi-\pi)/(r\sin\theta)$

11.25 (c) $G''(x) = \delta(x) + 5\delta'(x)$

12.2 $y = (\sin\omega t - \omega t\cos\omega t)/(2\omega^2)$

12.7 $y = [a(\cosh at - e^{-t}) - \sinh at]/[a(a^2-1)]$

12.11 $y = -\frac{1}{3}\sin 2x$

12.13 $y = \begin{cases} x - \sqrt{2}\,\sin x, & x < \pi/4 \\ \frac{1}{2}\pi - x - \sqrt{2}\,\cos x, & x > \pi/4 \end{cases}$

12.16 $y = -x\ln x - x - x(\ln x)^2/2$

12.18 $y = x^2/2 + x^4/6$

13.1 $y = -\frac{1}{3}x^{-2} + Cx$

13.3 $y = A + Be^{-x}\sin(x+\gamma)$

13.5 $x^2 + y^2 - y\sin^2 x = C$

13.7 $3x^2y^3 + 1 = Ax^3$

13.8 $y = x(A + B\ln x) + \frac{1}{2}x(\ln x)^2$

13.10 $u - \ln u + \ln v + v^{-1} = C$

13.13 $y = Ae^{-2x}\sin(x+\gamma) + e^{3x}$

13.15 $y = (A + Bx)e^{2x} + 3x^2e^{2x}$

13.18 $x = (y + C)e^{-\sin y}$

13.20 $y = Ae^x\sin(2x+\gamma) + x + \frac{2}{5} + e^x(1 - x\cos 2x)$

13.22 $y = (A + Bx)e^{2x} + C\sin(3x+\gamma)$

13.24 $y^2 = ax^2 + b$

13.26 $y = x^2 + x$

13.28 $y^2 + 4(x-1)^2 = 9$

13.30 $y = g/3$, $v = 7g/12$, $a = 5g/12$

13.32 1:23 p.m.

13.33 In both (a) and (b), the temperature of the mixture at time t is given by the formula $T_a(1 - e^{-kt}) + (n + n')^{-1}(nT_0 + n'T_0')e^{-kt}$.

13.38 $\frac{1}{2}\ln[(a^2+p^2)/p^2]$

13.41 $\frac{1}{4}(\tanh 1 - \operatorname{sech}^2 1) = 0.0854$

13.43 $(\sin at + at\cos at)/(2a)$

13.46 For e^{-x}: $g_s(\alpha) = (2/\pi)^{1/2}\alpha/(1+\alpha^2)$, $g_c(\alpha) = (2/\pi)^{1/2}/(1+\alpha^2)$

13.47 $y = A\sin t + B\cos t + \sin t\ln(\sec t + \tan t) - 1$

Chapter 9

2.1 Parabola

2.2 Circle

2.3 $ax = \sinh(ay + b)$

2.6 $x + a = \frac{4}{3}(y^{1/2} - 2b)(b + y^{1/2})^{1/2}$

3.1 $dx/dy = C/\sqrt{y^3 - C^2}$

3.3 $x^4y'^2 = C^2(1 + x^2y'^2)^3$

3.6 $x = ay^{3/2} - \frac{1}{2}y^2 + b$

3.7 $y = K\sinh(x + C)$

3.9 $\cot\theta = A\cos(\phi - \alpha)$

3.12 $(x-a)^2 + y^2 = C^2$

3.15 $r\cos(\theta+\alpha) = C$ or, in rectangular coordinates, the straight line $x\cos\alpha - y\sin\alpha = C$

3.18 See Problem 3.9

4.6 Cycloid

5.2 $\begin{cases} m(\ddot{r} - r\dot{\theta}^2) = -\partial V/\partial r \\ m(r\ddot{\theta} + 2\dot{r}\dot{\theta}) = -(1/r)(\partial V/\partial\theta) \\ m\ddot{z} = -\partial V/\partial z \end{cases}$

Comment: These equations are in the form $m\mathbf{a} = \mathbf{F}$; recall from Chapter 6, equation (6.7), the polar coordinate form for $\mathbf{F} = -\boldsymbol{\nabla}V$.

5.4 $l\ddot{\theta} + g\sin\theta = 0$

5.6 $\begin{cases} a\ddot{\theta} - a\sin\theta\cos\theta\,\dot{\phi}^2 - g\sin\theta = 0 \\ (d/dt)(\sin^2\theta\,\dot{\phi}) = 0 \end{cases}$

5.8 $L = \frac{1}{2}m(2\dot{r}^2 + r^2\dot{\theta}^2) - mgr$
$2\ddot{r} - r\dot{\theta}^2 + g = 0,\ (d/dt)(r^2\dot{\theta}) = 0$

5.11 $L = \frac{1}{2}(m + Ia^{-2})\dot{z}^2 - mgz$
$(ma^2 + I)\ddot{z} + mga^2 = 0$

5.12 $L = \frac{1}{2}m(\dot{r}^2 + r^2\dot{\theta}^2) - \left[\frac{1}{2}k(r - r_0)^2 - mgr\cos\theta\right]$
$\ddot{r} - r\dot{\theta}^2 + \frac{k}{m}(r - r_0) - g\cos\theta = 0,\quad (d/dt)(r^2\dot{\theta}) + gr\sin\theta = 0$

5.14 $L = M\dot{x}^2 + Mgx\sin\alpha,\ 2M\ddot{x} - Mg\sin\alpha = 0$

5.16 $L = \frac{1}{2}m(l + a\theta)^2\dot{\theta}^2 - mg[a\sin\theta - (l + a\theta)\cos\theta]$
$(l + a\theta)\ddot{\theta} + a\dot{\theta}^2 + g\sin\theta = 0$

5.19 $x = y$ with $\omega = \sqrt{g/l}$; $x = -y$ with $\omega = \sqrt{3g/l}$

5.21 $2\ddot{\theta} + \ddot{\phi}\cos(\theta - \phi) + \dot{\phi}^2\sin(\theta - \phi) + \frac{2g}{l}\sin\theta = 0$
$\ddot{\phi} + \ddot{\theta}\cos(\theta - \phi) - \dot{\theta}^2\sin(\theta - \phi) + \frac{g}{l}\sin\phi = 0$

5.23 $\phi = 2\theta$ with $\omega = \sqrt{2g/(3l)}$; $\phi = -2\theta$ with $\omega = \sqrt{2g/l}$

6.1 Catenary 6.3 Circular cylinder 6.5 Circle

8.4 $dr/d\theta = Kr\sqrt{r^4 - K^2}$

8.6 $(x - a)^2 + (y + 1)^2 = C^2$

8.8 Intersection of $r = 1 + \cos\theta$ with $z = a + b\sin(\theta/2)$

8.10 Intersection of $y = x^2$ with $az = b[2x\sqrt{4x^2 + 1} + \sinh^{-1}2x] + c$

8.12 $e^y\cos(x - a) = K$

8.16 Hyperbola: $r^2\cos(2\theta + \alpha) = K$ or $(x^2 - y^2)\cos\alpha - 2xy\sin\alpha = K$

8.17 $K\ln r = \cosh(K\theta + C)$

8.18 Parabola: $(x - y - C)^2 = 4K^2(x + y - K^2)$

8.20 $m(\ddot{r} - r\dot{\theta}^2) + Kr^{-2} = 0,\ r^2\dot{\theta} =$ const.

8.22 $r^{-1}m(r^2\ddot{\theta} + 2r\dot{r}\dot{\theta} - r^2\sin\theta\cos\theta\,\dot{\phi}^2) = -r^{-1}(\partial V/\partial\theta) = F_\theta = ma_\theta$,
$a_\theta = r\ddot{\theta} + 2\dot{r}\dot{\theta} - r\sin\theta\cos\theta\,\dot{\phi}^2$

8.27 $dr/d\theta = r\sqrt{K^2(1 + \lambda r)^2 - 1}$

Chapter 10

4.6 $I = \begin{pmatrix} 9 & 0 & -3 \\ 0 & 6 & 0 \\ -3 & 0 & 9 \end{pmatrix}$; principal moments: $(6, 6, 12)$; principal axes along the vectors $(1, 0, -1)$ and any two orthogonal vectors in the plane $z = x$, say $(0, 1, 0)$ and $(1, 0, 1)$.

5.6 (a) 3 (c) 2 (e) -1

6.15 (c) vector

8.1 $h_r = 1, \quad h_\theta = r, \quad h_\phi = r\sin\theta$

$d\mathbf{s} = \mathbf{e}_r\,dr + \mathbf{e}_\theta\,r\,d\theta + \mathbf{e}_\phi\,r\sin\theta\,d\phi$

$dV = r^2\sin\theta\,dr\,d\theta\,d\phi$

$\mathbf{a}_r = \mathbf{i}\sin\theta\cos\phi + \mathbf{j}\sin\theta\sin\phi + \mathbf{k}\cos\theta = \mathbf{e}_r$

$\mathbf{a}_\theta = \mathbf{i}r\cos\theta\cos\phi + \mathbf{j}\,r\cos\theta\sin\phi - \mathbf{k}r\sin\theta = r\mathbf{e}_\theta$

$\mathbf{a}_\phi = -\mathbf{i}r\sin\theta\sin\phi + \mathbf{j}\,r\sin\theta\cos\phi = r\sin\theta\,\mathbf{e}_\phi$

8.3 $d\mathbf{s}/dt = \mathbf{e}_r\dot{r} + \mathbf{e}_\theta\,r\dot{\theta} + \mathbf{e}_\phi\,r\sin\theta\;\dot{\phi}$

$$d^2\mathbf{s}/dt^2 = \mathbf{e}_r(\ddot{r} - r\dot{\theta}^2 - r\sin^2\theta\;\dot{\phi}^2) + \mathbf{e}_\theta(r\ddot{\theta} + 2\dot{r}\dot{\theta} - r\sin\theta\cos\theta\;\dot{\phi}^2) + \mathbf{e}_\phi(r\sin\theta\;\ddot{\phi} + 2r\cos\theta\;\dot{\theta}\dot{\phi} + 2\sin\theta\;\dot{r}\dot{\phi})$$

8.5 $\mathbf{V} = \mathbf{e}_r\cos\theta - \mathbf{e}_\theta\sin\theta - \mathbf{e}_\phi\,r\sin\theta$

8.6 $h_u = h_v = (u^2+v^2)^{1/2}, \quad h_z = 1$

$d\mathbf{s} = (u^2+v^2)^{1/2}(\mathbf{e}_u\,du + \mathbf{e}_v\,dv) + \mathbf{e}_z\,dz$

$dV = (u^2+v^2)\,du\,dv\,dz$

$\mathbf{a}_u = \mathbf{i}u + \mathbf{j}v = (u^2+v^2)^{1/2}\mathbf{e}_u$

$\mathbf{a}_v = -\mathbf{i}v + \mathbf{j}u = (u^2+v^2)^{1/2}\mathbf{e}_v$

$\mathbf{a}_z = \mathbf{k} = \mathbf{e}_z$

8.9 $h_u = h_v = a(\cosh u + \cos v)^{-1}$

$d\mathbf{s} = a(\cosh u + \cos v)^{-1}(\mathbf{e}_u\,du + \mathbf{e}_v\,dv)$

$dA = a^2(\cosh u + \cos v)^{-2}\,du\,dv$

$\mathbf{a}_u = (h_u^2/a)[\mathbf{i}(1+\cos v\cosh u) - \mathbf{j}\sin v\sinh u] = h_u\mathbf{e}_u$

$\mathbf{a}_v = (h_v^2/a)[\mathbf{i}\sinh u\sin v + \mathbf{j}(1+\cos v\cosh u)] = h_v\mathbf{e}_v$

8.11 $d\mathbf{s}/dt = (u^2+v^2)^{1/2}(\mathbf{e}_u\dot{u} + \mathbf{e}_v\dot{v}) + \mathbf{e}_z\dot{z}$

$$d^2\mathbf{s}/dt^2 = \mathbf{e}_u(u^2+v^2)^{-1/2}[(u^2+v^2)\ddot{u} + u(\dot{u}^2-\dot{v}^2) + 2v\dot{u}\dot{v}] + \mathbf{e}_v(u^2+v^2)^{-1/2}[(u^2+v^2)\ddot{v} + v(\dot{v}^2-\dot{u}^2) + 2u\dot{u}\dot{v}] + \mathbf{e}_z\ddot{z}$$

8.14 $d\mathbf{s}/dt = a(\cosh u + \cos v)^{-1}(\mathbf{e}_u\dot{u} + \mathbf{e}_v\dot{v})$

$$d^2\mathbf{s}/dt^2 = \mathbf{e}_u a(\cosh u + \cos v)^{-2}[(\cosh u + \cos v)\ddot{u} + (\dot{v}^2-\dot{u}^2)\sinh u + 2\dot{u}\dot{v}\sin v] + \mathbf{e}_v a(\cosh u + \cos v)^{-2}[(\cosh u + \cos v)\ddot{v} + (\dot{v}^2-\dot{u}^2)\sin v - 2\dot{u}\dot{v}\sinh u]$$

9.10 Let $h = h_u = h_v = (u^2+v^2)^{1/2}$ represent the u and v scale factors.

$$\boldsymbol{\nabla}U = h^{-1}\left(\mathbf{e}_u\frac{\partial U}{\partial u} + \mathbf{e}_v\frac{\partial U}{\partial v}\right) + \mathbf{k}\frac{\partial U}{\partial z}$$

$$\boldsymbol{\nabla}\cdot\mathbf{V} = h^{-2}\left[\frac{\partial}{\partial u}(hV_u) + \frac{\partial}{\partial v}(hV_v)\right] + \frac{\partial V_z}{\partial z}$$

$$\nabla^2 U = h^{-2}\left(\frac{\partial^2 U}{\partial u^2} + \frac{\partial^2 U}{\partial v^2}\right) + \frac{\partial^2 U}{\partial z^2}$$

$$\boldsymbol{\nabla}\times\mathbf{V} = \left(h^{-1}\frac{\partial V_z}{\partial v} - \frac{\partial V_v}{\partial z}\right)\mathbf{e}_u + \left(\frac{\partial V_u}{\partial z} - h^{-1}\frac{\partial V_z}{\partial u}\right)\mathbf{e}_v + h^{-2}\left[\frac{\partial}{\partial u}(hV_v) - \frac{\partial}{\partial v}(hV_u)\right]\mathbf{e}_z$$

9.13 Same as 9.10 if $h = a(\cosh u + \cos v)^{-1}$ and terms involving either z derivatives or V_z are omitted. Note, however, that $\nabla \times \mathbf{V}$ has *only* a z component if $\mathbf{V} = \mathbf{e}_u V_u + \mathbf{e}_v V_v$ where V_u and V_v are functions of u and v.

9.15 $h_u = 1$, $h_v = u/\sqrt{1-v^2}$
$\mathbf{e}_u = \mathbf{i}v + \mathbf{j}\sqrt{1-v^2}$, $\mathbf{e}_v = \mathbf{i}\sqrt{1-v^2} - \mathbf{j}v$
$m[\ddot{u} - u\dot{v}^2/(1-v^2)] = -\partial V/\partial u = F_u$
$m[(u\ddot{v} + 2\dot{u}\dot{v})/(1-v^2)^{1/2} + uv\dot{v}^2/(1-v^2)^{3/2}] = -h_v^{-1}\partial V/\partial v = F_v$

9.16 r^{-1}, 0, 0, $r^{-1}\mathbf{e}_z$

9.19 $2\mathbf{e}_\phi$, $\mathbf{e}_r\cos\theta - \mathbf{e}_\theta\sin\theta$, 3

9.21 $2r^{-1}$, 6, $2r^{-4}$, $-k^2e^{ikr\cos\theta}$

Chapter 11

3.3 $9/10$

3.7 8

3.9 $\Gamma(5/4)$

3.11 1

3.14 $-\Gamma(4/3)$

3.17 $\Gamma(p)$

7.1 $\frac{1}{2}B(\frac{5}{2},\frac{1}{2}) = 3\pi/16$

7.3 $\frac{1}{3}B(\frac{1}{3},\frac{1}{2})$

7.5 $B(3,3) = 1/30$

7.7 $\frac{1}{2}B(\frac{1}{4},\frac{1}{2})$

7.11 $2B(\frac{2}{3},\frac{4}{3})/B(\frac{1}{3},\frac{4}{3})$

7.13 $I_y/M = 8B(\frac{4}{3},\frac{4}{3})/B(\frac{5}{3},\frac{1}{3})$

8.1 $B(\frac{1}{2},\frac{1}{4})\sqrt{2l/g} = 7.4163\sqrt{l/g}$
(Compare $2\pi\sqrt{l/g}$)

8.3 $t = \pi\sqrt{a/g}$

10.2 $\Gamma(p,x) \sim x^{p-1}e^{-x}[1 + (p-1)x^{-1} + (p-1)(p-2)x^{-2}\cdots]$

10.5 (a) $E_1(x) = \Gamma(0,x)$

10.6 (b) $\mathrm{Ei}(x)$

11.5 1

12.1 $K = F(\pi/2,k) = (\pi/2)\{1 + (\frac{1}{2})^2k^2 + [(1\cdot3)/(2\cdot4)]^2k^4\cdots\}$
$E = E(\pi/2,k) = (\pi/2)\{1 - (\frac{1}{2})^2k^2 - [1/(2\cdot4)]^2\cdot 3k^4$
$- [(1\cdot3)/(2\cdot4\cdot6)]^2\cdot 5k^6\cdots\}$

Caution: For the following answers, see the warning about elliptic integral notation just after equations (12.3) and in Example 1.

12.5 $E(1/3) \cong 1.526$

12.6 $\frac{1}{3}F(\frac{\pi}{3},\frac{1}{3}) \cong 0.355$

12.7 $5E(\frac{5\pi}{4},\frac{1}{5}) \cong 19.46$

12.10 $\frac{1}{2}F(\frac{\pi}{4},\frac{1}{2}) \cong 0.402$

12.11 $F(\frac{3\pi}{8},\frac{3}{\sqrt{10}}) + K(\frac{3}{\sqrt{10}}) \cong 4.097$

12.13 $3E(\frac{\pi}{6},\frac{2}{3}) + 3E(\arcsin\frac{3}{4},\frac{2}{3}) \cong 3.96$

12.16 $2\sqrt{2}\,E(1/\sqrt{2}) \cong 3.820$

12.23 $T = 8\sqrt{\frac{a}{5g}}\,K(1/\sqrt{5})$; for small vibrations, $T \cong 2\pi\sqrt{\frac{2a}{3g}}$

13.8 $\frac{1}{2}\sqrt{\pi}\,\mathrm{erf}(1)$

13.10 $\sqrt{2}\,K(1/\sqrt{2}) \cong 2.622$

13.11 $\frac{1}{5}F(\arcsin\frac{3}{4},\frac{4}{5}) \cong 0.1834$

13.13 $-\mathrm{sn}\,u\,\mathrm{dn}\,u$

13.15 $\Gamma(7/2) = 15\sqrt{\pi}/8$

13.17 $\frac{1}{2}B(\frac{5}{4},\frac{7}{4}) = 3\pi\sqrt{2}/64$

13.19 $\frac{1}{2}\sqrt{\pi}\,\mathrm{erfc}\,5$

13.21 $5^4B(\frac{2}{3},\frac{13}{3}) = (\frac{5}{3})^5(\frac{14\pi}{\sqrt{3}})$

13.24 $-2^{55}\sqrt{\pi}/109!!$

Chapter 12

1.2 $y = a_0 e^{x^3}$ 1.3 $y = a_1 x$

1.7 $y = Ax + Bx^3$ 1.9 $y = a_0(1 - x^2) + a_1 x$

2.4 $Q_0 = \frac{1}{2}\ln\frac{1+x}{1-x}$, $Q_1 = \frac{x}{2}\ln\frac{1+x}{1-x} - 1$

3.3 $(30 - x^2)\sin x + 12x\cos x$ 3.5 $(x^2 - 200x + 9900)e^{-x}$

5.3 $P_0(x) = 1$ $P_4(x) = (35x^4 - 30x^2 + 3)/8$

$P_1(x) = x$ $P_5(x) = (63x^5 - 70x^3 + 15x)/8$

$P_2(x) = (3x^2 - 1)/2$ $P_6(x) = (231x^6 - 315x^4 + 105x^2 - 5)/16$

$P_3(x) = (5x^3 - 3x)/2$

5.9 $2P_2 + P_1$ 5.11 $\frac{2}{5}(P_1 - P_3)$

5.12 $\frac{8}{5}P_4 + 4P_2 - 3P_1 + \frac{12}{5}P_0$

8.2 $N = \sqrt{\frac{2}{5}}$, $\sqrt{\frac{5}{2}}\,P_2(x)$ 8.4 $N = \pi^{1/4}$, $\pi^{-1/4}e^{-x^2/2}$

9.1 $\frac{3}{2}P_1 - \frac{7}{8}P_3 + \frac{11}{16}P_5 \cdots$

9.4 $\frac{1}{8}\pi(3P_1 + \frac{7}{16}P_3 + \frac{11}{64}P_5 \cdots)$

9.6 $P_0 + \frac{3}{8}P_1 - \frac{20}{9}P_2 \cdots$

9.8 $\frac{1}{2}(1-a)P_0 + \frac{3}{4}(1-a^2)P_1 + \frac{5}{4}a(1-a^2)P_2 + \frac{7}{16}(1-a^2)(5a^2-1)P_3 \cdots$

9.11 $\frac{8}{5}P_4 + 4P_2 - 3P_1 + \frac{12}{5}P_0$

9.12 $\frac{2}{5}(P_1 - P_3)$

9.14 $\frac{1}{2}P_0 + \frac{5}{8}P_2 = \frac{3}{16}(5x^2 + 1)$

10.5 $\frac{1}{2}(\sin\theta)(35\cos^3\theta - 15\cos\theta)$

11.2 $y = Ax^{-3} + Bx^3$ 11.4 $y = Ax^{-2} + Bx^3$

11.6 $y = Ae^{-x} + Bx^{2/3}[1 - 3x/5 + (3x)^2/(5\cdot 8) - (3x)^3/(5\cdot 8\cdot 11) + \cdots]$

11.8 $y = A(x^{-1} - 1) + Bx^2(1 - x + 3x^2/5 - 4x^3/15 + 2x^4/21 + \cdots)$

11.10 $y = A[1 + 2x - (2x)^2/2! + (2x)^3/(3\cdot 3!) - (2x)^4/(3\cdot 5\cdot 4!) + \cdots]$

$+ Bx^{3/2}[1 - 2x/5 + (2x)^2/(5\cdot 7\cdot 2!) - (2x)^3/(5\cdot 7\cdot 9\cdot 3!) + \cdots]$

11.11 $y = Ax^{1/6}[1 + 3x^2/2^5 + 3^2x^4/(5\cdot 2^{10}) + \cdots]$

$+ Bx^{-1/6}[x + 3x^3/2^6 + 3^2x^5/(7\cdot 2^{11}) + \cdots]$

16.1 $y = x^{-3/2}Z_{1/2}(x)$ 16.3 $y = x^{-1/2}Z_1(4x^{1/2})$

16.5 $y = xZ_0(2x)$ 16.7 $y = x^{-1}Z_{1/2}(x^2/2)$

16.9 $y = x^{1/3}Z_{2/3}(4\sqrt{x})$ 16.11 $y = x^{-2}Z_2(x)$

16.15 $y = Z_2(5x)$ 16.17 $y = Z_0(3x)$

17.7 (a) $y = x^{1/2}I_1(2x^{1/2})$. Note that the factor i does not need to be included, since *any* multiple of y is a solution.

18.11 1.7 m for steel.

20.1 1/6 20.3 $4/\pi$ 20.5 1/2

20.7 $h_n^{(1)}(x) \sim x^{-1}e^{i[x-(n+1)\pi/2]}$ 20.9 $h_n^{(1)}(ix) \sim -i^{-n}x^{-1}e^{-x}$

21.1 $y = Ax + B\left(x \sinh^{-1} x - \sqrt{x^2+1}\right)$
21.2 $y = A(1+x) + Bxe^{1/x}$
21.5 $y = A(x-1) + B[(x-1)\ln x - 4]$
21.7 $y = A\frac{x}{1-x} + B[\frac{x}{1-x}\ln x + \frac{1+x}{2}]$
21.8 $y = A(x^2+2x) + B[(x^2+2x)\ln x + 1 + 5x - x^3/6 + x^4/72 + \cdots]$

22.4 $H_0(x) = 1$ $\quad H_3(x) = 8x^3 - 12x$
$H_1(x) = 2x$ $\quad H_4(x) = 16x^4 - 48x^2 + 12$
$H_2(x) = 4x^2 - 2$ $\quad H_5(x) = 32x^5 - 160x^3 + 120x$

22.13 $L_0(x) = 1$
$L_1(x) = 1 - x$
$L_2(x) = \frac{1}{2}(2 - 4x + x^2)$
$L_3(x) = \frac{1}{6}(6 - 18x + 9x^2 - x^3)$
$L_4(x) = \frac{1}{24}(24 - 96x + 72x^2 - 16x^3 + x^4)$
$L_5(x) = \frac{1}{120}(120 - 600x + 600x^2 - 200x^3 + 25x^4 - x^5)$

Note: The factor $1/n!$ is omitted in most quantum mechanics books but is included as here in most reference books.

Chapter 13

2.12 $$T = \sum_{\text{odd } n} \frac{400}{n\pi \sinh 3n\pi} \sinh\frac{n\pi}{10}(30-y)\sin\frac{n\pi x}{10} + \sum_{\text{odd } n} \frac{400}{n\pi \sinh(n\pi/3)} \sinh\frac{n\pi}{30}(10-x)\sin\frac{n\pi y}{30}$$

2.14 For $f(x) = x - 5$: $T = -\frac{40}{\pi^2}\sum_{\text{odd } n}\frac{1}{n^2}\cos\frac{n\pi x}{10}e^{-n\pi y/10}$
For $f(x) = x$: add 5 to the answer just given.

3.9 $$u = 100 - \frac{400}{\pi}\sum_{n=0}^{\infty}\frac{(-1)^n}{2n+1}e^{-[(2n+1)\pi\alpha/4]^2 t}\cos\left(\frac{2n+1}{4}\pi x\right)$$

3.11 $E_n = n^2\hbar^2/(2m)$; $\Psi(x,t) = \frac{4}{\pi}\sum_{\text{odd } n}\frac{\sin nx}{n}e^{-iE_n t/\hbar}$

4.8 $$y = \frac{4l}{\pi^2 v}\left[\frac{1}{3}\sin\frac{\pi x}{l}\sin\frac{\pi vt}{l} + \frac{\pi}{16}\sin\frac{2\pi x}{l}\sin\frac{2\pi vt}{l} - \sum_{n=3}^{\infty}\frac{\sin n\pi/2}{n(n^2-4)}\sin\frac{n\pi x}{l}\sin\frac{n\pi vt}{l}\right]$$

4.9 Problem 2: $n = 2$, $\nu = v/l$
Problem 3: $n = 3$, $\nu = \frac{3}{2}v/l$ and $n = 4$, $\nu = 2v/l$ have nearly equal intensity.
Problem 5: $n = 1$, $\nu = \frac{1}{2}v/l$

5.1 (a) $u \cong 9.76$

5.4 $$u = 200\sum_{m=1}^{\infty}\frac{1}{k_m J_1(k_m)}J_0(k_m r/a)e^{-(k_m\alpha/a)^2 t}, \quad k_m = \text{zeros of } J_0$$

5.10 $u = \dfrac{6400}{\pi^3} \sum_{\text{odd } n} \sum_{\text{odd } m} \sum_{\text{odd } p} \dfrac{1}{nmp} \sin\dfrac{n\pi x}{l} \sin\dfrac{m\pi y}{l} \sin\dfrac{p\pi z}{l} e^{-(\alpha\pi/l)^2(n^2+m^2+p^2)t}$

5.11 $R = r^n,\ r^{-n},\ n \neq 0;\ R = \ln r$, const., $n = 0$.
$R = r^l,\ r^{-l-1}$.

5.13 $u = \dfrac{400}{\pi} \sum_{\text{odd } n} \dfrac{1}{n} \left(\dfrac{r}{10}\right)^{4n} \sin 4n\theta$

5.14 $u = \dfrac{50 \ln r}{\ln 2} + \dfrac{200}{\pi} \sum_{\text{odd } n} \dfrac{r^n - r^{-n}}{n(2^n - 2^{-n})} \sin n\theta$

6.5 $z = \dfrac{64 l^4}{\pi^6} \sum_{\text{odd } m} \sum_{\text{odd } n} \dfrac{1}{n^3 m^3} \sin\dfrac{n\pi x}{l} \sin\dfrac{m\pi y}{l} \cos\dfrac{\pi v(m^2+n^2)^{1/2} t}{l}$

6.8 $\Psi_{mn} = J_n(k_{mn} r) \left\{ \begin{matrix} \sin n\theta \\ \cos n\theta \end{matrix} \right\} e^{-iE_{mn}t/\hbar},\quad E_{mn} = \dfrac{\hbar^2 k_{mn}^2}{2ma^2}$

7.2 $u = \frac{2}{5} r P_1(\cos\theta) - \frac{2}{5} r^3 P_3(\cos\theta)$
7.5 $u = \frac{1}{2} P_0(\cos\theta) + \frac{5}{8} r^2 P_2(\cos\theta) - \frac{3}{16} r^4 P_4(\cos\theta) \cdots$
7.6 $u = \frac{1}{8}\pi[3r P_1(\cos\theta) + \frac{7}{16} r^3 P_3(\cos\theta) + \frac{11}{64} r^5 P_5(\cos\theta) \cdots]$
7.8 $u = 25[P_0(\cos\theta) + \frac{9}{4} r P_1(\cos\theta) + \frac{15}{8} r^2 P_2(\cos\theta) + \frac{21}{64} r^3 P_3(\cos\theta) \cdots]$
7.10 $u = \frac{1}{15} r^3 P_3^2(\cos\theta) \cos 2\phi - r P_1(\cos\theta)$
7.12 $u = \frac{3}{4} r P_1(\cos\theta) + \frac{7}{24} r^3 P_3(\cos\theta) - \frac{11}{192} r^5 P_5(\cos\theta) \cdots$
7.13 $u = E_0(r - a^3/r^2) P_1(\cos\theta)$

7.15 $u = 100 + \dfrac{200a}{\pi r} \sum_1^\infty \dfrac{(-1)^n}{n} \sin\dfrac{n\pi r}{a} e^{-(\alpha n\pi/a)^2 t}$

$= 100 + 200 \sum_1^\infty (-1)^n j_0(n\pi r/a) e^{-(\alpha n\pi/a)^2 t}$

7.19 $\Psi(r,\theta,\phi) = j_l(\beta r) P_l^m(\cos\theta) e^{\pm im\phi} e^{-iEt/\hbar}$, where

$\beta = \sqrt{2ME/\hbar^2},\quad \beta a =$ zeros of j_l, $\quad E = \dfrac{\hbar^2}{2Ma^2}$(zeros of $j_l)^2$

7.20 $\psi_n(x) = e^{-\alpha^2 x^2/2} H_n(\alpha x),\quad \alpha = \sqrt{m\omega/\hbar}$
7.21 Degree of degeneracy of E_n is $C(n+2, n) = (n+2)(n+1)/2$, $n = 0$ to ∞.
7.22 $\Psi(r,\theta,\phi) = R(r) Y_l^m(\theta,\phi),\ R(r) = r^l e^{-r/(na)} L_{n-l-1}^{2l+1}\left(\frac{2r}{na}\right),\ E_n = -\frac{Me^4}{2\hbar^2 n^2}$

8.4 Let $K =$ line charge per unit length. Then

$$V = -K \ln(r^2 + a^2 - 2ra\cos\theta) + K \ln a^2 - K \ln R^2$$
$$+ K \ln[r^2 + (R^2/a)^2 - 2(R^2/a) r \cos\theta]$$

8.5 K at $(a, 0)$, $-K$ at $(R^2/a, 0)$

9.2 $u = 200\pi^{-1} \int_0^\infty k^{-2}(1 - \cos 2k) e^{-ky} \cos kx \, dk$
9.7 $u(x,t) = 100 \operatorname{erf}[x/(2\alpha t^{1/2})] - 50 \operatorname{erf}[(x-1)/(2\alpha t^{1/2})] - 50 \operatorname{erf}[(x+1)/(2\alpha t^{1/2})]$

10.3 $T = \dfrac{1}{4}(2 - y) + \dfrac{4}{\pi^2} \sum_{\text{odd } n} \dfrac{1}{n^2 \sinh 2n\pi} \sinh n\pi(2 - y) \cos n\pi x$

10.4 $T = 20 + \frac{40}{\pi}\sum_{\text{odd } n}\frac{1}{n\sinh(3n\pi/5)}\sinh\frac{n\pi y}{5}\sin\frac{n\pi x}{5} + \frac{40}{\pi}\sum_{\text{odd } n}\frac{1}{n\sinh(5n\pi/3)}\sinh\frac{n\pi(5-x)}{3}\sin\frac{n\pi y}{3}$

10.6 $u = 20 - \frac{80}{\pi}\sum_0^\infty\frac{(-1)^n}{2n+1}e^{-[(2n+1)\pi\alpha/(2l)]^2 t}\cos\left(\frac{2n+1}{2l}\pi x\right)$

10.8 $u = 20 - x - \frac{40}{\pi}\sum_{\text{even } n}\frac{1}{n}e^{-(n\pi\alpha/10)^2 t}\sin\frac{n\pi x}{10}$

10.10 $u = \frac{1600}{\pi^2}\sum_{\text{odd } n}\sum_{\text{odd } m}\frac{1}{nmI_n(3m\pi/20)}I_n\left(\frac{m\pi r}{20}\right)\sin n\theta\sin\frac{m\pi z}{20}$

10.16 $v\sqrt{5}/(2\pi)$

10.18 ν_{mn}, $n = 3, 6, \cdots$; the lowest frequencies are:
$\nu_{13} = 2.65\,\nu_{10}$, $\nu_{23} = 4.06\,\nu_{10}$, $\nu_{16} = 4.13\,\nu_{10}$, $\nu_{33} = 5.4\,\nu_{10}$

10.20 $\nu = v\lambda_l/(2\pi a)$ where $\lambda_l =$ zeros of j_l, $a =$ radius of sphere, $v =$ speed of sound

10.22 $u = 1 - \frac{1}{2}rP_1(\cos\theta) + \frac{7}{8}r^3P_3(\cos\theta) - \frac{11}{16}r^5P_5(\cos\theta)\cdots$

10.26 $\nu = [v/(2\pi)][(k_{mn}/a)^2 + \lambda^2]^{1/2}$ where k_{mn} is a zero of J_n

Chapter 14

1.1 $u = x^3 - 3xy^2$, $v = 3x^2y - y^3$
1.3 $u = x$, $v = -y$
1.4 $u = (x^2+y^2)^{1/2}$, $v = 0$
1.7 $u = \cos y\cosh x$, $v = \sin y\sinh x$
1.9 $u = x/(x^2+y^2)$, $v = -y/(x^2+y^2)$
1.11 $u = 3x/[x^2+(y-2)^2]$, $v = (-2x^2-2y^2+5y-2)/[x^2+(y-2)^2]$
1.13 $u = \ln(x^2+y^2)^{1/2}$, $v = 0$
1.17 $u = \cos x\cosh y$, $v = \sin x\sinh y$
1.18 $u = \pm 2^{-1/2}[(x^2+y^2)^{1/2}+x]^{1/2}$, $v = \pm 2^{-1/2}[(x^2+y^2)^{1/2}-x]^{1/2}$, where the $\pm$ signs are chosen so that uv has the sign of y.
1.19 $u = \ln(x^2+y^2)^{1/2}$, $v = \arctan(y/x)$
[The angle is in the quadrant of the point (x, y).]

In 2.1–2.23, A = analytic, N = not analytic

2.1 A
2.3 N
2.4 N
2.7 A
2.9 A, $z \neq 0$
2.11 A, $z \neq 2i$
2.13 N
2.17 N
2.18 A, $z \neq 0$
2.19 A, $z \neq 0$
2.23 A, $z \neq 0$
2.34 $-z - \frac{1}{2}z^2 - \frac{1}{3}z^3\cdots$, $|z| < 1$
2.38 $-\frac{1}{2}i + \frac{1}{4}z + \frac{1}{8}iz^2 - \frac{1}{16}z^3\cdots$, $|z| < 2$
2.42 $z + z^3/3! + z^5/5!\cdots$, all z
2.48 Yes, $z \neq 0$
2.52 No
2.53 Yes, $z \neq 0$
2.54 $-iz$
2.56 $-iz^2/2$
2.59 e^z
2.60 $2\ln z$
2.63 $-i/(1-z)$

3.1 $\frac{1}{2} + i$
3.3 0
3.5 -1
3.7 $\pi(1-i)/8$
3.9 1
3.12 (a) $\frac{5}{3}(1+2i)$
3.17 (a) 0 (b) $i\pi$
3.19 $16i\pi$
3.23 $72i\pi$

4.4 For $0 < |z| < 1$: $-\frac{1}{4}z^{-1} - \frac{1}{2} - \frac{11}{16}z - \frac{13}{16}z^2\cdots$; $R(0) = -\frac{1}{4}$
For $1 < |z| < 2$: $\cdots + z^{-3} + z^{-2} + \frac{3}{4}z^{-1} + \frac{1}{2} + \frac{5}{16}z + \frac{3}{16}z^2\cdots$
For $|z| > 2$: $z^{-4} + 5z^{-5} + 17z^{-6} + 49z^{-7}\cdots$

4.6 For $0 < |z| < 1$: $z^{-2} - 2z^{-1} + 3 - 4z + 5z^2 \cdots$; $R(0) = -2$
For $|z| > 1$: $z^{-4} - 2z^{-5} + 3z^{-6} \cdots$

4.8 For $|z| < 1$: $-5 + \frac{25}{6}z - \frac{175}{36}z^2 \cdots$; $R(0) = 0$
For $1 < |z| < 2$: $-5(\cdots + z^{-3} - z^{-2} + z^{-1} + \frac{1}{6}z + \frac{1}{36}z^2 + \frac{7}{216}z^3 \cdots)$
For $2 < |z| < 3$: $\cdots + 3z^{-3} + 9z^{-2} - 3z^{-1} + 1 - \frac{1}{3}z + \frac{1}{9}z^2 - \frac{1}{27}z^3 \cdots$
For $|z| > 3$: $30(z^{-3} - 2z^{-4} + 9z^{-5} \cdots)$

4.9 (a) Regular (b) Pole of order 3
4.10 (b) Pole of order 2 (d) Essential singularity
4.11 (c) Simple pole (d) Pole of order 3
4.12 (b) Pole of order 2 (d) Pole of order 1

6.1 $z^{-1} - 1 + z - z^2 \cdots$; $R = 1$
6.3 $z^{-3} - z^{-1}/3! + z/5! \cdots$; $R = -\frac{1}{6}$
6.5 $\frac{1}{2}e[(z-1)^{-1} + \frac{1}{2} + \frac{1}{4}(z-1) \cdots]$; $R = \frac{1}{2}e$
6.7 $\frac{1}{4}[(z - \frac{1}{2})^{-1} - 1 + (1 - \pi^2/2)(z - \frac{1}{2}) \cdots]$; $R = \frac{1}{4}$
6.9 $-[(z-2)^{-1} + 1 + (z-2) + (z-2)^2 \cdots]$; $R = -1$
6.14 $R(-2/3) = 1/8$, $R(2) = -1/8$ 6.16 $R(0) = -2$, $R(1) = 1$
6.18 $R(3i) = \frac{1}{2} - \frac{1}{3}i$ 6.19 $R(\pi/2) = 1/2$
6.21 $R[\sqrt{2}\,(1+i)] = \sqrt{2}\,(1-i)/16$ 6.22 $R(i\pi) = -1$
6.27 $R(\pi/6) = -1/2$ 6.28 $R(3i) = -\frac{1}{16} + \frac{1}{24}i$
6.31 $R(0) = 9/2$ 6.33 $R(\pi) = -1/2$
6.35 $R(i) = 0$ 6.14′ $\pi i/4$
6.16′ $-2\pi i$ 6.18′ 0
6.19′ 0 6.27′ $-\pi i$
6.28′ $\pi i/4$ 6.31′ $9\pi i$
6.33′ 0 6.35′ 0

7.1 $\pi/6$ 7.3 $2\pi/3$
7.5 $\pi/(1-r^2)$ 7.7 $\pi/6$
7.9 $2\pi/|\sin\alpha|$ 7.11 $3\pi/32$
7.13 $\pi/10$ 7.15 $\pi e^{-4/3}/12$
7.17 $(\pi/e)(\cos 2 + 2\sin 2)$ 7.19 $\pi e^{-3}/54$
7.23 $\pi/8$ 7.24 π
7.26 $-\pi/2$ 7.28 $\pi/4$
7.30 $\pi/(2\sqrt{2})$ 7.32 $\frac{3}{16}\pi\sqrt{2}$
7.33 $\pi\sqrt{2}/2$ 7.36 $-\pi^2\sqrt{2}$
7.39 2 7.41 $(2\pi)^{1/2}/4$
7.45 One negative real, one each in quadrants I and IV
7.48 Two each in quadrants I and IV
7.50 Two each in quadrants II and III
7.52 πi 7.54 $8\pi i$
7.55 $\cosh t \cos t$ 7.57 $1 + \sin t - \cos t$
7.60 $t + e^{-t} - 1$ 7.61 $\left(\cosh 2t + 2\cosh t \cos t\sqrt{3}\,\right)/3$
7.63 $(\cosh t - \cos t)/2$ 7.65 $(\cos 2t + 2\sin 2t - e^{-t})/5$

8.3 Regular, $R = -1$ 8.5 Regular, $R = -1$
8.7 Simple pole, $R = -2$ 8.9 Regular, $R = 0$
8.11 Regular, $R = -1$ 8.14 $-2\pi i$

9.3 $u = x/(x^2 + y^2)$, $v = -y/(x^2 + y^2)$
9.4 $u = e^x \cos y$, $v = e^x \sin y$
9.7 $u = \sin x \cosh y$, $v = \cos x \sinh y$

10.6 $T = 100y/(x^2 + y^2)$; isothermals $y/(x^2 + y^2) = \text{const.}$; flow lines $x/(x^2 + y^2) = \text{const.}$
10.9 Streamlines $y - y/(x^2 + y^2) = \text{const.}$
10.12 $T = (20/\pi) \arctan[2y/(1 - x^2 - y^2)]$, arc tan between $\pi/2$ and $3\pi/2$
10.14 $\Phi = \frac{1}{2} V_0 \ln\{[(x+1)^2 + y^2]/[(x-1)^2 + y^2]\}$
$\Psi = V_0 \arctan\{2y/[1 - x^2 - y^2]\}$, arc tan between $\pi/2$ and $3\pi/2$
$V_x = 2V_0(1 - x^2 + y^2)/[(1 - x^2 + y^2)^2 + 4x^2y^2]$,
$V_y = -4V_0 xy/[(1 - x^2 + y^2)^2 + 4x^2y^2]$

11.2 $-i \ln(1 + z)$ 11.5 $R(i) = \frac{1}{4}(1 - i\sqrt{3})$, $R(-i) = -\frac{1}{2}$
11.8 $R(1/2) = 1/2$ 11.10 -1 11.12 $1/2$
11.14 (a) 2 (b) $-\sin 5$ (c) 1/16 (d) -2π 11.16 $-\pi/6$
11.18 $\frac{1}{4}\pi e^{-\pi/2}$ 11.20 $\frac{1}{2}\pi(e^{-1} + \sin 1)$
11.29 $\pi^3/8$. Caution: $-\pi^3/8$ is wrong.
11.32 One negative real, one each in quadrants II and III
11.34 Two each in quadrants I and IV, one each in II and III
11.41 $\pi^2/8$

Chapter 15

1.2 3/8, 1/8, 1/4 1.5 1/4, 3/4, 1/3, 1/2
1.6 27/52, 16/52, 15/52 1.8 9/100, 1/10, 3/100, 1/10

2.12 (a) 3/4 (b) 1/5 (c) 2/3 (d) 3/4 (e) 3/7
2.14 (a) 3/4 (b) 25/36 (c) 37, 38, 39, 40
2.17 (a) 3 to 9 with $p(5) = p(7) = 2/9$; others, $p = 1/9$. (c) 1/3

3.4 (a) 8/9, 1/2 (b) 3/5, 1/11, 2/3, 2/3, 6/13 3.5 1/33, 2/9
3.12 (a) 1/49 (b) 68/441 (c) 25/169 (d) 15 times (e) 44/147
3.14 $n > 3.3$, so 4 tries are needed. 3.16 9/23
3.17 (a) 39/80, 5/16, 1/5, 11/16 (b) 374/819 (c) 185/374
3.20 5/7, 2/7, 11/14 3.21 2/3, 1/3

4.1 (a) $P(10, 8)$ (b) $C(10, 8)$ (c) 1/45
4.4 1.98×10^{-3}, 4.95×10^{-4}, 3.05×10^{-4}, 1.39×10^{-5} 4.7 1/26
4.8 1/221, 1/33, 1/17 4.11 0.097, 0.37, 0.67; 13
4.17 MB: 16, FD: 6, BE: 10

5.1 $\mu = 0$, $\sigma = \sqrt{3}$ 5.3 $\mu = 2$, $\sigma = \sqrt{2}$
5.5 $\mu = 1$, $\sigma = \sqrt{7/6}$ 5.7 $\mu = 3(2p - 1)$, $\sigma = 2\sqrt{3p(1-p)}$

6.1 (c) $\bar{x} = 0$, $\sigma = 2^{-1/2}a$ 6.4 $\bar{x} = 0$, $\sigma = (2^{1/2}\alpha)^{-1}$
6.5 $f(t) = \lambda e^{-\lambda t}$, $F(t) = 1 - e^{-\lambda t}$, $\bar{t} = 1/\lambda$, half life $= \bar{t} \ln 2$

6.7 (a) $F(s) = 2[1 - \cos(s/R)]$, $f(s) = (2/R)\sin(s/R)$
(b) $F(s) = [1 - \cos(s/R)]/[1 - \cos(1/R)] \simeq s^2$,
$f(s) = R^{-1}[1 - \cos(1/R)]^{-1}\sin(s/R) \cong 2s$

	n	Exactly $7\ h$	At most $7\ h$	At least $7\ h$	Most probable number of h	Expected number of h
7.1	7	0.0078	1	0.0078	3 or 4	7/2
7.2	12	0.193	0.806	0.387	6	6

In the following answers, the first number is the binomial result and the second number is the normal approximation using whole steps at the ends as in Example 2.

8.12 0.03987, 0.03989
8.14 0.9546, 0.9546
8.17 0.0770, 0.0782
8.18 0.372, 0.376
8.20 0.462, 0.455

9.3

Number of particles	0	1	2	3	4	5
Number of intervals	406	812	812	541	271	108

9.5 $P_0 = 0.37$, $P_1 = 0.37$, $P_2 = 0.18$, $P_3 = 0.06$
9.8 3, 10, 3
9.11 Normal: 0.08, Poisson: 0.0729, (binomial: 0.0732)

10.8 $\bar{x} = 5$, $\bar{y} = 1$, $s_x = 0.122$, $s_y = 0.029$,
$\sigma_x = 0.131$, $\sigma_y = 0.030$, $\sigma_{mx} = 0.046$, $\sigma_{my} = 0.0095$,
$r_x = 0.031$, $r_y = 0.0064$,
$\overline{x+y} = 6$ with $r = 0.03$, $\overline{xy} = 5$ with $r = 0.04$,
$\overline{x^3 \sin y} = 105$ with $r = 2.00$, $\overline{\ln x} = 1.61$ with $r = 0.006$
10.10 $\bar{x} = 6$ with $r = 0.062$, $\bar{y} = 3$ with $r = 0.067$,
$\overline{e^y} = 20$ with $r = 1.3$, $\overline{x/y^2} = 0.67$ with $r = 0.03$

11.3 20/47
11.7 $\bar{x} = 1/4$, $\sigma = \sqrt{3}/4$
11.9 (d) $\bar{x} = 1/4$, $\sigma = \sqrt{31}/12$
11.13 30, 60
11.17 $\bar{x} = 2$ with $r = 0.073$, $\bar{y} = 1$ with $r = 0.039$, $\overline{x - y} = 1$ with $r = 0.08$,
$\overline{xy} = 2$ with $r = 0.11$, $\overline{x/y^3} = 2$ with $r = 0.25$

Index